The Protein Protocols Handbook

The Protein Protocols Handbook

Second Edition

Edited by

John M. Walker

University of Hertfordshire, Hatfield, UK

Humana Press   Totowa, New Jersey

999 Riverview Drive, Suite 208
Totowa, New Jersey 07512

humanapress.com

This publication is printed on acid-free paper. ∞

ANSI Z39.48-1984 (American Standards Institute) Permanence of Paper for Printed Library Materials.

Production Editor: Diana Mezzina

Cover design by Patricia F. Cleary.

For additional copies, pricing for bulk purchases, and/or information about other Humana titles, contact Humana at the above address or at any of the following numbers: Tel: 973-256-1699; Fax: 973-256-8341; E-mail: humana@humanapr.com, or visit our Website at www.humanapress.com

Printed in the United States of America. 10 9 8 7 6 5 4 3 2 1

Library of Congress Cataloging in Publication Data

The Protein Protocols Handbook: Second Edition / edited by John M. Walker.
p. cm.
ISBN 0-89603-940-4 (HB); 0-89603-941-2 (PB)
Includes bibliographical references and index.
1. Proteins--Analysis--Laboratory manuals. I. Walker, John M., 1948-

Qp551 .P697512 2002
572'.6--dc21

2001039829

Preface

The Protein Protocols Handbook, Second Edition aims to provide a cross-section of analytical techniques commonly used for proteins and peptides, thus providing a benchtop manual and guide for those who are new to the protein chemistry laboratory and for those more established workers who wish to use a technique for the first time.

All chapters are written in the same format as that used in the *Methods in Molecular Biology*™ series. Each chapter opens with a description of the basic theory behind the method being described. The Materials section lists all the chemicals, reagents, buffers, and other materials necessary for carrying out the protocol. Since the principal goal of the book is to provide experimentalists with a full account of the practical steps necessary for carrying out each protocol successfully, the Methods section contains detailed step-by-step descriptions of every protocol that should result in the successful execution of each method. The Notes section complements the Methods material by indicating how best to deal with any problem or difficulty that may arise when using a given technique, and how to go about making the widest variety of modifications or alterations to the protocol.

Since the first edition of this book was published in 1996 there have, of course, been significant developments in the field of protein chemistry. Hence, for this second edition I have introduced 60 chapters/protocols not present in the first edition, significantly updated a number of chapters remaining from the first edition, and increased the overall length of the book from 144 to 164 chapters. The new chapters particularly reflect the considerable developments in the use of mass spectrometry in protein characterization. Recognition of the now well-established central role of 2-D PAGE in proteomics has resulted in an expansion of chapters on this subject, and I have also included a number of new techniques for staining and analyzing protein blots. The section on glycoprotein analysis has been significantly expanded, and aspects of single chain antibodies and phage-displayed antibodies have been introduced in the section on antibodies.

We each, of course, have our own favorite, commonly used methods, be it a gel system, gel-staining method, blotting method, and so on; I'm sure you will find yours here. However, I have, as before, also described alternatives for some of these techniques; though they may not be superior to the methods you commonly use, they may nevertheless be more appropriate in a particular situation. Only by knowing the range of techniques that are available to you, and the strengths and limitations of these techniques, will you be able to choose the method that best suits your purpose. Good luck in your protein analysis!

John M. Walker

Contents

PART VII : ANTIBODY TECHNIQUES

Contributors

THOMAS E. ADRIAN • *Department of Surgery, Northwestern University Medical School, Chicago, IL*
F. JAVIER ALBA • *Departament de Bioquímica i Biologia Molecular, Universität Autònoma de Barcelona, Bellaterra (Barcelona), Spain*
ALASTAIR AITKEN • *Division of Biomedical and Clinical Laboratory Sciences, Membrane Biology Group, University of Edinburgh, Scotland, UK*
ROBERT E. AKINS • *Nemours Biomedical Research Program, A.I. duPont Hospital for Children, Wilmington, DE*
SALLY ANN AMERO • *Center for Scientific Review, National Institutes of Health, Bethesda, MD*
DOUGLAS A. ANDRES • *Department of Biochemistry, University of Kentucky, Lexington, KY*
SARAH M. ANDREW • *Chester College of Higher Education, UK*
NEBOJSA AVDALOVIC • *Dionex Corporation, Life Science Research Group, Sunnyvale, CA*
GRAHAM S. BAILEY • *Department of Biological Sciences, University of Essex, Colchester, UK*
PASCAL BAILON • *Department of Pharmaceutical and Analytical R & D, Hoffmann-LaRoche Inc., Nutley, NJ*
MALCOLM S. BALL • *Co-operative Research Centre for Eye Research Technology, Sydney, Australia*
SALVADOR BARTOLOMÉ • *Departament de Bioquímica i Biologia Molecular, Universität Autònoma de Barcelona, Bellaterra (Barcelona), Spain*
ANTONIO BERMÚDEZ • *Departament de Bioquímica i Biologia Molecular, Universität Autònoma de Barcelona, Bellaterra (Barcelona), Spain*
WOLFGANG BERTHOLD • *Division of Biopharmaceutical Sciences, IDEC Pharmaceuticals Corp., San Diego, CA*
MAHESH K. BHALGAT • *Molecular Probes, Inc., Eugene, OR*
KEITH C. BIBLE • *Division of Medical Oncology, Mayo Clinic, Rochester, MN*
LUCA BINI • *Department of Molecular Biology, University of Siena, Italy*
CHRISTOPHER R. BIRD • *Division of Immunobiology, National Institute for Biological Standards and Control, Potters Bar, UK*
NICK BIZIOS • *AGI Dermatics, Freeport, NY*
SCOTT A. BOERNER • *Division of Medical Oncology, Mayo Clinic, Rochester, MN*
DÉBORA BONENFANT • *Department of Biochemistry, Biozentrum der Universität Basel, Switzerland*
BRIAN K. BRANDLEY • *Glyko Inc., Navato, CA*
MICHAEL BRAUNAGEL • *Affitech, Oslo, Norway*

ROBERT BURNS • *Antibody Unit, Scottish Agricultural Science Agency, Edinburgh, UK*
FRANCA CASAGRANDA • *CSIRO Division of Biomolecular Engineering, Victoria, Australia; Present address: European Molecular Biology Laboratory, Heidelberg, Germany*
BRIAN T. CHAIT • *The Rockefeller University, New York, NY*
JUNG-KAP CHOI • *College of Pharmacy, Chonnam National University, Kwangju, Korea*
PHILIPP CHRISTEN • *Biochemisches Institut der Universität Zürich, Switzerland*
ANTONELLA CIRCOLO • *Maxwell Finland Lab for Infectious Diseases, Boston, MA*
JOHN COLYER • *Department of Biochemistry & Molecular Biology, University of Leeds, UK*
J. MICHAEL CONLON • *Department of Biomedical Sciences, Creighton University School of Medicine, Omaha, NE*
CATHERINE COPSE • *Amersham Biosciences, Amersham, UK*
ALBERTO CORSINI • *Department of Pharmacological Sciences, University of Milan, Italy*
PAUL L. COURCHESNE • *Amgen Inc., Thousand Oaks, CA*
DEAN C. CRICK • *Department of Biochemistry, University of Kentucky, Lexington, KY*
JOAN-RAMON DABAN • *Departament de Bioquímica i Biologia Molecular, Universität Autònoma de Barcelona, Bellaterra (Barcelona), Spain*
JAMES R. DAVIE • *Manitoba Institute of Cell Biology, Winnipeg, Canada*
GENEVIÈVE P. DELCUVE • *Manitoba Institute of Cell Biology, Winnipeg, Canada*
PETER L. DEVINE • *Proteome Systems Ltd., Sydney, Australia*
MONIQUE DIANO • *IBDM, Faculté des Sciences de Luminy, Marseille, France*
JOANNE DICKINSON • *Amersham Biosciences, Amersham Labs, UK*
CARL DOLMAN • *Division of Immunobiology, National Institute for Biological Standards and Control, Potters Bar, UK*
HEINZ DÖRSAM • *German Cancer Research Center, Heidelberg, Germany*
STEVEN F. DOWDY • *University of California Medical Center, San Francisco, CA*
MICHAEL J. DUNN • *Department of Neuroscience, Institute of Psychiatry, De Crespigny Park, London, UK*
GEORGE K. EHRLICH • *Department of Pharmaceutical and Analytical R & D, Hoffman-LaRoche Inc., Nutley, NJ*
SARAH C. R. ELGIN • *Department of Biology, Washington University in St. Louis, MO*
SERGEI A. EZHEVSKY • *Howard Hughes Medical Institute, Washington University School of Medicine, St. Louis, MO*
CHRISTOPHER C. FARNSWORTH • *Department of Protein Chemistry, IMMUNEX Corporation, Seattle, WA*
GIORGIO FASSINA • *Biopharmaceuticals, Tecnogen SCPA, Parco Scientifico, Piana di Monte Verna (CE), Italy*
JOSEPH FERNANDEZ • *Protein/DNA Technology Center, Rockefeller University, NY*
CARLOS FERNANDEZ-PATRON • *Department of Biochemistry, University of Alberta, Edmonton, Canada*
BRIAN S. FINLIN • *Department of Biochemistry, University of Kentucky, Lexington, KY*
ANGELIKI FOTINOPOULOU • *Department of Clinical Biochemistry, The Medical School, University of Newcastle, Newcastle upon Tyne, UK*
SUSAN J. FOWLER • *Amersham Biosciences, Amersham, UK*

RUTH R. FRENCH • *Lymphoma Research Unit, Tenovus Research Laboratory, Southhampton General Hospital, UK*
THOMAS D. FRIEDRICH • *Center for Immunology and Microbial Disease, Albany Medical College, NY*
JOHANNES FREUND • *Institute of Chemistry and Biochemistry, Immunology Group, University of Salzburg, Austria*
MICHAEL H. GELB • *Departments of Chemistry and Biochemistry, University of Washington, Seattle, WA*
ELISABETTA GIANAZZA • *Istituto di Scienze Farmacologiche, Universita di Milano, Italy*
JOHN A. GLOMSET • *Howard Hughes Medical Institute, University of Washington, Seattle, WA*
MOHAMMAD T. GOODARZI • *Department of Clinical Biochemistry, The Medical School, University of Newcastle, New Castle upon Tyne, UK*
MORAG A. GRASSIE • *Department of Biochemistry & Molecular Biology, Institute of Biomedical and Life Sciences, University of Glasgow, UK*
PATRICIA GRAVEL • *Triskel Integrated Services, Geneva, Switzerland*
SUNITA GULATI • *Maxwell Finland Lab for Infectious Diseases, Boston, MA*
PETER HAMMERL • *Institute of Chemistry and Biochemistry, Immunology Group, University of Salzburg, Austria*
ARNULF HARTL • *Institute of Chemistry and Biochemistry, Immunology Group, University of Salzburg, Austria*
ROSARIA P. HAUGLAND • *Molecular Probes Inc., Eugene, OR*
LARS HENNIG • *Swiss Federal Institute of Technology, Zürich, Switzerland*
HEE-YOUN HONG • *College of Pharmacy, Chonnam National University, Kwangju, Korea*
ANDREW HOOKER • *Sittingbourne Research Centre, Pfizer Ltd, Analytical Research and Development (Biologics), Sittingbourne, UK*
MARTIN HORST • *STRATEC Medical, Oberdorf, Switzerland*
ELIZABETH F. HOUNSELL • *School of Biological and Chemical Sciences, Birkbeck University of London, UK*
G. BRENT IRVINE • *School of Biology and Biochemistry, Queen's University of Belfast, UK*
DAVID C. JAMES • *Sittingbourne Research Centre, Pfizer Ltd, Analytical Research and Development (Biologics), Sittingbourne, UK*
THARAPPEL C. JAMES • *Dublin, Ireland*
PAUL JENÖ • *Department of Biochemistry, Biozentrum der Universität Basel, Switzerland*
OLE NØØRREGAARD JENSEN • *Department of Biochemistry and Molecular Biology, Odense University, Denmark*
KARI JOHANSEN • *Department of Virology, Swedish Institute For Infectious Disease Control, Sweden*
WILLIAM JORDAN • *Department of Immunology, ICSM, Hammersmith Hospital, London, UK*
RALPH C. JUDD • *Division of Biological Science, University of Montana, Missoula, MT*
SCOTT H. KAUFMANN • *Division of Oncology Research, Mayo Clinic, Rochester, MN*

SERGEY M. KIPRIYANOV • *Affimed Therapeutics AG, Ladenburg, Germany*
CHRISTIAN KLEIST • *Institute for Immunology, Heidelberg, Germany*
JOACHIM KLOSE • *Institut für Humangenetik Charité, Humboldt-Universität, Berlin, Germany.*
SUNIL KOCHHAR • *Nestlé Research Center, Lausanne, Switzerland*
NICHOLAS J. KRUGER • *Department of Plant Sciences, University of Oxford, UK*
JUDITH A. LAFFIN • *Department of Microbiology, Immunology, and Molecular Genetics, The Albany Medical College, Albany, NY*
WILLIAM J. LAROCHELLE • *Laboratory of Cellular and Molecular Biology, National Cancer Institute, National Institute of Health, Bethesda, MD*
ROBERT R. LATEK • *Howard Hughes Medical Institute, Washington University School of Medicine, St. Louis, MO*
JOHN M. LEHMAN • *Center for Immunology and Microbial Disease, Albany Medical College, NY*
MICHELE LEARMONTH • *Department of Biomedical Sciences, University of Edinburgh, Scotland*
ANDRÉ LE BIVIC • *IBDM, Faculté des Sciences de Luminy, Marseille, France*
YEAN KIT LEE • *Division of Medical Oncology, Mayo Clinic, Rochester, MN*
PETER LEMKIN • *LECB/NCI-FCRDC, Frederick, MD*
KRISTI A. LEWIS • *Laboratory of Cellular and Molecular Biology, National Cancer Institute, National Institute of Health, Bethesda, MD*
SABRINA LIBERATORI • *Department of Molecular Biology, University of Siena, Italy*
FAN LIN • *Department of Pathology, Temple University Hospital, Philadelphia, PA*
Mary F. Lopez • *Proteome Systems, Woburn, MA*
BARBARA MAGI • *Department of Molecular Biology, University of Siena, Italy*
NGUYEN THI MAN • *MRIC, North East Wales Institute, Deeside, Clwyd, UK*
SU-YAU MAO • *Department of Immunology and Molecular Genetics, Medimmune Inc., Gaithersburg, MD*
PHILIP N. MCFADDEN • *Department of Biochemistry and Biophysics, Oregon State University, Corvallis, OR*
PAUL MCGEADY • *Department of Chemistry, Clark Atlanta University, Georgia*
TONY MERRY • *Department of Biochemistry, The Glycobiology Institute, University of Oxford, UK*
GRAEME MILLIGAN • *Department of Biochemistry & Molecular Biology, Institute of Biomedical and Life Sciences, University of Glasgow, UK*
JONATHAN MINDEN • *Millennium Pharmaceuticals, Cambridge, MA*
THIERRY MINI • *Department of Biochemistry, Biozentrum der Universität Basel, Switzerland*
SHEENAH M. MISCHE • *Protein/DNA Technology Center, Rockefeller University, NY*
HOLGER J. MØLLER • *Department of Clinical Biochemistry, Aarhus University Hospital, Amtssygehuset, Aarhus, Denmark*
GLENN E. MORRIS • *MRIC, North East Wales Institute, Wrexhäm, UK*
BARBARA MOURATOU • *Biochemisches Institut der Universität Zürich, Switzerland*
DANIEL MOYNET• *INSERM, Bordeaux Cedex, France*
HIKARU NAGAHARA • *Howard Hughes Medical Institute, Washington University School of Medicine, St. Louis, MO*

STEFANIE A. NELSON • *Laboratory of Cellular and Molecular Biology, National Cancer Institute, National Institute of Health, Bethesda, MD*
TOSHIAKI OSAWA • *Yakult Central Institute for Microbiology Research, Tokyo, Japan*
NICOLLE PACKER • *Proteome Systems Ltd., Sydney, Australia*
MARK PAGE • *Apovia Inc., San Diego, CA*
VITALIANO PALLINI • *Department of Molecular Biology, University of Siena, Italy*
GIOVANNA PALOMBO • *Biopharmaceuticals, Tecnogen SCPA, Parco Scientifico, Piana di Monte Verna (CE), Italy*
SCOTT D. PATTERSON • *Celera Genomics, Rockville, MD*
WAYNE F. PATTON • *Molecular Probes Inc., Eugene, OR*
JERGEN H. POULSEN • *Department of Clinical Biochemistry, Aarhus University Hospital, Amtssygehuset, Aarhus, Denmark*
THIERRY RABILLOUD • *DBMS/BECP, CEA-Grenoble, Grenoble, France*
ROBERTO RAGGIASCHI • *Department of Molecular Biology, University of Siena, Italy*
MENOTTI RUVO • *Biopharmaceuticals, Tecnogen SCPA, Parco Scientifico, Piana di Monte Verna (CE), Italy*
F. ANDREW RAY • *Department of Biology, Hartwick College, Oneonta, NY*
JEFFREY ROHRER • *Dionex Corporation, Life Science Research Group, Sunnyvale, CA*
DOUGLAS D. ROOT • *Department of Biological Sciences, University of North Texas, Denton, TX*
KENNETH E. SANTORA • *Laboratory of Cellular and Molecular Biology, National Cancer Institute, National Institute of Health, Bethesda, MD*
ALEXANDER SCHWARZ • *Biosphere Medical Inc., Rockland, MA*
BRYAN JOHN SMITH • *Celltech, R&D, Slough, UK*
VIRGINIA SPENCER • *Manitoba Institute of Cell Biology, Manitoba, Canada*
WAYNE R. SPRINGER • *VA San Diego Healthcare System, CA*
CHRISTOPHER M. STARR • *Glyko Inc., Novato, CA*
KATHRYN L. STONE • *Yale Cancer Center Mass Spectrometry Resource and W. M. Keck Foundation Biotechnology Resource Laboratory, New Haven, CT*
RICHARD A. W. STOTT • *Department of Clinical Chemistry, Doncaster Royal Infirmary, South Yorkshire, UK*
LENNART SVENSSON • *Department of Virology, Swedish Institute For Infectious Disease Control, Sweden*
PATRICIA J. SWEENEY • *School of Natural Sciences, Hatfield Polytechnic, University of Hertfordshire, UK*
DAN S. TAWFIK • *Department of Biological Chemistry, the Weizman Institute of Science, Rehovot, Israel*
JOSEPH THALHAMER • *Institute of Chemistry and Biochemistry, Immunology Group, University of Salzburg, Austria*
JAMES R. THAYER • *Dionex Corporation, Life Science Research Group, Sunnyvale, CA*
GEORGE C. THORNWALL • *LECB/NCI-FCRDC, Frederick, MD*
ROBIN THORPE • *Division of Immunobiology, National Institute for Biological Standards and Control, Potters Bar, UK*
TSUTOMU TSUJI • *Hoshi Pharmaceutical College, Tokyo, Japan*
ROCKY S. TUAN • *Department of Orthopaedic Surgery, Thomas Jefferson University Philadelphia, PA*

Jeremy E. Turnbull • *School of Biosciences, University of Birmingham, UK*
Graham A. Turner • *Department of Clinical Biochemistry, The Medical School, University of Newcastle, New Castle upon Tyne, UK*
Mustafa Ünlü • *Millennium Pharmaceuticals, Cambridge, MA*
Antonio Verdoliva • *Biopharmaceuticals, Tecnogen SCPA, Parco Scientifico, Piana di Monte Verna (CE), Italy*
Yoshinao Wada • *Osaka Medical Center and Research Institute for Maternal and Child Health, Osaka, Japan*
Charles J. Waechter • *Department of Biochemistry, University of Kentucky, Lexington, KY*
Kuan Wang • *Department of Biological Sciences, University of North Texas, Denton, TX*
Rong Wang • *Department of Human Genetics, Mount Sinai School of Medicine, New York, NY*
John M. Walker • *Department of Biosciences, University of Hertfordshire, School of Natural Sciences, Hatfield, UK*
Malcolm Ward • *Proteome Sciences plc, Kings College, London, UK*
Jakob H. Waterborg • *Cell Biology & Biophysics, University of Missouri-Kansas City, Kansas City, MO*
Darin J. Weber • *Department of Biochemistry and Biophysics, Oregon State University, Corvallis, OR*
Michael Weitzhandler • *Dionex Corporation, Life Science Research Group, Sunnyvale, CA*
Martin Welschof • *Axaron Bioscience AG, Heidelberg, Germany*
Matthias Wilm • *Department of Biochemistry and Molecular Biology, Odense University, Denmark*
John F. K. Wilshire • *CSIRO Division of Biomolecular Engineering, Victoria, Australia*
G. Brian Wisdom • *School of Biology and Biochemistry, The Queen's University, Medical Biology Centre, Belfast, UK*
Gary E. Wise • *Department of Anatomy & Cell Biology, Louisiana State University School of Veterinary Medicine, Baton Rouge, LA*
Kenneth R. Williams • *Yale Cancer Center Mass Spectrometry Resource and W. M. Keck Foundation Biotechnology Resource Laboratory, New Haven, CT*
Nicky K. C. Wong • *Department of Biochemistry, University of Hong Kong, Pokfulam, Hong Kong*
Kazuo Yamamoto • *Department of Integrated Biosciences, Graduate School of Frontier Sciences, University of Tokyo, Japan*
Gyurng-Soo Yoo • *College of Pharmacy, Chonnam National University, Kwangju, Korea*
Wendy W. You • *Department of Biochemistry and Biophysics, Oregon State University, Corvallis, OR*

PART I

QUANTITATION OF PROTEINS

1

Protein Determination by UV Absorption

Alastair Aitken and Michèle P. Learmonth

1. Introduction

1.1. Near UV Absorbance (280 nm)

Quantitation of the amount of protein in a solution is possible in a simple spectrometer. Absorption of radiation in the near UV by proteins depends on the Tyr and Trp content (and to a very small extent on the amount of Phe and disulfide bonds). Therefore the A_{280} varies greatly between different proteins (for a 1 mg/mL solution, from 0 up to 4 [for some tyrosine-rich wool proteins], although most values are in the range 0.5–1.5 *[1]*). The advantages of this method are that it is simple, and the sample is recoverable. The method has some disadvantages, including interference from other chromophores, and the specific absorption value for a given protein must be determined. The extinction of nucleic acid in the 280-nm region may be as much as 10 times that of protein at their same wavelength, and hence, a few percent of nucleic acid can greatly influence the absorption.

1.2. Far UV Absorbance

The peptide bond absorbs strongly in the far UV with a maximum at about 190 nm. This very strong absorption of proteins at these wavelengths has been used in protein determination. Because of the difficulties caused by absorption by oxygen and the low output of conventional spectrophotometers at this wavelength, measurements are more conveniently made at 205 nm, where the absorbance is about half that at 190 nm. Most proteins have extinction coefficients at 205 nm for a 1 mg/mL solution of 30–35 and between 20 and 24 at 210 nm *(2)*.

Various side chains, including those of Trp, Phe, Tyr, His, Cys, Met, and Arg (in that descending order), make contributions to the A_{205} *(3)*.

The advantages of this method include simplicity and sensitivity. As in the method outlined in **Subheading 3.1.** the sample is recoverable and in addition there is little variation in response between different proteins, permitting near-absolute determination of protein. Disadvantages of this method include the necessity for accurate calibration of the spectrophotometer in the far UV. Many buffers and other components, such as heme or pyridoxal groups, absorb strongly in this region.

From: *The Protein Protocols Handbook, 2nd Edition*
Edited by: J. M. Walker © Humana Press Inc., Totowa, NJ

2. Materials

1. 0.1 *M* K_2SO_4 (pH 7.0).
2. 5 m*M* potassium phosphate buffer, pH 7.0.
3. Nonionic detergent (0.01% Brij 35)
4. Guanidinium-HCl.
5. 0.2-µm Millipore (Watford, UK) filter.
6. UV-visible spectrometer: The hydrogen lamp should be selected for maximum intensity at the particular wavelength.
7. Cuvets, quartz, for <215 nm.

3. Methods

3.1. Estimation of Protein by Near UV Absorbance (280 nm)

1. A reliable spectrophotometer is necessary. The protein solution must be diluted in the buffer to a concentration that is well within the accurate range of the instrument (*see* **Notes 1** and **2**).
2. The protein solution to be measured can be in a wide range of buffers, so it is usually no problem to find one that is appropriate for the protein which may already be in a particular buffer required for a purification step or assay for enzyme activity, for example (*see* **Notes 3** and **4**).
3. Measure the absorbance of the protein solution at 280 nm, using quartz cuvets or cuvets that are known to be transparent to this wavelength, filled with a volume of solution sufficient to cover the aperture through which the light beam passes.
4. The value obtained will depend on the path length of the cuvet. If not 1 cm, it must be adjusted by the appropriate factor. The Beer-Lambert law states that:

$$A \text{ (absorbance)} = \varepsilon\, c\, l \tag{1}$$

where ε = extinction coefficient, c = concentration in mol/L and l = optical path length in cm. Therefore, if ε is known, measurement of A gives the concentration directly, ε is normally quoted for a 1-cm path length.

5. The actual value of UV absorbance for a given protein must be determined by some absolute method, e.g., calculated from the amino acid composition, which can be determined by amino acid analysis *(4)*. The UV absorbance for a protein is then calculated according to the following formula:

$$A_{280} \text{ (1 mg/mL)} = (5690n_w + 1280n_y + 120n_c)/M \tag{2}$$

where n_w, n_y, and n_c are the numbers of Trp, Tyr, and Cys residues in the polypeptide of mass M and 5690, 1280 and 120 are the respective extinction coefficients for these residues (*see* **Note 5**).

3.2. Estimation of Protein by Far UV Absorbance

1. The protein solution is diluted with a sodium chloride solution (0.9% w/v) until the absorbance at 215 nm is <1.5 (*see* **Notes 1** and **6**).
2. Alternatively, dilute the sample in another non-UV-absorbing buffer such as 0.1 *M* K_2SO_4, containing 5 m*M* potassium phosphate buffer adjusted to pH 7.0 (*see* **Note 6**).
3. Measure the absorbances at the appropriate wavelengths (either A_{280} and A_{205}, or A_{225} and A_{215}, depending on the formula to be applied), using a spectrometer fitted with a hydrogen lamp that is accurate at these wavelengths, using quartz cuvets filled with a volume of solution sufficient to cover the aperture through which the light beam passes (details in **Subheading 3.1.**).

4. The A_{205} for a 1 mg/mL solution of protein ($A_{205}^{1\ mg/mL}$) can be calculated within ±2%, according to the empirical formula proposed by Scopes *(2)* (*see* **Notes 7–10**):

$$A_{205}^{1\ mg/mL} = 27 + 120\ (A_{280}/A_{205}) \quad (3)$$

5. Alternatively, measurements may be made at longer wavelengths *(5)*:

$$\text{Protein concentration } (\mu g/mL) = 144\ (A_{215} - A_{225}) \quad (4)$$

The extinction at 225 nm is subtracted from that at 215 nm; the difference multiplied by 144 gives the protein concentration in the sample in μg/mL. With a particular protein under specific conditions accurate measurements of concentration to within 5 μg/L are possible.

4. Notes

1. It is best to measure absorbances in the range 0.05–1.0 (between 10 and 90% of the incident radiation). At around 0.3 absorbance (50% absorption), the accuracy is greatest.
2. Bovine serum albumin is frequently used as a protein standard; 1 mg/mL has an A_{280} of 0.66.
3. If the solution is turbid, the apparent A_{280} will be increased by light scattering. Filtration (through a 0.2-μm Millipore filter) or clarification of the solution by centrifugation can be carried out. For turbid solutions, a convenient approximate correction can be applied by subtracting the A_{310} (proteins do not normally absorb at this wavelength unless they contain particular chromophores) from the A_{280}.
4. At low concentrations, protein can be lost from solution by adsorption on the cuvet; the high ionic strength helps to prevent this. Inclusion of a nonionic detergent (0.01% Brij 35) in the buffer may also help to prevent these losses.
5. The presence of nonprotein chromophores (e.g., heme, pyridoxal) can increase A_{280}. If nucleic acids are present (which absorb strongly at 260 nm), the following formula can be applied. This gives an accurate estimate of the protein content by removing the contribution to absorbance by nucleotides at 280 nm, by measuring the A_{260} which is largely owing to the latter *(6)*.

$$\text{Protein (mg/mL)} = 1.55\ A_{280} - 0.76\ A_{260} \quad (5)$$

Other formulae (using similar principles of absorbance differences) employed to determine protein in the possible presence of nucleic acids are the following *(7,8)*:

$$\text{Protein (mg/mL)} = (A_{235} - A_{280})/2.51 \quad (6)$$

$$\text{Protein (mg/mL)} = 0.183\ A_{230} - 0.075.8\ A_{260} \quad (7)$$

6. Protein solutions obey Beer-Lambert's Law at 215 nm provided the absorbance is <2.0.
7. Strictly speaking, this value applies to the protein in 6 *M* guanidinium-HCl, but the value in buffer is generally within 10% of this value, and the relative absorbances in guanidinium-HCl and buffer can be easily determined by parallel dilutions from a stock solution.
8. Sodium chloride, ammonium sulfate, borate, phosphate, and Tris do not interfere, whereas 0.1 *M* acetate, succinate, citrate, phthalate, and barbiturate show high absorption at 215 nm.
9. The absorption of proteins in the range 215–225 nm is practically independent of pH between pH values 4–8.
10. The specific extinction coefficient of a number of proteins and peptides at 205 nm and 210 nm *(3)* has been determined. The average extinction coefficient for a 1 mg/mL solution of 40 serum proteins at 210 nm is 20.5 ± 0.14. At this wavelength, a protein concentration of 2 μg/mL gives $A = 0.04$ *(5)*.

References

1. Kirschenbaum, D. M. (1975) Molar absorptivity and A1%/1 cm values for proteins at selected wavelengths of the ultraviolet and visible regions. *Analyt. Biochem.* **68,** 465–484.
2. Scopes, R. K. (1974) Measurement of protein by spectrometry at 205 nm. *Analyt. Biochem.* **59,** 277–282.
3. Goldfarb, A. R., Saidel, L. J., and Mosovich, E. (1951) The ultraviolet absorption spectra of proteins. *J. Biol. Chem.* **193,** 397–404.
4. Gill, S. C. and von Hippel, P. H. (1989) Calculation of protein extinction coefficients from amino acid sequence data. *Analyt. Biochem.* **182,** 319–326.
5. Waddell, W. J. (1956) A simple UV spectrophotometric method for the determination of protein. *J. Lab. Clin. Med.* **48,** 311–314.
6. Layne, E. (1957) Spectrophotornetric and turbidimetric methods for measuring proteins. *Meth. Enzymol.* **3,** 447–454.
7. Whitaker, J. R. and Granum, P. E. (1980) An absolute method for protein determination based on difference in absorbance at 235 and 280 nm. *Analyt. Biochem.* **109,** 156–159.
8. Kalb, V. F. and Bernlohr, R. W. (1977). A new spectrophotometric assay for protein in cell extracts. *Analyt. Biochem.* **82,** 362–371.

2

The Lowry Method for Protein Quantitation

Jakob H. Waterborg

1. Introduction

The most accurate method of determining protein concentration is probably acid hydrolysis followed by amino acid analysis. Most other methods are sensitive to the amino acid composition of the protein, and absolute concentrations cannot be obtained *(1)*. The procedure of Lowry et al. *(2)* is no exception, but its sensitivity is moderately constant from protein to protein, and it has been so widely used that Lowry protein estimations are a completely acceptable alternative to a rigorous absolute determination in almost all circumstances in which protein mixtures or crude extracts are involved.

The method is based on both the Biuret reaction, in which the peptide bonds of proteins react with copper under alkaline conditions to produce Cu^+, which reacts with the Folin reagent, and the Folin–Ciocalteau reaction, which is poorly understood but in essence phosphomolybdotungstate is reduced to heteropolymolybdenum blue by the copper-catalyzed oxidation of aromatic amino acids. The reactions result in a strong blue color, which depends partly on the tyrosine and tryptophan content. The method is sensitive down to about 0.01 mg of protein/mL, and is best used on solutions with concentrations in the range 0.01–1.0 mg/mL of protein.

2. Materials

1. Complex-forming reagent: Prepare immediately before use by mixing the following stock solutions in the proportion 100:1:1 (by vol), respectively:
   Solution A: 2% (w/v) Na_2CO_3 in distilled water.
   Solution B: 1% (w/v) $CuSO_4 \cdot 5H_2O$ in distilled water.
   Solution C: 2% (w/v) sodium potassium tartrate in distilled water.
2. 2 *N* NaOH.
3. Folin reagent (commercially available): Use at 1 *N* concentration.
4. Standards: Use a stock solution of standard protein (e.g., bovine serum albumin fraction V) containing 2 mg/mL protein in distilled water, stored frozen at –20°C. Prepare standards by diluting the stock solution with distilled water as follows:

Stock solution (µL)	0	2.5	5	12.5	25	50	125	250	500
Water (µL)	500	498	495	488	475	450	375	250	0
Protein conc. (µg/mL)	0	10	20	50	100	200	500	1000	2000

From: *The Protein Protocols Handbook, 2nd Edition*
Edited by: J. M. Walker © Humana Press Inc., Totowa, NJ

3. Method

1. To 0.1 mL of sample or standard (*see* **Notes 1–4**), add 0.1 mL of 2 *N* NaOH. Hydrolyze at 100°C for 10 min in a heating block or boiling water bath.
2. Cool the hydrolysate to room temperature and add 1 mL of freshly mixed complex-forming reagent. Let the solution stand at room temperature for 10 min (*see* **Notes 5** and **6**).
3. Add 0.1 mL of Folin reagent, using a vortex mixer, and let the mixture stand at room temperature for 30–60 min (do not exceed 60 min) (*see* **Note 7**).
4. Read the absorbance at 750 nm if the protein concentration was below 500 µg/mL or at 550 nm if the protein concentration was between 100 and 2000 µg/mL.
5. Plot a standard curve of absorbance as a function of initial protein concentration and use it to determine the unknown protein concentrations (*see* **Notes 8–13**).

4. Notes

1. If the sample is available as a precipitate, then dissolve the precipitate in 2 *N* NaOH and hydrolyze as described in **Subheading 3, step 1**. Carry 0.2-mL aliquots of the hydrolyzate forward to **Subheading 3, step 2**.
2. Whole cells or other complex samples may need pretreatment, as described for the Burton assay for DNA *(3)*. For example, the perchloroacetic acid (PCA)/ethanol precipitate from extraction I may be used directly for the Lowry assay, or the pellets remaining after the PCA hydrolysis step (**Subheading 3, step 3** of the Burton assay) may be used for Lowry. In this latter case, both DNA and protein concentration may be obtained from the same sample.
3. Peterson *(4)* has described a precipitation step that allows the separation of the protein sample from interfering substances and also consequently concentrates the protein sample, allowing the determination of proteins in dilute solution. Peterson's precipitation step is as follows:
   a. Add 0.1 mL of 0.15% deoxycholate to 1.0 mL of protein sample.
   b. Vortex-mix, and stand at room temperature for 10 min.
   c. Add 0.1 mL of 72% trichloroacetic acid (TCA), vortex-mix, and centrifuge at 1000–3000*g* for 30 min.
   d. Decant the supernatant and treat the pellet as described in **Note 1.**
4. Detergents such as sodium dodecyl sulfate (SDS) are often present in protein preparations, added to solubilize membranes or remove interfering substances *(5–7)*. Protein precipitation by TCA may require phosphotungstic acid (PTA) *(6)* for complete protein recovery:
   a. Add 0.2 mL of 30% (w/v) TCA and 6% (w/v) PTA to 1.0 mL of protein sample.
   b. Vortex-mix, and stand at room temperature for 20 min.
   c. Centrifuge at 2000*g* and 4°C for 30 min.
   d. Decant the supernatant completely and treat the pellet as described in **Note 1**.
5. The reaction is very pH dependent, and it is therefore important to maintain the pH between 10 and 10.5. Therefore, take care when analyzing samples that are in strong buffer outside this range.
6. The incubation period is not critical and can vary from 10 min to several hours without affecting the final absorbance.
7. The vortex-mixing step is critical for obtaining reproducible results. The Folin reagent is reactive only for a short time under these alkaline conditions, being unstable in alkali, and great care should therefore be taken to ensure thorough mixing.
8. The assay is not linear at higher concentrations. Ensure that you are analyzing your sample on the linear portion of the calibration curve.

9. A set of standards is needed with each group of assays, preferably in duplicate. Duplicate or triplicate unknowns are recommended.
10. One disadvantage of the Lowry method is the fact that a range of substances interferes with this assay, including buffers, drugs, nucleic acids, and sugars. (The effect of some of these agents is shown in Table 1 in Chapter 3.) In many cases, the effects of these agents can be minimized by diluting them out, assuming that the protein concentration is sufficiently high to still be detected after dilution. When interfering compounds are involved, it is, of course, important to run an appropriate blank. Interference caused by detergents, sucrose, and EDTA can be eliminated by the addition of SDS *(5)* and a precipitation step (*see* **Note 4**).
11. Modifications to this basic assay have been reported that increase the sensitivity of the reaction. If the Folin reagent is added in two portions, vortex-mixing between each addition, a 20% increase in sensitivity is achieved *(8)*. The addition of dithiothreitol 3 min after the addition of the Folin reagent increases the sensitivity by 50% *(9)*.
12. The amount of color produced in this assay by any given protein (or mixture of proteins) is dependent on the amino acid composition of the protein(s) (*see* Introduction). Therefore, two different proteins, each for example at concentrations of 1 mg/mL, can give different color yields in this assay. It must be appreciated, therefore, that using bovine serum albumin (BSA) (or any other protein for that matter) as a standard gives only an approximate measure of the protein concentration. The only time when this method gives an absolute value for protein concentration is when the protein being analyzed is also used to construct the standard curve. The most accurate way to determine the concentration of any protein solution is amino acid analysis.
13. A means of speeding up this assay using raised temperatures *(10)* or a microwave oven (*see* Chapter 5) has been described.

References

1. Sapan, C. V., Lundblad, R. L., and Price, N. C. (1999) Colorimetric protein assay techniques. *Biotechnol. Appl. Biochem.* **29,** 99–108.
2. Lowry, O. H., Rosebrough, N. J., Farr, A. L., and Randall, R. J. (1951) Protein measurement with the Folin phenol reagent. *J. Biol. Chem.* **193,** 265–275.
3. Waterborg, J. H. and Matthews, H. R. (1984) The Burton assay for DNA, in *Methods in Molecular Biology*, Vol. 2: *Nucleic Acids* (Walker, J. M., ed.), Humana Press, Totowa, NJ, pp. 1–3.
4. Peterson, G. L. (1983) Determination of total protein. *Methods Enzymol.* **91,** 95–121.
5. Markwell, M.A.K., Haas, S. M., Tolbert, N. E., and Bieber, L. L. (1981) Protein determination in membrane and lipoprotein samples. *Methods Enzymol.* **72,** 296–303.
6. Yeang, H. Y., Yusof, F., and Abdullah, L. (1998) Protein purification for the Lowry assay: acid precipitation of proteins in the presence of sodium dodecyl sulfate and other biological detergents. *Analyt. Biochem.* **265,** 381–384.
7. Chang, Y. C. (1992) Efficient precipitation and accurate quantitation of detergent-solubilized membrane proteins. *Analyt. Biochem.* **205,** 22–26.
8. Hess, H. H., Lees, M. B., and Derr, J. E. (1978) A linear Lowry-Folin assay for both water-soluble and sodium dodecyl sulfate-solubilized proteins. *Analyt. Biochem.* **85,** 295–300.
9. Larson, E., Howlett, B., and Jagendorf, A. (1986) Artificial reductant enhancement of the Lowry method for protein determination. *Analyt. Biochem.* **155,** 243–248.
10. Shakir, F. K., Audilet, D., Drake, A. J., and Shakir, K. M. (1994) A rapid protein determination by modification of the Lowry procedure. *Analyt. Biochem.* **216,** 232–233.

3

The Bicinchoninic Acid (BCA) Assay for Protein Quantitation

John M. Walker

1. Introduction

The bicinchoninic acid (BCA) assay, first described by Smith et al. *(1)* is similar to the Lowry assay, since it also depends on the conversion of Cu^{2+} to Cu^{+} under alkaline conditions (*see* Chapter 2). The Cu^{+} is then detected by reaction with BCA. The two assays are of similar sensitivity, but since BCA is stable under alkali conditions, this assay has the advantage that it can be carried out as a one-step process compared to the two steps needed in the Lowry assay. The reaction results in the development of an intense purple color with an absorbance maximum at 562 nm. Since the production of Cu^{+} in this assay is a function of protein concentration and incubation time, the protein content of unknown samples may be determined spectrophotometrically by comparison with known protein standards. A further advantage of the BCA assay is that it is generally more tolerant to the presence of compounds that interfere with the Lowry assay. In particular it is not affected by a range of detergents and denaturing agents such as urea and guanidinium chloride, although it is more sensitive to the presence of reducing sugars. Both a standard assay (0.1–1.0 mg protein/mL) and a microassay (0.5–10 μg protein/mL) are described.

2. Materials

2.1. Standard Assay

1. Reagent A: sodium bicinchoninate (0.1 g), $Na_2CO_3 \cdot H_2O$ (2.0 g), sodium tartrate (dihydrate) (0.16 g), NaOH (0.4 g), $NaHCO_3$ (0.95 g), made up to 100 mL. If necessary, adjust the pH to 11.25 with $NaHCO_3$ or NaOH (*see* **Note 1**).
2. Reagent B: $CuSO_4 \cdot 5H_2O$ (0.4 g) in 10 mL of water (*see* **Note 1**).
3. Standard working reagent (SWR): Mix 100 vol of regent A with 2 vol of reagent B. The solution is apple green in color and is stable at room temperature for 1 wk.

2.2. Microassay

1. Reagent A: $Na_2CO_3 \cdot H_2O$ (0.8 g), NaOH (1.6 g), sodium tartrate (dihydrate) (1.6 g), made up to 100 mL with water, and adjusted to pH 11.25 with 10 *M* NaOH.
2. Reagent B: BCA (4.0 g) in 100 mL of water.
3. Reagent C: $CuSO_4 \cdot 5H_2O$ (0.4 g) in 10 mL of water.

From: *The Protein Protocols Handbook, 2nd Edition*
Edited by: J. M. Walker © Humana Press Inc., Totowa, NJ

4. Standard working reagent (SWR): Mix 1 vol of reagent C with 25 vol of reagent B, then add 26 vol of reagent A.

3. Methods

3.1. Standard Assay

1. To a 100-µL aqueous sample containing 10–100 µg protein, add 2 mL of SWR. Incubate at 60°C for 30 min (*see* **Note 2**).
2. Cool the sample to room temperature, then measure the absorbance at 562 nm (*see* **Note 3**).
3. A calibration curve can be constructed using dilutions of a stock 1 mg/mL solution of bovine serum albumin (BSA) (*see* **Note 4**).

3.2. Microassay

1. To 1.0 mL of aqueous protein solution containing 0.5–1.0 µg of protein/mL, add 1 mL of SWR.
2. Incubate at 60°C for 1 h.
3. Cool, and read the absorbance at 562 nm.

4. Notes

1. Reagents A and B are stable indefinitely at room temperature. They may be purchased ready prepared from Pierce, Rockford, IL.
2. The sensitivity of the assay can be increased by incubating the samples longer. Alternatively, if the color is becoming too dark, heating can be stopped earlier. Take care to treat standard samples similarly.
3. Following the heating step, the color developed is stable for at least 1 h.
4. Note, that like the Lowry assay, response to the BCA assay is dependent on the amino acid composition of the protein, and therefore an absolute concentration of protein cannot be determined. The BSA standard curve can only therefore be used to compare the relative protein concentration of similar protein solutions.
5. Some reagents interfere with the BCA assay, but nothing like as many as with the Lowry assay (*see* **Table 1**). The presence of lipids gives excessively high absorbances with this assay *(2)*. Variations produced by buffers with sulfhydryl agents and detergents have been described *(3)*.
6. Since the method relies on the use of Cu^{2+}, the presence of chelating agents such as EDTA will of course severely interfere with the method. However, it may be possible to overcome such problems by diluting the sample as long as the protein concentration remains sufficiently high to be measurable. Similarly, dilution may be a way of coping with any agent that interferes with the assay (*see* **Table 1**). In each case it is of course necesary to run an appropriate control sample to allow for any residual color development. A modification of the assay has been described that overcomes lipid interference when measuring lipoprotein protein content *(4)*.
7. A modification of the BCA assay, utilizing a microwave oven, has been described that allows protein determination in a matter of seconds (*see* Chapter 5).
8. A method has been described for eliminating interfering compounds such as thiols and reducing sugars in this assay. Proteins are bound to nylon membranes and exhaustively washed to remove interfering compounds; then the BCA assay is carried out on the membrane-bound protein *(5)*.

Table 1
Effect of Selected Potential Interfering Compounds[a]

	BCA assay (μg BSA found)		Lowry assay (μg BSA found)	
Sample (50 μg BSA) in the following	Water blank corrected	Interference blank corrected	Water blank corrected	Interference blank corrected
50 μg BSA in water (reference)	50.00	—	50.00	—
0.1 *N* HCl	50.70	50.80	44.20	43.80
0.1 *N* NaOH	49.00	49.40	50.60	50.60
0.2% Sodium azide	51.10	50.90	49.20	49.00
0.02% Sodium azide	51.10	51.00	49.50	49.60
1.0 *M* Sodium chloride	51.30	51.10	50.20	50.10
100 m*M* EDTA (4 Na)	No color		138.50	5.10
50 m*M* EDTA (4 Na)	28.00	29.40	96.70	6.80
10 m*M* EDTA (4 Na)	48.80	49.10	33.60	12.70
50 m*M* EDTA (4 Na), pH 11.25	31.50	32.80	72.30	5.00
4.0 *M* Guanidine HCl	48.30	46.90	Precipitated	
3.0 *M* Urea	51.30	50.10	53.20	45.00
1.0%Triton X-100	50.20	49.80	Precipitated	
1.0% SDS (lauryl)	49.20	48.90	Precipitated	
1.0% Brij 35	51.00	50.90	Precipitated	
1.0% Lubrol	50.70	50.70	Precipitated	
1.0% Chaps	49.90	49.50	Precipitated	
1.0% Chapso	51.80	51.00	Precipitated	
1.0% Octyl glucoside	50.90	50.80	Precipitated	
40.0% Sucrose	55.40	48.70	4.90	28.90
10.0% Sucrose	52.50	50.50	42.90	41.10
1.0% Sucrose	51.30	51.20	48.40	48.10
100 m*M* Glucose	245.00	57.10	68.10	61.70
50 mM Glucose	144.00	47.70	62.70	58.40
10 m*M* Glucose	70.00	49.10	52.60	51.20
0.2 *M* Sorbitol	42.90	37.80	63.70	31.00
0.2 *M* Sorbitol, pH 11.25	40.70	36.20	68.60	26.60
1.0 *M* Glycine	No color		7.30	7.70
1.0 *M* Glycine, pH 11	50.70	48.90	32.50	27.90
0.5 *M* Tris	36.20	32.90	10.20	8.80
0.25 *M* Tris	46.60	44.00	27.90	28.10
0.1 *M* Tris	50.80	49.60	38.90	38.90
0.25 *M* Tris, pH 11.25	52.00	50.30	40.80	40.80
20.0% Ammonium sulfate	5.60	1.20	Precipitated	
10.0% Ammonium sulfate	16.00	12.00	Precipitated	
3.0% Ammonium sulfate	44.90	42.00	21.20	21.40
10.0% Ammonium sulfate, pH 11	48.10	45.20	32.60	32.80
2.0 *M* Sodium acetate, pH 5.5	35.50	34.50	5.40	3.30
0.2 *M* Sodium acetate, pH 5.5	50.80	50.40	47.50	47.60
1.0 *M* Sodium phosphate	37.10	36.20	7.30	5.30
0.1 *M* Sodium phosphate	50.80	50.40	46.60	46.60
0.1 *M* Cesium bicarbonate	49.50	49.70	Precipitated	

[a]Reproduced from **ref. *1*** with permission from Academic Press Inc.

9. A comparison of the BCA, Lowry and Bradford assays for analyzing gylcosylated and non-glycosylated proteins have been made *(6)*. Significant differences wee observed between the assays for non-glycosylated proteins with the BCA assay giving results closest to those from amino acid analysis. Glycosylated proteins were underestimated by the Bradford the method and overestimated by the BCA and Lowry methods. The results suggest a potential interference of protein glycosylation with colorimetric assays.
10. A modification of this assay for analysis complex samples, which involves removing contaminants from the protein precipitate with 1 *M* HCl has been reported *(7)*.

References

1. Smith, P. K., Krohn, R. I., Hermanson, G. T., Mallia, A. K., Gartner, F. H., Provenzano, M. D., Fujimoto, E. K., Goeke, N. M., Olson, B. J., and Klenk, D. C. (1985) Measurement of protein using bicinchoninic acid. *Analyt. Biochem.* **150,** 76–85.
2. Kessler, R. J. and Fanestil, D. D. (1986) Interference by lipids in the determination of protein using bicinchoninic acid. *Analyt. Biochem.* **159,** 138–142.
3. Hill, H. D. and Straka, J. G. (1988) Protein determination using bicinchoninic acid in the presence of sulfhydryl reagents. *Analyt. Biochem.* **170,** 203–208.
4. Morton, R. E. and Evans, T. A. (1992) Modification of the BCA protein assay to eliminate lipid interference in determining lipoprotein protein content. *Analyt. Biochem.* **204,** 332–334.
5. Gates, R. E. (1991) Elimination of interfering substances in the presence of detergent in the bicinchoninic acid protein assay. *Analyt. Biochem.* **196,** 290–295.
6. Fountoulakis, M., Juranville, J. F., and Manneberg, M. (1992) Comparison of the coomassie brilliant blue, bicinchoninic acid and lowry quantitation assays, using nonglycosylated and glycosylated proteins. *J. Biochem. Biophys. Meth.* **24,** 265–274.
7. Schoel, B., Welzel, M., and Kaufmann, S. H. E. (1995) Quantification of protein in dilute and complex samples–modification of the bicinchoninic acid assay. *J. Biochem. Biophys. Meth.* **30,** 199–206.

4

The Bradford Method for Protein Quantitation

Nicholas J. Kruger

1. Introduction

A rapid and accurate method for the estimation of protein concentration is essential in many fields of protein study. An assay originally described by Bradford *(1)* has become the preferred method for quantifying protein in many laboratories. This technique is simpler, faster, and more sensitive than the Lowry method. Moreover, when compared with the Lowry method, it is subject to less interference by common reagents and nonprotein components of biological samples (*see* **Note 1**).

The Bradford assay relies on the binding of the dye Coomassie Blue G250 to protein. Detailed studies indicate that the free dye can exist in four different ionic forms for which the pK_a values are 1.15, 1.82, and 12.4 *(2)*. Of the three charged forms of the dye that predominate in the acidic assay reagent solution, the more cationic red and green forms have absorbance maxima at 470 nm and 650 nm, respectively. In contrast, the more anionic blue form of the dye, which binds to protein, has an absorbance maximum at 590 nm. Thus, the quantity of protein can be estimated by determining the amount of dye in the blue ionic form. This is usually achieved by measuring the absorbance of the solution at 595 nm (*see* **Note 2**).

The dye appears to bind most readily to arginyl and lysyl residues of proteins *(3,4)*. This specificity can lead to variation in the response of the assay to different proteins, which is the main drawback of the method (*see* **Note 3**). The original Bradford assay shows large variation in response between different proteins *(5–7)*. Several modifications to the method have been developed to overcome this problem (*see* **Note 4**). However, these changes generally result in a less robust assay that is often more susceptible to interference by other chemicals. Consequently, the original method devised by Bradford remains the most convenient and widely used formulation. Two types of assay are described here: the standard assay, which is suitable for measuring between 10 and 100 μg of protein, and the microassay, which detects between 1 and 10 μg of protein. The latter, although more sensitive, is also more prone to interference from other compounds because of the greater amount of sample relative to dye reagent in this form of the assay.

From: *The Protein Protocols Handbook, 2nd Edition*
Edited by: J. M. Walker © Humana Press Inc., Totowa, NJ

2. Materials

1. Reagent: The assay reagent is made by dissolving 100 mg of Coomassie Blue G250 in 50 mL of 95% ethanol. The solution is then mixed with 100 mL of 85% phosphoric acid and made up to 1 L with distilled water (*see* **Note 5**).

   The reagent should be filtered through Whatman no. 1 filter paper and then stored in an amber bottle at room temperature. It is stable for several weeks. However, during this time dye may precipitate from solution and so the stored reagent should be filtered before use.
2. Protein standard (*see* **Note 6**). Bovine γ-globulin at a concentration of 1 mg/mL (100 μg/mL for the microassay) in distilled water is used as a stock solution. This should be stored frozen at –20°C. Since the moisture content of solid protein may vary during storage, the precise concentration of protein in the standard solution should be determined from its absorbance at 280 nm. The absorbance of a 1 mg/mL solution of γ-globulin, in a 1-cm light path, is 1.35. The corresponding values for two alternative protein standards, bovine serum albumin and ovalbumin, are 0.66 and 0.75, respectively.
3. Plastic and glassware used in the assay should be absolutely clean and detergent free. Quartz (silica) spectrophotometer cuvettes should not be used, as the dye binds to this material. Traces of dye bound to glassware or plastic can be removed by rinsing with methanol or detergent solution.

3. Methods

3.1. Standard Assay Method

1. Pipet between 10 and 100 μg of protein in 100 μL total volume into a test tube. If the approximate sample concentration is unknown, assay a range of dilutions (1, 1:10, 1:100, 1:1000). Prepare duplicates of each sample.
2. For the calibration curve, pipet duplicate volumes of 10, 20, 40, 60, 80, and 100 μL of 1 mg/mL γ-globulin standard solution into test tubes, and make each up to 100 μL with distilled water. Pipet 100 μL of distilled water into a further tube to provide the reagent blank.
3. Add 5 mL of protein reagent to each tube and mix well by inversion or gentle vortex-mixing. Avoid foaming, which will lead to poor reproducibility.
4. Measure the A_{595} of the samples and standards against the reagent blank between 2 min and 1 h after mixing (*see* **Note 7**). The 100 μg standard should give an A_{595} value of about 0.4. The standard curve is not linear, and the precise absorbance varies depending on the age of the assay reagent. Consequently, it is essential to construct a calibration curve for each set of assays (*see* **Note 8**).

3.2. Microassay Method

This form of the assay is more sensitive to protein. Consequently, it is useful when the amount of the unknown protein is limited (*see also* **Note 9**).

1. Pipet duplicate samples containing between 1 and 10 μg in a total volume of 100 μL into 1.5-mL polyethylene microfuge tubes. If the approximate sample concentration is unknown, assay a range of dilutions (1, 1:10, 1:100, 1:1000).
2. For the calibration curve, pipet duplicate volumes of 10, 20, 40, 60, 80, and 100 μL of 100 μg/mL γ-globulin standard solution into microfuge tubes, and adjust the volume to 100 μL with water. Pipet 100 μL of distilled water into a tube for the reagent blank.
3. Add 1 mL of protein reagent to each tube and mix gently, but thoroughly.

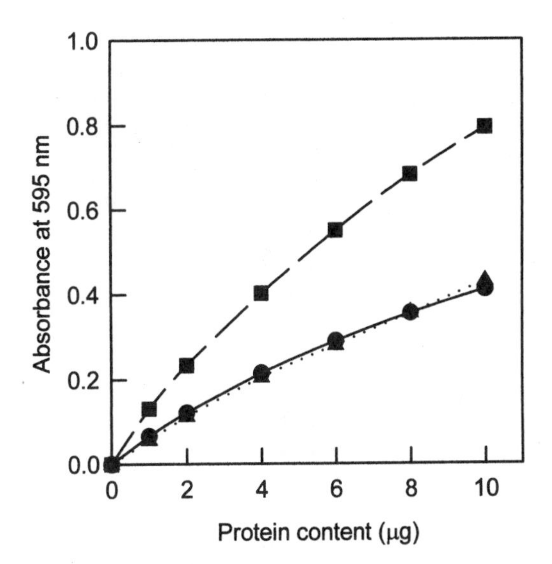

Fig. 1. Variation in the response of proteins in the Bradford assay. The extent of protein–dye complex formation was determined for bovine serum albumin (■), γ-globulin (●), and ovalbumin (▲) using the microassay. Each value is the mean of four determinations. For each set of measurements the standard error was <5% of the mean value. The data allow comparisons to be made between estimates of protein content obtained using these protein standards.

4. Measure the absorbance of each sample between 2 and 60 min after addition of the protein reagent. The A_{595} value of a sample containing 10 µg γ-globulin is 0.45. **Figure 1** shows the response of three common protein standards using the microassay method.

4. Notes

1. The Bradford assay is relatively free from interference by most commonly used biochemical reagents. However, a few chemicals may significantly alter the absorbance of the reagent blank or modify the response of proteins to the dye (**Table 1**). The materials that are most likely to cause problems in biological extracts are detergents and ampholytes *(3,8)*. These can be removed from the sample solution by gel filtration, dialysis, or precipitation of protein with calcium phosphate *(9,10)*. Alternatively, they can be included in the reagent blank and calibration standards at the same concentration as that found in the sample. The presence of base in the assay increases absorbance by shifting the equilibrium of the free dye toward the anionic form. This may present problems when measuring protein content in concentrated basic buffers *(3)*. Guanidine hydrochloride and sodium ascorbate compete with dye for protein, leading to underestimation of the protein content *(3)*.
2. Binding of protein to Coomassie Blue G250 may shift the absorbance maximum of the blue ionic form of the dye from 590 nm to 620 nm *(2)*. It might, therefore, appear more sensible to measure the absorbance at the higher wavelength. However, at the usual pH of the assay, an appreciable proportion of the dye is in the green form (λ_{max} = 650 nm) which interferes with absorbance measurement of the dye–protein complex at 620 nm. Measurement at 595 nm represents the best compromise between maximizing the absorbence due to the dye–protein complex while minimizing that due to the green form of the free dye (*2–4*; but *see also* **Note 9**).

Table 1
Effects of Common Reagents on the Bradford Assay

	Absorbance at 600 nm	
Compound	Blank	5 mg Immunoglobulin
Control	0.005	0.264
0.02% SDS	0.003	0.250
0.1% SDS	0.042a	0.059a
0.1% Triton	0.000	0.278
0.5% Triton	0.051a	0.311a
1 *M* 2-Mercaptoethanol	0.006	0.273
1 *M* Sucrose	0.008	0.261
4 *M* Urea	0.008	0.261
4 *M* NaCl	–0.015	0.207a
Glycerol	0.014	0.238a
0.1 *M* HEPES, pH 7.0	0.003	0.268
0.1 *M* Tris, pH 7.5	–0.008	0.261
0.1 *M* Citrate, pH 5.0	0.015	0.249
10 m*M* EDTA	0.007	0.235a
1 *M* $(NH_4)_2SO_4$	0.002	0.269

Data were obtained by mixing 5 µL of sample with 5 µL of the specified compound before adding 200 µL of dye reagent.

[a]Measurements that differ from the control by more than 0.02 absorbance unit for blank values or more than 10% for the samples containing protein.

Data taken from **ref. 7**.

3. The dye does not bind to free arginine or lysine, or to peptides smaller than about 3000 Da *(4,11)*. Many peptide hormones and other important bioactive peptides fall into the latter category, and the Bradford assay is not suitable for quantifying the amounts of such compounds.
4. The assay technique described here is subject to variation in sensitivity between individual proteins (*see* **Table 2**). Several modifications have been suggested that reduce this variability *(5–7,12)*. Generally, these rely on increasing either the dye content or the pH of the solution. In one variation, adjusting the pH by adding NaOH to the reagent improves the sensitivity of the assay and greatly reduces the variation observed with different proteins *(7)*. (This is presumably caused by an increase the proportion of free dye in the blue form, the ionic species that reacts with protein.) However, the optimum pH is critically dependent on the source and concentration of the dye (*see* ***Note 5***). Moreover, the modified assay is far more sensitive to interference from detergents in the sample.

   Particular care should be taken when measuring the protein content of membrane fractions. The conventional assay consistently underestimates the amount of protein in membrane-rich samples. Pretreatment of the samples with membrane-disrupting agents such as NaOH or detergents may reduce this problem, but the results should be treated with caution *(13)*. A useful alternative is to precipitate protein from the sample using calcium phosphate and remove contaminating lipids (and other interfering substances, *see* **Note 1**) by washing with 80% ethanol *(9,10)*.
5. The amount of soluble dye in Coomassie Blue G250 varies considerably between sources, and suppliers' figures for dye purity are not a reliable estimate of the Coomassie Blue G250 content *(14)*. Generally, Serva Blue G is regarded to have the greatest dye content

Table 2
Comparison of the Response of Different Proteins in the Bradford Assay

	Relative absorbance	
Protein	Assay 1	Assay 2
Myelin basic protein	139	—
Histone	130	175
Cytochrome *c*	128	142
Bovine serum albumin	100	100
Insulin	89	—
Transferrin	82	—
Lysozyme	73	—
α-Chymotrypsinogen	55	—
Soybean trypsin inhibitor	52	23
Ovalbumin	49	23
γ-Globulin	48	55
β-Lactoglobulin A	20	—
Trypsin	18	15
Aprotinin	13	—
Gelatin	—	5
Gramicidin S	5	—

For each protein, the response is expressed relative to that of the same concentration of BSA. The data for assays 1 and 2 are recalculated from **refs.** *5* and *7*, respectively.

and should be used in the modified assays discussed in **Note 4**. However, the quality of the dye is not critical for routine protein determination using the method described in this chapter. The data presented in **Fig. 1** were obtained using Coomassie Brilliant Blue G (C.I. 42655; product code B-0770, Sigma-Aldrich).

6. Whenever possible the protein used to construct the calibration curve should be the same as that being determined. Often this is impractical and the dye response of a sample is quantified relative to that of a "generic" protein. Bovine serum albumin (BSA) is commonly used as the protein standard because it is inexpensive and readily available in a pure form. The major argument for using this protein is that it allows the results to be compared directly with those of the many previous studies that have used bovine serum albumin as a standard. However, it suffers from the disadvantage of exhibiting an unusually large dye response in the Bradford assay, and thus, may underestimate the protein content of a sample. Increasingly, bovine γ-globulin is being promoted as a more suitable general standard, as the dye binding capacity of this protein is closer to the mean of those proteins that have been compared (**Table 2**). Because of the variation in response between different proteins, it is essential to specify the protein standard used when reporting measurements of protein amounts using the Bradford assay.
7. Generally, it is preferable to use a single new disposable polystyrene semimicrocuvette that is discarded after a series of absorbance measurements. Rinse the cuvette with reagent before use, zero the spectrophotometer on the reagent blank and then do not remove the cuvette from the machine. Replace the sample in the cuvette gently using a disposable polyethylene pipet.

8. The standard curve is nonlinear because of problems introduced by depletion of the amount of free dye. These problems can be avoided, and the linearity of the assay improved, by plotting the ratio of absorbances at 595 and 450 nm ***(15)***. If this approach is adopted, the absolute optical density of the free dye and dye–protein complex must be determined by measuring the absorbance of the mixture at each wavelength relative to that of a cuvette containing only water (and no dye reagent). As well as improving the linearity of the calibration curve, taking the ratio of the absorbances at the two wavelengths increases the accuracy and improves the sensitivity of the assay by up to 10-fold ***(15)***.
9. For routine measurement of the protein content of many samples the microassay may be adapted for use with a microplate reader ***(7,16)***. The total volume of the modified assay is limited to 210 µL by reducing the volume of each component. Ensure effective mixing of the assay components by pipetting up to 10 µL of the protein sample into each well before adding 200 µL of the dye reagent. If a wavelength of 595 nm cannot be selected on the microplate reader, absorbance may be measured at any wavelength between 570 nm and 610 nm. However, absorbance measurements at wavelengths other than 595 nm will decrease the sensitivity of response and may increase the minimum detection limit of the protocol.
10. For studies on the use of the Bradford assay in analyzing glycoproteins, *see* **Note 9** in Chapter 3.

References

1. Bradford, M. M. (1976) A rapid and sensitive method for the quantitation of microgram quantities of protein utilizing the principle of protein-dye binding. *Analyt. Biochem.* **72,** 248–254.
2. Chial, H. J., Thompson, H. B., and Splittgerber, A. G. (1993) A spectral study of the charge forms of Coomassie Blue G. *Analyt. Biochem.* **209,** 258–266.
3. Compton, S. J. and Jones, C. G. (1985) Mechanism of dye response and interference in the Bradford protein assay. *Analyt. Biochem.* **151,** 369–374.
4. Congdon, R. W., Muth, G. W., and Splittgerber, A. G. (1993) The binding interaction of Coomassie Blue with proteins. *Analyt. Biochem.* **213,** 407–413.
5. Friendenauer, S. and Berlet, H. H. (1989) Sensitivity and variability of the Bradford protein assay in the presence of detergents. *Analyt. Biochem.* **178,** 263–268.
6. Reade, S. M. and Northcote, D. H. (1981) Minimization of variation in the response to different proteins of the Coomassie Blue G dye-binding assay for protein. *Analyt. Biochem.* **116,** 53–64.
7. Stoscheck, C. M. (1990) Increased uniformity in the response of the Coomassie Blue protein assay to different proteins. *Analyt. Biochem.* **184,** 111–116.
8. Spector, T. (1978) Refinement of the Coomassie Blue method of protein quantitation. A simple and linear spectrophotometric assay for <0.5 to 50 µg of protein. *Analyt. Biochem.* **86,** 142–146.
9. Pande, S. V. and Murthy, M. S. R. (1994) A modified micro-Bradford procedure for elimination of interference from sodium dodecyl sulfate, other detergents, and lipids. *Analyt. Biochem.* **220,** 424–426.
10. Zuo, S.-S. and Lundahl, P. (2000) A micro-Bradford membrane protein assay. *Analyt. Biochem.* **284,** 162–164.
11. Sedmak, J. J. and Grossberg, S. E. (1977) A rapid, sensitive and versatile assay for protein using Coomassie Brilliant Blue G250. *Analyt. Biochem.* **79,** 544–552.
12. Peterson, G. L. (1983) Coomassie blue dye binding protein quantitation method, in *Methods in Enzymology,* vol. 91 (Hirs, C. H. W. and Timasheff, S. N., eds.), Academic Press, New York.

13. Kirazov, L. P., Venkov, L. G. and Kirazov, E. P. (1993) Comparison of the Lowry and the Bradford protein assays as applied for protein estimation of membrane-containing fractions. *Analyt. Biochem.* **208,** 44–48.
14. Wilson, C. M. (1979) Studies and critique of amido black 10B, Coomassie Blue R and Fast Green FCF as stains for proteins after polyacrylamide gel electrophoresis. *Analyt. Biochem.* **96,** 263–278.
15. Zor, T. and Selinger, Z. (1996) Linearization of the Bradford protein assay increases its sensitivity: theoretical and experimental studies. *Analyt. Biochem.* **236,** 302–308.
16. Redinbaugh, M. G. and Campbell, W. H. (1985) Adaptation of the dye-binding protein assay to microtiter plates. *Analyt. Biochem.* **147,** 144–147.

5

Ultrafast Protein Determinations Using Microwave Enhancement

Robert E. Akins and Rocky S. Tuan

1. Introduction

In this chapter, we describe modifications of existing protein assays that take advantage of microwave irradiation to reduce assay incubation times from the standard 15–60 min down to just seconds *(1)*. Adaptations based on two standard protein assays will be described:

1. The classic method of Lowry et al. (**ref. *2*** and *see* Chapter 2), which involves intensification of the biuret reaction through the addition of Folin-Ciocalteau phenol reagent; and
2. The recently developed method of Smith et al. *(3)*, which involves intensification of the biuret assay through the addition of bicinchoninic acid (BCA) (*see* Chapter 3).

Performing incubations in a 2.45–GHz microwave field (i.e., in a microwave oven) for 10–20 s results in rapid, reliable, and reproducible protein determinations. We have provided here background information concerning protein assays in general and the microwave enhanced techniques in particular. The use of microwave exposure as an aid in the preparation of chemical and biological samples is well established; interested readers are directed to Kok and Boon *(4)* for excellent discussions concerning the application of microwave technologies in biological research.

1.1. Microwave Assay

Household microwave ovens expose materials to nonionizing electromagnetic radiation at a frequency of about 2.45 GHz (i.e., 2.45 billion oscillations/s). Such exposures have been applied to a myriad of scientific purposes including tissue fixation, histological staining, immunostaining, PCR, and many others. The practical and theoretical aspects of many microwave techniques are summarized by Kok and Boon *(4)*, and the reader is directed to this reference for excellent discussions concerning general procedures and theoretical background.

Microwave ovens are conceptually simple and remarkably safe devices (*see* **Note 1**). Typically, a magnetron generator produces microwaves that are directed toward the sample chamber by a wave guide. The beam is generally homogenized by a "mode-stirrer," consisting of a reflecting fan with angled blades that scatter the beam as it

From: *The Protein Protocols Handbook, 2nd Edition*
Edited by: J. M. Walker © Humana Press Inc., Totowa, NJ

passes. The side walls of the chamber are made of a microwave reflective material, and the microwaves are thereby contained within the defined volume of the oven (*see* **Note 2**). Specimens irradiated in a microwave oven absorb a portion of the microwave energy depending on specific interactions between the constituent molecules of the sample and the oscillating field.

As microwaves pass through specimens, the molecules in that specimen are exposed to a continuously changing electromagnetic field. This field is often represented as a sine wave with amplitude related to the intensity of the field at a particular point over time and wavelength related to the period of oscillation. Ionically polarized molecules (or dipoles) will align with the imposed electromagnetic field and will tend to rotate as the sequential peaks and troughs of the oscillating "wave" pass. Higher frequencies would, therefore, tend to cause faster molecular rotations. At a point, a given molecule will no longer be able to reorient quickly enough to align with the rapidly changing field, and it will cease spinning. There exists, then, a distinct relationship between microwave frequency and the "molecular-size" of the dipoles that it will affect. This relationship is important, and at the 2.45–GHz frequency used in conventional microwave ovens, only small molecules may be expected to rotate; specifically, water molecules rotate easily in microwave ovens but proteins do not (*see* **Note 3**).

For most microwave oven functions, a portion of the rotational energy of the water molecules in a sample dissipates as heat. Since these molecular rotations occur throughout exposed samples, microwave ovens provide extremely efficient heating, and the effects of microwaves are generally attributed to changes in local temperature. At some as yet undetermined level, however, microwave exposure causes an acceleration in the rate of reaction product formation in the protein assays discussed here. This dramatic acceleration is independent of the change in temperature, and our observations have suggested that microwave-based heating is not the principal means of reaction acceleration.

1.2. Microwave Enhanced Protein Determinations

Modifications of the procedures of Lowry (Chapter 2) and Smith (Chapter 3) to include microwave irradiation result in the generation of linear standard curves. **Figure 1** illustrates typical standard curves for examples of each assay using bovine serum albumin (BSA). Both standard curves are linear across a practical range of protein concentrations.

One interesting difference between the two microwave enhanced assays concerns the relationship between irradiation time and assay sensitivity. In the *DC Protein Assay*, a modification of Lowry's *(2)* procedure supplied by Bio-Rad Laboratories, Inc. (Hercules, CA), illustrated in **Fig. 1**, a colorimetric end point was reached after 10 s of microwave irradiation; no further color development occurred in the samples. This end point was identical to that achieved in a 15-min room temperature control assay.

The *BCA* Protein Assay*, a version of Smith's assay *(3)* supplied by Pierce Chemical Co. (Rockford, IL), afforded some flexibility in assay sensitivity because the formation of detectable reaction product was a function of the duration of microwave exposure. **Figure 2** shows the rate of reaction product formation for three different concentrations of BSA as a function of microwave exposure in a BCA assay. Absorbance values increased for each BSA concentration as a second order function of irradiation time. A linear standard curve could be generated from BSA dilutions that were

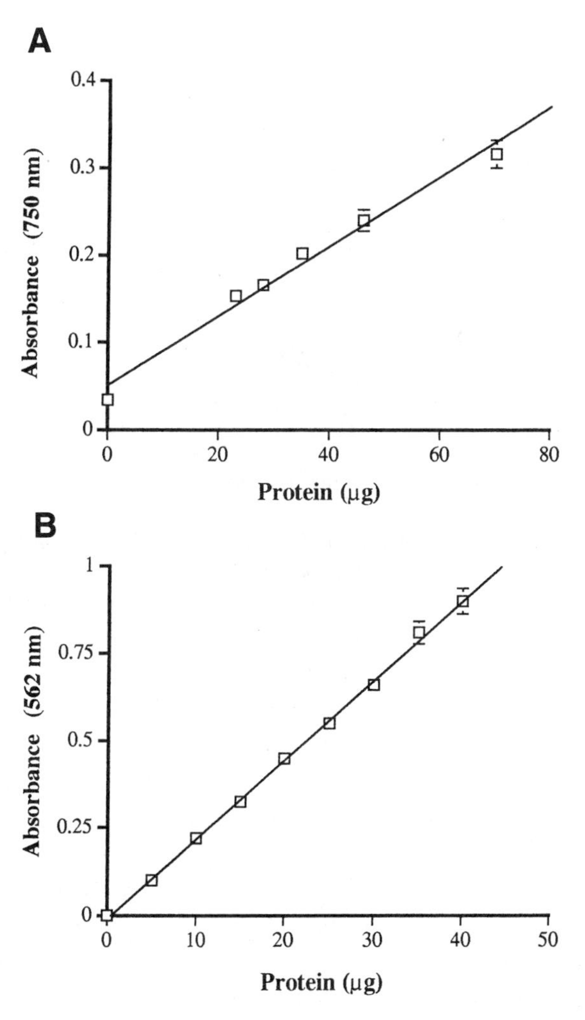

Fig. 1. Typical standard curves for microwave Lowry and BCA assays. Standard curves were generated with microwave protocols using BSA (Sigma) dissolved in water. BSA samples in 100-μL vol were prepared in triplicate for each assay and were combined with reagent as described in **Notes 1** and **2**. Tubes and a water load (total vol 100 mL, *see* **Note 5**) were placed in the center of a microwave oven. Samples for the Lowry assay were irradiated for 10 s; samples for the BCA assay were irradiated for 20 s. Results in **A** show a linear standard curve generated using a microwave Lowry assay. Results in **B** show a linear standard curve generated using a microwave BCA assay. Values presented are means ± SD.

irradiated for any specific time; longer irradiation times yielded more steeply sloped standard curves. In practice, then, the duration of microwave exposure can be selected to correspond to a desired sensitivity range with longer times being more suitable for lower protein concentrations. In contrast to the *DC Protein Assay*, the BCA microwave procedure described here is more sensitive than a standard, room temperature assay, and we have used it for most applications (*see* **Note 4**).

Figure 3 illustrates that the dramatic effects of microwave exposure cannot be mimicked by external heating. These results are surprising since microwave effects are

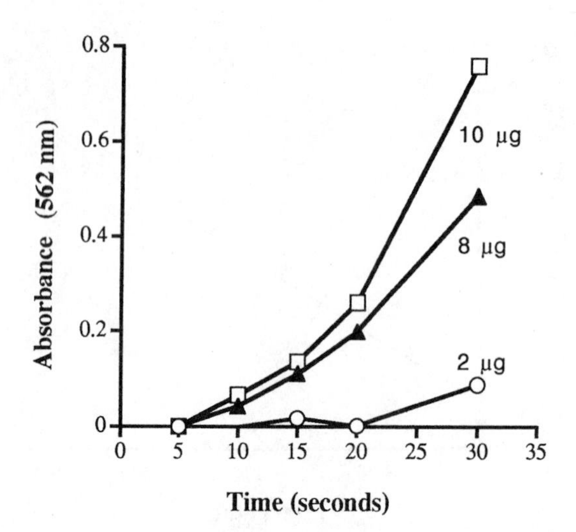

Fig. 2. Effect of Increasing microwave irradiation time on the BCA assay. Three amounts of BSA (Sigma) were prepared in water: 10 µg/tube (□), 8 µg/tube (▲), 2 µg/tube (○). Each time point was determined from triplicate samples in a single irradiation trial with the water load replaced between determinations. Each assay time resulted in the generation of a linear standard curve; the slope of each standard curve increased as a function of irradiation time. Values were normalized to the 5-s time point and presented as means ± SD.

generally attributed to increases in temperature. It is not clear at what level(s) microwaves interact with the biochemical processes involved in protein estimation; however, the acceleration is possibly related to an alteration in solvent/solute interactions. As the solvent water molecules rotate, specific structural changes may occur in the system such that interactions between solvent and solute molecules (or among the solvent molecules themselves) tend to enhance the chemical interactions between the protein and the assay components to accelerate the rate of product formation. For example, water molecules rotating in a microwave field may no longer be available to form hydrogen bonds within the solvent/solute structure. Clearly, the nature and mechanism of nonthermal microwave effects need to be studied further.

2. Materials

2.1. Lowry Assay (see Chapter 2)

Lowry reagents are available from commercial sources. Assay reagents were routinely purchased from Bio-Rad Laboratories in the form of a detergent-compatible Lowry kit (*DC Protein Assay*). Assay reactions are typically carried out in polystyrene Rohren tubes (Sarstedt, Inc., Newtown, NC). Tubes were placed in a plastic test tube rack at the center of a suitable microwave oven *(see the following)* along with a beaker containing approx 100 mL of H_2O (*see* **Note 5**).

2.2. BCA Assay (see Chapter 3)

BCA protein reagent is available from commercial sources. Assay reagents are routinely purchased from Pierce Chemical Co. in the form of a *BCA* Protein Assay* kit. As with the Lowry assay, reactions were typically carried out in polystyrene Rohren

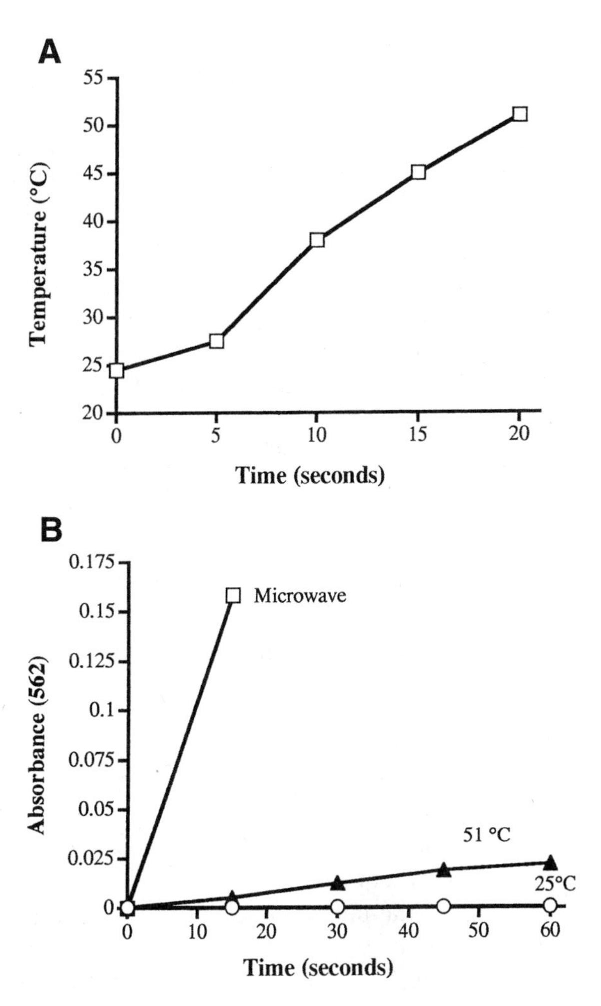

Fig. 3. Comparison of microwave irradiation with incubation at elevated temperature. **(A)** The change in temperature of BCA assay samples containing bovine serum albumin (BSA, Sigma). Temperatures reached 51°C during a typical 20-s irradiation. Since the assays are carried out in an open system, temperatures near 100°C would cause sample boil-over and should be avoided. **(B)** Development of reaction product under three different conditions: microwave irradiation (□), incubation at 51°C (▲), and incubation at 25°C (○). Incubation at 51°C, the maximal temperature reached during a 20-s irradiation, did not mirror microwave irradiation.

tubes (Sarstedt). Tubes were placed in a plastic test tube rack at the center of a suitable microwave oven *(see* **Subheading 3.2.**) along with a beaker containing approx 100 mL of H_2O (*see* **Note 5**).

3. Methods

3.1. Sample Preparation

Sample preparation should be carried out as specified by the manufacturers. Generally, samples are solubilized in a noninterfering buffer (*see* **Note 3**) so that the final protein concentration falls within the desired range (*see* **Note 4**). Samples should be either filtered or centrifuged to remove any debris prior to protein determination.

3.2. Selection of Microwave Oven

Microwave ovens differ substantially in their suitability for these assays. Desirable attributes include fine control of irradiation time, a chamber size large enough to easily accommodate the desired number of samples, and a configuration that results in a homogenous field of irradiation so that all samples within the central volume of the oven receive a uniform microwave dose (*see* **Note 4**). Samples should be placed in a nonmetallic test tube rack in the center of the oven. A volume of room temperature water is included in the oven chamber as well so that the total amount of fluid (samples + additional water) is constant from one assay to another (*see* **Note 5**).

3.3. Sample Irradiation

Once the samples are placed into the center of the microwave chamber, close the door and irradiate the samples. Using the *DC Protein Assay* , a 10-s irradiation was optimal as a replacement for the standard 15–min incubation. Using the *BCA* Protein Assay*, a 20-s irradiation has proved adequate in most situations. We have found it most convenient to use the highest setting on the microwave oven and to control exposure using an accurate timer.

3.4. Reading and Interpreting Assay Results

After irradiation, the absorbance of each sample and standard should be determined spectrophotometrically. Samples from Lowry assays should be read at 750 nm and samples from BCA assays should be read at 562 nm. A standard curve of absorbance values as a function of standard protein concentration can be generated easily and used to determine the protein levels in the unknown samples. It is recommended that standard curves be generated along with each assay to avoid any difficulties that may arise from differences in reagents or alterations in total microwave exposure.

3.5. Summary

In summary, the microwave BCA protein assay protocol is as follows:

1. Combine samples and BSA standards with BCA assay reagent in polystyrene tubes;
2. Place samples into an all plastic test tube rack in the center of a microwave oven along with a beaker containing a volume of room temperature water sufficient to make the total volume of liquid in the chamber 100 mL;
3. Irradiate samples for 20 s on the highest microwave setting; and
4. Measure A_{562} for each sample and determine protein concentrations based on a BSA calibration curve.

The microwave Lowry assay protocol is virtually identical:

1. Combine samples and Lowry assay reagents in polystyrene tubes;
2. Place samples into an all plastic test tube rack in the center of a microwave oven along with a beaker containing a volume of room temperature water sufficient to make the total volume of liquid in the chamber 100 mL;
3. Irradiate samples for 10 s on the highest microwave setting; and
4. Measure A_{750} for each sample and determine protein concentrations based on a BSA calibration curve.

Microwave protein assays are suitable for all situations where standard assays are presently used. The ability to determine accurate protein concentrations in so little time should greatly facilitate routine assays and improve efficiency when protocols require protein determination at multiple intermediate steps. Similar microwave techniques have also been applied as time-saving and efficiency-enhancing procedures by several authors (*see* **ref. *4***). We have used microwave assays to generate chromatograms during protein purification and for general protein determinations (e.g., before electrophoretic analysis). The assays consistently yield reliable results that are comparable to those obtained by standard protocols. The modifications we present here are very easily adapted to most commercially available microwave ovens and, in the case of the BCA assay, can be adjusted to cover a wide range of protein concentrations. Since the duration of the assays is so short, it is possible to try several irradiation times and water loads to select the specific conditions required by the particular microwave and samples to be used. Microwave enhanced protein estimations should prove to be extremely useful in laboratories currently doing standard Smith *(3)* or Lowry *(2)* based protein determinations.

4. Notes

1. Although microwave radiation is nonionizing, precautions should be taken to avoid direct irradiation of parts of the body. Microwaves can penetrate the skin and cause significant tissue damage in relatively short periods of time. Most contemporary microwave ovens are remarkably safe and leakage is unlikely; however, periodic assessment of microwave containment within the microwave chamber should be carried out. Perhaps more dangerous than radiation effects are potential problems caused by the rapid heating of irradiated samples. Care should be taken when removing samples from the microwave chamber to avoid getting burned, sealed containers should not be irradiated as they may explode, and metal objects should be excluded from the chamber to avoid sparking. Users should consult their equipment manuals and institute safety offices prior to using microwave ovens. A detailed discussion of microwave hazards is included in **ref. *4***.
2. Microwave ovens use a nominal frequency of 2.45 GHz. The energy put into the microwave chamber is actually a range of frequencies around 2.45 GHz. As the waves in the oven reflect off of the metal chamber walls, "hot" and "cold" spots may be set up by the constructive and destructive interference of the waves. The positions of these "hot" and "cold" spots is a function of the physical design of the oven chamber and the electrical properties of the materials contained in the oven. The unevenness of microwave fields can be minimized by mode stirring *(see* the preceeding*)*, appropriate oven configuration, or by rotating the specimen in the chamber during irradiation. If the design of a particular microwave oven does not provide a relatively uniform irradiation volume, it may not be useful for the assays outlined here.

   The suitability of a particular microwave oven may be tested easily using a number of tubes with known concentrations of protein. Uneven irradiation patterns will be detected by significant differences in the color development for a given protein concentration as a function of position within the oven chamber. We have had success with several microwave ovens including a 0.8 cu ft, 600W, General Electric oven, model JEM18F001 and a 1.3 cu ft, 650W, Whirlpool oven, model RJM7450.
3. The frequency used in household microwave ovens is actually below the resonance frequency for water. Above the 2.45 GHz used in household microwave ovens, water molecules are capable of rotating faster and of absorbing substantially more microwave

energy. Too much absorption, however, is undesirable. It is possible that the outer layers of a sample may absorb energy so efficiently that the interior portions receive substantially less energy. The resulting uneven exposure may have adverse effects.

It is important to note that molecules other than water may absorb microwave energy. If a compound added to a microwave enhanced assay absorbs strongly near 2.45 GHz, uneven sample exposures may result because of the presence of the compound. In addition, compounds that are degraded or converted during a microwave procedure, or compounds that have altered interactions with other assay components during irradiation, may substantially affect assay results. Although it may be possible to predict which materials would interfere with a given assay by considering the relevant chemical and electrical characteristics of the constituent compounds, potential interference is most easily assessed empirically by directly determining the effects of a given additive on the accuracy and sensitivity of the standard microwave assay.

4. Assay sensitivity may be improved by increasing irradiation time. The time required may be determined quickly by using BSA test solutions in the range of protein concentrations expected until desirable A_{562} values were obtained. By increasing microwave exposure time, it is possible to substantially increase assay sensitivity while keeping irradiation times below 60 s. The ease with which sensitivity may be adjusted within extremely short time frames places the microwave BCA assay among the quickest, most flexible assays available for protein determinations.
5. The addition of a volume of water to the microwave chamber, such that the total volume contained in the microwave chamber is constant from one assay to the next, allows irradiation conditions to be controlled easily from assay to assay. The additional water acts as a load on the oven and absorbs some of the microwave energy. Since the total amount of water remains constant from one assay to the next, the amount of energy absorbed also remains constant. The time of irradiation, therefore, becomes independent of the number of samples included in the assay.

Acknowledgments

This work was supported in part by funds from NASA-SBRA and AHA 9406244S (REA), as well as NIH HD 15822 and NIH HD 29937 (RST).

References

1. Akins, R. E. and Tuan, R. S. (1992) Measurement of protein in 20 seconds using a microwave BCA assay. *Biotechniques* **12,** 496–499.
2. Lowry, O. H., Rosebrough, N. J., Farr, A. L., and Randall, R. J. (1951) Protein measurement with the Folin phenol reagent. *J. Biol. Chem.* **193,** 265–275.
3. Smith, P. K., Krohn, R. I., Hermanson, G. T., Mallia, A. K., Gartner, F. H., Provenzano, M. D., Fujimoto, E. K., Goeke, N. M., Olson, B. J., and Klenk, D. C. (1985) Measurement of protein using bicinchoninic acid. *Analyt. Biochem.* **150,** 76–85.
4. Kok, L. P. and Boon, M. E. (1992) *Microwave Cookbook for Microscopists.* Coulomb, Leyden, Leiden, Netherlands.

6

The Nitric Acid Method for Protein Estimation in Biological Samples

Scott A. Boerner, Yean Kit Lee, Scott H. Kaufmann, and Keith C. Bible

1. Introduction

1.1. Background

The quantitation of protein in biological samples is of great importance and utility in many research laboratories. Protein measurements are widely utilized to ensure equal loading of samples on sodium dodecyl sulfide (SDS)-polyacrylamide gels and to provide a basis for comparison of enzyme activities and other analytes.

Several methods are commonly employed for the determination of protein content in biological samples, including the measurement of absorbance at 260 *(1)* or 205 nm (*2*, and Chapter 1); the method of Lowry et al. (*3*, and Chapter 2), which relies on the generation of a new chromophore on reaction of an alkaline protein hydrolysate with phosphotungstic-phosphomolybdic acid in the presence of Cu^{2+} *(4)*; the method of Smith et al. (*5*, and Chapter 3), which relies upon the reduction of Cu^{2+} to Cu^{+}, which then forms a colored complex with bicinchoninic acid (BCA); the method of Bradford (*6*, and Chapter 4), which relies on a change in absorbance on binding of Coomassie Blue to basic and aromatic amino acids under acidic conditions *(7)*; and the method of Böhlen et al. *(8)*, which relies on the reaction of fluorescamine with primary amines to generate a fluorescent product. Unfortunately, compounds that are constituents of biological buffers sometimes interfere with these methods, limiting their application.

Nucleic acids, neutral detergents of the polyoxyethylenephenol class, SDS, and urea interfere with spectrophotometric determinations at 260 and 205 nm *(9)*. Several components commonly used in biological buffers, including tris-hydroxymethyl-amino methane (Tris), N-2-hydroxyethylpiperazine-N-2-ethanesulfonic [acid] (HEPES), ethylenediaminetetraacetic acid (EDTA), neutral detergents, and reducing agents, interfere with the method of Lowry et al. *(10)*. In addition, Tris, ammonium sulfate, EDTA, and reducing agents interfere with the BCA method, although neutral detergents and SDS do not *(5)*. Finally, many commonly used reagents, including neutral detergents and SDS, interfere with the method of Bradford *(6)*. Details of compounds that interfere with these assays can be found in **Table 1**, Chapter 3.

From: *The Protein Protocols Handbook, 2nd Edition*
Edited by: J. M. Walker © Humana Press Inc., Totowa, NJ

Several approaches have been proposed to circumvent the aforementioned difficulties. Protein precipitation with deoxycholate and trichloroacetic acid will eliminate many interfering substances ***(10)***, but the inclusion of this step is laborious when multiple samples are being analyzed simultaneously. Alternatively, samples can be lysed in a buffer that is compatible with a particular assay, then diluted into a second buffer after analysis. For example, cells lysed in Triton X-100 can be assayed for protein and then diluted into SDS sample buffer in preparation for SDS-polyacrylamide gel electrophoresis (SDS-PAGE). This approach, however, can result in spurious results, as illustrated by the demonstration that caspase-3 activation in lysates of mitogen-stimulated lymphocytes reflects proteolytic cleavage occurring in the cell lysate rather than in the cells prior to lysis ***(11,12)***. Finally, protein can be estimated by solubilizing samples directly in SDS sample buffer, spotting them onto a solid support (e.g., glass or nitrocellulose), washing away interfering substances, reacting the immobilized polypeptides with Coomassie Blue, and eluting the Coomassie Blue to estimate the protein content ***(13,14)***, but this process that is potentially time and labor intensive.

As a result of these shortcomings, there is still a need for alternative methods of protein determination in biological samples. The nitric acid method reviewed in this chapter represents one such newly devised method ***(15)***. Studies dating to the 1800s have demonstrated that aromatic molecules undergo nitration when treated with nitric acid. In particular, treatment of tyrosine with nitric acid produces 3-nitrotyrosine ***(16)***, which is distinguished from the parent compound by the appearance of a new absorbance peak at 358 nm. The method detailed in the following subheading utilizes this chemistry to create a one-step method for protein determination in biological samples.

1.2. Nitric Acid Method for Protein Determination

1.2.1. Technique Derivation

Our laboratory has frequently examined the cellular accumulation of cytotoxic platinum analogues ***(17,18)***. Because of concern about the efflux of platinum analogues from cells during the course of manipulation (e.g., trypsinization and centrifugation), we sought a method for the assessment of protein content after cells were washed *in situ* and then solubilized directly in nitric acid, one of the matrices used for platinum determination.

During the course of solublizing cells in nitric acid in preparation for platinum determinations, we noted a brownish color change that appeared to vary in intensity with the quantity of cells treated. Examination of the absorption spectrum of these nitric acid lysates demonstrated a peak at 358 nm (**Fig. 1A**, inset) that was not present in cells solubilized in SDS sample buffer or in nitric acid itself. The absorbance at 358 nm increased linearly with cell number; comparison of two human leukemia cell lines, HL-60 (diameter 11 μm) and K562 (diameter 15 μm), revealed that the slope was twofold higher in the larger cell line (**Fig. 1A**). We hypothesized that the new absorbance peak at 358 nm reflected a reaction of nitric acid with one or more amino acids. While nitric acid solubilization of L-tyrosine produced a sharp peak at 358 nm and similar treatment of L-tryptophan produced a shoulder at the same wavelength, other amino acids, including L-serine, L-arginine, L-phenylalanine, L-threonine, L-alanine, L-proline, L-valine, L-histidine, L-methionine, L-isoleucine, L-glutamic acid, L-aspartic

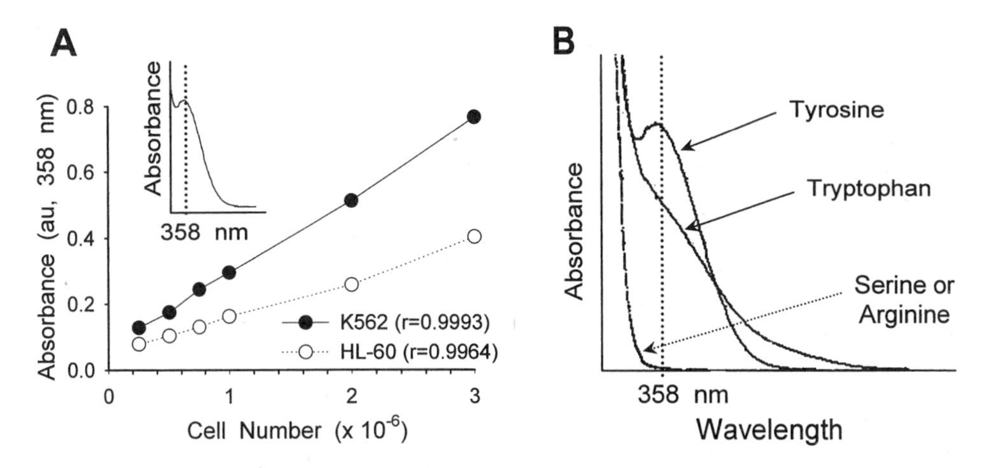

Fig. 1. Treatment of cells or L-tyrosine with nitric acid results in an increase in absorbance at 358 nm. **(A)** Relationship between cell number and absorbance at 358 nm in HL-60 and K562 cells solubilized in nitric acid for 24 h. *Inset*, absorption spectrum resulting from treatment of OV202 cells with nitric acid. **(B)** Absorption spectra resulting from treatment of L-tyrosine, L-tryptophan, L-serine, or L-arginine with nitric acid for 1 h. Amino acid concentrations in nitric acid were all 0.1 mg/mL.

acid and L-asparagine failed to yield products that absorbed at 358 nm (**Fig. 1B** and data not shown). Nucleotides (ATP, CTP, GTP, TTP, UTP), RNA, and DNA also failed to yield products that absorbed at 358 nm. These investigations led us to propose that the 358 nm absorption peak observed after nitric acid solubilization of cells likely primarily reflects the tyrosine content of cellular proteins.

To understand better the chemistry responsible for these observations, we solublized L-tyrosine in nitric acid for 1 h. A single reaction product that co-migrated with authentic 3-nitrotyrosine by high performance liquid chromatography (HPLC) and had an absorbance maximum at 358 nm was formed (**Fig. 2**). Incubation for prolonged periods of time resulted in disappearance of the 3-nitrotyrosine peak and appearance of additional peaks that are thought to reflect nitration at additional sites as well as oxidation (data not shown). As a consequence, the absorbance at 358 nm reached a peak between 1 and 6 h, and then subsequently declined.

To determine whether these observations could be utilized to devise a method for quantitating protein, bovine serum albumin (BSA) was solubilized in nitric acid and incubated for varying lengths of time. As indicated in **Fig. 3A**, absorbance at 358 nm measured after incubation for 0.5–70 h was a linear function of protein content. Examination of these data revealed that production of the chromophore was approx 88% complete within 30 min, reached a maximum at approx 24 h, and remained stable thereafter (**Fig. 3A**, inset). We hypothesize that the stability of the 358 nm peak in proteins treated for prolonged periods with nitric acid (in contrast to the disappearance of the same peak when L-tyrosine is subjected to prolonged nitric acid treatment) results from diminished susceptibility of tyrosyl residues in protein to reaction beyond 3-nitration.

To determine whether this approach could be utilized to quantitate protein under conditions compatible with SDS-PAGE, aliquots of BSA in SDS sample buffer

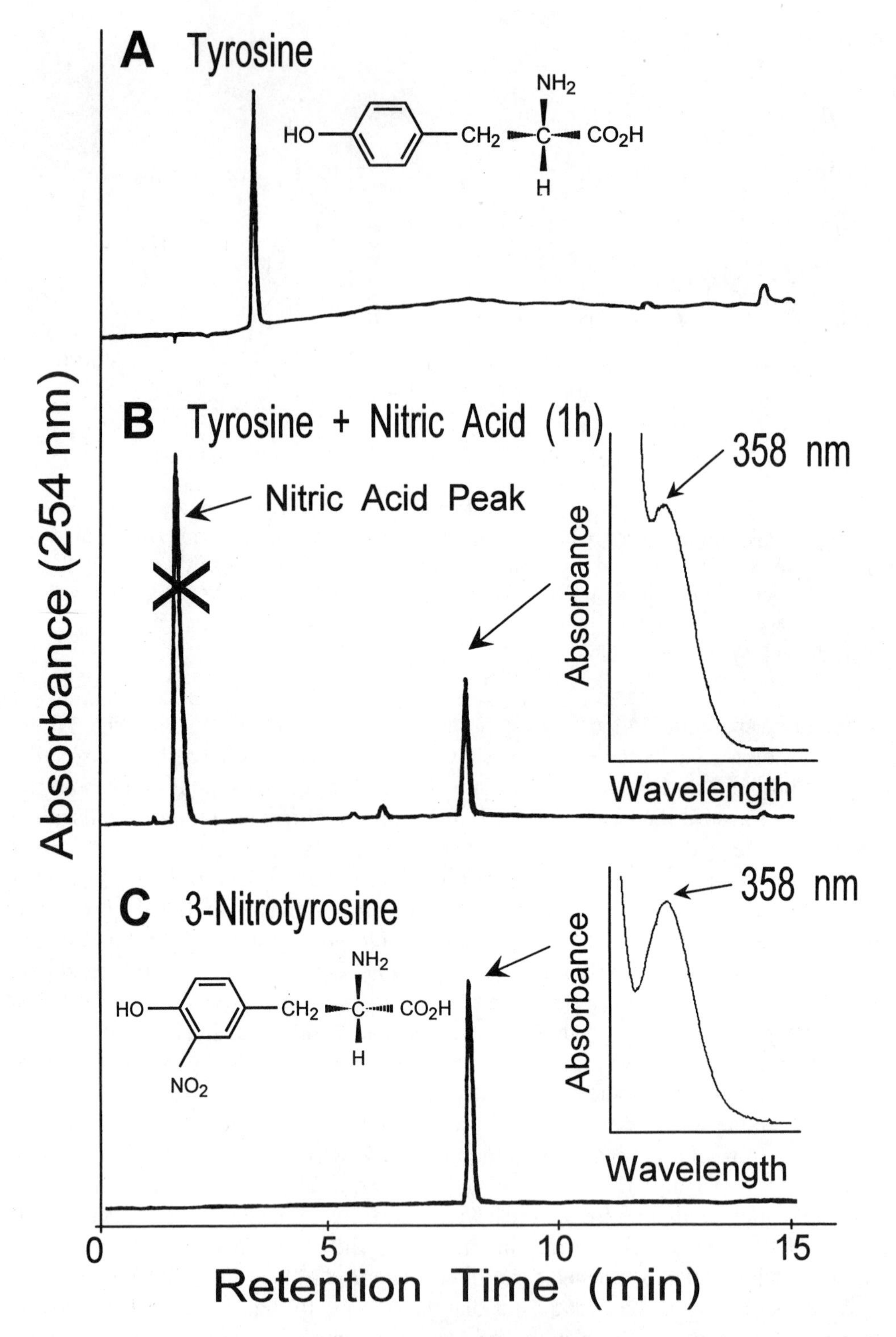

Fig. 2. 3-Nitrotyrosine results from the treatment of L-tyrosine with nitric acid. **(A)** HPLC chromatogram of L-tyrosine. **(B)** HPLC chromatogram of L-tyrosine reacted with nitric acid for 1 h. *Inset*, absorption spectrum of peak eluting at ~8 min. The prominent peak eluting at 2 min corresponds to a contaminant in the nitric acid. **(C)** HPLC chromatogram of 3-nitrotyrosine. *Inset*, absorption spectrum of peak eluting at ~8 min. HPLC was accomplished using a Beckman Ultrasphere obs column (4.6 × 250mm, 5 μm) as previously described ***(15)***.

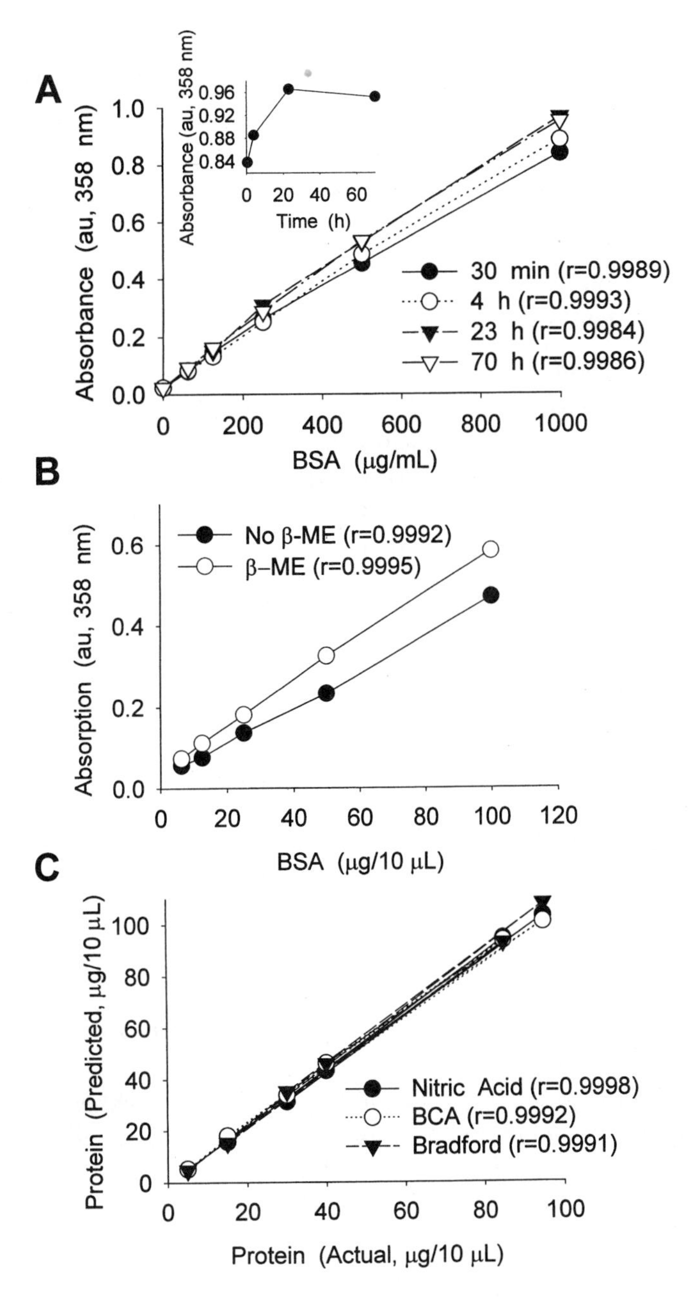

Fig. 3. Measurement of absorbance at 358 nm can be used to quantitate protein levels in samples treated with nitric acid. **(A)** Absorbance (at 358 nm) of varying concentrations of BSA incubated in nitric acid for 30 min, 4 h, 23 h, and 70 h. Inset, time-dependent increase of absorbance at 358 nm for a 1000 μg/mL solution of BSA dissolved in nitric acid. **(B)** Relationship between BSA concentration and absorbance at 358 nm for samples dissolved in electrophoresis sample buffer with or without 5% (v/v) β-mercaptoethanol. **(C)** Relationship between actual and measured BSA concentration in six blindly submitted BSA samples assessed by the nitric acid, BCA, and Bradford protein assays. Results in **B** and **C** represent the means of results from the assay of triplicate samples. Error bars (hidden by data points) represent ± 1 sample standard deviation.

containing Tris, SDS, urea, and β-mercaptoethanol were treated with nitric acid. As indicated in **Fig. 3B**, the presence of increasing amounts of protein resulted in increasing absorbance at 358 nm, which was again a linear function of protein content. To confirm that the present assay was useful for measuring polypeptides other than BSA, we assessed absorbance after reaction of five other polypeptides with nitric acid including bovine insulin, human fibrinogen, human fibronectin, chicken ovalbumin, and rat albumin. Each of these polypeptides reacted with nitric acid to produce species with absorption maxima at 358 nm. In each case, there was a strong correlation ($r > 0.99$) between protein content and absorbance (data not shown).

1.2.2. Interfering Substances

Urea, SDS, β-mercaptoethanol, 20% glycerol, and 50% saturated ammonium sulfate did not interfere with protein determination by the nitric acid method. In addition, the neutral detergents Tween 20 (1%), Brij 97 (1%), *n*-octyl α-D-glucopyranoside, and digitonin (10 μ*M*) did not interfere with this method. In contrast, the presence of 1% Triton X-100 (a phenol derivative) or trace amounts of phenol produced strong absorbance at 358 nm that interfered with the assay. Although 1% 3-([3-cholamidopropyl]dimethylammonio)-1-propanesulfonate (CHAPS) increased absorbance at 358 nm, it did not interfere with the assay as long as the protein samples used to generate standard curves were also solubilized in CHAPS.

1.2.3. Comparison of the Nitric Acid Method to Bradford and BCA Protein Assays

To assess the reliability of this technique, one of us blindly measured protein content in six coded "unknown samples" using the Bradford assay (Coomassie Blue binding), reaction with BCA, and the nitric acid method. Results of these experiments indicated that the three methods produce comparable results in assessing protein content in unknown samples (**Fig. 3C**). Interestingly, the correlation between actual and measured (predicted) protein content was slightly higher for the nitric acid method ($r = 0.9998$) than for the BCA assay ($r = 0.9992$) or Bradford assay ($r = 0.9991$).

1.2.4. Utility of Nitric Acid Method in Loading Extracts of Different Cell Types for SDS-PAGE

Over the course the past 3 yr, we have had considerable experience in using the nitric acid method to facilitate the equal loading of SDS-PAGE gels with cell extracts solublized in electrophoresis sample buffer. Most of our experience has involved using the method to facilitate equal loading of samples derived from single cell lines on a given gel (with respect to protein content).

When loading a single gel with samples derived from multiple different cell lines, however, we noted that the nitric acid method frequently gave the appearance of unequal loading as assessed by Western blotting for proteins typically used as loading controls (e.g., histone H1, β-actin). This observation led us to specifically examine whether the nitric acid method is suited to comparing protein levels between different cell lines. In assessing this possibility, we ran SDS-PAGE gels with extracts from five different human cancer cell lines loaded in accord with the amounts of protein in the extracts indicated by either the BCA or the nitric acid method. When we compared Western blots for a number of different proteins, we found that both protein assays facilitated similar gel loading (**Fig. 4**). Hence, the appearance of unequal gel loading

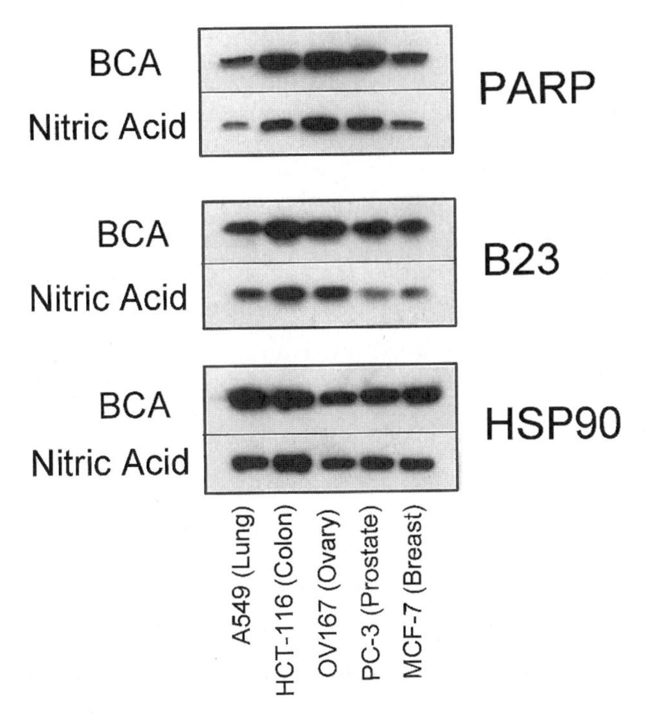

Fig. 4. BCA and nitric acid protein assays produce equivalent results when used as the basis for loading SDS-PAGE gels with extracts from different cell lines. Extracts from five human cancer cell lines (each containing 50 μg of total cellular proteins, as assessed by the BCA or the nitric acid method) were subjected to SDS-PAGE and probed for PARP, B23, and HSP90 via Western blotting. All shown pairs of BCA/nitric acid blots were derived from the same gel and have been aligned to facilitate comparison.

noted above is not, in fact, attributable to unequal gel loading, but instead attributable to differential expression of the probed proteins across the different cell lines. It should be noted, however, that the nitric acid method uniformly resulted in proportionately higher protein values (and hence proportionately less cell extract loaded per gel lane) in comparison to the BCA method.

2. Materials

2.1. Reagents/Equipment

All reagents are ACS grade unless otherwise indicated.

1. Nitric acid (J.T. Baker, Phillipsburg, NJ).
2. BSA, bovine insulin, human fibronectin, human fibrinogen, rat albumin, chicken ovalbumin (Sigma, St. Louis, MO).
3. Tris (Bio-Rad, Richmond, CA).
4. Triton X-100, Tween 20, Brij 97, *n*-octyl-glucoside, digitonin, glycerol, 3-nitrotyrosine, and various amino acids (Sigma, St. Louis, MO).
5. Urea (Aldrich, Madison, WI).
6. Electrophoresis grade SDS and β-mercaptoethanol (Bio-Rad, Richmond, CA).
7. Diode array UV-vis spectrophotometer (Model DU7400, Beckman, Palo Alto, CA).

2.2. Nitric Acid Protein Determination Method

1. SDS sample buffer: 2% (w/v) SDS; 4 *M* urea (prepared as an 8 *M* stock and deionized over Bio-Rad AG501-X8 mixed bed resin prior to use); 62.5 m*M* Tris-HCl, pH 6.8 at 20–22°C, and 1 m*M* EDTA with or without 5% (v/v) β-mercaptoethanol.
2. ACS reagent grade concentrated (70%) nitric acid.
3. BSA.

3. Methods

3.1. Nitric acid Protein Determination Method

The following is the recommended procedure for determining protein content in samples solubilized in SDS sample buffer (*see* **Notes 1** and **2**):

1. Remove an aliquot containing 5–100 µg of protein. Bring sample volume to 10 µL.
2. Add 140 µL 70% nitric acid.
3. Incubate at 20–22°C for 2 h (*see* **Note 3**).
4. Utilizing suitable microcuvettes, measure absorbance at 358 nm with H_2O as a blank (*see* **Note 4**).
5. Compare results to a standard curve constructed with varying amounts of BSA in 10 µL SDS sample buffer subjected to the same conditions (*see* **Notes 5** and **6**).

3.2. Nitric Acid Protein Determination Method Modification One: Whole Cells

Simple modifications allowed this procedure to be applied to other settings. When protein content in whole cells is to be determined:

1. Wash cells with serum-free buffer (*see* **Note 1**).
2. Solublize cells directly in 70% nitric acid.
3. Incubate at 22°C for 24 h (*see* **Note 3**).
4. Determine absorbance at 358 nm (dilution in 70% nitric acid may be necessary; *see* **Note 4**).
5. Compare results to a standard curve constructed with varying amounts of BSA subjected to the same assay conditions (*see* **Notes 5** and **6**).

3.3. Nitric Acid Protein Determination Method Modification Two: Perchloric or Trichloric Acid Precipitates

When perchloric acid or trichloric acetic acid precipitates are assayed for protein content (*see* **Note 1**):

1. Solubilize precipitates directly in nitric acid.
2. Incubate for 2 h at 22°C (*see* **Note 3**).
3. Determine absorbance at 358 nm (dilution in 70% nitric acid may be necessary; *see* **Note 4**).
4. Compare results to a standard curve constructed with varying amounts of BSA subjected to the same assay conditions (*see* **Notes 5** and **6**).

4. Notes

1. Prior to assay for protein, cells must be washed in serum-free medium or physiological buffer to remove (exogenous) protein and cross-reacting substances contained in culture media.
2. In contrast to existing assays, the nitric acid method is not adversely affected by urea, reducing agents, and most nonionic and ionic detergents. As a consequence, it is possible to solubilize samples directly in SDS sample buffer, remove an aliquot for protein deter-

mination, and then adjust the sample volume based on the result of this assay. One advantage of the nitric acid method, therefore, is that protein content can be assayed on samples solublized in SDS-PAGE sample buffer immediately prior to application of the samples to electrophoresis gels.

3. The assay of protein content in detergent-solubilized cells by the nitric acid method requires an incubation period of only 2 h at 22°C. In contrast, similar determinations on nitric acid-solubilized cells require incubation periods of at least 24 h for optimal detection of differences between different cell lines. Longer incubations may be required for nitric acid-solubilized cells because longer times may be required to complete protein nitration under these conditions.
4. Our experience indicates that water, rather than 70% nitric acid, should be used for a "blank" (or in the reference cell) for all spectrophotometric determinations.
5. Proteins with differing tyrosine/amino acid content would be expected to yield different slopes when utilized as standards for this assay. Although this can present a problem in experiments that require the precise quantitation of a single polypeptide species, it does not appear to be a limitation when this method is used to compare relative protein content in complex biological mixtures. Moreover, when the nitric acid assay was applied to a variety of polypeptides (including BSA, rat albumin, chicken ovalbumin, bovine insulin, human fibrinogen, and human fibronectin), the resulting slopes did not correlate with either tyrosine or tryptophan content of the polypeptides (data not shown), suggesting that tertiary structure or other factors might also influence the extent of nitration during nitric acid treatment. Similar differences in reactivity of various purified polypeptides have been previously observed with other protein estimation methods as well ***(5,6)***.
6. The nitric acid method has a limit of sensitivity of approx 5 μg when performed using 100-μL cuvettes. This sensitivity is slightly lower than the methods of Lowry et al. ***(3)***, Smith et al. ***(5)*** and Bradford ***(6)*** performed under similar semimicro conditions.

Acknowledgments

Preparation of this chapter was supported in part by grants from the NIH (CA 67818 and CA 69008), the American Cancer Society (RPG CCE 9918201), and the Jack Taylor Family Foundation. The secretarial assistance of Deb Strauss is gratefully acknowledged.

References

1. Warburg, O. and Christian, W. (1942) Isolierung und Kristallisation des Garungsferments Enolase. *Biochemi. Z.* **310,** 384–421.
2. Waddell, W. J. (1956) A simple ultraviolet spectrophotometric method for the determination of protein. *J. Clin. Lab. Med.* **48,** 311–314.
3. Lowry, O. H., Rosebrough, N. J., Farr, A. L., and Randall, R. J. (1951) Protein measurement with the Folin phenol reagent. *J. Biol. Chem.* **193,** 265–275.
4. Herbert, D., Phipps, P. J., and Strange, R. E. (1971) Determination of protein. *Methods Microbiol.* **5B,** 242–265.
5. Smith, P. K., Krohn, R. I., Hermanson, G. T., Mallia, A. K., Gartner, F. H., Provenzano, M. D., et al. (1985) Measurement of protein using bicinchoninic acid. *Analyt. Biochem.* **150,** 76–85.
6. Bradford, M. M. (1976) A rapid and sensitive method for the quantitation of microgram quantities of protein utilizing the principle of protein-dye binding. *Analyt. Biochem.* **72,** 248–254.
7. Compton, S. J. and Jones, C. G. (1985) Mechanism of dye response and interference in the Bradford Protein assay. *Analyt. Biochem.* **151,** 369–374.

8. Böhlen, P., Stein, S., Dairman, W., and Udenfriend, S. (1973) Fluorometric assay of proteins in the nanogram range. *Arch. Biochem. Biophys.* **155,** 213–220.
9. Simonian, M. H. and Smith, J. A. (1996) Quantitation of Proteins, *Current Protocols in Molecular Biology*, John Wiley & Sons, New York, Chanda, V. B. (ed.). Chapter 10, Suppl. 35, pp. 10.1.1–10.1.10.
10. Peterson, G. L. (1979) Review of the Folin phenol protein quantitation method of Lowry, Rosebrough, Farr and Randall. *Analyt. Biochem.* **100,** 201–220.
11. Miossec, C., Dutilleul, V., Fassy, F., and Diu-Hercend, A. (1997) Evidence for CPP32 activation in the absence of apoptosis during T lymphocyte stimulation. *J. Biol. Chem.* **272,** 13,459–13,462.
12. Zapata, J. M., Takahashi, R., Salvesen, G. S., and Reed, J. C. (1998) Granzyme release and caspase activation in activated human T-lymphocytes. *J. Biol. Chem.* **273,** 6916–6920.
13. McKnight, G. S. (1977) A colorimetric method for the determination of submicrogram quantities of protein. *Analyt. Biochem.* **78,** 86–92.
14. Winterbourne, D. J. (1993) Chemical assays for proteins: *Methods in molecular biology*, In: Biomembrane Protocols I., Graham, J. M. and Higgins, J. A. (eds.). Humana Press, Totowa, NJ, *Isolation and Analysis*, Vol. 19, pp. 197–202.
15. Bible, K. C., Boerner, S. A., and Kaufmann, S. H. (1999) A one-step method for protein estimation in biological samples: nitration of tyrosine in nitric acid. *Analyt. Biochem.* **267,** 217–221.
16. Waser, E. and Lewandowski, M. (1921) Untersuchungen in der Phenylalanin-Reihe I. Synthese des 1-3,4-Dioxy-phenylalanins. *Helv. Chim. Acta* **4,** 657–666.
17. Budihardjo, I. I., Walker, D. L., Svingen, P. A., Buckwalter, C. A., Desnoyers, S., Eckdahl, S., et al. (1998) 6-Aminonicotinamide sensitizes human tumor cell lines to cisplatin. *Clin. Cancer Res.* **4,** 117–130.
18. Bible, K. C., Boerner, S. A., Kirkland, K., Anderl, K. L., Bartelt, D., Jr., Svingen, P. A., et al. (2000) Characterization of an ovarian carcinoma cell line resistant to cisplatin and flavopiridol. *Clin. Cancer Res.* **6,** 661–670.

7

Quantitation of Tryptophan in Proteins

Alastair Aitken and Michèle Learmonth

1. Introduction

1.1 Hydrolysis Followed by Amino Acid Analysis

Accurate measurement of the amount of tryptophan in a sample is problematic, as it is completely destroyed under normal conditions employed for the complete hydrolysis of proteins. Strong acid is ordinarily the method of choice, and constant boiling 6 *M* hydrochloric acid is most frequently used. The reaction is usually carried out in evacuated sealed tubes or under nitrogen at 110°C for 18–96 h. Under these conditions, peptide bonds are quantitatively hydrolyzed (although relatively long periods are required for the complete hydrolysis of valine, leucine, and isoleucine). As well as complete destruction of tryptophan, small losses of serine and threonine occur, for which corrections are made. The advantages of amino acid analysis include the measurement of absolute amounts of protein, provided that the sample is not contaminated by other proteins. However, it may be a disadvantage if an automated amino acid analyzer is not readily available.

Acid hydrolysis in the presence of 6 *N* HCl, containing 0.5–6% (v/v) thioglycolic acid at 110°C for 24–72 h *in vacuo* will result in greatly improved tryptophan yields ***(1)***, although most commonly, hydrolysis in the presence of the acids described in **Subheading 3.1.** may result in almost quantitative recovery of tryptophan.

Alkaline hydrolysis followed by amino acid analysis is also used for the estimation of tryptophan. The complete hydrolysis of proteins is achieved with 2–4 *M* sodium hydroxide at 100°C for 4–8 h. This is of limited application for routine analysis, because cysteine, serine, threonine, and arginine are destroyed in the process and partial destruction by deamination of asparagine and glutamine to aspartic and glutamic acids occurs.

The complete enzymatic hydrolysis of proteins (where tryptophan would be quantitatively recovered) is difficult, because most enzymes attack only specific peptide bonds rapidly. Often a combination of enzymes is employed (such as Pronase), and extended time periods are required (*see* Chapter 82). A further complication of this method is possible contamination resulting from autodigestion of the enzymes.

From: *The Protein Protocols Handbook, 2nd Edition*
Edited by: J. M. Walker © Humana Press Inc., Totowa, NJ

1.2. Measurement of tryptophan content by UV

The method of Goodwin and Morton *(2)* is described in **Subheading 3.3.**

The absorption of protein solutions in the UV is due to tryptophan and tyrosine (and to a very minor, and negligible, extent phenylalanine and cysteine). The absorption maximum will depend on the pH of the solution, and spectrophotometric measurements are usually made in alkaline solutions. Absorption curves for tryptophan and tyrosine show that at the points of intersection, 257 and 294 nm, the extinction values are proportional to the total tryptophan + tyrosine content. Measurements are normally made at 294.4 nm, as this is close to the maximum in the tyrosine curve (where $\Delta\varepsilon/\Delta\lambda$, the change in extinction with wavelength, is minimal), and in conjunction with the extinction at 280 nm (where $\Delta\varepsilon/\Delta\lambda$ is minimal for tryptophan) the concentrations of each of the two amino acids may be calculated (*see* **Notes 1** and **2**).

2. Materials

1. 3 *M* *p*-toluenesulfonic acid.
2. 0.2% 3-(2-Aminoethyl) indole.
3. 3 *M* Mercaptoethanesulfonic acid (Pierce-Warriner).
4. 1 *M* NaOH.

3. Methods

3.1. Quantitation of Tryptophan by Acid Hydrolysis

1. To the protein dried in a Pyrex glass tube (1.2 × 6 cm or similar, in which a constriction has been made by heating in an oxygen/gas flame) is added 1 mL of 3 *M* *p*-toluenesulfonic acid, containing 0.2% tryptamine [0.2% 3-(2-aminoethyl) indole] *(3)*.
2. The solution is sealed under vacuum and heated in an oven for 24–72h at 110°C, *in vacuo*.
3. Alternatively, the acid used may be 3 *M* mercaptoethanesulfonic acid. The sample is hydrolyzed for a similar time and temperature *(4)*.
4. The tube is allowed to cool, and cracked open with a heated glass rod held against a horizontal scratch made in the side of the tube.
5. The acid is taken to near neutrality by carefully adding 2 mL of 1 *M* NaOH. An aliquot of the solution (which is still acid) is mixed with the amino acid analyzer loading buffer.
6. Following this hydrolysis, quantitative analysis is carried out for each of the amino acids on a suitable automated instrument.

3.2. Alkaline Hydrolysis

1. To the protein dried in a Pyrex glass tube (as described in **Subheading 3.1., step 1**), 0.5mL of 3 *M* sodium hydroxide is added (*See* **Notes 3**).
2. The solution is sealed under vacuum and heated in an oven for 4–8 h at 100°C, *in vacuo*.
3. After cooling and cracking open, *while one is wearing safety goggles*, the alkali is neutralized carefully with an equivalent amount of 1 *M* HCl. An aliquot of the solution is mixed with the amino acid analyzer loading buffer and analyzed (as described in **Subheading 3.1., step 6**).

3.3. Measurement of Tryptophan Content by UV

1. The protein is made 0.1 *M* in NaOH.
2. Measure the absorbance at 294.4 nm and 280 nm in cuvettes transparent to this wavelength (i.e. quartz) in a spectrometer (*see* **Note 4**).

3. The amount of tryptophan (w) is estimated from the relative absorbances at these wavelengths by the method of Goodwin and Morton *(2)* shown in **Eq. 1**.

where

x = total mole/L; w = tryptophan mole/L; $(x - w)$ = tyrosine mole/L.

e_y = Molar extinction of tyrosine in 0.1 M alkali at 280 nm = 1576.

e_w = Molar extinction of tryptophan in 0.1 M alkali at 280 nm = 5225.

Also, x is measured from $E_{294.4}$ (the molar extinction at this wavelength). This is 2375 for both Tyr and Trp (since their absorption curves intersect at this wavelength). An accurate reading of absorbance at one other wavelength is then sufficient to determine the relative amounts of these amino acids.

$$E_{280} = w\,\varepsilon_w + (x - {}_w)\varepsilon_y \quad (1)$$

Therefore,

$$w = (E_{280} - x\,\varepsilon_y)/(\varepsilon_w - \varepsilon_y)$$

4. An alternative method of obtaining the ratios of Tyr and Trp is to use the formulae (**Eq. 2**) derived by Beaven and Holiday *(5)*.

$$M_{Tyr} = (0.592\,K_{294} - 0.263\,K_{280}) \times 10^{-3}$$

$$= (0.263\,K_{280} - 0.170\,K_{294}) \times 10^{-3}$$

where M_{Tyr} and M_{Trp} are the moles of tyrosine and tryptophan in 1/ g of protein, and K_{294} and K_{280} are the extinction coefficients of the protein in 0.1 N alkali at 294 and 280 nm. Extinction values can be substituted for the K values to give the molar ratio of tyrosine to tryptophan according to the formula:

$$M_{Tyr} / M_{Trp} = (0.592\,E_{294} - 0.263\,E_{280})/(0.263\,E_{280} - 0.170\,E_{294}) \quad (2)$$

4. Notes

1. The extinction of nucleic acid in the 280 nm region may be as much as 10 times that of protein at the same wavelength and hence a few percent of nucleic acid can greatly influence the absorption.
2. In this analysis, the tyrosine estimate may be high and that of tryptophan low. If amino acid analysis indicates absence of tyrosine, tryptophan is more accurately determined at its maximum, 280.5 nm.
3. Absorption by most proteins in 0.1 M NaOH solution decreases at longer wavelengths into the region 330–450nm where tyrosine and tryptophan do not absorb. Suitable blanks for 294 and 280 nm are therefore obtained by measuring extinctions at 320 and 360 nm and extrapolating back to 294 and 280 nm.
4. In proteins, in a peptide bond, the maximum of the free amino acids is shifted by 1–3 nm to a longer wavelength and pure peptides containing tyrosine and tryptophan residues are better standards than the free amino acids. A source of error may be due to turbidity in the solution and if a protein shows a tendency to denature, it is advisable to treat with a low amount of proteolytic enzyme to obtain a clear solution.

References

1. Matsubara, H. and Sasaki, R. M. (1969) High recovery of tryptophan from acid hydrolysates of proteins. *Biochem. Biophys. Res. Commun.* **35,** 175–181.
2. Goodwin, T. W. and Morton, R. A. (1946) The spectrophotometric determination of tyrosine and tryptophan in proteins. *Biochem. J.* **40,** 628–632.
3. Liu, T.-Y. and Chang, Y. H. (1971) Hydrolysis of proteins with *p*-toluenesulphonic acid. *J. Biol. Chem.* **246,** 2842–2848.
4. Penke, B., Ferenczi, R., and Kovacs, K. (1974) A new acid hydrolyis method for determining tryptophan in peptides and proteins. *Analyt. Biochem.* **60,** 45–50.
5. Beaven, G. H. and Holiday, E. R. (1952) Utraviolet absorption spectra of proteins and amino acids. *Adv. Protein Chem.* **7,** 319.

8

Flow Cytometric Quantitation of Cellular Proteins

Thomas D. Friedrich, F. Andrew Ray, Judith A. Laffin, and John M. Lehman

1. Introduction

Quantitation of specific proteins in complex mixtures is simplified by the use of antibodies directed against the protein of interest. If the specific protein is differentially expressed within a population of cells, quantitation of the protein in cell lysates by immunoblotting will provide an average quantity of the protein per cell. As a result, when comparing lysates of different cell populations, large changes in the amount of a protein in a small percentage of cells cannot be distinguished from small changes in the amount of the protein in a large percentage of cells. Flow cytometric analysis solves this problem by providing a means to measure the amount of a protein within each individual cell in a population. One restriction is that a specific fluorescently labeled probe, usually an antibody, is required for detection of the protein. Cells that are reacted with the antibody are then passed through a flow cytometer where a single cell suspension is focused into a stream that intersects a laser beam. As each cell passes through the beam, the fluorescent probes in each cell are excited and photomultiplier tubes register the degree of fluorescence *(1)*.

Multivariate analysis, the simultaneous measurement of multiple parameters, is a powerful feature of flow cytometry. When determining the quantity of a specific protein in a cell, the quantity of other molecules in the same cell can be found by using fluorescent dyes or specific probes conjugated with fluors of different colors. Examples of second molecules measured simultaneously include DNA *(2–4)* and additional proteins *(5)*. In addition, measurement of light scatter can provide information about cell size and granularity. Most commercially available flow cytometers can measure at least two colors and light scatter, but specialized flow cytometers that are capable of measuring as many as nine colors have been described *(6)*.

In this chapter we describe techniques used to quantitate Simian Virus 40 (SV40) large T antigen (T Ag) in infected cells. By staining infected cells with antibodies against T Ag and with the DNA-binding dye propidium iodide it is possible to measure the amount of T Ag as a function of DNA content.

From: *The Protein Protocols Handbook, 2nd Edition*
Edited by: J. M. Walker © Humana Press Inc., Totowa, NJ

2. Materials

2.1. Cells and Antibodies

The CV-1 line of African Green monkey kidney cells was obtained from the American Type Culture Collection (ATTC, CCL-70). Confluent cultures of CV-1 were infected with SV40 strain RH-911 at 100 plaque-forming units per cell.

Pab101 is a monoclonal antibody specific to the carboxy (C)-terminus of SV40 T Ag. Hybridoma cells producing Pab101 were obtained from ATCC (no. TIB-117). Fluorescein isothiocyanate (FITC)-conjugated secondary antibodies were obtained from Antibodies Inc. (Davis, CA). Alexa conjugated secondary antibodies were obtained from Molecular Probes (Eugene, OR).

2.2. Solutions for Fixation and Staining

1. Ca^{2+}/Mg^{2+}-free phosphate-buffered saline (PBS). Dissolve 8.0 g of NaCl, 0.2 g of KCl, 1.2 g of Na_2HPO_4 and 0.2 g of KH_2PO_4 in 1 L of distilled water (dH_2O). Adjust the pH to 7.4, sterilize by passing through a 0.2-μm filter and store at room temperature.
2. Trypsin-ethylenediaminetetraacetic acid(EDTA): Dilute 10X trypsin/EDTA (Gibco, Rockville, MD) in PBS to give final concentrations of 0.05% trypsin, 0.53 m*M* EDTA. Filter-sterilize and store at 4°C.
3. Methanol: Store at –20°C.
4. Wash solution (WS): Heat inactivate 100 mL of normal goat serum (Gibco) at 56°C for 1 h. Mix with 900 mL of PBS, 200 μL of Triton X-100, and 1.0 g of sodium azide. Filter-sterilize and store at 4°C.
5. Propidium iodide (PI): Dissolve 1.0 mg of PI (Calbiochem) in 100 mL of PBS. Add 20 μL of Triton X-100 and 0.1 g of sodium azide. Protect solution from light and store at 4°C.
6. Ribonuclease: Dissolve 100 mg of RNase A (Sigma) in 100 mL of PBS. Boil for 1 h. Add 20 μL of Triton X-100 and 0.1 g of sodium azide. Store at 4°C.

3. Methods

3.1. Fixation

1. Remove culture media and rinse cells with PBS. If floating and/or mitotic cells are of interest, media and wash fractions should be saved, pelleted, and pooled with attached cells after trypsinization.
2. Prewarm trypsin/EDTA to 37°C and add 1 mL per 60-mm dish. (Adjust proportionally for larger culture areas). Tilt plate to thoroughly distribute solution and then remove.
3. Once cells have detached (*see* **Note 1**), add 1 mL of WS to plate and transfer cells to a 1.5-mL microcentrifuge tube. Remove a small volume for determination of cell number (*see* **Note 2**). Centrifuge remaining cells at 2000*g* for 15 s.
4. Discard supernatant, resuspend pellet in 1 mL of cold PBS, and centrifuge as in **step 3**.
5. Discard supernatant and thoroughly resuspend cell pellet in 0.1 mL of cold PBS.
6. Immediately add 0.9 mL of methanol (–20°C) (*see* **Notes 3**, **4**), mix, and store at –20°C (*see* **Note 5**).

3.2. Titration of Antibodies

It is important that the primary antibody (*see* **Note 6**) be present in excess to ensure quantitative measurement of the specific protein (*see* **Note 7**). Yet, the background of nonspecific antibody staining should be kept to a minimum. To determine the appropriate antibody concentration, stain parallel samples of cells with serial dilutions of antibody. If similar cells not expressing the protein of interest are available they can be stained at the same dilutions to determine levels of nonspecific staining.

3.3. Staining

1. Transfer 1.5×10^6 cells to a 1.5-mL microcentrifuge tube and centrifuge at 2000*g* for 15 s (*see* **Note 8**).
2. Discard supernatant and resuspend the cell pellet in 1 mL of cold PBS. Repeat centrifugation step.
3. Discard PBS and add 0.5-mL of diluted primary antibody. Mix gently.
4. Incubate at 4°C with gentle agitation (*see* **Note 9**).
5. Pellet at 4000*g* for 15 s and discard supernatant.
6. Resuspend pellet in 0.5 mL of WS and repeat **step 5**.
7. Resuspend pellet in 0.5-mL of diluted fluorescently labeled secondary antibody (*see* **Note 10**). Mix gently.
8. Incubate at 37°C for 2 h with gentle mixing; minimize exposure to light.
9. Add 0.5 mL of RNase and mix gently.
10. Incubate at 37°C for 30 min.
11. Add 0.5 mL of PI, bringing the total vol to 1 mL. Mix gently.
12. Filter each sample through a 53-μm mesh nylon grid (Nitex HC3-53, Tetko, Elmsford, NY) (*see* **Note 11**)

3.4. Flow Cytometry and Analysis

Most commercially available flow cytometers are capable of multiparameter analysis (*see* **Note 12**). Using 20 mW of power tuned to 488 nm, minimize the coefficient of variation (CV) for the green photomultiplier (*see* **Note 13**) to <2.0% by aligning the instrument with 2.0-μm fluorescent microspheres (Polysciences, Washington, PA). Minimize the CV for red fluorescence (*see* **Note 14**) to <6.0% with PI-stained lymphocytes (*see* **Note 15**).

Set data acquisition to trigger on a photomultiplier collecting unfiltered 90° light scatter. Set gates to collect data representing single cells, and to eliminate data representing cell debris and cell aggregates. Compare red fluorescence to light scatter and set gate 1 to eliminate subcellular debris (low red fluorescence) and clumped cells (excessive light scatter). Then compare the peak height and area of red fluorescence and eliminate doublets (off axis owing to biphasic peak height). Collect and display red fluorescence vs green fluorescence. Collect data from at least 10,000 cells. Include appropriate control samples, save data and analyze. Representative data are shown in **Fig. 1**. For absolute quantitation (number of molecules of the specific protein per cell), separate populations of cells with a known absolute quantities of the antigen can be used as standards (*see* **Note 16**).

4. Notes

1. It is best to monitor cell detachment microscopically. Cells should be treated to optimize the generation of a single-cell suspension.
2. Each sample should contain $1–1.5 \times 10^6$ cells, because cells will be lost during fixation and staining. Use polypropylene tubes and minimize tube size to reduce cell loss.
3. The cell concentration can be adjusted at the time of fixation by adding the appropriate amounts of PBS (10%) and then methanol (90%). The method of fixation can dramatically alter the results depending on the protein of interest.

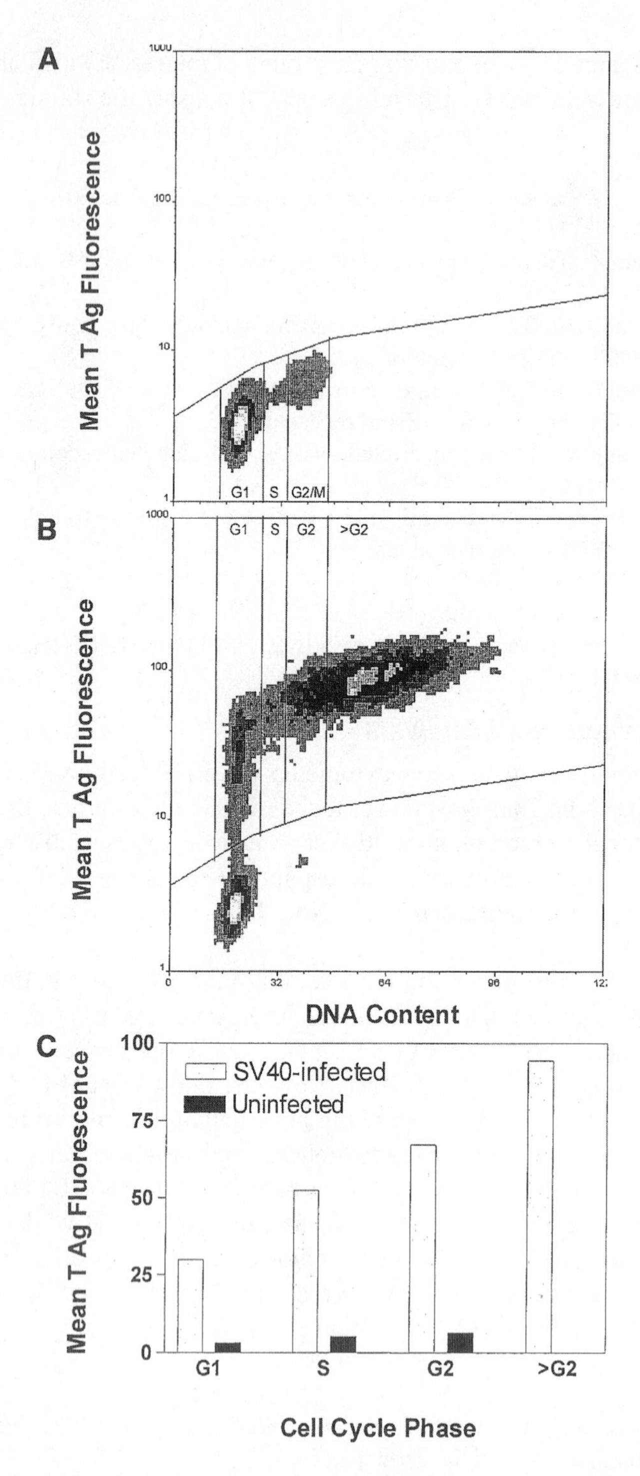

Fig. 1. Differential T antigen expression in SV40-infected monkey kidney cells. Cells in **A** and **B** were fixed and stained with propidium iodide (PI) and anti-T Ag as described in **Subheading 3.** PI staining (DNA content) is expressed on a linear scale and anti-T Ag fluorescence on a log scale. **(A)** An uninfected culture of confluent CV-1 cells. The cell cycle distribution, as determined by PI staining, was: 80% G_1 phase, 5% S phase, 12% G_2/M phase, and 3% >G_2 phase. **(B)** Confluent CV-1 cells infected with SV40 at a multiplicity of infection of 100 plaque-Fig. 1.

4. In some cases methanol fixation may result in loss of the protein antigen. In this case alternative fixation procedures should be compared. Fixation by 0.5% formaldehyde followed by permeabilization with 0.1% Triton X-100 has also been used.
5. Fixed cells have been stored at –20°C for >1 yr with negligible loss of T Ag. However, stability during storage must be evaluated for each antigen.
6. Monoclonal antibodies or affinity-purified polyclonal antibodies are preferred. Polyclonal antisera may have additional antibodies of unknown specificity that will increase background staining.
7. Antibody binding is actually a quantitative measure of the epitope recognized by the antibody. If this epitope is masked through protein conformation/association or as an artifact of fixation the epitope may not be detected.
8. Use of a swinging bucket microcentrifuge reduces cells loss.
9. Adjust time and temperature to maximize the signal over background. Incubation times generally range from 1 or 2 h to overnight. Gentle rocking is recommended.
10. Primary antibodies that are directly conjugated to fluors are commercially available. In addition, kits that allow conjugation of Alexa fluors to antibodies can be obtained from Molecular Probes. If particularly weak signals are encountered, the signal can amplified by additional layers of fluorescently tagged antibody. Cells stained with an FITC-labeled primary or secondary antibody are reacted with Alexa 488-labeled rabbit anti-fluorescein followed by Alexa 488-labeled goat anti-rabbit IgG.
11. Filtering of samples removes cell aggregates that may clog the tubing in the flow cytometer. If samples are stored and then reanalyzed, filtering should be repeated.
12. We use a Cytofluorograph Model 50 H-H with an air-cooled argon laser (model 532, Omnichrome, Chino, CA). The Cyclops analysis program (Cytomation, Fort Collins, CO) was used for data analysis.
13. Use a 535-nm band pass filter for this photomultiplier.
14. Use a 640-nm long pass filter for this photomultiplier.
15. Prepare a stock of spleen cells from a healthy mouse; fix and store in 90% methanol/10% PBS at –20°C. Prior to each run wash and stain 1×10^6 cells with PI only.
16. Flow cytometric standards can be developed following the procedure of Frisa et al. *(7)*. In brief, cell lines known to express different quantities of the protein are counted and divided into two portions. One is fixed for flow cytometry. The other is lysed and co-run on SDS-PAGE with known amounts purified protein in neighboring lanes. The immunoblot of the gel and the fixed cells are stained with the same antibody. On the immunoblot, comparison of the lysate band with the standards allows calculation of the average amount of protein per cell. This value can then be paired with mean fluorescence of the population to calculate a value for fluorescent units per protein molecule.

Fig. 1. (*continued*) forming units per cell. Cells were trypsinized and fixed at 48 h post-infection. The *sloped horizontal line* in each panel was set to divide T Ag positive from T Ag negative cells. *Vertical lines* indicate gates that were set on the basis of DNA content to discriminate cells within the different phases of the cell cycle. The cell cycle distribution of the T Ag expressing cells was: 27% G_1 phase, 9% S phase, 14% G_2/M phase, and 50% >G_2 phase. Levels of T Ag expression in G_1 phase cells cover a 10-fold range, but only cells expressing higher amounts of T Ag enter S phase. The cells in >G_2 phase are infected cells that are replicating viral DNA and re-replicating cellular DNA *(8)*. The T Ag negative G_1 phase cells serve as an internal negative control and have the same anti-T Ag fluorescence as the uninfected G_1 phase cells in **A**. **(C)** The relative quantities of T Ag within each cell cycle gate, as determined by anti-T Ag staining. The bar graph shows the average quantity of T Ag per cell in the T Ag expressing population.

References

1. Shapiro, H. M. (1995) *Practical Flow Cytometry.* Wiley-Liss, New York.
2. Jacobberger, J. W., Fogleman, D., and Lehman, J. M. (1986) Analysis of intracellular antigens by flow cytometry. *Cytometry* **7,** 356–364
3. Laffin, J. and Lehman, J. M. (1994) Detection of intracellular virus and virus products. *Meth. Cell Biol.* **41,** 543–557.
4. Ray, F. A., Friedrich, T. D., Laffin, J. and Lehman, J. M. (1996) Protein Quantitation using flow cytometry, in *The Protein Protocols Handbook* (Walker, J. M., ed.), Humana Press, Totowa, NJ, pp. 33–38.
5. Stewart, C. C. and Stewart, S. J. (1994) Multiparameter analysis of leukocytes by flow cytometry. *Meth. Cell Biol.* **41,** 61–79.
6. Bigos, M., Baumgarth, N., Jager, G. C., Herman, O. C., Nozaki, T., Stovel, R. T., et al. (1999) Nine color eleven parameter immunophenotyping using three laser flow cytometry. *Cytometry* **36,** 36–45.
7. Frisa, P. S., Lanford, R. E., and Jacobberger, J. W. (2000) Molecular quantification of cell cycle-related gene expression at the protein level. *Cytometry* **39,** 79–89.
8. Lehman, J. M., Laffin, J., and Friedrich, T. D. (2000) Simian virus 40 induces multiple S phases with the majority of viral DNA replication in the G2 and second S phase in CV-1 cells. *Exp. Cell Res.* **258,** 215–222.

9

Kinetic Silver Staining of Proteins

Douglas D. Root and Kuan Wang

1. Introduction

Silver staining methods have long been know to provide highly sensitive detection of proteins and nucleic acids following electrophoresis in agarose and polyacrylamide gels *(1,2)*. Silver staining technologies can be extended to other media such as blots, thin-layer chromatography (TLC), and microtiter plates. The quantification of proteins adsorbed to microtiter plate wells provides quantitative information for enzyme-linked immunosorbent assay (ELISA) and protein interaction assays. One nonradioactive procedure, copper iodide staining, is described in Chapter 51 *(3)*. The kinetic silver staining method for measuring the amount of adsorbed protein in a microtiter plate has been developed recently. The microtiter plate assay has a sensitivity similar to copper iodide staining (5–150 ng/well) but higher precision (<5%; *4*). When quantification is based on the time required for staining to reach a fixed optical density, very little protein-to-protein variation is observed, so a standard protein for calibration can be selected (such as bovine albumin) that is free of interfering substances to which the assay is sensitive *(4)*. Furthermore, this kinetic silver staining assay is found to be most sensitive for the detection of proteins on cellulose such as is commonly used for TLC (*see* Chapter 51 for a comparison with other solid-phase stains).

2. Materials

1. Polystyrene 96-well microtiter plates (e.g., Titertek 76-381-04, McLean, VA).
2. The kinetic silver staining reagent consists of
   Reagent A: 0.2% (w/v) $AgNO_3$, 0.2% (w/v) NH_4NO_3, 1% (w/v)tungstosilicic acid, and 0.3% (v/v) formaldehyde (from a 37% stock solution in water) in distilled water. Store in the dark at room temperature.
   Reagent B: 5% (w/v) Na_2CO_3 in distilled water.
3. A microtiter plate reader is required (e.g., EIA Autoreader, model EL310,Biotek Instruments, Burlington, VT).
4. Cellulose paper or TLC plates.

From: *The Protein Protocols Handbook, 2nd Edition*
Edited by: J. M. Walker © Humana Press Inc., Totowa, NJ

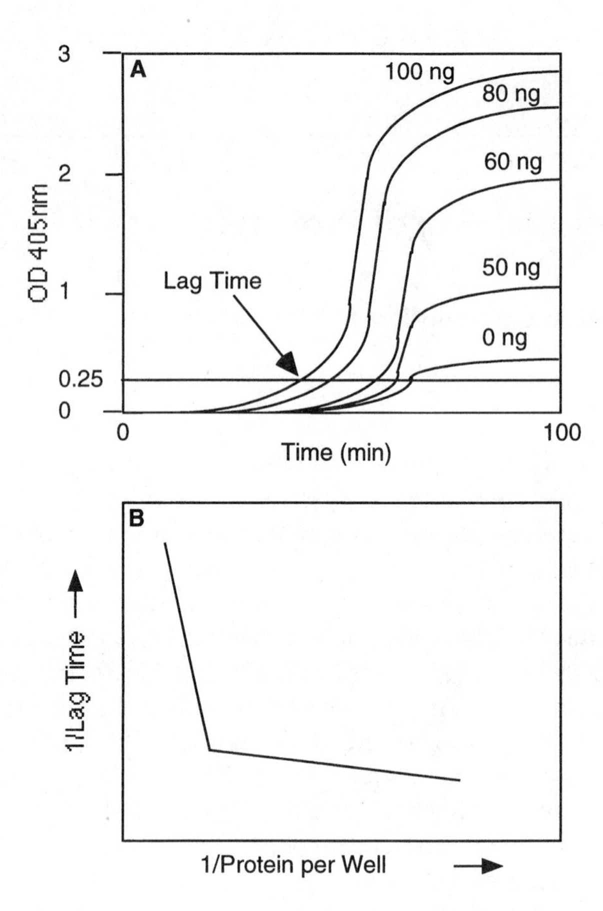

Fig. 1. Representative plots of standard curve data. **(A)** Raw data for varying standard protein mass per well.**(B)** Double-reciprocal plot of standard curve.

3. Methods

3.1. Microtiter Plate Assay

1. Adsorb the protein of interest to duplicate microtiter plates (*see* **Note 1**). Wash the microtiter plate profusely with distilled water after the adsorption (*see* **Note 2**). One of the microtiter plates is used for the kinetic silver staining assay and the other is for quantitative ELISA or binding experiments. The microtiter plate for the kinetic silver staining assay contains protein adsorbed to only a few of the wells (e.g., 4–16 wells).
2. Prepare a known concentration of the standard protein in distilled water (e.g., by dialysis) for use in a standard curve (*see* **Notes 2** and **3**).
3. Apply varying concentrations (e.g., 50–1000 ng/mL) of the standard protein to the blank wells on the microtiter plate at the same volume (50–200 μL) that was used to adsorb protein in **step 1**. Cover the microtiter plate with a tissue to avoid dust from settling in the wells and allow the protein to air dry for several days (*see* **Note 4**). Do not wash the plate after this step!
4. Mix equal volumes of reagents A and B immediately before use. A fixed volume (e.g., 100 μL) of the mixture is added quickly to the wells on the microtiter plate that were adsorbed with protein in **step 1**, **step 3**, and to blank control wells. The time for the addition of reagent to each well should be noted as time zero. All wells are filled in less than

Table 1
Compounds Known to Interfere with Kinetic Silver Staining

Compounds	Concentration in the staining reagent
TCA	>0.01 m*M*
Glucose	>1 m*M*
β-Mercaptoethanol	>0.1 m*M*
DTT	>0.01 m*M*
KOH	>0.01 m*M*
EDTA	>1 m*M*
Phosphate	>0.01 m*M*
SDS	> 1 m*M*
Triton X-100	>0.0005%
Ammonium sulfate	>1 m*M*
Urea	>1 m*M*
Imidazole	>0.01 m*M*
Tris	>1 m*M*
NaCl	>0.01 m*M*
KCl	>0.01 m*M*
Guanidine HCl	>0.01 m*M*

10 min from the time of mixing reagents A and B; otherwise, excessive silver development in the reagent may cause a high background.

5. Read optical densities of each well with a microtiter plate reader at 405 nm (*see* **Note 5**). The time elapsed from time zero for each well to reach 0.25 OD (lag time) is noted as the lag time.
6. The lag time is plotted against the mass per well of adsorbed protein in the standard curve (**step 3**), which typically yields an inverse sigmoidal shape curve. Comparison of the lag times of the sample wells to the standard curve allows the determination of the total mass of protein in the sample well. The standard curve may be linearized to a sharply biphasic shape curve by plotting 1/lag time vs 1/protein per well (*see* **Fig. 1**).

3.2. Cellulose Assay

1. Mix equal volumes of reagents A and B immediately before use.
2. Stain cellulose (*see* **Note 6**) with adsorbed proteins by immersion in the mixed reagents for at least 1 h.

4. Notes

1. The wells on the edge of microtiter plates should be avoided for quantitative measurements because they tend to yield less accurate numbers.
2. Washing of microtiter plates is essential, as residual buffer reagents may interfere with silver staining (*see* **Table 1**). The washing is performed by gently dipping in beakers of deionized water. Vigorous washing was avoided for fear of losing adsorbed protein.
3. Kinetic silver staining shows little protein-to-protein variation (<30% over six tested proteins); it is possible to estimate the total amount of adsorbed protein or mixture of proteins using a standard (e.g., bovine albumin) that can be easily dissolved in water and measured for concentration.

4. Kinetic silver staining detects proteins bound to polystyrene and apparently not protein in solution; thus the quantitative adsorption of the standard protein in distilled water (to avoid interference, *see* **Note 2**) to polystyrene by drying is necessary. As the binding capacity of polystyrene is exceeded (typically about 100-200 ng/well, but depends upon both the microtiter plate and the protein of interest), kinetic silver staining does not detect further increases in protein mass per well.
5. Kinetic silver staining is based on measurements of light scattering. Thus, other wavelengths may be used, but the corresponding optical density reading will be different and may require optimization.
6. The cellulose may require washing to remove solvents or buffers, if present prior to staining.

References

1. Somerville, L. L. and Wang, K. (1981) The ultrasensitive silver "protein" stain also detects nanograms of nucleic acids. *Biochem. Biophys. Res. Commun.* **102,** 53–58.
2. Gottlieb, M. and Chavko, M. (1987) Silver staining of native and denatured eucaryotic DNA in agarose gels. *Analyt. Biochem.* **165,** 33–37.
3. Root, D. D. and Reisler, E. (1990) Copper iodide staining and determination of proteins adsorbed to microtiter plates. *Analyt. Biochem.* **186,** 69–73.
4. Root, D. D. and Wang, K. (1993) Kinetic silver staining and calibration of proteins adsorbed to microtiter plates. *Analyt. Biochem.* **209,** 354–359.

Part II

Electrophoresis of Proteins and Peptides and Detection in Gels

10

Nondenaturing Polyacrylamide Gel Electrophoresis of Proteins

John M. Walker

1. Introduction

SDS-PAGE (Chapter 11) is probably the most commonly used gel electrophoretic system for analyzing proteins. However, it should be stressed that this method separates denatured protein. Sometimes one needs to analyze native, nondenatured proteins, particularly if wanting to identify a protein in the gel by its biological activity (for example, enzyme activity, receptor binding, antibody binding, and so on). On such occasions it is necessary to use a nondenaturing system such as described in this chapter. For example, when purifying an enzyme, a single major band on a gel would suggest a pure enzyme. However this band could still be a contaminant; the enzyme could be present as a weaker (even nonstaining) band on the same gel. Only by showing that the major band had enzyme activity would you be convinced that this band corresponded to your enzyme. The method described here is based on the gel system first described by Davis *(1)*. To enhance resolution a stacking gel can be included (*see* Chapter 11 for the theory behind the stacking gel system).

2. Materials

1. Stock acrylamide solution: 30 g acrylamide, 0.8 g *bis*-acrylamide. Make up to 100 mL in distilled water and filter. Stable at 4°C for months (*see* **Note 1**). **Care: Acrylamide Monomer Is a Neurotoxin.** Take care in handling acrylamide (wear gloves) and avoid breathing in acrylamide dust when weighing out.
2. Separating gel buffer: 1.5 *M* Tris-HCl, pH 8.8.
3. Stacking gel buffer: 0.5 *M* Tris-HCl, pH 6.8.
4. 10% Ammonium persulfate in water.
5. *N,N,N',N'*-tetramethylethylenediamine (TEMED).
6. Sample buffer (5X). Mix the following:
   a. 15.5 mL of 1 *M* Tris-HCl pH 6.8;
   b. 2.5 mL of a 1% solution of bromophenol blue;
   c. 7 mL of water; and
   d. 25 mL of glycerol.

   Solid samples can be dissolved directly in 1X sample buffer. Samples already in solution should be diluted accordingly with 5X sample buffer to give a solution that is 1X sample buffer. Do not use protein solutions that are in a strong buffer that is not near to pH

From: *The Protein Protocols Handbook, 2nd Edition*
Edited by: J. M. Walker © Humana Press Inc., Totowa, NJ

6.8 as it is important that the sample is at the correct pH. For these samples it will be necessary to dialyze against 1X sample buffer.

7. Electrophoresis buffer: Dissolve 3.0 g of Tris base and 14.4 g of glycine in water and adjust the volume to 1 L. The final pH should be 8.3.
8. Protein stain: 0.25 g Coomassie brilliant blue R250 (or PAGE blue 83), 125 mL methanol, 25 mL glacial acetic acid, and 100 mL water. Dissolve the dye in the methanol component first, then add the acid and water. Dye solubility is a problem if a different order is used. Filter the solution if you are concerned about dye solubility. For best results do not reuse the stain.
9. Destaining solution: 100 mL methanol, 100 mL glacial acetic acid, and 800 mL water.
10. A microsyringe for loading samples.

3. Method

1. Set up the gel cassette.
2. To prepare the separating gel (*see* **Note 2**) mix the following in a Buchner flask: 7.5 mL stock acrylamide solution, 7.5 mL separating gel buffer, 14.85 mL water, and 150 μL 10% ammonium persulfate.

   "Degas" this solution under vacuum for about 30 s. This degassing step is necessary to remove dissolved air from the solution, since oxygen can inhibit the polymerization step. Also, if the solution has not been degassed to some extent, bubbles can form in the gel during polymerization, which will ruin the gel. Bubble formation is more of a problem in the higher percentage gels where more heat is liberated during polymerization.
3. Add 15 μL of TEMED and gently swirl the flask to ensure even mixing. The addition of TEMED will initiate the polymerization reaction, and although it will take about 20 min for the gel to set, this time can vary depending on room temperature, so it is advisable to work fairly quickly at this stage.
4. Using a Pasteur (or larger) pipet, transfer the separating gel mixture to the gel cassette by running the solution carefully down one edge between the glass plates. Continue adding this solution until it reaches a position 1 cm from the bottom of the sample loading comb.
5. To ensure that the gel sets with a smooth surface, *very* carefully run distilled water down one edge into the cassette using a Pasteur pipet. Because of the great difference in density between the water and the gel solution, the water will spread across the surface of the gel without serious mixing. Continue adding water until a layer about 2 mm exists on top of the gel solution.
6. The gel can now be left to set. When set, a very clear refractive index change can be seen between the polymerized gel and overlaying water.
7. While the separating gel is setting, prepare the following stacking gel solution. Mix the following quantities in a Buchner flask: 1.5 mL stock acrylamide solution, 3.0 mL stacking gel buffer, 7.4 mL water, and 100 μL 10% ammonium persulfate. Degas this solution as before.
8. When the separating gel has set, pour off the overlaying water. Add 15 μL of TEMED to the stacking gel solution and use some (~2 mL) of this solution to wash the surface of the polymerized gel. Discard this wash, then add the stacking gel solution to the gel cassette until the solution reaches the cutaway edge of the gel plate. Place the well-forming comb into this solution and leave to set. This will take about 30 min. Refractive index changes around the comb indicate that the gel has set. It is useful at this stage to mark the positions of the bottoms of the wells on the glass plates with a marker pen.
9. Carefully remove the comb from the stacking gel, remove any spacer from the bottom of the gel cassette, and assemble the cassette in the electrophoresis tank. Fill the top reservoir with electrophoresis buffer ensuring that the buffer fully fills the sample loading wells,

and look for any leaks from the top tank. If there are no leaks, fill the bottom tank with electrophoresis buffer, then tilt the apparatus to dispel any bubbles caught under the gel.

10. Samples can now be loaded onto the gel. Place the syringe needle through the buffer and locate it just above the bottom of the well. Slowly deliver the sample (~5–20 μL) into the well. The dense sample solvent ensures that the sample settles to the bottom of the loading well. Continue in this way to fill all the wells with unknowns or standards, and record the samples loaded.
11. The power pack is now connected to the apparatus and a current of 20–25 mA passed through the gel (constant current) (*see* **Note 3**). Ensure that the electrodes are arranged so that the proteins are running to the anode (*see* **Note 4**). In the first few minutes the samples will be seen to concentrate as a sharp band as it moves through the stacking gel. (It is actually the bromophenol blue that one is observing, not the protein but, of course, the protein is stacking in the same way.) Continue electrophoresis until the bromophenol blue reaches the bottom of the gel. This will usually take about 3 h. Electrophoresis can now be stopped and the gel removed from the cassette. Remove the stacking gel and immerse the separating gel in stain solution, or proceed to step 13 if you wish to detect enzyme activity (*see* **Notes 5** and **6**).
12. Staining should be carried out, with shaking, for a minimum of 2 h and preferably overnight. When the stain is replaced with destain, stronger bands will be immediately apparent and weaker bands will appear as the gel destains. Destaining can be speeded up by using a foam bung, such as those used in microbiological flasks. Place the bung in the destain and squeeze it a few times to expel air bubbles and ensure the bung is fully wetted. The bung rapidly absorbs dye, thus speeding up the destaining process.
13. If proteins are to be detected by their biological activity, duplicate samples should be run. One set of samples should be stained for protein and the other set for activity. Most commonly one would be looking for enzyme activity in the gel. This is achieved by washing the gel in an appropriate enzyme substrate solution that results in a colored product appearing in the gel at the site of the enzyme activity (*see* **Note 7**).

4. Notes

1. The stock acrylamide used here is the same as used for SDS gels (*see* Chapter 11) and may already be availabe in your laboratory.
2. The system described here is for a 7.5% acrylamide gel, which was originally described for the separation of serum proteins *(1)*. Since separation in this system depends on both the native charge on the protein and separation according to size owing to frictional drag as the proteins move through the gel, it is not possible to predict the electrophoretic behavior of a given protein the way that one can on an SDS gel, where separation is based on size alone. A 7.5% gel is a good starting point for unknown proteins. Proteins of mol wt >100,000 should be separated in 3–5% gels. Gels in the range 5–10% will separate proteins in the range 20,000–150,000, and 10–15% gels will separate proteins in the range 10,000–80,000. The separation of smaller polypeptides is described in Chapter 13. To alter the acrylamide concentration, adjust the volume of stock acrylamide solution in **Subheading 3., step 2** accordingly, and increase/decrease the water component to allow for the change in volume. For example, to make a 5% gel change the stock acrylamide to 5 mL and increase the water to 17.35 mL. The final volume is still 30 mL, so 5 mL of the 30% stock acrylamide solution has been diluted in 30 mL to give a 5% acrylamide solution.
3. Because one is separating native proteins, it is important that the gel does not heat up too much, since this could denature the protein in the gel. It is advisable therefore to run the gel in the cold room, or to circulate the buffer through a cooling coil in ice. (Many gel

apparatus are designed such that the electrode buffer cools the gel plates.) If heating is thought to be a problem it is also worthwhile to try running the gel at a lower current for a longer time.

4. This separating gel system is run at pH 8.8. At this pH most proteins will have a negative charge and will run to the anode. However, it must be noted that any basic proteins will migrate in the opposite direction and will be lost from the gel. Basic proteins are best analyzed under acid conditions, as described in Chapters 16 and 17.
5. It is important to note that concentration in the stacking gel may cause aggregation and precipitation of proteins. Also, the pH of the stacking gel (pH 6.8) may affect the activity of the protein of interest. If this is thought to be a problem (e.g., the protein cannot be detected on the gel), prepare the gel without a stacking gel. Resolution of proteins will not be quite so good, but will be sufficient for most uses.
6. If the buffer system described here is unsuitable (e.g., the protein of interest does not electrophorese into the gel because it has the incorrect charge, or precipitates in the buffer, or the buffer is incompatible with your detection system) then one can try different buffer systems (without a stacking gel). A comprehensive list of alternative buffer systems has been published *(2)*.
7. The most convenient substrates for detecting enzymes in gels are small molecules that freely diffuse into the gel and are converted by the enzyme to a colored or fluorescent product within the gel. However, for many enzymes such convenient substrates do not exist, and it is necessary to design a linked assay where one includes an enzyme together with the substrate such that the products of the enzymatic reaction of interest is converted to a detectable product by the enzyme included with the substrate. Such linked assays may require the use of up to two or three enzymes and substrates to produce a detectable product. In these cases the product is usually formed on the surface of the gel because the coupling enzymes cannot easily diffuse into the gel. In this case the zymogram technique is used where the substrate mix is added to a cooled (but not solidified) solution of agarose (1%) in the appropriate buffer. This is quickly poured over the solid gel where it quickly sets on the gel. The product of the enzyme assay is therefore formed at the gel–gel interface and does not get washed away. A number of review articles have been published which described methods for detecting enzymes in gels *(3–7)*. A very useful list also appears as an appendix in **ref.** ***8***.

References

1. Davis, B. J. (1964) Disc electrophoresis II—method and application to human serum proteins. *Ann. NY Acad. Sci.* **121,** 404–427.
2. Andrews, A. T. (1986) *Electrophoresis: Theory, Techniques, and Biochem-ical and Clinical Applications.* Clarendon, Oxford, UK.
3. Shaw, C. R. and Prasad, R. (1970) Gel electrophoresis of enzymes—a compilation of recipes. *Biochem. Genet.* **4,** 297–320.
4. Shaw, C. R. and Koen, A. L. (1968) Starch gel zone electrophoresis of enzymes, in *Chromatographic and Electrophoretic Techniques,* vol. 2 (Smith, I., ed.), Heinemann, London, pp. 332–359.
5. Harris, H. and Hopkinson, D. A. (eds.) (1976) *Handbook of Enzyme Electrophoresis in Human Genetics.* North-Holland, Amsterdam.
6. Gabriel, O. (1971) Locating enymes on gels, in *Methods in Enzymology,* vol. 22 (Colowick, S. P. and Kaplan, N. O., eds.), Academic, New York, p. 578.
7. Gabriel, O. and Gersten, D. M. (1992) Staining for enzymatic activity after gel electrophoresis. I. *Analyt. Biochem.* **203,** 1–21.
8. Hames, B. D. and Rickwood, D. (1990) *Gel Electrophoresis of Proteins,* 2nd ed., IRL, Oxford and Washington.

11

SDS Polyacrylamide Gel Electrophoresis of Proteins

John M. Walker

1. Introduction

SDS-PAGE is the most widely used method for qualitatively analyzing protein mixtures. It is particularly useful for monitoring protein purification, and because the method is based on the separation of proteins according to size, the method can also be used to determine the relative molecular mass of proteins (*see* **Note 14**).

1.1. Formation of Polyacrylamide Gels

Crosslinked polyacrylamide gels are formed from the polymerization of acrylamide monomer in the presence of smaller amounts of *N,N'*-methylene-*bis*-acrylamide (normally referred to as "*bis*-acrylamide") (**Fig. 1**). Note that *bis*-acrylamide is essentially two acrylamide molecules linked by a methylene group and is used as a crosslinking agent. Acrylamide monomer is polymerized in a head-to-tail fashion into long chains, and occasionally a *bis*-acrylamide molecule is built into the growing chain, thus introducing a second site for chain extension. Proceeding in this way, a crosslinked matrix of fairly well-defined structure is formed (**Fig. 1**). The polymerization of acrylamide is an example of free-radical catalysis, and is initiated by the addition of ammonium persulfate and the base *N,N,N',N'*-tetramethylenediamine (TEMED). TEMED catalyzes the decomposition of the persulfate ion to give a free radical (i.e., a molecule with an unpaired electron):

$$S_2O_8^{2-} + e^- \rightarrow SO_4^{2-} + SO_4^{-\bullet} \qquad (1)$$

If this free radical is represented as $R^\bullet$ (where the dot represents an unpaired electron) and M as an acrylamide monomer molecule, then the polymerization can be represented as follows:

$$\begin{aligned} R^\bullet + M &\rightarrow RM^\bullet \\ RM^\bullet + M &\rightarrow RMM^\bullet \\ RMM^\bullet + M &\rightarrow RMMM^\bullet, \text{ and so forth} \end{aligned} \qquad (2)$$

In this way, long chains of acrylamide are built up, being crosslinked by the introduction of the occasional *bis*-acrylamide molecule into the growing chain. Oxygen "mops up" free radicals, and therefore the gel mixture is normally degassed (the solutions are briefly placed under vacuum to remove loosely dissolved oxygen) prior to addition of the catalyst.

From: *The Protein Protocols Handbook, 2nd Edition*
Edited by: J. M. Walker © Humana Press Inc., Totowa, NJ

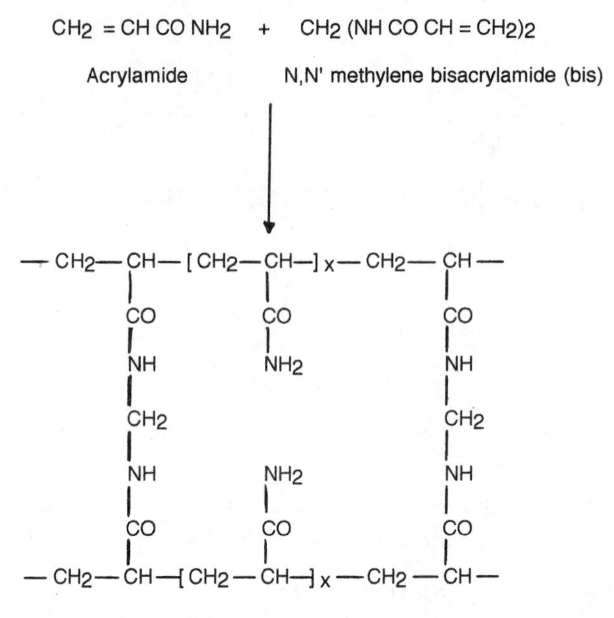

Fig. 1. Polymerization of acrylamide.

1.2. The Use of Stacking Gels

For both SDS and buffer gels samples may be applied directly to the top of the gel in which protein separation is to occur (the separating gel). However, in these cases, the sharpness of the protein bands produced in the gel is limited by the size (volume) of the sample applied to the gel. Basically the separated bands will be as broad (or broader, owing to diffusion) as the sample band applied to the gel. For some work, this may be acceptable, but most workers require better resolution than this. This can be achieved by polymerizing a short stacking gel on top of the separating gel. The purpose of this stacking gel is to concentrate the protein sample into a sharp band before it enters the main separating gel, thus giving sharper protein bands in the separating gel. This modification allows relatively large sample volumes to be applied to the gel without any loss of resolution. The stacking gel has a very large pore size (4% acrylamide) which allows the proteins to move freely and concentrate, or stack under the effect of the electric field. Sample concentration is produced by isotachophoresis of the sample in the stacking gel. The band-sharpening effect (isotachophoresis) relies on the fact that the negatively charged glycinate ions (in the reservoir buffer) have a lower electrophoretic mobility than the protein–SDS complexes. which in turn, have lower mobility than the Cl^- ions if they are in a region of higher field strength. Field strength is inversely proportional to conductivity, which is proportional to concentration. The result is that the three species of interest adjust their concentrations so that $[Cl^-]$ > [protein-SDS] > [glycinate]. There are only a small quantity of protein–SDS complexes, so they concentrate in a very tight band between the glycinate and Cl^- ion boundaries. Once the glycinate reaches the separating gel, it becomes more fully ionized in the higher pH environment and its mobility increases. (The pH of the stacking gel is 6.8 and that of the separating gel is 8.8.) Thus, the interface between glycinate and the Cl^- ions leaves behind the protein–SDS complexes, which are left to electrophorese at their own rates. A more detailed description of the theory of isotachophoresis and electrophoresis generally is given in **ref.** ***1***.

1.3. SDS-PAGE

Samples to be run on SDS-PAGE are first boiled for 5 min in sample buffer containing β-mercaptoethanol and SDS. The mercaptoethanol reduces any disulfide bridges present that are holding together the protein tertiary structure. SDS (CH_3-$[CH_2]_{10}$ - $CH_2OSO_3^-Na^+$) is an anionic detergent and binds strongly to, and denatures, the protein. Each protein in the mixture is therefore fully denatured by this treatment and opens up into a rod-shaped structure with a series of negatively charged SDS molecules along the polypeptide chain. On average, one SDS molecule binds for every two amino acid residues. The original native charge on the molecule is therefore completely swamped by the SDS molecules. The sample buffer also contains an ionizable tracking dye usually bromophenol blue that allows the electrophoretic run to be monitored, and sucrose or glycerol which gives the sample solution density, thus allowing the sample to settle easily through the electrophoresis buffer to the bottom when injected into the loading well. When the main separating gel has been poured between the glass plates and allowed to set, a shorter stacking gel is poured on top of the separating gel, and it is into this gel that the wells are formed and the proteins loaded. Once all samples are loaded, a current is passed through the gel. Once the protein samples have passed through the stacking gel and have entered the separating gel, the negatively charged protein–SDS complexes continue to move toward the anode, and because they have the same charge per unit length they travel into the separating gel under the applied electric field with the same mobility. However, as they pass through the separating gel the proteins separate, owing to the molecular sieving properties of the gel. Quite simply, the smaller the protein, the more easily it can pass through the pores of the gel, whereas large proteins are successively retarded by frictional resistance owing to the sieving effect of the gel. Being a small molecule, the bromophenol blue dye is totally unretarded and therefore indicates the electrophoresis front. When the dye reaches the bottom of the gel the current is turned off and the gel is removed from between the glass plates, shaken in an appropriate stain solution (usually Coomassie brilliant blue) for a few hours, and then washed in destain solution overnight. The destain solution removes unbound background dye from the gel, leaving stained proteins visible as blue bands on a clear background. A typical large format gel would take about 1 h to prepare and set, 3 h to run at 30 mA, and have a staining time of 2–3 h with an overnight destain. Minigels (e.g., Bio-Rad minigel) run at 200 V. Constant voltage can run in about 40 min, and require only 1 h staining. Most bands can be seen within 1 h of destaining. Vertical slab gels are invariably run since this allows up to 20 different samples to be loaded onto a single gel.

2. Materials

1. Stock acrylamide solution: 30% acrylamide, 0.8% *bis*-acrylamide. Filter through Whatman No. 1 filter paper and store at 4°C (*see* **Note 1**).
2. Buffers:
   a. 1.875 *M* Tris-HCl, pH 8.8.
   b. 0.6 *M* Tris-HCl, pH 6.8.

3. 10% Ammonium persulfate. Make fresh.
4. 10% SDS (*see* **Note 2**).
5. TEMED.
6. Electrophoresis buffer: Tris (12 g), glycine (57.6 g), and SDS (2.0 g). Make up to 2 L with water. No pH adjustment is necessary.
7. Sample buffer (*see* **Notes 3** and **4**):

0.6 *M* Tris-HCl, pH 6.8	5.0 mL
SDS	0.5 g
Sucrose	5.0 g
β-Mercaptoethanol	0.25 mL
Bromophenol blue, 0.5% stock	5.0 mL

   Make up to 50 mL with distilled water.
8. Protein stain: 0.1% Coomassie brilliant blue R250 in 50% methanol, 10% glacial acetic acid. Dissolve the dye in the methanol and water component first, and then add the acetic acid. Filter the final solution through Whatman No. 1 filter paper if necessary.
9. Destain: 10% methanol, 7% glacial acetic acid.
10. Microsyringe for loading samples. Micropipet tips that are drawn out to give a fine tip are also commercially available.

3. Method

The system of buffers used in the gel system described below is that of Laemmli *(2)*.

1. Samples to be run are first denatured in sample buffer by heating to 95–100°C for 5 min (*see* **Note 3**).
2. Clean the internal surfaces of the gel plates with detergent or methylated spirits, dry, then join the gel plates together to form the cassette, and clamp it in a vertical position. The exact manner of forming the cassette will depend on the type of design being used.
3. Mix the following in a 250-mL Buchner flask (*see* **Note 5**):

	For 15% gels	For 10% gels
1.875 *M* Tris-HCl, pH 8.8	8.0 mL	8.0 mL
Water	11.4 mL	18.1 mL
Stock acrylamide	20.0 mL	13.3 mL
10% SDS	0.4 mL	0.4 mL
Ammonium persulfate (10%)	0.2 mL	0.2 mL

4. "Degas" this solution under vacuum for about 30 s. Some frothing will be observed, and one should not worry if some of the froth is lost down the vacuum tube: you are only losing a very small amount of liquid (*see* **Note 6**).
5. Add 14 μL of TEMED, and gently swirl the flask to ensure even mixing. The addition of TEMED will initiate the polymerization reaction and although it will take about 15 min for the gel to set, this time can vary depending on room temperature, so it is advisable to work fairly quickly at this stage.
6. Using a Pasteur (or larger) pipet transfer this separating gel mixture to the gel cassette by running the solution carefully down one edge between the glass plates. Continue adding this solution until it reaches a position 1 cm from the bottom of the comb that will form the loading wells. Once this is completed, you will find excess gel solution remaining in your flask. Dispose of this in an appropriate waste container **not** down the sink.

7. To ensure that the gel sets with a smooth surface **very carefully** run distilled water down one edge into the cassette using a Pasteur pipet. Because of the great difference in density between the water and the gel solution the water will spread across the surface of the gel without serious mixing. Continue adding water until a layer of about 2 mm exists on top of the gel solution (*see* **Notes 7** and **8**).
8. The gel can now be left to set. As the gel sets, heat is evolved and can be detected by carefully touching the gel plates. When set, a very clear refractive index change can be seen between the polymerized gel and overlaying water.
9. While the separating gel is setting prepare the following stacking gel (4°C) solution. Mix the following in a 100-mL Buchner flask (*see* **Notes 8** and **9**):

0.6 *M* Tris-HCl, pH 6.8	1.0 mL
Stock acrylamide	1.35 mL
Water	7.5 mL
10% SDS	0.1 mL
Ammonium persulfate (10%)	0.05 mL

Degas this solution as before.
10. When the separating gel has set, pour off the overlaying water. Add 14 µL of TEMED to the stacking gel solution and use some (~2 mL) of this solution to wash the surface of the polymerized gel. Discard this wash, and then add the stacking gel solution to the gel cassette until the solution reaches the cutaway edge of the gel plate. Place the well-forming comb into this solution, and leave to set. This will take about 20 min. Refractive index changes around the comb indicate that the gel has set. It is useful at this stage to mark the positions of the bottoms of the wells on the glass plates with a marker pen to facilitate loading of the samples (*see* also **Note 9**).
11. Carefully remove the comb from the stacking gel, and then rinse out any nonpolymerized acrylamide solution from the wells using electrophoresis buffer. Remove any spacer from the bottom of the gel cassette, and assemble the cassette in the electrophoresis tank. Fill the top reservoir with electrophoresis buffer, and look for any leaks from the top tank. If there are no leaks fill the bottom tank with electrophoresis buffer, and then tilt the apparatus to dispel any bubbles caught under the gel.
12. Samples can now be loaded onto the gel. Place the syringe needle through the buffer and locate it just above the bottom of the well. Slowly deliver the sample into the well. Five- to 10-µL samples are appropriate for most gels. The dense sample buffer ensures that the sample settles to the bottom of the loading well (*see* **Note 10**). Continue in this way to fill all the wells with unknowns or standards, and record the samples loaded.
13. Connect the power pack to the apparatus, and pass a current of 30 mA through the gel (constant current) for large format gels, or 200 V (constant voltage) for minigels (Bio-Rad). Ensure your electrodes have correct polarity: all proteins will travel to the anode (+). In the first few minutes, the samples will be seen to concentrate as a sharp band as it moves through the stacking gel. (It is actually the bromophenol blue that one is observing not the protein, but of course the protein is stacking in the same way.) Continue electrophoresis until the bromophenol blue reaches the bottom of the gel. This will take 2.5–3.0 h for large format gels (16 µm × 16 µm) and about 40 min for minigels (10 µm × 7 µm) (*see* **Note 11**).
14. Dismantle the gel apparatus, pry open the gel plates, remove the gel, discard the stacking gel, and place the separating gel in stain solution.
15. Staining should be carried out with shaking, for a minimum of 2 h. When the stain is replaced with destain, stronger bands will be immediately apparent, and weaker bands will appear as the gel destains (*see* **Notes 12** and **13**).

4. Notes

1. Acrylamide is a potential neurotoxin and should be treated with great care. Its effects are cumulative, and therefore, regular users are at greatest risk. In particular, take care when weighing out acrylamide. Do this in a fume hood, and wear an appropriate face mask.
2. SDS come out of solution at low temperature, and this can even occur in a relatively cold laboratory. If this happens, simply warm up the bottle in a water bath. Store at room temperature.
3. Solid samples can be dissolved directly in sample buffer. Pure proteins or simple mixtures should be dissolved at 1–0.5 mg/mL. For more complex samples suitable concentrations must be determined by trial and error. For samples already in solution dilute them with an equal volume of double-strength sample buffer. Do not use protein solutions that are in a strong buffer, that is, not near pH 6.5, since it is important that the sample be at the correct pH. For these samples, it will be necessary to dialyze them first. Should the sample solvent turn from blue to yellow, this is a clear indication that your sample is acidic.
4. The β-mercaptoethanol is essential for disrupting disulfide bridges in proteins. However, exposure to oxygen in the air means that the reducing power of β-mercaptoethanol in the sample buffer decreases with time. Every couple of weeks, therefore, mercaptoethanol should be added to the stock solution or the solution remade. Similarly protein samples that have been prepared in sample buffer and stored frozen should, before being rerun at a later date, have further mercaptoethanol added.
5. Typically, the separating gel used by most workers is a 15% polyacrylamide gel. This give a gel of a certain pore size in which proteins of relative molecular mass (M_r) 10,000 move through the gel relatively unhindered, whereas proteins of 100,000 can only just enter the pores of this gel. Gels of 15% polyacrylamide are therefore useful for separating proteins in the range of 100,000–10,000. However, a protein of 150,000 for example, would be unable to enter a 15% gel. In this case, a larger-pored gel (e.g., a 10% or even 7.5% gel) would be used so that the protein could now enter the gel, and be stained and identified. It is obvious, therefore, that the choice of gel to be used depends on the size of the protein being studied. If proteins covering a wide range of mol-wt values need to be separated, then the use of a gradient gel is more appropriate (*see* Chapter 12).
6. Degassing helps prevent oxygen in the solution from "mopping up" free radicals and inhibiting polymerization although this problem could be overcome by the alternative approach of increasing the concentration of catalyst. However, the polymerization process is an exothermic one. For 15% gels, the heat liberated can result in the formation of small air bubbles in the gel (this is not usually a problem for gels of 10% or less where much less heat is liberated). It is advisable to carry out degassing as a matter of routine.
7. An alternative approach is to add a water-immiscible organic solvent, such as isobutanol, to the top of the gel. Less caution is obviously needed when adding this, although if using this approach, this step should be carried out in a fume cupboard, not in the open laboratory.
8. To save time some workers prefer to add the stacking gel solution directly and carefully to the top of the separating gel, i.e., the overlaying step (**step 7**) is omitted, the stacking gel solution itself providing the role of the overlaying solution.
9. Some workers include a small amount of bromophenol blue in this gel mix. This give a stacking gel that has a pale blue color, thus allowing the loading wells to be easily identified.
10. Even if the sample is loaded with too much vigor, such that it mixes extensively with the buffer in the well, this is not a problem, since the stacking gel system will still concentrate the sample.
11. When analyzing a sample for the first time, it is sensible to stop the run when the dye reaches the bottom of the gel, because there may be low mol-wt proteins that are running

close to the dye, and these would be lost if electrophoresis was continued after the dye had run off the end of the gel. However, often one will find that the proteins being separated are only in the top two-thirds of the gel. In this case, in future runs, the dye would be run off the bottom of the gel, and electrophoresis carried out for a further 30 min to 1 h to allow proteins to separate across the full length of the gel thus increasing the separation of bands.

12. Normally, destain solution needs to be replaced at regular intervals since a simple equilibrium is quickly set up between the concentration of stain in the gel and destain solution, after which no further destaining takes place. To speed up this process and also save on destain solution, it is convenient to place some solid material in with the destain that will absorb the Coomassie dye as it elutes from the gel. We use a foam bung such as that used in culture flasks (ensure it is well wetted by expelling all air in the bung by squeezing it many times in the destain solution), although many other materials can be used (e.g., polystyrene packaging foam).
13. It is generally accepted that a very faint protein band detected by Coomassie brilliant blue, is equivalent to about 0.1 μg (100 ng) of protein. Such sensitivity is suitable for many people's work. However if no protein bands are observed or greater staining is required, then silver staining (Chapter 33) can be further carried out on the gel.
14. Because the principle of this technique is the separation of proteins based on size differences, by running calibration proteins of known molecular weight on the same gel run as your unknown protein, the molecular weight of the unknown protein can be determined. For most proteins a plot of $\log_{10}$ molecular mass vs relative mobility provides a straight line graph, although one must be aware that for any given gel concentration this relationship is only linear over a limited range of molecular masses. As an approximate guide, using the system described here, the linear relationship is true over the following ranges: 15% acrylamide, 10,000–50,000; 10% acrylamide 15,000–70,000; 5% acrylamide 60,000–200,000. It should be stressed that this relationship only holds true for proteins that bind SDS in a constant weight ratio. This is true of many proteins but some proteins for example, highly basic proteins, may run differently than would be expected on the basis of their known molecular weight. In the case of the histones, which are highly basic proteins, they migrate more slowly than expected, presumably because of a reduced overall negative charge on the protein owing to their high proportion of positively-charged amino acids. Glycoproteins also tend to run anomalously presumably because the SDS only binds to the polypeptide part of the molecule.

    To determine the molecular weight of an unknown protein the relative mobilities (*Rf*) of the standard proteins are determined and a graph of log molecular weight vs *Rf* plotted.

    $$Rf = (\text{distance migrated by protein/distance migrated by dye}) \quad (3)$$

    Mixtures of standard mol-wt markers for use on SDS gels are available from a range of suppliers. The *Rf* of the unknown protein is then determined and the logMW (and hence molecular weight) determined from the graph. A more detailed description of protein mol-wt determination on SDS gels is described in **refs.** ***1*** and ***3***.

References

1. Deyl, Z. (1979) *Electrophoresis: A Survey of Techniques and Applications. Part A Techniques.* Elsevier, Amsterdam.
2. Laemmli, U. K. (1970) Cleavage of structural proteins during the assembly of the head of bacteriophage T4. *Nature* **227,** 680–685.
3. Hames, B. D. and Rickwood, D. (eds.) (1990) *Gel Electrophoresis of Proteins—A Practical Approach.* IRL, Oxford University Press, Oxford.

12

Gradient SDS Polyacrylamide Gel Electrophoresis of Proteins

John M. Walker

1. Introduction

The preparation of fixed-concentration polyacrylamide gels has been described in Chapters 10 and 11. However, the use of polyacrylamide gels that have a gradient of increasing acrylamide concentration (and hence decreasing pore size) can sometimes have advantages over fixed-concentration acrylamide gels. During electrophoresis in gradient gels, proteins migrate until the decreasing pore size impedes further progress. Once the "pore limit" is reached, the protein banding pattern does not change appreciably with time, although migration does not cease completely *(1)*. There are two main advantages of gradient gels over linear gels.

First, a much greater range of protein M_r values can be separated than on a fixed-percentage gel. In a complex mixture, very low-mol-wt proteins travel freely through the gel to begin with, and start to resolve when they reach the smaller pore size toward the lower part of the gel. Much larger proteins, on the other hand, can still enter the gel but start to separate immediately owing to the sieving effect of the gel. The second advantage of gradient gels is that proteins with very similar M_r values may be resolved, which otherwise cannot resolve in fixed percentage gels. As each protein moves through the gel, the pore size become smaller until the protein reaches its pore size limit. The pore size in the gel is now too small to allow passage of the protein, and the protein sample stacks up at this point as a sharp band. A similar-sized protein, but with slightly lower M_r, will be able to travel a little further through the gel before reaching its pore size limit, at which point it will form a sharp band. These two proteins, of slightly different M_r values, therefore separate as two, close, sharp bands.

The usual limits of gradient gels are 3–30% acrylamide in linear or concave gradients. The choice of range will of course depend on the size of proteins being fractionated. The system described here is for a 5–20% linear gradient using SDS polyacrylamide gel electrophoresis. The theory of SDS polyacrylamide gel electrophoresis has been decribed in Chapter 11.

2. Materials

1. Stock acrylamide solution: 30% acrylamide, 0.8% *bis*-acrylamide. Dissolve 75 g of acrylamide and 2.0 g of *N,N'*-methylene *bis*-acrylamide in about 150 mL of water. Filter and make the volume to 250 mL. Store at 4°C. The solution is stable for months.

From: *The Protein Protocols Handbook, 2nd Edition*
Edited by: J. M. Walker © Humana Press Inc., Totowa, NJ

2. Buffers:
   a. 1.875 *M* Tris-HCl, pH 8.8.
   b. 0.6 *M* Tris-HCl, pH 6.8.
   Store at 4°C.
3. Ammonium persulfate solution (10% [w/v]). Make fresh as required.
4. SDS solution (10% [w/v]). Stable at room temperature. In cold conditions, the SDS can come out of solution, but may be redissolved by warming.
5. *N,N,N',N'*-Tetramethylene diamine (TEMED).
6. Gradient forming apparatus (*see* **Fig. 1**). Reservoirs with dimensions of 2.5 cm id and 5.0 cm height are suitable. The two reservoirs of the gradient former should be linked by flexible tubing to allow them to be moved independently. This is necessary since although equal volumes are placed in each reservoir, the solutions differ in their densities and the relative positions of A and B have to be adjusted to balance the two solutions when the connecting clamp is opened (*see* **Note 3**).

3. Method

1. Prepare the following solutions:

	Solution A, mL	Solution B, mL
1.875 *M* Tris-HCl, pH 8.8	3.0	3.0
Water	9.3	0.6
Stock acrylamide, 30%	2.5	10.0
10% SDS	0.15	0.15
Ammonium persulfate (10%)	0.05	0.05
Sucrose	—	2.2 g (equivalent to 1.2 mL volume)

2. Degas each solution under vacuum for about 30 s and then, when you are ready to form the gradient, add TEMED (12 μL) to each solution.
3. Once the TEMED is added and mixed in, pour solutions A and B into the appropriate reservoirs (*see* **Fig. 1**.)
4. *With the stirrer stirring,* fractionally open the connection between A and B and adjust the relative heights of A and B such that there is no flow of liquid between the two reservoirs (easily seen because of the difference in densities). Do not worry if there is some mixing between reservoirs—this is inevitable.
5. When the levels are balanced, completely open the connection between A and B, turn the pump on, and fill the gel apparatus by running the gel solution down one edge of the gel slab. Surprisingly, very little mixing within the gradient occurs using this method. A pump speed of about 5 mL/min is suitable. If a pump is not available, the gradient may be run into the gel under gravity.
6. When the level of the gel reaches about 3 cm from the top of the gel slab, connect the pump to distilled water, reduce pump speed, and overlay the gel with 2–3 mm of water.
7. The gradient gel is now left to set for 30 min. Remember to rinse out the gradient former before the remaining gel solution sets in it.
8. When the separating gel has set, prepare a stacking gel by mixing the following:
   a. 1.0 mL 0.6 *M* Tris-HCl, pH 6.8;
   b. 1.35 mL Stock acrylamide;
   c. 7.5 mL Water;
   d. 0.1 mL 10% SDS;
   e. 0.05 mL Ammonium persulfate (10%).

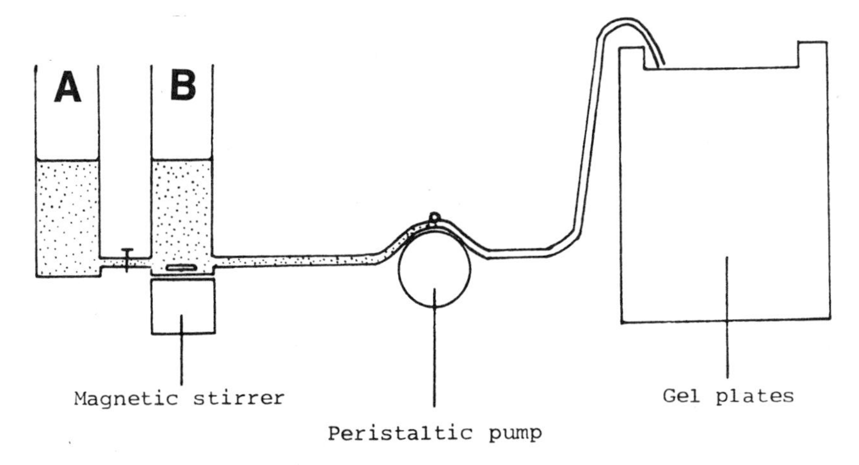

Fig. 1. Gradient forming apparatus.

9. Degas this mixture under vacuum for 30 s and then add TEMED (12 µL).
10. Pour off the water overlayering the gel and wash the gel surface with about 2 mL of stacking gel solution and then discard this solution.
11. The gel slab is now filled to the top of the plates with stacking gel solution and the well-forming comb placed in position (*see* Chapter 11).
12. When the stacking gel has set (~15 min), carefully remove the comb. The gel is now ready for running. The conditions of running and sample preparation are exactly as described for SDS gel electrophoresis in Chapter 11.

4. Notes

1. The total volume of liquid in reservoirs A and B should be chosen such that it approximates to the volume available between the gel plates. However, allowance must be made for some liquid remaining in the reservoirs and tubing.
2. As well as a gradient in acrylamide concentration, a density gradient of sucrose (glycerol could also be used) is included to minimize mixing by convectional disturbances caused by heat evolved during polymerization. Some workers avoid this problem by also including a gradient of ammonium persulfate to ensure that polymerization occurs first at the top of the gel, progressing to the bottom. However, we have not found this to be necessary in our laboratory.
3. The production of a linear gradient has been described in this chapter. However, the same gradient mixed can be used to produce a concave (exponential) gradient. This concave gradient provides a very shallow gradient in the top half of the gel such that the percentage of acrylamide only varies from about 5–7% over the first half of the gel. The gradient then increases much more rapidly from 7–20% over the next half of the gel. The shallow part of the gradient allows high-mol-wt proteins of similar size to sufficiently resolve while at the same time still allowing lower mol-wt proteins to separate lower down the gradient. To produce a concave gradient, place 7.5 mL of solution B in reservoir B, then tightly stopper this reservoir with a rubber bung. Equalize the pressure in the chamber by briefly inserting a syringe needle through the bung. Now place 22.5 mL of solution A in reservoir A, open the connector between the two chambers, and commence pouring the gel. The volume of reservoir B will be seen to remain constant as liquid for reservoir A is drawn into this reservoir and diluted.

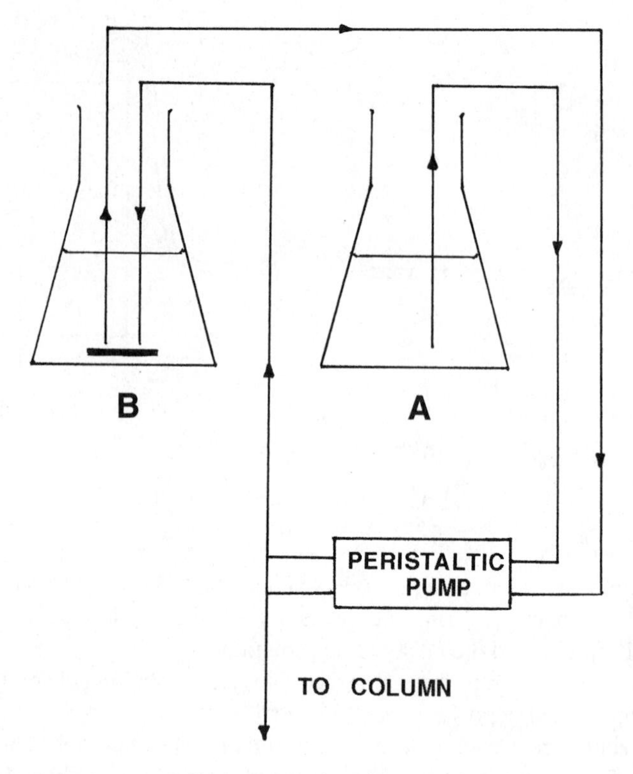

Fig. 2. Diagrammatic representation of a method for producing a gradient using a two-channel peristaltic pump. Reservoir B has the high percentage acrylamide concentration, reservoir A the lower.

4. We have described the production of a linear gradient using a purpose built gradient mixer. However, it is not necessary to purchase this since the simple arrangement, shown in **Fig. 2** usin1g just flasks or beakers, a stirrer, and a dual channel peristaltic pump, can just as easily be used.

Reference

1. Margolis, J. and Kenrick, K. G. (1967) *Nature (London),* **214,** 1334.

13

SDS-Polyacrylamide Gel Electrophoresis of Peptides

Ralph C. Judd

1. Introduction

Sodium dodecyl sulfate-polyacrylamide gel electrophoresis (SDS-PAGE) has proven to be among the most useful tools yet developed in the area of molecular biology. The discontinuous buffer system, first described by Laemmli *(1)*, has made it possible to separate, visualize, and compare readily the component parts of complex mixtures of molecules (e.g., tissues, cells). SDS-PAGE separation of proteins and peptides makes it possible to quantify the amount of a particular protein/peptide in a sample, obtain fairly reliable molecular mass information, and, by combining SDS-PAGE with immunoelectroblotting, evaluate the antigenicity of proteins and peptides. SDS-PAGE is both a powerful separation system and a reliable preparative purification technique *(2*; and *see* Chapter 11*)*.

Parameters influencing the resolution of proteins or peptides separated by SDS-PAGE include the ratio of acrylamide to crosslinker (*bis*-acrylamide), the percentage of acrylamide/crosslinker used to form the stacking and separation gels, the pH of (and the components in) the stacking and separation buffers, and the method of sample preparation. Systems employing glycine in the running buffers (e.g., Laemmli *[1]*, Dreyfuss et al. *[3]*) can resolve proteins ranging in molecular mass from over 200,000 Daltons (200 kDa) down to about 3 kDa. Separation of proteins and peptides below 3 kDa necessitates slightly different procedures to obtain reliable molecular masses and to prevent band broadening. Further, the increased use of SDS-PAGE to purify proteins and peptides for N-terminal sequence analysis demands that glycine, which interferes significantly with automated sequence technology, be replaced with noninterfering buffer components.

This chapter describes a modification of the tricine gel system of Schagger and von Jagow *(4)* by which peptides as small as 500 Daltons can be separated. This makes it possible to use SDS-PAGE peptide mapping (*see* Chapter 80), epitope mapping *(5)*, and protein and peptide separation for N-terminal sequence analyses *(6)* when extremely small peptide fragments are to be studied. Since all forms of SDS-PAGE are denaturing, they are unsuitable for separation of proteins or peptides to be used in functional analyses (e.g., enzymes, receptors).

From: *The Protein Protocols Handbook, 2nd Edition*
Edited by: J. M. Walker © Humana Press Inc., Totowa, NJ

2. Materials

2.1. Equipment

1. SDS-PAGE gel apparatus.
2. Power pack.
3. Blotting apparatus.

2.2. Reagents

1. Separating/spacer gel acrylamide (1X crosslinker): 48 g acrylamide, 1.5 g *N,N'*-methylene-*bis*-acrylamide. Bring to 100 mL, and then filter through qualitative paper to remove cloudiness (*see* **Note 1**).
2. Separating gel acrylamide (2X crosslinker): 48 g acrylamide, 3 g *N,N'*-methylenebis-acrylamide. Bring to 100 mL, and then filter through qualitative paper to remove cloudiness (*see* **Note 2**).
3. Stacking gel acrylamide: 30 g acrylamide, 0.8 g *N,N'*-methylene-*bis*-acrylamide. Bring to 100 mL, and then filter through qualitative paper to remove cloudiness.
4. Separating/spacer gel buffer: 3 *M* Trizma base, 0.3% sodium dodecyl sulfate (*see* **Note 3**). Bring to pH 8.9 with HCl.
5. Stacking gel buffer: 1 *M* Tris-HCl, pH 6.8.
6. Cathode (top) running buffer (10X stock): 1 *M* Trizma base, 1 *M* tricine, 1% SDS (*see* **Note 3**). Dilute 1:10 immediately before use. Do not adjust pH; it will be about 8.25.
7. Anode (bottom) buffer (10X stock): 2 *M* Trizma base. Bring to pH 8.9 with HCl. Dilute 1:10 immediately before use.
8. 0.2 *M* tetrasodium EDTA.
9. 10% ammonium persulfate (make fresh as required).
10. TEMED.
11. Glycerol.
12. Fixer/destainer: 25% isopropanol, 7% glacial acetic acid in dH_2O (v/v/v).
13. 1% Coomassie brilliant blue (CBB) (w/v) in fixer/destainer.
14. Sample solubilization buffer: 2 mL 10% SDS (w/v) in dH_2O, 1.0 mL glycerol, 0.625 mL 1 *M* Tris-HCl, pH 6.8, 6 mL dH_2O, bromphenol blue to color.
15. Dithiothreitol.
16. 2% Agarose.
17. Molecular-mass markers, e.g., low-mol-wt kit (Bio-Rad, Hercules, CA), or equivalent, and peptide molecular-mass markers (Pharmacia Inc., Piscataway, NJ), or equivalent.
18. PVDF (nylon) membranes.
19. Methanol.
20. Blotting transfer buffer: 20 m*M* phosphate buffer, pH 8.0: 94.7 mL 0.2 *M* Na_2HPO_4 stock, 5.3 mL 0.2 *M* NaH_2PO_4 stock in 900 mL H_2O.
21. Filter paper for blotting (Whatman No. 1), or equivalent.
22. Distilled water (dH_2O).

2.3. Gel Recipes

2.3.1. Separating Gel Recipe

Add reagents in order given (*see* **Note 4**): 6.7 mL water, 10 mL separating/spacer gel buffer, 10 mL separating/spacer gel acrylamide (1X or 2X crosslinker), 3.2 mL glycerol, 10 µL TEMED, 100 µL 10% ammonium persulfate.

2.3.2. Spacer Gel Recipe

Add reagents in order given (*see* **Note 4**): 6.9 mL water, 5.0 mL separating/spacer gel buffer, 3.0 mL separating/spacer gel acrylamide (1X crosslinker only), 5 μL TEMED, 50 μL 10% ammonium persulfate.

2.3.3. Stacking Gel Recipe

Add reagents in order given (*see* **Note 4**): 10.3 mL water, 1.9 mL stacking gel buffer, 2.5 mL stacking gel acrylamide, 150 μL EDTA, 7.5 μL TEMED, 150 μL 10% ammonium persulfate.

3. Methods

3.1. Sample Solubilization

1. Boil samples in sample solubilization buffer for 10–30 min. Solubilize sample at 1 mg/mL and run 1–2 μL/lane (1–2 μg/lane) (*see* **Note 5**). For sequence analysis, as much sample as is practical should be separated.

3.2. Gel Preparation/Electrophoresis

1. Assemble the gel apparatus (*see* **Note 6**). Make two marks on the front plate to identify top of separating gel and top of spacer gel (*see* **Note 7**). Assuming a well depth of 12 mm, the top of the separating gel should be 3.5 cm down from the top of the back plate, and the spacer gel should be 2 cm down from the top of the back plate, leaving a stacking gel of 8 mm (*see* **Note 8**).
2. Combine the reagents to make the separating gel, mix gently, and pipet the solution between the plates to lowest mark on the plate. Overlay the gel solution with 2 mL of dH_2O by gently running the dH_2O down the center of the inside of the front plate. Allow the gel to polymerize for about 20 min. When polymerized, the water–gel interface will be obvious.
3. Pour off the water, and dry between the plates with filter paper. Do not touch the surface of the separating gel with the paper. Combine the reagents to make the spacer gel, mix gently, and pipet the solution between the plates to second mark on the plate. Overlay the solution with 2 mL of dH_2O by gently running the dH_2O down the center of the inside of the front plate. Allow the gel to polymerize for about 20 min. When polymerized, the water–gel interface will be obvious.
4. Pour off the water, and dry between the plates with filter paper. Do not touch the surface of spacer gel with the paper. Combine the reagents to make the stacking gel and mix gently. Place the well-forming comb between the plates, leaving one end slightly higher than the other. Slowly add the stacking gel solution at the raised end (this allows air bubbles to be pushed up and out from under the comb teeth). When the solution reaches the top of the back plate, gently push the comb all the way down. Check to be sure that no air pockets are trapped beneath the comb. Allow the gel to polymerize for about 20 min.
5. When the stacking gel has polymerized, carefully remove the comb. Straighten any wells that might be crooked with a straightened metal paper clip. Remove the acrylamide at each edge to the depth of the wells. This helps prevent "smiling" of the samples at the edge of the gel. Seal the edges of the gel with 2% agarose.
6. Add freshly diluted cathode running buffer to the top chamber of the gel apparatus until it is 5–10 mm above the top of the gel. Squirt running buffer into each well with a Pasteur pipet to flush out any unpolymerized acrylamide. Check the lower chamber to ensure that no cathode running buffer is leaking from the top chamber, and then fill the bottom cham-

ber with anode buffer. Remove any air bubbles from the under edge of the gel with a bent-tip Pasteur pipet. The gel is now ready for sample loading.

7. After loading the samples and the molecular-mass markers, connect leads from the power pack to the gel apparatus (the negative lead goes on the top, and the positive lead goes on the bottom). Gels can be run on constant current, constant voltage, or constant power settings. When using the constant current setting, run the gel at 50 mA. The voltage will be between 50 and 100 V at the beginning, and will slowly increase during the run. For a constant voltage setting, begin the electrophoresis at 50 mA. As the run progresses, the amperage will decrease, so adjust the amperage to 50 mA several times during the run or the electrophoresis will be very slow. If running on constant power, set between 5 and 7 W. Voltage and current will vary to maintain the wattage setting. Each system varies, so empirical information should be used to modify the electrophoresis conditions so that electrophoresis is completed in about 4 h (*see* **Note 9**).
8. When the dye front reaches the bottom of the gel, turn off the power, disassemble the gel apparatus, and place the gel in 200–300 mL of fixer/destainer. Gently shake for 16 h (*see* **Note 10**). Pour off spent fixer/destainer, and add CBB. Gently shake for 30 min. Destain the gel in several changes of fixer/destainer until the background is almost clear. Then place the gel in dH_2O, and gently mix until the background is completely clear. The peptide bands will become a deep purple-blue. The gel can now be photographed or dried. To store the gel wet, soak the gel in 7% glacial acetic acid for 1 h, and seal in a plastic bag.

Figure 1 demonstrates the molecular mass range of separation of a 1X crosslinker tricine gel. Whole-cell (WC) lysates and 1X and 2X purified (*see* Chapter 80) 44 kDa proteins of *Neisseria gonorrhoeae*, Bio-Rad low-mol-wt markers (mw), and Pharmacia peptide markers (pep mw) were separated and stained with CBB. The top of the gel in this figure is at the spacer gel–separating gel interface. Proteins larger than about 100 kDa remained trapped at the spacer gel–separating gel interface, resulting in the bulging of the outside lanes. Smaller proteins all migrated into the gel, but many remained tightly bunched at the top of the separating gel. The effective separation range is below 40 kDa. Comparison of this figure with **Fig. 1** in Chapter 80, which shows gonococcal whole cells and the two mol-wt marker preparations separated in a standard 15% Laemmli gel *(1)*, demonstrates the tremendous resolving power of low-mol-wt components by this tricine gel system.

3.3. Blotting of Peptides

Separated peptides can be electroblotted to PVDF membranes for sequencing or immunological analyses (*see* **Note 11**).

1. Before the electrophoresis is complete, prepare enough of the 20 m*M* sodium phosphate transfer buffer, pH 8.0 (*see* **Note 12**) to fill the blotting chamber (usually 2–4 L). Degas about 1 L of transfer buffer for at least 15 min before use. Cut two sheets of filter paper to fit blotting apparatus, and cut a piece of PVDF membrane a little larger than the gel. Place the PVDF membrane in 10 mL of methanol until it is wet (this takes only a few seconds), and then place the membrane in 100 mL of degassed transfer buffer.
2. Following electrophoresis, remove the gel from the gel apparatus, and place it on blotting filter paper that is submersed in the degassed transfer buffer. Immediately overlay the exposed side of the gel with the wetted PVDF membrane, being sure to remove all air pockets between the gel and the membrane. Overlay the PVDF membrane with another piece of blotting filter paper, and place the gel "sandwich" into the blotting chamber using the appropriate spacers and holders.

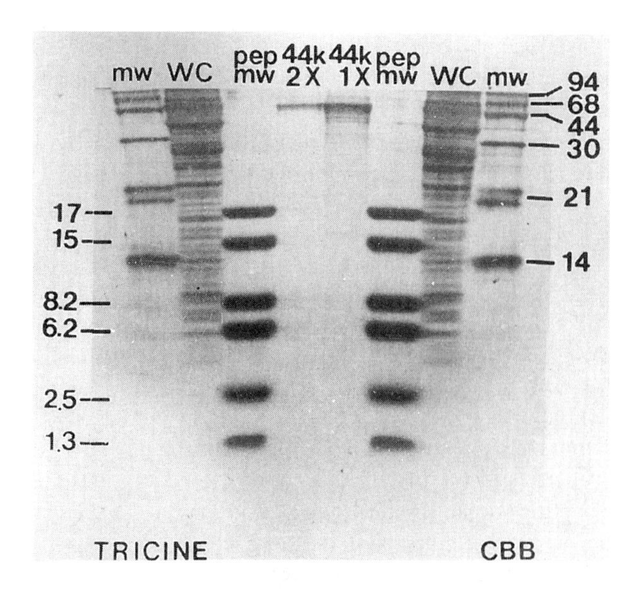

Fig. 1. WC lysates and 1X and 2X purified 44 kDa protein of *Neisseria gonorrhoeae* (*see* Chapter 79, **Subheading 3.1.**), Bio-Rad mw (1 µg of each protein), and Pharmacia pep mw (to which the 1.3-kDa protein kinase C substrate peptide [Sigma, St. Louis, MO] was added) (3 µg of each peptide), separated in a 1X crosslinker tricine gel, fixed, and stained with CBB. Molecular masses are given in thousands of daltons.

3. Connect the power pack electrodes to the blotting chamber (the positive electrode goes on the side of the gel having the PVDF membrane). Electrophorese for 16 h at 25 V, 0.8 A. Each system varies, so settings may be somewhat different than those described here.
4. Following electroblotting, disconnect the power, disassemble the blotting chamber, and remove the PVDF membrane from the gel (*see* **Note 13**). The PVDF membrane can be processed for immunological analyses or placed in CBB in fixer/destainer to stain the transferred peptides. Remove excess stain by shaking the membrane in several changes of fixer/destainer until background is white. Peptide bands can be excised, rinsed in dH_2O, dried, and subjected to N-terminal sequencing.

3.4. Modifications for Peptide Sequencing

Peptides to be used in N-terminal sequence analyses must be protected from oxidation, which can block the N-terminus. Several simple precautions can help prevent this common problem.

1. Prepare the separation and spacer gel the day before electrophoresing the peptides. After pouring, overlay the spacer gel with several milliliters of dH_2O, and allow the gel to stand overnight at room temperature (*see* **Note 14**).
2. On the next day, pour off the water, and dry between the plates with filter paper. Do not touch the gel with the filter paper. Prepare the stacking gel, but use half the amount of 10% ammonium persulfate (*see* **Note 15**). Pipet the stacking gel solution between the plates as described above, and allow the stacking gel to polymerize for at least 1 h. Add running buffers as described above, adding 1–2 mg of dithiothreitol to both the upper and lower chambers to scavenge any oxidizers from the buffers and gel.
3. Pre-electrophorese the gel for 15 min. then turn off the power, and load the samples and molecular-mass markers.

4. Run the gel as described above.
5. Blot the peptides to a PVDF membrane as described above, but add 1–2 mg of dithiothreitol to the blot transfer buffer.
6. Fix and stain as above, again adding 1–2 mg of dithiothreitol to the fixer/destainer, CBB, and dH_2O used to rinse the peptide-containing PVDF membrane.

4. Notes

1. Working range of separation about 40 kDa down to about 1 kDa.
2. Working range of separation about 20 kDa down to less than 500 Dalton.
3. Use electrophoresis grade SDS. If peptide bands remain diffuse, try SDS from BDH (Poole, Dorset, UK).
4. Degassing of gel reagents is **not** necessary.
5. Coomassie staining can generally visualize a band of 0.5 µg. This may vary considerably based on the properties of the particular peptide (some peptides stain poorly with Coomassie). Some peptides do not bind SDS well and may never migrate exactly right when compared to mol-wt markers. Fortunately, these situations are rare.
6. Protocols are designed for a standard 13 cm × 11 cm × 1.5 mm slab gel. Dimensions and reagent volumes can be proportionally adjusted to accommodate other gel dimensions.
7. Permanent marks with a diamond pencil can be made on the back of the back plate if the plate is dedicated to this gel system.
8. The depth of the spacer gel can be varied from 1 to 2 cm. Trial and error is the only way to determine the appropriate dimension for each system.
9. It is wise to feel the front plate several times during the electrophoresis to check for overheating. The plate will become pleasantly warm as the run progresses. If it becomes too warm, the plates might break, so turn down the power!
10. Standard-sized gels can be fixed in as little as 4 h with shaking.
11. It is best to blot peptides to PVDF membranes rather than nitrocellulose membranes, since small peptides tend to pass through nitrocellulose without binding. Moreover, peptides immobilized on PVDF membranes can be directly sequenced in automated instrumentation equipped with a "blot cartridge" *(6)*.
12. The pH of the transfer buffer can be varied from 5.7 to 8.0 if transfer is inefficient at pH 8.0 *(7)*.
13. Wear disposable gloves when handling membranes.
14. Do not refrigerate the gel. It will contract and pull away from the plates, resulting in leaks and poor resolution.
15. Do not pour the stacking gel the day before electrophoresis. It will shrink, allowing the samples to leak from the wells.

Acknowledgments

The author thanks Joan Strange for her assistance in developing this system, Pam Gannon for her assistance, and the Public Health Service, NIH, NIAID (grants RO1 AI21236) and UM Research Grant Program for their continued support.

References

1. Laemmli, U. K. (1970) Cleavage of structural proteins during the assembly of the head of bacteriophage T4. *Nature* **227,** 680–695.
2. Judd, R. C. (1988) Purification of outer membrane proteins of the Gram negative bacterium *Neisseria gonorrhoeae. Analyt. Biochem.* **173,** 307–316.

3. Dreyfuss, G., Adam, S. A., and Choi, Y. D. (1984) Physical change in cytoplasmic messenger ribonucleoproteins in cells treated with inhibitors of mRNA transcription. *Mol. Cell. Biol.* **4,** 415–423.
4. Schagger, H. and von Jagow, G. (1987) Tricine-sodium dodecyl sulfate-polyacrylamide gel electrophoresis for the separation of proteins in the range from 1 to 100 kDa. *Analyt. Biochem.* **166,** 368–397.
5. Judd, R. C. (1986) Evidence for N-terminal exposure of the PIA subclass of protein I of *Neisseria gonorrhoeae. Infect. Immunol.* **54,** 408–414.
6. Moos, M., Jr. and Nguyen, N. Y. (1988) Reproducible high-yield sequencing of proteins electrophoretically separated and transferred to an inert support. *J. Biol. Chem.* **263,** 6005–6008.
7. Stoll, V. S. and Blanchard, J. S. (1990) Buffers: principles and practice, in *Methods in Enzymology,* vol. 182, *A Guide to Protein Purification* (Deutscher, M. P., ed.), Academic, San Diego, CA, pp. 24–38.

14

Identification of Nucleic Acid Binding Proteins Using Nondenaturing Sodium Decyl Sulfate Polyacrylamide Gel Electrophoresis (SDecS-PAGE)

Robert E. Akins and Rocky S. Tuan

1. Introduction

Methods for the identification and characterization of nucleic acid binding proteins, such as DNA binding transcription factors, typically involve gel retardation assays *(1,2)* or Southwestern analysis *(3)*. Gel retardation assays allow the detection of DNA-binding factors by assessing the degree to which protein binding affects the electrophoretic mobility of specific DNA sequences. One drawback of gel retardation assays is that only the presence of the protein is indicated; specific information concerning protein molecular weight, or other characteristics, is obtained only through additional methods. Southwestern analysis detects DNA binding proteins through the use of nucleic acid probes applied to protein blots prepared from sodium dodecyl sulfate (SDS) gels. The Southwestern technique relies on the limited ability of proteins to renature after SDS-gel electrophoresis and does not specifically identify binding by protein complexes, which are dissociated during sample processing.

In this chapter, we describe a novel method for the identification of nucleic acid binding proteins and protein complexes. The method is based on a nondenaturing gel electrophoresis system. The system allows the separation of proteins and protein complexes as a function of log M_r with the maintenance of DNA oligomer binding activity. The use of the anionic detergent sodium **decyl** sulfate (SDecS) (*see* **Note 1**) in the gel electrophoresis system allows the identification of specific DNA binding proteins or protein complexes based on molecular size.

The SDecS-polyacrylamide gel electrophoresis (SDecS-PAGE) technique is straightforward. In brief, samples solubilized in a SDecS buffer are electrophoresed in a discontinuous gel *(4,5)* comprised of acrylamide with SDecS in a Tris-HCl buffer and using a Tris-glycine running buffer that also contains SDecS. The resulting formulation is similar to standard sodium **dodecyl** sulfate (SDS) electrophoresis systems with SDecS directly substituted for SDS.

From: *The Protein Protocols Handbook, 2nd Edition*
Edited by: J. M. Walker © Humana Press Inc., Totowa, NJ

After electrophoresis at 4°C, gels are removed from the apparatus and cut into several slices. The slices are stained for total protein or probed with radiolabeled cDNAs. The resulting staining and binding profiles can then be used to identify protein bands that specifically bind the cDNA probe. The SDecS system allows the rapid identification of nucleic acid binding moieties and allows the assessment of changes in binding level or in the M_r of the binding complex, for example, when additional moieties associate with or dissociate from the binding protein.

The SDecS system is conceptually related to the CAT-gel system, which is described in Chapter 15 of this volume. The reason for choosing SDecS- or cetyltrimethylammonium bromide- (CTAB-) based methods as substitutes for SDS electrophoresis systems is related to how the side chain and head group of a detergent affect the condensation of protein–detergent mixed micelles. The conformation of these mixed micelles affects protein structure and, hence, protein denaturation. In reports of electrophoretic systems that allow the separation of proteins based on M_r with the retention of native activity ***(6–8)***, the use of CTAB instead of SDS for the analysis of proteins has been discussed. These reports suggest that CTAB-based methods may be suitable for the identification of nucleic acid binding proteins; unfortunately, CTAB is a cationic detergent, which tends to precipitate anionic nucleic acids, and therefore is not useful in the present application. SDecS, on the other hand, is an anionic detergent, which does not precipitate nucleic acids. In the authors' laboratories, proteins solubilized in SDecS retain detectable levels of native activity, including nucleic acid binding activity.

The SDecS system allows protein mixtures to be fractionated under relatively nondenaturing conditions and to be subsequently probed with specific nucleic acid oligomers. After rinsing, probe binding is detected, and bands are assigned M_r values based on the relative distance migrated. M_r values for binding proteins may be affected by a variety of conditions, including gene product truncation during expression, posttranslational modification, protein multimerization, and assembly of multiprotein complexes. The SDecS system can be used to analyze the relationships between these conditions and the appearance of nucleic acid binding in different samples, for example, before and after drug treatment. The SDecS system should prove useful for the analysis of previously identified nucleic acid binding proteins and for the identification of novel binding proteins or protein complexes.

2. Materials (*See* Note 2)

1. 40% Acrylamide (Amresco, Solon, OH).
2. X-AR Autoradiography film (Kodak, Rochester, NY).
3. KinAce-It Kinasing Kit (Stratagene, LaJolla, CA) (*see* **Note 3**).
4. γ-^{32}PATP (ICN, Irvine, CA).
5. Protein M_r markers (Sigma).
6. 2X SDecS sample buffer: 2% SDecS, 20 m*M* Tris-HCl, pH 8.0; 2 m*M* EDTA, and 10% glycerol.
7. Separating gel solution: 8% T acrylamide; 375 m*M* Tris-HCl, pH 8.8, 0.1% SDecS.
8. Stacking gel solution: 4% T acrylamide; 125 m*M* Tris-HCl, pH 6.8, 0.1% SDecS.
9. Coomassie Brilliant Blue R-250 (CBB) stain: 0.25% CBB, 10% acetic acid, 50% methanol.
10. CBB destain: 7.5% acetic acid, 5% methanol.
11. 5X Rinsing buffer: 50 m*M* Tris-HCl, 50 m*M* phosphate buffer, 50 m*M* NaCl, 5 m*M* $MgCl_2$, 5 μ*M* $ZnCl_2$, and 2.5 m*M* DTT, pH 7.4.

12. 1× Blocking buffer: 1× Rinsing buffer plus 25 μg/mL of herring sperm DNA.
13. 1× Binding buffer: 1× Blocking buffer plus 100 ng of labeled probe.

3. Methods

The first part of the SDecS method is nearly identical to that described in Chapter 11, *SDS Polyacrylamide Gel Electrophoresis of Proteins*. The reader is referred to that chapter for details on the preparation of polyacrylamide gels. There are four main differences between the method described in Chapter 11 and the SDecS method: (1) SDecS replaces SDS in all solutions (*see* **Note 1**), (2) no reducing agent is added to the sample buffer (*see* **Note 2**), (3) samples are NOT boiled prior to loading the gel, and (4) the system is operated in the cold (4°C). The SDecS-PAGE technique is outlined in **Fig. 1**.

3.1. Sample Preparation

Samples are solubilized by combining nuclear extract or cell homogenate 1:1 with 2× SDecS sample buffer. Insoluble material is centrifuged in a microfuge immediately prior to loading the gel.

3.2. Electrophoresis

1. Samples are loaded onto discontinuous polyacrylamide minigels consisting of 8% T acrylamide separating gels polymerized in 375 m*M* Tris-HCl, pH 8.8 with 0.1% SDecS and 4% T acrylamide stacking gels polymerized in 125 m*M* Tris-HCl, pH 6.8 with 0.1% SDecS.
2. Gels are run at 4°C in a Bio-Rad Mini-Protean II apparatus by applying 100 V through the stacking gel and 150 V through the separating gel at constant voltage until the discontinuous buffer front approaches the bottom of the gel.

3.3. Staining/Probing of Gel Slices

1. After electrophoresis, the gel is removed from the apparatus and cut into three sections.
2. The first section, containing protein samples and M_r markers, can be stained using a solution of 0.25% CBB R-250, 10% acetic acid, 50% methanol, and destained using a solution of 7.5% acetic acid and 5% methanol until bands are visible. A plot of distance migrated vs known M_r of the marker proteins approximates a linear standard curve that can be used for the assignment of M_r values to sample proteins. An example of such a plot is given in **Fig. 2**.
3. The second and third sections cut from the gel are rinsed for 10 min. in DNA blocking buffer, then separated and incubated with DNA binding buffer (*see* **Note 3**) containing either experimental or control ^{32}P-end-labeled oligo probe.
4. Binding is allowed to occur for 30 min at room temperature after which the gel slices are rinsed with DNA rinsing buffer to remove any unbound probe (*see* **Note 4**).
5. The gel slices are then stained with CBB and destained for 10 min, covered with plastic wrap, and exposed to X-ray film at room temperature. In cases where a low amount of probe is found, gels may be dried down and exposed to X-ray film at –70°C for extended periods of time.

4. Notes

1. SDecS and SDS are similar detergents. The only difference between the two pure chemicals is in the length of the carbon side chain associated with the sulfate group (10 carbons

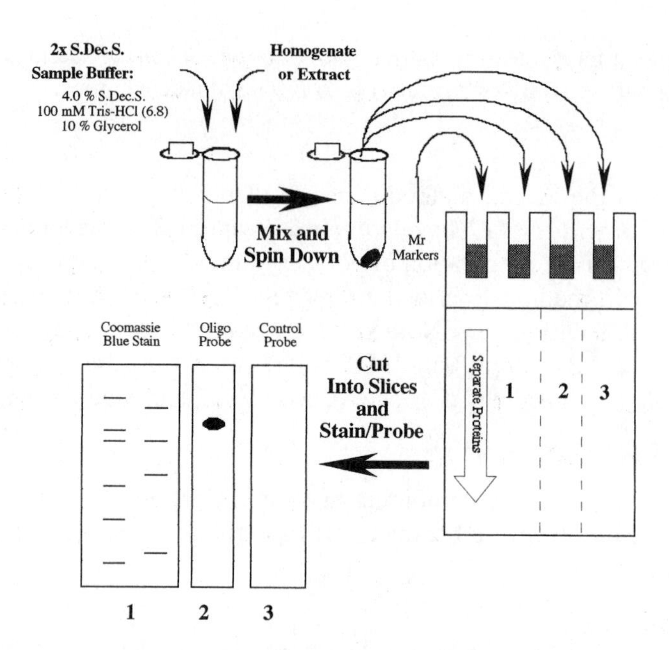

Fig. 1. Diagram of the use of an SDecS Gel System. The cell homogenate, nuclear extract, or another mixture of proteins is combined in a 1:1 ratio (v:v) with 2× Sample Buffer. After solubilization, any debris is centrifuged, and the sample is separated by electrophoresis on a polyacrylamide gel. The gel is then cut into strips, and the protein bands are identified by CBB staining or binding to labeled probe.

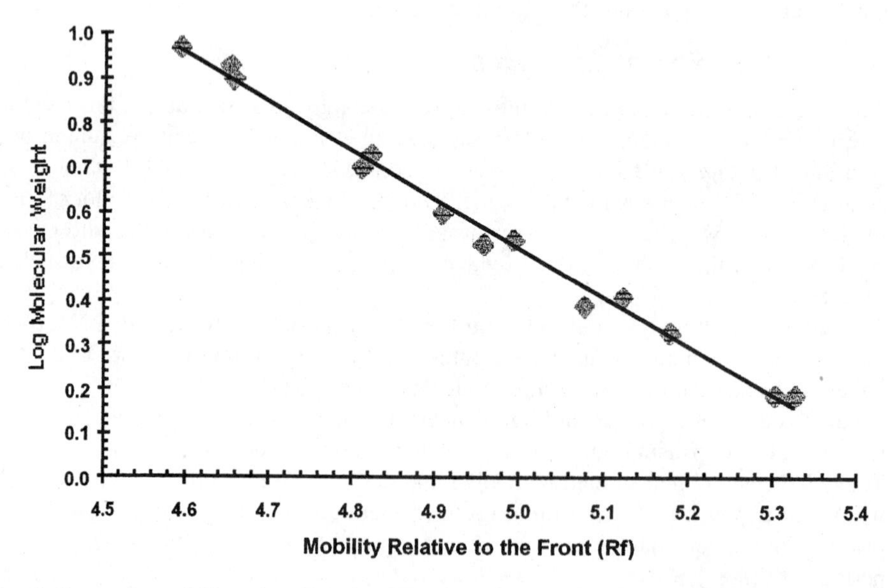

Fig. 2. Example mobility plot. Graph of the distance migrated in an SDecS gel relative to the salt front as a function of protein molecular weight. The proteins (M_r in kDa) included: α-macroglobulin (211); β-amylase (200); alcohol dehydrogenase (150); β-galactosidase (119); fructose-6-phosphate kinase (98); pyruvate kinase (80.6); bovine serum albumin (66 and 132); fumarase (64.4); ovalbumin (45 and 90); lactate dehydrogenase (44.6); triosephophate isomerase (38.9). $R^2 = 0.99$.

for SDecS vs 12 for SDS). In practice, preparations of detergents actually contain mixtures of detergents with different side chain lengths. The contribution of these multiple moieties to the formation of mixed micelles may affect protein structure and, therefore, function. Protein–detergent mixtures form mixed micelles in which the relative association of detergent and protein is balanced by a concentration of detergent that remains as monomer. The composition of these mixed micelles affects both protein solubility and protein structure/activity. A small amount of SDS (or other detergent) in mixed micelles with SDecS and protein may improve the overall solubility of the desired protein while maintaining native activity. Thus, it is advisable to compare different preparations of SDecS in terms of the retention of the desired native activity.

It should also be noted that sodium dodecyl **sulfonate**, which has a different polar head group than SDS, has been successfully used by the authors in the current methodology. Alkylsulfonates, alone or in combination with alkylsulfates, may represent additional options for the analysis of nucleic acid binding proteins. In most cases, depending on solubility, critical micelle concentration, and so forth, anionic detergents and mixtures of anionic detergents can be substituted directly for SDecS in the methods given in this chapter.

2. Specific components of the buffers described in this chapter can be modified to arrive at an acceptable level of DNA-binding activity. In the sample buffer, which is used to solubilize the protein mixture prior to electrophoresis, the amount of SDecS, ethylenediaminetetraacetic acid (EDTA), and glycerol can be modified to improve solubilization. Alternatively, the ratio of sample buffer to sample, which is 1:1 in the standard technique, can be adjusted, or a reducing agent such DTT or β-mercaptoethanol can be added to arrive at the desired formulation for the specific binding activity. The buffers described here are recommended for general use.
3. It is important to note that nuclear extracts and cell homogenates contain numerous proteins that may bind to single-stranded (ss) or double-stranded (ds) nucleic acid probes in a nonspecific manner. It is, therefore, important to block these interactions by including nonlabeled carrier ss- or ds-nucleic acid. In many cases, sheared herring sperm DNA provides an excellent blocking agent. Alternatives include other polyanionic polymers or small amounts of mild surfactant. The inclusion of any compound must be weighed against potential negative effects on the activity to be measured.
4. The use of radiolabeling is not a necessity in the SDecS system. Although we have successfully used ^{32}P-based detection systems, other methods of detection, for example, those based on digoxygenin or biotin tags, should work equally as well. The main concern to keep in mind is that the labeling should not affect the specificity or affinity of the probe.

References

1. Fried, M. and Crothers, D. M. (1981) Equilibria and kinetics of lac repressor-operator interactions by polyacrylamide gel electrophoresis. *Nucleic Acids. Res.* **9,** 6505–6525.
2. Garner, M. M. and Revzin, A. (1981) A gel electrophoresis method for quantifying the binding of proteins to specific DNA regions: application to components of the *Escherichia coli* lactose operon regulatory system. *Nucleic Acids Res.* **9,** 3047–3060.
3. Bowen, B., Steinberg, J., Laemmli, U. K., and Weintraub, H. (1980) The detection of DNA-binding proteins by protein blotting. *Nucleic Acids Res.* **8,** 1–20.
4. Ornstein, L. (1964) Disc electrophoresis I: Background and theory. *Ann. NY Acad. Sci.* **121,** 321–349.
5. Davis, B. J. (1964) Disc electrophoresis II: Method of application to human serum proteins. *Ann. NY Acad. Sci.* **121,** 404–427.

6. Akins, R. E. and Tuan, R. S. (1994) Separation of proteins using cetyltrimethylammonium bromide discontinuous gel electrophoresis. *Mol. Biotechnol.* **1,** 211–228.
7. Akins, R. E., Levin, P. M., and Tuan, R. S. (1992) Cetyltrimethylammonium bromide discontinuous gel electrophoresis: M_r-based separation of proteins with retention of enzymatic activity. *Analyt. Biochem.* **202,** 172–178.
8. Akin, D., Shapira, R., and Kinkade, J. M. (1985) The determination of molecular weights of biologically active proteins by cetyltrimethyl ammonium bromide-polyacrylamide gel electrophoresis. *Analyt. Biochem.* **145,** 170–176.

15

Cetyltrimethylammonium Bromide Discontinuous Gel Electrophoresis of Proteins

M_r-*Based Separation of Proteins with Retained Native Activity*

Robert E. Akins and Rocky S. Tuan

1. Introduction

This chapter describes a novel method of electrophoresis that allows the fine separation of proteins to be carried out with the retention of native activity. The system combines discontinuous gel electrophoresis in an arginine/*N*-Tris (hydroxymethyl) methylglycine) (Tricine) buffer with sample solubilization in cetyltrimethylammonium bromide (CTAB). Because the components that distinguish this system are **C**TAB, **a**rginine, and **T**ricine and because CTAB is a **cat**ionic detergent, we refer to this method as CAT gel electrophoresis *(1,2)*. Proteins separated on CAT gels appear as discrete bands, and their mobility is a logarithmic function of M_r across a broad range of molecular weights. After CAT electrophoresis, many proteins retain high enough levels of native activity to be detected, and gel bands may be detected by both M_r and protein-specific activities. In this chapter, we provide a description of the procedures for preparing and running CAT gels. We also provide some technical background information on the basic principles of CAT gel operation and some points to keep in mind when considering the CAT system.

1.1. Technical Background

The electrophoretic method of Laemmli *(3)* is among the most common of laboratory procedures. It is based on observations made by Shapiro et al. *(4)* and Weber and Osborn *(5)*, which showed that sodium dodecyl sulfate (SDS) could be used for the separation of many proteins based on molecular size. In Laemmli's method, SDS solubilization was combined with a discontinuous gel system using a glycine/Tris buffer, as detailed by Ornstein *(6)* and Davis *(7)* (*see* Chapter 11). Typically, SDS-discontinuous gel electrophoresis results in the dissociation of protein complexes into denatured subunits and separation of these subunits into discrete bands. Since the mobility of proteins on SDS gels is related to molecular size, many researchers have come to rely on SDS gels for the convenient assignment of protein subunit M_r.

Unfortunately, it is difficult to assess the biological activity of proteins treated with SDS: proteins prepared for SDS gel electrophoresis are dissociated from native complexes and are significantly denatured. Several proteins have been shown to renature to

From: *The Protein Protocols Handbook, 2nd Edition*
Edited by: J. M. Walker © Humana Press Inc., Totowa, NJ

an active form after removal of SDS *(8,9)*; however, this method is inconvenient and potentially unreliable. A preferred method for determining native protein activity after electrophoresis involves the use of nonionic detergents like Triton X 100 (Tx-100) *(10)*; however, proteins do not separate based on molecular size. The assignment of M_r in the nonionic Tx-100 system requires the determination of mobilities at several different gel concentrations and "Ferguson analysis" *(11–13)*. The CAT gel system combines the most useful aspects of the SDS and Tx-100 systems by allowing the separation of proteins based on M_r with the retention of native activity.

Previous studies have described the use of CTAB and the related detergent tetradecyltrimethylammonium bromide (TTAB), in electrophoretic procedures for the determination of M_r *(14–18)*. In addition, as early as 1965, it was noted that certain proteins retained significant levels of enzymatic activity after solubilization in CTAB *(19)*. A more recent report further demonstrated that some proteins even retained enzymatic activity after electrophoretic separation in CTAB *(14)*. Based on the observed characteristics of CTAB and CTAB-based gel systems, we developed the CAT gel system.

In contrast to previous CTAB-based gel methods, the CAT system is discontinuous and allows proteins to be "stacked" prior to separation (*see **refs.** 6* and *7*). CAT gel electrophoresis is a generally useful method for the separation of proteins with the retention of native activity. It is also an excellent alternative to SDS-based systems for the assignment of protein M_r (*see* **Note 1**).

1.2. Basic Principles of CAT Gel Operation

The CAT gel system is comprised of two gel matrices and several buffer components in sequence. A diagram of the CAT gel system is shown in **Fig. 1**. In an applied electric field, the positive charge of the CTAB–protein complexes causes them to migrate toward the negatively charged cathode at the bottom of the system. The arginine component of the tank buffer also migrates toward the cathode; however, arginine is a *zwitterion*, and its net charge is a function of pH. The arginine is positively charged at the pH values used in the tank buffer, but the pH values of the stacking gel and sample buffer are closer to the pI of arginine, and the arginine will have a correspondingly lower net positive charge as it migrates from the tank buffer into these areas. Therefore, the interface zone between the upper tank buffer and the stacking gel/sample buffer contains a region of high electric field strength where the sodium ions in the stacking gel/sample buffer (Tricine-NaOH) move ahead of the reduced mobility arginine ions (Tricine-arginine). In order to carry the electric current, the CTAB-coated proteins migrate more quickly in this interface zone than in the sodium-containing zone just below. As the interface advances, the proteins "stack," because the trailing edge of the applied sample catches up with the leading edge. When the cathodically migrating interface zone reaches the separating gel, the arginine once again becomes highly charged owing to a drop in the pH relative to the stacking gel. Because of the sieving action of the matrix, the compressed bands of stacked proteins differentially migrate through the separating gel based on size.

Two features of CTAB-based gels set them apart from standard SDS-based electrophoretic methods. First, proteins separated in CTAB gels migrate as a function of log M_r across a much broader range of molecular weights than do proteins separated in

TANK BUFFER	--------ANODE--------- 25 mM Tricine pH 8.2 0.1% CTAB 14 mM Arginine Free Base
SAMPLE BUFFER	10 mM Tricine-NaOH pH 8.8 1% CTAB 10% Glycerol
STACKING GEL	0.7% Agarose 125 mM Tricine-NaOH pH 9.96 0.1% CTAB
SEPARATING GEL	6% Polyacrylamide 375 mM Tricine-NaOH pH 7.96
TANK BUFFER	25 mM Tricine pH 8.2 0.1% CTAB 14 mM Arginine Free Base -------CATHODE-------

Fig. 1. Diagram of a CAT gel. CAT gels begin at the top with the anode immersed in tank buffer and end at the bottom with the cathode immersed in additional tank buffer. The tank buffer solution contains CTAB, Arginine, and Tricine. Between the tank buffers are the stacking gel and the separating gel. The gels are made up of acrylamide polymers in a Tricine-NaOH-buffered solution. Prior to electrophoresis, protein samples are solubilized in a sample buffer that contains CTAB, to solubilize the protein sample, Tricine-NaOH, to maintain pH, and carry current and glycerol, to increase specific gravity. Proteins solubilized in sample buffer are typically layered under the upper tank buffer and directly onto the stacking gel. *See* **Note 3** for a listing of some physical characteristics of the CAT gel components.

SDS gels. As shown in **Fig. 2**, a plot of relative migration distance, as a function of known log M_r of standard proteins, results in a straight line. Because of the consistent relationship between M_r and distance migrated, the relative molecular weights of unknown proteins can be determined. CAT gels may be especially useful for the assignment of M_r to small proteins or for the comparison of proteins with very different molecular weights. Second, the retention of significant levels of native activity in CAT gels allows electrophoretic profiles to be assessed *in situ* for native activities without additional steps to ensure protein renaturation (*see* **Note 2**). Taken together, these two characteristics of CAT gels make them an attractive alternative to standard electrophoretic systems.

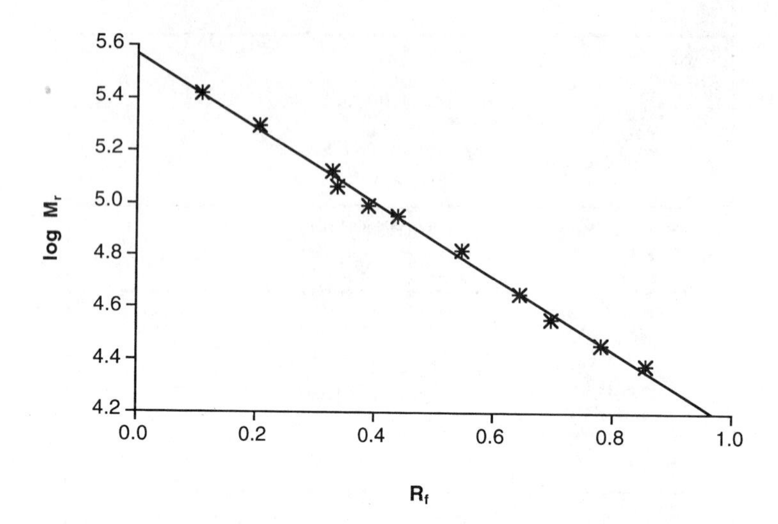

Fig. 2. Mobility of proteins in a CAT gel as a function of M_r. A mixture of proteins fractionated in a CAT gel with a 6% T acrylamide separator and a 0.7% agarose stacker was visualized by CBB R-250 staining. Relative mobilities (R_f) were calculated as distance, migrated divided by total distance to the salt/dye front and were plotted against the known M_r values for each protein band. The plot is linear across the entire range ($R^2 > 0.99$). Protein bands included trypsinogen (24 kDa), carbonic anhydrase (29 kDa), glyceraldehyde-3-phospate dehydrogenase (36 kDa), ovalbumin (45 kDa monomer and 90 kDa dimer), bovine serum albumin (66 kDa monomer and 132, 198, and 264 kDa multimers), phosphorylase-B (97.4 kDa), and β-galactosidase (116 kDa). *See* **Note 2** concerning the comparison of R_f values from different gels.

2. Materials

1. CAT tank buffer: One liter of 5X tank buffer may be prepared using CTAB, Tricine, and arginine free base. First, prepare 80 mL of a 1 *M* arginine free base solution by dissolving 13.94 g in distilled water. Next, dissolve 22.40 g of Tricine in 900 mL of distilled water; add 5 g of CTAB, and stir until completely dissolved. Using the 1 *M* arginine solution, titrate the Tricine/CTAB solution until it reaches pH 8.2. Approx 75 mL of 1 *M* arginine solution will be required/L of CAT tank buffer. Since Tricine solutions change pH with changes in temperature, the tank buffer should be prepared at the expected temperature of use (typically 10–15°C). Finally, add distilled water to 1000 mL. Store the CAT tank buffer at room temperature. Prior to use, prepare 1X tank buffer by diluting 200 mL of the 5X stock to 1000 mL using distilled water of the appropriate temperature (usually 10–15°C); filter the 1X tank buffer through #1 Whatman filter paper to remove any particulate material. The 1X CAT tank buffer may be stored cold, but it should not be reused (*see* **Note 5**). Note that CTAB is corrosive, and care should be taken when handling CTAB powder or CTAB solutions: avoid inhalation or skin contact as advised by the supplier.
2. CAT stacking gel buffer: Prepare a 500 m*M* Tricine-NaOH by dissolving 22.4 g of Tricine in 200 mL of distilled water. Add NaOH until the pH of the solution reaches 10.0. Bring the solution to a total volume of 250 mL using distilled water. As with all Tricine solutions, the pH of CAT stacking gel buffer should be determined at the expected temperature of use. The CAT stacking gel buffer should be stored at room temperature to avoid any precipitation that may occur during long-term cold storage.

3. CAT separating gel buffer: Prepare a 1.5 *M* Tricine-NaOH solution by dissolving 134.4 g of Tricine in 400 mL of distilled water. Add NaOH until the pH of the solution reaches 8.0. Bring the solution to a total volume of 500 mL using distilled water. As with all Tricine solutions, the pH of CAT separating gel buffer should be determined at the expected temperature of use. The CAT separating gel buffer should be stored at room temperature to avoid any precipitation that may occur during long-term cold storage.
4. CAT sample buffer: Dilute 0.67 mL of CAT separating gel buffer to approx 80 mL with distilled water; to this add 10 mL of glycerol and 1 g of CTAB. Mix the solution until all the components are dissolved, and adjust the pH to 8.8 using NaOH. Bring the solution to a final volume of 100 mL using distilled water. In some cases, it may be helpful to add a low-mol-wt cationic dye that will be visible during electrophoresis: 10 µL of a saturated aqueous solution of crystal violet may be added/mL of sample buffer. Note that CTAB is corrosive, and care should be taken when handling CTAB powder or CTAB solutions: avoid inhalation or skin contact as advised by the supplier. Store CAT sample buffer at room temperature to avoid precipitation of the components.
5. Acrylamide stock solution: A 40% acrylamide stock solution may be prepared by combining 38.93 g of ultrapure acrylamide with 1.07 g of *bis*-acrylamide in a total of 100 mL of distilled water. The final solution is 40%T (w/v) and 2.67%C (w/w). The "%T" and "%C" values indicate that the total amount of acrylamide in solution is 40 g/100 mL and that the amount of *bis*-acrylamide included is 2.67% of the total acrylamide by weight. The acrylamide stock solution should be stored in the refrigerator. Unpolymerized acrylamide is very toxic, and great care should be taken when handling acrylamide powders and solutions: Follow all precautions indicated by the supplier, including the wearing of gloves and a particle mask during preparation of acrylamide solutions.
6. Agarose stock solution: A ready-to-use agarose stacking gel solution may be prepared by combining 25 mL of CAT stacking gel buffer, 0.1 g CTAB, and 0.7 g of electrophoresis-grade agarose distilled to a final volume of 100 mL. Mix the components well, and, if necessary, adjust the pH to 10.0. Heat the solution in a microwave oven to melt the agarose, and swirl the solution to mix thoroughly. Divide the agarose stock solution into 10 aliquots, and store at 4°C until ready to use.
7. 10% Ammonium persulfate (AP): Dissolve 0.1 g of ammonium persulfate in 1 mL of distilled water. Make just prior to use.
8. Water saturated isobutanol: Combine equal volumes of isobutanol and distilled water. Mix well, and allow the two phases to separate: the water-saturated isobutanol will be the upper layer. Store at room temperature in a clear container so that the interface is visible.
9. CAT gel fixative: Combine 40 mL of distilled water, 10 mL of acetic acid, and 50 mL of methanol; mix well. Store CAT gel fixative in a tightly sealed container at room temperature.
10. Coomassie brilliant blue stain (CBB): Combine 40 mL of distilled water with 10 mL of acetic acid and 50 mL of methanol. Add 0.25 g of CBB R-250, and dissolve with stirring (usually overnight). Filter the solution through #1 Whatman paper to remove any particulate material. Store at room temperature in a tightly sealed container.
11. CBB Destain: Combine 437.5 mL of distilled water, 37.5 mL of acetic acid, and 25 mL of methanol. Mix well, and stored in a closed container at room temperature.
12. Electrophoresis apparatus: A suitable electrophoresis apparatus and power supply are required to run CAT gels. It is desirable to set aside combs, spacers, gel plates, and buffer tanks to use specifically with CAT gels; however, if the same apparatus is to be used alternately for CAT gels and SDS gels, it is necessary to clean it thoroughly between each use. Often, the first CAT gel run in an apparatus dedicated to SDS gels will have a smeared appearance with indistinct bands. This smearing is the result of residual SDS, and subse-

quent CAT electrophoretic runs will resolve protein bands distinctly. This smearing may be somewhat avoided by soaking the gel apparatus and gel plates in CAT tank buffer prior to a final rinse in distilled water at the final step in the cleaning process.

The selection of an electrophoresis apparatus to be used for CAT gels should be based on a consideration of the electrical configuration of the system. Because molecular bromine (Br_2) will form at the anode, the anode should be located away from the top of the gel (*see* **Note 5**). In addition, it is important to realize that CAT gels are "upside-down" relative to SDS gels: proteins migrate to opposite electrodes in the two systems. Some electrophoresis apparatus are intentionally designed for use with SDS, and the anode (usually the electrode with the red lead) may be fixed at the bottom of the gel, whereas the cathode (usually the electrode with the black lead) is fixed at the top of the gel. If such an apparatus is used, the red lead wire should be plugged into the black outlet on the power supply, and the black lead should be plugged into the red outlet on the power supply. Crossing the wires in this fashion ensures that the CTAB-coated proteins in the CAT system will run into the gel and not into the tank buffer.

3. Method

The methods for the preparation and running of CAT gels are similar to other familiar electrophoretic techniques. In this section, we will describe the basic methods for preparing samples, casting gels, loading and running gels, visualizing protein bands, and transferring proteins to nitrocellulose (or other) membranes. We will emphasize the differences between CAT gels and other systems. To provide the best results, the recommendations of the manufacturer should be followed concerning the assembly of the apparatus and the casting of discontinuous gels.

3.1. Preparing Samples

1. Protein samples should be prepared at room temperature immediately prior to loading the gel. Typically, tissue fragments, cells, or protein pellets are resuspended in 1.5-mL microfuge tubes using CAT sample buffer (*see* **Note 6**). CAT sample buffer may also be used to solubilize cultured cells or minced tissues directly. In each case, the samples should be spun in a microfuge for 0.5 min at 16,000*g* to pellet any debris or insoluble material prior to loading the gel. Good results have been obtained when the final concentration of protein in CAT sample buffer is between 1 and 5 mg/mL; however, the preferred concentration of protein will vary depending on the sample and the particular protein of interest. A series of protein dilutions should be done to determine the optimal solubilization conditions for a particular application.

3.2. Casting CAT Separating Gels

1. Assemble the gel plates and spacers in the gel casting stand as described by the manufacturer.
2. Prepare a separating gel solution by combining the 40%T acrylamide, CAT Separating gel buffer, and distilled water in the ratios indicated in Table 1. Mix the solution by swirling with the introduction of as little air as possible (oxygen inhibits the reactions necessary to accomplish acrylamide polymerization, *see* **Note 7**).
3. Degas the solution by applying a moderate vacuum for 5–10 min: the vacuum generated by an aspirator is generally sufficient.
4. Add 10% AP and TEMED to the solution as indicated in **Table 1**, and swirl the solution gently to mix. Note that insufficient mixing will result in the formation of a nonhomogeneous gel, but that vigorous mixing will introduce oxygen into the mixture.

Table 1
Preparation of Acrylamide Solutions for CAT Gels

Regent	4%T, mL	6%T, mL	8%T, mL	10%T, mL
40%T Acrylamide	1.00	1.50	2.00	2.50
Tricine buffer	2.50	2.50	2.50	2.50
Distilled water	6.39	5.89	5.39	4.89
Degas solution				
10% AP	0.10	0.10	0.10	0.10
TEMED	0.01	0.01	0.01	0.01

Volumes indicated are in milliliters required to prepare 10 mL of the desired solution. Solutions should be degassed prior to the addition of the crosslinking agents, AP and TEMED.

5. Carefully pour the gel mixture into the gel plates to the desired volume; remember to leave room for the stacking gel and comb.
6. Finally, layer a small amount of water-saturated isobutanol onto the top of the gel. The isobutanol layer reduces the penetration of atmospheric oxygen into the surface of the gel and causes the formation of an even gel surface. Allow polymerization of the separating gel to proceed for at least 60 min to assure complete crosslinking; then pour off the isobutanol, and rinse the surface of the separating gel with distilled water.

3.3. Casting CAT Stacking Gels

Two different types of gel stackers are routinely used with CAT gels. For gel histochemical analyses, or where subsequent protein activity assays will be performed, stacking gels made from agarose have provided the best results.

1. Slowly melt a tube of agarose stock solution in a microwave oven; avoid vigorous heating of the solution, since boiling will cause foaming to occur and may result in air pockets in the finished gel.
2. Insert the gel comb into the apparatus, and cast the stacking gel directly onto the surface of the acrylamide separating gel. Allow the agarose to cool thoroughly before removing the comb (*see* **Note 8**).
3. As an alternative to agarose stacking gels, low%T acrylamide stackers may also be used. To prepare an acrylamide stacking gel, combine the 40%T acrylamide stock, CAT separating gel buffer (0.5 *M* Tricine-NaOH, pH 10.0), and distilled water in the ratios indicated in **Table 1**. Typically, a 4%T stacking gel is used. Degas the solution by applying a moderate vacuum for 5–10 min. Next, add 10% AP and TEMED to the solution as indicated in **Table 1**, and swirl the solution gently to mix. Insert the gel comb and cast the stacking gel directly onto the surface of the acrylamide separating gel. Do not use water-saturated isobutanol with stacking gels! It will accumulate between the comb and the gel, and cause poorly defined wells to form. Allow the stacking gel to polymerize completely before removing the comb.

3.4. Loading and Running CAT Gels

1. After the stacking and separating gels are completely polymerized, add 1X CAT tank buffer to the gel apparatus so that the gel wells are filled with buffer prior to adding the samples.

2. Next, using a Hamilton syringe (or other appropriate loading device), carefully layer the samples into the wells. Add the samples slowly and smoothly to avoid mixing them with the tank buffer, and fill any unused wells with CAT sample buffer. The amount of sample to load on a given gel depends on several factors: the size of the well, the concentration of protein in the sample, staining or detection method, and so forth. It is generally useful to run several dilutions of each sample to ensure optimal loading. Check that the electrophoresis apparatus has been assembled to the manufacturer's specifications, and then attach the electrodes to a power supply. Remember that in CAT gels, proteins run toward the negative electrode, which is generally indicated by a black-colored receptacle on power supplies.
3. Turn the current on, and apply 100 V to the gel. For a single minigel (approx 80 mm across, 90 mm long, and 0.8 mm thick), 100 V will result in an initial current of approx 25 mA. Excessive current flow through the gel should be avoided, since it will cause heating.
4. When the front of the migrating system reaches the separating gel, turn the power supply up to 150 V until the front approaches the bottom of the gel. The total time to run a CAT gel should be around 45–60 min for minigels or 4–6 h for full-size gels.

3.5. Visualization of Proteins

1. As with any electrophoretic method, proteins run in CAT gels may be visualized by a variety of staining techniques. A simple method to stain for total protein may be carried out by first soaking the gel for 15 min in CAT gel fixative, followed by soaking the gel into CBB stain until it is thoroughly infiltrated. Infiltration can take as little as 5 min for thin (0.8 mm), low-percentage (6%T) gels or as long as 1 h for thick (1.5 mm), high-percentage (12%T) gels. When the gel has a uniform deep blue appearance, it should be transferred to CBB destain.
2. Destain the gel until protein bands are clearly visible (*see* **Note 9**). It is necessary to observe the gel periodically during the destaining procedure, since the destain will eventually remove dye from the protein bands as well as the background. Optimally, CBB staining by this method will detect about 0.1–0.5 µg of protein/protein track; when necessary, gels containing low amounts of total protein may be silver-stained (*see* **Note 10**). Note that the CBB stain may be retained and stored in a closed container for reuse.
3. In addition to total protein staining by the CBB method, enzyme activities may be detected by a variety of histochemical methods. The individual protocol for protein or enzyme detection will, of course, vary depending on the selected assay (*see* **Note 11**). Generally, when CAT gels are to be stained for enzyme activity, they should be rinsed in the specific reaction buffer prior to the addition of substrate or detection reagent. The CAT gel system provides an extremely flexible method for the analysis of protein mixtures by a variety of direct and indirect gel staining methods.

3.6. Electrophoretic Transfer Blotting

Similar to other methods of electroblotting (e.g., see Chapters 39 and 40), proteins can be transferred from CAT gels to polyvinylidene difluoride (PVDF) membranes. Blotting CAT gels is similar to the standard methods used for SDS-based gels with the notable difference that the current flow is reversed. We have successfully transferred proteins using the method described in Chapter 39 with the following changes: 1) A solution of 80% CAT gel running buffer and 20% MeOH was used as the transfer buffer, 2) the transfer membrane was PVDF, and 3) the polarity of the apparatus was reversed to account for the cationic charge of the CTAB. *(2)* As with electrophoresis, the tank and apparatus used for electroblotting should be dedicated to CTAB gels (*see* **Note 9**).

4. Notes

1. SDS vs CTAB: CTAB and SDS are very different detergents. CTAB is a cationic detergent, and proteins solubilized in CTAB are positively charged; SDS is an anionic detergent, and proteins solubilized in SDS are negatively charged. In terms of electrophoretic migration, proteins in CTAB gels migrate toward the cathode (black electrode), and proteins in SDS gels run toward the anode (red electrode). SDS is not compatible with the CAT gel system, and samples previously prepared for SDS-PAGE are not suitable for subsequent CAT gel electrophoresis. Also, the buffer components of the typical SDS-PAGE system, Tris and glycine, are not compatible with the CAT system: Tricine and arginine should be used with CAT gels.

   Although the detergents are different, protein banding patterns seen in CAT gels are generally similar to those seen when using SDS-PAGE. R_f values of proteins fractionated by CAT electrophoresis are consistently lower than R_f values determined on the same%T SDS-PAGE, i.e., a particular protein will run nearer the top of a CAT gel than it does in a similar %T SDS gel. As a rule of thumb, a CAT gel with a 4%T stacker and a 6%T separator results in electrophoretograms similar to an SDS gel with a 4%T stacker and 8%T separator. Differences between the protein banding patterns seen in CAT gels and SDS gels are usually attributable to subunit associations: multisubunit or self-associating proteins are dissociated to a higher degree in SDS than in CTAB, and multimeric forms are more commonly seen in CAT gels than in SDS gels.
2. Detergent solubilization and protein activity: Many proteins separated on CAT gels may be subsequently identified based on native activity, and under the conditions presented here, CTAB may be considered a nondenaturing detergent. Denaturants generally alter the native conformation of proteins to such an extent that activity is abolished; such is the case when using high levels of SDS. Sample preparation for SDS-based gels typically results in a binding of 1.4 g of SDS/1 g of denatured protein across many types of proteins *(20)*, and it is this consistent ratio that allows proteins to be electrophoretically separated by log M_r. Interestingly, at lower concentrations of SDS, another stable protein binding state also exists (0.4 g/1 g of protein) which reportedly does not cause massive protein denaturation *(20)*. In fact, Tyagi et al. *(21)* have shown that low amounts of SDS (0.02%) combined with pore-limit electrophoresis could be used for the simultaneous determination of M_r and native activity. The existence of detergent–protein complexes, which exhibit consistent binding ratios without protein denaturation, represents an exciting prospect for the development of new electrophoretic techniques: protein M_r and activity may be identified by any of a variety of methods selected for applicability to a specific system.
3. Comparing CAT gels: The comparison of different CAT gel electrophoretograms depends on using the same separating gel, stacking gel, and sample buffer for each determination. An increase in the acrylamide%T in the separating gel will cause bands to shift toward the top of the gel, and high%T gels are more suitable for the separation of low M_r proteins. In addition, the use of acrylamide or high-percentage agarose stackers will lead to the determination of R_f values that are internally consistent, but uniformly lower than those determined in an identical gel with a low-percentage agarose stacker. This effect is likely owing to some separation of proteins in the stacking gel, but, nonetheless, to compare R_f values among CAT gels, the stacker of each gel should be the same. Similarly, any changes in sample preparation (for example, heating the sample before loading to dissociate

protein subunits) or the sample buffer used (for example, the addition of salt or urea to increase sample solubilization) often precludes direct comparisons to standard CAT gel electrophoretograms.

4. Characteristics of system components: The CAT gel system is designed around the detergent CTAB. The other system components were selected based on the cationic charge of CTAB and the desire to operate the gel near neutral pH. Some of the important physical characteristics of system components are summarized here; the values reported are from information supplied by manufacturers and *Data for Biochemical Research* **(22)**.

   In solution, CTAB exists in both monomer and micelle forms. CTAB has a monomer mol wt of 365 Dalton. At room temperature and in low-ionic-strength solutions (<0.05 *M* Na^+), CTAB has a critical micelle concentration (CMC) of about 0.04% (≈1 m*M*). The CMC may be defined as the concentration of detergent monomer that may be achieved in solution before micellization occurs; it depends on temperature, ionic strength, pH, and the presence of other solutes. The actual CMC of CTAB in the solutions used in the CAT gel system is not known. In low-ionic-strength solutions (< 0.1 *M* Na^+), the mol wt of CTAB micelles is approx 62 kDa with about 170 monomers/micelle. The solubility of CTAB is also a function of ionic conditions and solution temperature, and CTAB will precipitate from CAT sample buffer at temperatures below 10–15°C.

   Arginine is an amino acid used in CAT tank buffer to carry current toward the cathode. Arginine is a zwitterion. It has a mol wt of 174.2 Dalton and contains three pH-sensitive charge groups (pK_a =1.8, 9.0, 12.5) with an isoelectric point (pI) near pH 10.8. Arginine was selected as the zwitterionic stacking agent because of its basic pI: at near-neutral pH levels, arginine will be positively charged and will migrate toward the cathode when an electric current is applied. Arginine free base is used in CAT tank buffer, and the proper pH of the CAT tank buffer is arrived at by mixing an acidic solution of Tricine with a basic solution of arginine.

   Tricine functions to maintain the desired pH levels throughout the system. Tricine has a mol wt of 179.2 Dalton, and a pK_a of 8.15 (there is also a second $pK_a \approx 3$). During electrophoresis, Tricine also functions as a counterion to carry current toward the anode during electrophoresis. Tricine-buffered solutions will tend to change pH as temperature changes (a drop in temperature from 25 to 4°C will result in a shift in pH of about 0.5 U). Also, at the pH used in CAT stacking gels and CAT sample buffer, Tricine has a relatively low buffering capacity; therefore, the pH of the stacker and sample buffer should be confirmed, especially when any additions or alterations are made to the standard recipes.

5. Bromine drip: CTAB is an ionic compound comprised of both cetyltrimethylammonium cations and bromide anions. During electrophoresis, bromide anions migrate toward the anode at the top of the CAT system. Since electrons are removed at the anode, molecular bromine (Br_2) is formed at the anode. Under standard conditions, Br_2 is a dense, highly reactive liquid. While running CAT gels, a small amount of Br_2 will drip from the anode, and, if the anode is located directly above the top of the gel, Br_2 may drip onto the samples. The "bromine drip" problem may be avoided if the apparatus used in CAT gel electrophoresis is configured so that the anode is away from (or even below) the top of the gel. Substitution of the bromide anion during CTAB preparation (perhaps with chloride to make CTACl) would be useful; however, the authors know of no high-quality commercial source. It should also be noted that the accumulation of reactive Br_2 during electrophoresis may preclude the reuse of tank buffer.

6. CAT sample buffer: The CTAB component of CAT sample buffer precipitates at low temperature (below 10–15°C). Samples should be prepared at room temperature immediately prior to use. Protease inhibitors, for example, phenylmethylsulfonyl fluoride (PMSF),

may be added to the sample buffer to inhibit endogenous protease activity. The potential effects of any sample buffer additives (including PMSF) on the enzyme of interest should be assessed in solution before using the additive. Also, to avoid contaminating samples with "finger proteins" from the experimenter's hands, gloves should be worn when handling samples or sample buffer.

7. Acrylamide CAT gels: The polymerization of acrylamide generally involves the production of acrylamide free radicals by the combined action of ammonium persulfate and TEMED. Oxygen inhibits acrylamide polymerization by acting as a trap for the ammonium persulfate and TEMED intermediate free radicals that are necessary to accomplish chemical crosslinking. It is important to degas acrylamide solutions prior to the addition of crosslinking agents. CTAB itself interferes with gel polymerization as well and should not be included with acrylamide solutions; in the case of agarose stacking gels, however, CTAB may be included. Also, we have noticed a slight increase in the time required for acrylamide polymerization in the presence of Tricine buffer, but Tricine does not apparently affect gel performance. Although gelation will occur in about 10–20 min, polymerization should be allowed to go to completion (1–2 h) before running the gel. Finally, to optimize gel performance and reproducibility, both degassing and polymerization should be done at room temperature: cooling a polymerizing gel does not accelerate gel formation, and warming a polymerizing gel may cause brittle matrices to form.

   The presence of residual crosslinking agents in the acrylamide matrix may result in the inhibition of some protein activities. An alternative method of gel formation that avoids the use of reactive chemical crosslinkers involves photoactivation of riboflavin to initiate acrylamide polymerization (*see* **ref. 23**). We have had success using acrylamide separating gels polymerized with ammonium persulfate and TEMED in combination with agarose stacking gels; however, in some instances, riboflavin polymerized gels may be useful.
8. Agarose CAT gels: When using agarose stackers, remove the comb carefully to avoid creating a partial vacuum in the wells. To avoid pulling the agarose stacker away from the underlying acrylamide separating gel, wiggle the comb so that air or liquid fills the sample wells as the comb is lifted out.
9. Smeared gels: During electrophoresis, proteins run as mixed micelles combined with CTAB and other solution components. The presence of other detergents or high levels of lipid in the sample solution may result in a heterogenous population of protein-containing micelles. In such cases, protein bands may appear indistinct or smeared. Samples that have been previously solubilized in another detergent or that contain high levels of lipid may not be suitable for subsequent CAT gel analysis. Also, precipitation may occur when CTAB is mixed with polyanions (e.g., SDS micelles or nucleic acids). Samples containing high levels of nucleic acid or SDS may not be suitable for CAT gel analysis owing to precipitation in the sample buffer. Furthermore, if the gel apparatus to be used for CAT gels is also routinely used for SDS-based gels, the apparatus should be thoroughly cleaned to remove traces of SDS prior to CAT gel analysis, especially from the gel plates. Interaction between SDS and CTAB during electrophoresis may cause the formation of heterogenous mixed micelles or the precipitation of the components, and will invariably result in smeared gels. Although it is not always possible, it is preferable to have a separate apparatus dedicated to CAT gel use in order to avoid this problem.
10. Staining gels and transfer membranes: In addition to the basic CBB staining of protein bands, CAT gels may be stained by a variety of other techniques. Silver staining *(24)* allows the detection of even trace amounts of proteins (1–10 ng), including any fingerprints or smudges that would be otherwise undetected. In silver staining it is, therefore, essential to wear gloves throughout the procedure, even when cleaning the glassware prior

to assembly of the apparatus. When the CAT gel is done, stain it with CBB to visualize proteins. (It is a good idea to take a photograph of the gel at this point, but be careful when handling it to avoid smudging the surface.) Next, place the gel into a 10% glutaraldehyde solution and soak in a fume hood for 30 min with gentle shaking. Rinse the gel well with distilled water for at least 2 h (preferably overnight) changing the water frequently; the gel will swell and become soft and somewhat difficult to handle while it is in the water. Combine 1.4 mL NH_4OH (concentrated solution ≈14.8*M*) and 21.0 mL 0.36% NaOH (made fresh); add approx 4.0 mL of 19.4% $AgNO_3$ (made fresh) with constant swirling. A brown precipitate will form as the $AgNO_3$ is added, but it will quickly disappear; if it does not, a small amount of additional NH_4OH (just enough to dissolve the precipitate) may be added. Soak the gel in the ammoniacal/silver solution for 10 min with shaking. Pour off the ammoniacal/silver solution and precipitate the silver with HCl. Transfer the gel to a fresh dish of distilled water and rinse for 5 min with two or three changes of water. Decant the water and add a freshly prepared solution of 0.005% citric acid and 0.019% formaldehyde (commercial preparations of formaldehyde are 37% solutions); bands will become visible at this point. Stop the silver-staining reaction by quickly rinsing the gel in water followed by soaking in a solution of 10% acetic acid and 20% methanol. Silver-staining takes a little practice to do well.

Transfer blots may be stained by a variety of conventional techniques. Two rapid and simple procedures are as follows: **Step (1)** Rinse the membrane briefly in phosphate-buffered saline containing 0.1% Tween 20 (PBS/Tw); soak the membrane in 0.1% solution of India ink in PBS/Tw until bands are detected (10 to 15 min.); rinse the blot in PBS/Tw. **Step (2)** Rinse the membrane briefly in phosphate-buffered saline (PBS); soak the blot in 0.2% Ponceau-S (3-hydroxy-4-(2-sulfo-4(sulfo-phenylazo)phenylazo)-2,7-naphthalene disulfonic acid) in 3% trichloroacetic acid and 3% sulfosalicylic acid for 10 min.; rinse the blot in PBS until bands appear. Ponceau S can be stored as a 10X stock solution and diluted with distilled water just prior to use. Staining with Ponceau-S is reversible, so blots should be marked at the position of mol-wt markers before continuing. When higher sensitivity is required to visualize bands, transfer membranes may be stained with ISS Gold Blot (Integrated Separation Systems, Natick, MA) (*see* Chapter 53).

11. Enzyme activities and protein banding patterns: The histochemical detection of proteins in CAT gels is generally a straightforward process of soaking the gel in reaction buffer, so that the necessary compounds penetrate the gel, followed by the addition of substrate; however, not all enzymes retain detectable levels of activity when solubilized in CTAB or when they are run on CAT gels. It is a good idea to check the relative activity of the protein of interest in a CTAB-containing solution vs standard reaction buffer prior to investing time and effort into running a CAT gel. Of the proteins that do retain detectable the levels of activity, there are substantial differences in the level of retained activity after solubilization in CTAB. There is also the possibility that the protein of interest will run anomalously in CAT gels. Anomalous migration may occur owing to differential CTAB binding, conformational differences in the protein/CTAB mixed micelle, or the presence of previously unrecognized subunits or associated proteins. Unexpected enzyme histochemistry patterns in samples with the expected CBB protein staining pattern may reflect the presence of cofactors or other protein/protein interactions that are necessary for activity and should be interpreted accordingly.

    It should also be pointed out that the intensity of a histochemical or binding assay does not necessarily reflect the actual level of enzyme in the sample. The rate of histochemical reaction within the gel matrix is not necessarily linear with respect to the amount of enzyme present: measured product also depends on the amount of substrate present and the ratio of product to substrate over time as well as the response of the detection apparatus. Reactions should be performed with an excess of substrate at the outset and should preferably be calibrated based on the varying level of a standard activity. Also, as with all detergents, CTAB may differentially solubilize proteins in a given sample, depending on the overall solution conditions. The detergent/protein solution is a complex equilibrium that may be shifted, depending on the level of other materials (proteins, lipids, salts, and so on) in the solution. When seeking to compare protein profiles on CAT gels, it is advisable to prepare samples from the same initial buffer and to solubilize at a uniform CTAB-to-protein ratio.

5. Conclusion

Gel electrophoresis of proteins is a powerful and flexible technique. There are many excellent references for general information concerning the theory behind electrophoresis. One good source of information is Hames and Rickwood *(25)*. The CAT gel electrophoresis system presented here efficiently stacks and separates a wide range of proteins as a function of M_r and preserves native enzymatic activity to such a degree that it allows the identification of protein bands based on native activity. The nature of the interaction between CTAB and native proteins allows the formation of complexes in which the amount of CTAB present is related to the size of the protein moiety. Based on characterizations of detergent/protein interactions that indicate consistent levels of detergent binding without massive denaturation ***(10,26–28)***, the retention of activity in certain detergent/protein complexes is expected, and CTAB is likely to represent a class of ionic detergents that allow the electrophoretic separation of native proteins by M_r. Since the level of retained activity after solubilization in CTAB varies depending on the protein of interest, and a given protein will retain varying amounts of measurable activity depending on the detergent used (*see* **ref. 1**, for example), a battery of detergents may be tried prior to selecting the desired electrophoresis system. In general, cationic detergents (e.g., the quartenary ammoniums like CTAB, TTAB, and so forth) may be used in the arginine/Tricine buffer system described above; anionic detergents (e.g., alkyl sulfates and sulfonates) may be substituted for SDS in the familiar glycine/Tris buffer system *(29)*.

The CAT gel system and its related cationic and anionic gel systems provide useful adjuncts to existing biochemical techniques. These systems allow the electrophoretic separation of proteins based on log M_r with the retention of native activity. The ability to detect native protein activities, binding characteristics, or associations and to assign accurate M_r values in a single procedure greatly enhances the ability of researchers to analyze proteins and protein mixtures.

Acknowledgment

This work has been supported in part by funds from Nemours Research Programs (to R. E. A.), NASA (SBRA93-15, to R. E. A.), the Delaware Affiliate of the American Heart Association (9406244S, to R. E. A.), and the NIH (HD29937, ES07005, and DE11327 to R. S. T.).

References

1. Akins, R. E., Levin, P., and Tuan, R. S. (1992) Cetyltrimethylammonium bromide discontinuous gel electrophoresis: M_r-based separation of proteins with retention of enzymatic activity. *Analyt. Biochem.* **202,** 172–178.
2. Akins, R. E. and Tuan, R. S. (1994) Separation of proteins using cetyltrimethylammonium bromide discontinuous gel electrophoresis. *Mol. Biotech.* **1,** 211–228.
3. Laemmli, U. K. (1970) Cleavage of structural proteins during the assembly of the head of bacteriophage T4. *Nature* **227,** 680–685.
4. Shapiro, A. L., Vinuela, E., and Maizel, J. V. (1967) Molecular weight estimation of polypeptide chains by electrophoresis in SDS-polyacrylamide gels. *Biochem. Biophys. Res. Commun.* **28,** 815–820.
5. Weber, K. and Osborn, M. (1969) The reliability of molecular weight determination by dodecyl sulfate-polyacrylamide electrophoresis. *J. Biol. Chem.* **244,** 4406–4412.
6. Ornstein, L. (1964) Disc electrophoresis I: Background and theory. *Ann. NY Acad. Sci.* **121,** 321–349.
7. Davis, B. J. (1964) Disc electrophoresis II: Method and application to human serum proteins. *Ann. NY Acad. Sci.* **121,** 404–427.
8. Manrow, R. E. and Dottin, R. P. (1980) Renaturation and localization of enzymes in polyacrylamide gels: Studies with UDP-glucose pyrophosphorylase of Dictyostelium. *Proc. Natl. Acad. Sci. USA* **77,** 730–734.
9. Scheele, G. A. (1982) Two-dimensional electrophoresis in basic and clinical research, as exemplified by studies on the exocrine pancreas. *Clin. Chem.* **28,** 1056–1061.
10. Hearing, V. J., Klingler, W. G., Ekel, T. M., and Montague, P. M. (1976) Molecular weight estimation of Triton X-100 solubilized proteins by polyacrylamide gel electrophoresis. *Analyt. Biochem.* **126,** 154–164.
11. Ferguson, K. (1964) Starch gel electrophoresis—Application to the classification of pituitary proteins and polypeptides. *Metabolism* **13,** 985–1002.
12. Hedrick, J. L. and Smith A. J. (1968) Size and charge isomer separation and estimation of molecular weights of proteins by disc gel electrophoresis. *Arch. Biochem. Biophys.* **126,** 154–164.
13. Tuan, R. S. and Knowles, K. (1984) Calcium activated ATPase in the chick embryonic chorioaliantoic membrane: Identification and topographic relationship with the calcium-biding protein. *J. Biol. Chem.* **259,** 2754–2763.
14. Akin, D., Shapira, R., and Kinkade, J. M. (1985) The determination of molecular weights of biologically active proteins by cetyltrimethylammonium bromide-polyacrylamide gel electrophoresis. *Analyt. Biochem.* **145,** 170–176.
15. Eley, M. H., Burns, P. C., Kannapell, C. C., and Campbell, P. S. (1979) Cetyltrimethylammonium bromide polyacrylamide gel electrophoresis: Estimation of protein subunit molecular weights using cationic detergents. *Analyt. Biochem.* **92,** 411–419.
16. Marjanen, L. A. and Ryrie, I. J. (1974) Molecular weight determinations of hydrophilic proteins by cationic detergent electrophoresis: Application to membrane proteins. *Biochem. Biophys. Acta* **37,** 442–450.
17. Panyim, S., Thitiponganich, R., and Supatimusro, D. (1977) A simplified gel electrophoretic system and its validity for molecular weight determinations of protein-cetyltrimethylammonium complexes. *Analyt. Biochem.* **81,** 320–327.
18. Schick, M. (1975) Influence of cationic detergent on electrophoresis in polyacrylamide gel. *Analyt. Biochem.* **63,** 345–349.
19. Spencer, M. and Poole, F. (1965) On the origin of crystallizable RNA from yeast. *J. Mol. Biol.* **11,** 314–326.

20. Reynolds, J. A. and Tanford, C. (1970) The gross conformation of protein-sodium dodedcyl sulfate complexes. *J. Biol. Chem.* **245,** 5161–5165.
21. Tyagi, R. K., Babu, B. R., and Datta, K. (1993) Simultaneous determination of native and subunit molecular weights of proteins by pore limit electrophoresis and restricted use of sodium dodecyl sulfate. *Electrophoresis* **14,** 826–828.
22. Dawson, R. M. C., Elliott, D. C., Elliott, W. H., and Jones, K. M. (1986) *Data for Biochemical Research.* Clarendon Press, Oxford.
23. Bio-Rad (1987) "Bio-Rad Technical Bulletin #1156: Acrylamide Polymerization—A Practical Approach." Bio-Rad Laboratories, Richmond, CA.
24. Oakley, B. R., Kirsch, D. R., and Morris, N. R. (1980) A simplified ultrasensitive silver stain for detecting proteins in polyacrylamide gels. *Analyt. Biochem.* **105,** 361–363.
25. Hames, B. D. and Rickwood, D. (1990) *Gel Electrophoresis of Proteins: A Practical Approach.* IRL, London.
26. Tanford, C., Nozaki, Y., Reynolds, J. A., and Makino, S. (1974) Molecular characterization of proteins in detergent solutions. *Biochemistry* **13,** 2369–2376.
27. Reynolds, J. A., Herbert, S., Polet, H., and Steinhardt, J. (1967) The binding of divers detergent anions to bovine serum albumin. *Biochemistry* **6,** 937–947.
28. Ray, A., Reynolds, J. A., Polet, H., and Steinhardt, J. (1966) Binding of large organic anions and neutral molecules by native bovine serum albumin. *Biochemistry* **5,** 2606–2616.
29. Akins, R. E. and Tuan, R. S. (1992) Electrophoretic techniques for the M_r-based separation of proteins with retention of native activity. *Mol. Biol. Cell* **3,** 185a.

16

Acetic Acid–Urea Polyacrylamide Gel Electrophoresis of Basic Proteins

Jakob H. Waterborg

1. Introduction

Panyim and Chalkley described in 1969 a continuous acetic acid–urea (AU) gel system that could separate very similar basic proteins based on differences in size and effective charge *(1)*. For instance, unmodified histone H4 can be separated from its monoacetylated or monophosphorylated forms *(2)*. At the acidic pH 3 of this gel system, basic proteins with a high isoelectric point will clearly have a net positive charge that will be the major determinant of electrophoretic mobility. If a single of these positive charges is removed, for example, by in vivo acetylation of one of the positively charged ε-amino lysine side chain residues in the small histone H4 protein (102 residues), a significant decrease in effective gel mobility is observed. Similarly, addition of a phosphate moiety decreases the net positive charge of the protein during gel electrophoresis by one. Separation between similarly sized and charged proteins, for example, the partially acetylated H2A, H2B, and H3 histones of most organisms, can typically be achieved only by inclusion of a nonionic detergent such as Triton X-100 (*see* Chapter 17).

In 1980, Bonner and co-workers introduced a discontinuous acetic acid–urea–Triton (AUT) variation that avoids the necessity for exhaustive preelectrophoresis *(3)*, prevents deformation of sample wells *(4)* and generally produces much sharper, straighter bands. Omission of Triton from this method creates the high capacity and high resolution AU gel electrophoresis protocol described in **Subheading 3. Figure 1** shows an example of the possibilities and limitations of the AU gel system. Yeast histones were extracted from four parallel cultures using a novel method that preserves all postsynthetic modifications *(5)*. The separation of single-charge differences, proven to be caused by acetylation of lysines, is clearly demonstrated for each histone species (**Fig. 1**). It is clear that, without fractionation, the patterns of H2B, H2A, and H3 histones would overlap, preventing quantitation of each protein and its acetylation. Protein band shapes are generally sharp and straight. Compression of band shapes, visible for some high-abundance nonhistone proteins with lower gel mobilities shown in **Fig. 1**, can be minimized by decreasing the protein concentration during gel stacking by increasing gel thickness.

From: *The Protein Protocols Handbook, 2nd Edition*
Edited by: J. M. Walker © Humana Press Inc., Totowa, NJ

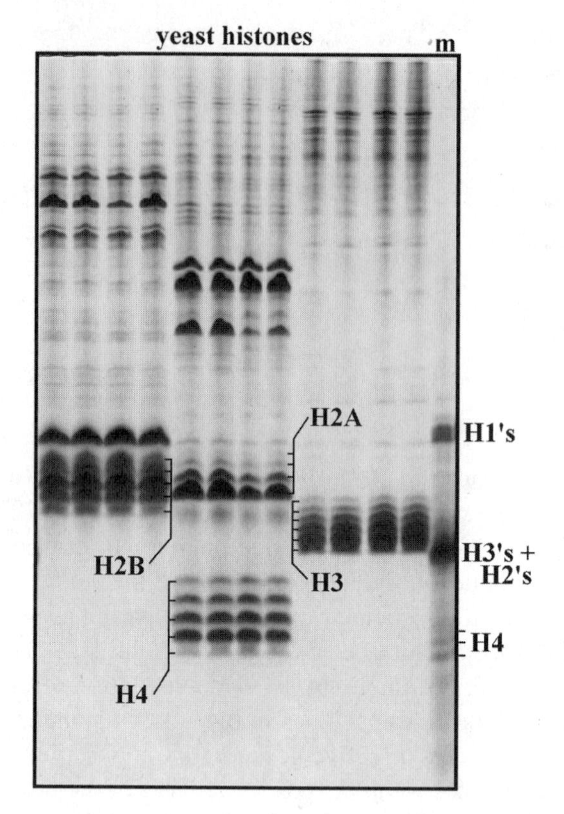

Fig. 1. Histones of the yeast *Saccharomyces cerevisiae* were extracted from crude nuclei, prepared in such a way that protein integrity and postsynthetic modifications were maintained ***(5)***. Histones from four parallel cultures (approx 2×10^{10} cells) were extracted, fractionated by reversed-phase HPLC into pools containing histone H2B (*left four lanes*), coeluting histone H2A and H4 (*center four lanes*) and histone H3 (*right four lanes*). In each histone species, the fastest mobility band represents nonacetylated histone. In each slower moving species one more lysine has been acetylated, neutralizing the positive charge of the lysine side chain, reducing the protein gel mobility by one charge. Virtually all low-mobility nonhistone proteins in this gel can be removed from yeast histone preparations by 100,000 mol wt ultrafiltration ***(5)***. The "m" indicates a marker lane with total calf thymus histones. Calf histone species are marked along the side. Note the closely overlapping pattern of H2A, H2B, and H3 histones. The long AU separating gel was 1 mm thick, 19 cm wide, and 27 cm long. The top of the gel is visible at the top of the figure. Electrophoresis was at 250 V constant voltage for 20 h with electrical current decreasing from 22 mA to 6 mA. The electrophoretic front, visualized using methylene blue, was below the lower edge of the figure, close to the lower margin of the gel. The gel was stained overnight in Coomassie, destained in 4 h with two aliquots of polyurethane foam, and digitized under standard conditions on a UMAX Powerlook II flatbed scanner with transilluminator.

AU gels are currently used for very dissimilar proteins with isoelectric points lower than those of histones but sufficiently above pH 3 to be positively charged during gel electrophoresis. Examples include neutrophil defensins ***(6)***, antimicrobiol nasal secretions ***(7)***, enzymes like tyrosinase (***8***), serum isoenzymes ***(9)***, chemokines ***(10)*** and basic protamines ***(11)***. Typically curved, somewhat diffuse protein bands, characteristic for native, non-stacking electrophoretic gel separations are obtained. The superior discontinuous gel system described here has only been used for histones, producing clearer bands in a shorter period of time than achievable in continuous AU gel modes ***(1–2)***.

2. Materials

1. Vertical gel apparatus for short (15 cm) or long (30 cm) slab gels. A gel electrophoresis apparatus that allows gel polymerization between the glass plates with spacers, without necessarily being assembled in the apparatus, is preferable. This facilitates the even and complete photopolymerization of the acrylamide gel. In this type of apparatus the glass–gel sandwich is typically clamped to the lower buffer reservoir, which acts as a stand, after which the upper buffer reservoir is clamped to the top of the gel assembly.

   Details of the procedure are described for a fairly standard and flexible gel apparatus that uses two rectangular glass plates (4 mm thick standard plate glass with sanded edges), 21 cm wide and 32.5 and 35.5 cm long, respectively. The Plexiglas bottom buffer reservoir with platinum electrode is 22.5 cm wide with three sides 5 cm high and one of the long sides 12.5 cm high. The glass plates are clamped to this side. The upper buffer reservoir with platinum electrode and a similar buffer capacity is 18 cm wide with one long side enlarged to measure 21 cm wide by 10 cm high. It contains a cutout of 18 cm wide and 3.5 cm high that allows access of the upper reservoir buffer to the top of the gel. The 21-cm-wide Plexiglas plate is masked with 5-mm-thick closed-cell neoprene tape (weather strip) and provides a clamping ridge for attachment to the top of the glass-gel sandwich.
2. Spacers and combs are cut from 1-mm Teflon sheeting. High-efficiency fluorography may benefit from 0.5-mm spacers. Teflon up to 3 mm thick is less easy to cut but yields very high capacity gels (*see* **Note 1**).

   Two side spacers (1.5 × 35 cm) and one bottom spacer (0.5 × 24 cm) are required. Added to the top of the side spacers is 3 cm adhesive, closed-cell neoprene tape (weather strip, 14 mm wide and 5 mm thick). This is not required if a more expensive glass plate with "rabbit ears" is used instead of the rectangular shorter plate.

   Combs have teeth 5–10 mm wide and 25–50 mm long, separated by gaps of at least 2.5 mm. For the detailed protocol described, a 15-cm-wide comb with 20 teeth of 5 × 30 mm is used.
3. Vaseline pure petroleum jelly.
4. Acrylamide stock solution: 60% (w/v) acrylamide, highest quality available, in water (*see* **Note 2**). The acrylamide is dissolved by stirring. Application of heat should be avoided, if possible, to prevent generating acrylic acid. The solution can be kept at least for 3 mo on the laboratory shelf at room temperature. Storage at 4°C can exceed 2 yr without detectable effects.
5. *N,N*′-methylene bis-acrylamide stock solution: 2.5% (w/v) in water (*see* **Note 2**).
6. Glacial acetic acid (HAc): 17.5 *M*.
7. Concentrated ammonium hydroxide: NH_4OH, 28–30%, approx 15 *M*.
8. *N,N,N*′,*N*′-Tetramethylenediamine (TEMED), stored at 4°C.

9. Riboflavin-5'-phosphate (R5P) solution: 0.006% (w/v) in water. This solution is stable for more than 6 mo if kept dark and stored at 4°C.
10. Urea, ultrapure quality.
11. Side-arm suction flasks with stoppers, magnetic stirrer, stirrer bar, and water-aspirator vacuum; measuring cylinders with silicon-rubber stoppers; pipets, and pipetting bulbs or mechanical pipetting aids; 1- and 5-mL plastic syringes, with 20-gauge needles.
12. Fluorescent light box with diffuser for even light output and with the possibility to stand vertically. Light intensity should equal or exceed 5 klx at a distance of 5–10 cm. A high-quality X-ray viewing light box with three 40-W bulbs typically will meet this specification.
13. Aluminum foil.
14. Electrophoresis power supply with constant voltage mode at 300–500 V with up to 50 mA current, preferably with a constant power mode option.
15. Urea stock solution: 8 *M* urea in water. An aliquot of 40 mL with 1 g of mixed-bed resin (Bio-Rad [Hercules, CA] AG 501-X8) can be used repeatedly over a period of months if refrozen and stored between use at –20°C (*see* **Note 3**).
16. Phenolphthalein indicator solution: 1% (w/v) in 95% ethanol, stored indefinitely at room temperature in a closed tube.
17. Dithiothreitol (DTT, Cleland's reagent) is stored at 4°C and is weighed freshly for each use.
18. Methylene blue running front indicator dye solution: 2% (w/v) in sample buffer (*see* **Subheading 3., step 22**).
19. Reference histones: Total calf thymus histones (Worthington, Freehold, NJ), stored dry at 4°C indefinitely or in solution at –80°C in 50-µL aliquots of 5 mg/mL in water for more than 1 yr (*see* **Subheading 3., step 23**).
20. Glass Hamilton microsyringe (100 µL) with Teflon-tipped plunger.
21. Electrophoresis buffer: 1 *M* acetic acid, 0.1 *M* glycine (*see* **Note 4**). This solution can be made in bulk and stored indefinitely at room temperature.
22. Destaining solution: 20% (v/v) methanol, 7% (v/v) acetic acid in water.
23. Staining solution: Dissolve a fresh 0.5 g of Coomassie Brilliant Blue R250 in 500 mL of destaining solution for overnight gel staining (*see* **Note 5**). For rapid staining within the hour the dye concentration should be increased to 1% (w/v). If the dye dissolves incompletely, the solution should be filtered through Whatman no. 1 paper to prevent staining artifacts.
24. Glass tray for gel staining and destaining.
25. Rotary or alternating table top shaker.
26. Destaining aids: Polyurethane foam for Coomassie-stained gels or Bio-Rad ion-exchange resin AG1-X8 (20–50-mesh) for Amido Black-stained gels.

3. Method

1. Assemble a sandwich of two clean glass plates with two side spacers and a bottom spacer, lightly greased with Vaseline to obtain a good seal, clamped along all sides with 2-in binder clamps. The triangular shape of these clamps facilitates the vertical, freestanding position of the gel assembly a few centimeters in front of the vertical light box.
2. Separating gel solution: Pipet into a 100-mL measuring cylinder 17.5 mL of acrylamide stock solution, 2.8 mL of *bis*-acrylamide stock solution, 4.2 mL of glacial acetic acid, and 0.23 mL of concentrated ammonium hydroxide (*see* **Notes 2** and **6**).
3. Add 33.6 g of urea and add distilled water to a total volume of 65 mL.
4. Stopper the measuring cylinder, and place on a rotary mixer until all urea has dissolved. Add water to 65 mL, if necessary.
5. Transfer this solution to a 200-mL sidearm flask with magnetic stir bar on a magnetic stirrer. While stirring vigorously, stopper the flask and apply water-aspirator vacuum.

Initially, a cloud of small bubbles of dissolved gas arises, which clears after just a few seconds. Terminate vacuum immediately to prevent excessive loss of ammonia.

6. Add 0.35 mL of TEMED and 4.67 mL of R5P (*see* **Note 7**), mix, and pipet immediately between the glass plates to a marking line 5 cm below the top of the shorter plate (*see* **Notes 2** and **8**).
7. Carefully apply 1 mL of distilled water from a 1-mL syringe with needle along one of the glass plates to the top of the separating gel solution to obtain a flat separation surface.
8. Switch the light box on, and place a reflective layer of aluminum foil behind the gel to increase light intensity and homogeneity (*see* **Note 1**). Gel polymerization becomes detectable within 2 min and is complete in 15–30 min.
9. Switch the light box off, completely drain the water from between the plates, and insert the comb 2.5 cm between the glass plates. The tops of the teeth should always remain above the top of the short glass plate.
10. Stacking gel solution, made in parallel to **steps 2–4**: Into a 25-mL measuring cylinder, pipet 1.34 mL of acrylamide stock solution, 1.28 mL of *bis*-acrylamide stock solution, 1.14 mL of glacial acetic acid, and 0.07 mL of concentrated ammonium hydroxide (*see* **Notes 2** and **6**).
11. Add 9.6 g of urea, and add distilled water to a total volume of 18.6 mL.
12. Stopper the measuring cylinder, and place on a rotary mixer until all urea has dissolved. Add water to 18.6 mL, if necessary.
13. Once the separating gel has polymerized, transfer the stacking gel solution to a 50-mL sidearm flask with stir bar and degas as described under **step 5**.
14. Add 0.1 mL of TEMED and 1.3 mL of R5P, mix, and pipet between the plates between the comb teeth. Displace air bubbles.
15. Switch the light box on, and allow complete gel polymerization in 30–60 min.
16. Prepare sample buffer freshly when the separation gel is polymerizing (**step 8**). The preferred protein sample is a salt-free lyophilisate (*see* **Note 9**). Determine the approximate volume of sample buffer required, depending on the number of samples.
17. Weigh DTT into a sample buffer preparation tube for a final concentration of 1 *M*, that is, 7.7 mg/mL.
18. Per 7.7 mg of DTT add 0.9 mL of 8 *M* urea stock solution, 0.05 mL of phenolphtalein, and 0.05 mL of NH_4OH to the tube to obtain the intensely pink sample buffer.
19. Add 0.05 mL of sample buffer/sample tube with lyophilized protein to be analyzed in one gel lane (*see* **Note 10**). To ensure full reduction of all proteins by DTT, the pH must be above 8.0. If the pink phenolphtalein color disappears owing to residual acid in the sample, a few microliters of concentrated ammonium hydroxide should be added to reach an alkaline pH.
20. Limit the time for sample solubilization and reduction to 5 min at room temperature to minimize the possibility of protein modification at alkaline pH by reactive urea side reactions, for example, by modification of cysteine residues by cyanate.
21. Acidify the sample by adding 1/20 volume of glacial acetic acid.
22. To each sample add 2 µL of methylene blue running front dye (*see* **Note 11**).
23. Prepare appropriate reference protein samples: To 2 and 6 µL of reference histone solution with 10 and 30 µg total calf thymus histones, add 40 µL of sample buffer (**step 18**), 2.5 µL of glacial acetic acid, and 2 µL of methylene blue.
24. When stacking gel polymerization is complete, remove the comb. Drain the wells completely, using a paper tissue as wick, to remove residual unpolymerized gel solution. At comb and spacer surfaces, gel polymerization is typically incomplete. The high urea concentration of unpolymerized gel solution interferes with the tight application of samples.

25. Remove the bottom spacer from the bottom of the gel assembly and use it to remove any residual Vaseline from the lower surface of the gel.
26. Clamp the gel assembly into the electrophoresis apparatus and fill the lower buffer reservoir with electrophoresis buffer.
27. Use a 5-mL syringe with a bent syringe needle to displace any air bubbles from the bottom of the gel.
28. Samples are applied deep into individual sample wells using a Hamilton microsyringe (rinsed with water between samples) (*see* **Note 12**). For the combination of comb and gel dimensions listed, 50-µL sample will reach a height of 1 cm (*see* **Note 8**). Samples can also be applied to sample wells by any micropipetter with plastic disposable tip. Pipet each sample solution against the long glass plate and let it run to the bottom of the well.
29. Apply reference samples in the outer lanes, which frequently show a slight loss of resolution due to edge effects. The threefold difference in reference protein amounts facilitates correct orientation of the gel following staining and destaining and obviates the need for additional markings. Optionally, apply 50 µL of acidified sample buffer to unused lanes.
30. Gently overlayer the samples with electrophoresis buffer, dispensed from a 5-mL syringe fitted with a 21-gauge needle until all wells are full.
31. Fill the upper buffer reservoir with electrophoresis buffer.
32. Attach the electrical leads between power supply and electrophoresis system: the + lead to the upper and the – lead to the lower reservoir. Note that this is opposite to the SDS gel electrophoresis configuration. Remember, basic proteins such as histones are positively charged and will move toward the cathode (negative electrode).
33. Long (30-cm) gels require 15–20 h of electrophoresis at 300 V in constant voltage mode. They are most easily run overnight. For maximum resolution and stacking capacity, the initial current through a 1-mm thick and 18-cm-wide gel should not exceed 25 mA. Gel electrophoresis is completed in the shortest amount of time in constant power mode with limits of 300 V, 25 mA, and 5 W. The current will drop towards completion of electrophoresis to 6 mA at 300 V.

    Short (15-cm) gels are run at 250 V in constant voltage mode with a similar maximal current, or in constant power mode starting at 25 mA. In the latter example, electrophoresis starts at 25 mA and 135 V, and is complete in 5.5 h at 13 mA and 290 V (*see* **Note 13**).
34. Electrophoresis is complete just before the methylene blue dye exits the gel. Obviously, electrophoresis may be terminated if lesser band resolution is acceptable, or may be prolonged to enhance separation of basic proteins with low gel mobilities, for example, histone H1 variants or phosphorylated forms of histone H1.
35. Open the glass–gel sandwich, and place the separating gel into staining solution, which is gently agitated continuously overnight on a shaker (*see* **Note 14**).
36. Decant the staining solution. The gel can be given a very short rinse in water to remove all residual staining solution.
37. Place the gel in ample destaining solution (*see* **Note 15**). Diffusion of unbound Coomassie dye from the gel is facilitated by the addition of polyurethane foam as an absorbent for free Coomassie dye. To avoid overdestaining and potential loss of protein from the gel (*see* **Note 14**), destaining aids in limited amounts are added to only the first and second destaining solutions. Final destaining is done in the absence of any destaining aids.
38. Record the protein pattern of the gel on film or on a flatbed digital scanner (*see* **Note 16**), possibly with quantitative densitometry ***(12)***. Subsequently the gel may be discarded, dried, eluted, blotted ***(6)***, or prepared for autoradiography or fluorography as required.

4. Notes

1. Owing to absorbance of the light that initiates gel polymerization, gels thicker than 1.5 mm tend to polymerize better near the light source and produce protein bands that are not perpendicular to the gel surface. For very thick gels, two high-intensity light boxes, placed at either side of the gel assembly, may be required for optimal gel polymerization and resolution.
2. All acrylamide and *bis*-acrylamide solutions are potent neurotoxins and should be dispensed by mechanical pipetting devices.
3. Storage of urea solutions at –20°C minimizes creation of ionic contaminants such as cyanate. The mixed-bed resin ensures that any ions formed are removed. Care should be taken to exclude resin beads from solution taken, for example, by filtration through Whatman no. 1 paper.
4. The stacking ions between which the positively charged proteins and peptides are compressed within the stacking gel during the initial phase of gel electrophoresis are NH_4^+ within the gel compartment and glycine$^+$ in the electrophoresis buffer. Chloride ions interfere with the discontinuous stacking system (*see* **Note 7**). This requires that protein samples should (preferably) be free of chloride salts, and that glycine base rather than glycine salt should be used in the electrophoresis buffer.
5. Amido Black is an alternate staining dye which stains less intensely and destains much slower than Coomassie, but is the better stain for peptides shorter than 30–50 residues.
6. The separating and stacking gels contain 15 vs 4% acrylamide and 0.1 vs 0.16% *N,N'*-methylene (*bis*-acrylamide), respectively, in 1 *M* acetic acid, 0.5% TEMED, 50 m*M* NH_4OH, 8 *M* urea, and 0.0004 % riboflavin-5'-phosphate.

   We have observed that 8M urea produces the highest resolution of histones in these gels when Triton X-100 is present (*see* **Chapter 17**). Equal or superior resolution of basic proteins has been reported for AU gels when the urea concentration is reduced to 5 *M*.
7. Acrylamide is photopolymerized with riboflavin or riboflavin 5'-phosphate as initiator, because the ions generated by ammonium persulfate initiated gel polymerization, as used for SDS polyacrylamide gels, interfere with stacking (*see* **Note 4**).
8. The height of stacking gel below the comb determines the volume of samples that can be applied and fully stacked before destacking at the surface of the separating gel occurs. In our experience, 2.5-cm stacking gel height suffices for samples that almost completely fill equally long sample wells. In general, the single blue line of completely stacked proteins and methylene blue dye should be established 1 cm above the separating gel surface. Thus, a 1–1.5 cm stacking gel height will suffice for small-volume samples.
9. Salt-free samples are routinely prepared by exhaustive dialysis against 2.5% (v/v) acetic acid in 3500 molecular weight cutoff dialysis membranes, followed by freezing at –70°C and lyophilization in conical polypropylene tubes (1.5-, 15-, or 50-mL, filled up to half of nominal capacity) with caps punctured by 21-gauge needle stabs. This method gives essentially quantitative recovery of histones, even if very dilute.

   Alternatively, basic proteins can be precipitated with trichloroacetic acid or acetone, acidified by hydrochloric acid to 0.02 *N*, provided that excess salt and acidity is removed by multiple washes with acetone.

   Solutions of basic proteins can be used directly, provided that the solution is free of salts that interfere with gel electrophoresis (*see* **Note 11**) and that the concentration of protein is high enough to compensate for the 1.8-fold dilution that occurs during sample preparation. Add 480 mg of urea, 0.05 mL of phenolphthalein, 0.05 mL of concentrated ammonium hydroxide, and 0.05 mL of 1 *M* DTT (freshly prepared) per milliliter sample. If not pink, add more ammonium hydroxide. Leave for 5 min at room temperature. Add 0.05 mL of glacial acetic acid. Measure an aliquot for one gel lane and continue at **step 22**.

10. The amount of protein that can be analyzed in one gel lane depends highly on the complexity of the protein composition. As a guideline, 5–50 µg of total calf thymus histones with five major proteins (modified to varying extent) represents the range between very lightly to heavily Coomassie-stained individual protein bands in 1-mm-thick, 30-cm-long gels using 5-mm-wide comb teeth.
11. Methylene blue is a single blue dye that remains in the gel discontinuity stack of 15% acrylamide separating gels (*see* **Note 4**). Methyl green is an alternate dye marker that contains methylene blue together with yellow and green dye components that remain together in discontinuous mode but that in continuous gel electrophoresis show progressively slower gel mobilities (*see* **Note 17**).
12. As an alternative to **step 28**, electrophoresis buffer is added to the upper buffer reservoir, and all sample wells are filled. Samples are layered under the buffer when dispensed by a Hamilton microsyringe near the bottom of each well. The standard procedure tends to prevent mixing of sample with buffer and thus retains all potential stacking capability.
13. Long gels used in overnight electrophoresis are made on the day that electrophoresis is started. The time for preparation of short gels may prevent electrophoresis on the same day. The nature of the stacking system of the gel (*see* **Note 4**) allows one to prepare a gel on day 1, to store it at room temperature overnight and to initiate electrophoresis on the morning of day 2. To prevent precipitation of urea, gels should not be stored in a refrigerator. The gels should not be stored under electrophoresis buffer as glycine would start to diffuse into the gel and destroy the stacking capability of the system. We routinely store short gels overnight once polymerization is complete (*see* **step 15**) and before bottom spacer and comb are removed. Saran Wrap® is used to prevent exposed gel surfaces from drying.
14. Be warned that small and strongly basic proteins such as histones are not fixed effectively inside 15% acrylamide gels in methanol–acetic acid without Coomassie. Comparison of identical gels, one fixed and stained as described and the other placed first in destain solution alone for several hours, followed by regular staining by Coomassie, reveals that 90% or more of core histones are lost from the gel. We speculate that the Coomassie dye helps to retain histones within the gel matrix. This is consistent with the observation that gradual loss of Coomassie intensity of histone bands is observed upon exhaustive removal of soluble dye.
15. Note that the gel increases significantly in size when transferred from the gel plates into the staining solution. The compositions of staining and detaining solvent mixtures are identical and changes in gel size are not observed upon destaining.
16. A standard method to record the protein staining pattern in Coomassie-stained polyacrylamide gels has been Polaroid photography, using an orange filter to increase contrast, with the gel placed on the fluorescent light box, covered by a glass plate to prevent Coomassie staining of the typical plastic surface. Polaroid negatives can be scanned but suffer from a nonlinear response of density, even within the range in which careful Coomassie staining leads to near-linear intensity of protein band staining.

    With the advent of 24+ bit color flatbed scanners with transmitted light capabilities and linear density capabilities in excess of three optical densities, it has become easy to record, and quantitate, intensity of protein staining patterns. Care should be taken to develop a standard scanning setup, using the full dynamic range (typically all three colors used at their full range, 0–255 for a 24-bit scanner, excluding automatic adjustments for density and contrast). A standard gamma correction value should be determined, using an optical density wedge (Kodak), to ensure that the density response is linear. Placing a detained polyacrylamide gel on the gel scanner, one should cover the top of the gel with an acetate film to prevent surface reflection abnormalities and prevent touching of the transilluminating light source surface to the film, avoiding moiré interference patterns. Recording gel patterns at 300 dpi in full color and saving loss-less tiff files facilitates faithful replication of experimental results (**Fig. 1**). Quantitative densitometry programs can use the image files.

17. Gel preelectrophoresis until the methylene blue dye, and thus all ammonium ions (*see* **Note 4**), have exited the separating gel converts this gel system into a continuous one. This option can be used to separate small proteins and peptides that do not destack at the boundary with the separating gel. Although this option is available, one should consider alternatives, such as increasing the acrylamide concentration of the separating gel. West and co-workers have developed a system with 40–50% polyacrylamide gels that is similar to the one described here and that has been optimized for the separation of small basic peptides *(13)*.

References

1. Panyim, S. and Chalkley, R. (1969) High resolution acrylamide gel electrophoresis of histones. *Arch. Biochem. Biophys.* **130,** 337–346.
2. Ruiz-Carrillo, A., Wangh, L. J., and Allfrey, V. G. (1975) Processing of newly synthesized histone molecules. Nascent histone H4 chains are reversibly phosphorylated and acetylated. *Science* **190,** 117–128.
3. Bonner, W. M., West, M. H. P., and Stedman, J. D. (1980) Two-dimensional gel analysis of histones in acid extracts of nuclei, cells, and tissues. *Eur. J. Biochem.* **109,** 17–23.
4. Paulson, J. R. and Higley, L. L. (1999) Acid-urea polyacrylamide slab gel electrophoresis of proteins: preventing distortion of gel wells during preelectrophoresis. *Analyt. Biochem.* **268,** 157–159.
5. Waterborg, J. H. (2000) Steady-state levels of histone acetylation in Saccharomyces cerevisiae. *J. Biol. Chem.* **275,** 13,007–13,011.
6. Wang, M. S., Pang, J. S., and Selsted, M. E. (1997) Semidry electroblotting of peptides and proteins from acid-urea polyacrylamide gels. *Analyt. Biochem.* **253,** 225–230.
7. Cole, A. M., Dewan, P., and Ganz, T. (1999) Innate antimicrobial activity of nasal secretions. *Infect. Immun.* **67,** 3267–3275.
8. Burzio, L. A., Burzio, V. A., Pardo, J., and Burzio, L. O. (2000) *In vitro* polymerization of mussel polyphenolic proteins catalyzed by mushroom tyrosinase. *Comp. Biochem. Physiol. B* **126,** 383–389.
9. Hansen, S., Thiel, S., Willis, A., Holmskov, U., and Jensenius, J. C. (2000) Purification and characterization of two mannan-binding lectins from mouse serum. *J. Immunol.* **164,** 2610–2618.
10. Krijgsveld, J., Zaat, S. A., Meeldijk, J., van Veelen, P. A., Fang, G., Poolman, B., et al. (2000) Thrombocidins, microbial proteins from human blood platelets, are C-terminal deletion products of CXC chemokines. *J. Biol. Chem.* **275,** 20,374–20,381.
11. Evenson, D. P., Jost, L. K., Corzett, M., and Balhorn, R. (2000) Characteristics of human sperm chromatin structure following an episode of influenza and high fever: a case study. *J. Androl.* **21,** 739–746.
12. Waterborg, J. H., Robertson, A. J., Tatar, D. L., Borza, C. M., and Davie, J. R. (1995) Histones of Chlamydomonas reinhardtii. Synthesis, acetylation and methylation. *Plant Physiol.* **109,** 393–407.
13. West, M. H. P., Wu, R. S., and Bonner, W. M. (1984) Polyacrylamide gel electrophoresis of small peptides. *Electrophoresis* **5,** 133–138.

17

Acid–Urea–Triton Polyacrylamide Gel Electrophoresis of Histones

Jakob H. Waterborg

1. Introduction

Acid–urea polyacrylamide gels are capable of separating basic histone proteins provided they differ sufficiently in size and/or effective charge (*see* Chapter 16). Separation between similarly sized and charged molecules, such as the histones H2A, H2B, and the H3 forms of most organisms, can typically not be achieved. Zweidler discovered that core histones but not linker histones (*see* **Note 1**) bind the nonionic detergent Triton *(1)*. Generally, Triton is added to an acetic acid–urea (AU) gel system to separate core histone sequence variants and histone species with overlapping AU gel patterns. This type of gel is known as an AUT or a TAU gel. To date, a single example is known where addition of Triton X-100 has allowed separation of a nonhistone primary sequence variation, the hydrophobic replacement variant of phenylalanine by leucine in fetal hemoglobin *(2)*.

The binding of Triton to a core histone increases the effective mass of the protein within the gel without affecting its charge, and thus reduces its mobility during electrophoresis. Separation between most or all core histone proteins of diverse species can virtually always be obtained by adjusting concentrations of Triton and of urea, which appears to act as a counteracting, dissociating agent *(3)*. Experimentally, an optimal balance can be determined by gradient gel electrophoresis with a gradient of urea *(4)* or Triton *(5)*. The Triton gradient protocol in the discontinuous gel system, developed by Bonner and co-workers *(6)*, is described in **Subheading 3.** It has a distinct advantage over the urea gradient protocol. Generally, it can identify a core histone protein band as belonging to histone H4, H2B, H3, or H2A. In this order the apparent affinities for Triton X-100 increase sharply *(5,7,8)*. An example of such a separation of a crude mixture of histones with nonhistone proteins from a tobacco callus culture is shown in **Fig. 1A**. In addition, a detailed working protocol for a long AUT gel at 9 m*M* Triton and 8 *M* urea is provided. It describes the protocol used extensively in my laboratory for the analysis of core histones, especially of histone H3, in dicot *(7)* and monocot plants *(8)*, in the green alga *Chlamydomonas* *(9)* and in budding yeast *Saccharomyces*

From: *The Protein Protocols Handbook, 2nd Edition*
Edited by: J. M. Walker © Humana Press Inc., Totowa, NJ

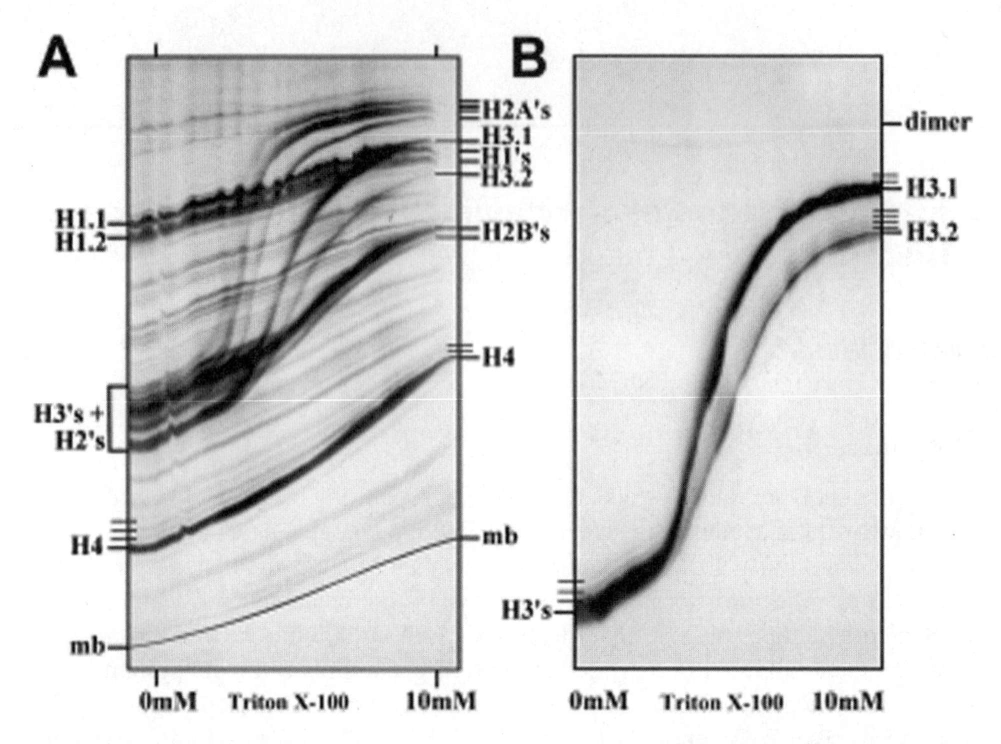

Fig. 1. Acid–urea gradient gel electrophoresis of tobacco histones. **(A)** A crude preparation of basic proteins, extracted from callus cultures of tobacco *(7)*, was electrophoresed on the gradient AUT gel and stained with Coomassie, as described. Marks at top and bottom indicate the joints between the gel compartments, from left to right, 0 m*M* Triton, 0–10 m*M* Triton gradient, and 10 m*M* Triton. The buffer front is marked by the methylene blue dye (mb). The identified histone bands are marked: two variants for histone H1, histone H4 with detectable mono- through triacetylation, two histone H2B variants, two histone H3 variants (H3.1 and the more highly acetylated histone H3.2) and, at least four histone H2A variant forms. Note that the Triton affinity of the core histones increases in the typical order of H4, H2B, H3, and H_2A. **(B)** Tobacco histone H3, purified by reversed-phase HPLC as a mixture of low acetylated histone H3.1 and highly acetylated histone H3.2, on a gradient AUT gel. The 0 and 10 m*M* Triton concentrations coincide with the edge of the figure. The presence of a small amount of histone H3 dimer is indicated.

cerevisiae ***(10)***. **Figure 1B** shows an example of the differentially acetylated histone H3 variant proteins of tobacco, purified by reversed-phase high performance liquid chromatography (HPLC) *(7)*. The protocol description parallels directly the acid–urea gel protocol described in Chapter 16, which also provides details for the use of different gel dimensions.

2. Materials

1. Vertical gel apparatus for long (30-cm) slab gels. A gel electrophoresis apparatus that allows gel polymerization between the glass plates with spacers, without being assembled in the apparatus, is required. This allows the even and complete photopolymerization of the acrylamide gel compartments in both orientations used. In this type of apparatus the glass–gel sandwich is typically clamped to the lower buffer reservoir, which acts as a stand, after which the upper buffer reservoir is clamped to the top of the gel assembly.

Details of the procedure are described for a fairly standard and flexible gel apparatus that uses two rectangular glass plates (4-mm-thick standard plate glass with sanded edges), 21 cm wide and 32.5 and 35.5 cm long, respectively. The Plexiglas bottom buffer reservoir with platinum electrode is 22.5 wide with three sides 5 cm high and one of the long sides 12.5 cm high. To this side the glass plates are clamped. The upper buffer reservoir with platinum electrode and with a similar buffer capacity is 18 cm wide with one long side enlarged to measure 21 cm wide by 10 cm high. It contains a cutout of 18 cm wide and 3.5 cm high that allows access of the upper reservoir buffer to the top of the gel. The 21-cm wide Plexiglas plate is masked with 5-mm-thick closed-cell neoprene tape (weather strip), and provides a clamping ridge for attachment to the top of the glass–gel sandwich.

2. Spacers and combs are cut from 1 mm Teflon sheeting. Required are two side spacers (1.5×35 cm), one bottom spacer (0.5×24 cm) and, for the gradient gel, one temporary spacer (1.5×21 cm). Added to the top of the side spacers is 3-cm adhesive, closed-cell neoprene tape (weather strip, 14 mm wide and 5 mm thick). A 15-cm-wide comb with 20 teeth of 5×30 mm, cut from a rectangle of 15×5 cm Teflon, is used with the teeth pointing down for the regular gel and with the teeth pointing up as a block comb for the gradient gel.
3. Vaseline pure petroleum jelly.
4. Acrylamide stock solution: 60% (w/v) acrylamide, highest quality available, in water (*see* **Note 2**). The acrylamide is dissolved by stirring. Application of heat should be avoided, if possible, to prevent generating acrylic acid. The solution can be kept at least for 3 mo on the laboratory shelf at room temperature. Storage at 4°C can exceed 2 yr without detectable effects.
5. *N,N'*-Methylene *bis*-acrylamide stock solution: 2.5% (w/v) in water (*see* **Note 2**).
6. Glacial acetic acid (HAc): 17.5 *M*.
7. Concentrated ammonium hydroxide: NH_4OH, 28–30%, approx 15 *M*.
8. Triton X-100 stock solution: 25% (w/v) in water (0.4 *M*) is used, as it is much easier to dispense accurately than 100%.
9. *N,N,N',N'*-Tetramethylenediamine (TEMED), stored at 4°C.
10. Riboflavin 5'-phosphate solutions: 0.006% (R5P) and 0.06% (R5P-hi) (w/v) in water. The lower concentration solution is used for the regular gel and for all stacking gels. It is stable for more than 6 mo if kept dark and stored at 4°C. Riboflavin 5'-phosphate readily dissolves at the higher concentration, which is used for the gradient gel formulation, but it cannot be stored for more than 1 d at room temperature in the dark, as precipitates form readily upon storage at 4°C.
11. Glycerol.
12. Urea, ultrapure quality.
13. Side-arm suction flasks with stoppers, magnetic stirrer, stirrer bar, and water-aspirator vacuum. Measuring cylinders with silicon-rubber stoppers. Pipets and pipetting bulbs or mechanical pipetting aids. Plastic syringes, 1 and 5 mL, with 20-gauge needles.
14. Fluorescent light box with diffuser for even light output and with the possibility to stand vertically. Light intensity should equal or exceed 5 klx at a distance of 5–10 cm. A high-quality X-ray viewing light box with three 40-W bulbs typically will meet this specification.
15. Aluminum foil.
16. Gradient mixer to prepare a linear concentration gradient with a volume of 30 mL. We use with success the 50-mL Jule gradient maker (Research Products International, Mt. Prospect, IL) with two 25-mL reservoirs.
17. Electrophoresis power supply with constant voltage mode at 300–500 V with up to 50 mA current, preferably with a constant power mode option.

18. Urea stock solution: 8 *M* urea in water. An aliquot of 40 mL with 1 g of mixed-bed resin (Bio-Rad [Hercules, CA] AG 501-X8) can be used repeatedly over a period of months if refrozen and stored between uses at –20°C (*see* **Note 3**).
19. Phenolphthalein indicator solution: 1% (w/v) in 95% ethanol, stored indefinitely at room temperature in a closed tube.
20. Dithiothreitol (DTT, Cleland's reagent) is stored at 4°C and is weighed freshly for each use.
21. Methylene blue running front indicator dye solution: 2% (w/v) in sample buffer (*see* **Subheading 3., step 16**).
22. Reference histones: Total calf thymus histones (Worthington, Freehold, NJ), stored dry at 4°C indefinitely or in solution at –80°C in 50-μL aliquots of 5 mg/mL in water for more than 1 yr (*see* **Subheading 3., step 17**).
23. Glass Hamilton microsyringe (100 μL) with Teflon-tipped plunger.
24. Electrophoresis buffer: 1 *M* acetic acid, 0.1 *M* glycine (*see* **Note 4**). This solution can be made in bulk and stored indefinitely at room temperature.
25. Destaining solution: 20% (v/v) methanol, 7% (v/v) acetic acid in water.
26. Staining solution: Dissolve a fresh 0.5 g of Coomassie Brilliant Blue R250 in 500 mL of destaining solution. If the dye dissolves incompletely, the solution should be filtered through Whatman no. 1 paper to prevent staining artifacts.
27. Glass tray for gel staining and destaining.
28. Rotary or alternating table top shaker.
29. Destaining aid: Polyurethane foam.

3. Method

1. For the AUT regular gel, follow **steps a–i** and continue at **step 3**.
   a. Assemble a sandwich of two clean glass plates with two side spacers and a bottom spacer, lightly greased with Vaseline to obtain a good seal, clamped along all sides with 2-in binder clamps. The triangular shape of these clamps facilitates the vertical, freestanding position of the gel assembly a few centimeters in front of the vertical light box (**Fig. 2A**).
   b. Separating gel solution: pipet into a 100-mL measuring cylinder 17.5 mL of acrylamide stock solution, 2.8 mL of *bis*-acrylamide stock solution, 4.2 mL of glacial acetic acid, and 0.23 mL of concentrated ammonium hydroxide solutions (*see* **Note 5**).
   c. Add 33.6 g of urea and add distilled water to a total volume of 63.5 mL.
   d. Stopper the measuring cylinder and place on a rotary mixer until all urea has dissolved. Add water to 63.5 mL, if necessary.
   e. Transfer this solution to a 200-mL sidearm flask with magnetic stir bar on a magnetic stirrer. While stirring vigorously, stopper the flask and apply water-aspirator vacuum. Initially, a cloud of small bubbles of dissolved gas arises, which clears after just a few seconds. Terminate vacuum immediately to prevent excessive loss of ammonia.
   f. Add 1.575 mL of Triton, 0.35 mL of TEMED and 4.67 mL of R5P (*see* **Note 6**), mix, and pipet immediately between the glass plates to a marking line 5 cm below the top of the shorter plate (*see* **Note 7**).
   g. Carefully apply 1 mL of distilled water from a 1-mL syringe with needle along one of the glass plates to the top of the separating gel solution to obtain a flat separation surface.
   h. Switch the light box on and place a reflective layer of aluminum foil behind the gel to increase light intensity and homogeneity. Gel polymerization becomes detectable within 2 min and is complete in 15–30 min.

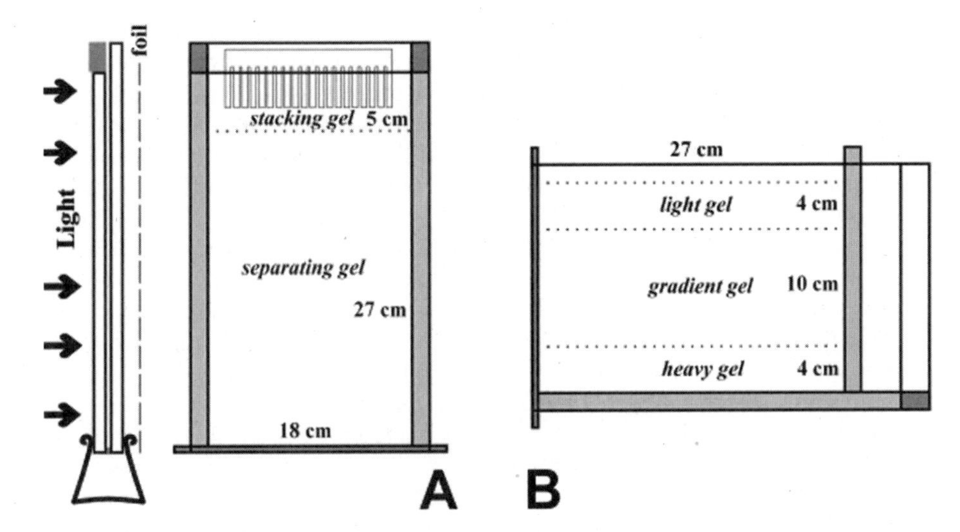

Fig. 2. **(A)** Side and front view of a long AUT gel assembly for 20 samples. The Teflon side spacers are shown in light gray, the bottom Teflon spacer and the neoprene blocks in dark gray. The fluorescent light source and reflecting aluminum foil are shown. One clamp is shown in the side view to demonstrate the independent vertical stand of this assembly and to represent the clamps all around the assembly at all spacer locations. **(B)** Front view of the AUT gradient assembly as used when the separation gel partitions are formed and polymerized.

i. Switch the light box off, completely drain the water from between the plates, and insert the comb 2.5 cm between the glass plates. The tops of the teeth should always remain above the top of the short glass plate.

2. For the Triton gradient gel, follow **steps a–q** and continue at **step 3**.
   a. Assemble a sandwich of two clean glass plates with one side spacer, a bottom spacer, and a temporary spacer, clamped with large binder clamps (**Fig. 2B**). The spacers are lightly greased with Vaseline to obtain a good seal but the amount of grease on the temporary spacer should be as low as possible. No grease should be present on the side of this spacer facing the buffer compartment to assure a flawless destacking surface. Place magic marker guidance lines at 4, 14, and 18 cm above the long side spacer (**Fig. 2B**). Place the assembly horizontally a few centimeters in front of the vertical light box and add a reflective layer of aluminum behind the gel.
   b. Heavy gel separating gel solution: Pipet into a 50-mL graduated, capped polypropylene tube, 4.0 mL of glycerol, 10 mL of acrylamide stock solution, 1.6 mL of *bis*-acrylamide stock solution, 2.4 mL of glacial acetic acid, and 0.13 mL of concentrated ammonium hydroxide. Add 19.2 g of urea. Add distilled water to a total volume of 38.5 mL (*see* **Note 5**).
   c. Light gel separating gel solution: Pipet into a 50-mL graduated, capped polypropylene tube, 10 mL of acrylamide stock solution, 1.6 mL of *bis*-acrylamide stock solution, 2.4 mL of glacial acetic acid, and 0.13 mL of concentrated ammonium hydroxide. Add 19.2 g of urea. Add distilled water to a total volume of 39.5 mL.
   d. Place the closed tubes on a rotary mixer until all urea has dissolved. Add water to the required final volume, if necessary.
   e. Transfer each solution to a 100-mL sidearm flask with magnetic stir bar on a magnetic stirrer. While stirring vigorously, stopper the flask and apply water-aspirator vacuum.

Initially, a cloud of small bubbles of dissolved gas arises, which clears after just a few seconds. Terminate vacuum immediately to prevent excessive loss of ammonia.

f. Return the gel solutions after degassing to the capped tubes.
g. Add to the heavy gel solution 1.0 mL of Triton stock solution.
h. Wrap each tube into aluminum foil to protect the gel solution from light (*see* **Note 6**).
i. Prepare freshly concentrated (0.06%) riboflavin 5'-phosphate (R5P-hi) solution (*see* **Subheading 2., step 10**).
j. Add to each gel solution 0.20 mL of TEMED and 0.27 mL of R5P-hi, and mix.
k. Under conditions of darkness (or very reduced light levels), pipet approx 12 mL of heavy gel solution between the plates, up to the 4-cm heavy gel surface mark (**Fig. 2B**).
l. Turn the light box on for 2 min only. This allows the heavy gel to initiate polymerization and gelling, but retains sufficient unpolymerized acrylamide to fuse this gel partition completely into the gel layered on top.
m. In darkness, set up the gradient maker with 15 mL of heavy gel and 15 mL of light gel solutions and create slowly, over a period of at least several minutes, a linear gradient that flows slowly and carefully between the glass plates. Empirically, we have observed that a 10% increase over the calculated volume of the 27 × 10 × 0.1 cm gradient partition will create a gradient that is complete near the marker line between gradient and light partition (**Fig. 2B**).
n. Turn the light box on for 2 min only.
o. In darkness, slowly add light gel solution on top of the partially polymerized gradient gel until the upper marker line.
p. Switch the light box on for complete gel polymerization in 15–30 min.
q. Remove the temporary spacer from the assembly. Insert a slightly greased spacer along the polymerized light gel. Reclamp the assembly. Insert the 20-well comb in the middle and upside down as a block comb for 2.5 cm below the edge of the shorter glass plate. Reposition the assembly vertically between light box and aluminum foil (**Fig. 2A**).

3. Stacking gel solution, made in parallel to **steps 1b–d** or **steps 2b–d**: Pipet into a 25-mL measuring cylinder 1.34 mL of acrylamide stock solution, 1.28 mL of *bis*-acrylamide stock solution, 1.14 mL of glacial acetic acid, and 0.07 mL of concentrated ammonium hydroxide (*see* **Note 5**).
4. Add 9.6 g of urea and add distilled water to a total volume of 18.6 mL.
5. Stopper the measuring cylinder and place on a rotary mixer until all urea has dissolved. Add water to 18.6 mL, if necessary.
6. Once the separating gel has polymerized, transfer the stacking gel solution to a 50-mL sidearm flask with stir bar and degas as in **step 1e**.
7. Add 0.1 mL of TEMED and 1.3 mL of R5P, mix and pipet between the plates between the comb teeth.
8. Displace air bubbles, especially in the gradient assembly where residual Vaseline may interfere with gel solution flow. In this case, all (mini) bubbles should be carefully removed from the separation gel surface that will act as the destacking boundary.
9. Switch the light box on and allow complete gel polymerization in 30–60 min.
10. Sample buffer is freshly prepared when the separation gel is polymerizing. The preferred protein sample is a salt-free lyophilisate (*see* **Note 8**). For regular gels, determine the approximate volume of sample buffer required, depending on the number of samples. For a gradient gel 0.5–1.5 mL sample buffer appears optimal.
11. Weigh DTT into a sample buffer preparation tube for a final concentration of 1 *M*, that is, 7.7 mg/mL.
12. Per 7.7 mg of DTT, add 0.9 mL of 8 *M* urea stock solution, 0.05 mL of phenolphthalein, and 0.05 mL of concentrated ammonium hydroxide to the tube to obtain the intensely pink sample buffer.

13. Add 0.05 mL of sample buffer per sample tube with lyophilized protein to be analyzed in one gel lane (*see* **Note 9**). Add 1 mL of sample buffer to sample for one gradient gel. To assure full reduction of all proteins by DTT, the pH must be above 8.0. If the pink phenolphthalein color disappears because of residual acid in the sample, a few microliters of concentrated ammonium hydroxide should be added to reach an alkaline pH.
14. Limit the time for sample solubilization and reduction to 5 min at room temperature to minimize the possibility of protein modification at alkaline pH by reactive urea side reactions, for example, by modification of cysteine residues by cyanate.
15. Acidify the sample by addition of 1/20 volume of glacial acetic acid.
16. Add methylene blue running front dye: 2 μL per gel lane or 50 μL for a gradient gel.
17. For a regular AUT gel, prepare reference histone samples: To 2 and 6 μL reference histone solution with 10 and 30 μg of total calf thymus histones, one adds 40 μL sample buffer (**step 12**), 2.5 μL of glacial acetic acid, and 2 μL of methylene blue.
18. When stacking gel polymerization is complete, remove the comb. Drain the wells completely, using a paper tissue as wick, to remove residual unpolymerized gel solution. At comb and spacer surfaces, gel polymerization is typically incomplete. The high urea concentration of unpolymerized gel solution interferes with the tight application of samples.
19. Remove the bottom spacer from the gel assembly gel assembly and use it to remove any residual Vaseline from the lower surface of the gel.
20. Clamp the gel assembly into the electrophoresis apparatus and fill the lower buffer reservoir with electrophoresis buffer.
21. Use a 5-mL syringe with a bent syringe needle to displace any air bubbles from the bottom of the gel.
22. For regular gel, follow **steps a–c** and continue at **step 24**.
    a. Samples are applied deep into individual sample wells by Hamilton microsyringe (rinsed with water between samples) (*see* **Note 10**). For the combination of comb and gel dimensions listed, a 50-μL sample will reach a height of 1 cm (*see* **Note 7**). Samples can also be applied to sample wells by any micropipetter with plastic disposable tip. Pipet each sample solution against the long glass plate and let it run to the bottom of the well.
    b. Apply reference samples in the outer lanes, which frequently show a slight loss of resolution due to edge effects. The threefold difference in reference protein amounts facilitates correct orientation of the gel following staining and destaining and obviates the need for additional markings. Optionally, apply 50 μL of acidified sample buffer to unused lanes.
    c. Gently overlayer the samples with electrophoresis buffer, dispensed from a 5-mL syringe fitted with a 21-gauge needle until all wells are full.
23. For gradient gel, follow **steps a–c** and continue at **step 24**.
    a. Fill the block well with electrophoresis buffer.
    b. Use a level to confirm that the bottom of the preparative well is exactly horizontal.
    c. Distribute the total sample evenly across the width of the well using a 250-μL Hamilton microsyringe. Limited mixing of sample and electrophoresis buffer will facilitate even loading and is easily dealt with by the strong stacking capability of the gel system (*see* **Note 11**).
24. Fill the upper buffer reservoir with electrophoresis buffer.
25. Attach the electrical leads between power supply and electrophoresis system: the + lead to the upper and the - lead to the lower reservoir. Note that this is opposite to the SDS gel electrophoresis configuration. Remember, basic proteins such as histones are positively charged and will move toward the cathode (negative electrode).

26. Long (30-cm) gels require 15–20 h of electrophoresis at 300 V in constant voltage mode. They are most easily run overnight. For maximum resolution and stacking capacity the initial current through a 1 mm thick and 18 cm wide gel should not exceed 25 mA. Gel electrophoresis is completed in the shortest amount of time in constant power mode with limits of 300 V, 25 mA, and 5 W. The current will drop toward completion of electrophoresis to 6 mA at 300 V.
    Gradient gel electrophoresis typically takes a few more hours. At 300 V constant voltage the electrophoretic current is reduced, in particular at the heavy side of the gel due to the reduced water concentration caused by the inclusion of glycerol. The ion front with methylene blue dye reflects this in a distinct curvature (**Fig. 1A**).
27. Electrophoresis is complete just before the methylene blue dye exits the gel. Obviously, electrophoresis may be terminated earlier if lesser band resolution is acceptable, or may be prolonged to enhance separation of histones with low gel mobilities, for example, histone H3 variants (**Fig. 1B**).
28. Open the glass–gel sandwich and place the separating gel into staining solution, which is gently agitated continuously overnight on a shaker (*see* **Note 12**).
29. Decant the staining solution. The gel can be given a very short rinse in water to remove all residual staining solution.
30. Place the gel in ample destaining solution. Diffusion of unbound Coomassie dye from the gel is facilitated by the addition of polyurethane foam as an absorbent for free Coomassie dye. To avoid overdestaining and potential loss of protein from the gel, polyurethane foam in limited amounts is added to only the first and second destaining solutions. Final destaining is done in the absence of any foam.
31. Record the protein pattern of the gel on film (**Fig. 1**) or on a flatbed digital scanner (*see* **Note 13**). Subsequently the gel may be discarded, dried, or prepared for autoradiography or fluorography as required. Blotting of AUT gels requires removal of Triton to allow successful histone transfer (*see* **Note 14**).

4. Notes

1. Core histones share a 65-basepair helix-turn-helix-turn-helix histone-fold motif which is the basis of the hydrophobic pairwise interaction between histones H2A and H2B and between histones H3 and H4. Our studies of histone H3 variant proteins from HeLa, *Physarum* ***(11)***, *Chlamydomonas* ***(9)*** and several plant species ***(5,7,8)*** support the notion that the characteristic differences in Triton X-100 affinity—Triton affinity: H2A > H3 > H2B > H4 (**Fig. 1**) —depend on sequence differences at residues whose side chains are mapped to the inside of the histone fold ***(12,13)***. To date, many proteins involved in DNA organization and transcription have been shown to share the histone fold, creating homo- and hetero-dimer complexes ***(14)***. However, none have yet been analyzed by AUT gradient gel electrophoresis to confirm the predicted interaction with Triton X-100.
2. All acrylamide and *bis*-acrylamide solutions are potent neurotoxins and should be dispensed by mechanical pipetting devices.
3. Storage of urea solutions at –20°C minimizes creation of ionic contaminants such as cyanate. The mixed-bed resin assures that any ions formed are removed. Care should be taken to exclude resin beads from solution taken, for example, by filtration through Whatman no. 1 paper.
4. The stacking ions between which the positively charged proteins and peptides are compressed within the stacking gel during the initial phase of gel electrophoresis are NH_4^+ within the gel compartment and glycine$^+$ in the electrophoresis buffer. Chloride ions interfere with the discontinuous stacking system. This requires that protein samples should (preferably) be free of chloride salts, and that glycine base rather than glycine salt should be used in the electrophoresis buffer.

5. The separating and stacking gels contain 15 vs 4% acrylamide and 0.1 vs 0.16% *N,N'*-methylene *bis*-acrylamide, respectively, in 1 *M* acetic acid, 0.5% TEMED, 50 m*M* NH_4OH, 8 *M* urea, and 0.0004% riboflavin 5'-phosphate. The concentration of Triton X-100 in the separating gel of the regular system is 9 m*M*, optimal for the separation desired in our research for histone H3 variant forms of plants and algae.
   In the gradient system, glycerol and Triton concentrations change in parallel: 10% (v/v) glycerol and 10 m*M* Triton in the heavy gel (**Fig. 2B**) and a gradient between 10 and 0% (v/v) glycerol and between 10 and 0 m*M* Triton in the gradient from heavy to light (**Fig. 2B**).
6. Acrylamide is photopolymerized with riboflavin or riboflavin 5'-phosphate as initiator because the ions generated by ammonium persulfate initiated gel polymerization, as used for SDS polyacrylamide gels, interfere with stacking (*see* **Note 4**).
7. The height of stacking gel below the comb determines the volume of samples that can be applied and fully stacked before destacking at the surface of the separating gel occurs. In our experience, a 2.5 cm stacking gel height suffices for samples that almost completely fill equally long sample wells. In general, the single blue line of completely stacked proteins and methylene blue dye should be established 1 cm above the separating gel surface. Thus, 1–1.5 cm stacking gel height will suffice for small volume samples.
8. Salt-free samples are routinely prepared by exhaustive dialysis against 2.5% (v/v) acetic acid in 3500 mol wt cutoff dialysis membranes, followed by freezing at –70°C and lyophilization in conical polypropylene tubes (1.5-, 15-, or 50-mL, filled up to half of nominal capacity) with caps punctured by 21-gauge needle stabs. This method gives essentially quantitative recovery of histones, even if very dilute. (For alternative methods, *see* Chapter 16).
9. The amount of protein that can be analyzed in one gel lane depends highly on the complexity of the protein composition. As a guideline, 5–50 µg of total calf thymus histones with five major proteins (modified to varying extent) represents the range between very lightly to heavily Coomassie-stained individual protein bands in 1-mm-thick, 30-cm-long gels using 5-mm-wide comb teeth.
   The optimal amount of protein for a gradient gel also depends on the number of protein species that must be analyzed. In general, a gradient using the block comb used, equivalent to the width of 30 gel lanes, should be loaded with 30 times the sample for one lane.
10. As an alternative to **step 22**, electrophoresis buffer is added to the upper buffer reservoir and all sample wells are filled. Samples are layered under the buffer when dispensed by a Hamilton microsyringe near the bottom of each well.
11. The preparative well of a gradient gel should not be loaded with sample prior to the addition of electrophoresis buffer. Uneven distribution of sample cannot be avoided when buffer is added.
12. For unknown reasons, the polyacrylamide gel below the buffer front tends to stick tightly to the glass in an almost crystalline fashion. This may cause tearing of a gradient gel at the heavy gel side below the buffer front. Since attachment to one of the gel plates is typically much stronger, one can release the gel without problems by immersing the glass plate, with gel attached, in the staining solution.
13. A standard method to record the protein staining pattern in Coomassie-stained polyacrylamide gels has been Polaroid photography, using an orange filter to increase contrast, with the gel placed on the fluorescent light box, covered by a glass plate to prevent Coomassie staining of the typical plastic surface. Polaroid negatives can be scanned but suffer from a nonlinear response of density, even within the range where careful Coomassie staining leads to near-linear intensity of protein band staining.
   With the advent of 24+ bit color flatbed scanners with transmitted light capabilities and linear density capabilities in excess of three optical densities, it has become easy to record,

and quantitate, intensity of protein staining patterns. Care should be taken to develop a standard scanning setup, using the full dynamic range (typically all three colors used at their full range, 0–255 for a 24-bit scanner; excluding automatic adjustments for density and contrast). A standard gamma correction value should be determined, using an optical density wedge (Kodak), to assure that the density response is linear. Placing a destained polyacrylamide gel on the gel scanner, one should cover the top of the gel with an acetate film to prevent surface reflection abnormalities and prevent touching of the transilluminating light source surface to the film, avoiding moiré interference patterns. Recording gel patterns at 300 dpi in full color and saving loss-less tiff files facilitates faithful replication of experimental results (**Fig. 1**). Quantitative densitometry programs can use the image files.

14. Triton X-100 interferes with native-mode electrotransfer of histones to nitrocellulose *(**15**)*. Exchange of Triton by SDS under acidic conditions allows Western blotting *(**16**)* (*see* Chapter 42).

References

1. Zweidler, A. (1978) Resolution of histones by polyacrylamide gel electrophoresis in presence of nonionic detergents. *Meth. Cell Biol.* **17,** 223–233.
2. Manca, L., Cherchi, L., De Rosa, M. C., Giardina, B., and Masala, B. (2000) A new, electrophoretically silent, fetal hemoglobin variant: Hb F-Calabria [Ggamma118 (GH1) Phe→Leu]. *Hemoglobin* **24,** 37–44.
3. Urban, M. K., Franklin, S. G., and Zweidler, A. (1979) Isolation and characterization of the histone variants in chicken erythrocytes. *Biochemistry* **18,** 3952–3960.
4. Schwager, S. L. U., Brandt, W. F., and Von Holt, C. (1983) The isolation of isohistones by preparative gel electrophoresis from embryos of the sea urchin *Parechinus angulosus. Biochim. Biophys. Acta* **741**, 315–321.
5. Waterborg, J. H., Harrington, R. E., and Winicov, I. (1987) Histone variants and acetylated species from the alfalfa plant *Medicago sativa. Arch. Biochem. Biophys.* **256,** 167–178.
6. Bonner, W. M., West, M. H. P., and Stedman, J. D. (1980) Two-dimensional gel analysis of histones in acid extracts of nuclei, cells, and tissues. *Eur. J. Biochem.* **109,** 17–23.
7. Waterborg, J. H. (1992) Existence of two histone H3 variants in dicotyledonous plants and correlation between their acetylation and plant genome size. *Plant Mol. Biol.* **18,** 181–187.
8. Waterborg, J. H. (1991) Multiplicity of histone H3 variants in wheat, barley, rice and maize. *Plant Physiol.* **96,** 453–458.
9. Waterborg, J. H., Robertson, A. J., Tatar, D. L., Borza, C. M., and Davie, J. R. (1995) Histones of *Chlamydomonas reinhardtii.* Synthesis, acetylation and methylation. *Plant Physiol.* **109,** 393–407.
10. Waterborg, J. H. (2000) Steady-state levels of histone acetylation in *Saccharomyces cerevisiae. J. Biol. Chem.* **275,** 13,007–13,011.
11. Waterborg, J. H. and Matthews, H. R. (1983) Patterns of histone acetylation in the cell cycle of *Physarum polycephalum. Biochemistry* **22,** 1489–1496.
12. Arents, G., Burlingame, R. W., Wang, B. C., Love, W. E., and Moudrianakis, E. N. (1991) The nucleosomal core histone octamer at 3.1 Å resolution. A tripartite protein assembly and a left-handed superhelix. *Proc. Natl. Acad. Sci. USA* **88,** 10,138–10,148.
13. Luger, K., Mäder, A. W., Richmond, R. K., Sargent, D. F., and Richmond, T. J. (1997) Crystal structure of the nucleosome core particle at 2.8Å resolution. *Nature* **389,** 251–260.
14. Sullivan, S. A., Aravind, L., Makalowska, I., Baxevanis, A. D., and Landsman, D. (2000) The histone database: a comprehensive WWW resource for histones and histone fold-containing proteins. *Nucleic Acids Res.* **28,** 320–322.

15. Waterborg, J. H. and Harrington, R. E. (1987) Western blotting from acid–urea–Triton– and sodium dodecyl sulfate-polyacrylamide gels. *Analyt. Biochem.* **162,** 430–434.
16. Delcuve, G. P. and Davie, J. R. (1992) Western blotting and immunochemical detection of histones electrophoretically resolved on acid–urea–Triton- and sodium dodecyl sulfate-polyacrylamide gels. *Analyt. Biochem.* **200,** 339–341.

18

Isoelectric Focusing of Proteins in Ultra-Thin Polyacrylamide Gels

John M. Walker

1. Introduction

Isoelectric focusing (IEF) is an electrophoretic method for the separation of proteins, according to their isoelectric points (pI), in a stabilized pH gradient. The method involves casting a layer of support media (usually a polyacrylamide gel but agarose can also be used) containing a mixture of carrier ampholytes (low-mol-wt synthetic polyamino-polycarboxylic acids). When using a polyacrylamide gel, a low percentage gel (~4%) is used since this has a large pore size, which thus allows proteins to move freely under the applied electrical field without hindrance. When an electric field is applied across such a gel, the carrier ampholytes arrange themselves in order of increasing pI from the anode to the cathode. Each carrier ampholyte maintains a local pH corresponding to its pI and thus a uniform pH gradient is created across the gel. If a protein sample is applied to the surface of the gel, where it will diffuse into the gel, it will also migrate under the influence of the electric field until it reaches the region of the gradient where the pH corresponds to its isoelectric point. At this pH, the protein will have no net charge and will therefore become stationary at this point. Should the protein diffuse slightly toward the anode from this point, it will gain a weak positive charge and migrate back towards the cathode, to its position of zero charge. Similarly diffusion toward the cathode results in a weak negative charge that will direct the protein back to the same position. The protein is therefore trapped or "focused" at the pH value where it has zero overall charge. Proteins are therefore separated according to their charge, and not size as with SDS gel electrophoresis. In practice the protein samples are loaded onto the gel before the pH gradient is formed. When a voltage difference is applied, protein migration and pH gradient formation occur simultaneously.

Traditionally, 1–2 mm thick isoelectric focusing gels have been used by research workers, but the relatively high cost of ampholytes makes this a fairly expensive procedure if a number of gels are to be run. However, the introduction of thin-layer isoelectric focusing (where gels of only 0.15 mm thickness are prepared, using a layer of electrical insulation tape as the "spacer" between the gel plate) has considerably reduced

From: *The Protein Protocols Handbook, 2nd Edition*
Edited by: J. M. Walker © Humana Press Inc., Totowa, NJ

the cost of preparing IEF gels, and such gels are therefore described in this chapter. Additional advantages of the ultra-thin gels over the thicker traditional gels are the need for less material for analysis, and much quicker staining and destaining times. Also, a permanent record can be obtained by leaving the gel to dry in the air, i.e., there is no need for complex gel-drying facilities. The tremendous resolution obtained with IEF can be further enhanced by combinations with SDS gel electrophoresis in the form of 2D gel electrophoresis. Various 2D gel systems are described in Chapters 22–26.

2. Materials

1. Stock acrylamide solution: acrylamide (3.88 g), bis-acrylamide (0.12 g), sucrose (10.0 g). Dissolve these components in 80 mL of water. This solution may be prepared some days before being required and stored at 4°C (*see* **Note 1**).
2. Riboflavin solution: This should be made fresh, as required. Stir 10 mg riboflavin in 100 mL water for 20 min. Stand to allow undissolved material to settle out (or briefly centrifuge) (*see* **Note 2**).
3. Ampholytes: pH range 3.5–9.5 (*see* **Note 3**).
4. Electrode wicks: 22 cm × 0.6 cm strips of Whatman No. 17 filter paper.
5. Sample loading strips: 0.5 cm square pieces of Whatman No. 1 filter paper, or similar.
6. Anolyte: 1.0 *M* H_3PO_4.
7. Catholyte: 1.0 *M* NaOH.
8. Fixing solution: Mix 150 mL of methanol and 350 mL of distilled water. Add 17.5 g sulfosalicylic acid and 57.5 g trichloroacetic acid.
9. Protein stain: 0.1% Coomassie brilliant blue R250 in 50% methanol, 10% glacial acetic acid. N.B. Dissolve the stain in the methanol component first.
10. Glass plates: 22 cm × 12 cm. These should preferably be of 1 mm glass (to facilitate cooling), but 2 mm glass will suffice.
11. PVC electric insulation tape: The thickness of this tape should be about 0.15 mm and can be checked with a micrometer. The tape we use is actually 0.135 mm.
12. A bright light source.

3. Methods

1. Thoroughly clean the surfaces of two glass plates, first with detergent and then methylated spirit. It is *essential* for this method that the glass plates are clean.
2. To prepare the gel mold, stick strips of insulation tape, 0.5 cm wide, along the four edges of one glass plate. Do not overlap the tape at any stage. Small gaps at the join are acceptable, but do *not* overlap the tape at corners as this will effectively double the thickness of the spacer at this point.
3. To prepare the gel solution, mix the following: 9.0 mL acrylamide, 0.4 mL ampholyte solution, and 60.0 μL riboflavin solution.

   N.B. Since acrylamide monomer is believed to be neurotoxic, **steps 4–6** must be carried out wearing protective gloves.
4. Place the glass mold in a spillage tray and transfer ALL the gel solution with a Pasteur pipet along one of the short edges of the glass mold. The gel solution will be seen to spread slowly toward the middle of the plate.
5. Take the second glass plate and place one of its short edges on the taped edge of the mold, adjacent to the gel solution. Gradually lower the top plate and allow the solution to spread across the mold. Take care not to trap any air bubbles. If this happens, carefully raise the top plate to remove the bubble and lower it again.

6. When the two plates are together, press the edges firmly together (NOT the middle) and discard the excess acrylamide solution spilled in the tray. Place clips around the edges of the plate and thoroughly clean the plate to remove excess acrylamide solution using a wet tissue.
7 Place the gel mold on a light box (*see* **Note 4**) and leave for at least 3 h to allow polymerization (*see* **Note 5**). Gel molds may be stacked at least three deep on the light box during polymerization. Polymerized gels may then be stored at 4°C for at least 2 mo, or used immediately. If plates are to be used immediately they should be placed at 4°C for ~15 min, since this makes the separation of the plates easier.
8. Place the gel mold on the bench and remove the top glass plate by inserting a scalpel blade between the two plates and *slowly* twisting to remove the top plate. (N.B. Protect eyes at this stage.) The gel will normally be stuck to the side that contains the insulation tape. Do *not* remove the tape. Adhesion of the gel to the tape helps fix the gel to the plate and prevents the gel coming off in the staining/destaining steps (*see* **Note 6**).
9. Carefully clean the underneath of the gel plate and place it on the cooling plate of the electrophoresis tank. Cooling water at 10°C should be passing through the cooling plate.
10. Down the full length of one of the longer sides of the gel lay electrode wicks, uniformly saturated with either 1.0 *M* phosphoric acid (anode) or 1.0 *M* NaOH (cathode) (*see* **Note 7**).
11. Samples are loaded by laying filter paper squares (Whatman No. 1, 0.5 cm × 0.5 cm), wetted with the protein sample, onto the gel surface. Leave 0.5 cm gaps between each sample. The filter papers are prewetted with 5–7 µL of sample and applied across the width of the gel (*see* **Notes 8–10**).
12. When all the samples are loaded, place a platinum electrode on each wick. Some commercial apparatus employ a small perspex plate along which the platinum electrodes are stretched and held taut. Good contact between the electrode and wick is maintained by applying a weight to the perspex plate.
13. Apply a potential difference of 500 V across the plate. This should give current of about 4–6 mA. After 10 min increase the current to 1000 V and then to 1500 V after a further 10 min. A current of ~4–6 mA should be flowing in the gel at this stage, but this will slowly decrease with time as the gel focuses.
14. When the gel has been running for about 1 h, turn off the power and carefully remove the sample papers with a pair of tweezers. Most of the protein samples will have electrophoresed off the papers and be in the gel by now (*see* **Note 11**). Continue with a voltage of 1500 V for a further 1 h. During the period of electrophoresis, colored samples (myoglobin, cytochrome c, hemoglobin in any blood samples, and so on) will be seen to move through the gel and focus as sharp bands (*see* **Note 12**).
15. At the end of 2 h of electrophoresis, a current of about 0.5 mA will be detected (*see* **Note 13**). Remove the gel plate from the apparatus and *gently* wash it in fixing solution (200 mL) for 20 min (overvigorous washing can cause the gel to come away from the glass plate).

    Some precipitated protein bands should be observable in the gel at this stage (*see* **Note 14**). Pour off the fixing solution and wash the gel in destaining solution (100 mL) for 2 min and discard this solution (*see* **Note 15**). Add protein stain and gently agitate the gel for about 10 min. Pour off and discard the protein stain and wash the gel in destaining solution. Stained protein bands in the gel may be visualized on a light box and a permanent record may be obtained by simply leaving the gel to dry out on the glass plate overnight (*see* **Note 16**).
16. Should you wish to stain your gel for enzyme activity rather than staining for total protein, immediately following electrophoresis gently agitate the gel in an appropriate substrate solution (*see* **Note 7**, Chapter 10).

4. Notes

1. The sucrose is present to improve the mechanical stability of the gel. It also greatly reduces pH gradient drift.
2. The procedure described here uses photopolymerization to form the polyacrylamide gel. In the presence of light, riboflavin breaks down to give a free radical which initiates polymerization (Details of acrylamide polymerization are given in the introduction to Chapter 11). Ammonium persulphate /TEMED can be used to polymerize gels for IEF but there is always a danger that artefactual results can be produced by persulfate oxidation of proteins or ampholytes. Also, high levels of persulfate in the gel can cause distortion of protein bands (*see* **Note 10**).
3. This broad pH range is generally used because it allows one to look at the totality of proteins in a sample (but note that very basic proteins will run off the gel). However, ampholytes are available in a number of different pH ranges (e.g., 4–6, 5–7, 4–8, and so on) and can be used to expand the separation of proteins in a particular pH range. This is necessary when trying to resolve proteins with very similar pI values. Using the narrower ranges it is possible to separate proteins that differ in their pI values by as little as 0.01 of a pH unit.
4. Ensure that your light box does not generate much heat as this can dry out the gel quite easily by evaporation through any small gaps at the joints of the electrical insulation tape. If your light box is a warm one, stand it on its side and stand the gels adjacent to the box.
5. It is not at all obvious when a gel has set. However, if there are any small bubbles on the gel (these can occur particularly around the edges of the tape) polymerization can be observed by holding the gel up to the light and observing a "halo" around the bubble. This is caused by a region of unpolymerized acrylamide around the bubble that has been prevented from polymerizing by oxygen in the bubble. It is often convenient to introduce a small bubble at the end of the gel to help observe polymerization.
6. If the gel stays on the sheet of glass that does not contain the tape, discard the gel. Although usable for electrofocusing you will invariably find that the gel comes off the glass and rolls up into an unmanageable "scroll" during staining/destaining. To ensure that the gel adheres to the glass plate that has the tape on it, it is often useful to siliconize (e.g., with trimethyl silane) the upper glass plate before pouring the gel.
7. The strips must be fully wetted but must not leave a puddle of liquid when laid on the gel. (Note that in some apparatus designs application of the electrode applies pressure to the wicks, which can expel liquid.)
8. The filter paper must be fully wetted but should have no surplus liquid on the surface. When loaded on the gel this can lead to puddles of liquid on the gel surface which distorts electrophoresis in this region. The most appropriate volume depends on the absorbancy of the filter paper being used but about 5 µL is normally appropriate. For pure proteins load approx 0.5–1.0 µg of protein. The loading for complex mixtures will have to be done by trial and error.
9. Theoretically, samples can be loaded anywhere between the anode or cathode. However, if one knows approximately where the bands will focus it is best not to load the samples at this point since this can cause some distortion of the band. Similarly, protein stability is a consideration. For example, if a particular protein is easily denatured at acid pH values, then cathodal application would be appropriate.
10. The most common problem with IEF is distortions in the pH gradient. This produces wavy bands in the focused pattern. Causes are various, including poor electrical contact between electrode and electrode strips, variations in slab thickness, uneven wetting of electrode strips, and insufficient cooling leading to hot spots. However, the most common cause is

excessive salt in the sample. If necessary, therefore, samples should be desalted by gel filtration or dialysis before running the gel.

11. Although not absolutely essential, removal of sample strips at this stage is encouraged since bands that focus in the region of these strips can be distorted if strips are not removed. Take care not to make a hole in the gel when removing the strips. Use blunt tweezers (forceps) rather than pointed ones. When originally loading the samples it can be advantageous to leave one corner of the filter strip slightly raised from the surface to facilitate later removal with tweezers.
12. It is indeed good idea to include two or three blood samples in any run to act as markers and to confirm that electrophoresis is proceeding satisfactorily. Samples should be prepared by diluting a drop of blood approx 1:100 with distilled water to effect lysis of the erythrocytes. This solution should be pale cherry in color. During electrophoresis, the red hemoglobin will be seen to electrophorese off the filter paper into the gel and ultimately focus in the central region of the gel (pH 3.5–10 range). If samples are loaded from each end of the gel, when they have both focused in the sample place in the middle of the gel one can be fairly certain that isoelectrofocusing is occurring and indeed that the run is probably complete.
13. Theoretically, when the gel is fully focused, there should be no charged species to carry a current in the gel. In practice there is always a slow drift of buffer in the gel (electro-endomosis) resulting in a small (~0.5 mA) current even when gels are fully focused. Blood samples (*see* **Note 9**) loaded as markers can provide additional confirmation that focusing is completed.
14. It is not possible to stain the IEF gel with protein stain immediately following electrophoresis since the ampholytes will stain giving a uniformly blue gel. The fixing step allows the separated proteins to be precipitated in the gel, while washing out the still soluble ampholytes.
15. This brief wash is important. If stain is added to the gel still wet with fixing solution, a certain amount of protein stain will precipitate out. This brief washing step prevents this.
16. If you wish to determine the isoelectric point of a protein in a sample, then the easiest way is to run a mixture of proteins of known pI in an adjacent track (such mixtures are commercially available). Some commercially available kits comprise totally colored compounds that also allows one to monitor the focusing as it occurs. However, it is just as easy to prepare ones own mixture from individual purified proteins. When stained, plot a graph of protein pI vs distance from an electrode to give a calibration graph. The distance moved by the unknown protein is also measured and its pI read from the graph. Alternatively, a blank track can be left adjacent to the sample. This is cut out prior to staining the gel and cut into 1 mm slices. Each slice is then homogenized in 1 mL of water and the pH of the resultant solution measured with a micro electrode. In this way a pH vs distance calibration graph is again produced.

19

Protein Solubility in Two-Dimensional Electrophoresis

Basic Principles and Issues

Thierry Rabilloud

1. Introduction

The solubilization process for two-dimensional (2-D) electrophoresis needs to achieve four parallel goals:

1. Breaking macromolecular interactions in order to yield separate polypeptide chains. This includes denaturing the proteins to break noncovalent interactions, breaking disulfide bonds, and disrupting noncovalent interactions between proteins and nonproteinaceous compounds such as lipids or nucleic acids.
2. Preventing any artefactual modification of the polypeptides in the solubilization medium. Ideally, the perfect solubilization medium should freeze all the extracted polypeptides in their exact state prior to solubilization, both in terms of amino acid composition and in terms of posttranslational modifications. This means that all the enzymes able to modify the proteins must be quickly and irreversibly inactivated. Such enzymes include of course proteases, which are the most difficult to inactivate, but also phosphatases, glycosidases, and so forth. In parallel, the solubilization protocol should not expose the polypeptides to conditions in which chemical modifications (e.g., deamidation of Asn and Gln, cleavage of Asp–Pro bonds) may occur.
3. Allowing the easy removal of substances that may interfere with 2-D electrophoresis. In 2-D electrophoresis, proteins are the analytes. Thus, anything in the cell but proteins can be considered as an interfering substance. Some cellular compounds (e.g., coenzymes, hormones) are so dilute they go unnoticed. Other compounds (e.g., simple nonreducing sugars) do not interact with proteins or do not interfere with the electrophoretic process. However, many compounds bind to proteins and/or interfere with 2-D electrophoresis and must be eliminated prior to electrophoresis if their amount exceeds a critical interference threshold. Such compounds mainly include salts, lipids, polysaccharides (including cell walls), and nucleic acids.
4. Keeping proteins in solution during the 2-D electrophoresis process. Although solubilization *stricto sensu* stops at the point where the sample is loaded onto the first dimension gel, its scope can be extended to the 2-D process *per se*, as proteins must be kept soluble

From: *The Protein Protocols Handbook, 2nd Edition*
Edited by: J. M. Walker © Humana Press Inc., Totowa, NJ

until the end of the second dimension. Generally speaking, the second dimension is a sodium dodecyl sulfate (SDS) gel, and very few problems are encountered once the proteins have entered the SDS-polyacrylamide gel electrophoresis (SDS-PAGE) gel. The one main problem is overloading of the major proteins when micropreparative 2-D electrophoresis is carried out, and nothing but scaling-up the SDS gel (its thickness and its other dimensions) can counteract overloading a SDS gel. However, severe problems can be encountered in the isoelectric fusing (IEF) step. They arise from the fact that IEF must be carried out in low ionic strength conditions and with no manipulation of the polypeptide charge. IEF conditions give problems at three stages:

a. During the initial solubilization of the sample, important interactions between proteins of widely different pI and/or between proteins and interfering compounds (e.g., nucleic acids) may occur. This yields poor solubilization of some components.
b. During the entry of the sample in the focusing gel, there is a stacking effect due to the transition between a liquid phase and a gel phase with a higher friction coefficient. This stacking increases the concentration of proteins and may give rise to precipitation events.
c. At, or very close to, the isoelectric point, the solubility of the proteins comes to a minimum. This can be explained by the fact that the net charge comes close to zero, with a concomitant reduction of the electrostatic repulsion between polypeptides. This can also result in protein precipitation or adsorption to the IEF matrix.

Apart from breaking molecular interactions and solubility in the 2-D gel which are common to all samples, the solubilization problems encountered will greatly vary from a sample type to another, owing to wide differences in the amount and nature of interfering substances and/or spurious activities (e.g. proteases). The aim of this outline chapter is not to give detailed protocols for various sample types, and the reader should refer to the chapters of this book dedicated to the type of sample of interest. I would rather like to concentrate on the solubilization rationale and to describe nonstandard approaches to solubilization problems. More detailed review on solubilization of proteins for electrophoretic analyses can be found elsewhere *(1,2)*.

2. Rationale of Solubilization-Breaking Molecular Interactions

Apart from disulfide bridges, the main forces holding proteins together and allowing binding to other compounds are non-covalent interactions. Covalent bonds are encountered mainly between proteins and some coenzymes. The noncovalent interactions are mainly ionic bonds, hydrogen bonds and "hydrophobic interactions." The basis for "hydrophobic interactions" is in fact the presence of water. In this very peculiar (hydrogen-bonded, highly polar) solvent, the exposure of nonpolar groups to the solvent is thermodynamically not favored compared to the grouping of these apolar groups together. Indeed, although the van der Waals forces give an equivalent contribution in both configurations, the other forces (mainly hydrogen bonds) are maximized in the latter configuration and disturbed in the former (solvent-binding destruction). Thus, the energy balance in clearly in favor of the collapse of the apolar groups together *(3)*. This explains why hexane and water are not miscible, and also that the lateral chain of apolar amino acids (L, V, I, F, W, Y) pack together and form the hydrophobic cores of the proteins *(4)*. These hydrophobic interactions are also responsible for some protein–

protein interactions and for the binding of lipids and other small apolar molecules to proteins.

The constraints for a good solubilization medium for 2-D electrophoresis are therefore to be able to break ionic bonds, hydrogen bonds, hydrophobic interactions, and disulfide bridges under conditions compatible with IEF, that is, with very low amounts of salt or other charged compounds (e.g., ionic detergents).

2.1. Disruption of Disulfide Bridges

Breaking of disulfide bridges is usually achieved by adding to the solubilization medium an excess of a thiol compound. Mercaptoethanol was used in the first 2-D protocols *(5)*, but its use does have drawbacks. Indeed, a portion of the mercaptoethanol will ionize at basic pH, enter the basic part of the IEF gel, and ruin the pH gradient in its alkaline part because of its buffering power *(6)*. Although its p*K* is around 8, dithiothreitol (DTT) is much less prone to this drawback, as it is used at much lower concentrations (usually 50 m*M* instead of the 700 m*M* present in 5% mercaptoethanol). However, DTT is still not the perfect reducing agent. Some proteins of very high cysteine content or with cysteines of very high reactivity are not fully reduced by DTT. In these cases, phosphines are very often an effective answer. First, the reaction is stoichiometric, which in turn allows use of very low concentration of the reducing agent (a few millimolar). Second, these reagents are not as sensitive as thiols to dissolved oxygen. The most powerful compound is tributylphosphine, which was the first phosphine used for disulfide reduction in biochemistry *(7)*. However, the reagent is volatile, toxic, has a rather unpleasant odor, and needs an organic solvent to make it water-miscible. In the first uses of the reagent, propanol was used as a carrier solvent at rather high concentrations (50%) *(7)*. It was found, however, that dimethyl sulfoxide (DMSO) and dimethyl formamide (DMF) are suitable carrier solvents, which enable the reduction of proteins by 2 m*M* tributylphosphine *(8)*. All these drawbacks have disappeared with the introduction of a watersoluble phospine, tris (carboxyethyl) phosphine (available from Pierce), for which 1 *M* aqueous stock solutions can be easily prepared and stored frozen in aliquots. The successful use of tributylphosphine in two-dimensional electrophoresis has been reported *(9)*. However, the benefits over DTT do not seem obvious in many cases *(10)*.

2.2. Disruption of Noncovalent Interactions

The perfect way to disrupt all types of noncovalent interactions would be the use of a charged compound that disrupts hydrophobic interactions by providing a hydrophobic environment. The hydrophobic residues of the proteins would be dispersed in that environment and not clustered together. This is just the description of SDS, and this explains why SDS has been often used in the first stages of solubilization *(11–14)*. However, SDS is not compatible with IEF, and must be removed from the proteins during IEF.

The other way of breaking most noncovalent interactions is the use of a chaotrope. It must be kept in mind that all the noncovalent forces keeping molecules together must be taken into account with a comparative view on the solvent. This means that the final energy of interaction depends on the interaction *per se* and on its effects on the solvent. If the solvent parameters are changed (dielectric constant, hydrogen bond formation, polarizability, etc.), all the resulting energies of interaction will change. Chaotropes, which alter all the solvent parameters, exert profound effects on all types of interactions. For example, by changing the hydrogen bond structure of the solvent, chaotropes disrupt hydrogen bonds but also decrease the energy penalty for exposure of apolar groups and therefore favor the dispersion of hydrophobic molecules and the unfolding of the hydrophobic cores of a protein ***(1,15)***. Unfolding the proteins will also greatly decrease ionic bonds between proteins, which are very often not very numerous and highly dependent of the correct positioning of the residues. As the gross structure of proteins is driven by hydrogen bonds and hydrophobic interactions, chaotropes decrease dramatically ionic interactions both by altering the dielectric constant of the solvent and by denaturing the proteins, so that the residues will no longer positioned correctly.

Nonionic chaotropes, such as those used in 2-D, however, are unable to disrupt ionic bonds when high charge densities are present (e.g., histones, nucleic acids) ***(16)***. In this case, it is often quite advantageous to modify the pH and to take advantage of the fact that the ionizable groups in proteins are weak acids and bases. For example, increasing the pH to 10 or 11 will induce most proteins to behave as anions, so that ionic interactions present at pH 7 or lower turn into electrostatic repulsion between the molecules, thereby promoting solubilization. The use of a high pH results therefore in dramatically improved solubilizations, with yields very close to what is obtained with SDS ***(17)***. The alkaline pH can be obtained either by addition of a few mM of potassium carbonate to the urea–detergent–ampholytes solution ***(17)***, or by the use of alkaline ampholytes ***(14)***, or by the use of a spermine–DTT buffer which allows better extraction of nuclear proteins ***(18)***.

For 2-D electrophoresis, the chaotrope of choice is urea. Although urea is less efficient than substituted ureas in breaking hydrophobic interactions ***(15)***, it is more efficient in breaking hydrogen bonds, so that its overall solubilization power is greater. It has been found that thiourea in addition to urea results in superior solubilization of proteins ***(19)***, and also in improved protein focusing. This may be explained by the fact that thiourea is superior to urea for protein denaturation, being inferior to only for guanidine ***(20)***. However, the solubility of thiourea in water is not sufficient to draw full benefits from its denaturing power, so that it must be used in admixture with urea. However, denaturation by chaotropes induces the exposure of the totality of the protein's hydrophobic residues to the solvent. This in turn increases the potential for hydrophobic interactions, so that chaotropes alone are often not sufficient to quench completely the hydrophobic interactions, especially when lipids are present in the sample. This explains why detergents, which can be viewed as specialized agents for hydrophobic interactions, are almost always included in the urea-based solubilization mixtures for 2-D electrophoresis. Detergents act on hydrophobic interactions by providing a stable dispersion of a hydrophobic medium in the aqueous medium, through the presence of micelles, for example. Therefore, the hydrophobic molecules (e.g., lipids) are no longer collapsed in the aqueous solvent but will disaggregate in the micelles, provided the amount of detergent is sufficient to ensure maximal dispersion of the hydrophobic molecules. Detergents have polar heads that are able to contract other types of noncovalent bonds (hydrogen bonds, salt bonds for charged heads, etc.). The action of detergents is the sum of the dispersive effect of the micelles on the hydrophobic parts of the molecules and the effect of their polar heads on the other types of bonds. This explains why various detergents show very ranging effects varying from a weak and often incomplete delipidation (e.g., Tweens) to a very aggressive action where the exposure of the hydrophobic core in the detergent-containing solvent is no longer energetically unfavored and leads to denaturation (e.g., SDS).

Of course, detergents used for IEF must bear no net electrical charge, and only nonionic and zwitterionic detergents may be used. However, ionic detergents such as SDS may be used for the initial solubilization, prior to isoelectric focusing, in order to increase solubilization and facilitate the removal of interfering compounds. Low amounts of SDS can be tolerated in the subsequent IEF ***(11)*** provided that high concentrations of urea ***(21)*** and nonionic ***(11)*** or zwitterionic detergents ***(22)*** are present to ensure complete removal of the SDS from the proteins during IEF. Higher amounts of SDS must be removed prior to IEF, by precipitation ***(11)*** for example. It must therefore be kept in mind that SDS will be useful only for solubilization and for sample entry, but will not eliminate isoelectric precipitation problems.

The use of nonionic or zwitterionic detergents in the presence of urea presents some problems due to the presence of urea itself. In concentrated urea solutions, urea is not freely dispersed in water but can form organized channels (see ***[18]***). These channels can bind linear alkyl chains, but not branched or cyclic molecules, to form complexes of undefined stoichiometry called inclusion compounds. These complexes are much less soluble than the free solute, so that precipitation is often induced upon formation of the inclusion compounds, precipitation being stronger with increasing alkyl chain length and higher urea concentrations. Consequently, many nonionic or zwitterionic detergents with linear hydrophobic tails ***(24,25)*** and some ionic ones ***(26)*** cannot be used in the presence of high concentrations of urea. This limits the choice of detergents mainly to those with nonlinear alkyl tails (e.g., Tritons, Nonidet P40, 3-[3 -cholamidopropyl) dimethlyammonio]-1-propanesulfonate [CHAPS]) or with short alkyl tails (e.g., octyl glucoside), which are unfortunately less efficient in quenching hydrophobic interactions. Thus, sulfobetaine detergents with long linear alkyl tails have received limited applications, as they require low concentrations of urea. However, good results have been obtained in certain cases for sparingly soluble proteins ***(27–29)***, although this type of protocol seems rather delicate owing to the need for a precise control of all parameters to prevent precipitation.

Apart from the problem of inclusion compounds, the most important problem linked with the use of urea is carbamylation. Urea in water exists in equilibrium with ammonium cyanate, the level of which increases with increasing temperature and pH ***(30)***. Cyanate can react with amines to yield substituted urea. In the case of proteins, this reaction takes place with the α-amino group of the amino-(N)-terminus and the ε-amino groups of lysines. This reaction leads to artefactual charge heterogeneity, N-terminus blocking, and adduct formation detectable in mass spectrometry. Carbamylation should therefore be completely avoided. This can be easily made with some simple precautions. The use of a pure grade of urea (p.a.) decreases the amount of cyanate present in the starting material. Avoidance of high temperatures (never heat urea-containing solutions above 37°C) considerably decreases cyanate formation. In the same trend, urea-containing solutions should be stored frozen (–20°C) to limit cyanate accumulation. Last but not least, a cyanate scavenger (primary amine) should be added to urea-containing solutions. In the case of isoelectric focusing, carrier ampholytes are perfectly suited for this task. If these precautions are correctly taken, proteins seem to withstand long exposures to urea without carbamylation *(31)*.

3. Solubility During IEF

Additional solubility problems often arise during the IEF at sample entry and solubility at the isoelectric point.

3.1. Solubility During Sample Entry

Sample entry is often quite critical. In most 2-D systems, sample entry in the IEF gel corresponds to a transition between a liquid phase (the sample) and a gel phase of higher friction coefficient. This induces a stacking of the proteins at the sample–gel boundary, which results in very high concentration of proteins at the application point. These concentrations may exceed the solubility threshold of some proteins, thereby inducing precipitation and sometimes clogging of the gel, with poor penetration of the bulk of proteins. Such a phenomenon is of course more prominent when high amounts of proteins are loaded onto the IEF gel. The sole simple but highly efficient remedy to this problem is to include the sample in the IEF gel. This process abolishes the liquid–gel transition and decreases the overall protein concentration, as the volume of the IEF gel is generally much higher than the one of the sample.

This process is, however, rather difficult for tube gels in carrier ampholyte-based IEF. The main difficulty arises from the fact that the thiol compounds used to reduce disulfide bonds during sample preparation are strong inhibitors of acrylamide polymerization, so that conventional samples cannot be used as such. Alkylation of cysteines and of the thiol reagent after reduction could be an answer, but many neutral alkylating agents (e.g., iodoacetamide, ethyl maleimide) also inhibit acrylamide polymerization. Owing to this situation, most workers describing inclusion of the sample within the IEF gel have worked with nonreduced samples *(32,33)*. Although this presence of disulfide bridges is not optimal, inclusion of the sample within the gel has proven of great but neglected interest *(32,33)*. It must, however, be pointed out that it is now possible to carry out acrylamide polymerization in an environment where disulfide bridges are reduced. The key is to use 2 mM tributylphosphine as the reducing agent in the sample and tetramethylurea as a carrier solvent. This ensures total reduction of disulfides and is totally compatible with acrylamide polymerization with the standard *N, N, N, N'*-tetramethylenediamine TEMED/persulfate initiator (T. Rabilloud, unpublished results). This modification should help the experimentators trying sample inclusion within the IEF gel when high amounts of proteins are to be separated by 2D.

The process of sample inclusion within the IEF gel is however much simpler for IPG gels. In this case, rehydration of the dried IPG gel in a solution containing the protein sample is quite convenient and efficient, provided that the gel has a sufficiently open structure to be able to absorb proteins efficiently *(18)*. Coupled with the intrinsic high capacity of IPG gels, this procedure enableseasy separation of milligram amounts of protein *(18,34)*.

3.2. Solubility at the Isoelectric Point

This is usually the second critical point for IEF. The isoelectric point is the pH of minimal solubility, mainly because the protein molecules have no net electrical charge. This abolishes the electrostatic repulsion between protein molecules, which maximizes in turn protein aggregation and precipitation.

The horizontal comet shapes frequently encountered for major proteins and for sparingly soluble proteins often arise from such a near-isoelectric precipitation. Such isoelectric precipitates are usually easily dissolved by the SDS solution used for the transfer of the IEF gel onto the SDS gel, so that the problem is limited to a loss of resolution, which, however, precludes the separation of high amounts of proteins.

The problem is however more severe for hydrophobic proteins when an IPG is used. In this case, a strong adsorption of the isoelectric protein to the IPG matrix seems to occur, which is not reversed by incubation of the IPG gel in the SDS solution. The result is severe quantitative losses, which seem to increase with the hydrophobicity of the protein and the amount loaded *(35)*. The sole solution to this serious problem is to increase the chaotropicity of the medium used for IEF, by using both urea and thiourea as chaotropes *(19)*.

The benefits of using thiourea–urea mixtures to increase protein solubility can be transposed to conventional, carrier ampholyte-based focusing in tube gels with minor adaptations. Thiourea strongly inhibits acrylamide polymerization with the standard TEMED/persulfate system. However, photopolymerization with methylene blue, sodium toluene sulfinate and diphenyl iodonium chloride *(36)* enables acrylamide polymerization in the presence of 2 *M* thiourea without any deleterious effect in the subsequent 2-D *(37)* so that higher amounts of proteins can be loaded without loss of resolution *(37)*.

3.3. The Epitome in Solubilization Problems: Membrane Proteins

Biological membranes represent a prototype of difficult samples for 2-D electrophoresis. They contain high amounts of lipids, which are troublesome compounds for 2-D electrophoresis. In addition, many membrane proteins are highly hydrophobic, and therefore very difficult to keep in aqueous solution, even with the help of chaotropes and detergents. Thus, membrane proteins have often been resistant to 2-D electrophoretic analysis *(38)*. In fact, it has been shown that typical membrane proteins with multiple transmembrane helices are resistant to solubilization by chaotrope–detergent mixtures, even using thiourea *(39)*. For such proteins, the development of new detergents proved necessary *(39,40)*. Even if some successes have been obtained *(39–42)*, it is still obvious that many membrane proteins are not properly solubilized and focused with the chemicals described up to now *(43)*. Therefore, further development is still needed to improve the representation of membrane proteins in 2-D maps.

4. Concluding Remarks

Although this outline chapter has dealt mainly with the general aspects of solubilization, the main concluding remark is that there is no universal solubilization protocol. Standard urea–reducer–detergent mixtures usually achieve disruption of disulfide bonds and noncovalent interactions. Consequently, the key issues for a correct solubilization is the removal of interfering compounds, blocking of protease action, and disruption of infrequent interactions (e.g., severe ionic bonds). These problems will strongly depend on the type of sample used, the proteins of interest, and the amount to be separated, so that the optimal solubilization protocol can vary greatly from a sample to another.

However, the most frequent bottleneck for the efficient 2-D separation of as many proteins as possible does not lie only in the initial solubilization but also in keeping the solubility along the IEF step. In both fields, the key feature is the disruption of hydrophobic interactions, which are responsible for most, if not all, of the precipitation phenomena encountered during IEF. This means improving solubility during denaturing IEF will focus on the quest of ever more powerful chaotropes and detergents. In this respect, the use of thiourea has proven to be one of the keys to increasing the solubility of proteins in 2-D electrophoresis. It would be nice to have other, even more powerful chaotropes. However, testing of the chatropes previously described as powerful ***(20)*** has not led yet to improvement in protein solubilization (T. Rabilloud, unpublished results). One of the other keys for improving protein solubilization is the use of as powerful a detergent or detergent mixtures as possible. In a complex sample, some proteins may be well denatured and solubilized by a given detergent or chaotrope, while other proteins will require another detergent or chaotrope (e.g. *see **[39]***). Consequently, the future of solubilization may still be to find mixtures of detergents and chaotropes able to cope with the diversity of proteins encountered in the complex samples separated by 2-D electrophoresis.

References

1. Rabilloud, T. (1996) Solubilization of proteins for electrophoretic analyses. *Electrophoresis* **17,** 813–829.
2. Rabilloud, T. and Chevallet, M. (1999) Solubilization of proteins in 2-D electrophoresis, in Proteome Research: Two-Dimensional Gel Electrophoresis, and Identification Methods Rabilloud, T., Ed., Springer-Verlag, Heidelberg, pp. 9–30.
3. Tanford, C. The Hydrophobic Effect, 2nd edit., John Wiley & Sons New York, 1980.
4. Dill, K. A. (1985) Theory for the folding and stability of globular proteins. *Biochemistry* **24,** 1501–1509.
5. O'Farrell P. H. (1975) High resolution two-dimensional electrophoresis of proteins. *J. Biol. Chem.* **250,** 4007–4021.
6. Righetti, P. G., Tudor, G., and Gianazza, E. (1982) Effect of 2-mercaptoethanol on pH gradients in isoelectric focusing. *J. Biochem. Biophys. Meth.* **6,** 219–227.
7. Ruegg, U. T. and Rüdinger, J. (1977) Reductive cleavage of cystine disulfides with tributylphosphine. *Meth. Enzymol.* **47,** 111–116.
8. Kirley, T. L. (1989) Reduction and fluorescent labeling of cyst(e)ine containing proteins for subsequent structural analysis. *Analyt. Biochem.* **180,** 231–236.
9. Herbert, B., Molloy, M. P., Gooley, A. A., Walsh, B. J., Bryson, W. G., and Williams, K. L. (1998) Improved protein solubility in two-dimensional electrophoresis using tributyl phosphine as a reducing agent. *Electrophoresis* **19,** 845–851.
10. Hoving, S., Voshol, H., and Van Oostrum, J. (2000) Towards high performance two-dimensional gel electrophoresis using ultrazoom gels *Electrophoresis* **21,** 2617–2621.
11. Wilson, D., Hall, M. E., Stone, G. C., and Rubin, R. W. (1977) Some improvements in two-dimensional gel electrophoresis of proteins. *Analyt. Biochem.* **83,** 33–44.

12. Hari, V. (1981) A method for the two-dimensional electrophoresis of leaf proteins. *Analyt. Biochem.*, **113,** 332–335.
13. Ames, G. F. L. and Nikaido, K.(1976) Two-dimensional electrophoresis of membrane proteins. *Biochemistry* **15,** 616–623.
14. Hochstrasser, D. F., Harrington, M. G., Hochstrasser, A. C., Miller, M. J., and Merril, C. R. (1988) Methods for increasing the resolution of two dimensional protein electrophoresis *Analyt. Biochem.* **173,** 424–435.
15. Herskovits, T. T., Jaillet, H., and Gadegbeku, B. (1970) On the structural stability and solvent denaturation of proteins. II. Denaturation by the ureas *J. Biol. Chem.* **245,** 4544–4550.
16. Sanders, M. M., Groppi, V. E., and Browning, E. T. (1980) Resolution of basic cellular proteins including histone variants by two-dimensional gel electrophoresis: evaluation of lysine to arginine ratios and phosphorylation. *Analyt. Biochem.* **103,** 157–165.
17. Horst, M. N., Basha, M. M., Baumbach, G. A., Mansfield, E. H., and Roberts, R. M.(1980) Alkaline urea solubilization, two-dimensional electrophoresis and lectin staining of mammalian cell plasma membrane and plant seed proteins *Analyt. Biochem.* **102,** 399–408.
18. Rabilloud, T., Valette, C., and Lawrence, J. J. (1994) Sample application by in-gel rehydration improves the resolution of two-dimensional electrophoresis with immobilized pH gradients in the first dimension. *Electrophoresis* **15,** 1552–1558.
19. Rabilloud, T., Adessi, C., Giraudel, A, and Lunardi, J. (1997) Improvement of the solubilization of proteins in two-dimensional electrophoresis with immobilized pH gradients. *Electrophoresis* **18,** 307–316.
20. Gordon, J. A. and Jencks, W. P. (1963) The relationship of structure to the effectiveness of denaturing agents for proteins. *Biochemistry* **2,** 47–57.
21. Weber, K. and Kuter, D. J. (1971) Reversible denaturation of enzymes by sodium dodecyl sulfate. *J. Biol. Chem.* **246,** 4504–4509.
22. Remy, R. and Ambard-Bretteville, F. (1987) Two-dimensional electrophoresis in the analysis and preparation of cell organelle polypeptides. *Meth. Enzymol.* **148,** 623–632.
23. March, J. (1977) Advanced Organic Chemistry, 2nd edit., McGraw-Hill London, pp. 83–84.
24. Dunn, M. J. and Burghes, A. H. M.(1983) High resolution two-dimensional polyacrylamide electrophoresi. I. Methodological procedures. *Electrophoresis* **4,** 97–116.
25. Rabilloud, T., Gianazza, E., Catto, N., and Righetti, P. G. (1990) Amidosulfobetaines, a family of detergents with improved solubilization properties: application for isoelectric focusing under denaturing conditions. *Analyt. Biochem.* **185,** 94–102.
26. Willard, K. E., Giometti, C., Anderson, N. L., O'Connor, T. E., and Anderson, N. G. (1979) Analytical techniques for cell fractions. XXVI. A two-dimensional electrophoretic analysis of basic proteins using phosphatidyl choline/urea solubilization. *Analyt. Biochem.* **100,** 289–298.
27. Clare Mills, E. N. and Freedman, R. B. (1983) Two-dimensional electrophoresis of membrane proteins. Factors affecting resolution of rat liver microsomal proteins *Biochim. Biophys. Acta* **734,** 160–167.
28. Satta, D., Schapira, G., Chafey, P., Righetti, P. G., and Wahrmann, J. P. (1984) Solubilization of plasma membranes in anionic, non ionic and zwitterionic surfactants for iso-dalt analysis: a critical evaluation. *J. Chromatogr.* **299,** 57–72.
29. Gyenes, T. and Gyenes, E. (1987) Effect of stacking on the resolving power of ultrathin layer two-dimensional gel electrophoresis. *Analyt. Biochem.* **165,** 155–160.
30. Hagel, P., Gerding, J. J. T., Fieggew, W., and Bloemendal, H. (1971) Cyanate formation in solutions of urea. I. Calculation of cyanate concentrations at different temperature and pH. *Biochim. Biophys. Acta* **243,** 366–373.
31. Bjellqvist,B., Sanchez, J. C., Pasquali, C., Ravier, F., Paquet, N., and Frutiger, S. (1993) Micropreparative two-dimensional electrophoresis allowing the separation of samples containing miiligram amounts of proteins. *Electrophoresis* **14,** 1375–1378.

32. Chambers, J. A. A., Degli Innocenti, F., Hinkelammert, K., and Russo, V. E. A (1985) Factors affecting the range of pH gradients in the isoelectric focusing dimension of two-dimensional gel electrophoresis: the effect of reservoir electrolytes and loading procedures. *Electrophoresis* **6,** 339–348.
33. Semple-Rowland, S. L., Adamus, G., Cohen, R. J., and Ulshafer, R. J. (1991) A reliable two-dimensional gel electrophoresis procedure for separating neural proteins. *Electrophoresis* **12,** 307–312.
34. Sanchez, J. C., Rouge, V., Pisteur, M., Ravier, F., Tonella, L., Moosmayer, M., et al. (1997) Improved and simplified in-gel sample application using reswelling of dry immobilized pH gradients. *Electrophoresis* **18,** 324–327.
35. Adessi, C., Miege, C., Albrieux, C., and Rabilloud, T. (1997) Two-dimensional electrophoresis of membrane proteins: a current challenge for immobilized pH gradients. *Electrophoresis* **18,** 127–135.
36. Lyubimova, T., Caglio, S., Gelfi, C., Righetti, P. G., and Rabilloud, T. (1993) Photopolymerization of polyacrylamide gels with methylene blue. *Electrophoresis* **14,** 40–50.
37. Rabilloud, T. (1998) Use of thiourea to increase the solubility of membrane proteins in two-dimensional electrophoresis. *Electrophoresis* **19,** 758–760.
38. Rubin, R. W. and Milikowski, C (1978) Over two hundred polypeptides resolved from the human erythrocyte membrane. *Biochim. Biophys. Acta 1978* **509,** 100–110.
39. Rabilloud, T., Blisnick, T., Heller, M., Luche, S., Aebersold, R., Lunardi, J., and Braun-Breton, C. (1999) Analysis of membrane proteins by two-dimensional electrophoresis: comparison of the proteins extracted from normal or *Plasmodium falciparum*-infected erythrocyte ghosts. *Electrophoresis* **20,** 3603–3610.
40. Chevallet, M., Santoni, V., Poinas, A., Rouquié, D., Fuchs, A., Kieffer, S., et al. (1998) New zwitterionic detergents improve the analysis of membrane proteins by two-dimensional electrophoresis. *Electrophoresis* **19,** 1901–1909.
41. Santoni, V., Rabilloud, T., Doumas, P., Rouquié, D., Mansion, M., Kieffer, S., et al. (1999) Towards the recovery of hydrophobic proteins on two-dimensional gels. *Electrophoresis* 1999. **20,** 705–711.
42. Friso, G. and Wikstrom, L. (1999) Analysis of proteins from membrane-enriched cerebellar preparations by two-dimensional gel electrophoesis and mass spectrometry *Electrophoresis* **20,** 917–927.
43. Santoni, V., Molloy, M. P., and Rabilloud, T. (2000) Membrane proteins and proteomics: un amour impossible? *Electrophoresis* **21,** 1054–1070.

20

Preparation of Protein Samples from Mouse and Human Tissues for 2-D Electrophoresis

Joachim Klose

1. Introduction

The protocol for extracting proteins from mouse and human tissues (organs) described in this chapter follows a strategy that is based on the intention to include all the various protein species of a particular tissue in a set of samples that are suitable for two-dimensional electrophoresis (2-DE), particularly for the large gel 2-DE technique *(1)*. The aim then is the resolution and visualization of all these protein species in 2-DE gels. This aim explains some features of our tissue extraction procedure for gaining the proteins. Fractionation of total tissue proteins was preferred to a one-step extraction of all proteins. Using the fractionation procedure, the many different protein species of a tissue can be distributed over several 2-DE gels, and this increases resolution. However, a postulate is that fractionation of tissue proteins results in fraction-specific proteins. To achieve this, cell fractionation is usually performed with the aim of isolating special cell organelles (nuclei, mitochondria) or cell structures (membranes). The proteins are then extracted from these natural fractions. This procedure, however, includes washing steps to purify the cell fractions, and the elimination of cell components that are not of interest and cell residues, which are rejected. Using this procedure, an uncontrolled loss of proteins is unavoidable.

The tissue fractionation procedure described in this chapter renounces the isolation of defined cell components. This allows us to avoid any selective loss of proteins. Mouse (human) tissues (liver, brain, heart) are fractionated into three fractions:

1. The "supernatant I + II" (SI + II) containing the proteins soluble in buffer (cytoplasmic proteins, nucleoplasmic proteins).
2. The "pellet extract" (PE) containing the proteins soluble in the presence of urea and CHAPS (proteins from membranes and other structures of the cells and cell organelles).
3. The "pellet suspension" (PS) containing proteins released by DNA digestion (histones and other chromosomal proteins).

The SI + II fraction is obtained by homogenization, sonication, and centrifugation of the tissue and the resulting pellet (I), and the combination of the two supernatants gained in this way. The solution thus obtained is the first protein sample. The pellet (II) that

From: *The Protein Protocols Handbook, 2nd Edition*
Edited by: J. M. Walker © Humana Press Inc., Totowa, NJ

remained is extracted with urea and CHAPS, and the homogenate is centrifuged. The supernatant is the PE fraction and gives the second protein sample. The final pellet (III) is suspended into buffer containing benzonase, a DNA-digesting enzyme. The PS (fraction) is the third protein sample. It is applied to 2-DE without further centrifugation.

Care is taken during the whole procedure to avoid any loss of material. In spite of that, some material may become lost, for example, by the transfer of the pulverized, frozen tissue from the mortar to the tube or by removing the glass beads from the sonicated homogenate. However, this does not lead to a preferential loss of certain protein species or protein classes.

It is evident that the best conditions for keeping proteins stable and soluble are given in the living cells *(2,3)*. Therefore, a principle of our tissue extraction procedure was to extract the so-called soluble proteins (SI + II) as far as possible under natural conditions. That means keeping the ionic strength of the tissue homogenate at 150–200 m*M*, the pH in the range of 7.0–7.5, the protein concentration high, and protecting the proteins against water by adding glycerol to the buffer. Generally, the best conditions for the first tissue extraction would be given if the addition of any diluent that disturbs the natural concentration and milieu of the cell proteins were avoided. We prepared a pure cell sap from a tissue by homogenizing the tissue without additives (except for protease inhibitor solutions added in small volumes) followed by high-speed centrifugation, and extracted the pellet that resulted successively in increasing amounts (0.5, 1, or 2 parts) of buffer. The series of protein samples obtained were separated by 2-DE and the patterns compared. The results showed that by increasing the dilution of the proteins, the number of spots and their intensities decreased in the lower part of the 2-DE patterns and increased in the upper part. The same phenomenon, but less pronounced, was observed even when the cell sap already extracted, was diluted successively. Apparently, low-mol-wt proteins are best dissolved in the pure cell sap and, presumably, tend to precipitate in more diluted extracts. High-mol-wt proteins, in contrast, become better dissolved in more diluted samples. This effect was most pronounced in protein patterns from the liver and not obvious in patterns from heart muscle. This is probably because of the high protein concentration in the liver cell sap that cannot be reached in extracts of other organs.

The dependency of the protein solubility on the molecular weight of the proteins is obscured in 2-DE patterns by another effect that leads to a similar phenomenon. The higher the concentration of the first tissue extract, the higher the activity of the proteases released by breaking the cellular structure by homogenization. Protein patterns from pure liver cell sap, extracted without protease inhibitors (but even with inhibitors) showed an enormous number of spots in the lower part of the gel and a rather depleted pattern in the upper part. In the pH range around 6.0, the protein spots disappeared almost completely, in the upper as well as in the lower part, suggesting that these proteins are most sensitive to the proteases. By extracting the tissue or first pellet with increasing amounts of buffer, the 2-DE pattern (spot number and intensity) shifted from the lower part to the upper part of the gel. Again, this observation was made particularly in liver.

The consequence of these observations for our protein extraction procedure was to gain the supernatant I and II at concentrations that keep all the soluble proteins in solution, but do not reach a level where proteases cannot be inhibited effectively enough. We introduced buffer factors that determine the concentration of the different

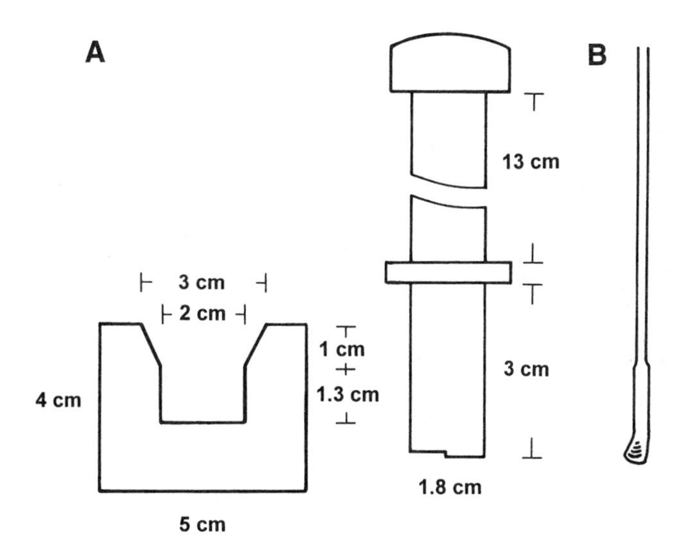

Fig. 1. Special equipment for the pulverization of frozen tissue (**A**) Glass mortar and plastic pestle. (**B**) Spatula used to transfer the frozen tissue from the mortar to the test tube. A regular spatula was formed to a small shovel.

Table 1
Buffer A

Components	Mixture	Final concentrations
Tris	0.606 g	50 m*M*
KCl	0.746 g	100 m*M*
Glycerol	20.000 g	20%
Buffer A	in 100 mL bidistilled water[a]	

[a]pH 7.1 (room temperature), adjusted with HCl.

extracts of each organ. The optimum concentrations were determined experimentally. The optimum was considered to be reached when a maximum of spots occurred in the upper as well as in the lower part of the 2-DE protein pattern. The region around pH 6.0 should not tend to become depleted, starting from the top.

The method described in the following was developed with mouse tissues, but was found to be applicable in the same manner for the corresponding human tissues.

2. Materials

2.1. Equipment

1. Sonicator for performing sonication in a water bath: A small apparatus is preferred (Transsonic 310 from Faust, D-78224 Singen, Germany).
2. Glass beads added to the tissue sample for sonication: The size (diameter) of the glass beads should be 2.0–2.5 mm. The factor 0.034 is calculated for this size of beads (*see* **Note 6**).
3. Mortar and pestle: Form and size of this equipment are shown in **Fig. 1**. Mortar and pestle are manufactured from achat or from glass (Spec. LAB, Breiter Weg 33, 12487 Berlin, Germany). Glass was found to be more stable in liquid nitrogen.

4. A small spatula was formed into a shovel by wrought-iron work (**Fig. 1**) and used to transfer tissue powder from the mortar to tubes.

2.2. Reagents

1. Buffer A: The composition is given in **Table 1**. The final solution is filtered, aliquoted into 150 μL portions, and stored at –70°C.
2. Buffer B: The composition is given in **Table 2**. The final solution was filtered, aliquoted precisely into 900-μL portions, and stored at –70°C. When used, 100 μL of an aqueous CHAPS solution are added. The CHAPS concentration in this aqueous solution is calculated in such a way (*see* **Table 6**) that the pellet II/buffer homogenate (*see* **Table 6**) contains 4.5% CHAPS. This concentration is found to be the best when 2-DE patterns are compared, which resulted from protein samples containing different CHAPS concentrations.
3. Buffer C: The composition is given in **Table 3**. The final solution is filtered, aliquoted into 1-mL portions, and stored at –70°C.
4. Protease inhibitor 1A: one tablet of Complete™ (Boehringer Mannheim, D-68305, Mannheim, Germany) is dissolved in 2 mL buffer A (according to the manufacturer's instructions), and the resulting solution aliquoted into 50-, 80-, and 100-μL portions. Inhibitor 1B was prepared in the same way, but with buffer B (900 μL buffer + 100 μL H_2O) and aliquoted into 30- and 50-μL portions. Inhibitor 2 is prepared as indicated in **Table 4** and aliquoted into 100-μL portions. The inhibitor solutions are stored at –70°C.
5. DTT-solution: 2.16 g DTT are dissolved in 10 mL bidistilled water. The solution is aliquoted into 100-μL portions and stored at –70°C.
6. Sample diluent: The composition is given in **Table 5**. The solution is aliquoted into 250-μL portions and stored at –70°C.

3. Methods

3.1. Extraction of Total Liver Proteins

3.1.1. Dissection of Mouse Liver

1. Kill the mouse by decapitation. Thereby, the body is allowed to bleed. The following steps are performed in the cold room.
2. Cut open the abdomen, cut through the vena femoralis on both sides, and perfuse the liver with 5 mL saline (0.9% NaCl solution).
3. Dissect the complete liver from the body, remove the gallbladder without injury, and cut the liver into its different lobes. The central part of each lobe, i.e., the region where the blood vessels enter the liver lobe, is cut off as well as remainders of other tissues (diaphragm, fascia).
4. Cut the liver lobes into two to four pieces, rinse in ice-cold saline and leave there. Immediately after this step, the next organ is prepared from the same animal, if desirable, and brought to the same stage of preparation as the liver.
5. Cut the liver pieces into smaller pieces (about 5 × 5 mm), and place each piece on filter paper, immerse into liquid nitrogen, and put into a screw-cap tube in which all pieces are collected. For the time of preparation, the tubes are kept in a box containing liquid nitrogen; then they are stored at –70°C.

3.1.2. Extraction of the Liver Proteins Soluble in Buffer (Supernatant I + II)

1. Fill the frozen liver pieces into a preweighed small plastic tube and weigh quickly without thawing. The weight of the liver pieces should be between 240 and 260 mg (*see* **Note 5**).
2. Place a mortar, a pestle, and a small metal spoon into a styrofoam box that contains liquid nitrogen up to a height not exceeding that of the mortar.

Table 2
Buffer B

Components	Mixture	Final concentrations[a]
KCl	1.491 g	0.2 *M*
Glycerol	20.000 g	20%
Phosphate buffer[b]	50.000 mL	0.1 *M*
Buffer B	in 90 mL bidistilled water	

[a]Concentrations in 100 mL of buffer B/CHAPS; *see* **Table 6**.
[b]Phosphate buffer: 33 mL 0.2 *M* NaH_2PO_4 solution + 67 mL 0.2 *M* Na_2HPO_4 solution; resulting pH: 7.1.

Table 3
Buffer C

Components	Mixture	Final concentrations
Tris	0.606 g	50 m*M*
$MgSO_4 \cdot 7H_2O$	0.049 g	[a]
Buffer C	in 100 mL bidistilled water[b]	

[a]If 1 part (mg) of pellet III is homogenized with 1 part (µL) of buffer C (*see* **Table 6**), the resulting concentration of $MgSO_4$ is 1 m*M*.
[b]pH 8.0 (room temperature), adjusted with HCl.

Table 4
Protease Inhibitor 2

Components	Stock solutions	Mixture
Pepstatin A	9.603 mg[a] in 100 mL ethanol	10 mL[c]
PMSF	1.742 g[b] in 100 mL ethanol	10 mL
Protease inhibitor 2		20 mL

[a]Concentration in the homogenate (*see* **Table 6**): 1.4 µ*M*.
[b]Concentration in the homogenate (*see* **Table 6**): 1.0 m*M*.
[c]The solution was incubated in a 37°C water bath to dissolve pepstatin A. The solution should be well-capped.

3. Put the frozen liver pieces into the mortar, and add buffer A, and inhibitor 1A and 2. The required volumes of each of the solutions are calculated as indicated in **Table 6**. The precise amount of each solution is pipeted as a drop onto the spoon that was kept in the N_2-box. The solution immediately forms an ice bead that can easily be transferred into the mortar.
4. Grind all the frozen components in the mortar to powder. Care should be taken that small pieces of the material do not jump out of the mortar when starting to break up the hard frozen material.
5. Transfer the powder into a 2-mL Eppendorf tube using a special spatula (**Fig. 1**). Forceps are used to freeze the tube briefly in N_2 and then to hold the tube near to the mortar in the N_2-box. Care is taken not to leave any powder in the mortar or at the pestle. For collection of this powder, always use the same type of plastic tube. This contributes to the reproducibility of the following sonication step. Compress the powder collected in the tube by knocking the tube against the mortar. The powder can be stored at –70°C or immediately subjected to sonication.

Table 5
Sample Diluent

Components	Mixture	Final concentrations
Urea	1.08 g (= 0.80 mL)	9.000 *M*
DTT solution[a]	0.10 mL	0.070 *M*
Servalyt, pH 2.0–4.0[b]	0.10 mL	2.000%
Bidistilled water	1.00 mL	50.000%
Sample diluent	2.00 mL	

[a] *See* **Subheading 2.2, item 5**.
[b] Serva (D-69115, Heidelberg, Germany).

6. For sonication, a calculated number of glass beads (*see* **Table 6** and **Note 6**) is given to the sample powder, and the powder is then thawed and kept in ice. Sonication is performed in a waterbath. The fill height of the water is critical for the sonication effect and should always be at the level indicated by the instruction manual of the apparatus. Furthermore, when dipping the sample tube into the water, it is important to do this at a "sonication center" visible on the concentric water surface motion and noticeable when holding the tube with the fingers into this center. We prefer a small sonication apparatus that forms only one sonication center (*see* **Subheading 2.1.**). The water must be kept ice-cold. Sonication is performed for 10 s. Immediately thereafter, the sample is stirred with a thin wire for 50 s with the tube still being in the ice water. The tube is then kept in ice for 1 min. Then the next sonication round is started, until a total of six 2-min rounds has been reached (*see* **Note 6**). After sonication, the glass beads are caught with fine forceps, cleaned as thoroughly as possible at the inner wall of the vial, and removed. The homogenate sticking on the wall is collected onto the bottom of the tube by a few seconds of spinning. The homogenate is then frozen in liquid nitrogen.
7. Detach the frozen homogenate in the tube from the wall by quickly knocking the top of the tube on the table. Transfer the frozen piece of homogenate into a centrifuge tube (before this, determine the weight of the tube) and thaw. Centrifuge the homogenate at 50,000 rpm (226,000*g* max.) for 30 min at 4°C.
8. Completely withdraw the supernatant (I) with a Pasteur pipet and fill into a small test tube the dead weight of which has been determined before. The centrifuge tube is kept on ice, and the pipet is put into the tube with the tip at the center of the bottom (the pellet sticks to the wall if a fixed-angle rotor was used). In this position, remainders of the supernatant of the bottom of the tube and inside the pipet accumulate in the pipet, and are added to the test tube. Then the supernatant is frozen in liquid nitrogen and stored at –70°C.
9. Weigh the pellet (I) left in the centrifuge tube on ice, and add buffer A at amounts calculated as indicated in **Table 6**. Mix the pellet and buffer by vortexing, collect the homogenate on the bottom by a short spin, freeze in liquid nitrogen, and store at –70°C or treat further immediately.
10. Grind the homogenate together with inhibitors 1A and 2 to powder as described above for the liver pieces. Transfer the homogenate frozen from the centrifuge tube to the mortar after detaching the frozen homogenate from the wall by knocking onto the bottom of the tube. Transfer the powder back into the used centrifuge tube, taking care that no powder remains in the mortar.
11. Thaw the powder, and slowly stir the homogenate for 45 min in the cold room.
12. Centrifuge the homogenate as described in **step 7**.

13. Completely withdraw the supernatant (II) from the pellet. This is done in such a way that a white layer, which partially covers the surface of the supernatant, sinks unaffected onto the pellet. Collect the remainders of the supernatant as mentioned in **step 8**. Add supernatant II to supernatant I, and thoroughly mix the two solutions. Measure the weight of the total supernatant.
14. Take a 50-μL aliquot from the total supernatant, and mix with urea, DTT solution, and ampholyte, pH 2.0–4.0, as indicated in **Table 6**. The final concentrations of these components are: 9 *M* urea, 70 m*M* DTT, and 2% ampholytes. These three components should be added to the supernatant in the order given here, and each component should be mixed with the supernatant before adding the next one. The final volume of this supernatant mixture is 100 μL. To this, add 100 μL sample diluent (*see* **Table 5**) and mix.
15. The resulting solution is the final sample (supernatant I + II, SI + II). Divide the sample into several portions, freeze each portion in liquid nitrogen, and store at –70°C. As a standard, 8 μL of the sample are applied to the IEF gel, if the large gel 2-DE technique *(3)* is used (*see* **Note 7**). Freeze the remaining portion of the pure supernatant and store.
16. Determine the weight of the pellet (II). Collect the pellet onto the bottom of the tube by a short spin, then freeze in liquid nitrogen, and store at –70°C.

3.1.3. Extraction of the Pellet Proteins Soluble in the Presence of Urea and CHAPS (Pellet Extract)

1. Grind pellet II, buffer B/CHAPS and inhibitor 1B to powder in a mortar placed in liquid nitrogen (*see* **Subheading 3.1.2., steps 2–4**). The calculation of the buffer and inhibitor volumes is given in **Table 6**. Place the powder back to the centrifuge tube, trying to leave no remainders in the mortar or on the pestle.
2. Thaw the powder mix, and stir slowly for 60 min in the cold room (CHAPS reaction).
3. Add urea (for the amount, *see* **Table 6**) to the homogenate, and stir the mixture for 45 min at room temperature (urea reaction). Some minutes after adding urea, a great part of the urea is dissolved. At this stage, add DTT solution (for the amount, *see* **Table 6**).
4. Remove the magnet rod from the homogenate. At this step, also avoid any loss of homogenate. Centrifuge the homogenate at 17°C for 30 min at 50,000 rpm (226,000*g* max.).
5. Completely withdraw the supernatant (III) with a Pasteur pipet, and fill into a small test tube the dead weight of which was determined. Collect the remainders of the supernatant as mentioned in **Subheading 3.1.2., step 8**. Measure the weight of the supernatant.
6. Add ampholytes, pH 2.0–4.0 (for the amount, *see* **Table 6**) to the supernatant, and immediately mix with this solution.
7. The resulting solution is the final sample (PE). Divide the sample into several portions, freeze each portion in liquid nitrogen, and store at –70°C. The standard volume of the sample applied per IEF gel is 8 μL, if the large gel 2-DE technique *(3)* is used (*see* **Note 7**).
8. Measure the weight of the pellet (III). Collect the pellet on the bottom of the tube by a short spin, then freeze in liquid nitrogen, and store at –70°C.

3.1.4. Suspension of the Remaining Pellet (Pellet Suspension)

1. Grind pellet III and buffer C to powder in a mortar as described in **Subheading 3.1.2., steps 2–4**. The buffer volume is calculated as indicated in **Table 6**. The powder is transferred into a test tube, thereby avoiding any loss of material.
2. Thaw the powder, and add benzonase (for the amount, *see* **Table 6**) in the form offered by the manufacturers (Merck, D-64271, Darmstadt, Germany). Slowly stir the homogenate for 30 min in the cold room (DNA digestion).

Table 6
Tissue Protein Extraction Protocol

SI + II fraction: liver			Brain	Heart
Liver pieces		250 mg[b] (A)[c]	240–260 mg fine pieces	100–130 mg (total heart)
Buffer A (liver mg × 1.5)[a]		375 µL	No buffer	←, Factor 1.0
	Σ_1	625		
Inhibitor 1A ($\Sigma_1 \times 0.08$)		50 µL	←	←
Inhibitor 2 ($\Sigma_1 \times 0.02$)		12.5 µL	←	←
	Σ_2	688		
Liver powder			←	←
Sonication, 6 × 10 s			No sonication	←, 12 × 10 s
Number of glass beads ($\Sigma 2 \times 0.034$)		23		
Centrifugation			←	←
Supernatant I store frozen			←	←
Pellet I weigh		110 mg[d]	←	←
Buffer A (pellet I mg × 2)		220 µL	← Factor 0.5	←, Factor 1.0
	Σ_3	330		
Inhibitor 1A ($\Sigma_3 \times 0.08$)		26.4 µL	←	←
Inhibitor 2 ($\Sigma_3 \times 0.02$)		6.6 µL	←	←
Pellet powder			←	←
			Sonication, 6 × 10 s	
			Number of glass beads ($\Sigma_3 + [\Sigma_3 \times 0.08] + [\Sigma_3 \times 0.02]) \times 0.034$	
Stirring			No stirring	Stirring
Centrifugation			←	←
Supernatant II add to I			←	←
Supernatant I + II weigh		628 mg[e] (B)[c]	←	←

SI + II fraction: liver			Brain	Heart
Aliquot of supernatant I + II (store rest of supernatant I + II frozen)		50 µL	←	←
Urea (50 µL × 1.08)		54 mg	←	←
DTT solution (50 µL × 0.1)		5 µL	←	←
Ampholyte pH 2–4 (50 µL × 0.1)		5 µL	←	←
Final volume of 50 µL supernatant plus additives		400 µL	←	←
Diluent		100 µL	No diluent	Diluent
Supernatant I + II, ready for use, store frozen in aliquots		200 µL	←	←
2-DE		8 µL/gel	9 µL/gel	6 µL/gel

PE fraction: liver			Brain	Heart
Pellet II weigh		92 mg[d] (C)[c]	←	←
Buffer B/CHAPS (pellet II mg × 1.6)		147 µL	←, Factor 1.4	←, Factor 2.2
900 µL buffer B			900 µL buffer B	900 µL buffer B
73 mg CHAPS (displace 69 µL)			77 mg CHAPS	65 mg CHAPS
31 µL bidistilled water			27 µL bidistilled water	38 µL bidistilled water
1000 µL buffer B/CHAPS			1000 µL buffer B/CHAPS	1000 µL buffer B/CHAPS
	Σ_4	239		
Inhibitor 1B ($\Sigma_4 \times 0.08$)		19 µL	←	←
	Σ_5	19 + 147[f] = 166		
Pellet powder			←	←
Stirring			←	←
Urea ([pellet II mg × 0.3] + Σ_5) × 1.08		207 mg	←, Pellet factor 0.56	←, Pellet factor 0.25
DTT solution ([Pellet II mg × 0.3) + Σ_5]) × 0.1		19 µL	←, Pellet factor 0.56	←, Pellet factor 0.25
Stirring			←	←

(continued)

Table 6 *(continued)*

Centrifugation			←	←
Supernatant III weigh		375 mg[e] (D)[c]	←	←
Ampholyte pH 2–4 (supernatant III mg × 0.0526)		20 µL	←	←
Pellet extract, ready for use, store frozen in aliquots			←	←
2-DE		8 µL/gel	8 µL/gel	7 µL/gel
PS fraction: liver			Brain	Heart
Pellet III		69 mg[d]	←	←
Buffer C (pellet III mg × 1.0)		69 µL	←	←
	Σ_6	138		
Pellet powder			←	←
Benzonase (Σ_6 × 0.025)		3.5 µL	←	←
Stirring			←	←
	Σ_7	3.5 + 69[g] = 73		
Urea (Σ_7 × 1.08)		79 mg	←	←
DTT solution (Σ_7 × 0.1)		7.3 µL	←	←
Stirring			←	←
Ampholyte pH 2.0–4.0 [(pellet III mg × 0.3) + Σ_7] × 0.1		9.4 µL	←, Pellet factor 0.56	←, Pellet factor 0.25
Pellet suspension, ready for use, store frozen in aliquots				
2-DE		9 µL/gel	8 µL/gel	8 µL/gel

[a]All factors used in this table are explained in **Note 3**.

[b]This figure is given as an example. The amount of the starting material may vary from 240 to 260 mg (*see* **Note 5**).

[c]B ÷ A = control value; D ÷ C = control value (*see* **Note 4**).

[d]This figure is given as an example and is not the result of a calculation. The pellet weight varies because slight losses of material are unavoidable, even during very precise work.

[e]*See* [d] for pellets; this also holds true for supernatants.

[f]Buffer B/CHAPS volume.

[g]Buffer C.

3. Add urea (for the amount, *see* **Table 6**), and stir the homogenate at room temperature for another 30 min. During this time, add DTT solution (for the amount, *see* **Table 6**) once the major part of urea is dissolved. At the end of this period, add ampholytes, pH 2.0–4.0 (for the amount, *see* **Table 6**), and quickly mix with the homogenate.
4. The resulting solution is the final sample (PS). The sample is frozen in liquid nitrogen and stored at –70°C. The standard volume of the sample applied per gel (large gel 2-DE; *[3]*) is 9 µL (*see* **Note 7**). The sample contains some fine, unsolved material and is therefore transferred to the gel with a thin Pasteur pipet instead of a microliter syringe.

3.2. Extraction of Total Brain Proteins

3.2.1. Dissection of Mouse Brain

1. Kill the mouse by decapitation. The following steps are performed in the cold room. If several organs have to be taken from the same animal, start with the brain.
2. Cut off the skin of the head, and open the cranium starting from the spinal canal proceeding in frontal direction. Break the cranial bones apart so that the brain is exposed. Take out the brain, including the two bulbi olfactorii and a short piece of the spinal cord. Place the brain into a Petri dish containing ice-cold saline. Remove any blood vessels and blood at the outside of the brain.
3. Cut the brain into four pieces, place the pieces on filter paper, and then individually immerse into liquid nitrogen and collect in a screw-cap tube. Keep the tubes in liquid nitrogen and finally store at –70°C.

3.2.2. Extraction of the Brain Proteins

The four pieces of a brain are put frozen into a mortar that was placed into a box containing liquid nitrogen and crushed with the pestle to fine pieces. The crushed material is transferred completely back to two test tubes in such a way that one of these tubes contains 240–260 mg of the frozen tissue (weigh the tube without thawing the tissue).

The weighed material is used to prepare the supernatant I + II, the pellet extract, and the pellet suspension. The procedure follows that of liver extraction with some exceptions, which are indicated in **Table 6**. One exception is that the tissue powder is produced without buffer and subjected to centrifugation without sonication. (A rather small amount of supernatant results.) Sonication is performed with the pellet I homogenate.

3.3. Extraction of Total Heart Proteins

3.3.1. Dissection of Mouse Heart

1. Kill the mouse by decapitation. The following steps are performed in the cold room.
2. Open the thorax, and remove the heart. Place the heart into a Petri dish containing ice-cold saline. Cut off the two atria, and open the ventriculi to remove any blood and blood clots.
3. Dry the heart on filter paper, freeze in liquid nitrogen, and store in a screw-cap tube at –70°C.

3.3.2. Extraction of the Heart Proteins

The supernatant I + II, the pellet extract, and the pellet suspension are prepared from a single heart. The procedure is as described for liver with some modifications. The modifications are indicated in **Table 6**.

3.4. Extraction Without Fractionation

In cases in which only very small amounts of tissue are available (e.g., 2–5 mg heart biopsy samples, 10–12 mg of two mouse eye lenses, early mouse embryos), a total protein extract is prepared instead of SI + II, PE, and PS fractions. The tissue is pulverized with buffer B/CHAPS containing 0.044 g $MgSO_4 \cdot 7\ H_2O$ in 90.00 mL (the buffer factor is less critical here than in fractionated extraction and should not be too low) and with inhibitors 1B and 2, sonicated, and stirred for 30 min at 4°C. After Λ5 min stirring Benzonase is added for DNA digestion: sample volume after sonication = ag; ag × 0.025 = bμL Benzonase (original solution). After the final Λ5 stirring the weight of the homogenate it is determined again, and then, accordingly, mixed with urea, DTT solution, and ampholyte solution (*see* **Note 3**), stirred for 45 min at room temperature, and used without centrifugation. Small plastic tubes and a glass rod with a rough surface at the well-fitting tip may serve as mortar and pestle. Total protein extraction was also preferred when a 2-DE pattern of low complexity was expected, for example, protein patterns from cultured human fibroblasts buffer fctor 1.25 was used in this case.

4. Notes

1. Maximum resolution of tissue protein fractions by 2-DE: **Fig. 2** shows the 2-DE patterns of the three protein fractions SI + II, PE, and PS of the mouse liver. The SI + II fraction reveals the highest number of protein spots. When the spots were counted visually, i.e., by placing the 2-DE gel on a light box and dotting each spot with a pencil *(3)*, about 9200 proteins were detected in this fraction. The SI + II pattern of the brain revealed about 7700 proteins, that of the heart being about 4800 protein spots. The high spot numbers reflect the high resolution of the large gel 2-DE *(1)*, which reveals many weak spots between the major spots. All these spots were counted precisely. The analysis of the large gel patterns by laser densitometry results in 24.5 million data. Treatment of the data by special computer programs for spot detection and correction of the computer pattern against the original gel pattern led us to the maximum number of spots we could obtain. For example, the SI + II pattern of brain evaluated in this way revealed about 8500 protein spots *(4)* compared to 7700 spots detected by visual inspection of this pattern.
2. Effect of fractionated extraction of tissue proteins: The purpose of fractionating the proteins of a certain tissue was to increase the number of proteins detectable in this tissue by 2-DE. This purpose, however, would only be fulfilled if each fraction contained a notable number of proteins that are strongly fraction-specific, so that the total tissue proteins can be distributed over several gels. Comparison of the 2-DE patterns from the SI + II and PE fractions of the liver (**Fig. 2**) showed that the PE pattern revealed about 2000 protein spots not detectable among the 9200 spots of the SI + II pattern. The PS pattern revealed only about 70 additional spots. The PS protein spots represent the class of the most basic proteins (**Fig. 2**) and belong mainly to the chromosomal proteins (e.g., histones). Therefore, the PS fraction is only of interest when the class of very basic proteins is subject of the investigation.

   Considering the 2-DE patterns of the SI + II and PE fraction in more detailed (**Fig. 3**), quite a number of very prominent spots can be observed that occur in one pattern, but do not occur, even in trace amounts, in the other pattern. At the same time, other spots revealing only low intensities are present in both patterns. This suggests that contamination of one fraction by the other one scarcely affects the fraction specificity of the SI + II and PE pattern, but, apparently, many proteins exist naturally in both the cytoplasm and the structural components of the cell.

   Taking all three fractions into account, the total liver proteins could be resolved into about 11,270 different proteins (polypeptide spots). This, however, does not mean that the protein sample preparation procedure described here in a 2-DE pattern, revealed all the

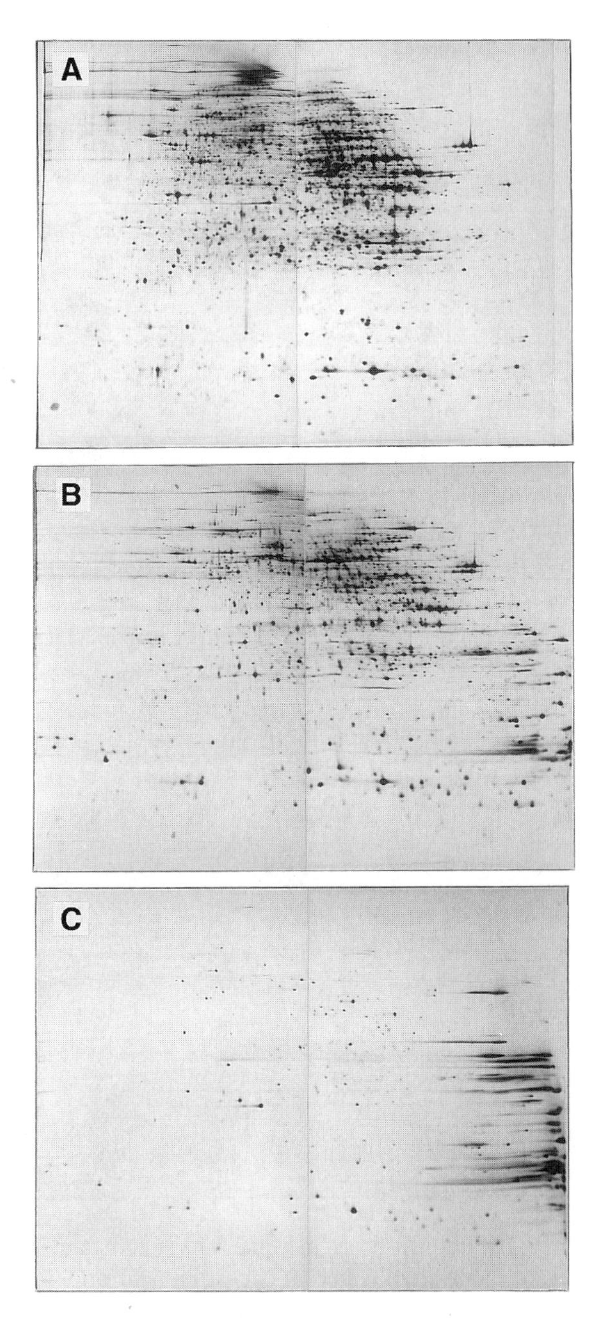

Fig. 2. 2-DE protein patterns from mouse liver. The tissue was fractionated in to supernatant I + II (**A**), pellet extract (**B**), and pellet suspension (**C**) as described in **Subheading 3.** The three fractions were subjected to the large gel 2-DE *(3)*. In pattern and vice versa. Pellet specific spots occur more in the basic half, and supernatant-specific spots more in the acid half of the pattern. The pellet suspension reveals the very basic protein of the tissue extracts. Because the IEF gels do not cover the entire basic pH range, these proteins cannot reach their isoelectric points. To prevent these proteins from accumulating at the end of the gels, the IEF run was shortened by 2 h at 1000 V. Consequently, the very basic proteins form streaks instead of focused spots. For the evaluation of the patterns in terms of number of spots, *see* **Notes 1** and **2**.

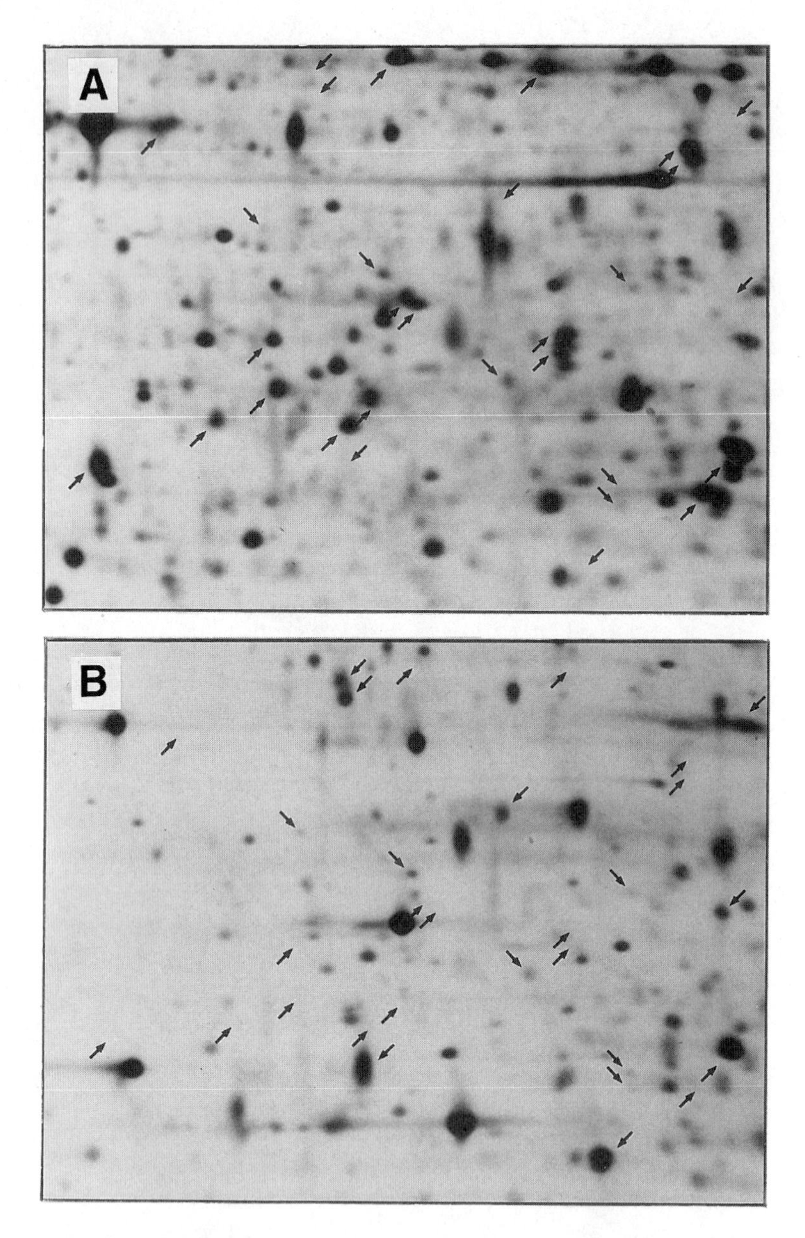

Fig. 3. Sections from 2-DE patterns shown in **Fig. 2**. The supernatant I + II (**A**) and the pellet extract (**B**) of the liver are compared. Some of the prominent protein spots present in the supernatant pattern (↗) but completely absent in the pellet extract pattern (↗), the reverse situation (↙), and some spots of low intensity present in both patterns (↘) are indicated. Other spots show high intensity in one pattern, but low intensity in the other one. These spots may reflect naturally occuring unequal distributions of proteins between the two different fractions rather than contaminations of one fraction by the other one (*see* **Note 2**).

different proteins of the liver in a 2-DE pattern. Many proteins may exist in a tissue in undetectable amounts, and special cell fraction procedures followed by protein concentration steps would be required to detect these proteins. We isolated cell nuclei from liver and brain, and extracted the nuclear pellet in a similar way to that for the tissues. The 2-DE patterns showed that the nuclear extracts add a large number of new proteins to those already known from the tissue extract patterns. However, protein spots present in both the nuclear and the tissue extracts occur as well, particularly in the acid halves of the supernatant patterns. In general, the proteins represented by the SI + II, PE, and PS patterns can be considered as the main population of protein species of a tissue to which further species can be added by analyzing purified and concentrated subfractions.

3. Explanation of factors used in **Table 6**: Factors were calculated to determine the amounts of urea, DTT solution, and ampholytes necessary to transmute any volume of a solution (theoretically water) into a mixture containing 9 *M* urea, 70 m*M* DTT, and 2% ampholytes. Calculations of the factors: 500 µL water (aqueous protein extract) + 540 mg urea (displaces 400 µL) + 50 µL DTT solution (*see* **Subheading 2.2., item 5**) + 50 µL ampholyte solution (commercial solutions that usually contain 40% ampholytes) = 1000 µL. If the volume of a protein solution to be mixed with urea, DTT, and ampholytes is n µL, the amounts of the components to be added are: $(540 \div 500) \times n = 1.08 \times n$ mg urea, $(500 \div 50) \times n = 0.1 \times n$ µL DTT solution and $0.1 \times n$ µL ampholyte solution.

   If the protein solution already contains urea and DTT (*see* **Table 6**: PE preparation), the factor 0.0526 is used to calculate the ampholyte volume for this solution. Calculation of the factor: (500 µL extract + 400 µL urea + 50 µL DTT solution) ÷ 50 µL ampholyte solution = 0.0526.

   If the protein solution of n µL includes a cell pellet, i.e., insoluble material, the n µL volume should be reduced by the volume of the insoluble material (theoretically by the volume of the dry mass of this material). For this reason, the pellet factor (e.g., 0.3) was introduced. This factor was determined experimentally using urea as an indicator. The factor reduces the volume n µL of a protein solution (containing a pellet) to the volume n' µL; n' µL × 1.08 results in an amount of urea that is added to the n µL of the solution, at the border of solubility, i.e., about 9 *M*. Note that pellet III contains urea and DTT by the foregoing steps. Therefore, in this case, the pellet volume was not taken into account when calculating the amounts for urea and DTT to be added to the final pellet suspension. In all these calculations, no distinction was made between values measured in volumes (µL) and values measured in weights (mg). This makes the calculation somewhat incorrect, but practicable and more reproducible.

   The inhibitor 2 solution was prepared as concentrated as possible to keep the volume of this solution small (1/50th of the homogenate volume, i.e., factor 0.02). This allowed us to ignore the error that resulted when the inhibitor solution was added to instead of included in the volume of the homogenate. The factor 0.08 for the inhibitor 1 solution was derived from the instructions of the manufacturers of the Complete™ tablets: 1 tablet should be dissolved in 2 mL buffer and this volume added to 25 mL of the homogenate, i.e., the volume of the inhibitor 1 solution to be added to n µL homogenate is $(2 \div 25) \times n = 0.08\ n$ µL.

   The volume of benzonase solution (ready-made solution of the manufacturer) necessary to digest the DNA in the pellet III suspension was determined experimentally. A chromatin pellet was prepared from isolated liver cell nuclei and found to change from a gelatinous clot to a fluid if treated as follows: 1 g chromatin pellet + 1 mL buffer + 0.050 mL benzonase solution (= 0.025 mL Benzonase/1 mL homogenate), stirred for 30 min at 4°C.

   The buffer factors used to calculate the volumes of the buffer added to the tissues or pellets are explained in **Subheading 1.** Note that the tissue to be homogenized must be free of any wash solutions. Otherwise, the calculated buffer volume becomes falsified.

4. Control values: Control values were calculated for each sample prepared by the fractionated extraction procedure to monitor the correctness and reproducibility of the prepara-

tion. The calculation of the control values is indicated in **Table 6**. In the following, an example is given from a real experiment. From a series of 73 individual mouse hearts, the SI + II fractions were prepared and the control values B ÷ A (*see* **Table 6**) were calculated: 60 samples showed values between 1.97 and 2.10, three samples between 1.94 and 1.96, and six samples between 2.11 and 2.13. Four samples with greater deviations (1.90, 1.91, 2.16, 2.24) were excluded from the investigation. The range 1.97 and 2.10 (mean 2.04 ± 0.04) was taken as standard control value for the preparation of mouse heart SI + II samples.

5. Amount of tissue used as starting material: The amount of tissues given in **Table 6** can be increased, but should not exceed 500 mg. However, the tissue amounts indicated give enough sample solutions to run many 2-DE gels, so that less rather than more material can be used.
6. Sonication: The conditions for sonicating mouse tissue homogenates (liver, brain, heart) were determined with the aim of breaking the membranes of all the cells and cell nuclei of the tissue. Three parameters were varied in the experiments: the length of the period of sonication, the number of repeats of sonication, and the number of glass beads added per volume of the homogenate. The effect of the various conditions was checked under the microscope by inspection of the sonicated material. During 10 s of sonication, the temperature of the homogenate increased from 0°C to 11–12°C. Therefore, sonication was not performed for more than 10 s. Glass beads are essential for breaking the cellular structures (membranes).

   Since most of the homogenates are rather thick fluids, the beads cannot flow freely. Consequently, for a given volume of homogenate, a certain number of glass beads is necessary to expose the homogenate evenly to the sonication effect. This number was standardized by using the factor 0.034 in calculating the number of glass beads for a given volume of homogenate.

   It follows from the above-mentioned three experimental parameters that the only parameter that can be varied to manipulate the effect of sonication was the number of repeats of the 10-s sonication period. Under the conditions described in **Subheading 3.**, the membranes of all cells were broken and no longer visible under the microscope. However, a certain number of intact nuclei were still detectable. We did not try to break even these nuclei by extending sonication. Sonication by using a metal tip cannot be recommended. We observed heavily disturbed 2-DE patterns as a result of employing this technique: many protein spots disappeared depending on the extent of sonication, and new spot series occurred in the upper part of the gel, apparently as a result of aggregation of protein fragments.
7. Amount of protein applied per gel: The protein samples applied to the IEF gels (*see* **Table 6**), contain about 100 µg protein. There is, however, no need to determine the protein concentration of each sample prepared in order to obtain protein patterns of reproducible intensities. The concept of the procedure described here for extracting tissues was to keep the volume of the extracts in strong correlation to the amount of the starting material (tissue or pellet) that was extracted. Therefore, by working precisely, the final sample should always contain nearly the same protein concentration. Accordingly, the reproducibility of the pattern intensity depends on the precise sample volume applied to the gel—and, of course, on the protein staining procedure.

   The sample volumes per gel given in **Table 6** are adapted for silver-staining protocols. A general guideline may be: decreasing the protein amount per gel and increasing the staining period is better than the other way around; diluted samples at a reasonable volume are better than concentrated protein samples at a small volume. But dilute, if necessary, the final sample (*see* **Table 5**), not the starting material.

Acknowledgments

The author appreciates critical comments and support from Marion Löwe and Michael Kastner.

References

1. Klose, J. and Kobalz, U. (1995) Two-dimensional electrophoresis of proteins: An updated protocol and implications for a functional analysis of the genome. *Electrophoresis* **16,** 1034–1059.
2. Scopes, R. K. (1987) *Protein Purification, Principles and Practice.* Springer-Verlag, New York, pp. 33–34.
3. Deutscher, M. P. (1990) Maintaining protein stability in *Methods of Enzymology*, vol. 182, *Guide to Protein Purification* (Deutscher, M. P., ed.), Academic, San Diego, pp. 85–86.
4. Gauss, C., Kalkum, M., Löwe, M., Lehrach, H., and Klose, J. (1999) Analysis of the mouse proteome. (I) Brain proteins: Separation by two-dimensional electrophoresis and identification by mass spectrometry and genetic variation. *Electrophoresis* **20,** 575–600.

21

Radiolabeling of Eukaryotic Cells and Subsequent Preparation for 2-Dimensional Electrophoresis

Nick Bizios

1. Introduction

Two-dimensional polyacrylamide gel electrophoresis (2-DE) provides not only the ability to resolve and quantify thousands of proteins, but it also gives those in research and industry the ability to monitor in-process protein purification, quickly and easily N. Bizios, unpublished date, (AGI Dermatics). 2-DE is also used to identify variability in protein expression in a variety of cell lines *(2,3)*. It is known that many parameters and laboratory conditions can influence the resolution of proteins on 2-DE, such as the pH range of carrier ampholytes used, the quality of reagents and equipment used, temperature, voltage, and the skill of the researcher or technician.

One of the biggest obstacles one may encounter prior to performing 2-DE is the proper documentation and validation of a Standard Operation Procedure (SOP). This falls under the category of current Good Laboratory Practices (cGLPs) and or current Good Manufacturing Practices (cGMPs). Without the proper documentation of reagents used, such as Certificates of Analysis (readily supplied by the supplier's Quality Control/Assurance department), adherence to expiration dates of commercially prepared reagents, implementation of expiration dates of reagents prepared in the laboratory, and the adherence of SOPs (*see* **Note 1**) the ability of generating repeatable and consistent results falls dramatically. However, the radiolabeling and preparation of eukaryotic cells for 2-D may be considered paramount in obtaining consistent results. Without proper radiolabeling and preparation, how would one properly find subsequent spots of interest and perform quantitative analysis?

This chapter describes a general method for labeling methionine-containing proteins, phosphorylation labeling, and subsequent lysate preparation for 2-DE that has been modified from Garrels *(3)* and Garrels and Franza *(4)*. The Jurkat T-lymphoblast cell line is used as an example. A large number of protocols have been published for the solubilization and sample preparation of eukaryotic cell lines and tissues for 2-DE. One of the best sources for additional protocols is found at the Geneva University Hospital's Electrophoresis laboratory, which can be accessed at http://expasy.hcuge.ch/ch2d/technical-info.html. (*See also* Chapters 19 and 20).

From: *The Protein Protocols Handbook, 2nd Edition*
Edited by: J. M. Walker © Humana Press Inc., Totowa, NJ

2. Materials

2.1. Equipment

1. 0.2 μm Filters.
2. Heat block or 100°C water bath.

2.2. Reagents

1. Complete culture medium: 90% RPMI-1640, 10% fetal bovine serum (FBS), streptomycin–penicillin: Mix 990 mL of RPMI-1640 with 100 mL of FBS, and add 10 mL of streptomycin–penicillin (100×). Cold-filter sterilize using a 0.2 μm filter. Store at 4°C. Maintain the Jurkat T lymphoblasts at a concentration of 10^5–10^6 cells/mL (*see* **Note 2**).
2. Methionine-free medium: 90% Methionine-free RPMI-1640, 10% dialyzed FBS (dFBS): Mix 990 mL of methionine-free RPMI-1640 with 100 mL of dFBS. Cold-filter sterilize using a 0.2μm filter. Store at 4°C.
3. Sodium phosphate-free medium: 90% sodium phosphate-free RPMI-1640, 10% FBS: Mix 90% sodium phosphate free RPMI with 10% FBS. Cold-filter-sterilize Store at 4°C.
4. ^{35}S-label (EXPRE^{35}S^{35}S[^{35}S] methionine/cysteine mix (New England Nuclear [NEN]).
5. ^{32}P orthophospa hte (NEN).
6. Phosphate-buffered saline (PBS): Mix 8 g of NaCl, 0.2 g of KCl, 1.44 g of Na_2HPO_4, and 0.24 g of KH_2PO_4 in 800 mL of dH_20, and adjust the pH to 7.4 with HCl. Add dH_2O to a final volume of 1000 mL and autoclave. Store at room temperature.
7. Dilute SDS (dSDS): 0.3% SDS, 1% β-mercaptoethanol (β-ME), 0.05 *M* Tris-HCl,pH 8.0. In a cold room, mix 3.0 g of SDS, 4.44 g of Tris-HCl, 2.65 g Tris base, and10 mL of β-ME in distilled water, and adjust the final volume to 1 L with dH_2O. Aliquot 500 mL into microcentrifuge tubes, and store at –70°C.
8. DNase/RNase solution: 1 mg/mL of DNase I, 0.5 mg/mL of RNase A, 0.5 *M* Tris-HCl, 0.05 *M* $MgCl_2$, pH 7.0. Thaw RNase, Tris, and $MgCl_2$ stocks, and thoroughly mix 2.5 mg of RNase A (Worthington Enzymes), 1585 μL of 1.5 *M* Tris-HCl, 80 μL of 1.5 *M* Tris base, 250 μL of 1.0 *M* $MgCl_2$, and 2960 μL of dH_2O. Mix the liquids and 5 mg of DNase I (Worthington Enzymes). Do not filter. Keep cool while dispensing into microcentrifuge tubes. Make 50 μL aliquots, and store at –70°C.
9. 1.5 *M* Tris-HCl solution: Weigh out 11.8 g of desiccated Tris-HCl, and add 41.4 g of dH_2O. Mix well and filter through a 0.2-μm filter. Aliquot into microcentrifuge tubes, and store at –70°C.
10. 1.5 *M* Tris base solution: Weigh out 9.09 g of Tris base, and add 41.4 g of dH_2O. Mix and filter through a 0.2-μm filter. Aliquot into microcentrifuge tubes, and store at –70°C.
11. 1.0 *M* $MgCl_2$ solution: Weigh out 30.3 g of $MgCl_2$, and add 85.9 g of dH_2O. Mix and filter through a 0.2-μm filter. Aliquot into microcentrifuge tubes, and store at –70°C.
12. Sample buffer solution (SB): 9.95 *M* urea, 4.0% Nonident P-40 (NP40) (Sigma), 2% pH 6.0–8.0 ampholytes, 100 m*M* dithiothreitol (DTT). Mix 59.7 g of urea, 44.9 g of dH_2O, 4.0 g of NP40, 5.5 g of pH 6.0 –8.0 ampholytes (AP Biotech), 1.54 g of DTT (Calbiochem) in this order in a 30–37° C waterbath just long enough to dissolve the urea. Filter through a 0.2-μm filter and aliquot 1 mL into microcentrifuge tubes. Snap-freeze in liquid nitrogen, and store at –70°C.
13. Sample buffer with SDS solution (SBS): 9.95 M urea, 4.0% NP40, 0.3% SDS, 2% pH 6.0–8.0 ampholytes, 100 m*M* DTT. Mix 59.7 g of urea, 44.9 g dH_2O, 4.0 g of NP40, 0.3 g of SDS, 5.5 g of pH 6.0–8.0 ampholytes, 1.54 g of DTT in this order in a 30–37°C waterbath just long enough to dissolve the urea. Filter through a 0.2-μm filter and aliquot 1 mL into microcentrifuge tubes. Snap-freeze in liquid nitrogen, and store at –70°C.

3. Methods

3.1. ^{35}S-Labeling (see Note 3)

1. Jurkat T lymphocytes are labeled for 3–24 h in methionine-free media containing 50–250 μCi/mL of ^{35}S.
2. Follow cell lysate protocol (**Subheading 3.3.**).

3.2. ^{32}P-Labeling (see Note 3)

1. Add 100 μCi/mL of ^{32}P for up to 3 h to cells that are in phosphate-free medium (*see* **Note 4**).

3.3. Whole-Cell Lysate Preparation

1. Wash cells with PBS three times in a microcentrifuge tube.
2. Add an equal volume of hot (100°C) dSDS solution to the pellet.
3. Boil tube (100°C) for 1–3 min.
4. Cool in an ice bath (*see* **Note 5**).
5. Add 1/10 volume of DNase/RNase solution.
6. Gently vortex mix for several minutes to avoid foaming. The sample should lose its viscosity, and the solution should look clear. If not, then add more dSDS and DNase/RNase solution (*see* **Note 6**).
7. Snap-freeze in liquid nitrogen, and store at –70°C. Samples may be kept for up to 6 mo at –70°C.

3.4. Preparaing the Sample for 2D-PAGE

3.4.1. Vacuum Drying

1. Lyophilize sample (frozen at –70°C) in a Speed Vac using no or low heat until dry.
2. Add SB solution to the sample equal to that of the original dSDS sample volume, and mix thoroughly.
3. Heat sample to 37°C for a short period if necessary (*see* **Note 7**).
4. Store at –70°C. samples can be kept for up to 6 mo at –70°C.
5. Radioisotope incorporation in the sample may now be determined by trichloroacetic acid (cTCA)precipitation.
6. Recommended first-dimension load is 500,000 dpm for ^{35}S-labeled proteins and 200,000 dpm for ^{32}P-labeled proteins.
7. If necessary, the sample is diluted and mixed thoroughly with SBS solution before loading onto the first-dimension gel (*see* **Note 7**).

4. Notes

1. cGLPs and cGMPs will help with all aspects in a research facility. Industry will have Quality Assurance/Control departments to oversee the implementation of these practices. Smaller research labs may wish to assign or hire a person to oversee that these practices are adhered to.
2. Maintain Jurkat T lymphocytes in complete culture medium supplemented with streptomycin/penicillin, in a humidified incubator with 95% air and 5% CO_2 at 37°C and at a concentration of 10^5–10^6 cells/mL.
3. It is extremely important that all radioactive work be performed with the utmost care, and according to your institutional and local guidelines.
4. It may be necessary to preincubate the cells. Preincubating the cells at a density of 1–10 × 10^6 cells/mL for 30 min in sodium-free, phosphate-free medium works well.

5. Sample preparation should be done quickly on ice to avoid degradation by proteases.
6. Keep salt concentrations as low as possible. high concentrations (>150 m*M*) of NaCl, KCl, and other salts cause streaking problems, as do lower concentrations of phosphate and charged buffers. Dialyzing samples to remove salts and other low-mol-weight substances is recommended.
7. After dissolving the sample in SB solution or diluting in SBS solution, it is imperative that the sample is not subjected to temperatures above 37°C. At extreme temperature (>40°C), the urea in SB and SBS will cause carbamylation. Charged isoforms will be generated by isocyanates formed by the decomposition of urea

References

1. Yu, L. R., Zeng, R., Shao, X. X., Wang, N., Xu, Y. H., and Xia, Q. C. (2000) Identification of diffentially expressed proteins betwen human hepatoma and normal liver cell lines by two-dimensional electrophoresis and liquid chromatography-ion trp mass spectrometry. *Electrophoresis* **14,** 3058–3068
2. Carroll, K., Ray, K., Helm, B., and Carey, E. (2000) Two-dimensional electrophoresis reveals differential protein expression in high- and low-secreting variants of the rat basophilic leukemia cell line. *Electrophoresis* **12,** 2476–2486.
3. Garrles, J. I. (1983) Quantitative two-dimensional gel electrophoresis of proteins, in *Methods of Enzymology*, Vol. 100 (Grossman, L., Moldave, K., and Wu, R., eds.), Academic Press, New York, pp. 411–423.
4. Garrels, J. I. and Franza, B. R., Jr. (1989) The REF52 protein database. *J. Biol. Chem.* **264,** 5283–5298.

22

Two-Dimensional Polyacrylamide Gel Electrophoresis of Proteins Using Carrier Ampholyte pH Gradients in the First Dimension

Patricia Gravel

1. Introduction

Two-dimensional polyacrylamide gel electrophoresis (2-D PAGE) is the only method currently available for the simultaneous separation of thousands of protein. This method separates individual proteins and polypeptide chains according to their isoelectric point and molecular weight. The 2-D PAGE technology can be used for several applications, including; separation of complex protein mixtures into their individual polypeptide components and comparison of protein expression profiles of sample pairs (normal vs transformed cells, cells at different stages of growth or differentiation, etc.).

This chapter describes the protocol for 2-D PAGE using carrier ampholyte pH gradient gel for the isoelectric focusing separation (pH gradient 3.5–10) and a polyacrylamide gradient gel for the second dimension ***(1,2)***.

For many years the 2-D PAGE technology relied on the use of carrier ampholytes to establish the pH gradient. This traditional technique has proven to be difficult in the hands of many because of the lack of reproducibility created by uncontrollable variations in the batches of ampholytes used to generate the pH gradients.

This problem was solved by using immobilized pH gradients (IPG) (*see* Chapter 23), in which the compounds used to set up the pH gradient are chemically immobilized and so the gradient is stable. Another advantage is that a larger amounts of protein could be used in the separation for micropreparative runs. The recent availability of commercial precast IPG gels in a variety of narrow and broad pH ranges, as well as precast sodium dodecyl sulfate (SDS)-PAGE gels has led also to major advances in protein separation, display, and protein characterisation ***(3)***. 2-D PAGE reference maps are now available over the World-Wide Web (***4***, and *see* Chapter 26).

2. Materials

2.1. Preparation of Samples

1. Lysis solution A: 10% (w/v) sodium dodecyl sulfate (SDS) and 2.32% (w/v) 1,4-dithioerythritol (DTE) in distilled water (dH_2O).

From: *The Protein Protocols Handbook, 2nd Edition*
Edited by: J. M. Walker © Humana Press Inc., Totowa, NJ

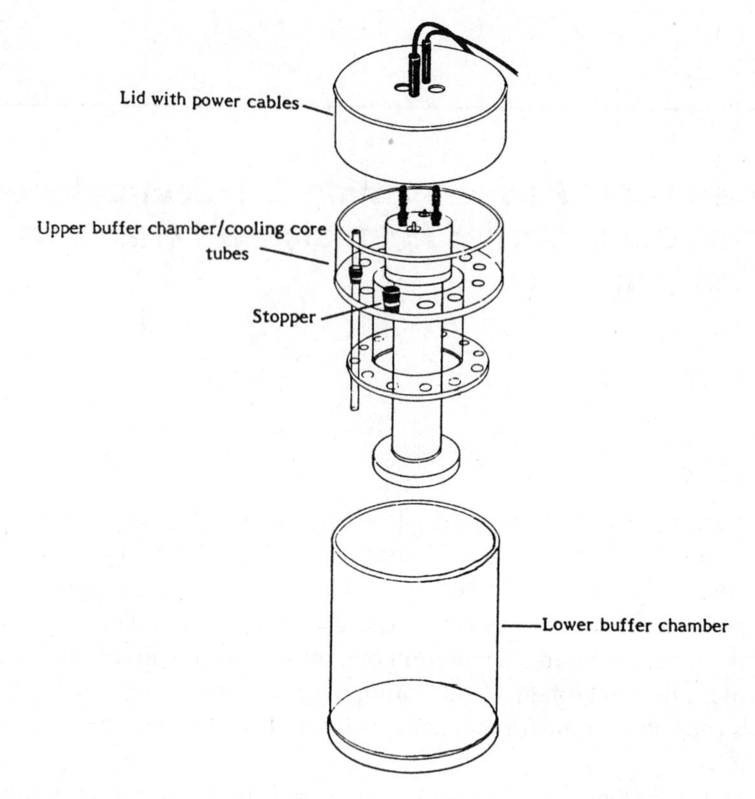

Fig. 1. Tube Cell Model 175 (Bio-Rad) for IEF with carrier ampholyte pH gradient capillary gels.

2. Lysis solution B: 5.4 g of 9.0 *M* urea (must be solubilized in water at warm temperature, around 35°C), 0.1 g of DTE (65 m*M*), 0.4 g of 65 m*M* 3-[3-cholamidopropyldimethylammonio]-1-propanesulfonate (CHAPS), 0.5 mL of Ampholines, pH range 3.5–10 (5% v/v), made to 10 mL with dH_2O.

These lysis solutions can be aliquoted and stored at –20°C for many months.

2.2. Isoelectric Focusing (IEF)

1. IEF is performed with the Tube Cell Model 175 (Bio-Rad) (*see* **Fig. 1**) and with glass capillary tubes (1.0–1.4 mm internal diameter and 210 mm long). Ampholytes pH 4–8 and Ampholytes pH 3.5–10 are from BDH (Poole, England).
2. Stock solution of acrylamide: 30% (w/v) acrylamide, 0.8% (w/v) piperazine diacrylamide (PDA) in dH_2O. This solution should be stored in the dark at 4°C for 1–2 mo.
3. Cathodic buffer: 20 m*M* NaOH in dH_2O. This solution should be made fresh.
4. Anodic buffer: 6 mM H_3PO_4 in dH_2O. This solution should be made fresh.
5. Ammonium persulfate (APS) stock solution: 10% (w/v) APS in dH_2O. This solution should be stored at 4°C, protected from light and made fresh every 2–3 wk.
6. SDS stock solution: 10% (w/v) SDS in dH_2O. This solution can be stored at room temperature.
7. Bromophenol blue stock solution: 0.05% (w/v) bromophenol blue in dH_2O. This solution can be stored at room temperature.
8. Capillary gel equilibration buffer:
   20 mL of 0.5 *M* Tris-HCl, pH 6.8

40 mL 10% (w/v) SDS stock solution
8 mL 0.05% (w/v) bromophenol blue stock solution
72 mL dH_2O

This solution can be stored at room temperature for 2 to 3 mo.

2.3. SDS-PAGE

1. The Protean II chamber (Bio-Rad) is employed by us for SDS-PAGE. The gels (160 × 200 ×1.5 mm) are cast in the Protean II casting chamber (Bio-Rad). The gradient former is model 395 (Bio-Rad).
2. Running buffer: 50 m*M* Tris base, 384 m*M* glycine, 0.1% (w/v) SDS in dH_2O. For 1 L: 6 g of Tris base, 28.8 g of glycine and 1 g of SDS. Do not adjust pH. Fresh solution is made up for the upper tank. The lower tank running buffer can be retained for more than 6 mo with the addition of 0.02% (w/v) sodium azide.
3. Sodium thiosulfate stock solution: 5% (w/v) sodium thiosulfate anhydrous in dH_2O. This solution should be stored at 4°C. (*See* **Note 1**).
4. Silver staining: Proteins in the 2-D gel are stained with silver. We used the ammoniacal silver nitrate method described by Oakley et al. *(5)* and modified by Hochstrasser et al. *(6)* and Rabilloud *(7)*. (*See also* Chapter 33.)

3. Method

3.1. Preparation of Protein Samples

Pellets of cells or tissue should be resuspended in 100 µL of 10 m*M* Tris-HCl, pH 7.4, and sonicated on ice for 30 s. Add one volume of lysis solution B and mix. The mixed solution should not be warmed up above room temperature. Samples can be loaded directly onto the gels or stored at –80°C until needed. (*See* **Note 2** for plasma sample preparation and **Note 3** for optimal sample loading.)

3.2. Isoelectric Focusing (IEF)

1. Draw a line at 16 cm on clean and dry glass capillary tubes (capillary should be cleaned with sulfochromic acid to eliminate all deposits). The remaining 0.5 cm of the capillary tubes is used to load the sample. Place each tube in a small glass test tube (tube of 5 mL) which will be filled with the isoelectric focusing gel solution. Connect the tops of each glass capillary tube to flexible plastic tubes joined together with a 1-mL plastic syringe.
2. At least 12 capillary gels can be cast with 11.5 mL of isoelectric focusing gel solution (800 µL of solution is needed per capillary).
   a. Prepare 1 mL of CHAPS 30% (w/v) and Nonidet P-40 (NP-40) 10% (w/v) (0.3 g of CHAPS and 0.1 g of NP-40) and degas for 5 min.
   b. Separately, prepare a second solution: 10 g of urea (7 mL of water is added to dissolve the urea at warm temperature, around 35°C), 2.5 mL of acrylamide stock solution, 0.6 mL of ampholytes, pH 4–8; 0.4 mL of ampholytes, pH 3.5–10; and 20 µL of *N,N,N'N'*-tetramethylethylenediamine (TEMED).
   c. Mix the first solution (CHAPS, NP-40) with the second one. Degas the mixture and add 40 µL of APS 10% (w/v) stock solution.
   Pipette 1 mL of this isoelectric focusing gel solution into the glass test tubes (along the side walls in order to prevent the formation of air bubbles in the solution). Fill up the capillary tubes by slowly pulling the syringe (up to the height of 16 cm).
   d. After 2 h of polymerization at room temperature, pull the capillary tubes out of the glass test tubes. Clean and gently rub the bottom of each capillary gel with parafilm.

3. Fill the lower chamber with the anodic solution. Wet the external faces of the capillary tubes with water and insert them in the isoelectric focusing chamber. Load the samples on the top of the capillary (cathodic side). Generally, 30–40 µL of the final diluted sample is loaded using a 25-µL Hamilton syringe (*see* **Notes 3** and **4**).
4. Lay down the cathodic buffer solution at the top of the sample in the capillary tube and then fill the upper chamber. Connect the upper chamber to the cathode, and the lower chamber to the anode.
5. Electrical conditions for IEF are 200 V for 2 h, 500 V for 5 h, and 1000 V overnight (16 h) at room temperature (*see* **Note 5**).
6. After IEF, remove the capillary tubes from the tank and force the gels out from the glass tube with a 1-mL syringe that is connected to a pipet tip and filled with water (*see* **Note 6**). Put the extruded capillary gels on the higher glass plate of the polyacrylamide gel (*see* **Subheading 3.3.**). Residual water around the capillary gel should be soaked up with a filter paper. Put 140 µL of IEF equilibration buffer down on the capillary gel. Immediatly push the capillary gels between the glass plates of the polyacrylamide gel using a small spatula. Place the cathodic side (basic end) of the capillary gel at the right side of the polyacrylamide gel. Care should be taken to avoid the entrapment of air bubbles between the capillary and the polyacrylamide gels. It is not necessary to seal the capillary gels with agarose solution but the contact between the capillary gel and the 9–16% polyacrylamide gradient gel should be very tight.

3.3. SDS-PAGE

1. To separate the majority of proteins, a 9–16% (w/v) polyacrylamide gradient gel is used for the second dimension. The gel is made as follows with 60 mL solution for a 9–16% (w/v) gel: 30 mL of 9% (w/v) acrylamide solution and 30 mL of 16% (w/v) acrylamide solution. Sixty milliliters are necessary to cast one gel of 1.5 mm × 200 mm × 160 mm (*see* **Note 7**).

   Light solution:

   9 mL of acrylamide stock solution (30% v/v)
   7.5 mL of 1.5 *M* Tris-HCl, pH 8.8 (25% v/v)
   150 µL of 5% sodium thiosulfate solution (0.5% v/v) (*see* **Note 1**)
   15 µL of TEMED (0.05% v/v)
   13.2 mL of water. This solution is degassed and then 150 µL of 10% APS solution (0.5% v/v) is added.

   Heavy solution:

   16 mL of acrylamide stock solution (53% v/v)
   7.5 mL of 1.5 *M* Tris-HCl pH 8.8 (25% v/v)
   150 µL of 5% sodium thiosulfate solution (0.5% v/v)
   15 µL of TEMED (0.05% v/v)
   6.2 mL of water. This solution is degassed and then 150 µL of 10% APS solution (0.5% v/v) is added.

   The gradient gel is formed using the Model 395 gradient Former (Bio-Rad) and a peristaltic pump. Immediately after the casting, the gels are gently overlayered with a water-saturated 2-butanol solution using a 1-mL syringe. This procedure avoids the contact of the gel with air and allows one to obtain a regular surface of the gel. Caution should be taken to avoid mixing the 2-butanol with the gel solution. The gels are stored overnight at room temperature to ensure complete polymerization.
2. Prior to the second dimension separation, wash extensively the top of the gels with deionized water to remove any remaining 2-butanol. Remove the excess water by suction with a syringe.

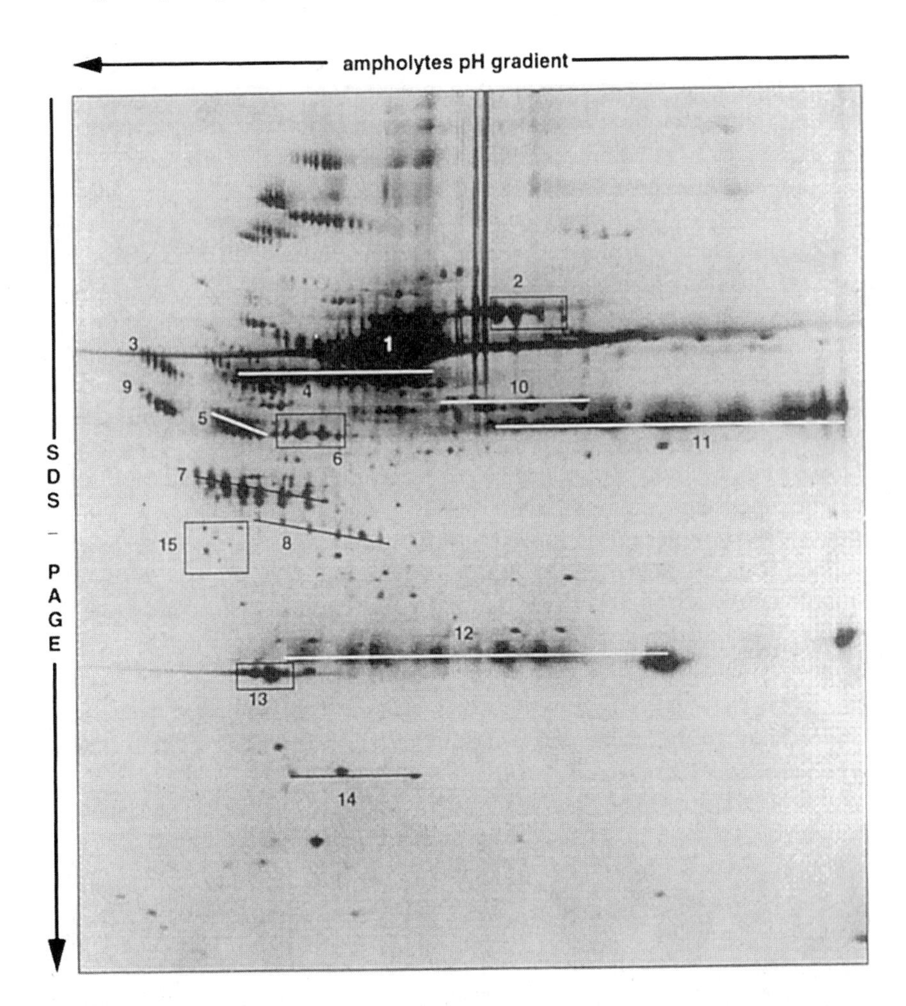

Fig. 2. Silver-stained plasma proteins separated by two-dimensional PAGE using carrier ampholyte pH gradient in the first dimension. *1*, albumin; *2*, transferrin; *3*, α_1-antichymotrypsin; *4*, IgA α-chain; *5*, α_1-antitrypsin; *6*, fibrinogen γ-chain; *7*, haptoglobin β-chain; *8*, haptoglobin cleaved β-chain; *9*, α_2-HS-glycoprotein; *10*, fibrinogen β-chain; *11*, IgG γ-chain; *12*, Ig light chain; *13*, Apolipoprotein A-1; *14*, haptoglobin α_2-chain; *15*,apolipoprotein J.

3. After the transfer of the capillary gel on top of the gradient polyacrylamide gel (*see* **Subheading 3.2., step 6**), fill the upper and the lower reservoirs with the running buffer and apply a constant current of 40 mA per gel. The separation usually requires 5 h. The temperature in the lower tank buffer is maintained at 10°C during the run.
4. At the end of the run, turn off the power, rinse the gel briefly in distilled water for a few seconds, and process for gel staining. A typical gel pattern of plasma proteins stained with the ammoniacal silver nitrate method is shown in **Fig. 2** *(6,7)*.

4. Notes

1. The addition of thiosulfate to the gel delays the appearance of background staining with the ammoniacal silver nitrate method *(6,7)*.
2. Human plasma proteins have been efficiently separated with the following sample preparation: 5 µL of plasma (containing approx 400 µg of proteins) are solubilized in 10 µL of

lysis solution A (SDS-DTE) and heated in boiling water for 5 min. After cooling for 2 min at room temperature, 485 μL of lysis solution B is added. For a silver-stained gel, 30–40 μL of the final diluted sample is loaded on the top (cathodic side) of the capillary gel.

3. The best separation of complex protein mixtures is performed when <100 μg of proteins are loaded onto the capillary gel. Overloading may cause streaking and inadequate resolution of spots. The use of a highly sensitive staining method (silver staining) allows application of a low amount of proteins onto the IEF gel. For a silver-stained gel, concentrations up to 25–40 μg per gel are enough.
4. During the sample loading, it is very important to avoid the formation of air bubbles between the gel and the sample. After the loading, the Hamilton syringe should be withdrawn slowly along the side wall.
5. If more than 20 kVh is applied during the IEF, more cathodic drift occurs and the protein pattern is not stationary.
6. Low pressure on the 1-mL syringe should be exerted to extrude the capillary gel. If too much pressure is applied, small lumps will be formed on the gel.
7. Precise determination of the volume of solution necessary to cast the gels should first be done by measuring the volume of distilled water required to fill one resolving gel or the casting chamber.

References

1. Hochstrasser, D. F., Harrington, M. G., Hochstrasser, A. C., Miller, M. J., and Merril, C. R. (1988) Methods for increasing the resolution of two-dimensional protein electrophoresis. *Analyt. Biochem.* **173,** 424–435.
2. Golaz, O. G., Walzer, C., Hochstrasser, D., Bjellqvist, B., Turler, H., and Balant, L. (1992) Red blood cell protein map: a comparison between carrier-ampholyte pH gradient and immobilized pH gradient, and identification of four red blood cell enzymes. *Appl. Theor. Electrophoresis* **3,** 77–82.
3. Herbert, B. R., Sanchez, J. C., and Bini, L. (1997) Two-dimensional electrophoresis: the state of the art and future directions, in *Proteomic Research: New Frontiers in Functional Genomics (Principles and Practice)* (Wilkins, M. R., Williams, K. L., Appel, R. D., and Hochstrasser, D. F., eds.), Springer-Verlag, Berlin, pp. 13–33.
4. Appel, R. D., Bairoch, A., and Hochstrasser, D. F. (1994) A new generation of information retrieval tools for biologists: the example of the expansy WWW server. *Trends Biochem. Sci.* **19,** 258–260.
5. Oakley, B. R., Kirsch, D. R., and Morris, N. R. (1990) A simplified ultrasensitive silver stain for detecting proteins in polyacrylamide gels. *Analyt. Biochem.* **105,** 361–363.
6. Hochstrasser, D. F. and Merril, C. R. (1988) Catalysts for polyacrylamide gel polymerization and detection of proteins by silver staining. *Appl. Theor. Electrophoresis* **1,** 35–40.
7. Rabilloud, T. (1992) A comparison between low background silver diammine and silver nitrate protein stains. *Electrophoresis* **13,** 429–439.

23

Casting Immobilized pH Gradients (IPGs)

Elisabetta Gianazza

1. Introduction

One of the main requirements for a 2-D protocol is reproducibility of spot position, and, indeed, the technique of isoelectric focusing on immobilized pH gradients (IPGs) is ideally suited to provide highly reproducible 1-D separations. IPGs are obtained through the copolymerization of acidic and basic acrylamido derivatives of different pKs within a polyacrylamide matrix *(1,2)* (**Fig. 1**). The pH gradient may be devised by computer modeling either with a linear or with an exponential course. IPGs are cast from two limiting solutions containing the buffering chemicals at concentrations adjusted to give the required pH course upon linear mixing. For consistent results, gradient pouring and polymerization are carried out under controlled conditions. The covalent nature of the chemical bonds formed during the polymerization step results in a permanent stability of the pH gradient within the matrix. Conflicting requirements during the focusing procedure prevent any effective use of IPGs into capillary tubes *(3)*: the need to buffer with carrier ampholytes (CAs) the pH extremes caused by the migration of the polymerization catalysts is contrasted by the adverse effects of the electroendosmotic flow brought about by the addition of CAs to the gel phase. The demand for the IPG gels to be backed by a binding support—they are usually cast on GelBond™ foils—results in dimensional stability between 1-D and 2-D as a further assistance to reproducibility.

Toward the aim of reproducibility, batch production and quality control as allowed by an industrial process give the commercially available IPG strips (Immobiline DryStrip™ from Pharmacia, Uppsala, Sweden) obvious advantages over homemade slabs. Moreover, the pH course of the commercial product is being carefully characterized in chemicophysical terms, by assessing the dissociation constants of the acrylamido buffers under the experimental conditions relevant to the 1-D run of an IPG-DALT *(4)*. The aim of this effort is to connect focusing position reliably with p*I* *(5)*; any discrepancy between computed and experimental values for known proteins would then hint at the occurrence of posttranslational modifications *(6)*.

From the above it seems this chapter could shrink to the statement: Use the ready-made IPG strips according to the manufacturer's instructions. However, there are reasons, and not only of economical order, for laboratories to cast their own slabs. The

From: *The Protein Protocols Handbook, 2nd Edition*
Edited by: J. M. Walker © Humana Press Inc., Totowa, NJ

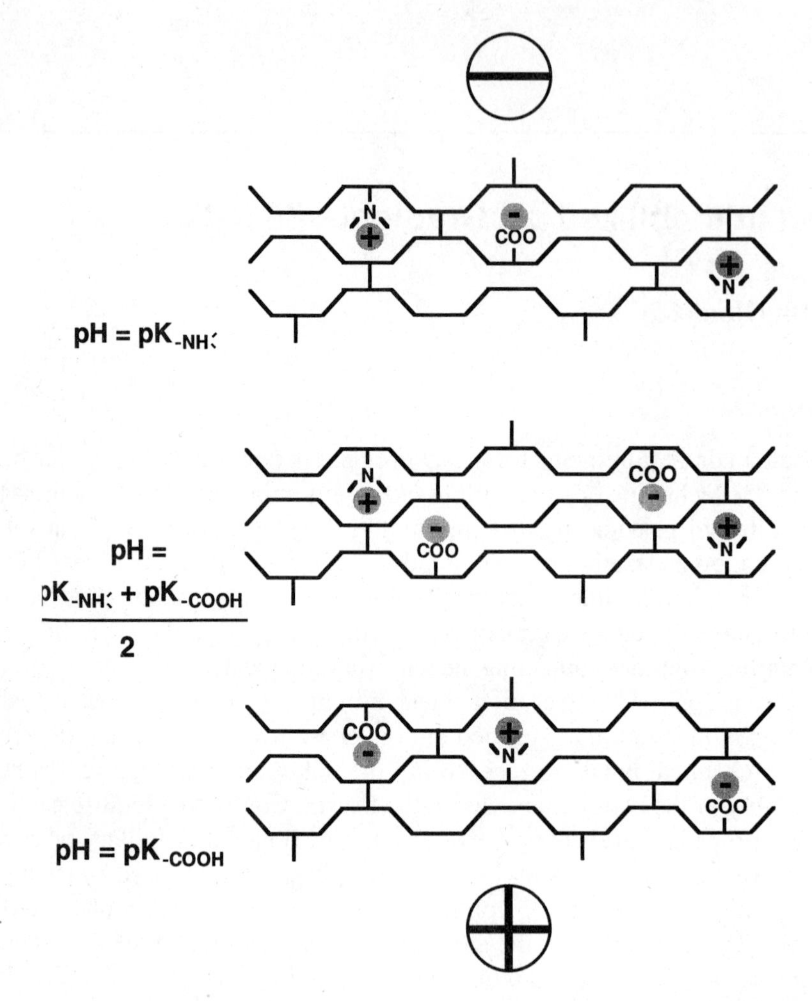

Fig. 1. Structure of the IPG matrix. Different regions of an IPG made up with a basic and an acidic Immobiline are schematically drawn from cathode to anode. In each region, pH depends from the concentration ratio between the basic and the acidic dissociating groups. From the Henderson-Hasselbalch equation, where the concentration of the basic monomer is twice the concentration of the acidic monomer, pH equals the pK of the base, where the concentration of the basic monomer is one-half the concentration of the acidic monomer pH equals pK of the acid, and where the two concentrations are alike, pH equals the mean between the two pKs.

main one is the variability of the analytical needs. 2-D protocols usually aim at the resolution of all peptide components of a complex mixture. Although the proteins in cellular extracts and biological fluids have mostly mildly acidic p*I* values, this is not true of all samples, and the optimal pH course should be devised accordingly. Moreover, after specific qualitative or quantitative variations have been detected for some spots as the result of a given experimental treatment, it is usually worth investigating the area of interest under conditions of maximal resolution. This is especially true if the spots have to be identified or characterized, e.g., by mass spectrometry (MS) techniques after a blotting step *(7)*. IPGs allow the tailoring of wide, narrow, or ultranarrow pH gradients, whereas migration on continuous or gradient PAA slabs of different %T in 2-D may further improve resolution of the spots of interest.

In order to optimize reproducibility and to increase throughput, laboratories handling a very large number of gels may consider a medium-scale production with gradient pouring through mechanical devices (computer-driven burettes) *(8,9)*. In any event, even at the typical laboratory level, the procedure of IPG casting is reliable and allows a remarkable reproducibility of results *(10)*. The following will try to convince the readers that it is also an easy one.

2. Materials

2.1. Equipment

1. Polymerization cassette:
   a. A molding plate with a 0.5-mm-thick permanent frame. Available from Pharmacia in either 12.5 × 25 cm, w × h (cat. no. 18-1013-74), or 25 × 12.5 cm, w × h, size (cat. no. 80-1106-89).
   b. Gel-supporting plate.
   c. Gel-binding foil, e.g., GelBond PAG™ (Pharmacia).
   d. Clamps.
2. Gradient-mixing device:
   a. A two-vessel chamber (e.g., gradient maker from Pharmacia, cat. no. 18-1013-72).
   b. Magnetic bars.
   c. Outlet to the mold, either a one-way silicone tubing (od 3 mm, id 1.5 mm, ca. 12 cm long, equipped with a 2-cm teflon tip) or multiple inlets cast by adapting the tips, ca. 6 cm long, of butterfly needles gauge 21 to a T-connector, available from Cole-Parmer (cat. no. K6365-70).
   d. A stirrer providing constant and even operation at medium to low rpm rating with minimal overheating.
   e. Screw-jack rising table.
3. Forced-ventilation oven at 50°C.
4. Shaking platform.
5. Fan.

2.2. Reagents

1. Acrylamide buffers: 0.2-*M* solutions (*see* **Table 1**), prepared either in water with 5 ppm of hydroquinone methylether as polymerization inhibitor for the acidic monomers or in *n*-propanol for the basic monomers. The chemicals are available from Pharmacia (as Immobiline™) or from Fluka (Buchs, Switzerland). Storage at 4°C; the expiration date is given by the manufacturer. The products are defined as irritant, whereas *n*-propanol is classified as highly flammable.
2. Acrylamide monomers stock solution: 30% T, 4% C *(11)*. Dissolve in water 1.2 g *bis*-acrylamide and 28.8 g acrylamide/100 mL final volume. Filter the stock through 0.8-mm membranes and store at 4°C for about 2 mo. The chemicals should be of the highest purity. Acrylamide has been recognized as a neurotoxin.
3. 40% w/v Ammonium persulfate solution: Dissolve 200 mg of ammonium persulfate in 440 µL of water. Store up to a week at 4°C.
4. TEMED: Store at 4°C for several months.
5. 87% v/v Glycerol.

Table 1
pK Values of Immobilines™

Chemicals	Water solution, 25°C[a]	PAA gel, 10°C[a]	8 M urea in PAA gel, 10°C[b]
Immobiline pK 3.6™	3.58	3.57	4.47
Immobiline pK 4.6™	4.61	4.51	5.31
Immobiline pK 6.2™	6.23	6.21	6.71
Immobiline pK 7.0™	6.96	7.06	7.53
Immobiline pK 8.5™	8.52	8.50	8.87
Immobiline pK 9.3™	9.27	9.59	9.94

[a]From LKB Application Note 324.
[b]From **ref. *42***.

6. 1 M Acetic acid: Dilute 57.5 mL of glacial acetic acid to 1 L.
7. Dimethyldichlorosilane solution: 2% v/v solution of dimethyldichlorosilane in 1,1,1,-trichloroethane. Both chemicals are available from Merck (Darmstadt, Germany), or a ready-made solution is marketed by Pharmacia as RepelSilane™.
8. Glycerol washing solution: 1% v/v glycerol.

3. Method

3.1. Selecting a Formulation

The formulations for narrow to wide-range IPGs have been reported in a series of papers ***(12–15)***, and the recipes for casting the gradients collected ***(2)***. Most of the formulations use the acrylamido buffers marketed as Immobiline™ from Pharmacia, since the pK distribution in their set is even and broad enough for all applications in the 4.0–10.0 pH region. For specific purposes—increased hydrophilicity, narrow ranges, pH extremes—other chemicals are available as acrylamido derivatives ***(16)*** or have been custom-synthesized ***(17,18)***. These chemicals are available from Fluka.

Computer programs for pH gradient modeling, with routines optimizing the concentrations of the acrylamido buffers required in the two limiting solutions in order to give the expected pH course, have been described in the literature ***(19–21)***. One of these programs is available from Pharmacia (Doctor pH™). For most practical applications, a wide pH gradient is required for 2-D mapping in order to resolve all components of a complex biological sample. A statistical examination of the p*I* distribution across all characterized proteins as well as the typical maps of most biological specimens show a prevalence of acidic to mildly alkaline values over basic proteins. A nonlinear pH course is then expected to give improved resolution of most proteins in a complex sample. An exponential gradient whose slope increases from pH 6.0 to 7.0 and to 8.0 optimally resolves most samples run in IPGs: one such recipe ***(14)*** is given in **Table 2** (I), whereas the gradient is plotted in **Fig. 2**.

The narrow ranges in **Table 2** (II–IV) have been interpolated from the sigmoidal gradient above, and provide a better resolution of the acidic, neutral, and basic protein components (**Fig. 3**) while maintaining the same slope ratios among the different pH regions. Alternatively, resolution may be improved by the parallel use of two linear, partly overlapping gradients, namely 4–7 and 6–10 ***(12)*** (**Table 3**).

Table 2
IPG 4-10 with a Nonlinear Course[a]

	Recipes for 5.1– + 5.1-mL Gradients							
	I: Whole range		II: Acidic region		III: Neutral region		IV: Basic region	
Chemicals	A[b] µL	B[c] µL	A µL	B µL	A µL	B µL	A µL	B µL
pK 0.8	135	16	135	62	103	45	86	14
pK 4.6	130	—	130	52	98	33	78	—
pK 6.2	153	46	153	89	126	73	110	46
pK 7.0	—	61	—	37	15	46	24	61
pK 8.5	—	23	—	14	6	17	9	23
pK 9.3	—	43	—	26	11	32	17	43
1 *M* Acetic acid	—	14	—	8.4	3.5	10.5	4.6	14
T30 C4	700	700	700	700	700	700	700	700
87% Glycerol	1000	—	1000	—	1000	—	1000	—
TEMED	3.1	2.7	3.1	2.9	3	2.8	2.9	2.7
40% APS[d]	4.75	4.75	4.75	4.75	4.75	4.75	4.75	4.75
pH at 25°C[e]	4.13	9.61	4.13	6.27	4.67	7.07	5.25	9.61

[a]From *(14)*; pH gradient course in **Fig. 2**.
[b]Acidic solution.
[c]Basic solution.
[d]To be added to the solutions after transferring to the gradient mixer and starting the stirrer.
[e]pH of the limiting solutions.

3.2. Preparing the Working Solutions

As an equilibrium technique, IPG technology freezes within the structure of the amphoteric matrix the actual composition of the acrylamido buffer mix. Since there is no adjustment once the gradients are cast, all steps in liquid handling require the careful use of properly calibrated measuring devices.

1. On two calibrated test tubes or 10-mL cylinders, mark the pH and the volume of the solutions to be prepared with a felt-tip (*see* **Note 1**).
2. Add the acrylamide. For reproducible measurements, all chemicals should be at room temperature (*see* **Note 2**).
3. After adding all the acrylamido buffers, fill the two solutions to one-half their final volume with water, and check their pH with a microelectrode. Refer to expected pH readings at 25°C in the liquid phase, not the values computed for 10°C in the gel phase (*see* **Tables 2** and **3**).
4. Add acetic acid to the basic solution to bring the pH to between 7.0 and 7.5 in order to prevent hydrolysis of the amide bonds during the polymerization step.
5. Add glycerol to the acidic solution to a final concentration of 15–20% v/v. Use a wide-mouth or a clipped pipet tip for dispensing the 87% glycerol stock.
6. The solutions are filled to their final volume with water and carefully mixed. Acrylamide is typically used at the final concentration of 4% T, 4% C. The concentration of TEMED is higher in the acidic than in the basic solution to counteract the effect of amine protonation. Ammonium persulfate is **not** added at this stage (*see* **Note 3**).

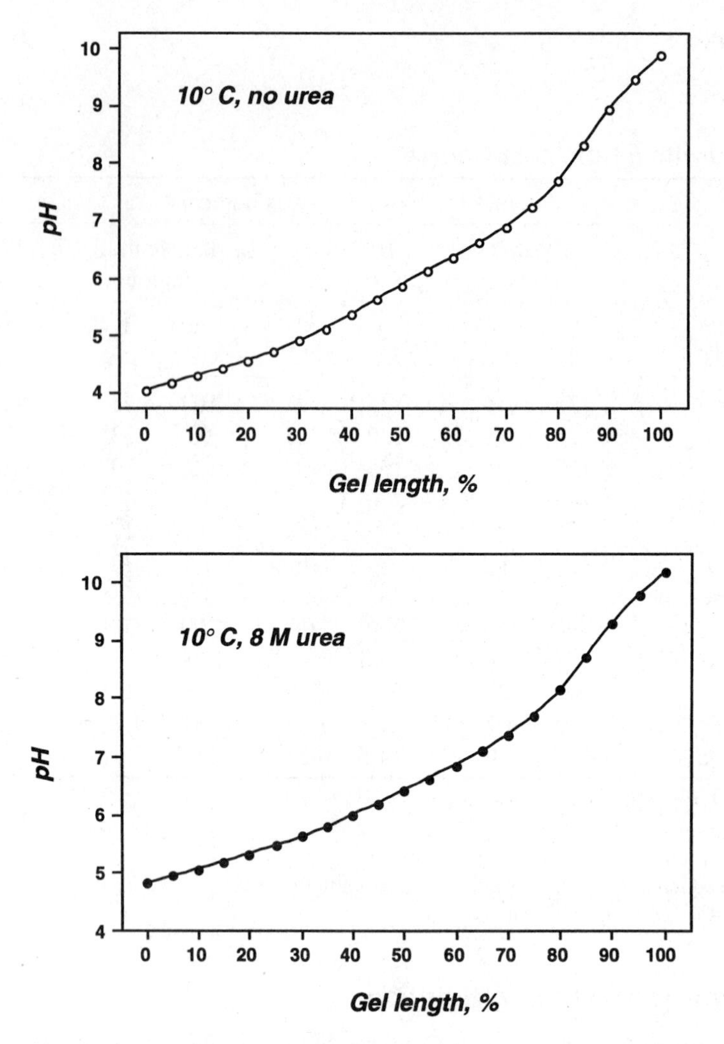

Fig. 2. Course of the gradient for the 4–10 nonlinear IPGs. The top panel shows the pH course computed for the gradient in **Table 2** *(14)* after grafting into a PAA matrix (pK values from **Table 1**, 2nd column); the bottom panel corresponds to the pH values corrected for the presence of 8 *M* urea (pK values from **Table 1**, 3rd column).

3.3. Assembling the Polymerization Cassette

To obtain a 16-cm high IPG slab that fits the 2-D SDS-PAGE gel (e.g., Protean™ from Bio-Rad [Hercules, CA]), the 12.5 × 25 cm w × h cassette is used (*see* **Note 4**).

1. Mark a GelBond PAG™ foil for polarity using a felt-tip marker (anode at the bottom, cathode at ⅔ height); additional annotations (pH gradient, date, even short reminders) may help further identification. The felt-tip marking does not interfere with electrophoretic procedures (*see* **Note 5**).
2. Wet the gel-supporting plate with distilled water.
3. Lay the GelBond PAG™ foil—hydrophobic side down, hydrophilic side up, still covered with the protective paper foil—on a gel-supporting plate. Adhere the foil to the plate by rolling it onto the glass. Care is taken to align the foil flush with the plate.
4. Discard the protecting sheet. Blot any water on the hydrophilic side of the foil with a paper tissue.

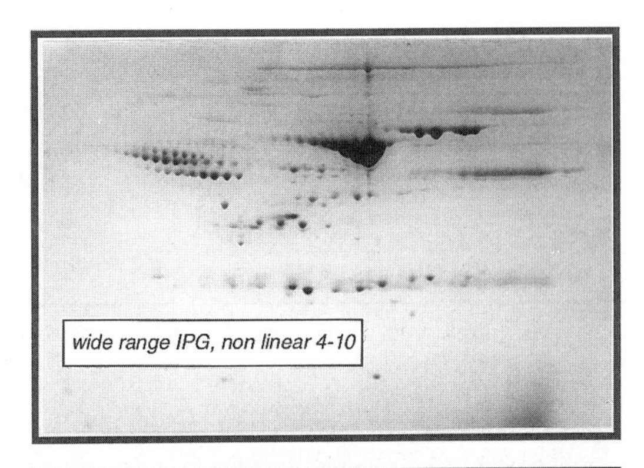

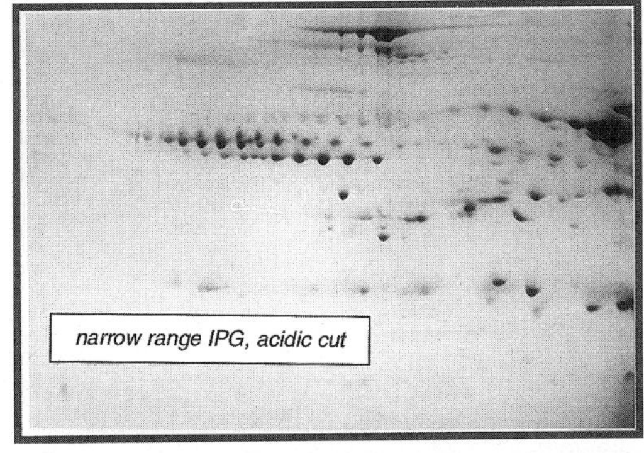

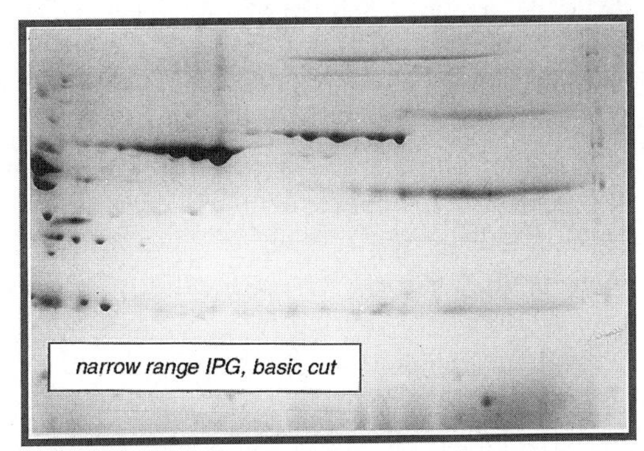

Fig. 3. Examples of IPG-DALT separations. Rat serum was run in 1-D on the 4–10 nonlinear IPG in **Table 2** (I) (top panel) and on its narrow cuts, also in **Table 2** (acidic region [II] in middle panel, basic region [IV] in bottom panel). The 2-D run was on a 7.5–17.5% T PAA gradient with the discontinuous buffer system of Laemmli ***(24)***; the samples from narrow-range IPGs were overrun in order to improve resolution along the *y*-axis (the above runs from: Haynes, P., Miller, I., Aebersold, R., Gemeiner, M., Eberini, I., Lovati, M. R., Manzoni, C., Vignati, M., Gianazza, E. (1998). Proteins of rat serum: I. Establishing a reference two-dimensional electrophoresis map by immunodetection and microbore high-performance liquid chromatography. *Electrophoresis* ***19***, 1484–1492).

5. Swab the inside of the molding plate with a paper towel moist with a dimethyldichlorosilane solution. It is recommended to wear gloves and to perform this treatment in a fume hood.

Table 3
IPG 4-7 and 6-10 with a Linear Course[a]

	Recipes for 5.1– + 5.1-mL Gradients			
	pH 4.0–7.0		pH 6.0–10.0	
Chemicals	A[b] μL	B[c] μL	A μL	B μL
pK 3.6	197	103	329	35
pK 4.6	37	251	—	—
pK 6.2	153	51	96	117
pK 7.0	—	91	85	127
pK 8.5	—	—	91	83
pK 9.3	—	298	99	114
1 *M* Acetic acid	—	—	—	40
T30 C4	700	700	700	700
87% Glycerol	1000	—	1000	—
TEMED	3.1	2.7	2.8	2.7
40% APS[d]	4.75	4.75	4.75	4.75
pH at 25°C[e]	4.02	6.93	5.99	9.72

[a]From **ref. *12***.
[b]Acidic solution.
[c]Basic solution.
[d]To be added to the solutions after transferring to the gradient mixer and starting the stirrer.
[e]pH of the limiting solutions.

6. After the solvent has dried out, rinse the surface of the glass with distilled water, and dry with a paper towel. Inspect the rubber gasket for any break or any adherent material (dried-out acrylamide, paper cuts, and so forth).
7. Invert the molding plate on the GelBond-glass assembly, and carefully align it. Check the marked polarity!
8. Fasten the cassette with clamps of adequate thickness and strength. If two strips from the 1-D IEF are to be aligned tail-to-head for the 2-D SDS-PAGE run on the same gel, the procedure detailed above applies to the 25 × 12.5-cm w × h mold. Pouring a gradient of the stated volume (**Table 2**) will result in an 8-cm-high gel.

3.4. Pouring the Gradient

1. Add the magnetic bar to each chamber of the gradient mixer. Set both connections between the two vessels as well as to the outlet tubing in the closed position. Arrange the mixer on a stirrer, about 5 cm above the mouth of the polymerization cassette.
2. Stick the outlet(s) (*see* **Subheading 2.1., item 2**) in the cassette taking extreme care to not miss the opening to reach instead the gap between GelBond and supporting glass plate.
3. Transfer the acidic, heavy solution into the mixing chamber of the gradient mixer (i.e., the vessel next to the outlet).
4. Briefly open the connection to the reservoir in order to purge air bubbles from the solution; return the solution that flowed into the reservoir during the purging operation to the mixing chamber with a Pasteur pipet.
5. Add the basic light solution to the reservoir.

6. Start the mixing on the stirrer at high speed. Slowly move the gradient mixer around the active area of the mixer in order to find a position in which the magnetic bars in both chambers turn evenly and synchronously. Reduce the stirring speed.
7. Add ammonium persulfate solution in the stated amounts (*see* **Table 2**) to the solutions in both chambers.
8. Open the outlet to the polymerization chamber; the liquid should easily flow by gravity: in case it does not, a light pressure (the palm of your hand over the mouth of the vessel) might solve the problem.
9. When the heavy solution is about to enter the polymerization cassette, open the connection between the two chambers of the mixer.
10. The in-flowing light solution should mix promptly with its heavy counterpart; if not (streams of liquid of different density are distinctly seen in the mixing chamber), the stirring speed should be increased accordingly (*see* **Note 6**).
11. When all solution has flown by gravity, force the few drops remaining in the outlet tubing and in the mixing chamber to the cassette by applying pressure *(see above)* first on the reservoir and then on the mixing chamber, while raising the gradient mixer.
12. Inspect the liquid within the cassette: no air bubbles should be present. It is possible to remove them, with **great caution**, with the help of a thin hook cut from GelBond foil.
13. Leave the gradient to stabilize for 5–10 min at room temperature before moving the cassette into the oven to start the polymerization step.
14. After use, promptly fill the gradient mixer with distilled water and flow the water through the outlet tubing; repeat this procedure twice.
15. Set aside the stirring bars; disassemble the outlet tubing and dry by suction with a pump; and blow the liquid still remaining inside the gradient mixer with compressed air, before drying the outside with a paper tissue.
16. Let the equipment dry to completion with all connections in the open position. Rinsing and drying the gradient mixer, and pouring a new gradient takes ~10 min, so no delay is actually required when processing a number of IPGs one after another.

3.5. Polymerization Conditions

1. Let the IPGs polymerize for 1 h at 50°C (*see* **Note 7**).
2. Gel setting occurs within approx 10 min, whereas polymerization slowly proceeds to plateau in the next 50 min. It is critical during the early period that the polymerization cassette is kept in an upright position **on a level surface**. Moreover, the oven should not be opened in order to avoid both shaking the cassette and causing the temperature to drop.

3.6. Washing, Drying, and Storing IPGs

1. At the end of the polymerization period, disassemble the cassette by removing the clamps.
2. Insert a scalpel blade between GelBond and glass plate to loosen the foil from its support.
3. Fill a plastic box in which the gel may lay flat with 1 L of glycerol washing solution.
4. Gently peel the IPG from the mold, and transfer (upside up!) into the box containing the glycerol solution.
5. Gently shake for 1 h at room temperature.
6. Lay the gel (upside up!) on a supporting glass plate partially or completely submerged in the washing solution in order to avoid trapping air bubbles (*see* **Note 8**).
7. Set the assembly upright, approx 30 cm in front of a (cool) fan, operating at high speed. The whole area should be as dustless as possible; glycerol smudge is better collected on a tray.
8. Evaluate the progression of the drying by sensing the cooling from water evaporation on the back of the glass plate and by looking for an even and matt appearance on the front of

the gel. Depending on the temperature and relative humidity, the drying step may take between 1 and 2 h, and should not be unnecessarily prolonged. It is acceptable, but not recommended to let the gels dry unattended, e.g., overnight, at room temperature.

9. For barely damp gels, the very last phase of drying may occur in the refrigerator, with the gel in an open box. If needed, the polymerized gels may be stored overnight in their cassettes at 4°C before washing.
10. The dry gels may be further processed right away or stored for later use. Dry IPGs may be kept in boxes at 4°C for several days. For longer periods of time *(22)*, they are sealed in plastic bags at –20°C.

4. Notes

1. All required stock solutions are orderly aligned on the bench. It is most helpful if their sequence corresponds to the order in which they are called for by the recipe. This is especially important for the acrylamido buffers, whose bottles look necessarily alike, whereas the pK label is in a relatively small print. It is strongly suggested that the recipe be marked with a ruler and that each chemical be put back in a box after its use. When a bottle is open, hold its cap upside down on a paper tissue. Make sure you are storing the open bottles on a flat, stable surface (like a tray): expensive Immobiline may be easily if inadvertently spilled.
2. Positive displacement measuring devices might be more accurate than air-displacement pipets. No liquid droplets must adhere to the outside of the pipet tips; for this purpose, it is usually sufficient to touch the neck of the bottle. For a quantitative transfer of the Immobiline chemicals, and especially of the basic monomers dissolved in an alcoholic solution, two precautions should be taken; namely, slow pipeting and rinsing the pipet tips twice with distilled water after each measurement.
3. It is not customary to prepare large batches of the limiting solutions—the storage conditions in the mixture are appropriate for none of the chemicals. However, leftover solutions might be used within a couple of days.
4. In the case where the commercial products were not available, an adequate substitution for the gel mold may be a plate of plain glass (polished and saturated with silane) together with a rubber gasket of approx 0.5-mm thickness. The latter may be cut from foils of *para*-, silicone, or nitrile rubber, should have a U-shape, and should be ~5 mm wide. Thick clamps from stationery stores may then be used, but care should be taken to apply the pressure on the gasket, not inside it.
5. A note of caution for 1-D experiments with radioactive samples. In some cases, the ink gives a strong positive signal; more commonly, the writing negatively interferes with film exposure. In these experiments, all marks should then be outside the area of sample application (where their disturbing appearance may turn into the definition of useful reference points).
6. The major points of caution will be stressed once more below. The flow between the different compartments must be unhindered by air bubbles between the two vessels of the mixer or by water droplets in the outlet. No backflow should occur. The pouring should not be too fast to allow proper mixing in the chamber and to avoid turbulence in the cassette (with a proper selection of the pressure drop). The solution should evenly flow along the hydrophilic wall of the cassette (lined with GelBond) and not fall dropwise along the hydrophobic one.
7. A forced ventilation is more appropriate to this purpose than a convection oven in order to provide even and controlled heating. It should be recalled that the selected temperature is the one allowing identical incorporation efficiency for all acrylamido buffers, which grafts into the gel matrix a pH gradient exactly matching the computed course *(23)*.
8. It is sensible to set aside some containers only for this purpose, i.e., to avoid recycling between gel washing, protein staining, and blot immunodetection. Owing to its high viscosity, 100% glycerol should be avoided as stock reagent in favor of the 87% preparation; ~12

mL of the latter may be measured in a small plastic beaker instead of in a graduated cylinder. Failure to include glycerol as a humidity-conditioning agent results in slab curling and easy peeling and tearing of the gel from its support. On washing, IPGs reswell to a different extent, depending on the pH of the matrix and on the course of the glycerol gradient.

Acknowledgment

The author is most thankful to Ivano Eberini for testing the content of this chapter as a teaching guide to the use of IPGs.

References

1. Bjellqvist, B., Ek, K., Righetti, P. G., Gianazza, E., Görg, A., Westermeier, R., et al. (1982) Isoelectric focusing in immobilized pH gradients: Principles, methodology and some applications. *J. Biochem. Biophys. Meth.* **6,** 317–339.
2. Righetti, P. G. (1990) *Immobilized pH Gradients:* Theory and Methodology. Elsevier, Amsterdam.
3. Gianazza, E., Tedesco, G., Cattò, N., Bontempi, L., and Righetti, P. G. (1988) Properties of thin-rod immobilized pH gradients. *Electrophoresis* **9,** 172–182.
4. Bjellqvist, B., Pasquali, C., Ravier, F., Sanchez, J.-C., and Hochstrasser, D. (1993) A non-linear wide-range immobilized pH gradient for two-dimensional electrophoresis and its definition in a relevant pH scale. *Electrophoresis* **14,** 1357–1365.
5. Bjellqvist, B., Hughes, G. J., Pasquali, C., Paquet, N., Ravier, F., Sanchez, J.-C., et al. (1993) The focusing position of polypeptides in immobilized pH gradients can be predicted from their amino acid sequences. *Electrophoresis* **14,** 1023–1031.
6. Gianazza, E. (1995) Isoelectric focusing as a tool for the investigation of post-translational processing and chemical modifications of proteins. *J. Chromatogr. A* **705,** 67–87.
7. Jungblut, P., Thiede, B., Zimny-Arndt, U., Müller, E.-C., Scheler, C., Wittmann-Liebold, B., et al. (1996) Resolution power of two-dimensional electrophoresis and identification of proteins from gels. *Electrophoresis* **17,** 839–847.
8. Altland, K. and Altland, A. (1984) Pouring reproducible gradients in gels under computer control: New devices for simultaneous delivery of two independent gradients, for more flexible slope and pH range of immobilized pH gradients. *Clin. Chem.* **30,** 2098–2103.
9. Altland, K., Hackler, R., Banzhoff, A., and Eckardstein, V. (1985) Experimental evidence for flexible slope of immobilized pH gradients poured under computer control. *Electrophoresis* **6,** 140–142.
10. Gianazza, E., Astrua-Testori, S., Caccia, P., Giacon, P., Quaglia, L., and Righetti, P. G. (1986) On the reproducibility of band position in electrophoretic separations. *Electrophoresis* **7,** 76–83.
11. Hjertén, S. (1962) "Molecular sieve" chromatography on polyacrylamide gels, prepared according to a simplified method. *Arch. Biochem. Biophys.* **Suppl. I,** 147–151.
12. Gianazza, E., Celentano, F., Dossi, G., Bjellqvist, B., and Righetti, P. G. (1984) Preparation of immobilized pH gradients spanning 2-6 pH units with two-chamber mixers: Evaluation of two experimental approaches. *Electrophoresis* **5,** 88–97.
13. Gianazza, E., Astrua-Testori, S., and Righetti, P. G. (1985) Some more formulations for immobilized pH gradients. *Electrophoresis* **6,** 113–117.
14. Gianazza, E., Giacon, P., Sahlin, B., and Righetti, P. G. (1985) Non-linear pH courses with immobilized pH gradients. *Electrophoresis* **6,** 53–56.
15. Gianazza, E., Celentano, F., Magenes, S., Ettori, C., and Righetti, P. G. (1989) Formulations for immobilized pH gradients including pH extremes. *Electrophoresis* **10,** 806–808.
16. Righetti, P., Chiari, M., Sinha, P., and Santaniello, E. (1988) Focusing of pepsin in strongly acidic immobilized pH gradients. *J. Biochem. Biophys. Meth.* **16,** 185–192.

17. Chiari, M., Pagani, L., Righetti, P. G., Jain, T., Shorr, R., and Rabilloud, T. (1990) Synthesis of an hydrophilic, pK 8.05 buffer for isoelectric focusing in immobilized pH gradients. *J. Biochem. Biophys. Meth.* **21,** 165–172.
18. Chiari, M., Righetti, P. G., Ferraboschi, P., Jain, T., and Shorr, R. (1990) Synthesis of thiomorpholino buffers for isoelectric focusing in immobilized pH gradients. *Electrophoresis* **11,** 617–620.
19. Celentano, F., Gianazza, E., Dossi, G., and Righetti, P. G. (1987) Buffer systems and pH gradient simulation. *Chemometrics Intell. Lab. Syst.* **1,** 349–358.
20. Altland, K. (1990) IPGMAKER: A program for IBM-compatible personal computers to create and test recipes for immobilized pH gradients. *Electrophoresis* **11,** 140–147.
21. Giaffreda, E., Tonani, C., and Righetti, P. G. (1993) pH gradient simulator for electrophoretic techniques in a Windows environment. *J. Chromatogr. A* **630,** 313–327.
22. Altland, K. and Becher, P. (1986) Reproducibility of immobilized pH gradients after seven months of storage. *Electrophoresis* **7,** 230–232.
23. Righetti, P. G., Ek, K., and Bjellqvist, B. (1984) Polymerization kinetics of polyacrylamide gels containing immobilized pH gradients for isoelectric focusing. *J. Chromatogr.* **291,** 31–42.
24. Laemmli, U. K. (1970) Cleavage of structural proteins during the assembly of the head of bacteriophage T4. *Nature* **227,** 680–685.

24

Nonequilibrium pH Gel Electrophoresis (NEPHGE)

Mary F. Lopez

1. Introduction

Nonequilibrium pH gel electrophoresis (NEPHGE) is a technique developed to resolve proteins with extremely basic isoelectric points (pH 7.5–11.0) *(1,2)*. These proteins are difficult to resolve using standard IEF, because the presence of urea in IEF gels has a buffering effect and prevents the pH gradient from reaching the very basic values (with a pH above 7.3–7.6) *(3)*. In addition, cathodic drift causes many very basic proteins to run off the end of the gel. During NEPHGE, proteins are not focused to their isoelectric point, but instead move at different rates across the gel owing to charge. For this reason, the accumulated volt hours actually determine the pattern spread across the gel. It is therefore imperative that volt hours be consistent to assure reproducible patterns.

2. Materials

2.1. Electrophoresis Equipment and Reagents

The procedures and equipment required for running NEPHGE are very similar to those for IEF. Please refer to Chapter 22 for the necessary electrophoresis equipment and reagents. In order to run the NEPHGE procedure, it will be necessary to reverse the polarity in the 1-D gel running tank. This is usually achieved with an adapter that attaches to the tank or power supply.

3. 2-D NEPHGE Protocol

3.1. Casting IEF and 2-D Slab Gels

Follow the instructions given in the Chapter 22 to cast the tube gels and the second-dimension slab gels. Use a wide-range ampholyte mixture, such as 3–10, for best results (*see* **Note 1**).

3.2. Running the NEpHGE Gels

1. Prepare anode and cathode solutions as described in **Subheading 2.** of Chapter 22.
2. Pour 2 L of degassed cathode solution into the lower reservoir of the 1-D running tank and lower the 1-D running rack into the cathode solution (*see* **Note 2**).

From: *The Protein Protocols Handbook, 2nd Edition*
Edited by: J. M. Walker © Humana Press Inc., Totowa, NJ

3. Slide a grommet onto each IEF tube so that the grommet is at the top of the tube. Slide the tube through a guide hole in the rack and press down on the grommet to seal.
4. Install as many IEF tubes into the 1-D running system as you wish to run. All other empty analytical and preparative holes should be plugged.
5. Fill the upper reservoir with 800 mL of anode solution (*see* **Note 2**).
6. Debubble the tubes by filling a 1-mL syringe with a long, fine needle attached with anode solution. Lower the needle to the gel surface of a tube, and expel the anode solution until all the bubbles are released.
7. Using a Hamilton syringe with a fine needle, load 1–50 μL sample in each tube. Optimum results for silver-stained gels are obtained when loading approx 60–100 μg protein. Samples should be salt-free and prepared as described in Chapter 22.
8. Install the cover on the 1-D running system.
9. Program the power supply with the following parameters to prerun the gels:
   a. Maximum voltage: 1500.
   b. Duration: 2 h.
   c. Maximum current (mA) per gel: 110.
10. Prerun the gels until the voltage reaches 1500 V. This should occur within 2 h (*see* **Note 3**).
11. Program the power supply with the following parameters to run the gels:
    a. Maximum voltage: 1000.
    b. Duration: 4 h.
    c. Maximum current (mA) per gel: 110.
    d. Total volt-hours: 4000.
12. Cast and prepare the second dimension gels as described in Chapter 22.
13. After running the NEPHGE IEF gels, proceed to extrude them and load onto the second dimension gels as described in Chapter 22.
14. Run the second-dimension gels as described in Chapter 22.

4. Notes

1. Ampholyte mixtures are usually provided as a 40% aqueous solution. Although the NEPHGE procedure separates proteins with basic pIs, it is best to use a wide-range ampholyte mixture. A basic mixture of ampholytes will crowd the acidic proteins in a narrow region and potentially obscure some of the basic proteins.
2. For the NEPHGE procedure, the positions of the anode and cathode buffers are reversed. This results in a better separation of the basic proteins.
3. During NEPHGE, the IEF gels are not actually "focused" to their pIs. The separation is based on the migration rates of the differentially charged polypeptides as they move across the gel. Therefore, it is necessary to pay strict attention to accumulated volt-hours during the run to assure reproducible patterns in subsequent separations. During the prerun, a pH gradient is set up, and the focusing voltage is reached. The resistance in the gels is such that this voltage should be achieved in <2 h. If it takes longer, the samples may have high conductivity or the ampholyte quality may be inferior.
4. The conditions for optimum polypeptide separation will most probably have to be empirically determined for the NEPHGE gels. This is because the proteins are not focused to their respective p*I*s as in standard IEF. Therefore, a series of test runs with different accumulated volt-hours should be compared to optimize for each sample.

References

1. O'Farrell, P. Z., Goodman, H. M., and O'Farrell, P. H. (1977) High resolution two-dimensional electrophoresis of basic as well as acidic proteins. *Cell* **12,** 1133–1142.

2. Dunn, M. J. (1993) Isoelectric focusing, in Gel *Electrophoresis: Proteins.* Bios Scientific Publishers, Oxford, UK, pp. 65–85.
3. Investigator 2-D Electrophoresis System Operating and Maintenance Manual, ESA, Chelmsford, MA 01824, cat. no. 70-3205.

25

Difference Gel Electrophoresis

Mustafa Ünlü and Jonathan Minden

1. Introduction

Proteomics is a relatively newly coined term. It refers to the field of study of the proteome, which was defined for the first time in 1994 as "the PROTEin products of a genOME" *(1)*. Part of the attractiveness of proteomics derives from its promise to uncover changes in global protein expression accompanying many biologically relevant processes, such as development, tumourigenesis, and so forth. Because proteins are the effector molecules that carry out most cellular functions, studying proteins directly has clear advantages over and (at least in theory) achieves results that go beyond genomic analyses. Whereas the information contained in the genome is almost always static, spatiotemporal patterns of protein expression are very complex and dynamic due to fluctuations in abundance. This complexity is increased several-fold by proteins' ability to be modified functionally through posttranslational modifications.

To detect these changes in global protein expression, there is a clear requirement that the methodology employed be able to generate and compare snapshots of the entire protein component of any organism, cell, or tissue type. Hence, as the method that offers the highest practical resolution in protein fractionation, two-dimensional gel electrophoresis (2-DE) is the most commonly used separation technique in proteomics. Its resolution power derives from orthogonally combining two separations based on two independent parameters: isoelectric focussing (IEF) separates based on charge and sodium dodecyl sulfate polyacrylamide gel electrophoresis (SDS-PAGE) separates by size. With 2-DE, each proteome snapshot becomes a protein map, a two-dimensional gel on which hundreds to a few thousand proteins are fractionated and displayed. Comparison of two or more gels then allows for the detection of the protein species differing between the compared states of the organism, cell, or tissue type under investigation.

2-DE was first described simultaneously by several groups in 1975 *(2–4)*. Despite the quite substantial advances in technology since its launch, the most notable of these being the introduction of immobilized gradients in the first dimension *(5)*, some of the more significant systemic shortcomings have remained unsolved. The most troublesome of these is the inherent lack of reproducibility between gels, despite the fact that immobilized IEF gradients have alleviated this problem to some extent. Efforts to surmount this limitation have focused on developing methods that have increasingly

From: *The Protein Protocols Handbook, 2nd Edition*
Edited by: J. M. Walker © Humana Press Inc., Totowa, NJ

utilized computational tools, applying statistical analyses to several experiments to distill the true differences away from those induced by inter-gel fluctuations. However, these efforts have been at best partially successful, since running different gels in order to compare different samples has been a requirement given that all conventional protein detection methods are nonspecific in nature. At the same time, the material and time investment required in a technique that is already considered to be difficult has increased.

Difference gel electrophoresis (DIGE) was developed to overcome the irreproducibility problem in 2-DE methodology by labeling two samples with two different fluorescent dyes prior to running them on the same gel *(6)*. The fluorescent dyes used in DIGE, Cy3 and Cy5 (**Fig. 1**), are cyanine based, molecular weight matched, amine reactive, and positively charged. These characteristics, coupled with substoichiometric labeling, result in no electrophoretic mobility shifts arising between the two differentially labeled samples when coelectrophoresed. Thus, in DIGE, every identical protein in one sample superimposes with its differentially labeled counterpart in the other sample, allowing for more reproducible and facile detection of differences. Furthermore, DIGE is a sensitive technique, capable of detecting a single protein between two otherwise identical samples at levels as low as 0.01% of total protein and having an overall detection sensitivity equal to that of silver staining.

2. Materials

All references to H_2O should be read as double-distilled H_2O unless stated otherwise.

In recipes, the information given in each line corresponds to final concentration, the name of ingredient, and the amount used, in that order.

2.1. Sample Solubilization and Labeling

1. 40% Methylamine in water, urea, thiourea, dithiothreitol (DTT) and 3-[3-cholamidopropyl) dimethyl ammonio]-1-propanesulfonate (CHAPS) are available from Sigma-Aldrich (Milwaukee, WI). N-[2-hydroxyethyl] piperazine *N*'-2'-ethanesulfonic acid (HEPES) was from Fisher Scientific (Pittsburgh, PA).
2. Lysis buffer:

   8 *M* Urea 24.0 g

(Alternatively, 6 *M* urea and 2 *M* thiourea may be used. *See* **Note 1**)

   6 *M* Urea 18.0 g

   2 *M* Thiourea 7.6 g

Make up to 40 mL with HPLC quality H_2O and dissolve the urea.

   2% CHAPS (Sigma-Aldrich) 1.0 g

   10 m*M* DTT 500 µL of 1 *M* stock or 0.077 g of solid

   10 m*M* NaHEPES, pH 8.0 5.0 mL of 100 m*M* stock (Add last, *see* **Notes 2** and **3**)

Make up to 50.0 mL with HPLC quality H_2O and store at –80°C in 1- to 1.5-mL aliquots.

3. Labeling solution: We typically label samples in lysis buffer with no further modification. Note that we do not add Pharmalytes to this solution (*see* **Note 4**).
4. Quenching solution: 5 *M* methylamine in 100 m*M* NaHEPES, pH 8.0. Dissolve 2.38 g of HEPES in 38.8 mL of 40% methylamine aqueous solution. Slowly add approx 60 mL concentrated HCl with stirring. Cool the solution on ice and measure the pH as the HCl is added until the pH reaches 8.0.

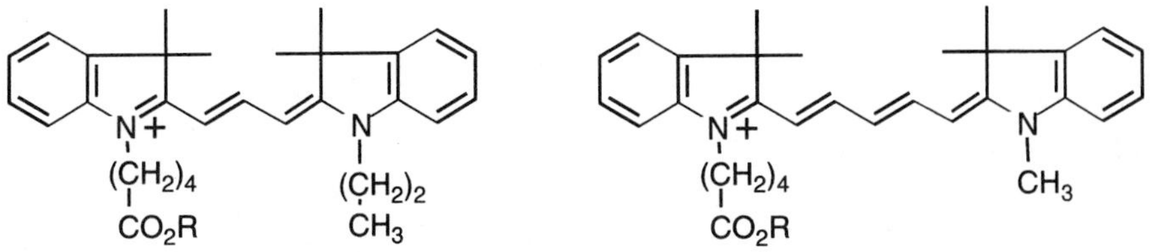

Cy-3

Cy-5

R = H, Su

Fig 1. The two cyanine dyes used in DIGE.

2.2. IEF

1. The IEF equipment is a Multiphor II from Amersham-Pharmacia (Peapack, NJ), with the DryStrip kit installed. The Teflon membranes are from the YSI (Yellow Springs, OH) model 5793 standard membrane kit for oxygen electrodes. Light paraffin oil was from obtained from Amersham-Pharmacia.
2. Rehydration buffer:

2% CHAPS	0.4 g
8 *M* Urea	9.6 g (same composition as lysis buffer; *see* **Note 1**)
6 *M* Urea	7.2 g
2 *M* Thiourea	3.0 g
2 m*M* Acetic acid	2.7 µL of glacial acetic acid (~17 *M*)
10 m*M* DTT	200 µL of 1 *M* stock or 0.031 g solid
1% Pharmalyte	500 µL of 40% Pharmalyte stock solution*

Make up to 20 mL with H_2O; store at 4°C (*see* **Notes 5** and **6**).

*Use the appropriate IPG solution which corresponds to the pH range of the IEF gels. *See* **Subheading 3.2.**

3. Equilibration buffer I:

1× Stacking gel buffer	25 mL of 4× solution
1% SDS	20 mL of 10% SDS solution
8.7% Glycerol	10 mL of 87% solution
5 m*M* DTT	500 µL of 1 *M* solution or 0.076 g

Bring up to 100 mL with H_2O *see* (**Note 5**).

4. Equilibration buffer II:
Same as above, except 2% iodoacetamide (add 2.0 g) replaces DTT, and a trace amount of bromophenol blue is added.

2.3. SDS-PAGE

1. SDS is available from Fisher Scientific. The gradient maker, acrylamide and *bis*-acrylamide of the highest purity are available from Bio-Rad (Hercules, CA). The gel equipment was a Hoefer SE-660 18 × 24 cm apparatus from Amersham-Pharmacia. The 0.2-µm filters are available from Nalgene (Rochester, NY).
2. 4× Resolving gel buffer: Dissolve 36.3 g Tris in 150 mL H_2O. Adjust to pH 8.6 with 6 *N* HCl then make up to 200 mL with H_2O. Filter sterilize through a 0.2-µm filter and store at 4°C.
3. 30% Monomer solution: 60.0 g of acrylamide, 1.6 g *bis*-acrylamide made up to 200 mL with H_2O. Filter sterilize and store as described.

4. 4× Stacking gel buffer:
   3.0 g of Tris.
   Add 40 mL of H_2O.
   Adjust pH to 6.8 with 6 *N* HCl.
   Made up to 50 mL with H_2O.
   Filter sterilize as described previously and store at –20°C.
5. Light gradient gel solution:

10% acrylamide	8.25 mL of 30% monomer solution
0.375 *M* Tris	6.25 mL of 4× resolving gel buffer
H_2O	10 mL
0.1% SDS	250 µL of 10% SDS solution

   Add these immediately pouring the gel:

APS	82.5 µL of 10% stock solution
TEMED	8.25 µL

6. Heavy gradient gel solution:

15% Acrylamide	12.25 mL of 30% monomer solution
0.375 *M* Tris	6.25 mL of 4× resolving gel buffer
H_2O	3.8 mL
0.1% SDS	250 µL of 10% SDS solution
Sucrose	3.75 g

   Add these immediately before pouring the gel:
   Ammonium pensulfate (APS) 82.5 µL of 10% stock solution
   N,N,N',N'-tetramethylethylenediamine (TEMED) 8.25 µL
7. Stacking gel solution:

3.5 % Acrylamide	400 µL of 30% monomer solution
0.175 *M* Tris	800 µL of 4× stacking gel solution
H_2O	2.0 mL
0.1% SDS	33 µL of 10% SDS solution

   Add these two just before pouring the gel:

APS	16.7 µL of 10% stock solution
TEMED	1.7 µL

2.4. Image Acquisition and Analysis

1. The cooled CCD camera is a 16-bit, series 300 model purchased from Photometrics/Roper Scientific (Tucson, AZ). It is fitted with a standard 105 mm macro lens from Nikon, available from most photographical suppliers. Two 250 W quartz–tungsten–halogen lamps from Oriel (Stratford, CT) are used for illumination. Single-bandpass excitation filters 2.5 cm diameter from Chroma Technology (Brattleboro, VT), are used to excite 545 ± 10 nm and 635 ± 15 nm for Cy3 and Cy5, respectively. A multiwavelength bandpass emission filter from Chroma Technology is used to image the gels at 587.5 ± 17.5 nm and 695 ± 30 nm for Cy3 and Cy5, respectively. The imager housing was constructed in-house from black Plexiglas. Image acquisition is semiautomated and is controlled by an Silicon Graphics. (Mountain View, CA) O_2 computer workstation.
2. Destain solution:

1% Acetic acid	10 mL of glacial acetic acid
40% Methanol	400 mL
H_2O	590 mL

3. Method

3.1. Sample Solubilization

As the number of different cells or tissue types that typically will be analyzed by 2-DE is quite large, it is difficult to describe a single protocol that will be applicable for all cases. Most samples have merely needed lysis buffer to be added to extract protein. Some examples of this kind are almost all prefractionated or prepurified proteins, human brain slices, *Drosophila* testes, mouse embryonic genital ridges, and pellets of *E. coli*. Some preparations have required only slightly more vigorous disruption, that is, with a ground glass homogenizer. Other preparations have required even more severe extractions, i.e. using sonication or glass beads for efficient disruption. Yeast and some cultured cells are examples of the latter. Each sample, if it does not simply lyse upon buffer addition, will require an empirical approach to determine the most efficient preparation method.

In general, all samples should be kept as cold as possible. All steps leading to and including lysis should be performed on ice. As soon as lysis is complete, samples may be stored indefinitely at –80°C. Repeated thawing and refreezing does not seem to have a deleterious effect. Since the lysis buffer contains urea, provisions of **Note 2** apply. Also *see* **Note 7**. Below are three sample extraction protocols that may be used as a basis for developing further protocols:

3.1.1. Tissue Cultured 3T3 Mouse Fibroblasts

1. Grow one 150-mm dish of cells per condition.
2. Wash 3× with 5 mL of Dulbecco's modified Eagle medium (DMEM) culture medium without serum.
3. Add 2 mL medium to plate.
4. Remove cells by scraping and transfer to a 15-mL conical tube.
5. Repeat the above with an additional 2 mL, transfer to the same conical tube.
6. Centrifuge for 5 min at 5000*g*.
7. Discard supernatant and resuspend cells in 1 mL medium.
8. Count cells.
9. Transfer cells to a 1.5-mL Eppendorf tube on ice.
10. Centrifuge for 5 s at maximum speed in a tabletop nanofuge.
11. Vortex-mix lightly to loosen pellet.
12. Add 50 µL of lysis buffer for every 10^6 cells.
13. Stir with pipet tip to remove clumps, if any.
14. Centrifuge in the cold room for 5–10 min at 13,000*g*.
15. Store at –80°C.
16. This extract is expected to have between 4 and 8 mg/mL of protein.

3.1.2. Drosophila *Embryos*

1. Embryos are dechorinoted and observed in eggwash solution under a dissection microscope to determine their age. Embryos that are at the right stage are removed and rinsed once with ethanol to remove the eggwash solution. They are immediately transferred into ethanol over dry ice. Several days of collections may be accumulated, with the embryos being kept in ethanol at –80°C.

2. After accumulating enough embryos (assume a yield of 1–1.5 μg of protein yield per embryo), transfer the embryos into a cold ground-glass homogenizer on ice, add 1 μL of lysis buffer per embryo and briefly homogenize.
3. Centrifuge in a tabletop microfuge for 5–10 s, take the supernatant and discard any solid material. Store at –80°C. This extract should have between 1–2 mg/mL of protein.

3.1.3. *S. cerevisiae*

1. Obtain a 250-mL yeast culture of 0.5 OD_{600}.
2. Centrifuge the culture at 5000*g* for 15 min at 4°C.
3. Decant the supernatant.
4. Wash the cells as follows:
   a. Suspend the cells in 5 mL of 100 m*M* NaHEPES, pH 8.0.
   b. Centrifuge the cells for 10 min at 5000*g* at 4°C (*see* **Note 15**).
   c. Remove the supernatant using an aspirator.
   d. Resuspend in 5 mL of 100 m*M* NaHEPES, pH 8.0, and repeat the preceding steps.
5. Suspend the pellet of washed cells in 750 μL of lysis buffer
6. Transfer to a 15-mL conical tube containing 1 g of acid-washed glass beads.
7. Vortex-mix for 3 min in the cold room
8. Centrifuge and transfer the supernatant to fresh tubes.
9. Store the extract at –80°C. This extract has an expected protein concentration between 1.5 and 4 mg/mL.

3.2. Sample Labeling

Measure the protein concentration in the extracts with the method of choice (Bradford, etc.). To obtain matched fluorescence images, use equal amounts of protein for each sample. Anywhere between 50 and 250 μg of protein per sample may be loaded on a single IEF strip, making the total maximum load 0.5 mg when two samples are being compared. There is very little to no loss of resolution even at the highest level of loading, and there is no reason not to try to achieve it. In fact, the greater load will allow for the detection of lower abundance proteins and permit a better chance of success in the eventual identification of the spots of interest. It is best to run two gels for each comparison where the order of labeling is reversed. This ensures that the observed differences are sample dependent and not dye dependent. Dye-dependent changes are rarely seen when the labeling reaction is done correctly. However, errors in labeling or quenching are often indicated by many dye-dependent changes.

1. Bring the desired amount of each sample to up to 48 μL with lysis buffer (*see* **Note 8**).
2. Add 1 μL of the appropriate dye to each sample (*see* **Note 9**).
3. Incubate on ice for 15 min.
4. Add 1 μL of quenching solution, followed by 0.5 μL of the appropriate Pharmalyte solution.
5. Incubate on ice for 30 min.
6. Immediately load and run on the first dimension.

3.3. IEF

This is arguably the most complex and problematic step in the whole procedure. Notable difficulties include, but are not limited to: sample leakage, gel sparking, and insufficient or incomplete focussing. The first dimension gels are purchased as dry strips and come in a variety of sizes and pH ranges. In the case of Pharmacia Dry

Strips, the instructions that come with the gels should be followed in general; however, do note that some of the solutions and procedures recommended by Pharmacia have been modified here. When there's a conflict, use this version for the best results.

3.3.1. Rehydration

1. The rehydration cassette is made up of two glass plates, one with a 0.5 mm rubber U-gasket, the other plain. Wash both plates with H_2O, then 95% EtOH and air dry.
2. Hold gel strip at either end and peel off protecting thin layer of plastic. The thicker piece has the gel cast on it. Discard the protecting plastic. The gel is bonded to special plastic called polyacrylamide (PAG) bond plastic. The back side of the strip opposite the gel is hydrophobic. Drop just a few drops of water on this hydrophobic side (*see* **Note 10**). Lay the plain glass plate face down and place the gel on it, wet (hydrophobic) side down. Make sure the acidic (pointed) end is near the bottom of plate.
3. After all gels are in place, roll them flat with a Teflon roller. There is no need to press hard; all that is needed is that the strips don't fall down when the glass plate is inverted. Remove any excess water that might have seeped out.
4. Make sure gasket around the edge of top glass plate is clean and dry, then place over bottom plate and clamp two plates together (two clips on each long side and one at bottom).
5. Using the special 25-mL syringe included with the kit, fill cassette with rehydration buffer by injecting from the bottom until the top of the gels are just covered. Try to avoid air bubbles. Lay the cassette assembly flat and rehydrate overnight (*see* **Note 11**).

3.3.2. Setting Up and Running the First Dimension

1. Remove the rehydration buffer from the IEF strips and save at 4°C for future reuse (*see* **Note 6**).
2. To make the contact wicks that will go between the gel and the electrodes, cut the filter paper supplied with the Immobiline apparatus to approx 0.5 × 0.3 mm rectangles (*see* **Note 12**). The wicks are used for ensuring good contact between the electrodes and the gel. They also act as a buffer space for salts that migrate to the electrodes. Cut enough wicks so that there are a few more than twice the number of strips. Immerse all wicks in HPLC quality H_2O for about a minute. Then place all wicks on a paper towel so that they cluster around each other to create a damp patch on the paper. They should be placed close to but not overlapping with each other. Then start placing the IEF strips on the gel apparatus. By the time the gels are in place and ready to receive the wicks, they are usually at the right dampness (*see* **Note 13**).
3. Lay down a thin line of light paraffin oil into each groove on the strip aligner that will receive an IEF strip. Make sure to use a minimal amount of oil. It is not desirable to have any excess oil on top of the gel, especially over the location of the sample cup. Place one IEF strip over each oil streak with the acidic end pointing toward the positive, red electrode, and making sure that it is parallel to and aligned with the other gels using the lines on the flatbed.
4. Apply one wick at each (basic and acidic) end of each gel. Make sure all the wicks are lined up with each other if there is more than one IEF strip on the plate. Place both electrodes over their respective wicks. Take care to orient the electrodes correctly. They are color coded and the instructions are included in the kit. Wrong orientation of the electrodes will result in no electrophoresis. Samples will be applied close to the anode—so the sample cup holder should next be placed as close to the positive electrode as possible.
5. Cover the gel strips with Teflon membranes, except the areas where the sample cups are going to be applied (*see* **Note 14**). Dab gel area around prospective sample cup location with a paper towel if there's liquid over it; otherwise leave it alone. After putting the cups

on, lightly press on the cup to make sure it sits flatly on the gel. This is also a quite tricky step. Any misalignment of the cups will cause the sample to leak out. Observe the cups from several angles to make sure that they are seated correctly (*see* **Note 16**). Apply sample, taking care not to disturb the sample cup.

Start the run at between 300 and 1000 V for 1 h or as long as the current stays above 0.5 mA. It is important to not allow the current to rise too high, especially during the initial stages of the electrophoresis. As the current drops through the run, slowly raise the voltage up to maximum allowed by your power supply. At least 2000 V should be achieved—the higher, the better focussing. The total amount of volt hours delivered over the run time should be 30 kVh to 50 kVh (*see* **Note 17**).

3.3.3. Equilibration of IEF Gels

1. At the end of the IEF run, equilibrate the gels in equilibration buffer I for 15 min, rinse briefly with H_2O to remove DTT, and equilibrate in buffer II for another 15 min.
2. After equilibration, gels should be either run immediately on the second dimension or stored at –80°C. They will last almost indefinitely at –80°C (*see* **Note 18**).

3.4. SDS-PAGE

3.4.1. Assembling the Gel Cassettes

1. Rinse the glass plates with H_2O, then 95% EtOH and air dry. Rinse clamps and spacers in H_2O and dry. Make sure the insides of clamps are dry and the spacers are free of particulate debris.
2. Align glass plates and spacers so that all edges are flush, especially at the bottom. Place a large thumbscrew clamp toward the top and a small one at the bottom on each of the long sides of the glass plates. Make sure the bottom of the cassette is sticking out slightly (1–2 mm) from the clamps. This drives the gel cassette into the stand and seals the bottom when the cams are inserted and twisted in opposite directions. If desired, check seal by squirting water on the outside of sandwich assembly.
3. Assemble the equipment for pouring, with the gradient maker on a stir plate, over the cassette(s) (*see* **Note 19**).

3.4.2. Pouring the 10–15% Gradient Gels

1. Pour the heavy solution in front chamber of gradient maker, add stirbar.
2. Pour light solution in back chamber of gradient maker, add stirbar balancer.
3. Open the mixing channel and the front stopcock at the same time.
4. Pour gel and overlay with *n*-butanol.
5. Allow gel to polymerize (about 30 min).
6. Pour off *n*-butanol, rinse top of gel with H_2O twice, and pour about 0.5 cm of stacking gel.
7. Overlay with *n*-butanol again.
8. Allow to polymerize for at least 8 h (*see* **Note 20**).

3.4.3. Setting Up and Running the Second Dimension

1. Microwave 10 mL of autoclaved 1% low melting agarose in 1× stacking gel buffer until it starts to boil. (We also add a trace amount of bromophenol blue.) Place an IEF strip on top of the second dimension gel plastic side down, acidic side facing left. Then rotate the IEF strip so that the plastic backing is parallel to the face of the back gel plate and slide it down vertically while making sure the plastic side, not the gel side, is contacting to the glass plate (*see* **Note 21**).

2. Push the strip down until it is in firm contact with the top of the SDS gel. Add the agarose until it barely covers the top of the strip. Make sure to add evenly from both sides and to avoid air bubbles. Burst bubbles with gel loading pipettor tips after the agarose hardens. Make sure to mark the second dimension gel with sample ID.
3. Electrophorese in the cold room. For a run of about 8–10 h, use 20 mA per gel at constant voltage. For a run of about 16 h, use 8–10 mA per gel.

3.5. Image Acquisition and Analysis

The layout of the imaging system is vital for the success of DIGE. The technique has unique requirements, so that all the hardware and most of the software for image acquisition had to be built from scratch. Our system is experimental and transitory in nature. At the time of the writing, the apparatus is under constant revision and development and is not commercially available in its current form. Thus, rather than give the exact minutiae of the image acquisition and analysis process, we list the nature and aim of the operations that are performed to arrive at the final result. Hopefully, this will assist those who are interested in either building or acquiring their own imaging system.

At the writing of this manuscript, commercially available DIGE imagers are just beginning to be developed. For those who might consider either building or acquiring an imager, we list the absolute minimum requirements that the hardware must meet:

1. In addition to the obvious requirement that the gel not physically move during imaging, no changes in protein spot position due to optically induced deformations should occur while switching between channels. Thus, a multiple bandpass emission filter must be employed.
2. The imaging hardware needs to be sensitive enough to detect minimally labeled proteins. Since a cooled, coupled charge device (CCD) had been used previously to image in gel fluorescently labeled proteins *(7)*, we decided to use a scientific grade CCD camera. CCD cameras without cooling do not have the requisite sensitivity of low noise capabilities.
3. The imaging cabinet must be light-tight and illumination must be filtered through the appropriate filter sets.

In the current incarnation of the DIGE imager, the gels are placed flat on a black Plexiglas surface at the bottom of the cabinet. The camera is mounted vertically over the gel at about 30 cm away. Illumination is provided by two halogen lamps with fiberoptic leads mounted on top of the cabinet at ±60° incident angles to the bottom to provide an even field of illumination. A schematic of the hardware is shown in **Fig. 2**.

The standard image acquired by the imager is a 4×4 cm square made up from 256×256 pixels, each storing 65,000 gray levels as unsigned short integers.

3.5.1. Image Acquisition

1. After the second dimension run, the gel are removed from their cassettes and incubated in destain solution for a minimum of 1 h. Gels are placed under either destain solution or in 1% acetic acid in H_2O during imaging.
2. Two images are acquired from the central region of the gel, one with each excitation filter. A few spot intensities are compared and used to normalize the acquisition times for the two channels in tile mode. Tile mode is where the entire gel is imaged into a single file by stitching together thirty 4×4 cm squares to generate one 20×24 cm image. Two such images are generated, one for each channel. Each image corresponds to one of dyes and

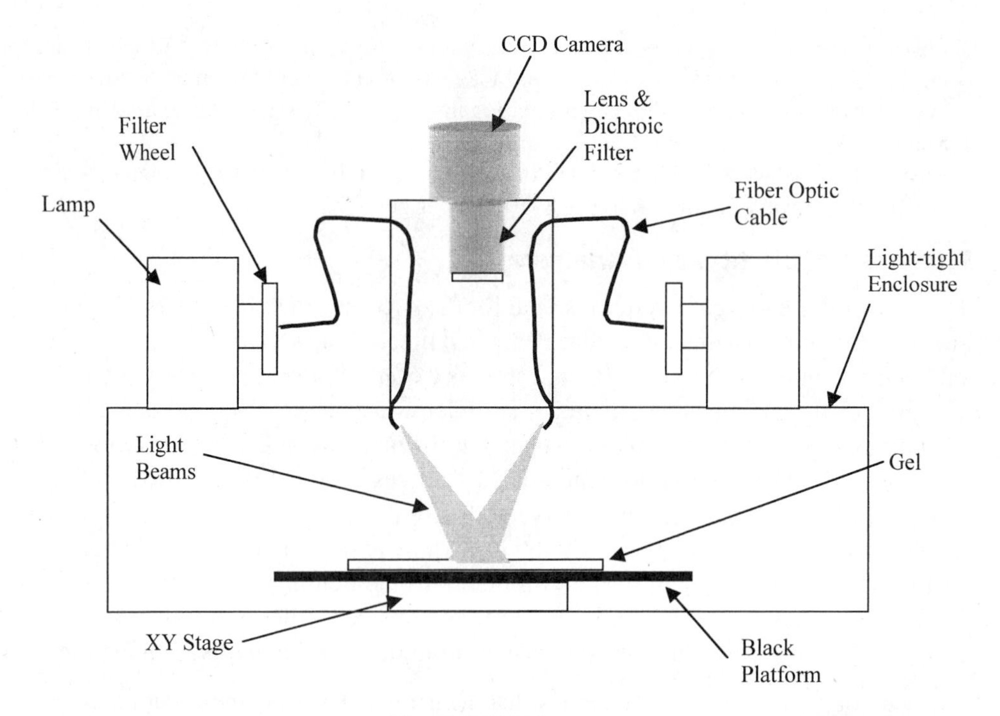

Fig. 2. DIGE Imager.

thus represents one of the samples which were run on the gel. Acquisition times typically vary between 10 and 180 s per square, which translates to 10–180 min total acquisition time per gels.

3.5.2. Image Analysis

1. The two images from a single gel are inverted and then normalized to each other so that the most abundant spots appear at the same level of intensity. The images are then converted to byte format and placed into a two-frame quicktime movie. Playing this movie in a continuous loop allows for the visual detection of differences. The images are normalized at several different grayscale levels. Normalizing for high values allows for the detection of the abundant protein changes and normalizing at lower values covers the lower-abundant protein changes.

4. Notes

1. A mixture of urea and thiourea has been reported to aid in the solubilization of membrane proteins ***(8)***.
2. Never warm a protein lysate in urea; always endeavour to keep at least on ice, if not frozen. Minimize the time samples spend away from –80°C. This is because at high temperature and pH urea spontaneously breaks down to yield cyanate, which modifies lysine residues and leads to carbamylated charge trains in the IEF dimension. Low temperature slows down but does not stop this process ***(9)***.
3. Usually the pH of HEPES is adjusted with KOH; however, if SDS is used in any subsequent step (such as running lysate directly on SDS-PAGE), KDS will be formed, which is insoluble in water. Use NaOH. When making up this buffer, make sure to add the HEPES last, *see* **Note 2**.

4. The presence of Pharmalytes in the lysis buffer interferes with labeling as some contain amines which react with and inactivate the dye. Similarly, the presence of any other primary amine containing compound (such as Tris) should be avoided.
5. As in **Note 2**, there's a danger of cyanate formation in this buffer. It is thus preferable to add the acetic acid at the same time as the urea. The lower pH will slow down the breakdown process.
6. This buffer may be reused up to three times or until the DTT is no longer detectable by its smell, whichever comes first.
7. We typically do not use protease inhibitors. The combination of the lysis buffer with its reducing ability, the chaotropic effects of the urea and the surfactant, and the cold temperature seems to inactivate proteolytic activity. We also do not perform any steps requiring room temperature or protein activity (such as the DNAse/RNAse treatment found in some protocols). Furthermore, the presence of the inhibitors may sometimes interfere with the fluorescent labeling.
8. The sample cups (*see* **Subheading 3.2.**) on the IEF gel have about 100 μL maximum capacity. However, if necessary, more volume can be handled by ordering more sample cup holder bars separately from the Dry strip kit and spreading one sample between several cups. As IEF is a focusing technique, sample does not necessarily be all applied in exactly the same spot.
9. The dye synthesis is detailed elsewhere *(6)*. The dyes are not commercially available as of the time of writing.
10. Try not to get any water on the gel side as this makes the gel very sticky, which causes problems when the strip is being rolled onto the glass plate. If the gel side does get wet, sometimes the situation can be saved by dabbing the water gently with a paper towel. Take care to not touch the paper to the gel. If you know you have sticky gels, then hold down the strips at the basic end while rolling.
11. Rehydration should continue for a minimum of 8 h. The gels should not be used after 24 h in the rehydration solution.
12. Getting the wicks to the exact size is not that important—eyeball accuracy is good enough. Also, even though Pharmacia supplies and sells "special" wick paper, we have found that 3MM Whatman chromatography paper will work just as well.
13. This is probably one of the more tricky aspects of the IEF procedure. It is important not to get the wicks too dry or to leave them too wet. Sparking is usually caused by the wicks being at the wrong dampness. Following these steps usually works, but work pace can play a role, as can the ambient humidity. One way of checking to see if the wicks are "correct" is to remove one from the damp patch and touch it lightly to a dry area on the paper. If it leaves behind a slight imprint, then it is most likely at the right dampness. If too wet, place firmly and dry a little more before placing on the gel. If too dry, rehydrate the wicks and repeat, waiting for half the time as the first try.
14. The Teflon membranes replace the paraffin oil which is normally used to cover the gels to isolate and keep the gels dry. We have found that using the oil produces a host of problems of its own. The membranes work just as well and are easy to use and remove without the messiness of the oil.
15. It is preferable to use a swingout rotor to get a more compact pellet and minimize loss of yeast cells. We use a benchtop centrifuge with a swingout "TechnoSpin" rotor from Sorvall Instruments.
16. Do not press on the sample cups too hard. The gel is quite delicate and making a dent or hole in it is as deleterious as not having the sample cup seated properly. Even if the sample does leak, this is not as big a tragedy as it might seem at first. As long as the sample stays

in the same groove as the gel, and the current is started promptly, most of the sample will still go in.

17. For 13 cm long dry strips, 20 kVh total also produces acceptable results in most cases.
18. A small disposable Petri dish makes a great receptacle for equilibration. Wrap the gel around the inside of the dish, with the gel side pointing in. The dishes may also be marked to easily keep track of the identities of multiple gels.
19. Before starting to pour the gradient gels, make sure the gradient maker chamber is free of polyacrylamide pieces. The tube and mixing channel can easily be blocked by small pieces of gel. Working fast, four gels can be poured one after the other. No washing of the chamber in between gels is necessary.
20. The 8 h is to allow the polymerization reaction to go to completion. In a crunch, this time may be shortened to as little as 1 h, but be aware that side chain and N-terminal modification by acrylamide then becomes more of a problem. Gels may be allowed to polymerize as long as overnight provided that the butanol layer is increased so that it does not dry out. We have also stored gels at 4°C for up to 24 h.
21. If the IEF gel has dried, this may cause a problem as the gel has a higher tendency to stick to the glass and getting torn. Should this start happening, try wetting the IEF strip a little with equilibration buffer II.

References

1. Wasinger, V. C., Cordwell S. J., Cerpa-Poljak, A., Yan, J. X., Gooley, A. A., Wilkins, M. R., et al. (1995) Progress with gene-product mapping of the Mollicutes: *Mycoplasma genitalium. Electrophoresis* **16,** 1090–1094.
2. Klose, J. (1975) Protein mapping by combined isoelectric focusing and electrophoresis in mouse tissues. A novel approach to testing for induced point mutations in mammals. Humangenetik 26, 231–243.
3. O'Farrell, P. H. (1975) High resolution two-dimensional electrophoresis of proteins. *J. Biol. Chem.* **250,** 4007–4021.
4. Scheele, G. A. (1975) Two dimensional gel analysis of soluble proteins. Characterization of guinea pig exocrine pancreatic proteins. *J. Biol. Chem.* **250,** 5375–5385.
5. Görg, A., Postel, W., and Günther, S. (1988) The current state of two-dimensional electrophoresis with immobilized pH gradients. *Electrophoresis* **9,** 531–546.
6. Ünlü, M., Morgan, M. E.,and Minden, J. S. (1997) Difference gel electrophoresis: A single gel method for detecting changes in protein extracts. *Electrophoresis* **18,** 2071–2077.
7. Urwin, V. E. and Jackson, P. (1993) Two-dimensional polyacrylamide gel electrophoresis of proteins labeled with the fluorophore monobromobimane prior to the first-dimensional isoelectric focusing: imaging of the fluorescent protein spot patterns using a cooled charge-coupled device. *Analyt. Biochem.* **209,** 57–62.
8. Rabilloud, T., Adessi, C., Giraudel, A., and Lunardi, J. (1997) Improvement of the solubilization of proteins in two-dimensional electrophoresis with immobilized pH gradients. *Electrophoresis* **18,** 307–316.
9. Hagel, P., Gerding, J. J. T., Fieggen, W., and Bloemendal, H. (1971) Cyanate formation in solutions of urea. *Biochim. Biophys. Acta* **243,** 366–379.

26

Comparing 2-D Electrophoresis Gels Across Internet Databases

Peter F. Lemkin and Gregory C. Thornwall

1. Introduction

In **refs.** *1–3* and in the previous edition of this book, we described a computer-assisted visual method, Flicker, for comparing two two-dimensional (2-D) protein gel images across the Internet, http://www.lecb.ncifcrf.gov/flicker/. This approach may be useful for comparing similar samples created in different laboratories to help putatively identify or suggest protein spot identification. Two-dimension gels and associated databases are increasingly appearing on the Internet *(4–17)* in World Wide Web (Web) servers *(18)* and through federated databases (DB) *(19)*. As this is an update on Flicker, we will not review the recent literature on 2-D gel databases or 2-D gel analysis systems here. **Table 1** lists some Web URL addresses for a number of 2-D protein gel databases that contain 2-D gel images with many identified proteins. This opens up the possibility of comparing one's own experimental 2-D gel image data with gel images of similar biological material from remote Internet databases in other laboratories. The image analysis method described here allows scientists to more easily collaborate and compare gel image data over the Web.

When two 2-D gels are to be compared, simple techniques may not suffice. There are a few ways to compare two images: (1) slide one gel (autoradiograph or stained gel) over the other while back lighted; or (2) build a 2-D gel quantitative computer database from both gels after scanning and quantitatively analyzing these gels using publicly available research (*[20–27]*) or commercial systems (recent systems are not reviewed here as the primary goal of this revision is the update on Flicker). These methods may be impractical for many investigators because the first case the physical gel or autoradiograph from another lab may not be locally available. In the latter case, the method may be excessive if only a single visual comparison is needed because of the costs (labor and equipment) of building a multigel database solely to answer the question of whether one spot is probably the same spot in the two gels.

This distributed Flicker gel comparison program runs on any World Wide Web-connected computer. It is a Java applet invoked from the user's Java-capable Web browser where it is then loaded from the NCI Flicker Web server. Alternatively, it may be downloaded to the user's computer and run locally (*see* **Subheading 2**, **step 26**).

From: *The Protein Protocols Handbook, 2nd Edition*
Edited by: J. M. Walker © Humana Press Inc., Totowa, NJ

Table 1
Partial List of World Wide Web 2-D Electrophoretic Gel Databases
Individual gel images with identified proteins are available within these databases. An on-line version of this table is available at **http://www.lecb.ncifcrf.gov/EP/table2Ddatabases.html**
The WORLD-2DPAGE Index to 2-D PAGE databases is available at **http://expasy.cbr.nrc.ca/ch2d/2d-index.html.**

SWISS-2DPAGE
Liver, plasma, HepG2, HepG2SP, RBC, lymphoma, CSF, macrophage-CL, erythroleukemia-CL, platelet, yeast, E.coli, colorectal, kidney, muscle, macrophage-like-CL, Pancreatic Islets, Epididymus, dictyostelium
http://www.expasy.ch/

Argonne Protein Mapping Group
Mouse liver, human breast cell lines, pyrococcus
http://www.anl.gov/BIO/PMG/

Danish Centre for Human Genome Research
Human: primary keratinocytes, epithelial, hematopoietic, mesenchymal, hematopoietic, tumors, urothelium, amnion fluid, serum, urine, proteasomes, ribosomes, phosphorylations.
Mouse: epithelial, new born (ear, heart, liver, lung)
http://biobase.dk/cgi-bin/celis/

Joint Protein Structure Lab
Human Colorectal-CL, Placental lysosomes
http://www.ludwig.edu.au/jpsl/jpslhome.html

UCSF 2D PAGE
A375 melanoma cell line
http://rafael.ucsf.edu/2DPAGEhome.html

ECO2DBASE
E.coli
http://pcsf.brcf.med.umich.edu/eco2dbase/
or ftp://ncbi.nlm.nih.gov/repository/ECO2DBASE/

PROTEOME Inc
Yeast
http://www.proteome.com/

Yeast 2D gel DB, Bordeaux
Yeast
http://www.ibgc.u-bordeaux2.fr/YPM

HSC-2DPAGE, Heart Science Centre, Harefield Hospital
Human, rat and mouse heart
http://www.harefield.nthames.nhs.uk/

HEART-2DPAGE, German Heart Inst. Berlin
Human heart
http://www.chemie.fu-berlin.de/user/pleiss/

HP-2DPAGE, MDC, Berlin
Human heart
http://www.mdc-berlin.de/~emu/heart/

Immunobiology, Univ. Edinburgh
Embryonal stem cells
http://www.ed.ac.uk/~nh/2DPAGE.html

Table 1 (*continued*)

Large Scale Biology Corp
Rat, mouse, human liver, corn, wheat
http://www.lsbc.com/
Maize Genome Database, INRA
Maize
http://moulon.moulon.inra.fr/imgd/
Univ. Greifswald
Bacillus subtilis
http://pc13mi.biologie.uni-greifswald.de/
IPS/LECB, NCI/FCRDC
Phosphoprotein, prostate, phosphoprotein, breast cancer drug screen, FAS (plasma), Cd toxicity (urine), leukemia
http://www.lecb.ncifcrf.gov/ips-databases.html
Washington Univ. Inner Ear Protein Database
Human: Inner Ear
http://oto.wustl.edu/thc/innerear2d.htm
Protein Project of Cyanobacteria
Cyano2Dbase - Synechocystis sp. PCC6803
http://www.kazusa.or.jp/cyano/cyano2D/
2-D PAGE Aberdeen
Haemophilus influenzae & Neisseria meningitidis
http://www.abdn.ac.uk/~mmb023/2dhome.htm
Lab. de Biochimie et Tech. des Proteines, Bobigny
Human leukemia cell lines
http://www-smbh.univ-paris13.fr/lbtp/biochemistry/biochimie/bque.htm
Max-Planck-Institut f. Infektionsbiologie
Mycobacterium tuberculosis, vaccine strain M. bovis BCG,
http://www.mpiib-berlin.mpg.de/2D-PAGE/
ToothPrint DB
Dental tissue in rat
http://bioc111.otago.ac.nz:8001/tooth/home.htm
Siena 2D-PAGE
Chlamydia trachomatis L2, Caenorhabditis elegans, Human breast ductal carcinoma and Histologically normal tissue, Human amniotic fluid
http://www.bio-mol.unisi.it/2d/2d.html
PHCI-2DPAGE
Parasite host cell interaction, IFN-gamma induced HeLa cells
http://www.gram.au.dk/
PMMA-2DPAGE
Human colorectal carcinoma
http://www.pmma.pmfhk.cz/
BALF 2D_AGE
Mouse, Human broncho-alveolar lavage fluid
http://www.umh.ac.be/~biochim/BALF2D.html
Mito-pick
Human Mitochondria
http://www-dsv.cea.fr/thema/MitoPick/Mito2D.html

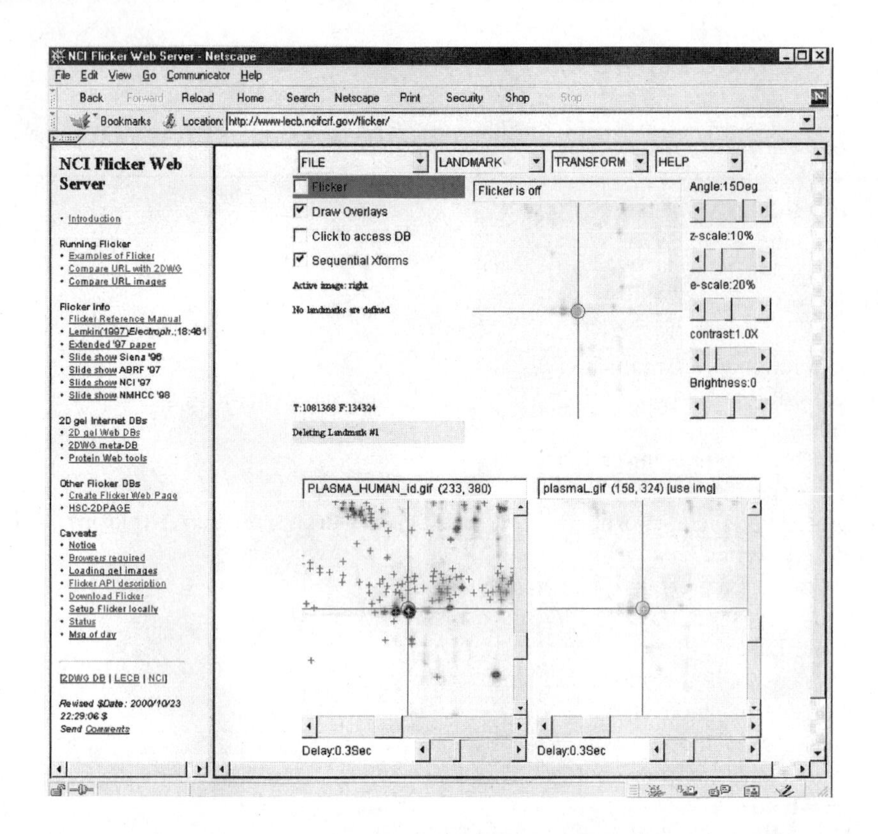

Fig. 1. Screen view of Netscape running the Flicker Java applet. Control-menus at the top invoke file operations, landmarking, image transforms. A set of scroll bars on the right determine various parameters used in the transforms. File menu options includes resetting images after a transform, aborting the current transform in progress, and help. Check-boxes on the left activate flickering and control display options. The "click to access DB" checkbox is available if one of the gel images is linked to a federated 2-D gel database. When this is set, clicking on a spot will get a pop-up browser report on that protein from the federated DB. A set of status lines appear below the check-boxes and indicate the state of operation. The flicker image is in the upper middle of the frame when it is enabled. The two labeled human blood plasma gel images are shown in the bottom scrollable windows that may be positioned to the region of interest. These windows also have associated flicker time-delays used when flickering. Image plasmaH (or PLASMA_HUMAN_id) is an IPG non-linear gradient gel from SWISS-2DPAGE in Geneva and plasmaL is a carrier-ampholyte linear gradient gel from the Merril Lab at NIMH. Transformed image results are shown in the same scrollable windows.

One gel image is read from any Internet 2-D gel database (e.g., SWISS-2DPAGE, etc.), the other may reside on the investigator's Web server where they were scanned or copied; or the two gel images may be from either Web server source. Portions of this paper were derived from **ref. *3***.

Flicker is also capable of interacting with federated 2-D gel databases to retrieve data on individual protein spots. Once gels are aligned in Flicker, you enable federated DB access; click on a spot in the gel belonging to the federated DB (*see* **Fig. 1**). This causes a Web page to pop up with information from the federated server describing that protein. We have set up a Web page to let users compare a gel from the Web (or their Web server if the gel image was copied there) to some of the SWISS-2DPAGE gels, http://www.lecb.ncifcrf.gov/flicker/swissProtIdFlkPair.html. First, select the resolution you want to use (it defaults to 1×). Second, select the SWISS-2DPAGE gel image to use using the pull-down menu (e.g., select Human Plasma). Third, enter the URL of the gel on the Web that you wish to compare. As a convenience, we show a sample URL of a human plasma gel from a NIMH database to illustrate how you should enter the URL. You could use this gel to demonstrate how it works. Enter the URL, after which press "Go Flicker." This methodology could be replicated with any federated 2-D gel server and is discussed in more detail in the on-line Flicker API description available on the main Flicker Web page.

The Flicker program is written in Java, a general purpose, object-oriented programming language developed by Sun Microsystems *(28)* http://java.sun.com/. Java has become a standard for portable Internet Web applications. A Java "applet" is the name Sun gave for a mini-application that runs inside of a Web browser. When a user accesses a Web page containing an applet reference, it automatically loads the applet into the user's Java-capable.

Normally, users interact with the NCI Flicker Web server located at http://www.lecb.ncifcrf.gov/flicker, using the client/server paradigm shown in **Fig. 2**. The Flicker program may be thought of as a client that makes requests of a 2-D gel database Web server. Because it runs on the user's computer, Java now gives us the ability to perform real-time comparisons of local 2-D gel image data with gel images residing in various remote databases on the Internet. Then Flicker will load two images regardless of their respective source. Sources include: data from the Flicker Web server, other Web servers, or locally. The latter case applies when the applet file is copied to your local computer, you can create a HTML document to start Flicker on your local data (discussed in **Subheading 2, step 2b**).

Although the original images may be compared directly, this may be more easily achieved by first applying spatial warping or other image enhancement transforms. For gels with a lot of geometric distortion, it is useful to adjust one gel so that the geometry of the local regions being compared match that of the other gel. By local geometry, we mean the relative positions, distances, and angles of a set of spots in corresponding regions. One technique to do this is called spatial warping. When doing spatial warping, regions of interest are (1) landmarked with several corresponding points in each gel image in the region of interest, and (2) then one gel image is warped to the geometry of the other gel (*see* **Eqs. 1** and **2**). A landmark is a corresponding spot that is present in both gels. Spatial warping does not change the underlying grayscale values of the synthesized warped image to the extent that cause local structural objects to would appear and disappear and thus spot artifacts might be created. Instead, it samples pixels from the original image to be transformed and places them in the output image according to the geometry of the other input image.

Gels are then compared by flickering them rapidly by alternately displaying them in a third "flicker" display window. Using the mouse, the user may drag one gel image over the other to visually align corresponding spots by matching local morphology.

1.1. Image Flickering

The basic concept of using flickering as a dynamic visualization technique is simple. If two images may be perfectly aligned then one could simply align them by overlaying one over the other and shifting one image until they line up. However many images

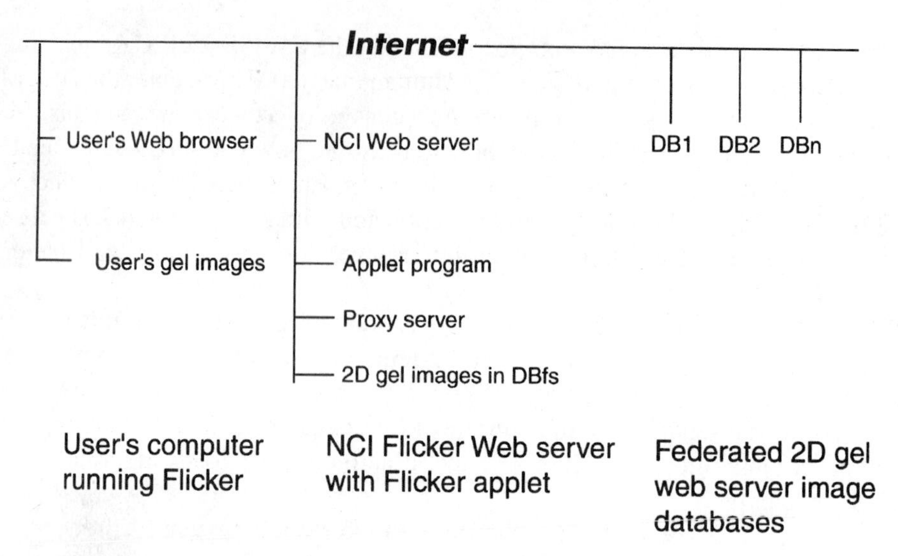

Fig. 2. This illustrates the client–server relationship between the user's Web browser with local 2-D gel images, the Web server that contains the Flicker program, and the 2-D gel image Web databases (DB) on other Web servers. Two gels to be compared may come from the Internet Web databases or from the user's local file system. The images may be from either the NCI Flicker 2-D gel image DB Web server DB_{fs} or from other 2-D gel image Web databases DB_1, DB_2, ..., DB_n. For example, DB_1 might be the SWISS-2DPAGE database, DB_2 might be the Danish Keratinocyte database, DB_3 might be the Cambridge heart database, DB_4 might be the Argonne breast cell line database, etc. For gel images on non-NCI servers, the request goes to the NCI server which contacts the other servers to get the image data and then sends it back to the user's browser.

such as 2-D polyacrylamide gel electrophoresis (PAGE) gels have "rubber-sheet" distortion (i.e., local translation, rotation, and magnification). This means there is more distortion in some parts of the image than in others. Although it is often impossible to align the two whole images at one time, they may be locally aligned piece-by-piece by matching the morphology of local regions.

If a spot and the surrounding region do match, then one has more confidence that the objects are the same. This putative visual identification is our definition of matching when doing a comparison. Full identification of protein spots requires further work such as cutting spots out of the gels and subjecting them to sequence analysis, amino acid composition analysis, mass spectrometry, testing them with monoclonal antibodies, or other methods.

1.2. Image Enhancement

It is well known that 2-D gels often suffer from local geometric distortions making perfect overlay impossible. Therefore, making the images locally morphologically similar while preserving their grayscale data may make them easier to compare. Even when the image sub regions are well aligned, it is still sometimes difficult to compare images that are quite different. Enhancing the images using various image transforms before flickering may help. Some of these transforms involve spatial warping, which maps a local region of one image onto the geometry of the local region of another image while preserving its grayscale values. Other transforms include image sharpening and contrast enhancement. Image sharpening is performed using edge enhancement techniques such as adding a percentage of the gradient or Laplacian edge detection functions to the original grayscale image. The gradient and Laplacian have higher values at the edges of objects. Another useful operation is contrast enhancement that helps when comparing light or dark regions by adjusting the dynamic range of image data to the dynamic range of the computer display. In all cases, the transformed image replaces the image previously displayed. Other functionality is available in Flicker and is described in the Flicker on-line "help," **Subheading 3.** of this chapter, and in **refs.** ***1,3***.

1.3. Image Processing Transforms

As mentioned, there are a number of different image transforms that may be invoked from the control panel.

1.3.1. Affine Spatial Warping Transform

The spatial warping transforms require defining several corresponding landmarks in both gels. As we mentioned, one gel image can be morphologically transformed to the geometry of the other using the affine or other spatial warping transformations. These transforms map the selected image to the geometry of the other image. It does not interpolate the gray scale values of pixels — just their position in the transformed image. As described in **refs.** ***1,3***, this might be useful for comparing gels that have some minor distortion, comparing local regions, gels of different sizes or gels run under slightly different conditions. Flicker uses the affine transform as an inverse mapping as described in ***(29)***. Let $(u_{xy},v_{xy}) = f(x, y)$ where (x, y) are in the output image, and (u, v) are in the input image. Then, in a raster sweep through the output image, pixels are copied from in the input image to the output image. The affine transformation is given in **Eqs. 1–2**:

$$u_{xy} = ax + by + c \quad (1)$$

$$v_{xy} = dx + ey + f \quad (2)$$

When the affine transform is invoked, Flicker solves the system of six linear equations for coefficients (a, b, c, d, e, f) using three corresponding landmarks in each gel.

1.3.2. Pseudo-3-D Transform

As described in **refs.** ***1,3***, the pseudo-3-D transform is a forward mapping that generates a pseudo 3-D relief image to enhance overlapping spots with smaller spots seen as side peaks. The gel size is width by height pixels. The gray value determines the amount of y shift scaled by a percentage z_{scale} (in the range of 0–50%). Pseudo-perspective is created by rotating the image to the right (left) by angle theta (in the range of –45 to +45 degrees). The transform is given in **Eqs. 3–5** for image of size width X height, shift in the horizontal dimension computed as d_x.

$$d_x = \text{width} \sin(\theta) \quad (3)$$

$$x' = [d_x (\text{height} - y)/\text{height}] + x \quad (4)$$

$$y' = y - z_{scale} * g(x, y) \quad (5)$$

where $g(x, y)$ is in the original input image and (x', y') is the corresponding position in the output mapped image. Pixels outside of the image are clipped to white. The pseudo-3-D transform is applied to both images so that one can flicker the transformed image.

1.3.3. Edge Sharpening

Edge sharpening may be useful for sharpening the edges of fuzzy spots. The sharpened image function $g'(x, y)$ is computed by adding a percentage of a 2-D edge function of the image to original image data $g(x, y)$ as shown in **Eq. 6**. The edge function increases at edges of objects in the original image and is computed on a pixel by pixel basis. Typical "edge" functions include the eight-neighbor gradient and Laplacian functions that are described in **ref. *1*** in more detail. The e_{scale} value (in the range of 0–50%) is used to scale the amount of edge detection value added.

$$g'(x, y) = (e_{scale} * \text{edge}(x, y) + (100 - e_{scale}) * g(x, y)/100 \quad (6)$$

2. Materials

The following lists all items necessary for carrying out the technique. Since it is a computer technique the materials consist of computer hardware, software and an Internet connection. We assume the user has some familiarity with computers and the World Wide Web.

1. A computer with a Java-compatible browser and an Internet connection is required. The actual computer could be a Windows-PC, Macintosh or Unix X-window system. The computer should have a minimum of 16 Mbytes of memory or more since intermediate images are held in memory when image transforms occur. If there is not enough memory, it will be unable to load the images, the transforms may crash the program or other problems may occur. Because a lot of computation is being performed, a computer with at least the power of an Intel 486/66 PC or better is suggested. A Pentium class machine with 32 Mb is more than adequate.

2a. To use Flicker directly from the NCI Flicker server, each time it is invoked it loads the applet into your Java-compatible Web browser. Browsers such as Netscape version 4.6 or later or Internet Explorer version 4.0 or later are required.

2b. Alternatively, if you decide to run Flicker locally, then you will need to download the Flicker Jar file and copy it to a directory with your GIF images. You will then need to edit an HTML file that invokes Flicker and also indicates that two GIF images you want to use. You may download the README, FlkJ2.jar, sample HTML, and other files from the http://www.lecb.ncifcrf.gov/FlkMirror Web site. There is a "tar" file at this site which bundles these files. You might also read the Flicker application programming interface (API) document located at http://www.lecb.ncifcrf.gov/flicker/flkParamList.html.

3. You will need a list of specific GIF image URLs from Internet 2-D gel image databases and/or copies of locally scanned gel images in GIF format. You can use the list of 2-D gel Web databases in **Table 1** as a starting point for finding gel GIF images you could download with Flicker. You might also investigate the ExPASy 2-D-hunt Web page which is a

search engine for 2-D gel electrophoresis Web sites at http://expasy.cbr.nrc.ca/ch2d/2DHunt/.

4. If the investigator will be using his or her own scanned gels, they will need to either have the gel scanner on the machine where Flicker will run or arrange to transfer the image files to that machine. In addition, the gel images may need to be converted to GIF images required by Flicker software (e.g., many scanners generate TIFF formatted images). Image format conversion software may be part of your scanner software. If not, there are a number of image file format converters available as part of various desktop publishing packages. Also some converter software is available free from the Internet (use a Web search engine such as Alta Vista to help find it).

3. Method

We now describe the operation of the Flicker applet from the point of view of the user. You first start up Flicker. This may include the specification of particular images from the NCI Flicker server. Otherwise, you need to specify the gel images to load once Flicker is running. Then you simply flicker the gel images or use image enhancement transforms first and then flicker them.

3.1. Using Flicker with the NCI Gel Image Proxy Server

3.1.1. Using Flicker Under a Web browser — No Installation of Any Software is Needed!

Assuming you have a working Web browser on your Internet-connected computer, there is nothing to do since Flicker is automatically downloaded into your browser every time you invoke its URL on the NCI Web server.

Using the NCI proxy server, you can use the following Web page to request two image URLs that you enter (http://www.lecb.ncifcrf.gov/flicker/urlFlkPair.html) and it will get the images and start Flicker in your Web browser.

Alternatively, you can compare a Web image URL against one of the images in the 2DWG gel image database *(2)* (http://www.lecb.ncifcrf.gov/2dwgDB). Search the 2DWG. Select ONE gel image you want to compare with one from the Web. Then scroll down to the bottom of the search results page and type in the image URL on the Web. Finally, press the "Go Flicker" button.

For those who are interested in the details on how Flicker is invoked, you can see some of the HTML (HyperText Markup Language) examples of how to start Flicker with pre-specified images from HTML in files linked from the Flicker home page at http://www.lecb.ncifcrf.gov/flicker/# Flicker-examples. There are additional examples given in the Flicker API. These use the HTML <APPLET> and </APPLET> tags. The following is an example of the <APPLET> HTML code required to start Flicker with two images.

```
<APPLET
CODEBASE=http://www.lecb.ncifcrf.gov/flicker/
CODE=FlkJ2.class ARCHIVE=FlkJ2.jar
WIDTH=650 HEIGHT=700 ALIGN=absmiddle
ALT="A java-enabled browser is needed to view Flicker applet.">
<PARAM NAME=image1 VALUE=plasmaH.gif>
<PARAM NAME=image2 VALUE=PlasmaL.gif>
</APPLET>
```

Note that in the example, CODEBASE was set to the NCI Flicker server. If you run it on your server, simply use the URL where the Flicker FlkJ2.jar file and GIF images reside. If your are running this applet with your browser on a local disk, then omit the CODEBASE line and change the image1 and image2 VALUE's to the names of your GIF image files.

3.2. Graphical User Interface for Flickering

Figure 1 shows the screen of the Flicker applet as seen from a Netscape browser. Control pull-down menus at the top invoke file operations, landmarking, image transforms. Scroll bars on the side determine various parameters used in the transforms. The two images to be compared are loaded into the lower scrollable windows. A flicker window appears in the upper middle of the screen. Check-boxes on the left activate flickering and control display options. A group of status lines below the check-boxes indicate the state of operations.

Only part of an image is visible in a scrollable window. This subregion is determined by setting horizontal and vertical scroll bars. Another, preferred, method of navigating the scrollable images is to click on the point of interest while the CONTROL key is pressed. This will recenter the scrollable image around that point. This lets the user view any sub-region of the image at high resolution. These images may be navigated using either the scroll bars or by moving the mouse with the button pressed in the scrollable image window. Then, each image in the flicker window is centered at the point last indicated in the corresponding scrollable image window.

A flicker window is activated in the upper-middle of the screen when the "Flicker" check-box is selected. Images from the left and right scrollable images are alternatively displayed in the flicker window. The flicker delay for each image is determined by the adjusting the scroll bar below the corresponding scrollable image window. Various graphic overlays may be turned on and off using the "Overlays" check-box.

Clicking on either the left or right image selects it as the image to use in the next transform. However, clicking on the flicker image window indicates the transform should be applied to both left and right images.

3.3. Loading Images

When Flicker is running under a Web browser, the names of the images are fixed and are specified in the HTML (as shown earlier).

3.4. Flickering

3.4.1. Use of Flicker for Comparing Images

When flickering two images with the computer, one aligns putative corresponding subregions of the two rapidly alternating images. The flicker-display overlays the same space on the screen with the two images and is aligned by interactively moving one image relative to the other using the cursor in either or both of the lower images. Using the mouse, the user initially selects what they suspect is the same prominent spot or object in similar morphologic regions in the two gel images. The images are then centered in the flicker window at these spots. When these two local regions come into alignment, they appear to pulse and the images fuse together. At this point, differences are more apparent and it is fairly easy to see which spots or objects correspond, which are different, and how they differ. We have found that the user should be positioned fairly close to the flicker window on the screen to optimize this image-fusion effect.

3.4.2. Selecting the Proper Time Delays When Flickering

The proper flicker delays, or time each image is displayed on the screen, is critical for the optimal visual integration of image differences. We have also found that optimal flicker rates are dependent on a wide variety of factors including: amount of distortion, similarity of corresponding subregions, complexity and contrast of each image, individual viewer differences, phosphor decay-time of the display, ambient light, distance from the display, and so forth. We have found the process of flickering images is easier for some people than for others.

When comparing a light spot in one gel with the putative paired darker spot in the other gel one may want to linger longer on the lighter spot to make a more positive identification. Because of this, we give the user the ability to set the display times independently for the two images (typically in the range of 0.01 s to 1.0 s with a default of 0.20 s) using separate "Delay" scroll bars located under each image. If the regions are complex and have a lot of variation, longer display times may be useful for both images. Differential flicker delays with one longer than the other are also useful for comparing light and dark sample gels.

3.5. Image Processing Methods

As mentioned, there are a number of different image transforms that may be invoked from the menus. These are useful for changing the geometry, sharpness, or contrast making it easier to compare potentially corresponding regions. As we go through the transforms we will indicate how they may be used. Some affect one image while some affect both. Flickering is deactivated during image transforms to use most computational power for doing the transforms.

The TRANSFORM menu has a number of selections that include warping, grayscale transforms and contrast functions. The two warp method selections: "Affine Warp and "Poly Warp" are performed on only one image (the last one selected by clicking on an image). Unlike the warp transforms, the grayscale transforms are performed on both images. These include: "Pseudo 3-D," "SharpenGradient," "SharpenLaplacian," "Gradient," "Laplacian," "Average." The contrast functions are "Complement" and "ContrastEnhance."

3.5.1. Landmarks: Trial and Active

The affine transform requires three active landmarks to be defined before it can be invoked. A trial landmark is defined by clicking on an object's center anywhere in a scrollable image window. This landmark would generally be placed on a spot. Clicking on a spot with or without the CONTROL key pressed still defines it as a trial landmark. After defining the trial landmark in both the left and right windows, selecting the "Add Landmark" option in the Landmark menu defines them as the next active landmark pair and identifies them with a red letter label in the two scrollable image windows. Selecting the "Delete Landmark" option deletes the last active landmark pair defined.

A

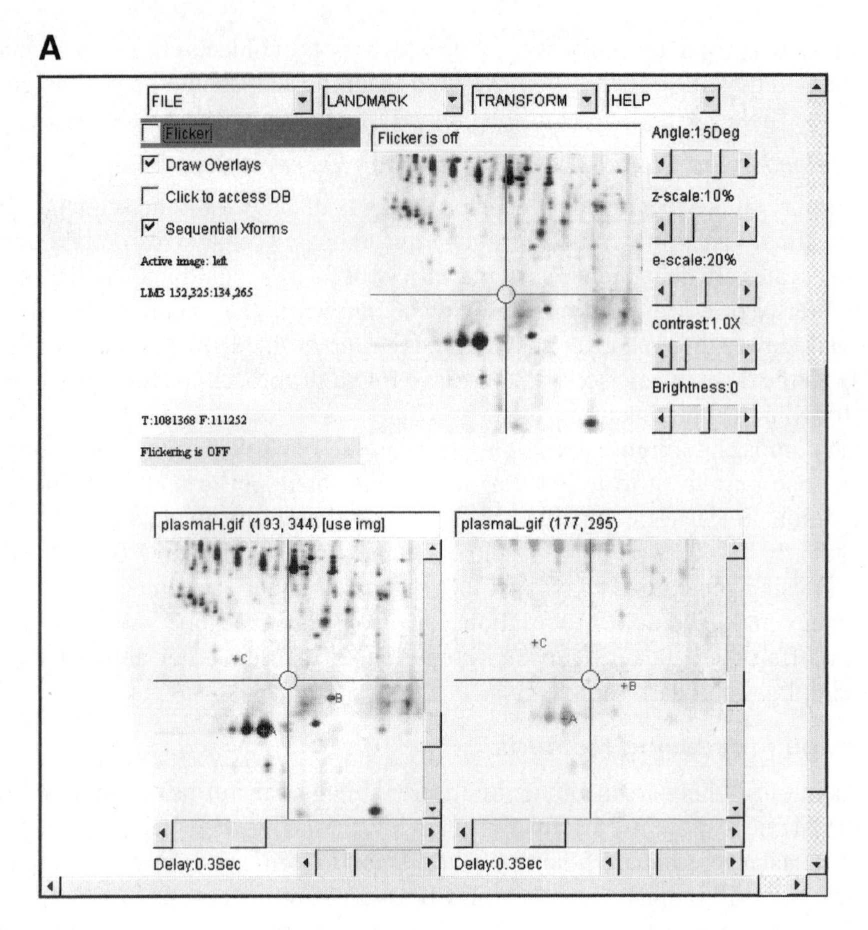

Fig. 3. Screen views of affine transform of human plasma gel image. The transform warps the geometry of a local region defined by the three landmarks so it more closely resembles the geometry of the corresponding local region in the other gel. **(A)** Scrollable image windows with three "active" landmarks defined in both gel images that were selected interactively in preparation for doing the affine image transform. Corresponding landmark spots are picked so as to be

3.5.2. Affine Transform

The two warping transforms, affine (*see* **Eqs. 1** and **2**) and polynomial, require three and six landmarks respectively. Attempting to run the transform with insufficient landmarks will cause Flicker to notify you that additional landmarks are required. The image to be transformed is the one last selected. You must select either the left or right image. **Figure 3** shows the landmarks the user defined in the two gels before the affine transform. **Figure 3b** shows the affine transform done on the right gel image.

3.5.3. Pseudo-3-D Transform

As described in (1) and as shown in **Eqs. 3–5**, the pseudo-3-D transform generates a pseudo-3-D relief image to enhance overlapping spots with smaller spots seen as side peaks. The gray value determines the amount of "*y*" shift scaled by a percentage (set by scroll bar z_{scale} (in the range of 0–50%). Pseudo perspective is created by shifting the image to the right or left by setting by scroll bar "angle" degrees (in the range of –45 to +45 degrees).

B

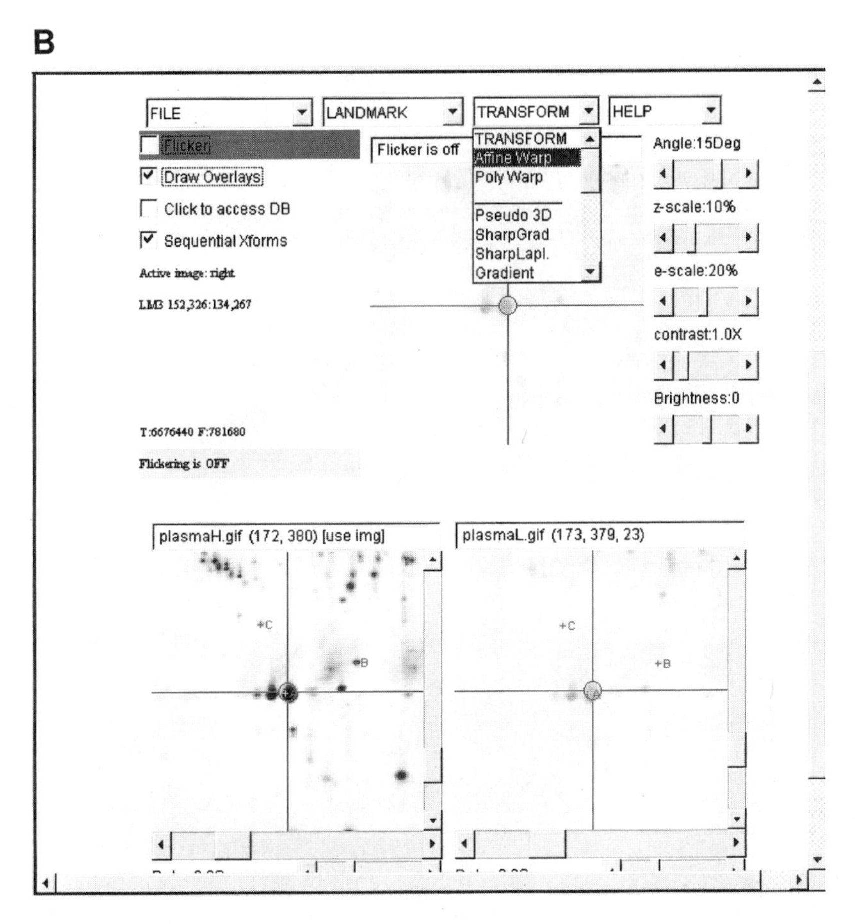

Fig. 3. (*continued from opposite page*) unambiguously defined in both gel images. **(B)** Scrollable image windows after the affine warp transform of the right (plasmaL, non-IPG) image to the geometry of the left (plasmaH, IPG gel) image. The plasmaH image is the same gel as PLASMA_HUMAN_id but without the graphic overlays.

Negative angles shift it to the right and positive angles to the left. The image to be transformed is the one last selected. If neither was selected (i.e., you clicked on the flicker-window), then both images are transformed. **Figure 4** shows the results of applying the pseudo-3-D transform to both images.

3.5.4. Edge Sharpening

Edge sharpening may be useful for improving the visibility of the edges of fuzzy spots. You can select either a Gradient or Laplacian edge sharpening function using the "SharpenGradient" or "SharpenLaplacian" operation in the "TRANSFORM" menu where the image to be transformed is the one last selected. The Laplacian filter generates a "softer" edge than the Gradient. You can set the scroll bar e_{scale} value (in the range of 0–50%) to scale the amount of edge detection value added. The image to be transformed is the one last selected. If neither was selected (i.e. you clicked on the flicker-window), then both images are transformed. **Figure 5** shows the results of applying the image-sharpening Laplacian transform to both images.

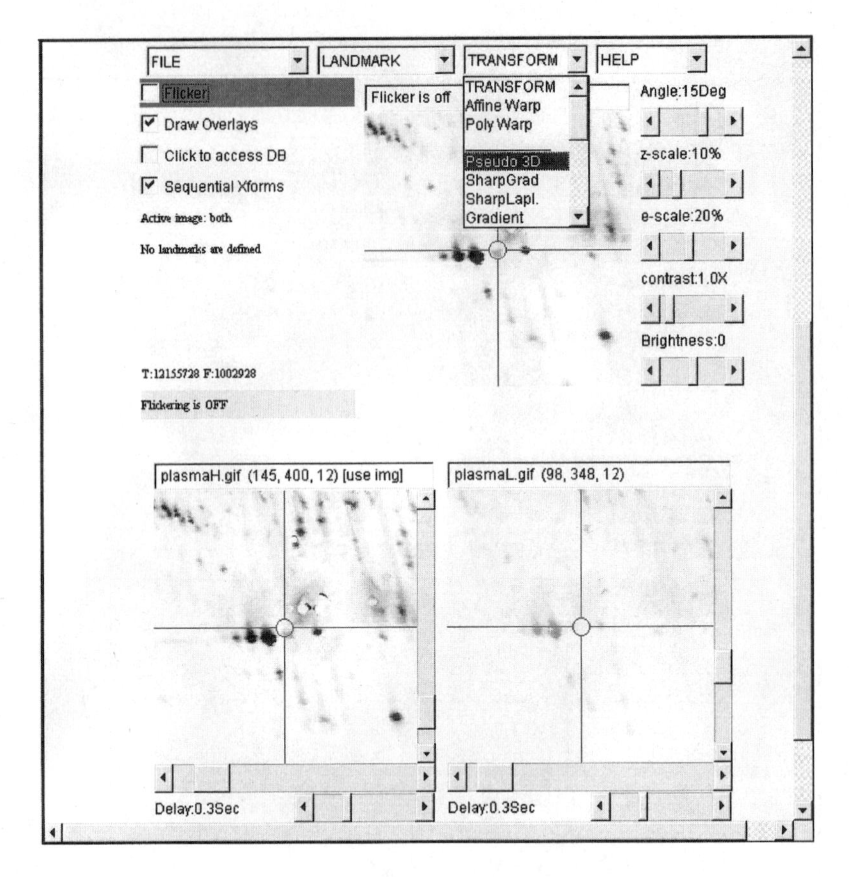

Fig. 4. Human plasma gel images after the pseudo-3-D transform was applied to both gel images. The parameter settings for angle and z_{scale} were 15 degrees and 20% respectively.

3.5.5. Other Image Transforms

There are a number of other image transforms which can be invoked. Like the edge sharpening transforms, the image to be transformed is the one last selected and if neither was selected (i.e., you clicked on the flicker-window), then both images are transformed.

4. Notes

1. *On installing Flicker on your computer*
   There are several advantages of running Flicker directly from the NCI Flicker server. Software updates are completely invisible to users since they don't have to waste time or space installing them on their computers. The technique uses existing low cost Web browser technology that requires little user effort. In addition, it saves time over the alternative ways that scientists might use to compare 2-D gels and other data. However, if you are far from our server, want to run Flicker behind a firewall, or want to use it exclusively on your own data, then it may be advantageous to download Flicker and install it on your computer (*see* **Subheading 2, step 2b**).

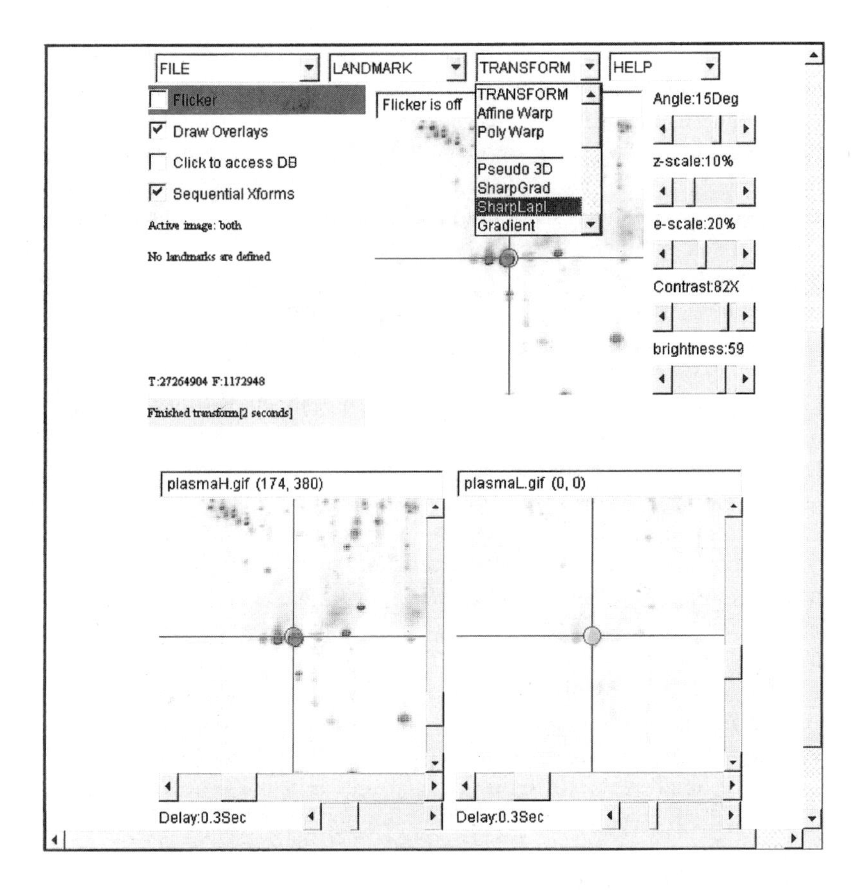

Fig. 5. Human plasma gel images after the SharpenLaplacian transform was applied to both gel images. The parameter settings for e_{scale} was 40%.

2. *On loading images*
   There are several problems with using Java with current Web browsers because of restrictions due to applet security restrictions. Because of fears of security breaches, Netscape and other Web browser providers have disabled Java applets running on their browsers from reading or writing local files. They also restrict access to Web URLs other than the host computer where the Java applet originated (i.e., in this case, the NCI Flicker Web server). Unfortunately, this prevents the Flicker applet from loading your local image files or other 2-D gel image databases not the Flicker Web server. It thus prevents you from comparing data from different sources.
   However, there are two ways to get around this security problem: (1) Use a Web proxy server (such as the NCI Flicker proxy server) to get the image and then pass it back to your browser as if the data came from the Flicker server. (2) Run Flicker as an applet in your Web browser using a local Web page that points to the data on your local computer or your Web server.
   Another restriction is that Flicker itself currently only handles GIF and JPEG image formats. Images in other formats such as TIFF currently need to be converted to GIF format. We do this on our NCI Flicker proxy server. If you are running a local copy, you

could convert your images to GIF using programs such as PhotoShop, and so forth. Because we are doing image pixel processing with the Flicker program, it requires more memory for intermediate images than programs that only manipulate text and so may require a more powerful CPU with more memory than some users currently have available.

3. *On flickering*

   There are some disadvantages in comparing gels this way. It is only good for doing a rough comparison and there is currently no simple way available to do quantitative comparison (as can be done with existing 2-D gel computer database systems *[20–27]*) — although we are working on the latter and will announce it on the Flicker Web site. One should keep these limitations in mind when using the technique.

   The intent of applying image transforms is to make it easier to compare regions having similar local morphologies but with some different objects within these regions. Image warping prior to flickering is intended to spatially warp and rescale one image to the "shape" of the other image so that we can compare them at the same scale. This should help make flickering of some local regions on quite different gels somewhat easier. Of the two warping transforms, affine and polynomial, the latter method handles nonlinearities better. For those cases where the gels are similar, the user which may be able to get away with using the simpler (affine) transform.

   In cases where there is a major difference in the darkness or lightness of gels, or where one gel has a dark spot and the other a very faint corresponding spot, it may be difficult to visualize the light spot. By differentially setting the flicker display-time delays, the user can concentrate on the light spot using the brief flash of the dark spot to indicate where they should look for the light spot. We have found differential-flicker to be very helpful for deciding difficult cases. Changing image brightness and contrast also is useful when flickering and the Flicker program has provision for interactively changing these parameters as well.

4. *On image transforms*

   Other transforms including image sharpening may be useful in cases where spots are very fuzzy, as might be the case when comparing Southern blots. When two corresponding local regions of the two images are radically different so the local morphologies are not even slightly similar (e.g., when high molecular weight regions of gels that are run differently as: IPG vs non-IPG, gradient vs non-gradient sodium dodecyl sulfate [SDS]), then even using these transforms may not help that much.

5. *The current status of Flicker*

   Of the features and operations we have mentioned, some are not fully functional and we are working to resolve this. The current state of Flicker is documented in the Flicker Reference Manual http://www.lecb.ncifcrf.gov/flicker/flkInfo.html. A future release of Flicker will contain quantification facility and be able to be run as a stand-alone Java application.

5. References

1. Lemkin, P. F. and Thornwall, G. (1999) Flicker image comparison of 2-D gel images for putative protein identification using the 2DWG meta-database. *Mol. Biotechnol.* **12,** 159–172.
2. Lemkin, P. F. (1997) 2DWG meta-database of 2D electrophoretic gel images on the Internet. *Electrophoresis* **18,** 2759–2773.
3. Lemkin, P. F. (1997) Comparing two-dimensional electrophoretic gels across the Internet. *Electrophoresis* **18,** 461–470.

4. Appel, R. D., Sanchez, J.-C., Bairoch, A., Golaz, O., Miu, M., Vargas, J. R., and Hochstrasser, D. F. (1993) SWISS-2DPAGE: a database of two-dimensional gel electrophoresis images. *Electrophoresis* **14,** 1232–1238.
5. Appel, R. D., Sanchez, J.-C., Bairoch, A., Golaz, O., Rivier, F., Pasquali, C., et al. (1994) SWISS-2DPAGE database of two-dimensional polyacrylamide gel electrophoresis. *Nucleic Acids Res.* **22,** 3581–3582.
6. Sanchez, J.-C., Appel, R. D., Golaz, O., Pasquali, C., Rivier, F., Bairoch, A., and Hochstrasser, D. F. (1995) Inside SWISS-2DPAGE database. *Electrophoresis* **16,** 1131–1151.
7. Jungblut, P., Thiede, B., Zinny-Arundl, U., Muller, E.-C., Scheler, C., Whittmann-Liebold, B., and Otto, A., (1996) Resolution power of two-dimensional electrophoresis and identification of proteins from gels. *Electrophoresis* **17,** 839–846.
8. Wilkins, M. R., Sanchez, J.-C., Williams, K. L., and Hochstrasser, D. F. (1996) Current challenges and future applications for protein maps and post translational vector maps in proteome projects. *Electrophoresis* **17,** 830–838.
9. Celis, J. E. (ed.). (1992) Special issue: Two-dimensional electrophoresis protein databases. *Electrophoresis* **13,** 891–1062.
10. Celis, J. E. (ed.). (1993) Special issue: Two-dimensional electrophoresis protein databases. *Electrophoresis* **14,** 1089–1240.
11. Celis, J. E. (ed.), (1994) Special issue: electrophoresis in cancer research. *Electrophoresis* **15,** 305–556.
12. Celis, J. E. (ed.), (1995) Special issue: two-dimensional electrophoresis protein databases. *Electrophoresis* **16,** 2175–2264.
13. Bjellqvist, B., Hughes, G. J., Pasquali, C., Paquet, N., Ravier, F., Sanchez, J.-C., et al. (1993) *Electrophoresis* **14,** 1023–1031.
14. Pallini, V., Bini, L., and Hochstrasser, D. (1994) Proceedings: 2D Electrophoresis: From Protein Maps to Genomes. University of Siena, Italy, Sept 5–7.
15. Pallini, V., Bini, L., and Hochstrasser, D. (1996) Proceedings: 2nd Siena 2D Electrophoresis Meeting: From Protein Maps to Genomes. University of Siena, Italy, Sept 16–18.
16. Pallini, V., Bini, L., and Hochstrasser, D. (1996) Proceedings: 3rd Siena 2D Electrophoresis Meeting: From Protein Maps to Genomes. University of Siena, Italy, Aug 31–Sept 3.
17. Pallini, V., Bini, L., and Hochstrasser, D. (2000) Proceedings: 4th Siena 2D Electrophoresis Meeting: From Protein Maps to Genomes. University of Siena, Italy, Sept 4–7.
18. Berners-Lee, T. J., Cailliau, R., Luotonen, A., Henrick, F., and Secret, A. (1994) World Wide Web. *Comm. Assoc. Comp. Mach.* **37,** 76–82.
19. Appel, R. D., Bairoch, A., Sanchez, J.-C., Vargas, J. R., Golaz, O., Pasquali, C., and Hochstrasser, D. F. (1996) Federated two-dimensional electrophoresis database: a simple means of publishing two-dimensional electrophoresis data. *Electrophoresis* **17,** 540–546.
20. Taylor, J., Anderson, N. L. Scandora, A. E., Willard, K. E., and Anderson, N. (1982) Design and implementation of a prototype human protein index. *Clin.Chem.* **28,** 861–866.
21. Lipkin, L. E. and Lemkin, P. F. (1980) Database techniques for multiple PAGE (2D gel) analysis. *Clin.Chem.* **26,** 1403–1413.
22. Lemkin, P. F. and Lester, E. P. (1989) Database and search techniques for 2D gel protein data: a comparison of paradigms for exploratory data analysis and prospects for biological modeling. *Electrophoresis* **10,** 122–140.
23. Appel, R. D., Hochstrasser, D. F., Funk, M., Vargas, J. R., Pellegrini, C., Muller, A. F., and Scherrer, J.-R. (1991) The MELANIE project from biopsy to automatic protein map interpretation by computer. *Electrophoresis* **12,** 722–735.

24. Olson, A. D. and Miller, M. J. (1988) Elsie 4: quantitative computer analysis of sets of two dimensional gel electrophoretograms. *Analyt. Biochem.* **169,** 49–70.
25. Garrels, J. I., Farrar, J. T., and Burwell, IV, C. B. (1984) Analyt. in Two-Dimensional Gel Electrophoresis of Proteins (Celis, J. E. and Bravo, R., eds.), Academic Press, New York, pp. 37–91.
26. Vincens, P. and Rabilloud, T. (1986) in Recent Progress in Two-Dimensional Electrophoresis' (Galteau, M. M. and Siest, G., eds.), University Press of Nancy, France, pp. 121–130.
27. Pleisner, K.-P., Hoffmann, F., Kriegel, K., Wenk, C., and Wegner, S. (1999) New algorithmic approaches to protein spot detection and pattern matching in two-dimensional electrophoresis gel databases. *Electrophoresis* **20,** 755–765.
28. Arnold, K. and Gosling, J. (1996) The Java Programming Language. Addison-Wesley, Reading, MA, pp. 1–704.
29. Wolberg, G. (1990) Digital Image Warping. IEEE Computer Press Monograph, Los Alamitos, CA, 1–318.

27

Immunoblotting of 2-D Electrophoresis Separated Proteins

Barbara Magi, Luca Bini, Sabrina Liberatori, Roberto Raggiaschi, and Vitaliano Pallini

1. Introduction

Electrotransfer (-blotting) of protein bands separated by sodium dodecyl sulfate-polyacrylamide gel electrophoresis (SDS-PAGE) onto nitrocellulose membranes allowed H. Towbin et al. *(1)* to exploit the specificity of the reaction between antibodies and protein epitopes, avoiding the interference of diffusion and denaturing reagents. Immunoreactive bands were detected by labeled "second antibody" or Protein A.

Devised at a time when one-dimensional SDS-PAGE was the dominant electrophoretic technique, immunoblotting (also called Western blotting *[2]*) is now widely used in conjunction with 2-dimensional (2-D) PAGE (electrofocusing/SDS-PAGE), the widespread use of which has been favored by its long-awaited standardization ***(3,4)*** and by the coming of age of proteome science and technology ***(5,6)***.

In the 2-D age, immunoblotting is still used for traditional goals, namely immunoaffinity identification of proteins *(7)* and analysis of immune responses ***(8,9)***, and also as a genome-proteome interface technique. In fact, specific gene products can be identified in 2-D protein maps using antibodies prepared with the help of modern biotechnology on the basis of gene and cDNA sequences. Today, techniques for antibody production have been improved so as not only to include immunization with recombinant proteins and synthetic oligopeptides, but also selection from complementarity determining region (CDR)-expressing phage libraries ***(10,11)*** without using animal immunization. When Enhanced Chemiluminescence (ECL) detection is applied to immunoblotting, even low abundance proteins undetectable by silver staining, such as oncogene products and cell cycle proteins, can be monitored (*see*, e.g., ***12***).

Immunoblotting results also complement with pI and M_r information to characterize posttranslational modifications of proteins. Protein isoforms are generally cross-reactive, especially when polyclonal antibodies are used, and primary gene products and posttranslational modifications are then distinguished by comparing experimentally determined electrophoretic parameters against values predicted from aminoacid sequences (*see*, e.g., ***3,13***). Protein polymorphism due to alternative splicing can be studied in the same way (*see*, e.g., ***14***).

From: *The Protein Protocols Handbook, 2nd Edition*
Edited by: J. M. Walker © Humana Press Inc., Totowa, NJ

Protein targets can be identified among thousands of non-modified proteins, by means of antibodies specific for epitopes generated by post-translational modification such as phosphotyrosine ***(15,16)***, carbonyl groups ***(17)***, nitrotyrosine ***(18)***, and covalent adducts with drugs ***(19)***. Separation of whole electrophoretic patterns into subsets of immunologically related spots may be a useful strategy for proteome research, as the high sensitivity of immunodetection can extend the analysis to low copy number proteins, because the sensitivity of immunostaining is often greater than that of silver staining ***(12)***. The extent of modification can be assessed by the number and pI shift of immunoreactive spots in the series, if the net charge of the protein is affected ***(20)***.

A modification of the immunoblotting procedures, called far-Western analysis, was developed to detect different kinds of protein–protein interaction ***(21–23)*** and is now also used for proteins separated by two-dimensional electrophoresis ***(24–26)***. The protocol is similar to a western blot, except that a labelled protein is used as a probe to detect specific protein–protein interactions. The probe can be directly biotinylated or labelled with ^{32}P, ^{125}I or ^{35}S, or can be indirectly detected by a labelled antibody to the protein used as the probe. Studies of protein–protein interaction can be useful to link a protein with a specific function in known cellular processes ***(27)***.

2-D immunoblotting maintains the traditional steps consisting in:

1. Electrotransfer of proteins from polyacrylamide gel onto chemically resilient membrane.
2. Chemical staining of protein pattern on the transblotted membrane.
3. Saturation of the membrane.
4. Application of the primary, and then of the secondary, labeled antibody.
5. Detection of immunoreactive proteins by the label on the secondary antibody, for example, radioactivity, enzyme activity.

In all steps, procedures developed for one-dimensional immunoblotting maintain their significance and utility. However, matching immunoreactive spots to silver-stained spots in complex, high-resolution 2-DE patterns has become a far more difficult task than matching bands in one-dimensional separations. As a consequence, chemical staining of total protein pattern on the transblotted nitrocellulose or on equivalent membrane plays a crucial role in 2-D immunoaffinity identification. Matching itself is better carried out on digitized images with adequate computer and software than by physically superimposing the chemiluminescent film or the membrane to the silver-stained gel (cf. **Subheading 3.5.**).

The cost of reagents per sample is also higher than with the one-dimensional procedure, as all volumes and surfaces are larger. As a preliminary step, it is convenient to find out antibody working dilution by one-dimensional immunoblotting. In the same way one can assess the need for a blank incubation with secondary antibody, omitting the primary antibody. Costs for 2-D gel run and electrotransfer reagents can be also reduced by reusing the electroblotted membrane consecutively with different antibodies (cf. **Subheading 3.4.**). All reagents can be saved if "multiple immuno-2-D blotting," that is, simultaneous application of several primary antibodies, can be performed. In this procedure, unambiguous immunodetection is allowed by different 2-D electrophoretic parameters of proteins ***(12)***.

2. Materials

2.1. Equipment

1. Blotting apparatus: Transfer cell, gel holder, magnetic stirrer, refrigerated thermostatic circulator unit.
2. Power supply.
3. Rocking agitator.
4. Computing densitometer.
5. Workstation with a computer program for 2-D gel analysis.

2.2. Reagents

1. Distilled water.
2. Nitrocellulose membrane.
3. Transfer buffer: 25 m*M* Tris, 192 m*M* glycine, 20% (v/v) methanol. Do not adjust pH; it is about 8.3.
4. Filter paper for blotting (Whatman 17 Chr).
5. Ponceau S solution: 0.2% (w/v) Ponceau S in 3% (w/v) trichloroacetic acid (TCA).
6. Phosphate-buffered saline (PBS): 0.15 *M* NaCl, 10 m*M* NaH_2PO_4, bring to pH 7.4 with NaOH.
7. Blocking solution: 3% (w/v) nonfat dry milk in PBS, Triton X-100 0.1% (w/v).
8. Primary antibody solution (primary antibody, appropriately diluted in blocking solution).
9. Secondary antibody solution (secondary antibody, appropriately diluted in blocking solution).
10. Washing solution: Triton X-100 0.5% (w/v) in PBS.
11. 0.05 *M* Tris-HCl, pH 6.8.
12. Amersham ECL (enhanced chemiluminescence) kit, cat. no. RPN 2106.
13. Saran Wrap or other cling films.
14. X-ray films, 18 cm × 24 cm (Amersham Hyper film ECL, cat. no. RPN 3103).
15. Developer and fixer for X-ray film (Developer replenisher; fixer and replenisher, 3M, cat. nos. XAF 3 and XAD 3) (3M Italia S.p.A., Segrate, Italy).
16. Stripping buffer: 100 m*M* 2-mercaptoethanol, 2% (w/v) SDS, 62.5 m*M* Tris-HCl pH 6.7.

3. Methods

3.1. Transfer

To avoid membrane contamination wear gloves during all the steps of the experiment.

1. Prepare the transfer buffer and cool it to 4°C before the end of the electrophoretic run (*see* **Notes 1–3**).
2. Cut to the dimensions of the gel, two pieces of filter paper and one piece of nitrocellulose/gel (*see* **Note 4**).
3. Following electrophoresis, wash the gel in distilled water and then equilibrate it in transfer buffer. The ideal time for 1.5-mm gels is 10–15 min. (*see* **Note 5**).
4. Soak the nitrocellulose membrane for 15–20 min in transfer buffer. Also wet two "Scotch-Brite" pads/gel and filter papers in transfer buffer.
5. Assemble the "sandwich" for transfer in this order: fiber pad, filter paper, nitrocellulose, gel, filter paper, fiber pad. Remove all air bubbles between membrane and gel and between paper and gel.
6. Put the blot sandwich in the gel holder and hold it firmly, to ensure a tight contact between gel and membrane (*see* **Note 6**).

7. Fill the cell with transfer buffer and place a stir bar inside the transfer cell, so that the buffer is stirred during electrotransfer (*see* **Note 7**).
8. Place the gel holder in the transfer cell with the sandwich oriented as follows: ANODE / fiber pad, filter paper, nitrocellulose, gel, filter paper, fiber pad/ CATHODE.
9. Carry out blotting at a constant current until it is reached a total of 1.5–2.0 A° (*see* **Note 8**), refrigerating the buffer to 4°C (*see* **Note 9**).
10. After electrotransfer, disassemble the blotting apparatus and remove the nitrocellulose membrane. To mark the orientation of the membrane, cut away the lower right corner, corresponding to low M_r, high pH.

The membrane can be processed immediately for immunoblotting or can be air-dried and stored at –20°C, within parafilm sheets for extended periods *(28)*.

3.2. Staining of Total Protein Pattern on Membrane

1. Before the immunodetection, stain the nitrocellulose membrane in 0.2% (w/v) Ponceau S in 3% (w/v) trichloroacetic acid for 3 min *(29)* (*see* **Note 10**).
2. Destain with several changes of distilled water to diminish background color. Because the red spots will disappear in the blocking step, circle with a waterproof pen some spots which will be used as landmarks to match total protein pattern on nitrocellulose vs immunoreactive pattern and vs silver stained polyacrylamide gel pattern (*see* **Note 10**).

3.3. Immunodetection

3.3.1. Incubation with Antibodies

All steps are carried out at room temperature and with gentle agitation on a rocking agitator.

1. Block nonspecific binding sites in the membrane with three washing steps, each 10 min in duration, in blocking solution (*see* **Notes 11** and **12**).
2. Incubate overnight in the primary antibody solution at the suitable dilution (*see* **Note 13**) in blocking solution.
3. Wash 3 × 10 minutes in blocking solution.
4. Incubate for 2 h in the secondary antibody solution (*see* **Note 14**).
5. Wash 3 × 10 min in blocking solution.
6. Wash 30 min in washing solution.
7. Wash 2 × 30 min in 0.05 *M* Tris-HCl, pH 6.8.

After this step one can go forward with ECL detection (*see* **Note 15**). Alternatively, one can choose detection with the chromogenic substrate (*see* **Note 16**).

3.3.2. Enhanced Chemiluminescent Detection

To detect the immunoreactive spot(s) with Amersham ECL kit, it is necessary to work in a darkroom and to wear gloves to prevent hand contact with film.

1. Mix equal volumes of detection reagent 1 and detection reagent 2 from the Amersham ECL kit and immerse the membrane in this solution for 1 min, ensuring that all the surface of the membrane is covered with solution (*see* **Note 17**).
2. Place the membrane on a glass and cover it with a layer of Saran Wrap.
3. Cut away a corner from a piece of autoradiography film to define its orientation (*see* **Subheading 3.1., step 10**). Superimpose the autoradiography film on the nitrocellulose membrane beginning from the upper left corners. Nitrocellulose membrane and X-ray

film may have different dimensions. Superimposing at the upper left corner for ECL impression will allow subsequent matching of images.
4. Expose the film for a time variable from 5 s to several minutes. It is convenient to begin with short exposure, develop the film, and then try longer exposures, if necessary.
5. Develop the film with the suitable reagents (*see* **Note 16**).

3.4. Stripping

At the end of a cycle of immunodetection, it is possible to strip the membrane with indicated solution and to carry out subsequent cycles incubating with different primary antibodies (*see* **Note 18**). The procedure for the stripping we use is:

1. Incubate the membrane in stripping buffer at 70°C for 30 min, with occasional shaking.
2. Wash the nitrocellulose 2 × 10 min in large volumes of washing solution at room temperature.
3. Block the membrane and perform immunodetection as described in **Subheading 3.3.**

3.5. Matching

For an accurate matching process we use a computer program (*see* **Note 19**) that permits matching the digitized images, using as landmarks the spots stained with Ponceau S (*see* **Note 20**).

To perform this operation we suggest the following procedure (cf. **Fig. 1**):

1. Scan the ECL-developed film, the Ponceau S-stained nitrocellulose membrane, and the silver-stained gel of the same sample with a computing densitometer with a sufficient resolution (*see* **Notes 21** and **22**).
2. Rotate left–right the nitrocellulose membrane, with an appropriate program, in order to have the three images with the cut lower corner on the right. In fact, the nitrocellulose membrane has the spots only in one face and the scanning process generates an image with the cut lower corner placed on the left.
3. Stack together the film and nitrocellulose membrane images, aligning the upper left corners and the two corresponding borders and placing the cut lower right corner in the same orientation for both.
4. Add "manually," with appropriate software tool, the Ponceau S spots chosen as landmarks onto the image of the ECL film.
5. Find the spots on the gel corresponding to landmarks on the film (*see* **Note 23**) and modify the size of the silver nitrate image adjusting it to the smaller one of film by the mean of adequate software. Actually the gel is larger than the film due to silver staining procedure.
6. Stack together the equalized ECL film and gel images, superimpose the landmarks carefully, and run the automatic match program. This operation permits automatically highlighting the silver-stained spots paired with the immunoreactive ones present on the ECL film.

4. Notes

1. We perform electrotransfer from gels to nitrocellulose membranes by a "wet" procedure. In a tank or "wet" apparatus the gel is submerged in a large volume of buffer during the transfer. We use a Bio-Rad transfer cell with 3 L of transfer buffer or an ISODALT cell (Hoefer Scientific Instruments) with 20 L of transfer buffer. ISODALT cells enable the simultaneous transfer of five gels. The transfer buffer can be used several times, if stored at 4°C.

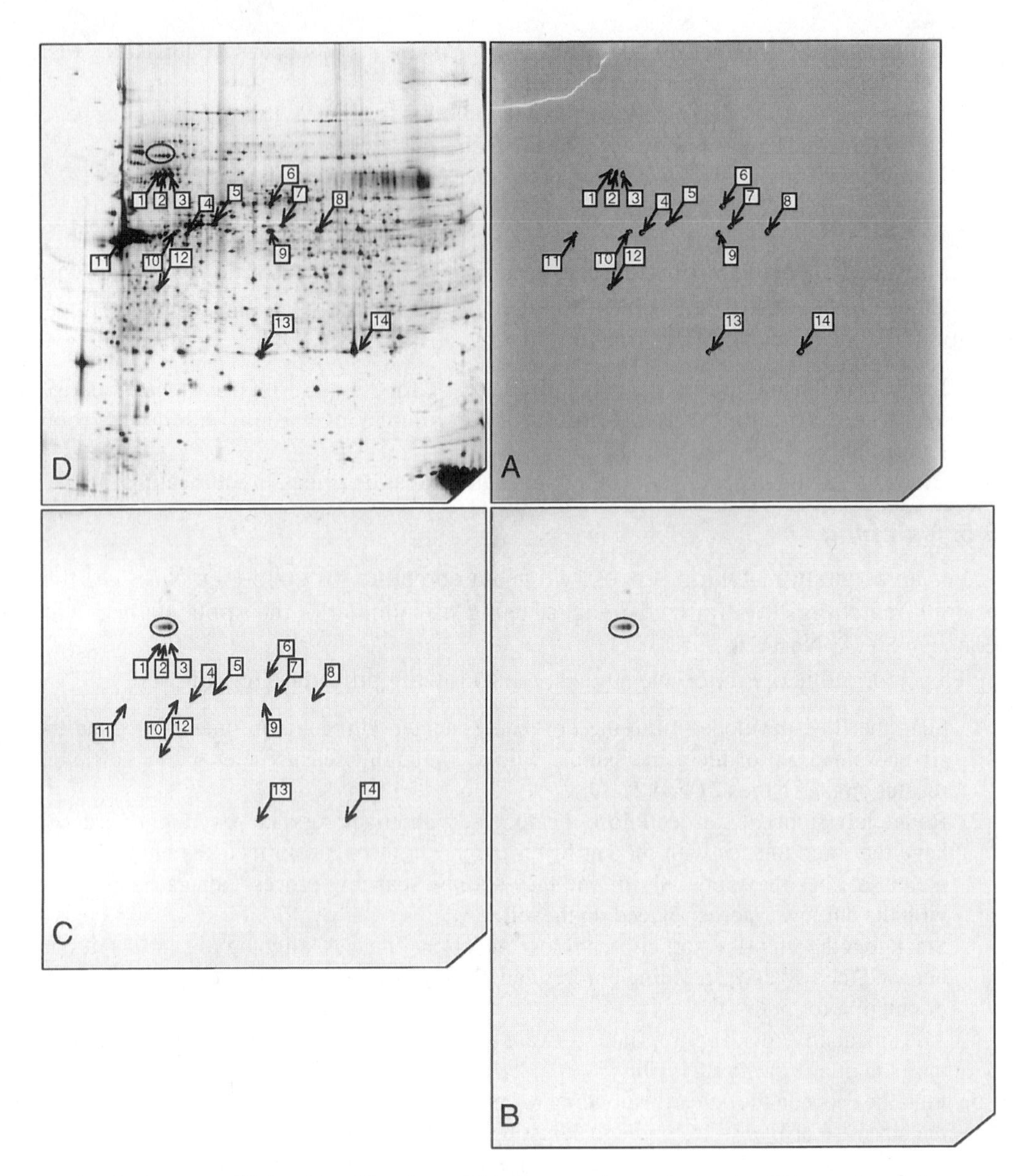

Fig. 1. Immunoaffinity identification of S1 ribosomal protein in the *Chlamydia trachomatis* serovar L2 protein map, comprised within pI 3.5–9, M_r 10–200 kDa window. Electrophoretic conditions as in **ref. *9***. Immunoblotting and ECL detection were reported (cf. **Subheading 3.**). (*Chlamydia* Elementary Bodies were provided by Dr. G. Ratti, Chiron-Biocine Research Centre, Siena, Italy; monoclonal antibody to S1 ribosomal protein was provided by Dr. S. Birkelund, Department of Medical Microbiology and Immunology, Aarhus University, Aarhus, Denmark). **(A)** Digitized image of Ponceau S-stained nitrocellulose membrane. *Arrows* and *number* indicate spots chosen as landmarks, being recognizable both in the transblotted nitrocellulose and in the silver-stained gels. **(B)** Digitized image of impressed ECL film. Immunoreactive isoelectric series is *circled*. **(C)** Image constructed in order to adding landmarks in **A** onto **B**. The process of image-stacking being at the *upper left corner*. Exceeding surface in the ECL films is cut by computer. **(D)** Digitized image of silver-stained gels. Size is equalized to that of image **C**. Recognition of landmarks allows recognition of immunoreactive spots.

"Semidry" electroblotters require smaller volumes of buffer, since only the membrane and filter paper have to be wet, and the procedure is faster. However, the "wet" method is recommended when antigen is present in small quantities (such as low abundance spots in 2-D gels) and/or its molecular weight is high *(30)*. It offers more options, such as temperature, time and voltage control.

2. The transfer buffer we use was first described by Towbin et al. *(1)*. Methanol is toxic and it can be omitted *(31–33)*. Still, we use it to reduce swelling of the gel during transfer and to increase the binding of proteins to nitrocellulose *(2,31,34)*. When working with high molecular weight proteins, elimination of the methanol results in a significant increase in protein transfer efficiency. Some recipes recommend the addition of low concentration of SDS to the buffer to help the transfer of high molecular weight proteins *(29)* and to improve the transfer of a variety of proteins *(31)*. However SDS reduces the amount of protein bound to the membrane *(2)* and may adversely affect immunoreactivity by inhibiting renaturation of antigenic sites *(35)*.
   For semidry blotting it is possible to use discontinuous buffer and/or "elution promoting" buffer on the gel side and a "retention promoting" buffer on the membrane side *(36)*.
3. Reagent grade methanol only is to be used because trace impurities in methanol can increase the conductivity of transfer buffer and decrease transfer efficiency.
4. Polyvinylidene difluoride (PVDF) may also be used *(37,38)*. Remember that unlike nitrocellulose, PVDF is a hydrophobic membrane and it must to be prewetted in methanol before use with aqueous solution. The buffer generally used to transfer proteins to PVDF is 10 m*M* 3-[cyclohexylamino]-1-propanesulfonic acid, 10% (v/v) methanol, pH 11 *(39)*, although it is possible also use the buffer described in Towbin et al. *(1)*.
5. Electrotransfer is usually carried out immediately after the electrophoretic run from unstained gels. However, transfer of proteins from polyacrylamide gels after Coomassie Blue or silver staining, has also been reported *(40,41)*. Proteins can also be transferred for immunodetection from gels previously stained in a reverse (negative) way, for example, with imidazole-zinc salts *(42)*.
   In these procedures, immunoreactivity pattern on the membrane, and total protein pattern can be obtained from the same gel from which spots have been transblotted, facilitating matching even with not easily reproducible 2-D separations.
6. It is important that the layers of sandwich are firmly held together to have a good transfer, without distortion of the proteins spots.
7. The stirring of the buffer ensures uniform temperature and conductivity during electrotransfer.
8. The transfer efficiency is adversely affected by high molecular weight and basic pI of proteins. Therefore, while attempting to transfer slow proteins, it may happen that fast proteins cross the nitrocellulose membrane and are lost. In these cases, one can use two stacked membranes or membranes with smaller pore diameter which would also prevent loss of small polypeptides during membrane manipulation *(43,44)*. *See* **ref. *45*** for information on blotting on various membranes.
   Some low molecular weight, basic proteins, such as histones, lysozymes, cytochromes and so forth, do not transfer well because they may be near their pI in currently used buffers, as SDS is lost during the transfer in methanol. Transfer of these proteins can be improved, without impairing transfer of other proteins, by introducing a more basic transfer buffer and/or omitting the equilibration (**Subheading 3.1., step 3**) *(46)*. Alternative buffers have also been proposed (e.g., *[47]*).
9. When the transfer is conducted at high voltage it is necessary to refrigerate the transfer tank with a thermostatic circulator. If possible, avoid transfer in the cold room.
10. Chemical staining of proteins patterns transblotted onto the nitrocellulose or other membrane plays an important role in 2-D immunoaffinity identification, since it provides

"landmark" spots to match immunoreactivity patterns to silver-staining patterns (cf. **Subheading 3.5.** and **Note 19**). Several staining procedures can be chosen. This step is usually carried out before the incubation of transblotted membranes with antibodies, employing dyes (e.g., Ponceau S, Fast Green, Amido Black) or metalchelates, which do not interfere with protein immunoreactivity ***(43,48–51)***. Staining with substances such as Ponceau S, Fast Green, and metal-chelates is reversible, eliminating interference in the immunoreactivity pattern obtained with chromogenic substrates, but it is not very sensitive. A dye-based staining method, using Direct Blue 71, was recently developed. It is reversible, compatible with immunodetection and is 10 times more sensitive than Ponceau S ***(52)***. Permanent staining, for example, with Amido Black, can also be used if the immunoreactivity pattern is collected from ECL-impressed films, but the stain must not interfere with the immunoreaction.

Fluorescent dyes were also recently introduced for membrane staining. For example, SYPRO Ruby protein blot stain is a new, luminescent metal chelate stain composed of ruthenium in an organic complex that interacts noncovalently with proteins. This stain is more sensitive than Ponceau, Coomassie Blue, Amido Black, or India ink and nearly as sensitive as colloidal gold stain procedures. This fluorescent stain is fully compatible with immunoblotting ***(53)***.

When radioactive labeling is possible, most accurate total protein patterns can be collected from transblotted membranes by phosphor imaging (*see*, e.g., **ref. *16***). The two images have the same dimensions so that general alignment and recognition of immunoreactive spots in the total protein pattern is readily achieved. However, the need for radioactive proteins is a huge limitation. Perfect alignment of immunoreactive spots to total protein pattern can be obtained, at least with PVDF membranes, by the conjunction of colloidal gold staining for total protein detection and ECL for immunoreactivity on the same membrane ***(54)***. This procedure produce an ECL-impressed film with low exposure allowing the detection of immunoreactive spots and a, ECL-impressed film with a strong exposure that produces a background pattern. The final result is a single image where immunoreactive spots appears as dark black spots and general protein pattern appears as light grey spots.

Colloidal gold ***(55,56)*** and India Ink staining ***(57)*** can be applied also after immunodetection. The latter approach is possible if membrane saturation is achieved by Tween 20, omitting proteins ***(58,59)***.

Finally, another method has been proposed by Zeindl-Eberhart ***(60)*** to localize antigen easily on 2-D gels: proteins are transferred to PVDF membrane, immunostained with specific antibodies, using Fast Red or 5-bromo-4-chloro-3-indolyl phosphate/nitroblue tetrazolium as a detection system, and then counterstained with Coomassie Brilliant Blue. The membrane appears with immunostained spots colored in red or black and the total protein pattern in blue. The blocking proteins are removed during staining with Coomassie and so do not create a background staining.

In all these methods all proteins are transferred onto one membrane and both total protein staining and immunostaining are performed on the same membrane. "Double replica" blotting methods were also developed, in order to obtain a membrane that has all the proteins stained and is an almost identical copy of the immunostained one. The first of its kind was described by Johansson ***(61)*** who found that by changing the direction of the blotting current, the proteins could be transferred simultaneously from one gel to two membranes on either side of the gel. A second method, described by Neumann and Mullner ***(62)***, combines the usual electroblotting procedure with the generation of a "contact copy" from a gel. Both systems enable one membrane to be immunostained while the second membrane is stained by highly sensitive total protein staining methods. Protein identification is then carried out by comparing the signals of both matrices. Similarly, a fast and simple method to produce print-quality like Ponceau replicas from blots was recently described ***(63)***. The positive replicas are the same size as the blots and can be stored without loss of intensity. This makes them useful for localizing immunoreactive spots in complex 2-D electrophoretograms.

11. For a 16 × 18-cm membrane, we use 50 mL of solution in each washing and incubation step. Volumes can be proportionally adjusted to other membrane dimension. It is important that the membrane is entirely soaked in solution during the washing and incubation step.

    We perform all steps of immunodetection in a flat glass vessel. Rotating glass cylinders are also convenient. Do not incubate membrane in sealed plastic envelopes, because it results in a very high background in ECL.
12. Our blocking procedure is suitable for routine use. However, special conditions and reagents are required for immunoblotting with some antibodies, such as antiphosphotyrosine antibodies ***(16,64,65)***. In this case, the presence of phosphorylated proteins in the blocking solution could determine high immunochemical background staining. Information on different blocking conditions can be found in the **refs. *66–69***. Chemicon have developed a new blocking agent, composed of nonanimal proteins that ensure uniform blocking, without nonspecific binding, eliminating all cross-reactivity with animal source antigens, primary antibodies and secondary antibodies. Blocking with a nonionic detergent such as Tween 20, without added protein, has also been used with the advantage that after immunodetection the blot can be stained for total protein pattern ***(58,59,70)*** (*see* **Note 10**). On the other hand, it has been found that blocking with detergent alone may cause loss of transblotted proteins ***(71,72)***. Using PVDF membranes it is possible to employ a nonblock technique: this method involves three cycles of methanol–water hydration of the membrane, allowing multiple erasure and probing of the same blot with little or no loss of signal ***(73)***.
13. Optimal dilution of the primary and secondary antibody should be determined by immunoblotting of one-dimensional gels or dot blot analysis can be used. Working solutions of antibodies can be stored at –20°C and used several times ***(74)***.
14. Secondary antibodies may often give problems of cross-reactivity, especially in the analysis of samples containing antibodies (such as immunoprecipitates, immune tissue, plasma, etc.), even when antibodies from different species are used. Bhatt et al. proposed a method to avoid this problem, suppressing extraneous signals in immunoblots. This method consist in preconjugating the primary with the secondary antibody before incubation. By this procedure, signals from secondary antibodies are completely quenched ***(75)***.
15. ECL detects horseradish peroxidase conjugated antibodies through oxidation of luminol in the presence of hydrogen peroxide and a phenolic enhancer under alkaline conditions. ECL reagents are capable of detecting 1–10 pg of protein antigen. Alternative enhancers that extend the duration of light emission and enable detection of even smaller quantities of target protein (in the order of femtograms) have recently been developed. One such system is ECL plus (Amersham Pharmacia Biotech) ***(76)***. These systems are suitable for use with charge-coupled device (CCD) camera that require longer exposure times for good quantification of immunoreaction.
16. In case chemiluminescence is too strong or background is too high, one can switch to detection with a chromogenic substrate.

    We use 4-chloro-1-naphthol ***(77)*** as chromogenic substrate, according to the following protocol:
    a. After ECL detection (or after **step 7** of **Subheading 3.3.1.**), wash the membrane briefly with 0.05 *M* Tris-HCl, pH 6.8.
    b. Soak it in developing solution (20 mL of 0.05 *M* Tris-HCl, pH 6.8; 7 µL of H_2O_2 30% (v/v); 5 mL 4-Chloro-1-naphthol 0.3% (w/v) in methanol) until the color appears.

c. Stop the reaction with washes in distilled water.
d. Air-dry the membrane and photograph it as soon as possible, because the color fades with time.

17. For a 16 cm × 18 cm membrane use 7.5 mL of reagent 1 and 7.5 mL of reagent 2.
18. It is also possible to perform stripping with kits as the CHEMICON Re-Blot™ Western blot recycling kit. Stripping of antibodies also elutes antigens from the membrane and signal intensity decreases in successive cycles. It is important therefore to remember that stripping should be used only for qualitative purpose.

    As alternatives to stripping, one can use: (a) different chromogenic substrates for peroxidase at each cycle (rainbow blotting, *[74]*), (b) ECL and inactivating peroxidase after each cycle ***(74)***, (c) different labels and detection methods at each cycle ***(78)***.
19. To perform the matching process we use the software Melanie II release 1.2, from Bio-Rad.
20. Matching can also be done by simple eye inspection directly on nitrocellulose and ECL film when the sample contains relatively few spots, all of them detectable by chemical staining of nitrocellulose. In the majority of cases samples are very complex, and many low abundance proteins occur. In these cases, matching by computer is mandatory in order to identify immunoreactive spots in silver-stained patterns. The following manual procedure is suggested:
    a. Match the exposed film with nitrocellulose membrane, aligning the upper left corner and the two corresponding borders and placing the cut lower right corner in the same orientation for both.
    b. Using a waterproof pen, mark the other two borders of the nitrocellulose on the film and transfer the chemically stained spot present on nitrocellulose on the ECL film in order to use them as landmarks for the next matching with the silver nitrate stained gel.
    c. Nitrocellulose membrane and film maintain the initial size, but the size of the gel increases after silver staining. Size equalization can be obtained by photographic or photocopy procedures.
    d. On a transilluminator match all the landmarks with the corresponding spots on the silver nitrate stained gel to identify the immunoreactive spots.
21. We use a computing densitometer 300 S from Molecular Dynamics with a resolution of 4000 × 5000 pixels, 12 bits/pixel, which generate 40 megabytes images on 16 bits.
22. The silver stained image used for matching can be taken from your archive of files or from images available in Internet, provided that the identical electrophoretic procedures have been applied. The possibility to match images deriving from different 2-DE procedures has been studied by Lemkin P. ***(79)***.
23. This step may be difficult if the "landmark " spots stained by Red Ponceau on the nitrocellulose membrane are few. To aid recognition of the spots chosen as landmarks on the silver stained gel, we suggest also staining with silver the gel from which the proteins are transferred, in which the amount of proteins loaded must be twice that used for a silver-stained gel. Most spots will be still visible on the transferred gel and can be used for a first matching with the membrane. Using the gel from which the membrane was obtained, the landmarks can be localized correctly. The landmarks are then easily transferred to the silver stained gel by computer matching.

References

1. Towbin, H., Staehlin, T., and Gordon, J. (1979) Electrophoretic transfer of proteins from polyacrylamide gels to nitrocellulose sheets: procedure and some applications. *Proc. Natl. Acad. Sci. USA* **76,** 4350–4354.

2. Burnette, W. N. (1981) "Western blotting": electrophoretic transfer of proteins from sodium dodecyl sulfate-polyacrylamide gels to unmodified nitrocellulose and radiographic detection with antibody and radioiodinated protein A. *Analyt. Biochem.* **112,** 195–203.
3. Bjellqvist, B., Hughes, G. J., Pasquali, C., Paquet, N., Ravier, F., Sanchez, J.-C., et al. (1993) The focusing positions of polypeptides in immobilized pH gradients can be predicted from their amino acid sequences. *Electrophoresis* **14,** 1023–1031.
4. Corbett, J. M., Dunn, M. J., Posh, A., and Görg, A. (1994) Positional reproducibility of protein spots in two-dimensional polyacrylamide gel electrophoresis using immobilized pH gradient isoelectric focusing in the first dimension: an interlaboratory comparison. *Electrophoresis* **15,** 1205–1211.
5. Kahn, P. (1995) From genome to proteome: looking at a cell's proteins. *Science* **270,** 369–370.
6. Wilkins, M. R., Sanchez, J.-C., Williams, K. L., and Hochstrasser, D. F. (1996) Current challenges and future applications for protein maps and post-translational vector maps in proteome projects. *Electrophoresis* **17,** 830–838.
7. Goldfarb, M. (1999) Two-dimensional electrophoresis and computer imaging: quantitation of human milk casein. *Electrophoresis* **20,** 870–874.
8. Pitarch, A., Pardo, M., Jimenez, A., Pla, J., Gil, C., Sanchez, M., and Nombela, C. (1999) Two-dimensional gel electrophoresis as analytical tool for identifying *Candida albicans* immunogenic proteins. *Electrophoresis* **20,** 1001–1010.
9. Sanchez-Campillo, M., Bini, L., Comanducci, M., Raggiaschi, R., Marzocchi, B., Pallini, V., and Ratti, G. (1999) Identification of immunoreactive proteins of *Chlamydia trachomatis* by Western blot analysis of a two-dimensional electrophoresis map with patient sera. *Electrophoresis* **20,** 2269–2279.
10. Pini A., Viti F., Santucci A., Carnemolla B., Zardi L., Neri P., and Neri D. (1998) Design and use of a phage display library. *J. Biol. Chem.* **273,** 21,769–21,776.
11. Ravn P., Kjaer S., Jensen K. H., Wind T., Jensen K. B., Kristensen, P., et al. (2000) Identification of phage antibodies toward the Werner protein by selection on Western blots. *Electrophoresis* **21,** 509–516.
12. Sanchez, J. C., Wirth, P., Jaccoud, S., Appel, R. D., Sarto, C., Wilkins, M. R., and Hochstrasser, D. F. (1997) Simultaneous analysis of cyclin and oncogene expression using multiple monoclonal antibody immunoblots. *Electrophoresis* **18,** 638–641.
13. Magi, B., Bini, L., Liberatori, S., Marzocchi, B., Raggiaschi, R., Arcuri, F., et al. (1998) Charge heterogeneity of macrophage migration inhibitory factor in human liver and breast tissue. *Electrophoresis* **19,** 2010–2013.
14. Janke, C., Holzer, M., Klose, J., and Arendt, T. (1996) Distribution of isoforms of the microtubule-associated protein tau in grey and white matter areas on human brain: a two dimensional gel electrophoretic analysis. *FEBS Lett.* **379,** 222–226.
15. Oda, T., Heaney, C., Hagopian, J. R., Griffin, J. D., and Druker, B. J. (1994) Crkl is the major tyrosine-phosphorylated protein in neutrophils from patients with chronic myelogenous leukemia. *J. Biol. Chem.* **269,** 22,925–22,928.
16. Birkelund, S., Bini, L., Pallini, V., Sanchez-Campillo, M., Liberatori, S., Clausen, J. D., et al. (1997) Characterization of *Chlamydia trachomatis* L2 induced tyrosine phosphorylated HeLa cell proteins by two-dimensional gel electrophoresis. *Electrophoresis* **18,** 563–567.
17. Reinhekel, T., Körn, S., Möhring, S., Augustin, W., Halangk, W., and Schild L. (2000) Adaption of protein carbonyl detection to the requirements of proteome analysis demonstrated for hypoxia/reoxygenation in isolated rat liver mitochondria. *Arch. Biochem. Biophys.* **376,** 59–65.

18. Strong, M. J., Sopper, M. M., Crow, J. P., Strong, W. L., and Beckman, J. S. (1998) Nitration of the low molecular weight neurofilament is equivalent in sporadic amyotrophic lateral sclerosis and control cervical spinal cord. *Biochem. Biophys. Res. Commun.* **248,** 157–164.
19. Magi, B., Marzocchi, B., Bini, L., Cellesi, C., Rossolini, A., and Pallini, V. (1995) Two-dimensional electrophoresis of human serum proteins modified by ampicillin during therapeutic treatment. *Electrophoresis* **16,** 1190–1192.
20. Marzocchi, B., Magi, B., Bini, L., Cellesi, C., Rossolini, A., Massidda, O., and Pallini, V. (1995) Two-dimensional gel electrophoresis and immunoblotting of human serum albumin modified by reaction with penicillins. *Electrophoresis* **16,** 851–853.
21. Arthur, T. M. and Burgess, R. R. (1998) Localization of a sigma70 binding site on the N terminus of the *Escherichia coli* RNA polymerase beta' subunit. *J. Biol. Chem.* **273,** 31,381–31,387.
22. Faust, M., Schuster, N., and Montenarh, M. (1999) Specific binding of protein kinase CK2 catalytic subunits to tubulin. *FEBS Lett.* **462,** 51–56.
23. Nakatani Y., Tanioka T., Sunaga S., Murakami, M., and Kudo, I. (2000) Identification of a cellular protein that functionally interacts with the C2 domain of cytosolic phospholipase A_2a. *J. Biol. Chem.* **275,** 1161–1168.
24. Bouvet, P., Diaz J.-J., Kindbeiter, K., Madjar, J.-J. and Amalric, F. (1998) Nucleolin interacts with several ribosomal proteins through its RGG domain. *J. Biol. Chem.* **273,** 19,025–19,029.
25. Makino, Y., Yoshida, T., Yogosawa, S., Tanaka, K., Muramatsu, M., and Tamura, T. A. (1999) Multiple mammalian proteasomal ATPases, but not proteasome itself, are associated with TATA-binding protein and a novel transcriptional activator, TIP120. *Genes Cells* **4,** 529–539.
26. Fouassier, L., Yun, C. C., Fitz, J. G., and Doctor, R. B. (2000) Evidence for ezrin-radixin-moesin-binding phosphoprotein 50 (EBP50) self-association through PDZ-PDZ interactions. *J. Biol. Chem.* **275,** 25,039–25,045.
27. Pasquali, C., Vilbois, F., Curchod, M.-L., van Huijsduijnen, R. H., and Arigoni, F. (2000) Mapping and identification of protein–protein interactions by two-dimensional far-western immunoblotting. *Electrophoresis* **21,** 3357–3368.
28. Bernstein, D. I., Garraty, E., Lovett, M. A., and Bryson, Y. J. (1985) Comparison of western blot analysis to microneutralization for the detection of type-specific antibodies to herpes simplex virus antibodies. *J. Med. Virol.* **15,** 223–230.
29. Sanchez, J. C., Ravier, F., Pasquali, C., Frutiger, S., Paquet, N., Bjellqvist, B, et al. (1992) Improving the detection of proteins after transfer to polyvinylidene difluoride membranes. *Electrophoresis* **13,** 715–717.
30. Okamura, H., Sigal, C. T., Alland, L., and Resh, M. D. (1995) Rapid high-resolution Western blotting. *Meth. Enzymol.* **254,** 535–550.
31. Nielsen, P. J., Manchester, K. L., Towbin, H., Gordon, J., and Thomas, G. (1982) The phosphorylation of ribosomal protein S6 in rat tissues following cycloheximide injection, in diabetes, and after denervation of diaphragm. *J. Biol. Chem.* **257,** 12,316–12,321.
32. Gershoni, J. M. and Palade, G. E. (1983) Protein blotting: principles and applications. *Analyt. Biochem.* **131,** 1–15.
33. Gershoni, J. M. (1988) Protein blotting: a manual. *Meth. Biochem. Analyt.* **33,** 1–58.
34. Gershoni, J. M. and Palade, G. E. (1982) Electrophoretic transfer of proteins from sodium dodecyl sulfate-polyacrylamide gels to a positively charged membrane filter. *Analyt. Biochem.* **124,** 396–405.

35. Birk, H.-W. and Koepsell, H. (1987) Reaction of monoclonal antibodies with plasma membrane proteins after binding on nitrocellulose: renaturation of antigenic sites and reduction of nonspecific antibody binding. *Analyt. Biochem.* **164,** 12–22.
36. Lauriere, M. (1993) A semidry electroblotting system efficiently transfers both high- and low-molecular weight proteins separated by SDS-PAGE. *Analyt. Biochem.* **212,** 206–211.
37. Pluskal, M. F., Przekop, M. B., Kavonian, M. R., Vecoli, C., and Hicks, D.A. (1986) Immobilon^{TM} PVDF transfer membrane. A new membrane substrate for Western blotting of proteins. *BioTechniques* **4,** 272–282.
38. Gultekin, H. and Heermann, K. H. (1988) The use of polyvinylidenedifluoride membranes as a general blotting matrix. *Analyt. Biochem.* **172,** 320–329
39. Matsudaira, P. (1987) Sequence from picomole quantities of proteins electroblotted onto polyvinylidene difluoride membranes. *J. Biol. Chem.* **262,** 10,035–10,038.
40. Ranganathan, V. and De, P. K. (1995) Western blot of proteins from Coomassie-stained polyacrylamide gels. *Analyt. Biochem.* **234,** 102–104.
41. Wise, G. E. and Lin, F. (1991) Transfer of silver-stained proteins from polyacrylamide gels to polyvinylidene difluoride membranes. *J. Biochem. Biophys. Meth.* **22,** 223–231.
42. Fernandez-Patron, C., Castellanos-Serra, L., and Rodriguez, P. (1992) Reverse staining of sodium dodecyl sulfate polyacrylamide gels by imidazole-zinc salts: sensitive detection of unmodified proteins. *BioTechniques* **12,** 564–573.
43. Lin, W. and Kasamatsu, H. (1983) On the electrotransfer of polypeptides from gels to nitrocellulose membrane. *Analyt. Biochem.* **128,** 302–311.
44. Polvino, W. J., Saravis, C. A., Sampson, C. E., and Cook, R. B. (1983) Improved protein analysis on nitrocellulose membrane. *Electrophoresis* **4,** 368–369.
45. Eckerskorn, C. and Lottspeich, F. (1993) Structural characterization of blotting membranes and the influence of membrane parameters for electroblotting and subsequent aminoacid sequence analysis of proteins. *Electrophoresis* **14,** 831–838.
46. Szewczyk, B. and Kozloff, L. M. (1985) A method for the efficient blotting of strongly basic proteins from sodium dodecyl sulphate-polyacrylamide gels to nitrocellulose. *Analyt. Biochem.* **50,** 403–407.
47. Dunn, S. D. (1986) Effects of the modification of transfer buffer composition and the renaturation of proteins in gels on the recognition of proteins on Western blots by monoclonal antibodies. *Analyt. Biochem.* **157,** 144–153.
48. Salinovich, O. and Montelaro, R. C. (1986) Reversible staining and peptide mapping of proteins transferred to nitrocellulose after separation by sodium dodecylsulfate-polyacrylamide gel electrophoresis. *Analyt. Biochem.* **156,** 341–347.
49. Reinhart, M. P. and Malamud, D. (1982) Protein transfer from isoelectric focusing gels: the native blot. *Analyt. Biochem.* **123,** 229–235.
50. Patton, W. F., Lam, L., Su, Q., Lui, M., Erdjument-Bromage, H., and Tempst, P. (1994) Metal chelates as reversible stains for detection of electroblotted proteins: application to protein microsequencing and immunoblotting. *Analyt. Biochem.* **220,** 324–335.
51. Root, D. D. and Reisler, E. (1989) Copper iodide staining of protein blots on nitrocellulose membranes. *Analyt. Biochem.* **181,** 250–253.
52. Hong, H.-Y., Yoo, G.-S., and Choi, J.-K. (2000) Direct Blue 71 staining of proteins bound to blotting membrane. *Electrophoresis* **21,** 841–845.
53. Berggren, K., Steinberg T. H., Lauber, W. M., Carroll, J. A., Lopez, M. F., Chernokalskaya, E., et al. (1999) A luminescent ruthenium complex for ultrasensitive detection of proteins immobilized on membrane supports. *Analyt. Biochem.* **276,** 129–143.
54. Chevallet, M., Procaccio, V., and Rabilloud, T. (1997) A nonradioactive double detection method for the assignment of spots in two-dimensional blots. *Analyt. Biochem.* **251,** 69–72.

55. Moeremans, M., Daneels, G., and De Mey, J. (1985) Sensitive colloidal metal (gold or silver) staining of protein blots on nitrocellulose membranes. *Analyt. Biochem.* **145,** 315–321.
56. Daneels, G., Moeremans, M., De Raeymaeker, M., and De Mey, J. (1986) Sequential immunostaining (gold/silver) and complete protein staining (Aurodye) on Western blots. *J. Immunol. Meth.* **89,** 89–91.
57. Glenney, J. (1986) Antibody probing of Western blots which have been stained with India ink. *Analyt. Biochem.* **156,** 315–319
58. Batteiger, B., Newhall, W. J. V., and Jones, R. B. (1982) The use of Tween 20 as a blocking agent in the immunological detection of proteins transferred to nitrocellulose membrane. *J. Immunol. Meth.* **55,** 297–307.
59. Fultz, C. D. and Witzmann, F. A. (1997) Locating Western blotted and immunostained proteins within complex two-dimensional patterns. *Analyt. Biochem.* **251,** 288–291.
60. Zeindl-Eberhart, E., Jungblut, P. R, and Rabes, H. M. (1997) A new method to assign immunodetected spots in the complex two-dimensional electrophoresis pattern. *Electrophoresis* **18,** 799–801.
61. Johansson, K. E. (1986) Double replica electroblotting: a method to produce two replicas from one gel. *J. Biochem. Biophys. Meth.* **13,** 197–203.
62. Neumann, H. and Mullner, S. (1998) Two replica blotting methods for fast immunological analysis of common proteins in two-dimensional electrophoresis. *Electrophoresis* **19,** 752–757.
63. Gotzmann, J. and Gerner, C. (2000) A method to produce Ponceau replicas from blots: application for Western analysis. *Electrophoresis* **21,** 523–525.
64. Kamps, M. P. (1991) Generation of anti-phosphotyrosine antibodies for immunoblotting. *Meth. Enzymol.* **201,** 101–110.
65. Michalewski, M. P., Kaczmarski, W., Golabek, A., Kida, E., Kaczmarski, A., and Wisniewski, K. E. (1999) Immunoblotting with antiphosphoaminoacid antibodies: importance of the blocking solution. *Analyt. Biochem.* **276,** 254–257.
66. Towbin, H. and Gordon, J. (1984) Immunoblotting and dot immunobinding. Current status and outlook. *J. Immunol. Meth.* **72,** 313–340.
67. Stott, D. I. (1989) Immunoblotting and dot blotting. *J. Immunol. Meth.* **119,** 153–187.
68. Poxton, I. R. (1990) Immunoblotting techniques. *Curr. Opin. Immunol.* **2,** 905–909.
69. Harper, D. R., Ming-Liu, K., and Kangro, H. O. (1990) Protein blotting: ten years on. *J. Virol. Meth.* **30,** 25–40.
70. Mohammad, K. and Esen, A. (1989) A blocking agent and a blocking step are not needed in ELISA, immunostaining dot-blots and Western blots. *J. Immunol. Meth.* **117,** 141–145.
71. Flanagan, S. D. and Yost, B. (1984) Calmodulin-binding proteins: visualization by ^{125}I-calmodulin overlay on blots quenched with Tween 20 or bovine serum albumin and poly(ethylene oxide). *Analyt. Biochem.* **140,** 510–519.
72. Hoffman, W. L. and Jump, A. A. (1986) Tween 20 removes antibodies and other proteins from nitrocellulose. *J. Immunol. Meth.* **94,** 191–197.
73. Sadra, A., Cinek, T., and Imboden, J. B. (2000) Multiple probing of an immunoblot membrane using a non-block technique: advantages in speed and sensitivity. *Analyt. Biochem.* **278,** 235–237.
74. Krajewski, S., Zapata, J. M., and Reed, J. C. (1996) Detection of multiple antigens on Western blots. *Analyt. Biochem.* **236,** 221–228.
75. Bhatt, T. R., Taylor III, P. A., and Horodyski, F. M. (1997) Suppression of irrelevant signals in immunoblots by preconjugation of primary antibodies. *BioTechniques* **23,** 1006–1010.
76. Wilkinson, D. (2000) Chemiluminescent techniques for Western blot detection let researcher shed their lead aprons. *The Scientist* **14,** 29–32.

77. Hawkes, R., Niday, E., and Gordon, J. (1982) A dot-immunobinding assay for monoclonal and other antibodies. *Analyt. Biochem.* **119,** 142–147.
78. Steffen, W. and Linck, R. W. (1989) Multiple immunoblot: a sensitive technique to stain proteins and detect multiple antigens on a single two-dimensional replica. *Electrophoresis* **10,** 714–718.
79. Lemkin, P. (1996) Matching 2-D gels on the Internet, in the Abstract Book of *2nd Siena 2-D Electrophoresis Meeting*, **57,** Siena, Sept. 16–18.

28

Quantification of Radiolabeled Proteins in Polyacrylamide Gels

Wayne R. Springer

1. Introduction

Autoradiography is often used to detect and quantify radiolabeled proteins present after separation by polyacrylamide gel electrophoresis (PAGE) (*see* Chapter 38). The method, however, requires relatively high levels of radioactivity when weak β-emitters, such as tritium, are to be detected. In addition, lengthy exposures requiring the use of fluorescent enhancers are often required. Recent developments in detection of proteins using silver staining *(1,2)* have added to the problem because of the fact that tritium emissions are quenched by the silver *(2)*. Since for many metabolic labeling studies, tritium labeled precursors are often the only ones available, it seemed useful to develop a method that would overcome these drawbacks. The method the author developed involves the use of a cleavable crosslinking agent in the polyacrylamide gels that allows the solubilization of the protein for quantification by scintillation counting. Although developed for tritium *(3)*, the method works well with any covalently bound label, as demonstrated here with ^{35}S. Resolution is as good as or better than autoradiography (*3* and **Fig. 1**), turnaround time can be greatly reduced, and quantification is more easily accomplished.

2. Materials

Reagents should be of high quality, particularly the sodium dodecyl sulfate (SDS) to obtain the best resolution, glycerol to eliminate extraneous bands, and ammonium persulfate to obtain proper polymerization. Many manufacturers supply reagents designed for use in polyacrylamide gels. These should be purchased whenever practical. Unless otherwize noted, solutions may be stored indefinitely at room temperature.

2.1. SDS-Polyacrylamide Gels

1. Acrylamide-DATD: Acrylamide (45 g) and *N,N'*-dialyltartardiamide (DATD 4.5 g) are dissolved in water to a final volume of 100 mL. Water should be added slowly and time allowed for the crystals to dissolve.
2. Acrylamide-*bis*: Acrylamide (45 g) and *N,N'*-methylenebis(acrylamide) (*bis*-acrylamide, 1.8 g), are dissolved as in item 1 in water to a final volume of 100 mL.
3. Tris I: 0.285 *M* Tris-HCl, pH 6.8.

From: *The Protein Protocols Handbook, 2nd Edition*
Edited by: J. M. Walker © Humana Press Inc., Totowa, NJ

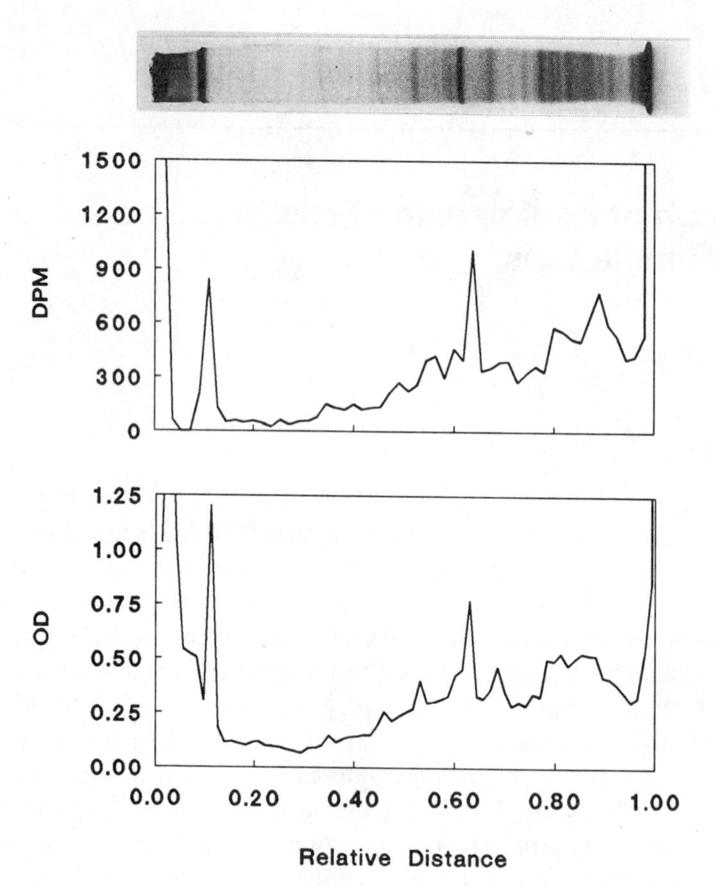

Fig. 1. Comparison of an autoradiogram, densitometry scan, and radioactivity in gel slices from a mixture of labeled proteins separated by PAGE. Identical aliquots of ^{35}S-methionine-labeled proteins from the membranes of the cellular slime mold, *D. purpureum,* were separated using the methods described in **Subheading 3.1.** One lane was fixed, soaked in Flouro-Hance (Research Products International Corp., Mount Prospect, IL), dried, and autoradiographed (photograph). Another lane was cut from the gel, sliced into 1-mm pieces, solubilized, and counted using a Tracor Mark III liquid scintillation counter with automatic quench correction (TM Analytic, Elk Grove Village, IL) as described in **Subheading 3.3.** The resultant disintigratiôns per minute (DPM) were plotted vs the relative distance from the top of the running gel (**top figure**). The autoradiogram (photograph) was scanned relative to the top of the running gel using white light on a Transidyne RFT densitometer (**bottom figure**).

4. Tris II: 1.5 *M* Tris-HCl, pH 8.8, 0.4% SDS.
5. Tris III: 0.5 *M* Tris-HCl pH 6.8, 0.4% SDS.
6. Ammonium persulfate (APS) solution: A small amount of ammonium persulfate is weighed out and water is added to make it 100 mg/mL. This should be made fresh the day of the preparation of the gel. The ammonium persulfate crystals should be stored in the refrigerator and warmed before opening.
7. *N,N,N',N'*-tetramethylethylenediamine (TEMED) is used neat as supplied by the manufacturer.
8. 10X Running buffer: Tris base (30.2 g), glycine (144.1 g), and SDS (10.00 g) are made up to 1 L with distilled water.

9. Sample buffer: The following are mixed together and stored at 4°C: 1.5 mL 20% (w/v) SDS, 1.5 mL glycerol, 0.75 mL 2-mercaptoethanol, 0.15 mL 0.2% (w/v) Bromophenol blue, and 1.1 mL Tris I.

2.2. Silver Staining of Gels

All solutions must be made with good-quality water.

1. Solution A: methanol:acetic acid:water (50:10:40).
2. Solution B: methanol:acetic acid:water (5:7:88).
3. Solution C: 10% (v/v) glutaraldehyde.
4. 10X Silver nitrate: 1% (w/v) silver nitrate stored in brown glass bottle.
5. Developer: Just before use, 25 µL of 37% formaldehyde are added to freshly made 3% (w/v) sodium carbonate.
6. Stop bath: 2.3 *M* Citric acid (48.3 g/100 mL).

2.3. Solubilization and Counting of Gel Slices

1. Solubilizer: 2% sodium metaperiodate.
2. Scintillation fluid: Ecolume (+) (ICN Biomedical, Inc., Irvine, CA) was used to develop the method. *See* **Subheading 4. Notes**. for other scintillants.

3. Methods

The general requirements to construct, prepare, and run SDS-PAGE gels as described by Laemmli *(4)* are given in Chapter 11. Presented here are the recipes and solutions developed in the author's laboratory for this particular technique. For most of the methods described a preferred preparation of a standard gel can be substituted, except for the requirement of the replacement of DATD for bis at a ratio of 1:10 DATD: acrylamide in the original formulation.

3.1. SDS-Polyacrylamide Gels

Sufficient medium for one 8 × 10 × 0.75 to 1.0-cm gel can be made by combining stock solutions and various amounts of acrylamide-DATD and water to achieve the required percentage of acrylamide (*see* **Note 1**) by using the quantities listed in **Table 1**. The stacking gel is that described by Laemmli *(4)* and is formed by combining 0.55 mL of acrylamide-*bis* with 1.25 mL of Tris III, 3.2 mL of water, 0.015 mL of APS, and 0.005 mL of TEMED.

1. For both the running and the stacking gel, degas the solutions by applying a vacuum for approx 30 s before adding the TEMED.
2. After the addition of the TEMED, pour the gels, insert the comb in the case of the stacking gel, and overlay quickly with 0.1% SDS to provide good polymerization.
3. Prepare protein samples by dissolving two parts of the protein sample in one part of sample buffer and heating to 100°C for 2 min.
4. Run gels at 200 V constant voltage for 45 min to 1 h or until the dye front reaches the bottom of the plate.

3.2. Silver Staining

The method used is that of Morrissey *(1)*.

1. Remove gels from the plates, and immerse in solution A for 15 min.
2. Transfer to solution B for 15 min and then solution C for 15 min, all while gently shaking.

Table 1
Recipe for Various Percentages of SDS-PAGE Gels

	Percentage of acrylamide		
	7.5	10	12.5
Stock solution			
Acrylamide-DATD, mL	0.96	1.33	1.66
Tris II, mL	1.50	1.50	1.50
Water, mL	3.50	3.13	2.80
APS, mL	0.03	0.03	0.03
TEMED, mL	0.003	0.003	0.003

3. At this point, the gel can be rinsed in glass-distilled water for 2 h to overnight (*see* **Note 2**).
4. After the water rinse, add fresh water and enough crystalline dithiotreitol (DTT) to make the solution 5 µg/mL.
5. After 15 min, remove the DTT, and add 0.1% silver nitrate made fresh from the 1% stock solution. Shake for 15 min.
6. Rapidly rinse the gel with a small amount of water followed by two 5–10 mL rinses with developer followed by the remainder of the developer.
7. Watch the gel carefully, and add stop bath to the gel and developer when the desired darkness of the bands is reached.
8. Store the stopped gel in water until the next step.

3.3. Slicing and Counting of Gel Slices

1. Remove individual lanes from the gel for slicing by cutting with a knife or spatula.
2. Cut each lane into uniform slices, or cut identified bands in the gel (*see* **Note 3**).
3. Place each slice into a glass scintillation vial, and add 0.5 mL of 2% sodium metaperiodate solution.
4. Shake the vials for 30 min to dissolve the gel.
5. Add a 10-mL aliquot of scintillation fluid to the vial, cool the vial, and count in a refrigerated scintillation counter (*see* **Notes 4–6**).

4. Notes

4.1. SDS-Polyacrylamide Gels

1. Acrylamide-DATD gels behave quite similarly to acrylamide-bis gels, except for the fact that for a given percentage of acrylamide, the relative mobility of all proteins are reduced in the DATD gel. In other words, a DATD gel runs like a higher percentage acrylamide-*bis* gel.

4.2. Silver Staining

2. The water rinse can be reduced to 1 h, if the water is changed at 10–15 min intervals. The author found in practice that the amount of DTT was not particularly critical, and routinely added the tip of a microspatula of crystals to approx 30 mL of water.

4.3. Slicing and Counting of Gel Slices

3. The method seems relatively insensitive to gel volume, as measured by changes in efficiency *(3)* over the range of 5–100 mm^3 for tritium and an even larger range for ^{14}C or ^{35}S.

Larger pieces of gel can be used if one is comparing relative amounts of label, such as in a pulse chase or other timed incorporation, but longer times and more metaperiodate may be required to dissolve the gel. Recovery of label from the gel is in the 80–90% range for proteins from 10–100 kDa *(3)*. In most cases, the label will remain with the protein, but in the case of periodate-sensitive carbohydrates on glycoproteins, it may be released from the protein. This, however, does not prevent the quantification. It just does not allow the solubilized labeled protein to be recovered for other manipulations.

4. It is necessary to cool the vials in the counting chamber before counting in order to eliminate occasional chemiluminescence. The cause of this phenomenon was not explored, but one should determine whether this occurs when using other scintillants or counters.
5. **Figure 1** shows the results of a typical experiment using ^{35}S-methionine to label proteins metabolically from the cellular slime mold, *Dictyostelium purpureum,* and quantify them using the method described or by autoradiography. As can be seen, the resolution of the method is comparable to that of the autoradiogram (*see* **ref. *3*** for similar results using tritium). The time to process the lane by the method described here was approx 8 h, whereas, the results of the autoradiogram took more than 3 d to obtain. Examination of the stained gel suggests that, if one were interested in a particular protein, it would be fairly easy to isolate the slice of gel containing that protein for quantification. This makes this method extremely useful for comparing incorporation into a single protein over time as in pulse/chase experiments *(3)*.
6. The author has found that as little as 400 dpm of tritium associated with a protein could be detected *(3)*, which makes this method particularly useful for scarce proteins or small samples.

References

1. Morrissey, J. H. (1981) Silver stain for proteins in polyacrylamide gels: a modified procedure with enhanced uniform sensitivity. *Analyt. Biochem.* **117,** 307–310.
2. Van Keuren, M. L. Goldman, D., and Merril, C. R. (1981) Detection of radioactively labeled proteins is quenched by silver staining methods: quenching is minimal for ^{14}C and partially reversible for ^{3}H with a photochemical stain. *Analyt. Biochem.* **116,** 248–255.
3. Springer, W. R. (1991) A method for quantifying radioactivity associated with protein in silver-stained polyacrylamide gels. *Analyt. Biochem.* **195,** 172–176.
4. Laemmli, U. K. (1970) Cleavage of structural proteins during assembly of the head of bacteriophage T4. *Nature* **227,** 680–685.

29

Quantification of Proteins on Polyacrylamide Gels

Bryan John Smith

1. Introduction

Quantification of proteins is a common challenge. There are various methods described for estimation of the amount of protein present in a sample, for example, amino acid analysis and the bicinchoninic acid (BCA) method (*see* Chapters 83 and 3). Proteins may be quantified in sample solvent prior to electrophoresis and can provide an estimate of total amount of protein in a sample loaded onto a gel ***(1,2)***. These methods quantify total protein present but not of one protein in a mixture of several. For this, the mixture must be resolved. Liquid chromatography achieves this, and the various proteins may be quantified by their absorbancy at 220 nm or 280 nm. Microgram to submicrogram amounts of protein can be analyzed in this way, using microbore high-performance liquid chromatography (HPLC). Problems may occur if particular species chromatograph poorly (such as hydrophobic polypeptides on reverse-phase chromatography).

An alternative is polyacrylamide gel electrophoresis (as the mixture-resolving step) followed by protein staining and densitometry. Exceptions are small peptides that are not successfully resolved and stained in gels. Microgram to submicrogram amounts of protein (of more than a few kilodaltons in size) may be quantified in this way. Not every protein stain is best suited to quantification in gels, however. The popular noncolloidal Coomassie gel stains such as PAGE blue '83 are not suitable. As discussed by Neuhoff et al. ***(3)***, staining by Coomassie dye is difficult to control—it may not fully penetrate and stain dense bands of protein, it may demonstrate a metachromatic effect (whereby the protein–dye complex may show any of a range of colors from blue through purple to red), and it may be variably or even completely decolorized by excessive destaining procedures (because the dye does not bind covalently). As a consequence, stoichiometric binding of Coomassie dye to protein is commonly not achieved. Sensitive silver stains are commonly used to detect rare proteins at levels at levels of the order of 0.05 ng/mm^2 in two-dimensional (2-D) gels, for example, but they, too, are not suitable for quantification. For various reasons, proteins vary widely in their stainability by silver—some do not stain at all, others may show a metachromatic effect. Use of silver staining for quantification is complicated by other factors,

From: *The Protein Protocols Handbook, 2nd Edition*
Edited by: J. M. Walker © Humana Press Inc., Totowa, NJ

too—such as the difficulty in obtaining staining throughout the gel, not just at the surface. This subject is reviewed at greater length in **ref. *4***, but in summary it may be said that while silver stain methods have great sensitivity, they are extremely problematical for quantification purposes.

One quantitative staining method generates negatively stained bands ***(5)***. The technique generates a white background of precipitated zinc salt, against which clear bands of protein (where precipitation is inhibited). This may be viewed by dark field illumination. The gel may be scanned at this stage, the negative staining being approximately quantitative (for horse myoglobin) in a range from about 100 ng/mm^2 to 2 µg/mm^2 or more. Sensitivity may be improved by staining of the background by incubation with tolidine. The method was described for 12% and 15% T gels, but the degree of zinc salt precipitation and subsequent toning is dependent upon %T and at closer to 20%T the toning procedure may completely destain the gel, restricting the usefulness of the method.

A number of other quantitative staining methods have been discussed in the literature. Notably, Neuhoff et al. ***(3)*** have made a thorough study of various factors affecting protein staining by Coomassie Brilliant Blue G-250 (0.1% [w/v] Color Index number 42655 in 2% [w/v] phosphoric acid, 6% [w/v] ammonium sulfate). This colloidal stain does not stain the background in a gel and so washing can give crystal-clear backgrounds. In my own laboratory I have used a similar colloidal Coomassie Brilliant Blue G250 perchloric acid staining procedure for quantification of proteins on gels. The method derives from that of Reisner et al. ***(6)***. The stain is 0.04% (w/v) Brilliant Blue G in 3.5% (w/v) perchloric acid, and is available commercially, ready to use. This simple method is described below.

O'Keefe *(7)* described a method whereby cysteine residues are labeled with the thiol-specific reagent monobromobimane. On transillumination (at λ = 302 nm) labeled bands fluoresce. However, the method requires quantitative reaction of cysteines, which may be totally lacking in some polypeptides, and a nominal detection limit of 10 pmol of cysteine is claimed. Better luminescent stains have been developed by Molecular Probes—the Sypro family of dyes. The Sypro Ruby stain is possibly the most sensitive of these, approaching the sensitivity of silver stains. The method for use of this stain is given below.

2. Materials

2.1. Colloidal Coomassie Brilliant Blue G Method

1. A suitable scanning densitometer, for example, the Molecular Dynamics Personal Densitometer SI with Image Quant software or the Bio-Rad Fluor S MultiImager with Multi Analyst software. Such equipment can scan transparent objects such as wet gels, gels dried between transparent films, and photographic film, and digitize the image that may then be analyzed.
2. Protein stain: Sigma, product no. B-8772:0.04% (w/v) Coomassie Brilliant Blue G (C.I. 42655) in 3.5% (w/v) perchloric acid (*see* **Note 1**). Stable for months at room temperature, in the dark. Beware the low pH of this stain. Wear protective clothing. Use fresh, undiluted stain, as supplied.
3. Destaining solution: Distilled water.

2.2. Sypro Ruby Method

1. Equipment for scanning fluorescent bands on gels, for example, the Bio-Rad Fluor S MultiImager with Multi Analyst software, which can be used to photographically record or scan wet gels, allowing repeated scanning procedures (as staining or destaining proceeds). Simple viewing of stained bands may be done under a hand-held 300-nm UV lamp or on a transilluminator. Protect eyes with UV-opaque glasses. The Sypro dye may be excited at 280 or 450 nm and emits at 610 nm.
2. Protein stain: Sypro Ruby gel stain (Molecular Probes, product number S-12000 or S-12001) (*see* **Note 2**). This stain is stable at room temperature in the dark. Exact details of the stain are not revealed but according to the manufacturer, the stain contains neither hazardous nor flammable materials. Use fresh, undiluted stain only.
3. Destain: Background may be reduced by rinsing in water.
4. Clean dishes, free of dust that might contribute to background staining.

3. Method

3.1. Colloidal Coomassie Brilliant Blue G Method

1. At the end of electrophoresis, wash the gel for a few minutes with several changes of water (*see* **Note 3**), then immerse the gel (with gentle shaking) in the colloidal Coomassie Brilliant Blue G. This time varies with the gel type (e.g., 1.0–1.5 h for a 1–1.5 mm thick sodium dodecyl sulfate [SDS] polyacrylamide gel slab), but cannot really be overdone. Discard the stain after use, for its efficacy declines with use.
2. At the end of the staining period, decolorize the background by immersion in distilled water, with agitation, and a change of water whenever it becomes colored. Background destaining is fairly rapid, giving a clear background after a few hours (*see* **Notes 4** and **5**).
3. Measure the extent of blue dye bound by each band by scanning densitometry. Compare the dye bound by a sample with those for standard proteins run and stained in parallel with the sample, on the same gel (*see* **Notes 6–8** and **11–17**).

3.2. Sypro Ruby Method

1. At the end of electrophoresis, rinse the gel in water briefly, put it into a clean dish and then cover it with Sypro Ruby gel stain solution. Gently agitate until staining is completed, which may take up to 24 h or longer (*see* **Note 9**). Overstaining will not occur during prolonged stained. Do not let the stain dry up on the gel during long staining procedures. Discard the stain after use, for it becomes less efficacious with use. During the staining procedure the gel may be removed from the stain and inspected under UV light to monitor progress. If the staining is insufficient, the gel may be replaced in the stain for further incubation.
2. Destain the background by washing the gel in a few changes of water for 15–30 min.
3. Measure the extent of dye bound, that is, the luminescence, by each band by scanning. Compare the dye bound by a sample with that for standard proteins run and stained in parallel with the sample, on the same gel (*see* **Note 10**).

4. Notes

1. The Coomassie Brilliant Blue G stain mixture may be readily made from the components. A commercially available alternative is the Gel Code blue stain reagent from Pierce (product no. 24590 or 24592). Details of the stain components are not divulged, other than they also include Coomassie (G250), but the stain is used in the same way as described for the Sigma reagent, gives similar results, and costs approximately the same.

2. Molecular Probes do not reveal the components of their reagent. Its cost is of the order of twice that of the Sigma Brilliant Blue G stain. Although more sensitive than the blue stain, it does require UV irradiation for detection.
3. The water wash that precedes staining by Coomassie Brilliant Blue G is intended to wash away at least some SDS from the gel, and so speed up destaining of the background. However, it should be remembered that proteins are not fixed in the gel until in the acidic stain mixture and consequently some loss of small polypeptides may occur in the wash step. Delete the wash step if this is of concern, or employ a fixing step immediately after electrophoresis, for example, methanol/glacial acetic acid/water: 50:7:43, by volume for 15–30 min, followed by water washing to remove the solvent and acid.
4. Destaining of the background may be speeded up by frequent changes of the water, and further by inclusion in this wash of an agent that will absorb free dye. Various such agents are commercially available (e.g., Cozap, from Amika Corp.), but a cheap alternative is a plastic sponge of the sort used to plug flasks used for microbial culture. The agent absorbs the stain and is subsequently discarded. The background can be made clear by these means, and the stained bands remain stained while stored in water for weeks. They may be restained if necessary.
5. The Coomassie Brilliant Blue G-stained gel may be dried between transparent sheets of dialysis membrane or cellophane (available commercially, e.g., from Novex) for storage and later scanning. Beware that bubbles or marks in the membrane may add to the background noise upon scanning.
6. Heavily loaded samples show up during staining with Coomassie Brilliant Blue G, but during destaining the blue staining of the proteins becomes accentuated. Bands of just a few tens of ng are visible on a 1 mm-thick gel (i.e., the lower limit of detection is < 10 ng/mm^2). Variability may be experienced from gel to gel, however. For example, duplicate loadings of samples on separate gels, electrophoresed and stained in parallel, have differed in staining achieved by a factor of 1.5, for reasons unknown. Furthermore, different proteins bind the dye to different extents: horse myoglobin may be stained twice as heavily as is bovine serum albumin, although this, too, is somewhat variable. While this Coomassie Brilliant Blue G is a good general protein stain, It is advisable to treat sample proteins on a case by case basis.
7. It is a requirement of this method that dye is bound stoichiometrically to polypeptide over a useful wide range of sample size. The staining by Coomassie Brilliant Blue G may be quantitative, or nearly so, from about 10 to 20 ng/mm^2 up to large loadings of 1–5 µg/mm^2. Concentrated protein solutions may be diluted to fall within the useful range.

   The stoichiometry of dye binding is subject to some variation, such that standard curves may be either linear or slightly curved, but even the latter case is acceptable provided standards are run on the same gel as samples.
8. Because performance of the Coomassie Brilliant Blue G stain is variable it essential to run and stain standards and samples on the same gel, and to do so in duplicate.
9. The Sypro Ruby method allows for periodic observation and further staining if required. It can be the case that for small loadings, however, prolonged staining (24 h or more) may be required to obtain best sensitivity. The stained gel may be dried for storage but can then lose luminescence.
10. The Sypro Ruby method is subject to the same sort of variation as the Coomassie Brilliant Blue G method, from gel to gel with standard curves being linear or slightly curved. Generally the Sypro Ruby method is more sensitive than the Coomassie Brilliant Blue G method, for instance about fivefold more so for bovine serum albumin. However, protein to protein variation may apparently be greater in the case of the Sypro Ruby stain. For

instance, horse myoglobin binds about 10-fold less dye than bovine serum albumin does. Thus in one experiment, the minimum amount of bovine serum albumin detectable after Sypro Ruby staining was about 5 ng/mm^2 (about four- to fivefold more sensitive than samples stained in parallel by the Coomassie Brilliant Blue G method), whereas the minimum amount of horse myoglobin detectable was about 50 ng/mm^2 (similar to that detectable by the Coomassie Brilliant Blue G stain). It is therefore recommended that samples and standards are run and stained in parallel on the same gel, in duplicate.

11. Best quantification is achieved after having achieved good electrophoresis. Adapt the electrophoresis as necessary to achieve good resolution, lack of any band smearing or tailing, and lack of retention of sample at the top of the gel by aggregation. For stains where penetration of dense bands may be a problem, avoid dense bands by reducing the size of the loaded sample and/or electrophoresing the band further down the length of the gel (to disperse the band further) and/or use lower %T gel. Thin gels, of 1 mm thickness or less, allow easier penetration of dye throughout their thickness. Gradient gels have a gradient of pore size that may cause variation of band density and background staining. Use nongradient gels if this is a problem.
12. For absolute quantification of a band on a gel, accurate pipetting is required, as is a set of standard protein solutions that cover the concentration range expected of the experimental sample. Ideally, adjust the concentrations of the various standards and a sample so that the volume of each that is loaded onto the gel is the same and the need for pipet adjustment is minimized. Run and stain these standard solutions at the same time and if possible on the same gel, as the experimental sample in order to reduce possible variations in band resolution or staining, background destaining and so on. Ideally the standard protein should be the same as that to be quantified but if, as is commonly the case, this is impossible then another protein may be used (while recognizing that this protein may bind a different amount of dye from that by the experimental protein, so that the final estimate obtained may be in error). The standard protein should have similar electrophoretic mobility to the proteins of interest, so that any effect such as dye penetration, due to pore size, is similar. Make the standard protein solution by dissolving a relatively large and accurately weighed amount of dry protein in water or buffer, and dilute this solution as required. If possible, check the concentration of this standard solution by alternative means (say amino acid analysis). Standardize treatment of samples in preparation for electrophoresis—treating of sample solutions prior to SDS-PAGE may cause sample concentration by evaporation of water for instance. To minimize this problem heat in small (0.5 mL) capped Eppendorf tubes, cool, and briefly centrifuge to take condensed water back down to the sample in the bottom of the tube.
13. When analyzing the results of scanning, construct a curve of absorbency vs protein concentration from standard samples and compare the experimental sample(s) with this. Construct a standard curve for each experiment.
14. If comparing the abundance of one protein species with others in the same sample, then standards are not required, provided that no species is so abundant that dye binding becomes saturated. Be aware that such relative estimates are approximate, as different proteins bind dye to different extents.
15. It is sometimes observed in electrophoresis that band shape and width are irregular—heavily loading a gel can generate a broad band that may interfere with the running of neighboring bands, for instance. It is necessary to include all of such a band for most accurate results.
16. Avoid damage to the gel—a crack can show artificially as an absorbing band (or peak) on the scan. Gels of low %T are difficult to handle without damage, but they may be made

tougher (and smaller) by equilibration in aqueous ethanol, say, 40% (v/v) ethanol in water, 1 h or so. Too much ethanol may cause the gel to become opaque, but if this occurs merely rehydrate the gel in a lower % (v/v) ethanol solution. A gradient gel may assume a slightly trapezoid shape upon shrinkage in ethanol solutions—this makes scanning tracks down the length of the gel more difficult. When scanning, eliminate dust, trapped air bubbles, and liquid droplets, all of which contribute to noisy baseline.

17. Methods have been described for quantification of submicrogram amounts of proteins that have been transferred from gels to polyvinylidene difluoride or similar membrane (e.g., *7*). A Sypro dye blot stain equivalent to the method described above for gels is available from Molecular Probes. Note, however, that transfer from gel to membrane need not (indeed, usually does not) proceed with 100% yield, so that results do not necessarily accurately reflect the content of the original sample.

References

1. Dráber, P. (1991) Quantification of proteins in sample buffer for sodium dodecyl sulfate-polyacrylamide gel electrophoresis using colloidal silver. *Electrophoresis* **12,** 453–456.
2. Henkel, A. W. and Bieger, S. C. (1994) Quantification of proteins dissolved in an electrophoresis sample buffer. *Analyt. Biochem.* **223,** 329–331.
3. Neuhoff, V., Stamm, R., Pardowitz, I., Arold, N., Ehhardt, W., and Taube, D. (1990) Essential problems in quantification of proteins following colloidal staining with Coomassie Brilliant Blue dyes in polyacrylamide gels, and their solution. *Electrophoresis* **11,** 101–117.
4. Syrovy, I. and Hodny, Z. (1991) Staining and quantification of proteins separated by polyacrylamide gel electrophoresis. *J. Chromatogr.* **569,** 175–196.
5. Ferreras, M., Gavilanes, J. G., and Garcia-Segura, J. M. (1993) A permanent Zn^{2+} reverse staining method for the detection and quantification of proteins in polyacrylamide gels. *Analyt. Biochem.* **213,** 206–212.
6. Reisner, A. H., Nemes, P., and Bucholtz, C. (1975) The use of Coomassie Brilliant Blue G250 perchloric acid solution for staining in electrophoresis and isoelectric focusing on polyacrylamide gels. *Analyt. Biochem.* **64,** 509–516.
7. O'Keefe, D. O. (1994) Quantitative electrophoretic analysis of proteins labeled with monobromobimane. *Analyt. Biochem.* **222,** 86–94.
8. Patton, W. F., Lam, L., Su, Q., Lui, M., Erdjument-Bromage, H., and Tempst, P. (1994) Metal chelates as reversible stains for detection of electroblotted proteins: application to protein microsequencing and immune-blotting. *Analyt. Biochem.* **220,** 324–335.

30

Rapid and Sensitive Staining of Unfixed Proteins in Polyacrylamide Gels with Nile Red

Joan-Ramon Daban, Salvador Bartolomé Antonio Bermúdez, and F. Javier Alba

1. Introduction

Sodium dodecyl sulfate-polyacrylamide gel electrophoresis (SDS-PAGE) is one of the most powerful methods for protein analysis *(1,2)*. However, the typical procedures for the detection of protein bands after SDS-PAGE, using the visible dye Coomassie Blue and silver staining, have several time-consuming steps and require the fixation of proteins in the gel. This chapter describes a rapid and very simple method for protein staining in SDS gels developed in our laboratory *(3–5)*. The method is based on the fluorescent properties of the hydrophobic dye Nile red (9-diethylamino-5*H*-benzo[α]phenoxazine-5-one; *see* **Fig. 1**), and allows the detection of < 10 ng of unfixed protein per band about 5 min after the electrophoretic separation. Furthermore, it has been shown elsewhere *(6,7)* that, in contrast to the current staining methods, Nile red staining does not preclude the direct electroblotting of protein bands and does not interfere with further staining, immunodetection and sequencing (*see* Chapter 50).

Nile red was considered a fluorescent lipid probe, because this dye shows a high fluorescence and intense blue shifts in presence of neutral lipids and lipoproteins *(8)*. We have shown that Nile red can also interact with SDS micelles and proteins complexed with SDS *(3)*. Nile red is nearly insoluble in water, but is soluble and shows a high increase in the fluorescence intensity in nonpolar solvents and in presence of pure SDS micelles and SDS–protein complexes. In the absence of SDS, Nile red can interact with some proteins in solution but the observed fluorescence is extremely dependent on the hydrophobic characteristics of the proteins investigated *(8,9)*. In contrast, Nile red has similar fluorescence properties in solutions containing different kinds of proteins associated with SDS, suggesting that this detergent induces the formation of structures having equivalent hydrophobic properties independent of the different initial structures of native proteins *(3)*. In agreement with this, X-ray scattering and cryoelectron microscopy results have shown that proteins having different properties adopt a uniform necklace-like structure when complexed with SDS *(10)*. In these

From: *The Protein Protocols Handbook, 2nd Edition*
Edited by: J. M. Walker © Humana Press Inc., Totowa, NJ

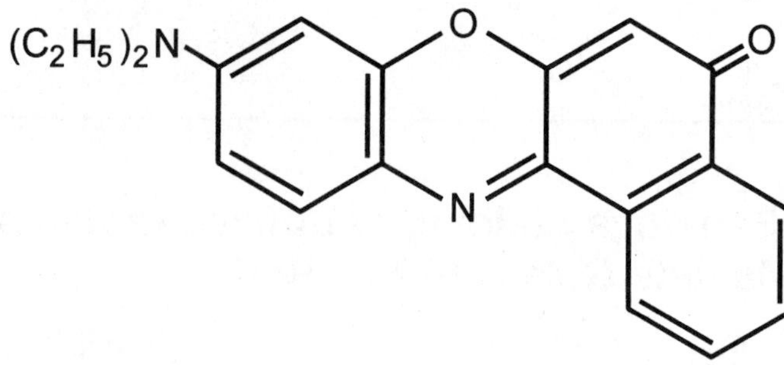

$(C_2H_5)_2N$ ... O ... =O ... N

Fig. 1. Structure of the noncovalent hydrophobic dye Nile red.

structures the polypeptide chain is mostly situated at the interface between the hydrocarbon core and the sulfate groups of the SDS micelles dispersed along the unfolded protein molecule.

The enhancement of Nile red fluorescence observed with different SDS–protein complexes occurs at SDS concentration lower than the critical micelle concentration of this detergent in the typical Tris–glycine buffer used in SDS-PAGE *(3)*. Thus, for Nile red staining of SDS-polyacrylamide gels *(4)*, electrophoresis is performed in the presence of 0.05% SDS instead of the typical SDS concentration (0.1%) used in current SDS-PAGE protocols. This concentration of SDS is high enough to maintain the stability of the SDS–protein complexes in the bands, but is lower than SDS critical micelle concentration and consequently precludes the formation of pure detergent micelles in the gel *(4,5)*. The staining of these modified gels with Nile red produces very high fluorescence intensity in the SDS–protein bands and low background fluorescence (*see* **Fig. 2**). Furthermore, under these conditions (*see* details in **Subheading 3.**), most of the proteins separated in SDS gels show similar values of the fluorescence intensity per unit mass.

2. Materials

All solutions should be prepared using electrophoresis-grade reagents and deionized water and stored at room temperature (exceptions are indicated). Wear gloves to handle all reagents and solutions and do not pipet by mouth. Collect and dispose all waste according to good laboratory practice and waste disposal regulations.

1. Nile red. Concentrated stock (0.4 mg/mL) in dimethyl sulfoxide (DMSO). This solution is stable for at least 3 mo when stored at room temperature in a glass bottle wrapped in aluminum foil to prevent damage by light. **Handle this solution with care, DMSO is flammable, and, in addition, this solvent may facilitate the passage of water-insoluble and potentially hazardous chemicals such Nile red through the skin.** Nile red can be obtained from Sigma Chemical (St. Louis, MO).
2. The acrylamide stock solution and the resolving and stacking gel buffers are prepared as described in Chapter 11.
3. 2× Sample buffer: 4% (w/v) SDS, 20% (v/v) glycerol, 10% (v/v) 2-mercaptoethanol, 0.125 *M* Tris-HCl, pH 6.8; bromophenol blue (0.05% [w/v]) can be added as tracking dye.
4. 10× Electrophoresis buffer: 0. 5% (w/v) SDS, 0.25 *M* Tris, 1.92 *M* glycine, pH 8.3 (do not adjust the pH of this solution).

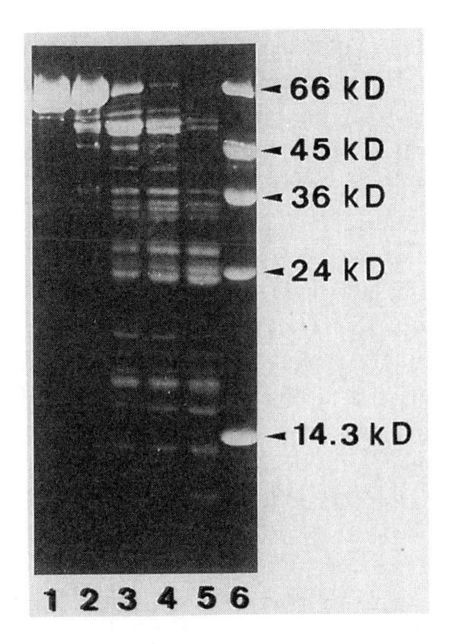

Fig. 2. Example of Nile red staining of different proteins and peptides in 0.05% SDS–15% polyacrylamide gels. The protein molecular weight markers (*lane 6*, from top to bottom: BSA, ovalbumin, glyceraldehyde-3-phosphate dehydrogenase, trypsinogen, and lysozyme), and BSA digested with increasing amounts of trypsin (*lanes 1–5*), were stained for 5 min with a solution of Nile red in water prepared by quick dilution of a stock solution of this dye in DMSO (*see* **Subheading 3.**).

5. Plastic boxes for gel staining. Use opaque polypropylene containers (e.g., 21 × 20 × 7.5 and 12 × 7.5 × 7 cm for large [20 × 16 × 0.15 cm] and small [8 × 6 × 0.075 cm] gels, respectively) with a close-fitting lid to allow intense agitation without spilling the staining solution.
6. Orbital shaker for the agitation of the plastic boxes during gel staining.
7. Transilluminator equipped with midrange ultraviolet (UV) bulbs (~300 nm) to excite Nile red *(3,4)*. A transilluminator with a cooling fan (e.g., Foto UV 300 [Fotodyne Inc., Harland, WI] or similar) is very convenient to prevent thermal damage of the gel when long exposures are necessary (*see* **Subheading 3., step 11**). UV light is dangerous to skin and particularly to eyes. UV-blocking goggles and a full face shield, and protective gloves and clothing, should be worn when the stained gel is examined using the transilluminator. For the photography, we place the transilluminator with the gel and the photographic camera inside a home-made cabinet with opaque (UV-blocking) curtains to prevent operator exposure to UV light. Alternatively, the gel can be transilluminated inside a compact darkroom of a documentation system provided with a charge-coupled device (CCD) camera (e.g., Gel Doc 1000 from Bio-Rad Laboratories [Hercules, CA]). To avoid the problems associated with UV light, we have constructed a transilluminator that works in the visible region (~540 nm; green-light) and can also be used for the excitation of Nile red stained bands *(11)*.
8. Photographic camera (e.g., Polaroid [Cambridge, MA] MP-4 camera) or a CCD camera.
9. Optical filters. With the Polaroid camera and UV transilluminator use the Wratten (Kodak [Rochester, NY]) filters number 9 (yellow) and 16 (orange) to eliminate the UV and visible light from the transilluminator. Place the two filters together in the filter holder of the camera so that filter 16 is on top of filter 9 (i.e., filter 16 should be facing towards the

camera lens). Store the filters in the dark and protect them from heat, intense light sources and humidity. With the CCD camera and UV transilluminator we have used the filter 590DF100 (Bio-Rad). Wratten filter number 26 (red) is required for both Polaroid and CCD cameras when the bands are visualized with the green-light transilluminator.

10. Photographic films. The following Polaroid instant films can be used for the photography of the red bands (maximum emission at ~640 nm *[3]*) seen after staining: 667 (3000 ASA, panchromatic, black-and-white positive film), 665 (80 ASA, panchromatic, black-and-white positive/negative film), and 669 (80 ASA, color, positive film). Store the films at 4°C.
11. Negative-clearing solution: 18% (w/v) sodium sulfite.
12. Wetting agent for Polaroid 665 negatives (Kodak Photo-Flo diluted at 1:600).
13. Densitometry: We have used a Shimadzu (Tokyo, Japan) CS-9000 densitometer for the scanning of photographic negatives; other densitometers available commercially can also be used. Alternatively, the images obtained with a CCD camera can be directly analyzed with the appropriate software (e.g., Molecular Analyst [Bio-Rad]).

3. Method

The method described in this part gives all the details for the staining SDS-polyacrylamide gels with Nile red. *See* **Note 1** for Nile red staining of gels without SDS. Unless otherwise indicated, all operations are performed at room temperature.

1. Typically we prepare 15% acrylamide–0.4% *bis*-acrylamide separating gel containing 0.05% SDS (*see* **Note 2**), 0.375 *M* Tris-HCl, pH 8.8 . The stacking gel contains 6% acrylamide–0.16% *bis*-acrylamide, 0.05% SDS (*see* **Note 2**), 0.125 *M* Tris-HCl, pH 6.8
2. Dissolve the proteins in water, add one volume of the 2× sample buffer (**Subheading 2.3.**; *see* **Note 3**), and incubate the resulting samples in a boiling-water bath for 3 min. The samples are kept at room temperature before loading them into the gel (*see* **Note 4**).
3. Place the gel sandwiched between the glass plates in the electrophoresis apparatus, fill the electrode reservoirs with 1× electrophoresis buffer (*see* **Note 2**), and rinse the wells of the gel with this buffer.
4. Load the protein samples (about 20 and 10 µL in large and small gels, respectively), carry out electrophoresis, and at the end of the run remove the gel sandwich from the electrophoresis apparatus and place it on a flat surface.
5. Wearing gloves, remove the upper glass plate (use a spatula) and excise the stacking gel and the bottom part of the separating gel (*see* **Note 5**).
6. Add quickly 2.5 mL of the concentrated Nile red (0.4 mg/mL) staining solution in DMSO to 500 mL of deionized water previously placed in a plastic box (*see* **Note 6**). These volumes are required for staining large gels; for small gels, use 0.25 mL of the concentrated Nile red solution in DMSO and 50 mL of water.
7. Immediately after the addition of concentrated solution of Nile red to water, agitate the resulting solution vigorously for 3 s (*see* **Note 6**).
8. Immerse the gel very quickly in the staining solution, put the lid on, and agitate vigorously (*see* **Note 6**) using an orbital shaker (at about 150 rpm) for 5 min (*see* **Note 7**).
9. Discard the staining solution and rinse the gel with deionized water (4×; about 10 s) to remove completely the excess Nile red precipitated during the staining of the gel (*see* **Note 8**).
10. Wearing gloves, remove the gel from the plastic box and place it on the UV or green-light transilluminator. Turn off the room lights, turn on the transilluminator and examine the protein bands, which fluoresces light red (*see* **Note 9**). Turn off the transilluminator immediately after the visualization of the bands (*see* **Note 10**).

11. Focus the camera (Polaroid or CCD) with the help of lateral illumination with a white lamp, place the optical filters described in **Subheading 2., item 9**, in front of the camera lens and, in the dark, turn on the transilluminator and photograph the gel (*see* **Note 10**). Finally, turn off the transilluminator.
12. Develop the different Polaroid films for the time indicated by the manufacturer (*see* **Note 11**). Spread the Polaroid print coater on the surface of the 665 positive immediately after development. (The positive prints of the 667 and 669 films do not require coating.) The 665 negatives can be stored temporary in water but, before definitive storage, immerse the negatives in 18% sodium sulfite and agitate gently for about 1 min, wash in running water for 5 min, dip in a solution of wetting agent (about 5 s), and finally dry in air (*see* **Note 12**).
13. Scan the photographic negatives to determine the intensity of the Nile red stained protein bands if a quantitative analysis is required. The quantitative analysis can also be performed directly from the image obtained with a CCD documentation system (*see* **Note 13**).

4. Notes

1. Isoelectric focusing gels do not contain SDS and should be treated with this detergent after the electrophoretic run to generate the hydrophobic SDS–protein complexes specifically stained by Nile red *(6)*. In general, for systems without SDS we recommend an extensive gel washing (20 min in the case of 0.75-mm thick isoelectric focusing gels) with 0.05% SDS, 0.025 *M* Tris, 0.192 *M* glycine, pH 8.3, after electrophoresis. The gel equilibrated in this buffer can be stained with Nile red following **steps 6–9** of **Subheading 3.**
2. To reduce the background fluorescence after the staining with Nile red it is necessary to preclude the formation of pure SDS micelles in the gel (*see* **Subheading 1.**). Thus, use 0.05% SDS to prepare both the separating and stacking gels and the electrophoresis buffer. This concentration is lower than the critical micelle concentration of this detergent (~0.1% *[3]*), but is high enough to allow the formation of the normal SDS–protein complexes that are specifically stained by Nile red *(4)*.
3. Use 2% SDS in the sample buffer to be sure that all protein samples are completely saturated with SDS. Lower concentrations of SDS in the sample buffer can produce only a partial saturation of proteins (in particular in highly concentrated samples), and consequently the electrophoretic bands could have anomalous electrophoretic mobilities. The excess SDS (uncomplexed by proteins) present in the sample buffer migrates faster than the proteins and form a broad band at the bottom of the gel (*see* **Note 5** and **ref.** *4*).
4. Storage of protein samples prepared as indicated in **Subheading 3.2.** at 4°C (or at lower temperatures) causes the precipitation of the SDS present in the solution. These samples should be incubated in a boiling-water bath to redissolve SDS before using them for electrophoresis.
5. The bottom part of the gel (i.e., from about 0.5 cm above the bromophenol blue band to the end) should be excised before the staining of the gel. Otherwise, after the addition of Nile red, the lower part of the gel produces a broad band with intense fluorescence. This band is presumably caused by the association of Nile red with the excess SDS used in the sample buffer (*see* **Note 3**). In the case of long runs, bromophenol blue and the excess SDS band diffuse into the buffer of the lower reservoir and it is not necessary to excise the gel bottom.
6. Nile red is very stable when dissolved in DMSO (*see* **Subheading 2.1.**), but this dye precipitates in aqueous solutions. Since the precipitation of Nile red in water is a rapid process and this dye is only active for the staining of SDS–protein bands before it is completely precipitated *(4,5)*, to obtain satisfactory results, it is necessary: (a) to perform all the agitations indicated in these steps (in order to favor as much as possible the disper-

sion of the dye); (b) to work very rapidly in the **steps 6–9** of **Subheading 3.** Furthermore, to obtain a homogeneous staining the gel has to be completely covered with the staining solution.

7. The staining time is the same for large and small gels.
8. After the water rinsing indicated in the **step 9** of **Subheading 3.**, the staining process is completely finished and the gel can be photographed immediately. It is not necessary, however, to examine and photograph the gel just after staining. Nile red stained bands are stable and the gel can be kept in the plastic box immersed in water for 1–2 h before photography.
9. About 100 ng of protein per band can be seen by naked eye (when using the green-light transilluminator it is necessary to place a Wratten filter number 26 in front of the eyes to visualize the bands). Faint bands that are not visible by direct observation of the transilluminated gel can be clearly seen in the photographic image. Very faint bands (containing as little as 10 ng of protein) can be detected by long exposure using Polaroid film. Long integration times with a CCD camera allow the detection of 5 ng of protein per band (using both UV and green-light transilluminators). To obtain this high sensitivity it is necessary to have sharp bands; broad bands reduce considerably the sensitivity.
10. Nile red is sensitive to intense UV irradiation *(3)*, but has a photochemical stability high enough to allow gel staining without being necessary to introduce complex precautions in the protocol (*see* **Subheading 3., steps 6–9**). For long-term storage, solutions containing this dye are kept in the dark. Transillumination of the gel for more than a few minutes produces a significant loss of fluorescence intensity. Thus, transillumination time must be reduced as much as possible both during visualization and photography. The relative short exposure times (4–12 s) required for the Polaroid film 667 and the CCD documentation system allow to make several photographs with different exposure times if necessary. With films (Polaroid 665 and 669) requiring longer exposures (4–5 min) only the first photograph from each gel shows the maximum intensity in the fluorescent bands.
11. The development time of Polaroid films is dependent on the film temperature. Store the films at 4°C (*see* **Subheading 2., item10**), but allow them to equilibrate at room temperature before use.
12. The relatively large area (7.3 × 9.5 cm) of the Polaroid 665 negative is very convenient for further densitometric measurements (*see* **Note 13**). Furthermore, this negative can be used for making prints with an adequate level of contrast (*see*, e.g., the photograph presented in **Fig. 2**). The images obtained with the CCD documentation system can be stored in the computer and printed using different printers (inkjet printers with glossy paper yield photo quality images).
13. In quantitative analyses care has to be taken to ensure that the photographic film or the CCD documentation system have a linear response for the amounts of protein under study. Use different amounts (in the same range as the analyzed protein) of an internal standard to obtain an exposure time producing a linear response. Furthermore, the internal standard is necessary to normalize the results obtained with different gels, under different electrophoretic conditions and with different exposure and development times. Nile red can be considered as a general stain for proteins separated in SDS gels *(4)*. However, proteins with prosthetic groups such as catalase, and proteins having anomalous SDS binding properties such as histone H5, show atypical values of fluorescence intensity after staining with Nile red *(4)*.

Acknowledgment

This work was supported in part by grants PB98-0858 (Dirección General de Investigación) and 2000SGRG5 (Generalitat de Catalunya).

References

1. Andrews, A. T. (1986) *Electrophoresis. Theory, Techniques, and Biochemical and Cinical Applications.* Oxford University Press, Oxford, UK.
2. Allen, R. C. and Budowle, B. (1999) *Protein Staining and Identification Techniques.* Eaton Publishing, Natick, MA.
3. Daban, J.-R., Samsó, M., and Bartolomé, S. (1991) Use of Nile red as a fluorescent probe for the study of the hydrophobic properties of protein–sodium dodecyl sulfate complexes in solution. *Analyt. Biochem.* **199,** 162–168.
4. Daban, J.-R., Bartolomé, S., and Samsó, M. (1991) Use of the hydrophobic probe Nile red for the fluorescent staining of protein bands in sodium dodecyl sulfate-polyacrylamide gels. *Analyt. Biochem.* **199,** 169–174.
5. Alba, F. J., Bermúdez, A., Bartolomé, S., and Daban, J.-R. (1996) Detection of five nanograms of protein by two-minute Nile red staining of unfixed SDS gels. *BioTechniques.* **21,** 625–626.
6. Bermúdez, A., Daban, J.-R., Garcia, J. R., and Mendez, E. (1994) Direct blotting, sequencing and immunodetection of proteins after five-minute staining of SDS and SDS-treated IEF gels with Nile red. *BioTechniques.* **16,** 621–624.
7. Alba, F. J. and Daban, J.-R. (1998) Rapid fluorescent monitoring of total protein patterns on sodium dodecyl sulfate-polyacrylamide gels and Western blots before immunodetection and sequencing. *Electrophoresis.* **19,** 2407–2411.
8. Greenspan, P. and Fowler, S. D. (1985) Spectrofluorometric studies of the lipid probe Nile red. *J. Lipid Res.* **26,** 781–789.
9. Sackett, D. L. and Wolff, J. (1987) Nile red as polarity-sensitive fluorescent probe of hydrophobic protein surfaces. *Analyt. Biochem.* **167,** 228–234.
10. Sansó, M., Daban, J.-R., Hansen, S., and Jones, G. R. (1995) Evidence for sodium dodecyl sulfate/protein complexes adopting a necklace structure. *Eur. J. Biochem.* **232,** 818–824.
11. Alba, F. J., Bermúdez, A., and Daban, J.-R. (2001) Green-light transilluminator for the detection without photodamage of proteins and DNA labeled with different fluorescent dyes. *Electrophoresis* **22,** 399–403.

31

Zinc-Reverse Staining Technique

Carlos Fernandez-Patron

1. Introduction

With the advent of "proteomics," many previously unidentified proteins will be isolated in gels for subsequent structural and functional characterization. It is therefore important that methods are available for detecting these proteins with minimal risk of modification. This chapter describes a "reverse" staining technique that facilitates the sensitive detection of unmodified proteins ***(1–9)***.

The Zn^{2+} reverse staining technique exploits the ability of biopolymers (in particular, proteins and protein–SDS complexes) to bind Zn^{2+} ***(9–11)*** and that of imidazole (ImH, $C_3N_3H_4$; *see* **Note 1**) to react with unbound Zn^{2+} to produce insoluble zinc imidazolate ($ZnIm_2$). Deposition of $ZnIm_2$ along the gel surface results in the formation of a deep white stained background, against which unstained biopolymer bands contrast (as biopolymer bands bind Zn^{2+}, they locally inhibit deposition of $ZnIm_2$ and are not stained; *see* **Notes 2** and **3**).

With regard to protein analysis, reverse staining is a very gentle method of detection that does not require strong acid solutions, organic dyes, chemical modifiers, or protein sensitizers. Protein interactions with Zn^{2+}, established during the reverse staining step, are abrogated by metal chelation. As such, the risk of protein modification during reverse staining is minimal and reverse stained proteins can be efficiently eluted and used in biological and enzymatic assays (*see* **Note 4** and **refs. *7–9,15***). Proteins can also be processed for microsequencing or mass spectrometric analysis at any time after detection (*see* **Note 5** and **refs. *5–9,13,14***).

Another benefit of the reverse staining technique is its high sensitivity of detection (~10 ng protein/band in sodium dodecyl sulfate-polyacrylamide gel electrophoresis [SDS-PAGE] gels), which is greater than that of the Coomassie Blue stain (~100 ng protein/band) and approaches the sensitivity of the silver stains (1–10 ng protein/band). Indeed, reverse staining often reveals many proteins that display low affinity for Coomassie Blue and are thus not detected with this stain (*see* **Note 6** and **ref. *11***).

Other advantages of the reverse staining technique include the speed of procedure, as it consumes ~ 15 min, thus being significantly faster than the Coomassie Blue (> 1 h) and silver (> 2 h) stains. The reverse stained gels can be kept in water for several hours

From: *The Protein Protocols Handbook, 2nd Edition*
Edited by: J. M. Walker © Humana Press Inc., Totowa, NJ

to years without loss of image or sensitivity of detection. Similar to other stains, reverse stained patterns can be analyzed densitometrically, and a gel "toning" procedure has been recently developed to preserve the image upon gel drying *(12)*.

The Zn^{2+} reverse staining technique can be applied to detect virtually any gel-separated biopolymer that binds Zn^{2+} (i.e., proteins/peptides, glycolipids, oligonucleotides, and their multimolecular complexes) (**ref.** ***9*** and references therein). Moreover, these biopolymers are detected regardless of the electrophoresis system used for their separation (*see* **Notes 7–9** and **ref.** ***9***), which was not possible with the previously developed metal salt stains *(1–3)*. Therefore, Zn^{2+} reverse staining is a widely applicable detection method.

The SDS-PAGE method of protein electrophoresis is probably the most popular. Thus, the author chose to describe a standard reverse staining method that works very well with SDS-PAGE. The detection of proteins, peptides and their complexes with glycolipids in less commonly used gel electrophoresis systems is addressed as well (*see* **Notes 7–9**).

2. Materials

Reagent- and analytical-grade zinc sulfate, imidazole, acetic acid, sodium carbonate are obtained from Sigma (St. Louis, MO).

1. Equilibration solution (1×): 0.2 *M* imidazole, 0.1% (w/v) SDS (*see* **Note 1**).
2. Developer (1×): 0.3 *M* zinc sulfate.
3. Storage solution for reverse stained gels (1×): 0.5 % (w/v) sodium carbonate.

All solutions are prepared as 10× concentrated stocks, stored at room temperature, and diluted (1:10) in distilled water, to yield the working concentration (1×), just before use.

3. Methods

3.1. PAGE

SDS-PAGE is conducted following the method of Laemmli (***16*** and *see* Chapter 11). Native PAGE is conducted following the protocol of Laemmli, except that SDS is not included in the gel and electrophoresis solutions (*see* Chapter 10). Conventional agarose gel electrophoresis is conducted in 0.8% agarose gels and using Tris-acetate, pH 8.0; as gel and running buffer *(17)*.

3.2. Standard Reverse Staining of SDS-PAGE and Native PAGE Gels

The following reverse staining method detects proteins in standard polyacrylamide gels (*see* **Note 2**, **Fig. 1**). All incubations are performed under continuous gentle agitation in a plastic or glass tray with a transparent bottom. The volume of the corresponding staining/storage solutions must be enough to cover the gel (typically 50 mL for 1 minigel [10 cm × 7 cm × 0.75 mm]).

1. Following electrophoresis, the gel is incubated for 15 min in the equilibration solution (*see* **Note 2**).
2. To develop the electropherogram, the imidazole solution is discarded and the gel soaked for 30–40 s in developer solution. **Caution!:** This step must not be extended longer than

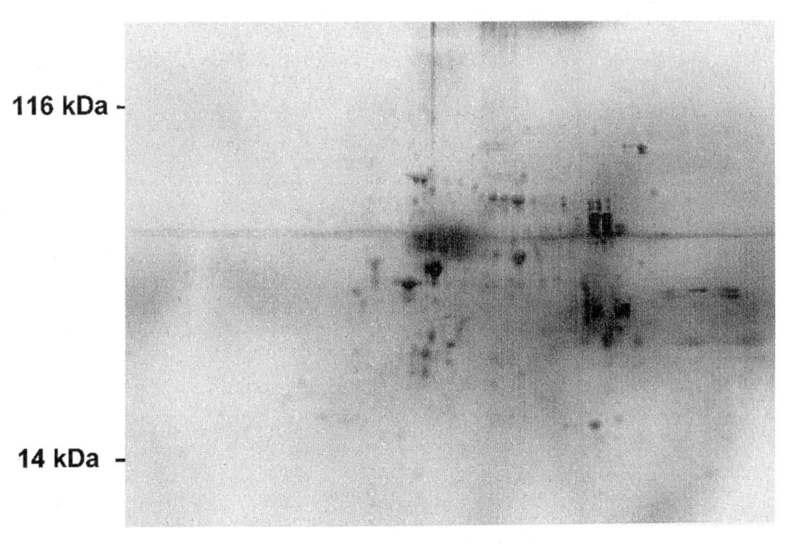

Fig. 1. Reverse staining of rat brain homogenate proteins (70 μg load) after 2-D-PAGE. Proteins in a wide range of molecular weights and isoelectric points are detected as transparent spots. These spots contain protein-SDS complexes that bind Zn^{2+} and thereby inhibit the precipitation of $ZnIm_2$ locally.

45 s or band overstaining and loss of the image will occur. Overstaining is prevented by pouring off the developer solution and rinsing the gel 3–5× (~1 min) in excess water (*see* **Note 3**).

At this point, the reverse stained gel can be photographed (**Fig. 1**). Photographic recording is best conducted with the gel placed on a glass plate held a few centimeters above a black underground and under lateral illumination.

While SDS-PAGE is a popular, high-resolution method for separating complex protein mixtures, sometimes it is desired to avoid either protein denaturation or disruption of macromolecular complexes during electrophoresis. In this case, the proteins are separated in the absence of SDS (native PAGE or agarose gel electrophoresis). Therefore, it is desirable to avoid the use of SDS during the reverse staining step. Two procedures have been developed to resolve this problem (*see* **Notes 8** and **9**; **Figs. 2** and **3**).

4. Notes

1. Imidazole is a five-membered heterocyclic ring containing a tertiary ("pyridine") nitrogen at position 3, and a secondary ("pyrrole") nitrogen at position 1. It is a monoacidic base whose basic nature is due to the ability of pyridine nitrogen to accept a proton. The pyrrole nitrogen can lose the hydrogen atom producing imidazolate anion (Im^-), at high pH values (pK_a ~ 14.2). However, deprotonation of imidazole's pyrrolic nitrogen may also occur at lower pH values, upon complexation of Zn^{2+} at the pyridinium nitrogen (**ref. *9*** and references therein). As a result, $ZnIm_2$ can form and precipitate at pH > 6.2. Upon treatment of

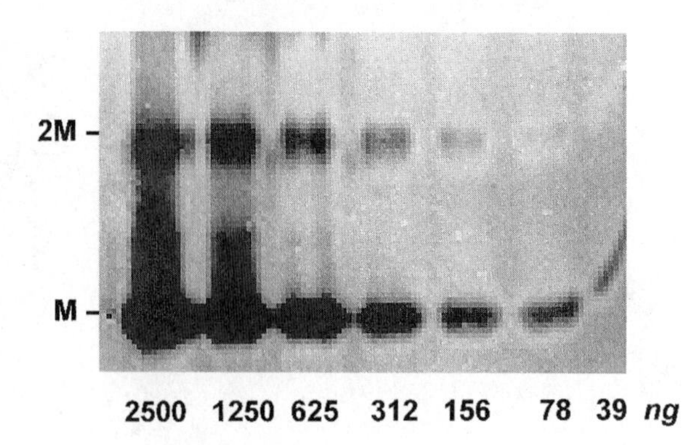

Fig. 2. Reverse staining of serial dilutions of human serum albumin (HSA) after native PAGE. HSA migrates yielding two main bands corresponding to its monomer (M) and its dimer (2M) and is detected under native conditions (no SDS) due to its natural ability to complex with Zn^{2+}.

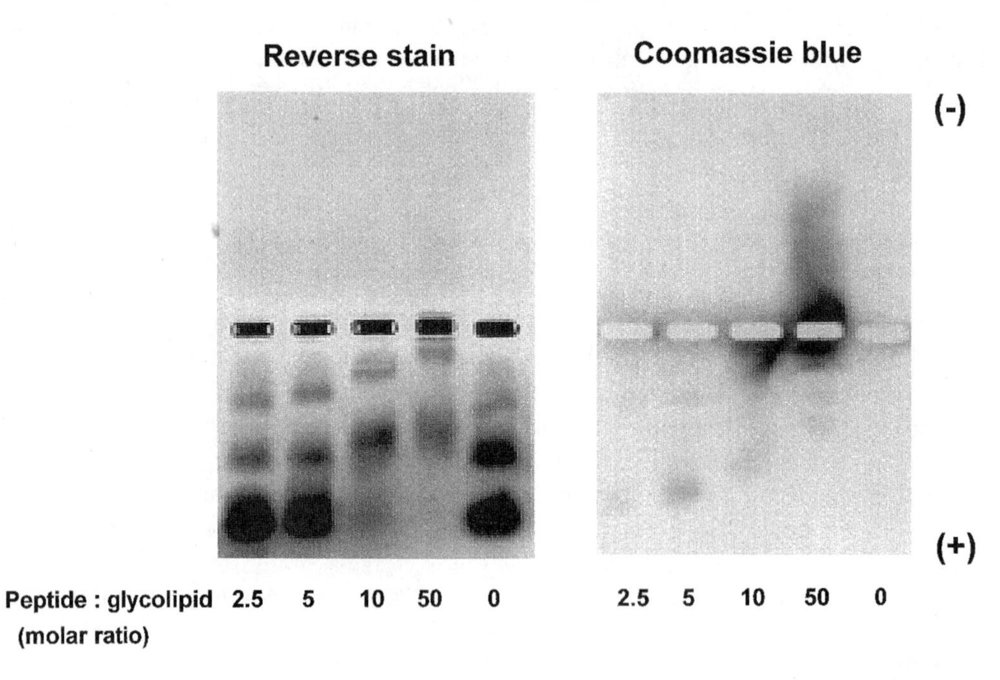

Fig. 3. Reverse staining of complexes between a synthetic cationic peptide with lipopolysaccharide binding properties and the glycolipid of a *N. meningitidis* strain. Formation of complexes was promoted by incubating (37°C) peptide and glycolipid for 30 min in 25 m*M* Tris-HCl, 0.1% Triton X-100, pH 8.0; at indicated peptide : glycolipid molar ratios. Reverse staining revealed bands of both: peptide.glycolipid complexes and uncomplexed glycolipid. Coomassie Blue stained mainly the uncomplexed peptide migrating toward the negative (–) electrode. Coomassie Blue failed to stain the glycolipid bands and stained peptide–glycolipid complexes very weakly.

a polyacrylamide or agarose electrophoresis gel with salts of zinc(II) and imidazole, a complex system is generated. Complexity is due to the presence of amide groups in the polyacrylamide matrix and sulfate groups in the agarose gel as well as Tris, glycine, dodecyl sulfate, and hydroxyl and carbonate anions in the electrophoresis buffers. These groups can coordinate with Zn(II), act as counteranions in the complexes of zinc with imidazole, and lead to the formation of complex salts and hydroxides. Diffusion phenomena (reflected in the times required for optimal gel "equilibration" and "development" during reverse staining) also critically influence the reverse staining reactions between Zn^{2+} and imidazole (*see* **Notes 2** and **3**). Nevertheless, when the protocol described in this chapter is followed, $ZnIm_2$ is the major component of the precipitate that stains the gels treated with zinc sulfate and imidazole ***(9)***.

2. This reverse staining protocol is optimized for use with any standard PAGE system regardless of whether PAGE is the first (1-D-PAGE) or the second (2-D-PAGE) dimension in the separation strategy. The equilibration step ensures that proteins in the gel are all uniformly coated with SDS. Therefore, protein's ability to bind Zn^{2+} is modulated by the protein–SDS complex and limits of detection (~10 ng protein/band) are similar for gels cast with and without SDS (i.e., SDS-PAGE and native PAGE, respectively). The larger the gel thickness or acrylamide concentration, the longer the equilibration step. A 15-min-long equilibration is enough for gels with ≤15% acrylamide and ≤1mm thickness. Preparative gels are often as thick as 3–5 mm; these gels should be equilibrated for 30–60 min. Insufficient equilibration may result in faint reverse stained patterns, which may fade upon prolonged storage.
3. Development time must be between 30 and 45 s as insufficient development results in pale background staining and overdevelopment causes overstaining. An overstained gel can be re-stained. For this, the gel is treated with 10 m*M* EDTA or 100 m*M* glycine for 5–10 min to redissolve the white $ZnIm_2$ precipitate that has deposited on the gel surface, rinsed in water (30 s) and, finally, soaked in the "storage" sodium carbonate solution. If the reverse stained pattern does not restore in ~5 min, the reverse staining procedure can be repeated as indicated in **Subheading 3.** Usually, the reverse stained pattern will restore during the equilibration step due to the precipitation of traces of Zn^{2+} already present in the gel with the imidazole from the equilibration solution. If the above suggestions do not lead to a homogenous reverse stained pattern of suitable quality, the gel can be "positively" restained with Coomassie Blue or silver ***(5)***. In this case, it is recommended that the reverse-stained gel be treated with EDTA (50 m*M*, pH 8.0; 30-min incubation) to free proteins of Zn^{2+}. Then, the gel can be processed with Coomassie Blue or silver stains.
4. Protein elution from reverse stained gels can be performed following any conventional method such as electroelution or passive elution; however, a highly efficient procedure has been developed ***(7,8)***. The reverse stained band of interest is excised, placed in a (1 mL) plastic vial and incubated with EDTA (50 m*M*, 2 × 5 min and 10 m*M*, 1 × 5 min) to chelate protein bound zinc ions. Supplementation of EDTA with nonionic detergents is optional but convenient if protein is to be in-gel refolded; for example, Triton X-100 was useful when refolding proteins that were to be bioassayed ***(9,15)***. EDTA (or EDTA, detergent mixture) is replaced by an appropriate "assay" buffer (e.g., phosphate saline solution), in which the protein band is equilibrated. Finally, the band is homogenized and protein is eluted into an appropriate volume of the assay buffer ***(7,8,14)***. The slurry is centrifuged and the supernatant filtered to collect a clean, transparent solution of protein ready for subsequent analysis ***(7–9,14,15)***.
5. An important application of reverse staining is in structural/functional proteomics ***(13,14)***. Identification of proteins separated by electrophoresis is a prerequisite to the construction of protein databases in proteome projects. Matrix assisted laser desorption/ionization-mass

spectrometry (MALDI-MS) has a sensitivity for peptide detection in the lower femtomole range. In principle, this sensitivity should be sufficient for an analysis of small amounts of proteins in silver-stained gels. However, a variety of known factors (e.g., chemical sensitizers such as glutaraldehyde) and other unknown factors modify the silver-stained proteins, leading to low sequence coverage ***(13)***. Low-abundance proteins have been successfully identified after $ZnIm_2$-reverse-staining ***(13,14)***.

6. Reverse staining of gels that have been already stained with Coomassie Blue reveals proteins undetected with Coomassie Blue, thus improving detection. Double staining can enhance sensitivity of detection in silver-stained gels as well. Before subjecting a Coomassie (or silver)-stained gel to double staining, the gel should be rinsed in water (2–3 × 10 min). This step assures a substantial removal of acetic acid from Coomassie and silver stains. Then, the gel is subjected to the standard reverse staining protocol. The resultant double stained patterns consist of the previously seen Coomassie Blue (brown, in the case of silver)-stained bands and new unstained (reverse-stained) bands that contrast against the deep-white, $ZnIm_2$, stained background.
7. Reverse staining can be used in combination with other "specific" protein stains to indicate the presence of both protein and glycolipid. As with proteins, the biological properties of the presumed glycolipid bands are testable in a functional experiment following elution from the reverse stained gel ***(9,18)***. After the electrophoresis step, the gel is rinsed for 30 min (2×, 15 min) in aqueous solution of methanol (40%, v/v). This step presumably removes excess SDS from glycolipid molecules that comigrate with the proteins in SDS-PAGE, making the glycolipid adopt a conformation with high affinity for Zn^{2+}. Next, the gel is reverse stained (Methods). Similar to proteins, glycolipid bands show up transparent and unstained. Sensitivity of detection is ~10 ng of glycolipid/band. If the same (or a parallel gel) is treated with Coomassie Blue (silver) stain, that does not detect glycolipids; protein and glycolipid bands can be distinguished by their distinct staining and migration properties ***(9,18)***. A replicate protein sample treated with proteinase K is a useful control, as this protease digests protein, but not glycolipid.
8. An alternative method of precipitating $ZnIm_2$ facilitates the reverse staining of native PAGE gels without the use of SDS ***(9)***. A *quick* and a *slow* reverse staining procedure were implemented ***(9)***. The *quick* version makes use of the characteristic neutral to basic pH of the gels immediately after electrophoresis. When the gel is incubated (3–6 min) in an slightly acidic solution containing zinc sulfate and imidazole (premixed to yield the molar ratio Zn(II) : ImH = 15 m*M* : 30 m*M*, pH ~5.0–5.5), the gel stains negatively as $ZnIm_2$ deposits along its surface. Blotchy deposits are prevented by intense agitation and resolubilized by lowering the pH of the zinc–imidazole solution; this also provides a basis for a *slow* version of this method. In the slow version (**Fig. 2**), the gel is incubated (10–20 min) in a solution of zinc sulfate (15 m*M*) and imidazole (30 m*M*) adjusted to pH 4.0. No precipitation occurs at this pH. The solution is poured off, the gel is rinsed with water (30 s); the electropherogram is developed by incubating the gel during ~5 min in 1% sodium carbonate. Sodium carbonate increases the pH first at the gel surface. Therefore, Zn^{2+} and imidazole, already present in the gel, react at the gel surface to form $ZnIm_2$. Again, protein bands are not stained as they complex with Zn^{2+} and locally prevent the precipitation of $ZnIm_2$.
9. Many proteins and peptides as well as glycolipids and their complexes with certain proteins/peptides separate well on agarose gels under nondenaturing conditions ***(9)***. To reverse stain an agarose gel ***(9)***; following electrophoresis, the gel is incubated for 25–30 min in a zinc sulfate–imidazole solution (Zn(II) : ImH = 15 m*M* : 30 m*M*, adjusted to pH 5.0 with glacial acetic acid). During this step, Zn^{2+} and imidazole diffuse into

agarose matrix. The gel is then rinsed in water for 5–8 min to remove any excess of the staining reagents from the gel surface. The reverse stained electropherogram is developed by incubating the gel for 5–8 min in 1% Na_2CO_3 (**Fig. 3**). Of note, due to poorly understood factors, agarose gels are not as amenable to reverse staining as polyacrylamide gels. Patterns of positive and negative bands are often seen in agarose gels.

Acknowledgment

The author is indebted to Drs. L. Castellanos-Serra and E. Hardy (Centre for Genetic Engineering and Biotechnology (CIGB), Havana, Cuba) for their contributions to the development of the reverse staining concept, to the CIGB for supporting this work and to Dr. D. Stuart (Dept. of Biochemistry, University of Alberta, Canada) for useful discussions. The author is a Fellow of the Canadian Institutes of Health Research and the Alberta Heritage Foundation for Medical Research, for support and to Dr. Stuart (Department of Biochemistry, University of Alberta, Canada) for useful comments on the manuscript.

References

1. Lee, C., Levin, A., and Branton, D. (1987) Copper staining: a five-minute protein stain for sodium dodecyl sulfate-polyacrylamide gels. *Analyt. Biochem.* **166,** 308–312.
2. Dzandu, J. K., Johnson, J. F., and Wise, G. E. (1988) Sodium dodecyl sulfate-gel electrophoresis: staining of polypeptides using heavy metal salts. *Analyt. Biochem.* **174,** 157–167.
3. Adams, L. D. and Weaver, K. M. (1990) Detection and recovery of proteins from gels following zinc chloride staining. *Appl. Theor. Electrophor.* **1,** 279–282.
4. Fernandez-Patron, C. and Castellanos-Serra, L. (1990) Abstract booklet from in *Eighth International Conference on Methods in Protein Sequence Analysis*, Kiruna, Sweden, July 1–6.
5. Fernandez-Patron, C., Castellanos-Serra, L., and Rodriguez P. (1992) Reverse staining of sodium dodecyl sulfate polyacrylamide gels by imidazole-zinc salts: sensitive detection of unmodified proteins. *Biotechniques* **12,** 564–573.
6. Fernandez-Patron, C., Calero, M., Collazo, P. R., Garcia, J. R., Madrazo, J., Musacchio, A., et al. (1995) Protein reverse staining: high-efficiency microanalysis of unmodified proteins detected on electrophoresis gels. *Analyt. Biochem.* **224,** 203–211.
7. Castellanos-Serra, L. R., Fernandez-Patron, C., Hardy, E., and Huerta, V. (1996) A procedure for protein elution from reverse-stained polyacrylamide gels applicable at the low picomole level: an alternative route to the preparation of low abundance proteins for microanalysis. *Electrophoresis* **17,** 1564–1572.
8. Castellanos-Serra, L. R., Fernandez-Patron, C., Hardy, E., Santana, H., and Huerta, V. (1997) High yield elution of proteins from sodium dodecyl sulfate-polyacrylamide gels at the low-picomole level: application to N-terminal sequencing of a scarce protein and to in-solution biological activity analysis of on-gel renatured proteins. *J. Protein Chem.* **16,** 415–419.
9. Fernandez-Patron, C., Castellanos-Serra, L., Hardy, E., Guerra, M., Estevez, E., Mehl, E., and Frank, R. W. (1998) Understanding the mechanism of the zinc-ion stains of biomacromolecules in electrophoresis gels: generalization of the reverse-staining technique. *Electrophoresis* **19,** 2398–2406.
10. *Bioinorganic Chemistry: Inorganic Elements in the Chemistry of Life* (1994) (Kaim W. and Schwederki, B., eds.), John Wiley & Sons, Chichester New York Brisbane Toronto Singapore.

11. Fernandez-Patron, C., Hardy, E., Sosa, A., Seoane, J., and Castellanos, L. (1995) Double staining of Coomassie Blue-stained polyacrylamide gels by imidazole-sodium dodecyl sulfate-zinc reverse staining: sensitive detection of Coomassie Blue-undetected proteins. *Analyt. Biochem.* **224,** 263–269.
12. Ferreras, M., Gavilanes, J. G., and Garcia-Segura, J. M. (1993) A permanent Zn^{2+} reverse staining method for the detection and quantification of proteins in polyacrylamide gels. *Analyt. Biochem.* **213,** 206–212.
13. Scheler, C., Lamer, S., Pan, Z., Li, X. P., Salnikow, J., and Jungblut, P. (1998) Peptide mass fingerprint sequence coverage from differently stained proteins on two-dimensional electrophoresis patterns by matrix assisted laser desorption/ionization-mass spectrometry (MALDI-MS). *Electrophoresis* **19,** 918–927.
14. Castellanos-Serra, L., Proenza, W., Huerta, V., Moritz, R. L., and Simpson, R. J. (1999) Proteome analysis of polyacrylamide gel-separated proteins visualized by reversible negative staining using imidazole-zinc salts. *Electrophoresis* **20,** 732–737.
15. Hardy, E., Santana, H., Sosa, A., Hernandez, L., Fernandez-Patron, C., and Castellanos-Serra, L. (1996) Recovery of biologically active proteins detected with imidazole-sodium dodecyl sulfate-zinc (reverse stain) on sodium dodecyl sulfate gels. *Analyt. Biochem.* **240,** 150–152.
16. Laemmli, U. K. (1970) Cleavage of structural proteins during the assembly of the head of bacteriophage T4. *Nature* **227,** 680–685.
17. *Molecular Cloning. A Laboratory Manual.* (1982) (Maniatis, T., Fritsch, E. F., and Sambrook, J., eds.), Cold Spring Harbor, New York.
18. Hardy, E., Pupo, E., Castellanos-Serra, L., Reyes, J., Fernandez-Patron, C. (1997) Sensitive reverse staining of bacterial lipopolysaccharides on polyacrylamide gels by using zinc and imidazole salts. *Analyt. Biochem.* **244,** 28–32.

32

Protein Staining with Calconcarboxylic Acid in Polyacrylamide Gels

Jung-Kap Choi, Hee-Youn Hong, and Gyurng-Soo Yoo

1. Introduction

Sodium dodecyl sulfate-polyacrylamide gel electrophoresis (SDS-PAGE) has become a highly reliable separation technique for protein characterization. Broad application of electrophoresis techniques has required the development of detection methods that can be used to visualize the proteins separated on polyacrylamide gels. A few of these methods are Coomassie brilliant blue (CBB) staining (*1* and *see* Chapter 11), silver staining (*2* and *see* Chapter 33), fluorescent staining (*3–5* and *see* Chapter 35), specific enzyme visualization *(6)*, and radioactive detection (*7* and *see* Chapters 28 and 38).

CBB staining is the most commonly used method owing to its proven reliability, simplicity, and economy *(8,9)*, but it lacks sensitivity compared with the silver-staining method. In addition, the staining/destaining process is time consuming *(10)*. Silver staining is the most sensitive nonradioactive protein detection method currently available, and can detect as little as 10 fg of protein *(11,12)*. However, it has several drawbacks, such as high-background staining, multiple steps, high cost of reagent, and toxicity of formaldehyde *(10)*.

In this chapter a protein staining method using calconcarboxylic acid (1-[2-hydroxy-4-sulfo-l-naphthylazo]-2-hydroxy-3-naphthoic acid, NN) is described *(13)*. This method can be performed by both simultaneous and postelectrophoretic staining techniques. Simultaneous staining using 0.01% of NN in upper reservoir buffer eliminates the poststaining step, and thus enables detection of the proteins more rapidly and simply. In poststaining, proteins can be stained by a 30-min incubation of a gel in 40% methanol/7% acetic acid solution of 0.05% NN and destaining in 40% methanol/7% acetic acid for 40 min with agitation. NN staining can detect as little as 10 ng of bovine serum albumin (BSA) by poststaining and 25 ng by simultaneous staining, compared to 100 ng detectable by CBB poststaining. These techniques produce protein-staining patterns identical to the ones obtained by the conventional CBB staining and also work well in nondenaturing PAGE, like in SDS-PAGE. The bands stained with NN present purple color. In addition, NN staining gives better linearity than CBB staining, although the slopes (band intensity/amount of protein) of it are somewhat lower (**Table 1**, **Fig. 1**). It suggests that NN staining is more useful than CBB staining for quantitative work of proteins.

From: *The Protein Protocols Handbook, 2nd Edition*
Edited by: J. M. Walker © Humana Press Inc., Totowa, NJ

Table 1
Linearity of CBB and NN Staining for Four Purified Proteins

	Slope[b]		y-intercept[b]		Correlation coefficient[b]	
Protein[a]	A	B	A	B	A	B
BSA	14.1	10.1	13.2	6.4	0.986	0.994
OVA	8.7	7.1	2.7	1.4	0.993	0.998
G-3-P DHase	8.0	5.8	5.3	2.2	0.997	0.999
CA	15.4	13.5	12.5	8.8	0.986	0.997

[a]Proteins were separated on 12.5% polyacrylamide gel, and densities and band area were determined with computerized densitometer. Some of the data are illustrated in **Fig. 1**. The range of amount of proteins was 0.25–12.5 μg. The number of points measured was six (0.25, 0.5, 1.0, 2.5, 5.0, and 12.5 μg).

[b]Slopes, y-intercepts, and correlation coefficients were determined by linear regression analysis. A, CBB staining; B, NN staining.

2. Materials

2.1. Equipment

1. Slab-gel apparatus.
2. Power supply (capacity 600 V, 200 mA).
3. Boiling water bath.
4. Gel dryer and vacuum pump.
5. Plastic container.
6. Rocking shaker.

2.2. Solutions

1. Destaining solution (1.0 L): Mix 530 mL distilled water with 400 mL methanol and 70 mL glacial acetic acid.
2. Simultaneous staining solution (1.0% [w/v] NN): Dissolve 1.0 g NN (pure NN* without K_2SO_4) in 100 mL reservoir buffer. Stir until fully dissolved at 50–60°C (stable for months at 4°C) (*see* **Note 7**).
3. Poststaining solution (0.05% [w/v] NN): Dissolve 0.05 g NN (pure NN*) in 100 mL destaining solution. Stir until dissolved thoroughly (store at room temperature).

3. Methods

3.1. Simultaneous Staining

The method is based on the procedure of Borejdo and Flynn ***(14)***, and is described for staining proteins in a 7.5% SDS-polyacrylamide gel.

1. Electrophorese the samples for 10 min to allow protein penetration into the upper gel phase.
2. Turn off the power, and then add 1% NN dissolved in reservoir buffer to the upper reservoir buffer to give a final concentration of 0.01–0.015% NN.
3. Stir the reservoir buffer sufficiently to ensure homogeneity.
4. Resume electrophoresis.

*NN diluted with 100- to 200-fold K_2SO_4 has been used as an indicator for the determination of calcium in the presence of magnesium with EDTA (*see* **Note 7**).

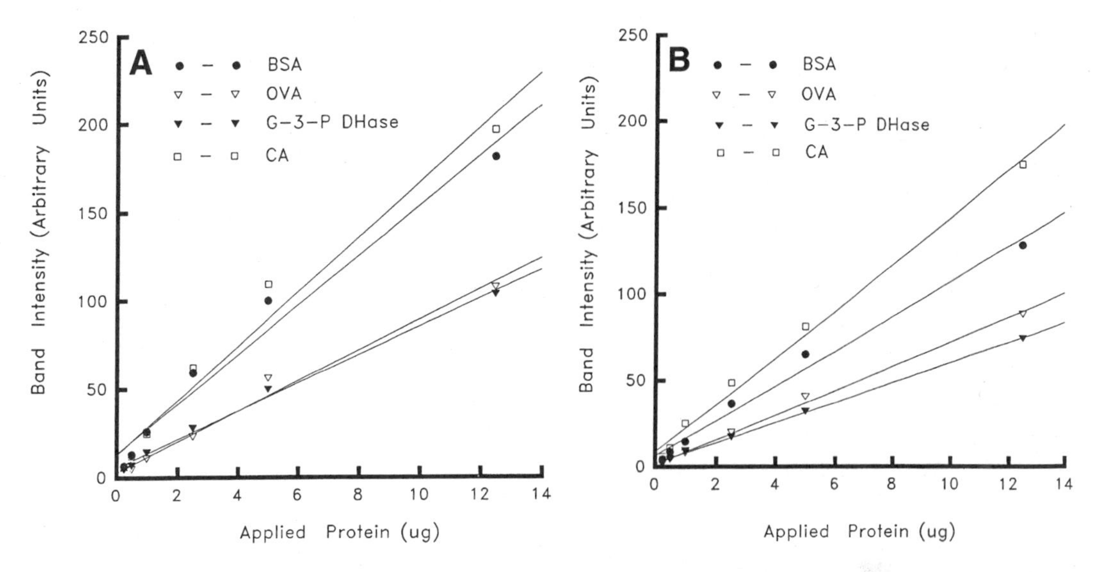

Fig. 1. Densitometric comparison of CBB and NN staining. Proteins were separated on 12.5% gels. **(A)** poststaining with 0.1% CBB; **(B)** simultaneous staining with 0.015% NN. Densitometric scanning was performed at 585 nm (CBB) and 580 nm (NN). The curves are fitted by the method of least squares. Each point represents the mean of three determinations. BSA, bovine serum albumin; OVA, ovalbumin; G-3-P DHase, glyceraldehyde-3-phosphate dehydrogenase; CA, carbonic anhydrase.

5. Immediately after electrophoresis, remove the stained gel from the apparatus.
6. Destain in 40% methanol/7% acetic acid for 30 min. To destain completely, change destaining solution several times and agitate (*see* **Notes 3–5,8**).

3.2. Postelectrophoretic Staining

1. Agitate the freshly run gel in 0.05% NN dissolved in destaining solution for 30 min.
2. Pour off the staining solution and rinse the gel with changes of destaining solution (two to three times). Staining solution can be reused several times.
3. Destain in 40% methanol/7% acetic acid for 40 min with agitation (*see* **Notes 3–5,8**).

4. Notes

1. The protocol for poststaining is the same as that for CBB staining, except for staining/destaining times and dye used.
2. The simultaneous staining method allows one to control the intensity of stained bands reproducibly by adjusting the concentration of the dye in the upper reservoir. More than 0.02% NN in the upper reservoir buffer does not increase sensitivity and requires greater destaining times.
3. Gel staining/destaining with NN is pH dependent. Intense staining occurs at pH 1.6–4.4, and weak staining with blue-purple color is observed at alkaline pH. In excessively strong acidic solution (<pH 1.0), however, the staining effect is markedly decreased, because the solubility of the dye is decreased and, thus replacement between dye anions and acetate ions may be suppressed (*see* **Fig. 2**). In destaining solution, the stained band diffuses, and its intensity is decreased significantly at pH higher than 4.4. Destaining in this pH range is rather slow compared with that in strongly acidic conditions.

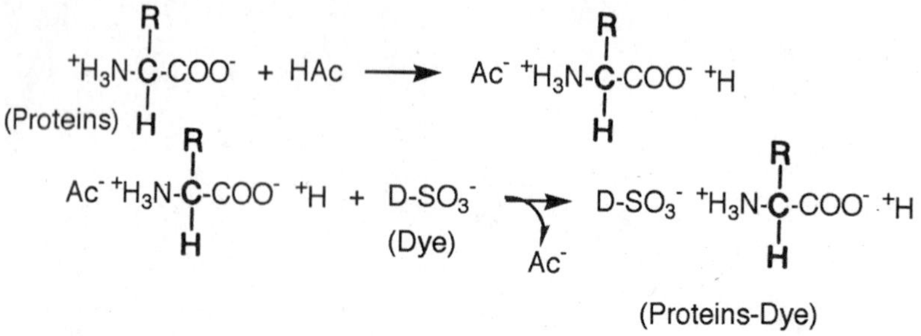

$$\underset{\text{(Proteins)}}{^{+}H_3N\text{-}\overset{R}{\underset{H}{C}}\text{-}COO^-} + HAc \longrightarrow Ac^-\ {}^{+}H_3N\text{-}\overset{R}{\underset{H}{C}}\text{-}COO^-\ {}^{+}H$$

$$Ac^-\ {}^{+}H_3N\text{-}\overset{R}{\underset{H}{C}}\text{-}COO^-\ {}^{+}H + \underset{\text{(Dye)}}{D\text{-}SO_3^-} \xrightarrow[Ac^-]{} \underset{\text{(Proteins-Dye)}}{D\text{-}SO_3^-\ {}^{+}H_3N\text{-}\overset{R}{\underset{H}{C}}\text{-}COO^-\ {}^{+}H}$$

Fig. 2. Mechanism of protein–dye interaction in acidic solution.

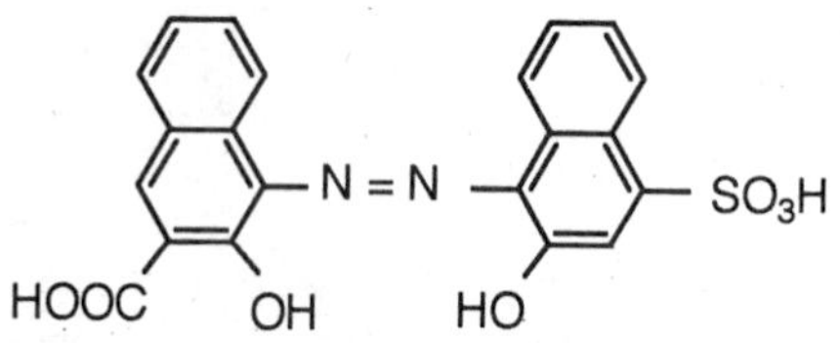

N = N
SO_3H
HOOC OH HO

Fig. 3. Structure of NN.

4. The rate of destaining speeds up with increasing methanol content; however, at high methanol content (>55%), gels are opaqued and shrunken. Addtionally, increasing the temperature of destaining solution is a great help in removing background (at 60–70°C, in 5 min), although sensitivity is a little reduced.
5. Destaining can be completed in 30 min in 7.5% polyacrylamide gels, but destaining time should be increased for 10 and 12.5% gels (50–60 min).
6. NN has several functional groups, such as hydroxyl, diazoic, carboxyl, and sulfonate groups (*see* **Fig. 3**). At acidic pH, NN probably forms electrostatic bonds with protonated amino groups, which are stabilized by hydrogen bonds and Van der Waals forces, as does CB *(1)*.
7. For maximal staining effect, the dye solution should be freshly prepared. The preparation of staining solution requires stirring and warming at 50–60°C since NN is poorly soluble.
8. Bands stained with NN are indefinitely stable when gels are stored in a refrigerator wrapped up in polyethylene films or dried on Whatmann No. 1 filter paper.
9. Throughout the staining/destaining processes, it is necessary to agitate the gel container using a shaker.

Acknowledgments

This work has been supported by a grant from KOSEF (981-0704-033-2) to J. K. Choi in 1988. This work has been supported by a Korean Research Foundation Grant (KRTF-2000-041-00301) to J. K. Choi in 2000.

References

1. Fazekas de St. Groth, S., Webster, R. G., and Datyner, A. (1963) Two new staining procedures for quantitative estimation of proteins on electrophoretic strips. *Biochim. Biophys. Acta* **71,** 377–391.

2. Merril, C. R., Goldman, D., Sedman, S., and Ebert, M. (1980) Ultrasensitive stain for proteins in polyacrylamide gels shows regional variation in cerebrospinal fluid proteins. *Science* **211,** 1437,1438.
3. Schetters, H. and McLeod, B. (1979) Simultaneous isolation of major viral proteins in one step. *Analyt. Biochem.* **98,** 329–334.
4. Jakowski, G. and Liew, C. C. (1980) Fluorescamine staining of nonhistone chromatin proteins as revealed by two-dimensional polyacrylamide gel electrophoresis. *Analyt. Biochem.* **102,** 321–325.
5. Weiderkamm, E., Wallach, D. F. H., and Fluckiger, R. (1973) A new sensitive, rapid fluorescence technique for the determination of proteins in gel electrophoresis and in solution. *Analyt. Biochem.* **54,** 102–114.
6. Gabriel, O. (1971) Locating enzymes on gels in *Methods in Enzymology,* vol. 22 (Colowick, S. P. and Kaplan, N. O., eds.), Academic, New York, pp. 363–367.
7. O'Farrell, P. H. (1975) High resolution two-dimensional electrophoresis of proteins. *J. Biol. Chem.* **250,** 4007–4021.
8. Diezel, W., Kopperschlager, G., and Hofmann, E. (1972) An improved procedure for protein staining in polyacrylamide gels with a new type of Coomassie Brilliant Blue. *Analyt. Biochem.* **48,** 617–620.
9. Zehr, B. D., Savin, T. J., and Hall, R. E. (1989) A one-step low background Coomassie staining procedure for polyacrylamide gels. *Analyt. Biochem.* **182,** 157–159.
10. Hames, B. D. and Rickwood, D. (1990) Analysis of gels following electrophoresis, in *Gel Electrophoresis of Proteins: A Practical Approach,* IRL, Oxford, UK, pp. 52–81.
11. Switzer, R. C., Merril, C. R., and Shifrin, S. (1979) A highly sensitive silver stain for detecting proteins and peptides in polyacrylamide gels. *Analyt. Biochem.* **98,** 231–237.
12. Ohsawa, K. and Ebata, N. (1983) Silver stain for detecting 10-femtogram quantities of protein after polyacrylamide gel electrophoresis. *Analyt. Biochem.* **135,** 409–415.
13. Hong, H. Y., Yoo, G. S., and Choi, J. K. (1993) Detection of proteins on polyacrylamide gels using calconcarboxylic acid. *Analyt. Biochem.* **214,** 96–99.
14. Borejdo, J. and Flynn, C. (1984) Electrophoresis in the presence of Coomassie Brilliant Blue R-250 stains polyacrylamide gels during protein fractionation. *Analyt. Biochem.* **140,** 84–86.

33

Detection of Proteins in Polyacrylamide Gels by Silver Staining

Michael J. Dunn

1. Introduction

The versatility and resolving capacity of polyacrylamide gel electrophoresis has resulted in this group of methods becoming the most popular for the analysis of patterns of gene expression in a wide variety of complex systems. These techniques are often used to characterize protein purity and to monitor the various steps in a protein purification process. Moreover, two-dimensional polyacrylamide gel electrophoresis (2-DE) remains the core technology of choice for separating complex protein mixtures in the majority of proteome projects (*1*, and *see* Chapter 22). This is due to its unrivalled power to separate simultaneously thousands of proteins and the relative ease with which proteins from 2-D gels can be identified and characterized using highly sensitive microchemical methods *(2)*, particularly those based on mass spectrometry *(3)*. Gel electrophoresis is now one of the most important methods of protein purification for subsequent identification and characterization.

Coomassie Brilliant Blue R-250 (CBB R-250) has been used for many years as a general protein stain following gel electrophoresis. However, the trend toward the use of thinner gels and the need to detect small amount of protein within single bands or spots resolved by one- or two-dimensional electrophoresis have necessitated the development of more sensitive detection methods *(4)*.

The ability of silver to develop images was discovered in the mid-17th century and this property was exploited in the development of photography, followed by its use in histological procedures. Silver staining for the detection of proteins following gel electrophoresis was first reported in 1979 by Switzer et al. *(5)*, resulting in a major increase in the sensitivity of protein detection. More than 100 publications have subsequently appeared describing variations in silver staining methodology *(4,6)*. This group of procedures is generally accepted to be around 100 times more sensitive than methods using CBB R-250, being able to detect low nanogram amounts of protein per gel band or spot.

All silver staining procedures depend on the reduction of ionic silver to its metallic form, but the precise mechanism involved in the staining of proteins has not been fully established. It has been proposed that silver cations complex with protein amino groups,

From: *The Protein Protocols Handbook, 2nd Edition*
Edited by: J. M. Walker © Humana Press Inc., Totowa, NJ

particularly the ε-amino group of lysine *(7)*, and with sulphur residues of cysteine and methionine *(8)*. However, Gersten and his colleagues have shown that "stainability" cannot be attributed entirely to specific amino acids and have suggested that some element of protein structure, higher than amino acid composition, is responsible for differential silver staining *(9)*.

Silver staining procedures can be grouped into two types of method depending on the chemical state of the silver ion when used for impregnating the gel. The first group is alkaline methods based on the use of an ammoniacal silver or diamine solution, prepared by adding silver nitrate to a sodium–ammonium hydroxide mixture. Copper can be included in these diamine procedures to give increased sensitivity, possibly by a mechanism similar to that of the Biuret reaction. The silver ions complexed to proteins within the gel are subsequently developed by reduction to metallic silver with formaldehyde in an acidified environment, usually using citric acid. In the second group of methods, silver nitrate in a weakly acidic (approx pH 6) solution is used for gel impregnation. Development is subsequently achieved by the selective reduction of ionic silver to metallic silver by formaldehyde made alkaline with either sodium carbonate or NaOH. Any free silver nitrate must be washed out of the gel prior to development, as precipitation of silver oxide will result in high background staining.

Silver stains are normally monochromatic, resulting in a dark brown image. However, if the development time is extended, dense protein zones become saturated and color effects can be produced. Some staining methods have been designed to enhance these color effects, which were claimed to be related to the nature of the polypeptides detected *(10)*. However, it has now been established that the colors produced depend on: (1) the size of the silver particles, (2) the distribution of silver particles within the gel, and (3) the refractive index of the gel *(11)*.

Rabilloud has compared several staining methods based on both the silver diamine and silver nitrate types of procedure *(12)*. The most rapid procedures were found to be generally less sensitive than the more time-consuming methods. Methods using glutaraldehyde pretreatment of the gel and silver diamine complex as the silvering agent were found to be the most sensitive. However, it should be noted that the glutaraldehyde and formaldehyde present in many silver staining procedures results in alkylation of α- and ε-amino groups of proteins, thereby interfering with their subsequent chemical characterization. To overcome this problem, silver staining protocols compatible with mass spectrometry in which glutaraldehyde is omitted have been developed *(13, 14)* but these suffer from a decrease in sensitivity of staining and a tendency to a higher background. This problem can be overcome using post-electrophoretic fluorescent staining techniques *(15)*. The best of these at present appears to be SYPRO Ruby (*see* Chapter 35), which has a sensitivity approaching that of standard silver staining and is fully compatible with protein characterization by mass spectrometry *(16)*.

The method of silver staining we describe here is recommended for analytical applications and is based on that of Hochstrasser et al *(17,18)*, together with modifications and technical advice that will enable an experimenter to optimize results. An example of a one-dimensional sodium dodecyl sulfate-polyacrylamide gel electrophoresis (SDS-PAGE) separation of the total proteins of human heart proteins stained by this procedure is shown in **Fig. 1**. The power of 2-DE combined with sensitive

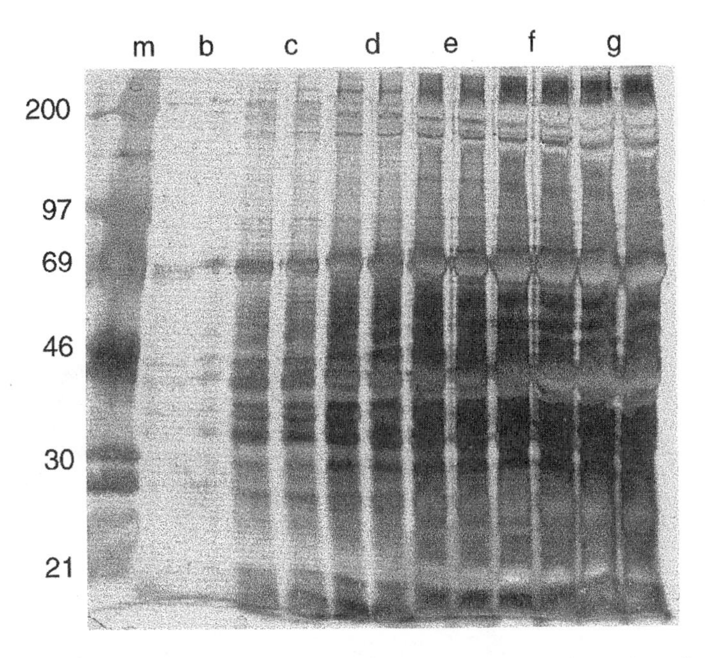

Fig. 1. One-dimensional 12%T SDS-PAGE separation of human heart proteins (*lanes b–g*) visualized by silver staining. Lane (m) contains the M_r marker proteins and the scale at the left indicates $M_r \times 10^{-3}$. The sample protein loadings were (**b**) 1 µg, (**c**) 5 µg, (**d**) 10 µg, (**e**) 25 µg, (**f**) 50 µg, (**g**) 100 µg.

detection by silver staining to display the complex protein profile of a whole tissue lysate is shown in **Fig. 2**.

2. Materials

1. All solutions should be freshly prepared, and overnight storage is not recommended. Solutions must be prepared using clean glassware and deionized, distilled water.
2. Gel fixation solution: Trichloroacetic acid (TCA) solution, 20% (w/v).
3. Sensitization solution: 10% (w/v) glutaraldehyde solution.
4. Silver diamine solution: 21 mL of 0.36% (w/v) NaOH is added to 1.4 mL of 35% (w/v) ammonia and then 4 mL of 20% (w/v) silver nitrate is added dropwise with stirring. When the mixture fails to clear with the formation of a brown precipitate, further addition of a minimum amount of ammonia results in the dissolution of the precipitate. The solution is made up to 100 mL with water. The silver diamine solution is unstable and should be used within 5 min.
5. Developing solution: 2.5 mL of 1% (w/v) citric acid, 0.26 mL of 36% (w/v) formaldehyde made up to 500 mL with water.
6. Stopping solution: 40% (v/v) ethanol, 10% (v/v) acetic acid in water.
7. Farmer's reducer: 0.3% (w/v) potassium ferricyanide, 0.6% (w/v) sodium thiosulfate, 0.1% (w/v) sodium carbonate.

3. Method

Note: *All incubations are carried out at room temperature with gentle agitation.*

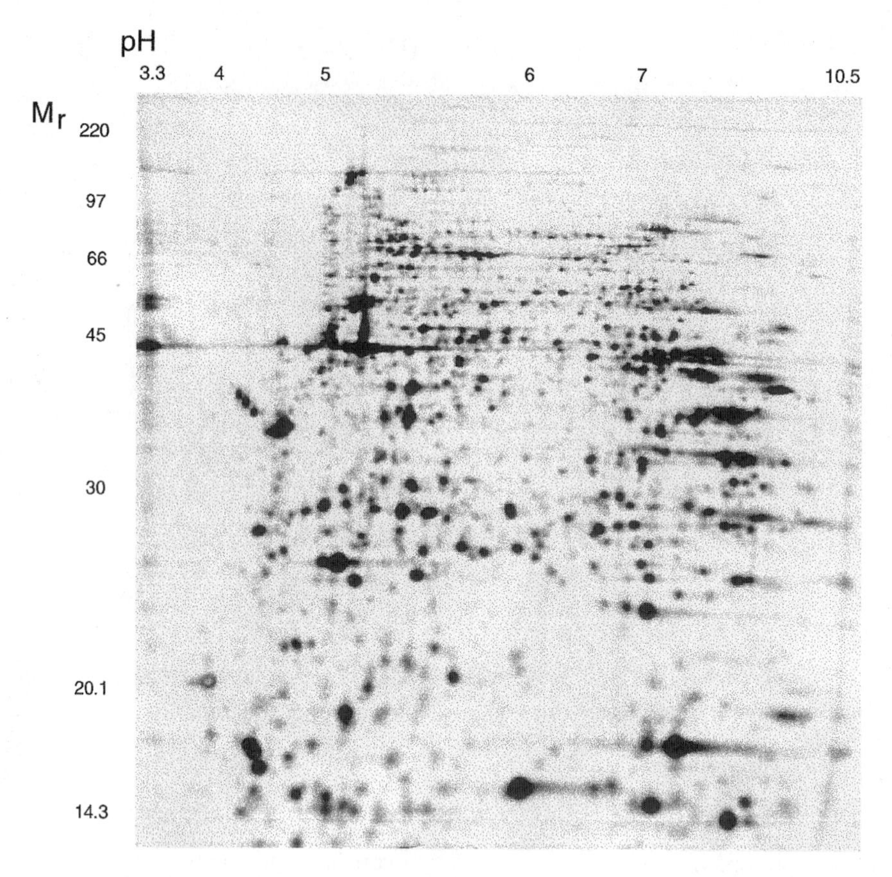

Fig. 2. Two-DE Separation of human heart proteins visualized by silver staining. A loading of 100 μg protein was used. The first dimension was pH 3–10 NL immobilized pH gradient (IPG) isoelectric focusing (IEF) and the second dimension was 12% T SDS-PAGE. The scale at the top indicates the nonlinear pH gradient used in the first IPG 3–10 NL IEF dimension, while the scale at the left indicates $M_r \times 10^{-3}$.

3.1. Fixation

1. After electrophoresis, fix the gel immediately (*see* **Note 1**) in 200 mL (*see* **Note 2**) of TCA (*see* **Note 3**) for a minimum of 1 h at room temperature. High-percentage polyacrylamide and thick gels require an increased period for fixation, and overnight soaking is recommended.
2. Place the gel in 200 mL of 40% (v/v) ethanol, 10% (v/v) acetic acid in water and soak for 2 × 30 min (*see* **Note 4**).
3. Wash the gel in excess water for 2 × 20 min, facilitating the rehydration of the gel and the removal of methanol. An indication of rehydration is the loss of the hydrophobic nature of the gel.

3.2. Sensitization

1. Soak the gel in a 10% (w/v) glutaraldehyde solution for 30 min at room temperature (*see* **Note 5**).
2. Wash the gel in water for 3 × 20 min to remove excess glutaraldehyde.

3.3. Staining

1. Soak the gel in the silver diamine solution for 30 min. For thick gels (1.5 mm), it is necessary to use increased volumes so that the gels are totally immersed. Caution should be exercised in disposal of the ammoniacal silver reagent, since it decomposes on standing and may become explosive. The ammoniacal silver reagent should be treated with dilute hydrochloric acid (1 *N*) prior to disposal.
2. Wash the gel (3 × 5 min) in water.

3.4. Development

1. Place the gel in developing solution. Proteins are visualized as dark brown zones within 10 min (*see* **Note 6**), after which the background will gradually increase (*see* **Note 7**). It is important to note that the reaction displays inertia, and that staining will continue for 2–4 min after removal of the gel from the developing solution. Staining times in excess of 20 min usually result in an unacceptable high background (*see* **Note 8**).
2. Terminate staining by immersing the gel in stopping solution.
3. Wash the stained gel in water prior to storage or drying.

3.5. Destaining

Partial destaining of gels using Farmer's reducing reagent is recommended for the controlled removal of background staining that obscures proper interpretation of the protein pattern.

1. Wash the stained gel in water for 5 min to remove the stop solution.
2. Place the gel in Farmer's reducer for a time dependent upon the intensity of the background.
3. Terminate destaining by returning the gel to the stop solution.

4. Notes

1. Gloves should be worn at all stages when handling gels, as silver staining will detect keratin proteins from the skin.
2. Volumes of the solutions used at all stages should be sufficient such that the gel is totally immersed. If the volume of solution is insufficient for total immersion, staining will be uneven and the gel surface can dry out.
3. A mixture of alcohol, acetic acid and water (9:9:2) is recommended for gel fixation in many published protocols, but TCA is a better general protein fixative and its use is compatible with silver staining provided that the gel is washed well after fixation to remove the acid.
4. In addition to removing TCA, the washing step also effectively removes reagents such as Tris, glycine, and detergents (especially SDS) which can bind silver and result in increased background staining.
5. Treatment of the gel with reducing agents such as glutaraldehyde prior to silver impregnation results in an increase in staining sensitivity by increasing the speed of silver reduction on the proteins.
6. If image development is allowed to proceed for too long, dense protein zones will become saturated and negative staining will occur, leading to serious problems if quantitative analysis is attempted. In addition, certain proteins stain to give yellow or red zones regardless of protein concentration, and this effect has been linked to the posttranslational modification of the proteins.

7. An inherent problem with the staining of gradient SDS-PAGE gels is uneven staining along the concentration gradient. The less concentrated polyacrylamide region develops background staining prior to the more concentrated region. A partial solution to this problem is to increase the time of staining in silver diamine (*see* **Subheading 3.3., step 1**).
8. Various chemicals used in one- and two-dimensional electrophoresis procedures can inhibit staining, whereas others impair resolution or produce artifacts. Acetic acid will inhibit staining and should be completely removed prior to the addition of silver diamine solution (*see* **Subheading 3.3., step 1**). Glycerol, used to stabilize SDS gradient gels during casting, and urea, used as a denaturing agent in isoelectric focusing (IEF), are removed by water washes. Agarose, often used to embed rod IEF gels onto SDS-PAGE gels in 2-D PAGE procedures, contains peptides that are detected by silver staining as diffuse bands and give a strong background. Tris, glycine and detergents (especially SDS) present in electrophoresis buffers can complex with silver and must be washed out with water prior to staining. The use of 2-mercaptoethanol as a disulfide bond reducing agent should be avoided since it leads to the appearance of two artifactual bands at 50 and 67 kDa on the gel *(19)*.
9. Radioactively labeled proteins can be detected by silver staining prior to autoradiography or fluorography for the majority of the commonly used isotopes (^{14}C, ^{35}S, ^{32}P, ^{125}I). In the case of ^{3}H, however, silver deposition will absorb most of the emitted radiation.

References

1. Dunn, M. J. and Görg, A. (2001) Two-dimensional polyacrylamide gel electrophoresis for proteome analysis, in *Proteomics, From Protein Sequence to Function* (Pennington, S. R. and Dunn, M. J., eds.), BIOS Scientific Publishers, Oxford, pp. 43–63.
2. Wilkins, M. R. and Gooley, A. (1997) Protein identification in proteome analysis, in *Proteome Research: New Frontiers in Functional Genomics* (Wilkins, M. R., Williams, K. L., Appel, R. D. and Hochstrasser, D. F., eds.), Springer-Verlag, Berlin, pp. 35–64.
3. Patterson, S. D., Aebersold, R., and Goodlett, D. R. (2001) Mass spectrometry-based methods for protein identification and phosphorylation site analysis, in *Proteomics, From Protein Sequence to Function* (Pennington, S. R. and Dunn, M. J., eds.), BIOS Scientific Publishers, Oxford, pp. 87–130.
4. Patton, W. F. (2001) Detecting proteins in polyacrylamide gels and on electroblot membranes, in *Proteomics, From Protein Sequence to Function* (Pennington, S. R. and Dunn, M. J., eds.), BIOS Scientific Publishers, Oxford, pp. 65–86.
5. Switzer, R. C., Merril, C. R., and Shifrin, S. (1979) A highly sensitive stain for detecting proteins and peptides in polyacrylamide gels. *Analyt. Biochem.* **98,** 231–237.
6. Rabilloud, T. (1990) Mechanisms of protein silver staining in polyacrylamide gels: a 10-year synthesis. *Electrophoresis* **11,** 785–794.
7. Dion, A. S. and Pomenti, A. A. (1983) Ammoniacal silver staining of proteins: mechanism of glutaraldehyde enhancement. *Analyt. Biochem.* **129,** 490–496.
8. Heukeshoven, J. and Dernick, R. (1985) Simplified method for silver staining of proteins in polyacrylamide gels and the mechanism of silver staining. *Electrophoresis,* **6,** 103–112.
9. Gersten, D. M., Rodriguez, L. V., George, D. G., Johnston, D. A., and Zapolski, E. J. (1991) On the relationship of amino acid composition to silver staining of protein in electrophoresis gels: II. Peptide sequence analysis. *Electrophoresis* **12,** 409–414.
10. Sammons, D. W., Adams, L. D., and Nishizawa, E. E. (1982) Ultrasensitive silver-based color staining of polypeptides in polycarylamide gels. *Electrophoresis* **2,** 135–141.
11. Merril, C. R., Harasewych, M. G., and Harrington, M. G. (1986) Protein staining and detection methods, in *Gel Electrophoresis of Proteins* (Dunn, M. J., ed.), Wright, Bristol, pp. 323–362.

12. Rabilloud, T. (1992) A comparison between low background silver diamine and silver nitrate protein stains. *Electrophoresis* **13,** 429–439.
13. Shevchenko, A., Wilm, M., Vorm, O., and Mann, M. (1996) Mass spectrometric sequencing of proteins from silver-stained polyacrylamide gels. *Analyt. Chem.* **68,** 85–858.
14. Yan, J. X., Wait, R., Berkelman, T., Harry, R., Westbrook, J. A. , Wheeler, C. H., and Dunn, M. J. (2000) A modified silver staining protocol for visualization of proteins compatible with matrix-assisted laser desorption/ionization and elctrospray ionization-mass spectrometry. *Electrophoresis* **21,** 3666–3672.
15. Patton, W. F. (2000) A thousand points of light: The application of fluorescence detection technologies to two-dimensional gel electrophoresis and proteomics. *Electrophoresis* **21,** 1123–1144.
16. Yan, J. X., Harry, R. A., Spibey, C., and Dunn, M. J. (2000) Postelectrophoretic staining of proteins separated by two-dimensional gel electrophoresis using SYPRO dyes. *Electrophoresis* **21,** 3657–3665.
17. Hochstrasser, D. F., Patchornik, A., and Merril, C. R. (1988) Development of polyacrylamide gels that improve the separation of proteins and their detection by silver staining. *Analyt. Biochem.* **173,** 412–423.
18. Hochstrasser, D. F. and Merril, C. R. (1988) 'Catalysts' for polyacrylamide gel polymerization and detection of proteins by silver staining. *Appl. Theor. Electrophoresis* **1,** 35–40.
19. Guevarra, J., Johnston, D. A., Ramagli, L. S., Martin, B. A., Capitello, S., and Rodriguez, L. V. (1982) Quantitative aspects of silver deposition in proteins resolved in complex polyacrylamide gels. *Electrophoresis* **3,** 197–205.

34

Background-Free Protein Detection in Polyacrylamide Gels and on Electroblots Using Transition Metal Chelate Stains

Wayne F. Patton

1. Introduction

Electrophoretically separated proteins may be visualized using organic dyes such as Ponceau Red, Amido Black, Fast Green, or most commonly Coomassie Brilliant Blue *(1,2)*. Alternatively, sensitive detection methods have been devised using metal ions and colloids of gold, silver, copper, carbon, or iron *(3–12)*. Metal chelates form a third class of stains, consisting of transition metal complexes that bind avidly to proteins resolved in polyacrylamide gels or immobilized on solid-phase membrane supports *(13–27)*. In recent years, metal chelate stains have been designed and optimized specifically for compatibility with commonly used microchemical characterization procedures employed in proteomics. The metal chelate stains are simple to implement, and do not contain extraneous chemicals such as glutaraldehyde, formaldehyde, or Tween-20 that are well known to interfere with many downstream protein characterization procedures.

Metal chelates can be used to detect proteins on nitrocellulose, poly(vinylidene difluoride) (PVDF), and nylon membranes as well as in polyacrylamide and agarose gels. The metal complexes do not modify proteins, and are compatible with immunoblotting, lectin blotting, mass spectrometry, and Edman-based protein sequencing *(13–17,22–27)*. Metal chelate stains are suitable for routine protein measurement in solid-phase assays owing to the quantitative stoichiometry of complex formation with proteins and peptides *(15,16)*. Such solid phase protein assays are more sensitive and resistant to chemical interference than their solution-based counterparts *(15)*.

A variety of metal ions and organic chelating agents may be combined to form metal chelate stains but only a few have been evaluated extensively for protein detection in electrophoresis. Ferrene S has been utilized for the specific detection of iron-containing proteins, such as cytochrome *c*, transferrin, ferritin, and lactoferrin in polyacrylamide gels *(18)*. In this situation the metal complex only forms when the chelate interacts with the native metal ion bound within the protein itself. Copper phthalocyanine 4, 4',4'', 4'''-tetrasulfonic acid has been shown to stain total protein in electrophoresis gels and on nitrocellulose membranes *(19)*. In addition, we demonstrated the use of Ferrene S-ferrous, Ferrozine-ferrous, ferrocyanide-ferric, and Pyrogallol Red-molybdate com-

From: *The Protein Protocols Handbook, 2nd Edition*
Edited by: J. M. Walker © Humana Press Inc., Totowa, NJ

plexes for colorimetric detection of electrophoretically separated proteins immobilized on membranes ***(13–16)***. In 1978, a pink bathophenanthroline disulfonate-ferrous complex was reported as a nonspecific protein stain for polyacrylamide gel electrophoresis ***(20)***. The stain is rather insensitive and was later modified by substituting [^{59}Fe] into the complex in order to detect proteins by autoradiography ***(21)***. Although increasing sensitivity substantially, the hazards associated with working with radioactivity and the burden of license application for an infrequently used radioisotope have precluded routine utilization of bathophenanthroline disulfonate-[^{59}Fe] as a general protein stain. Measuring light emission is intrinsically more sensitive than measuring light absorbance, as the later is limited by the molar extinction coefficient of the colored complex ***(28)***. Thus, luminescent protein detection systems utilizing chelates complexed to transition metal ions such as europium, or ruthenium should offer greater sensitivity than their colorimetric counterparts without the accompanying hazards associated with radioactivity. The organic component of the complex absorbs light and transfers the energy to the transition metal ion, which subsequently emits light at longer wavelength.

This is demonstrated by substituting europium into the bathophenanthroline–disulfonate complex ***(17)***. This luminescent reagent has been commercialized as SYPRO Rose protein blot stain (Molecular Probes, Eugene, OR). The bathophenanthroline disulfonate-europium complex can detect as little as 8 ng of protein immobilized on nitrocellulose or PVDF membranes. By comparison, the original bathophenanthroline disulfonate–ferrous complex is capable of detecting 600 ng of protein, while the modification employing [^{59}Fe] is capable of detecting 10–25 ng of protein ***(20,21)***. The luminescent stain is readily removed by incubating blots in mildly alkaline solution, is highly resistant to photobleaching and is compatible with popular downstream biochemical characterization procedures including immunoblotting, lectin blotting, and mass spectrometry ***(17)***. Disadvantages of the bathophenanthroline disulfonate–europium stain are that the dye can only be adequately visualized using 302 nm UV-B epi-illumination and the dye exhibits intense 430 nm (blue) fluorescence emission as well as the desired red emission maxima of 595 and 615 nm.

Subsequently, SYPRO Rose Plus protein blot stain, an improved europium-based metal chelate stain roughly 10 times brighter than the original bathophenanthroline disulfonate-europium stain, was introduced ***(25,26)***. The intense blue fluorescence from uncomplexed ligand, observed in the original stain, was eliminated by employing a thermodynamically more stable europium complex. Due to improved absorption properties, the stain could now be readily visualized with UV-A UV-B or UV-C epi-illumination. Just as with the bathophenanthroline disulfonate–europium stain, SYPRO Rose Plus stain is easily removed by increasing solution pH. The stain is fully compatible with biotin–streptavidin and immunoblotting detection technologies that use a wide variety of visualization strategies. Neither of the europium-based stains is compatible with laser-based gel scanners as they lack visible excitation peaks. SYPRO Ruby dye is a proprietary ruthenium-based metal chelate stain developed to address the limitations of the SYPRO Rose and SYPRO Rose Plus dyes. SYPRO Ruby protein blot stain visualizes electroblotted proteins on nitrocellulose and PVDF membranes with a detection sensitivity of 0.25–1 ng of protein/mm^2 in slot-blotting applications. Approximately 2–8 ng of protein can routinely be detected by electroblotting, which side-by-side comparisons demonstrate is as sensitive as colloidal gold stain ***(22)***. While colloidal gold stain-

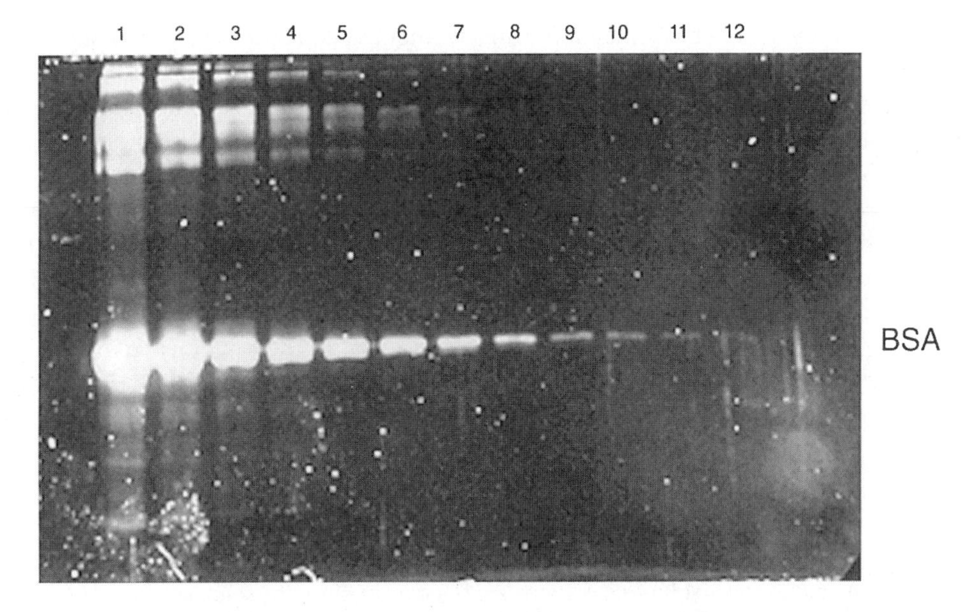

Fig. 1. Protein detection sensitivity in 1-D gels using SYPRO Ruby Protein gel stain: Twofold serial dilutions of solution-quantified bovine serum albumin (P-7656, Sigma Chemical, Saint Louis, MO) electrophoretically separated on 13% Duracryl SDS-polyacrylamide gels (Genomic Solutions, Ann Arbor, MI). After staining with SYPRO Ruby protein gel stain (Molecular Probes, Eugene, OR), gels were imaged using a Lumi-Imager-F1 scanner (Roche Biochemicals, Indianapolis, IN). The 302-nm UV-B transilluminator was used in conjunction with the system's 600-nm emission filter. As little as 250 pg of serum albumin is detectable, with a linearly increasing signal extending to 1000 ng.

ing requires 2–4 h, SYPRO Ruby dye staining is complete in 15 min. The linear dynamic range of SYPRO Ruby protein blot stain is vastly superior to colloidal gold stain, extending over a 1000-fold range. The dye can be excited using a standard 302 nm UV-B transilluminator or using imaging systems equipped with 450-, 473-, 488-, or even 532- nm lasers. Unlike colloidal gold stain, SYPRO Ruby stain does not interfere with mass spectrometry or immunodetection procedures *(22)*.

SYPRO Ruby protein gel stain and SYPRO Ruby IEF protein gel stains allow one-step, low background staining of proteins in polyacrylamide or agarose gels without resorting to lengthy destaining steps (*see* **Fig. 1**). The linear dynamic range of these dyes extends over three orders of magnitude, thus surpassing silver and Coomassie Blue stains in performance. An evaluation of 11 protein standards ranging in isoelectric point from 3.5 to 9.3 indicates that SYPRO Ruby IEF gel stain is 3–30 times more sensitive than highly sensitive silver stains *(24)*. Proteins that stain poorly with silver stain techniques are often readily detected by SYPRO Ruby dye *(27)*. Similar to colloidal Coomassie Blue stain but unlike silver stain, SYPRO Ruby dye stains are end point stains. Thus, staining times are not critical and staining can be performed over night without gels overdeveloping.

A potential disadvantage to detection of proteins using luminescent compared to colorimetric metal chelate stains is the requirement for ancillary equipment such as a laser gel scanner, UV light box, bandpass filters, and photographic or charge-coupled device (CCD) camera. The tremendous improvement in detection sensitivity and linear dynamic range certainly justifies the investment in equipment. Procedures for the detection of electrophoretically separated proteins utilizing colorimetric and luminescent metal chelate stains are presented in this chapter. Methods for the elution of the metal chelate stains are also presented. The procedures are applicable for detection of proteins or peptides in polyacrylamide or agarose gels as well as on nitrocellulose, PVDF, or nylon membranes. Owing to the electrostatic mechanism of the protein visualization methods, metal chelate stains are unsuitable for detection of proteins and peptides immobilized on cationic membranes.

2. Materials

2.1. Colorimetric Detection of Electroblotted Proteins on Membranes

1. Block buffer: 0.1% polyvinylpyrrolidone-40 (PVP-40) in 2% glacial acetic acid.
2. Ferrozine–ferrous stain (stable for at least 6 mo at room temperature): 0.75 m*M* 3-(2-pyridyl)-5,6-*bis* (4-phenylsulfonic acid)-1,2,4-triazine disodium salt (Ferrozine), 30 m*M* ferric chloride, 15 m*M* thioglycolic acid in 2% glacial acetic acid. Alternatively, commercially prepared stain solutions of Ferrozine–ferrous (Rev–Pro stain kit; Genomic Solutions, Ann Arbor, MI) or Pyrogallol Red-molybdenum (Microprotein-PR kit; Sigma Chemical Company, St. Louis, MO) can be used.
3. 2% Glacial acetic acid.
4. Ferrocyanide–ferric stain (stable for at least 6 mo at room temperature): 100 m*M* sodium acetate, pH 4.0; 100 m*M* potassium ferrocyanide, 60 m*M* ferric chloride.

2.2. Luminescent Detection of Electroblotted Proteins Using Bathophenanthroline Disulfonate-europium (SYPRO Rose Stain)

1. Formate buffer: 100 m*M* formic acid, pH 3.7, 100 m*M* sodium chloride.
2. Bathophenanthroline–europium blot stain (stable for at least 6 mo at room temperature): 1.5 m*M* bathophenanthroline disulfonic acid disodium salt, 0.5 m*M* europium chloride, and 0.2 m*M* EDTA (added from 1000× stock, pH 7.0).

2.3. Luminescent Detection of Electroblotted Proteins Using SYPRO Rose Plus Protein Blot Stain

1. SYPRO Rose Plus protein blot stain kit (Molecular Probes, Inc. cat. #S-12011) The kit contains the following components:
   SYPRO Rose Plus blot wash solution (component A), 200 mL
   SYPRO Rose Plus blot stain solution (component B), 200 mL
   SYPRO Rose Plus blot destain solution (component C), 200 mL
   The kit contains sufficient material for staining 10–40 minigel electroblots or four large-format electroblots (20 × 20 cm). The SYPRO Rose Plus solutions may be reused up to four times with little loss in sensitivity.

2.4. Luminescent Detection of Electroblotted Proteins Using SYPRO Ruby Protein Blot Stain

1. SYPRO Ruby protein blot stain (Molecular Probes, Inc.) is provided in a unit size of 200 mL. The 200-mL volume is sufficient for staining 10–40 minigel electroblots or four large-format electroblots (20 × 20 cm). SYPRO Ruby protein blot stain may be reused up to four times with little loss in sensitivity.
2. 7% Acetic Acid, 10% methanol.

2.5. Luminescent Detection of Proteins in SDS-Polyacrylamide Gels Using SYPRO Ruby Protein Gel Stain

1. SYPRO Ruby protein gel stain (Molecular Probes, Inc.) is provided ready to use, in either 200 mL volume (will stain ~four minigels) or 1 L volume (~20 minigels or 2–3 large-format gels).
2. 7% Acetic Acid, 10% methanol.

2.6. Luminescent Detection of Proteins in Isoelectric Focusing (IEF) Gels Using SYPRO Ruby IEF Protein Gel Stain

1. SYPRO Ruby IEF protein gel stain (Molecular Probes, Inc.) is supplied as a 400-mL ready-to-use staining solution, sufficient to stain ~10 IEF minigels or two standard flatbed IEF gels. Use caution when handling the SYPRO Ruby IEF protein gel stain as it contains a strong acid that can cause burns.

2.7. The Elution of Metal Chelate Stains

1. 50 m*M* Tris-HCl, pH 8.8; 200 m*M* NaCl; 20 m*M* EDTA (for the Ferrozine–ferrous stain).
2. 200 m*M* Sodium carbonate, 100 m*M* EDTA, pH 9.6 (for the Ferrozine–ferrous stain enhanced with the ferrocyanide–ferric stain and for the bathophenanthroline–europium blot stain).
3. 200 m*M* Sodium carbonate; 100 m*M* EDTA, pH 9.6 in 30% methanol (for the bathophenanthroline-europium gel stain).
4. SYPRO Rose Plus blot destain solution (component C) (for the SYPRO Rose Plus protein blot stain.)

3. Methods

3.1. Colorimetric Detection of Electroblotted Proteins on Membranes

The colorimetric metal chelate stains allow rapid visualization of proteins on solid-phase supports with detection sensitivities that are comparable to Coomassie Brilliant Blue staining ***(13–16)***. Detection sensitivity of the Ferrozine–ferrous stain can be enhanced to a level comparable to silver staining by further incubating membranes in ferrocyanide-ferric stain ***(13,15)***. The colorimetric metal chelate stains are fully reversible and compatible with Edman-based protein sequencing, lectin blotting, mass spectrometry and immunoblotting ***(13–16)***.

1. After electroblotting, vacuum slot blotting, or dot blotting, membranes are completely immersed in block buffer for 10 min. Blocking and staining steps are performed on a rotary shaker (50 rpm).
2. Thoroughly immerse membranes in Ferrozine-ferrous stain for 10–15 min until purple bands or spots appear.
3. Unbound dye is removed by several brief rinses in 2% glacial acetic acid until the membrane background appears white. Shaking can be performed manually using wash volumes roughly 2× greater than in the blocking and staining steps.
4. If increased sensitivity is desired, the blot can be double stained by subsequently incubating in the ferrocyanide-ferric stain for 10-15 min (*see* **Notes 1–3**).
5. Unbound dye is removed by several brief washes in 100 m*M* sodium acetate, pH 4.0 (*see* **Note 4**). Shaking can be performed manually using wash volumes roughly 2× greater than in the blocking and staining steps.
6. Stained proteins are visualized by eye and quantified using a CCD camera (*see* **Note 5**).

3.2. Luminescent Detection of Electroblotted Proteins Using Bathophenanthroline Disulfonate-Europium (SYPRO Rose Stain)

Bathophenanthroline disulfonate–europium complex (SYPRO Rose protein blot stain) is a medium sensitivity metal chelate stain that offers the same advantages as colorimetric stains, but with the additional benefits of a 500-fold linear dynamic range and detection sensitivity of <15–30 ng of protein *(17)*. The stain can be visualized using 302 nm UV epi-illumination and emits at 590 and 615 nm. There is also a blue 450 nm emission, however, due to uncomplexed ligand. Bathophenanthroline disulfonate–europium complex is readily removed by incubation of blots in mildly alkaline solution.

1. Following electroblotting, vacuum slot blotting, or dot blotting, membranes are washed 2× for 30 min with formate buffer followed by 4× for 30 min with deionized water (*see* **Note 6**). All washing and staining steps are performed on a rotary shaker (50 rpm).
2. Membranes are completely immersed in the bathophenanthroline-europium blot stain for 15 min.
3. Unbound dye is removed by washing 4–6× for 1 min in deionized water (*see* **Note 7**).
4. Membranes are dried at room temperature or in a 37°C drying oven (for quicker results) following the washing steps (*see* **Note 8**).
5. Stained proteins are visualized by reflective UV illumination (*see* **Note 9**).

3.3. Luminescent Detection of Electroblotted Proteins Using SYPRO Rose Plus Protein Blot Stain

SYPRO Rose Plus protein blot stain is an improved europium-based complex with a narrow emission maximum centered at 615 nm. The blue fluorescence observed with SYPRO Rose dye has been eliminated with the new stain and SYPRO Rose Plus dye is readily excited using UV-A, UV-B, or UV-C epi-illumination. Transillumination is not recommended. Typically 2–8 ng of electroblotted protein may be detected using the stain. SYPRO Rose Plus protein blot stain is easily removed by incubation in a mildly alkaline solution. The stain can not be visualized using a laser-based gel scanner, but is fully compatible with mass spectrometry and immunoblotting technologies.

1. Following electroblotting, vacuum slot blotting, or dot blotting, membranes are completely immersed in SYPRO Rose Plus protein blot wash solution and incubated at room temperature for 15 min in a small, polypropylene staining dish. Perform all washing and staining steps with continuous, gentle agitation (ideally, on an orbital shaker at 50 rpm).
2. Repeat the wash step.
3. Incubate the membrane in four changes of deionized water for 10 min each.
4. Completely immerse the membrane in SYPRO Rose Plus protein blot stain reagent for 15 min.
5. Wash the membrane 4–6× for 1 min in deionized water, to remove excess dye from the membrane. Stained membranes should be monitored using UV epi-illumination periodically to determine whether background luminescence has been washed away (*see* **Note 10**).
6. Blots treated with the SYPRO Rose Plus blot stain are best preserved by allowing membranes to air dry. After staining, wet membranes should not be touched, since residue found on latex gloves may destroy the staining pattern. Use forceps to handle wet blots. Once dry, membranes can be handled freely.
7. Stained proteins are visualized by reflective UV illumination (*see* **Note 11**).

3.4. Luminescent Detection of Electroblotted Proteins Using SYPRO Ruby Protein Blot Stain

SYPRO Ruby protein blot stain is suitable for visualizing proteins on nitrocellulose or PVDF membranes using UV transillumination, UV epi-illumination, or laser excitation at 450–532 nm *(22)*. Roughly 2–8 ng of protein can routinely be detected on electroblots and the linear dynamic range of detection is roughly 1000-fold. Unlike other sensitive detection technologies such as colloidal gold stain, SYPRO Ruby protein blot stain is fully compatible with mass spectrometry and Edman-based sequencing. The stain is fairly permanent, but is lost during protein blocking steps associated with Western blotting. Thus, the fluorescent pattern must be recorded and documented prior to immunodetection of a specific target on the blot.

3.4.1. Staining Proteins After Electroblotting to Nitrocellulose Membranes

1. Following electroblotting, vacuum slot blotting, or dot blotting to nitrocellulose, membranes are completely immersed in 7% acetic acid, 10% methanol and incubated at room temperature for 15 min in a small, polypropylene staining dish. Perform all washing and staining steps with continuous, gentle agitation (ideally, on an orbital shaker at 50 rpm).
2. Incubate the membrane in four changes of deionized water for 5 min each.
3. Completely immerse the membrane in SYPRO Ruby protein blot stain reagent for 15 min.
4. Wash the membrane 4–6× for 1 min in deionized water. This wash serves to remove excess dye from the membrane. Membranes stained with SYPRO Ruby protein blot stain should be monitored using UV epi-illumination periodically to determine if background luminescence has been washed away (*see* **Notes 12** and **13**).
5. Blots treated with the SYPRO Ruby protein blot stain can be preserved by allowing membranes to air-dry. After staining, wet membranes should not be touched, since residue found on latex gloves may destroy the staining pattern. Use forceps to handle wet blots. Once dry, membranes can be handled freely.
6. Stained proteins may be visualized by reflective or transmissive UV illumination or using a laser-based gel scanner (*see* **Notes 12** and **13**).

3.4.2. Staining Proteins After Electroblotting to PVDF Membranes

1. Following electroblotting, vacuum slot blotting, or dot blotting, to a sheet of PVDF membrane, allow the membrane to dry completely.
2. Float the membrane face down in 7% acetic acid, 10% methanol and incubate for 15 min. Perform all washing and staining steps with continuous, gentle agitation (e.g., on an orbital shaker at 50 rpm).
3. Incubate the membrane for 5 min each in four changes of deionized water.
4. Float the membrane face down in SYPRO Ruby protein blot stain reagent for 15 min.
5. Wash the membrane 2–3× for 1 min in deionized water. This wash serves to remove excess dye from the membrane. Membranes stained with SYPRO Ruby protein blot stain should be monitored using UV epi-illumination periodically to determine if background luminescence has been washed away (*see* **Notes 12** and **13**).
6. Blots treated with the SYPRO Ruby protein blot stain are best preserved by allowing the membranes to air-dry. After staining, wet membranes should not be touched, since residue found on latex gloves may destroy the staining pattern. Use forceps to handle wet blots. Once dry, the membranes can be handled freely.
7. Stained proteins may be visualized by reflective or transmissive UV illumination or using a laser-based gel scanner (*see* **Notes 12** and **13**).

3.5. Luminescent Detection of Proteins in Sodium Dodecyl Sulfate (SDS)-Polyacrylamide Gels Using SYPRO Ruby Protein Gel Stain

SYPRO Ruby protein gel stain is an ultrasensitive dye for detecting proteins separated by SDS-polyacrylamide or 2-D polyacrylamide gel electrophoresis (PAGE). The background fluorescence is low, the stain's linear dynamic range extends over three orders of magnitude, and shows low protein-to-protein variation. The stain is more sensitive than colloidal Coomassie Blue stain and equal in sensitivity to the best silver stains available. The stain is ready-to-use, and gels cannot over stain. Optimal staining requires about 3–4 h. Staining times are not critical, however, and staining can be performed overnight. SYPRO Ruby protein gel stain will not stain extraneous nucleic acids, and it is compatible with further downstream microchemical processing. The stain does not interfere with subsequent analysis of proteins by Edman-based sequencing or mass spectrometry. SYPRO Ruby protein gel stain can be used with many types of gels, including 2-D gels, Tris-glycine SDS gels, Tris-tricine precast SDS gels, and nondenaturing gels. SYPRO Ruby protein gel stain is also compatible with gels adhering to plastic backings, although the signal from the inherent blue fluorescence of the plastic must be removed with an appropriate emission filter. The stain is suitable for peptide mass profiling using MALDI-TOF mass spectrometry ***(25–27)***. Stained gels can be visualized with a 302-nm UV transilluminator, various laser scanners, or other blue light-emitting sources. The dye maximally emits at about 610 nm.

1. Prepare and run SDS-polyacrylamide or 2-D polyacrylamide gels according to standard protocols. Perform staining with SYPRO Ruby protein gel stain using continuous, gentle agitation (e.g., on an orbital shaker at 50 rpm). For small gels, use circular staining dishes on orbital shakers if possible.
2. Clean and *thoroughly rinse* the staining dishes before use (*see* **Note 14**).
3. Fix the gel in 10% methanol and 7% acetic acid in a plastic dish for 30 min. This step improves the sensitivity of the stain in 2-D gels, but is optional for 1-D SDS-PAGE gels.
4. Pour the staining solution into a small, clean plastic dish. For one or two standard-size minigels, use ~50–100 mL of staining solution; for larger gels, use 500–750 mL (*see* **Note 15**).
5. Place the gel into fresh staining solution (*see* **Note 16**). Protect the gel and staining solution from light at all times by covering the container with a lid or with aluminum foil.
6. Gently agitate the gel in stain solution at room temperature. The staining times range from 90 min to 3 h, depending on the thickness and percentage of polyacrylamide in the gel. Specific staining can be seen in as little as 30 min. However, a minimum of 3 h of staining is required for the maximum sensitivity and linearity. For convenience, gels may be left in the dye solution overnight or longer without over staining.
7. After staining, rinse the gel in deionized water for 30–60 min to decrease background fluorescence. To further decrease background fluorescence the gel can be washed in a mixture of 10% methanol and 7% acetic acid for 30 min instead of water (*see* **Note 17**). This is especially recommended for 2-D gels. The gel may be monitored periodically using UV illumination to determine the level of background fluorescence (*see* **Note 18**).
8. To dry the stained gel for permanent storage, incubate the gel in a solution of 2% glycerol for 30 min. Dry the stained gel using a gel dryer by standard methods. Note that proteins present at very low levels may no longer be detectable after gel drying.
9. Stained proteins in wet or dried gels may be visualized by reflective or transmissive UV illumination or using a laser-based gel scanner (*see* **Note 18**).

3.6. Luminescent Detection of Proteins in Gels Using SYPRO Ruby IEF Protein Gel Stain

SYPRO Ruby IEF protein gel stain is a ready-to-use, ultrasensitive, luminescent protein stain created especially for the analysis of proteins in IEF gels. This fluorescent stain attains comparable sensitivity to that of the best silver-staining techniques. Staining protocols are simple, the stain is ready-to-use, and it cannot overstain. It will not stain extraneous nucleic acids and the stain detects glycoproteins and other difficult-to-stain proteins. It does not interfere with subsequent analysis of proteins by Edman-based sequencing or mass spectrometry. SYPRO Ruby IEF protein gel stain is also compatible with agrose gels or polyacrylamide gels adhering to plastic backings.

1. Prepare and run IEF gels according to standard protocols. The staining technique is appropriate for carrier ampholyte isoelectric focusing or immobilized pH gradient (IPG) gels *(24)*. Perform SYPRO Ruby IEF gel staining with continuous, gentle agitation (e.g., on an orbital shaker at 50 rpm). For small gels, use circular staining dishes on orbital shakers if possible.
2. Clean and *thoroughly rinse* the staining dishes before use. (*see* **Note 20**).
3. Incubate the gel in the undiluted stain overnight. (*see* **Note 19**). **Caution: SYPRO Ruby IEF protein gel stain contains a caustic acid. The stain should be handled with care, using protective clothing, eye protection, and gloves.**
4. Rinse the gel in four changes of deionized water over 2 h to decrease background fluorescence. The gel may be monitored periodically using UV transillumination (*see* **Note 21**) to determine the level of background fluorescence. For IEF gels with plastic backings, the gel separates from the backing during this water wash. It is important that the backing be removed from the gels as it has high inherent fluorescence.
5. To dry the stained gel for permanent storage, incubate the gel in a solution of 2% glycerol for 30 min. Dry the stained gel using a gel dryer by standard methods. Note that proteins present at very low levels may no longer be detectable after gel drying.
6. Stained proteins in wet or dried gels may be visualized by reflective or transmissive UV illumination or using a laser-based gel scanner (*see* **Note 21**).

3.7. Elution of Metal Chelate Stains

The Ferrozine-ferrous, Pyrogallol Red–molybdate, Ferrocyanide–ferric, colorimetric metal chelate stains as well as the luminescent bathophenanthroline disulfonate–europium and SYPRO Rose Plus protein blot stains are readily removed from electroblotted proteins by incubation in mildly alkaline solution. SYPRO Ruby protein blot stain is not readily reversible by similar methods, although it is slowly removed from proteins during blocking and incubation steps commonly used in immunoblotting procedures. The gel stains are also not readily destained.

The Ferrozine–ferrous stain can be eluted by immersing the blots in 50 m*M* Tris-HCl, pH 8.8, 200 m*M* NaCl, 20 m*M* EDTA for 15 min on a rotary shaker (50 rpm). Blots treated with the Ferrozine–ferrous stain followed by enhancement with the ferrocyanide–ferric stain require harsher elution conditions. This stain is eluted by incubation in 200 m*M* sodium carbonate, 100 m*M* EDTA, pH 9.6 for 10 min. As the bathophenanthroline-europium and SYPRO Rose Plus stains can be observed only upon UV illumination, they do not require elution if subsequent colorimetric detection procedures are to be performed *(29)*. If concerned about interference with subsequent procedures, however, the stains can be eluted from blots by incubation in 200 m*M* sodium carbonate, 100 m*M*

EDTA, pH 9.6 or using SYPRO Rose Plus destain solution (component C). All the metal chelate stains mentioned above may be destained using the following protocol:

1. Incubate the membrane in appropriate destain reagent for 15 min with continuous gentle agitation.
2. Rinse twice for 1 min each in deionized water.

4. Notes

4.1. Colorimetric Detection of Proteins on Nitrocellulose, PVDF, or Nylon Membranes

1. Although enhancement of the Ferrozine–ferrous stain with ferrocyanide–ferric stain substantially increases detection sensitivity, the double stain is also more difficult to elute than the Ferrozine–ferrous stain alone (*see* **Subheading 3.7.** for methods of stain reversal).
2. The ferrocyanide-ferric stain may form a precipitate after long–term storage. The precipitate is easily resuspended by vigorous shaking or sonication.
3. If a dried nitrocellulose membrane is incubated in the ferrocyanide–ferric stain, a patchy background may result that is difficult to destain. Dry blots should be rehydrated briefly in deionized water prior to incubation in the ferrocyanide–ferric stain.
4. The number of washes necessary to remove background staining may vary slightly. The ferrocyanide–ferric stain may initially remain bound to the nitrocellulose membrane but the membrane background will become white with sufficient washing.
5. Membranes may be imaged transmissively or reflectively. Transmissive imaging is preferred over conventional reflective imaging as it improves signal detection while maintaining a white background. Furthermore, in some cases the protein sample penetrates through the membrane support to the reverse side (particularly with vacuum slot blotters). The metal chelate complexes stain proteins present throughout the thickness of the membrane and on both surfaces but reflective scanning only detects signal on the front surface, leading to a poorer linear dynamic range of quantitation. This problem is alleviated with transmissive scanning. Typically, a 45 W white light box is used in conjuction with a 450 ± 70 nm bandpass filter to enhance image contrast when imaging Ferrozine–ferrous or Pyrogallol Red-molybdenum stains. For blue stains such as Ferrozine-ferrous followed by the ferrocyanide-ferric stain, a 600 ± 70 nm bandpass filter may enhance the image contrast. Because the membrane support blocks a substantial amount of transmitted light, care must be taken to completely mask the sample with black cardboard for transmissive imaging.

4.2. Luminescent Detection of Electroblotted Proteins Using Bathophenanthroline Disulfonate-Europium (SYPRO Rose Stain)

6. Since formate ions may chelate europium ions, the washes with deionized water are crucial for complete removal of the formate buffer. Otherwise, the staining solution may be inactivated. Care should be taken to aspirate off all the solution between washes.
7. Removal of unbound bathophenathroline-europium stain can not be visually monitored as with the colorimetric stains. We have found that washing 4× for 1 min is effective. However, the number and duration of washes may vary from case to case (depending on the size of the membrane and the volume of each wash). Therefore, optimal washing should be determined empirically for each application.
8. After staining, wet membranes should not be touched since residue found on latex laboratory gloves may destroy the stain. Once dry, membranes can be handled freely. Since

water is known to quench europium luminescence, drying the membrane also serves to enhance the signal (about twofold).

9. Images are best obtained using a cooled CCD-camera by digitizing at about 1024 × 1024 picture elements (pixels) resolution with 12- or 16-bit gray scale levels assigned per pixel. To visualize the stain, the front face of membranes should be illuminated with the UV source: a hand-held, UV-B (302 nm) light, a UV light box placed on its side, or a top illuminating system such as the Bio-Rad Fluor-S imager. Direct transillumination through the blotting membrane yields unsatisfactory results. SYPRO Rose protein blot stain is best visualized using a 490-nm longpass filter such as a SYPRO protein gel photographic filter (Molecular Probes, Inc., cat. no. S-6656) or a 600-nm bandpass filter.

4.3. Luminescent Detection of Electroblotted Proteins Using SYPRO Rose Plus Protein Blot Stain

10. Removal of unbound bathophenathroline–europium stain cannot be visually monitored as with the colorimetric stains. We have found that washing 4× for 1 min is effective. However, the number and duration of washes may vary from case to case (depending on the size of the membrane and the volume of each wash). Therefore, optimal washing should be determined empirically for each application.
11. Proteins stained with SYPRO Rose Plus protein blot stain are readily visualized by eye using epi-illumination with a UV light source. Illuminate the front face of the membranes using a hand-held, UV-B (302 nm) light source, a UV light box placed on its side, or a top illuminating system such as the Bio-Rad Fluor-S™ imager. For greatest sensitivity and accurate quantitation, photograph the blot using Polaroid photography with a 490-nm longpass filter such as the SYPRO protein gel photographic filter (Molecular Probes, Inc., cat. no. S-6656) or use a computerized CCD camera-based image analysis system equipped with a 600-nm bandpass filter. Direct transillumination through the blotting membrane yields unsatisfactory results.

4.4. Luminescent Detection of Electroblotted Proteins Using SYPRO Ruby Protein Blot Stain

12. Proteins stained with SYPRO Ruby protein blot stain are readily visualized by eye using epi-illumination with UV light source. Illuminate the front face of the membranes using a hand-held, UV-B (302 nm) light source, a UV light box placed on its side, or a top illuminating system. Alternatively, use direct transillumination through the blotting membrane. For greater sensitivity and accurate quantitation, photograph the blot using Polaroid® photography or a computerized CCD camera-based image analysis system equipped with a 490-nm longpass filter such as the SYPRO protein gel photographic filter (Molecular Probes, Inc., cat no. S-6656) or a 600-nm bandpass filter. Laser-based imaging systems can also be used.
13. Images are best obtained using a cooled CCD-camera by digitizing at about 1024 × 1024 picture elements (pixels) resolution with 12- or 16-bit gray scale levels assigned per pixel. To visualize the stain, the front face of membranes should be illuminated with the UV source: a hand-held, UV-B (302 nm) light, a UV light box placed on its side, or a top illuminating system such as the Bio-Rad Fluor-S imager. Satisfactory results are obtained from direct transillumination through the blotting membrane as well. SYPRO Ruby protein blot stain is best visualized using a 490-nm longpass filter such as a SYPRO protein gel photographic filter (Molecular Probes, Inc., cat. no. S-6656) or a 600 nm bandpass

filter. Proteins stained with the dye can also be visualized using imaging systems equipped with 450-, 473-, 488-, or even 532-nm lasers.

4.5. Luminescent Detection of Proteins in SDS-Polyacrylamide Gels Using SYPRO Ruby Protein Gel Stain

14. Polypropylene dishes, such a Rubbermaid® Servin' Savers, are the optimal containers for staining because the high-density plastic adsorbs only a minimal amount of the dye. Clean and rinse the staining containers well before use, as detergent will interfere with staining. It is best to rinse the containers with ethanol before use. For small gels, circular staining dishes provide the best fluid dynamics on orbital shakers, resulting in less dye aggregation and better staining. For large format 2-D gels, polyvinyl chloride photographic staining trays, such as Photoquip Cesco-Lite 8 in. × 10 in. photographic trays also work well. Glass dishes are not recommended.
15. Minimal staining volumes for typical gel sizes are as follows:
    50 mL, for 8 cm × 10 cm × 0.75 mm gels (minigels)
    330 mL, for 16 cm × 20 cm × 1 mm gels
    500 mL, for 20 cm × 20 cm × 1 mm gels
    or ~10 times the volume of the gel for other gel sizes
16. Use only fresh staining solution for optimal sensitivity. Longer staining times result in greater sensitivity. Using too little stain will lower the sensitivity.
17. Always store gels in the dark to prevent photobleaching. When gels are stored in the staining solution, the signal decreases somewhat after several days; however, depending on the amount of protein in bands of interest, gels may retain a usable signal for many weeks.
18. The stained gel is best viewed on a standard 302 nm UV or a blue-light transilluminator. Gels may also be visualized using various laser scanners: 473 nm (SHG) laser, 488 nm argon-ion laser, 532 nm (YAG) laser. Alternatively, use a xeon arc lamp, blue fluorescent light, or blue light-emitting diode (LED) source. Gels may be photographed by Polaroid or CCD camera. Use Polaroid 667 black-and-white print film and the SYPRO protein gel stain photographic filter (Molecular Probes, Inc., cat. no. S-6656). Exposure times vary with the intensity of the illumination source; for an f-stop of 4.5, roughly 1–3 s should be required.

4.6. Luminescent Detection of Proteins in IEF Gels Using SYPRO Ruby IEF Protein Gel Stain

19. Minimal staining volumes for typical gel sizes are as follows:

    50 mL, for 6 cm × 9 cm × 1 mm gels (minigels)
    150 mL, for 22 cm × 22 cm × 1 mm (Multiphor II format gels, Amersham-Pharmacia Biotech)
    Or approx 10 times the volume of the gel for other gel sizes.
20. Polypropylene dishes, such as Rubbermaid® Servin' Savers, are the optimal containers for staining because the high-density plastic adsorbs only a minimal amount of the dye. Clean and rinse the staining containers well before use as detergent will interfere with staining. It is best to rinse the containers with ethanol before use. For small gels, circular staining dishes provide the best fluid dynamics on orbital shakers, resulting in less dye aggregation and better staining. For large-format 2-D gels, polyvinyl chloride photographic staining trays, such as Photoquip Cesco-Lite 8 in. × 10 in. photographic trays also work well. Glass dishes are not recommended.
21. The stained gel is best viewed on a standard 302-nm UV or a blue-light transilluminator. Gels may also be visualized using various laser scanners: 473-nm (SHG) laser, 488 nm

argon-ion laser, 532-nm (YAG) laser. Alternatively, use a xenon arc lamp, blue fluorescent light, or blue light-emitting diode (LED) source. Gels may be documented by conventional photography using the SYPRO protein gel stain photographic filter (Molecular Probes, Inc., cat. no. S-6656) or using a CCD camera equipped with 600-nm bandpass filter. Exposure times vary with the intensity of the illumination source; for an f-stop of 4.5, roughly 1–3 s should be required.

References

1. Merril, C. (1987) Detection of proteins separated by electrophoresis, in *Advances in Electrophoresis*, Vol. 1 (Chrambach, A., Dunn, M., and Radola B., eds.), VCH Press, Germany/Switzerland/Great Britian/New York, pp. 111–139.
2. Wirth, P. and Romano, A. (1995) Staining methods in gel electrophoresis, including the use of multiple detection methods. *J. Chromatogr.* A, **698,** 123–143.
3. Hancock, K. and Tsang, V. (1983) India ink staining of proteins on nitrocellulose paper. *Analyt. Biochem.* **133,** 157–162.
4. Moeremans, M., Daneels, G., and De Mey, J. (1985) Sensitive colloidal metal (gold or silver) staining of protein blots on nitrocellulose membranes. *Analyt. Biochem.* **145,** 315–321.
5. Moeremans, M., Raeymaeker, M., Daneels, G., and De Mey, J. (1986) Ferridye: colloidal iron binding followed by Perls' reaction for the staining of proteins transferred from sodium dodecyl sulfate gels to nitrocellulose and positively charged nylon membranes. *Analyt. Biochem.* **153,** 18–22.
6. Hunter, J. and Hunter, S. (1987) Quantification of proteins in the low nanogram range by staining with the colloidal gold stain Aurodye. *Analyt. Biochem.* **164,** 430–433.
7. Egger, D. and Bienz, K. (1987) Colloidal gold and immunoprobing of proteins on the same nitrocellulose blot. *Analyt. Biochem.* **166,** 413–417.
8. Yamaguchi, K. and Asakawa, H. (1988) Preparation of colloidal gold for staining proteins electrotransferred onto nitrocellulose membranes. *Analyt. Biochem.* **172,** 104–107.
9. Li, K., Geraerts, W., van Elk, R., and Joosse, J. (1988) Fixation increases sensitivity of india ink staining of proteins and peptides on nitrocellulose paper. *Analyt. Biochem.* **174,** 97–100.
10. Li, K., Geraerts, R., van Elk, R., and Joosse, J. (1989) Quantification of proteins in the subnanogram and nanogram range: comparison of the Aurodye, Ferridye, and India Ink staining methods. *Analyt. Biochem.* **182,** 44–47.
11. Root, D. and Reisler, E. (1989) Copper iodide staining of protein blots on nitrocellulose membranes. *Analyt. Biochem.* **181,** 250–253.
12. Root, D. and Wang, K. (1993) Silver-enhanced copper staining of protein blots. *Analyt. Biochem.* **209,** 15–19.
13. Patton, W., Lam, L., Su, Q., Lui, M., Erdjument-Bromage, H., and Tempst, P. (1994). Metal chelates as reversible stains for detection of electroblotted proteins: application to protein microsequencing and immunoblotting. *Analyt. Biochem.* **220,** 324–335.
14. Shojaee, N., Patton, W., Lim, M., and Shepro, D. (1996) Pyrogallol red-molybdate; a reversible, metal chelate stain for detection of proteins immobilized on membrane supports. *Electrophoresis* **17,** 687–695.
15. Lim, M., Patton, W., Shojaee, N., and Shepro, D. (1996) A solid-phase metal chelate assay for quantifying total protein; resistance to chemical interference. *Biotechniques* **21,** 888–897.
16. Lim, M., Patton, W., Shojaee, N., and Shepro, D. (1997) Comparison of a sensitive, solid-phase metal chelate protein assay with the bicinchoninic acid (BCA) assay. *Am. Biotech. Lab.* **15,** 16–18.

17. Lim, M., Patton, W., Shojaee, N., Lopez, M., Spofford, K., and Shepro, D. (1997) A luminescent europium complex for the sensitive detection of proteins and nucleic acids immobilized on membrane supports. *Analyt. Biochem.* **245,** 184–195.
18. Chung, M. (1985) A specific iron stain for iron-binding proteins in polyacrylamide gels: application to transferrin and lactoferrin. *Analyt. Biochem.* **148,** 498–502.
19. Bickar, D. and Reid, P. (1992) A high-affinity protein stain for Western blots, tissue prints, and electrophoresis gels. *Analyt. Biochem.* **203,** 109–115.
20. Graham, G., Nairn, R., and Bates, G. (1978) Polyacrylamide gel staining with iron (II)-bathophenanthroline sulfonate. *Analyt. Biochem.* **88,** 434–441.
21. Zapolski, E., Gersten, D., and Ledley, R. (1982) [59Fe] ferrous bathophenathroline sulfonate: a radioactive stain for labeling proteins in situ in polyacrylamide gels. *Analyt. Biochem.* **123,** 325–328.
22. Berggren, K., Steinberg, T., Lauber, W., Carroll, J., Lopez, M., Chernokalskaya, E., et al. (1999) A luminescent ruthenium complex for ultrasensitive detection of proteins immobilized on membrane supports. *Analyt. Biochem.* **276,** 129–143.
23. Steinberg, T., Lauber, W., Berggren, K. Kemper, C., Yue, S., and Patton, W. (200) Fluorescence detection of proteins in SDS-polyacrylamide Gels using environmentally benign, non-fixative, saline solution. *Electrophoresis* **21,** 497–508.
24. Steinberg, T., Chernokalskaya, E., Berggren, K., Lopez, M., Diwu, Z., Haugland, R., and Patton, W. (2000) Ultrasensitive fluorescence protein detection in isoelectric focusing gels using a ruthenium metal chelate stain. *Electrophoresis* **21,** 486–496.
25. Patton, W. (2000) A thousand points of light; the application of fluorescence detection technologies to two-dimensional gel electrophoresis and proteomics. *Electrophoresis* **21,** 1123–1144.
26. Patton, W. (2000) Making blind robots see; the synergy between fluorescent dyes and imaging devices in automated proteomics. *BioTechniques* **28,** 944–957.
27. Berggren, K., Chernokalskaya, E., Steinberg, T., Kemper, C., Lopez, M., Diwu, Z., et al. (2000) Background-free, high-sensitivity staining of proteins in one- and two-dimensional sodium dodecyl sulfate-polyacrylamide gels using a luminescent ruthenium complex. *Electrophoresis* **21,** 2509–2521.
28. Gosling, J. (1990) A decade of development in immunoassay methodology. *Clin. Chem.* **36,** 1408–1427.
29. Harlow, E. and Lane, D. (1988) Antibodies; A Laboratory Manual. Cold Spring Harbor Laboratory Press, Cold Spring Harbor, New York, pp. 505.

35

Detection of Proteins in Polyacrylamide Gels by Fluorescent Staining

Michael J. Dunn

1. Introduction

Techniques of polyacrylamide gel electrophoresis are often the method of choice for the analysis of patterns of gene expression in a wide variety of complex systems. In particular, two-dimensional polyacrylamide gel electrophoresis (2-DE) is the core technology for separating complex protein mixtures in the majority of proteome projects *(1)*. This is due to its unrivalled power to separate simultaneously thousands of proteins and the availability of sophisticated computer software for the qualitative and quantitative analysis of differential patterns of protein expression *(2)*. Once this analysis has indicated proteins of interest, it is now relatively straightforward to directly identify and characterize proteins from polyacrylamide gels using highly sensitive microchemical methods *(3)*, particularly those based on mass spectrometry *(4)*. This has resulted in gel electrophoresis becoming one of the most important methods of protein purification for subsequent identification and characterization.

The successful use of polyacrylamide gel electrophoresis as an analytical tool for investigating protein expression and as a micropreparative procedure for protein identification and characterization depends on the availability of detection methods that are both quantitative and sensitive. Coomassie Brilliant Blue R-250 (CBB R-250) has been used for many years as a general protein stain following gel electrophoresis. Its restricted dynamic range limits its suitability for accurate quantitative analysis, although it is compatible with most methods of chemical characterization. However, CBB R-250 has a limited sensitivity being capable of detecting microgram to submicrogram amounts of protein. The development of colloidal CBB staining using CBB G-250 *(5)* provides increased sensitivity at approx 10 ng of protein. However, the trend toward the use of thinner gels and the need to detect small amount of protein within single bands or spots resolved by one- or two-dimensional electrophoresis have necessitated the development of more sensitive detection methods *(6)*.

Silver staining for the detection of proteins following gel electrophoresis was first reported in 1979 by Switzer et al. *(7)*, resulting in a major increase in the sensitivity of protein detection. More than 100 publications have subsequently appeared describing

From: *The Protein Protocols Handbook, 2nd Edition*
Edited by: J. M. Walker © Humana Press Inc., Totowa, NJ

variations in silver staining methodology ***(6,8)***. This group of procedures is generally accepted to be approx 100 times more sensitive than methods using CBB R-250, being able to detect low nanogram amounts of protein per gel band or spot. Although silver staining has become the most widely used technique for high sensitivity visualization of gel separated proteins (*see* Chapter 33), its variable binding characteristics toward many proteins and its relatively restricted dynamic range limit the accuracy and reliability of quantitation. Moreover, saturated protein bands and spots tend to be negatively stained making their quantitation impossible. Standard methods of silver staining are also often not compatible with chemical characterization by mass spectrometry as the glutaraldehyde and formaldehyde present results in alkylation of α- and ε-amino groups of proteins. To overcome this problem, silver staining protocols compatible with mass spectrometry in which glutaraldehyde is omitted have been developed ***(9,10)*** but these suffer from a decrease in sensitivity of staining and a tendency to a higher background.

Many of the problems associated with silver staining can be overcome using detection methods based on the use of fluorescent compounds. This group of methods are highly sensitive and generally exhibit excellent linearity and a high dynamic range, making it possible to achieve good quantitative analysis, particularly if a suitable imaging device is used. In addition, these methods should in general have good compatibility with protein characterization methods. Two approaches can be used, the first being to couple the proteins with a fluorescently labeled compound prior to electrophoresis. Examples of such compounds are dansyl chloride ***(11)***, fluorescamine (4-phenyl-[furan-2(3H),1-phthalan]-3,3'-dione) ***(12)***, o-pththaldialdehyde (OPA) ***(13)***, and MDPF (2-methoxy-2,4-diphenyl-3([2H])-furanone) (MDPF) ***(14)***. The latter reagent has a reported sensitivity of 1 ng protein/band and is linear over the range 1–500 ng of protein/band.

The main disadvantage of preelectrophoretic staining procedures is that they can cause protein charge modifications; for example, fluorescamine converts an amino group to a carboxyl group when it reacts with proteins. Such modifications usually do not compromise sodium dodecyl sulfate-polyacrylamide gel electrophoresis (SDS-PAGE) unless a large number of additional charged groups are introduced into the protein. However, they result in altered mobility during other forms of electrophoresis, resulting in altered separations by native PAGE, isoelectric focussing (IEF), and 2-DE. Recently, compounds that react with cysteine or lysine residues have been described and used successfully for 2-DE separations. The cysteine-reactive reagent monobromobimane ***(15)*** has been used to label proteins prior to analysis by 2-DE ***(16)***. Using a cooled CCD camera to measure fluorescence , the limit of detection was found to be 1 pg protein/spot ***(17)***.

In an alternative approach, two amine-reactive dyes (propyl Cy3 and methyl Cy5) have been synthesized and used to label *E. coli* proteins prior to electrophoresis ***(18)***. These cyanine dyes have an inherent positive charge, which preserves the overall charge of the proteins after dye coupling. The two dyes have sufficiently different fluorescent spectra that they can be distinguished when they are present together. This allowed two different protein samples, each labeled with one of the dyes, to be mixed together and subjected to 2-DE on the same gel. This method, which has been termed difference gel electrophoresis (DIGE) *(18)*, has great potential for improving the efficiency of detection of differences in 2-DE protein profiles between different samples *(19)*.

For 2-DE, one approach to overcoming the problems associated with charge modification during the IEF dimension is to label the proteins while present in the first dimension gel after IEF, prior to the second dimension separation by SDS-PAGE. Two fluorescent labels that have been used in this way are MDPF *(20)* and a fluorescent maleimide derivative *(21)*.

The alternative approach, which also overcomes the problem of protein charge modifications, is to label the proteins with fluorescent molecules such as 1-aniline-8-naphthalene sulpfonate (ANS) *(22)* and OPA after the electrophoretic separation has been completed. However, these two methods suffer the disadvantage of relative insensitivity. Two additional post-electrophoretic fluorescent staining reagents, SYPRO Orange and Red, have been described *(23,24)*. These stains have a high sensitivity (4–10 ng of protein/band) and excellent linearity with a high dynamic range. Using a fluorescent imaging device, the SYPRO dyes have been shown to be linear over three orders of magnitude in protein quantity *(24)*. The other advantage of this method is that staining can be achieved in only 30 min, compared with staining with silver and CBB R-250 which can take from 2 h to overnight. Gels can be stained without fixation so that they can be subjected to subsequent Western blotting procedures. However, staining with these reagents requires that the proteins are complexed with SDS, so that if the gels are fixed prior to staining or electrophoresis is carried out in the absence of this detergent, then the gels must be incubated in a solution of SDS prior to staining. Moreover, these dyes require 7% acetic acid and organic solvent in the staining solution, which can cause problems in electroblotting, electroelution, and measurement of enzyme activity. The dye SYPRO Tangerine was subsequently developed to overcome these shortcomings *(25)*.

Recently, a ruthenium-based metal chelate stain, SYPRO Ruby, has been developed *(26,27)*. The advantage of this dye over the other SYPRO stains is that the presence of SDS is not required for binding so that gels can be fixed in the normal way and then simply soaked in the SYPRO Ruby staining solution for a minimum of 3 h. In addition, SYPRO Ruby has a sensitivity similar to the most sensitive silver stains *(26,28)*, and it has been reported that some proteins that stain poorly with silver techniques are readily detected by SYPRO Ruby *(27)*. In addition, staining times are not critical and staining can be performed overnight without gels overdeveloping. Together with its very high dynamic range of more than 1000-fold, these properties make SYPRO Ruby an excellent choice for quantitative applications. It also has been shown to have excellent compatibility with chemical characterization methods including Edman degradation *(29)* and MALDI-TOF MS *(30)*. A 2-DE gel separation of rat heart proteins visualized with SYPRO Ruby is shown in **Fig. 1**.

2. Materials

All solutions should be freshly prepared, and overnight storage is not recommended.

1. Fixation solution: 10% (v/v) methanol, 7% (v/v) acetic acid.
2. SYPRO Ruby staining solution (*see* **Note 1**).
3. Washing solution: 10% (v/v) methanol, 7% (v/v) acetic acid.

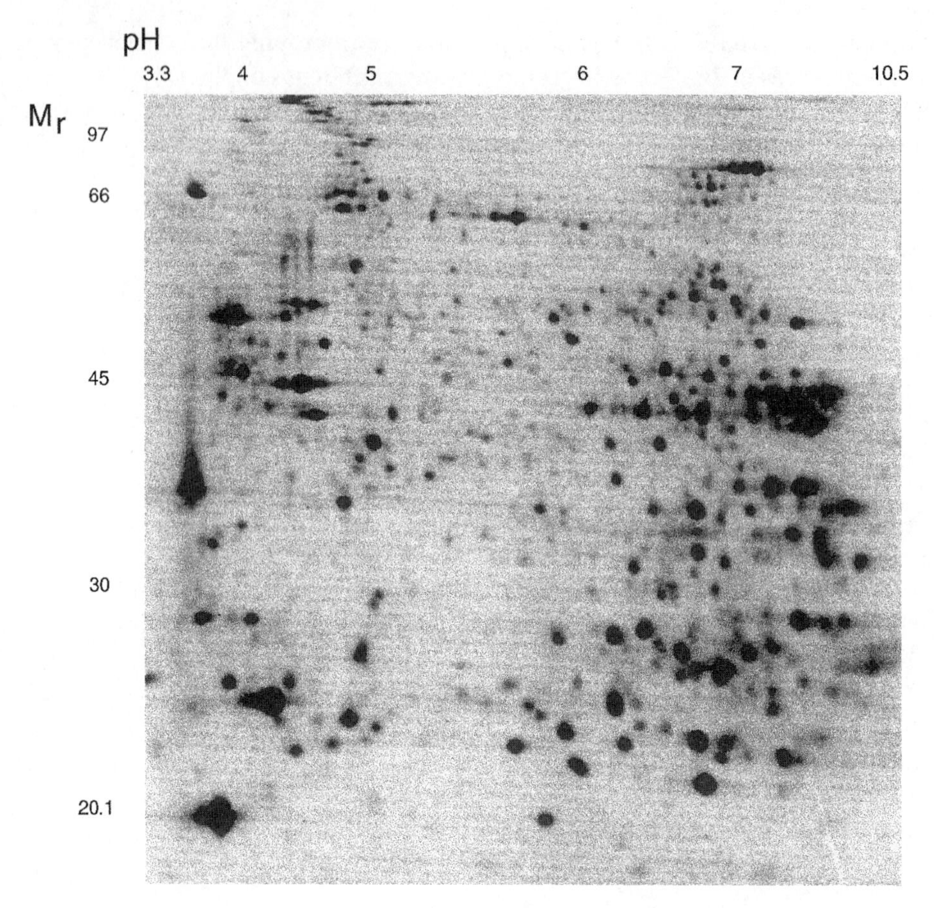

Fig. 1. A 2-DE separation of rat heart proteins visualized by SYPRO Ruby staining. Note that the fluorescent image is shown as an inverted image to show black spots on a white background for convenient viewing. The first dimension was pH 3–10 NL immobilized pH gradient (IPG) IEF and the second dimension was 12% T SDS-PAGE. The scale at the top indicates the nonlinear pH gradient used in the first IPG 3–10 NL IEF dimension, while the scale at the left indicates $M_r \times 10^{-3}$.

3. Method

***Note:** All incubations are carried out at room temperature with gentle agitation.*

1. After electrophoresis, fix the gel immediately (*see* **Note 2**) in 500 mL (*see* **Note 3**) of fixation solution for a minimum of 1 h at room temperature (*see* **Note 4**).
2. Place the gel into 500 mL of SYPRO Ruby protein gel stain solution for a minimum of 3 h (*see* **Note 5**).
3. Wash the gel in 500 mL of washing solution for 10 min (*see* **Note 6**).
4. Image the gel (*see* **Note 7**).

4. Notes

1. SYPRO Ruby protein gel stain is available only as a premade solution from the manufacturers (Molecular Probes, Eugene, OR, USA).

2. Gloves should be worn at all stages when handling gels, particularly when they will be used for subsequent chemical characterization, to prevent contamination with keratin proteins from the skin.
3. Volumes of the solutions used at all stages should be sufficient such that the gel is totally immersed. If the volume of solution is insufficient for total immersion, staining will be uneven and the gel surface can dry out. The volumes given here are suitable for a 20 × 20 cm 2-D gel.
4. High-percentage polyacrylamide and thick gels require an increased period for fixation, and overnight soaking is recommended.
5. The staining time is not critical and staining can be performed overnight without gels overdeveloping. In this case, the containers used for staining should be wrapped with aluminum foil to avoid light exposure.
6. Gels stained with SYPRO Ruby can be stored in the washing solution for a period of time (several days to a week) before image acquisition. In this case, the containers used for staining should be wrapped with aluminum foil to avoid light exposure.
7. SYPRO Ruby dye can be excited using a standard 300 nm UV trans-illuminator, so that gel images can be documented using any suitable camera system. However, for quantitative analysis the gel must be digitized using a suitable imaging system such as laser fluorimager (equipped with 450-, 473-, 488-, or even 532-nm lasers) or a multiwavelength coupler charge device (CCD)-based fluorimaging system *(28)*.

References

1. Dunn, M. J. and Görg, A. (2001) Two-dimensional polyacrylamide gel electrophoresis for proteome analysis, in *Proteomics, From Protein Sequence to Function* (Pennington, S. R. and Dunn, M. J., eds.), BIOS, Oxford, pp. 43–63.
2. Pleissner, K. P., Oswald, H., and Wegner, S. (2001) Image analysis of two-dimensional gels, in *Proteomics, From Protein Sequence to Function* (Pennington, S. R. and Dunn, M. J., eds.), BIOS, Oxford, pp. 131–149.
3. Wilkins, M. R. and Gooley, A. (1997) Protein identification in proteome analysis, in *Proteome Research: New Frontiers in Functional Genomics* (Wilkins, M. R., Williams, K. L., Appel, R. D. and Hochstrasser, D. F., eds.), Springer-Verlag, Berlin, pp. 35–64.
4. Patterson, S. D., Aebersold, R., and Goodlett, D. R. (2001) Mass spectrometry-based methods for protein identification and phosphorylation site analysis, in *Proteomics, From Protein Sequence to Function* (Pennington, S. R. and Dunn, M. J., eds.), BIOS, Oxford, pp. 87–130.
5. Neuhoff, V., Arold, N., Taube, D., and Erhardt, W. (1988) Improved staining of proteins in polyacrylamide gels including isoelectric focusing gels with clear background at nanogram sensitivity using Coomassie Brilliant Blue G-250 and R-250. *Electrophoresis* **9,** 255–262.
6. Patton, W. F. (2001) Detecting proteins in polyacrylamide gels and on electroblot membranes, in *Proteomics, From Protein Sequence to Function* (Pennington, S. R. and Dunn, M. J., eds.), BIOS, Oxford, pp. 65–86.
7. Switzer, R. C., Merril, C. R., and Shifrin, S. (1979) A highly sensitive stain for detecting proteins and peptides in polyacrylamide gels. *Analyt. Biochem.* **98,** 231–237.
8. Rabilloud, T. (1990) Mechanisms of protein silver staining in polyacrylamide gels: A 10-year synthesis. *Electrophoresis* **11,** 785–794.
9. Shevchenko, A., Wilm, M., Vorm, O., and Mann, M. (1996) Mass spectrometric sequencing of proteins from silver-stained polyacrylamide gels. *Analyt. Chem.* **68,** 850–858.
10. Yan, J. X., Wait R, Berkelman T., Harry R., Westbrook, J. A., Wheeler, C. H., Dunn, M. J. (2000) A modified silver staining protocol for visualization of proteins compatible with matrix-assisted laser desorption/ionization and electrospray ionization-mass spectrometry. *Electrophoresis* **21,** 3666–3672.

11. Stephens, R. E. (1975) High-resolution preparative SDS-polyacrylamide gel electrophoresis: fluorescent visualization and electrophoretic elution-concentration of protein bands. *Analyt. Biochem.* **65,** 369–379.
12. Eng, P. R., and Parkes, C. O. (1974) SDS electrophoresis of fluorescamine-labeled proteins. *Analyt. Biochem.* **59,** 323–325.
13. Weidekamm, E., Wallach, D. F., and Fluckiger, R. (1973) A new sensitive, rapid fluorescence technique for the determination of proteins in gel electrophoresis and in solution. *Analyt. Biochem.* **5,** 102–114.
14. Barger, B. O., White, R. C., Pace, J. L., Kemper, D. L., Ragland, W. L. (1976) Estimation of molecular weight by polyacrylamide gel electrophoresis using heat stable fluors. *Analyt. Biochem.* **70**, 327–335.
15. O'Keefe, D. O. (1994) Quantitative electrophoretic analysis of proteins labeled with monobromobimane. *Analyt. Biochem.* **222,** 86–94.
16. Urwin, V. E. and Jackson, P. (1993) Two-dimensional polyacrylamide gel electrophoresis of proteins labeled with the fluorophore monobromobimane prior to first-dimensional isoelectric focusing: imaging of the fluorescent protein spot patterns using a charge-coupled device. *Analyt. Biochem.* **209,** 57–62.
17. Fey, S. J., Nawrocki, A., Larsen, M. R., Görg, A., Roepstorff, P., Skews, G. N., et al. (1997) Proteome analysis of *Saccharomyces cerevisia*: A methodological outline. *Electrophoresis* **18,** 1361–1372.
18. Ünlü, M., Morgan, M. E., and Minden, J. S. (1997) Difference gel electrophoresis: a single gel method for detecting changes in protein extracts. *Electrophoresis* **18,** 2071–2077.
19. Tonge, R., Shaw, J., Middleton, B., Rowlinson, R., Rayner S., Young, J., et al. (2001) Validation and development of fluorescence two-dimensional differential gel electrophoresis proteomics technology. *Proteomics* **1,** 377–396.
20. Jackson, P., Urwin, V. E., and Mackay, C. D. (1988) Rapid imaging, using a cooled charge-coupled-device, of fluorescent two-dimensional polyacrylamide gels produced by labelling proteins in the first-dimensional isoelectric focusing gel with the fluorophore 2-methoxy-2,4-diphenyl-3(2*H*)furanone. *Electrophoresis* **9,** 330–339.
21. Herbert, B. R., Molloy, M. P., Gooley, A. A., Walsh, B. J., Bryson, W. G., and Williams, K. L. (1998) Improved protein solubility in 2-D electrophoresis using tributyl phopshine as the reducing agent. *Electrophoresis* **19,** 845–851.
22. Hartman, B. K. and Udenfriend, S. (1969) A method for immediate visualization of proteins in acrylamide gels and its use for preparation of antibodies to enzymes. *Analyt. Biochem.* **30,** 391–394.
23. Steinberg, T. H., Jones, L. J., Haugland, R. P., and Singer, V. L. (1996) SYPRO orange and SYPRO red protein gel stains: one-step fluorescent staining of denaturing gels for detection of nanogram levels of protein. *Analyt. Biochem.* **239,** 223–237.
24. Steinberg, T. H., Haugland, R. P., and Singer, V. L. (1996) Applications of SYPRO orange and SYPRO red protein gel stains. *Analyt. Biochem.* **239,** 238–245.
25. Steinberg, T. H., Lauber, W. M., Berggren, K., Kemper, C., Yue, S., and Patton, W. F. (1999) Fluorescence detection of proteins in sodium dodecyl sulfate-polyacrylamide gels using environmentally benign, nonfixative, saline solution *Electrophoresis* **20,** 497–508.
26. Steinberg, T., Chernokalskaya, E., Berggren, E., Lopez, M., Diwu, Z., Haugland R., and Patton, W. (2000) Ultrasensitive fluorescence protein detection in isoelectric focusing gels using a ruthenium metal chelate stain. *Electrophoresis* **21,** 486–496.
27. Patton, W. F. (2002) A thousand points of light: the application of fluorescence detection technologies to two-dimensional gel electrophoresis and proteomics. *Electrophoresis* **21,** 1123–1144.

28. Yan, J. X., Harry, R. A., Spibey, C., and Dunn, M. J. (2000) Post-electrophoretic staining of proteins separated by two-dimensional gel electrophoresis using SYPRO dyes. *Electrophoresis* **21,** 3657–3665.
29. Berggren, K., Steinberg, T. H., Lauber, W. M., Carroll, J. A., Lopez, M. F., Chernokalskaya, E., et al. (1999) A luminescent ruthenium complex for ultrasensitive detection of proteins immobilized on membrane supports. *Analyt. Biochem.* **276,** 129–143.
30. Berggren, K., Chernokalskaya, E., Steinberg, T. H., Kemper, C., Lopez, M. F., Diwu, Z., et al. (2000) Background-free, high sensitivity staining of proteins in one- and two-dimensional sodium dodecyl sulfate-polyacrylamide gels using a luminescent ruthenium complex. *Electrophoresis.* **21,** 2509–2521.

36

Detection of Proteins and Sialoglycoproteins in Polyacrylamide Gels Using Eosin Y Stain

Fan Lin and Gary E. Wise

1. Introduction

A rapid, sensitive, and reliable staining technique is essential in detection of proteins in polyacrylamide gels. Coomassie brilliant blue R-250 (CBB) is the stain that meets these criteria except for sensitivity; i.e., CBB staining requires relatively large amounts of proteins. It has been reported that the sensitivity for CBB stain in polyacrylamide gels is 0.1–0.5 μg/protein band *(1)*. This problem of relatively low staining sensitivity is often circumvented by employing silver staining techniques *(2–5)*. However, it is difficult to transfer silver stained proteins to transfer membranes unless they either are negatively stained by silver *(6)* or the positively silver-stained proteins are treated with 2X SDS sample buffer prior to transfer *(7)*. In addition, sialoglycoproteins cannot be detected by CBB and thus have to be visualized by other stains, such as the periodic acid-Schiff (PAS) reagent *(8)*, silver stains *(9)*, or silver/Coomassie blue R-250 double staining technique *(10)*.

To circumvent the deficiencies of the above staining techniques, we have developed an eosin Y staining technique *(11)*. This staining method allows one to detect proteins more rapidly than most CBB and silver staining methods. It detects a variety of proteins in amounts as little as 10 ng in polyacrylamide gels, including membrane sialoglycoproteins, and has the added advantage of the antigenicity of the stained proteins being retained. The precise mechanism by which eosin Y stains both proteins and sialoglycoproteins is not fully understood.

However, in the staining protocol described here, the fixing and developing solution of 10% acetic acid/40% methanol (pH 2.5) is strongly acidic. Under such conditions, eosin Y might be converted into its precursor form of the dihydrofluoran. Thus, a protein might be stained by means of both hydrophobic interaction between aromatic rings of eosin Y and hydrophobic sites of the protein and by hydrogen bonding between hydroxyl groups of eosin Y and the backbone of a protein. Here we describe this detailed staining protocol and technical advice in order to enable others to obtain optimal staining results.

From: *The Protein Protocols Handbook, 2nd Edition*
Edited by: J. M. Walker © Humana Press Inc., Totowa, NJ

2. Materials

1. Eosin Y staining solution: A stock solution of 10% eosin Y (w/v) is prepared. This solution is stable at room temperature for at least 6 mo. Each 100 mL of staining solution contains 10 mL of 10% eosin Y solution, 40 mL of 100% methanol, 49.5 mL distilled deionized water, and 0.4–0.5 mL of full strength glacial acetic acid. The staining solution is made and filtered prior to use.
2. Gel fixation solution: 10% glacial acetic acid/40% methanol.
3. Gel developing solution: distilled-deionized water, 10% glacial acetic acid/40% methanol.
4. A black plastic board and a transilluminated fluorescent white light box.

3. Methods

3.1. Staining of Various Protein in SDS-PAGE

1. Immediately following electrophoresis, the SDS-polyacrylamide gel containing given proteins is fixed in 5 gel volumes of 10% glacial acetic acid/40% methanol for 10 min at room temperature with shaking and then rinsed with distilled water twice.
2. The gel is immersed with 200 mL (5–6 gel volumes) of the 1% eosin Y staining solution for 15 min at room temperature with shaking.
3. The gel is transferred to a clean glass container, quickly rinsed with distilled water and then washed with distilled water for 3 min (*see* **Note 2**).
4. The stained bands of the gel are developed by placing the gel in 10% acetic acid/40 methanol for about 15 s (*see* **Note 3**).
5. The development is stopped by immersing the gel in distilled water.
6. The gel can be kept in distilled water for at least 1 mo without fading.
7. The stained gel can be viewed either by using transilluminated fluorescent white light or by placing the gel on a black plastic board with top light illumination.

3.2. Staining of Membrane Sialoglycoproteins in SDS-PAGE

1. Immediately following electrophoresis, the SDS-polyacrylamide gel is placed in 200 mL (5–6 gel volumes) eosin Y staining solution for 45 min at room temperature with shaking.
2. The gel is quickly rinsed with distilled water and then washed with distilled water for 3 changes of 5 min each. Protein bands are visualized at the end of washing step (*see* **Note 4**).
3. The gel is then developed in 10% acetic acid/40% methanol for about 2 min.
4. The development of staining is terminated by washing the gel in distilled water twice.
5. The gel can be kept in distilled water for at least 1 mo without fading.
6. The stained gel can be viewed either by using transilluminated fluorescent white light or by placing the gel on a black plastic board with table top light illumination.

4. Notes

1. The 1% eosin Y staining solution needs to be made fresh and filtered prior to use. Eosin Y should be soluble in both water and methanol. Acetic acid must be added last. The final concentration of acetic acid in this staining solution is critical for staining background and sensitivity. The acetic acid should be added to solution with stirring. After adding acetic acid, the staining solution should appear to be cloudy but should not precipitate. If any precipitation occurs, it indicates that the acetic acid concentration is too high.
2. In **step 3** of staining various proteins, some orange precipitation may cover a SDS-gel surface. One may gently clean up the precipitation by wiping off the gel surface with a latex glove or by using Kimwipe tissue, which will reduce the background and improve the staining sensitivity.

3. An appropriate development time will ensure yellow–orange staining bands. Prolonging development in 10% acetic acid/40% methanol often results in yellow–orange bands becoming brown bands which, in turn, will decrease the staining sensitivity.
4. It should be noted that at the end of the washing step, the proteins should appear to be yellow-orange and the sialoglycoproteins appear to be light yellow. If the protein bands are still not visualized at this point, one may prolong the washing step for another 5–10 min.
5. The eosin Y stained gel can be stored in distilled water for at least 1 mo without fading. It is not recommended to dry and store the stained gels because the intensity and resolution of protein bands are greatly decreased.
6. The eosin Y stained proteins in SDS-gels can be transferred to immobilon-P membrane without additional treatment. The antigenicity of a given protein is usually not affected by the eosin Y stain *(11)*.

References

1. Harlow, E. and Lane, D. (1988) *Antibodies*, Cold Spring Harbor Laboratory, Cold Spring Harbor, NY, pp. 649–653.
2. Switzer, R. C., Merril, C. R., and Shifrin, S. (1979) A highly sensitive silver stain for detecting proteins and peptides in polyacrylamide gels. *Analyt. Biochem.* **98,** 231–237.
3. Merril, C. R., Switzer, R. C., and Van Keuren, N. L. (1979) Trace polypeptides in cellular extracts and human body fluids detected by two-dimensional electrophoresis and a highly sensitive silver stain. *Proc. Natl. Acad. Sci. USA* **76,** 4335–4339.
4. Oakley, B. R., Kirsch, D. R., and Morris, N. R. (1980) A simplified ultrasensitive silver stain for detecting proteins in polyacrylamide gels. *Analyt. Biochem.* **105,** 361–363.
5. Wray, W., Boulikas, T., Wray, V. P., and Hancook, R. (1981) Silver staining of proteins in polyacrylamide gels. *Analyt. Biochem.* **118,** 197–203.
6. Nalty, T. J. and Yeoman, L. C. (1988) Transfer of proteins from acrylamide gels to nitrocellulose paper after silver detection. *Immunol. Meth.* **107,** 143–149.
7. Wise, G. E. and Lin, F. (1991) Transfer of silver-stained proteins from polyacrylamide gels to polyvinylidene difluoride membranes. *J. Biochem. Biophys. Meth.* **22,** 223–231.
8. Glossmann, H. and Neville, D. M. (1971) Glycoproteins of cell surfaces. *J. Biol. Chem.* **246,** 6339–6346.
9. Mueller, T. J., Dow, A. W., and Morrison, M. (1976) Heterogeneity of the sialoglycoproteins of the normal human erythrocyte membrane. *Biochem. Biophys. Res. Commun.* **72,** 94–99
10. Dzandu, J. K., Deh, M. E., Barratt, D. L., and Wise, G. E. (1984) Detection of erythrocyte membrane proteins, sialoglycoproteins, and lipids in the same polyacrylamide gel using a double-staining technique. *Proc. Natl. Acad. Sci. USA* **81,** 1733–1737.
11. Lin, F., Fan, W., and Wise, G. E. (1991) Eosin Y staining of proteins in polyacrylamide gels. *Analyt. Biochem.* **196,** 279–283.

37

Electroelution of Proteins from Polyacrylamide Gels

Paul Jenö and Martin Horst

1. Introduction

Understanding the function of proteins requires determination of their structures. Advances in chemical technology make it possible to obtain a picture of global protein expression in model organisms such as the yeast *Saccharomyces cerevisiae*. Instrumental to these breakthroughs was the development of high-resolution two-dimensional gel electrophoresis and mass spectrometric means for rapid detection and analysis of peptides and proteins. While sequence information on single spots separated by two-dimensional electrophoresis can be quickly obtained by today's technology, structural characterization of proteins expressed in low abundance is still a difficult task. For example, in our attempts to isolate components of the import machinery of yeast mitochondria, some proteins turned out to be expressed at extremely low levels, and it soon became evident that conventional purification techniques were impractical to obtain these proteins in amounts sufficient for structural analysis. Furthermore, some proteins of the mitochondrial import machinery were rapidly degraded by proteases, and they could be obtained intact only by first denaturing mitochondria with trichloroacetic acid and then solubilizing them in boiling sample buffer for sodium dodecyl sulfate-polyacrylamide gel electrophoresis (SDS-PAGE) *(1)*. This in turn made the separation of several milligrams of denatured proteins necessary, which was best achieved by preparative SDS-PAGE followed by electroelution of individual proteins.

Proteins isolated by electroelution usually contain large amounts of salts and SDS, which interfere with enzymatic digestion or Amino-D (N)-terminal sequencing. A number of methods are available to remove SDS from proteins, including precipitation of the detergent by organic solvents *(2,3)*, and solvent extraction with *(4)* or without ion-pairing reagents *(5)*. However, they all suffer from the disadvantage that once the detergent is removed, many proteins become virtually insoluble in buffers lacking SDS. Alternatively, chromatographic methods can be used to remove SDS from proteins. Simpson et al. *(6)* described desalting of electroeluates based on the finding that certain reverse-phase matrices retain proteins at high organic modifier concentrations, whereas small molecular weight compounds are not. This allows bulk separation of salts, SDS,

From: *The Protein Protocols Handbook, 2nd Edition*
Edited by: J. M. Walker © Humana Press Inc., Totowa, NJ

and Coomassie Blue staining components from protein material. Based on a different stationary phase, a similar approach was developed *(7)* to separate protein from contaminants originating from the electroelution process. In this method, electroeluted proteins are applied to a poly (2-hydroxyethyl) aspartamide-coated silica, which provides a polar medium for binding of proteins when equilibrated at high organic solvent concentration *(8)*. Bound proteins are then eluted with a decreasing gradient of organic modifier, allowing recovery of protein free of SDS and buffer salts. Although the methodology is similar to the one described by Simpson et al. *(6)*, proteins tend to adsorb to the poly (hydroxyethyl) aspartamide matrix at lower *n*-propanol concentrations than to reverse-phase matrices, therefore minimizing the danger of irreversible protein precipitation on the stationary phase.

In this chapter, we describe an electroelution procedure that has worked well with proteins in the molecular weight range of 20–100 kDa isolated from yeast mitochondrial membranes. In addition, a method is described which allows the desalting of these proteins into volatile buffer systems by hydrophilic-interaction chromatography.

2. Materials

1. Electrophoresis apparatus: Preparative electrophoresis is carried out on 16 × 10 cm separating and 16 × 2 cm stacking gels of 1.5 mm thickness with the buffer system described by Laemmli *(9)*. For sample application, a preparative sample comb of 2 cm depth and 14 cm width is used. Up to 2.5 mg total protein is applied onto one preparative slab gel. Samples are electrophoresed at 15 V for 14 h. Chemicals used for electrophoresis (acrylamide, *N,N'*-methylene *bis*-acrylamide, ammonium persulfate, *N,N,N',N'*-tetramethylenediamine, SDS, and Coomassie Blue R250) are electrophoresis-grade and are purchased from Bio-Rad (Hercules, CA). Methanol and acetic acid used for staining are *pro analysis* (p.a.) grade from Merck (Darmstadt, Germany). Unless stated, all other chemicals used are of the highest grade available.
2. Staining solution: 0.125% (w/v) Coomassie Brilliant Blue R250, 50% (v/v) methanol, 10% acetic acid. Filter the staining solution over 320-µm filters (Schleicher and Schuell, Dassel, Germany) before use.
3. Destaining solution: 50% methanol, 10% acetic acid.
4. Electroelution apparatus: BIOTRAP from Schleicher and Schuell (Dassel, Germany, "Elutrap" trademark in the US and Canada).
5. BT1 and BT2 membranes for electroelution (Schleicher and Schuell).
6. Electroelution buffer: 25 m*M* Tris, 192 m*M* glycine, 0.1% SDS. This buffer is prepared with NANOpure water from a water purification system (Barnstead, Dubuque, IA, USA) and electrophoresis-grade SDS (Bio-Rad).
7. Electrodialysis buffer: 15 m*M* NH_4HCO_3, 0.025% SDS. Prepare the buffer with NANOpure water and electrophoresis-grade SDS (Bio-Rad).
8. High-pressure liquid chromatography (HPLC) equipment: we are using a Hewlett Packard HP1090M (Palo Alto, CA) liquid chromatograph connected to a diode array UV detector (Hewlett Packard, Palo Alto, CA) for operating columns of 4.6 × 200 mm or 2.1 mm diameter.
9. Poly (hydroxyethyl) aspartamide (PHA) columns: 5 µm particle size, 20 nm pore size, 4. × 200 mm or 2.1 × 200 mm (PolyLC, Columbia, MD).
10. Solvents for hydrophilic interaction chromatography: solvent A: 100% *n*-propanol (HPLC grade, Merck), 50 m*M* formic acid (Merck, analytical grade); solvent B: 50 m*M* formic acid in water (NANOpure).

3. Method

1. After electrophoresis, stain the gel for 15 min with Coomassie Blue. To prevent irreversible fixation of the protein, destaining is observed on a light box. As soon as the protein of interest becomes visible, the band is cut out with a razor blade. The gel piece is washed once with 10-mL of 1 *M* Tris-HCl, pH 8.0 for 5 min, followed by three 10-mL washes with water for 5 min each. Cut the gel piece with a razor blade into small cubes. Equilibrate them in 10-mL of electroelution buffer for 10 min with occasional shaking. In the meantime, assemble the electroelution apparatus.
2. The electroelution device is a block of polycarbonate (160 × 30 × 30 mm) that has an open channel along its axis (**Fig. 1**). An elution chamber that holds the polyacrylamide pieces is formed with trap inserts between points C and F of the body. During the elution process, the protein is trapped into a chamber formed between points F and G (**Fig. 1**). The device works with two types of membranes having different ion permeabilities: the BT1 membrane retains all charged macromolecules larger than 5 kDa, whereas buffer ions can freely permeate under the influence of an electric field. The macroporous BT2 membrane acts as a barrier that prevents particulate matter from entering the trap. It also keeps the buffer from flowing into the trap when the electric field is switched off, preventing dilution of the protein in the trap.
   Slide BT1 membranes between the clamping plates and the trap inserts at positions A and G of the BIOTRAP apparatus (**Fig. 1**). Since the BT1 membrane is an asymmetric membrane with two different surfaces, make sure that they are mounted in the proper orientation. The BT1 membrane is delivered moist and should not dry out. Buffer should be added within 5 min after insertion of the membranes. Insert a BT2 membrane at points C and F. If necessary, smaller elution chambers can be formed by inserting the BT2 membrane at positions D or E. Tighten the pressure screws to hold the membranes in place. Transfer the gel pieces with a spatula into the elution chamber formed between the two BT2 membranes inserted at positions C and F. Carefully overlay the gel pieces with electroelution buffer until the level of the liquid is approx 5 mm above the gel pieces. After some minutes, the trap is filled with buffer by seeping through membrane BT2. Make sure enough liquid is in the elution chamber so that the gel pieces remain completely immersed. Place the electroelution device into a horizontal electrophoresis chamber with the + mark directed toward the anode of the electrophoresis chamber. The dimensions of the horizontal electrophoresis tank are as follows: 30 cm length, 20 cm width, and 7 cm depth. The T-shaped table for agarose gels is 3 cm from the bottom. Add enough electroelution buffer to the electrophoresis chamber to fill half of the BIOTRAP (approx 3 L). Electroelute the protein for 18 h at 100 V (the current will be in the range of 70–90 mA). The volume into which the eluted protein is recovered depends on the buffer level inside the BIOTRAP and ranges from 200–800 µL.
3. Replace the electroelution buffer with 3 L of electrodialysis buffer and electrodialyze the sample for 6 h at 40 V against 15 m*M* NH_4HCO_3, 0.025% SDS.
4. Remove the eluted protein from the trap. Be careful not to perforate the BT1 membrane with the pipet tip! Rinse the trap twice with 100 µL of fresh electrodialysis buffer. Combine the dialysate and the washes. The solution is dried in a Speed Vac concentrator and stored at –20°C.
5. For desalting of the electroeluted protein, equilibrate the PHA column with solvent A (65% *n*-propanol, 50 m*M* formic acid). Electroeluates containing > 5 µg of protein are desalted on 4.6 mm internal diameter columns, which are operated at a flow rate of 0.5 mL/min. Less than 5 µg protein are chromatographed on 2.1 mm internal diameter columns at 75 µL/min. The effluent is monitored at 280 nm.

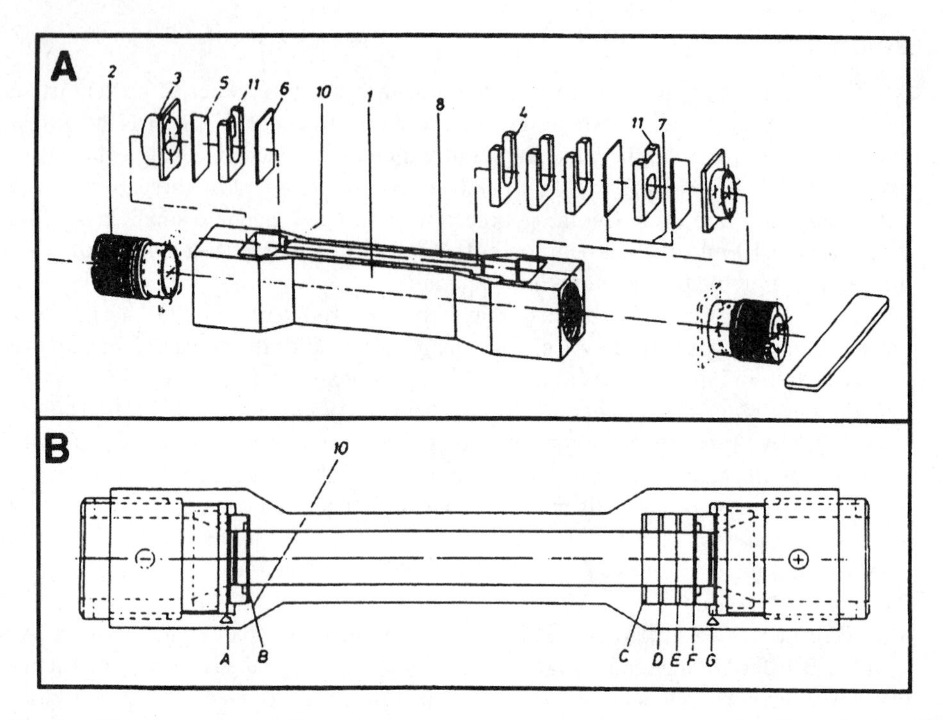

Fig. 1. Side **(A)** and top **(B)** view of the electroelution device. The apparatus is assembled by inserting a BT1 membrane at points A and G. The trap that collects the protein during the elution process is formed between points F and G. The chamber that holds the gel pieces is formed by inserting a BT2 membrane at point C. Smaller elution chambers can be made by inserting the BT2 at point D or E. The trap inserts and membranes are fixed by clamping plates, which press the trap inserts against the cell body. *1*, Cell body; *2*, pressure screw; *3*, clamping plate; *4*, trap inserts; *5*, membrane BT1; *6*, membrane BT2; *7*, trap chamber; *8*, elution chamber; *10*, mark for correct orientation of membrane BT1; *11*, trap insert for membranes BT1 (modified with permission from Schleicher and Schuell).

6. Dissolve the dried protein in a small volume of water (50–100 µL). The dried SDS efficiently solubilizes the protein. *n*-propanol is added to 65% final concentration. The sample is then applied in 50-µL aliquots onto the PHA column. After each injection, a number of UV-absorbing peaks, caused by Coomassie blue components, elute from the column. It is important that these components are completely washed out before the next aliquot is injected. With this procedure, the protein is efficiently concentrated on the column inlet. After the entire sample has been applied, the gradient is initiated, which is developed in 10 min from 65% *n*-propanol/50 m*M* formic acid to 50 m*M* formic acid. The protein elutes at the end of the gradient and is now devoid of any salt or SDS (**Fig. 2**).
7. The protein is now ready for further protein structural characterization. It can be directly subjected to automated Edman degradation. For enzymatic fragmentation, residual *n*-propanol has to be removed in the Speed Vac prior to adding the protease.

4. Notes

1. To locate the protein of interest in the gel, a staining method has to be chosen so that maximal sensitivity with minimal fixation is obtained. A number of methods exist to visualize proteins in the polyacrylamide matrix, such as formation of insoluble

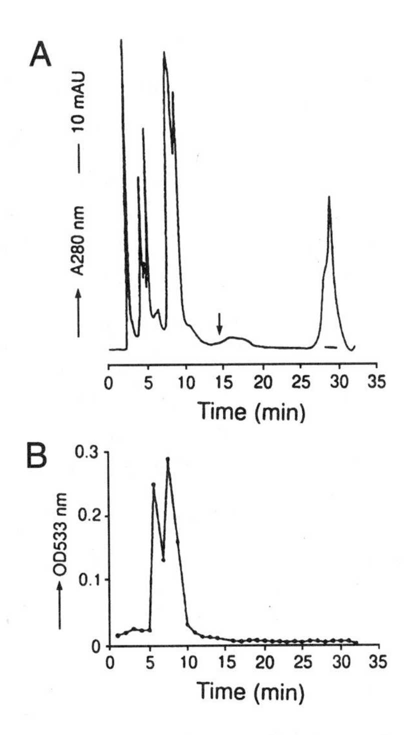

Fig. 2. Removal of SDS from an electroeluate. **(A)** 20 µg of a 45-kDa mitochondrial outer membrane protein in 50 µL was injected onto a PHA column (4.6 × 200 mm) that had been equilibrated in 70% *n*-propanol–50 m*M* formic acid. After the baseline had stabilized, the gradient was initiated (marked with an arrow). Bound protein was eluted with a linear 10-min gradient from 70% *n*-propanol–50 m*M* formic acid to 50 m*M* formic acid at a flow rate of 0.5 mL/min. The protein eluted between 27 and 30 min. **(B)** Fractions of 500 µL were collected and tested with Fuchsin red for the presence of SDS. (Modified with permission from **ref.** *7*).

protein–SDS complexes with potassium *(10)*, or precipitation of SDS by 4 *M* sodium acetate *(11)*. We found staining of complex protein patterns with these methods difficult, as they tend to produce diffuse bands. Staining the gel for 15 min with Coomassie Blue is sufficient to visualize also faint bands without fixing the protein irreversibly. To minimize fixation of proteins, destaining is carried out on a light box, so that the band of interest can be sliced out of the gel as soon as it becomes visible.

2. The electroelution apparatus routinely used in our laboratory was originally described by Jacobs and Clad *(12)* and is commercially available from Schleicher and Schuell. We found this type of apparatus very reliable for routine use. The volume of the elution chamber can be adjusted depending on the volume of the gel pieces used. The volume can be increased or decreased by varying the position of the BT2 membranes between positions C and F (*see* **Fig. 1**). By forming the smallest possible elution chamber, one can process Coomassie Blue-stained bands from a single one-dimensional analytical PAGE. With the larger elution chamber, up to five preparative gels can be processed at a time. However, other suitably constructed devices will give identical results.
3. So far we have eluted proteins between 20 and 100 kDa with protein amounts ranging from 10 to 50 µg per band. After elution and dialysis, proteins are typically recovered in volumes of between 300 and 800 µL. However, elution of <10 µg of protein becomes difficult owing to the large volume of the trap, nonspecific adsorption of proteins to the

BT1 membrane, or microleaks in the trap. In such cases, we prefer to run several preparative gels in parallel and pool multiple protein slices until the required amount of protein for electroelution is obtained. For low amounts of protein, several commercially available devices can be used which elute proteins into smaller volumes than the BIOTRAP apparatus.

4. Electroelution of proteins into ammonium hydrogen carbonate would be preferable, since NH_4HCO_3 can be removed by lyophilization. However, owing to the low buffering capacity of NH_4HCO_3, the pH of the buffer drops after 4 h, rendering electroelution of high molecular weight proteins difficult due to their slow elution from the gel pieces. Therefore, it is preferable to electroelute proteins into Tris–glycine–SDS followed by electrodialysis into NH_4HCO_3-containing buffers.
5. When using 0.1% SDS in the electroelution buffer, the micelles formed in front of the BT1 membrane lead to massive accumulation of SDS in the trap. The detergent and low molecular weight contaminants can be easily removed by hydrophilic interaction chromatography with simultaneous desalting of the protein into a volatile buffer *(7)*. Alternatively, the procedure devised by Simpson et al. *(6)* can be used to desalt the electroeluted protein.
6. Hydrophilic interaction chromatography of proteins requires careful control of the solvent composition. To test column performance, we use a test mixture consisting of cytochrome *c*, ovalbumin, and bovine serum albumin. Use of this test mixture allows finding the minimal *n*-propanol concentration at which the electroeluted protein binds to the stationary phase without the risk of precipitation by the organic solvent. Protein binding usually occurs at *n*-propanol concentrations between 60% and 65%. When performing electroelution with Tris–glycine buffers, direct application of the electroeluate is not possible due to precipitation of buffer salts above 50% *n*-propanol concentration. In such cases, buffer exchange into 0.1 *M* NH_4HCO_3, 0.01% SDS is carried out by electrodialysis in the electroelution apparatus. Ammonium bicarbonate can be removed by Speed Vac drying. The residual SDS facilitates solubilization of the dried protein with water. No salt precipitation is observed when adding *n*-propanol to 65% final concentration

   The solvent system used to elute bound proteins contains 50 m*M* formic acid. Unfortunately, this precludes detection of proteins at wavelengths below 250 nm. Since most proteins contain tyrosine and tryptophane residues, detection of eluting proteins is done at 280 nm. For proteins lacking tyrosines or tryptophanes, 0.05% trifluoroacetic acid (TFA) can be used instead of formic acid. However, exposure of the stationary phase to TFA should be kept to an absolute minimum, since TFA greatly reduces the lifetime of the column.

   Efficient removal of small molecular weight contaminants such as SDS and Coomassie Blue requires small injection volumes of the electroeluted protein. When using 4.6 mm i.d. columns, 50 µL injections were found to be optimal; for 2.1 mm i.d. columns, the injection volume is reduced to 20 µL. Larger volumes are concentrated on the column inlet with multiple injections. In such cases, the column should be allowed to reequilibrate between individual injections, otherwise loss of protein in the breakthrough volume occurs.

   Electroeluted proteins chromatographed by hydrophilic interaction often display unsymmetrical peaks. This may indicate the presence of several different proteins in the eluate, which are partially resolved by the stationary phase, or heterogeneity of a single protein, which became modified during electrophoresis. Since the main purpose is to free the electroeluted protein from SDS and Coomassie Blue, we tend to elute the protein into one single peak by running very steep gradients, rather than trying to separate the eluting material into single components. This can be subsequently achieved by two-dimensional gel electrophoresis, or by reverse-phase chromatography after removal of residual *n*-propanol.

References

1. Horst, M., Jenö, P., Kronidou, N. G., Bolliger, L., Oppliger, W., Scherer, P., et al. (1993) Protein import into yeast mitochondria: the inner membrane import site ISP45 is the *MPI1* gene product. *EMBO J.* **12,** 3035–3041.
2. Wessel, D. and Flügge, U. I. (1984) A method for the quantitative recovery of protein in dilute solution in the presence of detergent and lipids. *Analyt. Biochem.* **138,** 141–143.
3. Stearne, P. A., van Driel, I. R., Grego, B., Simpson, R. J., and Goding, J. W. (1985) The murine plasma cell antigen PC-1: purification and partial amino acid sequence. *J. Immunol.* **134,** 443–448.
4. Konigsberg, W. H. and Henderson, L. (1983) Removal of sodium dodecylsulfate from proteins by ion-pair extraction. *Meth. Enzymol.* **91,** 254–259.
5. Bosserhoff, A., Wallach, J., and Frank, R. (1989) *J. Chromatogr.* **437,** 71–77.
6. Simpson, R. J., Moritz, R. L., Nice, E. E., and Grego, B. (1987) A high-performance liquid chromatography procedure for recovering subnanomole amounts of protein from SDS-gel electroeluates for gas-phase sequence analysis. *Eur. J. Biochem.* **165,** 292–298.
7. Jenö, P., Scherer, P., Manning-Krieg, U., and Horst, M. (1993) Desalting electroeluted proteins with hydrophilic chromatography. *Analyt. Biochem.* **215,** 292–298.
8. Alpert, A. J. (1990) Hydrophilic-interaction chromatography for the separation of peptides, nucleic acids and other polar compounds. *J. Chromatogr.* **215,** 292–298.
9. Laemmli, M. K. (1974) Cleavage of structural proteins during the assembly of the head of bacteriophage T4. *Nature.* **227,** 680–685.
10. Hager, D. A. and Burgess, R. (1980) Elution of proteins from sodium dodecyl sulfate polyacrylamide gels, removal of sodium dodecyl sulfate, and renaturation of enzymatic activity: results with sigma subunit of *Escherichia coli* RNA polymerase, wheat germ DNA topoisomerase, and other enzymes. *Analyt. Biochem.* **109,** 76–86.
11. Higgins, R. C. and Dahmus, M. E. (1979) Rapid visualization of protein bands in preparative SDS-polyacrylamide gels. *Analyt. Biochem.* **93,** 257–260.
12. Jacobs, E. and Clad, A. (1986) Electroelution of fixed and stained membrane proteins from preparative sodium dodecyl sulfate polyacrylamide gels into a membrane trap. *Analyt. Biochem.* **154,** 583–589.

38

Autoradiography and Fluorography of Acrylamide Gels

Antonella Circolo and Sunita Gulati

1. Introduction

Autoradiography detects the distribution of radioactivity on gels or filters by producing permanent images on photographic film. It is frequently used in a variety of experimental techniques ranging from Southern and Northern blot analysis *(1)*, to visualization of radioactive proteins separated in a sodium dodecyl sulfate (SDS)-Polyacrylamide gels *(2)*, to detection of nuclear factors bound to a labeled DNA probe in gel shift analysis *(3)*, and to localization of DNA bands in sequencing gels *(4)*.

Autoradiographic images are formed when particles emitted by radioactive isotopes encounter the emulsion of an X-ray film and cause emission of electrons from silver halide crystals that, in turn, react with positively charged silver ions, resulting in the precipitation of silver atoms and the formation of an image *(5)*.

^{35}S and ^{32}P isotopes are the most commonly used isotopes for autoradiography. ^{35}S is a β-emitter of relatively low energy (0.167 MeV). Thus ^{35}S particles penetrate a film to a depth of 0.22 mm, generally sufficient to interact with the emulsion in the film, as long as care is taken to assure that the film and the source of radioactivity are in direct contact, and that no barriers are posed between the film and the gel. In addition, gels must be completely dry before autoradiography. ^{32}P is a β-emitter with an energy of 1.71 MeV. Therefore, its particles penetrate water or other materials to a depth of 6 mm, passing completely through a film. In this case, gels or filters do not need to be dry, since water will not block particles of this energy and may be covered with a clear plastic wrap before autoradiography. The efficiency of ^{32}P-emitted β-particles is enhanced when an intensifying screen is placed behind the X-ray film, because radioactive particles that pass through the film cause the screen to emit photons that sensitize the film emulsion. The use of intensifying screens results in a fivefold increased enhancement of the autoradiographic image when the exposure is performed at low temperature (–70°C). In general, calcium tungstate screens are the most suitable because they emit blue light to which X-ray films are very sensitive *(6)*.

Radiation of sufficiently high energy (e.g., ^{32}P and ^{35}S) can be detected by simple autoradiography, but low energy emissions may not penetrate the coating of the film, and the most sensitive fluorography procedure is used in these cases. In fluorography, the use of fluorescent chemicals increases about 10-fold the sensitivity of detection of

From: *The Protein Protocols Handbook, 2nd Edition*
Edited by: J. M. Walker © Humana Press Inc., Totowa, NJ

weak β-emitters (^{35}S and ^{14}C), and permits detection of radioactivity from ^{3}H, virtually undetectable with simple autoradiography. Gels are impregnated with scintillant or fluors, which come in direct contact with the low-energy particles emitted by ^{35}S, ^{14}C, and ^{3}H. In response to radiation, fluors emit photons that react with the silver halide crystals in the emulsion of the film. Because the wavelength of the emitted photons depends on the fluorescent chemical and not on the radioactive emission of the isotope, the same type of film can be used to detect radioactivity from isotopes of varying energies *(7)*.

In this chapter, we describe the most common techniques of autoradiography and fluorography, and a double silver-staining method for acrylamide gels that approaches the sensitivity of autoradiography and that may be used for those experiments in which radiolabeling of proteins is not easily obtained (*see* **Note 1**). Autoradiography of wet and dry gels is described in **Subheading 3.1.**, and protocols for fluorography and methods for quantification of radioactive proteins in polyacrylamide gels are given in **Subheading 3.2.** Double silver staining is described in **Subheading 3.3.**

2. Materials

2.1. Autoradiography

1. X-ray film (Kodak XR [Rochester, NY], Fuji RX [Pittsburgh, PA], Amersham [Arlington Heights, IL], or equivalent).
2. Plastic wrap (e.g., Saran Wrap).
3. 3MM Whatman paper.
4. SDS-PAGE fixing solution: 46% (v/v) methanol, 46% (v/v) water, 8% (v/v) acetic acid glacial. Store in an air-tight container. It is stable for months at room temperature.

2.2. Fluorography

1. X-ray blue-sensitive film (Kodak XAR-5, Amersham, or equivalent).
2. Plastic wrap.
3. 3MM Whatman paper.
4. SDS-PAGE fixing solution (as in **Subheading 2.1.**).
5. Coomassie blue staining solution: 0.125% Coomassie brilliant blue R250 in SDS-PAGE fixing solution. Stir overnight to dissolve, filter through a Whatman paper, and store at room temperature protected from light. It is stable for several weeks. For longer storage, dissolve the Coomassie blue in 46% water and 46% methanol only, and add 8% acetic acid glacial just before use.
6. Commercially available autoradiography enhancers (acidic acid-based: En^{3}Hance from Dupont [Mount Prospect, IL], water-soluble: Fluoro-Hance from RPI [Dupont NEN, Boston, MA], or equivalent).
7. Hydrogen peroxide (15% solution).
8. Scintillation fluid.

2.3. Double Silver Staining

1. Silver stain kit (Bio-Rad, Hercules, CA).
2. Methanol, ethanol, acetic acid glacial.
3. Solution A: Sodium thiosulfate 436 g/L.
4. Solution B: Sodium chloride 37 g, cupric sulfate 37 g, ammonium hydroxide 850 mL to 1 L of dd H_2O. Store these solutions at room temperature.

3. Methods

3.1. Autoradiography

3.1.1. Wet Gels

When radioactive proteins have to be recovered from the gel, the gel should not be fixed and dried before autoradiography. However, only high-energy isotopes (^{32}P or ^{125}I) should be used.

1. At the end of the electrophoresis, turn off the power supply and disassemble the gel apparatus. With a plastic spatula, pry apart the glass plates, cut one corner of the gel for orientation, carefully remove the gel, place it over two pieces of Whatman paper, and cover with a plastic wrap, avoiding the formation of bubbles or folds.
2. In a dark room, place the gel in direct contact with the X-ray film, and place an intensifying screen over the film. Expose for the appropriate length of time at room temperature or freeze at –70°C.
3. Develop the film using an automatic X-ray processor or a commercially available developer as following: immerse the gel for 5 min in the X-ray developer, rinse in water for 30 s, transfer into the fixer for 5 min, rinse in running water for 10–15 min, and let dry. All the solutions should be at 18–20°C.

   If the exposure of the gel is made at –70°C, allow the cassette to warm to room temperature and wipe off any condensation before opening. Alternatively, develop the film immediately as soon as the cassette is removed from the freezer, before condensation forms. If an additional exposure is needed, allow the cassette to dry completely before reusing.

   Cardboard exposure holders may work better with wet gel, since metal cassettes may compress the gel too tightly. When an intensifying screen is used, the film should be exposed at –70°C. The screen enhances the detection of radioactivity up to 10-fold, but may decrease the resolution.

3.1.2. Dry Gels

Gels containing urea should always be fixed to remove urea crystals. When high resolution is required, gels should always be dried before exposure to film. Gels should also be dried when ^{35}S or other low-energy β-emitters are used as radioactive tracers. For improved sensitivity and resolution, gels containing ^{32}P should also be fixed and dried before autoradiography.

1. At the end of the electrophoresis, turn off the power supply, disassemble the gel apparatus, and carefully pry open the two glass plates with a plastic spatula. (To assure that the gel will adhere to one glass plate only, one of the glass plates should be treated with silicon before casting the gel.)
2. Cut one corner of the gel for orientation, place the gel with the supporting glass plate into a shallow tray containing a volume of fixing solution sufficient to cover the gels, and fix for 30 min (the time necessary for fixation varies according to the thickness of the gel, but longer times do not have deleterious effects).
3. After fixation, carefully remove the plate from the tray, taking care not to float away the gel, or with a pipet connected to a vacuum pump, remove the solution from the tray and carefully from the glass plate. Place a wet piece of Whatman paper over the gel, being careful to avoid formation of bubbles and folding of the gel. Blot dry the Whatman paper with dry paper towels, applying gentle pressure.

4. Flip over the plate and maintain it exactly over the tray of a gel dryer. Carefully begin to detach the gel from one corner of the glass plate. The gel will remain adherent to the Whatman paper and will detach easily from the plate. Carefully cover the gel with plastic wrap, avoiding folding, and then dry the gel in a slab gel vacuum dryer for 1 h at 80°C (*see* **Notes 2–6**).
5. When the gel is dry, remove the plastic wrap and expose to X-ray film. After an appropriate length of time, develop the film as described in **Subheading 3.1.1., step 3**.

To increase the resolution of the radiogaphy, gels in which several bands of similar molecular mass are visualized, should be exposed at room temperature, without an intensifying screen. The use of a screen will result in increased intensity of the radioactive signal, but in decreased sharpness of the image. SDS-PAGE or other gels in which fewer bands need to be resolved may be exposed with an intensifying screen.

3.2. Fluorography

Gels containing weak β-emitters (^{35}S, ^{14}C, and ^{3}H) should be fixed, impregnated with autoradiography enhancers, and dried to reduce the film exposure time necessary for visualization of radioactive bands. SDS-PAGE should also be stained if nonradiolabeled mol-wt markers are used and for quantitative experiments (e.g., immunoprecipitation, detection of cell-free translated products, and so forth) where the radioactivity contained in specific proteins is to be measured in bands cut out from the gel *(8,9)*.

1. Turn off the power supply, remove the gel from the mold, cut a corner for orientation, and place the gel on a tray containing Coomassie blue-staining solution (the volume of the solution should always be adequate to cover the gel, so that it can float freely). Incubate for 45 min at room temperature with gentle shaking.
2. Remove the staining solution, and replace with SDS-PAGE fixing solution (the staining solution can be filtered through filter paper and reused as long as the radioactivity on it remains low, or until the color changes from blue to purple).
3. Incubate overnight at room temperature with gentle shaking, and replace the fixing solution at least once to accelerate the destaining. Gels should be destained until the mol-wt markers are clearly visible and the background is clear (*see* **Note 7**).
4. Discard the fixing solution in accordance with radioactive liquid waste disposal procedures, and add the autoradiography enhancer.
5. If the enhancer used is based on acetic acid (e.g., En^{3}Hance from Dupont or its equivalent), the gel can be soaked in the enhancer without rinsing, and **steps 6–8** should be followed (*see* **Note 2**).
6. Allow the gel to impregnate with enhancer for 1 h with gentle shaking. Initially, a white precipitate may form on the surface of the gel, but it will disappear within the first 15 min of impregnation. Following impregnation, discard the used enhancer solution (do not mix with waste containing NaOH, $NaHCO_3$, and so forth). Add cold tap water to the gel to precipitate the fluorescent material and incubate the gel in water for 30 min. At this stage, the gel should appear uniformly opaque.
7. After the precipitation step, carefully place the gel over two pieces of wet filter paper, cover with plastic wrap, and dry under heat (60 –70°C) and vacuum for 1–2 h on a slab gel dryer.
8. Remove the plastic wrap, tape the gel on a rigid support, and place it against a suitable blue-sensitive X-ray film, with an intensifying screen. Expose at –70°C for an appropriate length of time. Do not store the gel at room temperature for >48 h before exposure, since evapora-

tion of the fluors may occur, resulting in reduced sensitivity. For a longer period of storage prior to exposure, freeze the gel at –70°C.

9. If water-soluble fluorography solutions are used (e.g., Fluoro-Hance from R.P.I. or equivalent), the gel must be equilibrated in water after destaining, and steps 10 and 11 should be followed.
10. Discard the destaining solution, and wash the gel in distilled water for 30 min at room temperature, with shaking.
11. Discard the water and impregnate the gel with the enhancer for 30 min at room temperature with shaking. Remove the enhancer, place the gel over two wet pieces of filter paper, cover with a plastic wrap, and dry under vacuum with heat (60–80°C) for 2 h (*see* **Note 3**). Expose the gel as described above. (Flouro-Hance can be reused, but should be discarded as soon as the solution shows sign of discoloration) (*see* **Notes 2–7**).

    If the gels are not stained with Coomassie blue, after electrophoresis, place the gel in SDS-PAGE fixing solution, and incubate for 45 min at room temperature, with gentle shaking. After incubation, discard the fixing solution and impregnate the gel with enhancer as described (**steps 5** or **9**).
12. Radioactivity incorporated into specific proteins can be determined by cutting out the radioactive bands from the dried fluorographed gel.
13. Precisely position the film over the gel. With a sharp blade or scalpel, cut out the area of the gel corresponding to the band on the film. Also cut out an area of the gel free from radioactivity immediately below (or above) the radioactive band, for subtraction of background.
14. Place each gel slice into a scintillation vial (detaching the filter paper from the slice is not necessary), add 1 mL of a 15% solution of hydrogen peroxide, and incubate overnight in a water bath at 60°C to digest the gel and release the radioactivity.
15. After incubation, allow the vials to cool down to room temperature, add scintillation fluid, and measure the radioactivity in a scintillation counter. Alternatively, autoradiographic images can be quantitated by densitometric scanning of different exposures of the film (*see* **Note 8**).

3.3. Double Silver Staining (see Note 9)

1. After electrophoresis, transfer the gel in a glass tray containing 40% methanol and 10% acetic acid, and incubate at room temperature for at least 30 min (longer periods of time have no detrimental effect).
2. Discard this solution, and incubate for 15 min in 10% ethanol, 5% acetic acid.
3. Repeat **step 2** one more time.
4. Add Oxidizer (diluted according to the manufacturer's protocol) and incubate 5 min, taking care that the gel is completely submerged in the solution.
5. Rinse the gel twice in distilled water.
6. Incubate 15 min in double-distilled water.
7. Repeat step 6 until the yellow color is completely removed from the gel.
8. Add Silver reagent (diluted according to the manufacturer's protocol), and incubate for 15 min.
9. Wash the gel once in double-distilled water.
10. Add developer (prepared according to the manufacturer's protocol), swirl the gel for 30 s, discard the solution, and wash once in double-distilled water.
11. Repeat **step 10** and develop the gel until bands appear and the mol-wt markers become clearly visible.
12. Stop the reaction with 5% acetic acid, incubate 5 min, and wash with double-distilled water.
13. Add 3.5% solution A and 3.5% solution B, and incubate 5–10 min, or until the gel is clear.

14. Incubate two times in 10% acetic acid, 30 min each time.
15. Restain the gel by repeating **steps 1–12**.
16. Dry the gel in a slab dryer with heat and vacuum, or use a gel rap (Bio-Rad or equivalent), and dry overnight.

4. Notes

1. Autoradiography of SDS-PAGE is a powerful technique that permits detection of very low amounts of protein. However, in some instances, radioactive protein labeling cannot be easily accomplished. For example, metabolic labeling requires active protein synthesis ***(12)***; thus, proteins present in body fluids or in tissue biopsy cannot be labeled ***(13)***. Moreover, in in vivo animal experiments, it is often difficult to obtain radiolabeled proteins with high specific activity. In this case, silver staining of gels can be used as an alternative method, since it approaches the sensitivity of autoradiography ***(14)***. We have developed a double silver-staining technique that is about 10-fold more sensitive than the conventional silver staining. This method has not been previously published in detail, except for figures presented elsewhere ***(15)*** (*see* **Subheading 3.3.**).
2. Gels of high polyacrylamide concentration (>10%) or gradient gels (acrylamide concentration 5–15%) may crack when dried. This problem is reduced by adding glycerol (1–5%) before drying. When acid-based fluorography enhancers are used, the glycerol should be added during the fluors precipitation step in water after removal of the enhancer. When water-soluble enhancers are used, the glycerol is added during equilibration of the gel in water, before addition of the enhancer. If the concentration of glycerol is too high, gels are difficult to dry, and they may stick to the film. Addition of the enhancer does not increase the cracking. We currently use water-based enhancers for our experiments, because they give sharp autoradiography images and good sensitivity, and can be reused for several gels. Fluorography with commercially available enhancers is simpler and less tedious than the traditional method with PPO-DMSO, and the results are as good as, or better than, those obtained with this method.
3. Enhancers are also used to increase the sensitivity of the autoradiograpy of DNA and RNA of agarose, acrylamide, or mixed gels. When enhancers are used, the gels must be dried at the suggested temperature, because excessive heat will cause damage of the fluors crystal of the enhancer and formation of brown spots on the surface of the gel.
4. Cracking may also be owing to the formation of air bubbles between the gel and the rubber cover of the dryer. This is generally caused by a weak vacuum that is insufficient to maintain the gel well adherent to the paper filter and to the dryer's tray. Air bubbles can be eliminated by rolling a pipet over the rubber cover while the vacuum is being applied.
5. Also, excessive stretching of the gel during the transfer to the filter paper may contribute to cracking, particularly for gels that contain high acrylamide concentration. A filter paper should be placed under the gel when removing it from the solution, and the method described in **Subheading 3.1.2.** should be used for larger gels.
6. It is always necessary to cover the gel with a plastic wrap to prevent sticking to the cover of the dryer. In addition, the vacuum should never be released during the drying procedure until the gel is completely dry, since this will cause the gel to shatter.
7. Staining of the gels with Coomassie blue before fluorography may quench the effect of the enhancer, particularly when low amounts of radioactivity are used. Therefore, the gels must be destained thoroughly, until the background is clear and only the protein bands are stained.

8. X-ray films that are not pre-exposed to light respond to radiation in a sigmoidal fashion, because the halide crystals of the emulsion are not fully activated *(10)*. In contrast, in a pre-exposed film, the response becomes linear and proportional to the amount of radioactivity, therefore, allowing precise quantitative measurement of radioactivity by scanning the autoradiography *(11)*. In addition, pre-exposure (preflashing) of the film results in a two- to threefold increase in sensitivity, for levels of radioactivity near the minimum level of detectability *(7)*, enabling autoradiography of gels containing ^{14}C and ^{3}H radioisotopes to be performed at room temperature, instead of –70°C. To pre-expose a film, a stroboscope or a flash of light (<1 ms) from an electronic flash unit can be used, but it is necessary to reduce the light emission with a Deep Orange Kodak Wratten No. 22, or an Orange Kodak Wratten No. 21 filter, and to diffuse the image of the bulb with two pieces of Whatman No. 1 filter paper placed over the light-emitting lens. The distance between the film and the light source should be determined empirically *(11)*.
9. In silver staining, surface artifacts can be caused by pressure, fingerprints, and surface drying. Gloves should always be worn when handling the gel. A gray precipitate on the gel may be owing to insufficient washing.

Acknowledgment

We thank Peter A. Rice for support and for critical review of the manuscript.

References

1. Thomas, P. (1980) Hybridization of denatured RNA and small DNA fragments to nitrocellulose. *Proc. Natl. Acad. Sci. USA* **77,** 5201–5205.
2. Bonner, W. M. and Laskey, R. A. (1974) A film detection method for Tritium-labeled proteins and nucleic acid. *Eur. J. Biochem.* **46,** 83–88.
3. Garnier, G., Ault, B., Kramer, M., and Colten, H. R. (1992) *cis* and *trans* elements differ among mouse strains with high and low extrahepatic complement factor B gene expression. *J. Exp. Med.* **175,** 471–479.
4. Sanger, F., Nicklen, S., and Coulson, A. R. (1977) DNA sequencing with chain termination inhibitors. *Proc. Natl. Acad. Sci. USA* **74,** 5463–5467.
5. Sambrook, J., Fritsch, E. F., and Maniatis, T. (1989) Autoradiography, in *Molecular Cloning, A Laboratory Manual* (Nolan, C., ed.), Cold Spring Harbor Laboratory, Cold Spring Harbor, NY, pp. E.21–E.24.
6. Swanstrom, R., and Shank, P. R. (1978) X-ray intensifying screens greatly enhance the detection of radioactive isotopes ^{32}P and ^{125}I. *Analyt. Biochem.* **86,** 184–192.
7. Bonner, W. M. (1984) Fluorography for the detection of radioactivity in gels. *Meth. Enzymol.* **104,** 460–465.
8. Circolo, A., Welgus, H. G., Pierce, F. G., Kramer, J., and Strunk, R. C. (1991) Differential regulation of the expression of proteinases/antiproteinases in human fibroblasts: effects of IL-1 and PDGF. *J. Biol. Chem.* **266,** 12,283–12,288.
9. Garnier, G., Circolo, A., and Colten, H. R. (1995) Translational regulation of murine complement factor B alternative transcripts by upstream AUG codons. *J. Immunol.* **154,** 3275–3282.
10. Laskey, R. A. (1977) Enhanced autoradiographyc detection of ^{32}P and ^{125}I using intensifying screens and hypersensitized films. *FEBS Lett.* **82,** 314–316.
11. Laskey, R. A. and Mills, A. D. (1975) Quantitative film detection of ^{3}H and ^{14}C in polyacrylamide gels by fluorography. *Eur. J. Biochem.* **56,** 335–341.

12. Switzar, R. C. III, Merril, C. R., and Shifrin, S. (1979) A highly sensitive silver stain for detecting proteins and peptides in polyacrylamide gels. *Analyt. Biochem.* **98,** 231–237.
13. Irie, S., Sezaki, M., and Kato, Y. (1982) A faithful double stain of proteins in the polyacrylamide gel with Coomassie Blue and silver. *Analyt. Biochem.* **126,** 350–354.
14. Berry, M. J. and Samuel, C. E. (1982) Detection of subnanogram amounts of RNA in polyacrylamide gels in the presence and absence of proteins by staining with silver. *Analyt. Biochem.* **124,** 180–184.
15. Densen, P., Gulati, S., and Rice, P. A. (1987) Specificity of antibodies against *Neisseria gonorrhoeae* that stimulate neutrophil chemotaxis. *J. Clin. Invest.* **80,** 78–87.

Part III

Blotting and Detection Methods

39

Protein Blotting by Electroblotting

Mark Page and Robin Thorpe

1. Introduction

Identification of proteins separated by gel electrophoresis or isoelectric-focusing is often compounded by the small pore size of the gel, which limits penetration by macromolecular probes. Overcoming this problem can be achieved by blotting the proteins onto an adsorbent porous membrane (usually nitrocellulose or diazotized paper), which gives a mirror image of the gel *(1)*. A variety of reagents can be incubated with the membrane specifically to detect and analyze the protein of interest. Antibodies are widely used as detecting reagents, and the procedure is sometimes called Western blotting. However, protein blotting or immunoblotting is the most descriptive.

Electroblotting is usually preferred for immunoblotting in which proteins are transferred to the membrane support by electrophoresis *(2)*. A possible exception to this is in transfer from isoelectric focusing gels, where the proteins are at their isoelectric points and uncharged. Therefore, there is considerable delay before the proteins start to migrate in an electric field; also they can migrate in different directions and their rate of transfer can vary. For these reasons, transfer from isoelectric focusing gels by capillary blotting (*see* Chapter 41) may be preferable, particularly if very thin gels are used.

Electroblotting has the advantage that transfer takes only 1–4 h, and lateral diffusion of proteins (which causes diffuse bands) is reduced. Overall, nitrocellulose membranes are recommended. These are efficient protein binders and do not require activation, but are fragile and need careful handling. Nitrocellulose membranes, such as Hybond-C extra, are more robust. If the antigens of interest do not bind efficiently or if the blot is to be reused, then diazotized paper, or possibly nylon membranes, can be used. A suitable electrophoretic transfer chamber and power pack are required; these are available commercially or can be made in a laboratory work shop. The apparatus consists of a tank containing buffer, in which is located a cassette, clamping the gel and membrane tightly together. A current is applied from electrodes situated at either side of the cassette. The buffer is often cooled during transfer to avoid heating effects.

2. Materials

1. Transfer apparatus.
2. Orbital shaker.
3. Nitrocellulose sheet, 0.45-µm pore size, e.g., Hybond-C extra.

From: *The Protein Protocols Handbook, 2nd Edition*
Edited by: J. M. Walker © Humana Press Inc., Totowa, NJ

4. Filter paper, Whatman 3MM (Maidstone, UK).
5. Plastic box large enough to hold the blot and allow movement on agitation.
6. Transfer buffer: 0.025 *M* Tris, 0.052 *M* glycine, 20% methanol.
7. Blocking buffer: PBS (0.14 *M* NaCl, 2.7 m*M* KCl, 1.5 m*M* KH_2PO_4, 8.1 m*M* Na_2HPO_4) containing 5% dried milk powder.

3. Method

1. Assemble transfer apparatus and fill the tank with transfer buffer (*see* **Note 1**). If a cooling device is fitted to the apparatus, switch on before it is required to allow the buffer to cool down sufficiently (to ~8–15°C).
2. Cut two pieces of filter paper to the size of the cassette clamp, soak in transfer buffer, and place one on the cathodal side of the cassette on top of a wetted sponge pad (*see* **Note 2**).
3. Place the gel on the filter paper covering the cathodal side of the cassette (*see* **Note 3**). Keep the gel wet at all times with transfer buffer. Soak the nitrocellulose sheet (cut to the same size as the gel) in transfer buffer, and place it on the gel, i.e., on the anodal side (*see* **Note 4**). Avoid trapping air bubbles throughout the process.
4. Place the remaining filter paper over the nitrocellulose, and expel all air bubbles between the nitrocellulose and gel. This is achieved by soaking the gel/nitrocellulose/filter paper assembly liberally with transfer buffer and then pressing with a small hand roller.
5. Finally, place a wetted sponge pad on top of the filter paper, and clamp securely in the cassette. This should be a tight fit (*see* **Note 5**).
6. Place the cassette in the tank and fit the lid. Recirculate the transfer buffer either by a recirculating pump or a magnetic stirrer.
7. Electrophorese for 1–4 h at 0.5 A (*see* **Note 6**).
8. Turn off power, remove nitrocellulose sheet, and agitate in 50–200 mL of PBS containing 5% dried milk powder (*see* **Note 7**).
9. Process nitrocellulose sheet as required (*see* Chapters 50–58).

4. Notes

1. Methanol prevents polyacrylamide gels, removes SDS from polypeptides, and enhances the binding of proteins to the membrane, but it reduces the efficacy of transfer of larger proteins. It can be omitted from the transfer buffer with no adverse effects, but this has to be established empirically. Methanol is not necessary for non-SDS gels or isoelectric focusing gels.
2. Cassette clamps are normally provided with two sponge pads that fit either side of the gel/membrane/filter paper assembly to fill any dead space to squeeze the gel and membrane tightly together during electrophoresis.
3. Handle the gel and nitrocellulose sheet with gloved hands to prevent contamination with finger-derived proteins.
4. Cut a small piece of the bottom corner of the gel and nitrocellulose to orient the assembly.
5. If the gel and nitrocellulose are not clamped tightly together, then the proteins migrating from the gel can move radially and give a smeared blot.
6. Higher current can be used, but may result in uneven heating effects and blurred or distorted blots. The use of lower currents is not recommended, since transfer efficiency is reduced and poor quality blots are obtained. Overnight transfer can be used, but is not generally recommended. The time required for efficient transfer depends on the acrylamide concentration, gel thickness, gel buffer system, and the molecular size and shape of the proteins. Most proteins will pass through the nitrocellulose sheet if transfer is contin-

ued for too long. Proteins migrate faster from SDS gels (they are coated with SDS and highly charged), and transfer from non-SDS gels takes longer (around 4–5 h).

7. Other proteins can be used for blocking (e.g., albumin, hemoglobin), but dried milk powder is usually the best option. Thirty-minute incubation with blocking protein is sufficient to saturate all the protein binding sites on the blot. Longer times and overnight incubation can be used if this is more convenient. Protein blocked sheets can be stored frozen for long periods. For this, drain excess blocking solution and place in a plastic bag when required, wash with blocking buffer for 10 min, then continue with the next processing steps.

References

1. Towbin, H., Staehelin, T., and Gordon, J. (1979) Electrophoretic transfer of proteins from polyacrylamide gels to nitrocellulose sheets: procedure and some applications. *Proc. Natl. Acad. Sci. USA* **76,** 4350–4354.
2. Johnstone, A. and Thorpe, R. (1996) Immunochemistry in Practice, 3rd ed. Blackwell Scientific, Oxford, UK.

40

Protein Blotting by the Semidry Method

Patricia Gravel

1. Introduction

Protein blotting involves the transfer of proteins to an immobilizing membrane. The most widely used blotting method is the electrophoretic transfer of resolved proteins from a polyacrylamide gel to a nitrocellulose or polyvinylidene difluoride (PVDF) sheet and is often referred to as "Western blotting." Electrophoretic transfer uses the driving force of an electric field to elute proteins from gels and to immobilize them on a matrix. This method is fast, efficient, and maintains the high resolution of the protein pattern *(1)*. There are currently two main configurations of electroblotting apparatus: (1) tanks of buffer with vertically placed wire (*see* Chapter 39) or plate electrodes and (2) semidry transfer with flat-plate electrodes.

For semidry blotting, the gel and membrane are sandwiched horizontally between two stacks of buffer-wetted filter papers that are in direct contact with two closely spaced solid-plate electrodes. The name semidry refers to the limited amount of buffer that is confined to the stacks of filter paper. Semidry blotting requires considerably less buffer than the tank method, the transfer from single gels is simpler to set up, it allows the use of multiple transfer buffers (i.e., different buffers in the cathode and anode electrolyte stacks) and it is reserved for rapid transfers because the amount of buffer is limited and the use of external cooling is not possible. Nevertheless, both techniques have a high efficacy and the choice between the two types of transfer is a matter of preference.

Once transferred to a membrane, proteins are more readily and equally accessible to various ligands than they were in the gel. The blot (i.e. the immobilizing matrix containing the transferred proteins) is therefore reacted with different probes, such as antibody for the identification of the corresponding antigen, lectin for the detection of glycoproteins, or ligand for the detection of blotted receptor components, as well as for studies of protein–ligand associations (*see* **ref. *2***; and Chapters 54–55 and 106). The blot is also widely used with various techniques of protein identification, from which the measurement of protein mass (by mass spectrometry using nitrocellulose or PVDF membrane), or the determination of the protein sequence (amino-[N]-terminal Edman degradation, carboxy-[C]-terminal sequence or amino acid analysis). Recently, a novel approach has been described for the sensitive analysis and identification of proteins, separated by 2-D PAGE, transferred onto a PVDF membrane, incubated with succinic acid

From: *The Protein Protocols Handbook, 2nd Edition*
Edited by: J. M. Walker © Humana Press Inc., Totowa, NJ

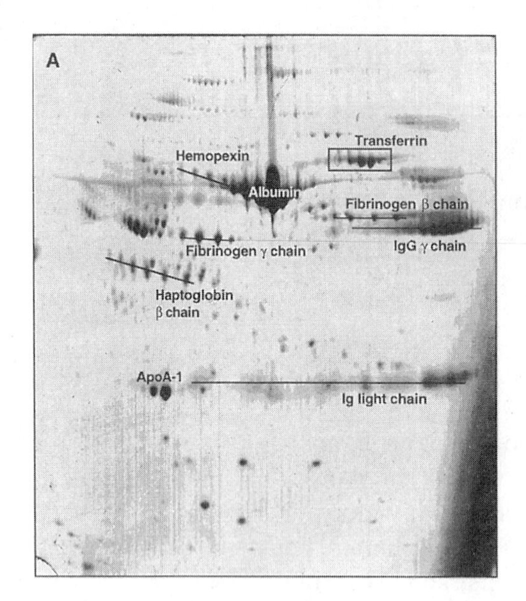

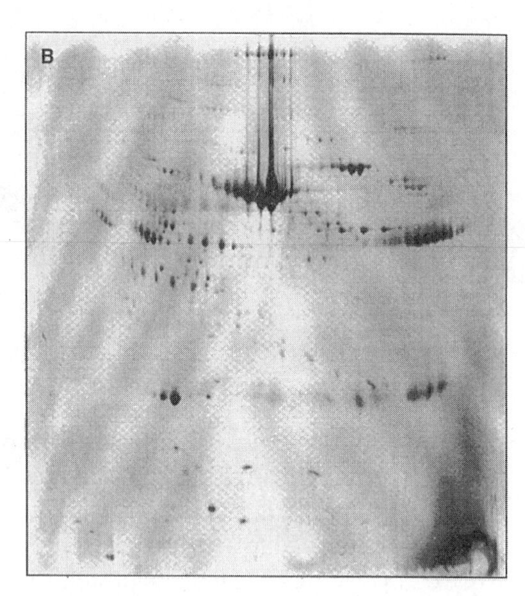

Fig. 1. Plasma proteins separated by two-dimensional polyacrylamide gel electrophoresis and (**A**) stained with Coomasie Brilliant Blue R250 or (**B**) transferred to PVDF membrane using the semidry system (2 h, 15 V) with Towbin buffer diluted 1:2 in water and stained with Coomasie Blue.

and then scanned with IR-MALDI-MS (infrared matrix-assisted laser desorption/ ionization-mass spectrometer). They demonstrated that the sensitivity for protein detection was comparable if not better than that of sensitive silver-stained gels *(3)*.

The blot analysis generally requires small amounts of reagents, the transferred proteins on membrane can be stored for many weeks prior to their use and the same blot can be used for multiple successive analyses. For reviews on the basic principles involved in performing protein blotting and for an overview of some possible applications, see articles by Garfin et al. *(1)*, Beisiegel *(4)*, Gershoni et al. *(5)*, Towbin et al. *(6)*, and Wilkins et al. *(7)*.

In the following sections, we describe a protocol for semidry blotting that uses a simple buffer system *(8)*. The efficacy of this method is illustrated in **Fig. 1** with human plasma proteins separated by two-dimensional polyacrylamide gel electrophoresis (2-D PAGE), transferred on PVDF membrane and stained with Coomasie Blue. The blot pattern is compared to the Coomasie Blue staining of the same protein sample before transfer from 2-D PAGE. The resolution, shape, and abundance of protein spots on membrane are comparable to the 2-D polyacrylamide gel pattern. This blotting procedure allows a good and almost complete elution of proteins from the gel and their immobilization on the membrane.

We also review in **Subheading 4.** the principal types of transfer matrix and the different discontinuous or continuous buffers that can be used with this technique.

2. Materials

1. Buffer: The transfer and equilibration buffer is the Towbin buffer diluted 1:2 with distilled water: 12.5 m*M* Tris, 96 m*M* glycine, and 10% (v/v) methanol *(9)* (*see* **Note 1** and **Table 1**).

2. Membranes and filter papers: PVDF (0.2 µm, 200 × 200 mm, Bio-Rad) or nitrocellulose (0.45 µm, Schleicher & Shuell). Filter papers are chromatography papers (grade 3 mm CHr, Whatman). Thicker blotting papers can also be used (Pharmacia-LKB, filter paper for blotting, 200 × 250 mm).
3. Electroblotting apparatus: Proteins are transferred with a Trans-Blot SD semidry cell (Bio-Rad). The anode of the apparatus is made of platinum-coated titanium and the cathode is made of stainless steel. The maximum gel size that can be used with this apparatus is 25 cm × 18.5 cm.
4. Coomasie staining: Proteins in the 2-D gel and on the blot (**Fig. 1**) are stained with a 0.025% (w/v) solution of Coomasie Brilliant Blue R250 solubilized in 43% (v/v) methanol and 7% (v/v) acetic acid. The destaining solution contains 30% (v/v) methanol and 7% (v/v) acetic acid.
   All chemicals and methanol are of analytical reagent grade. Metallic contaminants in low-grade methanol normally deposit on the electrodes.

3. Method

Note: *To avoid membrane contamination, always use forceps or wear gloves when handling membranes.*

3.1. Preparation for Semidry Blotting

1. Prepare the transfer buffer the day preceding the blotting experiments and store it at 4°C.
2. After the separation of proteins by sodium dodecyl sulfate (SDS)-PAGE or 2-D PAGE, the gel is briefly rinsed in distilled water for a few seconds and then equilibrated in transfer buffer for 10 min under gentle agitation (*see* **Note 2**).
3. During the equilibration, the filter paper and the transfer membrane are cut to the dimensions of the gel. Six pieces of filter paper per gel are needed for each gel–membrane sandwich (or two pieces of thick filter papers).
4. If the hydrophobic PVDF membrane is used, it should be prewetted prior to equilibration in transfer buffer. Immerse the membrane in a small volume of 100% methanol for a few seconds, until the entire membrane is translucent, rinse it in deionized water and then equilibrate it in transfer buffer for 3–5 min. It is important to keep the membrane wet at all times since proteins will not bind to the dried PVDF membrane (*see* **Note 3**).
   For hydrophilic nitrocellulose membrane, wet it in transfer buffer directly and allow it to soak for approx 5 min (*see* **Notes 4** and **5** and **Table 2** for the description of the different transfer membranes).
5. If multiple full-size gels are to be transferred at one time, there is a necessity to interleave a dialysis membrane between the gel-membrane pairs to prevent proteins being driven through membranes into subsequent stacks.
   Cut a piece of dialysis membrane with the appropriate molecular weight cutoff to the dimensions of the gel and soak it in the transfer buffer (*see* **Note 6**).

3.2. Assembly of the Semidry Unit

The assembly of a semidry electroblotting apparatus is represented in **Fig. 2**. Four mini gels can also be transferred at the same time by placing them side-by-side on the anode platform.

1. The filter papers are briefly soaked in transfer buffer for a few seconds.
   Place a prewetted filter paper onto the anode. Use a pipette or a painter roller to eliminate all air bubbles by pressing firmly all over the area of the paper (*see* **Note 7**). If a thin filter paper is used, add two more sheets and remove air bubbles between each layer.

Table 1
Examples of Transfer Buffers Used for Semidry Blotting of Proteins from Polyacrylamide Gels to Immobilized Matrices

Buffer		pH	Remarks	Ref.
Continuous buffer system				
Towbin buffer	25 m*M* Tris 192 m*M* Glycine 20 % (v/v) Methanol	8.3	First practical method for the transfer of proteins in an electric field from gel electropherograms to nitrocellulose matrix.	***10***
Towbin buffer with addition of SDS	25 m*M* Tris 192 m*M* Glycine 20% (v/v) Methanol 0.1 % (w/v) SDS	8.3	Facilitates the elution of high molecular weight proteins.	***11***
Towbin buffer diluted 1:2 with water	12.5 m*M* Tris 96 m*M* Glycine 10% (v/v) Methanol	8.3		***9***
Dunn carbonate buffer	10 m*M* $NaHCO_3$ 3 m*M* $Na2CO_3$ 20% (v/v) Methanol	9.9	Carbonate buffer enhances immunochemical recognition and therefore the sensitivity of Western blots by increasing the retention of small proteins and restoration of a more nearly native state for the larger proteins.	***12***
Bjerrum and Schafer–Nielsen buffer	40 m*M* Tris 39 m*M* Glycine 20% (v/v) Methanol	9.2	This buffer and the Dunn carbonate buffer have a higher pH and a lower conductivity than the Towbin buffer and are recommended for semidry transfers.	***13***
Buffer without methanol for transferring SDS–protein complexes using nylon membrane	All previous buffers without the addition of methanol	–	Without methanol, the SDS from SDS gels are still complexed with proteins and confers similar negative charge to mass ratio for all proteins. Therefore, electrostatic interactions between the nylon matrix (positively charged) and the proteins can occur.	***14***
CAPS[a] buffer	10 m*M* CAPS 10% (v/v) Methanol	11	This buffer without glycine is useful to perform a complete amino acid composition analysis.	***15,16***

Discontinuous buffer systems				
Tris-glycine buffer with addition of SDS at the cathodic side and methanol at the anodic side (Svoboda buffer)	Cathodic side: 0,1% (w/v) SDS 25 m*M* Tris Glycine 192 m*M* (no methanol)	8.3	The presence of 20% methanol at the anodic side facilitates the dissociation of protein–SDS complexes and the proteinbinding to the negative charge matrix, while the presence of 0.1% SDS at the cathodic side increases the rate of transfer of cationic and/or hydrophobic at the cathodic side increases the rate of transfer of cationic and/or hydrophobic proteins.	***17***
	Anodic side: 25 m*M* Tris 192 m*M* Glycine 20% (v/v) Methanol	8.3		
Kyhse–Andersen buffer	Cathodic side: 40 m*M* 6-amino-*n*-hexanoic acid 25 m*M* Tris 20% (v/v) Methanol	9.4	This system uses isotachophoresis to elute proteins from from the gel. SDS, glycine, and SDS–protein complexes from the gel move toward the anode according to their net mobilities in the stated order. At the cathodic side, 6-amino-*n*-hexanoic acid is the terminating ion and thereby, carries away the proteins from the polyacrylamide gel.	***18***
	Anodic side: 300 m*M* Tris 20% (v/v) Methanol (in contact with the anode)	10.4		
	25 m*M* Tris 20% (v/v) Methanol (in contact with the membrane)	10.4		
Laurière buffer	Cathodic side: 0.4% (w/v) SDS 60 m*M* lactic acid 100 m*M* Tris (in contact with the cathode)	8.4	This system of buffers introduces a difference of pH which remains stable during the entire electro-transfer time period and an asymetrical disposition of methanol and SDS on each side of the gel–membrane pair.	***19***

(*continued*)

Table 1 (*continued*)
Exampled of Transfer Buffers Used for Semidry Blotting of Proteins from Polyacrylamide Gels to Immobilized Matrices

Buffer	pH	Remarks	Ref.
0.1% (w/v) SDS 15 m*M* lactic acid 25 m*M* Tris (in contact with the gel)	8.4		
Anodic side: 20% (v/v) Methanol 60 m*M* Lactic acid 20 m*M* Tris (in contact with the membrane)	3.8	Under these conditions, semidry blotting allows quantitative transfer of proteins from SDS gels to membranes almost regardless of their molecular weight and solubility.	
20% (v/v) Methanol 100 m*M* Tris (in contact with the anode)	10.4		

[a]CAPS: 3-[cyclohexylamino]-1-propanesulfonic acid.

Table 2
Membranes Used for Electroblotting Proteins from Polyacrylamide Gels

Transfer matrix	Charge of the matrix	Mechanism of protein binding[a]	Capacity of protein binding ($\mu g/cm^2$)[b]	Advantages	Disadvantages
Nitrocellulose	Negative	Hydrophobic and electrostatic forces	249	High capacity	They are mechanically fragile
				Low cost	The presence of cellulose acetate in nitrocellulose membranes seems to reduce their capacity to bind proteins. Pure cellulose nitrate has the highest binding capacity ***(14)***
(PVDF) Polyvinylidene difluoride	Negative	Hydrophobic forces	172	High capacity	High cost
				Ideal for N-terminal micro-sequencing, amino acid composition analysis (resists to acidic and organic solvent)	
				Protein transfer efficacy can be visualized without staining (*see* **Note 4**)	
Modified nylon (made by incorporating tertiary and quaternary amines)	Positive	Electrostatic forces	149	Multiple reprobing possible	Protein staining is difficult because common anionic dyes cannot be used (Coomasie Blue, Amido Black, Colloidal Gold, Ponceau S)

(*continued*)

Table 2 (*continued*)
Membranes Used for Electroblotting Proteins from Polyacrylamide Gels

Transfer matrix	Charge of the matrix	Mechanism of protein binding[a]	Capacity of protein binding ($\mu g/cm^2$)[b]	Advantages	Disadvantages
					Blocking of unoccupied binding sites is difficult due to the high charge density of the matrices
Carboxymethyl cellulose	Negative	Ionic interactions	2500 for histones (probably due to the high affinity of histones for anionic groups)	Helpful for the transfer and microsequencing of basic proteins or peptides ***(22)***	Most acidic proteins poorly transferred An elution step is required before microsequencing

[a]Refers to the mechanism of adsorption between the surface of the membrane and structures on the protein.
[b]Reference 22.

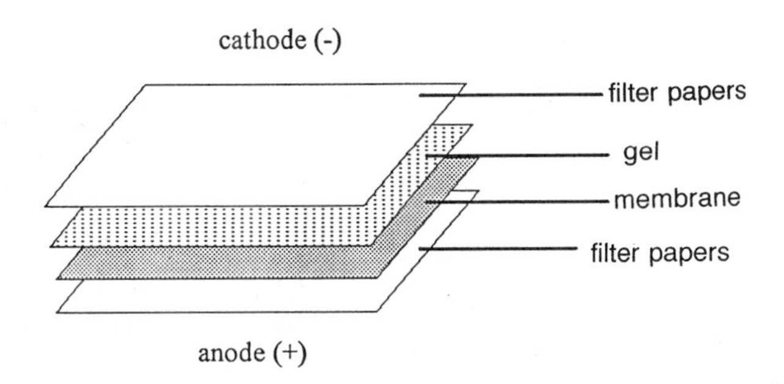

Fig. 2: Assembly of a horizontal electroblotting apparatus.

2. Place the preequilibrated transfer membrane on top of the filter paper.
3. Place the equilibrated gel on top and on the center of the membrane.
4. Place another wetted sheet of filter paper on top of the gel. If a thin filter paper is used, add two more sheets. Roll out air bubbles.
5. Place the cathode onto the stack.
6. The blotting unit is then connected to a power supply and proteins are transferred for 2 h at a constant voltage of 15 V and at room temperature.
7. After protein blotting, the membrane is rinsed (3 × 5 min) with distilled water and is then ready for blot analysis. Following transfer, the first step for blot analysis is to block unoccupied binding sites on the membrane to prevent nonspecific binding of probes, most of which are proteins (antibody, lectin). *See* **Note 8** and **Table 3** for a description of the blocking procedure.
8. The blots can be stored for many weeks prior to their use. *See* **Note 9** for storage conditions.

4. Notes

1. Since the introduction of western blotting in 1979 by Towbin et al. ***(10)***, many other buffer systems have been developed in order to improve electrophoretic transfer of proteins. The most common systems used with semidry apparatus are listed in **Table 1**.

   The two critical factors during transfer are the elution efficacy of proteins out of the gel and the binding capacity of the matrix for proteins. The elution efficacy is mainly determined by the acrylamide concentration of the gel, the ionic strength, the pH of the buffer and additional constituents of the transfer buffer such as (SDS) and methanol.

   The binding capacity is mainly determined by the character of the membrane but also by the transfer buffer composition ***(5)***.

   Alkaline pH and SDS favor the elution of the proteins from the gel, whereas acidic pH and methanol favor their adsorption on the negatively charged membrane ***(19)***. Methanol increases the binding capacity of matrix presumably by exposing hydrophobic protein domains, so that they can interact with the matrix. Also, methanol decreases the elution efficacy by fixing the proteins in the gel and by reducing the gel pore size ***(1,5)***. When there is SDS in transfer buffer (up to about 0.1% w/v), the proteins are negatively charged and elute efficiently from the gel.

   Semidry conditions for blotting allow the use of multiple transfer buffers (discontinuous buffer systems) to ensure a faster and better electrotransfer ***(18)***. Examples of discontinuous buffer systems are listed in **Table 1**. However, Bjerrum and Schafer-Nielsen ***(13)*** showed that there is no advantage in using different buffers in the cathode and anode electrolyte stacks. They found comparable transfer efficiencies for semidry blots performed in continuous and discontinuous buffers. We tested different buffer systems (Towbin buffer;

Table 3
Blocking Agents for Blots

Blocking agent	Remarks
Proteins: Should not be used in excess of 3%.	
Bovine serum albumin (BSA)	Can contain carbohydrate contaminants that can increase background when lectin probes are used ***(1,24)***. For nylon membranes, high concentration of BSA is required because of their very high protein binding capacity. A blocking solution made of 10% BSA in PBS (phosphate buffered saline) is used for at least 12 h at 45–50°C for satisfactory quenching ***(14)***.
Non-fat dried milk	Inexpensive and easy to use Can contain competing reactants such as biotin, which can cause a diminution of the signal intensity ***(1)***.
Gelatin	The use of gelatin as a blocking agent is contested: it may mask blotted proteins ***(1)***, it leads to strong unspecific reactions with peroxidase-labelled antibodies ***(5)*** and when used alone, gelatin is not a satisfactory blocking agent ***(6)***. Fish skin gelatin gives lower background than mammalian gelatin and does not have to be heated to dissolve (can be used at 4°C) ***(25)***.
Nonionic detergents Should not be used in excess of 0.3–0.5%	
Tween-20	Simple and effective blocker. Widely used at 0.05% for immunoblotting on nitrocellulose or PVDF ***(1,26,27)***. Tween-20 minimizes non specific protein-protein and protein-matrix adsorption by disrupting the underlying non covalent/hydrophobic interactions, while leaving antigen-antibody interactions relatively unaffected ***(27)***. However, some studies reported proteins and antibodies removal from the membrane after Tween-20 incubation ***(28)***.
NP-40, Triton X-100	These agents carry slight charges and should be avoided since they displace proteins from nitrocellulose at a higher extent than Tween-20 ***(28)***.

(*continued*)

Table 3 (*continued*)
Blocking Agents for Blots

Blocking agent	Remarks
Other:	
Polyvinylpyrolidone (PVP-40, 40,000 M_r)	As adjunct to Tween-20, PVP-40 produces backgrounds lower than those of Tween-20 alone, approximately doubling the signal-to-background ratio (without decreasing specific immunoreactivity in Western blots) Should be used at a concentration of 1% (w/v) and added to 0.05% (w/v) Tween-20 *(27)*.

Towbin buffer diluted 1:2 with water and Laurière, buffer) for transferring plasma proteins from 2-D PAGE by semidry method and we similarly found no advantage in using discontinuous buffer systems (unpublished results).

2. Rinsing and equilibrating the gel facilitate the removal of electrophoresis buffer salt and excess of detergent. If salts are still present, the conductivity of the transfer buffer increases and excessive heat is generated during the transfer. Also, an equilibration period allows the gel to adjust to its final size prior to electroblotting because the gel shrinks in methanol-containing transfer buffer.
   The duration of equilibration depends on the gel thickness. For a 1.5 mm thick gel, 10 min of equilibration are enough.
   Recently, poly(ethylene glycol) polymers (PEG 1000–2000) have been used to complete electroblotting, in order to obtain better resolution and enhancement of sensitivity of proteins on membrane. PEG reduces background, raises signal-to-noise ratio and sharpens protein band *(20)*. After polyacrylamide gel electrophoresis, 30% (w/v) PEG (solubilized in transfer buffer: 12.5 m*M* Tris, 96 m*M* glycine and 15% [v/v] methanol) is applied to reversibly fix proteins within the gel. The intragel proteins can then be electroblotted directly onto membranes using the same transfer buffer containing PEG.
3. If the PVDF membrane does dry out during use it can be rewet in methanol. Membranes that contain adsorbed proteins and that have been allowed to dry can also be rewet in methanol. In our experience, we have not seen any difficulty in protein staining, glycoprotein detection and immunostaining after this rewetting procedure in methanol.
4. Nitrocellulose membrane was the first matrix used in electroblotting *(10)* and is still the support used for most protein blotting experiments. It has a high binding capacity, the nonspecific protein binding sites are easily and rapidly blocked, allowing low background staining and it is not expensive. For blotting with mixtures of proteins, standard nitrocellulose with a pore size of 0.45 μm should be used first. However, membranes having 0.1 μm and 0.2 μm pore sizes should be tried if some low molecular weight proteins do not bind efficiently to the standard membrane *(1)*. After staining, the nitrocellulose membranes become transparent simply by impregnating the membrane with concentrated Triton X-114 *(21)*. The blot can thus be photographed by transillumination or scanned with a densitometer for quantitative analysis. Long-term stability (several months) of transparent stained blots is possible if they are stored at –20°C.

PVDF membranes are more expensive but have high mechanical strength, high protein binding capacity, and are compatible with most commonly used protein stains and immunochemical detection systems. The chemical structure of the membrane offers excellent resistance to acidic and organic solvents. This makes PVDF membrane an appropriate support for N-terminal protein sequencing and amino acid composition analysis. Another interesting advantage of PVDF matrix over the nitrocellulose is the possibility to visualize the protein pattern on the blot without staining. After blotting, the PVDF membrane should be placed on top of a vessel containing distilled water. The immersion of the membrane in water should be avoided. It should be laid down at the surface. The protein spots (or bands) contrast with the remainder of the membrane and can be visualized. To obtain a clear image of the protein pattern, the surface of the membrane should be observed from different angles and under appropriate lighting. This procedure, which was found unintentionally, is easy to perform and allows rapid and good evaluation of the transfer quality. **Table 2** summarizes the most common matrices that can be used for transferring proteins from polyacrylamide gels.

5. Whatever the membrane used, exceeding its binding capacity tends to reduce the signal eventually obtained on blots. It can be assumed that excess protein, weakly associated with the membrane, may be readily accessible to react with the probe in solution, but the probe–protein complexes formed may then be easily washed off during the further processing of the membrane ***(5)***. This situation does not occur if the proteins are initially in good contact with the membrane.
   For 2-D PAGE, the best recovery and resolution of proteins are obtained when loading 120 µg of human plasma or platelet proteins. When 400 µg of protein are separated by 2-D PAGE and transferred on membrane, the spots are diffused and the basic proteins poorly transferred (not shown). This could be attributed to overloading which prevents a good separation of proteins and an adequate binding to transfer matrix.
6. It is very difficult to form large stacks of gel–membrane pairs. Even two pairs are often associated with the introduction of air bubbles. We prefer to use a semidry unit to transfer proteins from a single gel only.
7. Air bubbles create points of high resistance and this results in spot (or band) areas of low efficacy transfer and spot (or band) distortion.
8. The quality or extent of the blocking step determines the level of background interference. It has been recognized that the blocking step may also promote renaturation of epitopes ***(23)***. This latter aspect is particularly important when working with monoclonal antibodies (which often fail to recognize the corresponding antigenic site after electroblotting). Hauri and Bucher ***(23)*** suggest that monoclonal antibodies may have individual blocking requirements, probably due to different degrees of epitope renaturation and/or accessibility of antibody under the various blocking conditions. Some common blocking agents are listed in **Table 3**.
   For immunoblotting on nitrocellulose membrane we obtain good results by using a blocking solution made of 0.5% (w/v) BSA, 0.2% (w/v) Tween 20 and 5% (w/v) nonfat dried milk in PBS buffer (137 m*M* NaCl, 27 m*M* KCl, 10 m*M* Na2HPO$_4$, 1.8 m*M* KH_2PO_4, pH 7.4) ***(29)***. For lectin blotting on PVDF membrane, we use 0.5% (w/v) Tween 20 in PBS buffer ***(8)***. (*See also* Chapter 106).
9. We store the blots at 4°C in PBS containing 0.005% (w/v) sodium azide. To evaluate the effect of storage on blotted proteins, we stained with colloidal gold a nitrocellulose blot of platelet proteins stored for a period of 4 mo in PBS–azide at 4°C. We observed a protein pattern identical to the same blot stained immediately after western blotting (not shown) On the other hand, when we dried a nitrocellulose blot at room temperature and stained

24 h later the proteins with colloidal gold, we observed important contaminating spots on the blots.

Only one point should be kept in mind when storing blots in PBS–azide. Sodium azide inhibits peroxidase activity and therefore good washing of the membrane with PBS is necessary before blot analysis using probe labeled with peroxidase.

As an alternative to blot storage, the gel can be frozen at –80°C. Immediately after electrophoresis, the gel should be rinsed in distilled water for a few seconds and placed in a plastic bag between two precooled glass plates. It is important to precool glass plates at –80°C otherwise the gel will crack when thawed. These frozen gels can be stored at –80°C for at least 3 mo *(19)*. Before the transfer procedure, the frozen gels are thawed in their plastic bags and then equilibrated as described in **Subheading 3.**

References

1. Garfin, D. E. and Bers, G. (1982) Basic aspects of protein blotting, in *Protein Blotting* (Baldo B. A. and Tovey, E. R., eds.), Karger, Basel, Switzerland, pp. 5–42.
2. Gershoni, J. M. (1985) Protein blotting: developments and perspectives. *TIBS* **10,** 103–106.
3. Eckerskorn, C., Strupat, K., Schleuder, D., Hochstrasser, D. F., Sanchez, J. C., Lottspeich, F., and Hillenkamp, F. (1997) Analysis of proteins by direct scanning-infrared-MALDI mass spectrometry after 2-D PAGE separation and electroblotting. *Analyt. Chem.* **69,** 2888–2892.
4. Beisiegel, U. (1986) Protein blotting. *Electrophoresis* **7,** 1–18.
5. Gershoni, J. M. and Palade, G. E. (1983) Protein blotting: principles and applications. *Analyt. Biochem.* **131,** 1–15.
6. Towbin, H. and Gordon, J. (1984) Immunoblotting and dot immunobinding: current status and outlook. *J. Immunol. Meth.* **72,** 313–340.
7 Wilkins, M. R. and Gooley, A. A. (1997) Protein identification in proteome projects, in *Proteomic Research: New Frontiers in Functional Genomics (Principles and Practice)* (Wilkins, M. R., Williams, K. L., Appel, R. D., Hochstrasser, D. F., eds.), Springer Verlag, Berlin, pp. 35–64.
8. Gravel, P., Golaz, O., Walzer, C., Hochstrasser, D. F., Turler, H., and Balant, L. P. (1994) Analysis of glycoproteins separated by two-dimensional gel electrophoresis using lectin blotting revealed by chemiluminescence. *Analyt. Biochem.* **221,** 66–71.
9. Sanchez, J. C., Ravier, F., Pasquali, C., Frutiger, S., Bjellqvist, B., Hochstrasser, D. F., and Hughes, G. J. (1992) Improving the detection of proteins after transfer to polyvinylidene difluoride membranes. *Electrophoresis* **13,** 715–717.
10. Towbin, H., Staehelin, T., and Gordon, J. (1979) Electrophoretic transfer of proteins from polyacrylamide gels to nitrocellulose sheets: procedure and some applications. *Proc. Natl. Acad. Sci. USA* **76,** 4350–4354.
11. Erickson, P. F., Minier, L. N., and Lasher, R. S. (1982) Quantitative electrophoretic transfer of polypeptides from SDS polyacrylamide gels to nitrocellulose sheets: a method for their re-use in immunoautoradiographic detection of antigens. *J. Immunol. Meth.* **51,** 241–249.
12. Dunn, S. D. (1986) Effects of the modification of transfer buffer composition and the renaturation of proteins in gels on the recognition of proteins on Western blots by monoclonal antibodies. *Analyt. Biochem.* **157,** 144–153.
13. Bjerrum, O. J. and Schafer-Nielsen, C. (1986). Buffer systems and transfer parameters for semidry electroblotting with a horizontal apparatus, in *Electrophoresis 1986* (Dunn, M. J., ed.) VCH, Weinheim, pp. 315–327.
14. Gershoni, J. M. and Palade, G. E. (1982) Electrophoretic transfer of proteins from sodium dodecyl sulfate-polyacrylamide gels to a positively charged membrane filter. *Analyt. Biochem.* **124,** 396–405.

15. Jin, Y. and Cerletti, N. (1992) Western blotting of transforming growth factor b2. Optimization of the electrophoretic transfer. *Appl. Theor. Electrophoresis* **3,** 85–90.
16. Matsudaira, P. J. (1987) Sequence from picomole quantities of proteins electroblotted onto polyvinylidene difluoride membranes. *J. Biol. Chem.* **21,** 10,035–10,038.
17. Svoboda, M., Meuris, S., Robyn, C., and Christophe, J. (1985) Rapid electrotransfer of proteins from polyacrylamide gel to nitrocellulose membrane using surface-conductive glass as anode. *Analyt. Biochem.* **151,** 16–23.
18. Kyhse-Andersen, J. J. (1984) Electroblotting of multiple gels: a simple apparatus without buffer tank for rapid transfer of proteins from polyacrylamide to nitrocellulose. *J. Biochem. Biophys. Meth.* **10,** 203–209.
19. Laurière, M. (1993) A semidry electroblotting system efficiently transfers both high and low molecular weight proteins separated by SDS-PAGE. *Analyt. Biochem.* **212,** 206–211.
20. Zeng, C., Suzuki, Y., and Alpert, E. (1990) Polyethylene glycol significantly enhances the transfer of membrane immunoblotting. *Analyt. Biochem.* **189,** 197–201.
21. Vachereau, A. (1989) Transparency of nitrocellulose membranes with Triton X-114. *Electrophoresis* **10,** 524–527.
22. Alimi, E., Martinage, A., Sautière, P., and Chevaillier, P. (1993) Electroblotting proteins onto carboxymethylcellulose membranes for sequencing. *BioTechniques* **15,** 912–917.
23. Hauri, H. P. and Bucher, K. (1986) Immunoblotting with monoclonal antibodies: importance of the blocking solution. *Analyt. Biochem.* **159,** 386–389.
24. Rohringer, R. and Holden, D. W. (1985) Protein blotting: detection of proteins with colloidal gold, and of glycoproteins and lectins with biotin-conjugated and enzyme probes. *Analyt. Biochem.* **144,** 118–127.
25. Saravis, C. A. (1984) Improved blocking of nonspecific antibody binding sites on nitrocellulose membranes. *Electrophoresis* **5,** 54–55.
26. Batteiger, B., Newhall, W. J., and Jones, R. B. (1982) The use of tween-20 as a blocking agent in the immunological detection of proteins transferred to nitrocellulose membranes. *J. Immunol. Meth.* **55,** 297–307.
27. Haycock, J. W. (1993) Polyvinylpyrrolidone as a blocking agent in immunochemical studies. *Analyt. Biochem.* **208,** 397–399.
28. Hoffman, W. L. and Jump, A. A. (1986) Tween-20 removes antibodies and other proteins from nitrocellulose. *J. Immunol. Meth.* **94,** 191–196.
29. Gravel, P., Sanchez, J. C., Walzer, C., Golaz, O., Hochstrasser, D. F., Balant, L. P., et al. (1995) Human blood platelet protein map established by two-dimensional polyacrylamide gel electrophoresis. *Electrophoresis* **16,** 1152–1159.

41

Protein Blotting by the Capillary Method

John M. Walker

1. Introduction

The ability to transfer (blot) separated proteins from a polyacrylamide gel matrix onto a sheet of nitrocellullose paper (where the proteins bind to the surface of the paper) has provided a powerful tool for protein analysis. Once immobilized on the surface of the nitrocellulose sheet, a variety of analytical procedures may be carried out on the proteins that otherwise would have proven difficult or impossible in the gel. Such procedures may include hybridization with labeled DNA or RNA probes, detection with antibodies (probably the most commonly used procedure), detection by specific staining procedures, autoradiographic assay, and so forth. The most commonly used, and indeed most efficient, methods for transferring proteins from gels to nitrocellulose (blotting) are by electrophoresis, and these methods are described in Chapters 39 and 40. An alternative method, capillary blotting, is described here. Although this method takes longer than electroblotting methods (it takes overnight) and transfer of proteins from the gel is not as complete as it is for electroblotting (although sufficient protein is transferred for most purposes) the method does have its uses. It is of course ideal for those who only wish to carry out occasional blotting experiments and therefore do not wish to commit themselves to the purchase or the purpose-built apparatus (plus power pack) needed for electroblotting. Second, this method is particularly useful for blotting isoelectric focusing gels where the proteins are at their isoelectric point (i.e., they have zero overall change) and are not easy therefore to transfer by electrophoresis. The method simply involves placing the gel on filter paper soaked in buffer. Buffer is drawn through the gel by capillary action by placing a pad of **dry** absorbent material on top of the gel. As the buffer passes through the gel, it carries the protein bands with it, and these bind to the nitrocellulose sheet that is placed between the top of the gel and the dry absorbent material.

2. Materials

1. Blotting buffer: 20 m*M* Tris, 150 m*M* glycine, 20% methanol, pH 8.3. Dissolve 4.83 g of Tris base, 20.5 g of glycine, and 400 mL of methanol in 2 L of distilled water. The solution is stable for weeks at 4°C.
2. Nitrocellulose paper.

From: *The Protein Protocols Handbook, 2nd Edition*
Edited by: J. M. Walker © Humana Press Inc., Totowa, NJ

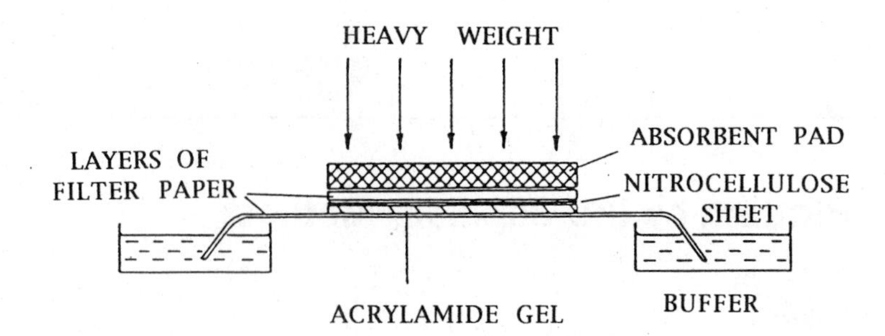

Fig. 1. A typical arrangement for capillary blotting.

3. Whatman 3MM filter paper
4. Absorbent material: e.g., a wad of filter paper or paper towels; this author uses baby diapers.

3. Method

1. Place two sheets of 3MM filter paper on a glass plate, and thoroughly soak them in blotting buffer. The ends should be dipped in reservoirs containing blotting buffer to ensure this filter paper pad remains wet overnight (*see* **Fig. 1**).
2. Place the gel to be blotted on top of this filter paper bed. Make sure the bed is fully wetted, and that no bubbles are trapped between the gel and filter paper. Thoroughly wet the top of the gel with blotting buffer.
3. Cut a piece of nitrocellulose paper to the size of the gel to be blotted. Wet the paper by dipping it in blotting buffer. **(Care: Use gloves; there are more proteins on your fingers than you are blotting from the gel!)** Then lay the nitrocellulose sheet on the gel surface. Take great care to ensure no air bubbles are trapped (buffer cannot pass through an air bubble).
4. Now dry material must be placed on top of the nitrocellulose. Start with three sheets of 3MM filter paper, cut to the same size as the gel, and then place on top of this a pad of your absorbent material (*see* **item 4**, **Subheading 2.**).
5. Finally, place a heavy weight on top of the absorbent rnaterial, e.g., a sheet of thick glass that supports a 2–3 L flask filled with water (*see* **Note 1**).
6. Allow blotting to take place overnight (preferably for 24 h). The setup may then be dismantled. the nitrocellulose sheet stained for protein to confirm that transfer has been achieved (e.g., with 0.2% Ponceau S in 10% acetic acid), and then the nitrocellulose sheet blocked and probed using any of the methods described in Chapters 54–58, and 106 .

4. Notes

1. The only error you can make with this method is to have any of your dry material (e.g., filter paper or absorbance pad) overhanging the gel and making contact with the wet base of filter paper. In this case, buffer preferentially travels around the gel into the absorbant pad, rather than through the gel. When the heavy weight is applied to the setup check that this is not happening (the absorbent pad often "sags" quite easily). If there is overlap, simply trim with scissors. Do not expect the absorbent pad to be particularly wet after an overnight blot. It should be barely damp after an overnight run. If it is soaking wet, then this indicates that buffer has traveled around the gel. However, even if this is the case, it is probably worth proceeding since some protein will have transferred nevertheless and this can probably still be detected with your probe.

42

Protein Blotting of Basic Proteins Resolved on Acid-Urea-Triton-Polyacrylamide Gels

Geneviève P. Delcuve and James R. Davie

1. Introduction

The electrophoretic resolution of histones on acetic acid-urea-Triton (AUT) polyacrylamide gels is the method of choice to separate basic proteins, such as histone variants, modified histone species, and high-mobility group proteins 14 and 17 (***1–6*** and *see* Chapters 16 and 17). Basic proteins are resolved in this system on the basis of their size, charge, and hydrophobicity. In previous studies, we analyzed the abundance of ubiquitinated histones by resolving the histones on two-dimensional (AUT into SDS) polyacrylamide gels, followed by their transfer to nitrocellulose membranes, and immunochemical staining of nitrocellulose membranes with an antiubiquitin antibody ***(7–9)***. However, transfer of the basic proteins directly from the AUT polyacrylamide gel circumvents the need to run the second-dimension SDS gel and accomplishes the analysis of several histone samples. We have described a method that efficiently transfers basic proteins from AUT polyacrylamide gels to nitrocellulose membranes ***(10)***. This method has been used in the immunochemical detection of modified histone, isoforms, and histone H1 subtypes ***(6,11–13)***.

To achieve satisfactory transfer of basic proteins from AUT gels to nitrocellulose, polyacrylamide gels are submerged in 50 m*M* acetic acid, and 0.5% SDS (equilibration buffer 1; 2 × 30 min) to displace the Triton X-100, followed by a 30-min incubation in a Tris-SDS buffer (equilibration buffer 2). The transfer buffer is an alkaline transfer buffer (25 m*M* CAPS, pH 10.0, 20% [v/v] methanol). Szewczyk and Kozloff ***(14)*** reported that alkaline transfer buffers increase the efficiency of transfer of strongly basic proteins from SDS gels to nitrocellulose membranes. We reasoned that this transfer buffer would improve the transfer of histones from AUT gels that had been treated with SDS.

2. Materials

Buffers are made from analytical-grade reagents dissolved in double-distilled water.

1. Equilibration buffer 1: 0.575 mL of glacial acetic acid (50 m*M*), 10 mL of 10% (10 g in 100 mL of water) SDS (0.5%), and water to 200 mL.
2. Equilibration buffer 2: 6.25 mL of 1 *M* Tris-HCl (62.5 m*M*), pH 6.8, 23 mL of 10% (w/v) SDS (2.3%), 5 mL of β-mercaptoethanol (5%), and water to 100 mL. 1 *M* Tris-HCl: 121 g

From: *The Protein Protocols Handbook, 2nd Edition*
Edited by: J. M. Walker © Humana Press Inc., Totowa, NJ

of Tris base in 800 mL of water and adjusted to pH 6.8 with hydrochloric acid. The buffer is made up to 1000 mL.

3. Transfer membrane: Nitrocellulose membranes (0.45-µm pore size, Schleicher & Schuell [Keene, NH], BA85) are used.
4. Transfer buffer: 187 mL of Caps (3-[cyclohexylamino]-1-propanesulfonic acid) stock (16X) solution (final concentration, 25 m*M*), 600 mL of methanol (20%), and water to 3 L. Caps stock solution (400 m*M*): 88.5 g of Caps dissolved in 800 mL of water and adjusted to pH 10.0 with 10 *M* NaOH. The buffer is made up to 1000 mL.
5. Electroblotting equipment: Proteins are transferred with the Bio-Rad (Hercules, CA) Trans-Blot transfer cell containing a super cooling coil (Bio-Rad). Cooling is achieved with a Lauda circulating bath.
6. Protein stain: Proteins in the AUT or SDS polyacrylamide gel are stained with Coomassie blue (Serva Blue G). Proteins on the membrane are stained with Indian ink (Osmiroid International Ltd.) (0.01% v/v in TBS-TW). TBS-TW: 20 m*M* Tris-HCl, pH 7.5, 500 m*M* NaCl, and 0.03% (v/v) Tween-20.

3. Methods

3.1. Protein Blotting

1. Nitrocellulose membranes are cut to size of gel at least a day before transfer and stored in water at 4°C.
2. The proteins are electrophoretically resolved on AUT- or SDS-polyacrylamide minislab gels (6 mm long, 10 mm wide, 0.8–1.0 mm thick).
3. Following electrophoresis, the AUT polyacrylamide gel is gently shaken in 100 mL of equilibration buffer 1 for 30 min at room temperature. This solution is poured off, and another 100 mL of equilibration buffer 1 are added. The gel is again shaken for 30 min. The solution is discarded and replaced with equilibration buffer 2. The gel is agitated in this solution for 30 min.
4. Onto the gel holder, place the porous pad that is equilibrated with transfer buffer. Three sheets of 3MM paper soaked in transfer buffer are placed on top of the porous pad. The treated AUT or SDS slab gel is placed onto the 3MM paper sheets. Nitrocellulose membrane is placed carefully on top of the gel, avoiding the trapping of air between the gel and nitrocellulose membrane. One sheet of 3MM paper soaked in transfer buffer is placed on top of the nitrocellulose membrane, followed by the placement of a porous pad that has been equilibrated with transfer buffer.
5. The gel holder is put into the transblot tank with the polyacrylamide gel facing the cathode and the nitrocellulose membrane facing the anode. The tank is filled with transfer buffer. Protein transfer is carried out at 70 V for 2 h and/or at 30 V overnight with cooling at 4°C.
6. Following transfer, the nitrocellulose membrane is placed onto a sheet of 3MM paper and allowed to dry for 30 min at room temperature. The gel is stained with Coomassie blue. The air-dried nitrocellulose membrane is placed between two sheets of 3MM paper. This is wrapped with aluminum foil and baked at 65°C for 30 min. The proteins transferred onto the nitrocellulose membrane may be visualized by staining the nitrocellulose membrane with India ink (*see* **Subheading 3.2.2.**).

3.2. Protein Staining

1. The nitrocellulose membrane is agitated in TBS-TW (0.7 mL per cm^2) for 10 min at room temperature in a sealed plastic box. The solution is discarded, and fresh TBS-TW is added. These steps are repeated twice more for a total of four washes of the membrane in the TBS-TW solution.

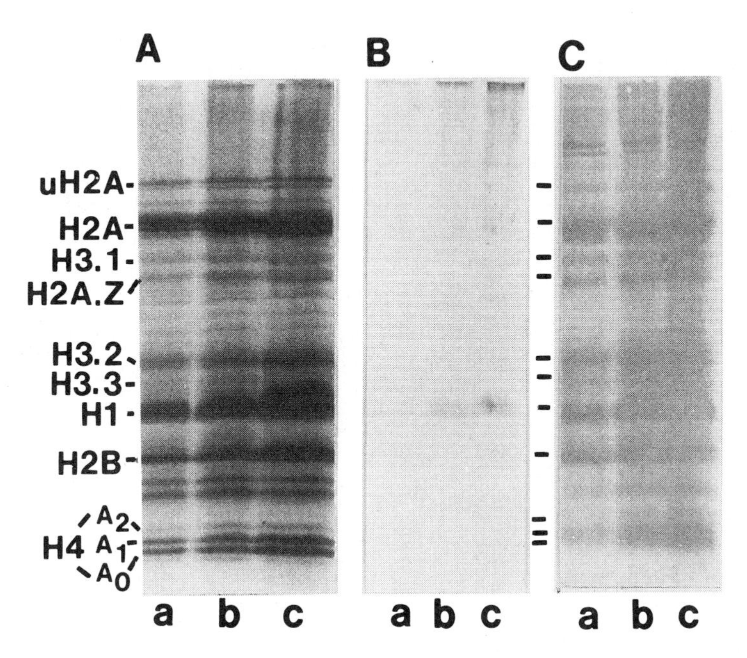

Fig. 1. Electrophoretic transfer of histones from AUT minislab gels. **(A)** Histones (9, 18, and 36 μg in lanes a, b, and c, respectively) isolated from T-47D-5 human breast cancer cells were electrophoretically resolved on AUT minislab gels. The gel was stained with Coomassie blue. **(B)** The Coomassie blue-stained AUT minislab gel pattern of histones remaining after transfer (30 V overnight) to nitrocellulose is shown. **(C)** The India ink-stained nitrocellulose pattern of histones transferred from the AUT minislab gel in B is shown. A_0, A_1, and A_2 correspond to the un-, mono-, and diacetylated species of histone H4, respectively. The ubiquitin adduct of histone H2A is denoted as uH2A. Reprinted with permission from **ref. *10*** (copyright by the Academic Press).

2. The India ink stain is added to the nitrocellulose membrane (0.56 mL/cm^2) which is agitated for 30 min to 2 h at room temperature.
3. The stain is discarded, and the nitrocellulose membrane is shaken in water for 5 min at room temperature.
4. The nitrocellulose membranes are dried and stored.

4. Notes

1. Nitrocellulose membranes have been used in the majority of our studies. However, these membranes are fragile and must be handled with care. An alternate membrane, which is stronger than nitrocellulose, is PVDF (Bio-Rad). The PVDF membranes are wetted with 100% methanol for 3 min and then equilibrated with transfer buffer for 3 min.
2. The efficiency of transfer of basic proteins (histones) from AUT polyacrylamide slab gels to nitrocellulose membranes is shown in **Fig. 1**. Most of the histones were efficiently transferred. The efficiency of elution was poorest for histone H1. **Figure 1C** demonstrates that the histone variants of histone H2A (H2A, H2A.Z) and of histone H3 (H3.1, H3.2, H3.3) and the modified histone species (e.g., ubiquitinated histone H2A, acetylated histone H4) were transferred. Densitometric tracings of the gel patterns before and after transfer demonstrated that >90% of the histones H2A, H2B, H3, and H4, and approx 80% of histone H1 were eluted from the AUT gel.

3. The transfer efficiency of basic proteins from AUT minislab polyacrylamide gels to nitrocellulose membranes was poor when a Tris-glycine-methanol (25 m*M* Tris, 192 m*M* glycine, 20% [v/v] methanol, and 0.1% SDS) transfer buffer was used.
4. With the Bio-Rad Trans-Blot cassette, four minislab gels can be easily accommodated.
5. Following transfer and baking, the nitrocellulose membrane may be stored for several weeks at room temperature before proceeding with immunochemical staining.
6. Leaving the nitrocellulose membrane in water for too long after staining with India ink will result in removal of the stain.
7. We have used this alkaline transfer buffer to transfer histones from SDS slab gels to nitrocellulose membranes. Pretreatment of the SDS slab gel is not required. However, we have found that washing the SDS slab gel in equilibration buffer 2 for 30 min improved the efficiency of elution of the histones from the SDS gel.

Acknowledgments

This work was supported by grants from the Medical Research Council of Canada (MT-9186, MT-12147, MA-12283, PG-12809) and the University of Manitoba Research Development Fund.

References

1. Urban, M. K., Franklin, S. G., and Zweidler, A. (1979) Isolation and characterization of the histone variants in chicken erythrocytes. *Biochemistry* **18,** 3952–3959.
2. Strickland, M., Strickland, W. N., and Von Holt, C. (1981) The occurrence of sperm isohistones H2B in single sea urchins. FEBS *Lett.* **135,** 86–88.
3. Meistrich, M. L., Bucci, L. R., Trostle-Weige, P. K., and Brock, W. A. (1985) Histone variants in rat spermatogonia and primary spermatocytes. *Dev. Biol.* **112,** 230–240.
4. Waterborg, J. H. (1990) Sequence analysis of acetylation and methylation in two histone H3 variants of alfalfa. *J. Biol. Chem.* **265,** 17,157–17,161.
5. Davie, J. R. and Delcuve, G. P. (1991) Characterization and chromatin distribution of the H1 histones and high-mobility-group non-histone chromosomal proteins of trout liver and hepatocellular carcinoma. *Biochem. J.* **280,** 491–497.
6. Li, W., Nagaraja, S., Delcuve, G. P., Hendzel, M. J., and Davie, J. R. (1993) Effects of histone acetylation, ubiquitination and variants on nucleosome stability. *Biochem. J.* **296,** 737–744.
7. Nickel, B. E., Allis, C. D., and Davie, J. R. (1989) Ubiquitinated histone H2B is preferentially located in transcriptionally active chromatin. *Biochemistry* **28,** 958–963.
8. Davie, J. R. and Murphy, L. C. (1990) Level of ubiquitinated histone H2B in chromatin is coupled to ongoing transcription. *Biochemistry* **29,** 4752–4757.
9. Davie, J. R., Lin, R., and Allis, C. D. (1991) Timing of the appearance of ubiquitinated histones in developing new macronuclei of *Tetrahymena thermophila. Biochem. Cell Biol.* **69,** 66–71.
10. Delcuve, G. P. and Davie, J. R. (1992) Western blotting and immunochemical detection of histones electrophoretically resolved on acid-urera-triton- and sodium dodecyl sulfate-polyacrylamide gels. *Analyt. Biochem.* **200,** 339–341.
11. Lee, D. Y., Hayes, J. J., Pruss, D., and Wolffe, A. P. (1993) A positive role for histone acetylation in transcription factor access to nucleosomal DNA. *Cell* **72,** 73–84.
12. Davie, J. R. and Murphy, L. C. (1994) Inhibition of transcription selectively reduces the level of ubiquitinated histone H2B in chromatin. *Biochem. Biophys. Res. Commun.* **203,** 344–350.

13. Nagaraja, S., Delcuve, G. P., and Davie, J. R. (1995) Differential compaction of transcriptionally competent and repressed chromatin reconstituted with histone H1 subtypes. *Biochim. Biophys. Acta* **1260,** 207–214.
14. Szewczyk, B. and Kozloff, L. M. (1985) A method for the efficient blotting of strongly basic proteins from sodium dodecyl sulfate-polyacrylamide gels to nitrocellulose. *Analyt. Biochem.* **150,** 403–407.

43

Alkaline Phosphatase Labeling of IgG Antibody

G. Brian Wisdom

1. Introduction

Alkaline phosphatase (EC 3.1.3.1) from bovine intestinal mucosa is a valuable label. It is stable, has a moderate size (140 kDa), a high turnover number, and can be assayed using a variety of different substrates which are measured by changes in absorbance, fluorescence, or luminescence. The most common method of labeling immunoglobulin G (IgG) antibody with this enzyme uses the homobifunctional reagent glutaraldehyde.

The chemistry of glutaraldehyde is complex. It reacts with the amino and, to a lesser extent, the thiol groups of proteins and when two proteins are mixed in its presence, stable conjugates are produced without the formation of Schiff bases. Self-coupling can be a problem unless the proteins are at appropriate concentrations. In the method *(1)* described there is usually little self-coupling of the enzyme or the IgG antibody, however, the size of the conjugate is large ($>10^6$ Da) as several molecules of each component are linked. This is the most simple labeling procedure to carry out and, although the yields of enzyme activity and immunoreactivity are small, the conjugates obtained are stable and practical reagents.

Alkaline phosphatase may also be coupled using heterobifunctional reagents containing the *N*-hydroxysuccinimide and maleimide groups, for example, succinimidyl 4-(*N*-maleidomethyl)-cyclohexane-1-carboxylate. However, because the enzyme has no free thiol groups this approach is usually used for the labeling of $F(ab')_2$ fragments of IgG via their thiols *(2)*.

2. Materials

1. Alkaline phosphatase from bovine intestinal mucosa, 2000 U/mg or greater (with 4-nitrophenyl phosphate as substrate); this is usually supplied at a concentration of 10 mg/mL. If ammonium sulfate, Tris, or any other amine is present, it must be removed (*see* **Note 1**). There are numerous commercial sources of labeling-grade enzyme.
2. IgG antibody. This should be the pure IgG fraction or, better, affinity-purified antibody from an antiserum or pure monoclonal antibody (*see* Chapters 137–144).
3. Phosphate buffered saline (PBS): 20 m*M* sodium phosphate buffer, pH 7.2, containing 0.15 *M* NaCl.

From: *The Protein Protocols Handbook, 2nd Edition*
Edited by: J. M. Walker © Humana Press Inc., Totowa, NJ

4. Glutaraldehyde.
5. 50 m*M* Tris-HCl buffer, pH 7.5, containing 1 m*M* $MgCl_2$, 0.02% NaN_3, and 2% bovine serum albumin (BSA).

3. Methods

1. Add 0.5 mg of IgG antibody in 100 µL of PBS to 1.5 mg of alkaline phosphatase.
2. Add 5% glutaraldehye (about 10 µL) to give a final concentration of 0.2% (v/v), and stir the mixture for 2 h at room temperature.
3. Dilute the mixture to 1 mL with PBS and dialyze against PBS (2 L) at 4°C overnight (*see* **Note 1**).
4. Dilute the solution to 10 mL with 50 m*M* Tris-HCl buffer, pH 7.5, containing 1 m*M* $MgCl_2$, 0.02% NaN_3, and 2% BSA, and store at 4°C (*see* Notes 2,–4).

4. Notes

1. Dialysis of small volumes can be conveniently done in narrow dialysis tubing by placing a short glass tube, sealed at both ends, in the tubing so that the space available to the sample is reduced. Transfer losses are minimized by carrying out the subsequent steps in the same dialysis bag. There are also various microdialysis systems available commercially.
2. The conjugates are stable for several years at 4°C as the NaN_3 inhibits microbial growth and the BSA minimizes denaturation and adsorption losses. These conjugates should not be frozen.
3. Purification of the conjugates is usually unnecessary; however, if there is evidence of the presence of free antibody it can be removed by gel filtration in Sepharose CL-6B (Amersham Pharmacia Biotech, Uppsala, Sweden) or a similar medium with PBS as solvent.
4. The efficacy of the enzyme-labeled antibody may be tested by immobilizing the appropriate antigen on the wells of a microtiter plate or strip, incubating various dilutions of the conjugate for a few hours, washing the wells, adding substrate, and measuring the amount of product formed. This approach may also be used for monitoring conjugate purification in chromatography fractions.

References

1. Engvall, E. and Perlmann, P. (1972) Enzyme-linked immunosorbent assay, ELISA. III. Quantitation of specific antibodies by enzyme-labelled anti-immunoglobulin in antigen-coated tubes. *J. Immunol.* **109,** 129–135.
2. Mahan, D. E., Morrison, L., Watson, L., and Haugneland, L. S. (1987) Phase change enzyme immunoassay. *Analyt. Biochem.* **162,** 163–170.

44

β-Galactosidase Labeling of IgG Antibody

G. Brian Wisdom

1. Introduction

The *Escherichia coli* β-galactosidase (EC 3.2.1.23) is a large enzyme (465 kDa) with a high turnover rate and wide specificity. Unlike several other enzyme labels it is not found in mammalian tissues, hence background contributions from these sources are negligible when this label is measured at a neutral pH.

The heterobifunctional reagent, *m*-maleimidobenzoyl-*N*-hydroxysuccinimide ester (MBS), is of value when one of the components of a conjugate has no free thiol groups, for example, immunoglobulin G (IgG). In this method *(1)* the IgG antibody is first modified by allowing the *N*-hydroxysuccinimide ester group of the MBS to react with amino groups in the IgG; the β-galactosidase is then added and the maleimide groups on the modified IgG react with thiol groups in the enzyme to form thioether links. This procedure produces conjugates with molecular weights in the range $0.6–1 \times 10^6$.

Many other heterobifunctional crosslinking reagents similar to MBS allow the coupling of the enzyme's thiols with the amino groups on IgG *(2)*.

2. Materials

1. β-Galactosidase from *E. coli*, 600 U/mg or greater (with 2-nitrophenyl-β-galactopyranoside as substrate) (*see* **Note 1**).
2. IgG antibody. This should be the pure IgG fraction or, better, affinity-purified antibody from an antiserum or pure monoclonal antibody (*see* Chapters 137–144).
3. 0.1 *M* Sodium phosphate buffer, pH 7.0, containing 50 m*M* NaCl.
4. MBS.
5. Dioxan.
6. Sephadex G25 (Amersham Pharmacia Biotech, Uppsala, Sweden) or an equivalent gel filtration medium.
7. 10 m*M* Sodium phosphate buffer, pH 7.0, containing 50 m*M* NaCl and 10 m*M* $MgCl_2$.
8. 2-Mercaptoethanol.
9. DEAE-Sepharose (Amersham Pharmacia Biotech) or an equivalent ion-exchange medium.
10. 10 m*M* Tris-HCl buffer, pH 7.0, containing 10 m*M* $MgCl_2$ and 10 m*M* 2-mercaptoethanol.
11. Item 10 containing 0.5 *M* NaCl.
12. Item 10 containing 3% bovine serum albumin (BSA) and 0.6% NaN_3.

From: *The Protein Protocols Handbook, 2nd Edition*
Edited by: J. M. Walker © Humana Press Inc., Totowa, NJ

3. Methods

1. Dissolve 1.5 mg of IgG in 1.5 mL of 0.1 *M* sodium phosphate buffer, pH 7.0, containing 50 m*M* NaCl.
2. Add 0.32 mg of MBS in 15 µL of dioxan, mix, and incubate for 1 h at 30°C.
3. Fractionate the mixture on a column of Sephadex G25 (approx 0.9 × 20 cm) equilibrated with 10 m*M* sodium phosphate buffer, pH 7.0, containing 50 m*M* NaCl and 10 m*M* $MgCl_2$, and elute with the same buffer. Collect 0.5-mL fractions, measure the A_{280} and pool the fractions in the first peak (about 3 mL in volume).
4. Add 1.5 mg of enzyme, mix, and incubate for 1 h at 30°C.
5. Stop the reaction by adding 1 *M* 2-mercaptoethanol to give a final concentration of 10 m*M* (about 30 µL).
6. Fractionate the mixture on a column of DEAE-Sepharose (approx 0.9 × 15 cm) equilibrated with 10 m*M* Tris-HCl buffer, pH 7.0, containing 10 m*M* $MgCl_2$ and 10 m*M* 2-mercaptoethanol; wash the column with this buffer (50 mL) followed by the buffer containing 0.5 *M* NaCl (50 mL). Collect 3-mL fractions in tubes with 0.1 mL of Tris buffer containing 3% BSA and 0.6% NaN_3. Pool the major peak (this is eluted with 0.5 *M* NaCl), and store at 4°C (*see* **Notes 2** and **3**).

4. Notes

1. The thiol groups of β-galactosidase may become oxidized during storage, thus diminishing the efficacy of the labeling. It is relatively easy to measure these groups using 5,5'-dithiobis(2-nitrobenzoic acid) (*see* Chapter 70); about 10 thiol groups per enzyme molecule allow the preparation of satisfactory conjugates.
2. The conjugates are stable for a year at 4°C as the NaN_3 inhibits microbial growth and the BSA minimizes denaturation and adsorption losses.
3. The activity of the conjugate can be checked by the method described in **Note 4** of Chapter 43.

References

1. O'Sullivan, M. J., Gnemmi, E., Morris, D., Chieregatti, G., Simmonds, A. D., Simmons, M., et al. (1979) Comparison of two methods of preparing enzyme-antibody conjugates: application of these conjugates to enzyme immunoassay. *Analyt. Biochem.* **100,** 100–108.
2. Hermanson, G. T. (1996) *Bioconjugate Techniques,* Academic Press, San Diego, pp. 229–248.

45

Horseradish Peroxidase Labeling of IgG Antibody

G. Brian Wisdom

1. Introduction

Horseradish peroxidase (HRP; EC 1.11.1.7) is probably the most widely used enzyme label. This protein is relatively small (44 kDa), stable, and has a broad specificity which allows is to be measured by absorption, fluorescence and luminescence.

The most popular method *(1)* for labeling IgG antibody molecules with HRP exploits the glycoprotein nature of the enzyme. The saccharide residues are oxidized with sodium periodate to produce aldehyde groups that can react with the amino groups of the IgG molecule, and the Schiff bases formed are then reduced to give a stable conjugate of high molecular weight ($0.5–1 \times 10^6$). The enzyme has few free amino groups so self-coupling is not a significant problem.

IgG may also be labeled with HRP using glutaraldehyde in a two-step procedure *(2)*.

2. Materials

1. HRP, 1000 U/mg or greater (with 2,2'-azino-*bis*-[3-ethylbenzthiazoline-6-sulfonic acid] as substrate) (*see* **Note 1**).
2. IgG antibody. This should be the pure IgG fraction or, better, affinity purified antibody from an antiserum or pure monoclonal antibody (*see* Chapters 137–148, 163).
3. 0.1 *M* Sodium periodate.
4. 1 m*M* Sodium acetate buffer, pH 4.4.
5. 10 m*M* Sodium carbonate buffer, pH 9.5.
6. 0.2 *M* Sodium carbonate buffer, pH 9.5.
7. Sodium borohydride, 4 mg/mL (freshly prepared).
8. Sepharose CL-6B (Amersham Pharmacia Biotech, Uppsala, Sweden) or a similar gel filtration medium.
9. Phosphate-buffered saline (PBS): 20 m*M* sodium phosphate buffer, pH 7.2, containing 0.15 *M* NaCl.
10. Bovine serum albumin (BSA).

3. Method

1. Dissolve 2 mg of peroxidase in 500 µL of water.
2. Add 100 µL of freshly prepared 0.1 *M* sodium periodate, and stir the solution for 20 min at room temperature. (The color changes from orange to green.)

From: *The Protein Protocols Handbook, 2nd Edition*
Edited by: J. M. Walker © Humana Press Inc., Totowa, NJ

3. Dialyze the modified enzyme against 1 m*M* sodium acetate buffer, pH 4.4 (2 L) overnight at 4°C (*see* **Note 2**).
4. Dissolve 4 mg of IgG in 500 µL of 10 m*M* sodium carbonate buffer, pH 9.5.
5. Adjust the pH of the dialyzed enzyme solution to 9.0–9.5 by adding 10 µL of 0.2 *M* sodium carbonate buffer, pH 9.5, and immediately add the IgG solution. Stir the mixture for 2 h at room temperature.
6. Add 50 µL of freshly prepared sodium borohydride solution (4 mg/mL), and stir the mixture occasionally over a period of 2 h at 4°C.
7. Fractionate the mixture by gel filtration on a column (approx 1.5 × 85 cm) of Sepharose CL-6B in PBS. Determine the A_{280} and A_{403} (*see* **Note 3**).
8. Pool the fractions in the first peak (both A_{280} and A_{403} peaks coincide), add BSA to give a final concentration of 5 mg/mL, and store the conjugate in aliquots at –20°C (*see* **Notes 4** and **5**).

4. Notes

1. Preparations of HRP may vary in their carbohydrate content and this can affect the oxidation reaction. Free carbohydrate can be removed by gel filtration. Increasing the sodium periodate concentration to 0.2 *M* can also help, but further increases lead to inactivation of the peroxidase.
2. Dialysis of small volumes can be conveniently done in narrow dialysis tubing by placing a short glass tube, sealed at both ends, in the tubing so that the space available to the sample is reduced. There are also various microdialysis systems available commercially.
3. The absorbance at 403 nm is caused by the peroxidase's heme group. The enzyme is often specified in terms of its *RZ* value; this is the ratio of A_{280} and A_{403}, and it provides a measure of the heme content and purity of the preparation. Highly purified peroxidase has an *RZ* of about 3. Conjugates with an *RZ* of 0.4 perform satisfactorily.
4. BSA improves the stability of the conjugate and minimizes loses due to adsorption and denaturation. NaN_3 should not be used with peroxidase conjugates because it inhibits the enzyme. If an antimicrobial agent is required, 0.2% sodium merthiolate (thimerosal) should be used.
5. The activity of the conjugate can be checked by the method described in **Note 4** of Chapter 43.

References

1. Wilson, M. B. and Nakane, P. P. (1978) Recent developments in the periodate method of conjugating horse radish peroxidase (HRPO) to antibodies, in *Immunofluorescence and Related Staining Techniques* (Knapp, W., Holubar, K., and Wick, G., eds.), Elsevier/North Holland Biomedical, Amsterdam, pp. 215–224.
2. Avrameas, S. and Ternynck, T. (1971) Peroxidase labeled antibody and Fab conjugates with enhanced intracellular penetration. *Immunochemistry* **8,** 1175–1179.

46

Digoxigenin (DIG) Labeling of IgG Antibody

G. Brian Wisdom

1. Introduction

Digoxigenin (DIG) is a plant steroid (390 Da) that can be used as a small, stable label of immunoglobulin G (IgG) molecules. It is a valuable alternative to biotin as the biotin–streptavidin system can sometimes give high backgrounds due, for example, to the presence of biotin-containing enzymes in the sample. There is a range of commercially available mouse and sheep anti-DIG Fab antibody fragments labeled with various enzymes and fluorescent molecules for the detection of the DIG-labeled IgG antibody in many applications *(1)*.

The IgG molecule is labeled, via its amino groups, with an *N*-hydroxysuccinimide ester derivative of the steroid containing the 6-aminocaproate spacer.

2. Materials

1. Digoxigenin-3-*O*-succinyl-ε-aminocaproic acid-*N*-hydroxysuccinimide ester (Roche, Indianapolis, IL).
2. IgG antibody. This should be the pure IgG fraction or, better, affinity purified antibody from an antiserum or pure monoclonal antibody (*see* Chapters 137–144).
3. Dimethyl sulfoxide (DMSO).
4. PBS: 20 m*M* sodium phosphate buffer, pH 7.2, containing 0.15 *M* NaCl.
5. 0.1 *M* ethanolamine, pH 8.5.
6. PBS containing 0.1% bovine serum albumin (BSA).
7. Sephadex G-25 (Amersham Pharmacia Biotech, Uppsala, Sweden) or similar gel filtration medium (*see* **Note 1**).

3. Method

1. Dissolve 1 mg of the antibody in 1 mL of PBS.
2. Prepare the digoxigenin-3-*O*-succinyl-ε-aminocaproic acid-*N*-hydroxysuccinimide ester immediately prior to use by dissolving it at a concentration of 2 mg/mL in DMSO.
3. Add 24 μL of the DIG reagent to the antibody solution slowly with stirring and incubate at room temperature for 2 h.
4. Terminate the reaction by adding 0.1 mL of 0.1 *M* ethanolamine and incubate for 15 min.
5. Remove the excess DIG reagent by gel filtration in a small column of Sephadex G-25 equilibrated with PBS containing 0.1% BSA. The IgG is in the first A_{280} peak.

From: *The Protein Protocols Handbook, 2nd Edition*
Edited by: J. M. Walker © Humana Press Inc., Totowa, NJ

6. Store the labeled antibody at 4°C with 0.05% NaN_3 or in aliquots at –20°C or lower (*see* **Note 2**).

4. Notes

1. Suitable ready-made columns of cross-linked dextran are available from Amersham Pharmacia Biotech (PD-10 or HiTrap columns) and from Pierce (Presto and Kwik columns).
2. The activity of the conjugate can be checked by the method described in **Note 4** of Chapter 43.

Reference

1. Kessler, C. (1991) The digoxigenin-anti-digoxigenin (DIG) technology: a survey of the concept and realization of a novel bioanalytical indicator system. *Mol. Cell. Probes* **5,** 161–205.

47

Conjugation of Fluorochromes to Antibodies

Su-Yau Mao

1. Introduction

The use of specific antibodies labeled with a fluorescent dye to localize substances in tissues was first devised by A. H. Coons and his associates. At first, the specific antibody itself was labeled and applied to the tissue section to identify the antigenic sites (direct method) *(1)*. Later, the more sensitive and versatile indirect method *(2)* was introduced. The primary, unlabeled, antibody is applied to the tissue section, and the excess is washed off with buffer. A second, labeled antibody from another species, raised against the IgG of the animal donating the first antibody, is then applied. The primary antigenic site is thus revealed. A major advantage of the indirect method is the enhanced sensitivity. In addition, a labeled secondary antibody can be used to locate any number of primary antibodies raised in the same animal species without the necessity of labeling each primary antibody.

Four fluorochromes are commonly used: fluorescein, rhodamine, Texas red, and phycoerythrin *(3)*. They differ in optical properties, such as the intensity and spectral range of their absorption and fluorescence. Choice of fluorochrome depends on the particular application. For maximal sensitivity in the binding assays, fluorescein is the fluorochrome of choice because of its high quantum yield. If the ligand is to be used in conjunction with fluorescence microscopy, rhodamine coupling is advised, since it has superior sensitivity in most microscopes and less photobleaching than fluorescein. Texas red *(4)* is a red dye with a spectrum that minimally overlaps with that of fluorescein; therefore, these two dyes are suitable for multicolor applications. Phycoerythrin is a 240-kDa, highly soluble fluorescent protein derived from cyanobacteria and eukaryotic algae. Its conjugates are among the most sensitive fluorescent probes available *(5)* and are frequently used in flow cytometry and immunoassays *(6)*. In addition, the newly introduced Alexa fluorochromes are a series of fluorescent dyes with excitation/emission spectra similar to those of commonly used ones, but are more fluorescent and more photostable *(7)*.

Thiols and amines are the only two groups commonly found in biomolecules that can be reliably modified in aqueous solution. Although the thiol group is the easiest functional group to modify with high selectivity, amines are common targets for modifying proteins. Virtually all proteins have lysine residues, and most have a free amino

From: *The Protein Protocols Handbook, 2nd Edition*
Edited by: J. M. Walker © Humana Press Inc., Totowa, NJ

terminus. The ε-amino group of lysine is moderately basic and reactive with acylating reagents. The concentration of the free-base form of aliphatic amines below pH 8.0 is very low. Thus, the kinetics of acylation reactions of amines by isothiocyanates, succinimidyl esters, and other reagents is strongly pH-dependent. Although amine acylation reactions should usually be carried out above pH 8.5, the acylation reagents degrade in the presence of water, with the rate increasing as the pH increases. Therefore, a pH of 8.5–9.5 is usually optimal for modifying lysines.

Where possible, the antibodies used for labeling should be pure. Affinity-purified, fluorochrome-labeled antibodies demonstrate less background and nonspecific fluorescence than fluorescent antiserum or immunoglobulin fractions. The labeling procedures for the isothiocyanate derivatives of fluorescein and sulfonyl chloride derivatives of rhodamine are given below ***(8)***. The major problem encountered is either over- or undercoupling, but the level of conjugation can be determined by simple absorbance readings.

2. Materials

1. IgG.
2. Borate buffered saline (BBS): 0.2 *M* boric acid, 160 m*M* NaCl, pH 8.0.
3. Fluorescein isothiocyanate (FITC) or Lissamine rhodamine B sulfonyl chloride (RBSC).
4. Sodium carbonate buffer: 1.0 *M* $NaHCO_3$-Na_2CO_3 buffer, pH 9.5, prepared by titrating 1.0 *M* $NaHCO_3$ with 1.0 *M* Na_2CO_3 until the pH reaches 9.5.
5. Absolute ethanol (200 proof) or anhydrous dimethylformamide (DMF).
6. Sephadex G-25 column.
7. Whatman DE-52 column.
8. 10 m*M* Sodium phosphate buffer, pH 8.0.
9. 0.02% Sodium azide.
10. UV spectrophotometer.

3. Methods

3.1. Coupling of Fluorochrome to IgG

1. Prior to coupling, prepare a gel-filtration column to separate the labeled antibody from the free fluorochrome after the completion of the reaction. The size of the column should be 10 bed volumes/sample volume (*see* **Note 1**).
2. Equilibrate the column in phosphate buffer. Allow the column to run until the buffer level drops just below the top of bed resin. Stop the flow of the column by using a valve at the bottom of the column.
3. Prepare an IgG solution of at least 3 mg/mL in BBS, and add 0.2 vol of sodium carbonate buffer to IgG solution to bring the pH to 9.0. If antibodies have been stored in sodium azide, the azide must be removed prior to conjugation by extensive dialysis (*see* **Note 2**).
4. Prepare a fresh solution of fluorescein isothiocyanate at 5 mg/mL in ethanol or RBSC at 10 mg/mL in DMF immediately before use (*see* **Note 3**).
5. Add FITC at a 10-fold molar excess over IgG (about 25 μg of FITC/mg IgG). Mix well and incubate at room temperature for 30 min with gentle shaking. Add RBSC at a 5-fold molar excess over IgG (about 20 μg of RBSC/mg IgG), and incubate at 4°C for 1 h.
6. Carefully layer the reaction mixture on the top of the column. Open the valve to the column, and allow the antibody solution to flow into the column until it just enters the bed resin. Carefully add phosphate buffer to the top of the column. The conjugated antibody elutes in the excluded volume (about one-third of the total bed volume).

7. Store the conjugate at 4°C in the presence of 0.02% sodium azide (final concentration) in a light-proof container. The conjugate can also be stored in aliquots at –20°C after it has been snap-frozen on dry ice. Do not refreeze the conjugate once thawed.

3.2. Calculation of Protein Concentration and Fluorochrome-to-Protein Ratio

1. Read the absorbance at 280 and 493 nm. The protein concentration is given by **Eq. 1**, where 1.4 is the optical density for 1 mg/mL of IgG (corrected to 1-cm path length).

$$\text{Fluorescein-conjugated IgG conc. (Fl IgG conc.) (mg/mL)} = (A_{280\text{ nm}} - 0.35 \times A_{493\text{ nm}})/1.4 \quad (1)$$

The molar ratio (F/P) can then be calculated, based on a molar extinction coefficient of 73,000 for the fluorescein group, by **Eq. 2** (*see* **Notes 4** and **5**).

$$\text{F/P} = (A_{493\text{ nm}}/73{,}000) \times (150{,}000/\text{Fl IgG conc.}) \quad (2)$$

2. For rhodamine-labeled antibody, read the absorbance at 280 and 575 nm. The protein concentration is given by **Eq. 3**.

$$\text{Rhodamine-conjugated IgG conc. (Rho IgG conc.) (mg/mL)} = (A_{280\text{ nm}} - 0.32 \times A_{575\text{ nm}})/1.4 \quad (3)$$

The molar ratio (F/P) is calculated by **Eq. 4**.

$$\text{F/P} = (A_{575\text{ nm}}/73{,}000) \times (150{,}000/\text{Rho IgG conc.}) \quad (4)$$

4. Notes

1. Sephadex G-25 resin is the recommended gel for the majority of desalting applications. It combines good rigidity, for easy handling and good flow characteristics, with adequate resolving power for desalting molecules down to about 5000 Dalton mol wt. If the volume of the reaction mixture is <1 mL, a prepacked disposable Sephadex G-25 column (PD-10 column from Pharmacia, Piscataway, NJ) can be used conveniently.
2. When choosing a buffer for conjugation of fluorochromes, avoid those containing amines (e.g., Tris, azide, glycine, and ammonia), which can compete with the ligand.
3. Both sulfonyl chloride and isothiocyanate will hydrolyze in aqueous conditions; therefore, the solutions should be made freshly for each labeling reaction. Absolute ethanol or dimethyl formamide (best grade available, stored in the presence of molecular sieve to remove water) should be used to dissolve the reagent. The hydrolysis reaction is more pronounced in dilute protein solution and can be minimized by using a more concentrated protein solution. Caution: DMSO should not be used with sulfonyl chlorides, because it reacts with them.
4. An F/P ratio of two to five is optimal, since ratios below this yield low signals, whereas higher ratios show high background. If the F/P ratios are too low, repeat the coupling reaction using fresh fluorochrome solution. The IgG solution needs to be concentrated prior to reconjugation (e.g., Centricon-30 microconcentrator from Amicon Co., Beverly, MA, can be used to concentrate the IgG solution).
5. If the F/P ratios are too high, either repeat the labeling with appropriate changes or purify the labeled antibodies further on a Whatman DE-52 column (diethylaminoethyl microgranular preswollen cellulose, 1-mL packed column/1–2 mg of IgG). DE-52 chromatography removes denatured IgG aggregates and allows the selection of the fraction of the

conjugate with optimal modification. Equilibrate and load the column with 10 m*M* phosphate buffer, pH 8.0. Wash the column with equilibrating buffer and elute with the same buffer containing 100 m*M* NaCl (first) and 250 m*M* NaCl (last). Measure the F/P ratios of each fraction, and select the appropriate fractions.

6. Alexa fluorochromes are available only as a protein labeling kit from Molecular Probes, Inc. (Eugene, Oregon, USA; www.probes.com). The reactive dye has a succinimidyl ester moiety that reacts with primary amines of proteins. The conjugation steps are similar to those for fluorescein isothiocyanate.

References

1. Coons, A. H., Creech, H. J., and Jones, R. N. (1941) Immunological properties of an antibody containing a fluorescent group. *Proc. Soc. Exp. Biol. Med.* **47,** 200–202.
2. Coons, A. H., Leduc, E. H., and Connolly, J. M. (1955) Studies on antibody production I. A method for the histochemical demonstration of specific antibody and its application to a study of the hyperimmune rabbit. *J. Exp. Med.* **102,** 49–60.
3. Mullins, J. M. (1999) Overview of fluorochromes, in Methods in Molecular Biology, vol. 115 (Javois, L. C., ed), Humana Press, USA, pp. 97–105.
4. Titus, J. A., Haugland, R., Sharrow, S. O., and Segal, D. M. (1982) Texas Red, a hydrophilic, red-emitting fluorophore for use with fluorescein in dual parameter flow microfluorometric and fluorescence microscopic studies. *J. Immunol. Meth.* **50,** 193–204.
5. Oi, V. T., Glazer, A. N., and Stryer, L. (1982) Fluorescent phycobiliprotein conjugates for analyses of cells and molecules. *J. Cell Biol.* **93,** 981–986.
6. Bochner, B. S., McKelvey, A. A., Schleimer, R. P., Hildreth, J. E., and MacGlashan, D. W., Jr. (1989) Flow cytometric methods for the analysis of human basophil surface antigens and viability. *J. Immunol. Meth.* **125,** 265–271.
7. Panchuk-Voloshina, N., Haugland, R. P., Bishop-Stewart, J., Bhalgat, M. K., Millard, P. J., Mao, F., Leung, W. Y., and Haugland, R. P. (1999) Alexa dyes, a series of new fluorescent dyes that yield exceptionally bright, photostable conjugates. *J. Histochem. Cytochem.* **47,** 1179–1188.
8. Schreiber, A. B. and Haimovich, J. (1983) Quantitative fluorometric assay for detection and characterization of Fc receptors. *Meth. Enzymol.* **93,** 147–155.

48

Coupling of Antibodies with Biotin

Rosaria P. Haugland and Wendy W. You

1. Introduction

The avidin–biotin bond is the strongest known biological interaction between a ligand and a protein ($K_d = 1.3 \times 10^{-15}$ *M* at pH 5) ***(1)***. The affinity is so high that the avidin–biotin complex is extremely resistant to any type of denaturing agent ***(2)***. Biotin (**Fig. 1**) is a small, hydrophobic molecule that functions as a coenzyme of carboxylases ***(3)***. It is present in all living cells. Avidin is a tetrameric glycoprotein of 66,000–68,000 mol wt, found in egg albumin and in avian tissues. The interaction between avidin and biotin occurs rapidly, and the stability of the complex has prompted its use for *in situ* attachment of labels in a broad variety of applications, including immunoassays, DNA hybridization ***(4–6)***, and localization of antigens in cells and tissues ***(7)***. Avidin has an isoelectric point of 10.5. Because of its positively charged residues and its oligosaccharide component, consisting mostly of mannose and glucosamine ***(8)***, avidin can interact nonspecifically with negative charges on cell surfaces and nucleic acids, or with membrane sugar receptors. At times, this causes background problems in histochemical and cytochemical applications. Streptavidin, a near-neutral, biotin binding protein ***(9)*** isolated from the culture medium of *Streptomyces avidinii,* is a tetrameric nonglycosylated analog of avidin with a mol wt of about 60,000. Like avidin, each molecule of streptavidin binds four molecules of biotin, with a similar dissociation constant. The two proteins have about 33% sequence homology, and tryptophan residues seem to be involved in their biotin binding sites ***(10,11)***. In general, streptavidin gives less background problems than avidin. This protein, however, contains a tripeptide sequence Arg-Tyr-Asp (RYD) that apparently mimics the binding sequence of fibronectin Arg-Gly-Asp (RGD), a universal recognition domain of the extracellular matrix that specifically promotes cell adhesion. Consequently, the streptavidin–cell-surface interaction causes high background in certain applications ***(12)***.

As an alternative to both avidin and streptavidin, a chemically modified avidin, NeutrAvidin™ (NeutrAvidin™ is a trademark of Pierce Chemical Company, Rockford, IL) and its conjugates with enzymes or fluorescent probes are available from both Molecular Probes and Pierce. NeutrAvidin™ consists of chemically deglycosylated avidin, which has been modified to reduce the isoelectric point to a neutral value, without loss of its biotin binding properties and without significant change in the lysines avail-

From: *The Protein Protocols Handbook, 2nd Edition*
Edited by: J. M. Walker © Humana Press Inc., Totowa, NJ

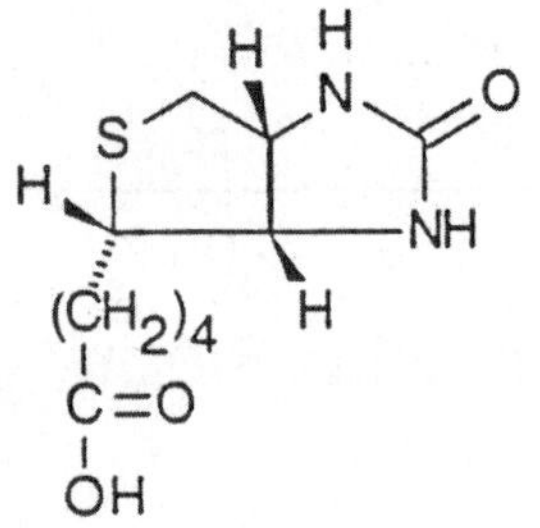

Biotin MW 244.31

Fig. 1. Structure of biotin.

able for derivatization *(13)*. (Fluorescent derivatives and enzyme conjugates of NeutraLite avidin, as well as the unlabeled protein, are available from Molecular Probes [Eugene, OR].)

As shown in **Fig. 1**, biotin is a relatively small and hydrophobic molecule. The addition to the carboxyl group of biotin of one (*X*) or two (*XX*) aminohexanoic acid "spacers" greatly enhances the efficiency of formation of the complex between the biotinylated antibody (or other biotinylated protein) and the avidin–probe conjugate, where the probe can be a fluorochrome or an enzyme *(14,15)*. Each of these 7- or 14-atom spacer arms has been shown to improve the ability of biotin derivatives to interact with the binding cleft of avidin. The comparison between streptavidin binding activity of proteins biotinylated with biotin-*X* or biotin-*XX* (labeled with same number of moles of biotin/mol of protein) has been performed in our laboratory (**Fig. 2**). No difference was found between the avidin or streptavidin–horseradish peroxidase conjugates in their ability to bind biotin-*X* or biotin-*XX*. However, biotin-*XX* gave consistently higher titers in enzyme-linked immunosorbent (ELISA) assays, using biotinylated goat antimouse IgG (GAM), bovine serum albumin (BSA), or protein A (results with avidin and with protein A are not presented here). Even nonroutine conjugations performed in our laboratory have consistently yielded excellent results using biotin-*XX*.

Biotin, biotin-*X,* and biotin-*XX* have all been derivatized for conjugation to amines or thiols of proteins and aldehyde groups of glycoproteins or other polymers. The simplest and most popular biotinylation method is to label the ε-amino groups of lysine residues with a succinimidyl ester of biotin. Easy-to-use biotinylation kits are commercially available that facilitate the biotinylation of 1–2 mg of protein or oligonucleotides *(16)*. One kit for biotinylating smaller amounts of protein (0.1–3 mg) utilizes biotin-*XX* sulfosuccinimidyl ester *(17)*. This compound is water-soluble and allows for the efficient labeling of dilute protein samples. Another kit uses biotin-*X* 2,4-dinitrophenyl-X-lysine succinimidyl ester (DNP-biocytin) as the biotinylating reagent. DNP-biocytin was developed by Molecular Devices (Menlo Park, CA) for their patented Threshold-Immunoligand System *(18)*. DNP-biocytin permits the direct measurement of the degree of biotinylation of the reaction product by using the molar extinction coefficient of DNP (15,000 M^{-1} cm^{-1} at 364 nm). Conjugates of DNP-biocytin can be probed separately or simultaneously using either anti-DNP antibodies or avidin/streptavidin; this flexibility is useful when combining techniques such as fluorescence and electron

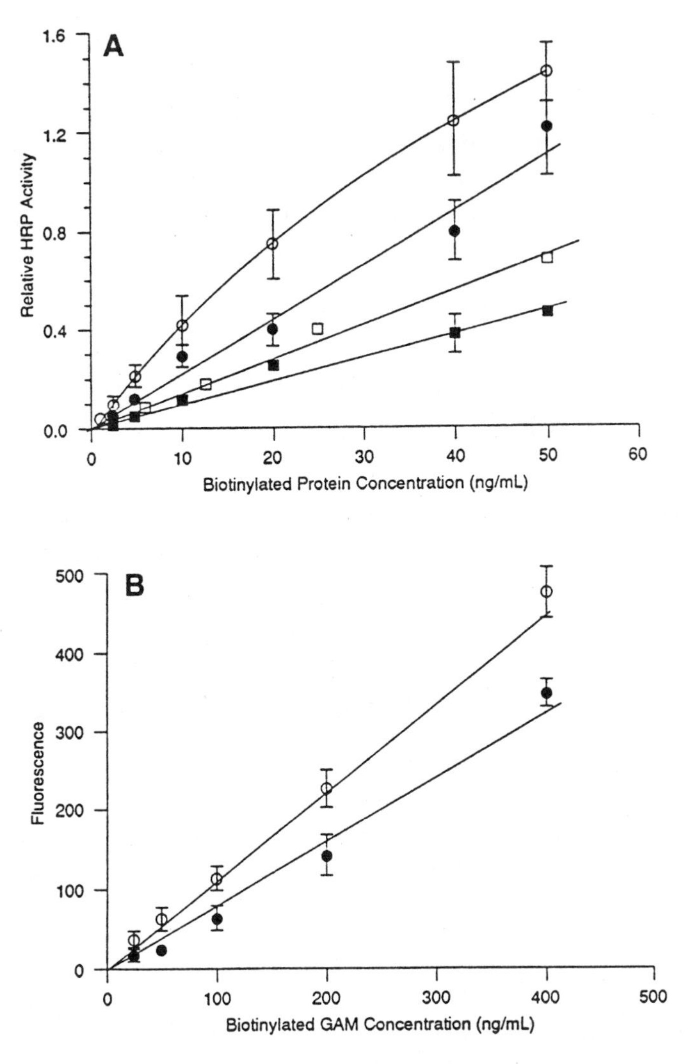

Fig. 2. **(A)** ELISA-type assay comparing the binding capacity of BSA and GAM biotinylated with biotin-*X* or biotin-*XX*. The assay was developed using streptavidin-HRP conjugate (0.2 µg/mL) and *o*-phenylenediamine dihydro-chloride (OPD). The number of biotin/mol was: 4.0 biotin-*X*/GAM (●), 4.4 biotin-*XX*/GAM (○), 6.7 biotin-*X*/BSA (■), and 6.2 biotin-*XX*/BSA (□). Error bars on some data points have been omitted for clarity. **(B)** Similar assay using GAM biotinylated with biotin-*X* (●) or biotin-*XX* (○). The assay was developed with streptavidin–R-phycoerythin conjugate (25 µg/mL using a Millipore CytoFluor™ fluorescence microtiter plate reader).

microscopy. Biotin iodoacetamide or maleimide, which could biotinylate the reduced sulfhydryls located at the hinge region of antibodies, is not usually used for this purpose. More examples in the literature describe biotinylation of antibodies with biotin hydrazide at the carbohydrate prosthetic group, located in the Fc portion of the molecule, relatively removed from the binding site. Conjugation of carbohydrates with hydrazides requires the oxidation of two adjacent hydroxyls to aldehydes and optional stabilization of the reaction with cyanoborohydride ***(19)***.

Because of its strength, the interaction between avidin and biotin cannot be used for preparing matrices for affinity column purification, unless columns prepared with avidin monomers are used *(20)*. The biotin analog, iminobiotin, which has a lower affinity for avidin, can be used for this purpose *(21,22)*. Iminobiotin in reactive form is commercially available, and the procedure for its conjugation is identical to that used for biotin. Detailed, practical protocols for biotinylating antibodies at the lysine or at the carbohydrate site, and a method to determine the degree of biotinylation are described in detail in this chapter (*see* **Notes 1–10** for review of factors that affect optimal conjugation and yield of biotinylated antibodies).

2. Materials

2.1. Conjugation with Amine-Reactive Biotin

1. Reaction buffer: 1 *M* sodium bicarbonate, stable for about 2 wk when refrigerated. Dissolve 8.3 g of $NaHCO_3$ in 100 mL of distilled water. The pH will be about 8.3. Dilute 1:10 before using to obtain a 0.1 *M* solution. Alternate reaction buffer: 0.1 *M* sodium phosphate, pH 7.8. Dissolve 12.7 g Na_2HPO_4 and 1.43 g NaH_2PO_4 in 800 mL of distilled water. Adjust pH to 7.8 if necessary. Bring the volume to 1000 mL. This buffer is stable for 2 mo when refrigerated.
2. Anhydrous dimethylformamide (DMF) or dimethyl sulfoxide (DMSO).
3. Phosphate-buffered saline (PBS): Dissolve 1.19 g of K_2HPO_4, 0.43 g of $KH_2PO_4 \cdot H_2O$ and 8.8 g NaCl in 800 mL of distilled water, adjust the pH to 7.2 if necessary or to the desired pH, and bring the volume to 1000 mL with distilled water.
4. Disposable desalting columns or a gel-filtration column: Amicon GH-25 and Sephadex G-25 or the equivalent, equilibrated with PBS or buffer of choice.
5. Good-quality dialysis tubing as an alternative to the gel-filtration column when derivatizing small quantities of antibody.
6. Biotin, biotin-*X* or biotin-*XX* succinimidyl ester: As with all succinimidyl esters, these compounds should be stored well desiccated in the freezer.

2.2. Conjugation with Biotin Hydrazide at the Carbohydrate Site

1. Reaction buffer: 0.1 *M* acetate buffer, pH 6.0. Dilute 5.8 mL acetic acid in 800 mL distilled water. Bring the pH to 6.0 with 5 *M* NaOH and the volume to 1000 mL. The buffer is stable for several months when refrigerated.
2. 20 m*M* Sodium metaperiodate: Dissolve 43 mg of $NaIO_4$ in 10 mL of reaction buffer, protecting from light. Use fresh.
3. Biotin-*X* hydrazide or biotin-*XX* hydrazide.
4. DMSO.
5. Optional: 100 m*M* sodium cyanoborohydride, freshly prepared. Dissolve 6.3 mg of $NaBH_3CN$ in 10 mL of 0.1 m*M* NaOH.

2.3. Determination of the Degree of Biotinylation

1. 10 m*M* 4'-Hydroxyazobenzene-2-carboxylic acid (HABA) in 10 m*M* NaOH.
2. 50 m*M* Sodium phosphate and 150 m*M* NaCl, pH 6.0. Dissolve 0.85 g of Na_2HPO_4 and 6.07 g of NaH_2PO_4 in 800 mL of distilled water. Add 88 g of NaCl. Bring the pH to 6.0 if necessary and the volume to 1000 mL.
3. 0.5 mg/mL Avidin in 50 m*M* sodium phosphate and 150 m*M* NaCl, pH 6.0.
4. 0.25 m*M* Biotin in 50 m*M* sodium phosphate, and 150 m*M* NaCl, pH 6.0.

3. Methods

3.1. Conjugation with Amine-Reactive Biotin

1. Calculate the amount of a 10 mg/mL biotin succinimidyl ester solution (biotin-NHS) needed to conjugate the desired quantity of antibody at the chosen biotin/antibody molar ratio, according to the following formula:

$$\text{(mL of 10 mg/mL biotin-SE)} = \{[(\text{mg antibody} \times 0.1)/\text{mol wt of antibody}] \times R \times \text{mol wt of biotin-SE})\} \quad (1)$$

where R = molar incubation ratio of biotin/protein. For example, using 5 mg of IgG and a 10:1 molar incubation ratio of biotin-*XX*-SE, **Eq. (1)** yields:

$$\text{(mL of 10 mg/mL biotin-}XX\text{-SE)} = \{[(5 \times 0.1)/145{,}000] \times (10 \times 568)\} = 0.02 \text{ mL} \quad (2)$$

2. Dissolve the antibody, if lyophilized, at approx 5–15 mg/mL in either of the two reaction buffers described in **Subheading 2.1.** If the antibody to be conjugated is already in solution in 10–20 m*M* PBS, without azide, the pH necessary for the reaction can be obtained by adding 1/10 vol of 1 *M* sodium bicarbonate. IgM should be conjugated in PBS, pH 7.2 (*see* **Note 3**).
3. Weigh 3 mg or more of the biotin-SE of choice, and dissolve it in 0.3 mL or more of DMF or DMSO to obtain a 10 mg/mL solution. **It is essential that this solution be prepared immediately before starting the reaction,** since the succinimidyl esters or any amine-reactive reagents hydrolyze quickly in solution. Any remaining solution should be discarded.
4. While stirring, slowly add the amount of 10 mg/mL solution, calculated in **step 1**, to the antibody prepared in **step 2**, mixing thoroughly.
5. Incubate this reaction mixture at room temperature for 1 h with gentle stirring or shaking.
6. The antibody conjugate can be purified on a gel-filtration column or by dialysis. When working with a few milligrams of dilute antibody solution, care should be taken not to dilute the antibody further. In this case, dialysis is a very simple and effective method to eliminate unreacted biotin. A few mL of antibody solution can be effectively dialyzed in the cold against 1 L of buffer with three to four changes. Small amounts of concentrated antibody can be purified on a prepackaged desalting column equilibrated with the preferred buffer, following the manufacturer's directions. Five or more milligrams of antibody can be purified on a gel-filtration column. The dimensions of the column will have to be proportional to the volume and concentration of the antibody. For example, for 5–10 mg of antibody in 1 mL solution, a column with a bed vol of 10 × 300 mm will be adequate. To avoid denaturation, dilute solutions of biotinylated antibodies should be stabilized by adding BSA at a final concentration of 0.1–1%.

3.2. Conjugation with Biotin Hydrazide at the Carbohydrate Site

1. It is essential that the entire following procedure be carried out with the sample completely protected from light (*see* **Note 9**).
2. Dissolve antibody (if lyophilized) or dialyze solution of antibody to obtain a 2–10 mg/mL solution in the reaction buffer described in **Subheading 2.1., item 1**. Keep at 4°C.
3. Add an equal volume of cold metaperiodate solution. Incubate the reaction mixture at 4°C for 2 h in the dark.
4. Dialyze overnight against the same buffer protecting from light, or, if the antibody is concentrated, desalt on a column equilibrated with the same buffer. This step removes the iodate and formaldehyde produced during oxidation.

5. Dissolve 10 mg of the biotin hydrazide of choice in 0.25 mL of DMSO to obtain a 40 mg/mL solution, warming if needed. This will yield a 107 m*M* solution of biotin-*X* hydrazide or an 80 m*M* solution of biotin-*XX* hydrazide. These solutions are stable for a few weeks.
6. Calculate the amount of biotin hydrazide solution needed to obtain a final concentration of approx 5 m*M,* and add it to the oxidized antibody. When using biotin-*X* hydrazide, 1 vol of hydrazide should be added to 20 vol of antibody solution. When using biotin-*XX* hydrazide, 1 vol of hydrazide should be added to 15 vol of antibody solution.
7. Incubate for 2 h at room temperature with gentle stirring.
8. This step is optional. The biotin hydrazone–antibody conjugate formed in this reaction (**steps 6** and **7**) is considered by some researchers to be relatively unstable. To reduce the conjugate to a more stable, substituted hydrazide, treat the conjugate with sodium cyanoborohydride at a final concentration of 5 m*M* by adding a 1/20 vol of a 100-m*M* stock solution. Incubate for 2 h at 4°C (*see* **Note 5**).
9. Purify the conjugate by any of the methods described for biotinylating antibodies at the amine site (*see* **Subheading 3.1., step 6**).

3.3. Determination of the Degree of Biotinylation

The dye HABA interacts with avidin yielding a complex with an absorption maximum at 500 nm. Biotin, because of its higher affinity, displaces HABA, causing a decrease in absorbance at 500 nm proportional to the amount of biotin present in the assay.

1. To prepare a standard curve, add 0.25 mL of HABA reagent to 10 mL of avidin solution. Incubate 10 min at room temperature and record the absorbance at 500 nm of 1 mL avidin–HABA complex with 0.1 mL buffer, pH 6.0. Distribute 1 mL of the avidin–HABA complex into six test tubes. Add to each the biotin solution in a range of 0.005–0.10 mL. Bring the final volume to 1.10 mL with pH 6.0 buffer, and record the absorbance at 500 nm of each concentration point. Plot a standard curve with the nanomoles of biotin vs the decrease in absorbance at 500 nm. An example of a standard curve is illustrated in **Fig. 3**.
2. To measure the degree of biotinylation of the sample, add an aliquot of biotinylated antibody of known concentration to 1 mL of avidin–HABA complex. For example, add 0.05–0.1 mL of biotinylated antibody at 1 mg/mL to 1 mL of avidin–HABA mixture. Bring the volume to 1.10 mL, if necessary, incubate for 10 min, and measure the decrease in absorbance at 500 nm.
3. Deduct from the standard curve the nanomoles of biotin corresponding to the observed change in absorbance. The ratio between nanomoles of biotin and nanomoles of antibody used to displace HABA represents the degree of biotinylation, as seen from the following equation:

$$[(\text{nmol biotin} \times 145{,}000 \times 10^{-6})/(\text{mg/mL antibody} \times 0.1\text{ mL})] = (\text{mol of biotin/mol of antibody}) \quad (3)$$

where 145,000 represents the mol wt of the antibody and 0.1 mL is the volume of 1 mg/mL of biotinylated antibody sample.

4. Notes

4.1. Factors that Influence the Biotinylation Reaction

1. Protein concentration: As in any chemical reaction, the concentration of the reagents is a major factor in determining the rate and the efficiency of the coupling. Antibodies at a concentration of 5–20 mg/mL will give better results; however, it is often difficult to have such concentrations or even such quantities available for conjugation. Nevertheless, the

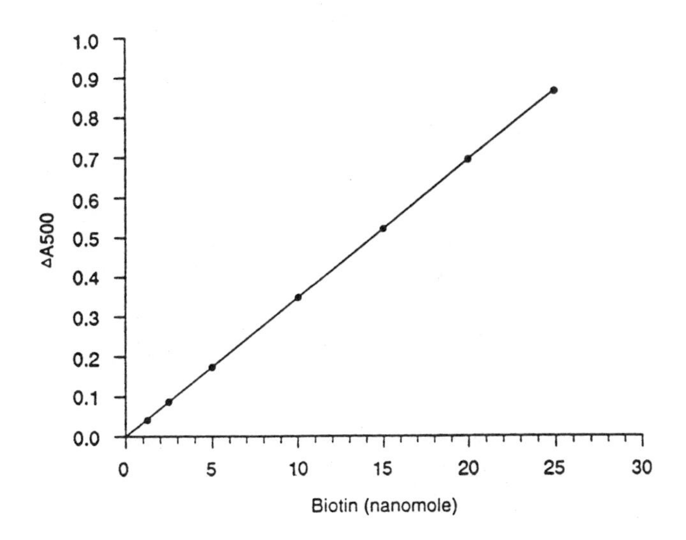

Fig. 3. Examples of standard curve for biotin assay with avidin-HABA reagent, obtained as described in **Subheading 3.3.**

antibody should be as concentrated as possible. In the case of solutions of antibody <2–3 mg/mL, the molar ratio of biotinylating reagent (or of both the oxidizing and biotinylating reagent, in the case of labeling the carbohydrate region) should be increased. It is also essential that the antibody solutions do not contain gelatin or BSA, which are often added to stabilize dilute solutions of antibodies. These proteins, generally present at a 1% concentration, will also react with biotinylating reagents.

2. pH: The reactivity of amines increases at basic pH. Unfortunately, so does the rate of hydrolysis of succinimidyl esters. We have found that the best pH for biotinylation of the ε-amino groups of lysines is 7.5–8.3. IgM antibodies, which denature at basic pH, can be biotinylated at pH 7.2 by increasing the molar ratio of the biotinylating reagent to antibody to at least 20. The optimum pH for oxidation and conjugation with hydrazides is 5.5–6.0.
3. Buffer: Bicarbonate or phosphate buffers are suitable for biotinylation. Organic buffers, such as Tris which contain amines, should be avoided, because they react with amino-labeling reagents or interfere with the reaction between aldehydes and hydrazides. However, HEPES and EPPS, which contain tertiary amines, are suitable. Antibodies dissolved in 10–20 m*M* PBS can be readily prepared for conjugation at the lysine site by adding 1/10–1/5 of the volume of 1 *M* sodium bicarbonate. As noted, because IgM antibodies are unstable in basic solution, biotinylation at the ε-amino group of lysines should be attempted in PBS or equivalent buffer at pH 7.2. Reactions of antibodies with periodate and biotin hydrazide can be performed in PBS at pH 7 or in acetate buffer, pH 6.0 (*see* **Subheading 2.2.**).
4. Temperature: Biotinylations at the amino group sites are run at room temperature, at the carbohydrate site at 0–4°C.
5. Time: Succinimidyl ester derivatives will react with a protein within 1 h. Periodate oxidation will require 2 h at pH 6.0. Reaction with biotin hydrazide can be performed in a few hours. Stabilization with cyanoboro-hydride requires <2 h.
6. Desired degree of biotinylation and stability of the conjugate: Reaction of an antibody with biotin does not significantly alter the size or charge of the molecule. However, because of the size of avidin or its analogs (mol wt = 60,000–68,000), an increase in the number of biotins per antibody will not necessarily increase the number of avidins capable of react-

ing with one antibody molecule. Because biotin, biotin-*X*, and biotin-*XX* are very hydrophobic molecules, a high degree of biotinylation might increase the background or might destabilize the antibody. To obtain a degree of biotinylation of about 3–7 biotins/IgG, generally a molar ratio of 15 mol of amino biotinylating reagent/mol of protein is used. When the concentration of the antibody is <3 mg/mL, this ratio should be increased. The amount of increase should be determined experimentally, because the reactivity of the lysines available for conjugation varies for each antibody (Ab). This could become a significant factor, especially at low antibody concentrations.

The succinimidyl esters or hydrazides of biotin, biotin-*X*, and biotin-*XX* exhibit similar degrees of reactivity, and the choice is up to the researcher. In general, the longer spacer arm in biotin-*XX* should be advantageous (**Fig. 2**). The overall stability of biotinylated MAbs derivatized with a moderate number of biotin should be similar to the stability of the native antibody, and the storage conditions also should be the same.

4.2. Factors that Affect Antibodies

7. Most Abs can withstand biotinylation with minimal change in activity and stability, especially if the degree of biotinylation is about 3–6 biotins/mol.
8. Biotin or any of its longer chain derivatives do not contribute to the absorbance of the antibody at 280 nm. Consequently, the concentration of the antibody can be measured by using $A^{1\%}_{1\,cm} = 14$ at 280 nm.
9. It is essential that the entire procedure for biotinylation of antibodies at the carbohydrate site (**Subheading 3.2.**) be performed in the dark, protected from light.
10. It should be noted that dry milk, serum, and other biological fluids contain biotin and, consequently, they should not be used as blocking agents in systems where blocking is required.

References

1. Green, N. M. (1963) Avidin. 3. The nature of the biotin binding site. *Biochem. J.* **89,** 599–609.
2. Green, N. M. (1963) Avidin. 4. Stability at extremes of pH and dissociation into subunits by guanidine hydrochloride. *Biochem. J.* **89,** 609–620.
3. Knappe, J. (1970) Mechanism of biotin action. *Annu. Rev. Biochem.* **39,** 757–776.
4. Wilchek, M. and Bayer, E. A. (1988) The avidin–biotin complex in bioanalytical applications. *Analyt. Biochem.* **171,** 1–32.
5. Wilchek, M. and Bayer, E. A. (1990) Avidin–biotin technology, in *Methods in Enzymology,* vol. 184, Academic, New York, pp. 213–217.
6. Levi, M., Sparvoli, E., Sgorbati, S., and Chiantante, D. (1990) Biotin–streptavidin immunofluorescent detection of DNA replication in root meristems through Brd Urd incorporation: cytological and microfluorimetric applications. *Physiol. Plantarum* **79,** 231–235.
7. Armstrong, R., Friedrich, V. L., Jr., Holmes, K. V., and Dubois-Dalcq, M. (1990) *In vitro* analysis of the oligodendrocyte lineage in mice during demyelination and remyelination. *J. Cell Biol.* **111,** 1183–1195.
8. Bruch, R. C. and White, H. B. III (1982) Compositional and structural heterogeneity of avidin glycopeptides. *Biochemistry* **21,** 5334–5341.
9. Hiller, Y., Gershoni, J. M., Bayer, E. A., and Wilchek, M. (1987) Biotin binding to avidin: oligosaccharide side chain not required for ligand association. *Biochem. J.* **248,** 167–171.
10. Green, N. M. (1975) Avidin, in *Advances in Protein Chemistry,* vol. 29 (Anfinsen, C. B., Edsall, J. T., and Richards, F. M., eds.), Academic, New York, pp. 85–133.
11. Chaiet, L. and Wolf, F. J. (1964) The properties of streptavidin, a biotin-binding protein produced by *Streptomycetes. Arch. Biochem. Biophys.* **106,** 1–5.

12. Alon, R., Bayer, E. A., and Wilcheck, M. (1990) Streptavidin contains an Ryd sequence which mimics the RGD receptor domain of fibronectin. *Biochem. Biophys. Res. Commun.* **170,** 1236–1241.
13. Wilchek, M. and Bayer, E. A. (1993) Avidin–biotin immobilization systems, in *Immobilized Macromolecules: Application Potentials* (Sleytr, U. B., ed.), Springer-Verlag, New York, pp 51–60.
14. Gretch, D. R., Suter, M., and Stinski, M. F. (1987) The use of biotinylated monoclonal antibodies and streptavidin affinity chomatography to isolate Herpes virus hydrophobic proteins or glycoproteins. *Analyt. Biochem.* **163,** 270–277.
15. Hnatowich, D. J., Virzi, F., and Rusckowski, M. (1987) Investigations of avidin and biotin for imaging applications. *J. Nucl. Med.* **28,** 1294–1302.
16. Haugland, R. P. (1996) Biotin derivatives, in *Handbook of Fluorescent Probes and Research Chemicals,* 6th ed. (Spence, M., ed.), Molecular Probes, Inc., Eugene, OR, Chapter 4.
17. LaRochelle, W. J. and Froehner, S. C. (1986) Determination of the tissue distributions and relative concentrations of the postsynaptic 43-kDa protein and the acetylcholine receptor in *Torpedo. J. Biol. Chem.* **261,** 5270–5274.
18. Briggs, J. and Panfili, P. R. (1991) Quantitation of DNA and protein impurities in biopharmaceuticals. *Analyt. Chem.* **63,** 850–859.
19. Wong, S. S. (1991) Reactive groups of proteins and their modifying agents, in *Chemistry of Protein Conjugation and Crosslinking,* CRC, Boston, MA, pp. 27–29.
20. Kohanski, R. A. and Lane, M. D. (1985) Receptor affinity chomatography. *Ann. NY Acad. Sci.* **447,** 373–385.
21. Orr, G. A. (1981) The use of the 2-iminobiotin-avidin interaction for the selective retrieval of labeled plasma membrane components. *J. Biol. Chem.* **256,** 761–766.
22. Hoffmann, K., Wood, S. W., Brinton, C. C., Montibeller, J. A., and Finn, F. M. (1980) Iminobiotin affinity columns and their application to retrieval of streptavidin. *Proc. Natl. Acad. Sci.* **77,** 4666–4668.

49

Preparation of Avidin Conjugates

Rosaria P. Haugland and Mahesh K. Bhalgat

1. Introduction

The high-affinity avidin–biotin system has found applications in different fields of biotechnology including immunoassays, histochemistry, affinity chromatography, and drug delivery, to name a few. A brief description of avidin and avidin-like molecules, streptavidin, deglycosylated avidin, and NeutrAvidin™ avidin, is presented in Chapter 48. With four biotin binding sites per molecule, the avidin family of proteins is capable of forming tight complexes with one or more biotinylated compounds *(1)*. Typically, the avidin–biotin system is used to prepare signal-amplifying "sandwich" complexes between specificity reagents (e.g., antibodies) and detection reagents (e.g., fluorophores, enzymes, and so on). The specificity and detection reagents are independently conjugated, one with avidin and the other with biotin, or both with biotin, providing synthetic flexibility *(2)*.

Avidin conjugates of a wide range of fluorophores, phycobiliproteins, secondary antibodies, microspheres, ferritin, and enzymes commonly used in immunochemistry are available at reasonable prices, making their small scale preparation impractical and not cost effective (*see* **Note 1**). However, conjugations of avidin to specific antibodies, to uncommon enzymes, and to other proteins and peptides are often performed onsite. A general protocol for the conjugation of avidin to enzymes, antibodies, and other proteins is described in this chapter.

Avidin conjugates of oligodeoxynucleotides are hybrid molecules that not only provide multiple biotin binding sites, but can also be targeted to complementary DNA or RNA sequences, by annealing interactions. Such conjugates are useful for the construction of macromolecular assemblies with a wide variety of constituents *(3)*. The protocol outlined in **Subheading 3.1.** can be modified (*see* **Note 2**) for the conjugation of oligonucleotides to avidin.

Streptavidin conjugates are also being evaluated for use in drug delivery systems. A two-step imaging and treatment protocol has been developed that involves injection of a suitably prepared tumor-specific monoclonal antibody, followed by a second reagent that carries an imaging or therapeutic agent, capable of binding to the tumor-targeted antibody *(4)*. Owing to complications associated with the injection of radiolabeled biotin *(5)*, conjugation of the imaging or therapeutic agent to streptavidin is being con-

From: *The Protein Protocols Handbook, 2nd Edition*
Edited by: J. M. Walker © Humana Press Inc., Totowa, NJ

sidered, instead. A protocol for radioiodination of streptavidin using IODO-BEADS® *(6)* is described in **Subheading 3.2.** Some other methods that have been developed include the iodogen method *(7,8)*, *see also* Chapter 115, the Bolton–Hunter reagent method ***(9)***, and a few that do not involve direct iodination of tyrosine residues ***(10–13)***. Streptavidin–drug conjugates are also candidates for therapeutic agents. Synthesis of a streptavidin-drug conjugate involves making a chemically reactive form of the drug followed by its conjugation to streptavidin. The synthetic methodology thus depends on the structure of the specific drug to be conjugated ***(14–16)***.

The avidin-biotin interaction can also be exploited for affinity chromatography; however, there are limitations to this application, because a biotinylated protein captured on an avidin affinity matrix would likely be denatured by the severe conditions required to separate the high-affinity avidin–biotin complex. On the other hand, an avidin affinity matrix may find utility in the removal of undesired biotinylated moieties from a mixture or for the purification of compounds derivatized with 2-iminobiotin. The biotin derivative 2-iminobiotin has reduced affinity for avidin, and its moderate binding to avidin at pH 9.0 is greatly diminished at pH 4.5 ***(17)***. Another approach to reducing the affinity of the interaction is to denature avidin to its monomeric subunits. The monomeric subunits have greatly reduced affinity for biotin ***(18)***. We describe here a protocol for preparing native ***(19)*** and monomeric avidin matrices ***(20)***. Modified streptavidins, hybrids of native and engineered subunits with lower binding constants, have been prepared that may also be suitable for affinity matrices ***(21)***. A new form of monovalent avidin, nitrated at the tyrosine located at three of the four biotin binding sites, has recently been available from Molecular Probes. The binding affinity of this modified avidin ***(22)***, called CaptAvidin™ biotin-binding protein, is lower than for the native protein. At pH 4 CaptAvidin biotin-binding protein associates with biotin with a K_a of ~10^9 M^{-1}; if needed, the complex can be completely dissociated at pH 10.0. This property makes CaptAvidin biotin-binding protein the ideal ligand for affinity matrices suitable for isolation and purification of biotinylated compounds. A nitration protocol is described in Chapter 64.

2. Materials

2.1. Conjugation to Antibodies, Enzymes or Oligonucleotides

1. Avidin (mol wt = 66,000).
2. Antibody, enzyme, peptide, protein, or thiolated oligonucleotide to be conjugated to avidin.
3. Succinimidyl 3-(2-pyridyldithio)propionate (SPDP; mol wt = 312.36) (*see* **Note 3**).
4. Succinimidyl *trans*-4-(*N*-maleimidylmethyl)cyclohexane-1-carboxylate (SMCC; mol wt = 334.33).
5. Dithiothreitol (DTT; mol wt = 154.24).
6. Tris-(2-carboxyethyl) phosphine (TCEP; mol wt = 286.7).
7. *N*-ethylmaleimide (NEM; mol wt = 125.13)
8. Anhydrous dimethyl sulfoxide (DMSO) or anhydrous dimethylformamide (DMF).
9. 0.1 *M* Phosphate buffer: Contains 0.1 *M* sodium phosphate, 0.1 *M* NaCl at pH 7.5. Dissolve 92 g of Na_2HPO_4, 21 g of $NaH_2PO_4 \cdot H_2O$, and 46.7 g of NaCl in approx 3.5 L of distilled water and adjust the pH to 7.5 with 5 *M* NaOH. Dilute to 8 L. Store refrigerated.
10. 1 *M* Sodium bicarbonate (*see* **Note 4**). Dissolve 8.4 g in 90 mL of distilled water and adjust the volume to 100 mL. A freshly prepared solution has a pH of 8.3–8.5.

11. Molecular exclusion matrix with properties suitable for purification of the specific conjugate. Sephadex G-200 (Pharmacia Biotech, Uppsala, Sweden), Bio-Gel® A-0.5m or Bio-Gel® A-1.5 m (Bio-Rad Laboratories, Hercules, CA) are useful for relatively small to large conjugates, respectively.
12. Sephadex G-25 (Pharmacia Biotech) or other equivalent matrix

2.2. Radioiodination Using IODO-BEADS

1. Streptavidin (mol wt = 60,000).
2. $Na^{131}I$ or $Na^{125}I$, as desired.
3. IODO-BEADS (Pierce Chemical, Rockford, IL).
4. Phosphate-buffered saline (PBS), pH 7.2 : Dissolve 1.19 g of K_2HPO_4, 0.43 g of KH_2PO_4, and 9 g of NaCl in 900 mL of distilled water. Adjust the pH to 7.2 and dilute to 1 L with distilled water.
5. Saline solution: 9 g of NaCl dissolved in 1 L of distilled water.
6. 0.1% Bovine serum albumin (BSA) solution in saline: 0.1 g of bovine serum albumin dissolved in 100 mL of saline solution.
7. Trichloroacetic acid (TCA), 10% w/v solution in saline: Dissolve 1 g of TCA in 10 mL of saline solution.
8. Bio-Gel® P-6DG Gel (Bio-Rad Laboratories).

2.3. Avidin Affinity Matrix

1. 50–100 mg of avidin.
2. Sodium borohydride.
3. 1,4-Butanediol-diglycidyl ether.
4. Succinic anhydride.
5. 6 *M* Guanidine·HCl in 0.2 *M* KCl/HCl, pH 1.5: Dissolve 1.5 g of KCl in 50 mL of distilled water. Add 57.3 g of guanidine·HCl with stirring. Adjust the pH to 1.5 with 1 *M* HCl. Adjust the volume to 100 mL with distilled water.
6. 0.2 *M* Glycine·HCl pH 2.0: Dissolve 22.3 g of glycine·HCl in 900 mL of distilled water. Adjust the pH to 2.0 with 6*M* HCl and the volume to 1 L with distilled water.
7. PBS: *See* **Subheading 2.2.4.**
8. 0.2 *M* Sodium carbonate, pH 9.5: Dissolve 1.7 g of sodium bicarbonate in 80 mL of distilled water. Adjust the pH to 9.5 with 1 *M* NaOH and the volume to 100 mL with distilled water.
9. 0.2 *M* Sodium phosphate, pH 7.5: Weigh 12 g of Na_2HPO_4 and 2.5 g of $NaH_2PO_4 \cdot H_2O$ and dissolve in 900 mL of distilled water. Adjust the pH to 7.5 with 5 *M* NaOH and the volume to 1 L with distilled water.
10. 20 m*M* Sodium phosphate, 0.5 *M* NaCl, 0.02% sodium azide pH 7.5: Dilute 100 mL of the buffer described in **Subheading 2.3., item 9**) to 900 mL with distilled water. Add 28 g of NaCl and 200 mg of sodium azide. Adjust pH if necessary, and dilute to 1 L with distilled water.
11. Sepharose 6B (Pharmacia Biotech) or other 6% crosslinked agarose gel.

3. Methods

3.1. Conjugation to Antibodies or to Enzymes

3.1.1. Avidin Thiolation

An easy-to-use, protein-to-protein crosslinking kit is now commercially available (Molecular Probes, Eugene, OR). This kit allows predominantly 1:1 conjugate formation between two proteins (0.2–3 mg) through the formation of a stable thioether bond

(23), with minimal generation of aggregates. A similar protocol is described here for conjugation of 5 mg of avidin to antibodies or enzymes. Modifications of the procedure for conjugation of avidin to thiolated oligonucleotides and peptides are described in **Notes 2** and **5**, respectively. Although the protocol described in this section uses avidin for conjugation, it can be applied for the preparation of conjugates using either avidin, streptavidin, deglycosylated avidin, or NeutrAvidin avidin.

1. Dissolve 5 mg of avidin (76 nanomol) in 0.5 mL of 0.1 *M* phosphate buffer to obtain a concentration of 10 mg/mL.
2. Weigh 3 mg of SPDP and dissolve it in 0.3 mL of DMSO to obtain a 10 mg/mL solution. This solution must be prepared **fresh** immediately before using. Vortex-mix or sonicate to ensure that the reagent is completely dissolved.
3. Slowly add 12 µL (380 nanomoles) of the SPDP solution (*see* **Note 3**) to the stirred solution of avidin. Stir for 1 h at room temperature.
4. Purify the thiolated avidin on a 7 × 250 mm size exclusion column, such as Sephadex G-25 equilibrated in 0.1 *M* phosphate buffer.
5. Determine the degree of thiolation (optional):
   a. Prepare a 100 m*M* solution of DTT by dissolving 7.7 mg of the reagent in 0.5 mL of distilled water.
   b. Transfer the equivalent of 0.3–0.4 mg of thiolated avidin (absorbance at 280 nm of a 1.0 mg/mL avidin solution = 1.54) and dilute to 1.0 mL using 0.1*M* phosphate buffer. Record the absorbance at 280 nm and at 343 nm.
   c. Add 50 mL of DTT solution. Mix well, incubate for 3–5 min at room temperature and record the absorbance at 343 nm.
   d. Using the extinction coefficient at 343 nm of 8.08×10^3/cm/M *(24)*, calculate the amount of pyridine-2-thione liberated during the reduction, which is equivalent to the number of thiols introduced on avidin, using the following equation along with the appropriate extinction coefficient shown in **Table 1**:

$$\text{Number of thiols/avidin} = [\Delta A_{343}/(8.08 \times 10^3)] \times (A_{280}\text{–}0.63\Delta A_{343})] \quad (1)$$

   where ΔA_{343} = change in absorbance at 343 nm; $E^M{}_{avidin}$ = molar extinction coefficient; and $0.63\Delta A_{343}$ = correction for the absorbance of pyridyldithiopropionate at 280 nm *(24)*.
6. **Equation 1** allows the determination of the average number of moles of enzyme or antibody that can be conjugated with each mole of avidin (*see* **Note 6**). For a 1:1 protein-avidin conjugate, avidin should be modified with 1.2–1.5 thiols/mol. Thiolated avidin prepared by the above procedure can be stored in the presence of 2 m*M* sodium azide at 4°C for 4–6 wk.

3.1.2. Maleimide Derivatization of the Antibody or Enzyme

In this step, which should be completed prior to the deprotection of thiolated avidin, some of the amino groups from the antibody or enzyme are transformed into maleimide groups by reacting with the bifunctional crosslinker, SMCC. (*see* **Note 7**).

1. Dissolve or, if already in solution, dialyze the protein in 0.1 *M* phosphate buffer to obtain a concentration of 2–10 mg/mL. If the protein is an antibody, 11 mg are required to obtain an amount equimolar to 5 mg of avidin (*see* **Note 6**).
2. Prepare a **fresh** solution of SMCC by dissolving 5 mg in 0.5 mL of dry DMSO to obtain a 10 mg/mL solution. Vorte-mix or sonicate to assure that the reagent is completely dissolved.
3. While stirring, add an appropriate amount of SMCC solution to the protein solution to obtain a molar ratio of SMCC to protein of approx 10. (If 11 mg of an antibody is the protein used, 30 µL of SMCC solution is required.)

Table 1
Molar Extinction Coefficients at 280 nm and Molecular Weights of Avidin and Avidin-Like Proteins

Protein	Molecular weight	E^{M}_{Avidin}/cm/M
Avidin	66,000	101,640
Deglycosylated avidin/NeutrAvidin	60,000	101,640
CaptAvidin biotin-binding protein	66,000	118,800
Streptavidin	60,000	180,000

4. Continue stirring at room temperature for 1 h.
5. Dialyze the solution in 2 L of 0.1 *M* phosphate buffer at 4°C for 24 h, with four buffer changes using a membrane with a suitable molecular weight cutoff.

3.1.3. Deprotection of the Avidin Thiol Groups

This procedure is carried out immediately before reacting thiolated avidin with the maleimide derivative of the antibody or enzyme prepared in **Subheading 3.1.2.**

1. Dissolve 3 mg of TCEP in 0.3 mL of 0.1 *M* phosphate buffer.
2. Add 11 µL of TCEP solution to the thiolated avidin solution. Incubate for 15 min at room temperature.

3.1.4. Formation and Purification of the Conjugate

1. Add the thiolated avidin–TCEP mixture dropwise to the dialyzed maleimide derivatized protein solution with stirring. Continue stirring for 1 h at room temperature, followed by stirring overnight at 4°C.
2. Stop the conjugation reaction by capping residual sulfhydryls with the addition of NEM at a final concentration of 50 µ*M*. Dissolve 6 mg of NEM in 1 mL of DMSO and dilute 1:1000 in the conjugate reaction mixture. Incubate for 30 min at room temperature or overnight at 4°C (*see* **Note 8**). The conjugate is now ready for final purification.
3. Concentrate the avidin–protein conjugate mixture to 1–2 mL in a Centricon®-30 (Amicon, Beverly, MA) or equivalent centrifuge tube concentrator.
4. Pack appropriate size columns (e.g., 10 × 60 mm for approx 15 mg of final conjugate) with a degassed matrix suitable for the isolation of the conjugate from unconjugated reagents. If the protein conjugated is an antibody, a matrix such as Bio-Gel A-0.5 m is suitable. For other proteins, Sephadex G-200 or a similar column support may be appropriate, depending on the size of the protein–avidin conjugate.
5. Collect 0.5–1-mL fractions. The first protein peak to elute contains the conjugate, however the first or second fraction may contain some aggregates. Analyze each fraction absorbing at 280 nm for biotin binding and assay it for the antibody or enzyme activity. High-performance liquid chromatography (HPLC) may be also be performed for further purification, if necessary.

3.2. Radioiodination Using IODO-BEADS

The radioiodination procedure (*see* **Note 9**) described here uses IODO-BEADS, which contain the sodium salt of *N*-chlorobenzenesulfonamide immobilized on nonporous, polystyrene beads. Immobilization of the oxidizing agent allows for easy separation of the latter from the reaction mixture. This method also avoids the use of reducing agents.

1. Wash six to eight IODO-BEADS twice with 5 mL of PBS. Dry the beads by rolling them on a clean filter paper.
2. Add 500 µL of PBS to the supplier's vial containing 8–10 mCi of carrier-free $Na^{125}I$ or $Na^{131}I$. Place the beads in the same vial and gently mix the contents by swirling. Allow the mixture to sit for 5 min at room temperature with the vial capped.
3. Dissolve or dilute streptavidin in PBS to obtain a final concentration of 1 mg/mL. Add 500 µL of streptavidin solution to the vial containing sodium iodide. Cap the vial immediately and mix the contents thoroughly. Incubate for 20–25 min at room temperature, with occasional swirling (*see* **Note 10**).
4. Carefully remove and save the liquid from the reaction vessel; this is the radioiodinated streptavidin solution. Wash the beads by adding 500 µL of PBS to the reaction vial. Remove the wash solution and add it to the radioiodinated streptavidin.
5. For purification, load the reaction mixture onto a 9 × 200 mm Bio-Gel P-6DG column packed in PBS (0.1% BSA may be added as a carrier to the PBS to reduce loss of streptavidin by adsorption to the column). Elute the column with PBS and collect 0.5-mL fractions. The first set of radioactive fractions (as determined by counting in a γ-ray counter) contain radioiodinated streptavidin, while the unreacted radioiodine elutes in the later fractions. Pool the radioiodinated streptavidin fractions.
6. Assessment of protein-associated activity with TCA acid precipitation:
   a. Dilute a small volume of the pooled radiolabeled streptavidin with saline solution such that 50 µL of the diluted solution has 10^4–10^6 cpm.
   b. Add 50 µL of the diluted streptavidin solution to a 12 mm × 75 mm glass tube, followed by 500 µL of a 0.1% BSA solution in saline.
   c. For precipitating the proteins, add 500 µL of 10% (w/v) TCA solution in saline.
   d. Incubate the solution for 30 min at room temperature and count the radioactivity of the solution for 10 min ("Total Counts").
   e. Centrifuge the tube at 500 g for 10 min and carefully discard the supernatant in a radioactive waste container.
   f. Resuspend the pellet in 1 mL of saline and count its radioactivity for 10 min ("Bound Counts").
   g. The percentage of radioactivity bound to streptavidin is determined using the following equation:

$$[(\text{Bound counts})/(\text{Total counts})] \times 100 = \%\ \text{of radioactivity bound to streptavidin} \quad (2)$$

3.3. Avidin Affinity Matrices

3.3.1. CaptAvidin Biotin-Binding Protein Affinity Matrix

CaptAvidin agarose is available from Molecular Probes. For the preparation of this column avidin has been nitrated at the tyrosine sites involved with biotin binding. Nitration is performed to the extent that three of the four active sites of avidin are modified and loose their binding activity for biotin. The fourth site allows binding of biotin at pH 4 with a K_a of 10^9 M^{-1} as reported by Morag et al. *(22)*. This monovalent form of modified avidin had been covalently attached to agarose to generate an affinity matrix that does not need the harsh eluting conditions necessary for the avidin or even the iminobiotin affinity columns. The biotinylated compounds are easily dissociated from the CaptAvidin, biotin-binding protein matrix by elution at pH 10.0.

3.3.2. Native Avidin Affinity Matrix

1. Wash 10 mL of sedimented 6% crosslinked agarose with distilled water on a glass or Buchner filter and remove excess water by suction.

2. Dissolve 14 mg of $NaBH_4$ in 7 mL of 1 *M* NaOH. Add this solution along with 7 mL of 1,4-butanediol-diglycidyl ether to the washed agarose, with mixing. Allow the reaction to proceed for 10 h or more at room temperature with gentle stirring.
3. Extensively wash the activated gel with distilled water on a supporting filter. The washed gel can be stored in water at 4°C, for up to 10 d.
4. Dissolve 50–100 mg of avidin in 10–20 mL of 0.2 *M* sodium carbonate, pH 9.5, and suspend the sedimented activated agarose gel in the same buffer to obtain a workable slurry.
5. Slowly drip the agarose slurry into the stirred protein solution and allow the binding to take place at room temperature for 2 d with continuous gentle mixing.
6. Wash the avidin–agarose mixture in PBS until the filtrate shows no absorbance at 280 nm. Store at 4°C in the presence of 0.02% sodium azide.

3.3.3. Monomeric Avidin Affinity Matrix

1. Filter the avidin–agarose matrix (from **Subheading 3.3.1., step 6**) on a glass or Buchner filter (or pack in a column) and wash 4× with two volumes of 6 *M* guanidine·HCl in 0.2 *M* KCl, pH 1.5, to dissociate the tetrameric avidin.
2. Thoroughly wash the gel with 0.2 *M* potassium phosphate, pH 7.5, and suspend in 10 mL of the same buffer.
3. Add 3 mg of solid succinic anhydride to succinylate the monomeric avidin and incubate for 1 h at room temperature with gentle stirring.
4. Wash the gel with 0.2 *M* potassium phosphate, pH 7.5, pack in a column, and saturate the binding sites by running through three volumes of 1 m*M* biotin dissolved in the same buffer.
5. Remove biotin from the low-affinity binding sites by washing the column with 0.2 *M* glycine·HCl, pH 2.0.
6. Store the column equilibrated in 20 m*M* sodium phosphate, 0.5 *M* NaCl, 0.02% sodium azide, pH 7.5. The column is now ready to use.
7. Load the column with the mixture to be purified. Elute any unbound protein by adding 20 m*M* sodium phosphate, 0.5 *M* NaCl, pH 7.5. Add biotin to the same buffer to obtain a final concentration of 0.8 m*M* to elute the biotinylated compound.
8. Regenerate the column after each run by washing with 0.2 *M* glycine·HCl, pH 2.0.

4. Notes

1. A detailed procedure for the conjugation of fluorophores to antibodies has been recently published ***(25)***. This protocol can be modified for conjugation of fluorophores to avidin or avidin-related proteins by using a dye to avidin molar ratio of 5–8:1.
2. The conjugation reaction for oligonucleotides synthesized with a disulfide containing a protecting group, should be performed under nitrogen or argon. Deprotect the disulfide of the oligonucleotide using DTT. Add 1 mg of DTT to 140 µL of a 6 µ*M* oligonucleotide (21–33 mer) solution in 0.1 *M* phosphate buffer containing 5 m*M* ethylenediami–netetraacetic acid. Stir the solution at 37°C for 0.5 h. Purify the reaction mixture using a disposable desalting column. Combine the oligonucleotide-containing fractions with thiolated avidin prepared as described in **Subheading 3.1.1.** It should be noted that, in this case, conjugation occurs through the formation of a disulfide bond instead of a thioether bond. Disulfides are sensitive to reducing agents; however, they make reasonably stable conjugates, useful in most applications ***(26)***. Purify the conjugate as outlined in **Subheading 3.1.4.**
3. Using a molar ratio of SPDP to avidin of 5 yields one or two protected sulfhydryls per molecule of avidin. This range of thiols per mole is found to produce the best yield of a 1:1 conjugate.
4. Buffer and pH: The entire procedure for preparation of conjugates through thioether bonds can be performed at pH 7.5. (**Note:** Organic buffers containing amines, such as Tris, are unsuitable.) Antibodies or enzymes in PBS can be prepared for reaction with SMCC by

adding 1/10 volume of 1 *M* sodium bicarbonate solution. This step eliminates dialysis and consequent dilution of the protein. The presence of azide at concentrations above 0.1% may interfere with the reaction of the protein with SMCC or of avidin with SPDP. Some gM antibodies denature above pH 7.2. They can, however, be conjugated in PBS at pH 7.0 by increasing the molar ratio of maleimide to antibody.

5. Peptides (20–25 amino acids) containing a single cysteine can also be conjugated to thiolated avidin by modifying the procedure described in **Subheading 3.1.** and performing the reaction under argon or nitrogen *(26)*. Peptide–avidin conjugate formation described here also involves the formation of a disulfide bond. For conjugation with 5 mg of avidin, dissolve 1.6 mg of a lyophilized cysteine-containing peptide in 900 μL of water–methanol (2:1 v/v) using 50 m*M* NaOH (a few microliters at a time) to improve solubility. Immediately prior to use, cleave any cystine-bridged homodimer that may be present by the addition of TCEP solution (10 mg/mL in 0.1 *M* phosphate buffer) to obtain a TCEP to peptide ratio of 3. Incubate for 15 min at room temperature. Purify the peptide–TCEP mixture using a disposable desalting column. Combine the peptide-containing fractions with thiolated avidin prepared as described in **Subheading 3.1.1.** Purify the conjugate as described in **Subheading 3.1.4.**
6. Avidin and antibody or enzyme concentration: The concentration of avidin as well as that of the protein to be conjugated should be 2–10 mg/mL. The crosslinking efficiency and, consequently, the yield of the conjugate decreases at lower concentrations of the thiolated avidin and maleimide-derivatized protein. To obtain 1:1 conjugates, equimolar concentrations of avidin and the protein are desirable. However, most methods of conjug`ation will generate conjugates of different sizes, following the Poisson distribution. The size range obtained with the method described here is much narrower because the number of proteins reacting with each mole of avidin can be regulated by the degree of thiolation of avidin.
7. It is essential that the procedure described in **Subheading 3.1.2.** be performed approx 24 h before the procedure described in **Subheading 3.1.3.**, because the deprotected thiolated avidin and the maleimide derivative of the protein are unstable. Purification of the maleimide-derivatized protein by size exclusion chromatography can be performed more rapidly than dialysis; however, the former leads to dilution of the protein and a decrease in the yield of the conjugate.
8. If the molecule being conjugated to avidin is β-galactosidase or other free thiol-containing oligonucleotide or protein, NEM treatment is not performed.
9. Radioiodination of streptavidin uses procedures similar to those used for stable nuclides. However, some distinct differences remain, since radioiodinations are performed in dilute solutions. Also, the radioiodination mixture contains minor impurities formed during the preparation and purification of the radionuclide. Thus, optimization of reaction parameters is essential for performing radioiodination. This reaction is carried out in small volumes; it is therefore essential to ensure adequate mixing at the outset of the reaction. Inadequate mixing is often responsible for poor radioiodination yield.
10. Specific activity using the method described in **Subheading 3.2.** is usually in the range of 10–50 mCi/mg and the protein-bound radioactivity obtained is >95%. Higher specific activity can be achieved by increasing the reaction time of **step 3** in **Subheading 3.2.**, by using more beads or by increasing the amount of radioiodine. However, one must bear in mind that at longer incubation times, the risk of damage to streptavidin or avidin is greater.
11. Storage and stability of avidin conjugates: Most avidin conjugates can be stored at 4°C or –20°C after lyophilization. Because of the variation in antibody structure, there is no general rule on the best method to store avidin–antibody conjugates, and the best conditions are determined experimentally. Aliquoting in small amounts and freezing is generally

satisfactory. Radiolabeled streptavidin is aliquoted (~100 μL/tube) and stored at 4°C or –20°C until use.

References

1. Green, N. M. (1975) Avidin, in *Advances in Protein Chemistry,* Vol. 29 Anfinsen, C. M., Edsall, J. T., and Richards, F. M., eds.), Academic Press, New York, pp. 85–133.
2. Bayer, E. A. and Wilchek, M. (1980) The use of the avidin-biotin complex as a tool in molecular biology. *Meth. Biochem. Analyt.* **26,** 1–45.
3. Niemeyer, C. M., Sano, T., Smith, C. L., and Cantor, C. R. (1994) Oligonucleotide-) directed self-assembly of proteins: semisynthetic DNA-streptavidin hybrid molecules as connectors for the generation of macroscopic arrays and the construction of supramolecular bioconjugates. *Nucleic Acids Res.* **22,** 5330–5339.
4. Paganelli, G., Belloni, C., Magnani, P., Zito, F., Pasini, A., Sassi, I., et al. (1992) Two-step tumor targetting in ovarian cancer patients using biotinylated monoclonal antibodies and radioactive streptavidin. *Eur. J. Nucl. Med.* **19,** 322–329.
5. van Osdol, W. W., Sung, C., Dedrick, R. L., and Weinstein, J. N. (1993) A distributed pharmacokinetic model of two-step imaging and treatment protocols: application to streptavidin-conjugated monoclonal antibodies and radiolabeled biotin. *J. Nucl. Med.* **34,** 1552–1564.
6. Markwell, M. A. K. (1982) A new solid-state reagent to iodinate proteins. I. Conditions for the efficient labeling of antiserum. *Analyt. Biochem.* **125,** 427–432.
7. Salacinski, P. R. P., McLean, C., Sykes, J. E. C., Clement-Jones, V. V., and Lowry, P. J. (1981) Iodination of proteins, glycoproteins, and peptides using a solid-phase oxidizing agent, 1,3,4,6,-tetrachloro-3α,6α-diphenyl glycoluril (iodogen). *Analyt. Biochem.* **117,** 136–146.
8. Mock, D. M. (1990) Sequential solid-phase assay for biotin based on ^{125}I-labeled avidin in *Meth. Enzymol.* **184**, 224–233.
9. Bolton, A. E. and Hunter, W. M. (1973) The labeling of proteins to high specific radioactivity by conjugation to an ^{125}I-containing acylating agent. Applications to the radioimmunoassay. *Biochem. J.* **133,** 529–539.
10. Vaidyanathan, G., Affleck, D. J., and Zalutsky, M. R. (1993) Radioiodination of proteins using *N*-succinimidyl 4-hydroxy-3-iodobenzoate. *Bioconjugate Chem.* **4,** 78–84.
11. Vaidyanathan, G. and Zalutsky, M. R. (1990) Radioiodination of antibodies via *N*-succinimidyl 2,4-dimethoxy-3-(trialkylstannyl)benzoates. *Bioconjugate Chem.* **1,** 387–393.
12. Hylarides, M. D., Wilbur, D. S., Reed, M. W., Hadley, S. W., Schroeder, J. R. and Grant, L. M. (1991) Preparation and *in vivo* evaluation of an *N*-(*p*-[^{125}I]iodophenethyl)maleimide-antibody conjugate. *Bioconjugate Chem.* **2,** 435–440.
13. Arano, Y., Wakisaka, K., Ohmomo, Y., Uezono, T., Mukai, T., Motonari, H., Shiono, et al. (1994) Maleimidoethyl 3-(tri-*n*-butylstannyl)hippurate: A useful radioiodination reagent for protein radiopharmaceuticals to enhance target selective radioactivity localization. *J. Med. Chem.* **37,** 2609–2618.
14. Willner, D., Trail, P. A., Hofstead, S. J., Dalton King, H., Lasch, S. J., Braslawsky, G. R., et al. (1993) (6-Maleimidocaproyl)hydrazone of doxorubicin – a new derivative for the preparation of immunoconjugates of doxorubicin. *Bioconjugate Chem.* **4,** 521–527.
15. Arnold, Jr., L. J. (1985) Polylysine-drug conjugates. *Meth. Enzymol.* **112,** 270–285.
16. Pietersz, G. A. and McKenzie, I. F. (1992) Antibody conjugates for the treatment of cancer. *Immunol. Rev.* **129,** 57–80.
17. Orr, G. A (1981) The use of the 2-iminobiotin-avidin interaction for the selective retrieval of labeled plasma membrane components. *J. Biol. Chem.* **256,** 761–766.
18. Dimroth, P. (1986) Preparation, characterization, and reconstitution of oxaloacetate decarboxylase from *Klebsiella aerogenes*, a sodium pump. *Meth. Enzymol.* **125,** 530–540.

19. Dean, P. D. G., Johnson, W. S., and Middle, F. S. (1985) Activation procedures. In *Affinity chromatography. A practical approach,* IRL Press, Washington, DC, 34–35.
20. Kohanski, R. A. and Lane, D. (1990) Monovalent avidin affinity columns. *Meth. Enzymol.* **184,** 194–220.
21. Chilkoti, A., Schwartz, B. L., Smith, R. D., Long, C. J. and Stayton, P. S. (1995) Engineered chimeric streptavidin tetramers as novel tools for bioseparations and drug delivery. *Bio/Technology* **13,** 1198–1204.
22. Morag, E., Bayer, E. A., and Wilchek, M. (1996) Immobilized Nitro-Avidin and Nitro-Streptavidin as Reusable Affinity Materials for Application in Avidin-Biotin Technology. *Analyt. Biochem.* **243,** 257–263.
23. Wong, S. S. (1991) Reactive groups of proteins and their modifying agents, in *Chemistry of Protein Conjugation and Crosslinking*, CRC, Boston, MA, 30.
24. Carlsson, J., Drevin, H., and Axen, R. (1978) Protein thiolation and reversible protein-protein conjugation. N-Succinimidyl 3-(2-pyridyldithio)propionate, a new heterobifunctional reagent. *Biochem. J.* **173,** 723–737.
25. Haugland, R. P. (1995) Coupling of monoclonal antibodies with fluorophores. In *Methods in Molecular Biology*, (Ed. Davis, W. C.) Humana, Totowa, NJ, **45,** 205–221.
26. Kronick, M. N. and Grossman, P. D. (1983) Immunoassay techniques with fluorescent phycobiliprotein conjugates. *Clin. Chem.* **29,** 1582–1586.
27. Bongartz, J.-P., Aubertin, A.-M., Milhaud, P. G. and Lebleu, B. Improved biological activity of antisense oligonucleotides conjugated to a fusogenic peptide. *Nucleic Acids Res.* **22,** 4681–4688.

50

MDPF Staining of Proteins on Western Blots

F. Javier Alba and Joan-Ramon Daban

1. Introduction

We describe a method for the detection of total protein patterns on polyvinylidene difluoride (PVDF) membranes using the fluorogenic dye 2-methoxy-2,4-diphenyl-3(2*H*)-furanone (MDPF) *(1)*. This method is based on the fluorescent properties of this dye *(2)*. As can be seen in **Fig. 1**, MDPF (**A**) and the hydrolysis product (**C**) are nonfluorescent; only the adduct **B** formed with the proteins is fluorescent. This makes unnecessary the destaining of the PVDF membrane after protein labeling. The whole process of staining with MDPF is completed in about 20 min. Wet membranes are translucent, allowing the visualization of MDPF labeled protein bands by transillumination with UV light (*see* **Fig. 2A**). Electrophoretic bands containing less than 10 ng of protein transferred to PVDF membranes can be detected after the reaction with MDPF.

This staining method is compatible with previous visualization of protein bands on the sodium dodecyl sulfate (SDS)-polyacrylamide gel with the noncovalent fluorescent dye Nile red (*see* Chapter 30). Thus, Nile red and MDPF staining can be performed sequentially. This allows the rapid monitoring of total protein patterns on both the electrophoretic gel and Western blot. In addition, MDPF staining allows further immunodetection of specific bands with polyclonal antibodies (*see* **Fig. 2B**). Finally, using the adequate conditions described in the Materials section, MDPF staining does not preclude the N-terminal sequence analysis of proteins in selected bands.

2. Materials

All solutions should be prepared using electrophoresis-grade reagents and deionized water and stored at room temperature. Wear gloves to handle all reagents and solutions and do not pipet by mouth. Collect and dispose all waste according to good laboratory practice and waste disposal regulations.

1. MDPF: Concentrated stock (35 m*M*) in dimethyl sulfoxide (DMSO). This solution is stable for 1 wk at room temperature in a glass bottle wrapped in aluminum foil. **Handle this solution with care, DMSO is flammable and, in addition, this solvent may facilitate the passage of hazardous chemicals such MDPF through the skin.** MDPF can be obtained from Fluka (Bunch, Switzerland).
2. PVDF membranes (Bio-Rad Laboratories [Hercules, CA]).

From: *The Protein Protocols Handbook, 2nd Edition*
Edited by: J. M. Walker © Humana Press Inc., Totowa, NJ

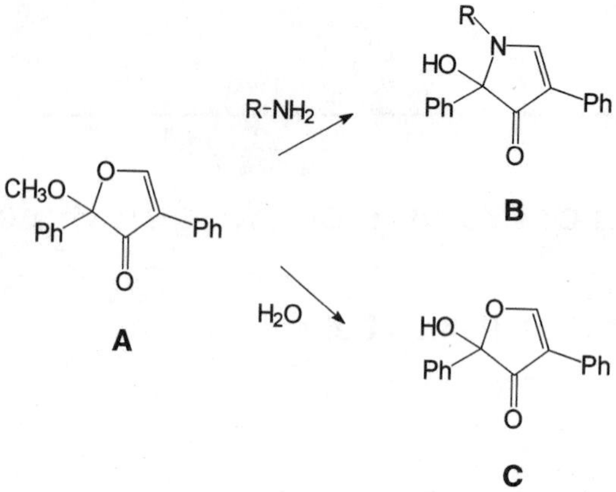

Fig. 1. According to Weigele et al. *(2)*, the reaction of MDPF **(A)** with primary amino groups of proteins produces the fluorescent adduct **(B)**; the excess reagent is hydrolyzed forming the nonfluorescent product **(C)**.

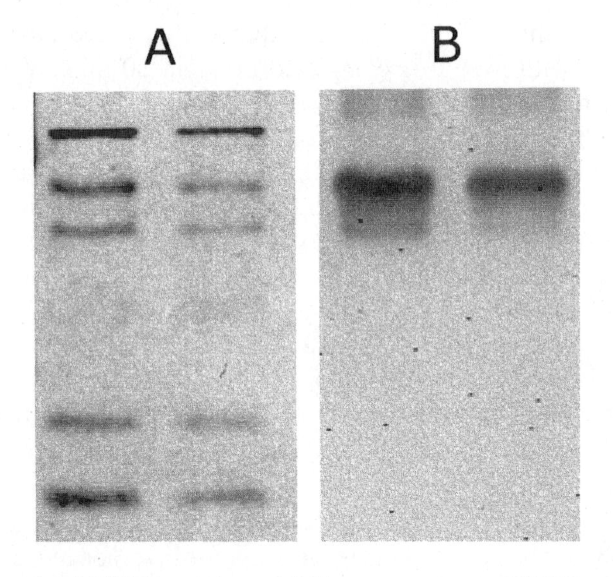

Fig. 2. (**A**) Example of MDPF staining of different proteins on PVDF membranes. The blot was equilibrated twice (5 min each time) in borate buffer and stained for 10 min with MDPF (*see* **Subheading 3.**). Proteins, from top to bottom, are: BSA, ovalbumin, glyceraldehyde-3-phosphate dehydrogenase, β-lactoglobulin, and α-lactalbumin; the amount of each protein loaded initially onto the SDS gel before electroblotting was 100 *(left lane)* and 50 *(right lane)* ng. (**B**) After staining with MDPF ovalbumin was immunodetected with the ECL system (Amersham).

3. Transfer apparatus (e.g., Mini-Trans-Blot Cell [Bio-Rad]).
4. Opaque plastic box for membrane equilibration and staining.
5. Orbital shaker.
6. Transilluminator equipped with midrange ultraviolet (UV) bulbs (~300 nm) to excite MDPF labeled proteins on blots (e.g., Foto UV 300 [Fotodyne Inc., Harland, WI] or the

transilluminator included in the Gel Doc 1000 system from Bio-Rad). *See* Chapter 30 for details about protection from UV during transillumination.

7. Photography: We have used the Polaroid (Cambridge, MA) MP-4 camera with Polaroid instant film 667 and the CCD camera of the Gel Doc 1000 system (Bio-Rad). With both cameras use the Wratten (Kodak [Rochester, NY]) filters numbers 3 (yellow) and 47 (blue). Place filter 47 on top of filter 3. Store the filters in the dark and protect them from heat, intense light sources, and humidity. Store the Polaroid film at 4°C.
8. Immunodetection: Ovalbumin antiserum developed in rabbit (Sigma [St. Louis, MO]); ECL detection kit, including horseradish peroxirase-labeled antirabbit antibodies (Amersham [Buckinghamshire, UK]).
9. Transfer buffer: 25 m*M* Tris, 192 m*M* glycine, pH 8.3.
10. Borate buffer: 10 m*M* sodium borate, pH 9.5.
11. Phosphate-buffered saline: 80 m*M* Na_2HPO_4, 20 m*M* NaH_2PO_4, 100 m*M* NaCl, pH 7.4.

3. Method

After SDS-polyacrylamide gel electrophoresis (SDS-PAGE), the gels were stained with Nile red (*see* Chapter 30). Protein bands stained with Nile red can be transferred to PVDF membranes without being necessary to perform the destaining of the gel *(1)*. The method described in this part gives all the details for the staining of the blotted proteins with MDPF. All operations are performed at room temperature. Handle blots by their edges using stainless steel forceps.

1. After electrophoresis and Nile red staining, equilibrate the gel ($8 \times 6 \times 0.075$ cm) in 100 mL of transfer buffer for 15 min. Immerse sequentially the PVDF membrane in 20 mL of methanol for 5 s, in 100 mL of water for 2–3 min, and finally in 100 mL of transfer buffer for 10 min.
2. Assemble the gel and the membrane in the blotting apparatus and fill the tank with transfer buffer. Perform electroblotting at 100 V for 1 h.
3. Following transfer, equilibrate the blot twice (for 5 min each time) in 100 mL of borate buffer. Use an orbital shaker at about 75 rpm.
4. Incubate the blots for 10 min in a staining solution containing 40 mL of borate buffer and 0.2 mL of the concentrated stock (35 m*M*) of MDPF in DMSO (*see* **Note 1**); use an orbital shaker (~75 rpm). The plastic box containing the blot should be covered with aluminum foil during staining. The staining conditions compatible with N-terminal sequencing of selected bands are indicated in **Note 2**.
5. After staining, rinse the blot briefly (for about 10 s) with borate buffer.
6. Place the wet membrane (*see* **Note 3**) on the UV transilluminator. Focus the camera (Polaroid or charge coupled device [CCD]) with the help of lateral illumination with a white lamp, place the optical filters indicated in **Subheading 2.7.** (*see* **Note 4**) in front of the camera lens and, in the dark, turn on the transilluminator and photograph the blot (*see* **Note 5**). Finally, turn off the transilluminator.
7. Develop the Polaroid film for the time indicated by the manufacturer (*see* **Note 6**). The images obtained with the CCD camera are stored directly in the computer and can be printed afterwards (*see* **Note 7**).
8. After photography, if specific bands have to be immunodetected (*see* **Note 8**), equilibrate the stained blot for 15 min in phosphate buffered saline containing 0.1% Tween 20, and then perform the ECL immunodetection according to the manufacturer's instructions. An example of immunodetection with the ECL system (*see* **Subheading 2.8.**) after MDPF staining is presented in **Fig. 2B.**

9. Mark with a soft pencil the stained protein bands (*see* **Note 2**) to be sequenced (*see* **Note 9**). Use a UV transilluminator to visualize the fluorescent bands. Cut out of the membrane the selected bands and apply them to the sequencer (we have used an LF3000 Automatic Sequencer [Beckman, Palo Alto, CA], *see* **Ref. *1***).

4. Notes

1. The reaction of MDPF with different proteins in solutions containing 10 m*M* sodium borate, pH 9.5, is completed in <10 min *(3)*.
2. When the protein bands have to be used for sequencing (*see* **Note 9**), use a lower concentration of MDPF (0.1 mL of the concentrated solution of MDPF in 100 mL of borate buffer) and reduce the incubation time to 2 min. The sensitivity obtained under these conditions is lower than that obtained using the normal staining conditions (*see* **Note 5**), but it is high enough to detect bands containing protein amounts suitable for microsequencing by Edman degradation.
3. Membranes must be wet during visualization and imaging. Wet membranes are highly translucent, but dry membranes are opaque and reduce dramatically the sensitivity *(1)*.
4. Wratten filters nos. 3 and 47 allow the visualization of the blue fluorescence emission from MDPF-labeled bands. In addition, these filters eliminate the light coming from the transilluminator and the background fluorescence produced by Nile red adsorbed on the surface of the PVDF membrane during the electrotransfer.
5. Electrophoretic bands containing 5–10 ng of protein can be seen in the images obtained with Polaroid film and the CCD system. However, considering that the yield of protein transfer from the gel to the membrane is relatively low, the actual sensitivity is presumably higher. In fact, we have observed that when proteins are transferred directly to the membrane using a slot-blotting device, 0.5 ng of protein per slot can be detected *(1)*.
6. The development time of Polaroid film is dependent on the film temperature. Store the film at 4°C (*see* **Subheading 2.7.**), but allow them to equilibrate at room temperature before use.
7. Ink jet printers with glossy paper yield images with photographic quality (*see* e.g., **Fig. 2A**).
8. Our results *(1)* have shown that the covalent modification of blotted proteins produced by the reaction with MDPF does not alter the antigenic properties that allow the binding of polyclonal antibodies.
9. Our results *(1)* have shown that the staining conditions indicated in **Note 2** do not preclude further sequencing reactions with a high yield. Since MDPF reacts with primary amino groups of proteins (*see* **Fig. 1**), probably including N-teminal groups, these results indicate that under these staining conditions the reaction with MDPF is not complete.

Acknowledgment

This work was supported in part by grants PB98-0858 (Dirección General de Investigación) and 2000SGR65 (Generalitat de Catalunya).

References

1. Alba, F. J. and Daban, J.-R. (1998) Rapid fluorescent monitoring of total protein patterns on sodium dodecyl sulfate-polyacrylamide gels and Western blots before immunodetection and sequencing. *Electrophoresis*, **19,** 2407–2411.
2. Weigele, M., DeBernardo, S., Leimgruber, W., Cleeland, R., and Grunberg, E. (1973) Fluorescent labeling of proteins. A new methodology. *Biochem. Biophys. Res. Commun.* **54,** 899–906.

3. Alba, F. J. and Daban, J.-R. (1997) Nonenzymatic chemiluminescent detection and quantitaion of total protein on Western and slot blots allowing subsequent immunodetection and sequencing. *Electrophoresis* **18,** 1960–1966.

51

Copper Iodide Staining of Proteins and Its Silver Enhancement

Douglas D. Root and Kuan Wang

1. Introduction

Copper iodide staining and silver-enhancement is designed to quantify proteins adsorbed to solid surfaces such as nitrocellulose, nylon, polyvinylidene difluoride (PVDF), silica, cellulose, and polystyrene ***(1–5)*** and has important applications in Western blotting and thin layer chromatography ***(3,6)***. The binding of cupric ions to the backbone of proteins under alkaline conditions and their reduction to the cuprous state is the basis of several protein assays in solution including the biuret, Lowry, and bicinchoninic acid methods (***1–3,7*** and *see* Chapters 2–4). In the case of copper iodide staining, the protein binds copper iodide under highly alkaline conditions. This protein assay demonstrates sensitivity, speed, reversibility, low cost, and the lack of known interfering substances (including nucleic acid; **refs.** ***4,5***). Copper iodide staining is sufficiently sensitive to permit the quantification of proteins adsorbed to microtiter plates ***(5)***. The information is particularly useful for the quantitative interpretation of enzyme-linked immunosorbent assay (ELISA) and protein binding experiments. The precision of the determination of protein adsorbed to the microtiter plate by copper iodide staining is typically about 10–15%. The high sensitivity of copper iodide staining (about 40 pg/μL) may be increased several fold by a silver-enhancement procedure that allows the detection of protein down to about 10 pg/μL, which is more sensitive than common solution-based assays ***(7)***. The sensitivity of the assay can be increased by repeated applications of the protein on a membrane to concentrate it. Protein concentrations may be estimated from copper iodide staining from very dilute protein solutions or when only small amounts of a precious protein are available such as for the analysis of chromatography fractions.

2. Materials

2.1. Copper Iodide Staining

1. Prepare the copper iodide staining reagent by mixing 12 g of $CuSO_4 \cdot 5H_2O$, 20 g of KI, and 36 g of potassium sodium tartrate with 80 mL of distilled water in a glass beaker (*see* **Note 1**). As the slurry is vigorously stirred, 10 g of solid NaOH is slowly added. The

From: *The Protein Protocols Handbook, 2nd Edition*
Edited by: J. M. Walker © Humana Press Inc., Totowa, NJ

suspension becomes warmer and changes color from brown to green to dark blue. After the NaOH is completely dissolved, the beaker is allowed to cool at room temperature for 30 min without stirring to allow the brownish-red precipitate to settle. Then 70 mL of solution is aspirated from the top to leave approx 50 mL of reagent with precipitate. The reagent is stable and may be stored in a sealed bottle at room temperature for at least 1 mo or at 4°C for at least 1 yr (*see* **Note 2**).

2. Prepare the copper iodide stain remover solution with 0.19 g of $Na_4EDTA \cdot H_2O$, 0.28 g of $NaH_2PO_4 \cdot H_2O$, and 2.14 g of $Na_2HPO_4 \cdot 7H_2O$ in 100 mL of deionized water.
3. Nitrocellulose (e.g., BA85, Schleicher & Schuell, Keene, NH), PVDF (e.g., PVDF, Millipore Corporation, Bedford, MA), or nylon blotting paper (e.g., Zeta probe, Bio-Rad Laboratories, Richmond, CA; *see* **Note 3**).

2.2. Silver-Enhanced Copper Staining (SECS)

1. Prepare the silver enhancing reagent just prior to use by dissolving 0.1 g of $AgNO_3$, 0.1 g of NH_4NO_3, and 7 µL of 5% (v/v in water) β-mercaptoethanol in 100 mL of distilled water. After the other components are dissolved and immediately before use add 2.5 g of Na_2CO_3.
2. Prepare the SECS stain remover by dissolving 33.2 g of KI in 100 mL of distilled water (final concentration, 2 *M* KI).

2.3. Copper Iodide Microtiter Plate Assay

1. Polystyrene 96-well microtiter plates (e.g., Nunc Immuno-Plate, Denmark; Titertek 76-381-04, McLean, VA; or Immulon 1 and 2, Dynatech Laboratories, Chantilly, VA).
2. Nitrocellulose membranes (e.g., BA85, Schleicher & Schuell, Keene, NH).
3. Household 3-in-one lubricating oil (Boyle-Midway, New York, NY).
4. A standard single hole puncher (6 mm diameter).
5. Some form of densitometer is required such as a flatbed scanner or video camera and framegrabber with image analysis software (*see* **Note 4**).

3. Methods

3.1. Copper Iodide Staining

1. Stir the copper iodide staining reagent vigorously at room temperature immediately prior to use. The reagent should be a fine slurry.
2. Rock the copper iodide staining reagent over the dried protein blot (or Western blot) for at least 2 min (but not more than 5 min; *see* **Note 5**) as the reddish-brown bands appear on the blot.
3. Gently dip the stained blot up and down in three beakers of deionized water and then allow the blot to dry (*see* **Note 6**).
4. The stained blot then may be quantified by densitometry (*see* **Note 7**) and photographed for documentation.
5. The staining pattern is stable at room temperature for at least 1 yr.
6. If greater sensitivity is desired, the blot may be used directly in the silver-enhanced copper staining procedure (*see* **subheading 3.2.**).
7. If subsequent immunostaining on the same blot is required, destain for 15 min with gentle agitation in the copper iodide stain remover prior to immunostaining.

3.2. Silver-Enhanced Copper Staining (SECS)

1. Rock a nitrocellulose protein blot stained with copper iodide (*see* **subheading 3.1.**) for 5 min in freshly prepared silver-enhancing reagent until the bands become dark black.

2. Dip the blot in a 1-L beaker of deionized water and stored in the dark to dry (to prevent background development).
3. The blot then may be quantified by densitometry and photographed for documentation.
4. If subsequent immunostaining on the same blot is required, destain for 30 min with gentle agitation in the SECS stain remover prior to immunostaining (*see* **Note 8**).

3.3. Copper Iodide Microtiter Plate Assay

1. Adsorb the protein of interest to duplicate microtiter plates and note the volume (V) that was used to adsorb the protein in each well (*see* **Note 9**). One of the microtiter plates is for copper iodide staining and the other is for quantitative ELISA or binding experiments. The microtiter plate for copper iodide staining contains protein adsorbed to only a few of the wells (e.g., 4–16 wells; see the schematic diagram and **Note 10**).
2. Create a standard curve by dot blotting approx 5–100 ng of the protein of interest per 5-μL drop onto nitrocellulose paper and allow to air dry. Blanks are dotted with equal volumes of buffers.
3. Stain both the nitrocellulose paper and the microtiter plate with adsorbed protein by the copper iodide staining procedure (*see* **Subheading 3.1.**).
4. Use the hole puncher to excise stained dots from the nitrocellulose membrane and place them stained side down into blank wells on the microtiter plate (*see* **Note 11**).
5. Use the hole puncher to excise blank nitrocellulose circles and place them in microtiter plate wells containing copper iodide-stained protein and also in some blank wells to determine the background density (*see* **Fig. 1**).
6. For transmittance densitometry, first add 5 μL per well of Household 3-in-one oil to make the nitrocellulose translucent, thus reducing the background. For reflectance densitometry, the bottom of the microtiter plate may be scanned directly (the stained sides of the nitrocellulose must all be face down in the wells).
7. Measure the mean optical density and total area of the known amounts of stained protein (adsorbed to nitrocellulose) and construct a standard curve of mean optical density vs ng/mm^2 of protein (*see* **Note 12**).
8. Measure the mean optical density of stained sample protein (adsorbed to microtiter plate) and compare to the standard curve to determine the concentration of protein (in ng/mm^2) on the microtiter plate well.
9. Calculate the total area of the stained sample protein (adsorbed to microtiter plate) on the microtiter plate well from the equation:

$$\text{Total Area} = 3.14r^2 + (2V/r)$$

in which r (in mm) is the radius of the cylindrical flat-bottom well, and V (in μL) is the volume that was used to adsorb protein to the well. The Total Area (typically 94.7 mm^2 for a 100-μL volume applied to a plate with a 3.25-mm well radius) multiplied by the surface density (in ng/mm^2) of the sample protein yields the amount of protein (in nonograms) adsorbed to each microtiter plate well.

4. Notes

1. Proportions of sodium and potassium ions in the copper iodide staining reagent are important. Thus potassium sodium tartrate (sodium potassium tartrate) should not be substituted with, for instance, sodium tartrate.
2. The copper iodide staining reagent can generally be reused two or three times but will eventually become less sensitive.
3. Copper iodide staining reagent stains proteins adsorbed on most solid phase adsorbents including nitrocellulose, nylon, PVDF, silica, cellulose, and polystyrene. PVDF requires that the blot be prewetted in 50% methanol immediately prior to staining, and some smear-

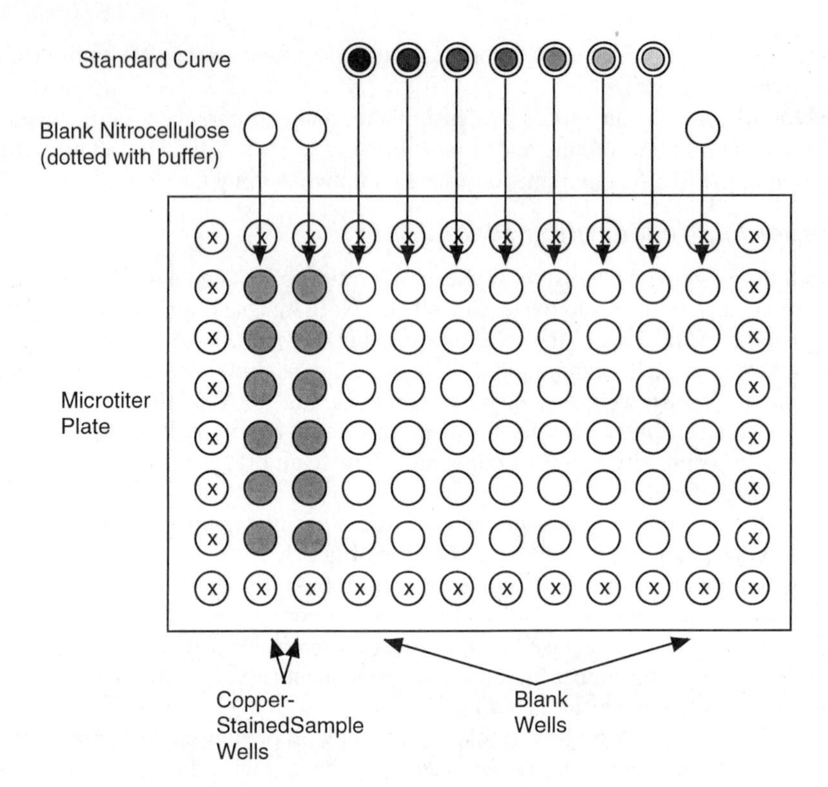

Fig. 1. Schematic diagram of the assembly of a microtiter plate for the copper iodide staining assay. Avoid wells marked with "x" due to possible distortions from edge effects.

ing of bands can occur at high loads of protein with PVDF. Copper iodide staining of silica leaves an uncharacteristic bluish background. Comparisons of the sensitivities of copper iodide staining and SECS with other stains on a variety of common supports are illustrated in **Table 1**.

4. Microtiter plate readers are to be avoided for quantitative measurements, because they are usually not sensitive enough (copper iodide staining yields OD ≤ 0.1) and do not sample a large enough area of the stained surface to detect any nonuniformity in the staining density.
5. Exceeding a staining time of 5 min can both damage nitrocellulose membranes and lead to solubilization of adsorbed protein. A staining time of 2 min is optimal.
6. Washing of microtiter plates should be handled gently by dipping in beakers of deionized water. Vigorous washing procedures often lead to nonuniform protein distribution and consequent uneven staining of microtiter plates.
7. An example of a low cost densitometer is a desktop flatbed scanner, color Apple Macintosh computer, and NIH Image software (public domain, by Wayne Rasband; further details are available by personal communication).
8. SECS may be removed by concentrations of KI that are less than 2 *M* but will require longer incubations (e.g., 90 min for 0.5 *M* KI).
9. The wells on the edge of microtiter plates should be avoided for quantitative measurements because they tend to yield less accurate numbers.
10. Nitrocellulose quantitatively binds most proteins that are dot blotted onto it and retains them well throughout copper iodide staining.

Table 1
Comparison of the Detection Limits (in ng/1μL) of Protein Stains[a]

Stain	Nitrocellulose	PVDF	Nylon	Silica	Cellulose
SECS	0.01	0.071	1	5	0.02
Copper iodide	0.04	0.014	5	20	0.04
Kinetic silver staining[b]	40	None	300	70	0.01
Coomassie[c]	9	150	None	40	300
Eu metal chelate[d]	9	150	2000	70	150

[a]Two-fold serial dilutions of bovine albumin were dotted in 1-μL spots on the indicated support, then dried, stained, and evaluated visually.
[b]Performed as described in Chapter 9.
[c]Performed as described in **ref. *8***.
[d]Performed as described in **ref. *9***. The Eu metal chelate stain was visualized over a UVP (Upland, CA) 302 nm UV transilluminator. When spots were excised and quantified by time-resolved fluorescence in an SLM Aminco Bowman II spectrometer, a detection limit of 3 ng/1 μL was observed on nitrocellulose consistent with reported values ***(9)***.

11. If there is a problem with static repulsion and transmission densitometry will be used, the Household 3-in-one oil (**Subheading 3.3., step 6**) may be first applied to the microtiter plate well to release the static charge.
12. Quantitative measurements of copper iodide staining should be done at least in triplicate because of the relatively high (10–15%) standard deviation of the results.

References

1. Jenzano, J. W., Hogan, S. L., Noyes, C. M., Featherstone, G. L., and Lundblad, R. L. (1986) Comparison of five techniques for the determination of protein content in mixed human saliva. *Analyt. Biochem.* **159,** 370–376.
2. Lowry, O. H., Rosebrough, N. J., Farr, A. L., and Randall, R. J. (1951) Protein measurement with the Folin phenol reagent. *J. Biol. Chem.* **193,** 265–275.
3. Smith, P. K., Krohn, R. I., Hermanson, G. T., Mallia, A. K., Gartner, F. H., Provenzano, M. D., et al. (1985) Measurement of protein using bicinchoninic acid. *Analyt. Biochem.* **150,** 76–85.
4. Root, D. D. and Reisler, E. (1989) Copper iodide staining of protein blots on nitrocellulose membranes. *Analyt. Biochem.* **181,** 250–253.
5. Root, D. D. and Reisler, E. (1990) Copper iodide staining and determination of proteins adsorbed to microtiter plates. *Analyt. Biochem.* **186,** 69–73.
6. Talent, J. M., Kong, Y., and Gracy, R. W. (1998) A double stain for total and oxidized proteins from two-dimensional fingerprints. *Analyt. Biochem.* **263,** 31–38.
7. Sapan, C. V., Lundblad, R. L., and Price, N. C. (1999) Colorimetric protein assay techniques. *Biotechnol. Appl. Biochem.* **29,** 99–108.
8. Christian, J. and Houen, G. (1992) Comparison of different staining methods for polyvinylidene difluoride membranes. *Electrophoresis* **13,** 179–183.
9. Lim, M. J., Patton, W. F., Lopez, M. F., Spofford, K. H., Shojaee, N., and Shepro, D. (1997) A luminescent europium complex for the sensitive detection of proteins and nucleic acids immobilized on membrane supports. *Analyt. Biochem.* **2,** 184–195.

52

Detection of Proteins on Blots Using Direct Blue 71

Hee-Youn Hong, Gyurng-Soo Yoo, and Jung-Kap Choi

1. Introduction

The visualization of proteins blotted onto transfer membranes has important applications for the localization of bands prior to further steps and for the comparison of individual proteins detected by immunostaining with total proteins *(1,2)*. Consequently, sensitive, convenient, and quantitative total protein staining on membranes is required *(3,4)*. Among transfer membranes available, nitrocellulose (NC), polyvinylidene difluoride (PVDF), and nylon, NC appears to be the material of choice because it is relatively inexpensive and requires a fast and simple step for blocking from nonspecific binding *(5)*. The widely used methods for staining of NC membranes are Amido Black *(2,6)*, Coomassie Blue *(7,8)*, India ink *(9,10)*, and Ponceau S *(11,12)* staining. None of them is as sensitive as silver stain on gels *(4)*. Metal stains (colloidal gold stain, silver stain) of membranes provide the highest sensitivity, but they have certain disadvantages: long incubations, troublesome preparation, an expensive metal, or irreversible protein staining *(4,13)*. Copper iodide staining is also sensitive, but complicated steps are required to prepare reagents. Further, this method employs strongly alkaline solution of pH 13.8, resulting in membrane impairment after incubation for more than 5 min *(14)*. Recently, reversible metal complex stains were introduced, which are compatible with subsequent evaluation *(15–17)*.

We have attempted to develop a dye-based staining method with sensitivity, convenience, and economy. Direct Blue 71 (DB71, C.I. 34140) produces blue color in water with an absorption maximum at 594 nm and possesses excellent light fastness (Product Information of Organic Dyestuffs Co., East Providence, RI). No hazard has been connected with the use of the dye under normal conditions. The structure of the dye consists of four biphenyls with a hydroxyl, a primary amine, three azo, and four sulfonate groups (*see* **Fig. 1**). The relatively strong hydrophobicity imparted by four biphenyl rings of DB71 is expected to improve the low sensitivity observed in Ponceau S staining. On the basis of this reasoning, DB71 appears to be a satisfactory candidate for protein staining on membranes *(18)*.

From: *The Protein Protocols Handbook, 2nd Edition*
Edited by: J. M. Walker © Humana Press Inc., Totowa, NJ

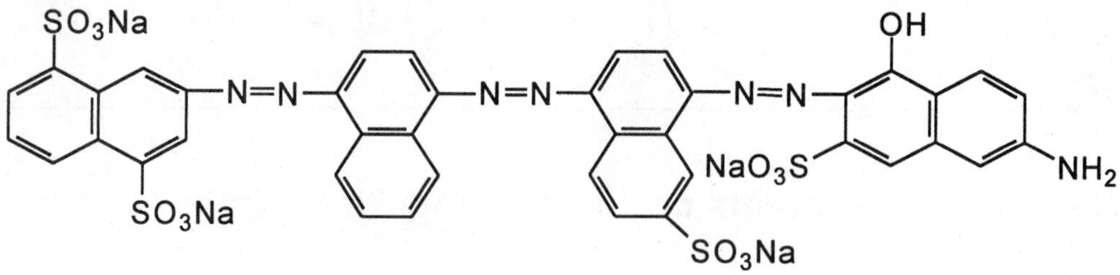

Fig. 1. Chemical structure of DB71.

2. Materials

2.1. General

1. Transfer apparatus.
2. Power supply (capacity 200 V, 2 A): Preferably capable of delivering constant current.
3. Nitrocellulose or PVDF membrane.
4. Filter papers (e.g., Whatmann no. 1).
5. Rocking shaker.
6. Plastic or glass container.
7. Transfer buffer: 25 m*M* Tris-HCl, pH 9.4, containing 20% methanol. Many different buffers can be used in transferring proteins from gels, depending on the efficacy of the blotting.
8. Dot or slot blot apparatus.
9. Phosphate buffered saline (PBS): Needed for dot or slot blotting. For 1 L of solution, use 0.2 g of KH_2PO_4, 1.15 g of Na_2HPO_4, 0.2 g of KCl, and 8.0 g of NaCl.

2.2. DB71 Staining

1. Stock dye solution (0.1% [w/v] DB71): Dissolve 0.1 g of DB71 (dye content, ~50%; from Aldrich) in 100 mL of distilled water. The stock solution is stable for months at room temperature.
2. Working dye solution (0.008% [w/v] DB71): Dilute 4 mL of stock dye solution to 50 mL of washing solution (*see* **ref. 3**). The working solution is stable for weeks at room temperature.
3. Washing solution: For 1 L of solution, mix 500 mL of distilled water with 400 mL of absolute ethanol and 100 mL of glacial acetic acid.
4. Destaining solution: For 1 L of solution, mix 350 mL of distilled water with 500 mL of absolute ethanol and 150 mL of 1 *M* sodium bicarbonate.

3. Methods

3.1. Electrophoresis and Blotting

All procedures refer to product information of the apparatus used and general recipes.

1. Electrophorese proteins diluted to the appropriate concentrations according to Laemmli *(19)* on 10% sodium dodecyl sulfate (SDS)-polyacrylamide gels.
2. Transfer proteins from gels onto membranes in a semi-dry transfer kit at a constant current of 1 mA/cm^2 for 1–2 h, in 25 m*M* Tris-HCl –20% methanol, pH 9.4.
3. *(Optionally)* Make dot or slot blots with a Bio-Dot® microfiltration apparatus (Bio-Rad Lab., Richmond, CA).

3.2. Membrane Staining

1. Prior to staining, wet protein transferred membranes, if the membranes had been dried: Wet NC membranes for a minute in distilled water, and wet PVDF membranes first in methanol, then rinse them in distilled water.
2. Immerse gently membranes in working dye solution for 5 min.
3. Rinse membranes briefly with washing solution for seconds.
4. Allow membranes to be air-dry or wrap them up with Saran Wrap to stay wet (*see* **Note 15**).
5. Keep membranes in a refrigerator for further uses.

3.3. Dye Removal

1. After staining, if necessary, incubate membranes for 5–10 min in destaining solution (*see* **Note 13**).
2. Rinse membranes briefly with distilled water.
3. Proceed to next steps (e.g., immunostaining).

4. Notes

1. DB71 staining is based on dye binding to proteins under acidic conditions similar to that in Amido Black ***(2,6)***, Coomassie Blue ***(7,8)***, and Ponceau S ***(11,12)*** staining methods.
2. DB71 possesses the properties of an ideal protein stain, including high, nonspecific affinity for protein; rapid staining; convenient application conditions compatible with matrix material and with protein; large molar absorptivity with the absorption maximum in blue region; and safety in use ***(20)***.
3. At concentrations < 0.002%, the staining intensity of protein bands is weak; at concentrations > 0.01%, unwanted background coloration is observed.
4. The binding of DB71 to proteins is favored under acidic condition: It is quite probable that in DB71 staining solution most carboxyl groups of proteins are protonated and thus there is no electrostatic repulsion between proteins and the dye molecule containing four sulfonate groups. In the acidic solutions, ionic interactions between maximally protonated amino moieties of proteins and anionic functional groups of the dye might primarily contribute to the stain. Additional forces for the staining could involve hydrophobic interaction, hydrogen bond, and Van der Waals forces which have been reported to be the binding mode of proteins to most anionic dyes ***(21)***.
5. For the visualization of proteins in gels and membranes, most of dye-based protein staining methods have generally employed methanol as a component of staining/destaining solution ***(2,6–9)***, despite its toxicity. DB71 can use ethanol instead of methanol, without any loss of staining intensity by the dye.
6. DB71 staining is influenced by the concentration of ethanol. Less than 20% ethanol causes background coloration; more than 60% ethanol not only reduces band intensity, but also causes distortion of the membranes.
7. In DB71 staining solution containing 0.008% DB71 in 40% ethanol–10% acetic acid, major protein bands appear in usually 1–2 min, with minor protein bands being detected within 5 min.
8. The detection limits of DB71 staining for slot blot and electroblot are 5–10 ng of protein on NC, which is 10-fold more sensitive than Ponceau S staining that detected down to 50–100 ng. DB71 can also stain proteins bound to PVDF; however, background is colored a little, possibly due to the relatively increased hydrophobicity in PVDF, comparing with NC. DB71 staining detects down to 10–20 ng of protein on PVDF, whereas Ponceau S staining does to 100 ng of protein.
9. The removal of DB71 from stained bands requires incubating membranes in ethanol–1 *M* sodium bicarbonate–water (10:3:7) for a few minutes. Complete destaining, however, may

Table 1
Comparison of Staining Procedures for Protein Blots[a]

Dyes	Conditions	Approx. detection limits (ng)
0.05% Coomassie Blue R-250	S: 50% MeOH–7% HAc (10 min) D: 10% MeOH–14% HAc (overnight at 65°C)	100[d]
0.1% Amido Black 10B	S: 45% MeOH–10% HAc (10 min) D: 10% HAc (15 min)	50[d]
0.1% Ponceau S	S: 5% HAc (3 min) W:Water[b]	100[d]
0.05% CPTS	S: 12 mM HCl (~1min) W: 12 mM HCl[b]	10[d]
0.008% DB71	S: 40% EtOH–10% HAc (5 min) W: 40% EtOH–10% HAc[c]	5–10 (our work)

[a]The staining procedures of Coomassie Blue *(9)*, Amido Black *(6)*, and CPTS *(20)* were performed as in references cited. Ponceau S staining followed the product information of Sigma Chemical.

[b]Membranes were briefly washed in several changes of the solutions to remove excess dye.

[c]Membranes were washed once in the solution, since background was almost clear.

[d]Approximate detection limits are from the references cited, where they were determined by comparing protein slot blots stained by these methods.

S, staining; D, destaining; W, washing; MeOH, methanol; HAc, acetic acid; EtOH, ethanol.

take up to an hour or more, depending on the amount or nature of proteins of interest. The membranes can be repeatedly stained and destained, with no apparent loss of sensitivity.

10. Both Coomassie Blue and Amido Black are inferior to DB71 stain in terms of requiring separate destaining solution and lengthy time (*see* **Table 1**), and giving lower sensitivity on PVDF.
11. DB71 is comparable with copper phthaloryanine-3,4',4",4'"-tetrasulfonic acid (CPTS) (*see* **Figs. 2C** and **G**) reported as the highest sensitive dye-binding staining method *(20)*, in its sensitivity, rapidity, and reversibility. The bluish violet color of DB71 stained proteins gives better band contrast than the turquoise blue CPTS stained ones, allowing easy photography without any filter system; moreover, DB71 is much cheaper than CPTS.
12. It is noticeable that among the methods tested, DB71 staining allows the highest band contrast on PVDF with a similar staining intensity as on NC (*see* **Figs. 2D** and **H**).
13. DB71 staining is applicable to a subsequent immunostaining without removing the dye from developed bands. However, we recommend the dye removal, prior to the immunodetection of small amounts of a particular protein.
14. DB71 staining can also be applied for quantification purposes with simplicity and convenience over the conventional spectrophotometric procedures. The relationships between peak area representing band intensity after DB71 staining and the protein amounts hold a linearity between 20 and 1000 ng, with a correlation coefficient of 0.991 (*see* **Fig. 3**).
15. A higher band contrast for photography is obtained when the stained membrane keeps wet rather than dry. The stain remains stable for several months in a refrigerator.

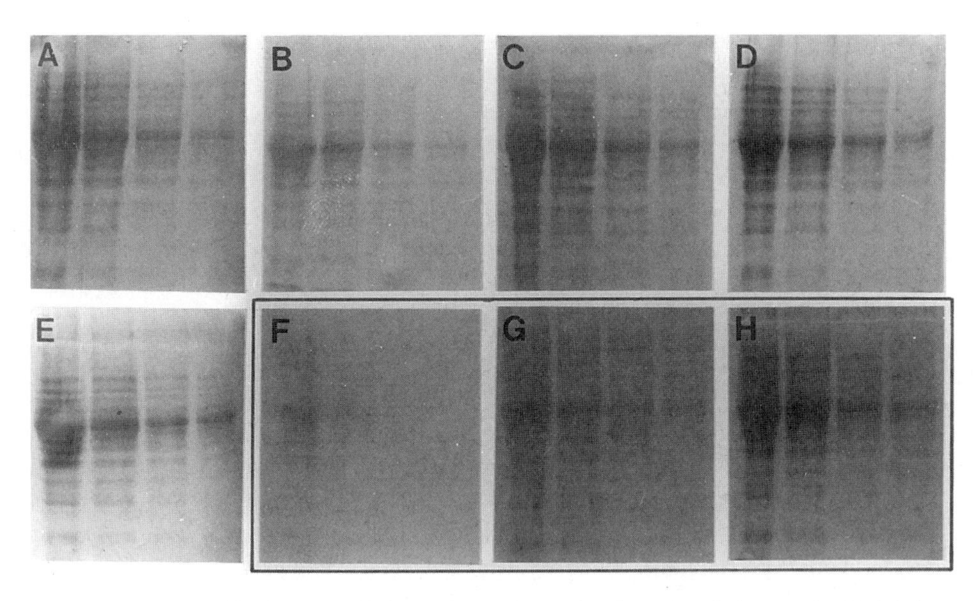

Fig. 2. Comparison of DB71 staining with various dye-based staining methods. Blots contained a whole cell lysate of HL-60, a human promyelocytic leukemia cell line, that 2 × 106 cells were lysed in 150 µL of lysis buffer. Four serial half dilutions of the lysates were separated on 10% SDS-PAGE and transferred onto two types of membranes (**A–E**, NC, **F–H**, PVDF). Transferred proteins were visualized by Coomassie Blue (**A**), Ponceau S (**B** and **F**), CPTS (**C** and **G**), DB71 (**D** and **H**), and Amido Black (**E**). All staining procedures were as described in **Table 1**.

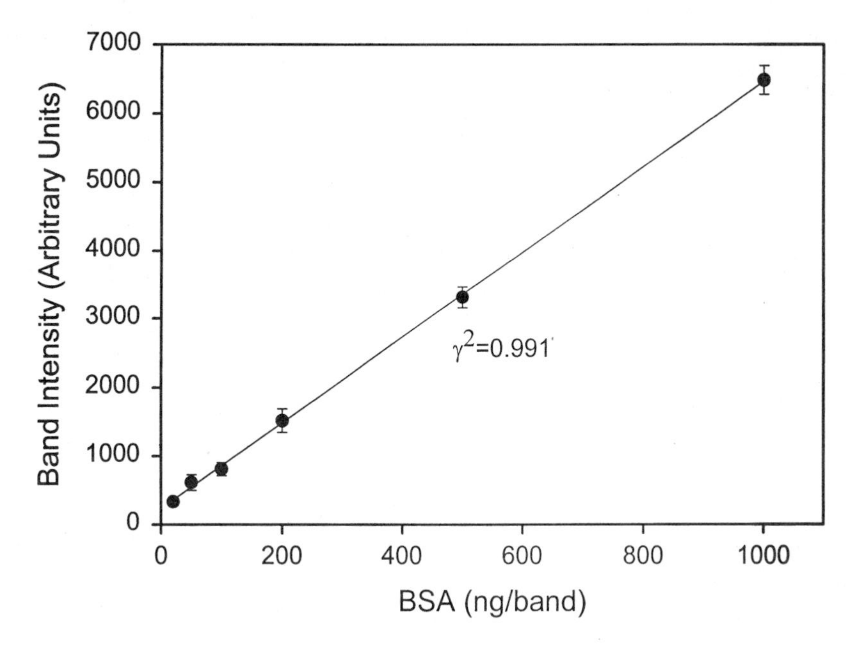

Fig. 3. Standard assay of BSA after DB71 staining. The peak areas of the densitometric measurements of BSA representing the relative band intensities were used to obtain a response curve. Electroblotted BSA was stained for 5 min in 0.008% DB71 solution and washed briefly with 40% ethanol–10% acetic acid, and quantified by using a TINA 2.09 software program. Data are expressed as the means ± SEM of the mean of triplicate of the experiments.

References

1. Towbin, H., Gordon, J. (1984) Immunoblotting and dot immunoblotting -current status and outlook. *J. Immunol. Meth.* **72,** 313–340.
2. Gershoni, J. M. and Palade, G. E. (1983) Protein blotting: principles and applications. *Analyt. Biochem.* **131,** 1–15.
3. Blake, M. S., Johnston, K. H., Russel-Jones, G. J., and Gotschlich, E. C. (1984) A rapid, sensitive method for detection of alkaline phosphatase conjugated anti-antibody on Western blots. *Analyt. Biochem.* **136,** 175–179.
4. Moeremans, M., Daneels, G., and De Mey, J. (1985) Sensitive colloidal metal (gold or silver) staining of protein blots on nitrocellulose membranes. *Analyt. Biochem.* **145,** 315–321.
5. Bollag, D. M. and Edelstein, S. J. (1991) Immunoblotting, in *Protein Methods*, Wiley-Liss, New York, pp. 181–211.
6. Soutar, A. K. and Wade, D. P. (1989) In *Protein Function; A Practical Approach* (Creighton, T. E., ed.), IRL Press, Oxford and Washington DC, p. 55.
7. Mitra, P., Pal, A. K., Basu, D., and Hati, R. N. (1994) A staining procedure using Coomassie Brilliant Blue G-250 in phosphoric acid for detection of protein bands with high resolved polyacrylamide gel and nitrocellulose membrane. *Analyt. Biochem.* **223,** 327–329.
8. Bio-Rad Laboratories, (1987) *Minitrans-Blot Electrophoretic Transfer Cell Manual*, Richmond, CA.
9. Hancock, K. and Tsang, V. C. (1983) India ink staining of proteins on nitrocellulose paper. *Analyt. Biochem.* **133,** 157–162.
10. Hughes, J. H., Mack, K., and Hamparian, V. V. (1988) India ink staining of proteins on nylon and hydrophobic membranes. *Analyt. Biochem.* **173,** 18–25.
11. Aebersold, R., Leavitt, J., Saavedra, R., Hood, L., and Kent, S. (1987) Internal amino acid sequence analysis of proteins separated by one- or two-dimensional gel electrophoresis after in situ protease digestion on nitrocellulose. *Proc. Natl. Acad. Sci. USA* **84,** 6970–6974.
12. Salinovich, O. and Montelaro, R. C. (1986) Reversible staining and peptide mapping of proteins transferred to nitrocellulose after separation by sodium dodecylsulfate-polyacrylamide gel electrophoresis. *Analyt. Biochem.* **156,** 341–347.
13. Wirth, P. and Romano, A. (1995) Staining methods in gel electrophoresis, including the use of multiple detection methods. *J. Chromatogr. A* **698,** 123–143.
14. Root, D. D. and Reisler, E. (1989) Copper iodide staining of protein blots on nitrocellulose membranes. *Analyt. Biochem.* **181,** 250–253.
15. Patton, W. F., Lam, L., Su, Q., Lui, M., Erdjument-Bromage, H., and Tempst, P. (1994) Metal chelates as reversible stains for detection of electroblotted proteins: application to protein microsequencing and immunoblotting. *Analyt. Biochem.* **220,** 324–335.
16. Shojaee, N., Patton, W. F., Lim, M. J., and Shepro, D. (1996) Pyrogallol red-molybdate: a reversible metal chelate stain for detection of proteins immobilized on membrane supports. *Electrophoresis* **17,** 687–693.
17. Lim, M. J., Patton, W. F., Lopez, M., Spofford, K., Shojaee, N., and Shepro, D. (1997) A luminescent europium complex for the sensitive detection of proteins and nucleic acids immobilized on membrane supports. *Analyt. Biochem.* **245,** 184–195.
18. Hong, H. Y., Yoo, G. S., and Choi J. K. (2000) Direct Blue 71 staining of proteins bound to blotting membranes, *Electrophoresis* **21,** 841–845.
19. Laemmli, U. K. (1970) Cleavage of structural proteins during the assembly of the head of bacteriophage T4. *Nature* **227,** 680–685.
20. Bickar, D. and Reid, P. D. (1992) A high-affinity protein stain for western blots, tissue prints, and electrophoretic gels. *Analyt. Biochem.* **203,** 109–115.
21. Fazekas de St. Groth, S., Webster, R. G., and Datyner, A. (1963) Two new staining procedures for quantitative estimation of proteins on electrophoretic strips. *Biochim. Biophys. Acta* **71,** 377–391.

53

Protein Staining and Immunodetection Using Immunogold

Susan J. Fowler

Introduction

Probes labeled with colloidal gold were originally used as electron-dense markers in electron microscopy *(1–3)* and as color markers in light microscopy *(4)*. Their application to immunoblotting was not examined until later *(5–7)*. The combination of gold-labeled antibodies and protein A was demonstrated to be suitable for the visualization of specific antigens on Western blots and dot blots *(5,6)*. When gold-labeled antibodies are used as probes on immunoblots, the antigen–antibody interaction is seen as a pinkish signal owing to the optical characteristics of colloidal gold *(5)*. Used on its own, the sensitivity of immunogold detection is equivalent to indirect peroxidase methods, and hence, only suitable for situations where there are higher levels of antigen. In addition, the signal produced is not permanent. In order to overcome this problem and to allow the technique to be used for more demanding applications, a way of amplifying the signal was subsequently developed using the capacity of gold particles to catalyze the reduction of silver ions *(8)*. This reaction results in the growth of the gold particles by silver disposition. A stable dark brown signal is produced on the blot, and sensitivity is increased 10-fold. The sensitivity achieved using immunogold silver staining (IGSS) is similar to that obtained with alkaline phosphatase using colorimetric detection and several times more sensitive than ^{125}I-labeled antibodies. However, unlike colorimetric detection, the result is stable and not prone to fading, and the chemicals used present no hazards. In addition, the signal-to-noise ratio of IGSS is usually very high, and there are none of the handling or disposal problems that are associated with ^{125}I-labeled antibodies.

The binding of the gold to antibodies is via electrostatic adsorption. It is influenced by many factors, including particle size, ionic concentration, the amount of protein added, and its molecular weight. Most importantly, it is pH dependent *(9)*. An additional feature is that the binding of the gold does not appear to alter the biological or immunological properties of the protein to which is attached. The colloidal gold particles used to label antibodies can be produced in different sizes ranging from 1 to 40 nm in diameter. For immunoblotting, Amersham Pharmacia Biotech Ltd. (Amersham, UK) supplies AuroProbe BL plus antibodies labeled with 10 nm particles.

From: *The Protein Protocols Handbook, 2nd Edition*
Edited by: J. M. Walker © Humana Press Inc., Totowa, NJ

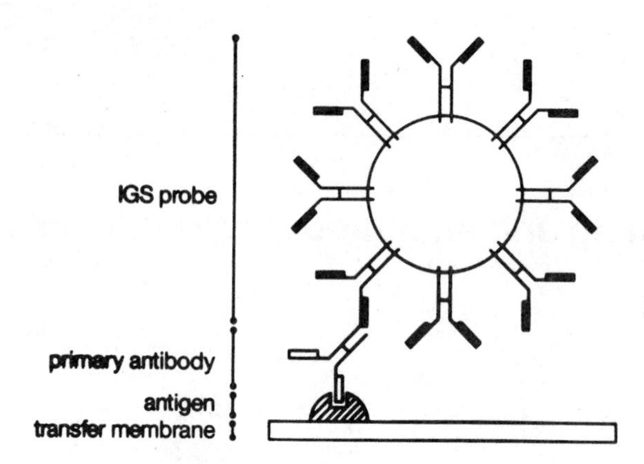

Fig. 1. Principle of the indirect visualization of antigens using immunogold probes. The primary antibodies bind to immobilized antigens on the blot and are in turn recognized by gold-labeled secondary antibodies. The above illustration of the binding pattern of secondary antibodies to gold particles represents the type of conjugate formed with gold particles of 10 nm or larger. The exact configuration adopted by the antibodies on the gold particle is not known.

For each 10 nm gold particle, there will be several antibodies bound, and the probe can be considered to be a gold particle coated with antibodies (*see* **Fig. 1**). For certain applications 1-nm gold particles allow increased labeling efficiency and can give greater sensitivity than larger particles ***(10)***. This may be owing in part to the larger number of intensifiable gold particles/unit of antigen ***(11)***. However, when using antibodies labeled with 1-nm gold particles, the small size of the gold particles means that visualization of the antigen–antibody interaction can be achieved only using silver enhancement.

Silver enhancement was first reported by Danscher ***(8)***, who used silver lactate and hydroquinone at a pH of 3.5. In the presence of the gold particles, that act as catalysts, the silver ions are reduced to metallic silver by the hydroquinone. The silver atoms formed are deposited in layers on the gold surface, resulting in significantly larger particles and a more intense macroscopic signal (*see* **Fig. 2**). This classical enhancement worked well, but had several disadvantages. The system was sensitive to light and chemical contamination. In addition, the components were not stable and were prone to self-nucleation, a phenomenon whereby the reduction of silver ions occurs spontaneously in solution to form silver particles that can be deposited and lead to high background.

More recently, silver-enhanced reagents have been developed that overcome these problems. IntenSE BL (Amersham Pharmacia Biotech) is light insensitive and has a neutral pH. It also exhibits delayed self-nucleation, which allows a fairly large time margin before it is necessary to stop the reaction (*see* **Fig. 3**). Using silver enhancement allows sensitivity to be increased 10-fold over immunogold detection.

In addition to the immunological detection of proteins, gold particles can also be used as a general stain for proteins on blots ***(12)***. AuroDye forte (Amersham Pharmacia Biotech Ltd) is a stabilized colloidal gold solution adjusted to a pH of 3.0. At this pH, the negatively charged gold particles bind very selectively to proteins by hydrophobic and ionic interactions. The proteins thus stained appear as dark red. The sensitivity

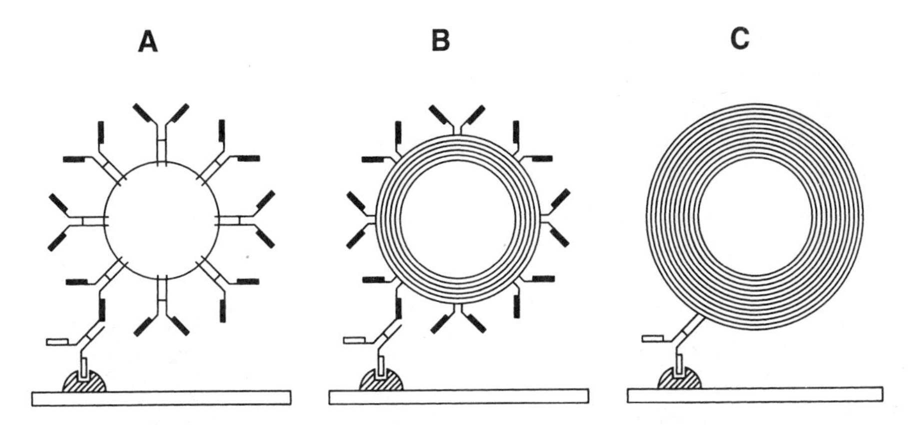

Fig. 2. Schematic representation of the silver-enhancement process. In the initial phase (**A**), the gold probe attaches to the primary antibody, which is bound to immobilized antigen. During the silver-enhancement process, layers of silver selectively precipitate on the colloidal gold surface (**B**). The result is a significantly larger particle and a silver surface that generates a more intense macroscopic signal (**C**).

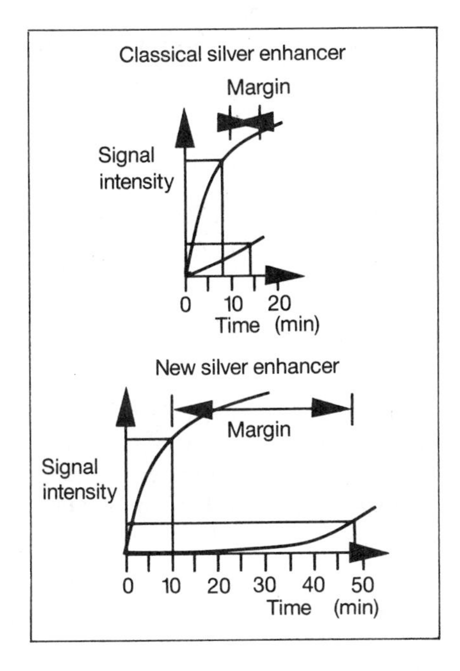

Fig. 3. Silver-enhancement time dependency for both the classical and IntenSE BL silver enhancers.

obtained is comparable to that of silver staining for polyacrylamide gels. Segers and Rabaey ***(13)*** found it detected more spots on transfers of two-dimensional (2-D) gels than silver straining of the gels. For applications demanding very high sensitivity, it is also possible to amplify the signal further by performing a silver-enhancement step using ItenSE BL. An additional feature of total protein staining with gold is that, as with India ink ***(14)***, the immunoreactivity of the proteins is not altered.

There are two approaches to allow combining of immunodetection and total protein staining on the same blot: the immune reaction is performed first and the total protein stain is performed afterwards (AuroDye method) or the blot is stained for total protein and then the immunodetection is performed ("citrate gold" method). Citrate gold ***(15)*** gives approx 30-fold lower sensitivity than AuroDye for protein staining but allows the use of protein based blocking agents prior to immunodetection. In contrast, AuroDye gives high sensitivity but only Tween 20 can be used to block prior to immunodetection, which can make background difficult to control for chemiluminescent detection methods. Thus, using citrate gold it is possible for specific proteins to be immunodetected with chemiluminescent ***(16)*** or colorimetric substrates after total protein staining ***(15,17)***. This dual detection has been used with blotted 2D gels ***(16,18)***, where it allows very precise mapping of the immunodetected protein against the background of total proteins. Alternatively, if proteins are omitted from block solutions, total protein staining can be performed after immunodetection using AuroDye ***(19)***.

In summary, immunogold silver staining is a highly sensitive detection method that provides a permanent record of results. Protocols are simple to use, and provided the silver-enhancement step is carefully monitored, background interference is negligible. Results are obtained without the prolonged autoradiographic exposures associated with radioactive methods, and the chemicals used do not present any hazards or disposal problems.

2. Materials

2.1. General

1. Nitrocellulose, nylon, or polyvinylidene fluoride (PVDF) membranes: AuroProbe BL plus can be used with any of the three types of membrane; AuroDye forte is compatible only with nitrocellulose and PVDF membranes.
2. Phosphate-buffered saline (PBS), pH 7.2, containing 0.02% azide: 8 g of NaCl, 0.2 g KCl, 1.4 g of $Na_2HPO_4 \cdot 2H_2O$, 0.2 g of KH_2PO_4, 0.2 g of sodium azide. Adjust pH to 7.2, and make up to 1000 mL with distilled water.
3. Gelatin: The gelatin used should be of high quality if it is to inhibit nonspecific binding of gold probes effectively. IGSS quality is supplied as a component of AuroProbe BLplus kits.
4. Analytical-grade chemicals should be used throughout, and water should be distilled and deionized. Where silver enhancement is used, it is important that the glassware and plastic containers used are scrupulously clean and are not contaminated with heavy metals or their salts.

2.2. Immunogold Silver Staining Using AuroProbe BLplus

The buffer system outlined below gives very clean backgrounds without the use of Tween 20. The use of Tween 20 in blocking, incubation, and/or washing can lead to nonspecific binding of gold probes to blotted proteins from certain types of sample, such as whole cultured cells and isolated nuclei extracts.

1. Wash buffer: 0.1% (w/v) bovine serum albumin (BSA) in PBS.
2. Block buffer for nitrocellulose or PVDF membranes: 5% (w/v) BSA in PBS.
3. Block buffer for nylon membranes: 10% (w/v) BSA in PBS.
4. Primary antibody diluent buffer: 1% normal serum v/v (from the same species as that in which the secondary antibody was raised) diluted in wash buffer.
5. Gelatin buffer: 1% (v/v) IGSS-quality gelatin in wash buffer (equivalent to a 1:20 dilution of the gelatin supplied with AuroProbe BLplus).

6. AuroProbe BLplus secondary antibody, 1:100 diluted in gelatin buffer; or biotinylated secondary antibody 2 µg/mL diluted in gelatin buffer, and AuroProbe BLplus streptavidin 1:100 diluted in gelatin buffer.
7. Enhancer solution: Ready-to-use component of the IntenSE BL kit.
8. Initiator solution: Ready-to-use component of the IntenSE BL kit.

2.3. General Staining of Blotted Proteins Using AuroDye Forte

1. Tween 20: Component of AuroDye forte kit (not all brands of Tween 20 give satisfactory results; it is important to use reagent that has been quality controlled for this purpose).
2. Wash and block buffer: 0.3% (v/v) Tween 20 in PBS.

3. Methods

3.1. General

Avoid skin contact with the transfer membrane. Wear gloves throughout the procedure. Handle blots by their edges using clean plastic forceps. Incubations and washes should be carried out under constant agitation. Plastic containers on an orbital shaker are ideal for this purpose. Alternatively, if it is necessary to conserve antibodies, antibody incubations can be performed in cylindrical containers on roller mixers ***(20)***.

3.2. Immunogold Silver Staining Using AuroProbe BLplus

1. After transfer of the proteins to this membrane, incubate the blot in block solution for 30 min at 37°C. A shaking water bath is suitable for this incubation. If nylon membranes are used, this period should be extended to overnight. All subsequent steps are performed at room temperature. It is important that there be enough block solution to cover the blot easily.
2. Remove excess block by washing the blot 3× for 5 min in wash buffer. As large a volume of wash buffer as possible should be used each time.
3. Prepare a suitable dilution of primary antibody in diluent buffer. If the antibody is purified, 1–2 µg/mL is a suitable concentration. If unpurified antiserum is used, a dilution >1:500 is recommended.
4. Incubate the blot in this solution for 1–2 h.
5. Wash the blot as described in **step 2**.
6. Prepare a 1:100 dilution of AuroProbe BLplus antibody in gelatin buffer. If the AuroProbe BLplus streptavidin system is being used, the biotinylated second antibody should be diluted to a concentration of 2mg/mL gelatin buffer.
7. Incubate the blot in second antibody for 2 h under constant agitation.
8. Wash blot as described in **step 2**.
9. If using the AuroProbe BLplus streptavidin system, incubate the blot in a 1:100 dilution of streptavidin-gold for 2 h.
10. Wash blot as described in **step 2**.
11. Wash the gold-stained blot twice for 1 min in distilled water. Do not leave the blot in distilled water for long periods, since this may lead to the release of gold particles from the surface. The result can be reviewed at this stage before going on to perform the silver-enhancement step.
12. Pour equal volumes of enhancer and initiator solutions into a plastic container (100 mL is sufficient for a 10 × 15 cm blot). Immediately add the gold-stained blot, and incubate under constant agitation for 15–40 min. The enhancement procedure can be monitored and interrupted or extended as necessary.
13. Wash the blot 3 × for 10 min in a large volume of distilled water.
14. Remove blot and leave to air-dry on filter paper (*see* **Fig. 4**).

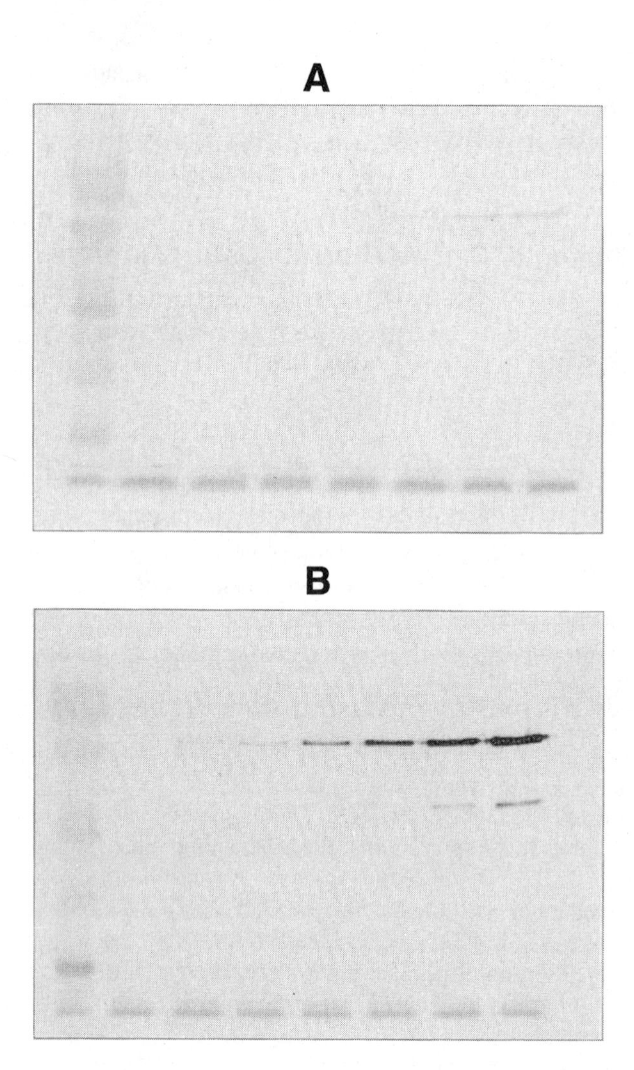

Fig. 4. Detection of bound proteins using colloidal gold-labeled antibodies with (**B**) and without (**A**) silver-enhancement. (**A**) Doubling dilutions of rat brain homogenate separated by 12% SDS-PAGE and transferred to nitrocellulose membrane, followed by immunodetection with mouse monoclonal anti-B-tubulin (1:1000) and AuroProbe BLplus GAM IgG (1:1000). (**B**) As for (**A**), but then subjected to silver-enhancement with IntenSE BL for 20 min.

3.3. General Protein Staining with AuroDye Forte

1. Incubate the blot in an excess of PBS containing 0.3% Tween 20 at 37°C for 30 min. Perform subsequent incubations at room temperature.
2. Further incubate the blot in PBS containing 0.3% Tween 20 3× for 5 min at room temperature.
3. Rinse the blot for 1 min in a large volume of distilled water.
4. Place the blot in AuroDye forte for 2-4 h. The staining can be monitored during this time.
5. When sufficient staining has been obtained, wash the blot in a large volume of distilled water, and leave to air-dry on a piece of filter paper.
6. In cases where extremely high sensitivity or contrast is required, the AuroDye signal can be further amplified with IntenSE BL as described in **Subheading 3.2., steps 12–14** (*see* **Fig. 5**).

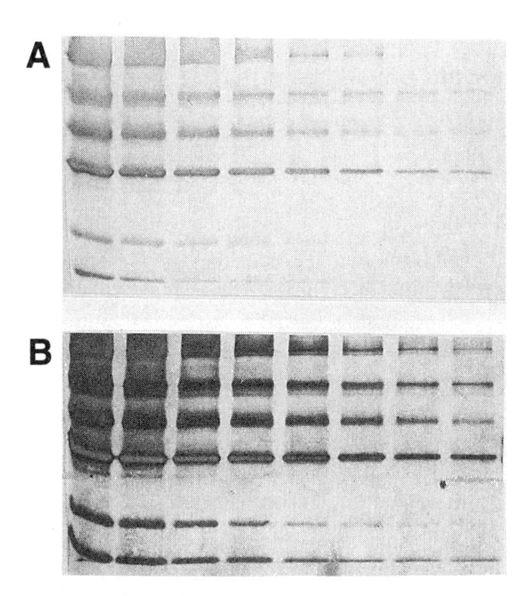

Fig. 5. Comparison of AuroDye forte total protein staining with AuroDye forte staining with silver enhancement. Doubling dilutions of molecular weight marker proteins: phosphorylase *b*, BSA, ovalbumin, carbonic anhydrase, trypsin inhibitor, and lysozyme from 200 to 0.375 ng were separated on a 12% gel and transferred to Hybond PVDF membrane (Amersham Pharmacia Biotech). Both blots were stained with AuroDye after which blot (**B**) was amplified with IntenSE BL for 15 min.

4. Notes

4.1. Electrophoresis and Electroblotting

1. During electrophoresis, care should be taken to ensure that no extraneous proteins are introduced. Glassware should be thoroughly cleaned, and all solutions should be prepared freshly. Low-ionic-strength transfer buffers are recommended for blotting (i.e., 25 m*M* Tris, 192 m*M* glycine, 20% methanol, pH 8.3).
2. For optimum total protein staining with AuroDye forte, it is important to place a piece of transfer membrane on both sides of the gel. The use of the extra transfer membrane at the cathodic side of the gel, combined with the use of high-quality filter paper, ensures high contrast staining with negligible background. For semidry blotting, the extra piece of transfer membrane is not necessary.

4.2. Immunogold Silver Staining with AuroProbe BLplus

3. When using a primary antibody for the first time, it is recommended that its concentration be optimized by performing a dot-blot assay. Antigen dot blots of a suitable concentration are prepared and air-dried. The blocking, washing and incubation conditions are as outlined in **Subheading 3.** A series of primary antibody dilutions are then made, and a dot blot incubated in each. The dilution of the gold-labeled second antibody is kept constant. The primary antibody dilution giving maximum signal with minimum nonspecific binding should be chosen.
4. It is essential to have gelatin in the incubation with the immunogold reagent to prevent nonspecific binding. The source of gelatin is extremely important. If gelatin other than the one supplied in the AuroProbe kits is used, the inclusion of a negative control is essential.

Table 1
Troubleshooting Immunogold Silver Staining

Observation	Probable cause	Remedy
Precipitation of silver enhancement mixture before indicated time interval	Glassware: chromic acid was used for rinsing and it was not washed away with HCl.	Rinse glassware several times with 0.1 *M* HCl and distilled water
	Glassware: Traces of metals originating from metal parts, e.g., cleaning brushes, or originating from previous experiments in which metallic compounds were used, e.g., silver staining.	Avoid contact with metallic objects; do not use glassware brush with metallic handle to clean glassware; use disposable plastics instead.
High background and non specific staining	Microprecipitates that are macroscopically invisible produce a high background when the stability time limit is reached.	Incubate for a shorter time in silver-enhancement mixture.
	Primary antibody is too concentrated.	Optimize dilution of primary antibody using dot-blot assay.
	Wrong type of gelatin gelatin used during incubation with AuroProbe	Use the type of gelatin prescribed.
No staining	Difficulties regarding the reactivity of the primary antibody with the immunogold reagents	Perform a dot-blot assay (*see* **Subheading 4.3.**)
	Inefficient transfer from gel to membrane	Optimize blotting conditions; silver-stain gel to see what remains.
	Error in handling: the steps were not performed in the right order or a step was omitted.	Repeat the procedure in the right order.
	Excessive dilution of primary antibody	Optimize primary antibody dilution using a dot-blot assay.
	Nonreactive primary antibody or a primary antibody that was destroyed by inappropriate storage conditions	Use a primary antibody of highest possible antibody quality, and repeat the procedure with a new batch of primary antibody.
	The gold probe may have been denatured owing to wrong storage conditions.	Repeat procedure with fresh gold probe.
Signal too weak	Excessive dilution of primary antibody	Optimize dilution of primary antibody using a dot-blot assay.
	Excessive dilution of AuroProbe reagent or too short an incubation time	Use AuroProbe reagent as recommended in **Subheading 3.2.**
	Silver-enhancement time too short	Rinse the membrane in distilled water and repeat the silver-enhancement in fresh reagent.

Table 2
Effect of Temperature on Enhancement Time and Enhancement Reagent Stability

Temperature (°C)	Typical enhancement time (min)	Typical stability time (min)
16	27–45	>80
18	22–38	>70
20	20–35	>55
22	18–33	>45
24	16–27	>40

5. If desired, the incubations in immunogold reagent can be extended to overnight when concentrations of 1:100 to 1:400 are used.
6. If, after performing the experiment, there is a complete absence of signal, the reactivity of the immunogold reagent with the primary antibody should be checked by performing a dot-blot assay. Prepare a dilution series of primary antibody, for example, from 250 to 0.5. ng/μL. Spot out 1 μL onto the membrane, and allow to air-dry. Proceed with the appropriate immunogold silver staining using the blocking, washing and incubation conditions as described in **Subheadings 3.2.** and **3.3.** (*see* **Table 1**).

4.3. Silver Enhancement

7. The enhancement reagents are extremely sensitive to the purity of the water used. Low quality water results in the formation of precipitates that reduce the reactivity of the enhancement and can lead to high backgrounds. In addition, glassware contaminated with heavy metals in elemental form or as heavy metal salts will decrease the performance of the enhancer reagents.
8. The silver enhancement reagents are prepared by mixing the enhancer and initiator solutions in equal quantities. The mixture is usable only over a defined time period (*see* **Table 2**), so it is important that the components be combined immediately before use. There is no need to shield the silver enhancer from normal daylight.
9. Both the enhancement time and the stability of the silver-enhancement mixture vary considerably with ambient temperature. A typical enhancement time for most blotting experiments, of 15–40 min at room temperature (22°C) is recommended. For some applications, it may be necessary to extend the enhancement time, and for others, to shorten it. At room temperature, there is a comfortable safety margin to enable maximum enhancement before there is any danger of self-nucleation of the enhancer reagent occurring.
10. When a very strong amplification signal is desired, it is possible to perform a second silver-enhancement step before self-nucleation starts. In this case, the blot is subjected to silver enhancement for 30 min at room temperature (22°C). It is then rinsed with distilled water and immersed in a freshly prepared silver-enhancement solution for another 20–30 min. The increase in signal slows down considerably with time. After enhancement, the blot should be washed in distilled water and dried.

4.4. General Protein Staining with AuroDye Forte

11. Owing to the high sensitivity of AuroDye forte, special care needs to be taken to avoid background staining. It is important to wear gloves when handling gels and blots. Where possible, handle blots by their edges with forceps, since gloves can leave smears. High quality chemicals should be used throughout.

Table 3
Troubleshooting General Protein Staining with AuroDye Forte

Observation	Probable cause	Remedy
AuroDye forte turns purplish during staining	Agglutination of gold particles by proteins released from the blot	Wash blot briefly in excess distilled water; replace AuroDye forte; if possible, use lower protein loads; this observed with PVDF membrane because it retains proteins better than nitrocellulose.
AuroDye forte turns colorless during staining.	Adsorption of all the gold particles by excess protein on the blot	Replace AuroDye forte and double the volume used per cm^2 of blot; if possible use lower protein loads.
Spotty background	Impurities released from filter paper adsorbed onto blot during transfer	Use high-quality filter paper during electroblotting procedure.
High background	Interference by proteinaceous contaminants	Use extra transfer membrane on cathodic side of the gel; use clean Scotch-Brite pads.
	Optional silver-enhancement time was too long	Use a shorter silver-enhancement time.
	Interference by chemical contaminants	Always use high-quality chemicals.
Smears on background	Incorrect handling	Handle blots by their edges using clean plastic forceps; avoid contact with gloves.

12. If large amounts of protein are loaded on the gel, when transferred to the membrane they will not only be heavily stained, but will leak off the membrane from saturated sites. Excessive protein leakage will cause an aggregation of the gold particles and destroy the AuroDye forte reagent. The problem is generally more severe with 1-D, than with 2-D gels, in which the proteins are more spread out. For 1-D gels, it is recommended that protein loads should be equivalent to amounts capable of giving resolvable bands after silver staining. Single bands in 1-D gels should not exceed 1000 ng of protein. Molecular weight standards should be loaded at approx 200 ng/band. In general, the use of lower protein loadings will give better separation, and samples will be conserved.
13. When staining 2-D gels, they should be thoroughly washed in several changes of excess transfer buffer after electrophoresis to remove any remaining ampholytes.
14. AuroDye forte is a stabilized gold colloid sol, adjusted to a pH of approx 3. At this low pH, the negatively charged gold particles bind very selectively to proteins by hydrophobic and ionic interactions. This may result in different staining intensities, depending on the isoelectric point of the proteins being stained. However, this has been reported to be a feature of other protein staining methods, such as silver staining and Coomassie Blue ***(21)***.

15. AuroDye forte is designed for use on nitrocellulose and PVDF membranes. It cannot be used on nylon membranes where its charge will result in staining of the whole membrane (*see* **Table 3**).

References

1. Faulk, W. P. and Taylor, G. M. (1971) An immunocolloid method for the electron microscope. *Immunochemistry* **8,** 1081.
2. Romano, E. L., Stolinski, C., and Hughes-Jones, N. C. (1974) An immunoglobulin reagent labelled with colloidal gold for use in electron microscopy. *Immunocytochemistry* **14,** 711–715.
3. Horisberger, M. and Rosset, J. (1977) Colloidal gold, a useful marker for transmission and scanning electron microscopy. *J. Histochem. Cytochem.* **25,** 295–305.
4. Roth, J. (1982) Applications of immunocolloids in light microscopy: preparation of protein A-silver and protein A-gold complexes and their applications for the localization of single and multiple antigens in paraffin sections. *J. Histochem. Cytochem.* **30,** 691–696.
5. Moeremans, M., Daneels, G., Van Dijck, A., Langanger, G., and De Mey, J. (1984) Sensitive visualization of antigen-antibody reactions in dot and blot immune overlay assays with immunogold and immunogold/silver staining. *J. Immunol. Meth.* **74,** 353–160.
6. Brada, D. and Roth, J. (1984) Golden Blot - detection of polyclonal and monoclonal antibodies bound to antigens on nitrocellulose by protein A-gold complexes. *Analyt. Biochem.* **142,** 79–83.
7. Hsu, Y. H. (1984) Immunogold for detection of antigen on nitrocellulose paper. *Analyt. Biochem.* **142,** 221–225.
8. Danscher, G. (1981) Histochemical demonstration of heavy metals, a revised version of the sulphide silver method suitable for both light and electron microscopy. *Histochemistry* **71,** 1–16.
9. Geoghegan, W. D. and Ackerman, G. A. (1977) Adsorption of horseradish peroxidase, ovomucoid and anti-immunoglobulin to colloidal gold for the indirect detection of concanavilin A, wheat germ agglutinin and goat antihuman immunoglobulin G on cell surfaces at the electron microscopic level: a new method, theory and application. *J. Histochem. Cytochem.* **25,** 1182–1200.
10. Western Blotting Technical Manual, Amersham International plc. 1991, Amersham UK.
11. Moeremans, M., Daneels, G., De Raeymaeker, M., and Leunissen, J. L. M. (1989) AuroProbe One in immunoblotting, in *Aurofile 02,* Janssen Life Sciences, Wantage, UK, pp. 4, 5.
12. Moeremans, M., Daneels, G., and DeMey, J. (1985) Sensitive colloid (gold or silver) staining of protein blots on nitrocellulose membrane. *Analyt. Biochem.* **145,** 315–321.
13. Segers, J. and Rabaey, M. (1985) Sensitive protein stain on nitrocellulose blots. *Protides Biol. Fluids* **33,** 589–591.
14. Glenney, J. (1986) Antibody probing of Western blots which have been stained with India ink. *Analyt. Biochem.* **156,** 315–319.
15. Chevallet, M., Procaccio V., and Rabilloud, R. (1997) A non-radioactive double detection method for the assignment of spots in two-dimensional blots. *Analyt. Biochem.* **251,** 69–72.
16. Egger, D. and Bienz, K. (1987) Colloidal gold staining and immunoprobing of proteins on the same nitrocellulose blot. *Analyt. Biochem.* **166,** 413–417.
17. Egger, D. and Bienz, K. (1992) Colloidal gold staining and immunoprobing on the same Western blot, in *Methods in Molecular Biology,* Vol. 10, *Immunochemical Protocols* (Manson, M., ed.). Humana Press, Totowa, NJ, pp. 247–253.

18. Schapira, A. H. V. (1992) Colloidal gold staining and immunodetection in 2D protein mapping, in *Methods in Molecular Biology,* Vol. 10, *Immunochemical Protocols* (Manson, M., ed.), Humana Press, Totowa, NJ, pp. 255–266.
19. Daneels, I. J., Moeremans, M., De Raemaeker, M., and De Mey, J. (1986) Sequential immunostaining (gold/silver) and complete protein staining (AuroDye) on Western blots. *J. Immunol. Meth.* **89,** 89–91.
20. Thomas, N., Jones, C. N., and Thomas, P. L. (1988) Low volume processing of protein blots in rolling drums. *Analyt. Biochem.* **170,** 393–396.
21. Jones, A. and Moeremans, M. (1988) Colloidal gold for the detection of proteins on blots and immunoblots, in *Methods in Molecular Biology,* Vol. 3, *New Protein Techniques* (Walker, J. M., ed.), Humana Press, Totowa, NJ, pp. 441–479.

54

Detection of Polypeptides on Immunoblots Using Enzyme-Conjugated or Radiolabeled Secondary Ligands

Nicholas J. Kruger

1. Introduction

Immunoblotting provides a simple and effective method for identifying specific antigens in a complex mixture of proteins. Initially, the constituent polypeptides are separated using sodium dodecyl sulfate-polyacrylamide gel electrophoresis (SDS-PAGE), or a similar technique, and are then transferred either electrophoretically or by diffusion onto a nitrocellulose filter. Once immobilized on a sheet of nitrocellulose, specific polypeptides can be identified using antibodies that bind to antigens retained on the filter and subsequent visualization of the resulting antibody-antigen complex. This chapter describes conditions suitable for binding antibodies to immobilized proteins and methods for locating these antibody-antigen complexes using appropriately labeled ligands. These methods are based on those of Blake et al. *(1)*, Burnette *(2)* and Towbin et al. *(3)*.

Although there are several different techniques for visualizing antibodies bound to nitrocellulose, most exploit only two different types of ligand. One is protein A conjugated to a marker enzyme, radiolabeled or otherwise tagged. The other ligand is an antibody raised against immunoglobulin G (IgG) from the species used to generate the primary antibody. Usually, this secondary antibody is either conjugated to a marker enzyme or linked to biotin. In the latter instance, the biotinylated antibody is subsequently detected using avidin (or streptavidin) linked to a marker enzyme.

Detection systems based on protein A are both convenient and sensitive. Protein A, from the cell wall of *Staphylococcus aureus*, specifically binds the Fc region of IgG from many mammals *(4)*. Thus, this compound provides a general reagent for detecting antibodies from several sources. Using this ligand, as little as 0.1 ng of protein may be detected, although the precise amount will vary with the specific antibody titer *(5)*. The principal disadvantage of protein A is that it fails to bind effectively to major IgG subclasses from several experimentally important sources, such as rat, mouse, goat, and sheep (*see* **Table 1**). For antibodies raised in such animals a similar method using derivatives of other bacterial immunoglobulin-binding proteins may be suitable (*see* **Note 1**). Alternatively, antibody bound to the nitrocellulose filter may be detected

From: *The Protein Protocols Handbook, 2nd Edition*
Edited by: J. M. Walker © Humana Press Inc., Totowa, NJ

Table 1
Variation in Species Specificity of Various Immunogobulin-Binding Proteins

Species	Immunoglobulin class	Affinity of binding by			
		Protein A	Protein G	Protein L	Protein LA
Human	IgG1[a]	+++	+++	+++	+++
	IgG2	+++	+++	+++	+++
	IgG3	–	+++	+++	+++
	IgG4	+++	+++	+++	+++
	IgM	–	–	+++	+++
	IgA	–	–	+++	+++
	IgE	–	–	+++	+++
	IgD	–	–	+++	+++
Mouse	IgG1[a]	+	+++	+++	+++
	IgG2a	+++	+++	+++	+++
	IgG2b	++	++	+++	+++
	IgG3	+	++	+++	+++
Rat	IgG1	–	+	+++	+++
	IgG2a	–	+++	+++	+++
	IgG2b	–	++	+++	+++
	IgG2c	+	++	+++	+++
Cow	IgG	++	+++	–	++
Cat	IgG	+++	–	n.d.	n.d.
Chicken	IgG	–	+	++	++
Dog	IgG	+++	+++	+	++
Goat	IgG	+/–	++	–	+/–
Guinea pig	IgG	+++	++	++	+++
Hamster	IgG	+	++	+++	+++
Horse	IgG	++	+++	+/–	++
Pig	IgG	++	++	+++	+++
Rabbit	IgG	+++	++	+	+++
Sheep	IgG	+/–	++	–	+/–

The binding affinities for immunoglobins from different sources is indicated as follows: –, no binding; +, low; ++, moderate; +++, high; n.d., not determined.

[a]Denotes major subclass of IgG.

Based on data in **refs.** ***4,8–10*** and references therein.

using a second antibody raised against IgG (or other class of immunoglobulin) from the species used to generate the primary antibody. The advantage of such secondary antibody systems is that they bind only to antibodies from an individual species. When combined with different marker enzymes, the specificity of secondary antibodies may be exploited to identify multiple polypeptides on a single nitrocellulose membrane ***(4,6)***.

The marker enzymes most commonly used for detection are alkaline phosphatase and horseradish peroxidase. Both enzymes can be linked efficiently to other proteins, such as antibodies, protein A, and avidin, without interfering with the function of the latter proteins or inactivating the enzyme. Moreover, a broad range of synthetic substrates have been developed for each of these enzymes. Enzyme activity is normally visualized by incubating the membrane with an appropriate chromogenic substrate that is converted to a colored, insoluble product. The latter precipitates onto the membrane in the area of enzyme activity, thus identifying the site of the antibody-antigen complex (*see* **Note 2**).

Both antigens and antisera can be screened efficiently by immunoblotting. Probing of a crude extract after fractionation by SDS-PAGE can be used to assess the specificity of an antiserum. The identity of the antigen can be confirmed using a complementary technique, such as immunoprecipitation of enzyme activity. This information is essential if the antibodies are to be used reliably. Once characterized, an antiserum may be used to identify antigenically related proteins in other extracts using the same technique (*see* **Note 17**). Examples of the potential of immunoblotting have been described by Towbin and Gordon *(7)*.

2. Materials

1. Electrophoretic blotting system, such as Trans-Blot, supplied by Bio-Rad.
2. Nitrocellulose paper: 0.45 µm pore size.
3. Protein A derivative.
   a. Alkaline phosphatase-conjugated protein A obtained from Sigma–Aldrich Co. Dissolve 0.1 mg in 1 mL of 50% (v/v) glycerol in water. Store at –20°C.
   b. Horseradish peroxidase-conjugated protein A obtained from Sigma–Aldrich Co. Dissolve 0.1 mg in 1 mL of 50% (v/v) glycerol in water. Store at –20°C.
   c. ^{125}I-labeled protein A, specific activity 30 m Ci/mg. Affinity-purified protein A, suitable for blotting, is available commercially (*see* **Note 3**). **^{125}I emits γ-radiation. Be sure that you are familiar with local procedures for safe handling and disposal of this radioisotope.**
4. Secondary antibody: A wide range of both alkaline phosphatase and horseradish peroxidase conjugated antibodies are available commercially. They are usually supplied as an aqueous solution containing protein stabilizers. The solution should be stored under the conditions recommended by the supplier. *Ensure that the enzyme-linked antibody is against IgG of the species in which the primary antibody was raised.*
5. Washing solutions: Phosphate-buffered saline (PBS): Make 2 L containing 10 m*M* NaH_2PO_4, 150 m*M* NaCl adjusted to pH 7.2 using NaOH. This solution is stable and may be stored at 4°C. It is susceptible to microbial contamination, however, and is usually made as required.
   The other washing solutions are made by dissolving the appropriate weight of bovine serum albumin or Triton X-100 in PBS. Dissolve bovine serum albumin by rocking the mixture gently in a large, sealed bottle to avoid excessive foaming. The "blocking" and "antibody" solutions containing 8% albumin may be stored at –20°C and reused several times. Microbial contamination can be limited by filter-sterilizing these solutions after use or by adding 0.05% (w/v) NaN_3 (but *see* **Note 4**). Other solutions are made as required and discarded after use.
6. Alkaline phosphatase substrate mixture:
   a. Diethanolamine buffer. Make up 100 m*M* diethanolamine and adjust to pH 9.8 using HCl. This buffer is usually made up as required, but may be stored at 4°C if care is taken to avoid microbial contamination.
   b. 1 *M* $MgCl_2$. This can be stored at 4°C.
   Combine 200 µL of 1 *M* $MgCl_2$, 5 mg of nitroblue tetrazolium, 2.5 mg of 5-bromo-4-chloroindolyl phosphate (disodium salt; *see* **Note 5**). Adjust the volume to 50 mL using 100 m*M* diethanolamine buffer. Make up this reaction mixture as required and protect from the light before use.

7. Horseradish peroxidase substrate mixture:
   a. Make up 50 m*M* acetic acid and adjust to pH 5.0 using NaOH. This buffer is usually made up as required, but may be stored at 4°C if care is taken to avoid microbial contamination.
   b. Diaminobenzidine stock solution of 1 mg/mL dissolved in acetone. Store in the dark at –20°C. **Caution: Diaminobenzidine is potentially carcinogenic; handle with care.**
   c. Hydrogen peroxide at a concentration of 30% (v/v). This compound decomposes, even when stored at 4°C. The precise concentration of the stock solution can be determined by measuring its absorbance at 240 nm. The molar extinction coefficient for H_2O_2 is 43.6 $M^{-1}cm^{-1}$ at this wavelength (*see* **Note 6**).

   Combine 50 mL of acetate buffer, 2 mL of diaminobenzidine stock solution, and 30 µL of hydrogen peroxide immediately before use. Mix gently and avoid vigorous shaking to prevent unwanted oxidation of the substrate. Protect the solution from the light.
8. Protein staining solutions: These are stable at room temperature for several weeks and the stains may be reused.
   a. Amido Black stain (100 mL): 0.1% (w/v) Amido Black in 25% (v/v) propan-2-ol, 10% (v/v) acetic acid.
   b. Amido Black destain (400 mL): 25% (v/v) propan-2-ol, 10% (v/v) acetic acid.
   c. Ponceau S stain (100 mL): 0.2% (w/v) Ponceau S, 10% (w/v) acetic acid.
   d. Ponceau S destain (400 mL): distilled water.

3. Methods

3.1. Immunodetection of Polypeptides

1. Following SDS-PAGE (*see* Chapter 11), electroblot the polypeptides from the gel onto nitrocellulose at 50 V for 3 h using a Bio-Rad Trans-Blot apparatus, or at 100 V for 1 h using a Bio-Rad Mini Trans-Blot system.
2. After blotting, transfer the nitrocellulose filters individually to plastic trays for the subsequent incubations. Ensure that the nitrocellulose surface that was closest to the gel is uppermost. Do not allow the filter to dry out, as this often increases nonspecific binding and results in heavy, uneven background staining. The nitrocellulose filter should be handled sparingly to prevent contamination by grease or foreign proteins. Always wear disposable plastic gloves, and only touch the edges of the filter.
3. If desired, stain the blot for total protein using Ponceau S as described in **Subheading 3.3.2.** (*see* **Note 7**).
4. Rinse the nitrocellulose briefly with 100 mL of PBS. Then incubate the blot at room temperature with the following solutions, shaking gently (*see* **Note 8**).
   a. 50 mL of PBS–8% bovine serum albumin for 30 min. This blocks the remaining protein-binding sites on the nitrocellulose (*see* **Note 9**).
   b. 50 mL of PBS–8% bovine serum albumin containing 50–500 mL of antiserum for 2–16 h (*see* **Note 10**).
   c. Wash the nitrocellulose at least 5×, each time using 100 mL of PBS for 15 min, to remove unbound antibodies.
   d. 50 mL of PBS–4% bovine serum albumin containing an appropriate ligand for 2 h (*see* **Note 11**). This is likely to be one of the following:
   i. enzyme-conjugated secondary antibody at the manufacturer recommended dilution (normally between 1:1000 and 1:10,000).
   ii. 5 µg of enzyme-conjugated protein A.
   iii. 1 µCi ^{125}I-labeled protein A.

e. Wash the nitrocellulose at least 5×, each time using 100 mL PBS–1% Triton X-100 for 5 min, to remove unbound protein A or secondary antibody.

To ensure effective washing of the filter, pour off each solution from the same corner of the tray and replace the lid in the same orientation.

3.2. Visualization of Antigen–Antibody Complex

3.2.1. Alkaline Phosphatase Conjugated Ligand

In this method the enzyme hydrolyzes 5-bromo-4-chloroindolyl phosphate to the corresponding indoxyl compound. The product tautomerizes to a ketone, oxidizes, and then dimerizes to form an insoluble blue indigo that is deposited on the filter. Hydrogen ions released during the dimerization reduce nitroblue tetrazolium to the corresponding diformazan. The latter compound is an insoluble intense purple compound that is deposited alongside the indigo, enhancing the initial signal.

1. Briefly rinse the filter twice, each time using 50 mL of diethanolamine buffer.
2. Incubate the filter with 50 mL of alkaline phosphatase substrate mixture until the blue-purple products appear, usually after 5–30 min.
3. Prevent further color development by removing the substrate mixture and washing the filter 3×, each time in 100 mL of distilled water. Finally, dry the filter thoroughly before storing (*see* **Note 12**).

3.2.2. Horseradish Peroxidase-Conjugated Ligand

Peroxidase catalyzes the transfer of hydrogen from a wide range of hydrogen donors to H_2O_2, and it is usually measured indirectly by the oxidation of the second substrate. In this method, soluble 3,3'-diaminobenzidine is converted to a red-brown insoluble complex that is deposited on the filter. The sensitivity of this technique may be increased up to 100-fold by intensifying the diaminobenzidine-based products using a combination of cobalt and nickel salts that produce a dense black precipitate (*see* **Note 13**).

1. Briefly rinse the filter twice, each time using 50 mL of sodium acetate buffer.
2. Incubate the filter with 50 mL of horseradish peroxidase substrate mixture until the red-brown insoluble products accumulate. Reaction times longer than about 30 min are unlikely to be effective owing to substrate-inactivation of peroxidase (*see* **Note 6**).
3. When sufficient color has developed, remove the substrate mixture and wash the filter 3× with 100 mL of distilled water. Then dry the filter and store it in the dark (*see* **Note 12**).

3.2.3. ^{125}I-Labeled Protein A

1. If desired, stain the blot for total protein as described below.
2. Allow the filter to dry. Do not use excessive heat since nitrocellulose is potentially explosive when dry.
3. Mark the nitrocellulose with radioactive ink to allow alignment with exposed and developed X-ray film.
4. Fluorograph the blot using suitable X-ray film and intensifying screens. Expose the film at –70°C for 6–72 h, depending on the intensity of the signal.

3.3. Staining of Total Protein

Either of the following stains is suitable for visualizing polypeptides after transfer onto nitrocellulose. Each can detect bands containing about 1 µg of protein. Coomassie

Brilliant Blue is unsuitable for nitrocellulose membranes, since generally it produces heavy background staining (*see* **Note 14**).

3.3.1. Amido Black

Incubate the filter for 2–5 s in 100 mL of stain solution. Transfer immediately to 100 mL of destain solution, and wash with several changes to remove excess dye. Unacceptably dark backgrounds are produced by longer incubation times in the stain solution.

3.3.2. Ponceau S

Incubate the filter with 100 mL of Ponceau S stain solution for 30 min. Wash excess dye off the filter by rinsing in several changes of distilled water. The proteins may be destained by washing the filter in PBS (*see* **Note 7**).

4. Notes

1. Protein G, a cell wall component of group C and group G streptococci, binds to the Fc region of IgG from a wider range of species than that recognized by protein A *(8)*. Therefore antibodies that react poorly with protein A, particularly those from rat, mouse, goat, and sheep, may be detected by a similar method using protein G derivatives. Natural protein G also contains albumin binding sites and membrane binding regions which can lead to nonspecific staining. However, these problems can be avoided by using protein G', a recombinant, truncated form of the enzyme which lacks both albumin and membrane binding sites and is thus more specific for IgG than the native form of the protein.
   Protein L from *Peptostreptococcus magus* has affinity for κ light chains from various species, and will bind to IgG, IgA, and IgM as well as Fab, $F(ab')_2$ and recombinant scFv fragments that contain κ light chains. It will also bind chicken IgG. However, species such as cow, goat, sheep, and horse whose immunoglobulins contain predominantly λ chains will not bind well, if at all, to protein L *(9)*. This problem has led to the development of protein LA, a recombinant fusion protein that combines Fc- and Fab-binding regions of protein A with κ light chain binding regions of protein L. This generates a molecule that combines the favorable binding properties of both proteins *(10)*.
   Currently, alkaline phosphatase and horseradish peroxidase conjugated and ^{125}I-labeled protein G are commercially available, while only peroxidase conjugates of protein L and protein LA are produced.
2. Several visualization systems have been developed for both alkaline phosphatase and horseradish peroxidase. Other colorimetric assay systems are described by Tijssen *(4)*. However, currently, greatest sensitivity is provided by chemiluminescent detection systems (*see* Chapter 56). The alkaline phosphatase system is based on the light emission that occurs during the hydrolysis of AMPPD (3-[2'-spiroadamantane]-4-methoxy-4-[3"-phosphoryloxy]-phenyl-1,2-dioxetane). The mechanism involves the enzyme catalyzed formation of the dioxetane anion, followed by fragmentation of the anion to adamantone and the excited state of methyl *m*-oxybenzoate. This latter anion is the source of light emission. The peroxidase detection system relies on oxidation of luminol (3-aminophthalhydrazine) by hydrogen peroxide, or other suitable substrates, to produce a luminol radical. This radical subsequently forms an endoperoxide that on decomposition, generates an electronically excited 3-aminophthalate dianion that emits light on decay to its ground state. Light emission from this system can be enhanced by the presence of 6-hydroxybenothiazole derivatives or substituted phenols, which act as electron-transfer mediators between peroxidase and luminol.

Recently, chemifluorescent detection systems have been introduced for both alkaline phosphatase and horseradish peroxidase (*see* Chapter 56). These sytems are based on the hydrolysis or oxidation of a fluorogenic substrate to yield an insoluble product that is visualized by fluorimetry using a charge coupled device CCD camera or other imaging system. The sensitivities of commercially developed systems such as FluoroBlot (Pierce Chemical) and Vistra ECF (Amersham Pharmacia Biotech) are claimed to be comparable to those of chemiluminescent systems. Relative to chemiluminescence, the potential benefits of fluorescence detection are linearity of response over a greater range of signal, adjustable sensitivity, and stability of signal. Because light emission is dependent on exposure to the excitation light, a single fluorochrome molecule can be excited repeatedly, and a fluorescent blot can be stored and rescanned after several weeks with little loss of signal. In contrast, using chemiluminescence detection, light emission peaks within minutes of substrate addition and decays within a few hours. Subsequent revisualization requires reapplication of substrate and relies on the extent to which the activity of phosphatase or peroxidase is retained during storage.

3. Iodination of protein A using Bolton and Hunter reagent (*see* Chapter 134) labels the ε-NH_2 group of lysine, which apparently is not involved directly in the binding of protein A to the Fc region of IgG. This method is preferable to others, such as those using chloramine T or iodogen, which label tyrosine. The only tyrosine residues in protein A are associated with Fc binding sites, and their iodination may reduce the affinity of protein A for IgG ***(11)***.
4. Many workers include up to 0.05% sodium azide in the antibody and washing buffers to prevent microbial contamination. However, azide inhibits horseradish peroxidase. Therefore, do not use buffers containing azide when using this enzyme.
5. In the original description of this protocol 5-bromo-4-chloroindolyl phosphate was made up as a stock solution in dimethylformamide. However, this is not necessary if the disodium salt is used since this compound dissolves readily in aqueous buffers.
6. Urea peroxide may be used instead of hydrogen peroxide as a substrate for peroxidase. The problems of instability, enzyme inactivation and possibility of caustic burns associated with hydrogen peroxide are eliminated by using urea peroxide. A 10% (w/v) stock solution of urea peroxide is stable for several months and is used at a final concentration of 0.1% in the peroxidase substrate mixture.
7. If desired, the nitrocellulose filter may be stained with Ponceau S immediately after electroblotting. This staining apparently does not affect the subsequent immunodetection of polypeptides, if the filter is thoroughly destained using PBS before incubation with the antiserum. In addition to confirming that the polypeptides have been transferred successfully onto the filter, initial staining allows tracks from gels to be separated precisely and probed individually. This is useful when screening several antisera.
8. Nonspecific binding is a common problem in immunoblotting. Several factors are important in reducing the resulting background.

   First, the filter is washed in the presence of an "inert" protein to block the unoccupied binding sites. Bovine serum albumin is the most commonly used protein, but others, such as fetal calf serum, hemoglobin, gelatin, and nonfat dried milk, have been used successfully. Economically, the latter two alternatives are particularly attractive.

   The quality of protein used for blocking is important, as minor contaminants may interfere with either antigen–antibody interactions or the binding of protein A to IgG. These contaminants may vary between preparations and can be sufficient to inhibit completely the detection of specific polypeptides. Routinely we use bovine serum albumin (fraction V) from Sigma–Aldrich (product no. A 4503), but no doubt albumin from other sources is equally effective. The suitability of individual batches of protein should be checked using antisera known to react well on immunoblots.

Second, the background may be reduced further by including nonionic detergents in the appropriate solutions. These presumably decrease the hydrophobic interactions between antibodies and the nitrocellulose filter. Tween 20, Triton X-100, and Nonidet P-40 at concentrations of 0.1–1.0% have been used. In my experience, such detergents may supplement the blocking agents described previously, but cannot substitute for these proteins. In addition, these detergents sometimes remove proteins from nitrocellulose (*see* **Note 9**).

Third, the nitrocellulose must be washed effectively to limit nonspecific binding. For this, the volumes of the washing solutions should be sufficient to flow gently over the surface of the filter during shaking. The method described in this chapter is suitable for 12 × 7 cm filters incubated in 14 × 9 cm trays. If the size of the filter is significantly different, the volumes of the washing solutions should be adjusted accordingly.

Finally, decreasing the incubation temperature to 4°C may greatly decrease the extent of nonspecific background binding *(12)*.

9. Protein desorption from the membrane during the blocking step and subsequent incubations can result in the loss of antigen and decrease the sensitivity of detection *(13,14)*. In some instances, this problem may be reduced by incubating the membrane in 0.1 *M* phosphate buffer, pH 2.0 for 30 min and then rinsing in PBS prior to treatment with the blocking agent. Such acid treatment is particularly effective when using non-denaturing gel blots, or SDS-PAGE blots transferred onto polyvinylidene difluoride rather than nitrocellulose membrane *(14)*.

   Alternatively, polyvinyl alcohol may be used as a blocking agent *(15)*. In comparative tests, PBS containing 1 mg/mL polyvinyl alcohol produced lower background staining than other commonly used blocking agents. Moreover, the blocking effect of polyvinyl alcohol is virtually instantaneous, allowing the incubation time to be reduced to 1 min and decreasing the opportunity for loss of protein from the membrane *(15)*.
10. The exact amount of antibody to use will depend largely on its titer. Generally it is better to begin by using a small amount of antiserum. Excessive quantities of serum tend to increase the background rather than improve the sensitivity of the technique. Nonspecific binding can often be reduced by decreasing the amount of antibody used to probe the filter. Also, *see* **Notes 15** and **16**.
11. Deciding which form of detection system to use is largely a personal choice. Detection using ^{125}I-labeled protein A is very sensitive. However, many researchers prefer to use nonradioactive methods, and the sensitivities of chemiluminescent and chemifluorescent systems approach that of radiolabeling. Comparison between the two enzymic detection systems is difficult because the reported sensitivity limits of both systems vary considerably, and most studies use different antigens, different primary antibodies, and different protocols. Despite these uncertainties, alkaline phosphatase is generally considered more sensitive than horseradish peroxidase when visualized using the standard colorimetric detection systems described in this chapter. For routine work I prefer to use alkaline phosphatase conjugated protein A.
12. The products of the peroxidase reaction are susceptible to photobleaching and fading. Consequently, the developed filters should be stored in the dark, and the results photographed as soon as possible. The products of the phosphatase reaction are reportedly stable in the light. However, I treat such filters in the same way—just in case!
13. To increase sensitivity of the diaminobenzidine-based staining protocol replace the standard substrate mixture with the following intensifying solution *(16)*. Dissolve 100 mg of diaminobenzidine in 100 mL of 200 m*M* phosphate buffer, pH 7.3. To this solution add,

dropwise and with constant stirring, 5 mL of 1% (w/v) cobalt chloride followed by 4 mL of 1% (w/v) nickel ammonium sulfate. Finally add 60 µL of 30% (v/v) hydrogen peroxide just before use.

14. Coomassie Brilliant Blue R-250 should be used in preference to Ponceau S to stain proteins blotted onto polyvinylidene difluoride because it is more sensitive and does not bind to this membrane, whereas Ponceau S appears to interact only weakly with polypeptides attached to this matrix. If greater sensitivity is required, as little as 2 ng of a polypeptide can be detected using SYPRO Ruby, a metal chelate stain containing ruthenium as part of an organic complex (developed by Molecular Probes), which works well on both nitrocellulose and polyvinylidene difluoride membranes ***(17)***. Note that SYPRO Red and Orange, which have been developed to stain proteins within polyacrylamide gels, react relatively poorly with polypeptides attached to polyvinylidene difluoride.
15. Particular care should be taken when attempting to detect antigens on nitrocellulose using monoclonal antibodies. Certain cell lines may produce antibodies that recognize epitopes that are denatured by detergent. Such "conformational" antibodies may not bind to the antigen after SDS-PAGE.
16. Even before immunization, serum may contain antibodies, particularly against bacterial proteins. These antibodies may recognize proteins in an extract bound to the nitrocellulose filter. Therefore, when characterizing an antiserum, control filters should be incubated with an equal amount of pre-immune serum to check whether such preexisting antibodies interfere in the immunodetection of specific proteins.
17. Quantitation of specific antigens using this technique is difficult and must be accompanied by adequate evidence that the amount of product or radioactivity bound to the filter is directly related to the amount of antigen in the initial extract. This is important, as polypeptides may vary in the extent to which they are eluted from the gel and retained by the nitrocellulose (*see* **Note 9**). In addition, in some tissues proteins may interfere with the binding of antigen to the filter or their subsequent detection ***(18)***. Therefore, the reliability of the technique should be checked for each extract.

    Perhaps the best evidence is provided by determining the recovery of a known amount of pure antigen. For this, duplicate samples are prepared, and to one is added a known amount of antigen comparable to that already present in the extract. Ideally the pure antigen should be identical to that in the extract. The recovery is calculated by comparing the antigen measured in the original and supplemented samples. Such evidence is preferable to that obtained from only measuring known amounts of pure antigen. The latter indicates the detection limits of the assay, but does not test for possible interference by other components in the extract.

    The other major problem in quantifying the level of antigen on immunoblots derives from the technical problems associated with relating densitometric measurements to the amount of antibody bound to the filter. A combined radiochemical-color method has been developed that circumvents these problems ***(19)***. The technique involves challenging the filter sequentially with alkaline phosphatase conjugated secondary antibody and ^{125}I-labeled protein A (which binds to the secondary antibody). The color reaction derived from the enzyme conjugate is used to localize the antibody–antigen complex. The appropriate region of the filter is then excised and the radioactivity derived from the protein A associated with the band is measured to provide a direct estimate of the amount of antigen. Alternatively, if ^{125}I-labeled protein A is used initially, the amount of radioactivity associated with a particular band may be quantified directly using a PhosphorImager ***(18)***. However, perhaps the most robust approach is to determine the amount of antigen using enzyme-linked immunosorbent (ELISA) techniques (*see* Chapter 55) after establishing the specificity of the antibody by immunoblotting as described in this chapter.

References

1. Blake, M. S., Johnson, K. H., Russell-Jones, G. J., and Gotschlich, E. C. (1984) A rapid, sensitive method for detection of alkaline phosphatase-conjugated anti-antibodies on Western blots. *Analyt. Biochem.* **136,** 175–179.
2. Burnette, W. N. (1981) "Western blotting": Electrophoretic transfer of proteins from sodium dodecyl sulfate-polyacrylamide gels to unmodified nitrocellulose and radiographic detection with antibody and radioiodinated protein A. *Analyt. Biochem.* **112,** 195–203.
3. Towbin, H., Staehelin, T., and Gordon, J. (1979) Electrophoretic transfer of proteins from polyacrylamide gels to nitrocellulose sheets: Procedure and some applications. *Proc. Natl. Acad. Sci. USA* **76,** 4350–4354.
4. Tijssen, P. (1985) *Practice and Theory of Enzyme Immunoassays.* Elsevier, Amsterdam.
5. Vaessen, R. T. M. J., Kreide, J., and Groot, G. S. P. (1981) Protein transfer to nitrocellulose filters. *FEBS Lett.* **124,** 193–196.
6. Hattori, S. and Fujisaki, H. (1996) Double immunodetection of proteins transferred onto a membrane using two different chemiluminescent reagents. *Analyt. Biochem.* **243,** 277–279.
7. Towbin, H. and Gordon, J. (1984) Immunoblotting and dot immunobinding - current status and outlook. *J. Immunol. Meth.* **72,** 313–340.
8. Akerstrom, B., Brodin, T., Reis, K., and Bjorck, L. (1985) Protein G: a powerful tool for binding and detection of monoclonal and polyclonal antibodies. *J. Immunol.* **135,** 2589–2592.
9. De Chateau, M., Nilson, B. H. K., Erntell, M., Myhre, E., Magnusson, C. G. M., Åkerström, B., and Björck, L. (1993) On the interaction between protein L and immunoglobulins of various mammalian species. *Scand. J. Immunol.* **37,** 399–405.
10. Svensson, S. G., Hoogenboom, H. R., and Sjöbring, U. (1998) Protein LA, a novel hybrid protein with unique single-chain Fv antibody- and Fab-binding properties. *Eur. J. Biochem.* **258,** 890–896.
11. Langone, J. J. (1980) ^{125}I-Labelled Protein A: Reactivity with IgG and use as a tracer in radioimmunoassay, in *Methods in Enzymology* Vol. 70 (Vunakis, H. V. and Langone, J. J., eds.), Academic Press New York, pp. 356–375.
12. Thean, E. T. and Toh, B. H. (1989) Western immunoblotting: temperature-dependent reduction in background staining. *Analyt. Biochem.* **177,** 256–258.
13. Den Hollander, N. and Befus, D. (1989) Loss of antigens from immunoblotting membranes. *J. Immunol. Meth.* **122,** 129–135.
14. Hoffman, W. L., Jump, A. A., and Ruggles, A. O. (1994) Soaking nitrocellulose blots in acidic buffers improves the detection of bound antibodies without loss of biological activity. *Analyt. Biochem.* **217,** 153–155.
15. Miranda, P. V., Brandelli, A., and Tezon, J. G. (1993) Instantaneous blocking for immunoblots. *Analyt. Biochem.* **209,** 376–377.
16. Adams, J. C. (1981) Heavy metal intensification of DAB-based HRP reaction product. *J. Histochem. Cytochem.* **29,** 775.
17. Breggren, K., Steinberg, T. H., Lauber, W. M., Carroll, J. A., Lopez, M. F., Chernokalskaya, E., et al.(1999) A luminescent ruthenium complex for ultrasensitive detection of proteins immobilized on membrane supports. *Analyt. Biochem.* **276,** 129–143.
18. O'Callaghan, J. P., Imai, H., Miller, D. B., and Minter, A. (1999) Quantitative immunoblots of proteins resolved from brain homogenates: underestimation of specific protein concentration and of treatment effects. *Analyt. Biochem.* **274,** 18–26.
19. Esmaeli-Azad, B. and Feinstein, S. C. (1991) A general radiochemical-color method for quantitation of immunoblots. *Analyt. Biochem.* **199,** 275–278.

55

Utilization of Avidin or Streptavidin-Biotin as a Highly Sensitive Method to Stain Total Protein on Membranes

Kenneth E. Santora, Stefanie A. Nelson, Kristi A. Lewis, and William J. LaRochelle

1. Introduction

Since the initial publication by Towbin and coworkers *(1)* on the preparation of replicas of sodium dodecyl sulfate (SDS)-polyacrylamide gel patterns, commonly called protein blots, the technique of transferring proteins from inaccessible gel matrices to accessible solid supports, such as nitrocellulose or nylon membranes, has become widely utilized *(2,3)*.

The detection of proteins on blots has ranged from the specific visualization of an individual protein of interest to the general staining of total protein *(4)*. Specific proteins are detected with probes, such as antibodies or toxins, that are either directly radiolabeled or conjugated to an enzyme *(5,6)*. Alternatively, bound and unlabeled antibodies or toxins are amplified by a secondary affinity probe similarly conjugated. The use of biotinylated antibodies or toxins followed by avidin or streptavidin enzyme conjugates is also gaining in popularity *(7)*.

Total protein detection is usually based on a chemical affinity staining method such as Coomassie Blue *(8)*, Amido Black *(9)*, India ink *(10)*, oxidation/ derivation of carbohydrate moieties *(11,12)*, silver-enhanced copper detection *(13)* or Ponceau S *(14)*. Other approaches often require the chemical modification of the polypeptide with hapten followed by detection with anti-hapten antibody and labeled secondary, protein A, or protein G *(15)*.

Here, we exploit the high-affinity and well-characterized interactions of biotin with either avidin or streptavidin *(16–18)*. Initially, proteins are resolved by sodium dodecyl sulfate-polyacrylamide gel electrophoresis (SDS-PAGE), transferred to nitrocellulose membrane, and the amino groups covalently derivatized *(19)* with sulfosuccinimidobiotin. Depending on the sensitivity required, either of two techniques illustrated in **Fig. 1** are used to stain the proteins, which appear as dark bands against an essentially white background *(19,20)*. The first method utilizes avidin or streptavidin conjugated to horseradish peroxidase (HRP) and detects <25 ng of protein in a single band. The second technique, although slightly more lengthy, requires streptavidin amplification with

From: The Protein Protocols Handbook, 2nd Edition
Edited by: J. M. Walker © Humana Press Inc., Totowa, NJ

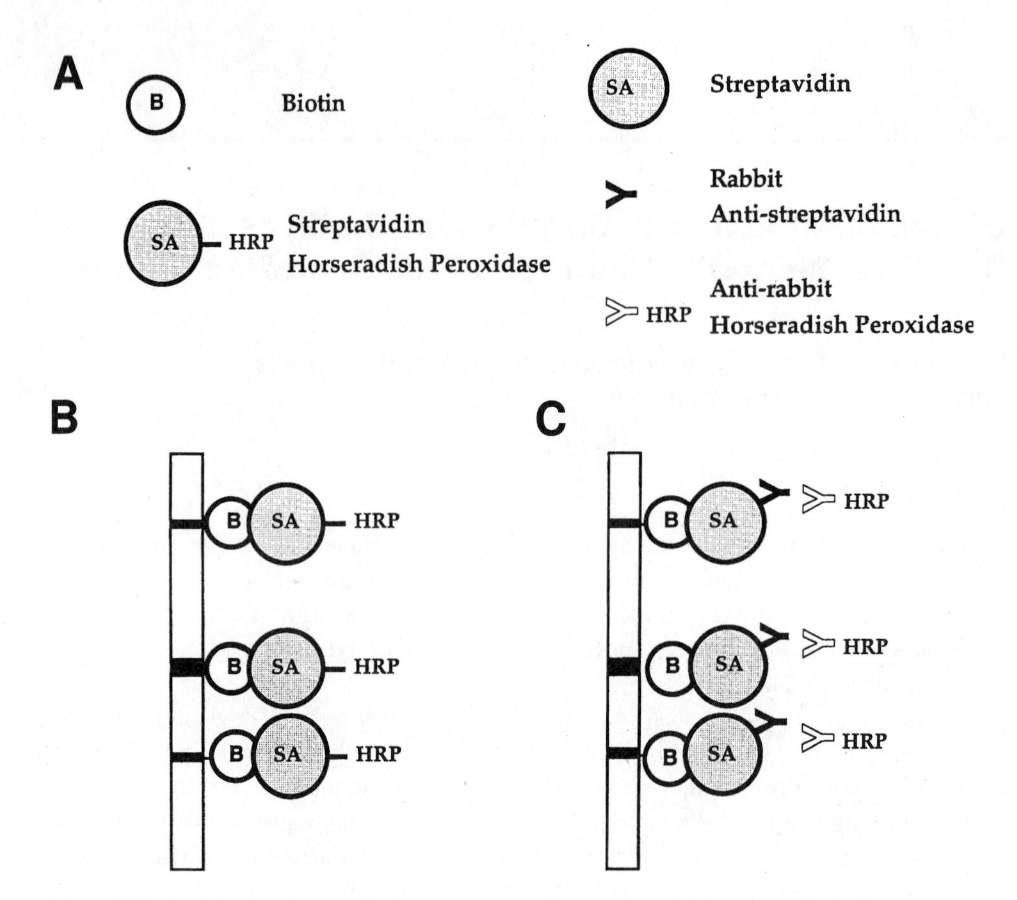

Fig. 1. Diagrammatic representation of streptavidin or amplified antistreptavidin staining of nitrocellulose replicas. Electrophoretically transferred proteins were biotinylated on a nitrocellulose replica depicted here as a strip blot. **(A)** B, Biotin; SA, streptavidin; HRP, horseradish peroxidase. **(B)** Schematic diagram of streptavidin staining method (**Subheading 3.1.**). Total protein, shown as individual dark bands, was biotinylated and detected with streptavidin conjugated to horseradish peroxidase followed by the α-chloronaphthol color reaction. **(C)** Schematic diagram of amplified antistreptavidin staining method (**Subheading 3.2.**). After electrophoretic transfer and biotinylation, streptavidin binding to biotinylated proteins was detected with rabbit antistreptavidin followed by goat antirabbit antibody conjugated to horseradish peroxidase as described in **Subheading 3.2.**

anti-streptavidin antisera followed by a secondary antibody conjugated to HRP. This procedure detects <5 ng of protein per band.

The methods described here *(19–20)* permit direct comparison of stained replicas with a duplicate blot that has been probed with antibody or ligand. Our approach is rapid and possesses greater sensitivity than the commonly used dyes. Our detection scheme is less costly and time consuming than the use of the metal stains, that in some instances possess greater sensitivity. Moreover, problems associated with gel shrinkage on drying or altered electrophoretic mobility of proteins caused by staining or derivatization prior to electrophoresis are avoided. Since our initial study *(19)*, this approach has also proven useful in detection and labeling of DNA on membrane supports *(22,23)*.

2. Materials

2.1. Avidin–Streptavidin–HRP

1. Distilled, deionized water.
2. Plastic trays rather than glass are preferred.
3. 0.45-μm Nitrocellulose membranes, (Bio-Rad, Richmond, CA).
4. Sulfosuccinimidobiotin (Pierce Chemical, Rockford, IL) solution: 10 m*M* in 100 m*M* sodium bicarbonate, pH 8.0.
5. Avidin conjugated to HRP (Cappel Laboratories, Malvern, PA).
6. Streptavidin conjugated to HRP (Gibco-BRL, Gaithersburg, MD).
7. Phosphate-buffered saline (PBS): 10 m*M* sodium phosphate, 150 m*M* sodium chloride, pH 7.4.
8. Block solution: PBS containing 5% v/v newborn calf serum (Gibco-BRL, Gaithersburg, MD) and 3% w/v bovine serum albumin (BSA, Fraction V, 98–99%, Sigma Chemical, St. Louis, MO).
9. Wash solution: PBS containing 0.05% w/v Tween-80 (Sigma Chemical, St. Louis, MO).
10. PBS containing 0.05% w/v Tween-80 and 1.0% w/v BSA.
11. 1.0*M* Glycine-HCl, pH 6.5.
12. α-Chloronaphthol solution: PBS containing 0.6 mg/mL of α-chloronaphthol (4-chloro-1-naphthol, Sigma) and 0.01% hydrogen peroxide.

2.2. Streptavidin/Antistreptavidin Amplified Staining

1. Materials listed in **Subheading 2.1.**, excluding avidin or streptavidin–HRP.
2. Streptavidin (Gibco-BRL, Gaithersburg, MD).
3. Antistreptavidin (Sigma Chemical, St. Louis, MO or Zymed Laboratories, South San Francisco, CA).
4. Goat antirabbit immunoglobulin conjugated to horseradish peroxidase (Cappel Laboratories).

3. Methods

3.1. Avidin or Streptavidin Horseradish Peroxidase Staining

1. After SDS-PAGE, transfer proteins to nitrocellulose membranes (*see* **Notes 1** and **2**, and Chapters 37–40). Rinse the nitrocellulose replicas three times and soak in 100 m*M* sodium bicarbonate, pH 8.0, for 5 min (0.25–0.50 mL/cm^2 nitrocellulose). Typically, a 10-mL volume is used for a minigel replica. The solution volume should permit the filter to move freely on agitation. Carry out all incubation and washing reactions using an orbital shaker or rocker platform at ambient temperature.
2. Transfer and submerge the replicas in the same volume of freshly prepared sulfosuccinimidobiotin solution for 45 min (*see* **Notes 3** and **4**).
3. Add 1 *M* glycine-HCl, pH 6.5, to a final concentration of 1 m*M* for approx 5 min to quench the derivatization reaction.
4. Wash the filters three times (5 min each wash) with PBS to remove free sulfosuccinimidobiotin. Incubate filters for 30 min with block solution.
5. Incubate the replicas for 1 h with either avidin conjugated to horseradish peroxidase (5 μg/mL) or streptavidin conjugated to horseradish peroxidase (1 μg/mL) diluted in PBS containing 1% BSA, 0.05% Tween-80 (*see* **Note 5**).
6. Wash filters three times for 15 min each time with the same volume of wash solution. Protein bands are visualized by immersing the replicas in α-chloronaphthol solution (*see* **Note 6**).

7. After allowing sufficient time for color development (usually 30 min), rinse the replicas with distilled water, and dry between two sheets of dialysis membrane. The replicas may also be dried, and stored in cellophane for future use.

3.2. Streptavidin/Antistreptavidin Amplified Staining

1. If desired, an alternative procedure to amplify fivefold the detection method described above is utilized. First, biotinylate and block the nitrocellulose filters as described in **Subheading 3.1., steps 1–4** of the avidin- or streptavidin-staining protocol.
2. Incubate blots with streptavidin (1 μg/mL) for 1 h in PBS containing 1% BSA, 0.05% Tween-80. Wash the replicas three times for 15 min each time with wash solution.
3. Next, dilute affinity-purified rabbit antistreptavidin IgG (0.5 μg/mL) in PBS containing 1% BSA, 0.05% Tween-80. Add solution to replica for 4 h or overnight if convenient. Wash replicas three times for 15 min each time with wash solution.
4. Incubate replicas with goat antirabbit immunoglobulin conjugated to horseradish peroxidase diluted (4 μg/mL) in PBS containing 1% BSA, 0.05% Tween-80 for 4 h. Wash replicas three times for 15 min each time with wash solution.
5. Protein bands are visualized by immersing the replicas in α-chloronaphthol solution for approx 30 min (*see* **Note 6**). The replicas are rinsed with distilled water and dried between two sheets of dialysis membrane.

4. Notes

1. This method is more sensitive for nitrocellulose membranes than for proteins transferred to Biodyne membranes by approx 10-fold owing to the higher background staining of Biodyne membranes.
2. Because of the sensitivity of the staining, care should be taken to avoid protein contamination of replicas with fingertips, and so forth. Gloves or forceps should be used.
3. This method derivatizes the free amino groups of proteins bound to filters. Amine containing compounds, such as Tris or glycine buffers, will compete for biotinylation with the sulfosuccinimidobiotin. Blot transfer buffers that contain free amino groups, such as Tris or glycine, may be used, but must be thoroughly removed by soaking and rinsing as indicated in **Subheading 3.1., step 1**.
4. The staining is highly dependent on the sulfosuccinimidobiotin concentration. Sulfosuccinimidobiotin concentrations of >10 μ*M* have resulted in a dramatic decrease of protein staining intensity ***(12)***. In some instances, it may be necessary to determine empirically the optimal concentration of sulfosuccinimidobiotin to use for staining particular proteins.
5. Little or no differences were observed when avidin conjugated to horseradish peroxidase was substituted for streptavidin conjugated to horseradish peroxidase. However, for some applications, streptavidin may present fewer problems owing to a neutral isoelectric point and apparent lack of glycosylation.
6. Sodium azide will inhibit horseradish peroxidase and, accordingly, must be removed before addition of the enzyme solution.
7. All reagent concentrations were determined empirically and were chosen to give maximum staining sensitivity. In principle, our procedure can be used in double-label experiments in which all proteins on the replica are biotinylated and the same blot is then probed with radioactive antibody or protein A.

Acknowledgments

The author thanks Stanley C. Froehner for helpful advice, discussions, and continued encouragement. This work was supported by grants to SCF from the NIH (NS-14781) and the Muscular Dystrophy Association.

References

1. Towbin, H. E., Staehelin, T., and Gordon, J. (1979) Electrophoretic transfer of proteins from polyacrylamide gels to nitrocellulose sheets: procedure and some applications. *Proc. Natl. Acad. Sci. USA* **76,** 4350–4354.
2. Gershoni, J. M. and Palade, G. E. (1983) Electrophoretic transfer of proteins from sodium dodecyl sulfate-polyacrylamide gels to a positively charged membrane filter. *Analyt. Biochem.* **124,** 396–405.
3. Bers, G. and Garfin, D. (1985) Protein and nucleic acid blotting and immunobiochemical detection. *Biotechniques* **3,** 276–288.
4. Moremans, M., Daneels, G., Van Dijck, A., Langanger, G., and De Mey, J. (1984) Sensitive visualization of antigen-antibody reactions in dot and blot immune overlay assays with immunogold and immunogold/silver staining. *J. Immunol. Meth.* **74,** 353–360.
5. Hsu, Y.-H. (1984) Immunogold for detection of antigen on nitrocellulose paper. *Analyt. Biochem.* **142,** 221–225.
6. Burnette, W. N. (1981) Western blotting: electrophoretic transfer of proteins from sodium dodecyl sulfate-polyacrylamide gels to unmodified nitrocellulose and radiographic detection with antibody and radioiodinated protein A. *Analyt. Biochem.* **112,** 195–203.
7. Harper, D. R., Liu, K.-M., and Kangro, H. O. (1986) The effect of staining on the immunoreactivity of nitrocellulose bound proteins. *Analyt. Biochem.* **157,** 270–274.
8. Hancock, K. and Tsang, V. C. W. (1983) India ink staining of proteins on nitrocellulose paper. *Analyt. Biochem.* **133,** 157–162.
9. Root, D. D. and Wang, K. (1993) Silver-enhanced copper staining of protein blots. *Analyt. Biochem.* **209,** 15–19.
10. Wojtkowiak, Z., Briggs, R. C., and Hnilica, L. S. (1983) A sensitive method for staining proteins transferred to nitrocellulose paper. *Analyt. Biochem.* **129,** 486–489.
11. Bayer, E. A., Wilchek, M., and Skutelsky, E. (1976) Affinity cytochemistry: the localization of lectin and antibody receptors on erythrocytes via the avidin-biotin complex. *FEBS Lett.* **68,** 240–244.
12. LaRochelle, W. J. and Froehner, S. C. (1986) Immunochemical detection of proteins biotinylated on nitrocellulose replicas. *J. Immunol. Meth.* **92,** 65–71.
13. LaRochelle, W. J. and Froehner, S. C. (1990) Staining of proteins on nitrocellulose replicas, in *Methods in Enzymology*, vol. 184 (Wilchek, M. and Bayer, E., eds.), Academic, San Diego, CA, pp. 433–436.
14. Bio-Rad Biotin-Blot Protein Detection Kit Instruction Manual (1989) Bio-Rad Laboratories Lit. No. 171., Richmond, CA, pp. 1–11.
15. Didenko, V. V. (1993) Biotinylation of DNA on membrane supports: a procedure for preparation and easy control of labeling of nonradioactive single-stranded nucleic acid probes. *Analyt. Biochem.* **213,** 75–78.

56

Detection of Proteins on Western Blots Using Chemifluorescence

Catherine Copse and Susan J. Fowler

1. Introduction

Immunodetection of specific proteins that have been immobilized on membrane supports by Western blotting is a widely used protein analysis technique. Traditionally, radioactively labeled antibodies or ligands (usually ^{125}I) were used to probe for specific antigens *(1–3)*. Autoradiographic detection on X-ray film, for example, reveals the location of the antigen with reproducibly high sensitivity and the results are easily quantifiable. However, the need for lengthy exposure times and concerns related to the handling and disposal of radioactive materials have led to the development of nonradioactive techniques for Western blot analysis, for example, using colorimetric and chemiluminescent detection.

In this case, probes can be labeled with reporter enzymes and when the appropriate substrates are added, the position and quantity of antigen are indicated by the presence and intensity of colored or luminescent products precipitated on the membrane at the site of the immobilized enzyme. The two systems that are commonly used to achieve this conversion are oxidation by horseradish peroxidase (HRP), or cleavage of a phosphate group by alkaline phosphatase (AP). These enzymes can be conjugated directly to an appropriate primary antibody or to a secondary antibody or ligand such as streptavidin that binds to biotinylated primary antibodies.

An alternative detection strategy involves the use of antibodies or ligands linked directly to fluorescent dyes, for example, fluorescein isothiocyanate (FITC) *(1)* or cyanine dyes *(4,5)*. The absorption of light energy by a fluorochrome causes excitation of electrons to a higher energy state. These excited electrons spontaneously decay back toward the ground state (lower energy level) within a few nanoseconds of excitation, and as the electrons decay, the fluorochrome emits light of a characteristic spectrum. Because the emitted light is always of longer wavelength (lower energy) than the excitation light, optical filters can be used to separate excitation light from emitted light. This allows the capture of the emitted light using imaging instruments that have the appropriate excitation wavelength and emission filters for the particular fluorophore.

A more recent advance has been the development of chemical fluorescence detection systems. Chemifluorescence relies on the enzymatic conversion of a nonfluorescent

From: *The Protein Protocols Handbook, 2nd Edition*
Edited by: J. M. Walker © Humana Press Inc., Totowa, NJ

substrate to a fluorescent product, or the production of a measurable change in the spectral properties of the substrate. As with chemiluminescence, this is usually achieved by oxidation or dephosphorylation by HRP or AP, respectively. This technique offers a signal amplification advantage over direct fluorescent labeling as each enzyme molecule can convert multiple substrate molecules into fluorescent products. The fluorescent product must also be immobilized at the site of enzyme activity on the membrane to ensure correct imaging of the protein sample. This is achieved by a combination of precipitation of the product and charge interaction with the membrane.

Fluorescence is detected by exciting the membrane with light of the appropriate wavelength, from a broad wavelength source (ultraviolet [UV] or xenon arc lamps) or using discrete wavelength excitation with a laser based fluorescence scanner. The emission can be captured directly using a fluorescence scanner, equipped with a photomultiplier tube (PMT), or, using a charge coupled device (CCD) camera. After the emitted light is captured, the analog signal from a PMT or CCD is converted into a digital signal. Alternatively, a Polaroid™ camera can be used to provide a hard copy of the results. Chemifluorescence can be used in place of standard chemiluminescence protocols but for optimal sensitivity it is important to use imaging equipment that closely matches the excitation and emission wavelengths of the substrate.

Currently available chemifluorescent *alkaline phosphatase substrates* include DDAO-phosphate (Molecular Probes) in which the phosphate group is cleaved to yield a product with an excitation maximum of 646 nm and an emission maximum of 660 nm. The ECF™ (Enhanced Chemifluorescence) substrate available from Amersham Pharmacia Biotech also has a phosphate group that can be cleaved by AP to yield a product with fluorescence maxima of 435 nm for excitation and 555 nm for emission. Kits containing the substrate and working buffer plus AP-conjugated secondary antimouse and antirabbit antibodies are available. **Figure 1** shows a typical Western blot detected using ECF substrate. Alternatively a three-tier system is offered in which the secondary antibody is conjugated to fluorescein. This allows direct visualization of the fluorescein fluorochrome (maximum excitation 494 nm and maximum emission 518 nm), but where greater sensitivity is required, tertiary AP-conjugated anti-fluorescein antibodies can be used and the ECF substrate applied to yield a chemifluorescent product.

Currently available *peroxidase substrates* yielding fluorescent products include FluoroBlot™ Peroxidase Substrate (Pierce Chemical) and ECL Plus™ (Amersham Pharmacia Biotech). The addition of the FluoroBlot working substrate to HRP, immobilized on a membrane, results in the generation of a blue fluorescent product (excitation maximum 325 nm, emission maximum 420 nm), which can be quantified using a CCD or Polaroid camera and UV excitation. Finally, ECL Plus is a highly sensitive chemiluminescent substrate system that utilizes a novel acridan-based chemistry ***(6)*** to generate a light signal. However, it also produces a stable fluorescent intermediate that can be detected with an excitation maximum of 430 nm and emission maximum of 503 nm on a membrane *(7)*. It can be visualized using the Storm™ (Amersham Pharmacia Biotech) fluorescence scanner with excitation at 450 nm.

1.1. Advantages of Chemifluorescent Detection

Chemifluorescence shares many of the advantages of chemiluminescence and has additional features:

1. Like chemiluminescence, chemifluorescence is based on the affinity binding of an enzyme-conjugated probe that acts to convert a substrate to product. Each enzyme conjugate molecule can generate many fluorochromes. Under appropriate conditions, enzymes can continue to convert substrate to product as long as the substrate is available. At the lower limits of detection, it can be beneficial to allow the signal to develop for longer, maximizing the signal from low abundance targets.
2. The signal does not decay in the same rapid time frame as chemiluminescence. Light emission from a fluorochrome occurs only during exposure to excitation light, but a single fluorochrome can be excited repeatedly and will emit light each time. As a consequence, blots can be scanned many times.
3. If an initial image is too weak or the pixels are saturated, the sensitivity of detection can be adjusted by changing the (PMT) voltage setting if acquired using a scanner, or by altering the length of exposure if using a CCD camera.
4. The use of film and associated volatile and unpleasant development and fixing solutions are avoided, together with the need for a dark room.
5. Scanning samples labeled with dyes excitable by the laser of an appropriate wavelength in a fluorescence scanner or CCD camera enables images to be captured directly in digital format, allowing for easier and more accurate quantification using image analysis software.

2. Materials

2.1. Solutions

1. Detection substrate:
   1–5 μ*M* DDAO-phosphate (Molecular Probes) in 10 m*M* Tris-HCl, 1 m*M* $MgCl_2$, pH 9.5.
2. Detection kits:
   ECF Western blotting kits (Amersham Pharmacia Biotech).
   Pro-Q™ Western Blot Stain Kit (Molecular Probes) DDAO-phosphate based (*see* **Note 18**).
   FluoroBlot Peroxidase Substrate (Pierce Chemical).
   ECL Plus (Amersham Pharmacia Biotech).
3. Phosphate-buffered saline (PBS), pH 7.5: 11.5 g of disodium hydrogen orthophosphate anhydrous (80 m*M*), 2.96 g of sodium dihydrogen orthophosphate (20 m*M*), 5.84 g sodium chloride, dilute to 1000 mL with distilled water and check pH.
4. Tris-buffered saline (TBS), pH 7.6: 20 mL of 1 *M* Tris-HCl, pH 7.6 (20 m*M*), 8 g of sodium chloride (137 m*M*), dilute to 1000 mL with distilled water and check pH.
5. 0.1% PBS– and TBS–Tween™ 20 (PBST and TBST): Dilute the appropriate volume of Tween 20 in PBS or TBS.
6. Blocking agent: PBST or TBST and blocking agent. Weigh out the appropriate amount of blocking agent and dissolve in PBST or TBST as appropriate. The blocking agent can be BSA, nonfat dried milk (5%), casein, fish gelatin, at between 1% and 5%, or various cocktails thereof.
7. Stripping buffer: 100 m*M* 2-mercaptoethanol, 2% (w/v) sodium dodecyl sulfate, 62.4 m*M* Tris-HCl, pH 6.7 *(8)*.
   Prepare all solutions using high-quality reagents and deionized water (18.2 MΩ resistivity).

2.2. Equipment

1. Membrane: Polyvinylidene fluoride (PVDF), for example, Immobilon-P™ (Millipore), Hybond™-P (Amersham Pharmacia Biotech), or nitrocellulose, for example, Hybond ECL (Amersham Pharmacia Biotech).

2. Orbital shaker/mixer platform.
3. Timer.
4. Forceps.
5. Transparent plastic, for example, SaranWrap™ (Dow Chemical Co.), Mylarplastic™ (DuPont).
6. *Imaging Systems*: Fluorescence scanners, for example, Storm 840/ 860 (Amersham Pharmacia Biotech), FluorImager™ 595 (Amersham Pharmacia Biotech), Typhoon™ (Amersham Pharmacia Biotech), Molecular Imager FX System (Bio-Rad Laboratories). *CCD cameras with UV light sources*, for example the Imagemaster™ VDS-CL (Amersham Pharmacia Biotech), Fluor-S™ (Bio-Rad Laboratories), ChemiImager™ (Alpha Innotech Corp.) Alternatively, a UV epi-illumination source in conjunction with a Polaroid or digital camera could be used for dyes excited at shorter wavelengths.
7. *Image analysis software*, for example, ImageQuant™ (Amersham Pharmacia Biotech).

3. Methods

3.1. Blotting and Probing the Membrane

The following standard protocol is suitable for use with ECF substrate, but can be adapted for use with any of the other systems. It should be optimized for the user's particular experimental requirements. TBST can be used in place of PBST where necessary. Unless stated otherwise, constant agitation should be maintained during each stage, using a mixer platform and steps should be performed at room temperature. Blots should be handled directly using only gloves (*see* **Note 17**).

1. After separation of the protein mixture by electrophoresis, they are transferred to a membrane (*see* **Note 16**) by electroblotting (*see* **Note 1**). These blots may be air-dried (*see* **Note 2**) and stored at 2–8°C in sealed bags, or used immediately.
2. Air-dried PVDF blots should be prewetted using methanol (*see* **Note 3**), followed by equilibration with distilled water. This step is not necessary for nitrocellulose membranes.
3. The blocking of nonspecific binding sites on the membrane can be achieved by incubation with PBST containing blocking agent (*see* **Subheading 2.1.**) for 1 h at room temperature or overnight at 2–8°C (*see* **Notes 4–6**). This step is particularly important in achieving good levels of specific signal against background.
4. Rinse the membrane twice with PBST.
5. Dilute the primary antibody using PBST and incubate with the membrane for 1 h at room temperature. Alternatively, PBST and blocking agent may be used to achieve this dilution (*see* **Note 7**). Typical dilutions in the range of 1:500–1:5000 are commonly used, but the dilution that gives optimal results will vary from application to application, depending on the antibody concentration and its affinity for the antigen (*see* **Notes 8** and **9**).
6. Rinse the membrane twice with PBST.
7. Wash 1× 15 min and 3× 5 min with PBST (*see* **Note 10**).
8. The secondary AP-conjugated antibody or ligand should be diluted in PBST. Typical dilutions of 1:10,000 for antimouse AP can be used with the ECF system. However, with streptavidin-AP higher dilutions may be necessary.
9. The rinsing and washing protocols (**steps 6** and **7** above) should be repeated.
10. The wet blot should then be drained and the ECF substrate solution pipetted gently over the surface, ensuring that the surface is completely covered and free of bubbles. Exposure to the detection reagent should be timed and this may also require optimization. Exposure to the ECF substrate for 5 min should be adequate.

11. Excess reagent should be drained from the blot and the blot placed in clear plastic wrap (e.g., SaranWrap) (*see* **Note 11, 12,** and **15**).
12. The blots can then be transferred to the image acquisition instrument, and scanned using the appropriate excitation light and emission filters. For ECF, this is usually achieved by scanning face down on the glass platen of the Storm scanner (*see* **Note 19**) using the blue chemifluorescence mode (450 nm excitation) or on the platform of the CCD camera.
13. Images should be analyzed immediately, to ensure that the pixel values are not saturated. If saturated data is present the PMT voltage should be reduced (or the CCD camera exposure time).
14. The membrane can be dried and stored in the refrigerator for several days before further treatment.

3.2. Stripping and Reprobing of Western Blots

The complete removal of primary and secondary antibodies from dried membranes is possible by following the method outlined below. Because a certain amount of protein will be lost from the membrane during this procedure it is recommended that lower abundance antigens be detected first.

1. Remove the ECF reaction product by submerging the membrane in methanol and incubation at room temperature with agitation for 30 min.
2. Rinse the membrane twice in PBS or TBS to remove all methanol.
3. Remove antibodies by submerging the membrane in stripping buffer at 50°C for 30 min with occasional agitation.
4. The membrane should then be rinsed and washed, including two 15-min incubations in large volumes of PBST followed by a further three 5-min washes.
5. The membrane should be blocked according to **step 3** above.
6. Immunodetection can now be repeated.

3.3. Quantification of Proteins on Western Blots Using Chemifluorescence

Quantification of proteins on Western blots detected directly with anti-mouse-AP or anti-rabbit-AP antibodies using a Molecular Dynamics FluorImager has been shown to give a linear response within a 25- to 50-fold dilution range. This linear relationship between the fluorescent signal and the amount of protein can be used for accurate protein quantification. The following description assumes the ability to acquire digital images of the blot. Where this is not possible, Polaroid film images can be analyzed using a laser densitometer.

1. To quantify the amount of a specific antigen in a sample, a set of standards (known concentrations of the same antigen) should be run on the same gel.
2. Immunodetection is performed as described in the **Note 13**, **Subheading 3.1.**
3. After detection, digitally stored images can be analyzed using software to quantify the fluorescence intensity over a defined area, to include the band representing the antigen. This is usually corrected for background by subtracting a background fluorescence value over an identical area a short distance from the band of interest.
4. The fluorescence volume report for the protein standards can be plotted against protein concentration and the amount of antigen in the unknown sample can be calculated from this relationship provided its fluorescence volume falls within the range of the standard data.

4. Notes

4.1. Sensitivity Optimization

1. Transfer of proteins from the gel to the membrane must be satisfactory. The efficiency of transfer can be determined by staining the blot with Ponceau S solution and the gel with silver stain. Rainbow™ molecular weight markers (Amersham Pharmacia Biotech) give some indication of transfer efficiency, and are useful where the protein bands of interest are not visible until probed.
2. The efficiency of antigen binding improves considerably after air-drying of the blot *(9)*. This is because the drying process forces unbound molecules further into the pore structure of the membrane and proteins with limited membrane contact bind more tightly to the surface *(10)*.
3. It is important to note that soaking blots in methanol will denature proteins, and this step should be avoided if the native conformation of the protein is required subsequently. This means native protein blots on PVDF membrane must be processed immediately after blotting.
4. The detergent content of the washing buffer and the incubation times can be adjusted to improve background, but it should be noted that increasing these parameters to the extremes (e.g., >0.2%) could result in loss of protein from the membrane.
5. Blocking agents should be prepared as required, because of the likelihood of bacterial contamination.
6. Optimization of the blocking step may involve the choice of blocking agent and the incubation period, both of which can affect background significantly. Incubations in blocking agent longer than those specified should be avoided, as this may result in leaching of proteins from the membrane, thereby reducing the protein concentration by an unknown quantity and therefore the reliability/sensitivity of the technique.
7. When considering the use of blocking agent for antibody dilution, it is important to note that nonfat dried milk should **not** be included in any incubation using streptavidin HRP or AP. This is because the endogenous biotin content of milk will mop up the streptavidin conjugate, resulting in lower sensitivity and higher background *(11)*.
8. The amount of primary and secondary antibodies used may need to be optimized to get the best signal-to-noise ratio. This is best performed using dot blots with a range of primary and secondary antibody dilutions. The incubation times may also be adjusted to suit.
9. The primary antibody should be able to detect denatured antigens when used in conjunction with sodium dodecyl sulfate-polyacrylamide gel electrophoresis (SDS-PAGE) blots, although some renaturation of the protein can take place, allowing antibodies that detect nonlinear epitopes to be used sometimes.
10. The washing volume should be as large as possible. A minimum of 4 mL/cm^2 of membrane is recommended.
11. Following removal of excess substrate, the plastic wrapped blot can be left to incubate in ECF substrate for up to 120 min for increased sensitivity although some substrate diffusion may occur after this time.
12. The membrane can be dried before image acquisition, as this will stop the reaction proceeding further and allow scanning at a later time. However, with PVDF membrane, drying can increase the background signal, resulting in reduced sensitivity. Drying of nitrocellulose membranes is not recommended as they become very brittle and break easily.
13. For accurate quantification of unknowns, it is suggested that at least five different dilutions of standard antigen be used over a 50-fold range. It is also important that the concentration of the unknown lies within this range; therefore it may be advantageous to load several dilutions of the unknown.

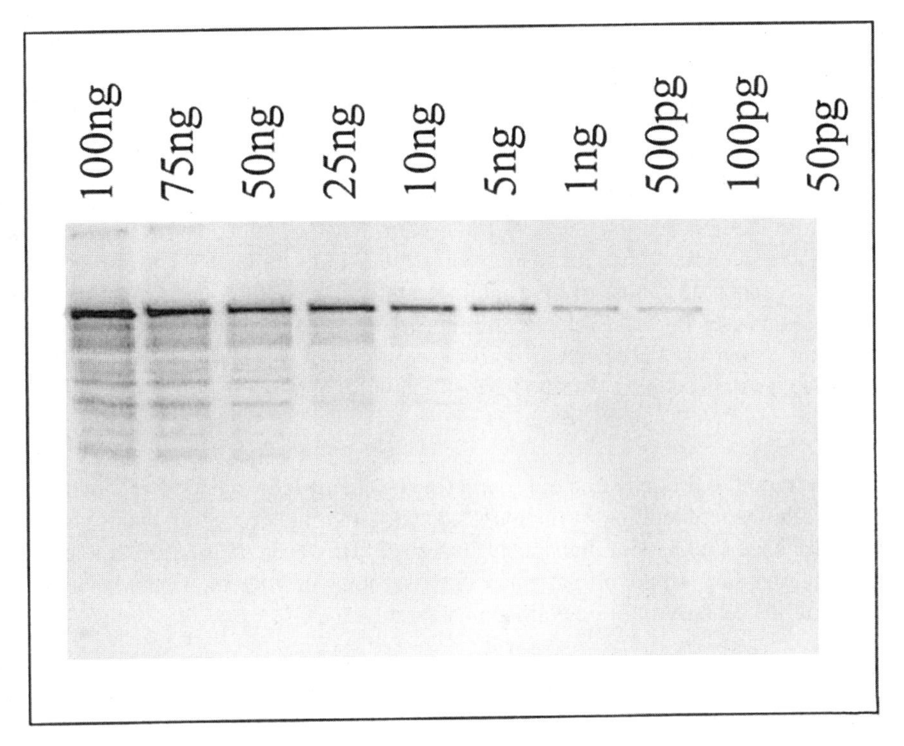

Fig. 1. β-Galactosidase was separated on a 10% SDS-PAGE gel. The gel was blotted onto PVDF membrane and blocking and immunodetection was performed as described in the text. The primary antibody was anti-β-galactosidase (from rabbit), diluted 1:32,000. The secondary antibody was antirabbit antibody conjugated to AP, diluted 1:2000. Detection was performed using ECF substrate and the image acquired using the Storm 860 (Amersham Pharmacia Biotech) in the blue chemifluorescence mode (450 nm) with a PMT setting of 600V. Protein is detectable down to 100 pg.

4.2. Background Fluorescence

14. Alkaline phosphatases are fairly ubiquitous, being endogenous to many animal and plant tissues and bacterial cells. This can contribute to background staining, making it particularly important to expose blots only to solutions, glassware and surfaces that are free of bacteria. Autoclaving washing and blocking solutions and filtering reagents can reduce background from ambient alkaline phosphatases.
15. It is also important to select a plastic wrap with low intrinsic fluorescence (*see* **Subheading 2.2.**), as this too will contribute to the background fluorescence and may affect the sensitivity of detection.
16. Intrinsic membrane fluorescence is also an important consideration and may affect the choice of membrane for a particular detection system. Membranes should be tested to determine their fluorescence properties before being used in an experiment. It is advisable to incubate blocked and washed blank membrane with detection reagent to determine the background membrane fluorescence likely to be measured. Choosing a membrane with low fluorescence will improve the limit and dynamic linear range of detection.
17. The use of powdered gloves should be avoided, as powder is fluorescent and can scatter light, contributing toward poor background. It is also advisable to rinse gloves with distilled water before directly handling the membrane.

18. DDAO-phosphate should not be used with diethanolamine buffer, as it causes hydrolysis of the substrate even in the absence of alkaline phosphatase.
19. The use of water between the scanning bed and the plastic wrap can reduce optical refraction and give a clearer image.

4.3. Molecular Weight Markers

20. Protein molecular weight markers for use in Western blotting systems are usually biotinylated, such that when incubated with streptavidin conjugated with AP or HRP, they provide a ladder of bands, to give internal molecular weight standards. An alternative system to this uses Perfect Protein™ Markers (Novagen), in which each protein contains an S-Tag™ peptide that enables their detection on a blot using S-protein HRP or AP conjugates in the secondary antibody or streptavidin incubation.

4.4. Time

21. The time required for detection of proteins by the chemifluorescence protocol described is approx 5 h. Direct handling of the blots is usually minimal, meaning that it is possible to run other experiments simultaneously if desired. However, if it is preferable to reduce handling time further, the primary antibody itself may be HRP or AP conjugated, but this may limit the sensitivity of detection considerably.

References

1. Towbin, H., Staehelin, T., and Gordon, J. (1979) Electrophoretic transfer of proteins from polyacrylamide gels to nitrocellulose sheets: procedure and some applications. *Proc. Natl. Acad. Sci USA* **76,** 4350–4354.
2. Renart, J., Reiser, J., and Stark, G. R. (1979) Transfer of proteins from gels to diazobenzyloxymethyl-paper and detection with antisera: a method for studying antibody specificity and antigen structure. *Proc. Natl. Acad. Sci. USA* **76,** 3116–3120.
3. Bowen, B., Steinberg, J., Laemmli, U. K., and Weintraub, H. (1980) The detection of DNA-binding proteins by protein blotting. *Nucleic Acids Res.* **8,** 1–20.
4. Gingrich, J. C., Davis, D. R., and Nguyen, Q. (2000) Multiplex detection and quantitation of proteins on Western blots using fluorescent probes. *BioTechniques* **29,** 636–642.
5. Fradelizi, J., Friederich, E., Beckerle, M. C., and Golsteyn, R. M. (1999) Quantitative measurement of proteins by Western blotting with Cy™5-coupled secondary antibodies. *BioTechniques* **26,** 484–494.
6. Akhavan-Tafti, H., DeSilva, R., Arghavani, Z., Eikholt, R. A., Handley, S., Schoenfelner, B. A., et al. (1998) Characterization of acridancarboxylic acid derivatives of chemiluminescent peroxidase substrates. *J. Org. Chem.* **63,** 930–937.
7. Appendix 2, *Fluorescence Imaging: Principles and Methods* (2000) Amersham Pharmacia Biotech, publisher, state, pp. xx–xx,
8. Kaufman, S., Ewing, C., and Shaper, J. (1987) The erasable Western blot. *Analyt. Biochem.* **161,** 89–95.
9. Van Dam, A. (1994) Transfer and blocking conditions in immunoblotting, in *Protein Blotting–A Practical Approach* (Dunbar, B. S., ed.), Oxford University Press, New York, pp. 73–85.
10. Mansfield, M. A. (1994) Transfer and blocking conditions in immunoblotting, *in Protein Blotting–A Practical Approach* (Dunbar, B. S., ed.), Oxford University Press, New York,
11. Hoffman, W. L and Jump, A. A. (1989) Inhibition of the streptavidin–biotin interaction by milk. *Analyt. Biochem.* **181,** 318–320.

57

Quantification of Proteins on Western Blots Using ECL

Joanne Dickinson and Susan J. Fowler

Introduction

The technique of Western blotting, the electrophoretic transfer of proteins from sodium dodecyl sulfate (SDS)-polyacrylamide gels to membranes, is a core technique in molecular and cell biology, *see* **ref. *1*** and Chapters 39–42. Western blots are detected with antibodies specific to the target protein which are known as primary antibodies. A labeled secondary antibody is then added, which binds to the primary antibody. The label may be radioactive, colorimetric, fluorescent, or chemiluminescent. The major application of Western blotting has been the detection of individual proteins in complex mixtures such as cell lysates. Information can be obtained about the molecular weight of the protein of interest and its relative abundance in the sample. However, it is also possible to use Western blotting to provide accurate quantification of proteins in samples, to assess changes in protein expression levels *(2)*.

The success of measuring proteins by Western blotting depends on calibrating signals from the chosen labels in a simple and reproducible manner over a usable range of protein concentrations. The method used for quantification depends on the type of detection reagent employed and the instrumentation available. In the past, radiolabels, particularly ^{125}I, were used to detect proteins on Western blots. This method is sensitive and gives a hard copy image on X-ray film, but long exposure times and restrictions over the use of radioactivity have made this technique less popular. However, although not discussed in this chapter, when used with phosphorimaging instrumentation, sensitivity is increased, exposure times can be significantly reduced and the signal is detected over a wide dynamic range (10^5), enabling more accurate quantification compared to film which has a dynamic range of only 5×10^2.

More recently nonradioactive approaches have increased in popularity. Colorimetric approaches followed by densitometry scanning have been used for quantification *(3)* but offer limited sensitivity and the blots are not easily reprobed. Alternative systems most favored for the visualization and quantification of proteins are based on fluorescence, chemifluorescence, or chemiluminescence. The direct detection of fluorescently labeled antibodies has been employed to measure protein concentrations on Western blots in conjunction with a fluorescent laser scanner *(4)*. This approach has

From: *The Protein Protocols Handbook, 2nd Edition*
Edited by: J. M. Walker © Humana Press Inc., Totowa, NJ

the advantage that the signal is read directly from the bound antibody but offers lower sensitivity *(5)* than chemifluorescence or chemiluminescence. Chemifluoresence *(6)* is analogous to chemiluminescence in utilizing an enzyme to cleave a substrate but generating a fluorescent product rather than light in the process.

Light emitting chemiluminescence detection has become extremely popular over the last 10 yr and many reagents are now available. Utilization of substrates for alkaline phosphatase produces hard copy on film with good sensitivity, but can sometimes suffer from background problems due to bacterial alkaline phosphatase contamination of buffers. Chemiluminescent systems based on horseradish peroxidase (HRP) offer rapid highly sensitive results with excellent signal-to-noise ratio and have become the predominantly used detection method. There are two different substrate types available, those based on the oxidation of luminol by HRP–hydrogen peroxide in the presence of phenolic enhancers *(7)* (e.g., ECL™), and more recently, those based on the oxidation of acridan substrates *(8)* (e.g., ECL Plus™). During the course of the ECL Plus reaction, fluorescent intermediates are generated, allowing signal to be measured using fluorescence scanners. As an example, ECL Plus detected using three imaging techniques—film plus densitometry, charge coupled device (CCD) camera detection, and fluorescence scanning—is described in this chapter.

Western blots detected with chemiluminescent substrates are typically quantified using images generated by autoradiography film followed by digitization on a densitometer. Although film itself has a dynamic range of 5×10^2, in practice the linear range over which proteins can be quantified has been estimated to be just over one order of magnitude *(9)*. Film-based detection suffers several drawbacks. At low levels of target, film exhibits reciprocity failure, which means it requires a threshold level of signal before an image is generated. Preflashing of the film to the detection threshold can overcome this to some extent and improve linearity *(10)*. At high target levels, saturation effects mean film darkening is again not linear with respect to the amount of light produced. For this reason, when using film, it is important that several exposures are taken to ensure the signal from the blot falls within the linear range of the film.

An alternative approach is to use a CCD camera to capture the chemiluminescent light output directly. CCD cameras work by converting the light emitted from the blot into an electrical signal which is then subsequently converted to a digital signal to obtain a quantifiable value *(11)*. For low level light imaging, cooled CCD cameras are necessary to reduce background noise to obtain improved signal to noise and allow longer exposures. Typically Peltier cooling is used, reducing the camera operating temperature to –20°C to –60°C below ambient. Some cameras are capable of providing equal or better sensitivity than that obtained with film. In addition, CCD cameras can offer excellent dynamic range (10^3–10^5) allowing accurate quantification.

An equally suitable alternative imaging method for quantification with ECL Plus is to detect the fluorescent signal. This can be achieved using a flat bed laser scanner or CCD camera coupled with a suitable excitation light source. Laser based fluorescence scanners also offer excellent dynamic range (10^5) and the convenience of direct digitization of signal. When imaging fluorescent blots, to obtain maximum sensitivity, it is important the excitation light of the imager is well matched with the excitation wavelength of the fluorescent product and that appropriate emission filters are used.

In this chapter, quantification of ECL Plus Western blots is used as the example since with this substrate there is a choice of imaging methods that can be used for quantification. However, the principle remains the same for any substrate. A known dilution series of the protein to be measured is prepared along with dilutions of the same protein to be measured. Following immunodetection and imaging, the results can be analyzed using a suitable computer program to measure band intensities. This data is then used to plot a standard curve from which unknown samples that fall within the linear range of the assay can be quantified. The dilution range over which samples can be quantified with CCD cameras or fluorescence scanners will be greater than film due to their increased dynamic range. However, in practice, saturation of membrane binding capacity and steric hindrance of antibody binding to target usually means signal linearity rarely extends over more than a 10^2–10^3 sample dilution range.

2. Materials

2.1. General Equipment and Materials

1. Electrophoresis equipment.
2. Electroblotting equipment.
3. Power pack.
4. Transfer membrane, for example, polyvinylidene fluoride (PVDF): Hybond™ P (Amersham Pharmacia Biotech) or nitrocellulose: Hybond ECL (Amersham Pharmacia Biotech).
5. SaranWrap™ (Dow Chemical Company) or low fluorescence plastic document wallets.
6. Autoradiography film, for example, Hyperfilm™ ECL (Amersham Pharmacia Biotech).
7. Densitometer, Personal Densitometer™ SI (PDSI) (Amersham Pharmacia Biotech).
8. Fluorescence Scanner e.g., Storm™, Imager, Model 840/ 860 Typhoon™ Imager, model 9400/9410 (Amersham Pharmacia Biotech).
9. CCD camera, for example, ImageMaster™ VDS-CL (Amersham Pharmacia Biotech).
10. Image analysis software, for example, ImageQuant™ (Amersham Pharmacia Biotech).

2.2. Preparation of Protein Blots

1. Standard protein for preparation of calibration curve.
2. Test protein samples.
3. Two 8 × 10 × 0.75 cm gels of 10–12% SDS-polyacrylamide prepared according to method of Laemmli ***(12)***.
4. Sample loading buffer: 2% (w/v) SDS; 10% (v/v) mercaptoethanol, 62.5 m*M* Tris-HCl, pH 6.8, 10% (v/v) glycerol, 0.01% (w/v) bromophenol blue.
5. Running buffer: 25 m*M* Tris base, 192 m*M* glycine, 0.1% (w/v) SDS in dH_2O. For 1 L: 3 g of Tris base, 14.4 g of glycine, and 1 g of SDS. Make up to 1 L with deionized water. Do not adjust pH.
6. Transfer buffer: 25 m*M* Tris base, 192 m*M* glycine, 20% (v/v) methanol. For 1 L: 3 g of Tris base, 14.4 g of glycine, 200 mL of methanol. Make up to 1 L with deionized water. Do not adjust pH.

2.3. Immunodetection

1. Phosphate buffered saline, pH 7.5, 0.1% (v/v) Tween 20 (PBST). For 1 L 11.5 g of disodium hydrogen orthophosphate anhydrous (80 m*M*), 2.96 g of 20 m*M* sodium dihydrogen orthophosphate (20 m*M*), 5.84 g of sodium chloride (100 m*M*), 1 mL of Tween 20. Make up to 1 L with deionized water and check pH.

2. Membrane blocking agent, for example 5% nonfat dried milk powder in PBST.
3. Primary antibody diluted in PBST.
4. Secondary anti-species HRP-labeled antibody diluted in PBST.

2.4. Detection

1. ECL Plus Western blotting reagents (Amersham Pharmacia Biotech).

3. Method

3.1. Preparation of Protein Blots

1. The sample containing the protein to be quantified plus a dilution series containing a known amount of the same protein are used for the Western blot. It is suggested that 5–10 standard dilutions are prepared, sufficient to load onto two gels. It is important that the concentration of the protein in the sample to be quantified lies within the standard range (*see* **Note 1**). It is therefore worthwhile preparing more than one concentration of the test sample, again sufficient to load onto two gels.
2. Equal volumes of the protein dilutions are mixed with sample loading buffer and heated to 95°C for 4 min. Ten microliters of each dilution is then loaded into the wells of each of the SDS-polyacrylamide gels.
3. Following electrophoresis, the gels are electroblotted onto PVDF or nitrocellulose membrane overnight using standard tank electroblotting methodology *(13)*.

3.2. Immunodetection (See also *Chapter 54*)

Steps 1–6 should be carried out with gentle shaking. Do not allow the blot to dry out between steps (*see* **Note 2**).

1. The blots are blocked in 5% nonfat dried milk for 60 min at room temperature.
2. The blot is rinsed briefly with two changes of PBST.
3. The blot is incubated for 60 min at room temperature with primary antibody at the appropriate dilution. In the example shown in **Fig. 1** an anti-actin monoclonal was diluted 1:5,000 with PBST.
4. The blot is washed in PBST for 15 min followed by three repeat washes of 5 min in fresh PBST.
5. The blot is incubated for 60 min at room temperature with secondary anti-species HRP-labeled antibody at the appropriate dilution. In the example shown in **Fig. 1**, anti-mouse HRP was diluted 1:75,000 with PBST.
6. The washing **step 4** is repeated.
7. The blot is incubated with ECL Plus reagents (0.1 mL/cm^2 membrane) for 5 min. Excess reagent is drained off and the blot is placed between Saran Wrap or in a clear plastic sandwich wrap.

3.3. Detection Using Film (See *Fig. 1*)

1. The blot is exposed to autoradiography film for 1–5 min shown in **Fig. 2**. If desired, the film to be used can be preflashed. This is performed using a modified flash unit that has been calibrated (by adjusting its distance from the film) to raise the film optical density 0.1–0.2 OD units above that of the standard film. The flash duration should be in the region of 1 ms.
2. For quantification to be accurate, it is important that the light produced is in the linear dynamic range of the film (*see* **Notes 2** and **3**). It is therefore necessary to make several exposures of different lengths of time. The film on which the standard protein of the low-

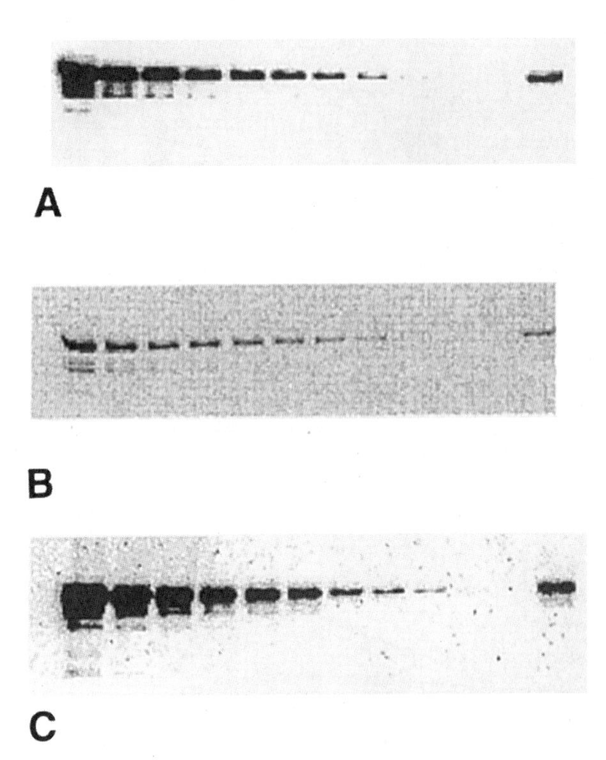

Fig. 1. Western blots of actin dilutions detected with ECL Plus. Actin concentrations from the *left*: 400, 200, 150, 100, 75, 50, 35, 25, 12.5, 6.25 ng, blank, "unknown." The "unknown" was a test sample whose concentration was calculated from the standard curve. **(A)** Film autoradiograph, exposure time 5 min. The film was digitized on a Personal Densitometer SI (PDSI). **(B)** Fluorescence scanner (Storm, Model 860) image (same blot as in **a**). **(C)** CCD camera (ImageMaster VDS-CL) image, 5 min exposure.

est concentration is only just visible should have the rest of the standards in the linear range of the film.

The resultant film is then digitized with a densitometer and saved as a 12 bit file. The band intensities can then be quantified using an appropriate image analysis program.

3.4. Detection Using a Fluorescence Scanner (See Fig. 2)

1. Ensure that the plastic used to encase the blot does not exhibit high fluorescence.
2. Blots should be scanned between 5 min and 1 h after addition of ECL Plus substrate reagents.
3. The blot is scanned face down at 450 nm on the Storm scanner using blue fluorescence/chemifluorescence scan mode. Set the photomultiplier tube (PMT) voltage between 500 and 900 for maximum linearity (*see* **Note 4**).
4. The band intensities are then quantified using an appropriate image analysis program such as ImageQuant.

3.5. Detection Using a CCD Camera (See Table 1)

1. Exposure times will typically be between 30 s and 10 min depending on the sensitivity desired and the type of camera used (*see* **Note 5**).

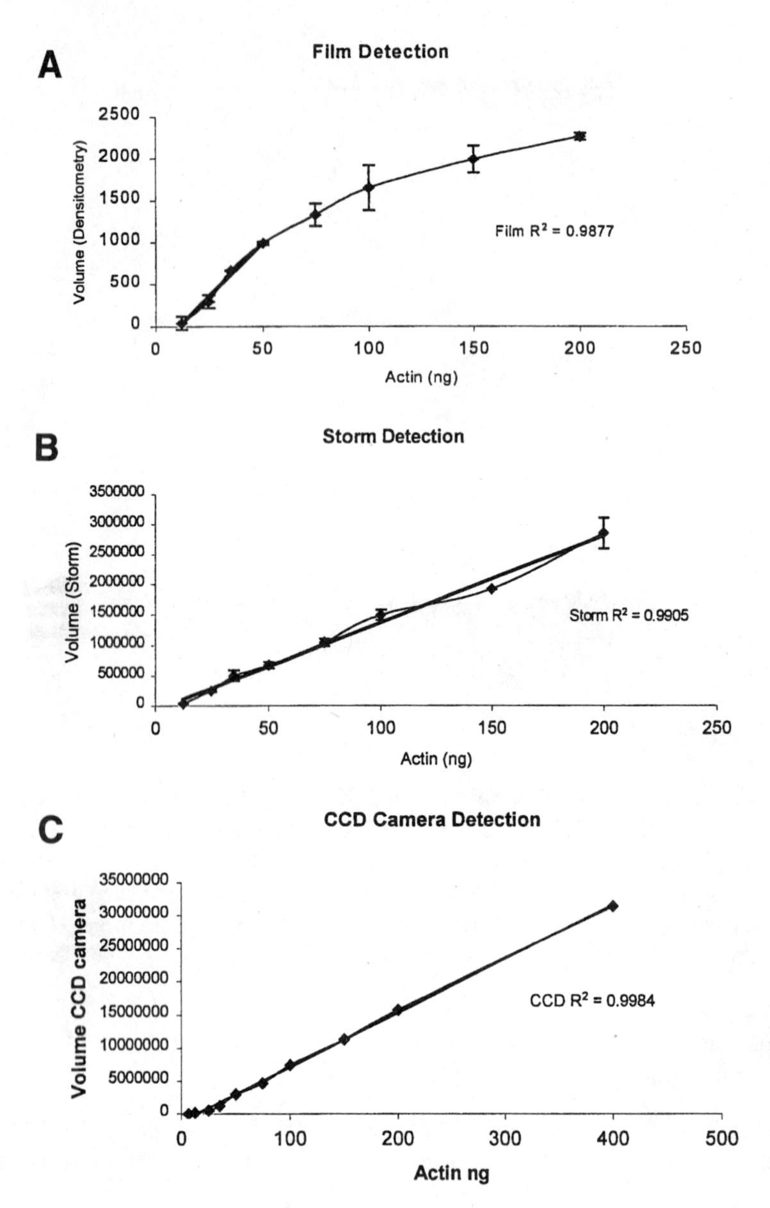

Fig. 2. **(A)** Quantification of film Western blot data. Signal intensities of actin bands were determined using volume analysis with object average background correction applied. Data were exported to Microsoft Excel to generate the plot. The background corrected volume of each band (except 400 ng) is shown versus ng actin. A linear fit to data points 50–12.5 ng is shown with the corresponding R^2 value. **(B)** Quantification of Storm Western blot data. Signal intensities of actin bands were determined using volume analysis with object average background correction applied. Data were exported to Microsoft Excel to generate the plot. The background corrected volume of each band (except 400 ng) is shown versus ng actin. A linear fit to data points 200–12.5 ng is shown with the corresponding R^2 value. **(C)** Quantification of CCD camera Western blot data. Signal intensities of actin bands were determined using volume analysis with object average background correction applied. Data were exported to Microsoft Excel to generate the plot. The background corrected volume of each band is shown versus (ng) actin. A linear fit to data points 400–12.5 ng is shown with the corresponding R^2 value.

Table 1
Calculation of Unknown Actin Concentrations Using Data from Images Generated by Storm and ImageMaster VDS-CL. For images created on film and analyzed by the PDSI the unknown was outside the linear range of the standard curve

	Volume rfu	Calculated ng	Actual ng
Storm:			
Sample			
Actin unknown	938162	61	54
CCD camera:			
Sample			
Actin unknown	3582848	56	54

2. As the optimum exposure time may not be known, it is recommended to take several exposures of increasing time in rapid succession. This avoids losing signal while appropriate exposure times are estimated.
3. Quantification can then be performed by exporting the image as a 12–16-bit file into a suitable image analysis program to measure band intensities.

4. Notes

4.1. Preparation of Protein Blots

1. For accurate quantification a dilution series of protein standards is required. This is easily done if the purified protein is commercially available; otherwise cloning and expression may be required.
2. If using film to image the Western blots, the dilution series should extend over a 10–20-fold range. For detection using fluorescence scanning or a CCD camera, a dilution range of up to 50-fold is recommended owing to the increased linear dynamic range achievable.
3. For greater accuracy it is recommended that duplicate gels be used for analysis. The running of several concentrations of the test sample will also help to increase the accuracy of quantification.
4. Overnight electroblotting is generally more reproducible than shorter blotting at higher voltages. However, the optimum procedure will depend on the particular protein being analyzed. Blots may be used immediately or air-dried and stored in a desiccator in a refrigerator (2–8°C) for up to 3 mo.

4.2. Immunodetection

5. PVDF membranes will require rewetting in methanol if they have been allowed to dry out after electroblotting.
6. The use of 5% nonfat dried milk in PBST is suitable for most applications but, for certain antigen antibody systems, alternative blocking conditions may need to be used.
7. The dilution of the primary antibody required to give optimum results will vary and should be determined for each antibody used. These optimization experiments may be performed by dot blot analysis.
8. Where backgrounds are still high, the inclusion of 0.25–0.5% blocking reagent in the antibody incubation may sometimes improve backgrounds.
9. As a general rule, as large a volume as possible of washing buffer should be used each time.
10. Due to the sensitivity of the detection system, the dilution of the HRP-labeled second antibody should be optimized to give the highest signal with minimum background. A

biotinylated antibody can be used in place of the HRP-labeled antibody in **Subheading 3.2., step 5**. This is then detected using a streptavidin–HRP complex. This system may result in more sensitive detection.

11. Drain off excess reagent by holding the membrane vertically and touching the edge of the membrane against tissue paper. Gently place the membrane, protein side down onto the Saran Wrap and lay a second piece on top, smoothing out any wrinkles. Alternatively, if using a fluorescent scanner, a low fluorescence plastic document folder may be used.

4.3. Detection Using Film

12. Preflashing raises the film to the threshold of response so that the number of photons of light hitting the film is more proportional to the number of silver grains converted.
13. Blots should be exposed to film immediately. The exact exposure time will have to be determined empirically. If the exposure time required to view all standards is very short, for example, <15 s, it may be advisable to use more dilute antibody solutions.
14. Laser densitometers are recommended for most accurate quantification. Some researchers have reported that document scanners can be used although correction for nonlinearity in response may be necessary ***(14)***.

4.4. Detection Using a Fluorescent Scanner

15. ECL Plus detection is approximately five times more sensitive with the Storm imager (450 nm excitation) than with a scanner utilizing a 488 nm laser. This is because the excitation wavelength of the Storm imager more closely matches that of the ECL Plus fluorescent intermediate allowing optimal excitation.
16. A small puddle of water may be used on the surface of the glass to help form a tight interface between the plastic and the glass.
17. Blots can be air dried and imaged directly on the glass plate, although sensitivity may be compromised by around twofold.
18. Leaving the blots for longer than 1 h after ECL Plus substrate addition may also result in reduced sensitivity of detection.

4.5. Detection Using a CCD Camera

19. As for film, to avoid losing signal, it is important to start exposures as soon as the incubation in substrate has been completed.
20. Care needs to be taken that saturation of the camera does not occur in regions where light intensity is high. This can be avoided by checking the peak pixel values and decreasing exposure times if their values are at maximal levels.

References

1. Bers, G. and Garfin, D. (1985) Protein and nucleic acid blotting and immunobiochemical detection. *Biotechniques* **3,** 276–287.
2. Holmes, J. L. and Pollenz, R. S. (1997) Determination of aryl hydrocarbon receptor nuclear translocator protein concentration and subcellular localization in hepatic and nonhepatic cell culture lines: Development of quantitative Western blotting protocols for calculation of aryl hydrocarbon receptor and aryl hydrocarbon receptor nuclear translocator protein in total cell lysates. *Mol. Pharmacol.* **52,** 202–211.
3. Gillespie, P. and Hudspeth, A. (1991) Chemiluminescence detection of proteins from single cells. *Proc. Nat Acad. Sci. USA* **88,** 2563–2567.

4. Fradelizi, J., Friederich, E., Beckerle, M. C., and Goldsteyn, R. M. (1998) Quantitative measurement of proteins by Western blotting with Cy5-coupled secondary antibodies. *Biotechniques* **26,** 484–494.
5. Gingrich, J. C., Davis, D. R., and Nguyen, Q. (2000) Multiplex detection and quantitation of proteins on Western blots using fluorescent probes. *Biotechniques* **2,** 636–642.
6. Worley, J. M., Powell, R. I., and Mansfield, E. S. (1994) Molecular Dynamics Application Note **#57**, Molecular Dynamics Inc.
7. Durrant, I. (1990) Light-based detection of biomolecules. *Nature* **346,** 297.
8. Akhavan-Tafti, H., DeSilva, R., Arghavani, Z., Eickholt, R. A., Handley, S., Schoenfelner, B. A., Sugioka, K., Sugioka, Y., and Schaap, A. P. (1998) Characterization of acridancarboxylic acid derivatives as chemiluminescent peroxidase substrates. *J. Org. Chem.* **63,** 930–937.
9. Heinicke, E., Kumar, U., and Munoz, D. G. (1992) Quantitative dot-blot assay for proteins using enhanced chemiluminescence. *J. Immunol. Meth.* **152,** 227–236.
10. Laskey, R. A. (1980) The use of intensifying screens or organic scintillators for visualizing radioactive molecules resolved by gel electrophoresis. *Meth. Enzymol.* **65,** 363–371.
11. Boniszewski, Z. A. M., Comley, J., Hughes, B., and Read, C. (1990) The use of charge coupled devices in the quantitative evaluation of images, on photographic film or membranes, obtained following electrophoretic separation of DNA fragments. *Electrophoresis* **11,** 432–440.
12. Laemmli, U. K. (1970) Cleavage of structural proteins during assembly of the head of bacteriophage T4. *Nature* **227,** 680–685.
13. Towbin, H., Staehelin, T., and Gordon, J. (1979) Electrophoretic transfer of proteins from polyacrylamide gels to nitrocellulose sheets: Procedure and some appliactions, *Proc. Natl. Acad. Sci. USA* **76,** 4350–4354.
14. Tarlton, J. F. and Knight, P. J. (1996) Comparison of reflectance and transmission densitometry, using document and laser scanners, for quantitation of stained western blots. *Analyt. Biochem.* **237,** 123–128.

58

Reutilization of Western Blots After Chemiluminescent Detection or Autoradiography

Scott H. Kaufmann

1. Introduction

Western blotting (also called immunoblotting) is a widely utilized laboratory procedure that involves formation and detection of antibody–antigen complexes between antibodies that are initially in solution and antigens that are immobilized on derivatized paper (reviewed in **refs.** ***1–3***). This procedure is most commonly performed by sequentially subjecting a complex mixture of polypeptides to three manipulations: (1) electrophoretic separation, usually through polyacrylamide gels in the presence of sodium dodecyl sulfate (SDS); (2) electrophoretic transfer of the separated polypeptides to thin sheets of nitrocellulose or polyvinylidene fluoride (PVDF); and (3) reaction of the sheets sequentially with one or more antibody-containing solutions. Because these are laborious, low-throughput procedures, there has been considerable interest over the past 15 yr in increasing the information gained from immunoblots in various ways. The approaches to this problem have included removal of bound antibodies so that immobilized polypeptides can be reprobed with additional antisera (last reviewed in **ref.** ***4***), the sequential ***(5)*** or simultaneous ***(6)*** use of multiple antisera to detect different antigens, and the development of higher density filter manifolds to permit more spots to be placed in the same area when purified proteins are applied to filters by spot adsorption. After a brief review of critical parameters for successful immunoblotting, some of these approaches are briefly discussed and illustrated.

Because of its versatility, immunoblotting is widely employed in biological studies. Critical to the success of this procedure is the quality of the immunological reagents utilized. With a high quality antiserum or monoclonal antibody, this approach can be utilized to determine whether an antigen of interest is present in a particular biological sample, to monitor the purification of the antigen, or to assess the location of epitopes within the antigen after chemical or enzymatic degradation. With suitable immunological probes, this same approach can be utilized to search for proteins that bear a particular physiological or pathological posttranslational modification, for example, phosphorylation ***(7)*** or nitrosylation of tyrosine ***(8)***, covalently attached glycosylphosphatidylinositol ***(9)***, or modification by D-penicillamine metabolites ***(10)***. Conversely, immunoblotting can also be utilized to determine whether antibodies that recognize a

From: *The Protein Protocols Handbook, 2nd Edition*
Edited by: J. M. Walker © Humana Press Inc., Totowa, NJ

particular antigen are present in a sample of biological fluid. Western blotting has become a method of choice, for example, for confirming that antibodies to the human immunodeficiency virus-1 ***(11)*** or *B. burgdorferi* ***(12)*** are present in patients suspected of having Acquired Immunodeficiency Syndrome (AIDS) or Lyme disease, respectively.

There are certain circumstances in which it is convenient to be able to probe immunoblots for the presence of multiple antigens. If, for example, an immunoblotting experiment gives an unexpected result regarding the subcellular distribution of a polypeptide, it is important to reprobe the blot with a reagent that recognizes a second polypeptide to confirm that the samples have been properly prepared, loaded, and transferred. Likewise, if the polypeptides being analyzed are derived from a precious source (e.g., pathological tissue, biological fluid, or organism that is not readily available), it is sometimes convenient to reutilize the blots.

1.1. Reutilization of Blots After Removal of Antibodies

One method for reutilizing blots involves removing the bound antibodies before probing with a new immunological reagent. This approach is patterned after methods for reutilizing Southern and Northern blots ***(13)***. In each case, the original idea was to remove radioactive probe from the blots without removing the target of interest. Older methods that permitted this approach with protein blots (reviewed in **ref.** ***4***) involved the covalent binding of target polypeptides to derivatized paper so that subsequently bound antibodies could be solubilized under denaturing conditions without eluting the antigens of interest. These older techniques never gained widespread acceptance because of the lack of durability of the derivatized supports.

Polypeptides are more commonly immobilized on nitrocellulose or PVDF, solid supports that are thought to bind macromolecules noncovalently. The critical issue then becomes the identification of conditions that solubilize antibodies without solubilizing the target polypeptides. Some reports have indicated that treatment of nitrocellulose blots with glycine at pH 2.2 ***(14)*** or with 8 *M* urea at 60°C ***(15)*** can remove antibodies and permit reutilization of blots. Other studies have indicated that these techniques are effective at disrupting low-affinity antibody-antigen interactions but not high-affinity interactions ***(14,16)***.

Two subsequent observations allowed the development of a more widely applicable technique for the removal of antibodies from Western blots. First, treatment of nitrocellulose with acidic solutions of methanol was observed to "fix" transferred polypeptides to nitrocellulose ***(17,18)***. Polypeptides treated in this fashion remained bound to nitrocellulose even during treatment with SDS at 70–100°C under reducing conditions ***(16,17)***. Second, treatment of blots with a large excess of irrelevant protein immediately prior to drying and autoradiography was observed to facilitate subsequent removal of radiolabeled antibodies from nitrocellulose ***(16)***.

Based on these observations, a technique that allows the reutilization of Western blots after reaction with a wide variety of radiolabeled antibodies or lectins was developed ***(16)***. This technique involves the initial immobilization of polypeptides on nitrocellulose by staining with dye dissolved in an acidic solution of methanol. After unoccupied binding sites have been saturated with irrelevant protein, the nitrocellulose is treated sequentially with unlabeled primary antibodies and radiolabeled secondary antibodies. Prior to drying, the blot is briefly incubated in a protein-containing buffer. After subsequent drying and autoradiography, the antibodies are removed by treating the nitrocellulose with SDS at 65–70°C under reducing conditions. Minor modifications of this procedure permit removal of peroxidase-coupled antibodies after detection of antigens by enhanced chemiluminescence. These procedures are described in detail in **Subheadings 2–4.** of this chapter.

1.2. Reutilization of Blots Without Removal of Antibodies

The preceding techniques were developed in an era when radiolabeled antibodies or lectins were widely utilized for blotting ***(16)***. At the present time, enzyme-coupled secondary antibodies are more widely utilized for immunoblotting; and fluorochrome-coupled antibodies are also being investigated ***(6)***. The use of these nonradioactive reagents raises the possibility of detecting multiple antigens by either simultaneous or sequential probing of blots without intervening dissociation of bound antibodies.

When two antigens of interest can be readily distinguished by size, blots have been probed with two different immunological reagents simultaneously. Bound primary antibodies are then visualized by probing with one or more radiolabeled or enzyme-coupled secondary antibodies as needed. Interpretation of results obtained in this manner can be difficult, however, if one or more of the immunological reagents cross-reacts with multiple species on the blot or one of the polypeptides of interest exists as multiple species, for example, splice variants or posttranslationally modified forms. A recently described alternative to this procedure involves the use of primary antisera raised in several different species followed species-specific secondary antibodies coupled to fluorochromes with distinguishable spectral properties ***(6)***. This approach has the advantage of allowing unequivocal detection of the species that react with each primary antibody. Its disadvantages, however, include the need for primary antisera raised in different species and the current lack of widespread availability of suitable imaging equipment.

An alternative to simultaneous probing involves sequential probing of blots with different reagents. Krajewski et al. ***(5)*** reported that results obtained during sequential probing of blots with different primary antibodies and peroxidase-coupled secondary antibodies could be simplified by incubating blots with chromogenic peroxidase substrates after each chemiluminescent detection. In brief, the chromogenic substrates irreversibly inhibit the peroxidase after chemiluminescent detection is complete, insuring that the peroxidase bound to one secondary antibody is not active at the time subsequent secondary antibodies are bound. We have recently observed that the sodium azide added to blocking buffers as a preservative also inactivates horseradish peroxidase. As a consequence, blots can be sequentially reacted with multiple different secondary antibodies without interference provided that azide is present in the solution containing the primary antibodies ***(26)***. This process is illustrated in **Fig. 1**, where the same piece of nitrocellulose was sequentially reacted with immunological reagents raised in chickens, rabbits, and mice. With the successive use of reagents raised in different species, the signal obtained at each step is unaffected by the prior probings (**Fig. 1**, *lanes 1–3*) even though the previously bound antibodies remain attached to the

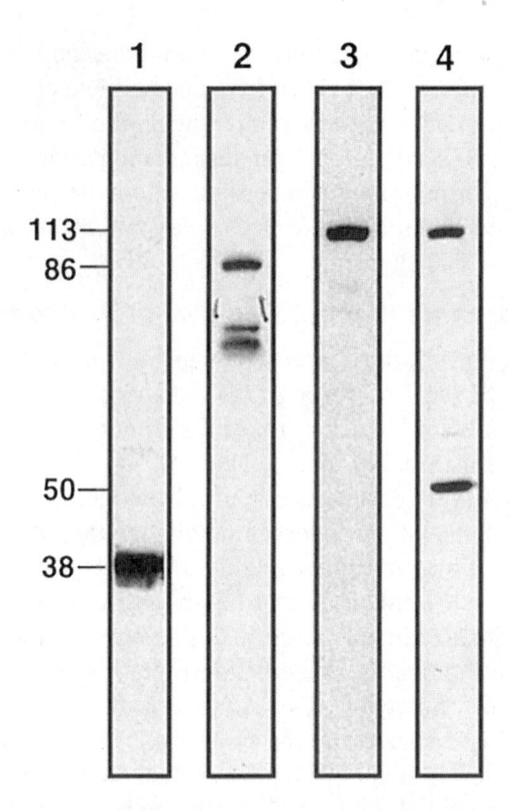

Fig. 1. Reutilization of immunoblots without intervening dissociation of antibodies. After staining with fast green FCF and coating with TSM (**Subheading 3.1., steps 8–11**), a single strip containing total cellular protein from 3×10^5 K562 human leukemia cells was sequentially subjected to immunoblotting using the following primary (secondary) reagents: chicken anti–B23 (peroxidase-coupled goat anti-chicken IgG), rabbit anti-protein kinase Cδ (peroxidase-coupled goat anti-rabbit IgG), mouse anti-poly(ADP-ribose) polymerase (peroxidase-coupled goat anti-mouse IgG), and mouse anti-procaspase-9 (peroxidase-coupled goat anti-mouse IgG). At each step, bound secondary antibodies were detected using ECL-enhanced chemiluminescence reagents (Amersham Pharmacia Biotech, Piscataway, NJ). At each step, primary antibody was added in TSM buffer, which contains sodium azide. Note that the peroxidase bound to the blot at each step is not catalytically active at subsequent steps (cf. *lanes 1–3*), whereas the previously bound primary antibodies react with fresh secondary antibodies at subsequent steps (cf. *lanes 3* and *4*). Numbers at *left*, molecular masses of the respective polypeptides in kilodaltons.

blot. It is important to stress, however, that each of these probings involves a different secondary antibody. When the same blot is subsequently probed with a second mouse immunoglobulin G (IgG) monoclonal antibody, the previously bound mouse IgG is detected by the peroxidase-coupled anti-mouse IgG (e.g., **Fig. 1**, *lanes 3* and *4*). Thus, sequential probing works well when multiple primary antibodies raised in different species are available or when antigens of different molecular weights can be unequivocally detected using antibodies from the same species. This approach can become problematic, however, if the antigens of interest have similar molecular weights or the immunological reagents cross-react with multiple species of different molecular weights. Under these circumstances, it might still be necessary to strip antibodies off the blots between sequential probing using the techniques described below.

2. Materials

1. Apparatus for transferring polypeptides from gel to solid support. (Design principles are reviewed in **refs. *2,3*.**)
   a. TE52 reservoir-type electrophoretic transfer apparatus (Hoefer Scientific, San Francisco, CA) or equivalent.
   b. Polyblot semidry blotter (Pharmacia, Piscataway, NJ) or equivalent.
2. Paper support for binding transferred polypeptides.
   a. Nitrocellulose.
   b. Nylon (e.g., Genescreen from New England Nuclear, Boston, MA, or Nytran from Schleicher and Schuell, Keene, NH).
   c. Polyvinylidene difluoride PVDF (e.g., Immobilon from Millipore, Bedford, MA).
3. Fast green FCF for staining polypeptides after transfer to solid support.
4. 10,000 U/mL Penicillin and 10 mg/mL of streptomycin.
5. Reagents for electrophoresis (acrylamide, *bis*-acrylamide, 2-mercaptoethanol, SDS) should be electrophoresis grade.
6. All other reagents (Tris, glycine, urea, methanol) are reagent grade.
7. Transfer buffer: 0.02% (w/v) SDS, 20% (v/v) methanol, 192 m*M* glycine-HCl, and 25 m*M* Tris base. Prepare enough buffer to fill the chamber of the transfer apparatus and a container for assembling cassette.
8. Fast green stain: 0.1% (w/v) Fast green FCF in 20% (v/v) methanol –5% (v/v) acetic acid. This stain is reusable. Prepare 50–100 mL per blot.
9. Fast green destain: 20% (v/v) Methanol in 5% (v/v) acetic acid.
10. TS buffer: 150 m*M* NaCl, 10 m*M* Tris-HCl, pH 7.4. This can be conveniently prepared as a 10X stock (1.5 *M* NaCl, 100 m*M* Tris-HCl, pH 7.4). The 10X stock can be stored indefinitely at 4°C and then used to prepare 1× TS buffer, TSM buffer, and the other buffers described below.
11. TSM buffer: TS buffer containing 10% (w/v) powdered milk, 100 U/mL penicillin, 100 µg/mL streptomycin, and 1 m*M* sodium azide. This buffer can be stored for several days at 4°C. Note that sodium azide is poisonous and can form explosive copper salts in drain pipes if not handled properly.
12. Phosphate-buffered saline (PBS): 137 m*M* NaCl, 2.7 m*M* KCl, 1.5 m*M* KH_2PO_4, 8 m*M* Na_2HPO_4, pH 7.4. This can be conveniently prepared as a 10-fold concentrated stock at pH 7.0 and stored at 4°C. Upon dilution the pH will be 7.4.
13. PBS containing 2 *M* urea and 0.05% (w/v) Nonidet P-40 (NP-40). Prepare 300 mL per blot by combining 0.15 g of NP-40, 30 mL of 10× PBS buffer, 75 mL of 8 *M* urea (freshly deionized over Bio-Rad AG1X-8 mixed bed resin to remove traces of cyanate), and 195 mL of water.
14. PBS buffer containing 0.05% (w/v) NP-40. Prepare 300 mL per blot.
15. PBS buffer containing 3% (w/v) nonfat powdered milk. Prepare 25 mL per blot.
16. Blot erasure buffer: 2% (w/v) SDS; 62.5 m*M* Tris-HCl, pH 6.8 at 21°C; and 100 m*M* 2-mercaptoethanol. The SDS–Tris-HCl solution is stable indefinitely at 4°C. Immediately prior to use, 2-mercaptoethanol is added to a final concentration of 6 µL/mL.
17. Primary antibody.
18. Reagents for detection by chemiluminescence. This approach requires enzyme-coupled secondary antibody and a substrate that becomes chemiluminescent as a consequence of enzymatic modification, for example, peroxidase-coupled secondary antibody and luminol

(Amersham Pharmacia Biotech [Piscataway, NJ] ECL enhanced chemiluminescence kit or equivalent).

19. In lieu of enzyme-coupled secondary antibody, ^{125}I-labeled secondary antibody can be used. Secondary antibodies can be labeled as previously described ***(16)*** or purchased commercially. Radiolabeled antibodies should only be used by personnel trained to properly handle radioisotopes.

3. Methods

3.1 Transfer of Polypeptides to Nitrocellulose

The following description is appropriate for transfer in a transfer reservoir. If a semidry transfer apparatus is to be used, follow the manufacturer's instructions (*see* **Note 1**).

1. Perform SDS-polyacrylamide gel electrophoresis (SDS-PAGE) using standard techniques (*see* **ref. 19** for description of this method).
2. Wear disposable gloves while handling the gel and nitrocellulose at all steps. This avoids cytokeratin-containing fingerprints.
3. Cut nitrocellulose sheets to a size slightly larger than the polyacrylamide gel (*see* **Note 2**).
4. Fill the transfer apparatus with transfer buffer (*see* **Note 3**).
5. Fill a container large enough to accommodate the transfer cassettes with transfer buffer. Assemble the cassette under the buffer in the following order:
   a. Back of the cassette.
   b. Two layers of filter paper.
   c. The gel.
   d. One piece of nitrocellulose—gently work bubbles out from between the nitrocellulose and gel by rubbing a gloved finger or glass stirring rod over the surface of the nitrocellulose.
   e. Two layers of filter paper—again, gently remove bubbles.
   f. Sponge or flexible absorbent pad.
   g. Front of the cassette.
6. Place the cassette in the transfer apparatus so that the front is oriented toward the POSITIVE pole.
7. Transfer at 4°C in a cold room with the transfer apparatus partially immersed in an ice-water bath. Power settings: 90 V for 5–6 h or 60 V overnight.
8. Place the fast green stain in a container with a surface area slightly larger than one piece of nitrocellulose (*see* **Notes 4** and **5**). After the transfer is complete, place all the pieces of nitrocellulose in the stain and incubate for 2–3 min with gentle agitation. Decant the stain solution, which can be reused.
9. Destain the nitrocellulose by rinsing it for 3–5 min in fast green destain solution with gentle agitation. Decant the destain, which can also be reused. Rinse the nitrocellulose four times (1–2 min each) with TS buffer (200 mL/rinse).
10. Mark the locations of lanes, standards, and any other identifying features by writing on the blot with a standard ball point pen.
11. Coat the remaining protein binding sites on the nitrocellulose by incubating the blot in TSM buffer (50–100 mL/blot) for 6–12 h at room temperature (*see* **Note 6**). Remove the blot from the TSM buffer. Wash the blot four times in quick succession with TS buffer (25–50 mL/wash) and dry the blot on fresh paper towels.

Either before or after coating of the unoccupied protein binding sites, blots can be dried and stored indefinitely in an appropriate container, for example, disposable resealable food storage bags (1–2 blots/bag).

3.2. Detection of Antibody–Antigen Complexes Using Peroxidase-Coupled Secondary Antibody

1. Place the nitrocellulose blot in an appropriate container for reaction with the antibody. A 15– or 20–lane sheet can be reacted with 15–20 mL of antibody solution in a resealable plastic bag or an open plastic dish of suitable size. A 1– or 2–lane strip can be reacted with 2–5 mL of antibody solution in a disposable 15–mL conical test tube.
2. If the nitrocellulose has been dried, rehydrate it by incubation for a few minutes in a small volume of TSM buffer.
3. Add an appropriate dilution of antibody to the TSM buffer and incubate overnight (10–15 h) at room temperature with gentle agitation (*see* **Notes 7** and **8**).
4. Remove the antibody solution and save for reuse (*see* **Note 9**).
5. Wash the nitrocellulose (100 mL/wash for each large blot or 15–50 mL/wash for each individual strip) with PBS containing 2 *M* urea and 0.05% NP-40 (three washes, 15 min each) followed by PBS buffer (two washes, 5 min each) (*see* **Note 10**).
6. Add freshly prepared PBS containing 3% milk to the nitrocellulose sheets or strips (*see* **Note 11**). For nitrocellulose sheets (or pooled strips) in resealable plastic bags it is convenient to use a 25 mL volume. Add peroxidase-labeled secondary antibody at a suitable dilution (*see* **Note 12**). Incubate for 90 min at room temperature.
7. Remove the secondary antibody and discard appropriately.
8. Wash the sheets (100 mL/wash for each large blot or each group of pooled strips) with PBS containing 0.05% NP-40 (two washes of 5 min each, two washes of 15 min each, and two washes of 5 min each).
9. Expose blots to enhanced chemiluminescent reagents according to the supplier's protocol.
   a. In the case of ECL, combine equal volumes of the two solutions (e.g., 5 mL of each for a gel containing 20 lanes).
   b. Remove the last wash solution from the blot.
   c. Incubate the blot with the combined ECL reagents for 60 s with gentle agitation.
   d. Remove ECL reagent from container (*see* **Note 13**). Remove excess reagent but do not dry the blot.
   e. Seal the bag containing the blot and a minimal amount of ECL reagent. Immediately expose to X-ray film (*see* **Notes 14** and **15**).
10. After exposure to X-ray film, rinse the blot three or four times with PBS (25 mL/wash) and store it dry at room temperature.

3.3. Dissociation of Bound Antibodies

1. Place the blot to be stripped in a resealable plastic bag.
2. Add 50 mL of erasure buffer, seal the bag, and incubate in a water bath at 65°C for 30 min with gentle agitation every 5–10 min (*see* **Notes 16–18**.)
3. Decant and discard the erasure buffer. Rinse the blot 3× with 50–100 mL of TS buffer to remove SDS.
4. To ensure that nonspecific binding sites on the blot are well coated, incubate with TSM buffer for at least 2 h at room temperature with gentle agitation.
5. The blot is ready to be dried and stored or to be incubated with a new antibody as described in **Subheading 3.2.**

3.4. Modification for Dissociation of Antibodies After Autoradiographic Detection

1. Perform **steps 1–8** as described in **Subheading 3.2.**, substituting 5–10 μCi ^{125}I-labeled secondary antibody for peroxidase-coupled antibody (*see* **Note 19**).

2. Rather than exposing the blot to ECL reagents (**Subheading 3.2., step 9**), follow the last wash in PBS/Tween with a 5-min incubation with TSM buffer. This incubation facilitates subsequent removal of antibody and reuse of the blot (*see* **Note 20**).
3. After incubating the blot with TSM buffer, immediately dry it between several layers of paper towels. After 5 min, move the blot to fresh paper towels to prevent the nitrocellulose from sticking to the paper towels. When the blot has dried thoroughly, tape it to a solid support (e.g. thin cardboard), wrap it with clear plastic wrap, and subject it to autoradiography (*see* **ref. *20*** for details).

4. Notes

4.1. Transfer of Polypeptides to Nitrocellulose

1. The method described is for transferring polypeptides after electrophoresis in SDS-containing polyacrylamide gels. Alternative methods have been described for transferring polypeptides after acid–urea gels and after isoelectric focusing (reviewed in **refs. *2,3***).
2. Choice of solid support for polypeptides: Various solid supports (nitrocellulose, nylon, or PVDF) can be used for Western blotting, stripped, and reprobed. Use of nitrocellulose and PVDF is illustrated in **Fig. 2**. Nitrocellulose is easy to use and is compatible with a wide variety of staining procedures. With multiple cycles of blotting and erasing, however, nitrocellulose becomes brittle. PVDF membranes are durable, are compatible with a variety of nonspecific protein stains, and are capable of being stripped and reprobed. PVDF membranes, however, require a prewetting step in methanol before protein can be transferred. Derivatized nylon has the advantage of greater protein binding capacity and greater durability, but avidly binds many nonspecific protein stains (reviewed in **refs. *1–3***). The higher binding capacity of nylon is said to contribute to higher background binding despite the use of blocking solutions containing large amounts of protein (reviewed in **ref. *3***). As illustrated below, however, some antibodies can be more easily dissociated from nylon than from nitrocellulose (*see* **Fig. 3**).
3. Various compositions of transfer buffer have been described (reviewed in **refs. *1–3***). Methanol is said to facilitate the binding of polypeptides to nitrocellulose, but to retard the electrophoretic migration of polypeptides out of the gel. In the absence of SDS, polypeptides with molecular masses above 116 kDa do not transfer efficiently. Low concentrations of SDS (0.01–0.1%) facilitate the transfer of larger polypeptides, but simultaneously increase the current generated during electrophoretic transfer, necessitating the use of vigorous cooling to prevent damage to the transfer apparatus.
4. Alternative staining procedures (reviewed in **refs. *1–3***) utilize Coomassie Blue, Ponceau S, Amido Black, India drawing ink, colloidal gold, or silver. A highly sensitive technique utilizing eosin Y has also been described ***(21)***.
5. A washing step in acidified alcohol is probably essential to immobilize the polypeptides on the nitrocellulose ***(3,17,18)***. The fast green staining procedure satisfies this requirement. Polypeptides are observed to elute from nitrocellulose under mild conditions if a wash in acidified alcohol is omitted ***(18,22)***.
6. Various proteins have been utilized to block unoccupied binding sites on nitrocellulose (reviewed in **refs. *2,3***). These include 5–10% (w/v) powdered dry milk, 3% (w/v) bovine serum albumin, 1% (w/v) hemoglobin, and 0.1% (w/v) gelatin. Although the choice of protein can affect antibody binding (e.g., **Fig. 4C**), antibodies have been successfully stripped from blots of coated with any of these solutions ***(16)***.

4.2. Formation of Antigen-Antibody Complexes

7. No guidelines can be provided regarding the appropriate dilution of antibody to use. Some antisera are useful for blotting at a dilution of >1:20,000. Other antisera yield detectable

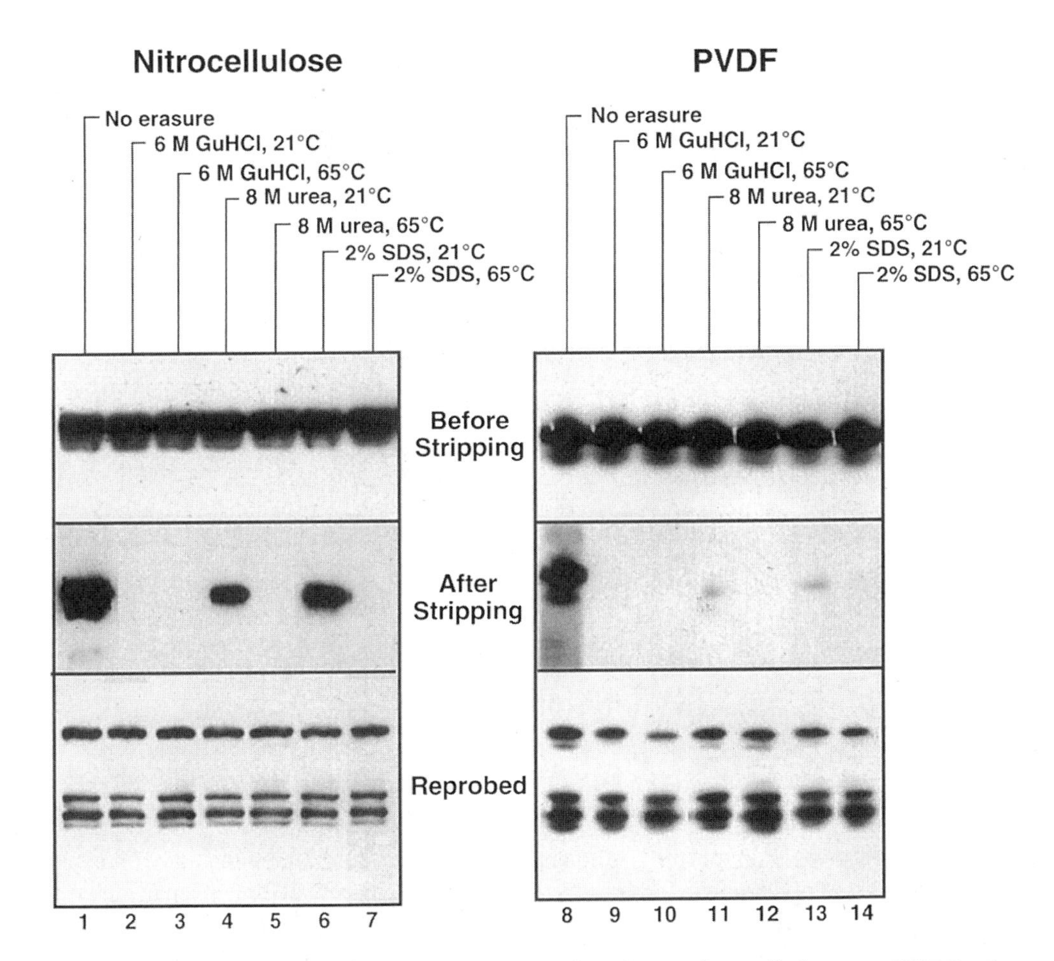

Fig. 2. Conditions for dissociating antibodies from nitrocellulose or PVDF after immunoblotting with enhanced chemiluminscence detection. Replicate samples containing whole cell lysates prepared from 3×10^5 K562 human leukemia cells were subjected to SDS-PAGE followed by transfer to nitrocellulose (*lanes 1–7*) or Immobilon PVDF (*lanes 8–14*). After unoccupied binding sites were blocked by incubation with TSM buffer, blots were incubated with a chicken polyclonal antiserum that reacts with the nucleolar protein B23 followed by peroxidase-labeled goat anti-chicken IgG. After detection with ECL reagents (*upper panels*), the blots were dried. The indicated strips were then incubated for 30 min at 21°C or 65°C with 62.5 m*M* Tris-HCl (pH 6.8 at 21°C) containing 100 m*M* 2-mercaptoethanol and 6 *M* guanidine hydrochloride (GuHCl), 8 *M* urea, or 2% (w/v) SDS as indicated. At the completion of the incubation, the strips were washed 4 times with PBS, recoated for 6 h in TSM buffer, and dried overnight. The strips were then reprobed with peroxidase-coupled goat anti-chicken IgG (**Subheading 3.2., steps 6–9**) to detect primary antibodies that were still bound (*middle panels*). After being dried again, strips were probed with rabbit anti-protein kinase Cδ followed by peroxidase-coupled goat anti-rabbit IgG to confirm that the cell lysate proteins remained bound to the strips (*lower panels*). Note that the efficacy of various treatments at removing antibodies varies depending upon the solid support, the buffer utilized for stripping, and the temperature applied. After ECL, all of these denaturing buffers remove the anti-B23 antibodies at 65°C. At 21°C, urea and SDS leave substantial amounts of antibody bound to nitrocellulose (*middle panel*, *lanes 4* and *6*) and small but detectable amounts of antibody bound to PVDF (*middle panel*, *lanes 11* and *13*). In each case, the stripping procedure has little or no effect on subsequent blotting with another antiserum (*bottom panels*).

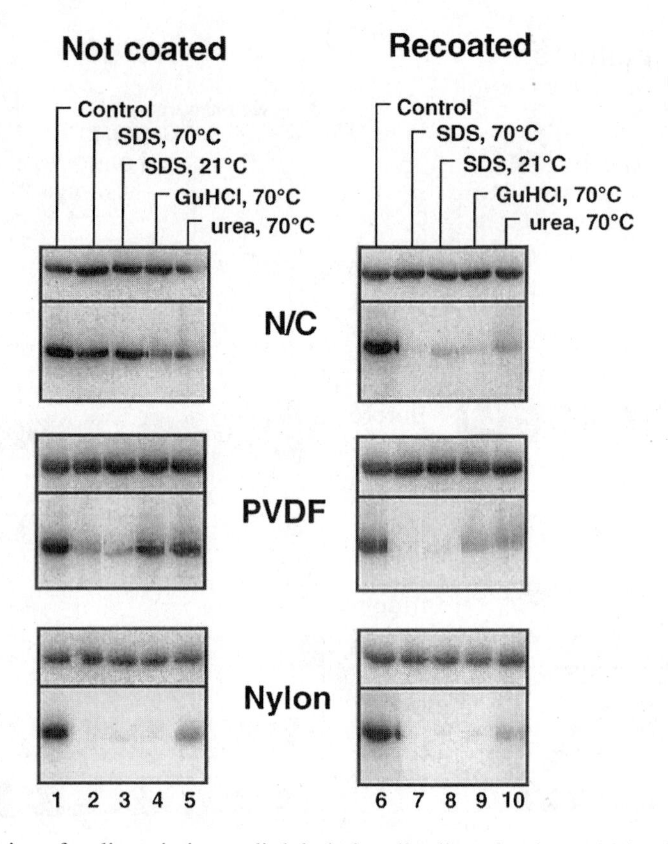

Fig. 3. Conditions for dissociating radiolabeled antibodies after immobilization of polypeptides on various solid supports. Replicate samples containing 2×10^6 rat liver nuclei were subjected to SDS-PAGE followed by transfer to nitrocellulose, PVDF, or Nytran nylon sheets as indicated (*see* **Note 2**). After staining with fast green FCF and treatment with TSM buffer to coat unoccupied protein binding sites (**Subheading 3.1., steps 8–11**), blots were incubated with chicken polyclonal antiserum that reacts with the nuclear envelope polypeptide lamin B_1 *(25)* followed by ^{125}I-labeled rabbit anti-chicken IgG (**Subheading 3.4., step 1**). Half of each blot (*lanes 6–10*) was recoated with milk-containing buffer for 5 min prior to drying (**Subheading 3.4., step 2**) and the other half (*lanes 1–5*) was dried without being recoated with protein. After autoradiography confirmed that the signals in all lanes of a given panel were comparable (*upper panels*), samples were incubated for 30 min at the indicated temperature with 50–62.5 m*M* Tris-HCl (pH 6.8) containing 100 m*M* 2-mercaptoethanol and one of the following denaturing agents: 2% (w/v) SDS, 6 *M* guanidine hydrochloride (GuHCl) or 8 *M* urea. Following this erasure procedure, strips were subjected to autoradiography again (*bottom panels*). In each case, untreated strips (*lanes 1* and *6*) served as controls. This analysis leads to several conclusions. First, recoating with milk prior to drying (**Subheading 3.4., step 2**) does not affect the amount of radiolabeled antibody initially bound to the blots (cf. *lanes 1* and *6*). Second, the efficacy of various treatments at removing these antibodies varies depending upon the solid support. For nitrocellulose or PVDF, coating of the blots with protein prior to drying facilitates the subsequent dissociation of the anti-lamin B antibodies. In both cases, SDS-containing buffer (*lanes 7 and 8*) is more effective than guanidine hydrochloride (*lane 9*) or urea (*lane 10*). For nylon, SDS is again slightly more effective than urea at dissociating the antibodies (cf. *lanes 2, 5,* and *7, 10*). Interestingly, when SDS-containing erasure buffer is used, it is not necessary to recoat nylon with protein prior to drying for autoradiography (cf. *lanes 2* and *7*). This is in contrast to nitrocellulose or PVDF, which were examined in the same experiment.

signals only when used at a dilution of 1:5 or 1:10. When attempting to blot with an antiserum for the first time, it is reasonable to try one or more arbitrary concentrations in the range of 1:10 to 1:500. If a strong signal is obtained at 1:500, further dilutions can be performed in subsequent experiments.

8. Different investigators incubate blots with primary antibodies for different lengths of time (reviewed in **ref. *3***). Preliminary studies with some of our reagents have indicated that the signal intensity is greater when blots are incubated with antibody overnight rather than 1–2 h at room temperature.
9. Most diluted antibody solutions can be reused multiple times. They should be stored at 4°C after additional aliquots of penicillin/streptomycin and sodium azide have been added. Some workers believe that the amount of nonspecific (background) staining on Western blots diminishes as antibody solutions are reutilized. Antibody solutions are discarded or supplemented with additional antibody when the intensity of the specific signal begins to diminish.
10. Choice of wash buffer after incubation with primary antibody: 2 *M* urea is included in the suggested wash buffer to diminish nonspecific binding. Alternatively, some investigators include a mixture of SDS and nonionic detergent (e.g., 0.1% (w/v) SDS and 1% [w/v] Triton X-100] in the wash buffers. For antibodies with low avidity (especially monoclonal antibodies and anti-peptide antibodies), the inclusion of 2 *M* urea or SDS might diminish the signal intensity. These agents are, therefore, optional depending upon the properties of the primary antibody used for blotting.
11. It is important to avoid sodium azide when using horse radish peroxidase-coupled antibodies, as the azide inhibits peroxidase ***(26)***. For this reason, adding milk to PBS immediately before use of each aliquot of this buffer is advisable.
12. The concentration of secondary antibody is determined empirically. Most suppliers recommend a dilution for their reagents, typically to the 0.1–0.5 µg/mL range.
13. The same ECL reagent can be transferred from one blot to another to develop several blots simultaneously.
14. The length of exposure required to give a good signal varies depending on the abundance of the antigen, the quality of the primary and secondary antibodies, and the nature of the solid support utilized. One convenient way to proceed is to expose each blot for an arbitrary time (e.g., 2 min) and then increase or decrease the exposure time based on the results of the trial exposure.
15. The signal obtained with some chemiluminescence reagents decays with a halftime of 30–60 min. With these reagents, it is important to expose film promptly after treating blots. Other chemiluminescence reagents continue to generate a strong signal for many hours.

4.3. Dissociation of Antibodies after Chemiluminescent Detection

16. Choice of erasure buffer: Experiments showing the effect of various erasure buffers on removal of primary antibodies are illustrated in **Fig. 2**. After detection by enhanced chemiluminescence, bound antibodies can be solubilized using 6 *M* guanidine hydrochloride, 8 *M* urea, or 2% SDS (*middle panel*, **Fig. 2**). None of these treatments elutes significant amounts of antigen from nitrocellulose (*lower panel*, **Fig. 2**). After enhanced chemiluminescent detection, antibodies appear to be more easily removed from PVDF than from nitrocellulose (cf. *lanes 6* and *13* in **Fig. 2**).

17a. Temperature of incubation: When blotting is performed after immobilization of polypeptides on nitrocellulose, complete removal of antibodies by either 8 *M* urea or 2% SDS requires heating to >50°C for 30 min (**ref. *16***; *see* also **Fig. 2**, *lanes 4–7*).

17b. Length of incubation. When blotting is performed on nitrocellulose, complete dissociation of radiolabeled antibodies at 70°C requires a minimum of 20 min incubation with erasure buffer *(16)*. Incubation times for removal of antibodies after chemiluminescent detection have not been systematically investigated.

18. Although most epitopes are resistant to the erasure procedure (**Figs. 2** and **4B**; *see also* **refs.** ***15,16***), epitopes recognized by an occasional monoclonal antibody are destroyed by erasure (**Fig. 4A,** *lane 2*). Observations from our laboratory also indicate that certain epitopes are lost upon prolonged storage of blots (**Fig. 4B**, *lane 3*). There does not appear to be any relationship between the loss of epitopes upon blot storage and the damage of epitopes during the erasure procedure.

4.4. Removal of Antibodies After Autoradiographic Detection

19. ^{125}I-labeled protein A can be substituted for radiolabeled secondary antibody. Protein A, however, can bind to the immunoglobulins present in milk, causing a high background on the blot. Therefore, when ^{125}I-labeled protein A is to be used, milk should not be utilized to block unoccupied binding sites (**Subheading 3.1., step 11**), nor as a diluent for antibodies (**Subheading 3.2., steps 2, 3,** and **6**). Instead, bovine serum albumin, hemoglobin, or gelatin should be considered (*see* **Note 6**).
20. The major modification suggested with radiolabled secondary antibodies is a reincubation of blots with protein-containing buffer prior to drying. For reasons that are unclear, this step appears to be essential for efficient dissociation of certain antibodies from nitrocellulose or PVDF after incubation with radiolabeled secondary antibodies (**Fig. 3**), but not after incubation with peroxidase-coupled antibodies (**Fig. 2**).

4.5. General Notes

21. Some blotting procedures involve deposition of chromogenic reaction products directly on immunoblots by enzyme-coupled secondary antibodies *(5,23,24)*. The technique described above is not useful for removing colored peroxidase reaction products (e.g., diaminobenzidine oxidation products) from blots.
22. A modification of the techniques described in **Subheading 3.2.** allows the detection of glycoproteins by radiolabeled lectins. For this application, blots would be coated with albumin or gelatin, reacted with radiolabeled lectin (**Subheading 3.2., steps 6–8**), and recoated with albumin or gelatin (**Subheading 3.4., steps 2** and **3**) prior to drying. After autoradiography, the radiolabeled lectin would be solubilized in warm SDS under reducing conditions (**Subheading 3.3.**)

Acknowledgments

The technical assistance of Sharon J. McLaughlin, Phyllis A. Svingen, and Timothy J. Kottke in the preparation of **Figs. 1–4**, as well as secretarial assistance of Deb Strauss, are gratefully acknowledged. Studies in my laboratory are supported by grants from the National Cancer Institute.

References

1. Gershoni, J. M. and Palade, G. E. (1983) Protein blotting: principles and applications. *Analyt. Biochem.* **131,** 1–15.
2. Beisiegel, U. (1986) Protein blotting. *Electrophoresis* **7,** 1–18.
3. Stott, D. I. (1989) Immunoblotting and dot blotting. *J. Immunol. Meth.* **119,** 153–187.
4. Kaufmann, S. H. and Kellner, U. (1998) Erasure of Western blots after autoradiographic or chemiluminescent detection, in *Methods in Molecular Biology,* Vol. 80, Humana Press, Totowa, NJ, pp. 223–235.

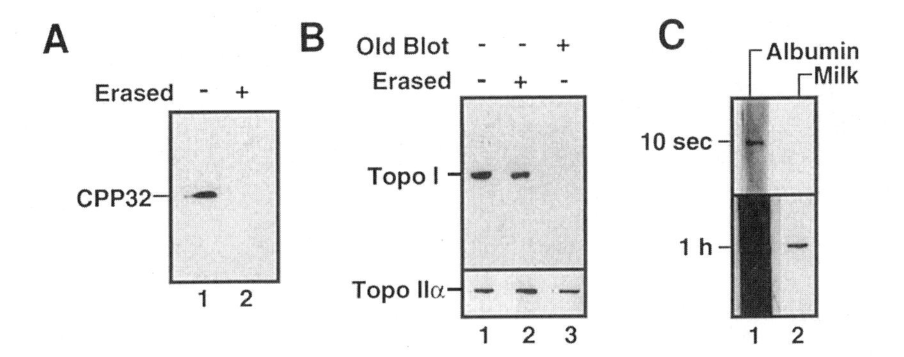

Fig. 4. Effect of various treatments on antigen recognition. **(A)**. Replicate aliquots containing polypeptides from 2×10^5 K562 cells were separated by SDS-PAGE, transferred to nitrocellulose, and stained with fast green FCF (**Subheading 3.1., steps 1–11**). Strip 2 was then treated as described in **Subheading 3.3. (steps 1–5)** to simulate an erasure procedure. Strips were blocked with TSM buffer and blotted with a mouse monoclonal IgG recognizing CPP32/procaspase-3. The epitope recognized by this antibody was destroyed by erasure (*lane 2*). In contrast, many other epitopes are not (e.g., **Fig. 2**, *lower panels*). **(B)**. Replicate gel lanes containing identical amounts of protein from a single batch of HL-60 leukemia cell lysate were transferred to nitrocellulose 1 w (strips 1 and 2) or 3 y (strip 3) prior to blotting. Strip 2 was treated as described in **Subheading 3.3., steps 1–5** to simulate an erasure procedure. After unoccupied binding sites were reblocked by incubation with TSM buffer, the strips were incubated with the monoclonal antibodies C-21, which recognizes DNA topoisomerase I (*upper panel*), or Ki-S1, which recognizes DNA topoisomerase IIα (*lower panel*). After washing and incubation with peroxidase-coupled secondary antibodies, blots were subjected to chemiluminescent detection. The epitope recognized by antibody C-21 has been damaged by prolonged storage (*lane 3, upper panel*) but not by the erasure procedure (*lane 2, upper panel*). In contrast, the epitope recognized by antibody Ki-S1 is unaffected by either manipulation. **(C)**. Effect of blocking solution on reactivity. Replicate gel lanes containing protein from 3×10^5 HL-60 cells were blocked with TS buffer containing 3% albumin (strip 1) or 10% milk (strip 2), then reacted with antibody Ki-S1 diluted 1:1000 in the corresponding protein solution. After reaction with peroxidase-coupled anti-mouse IgG diluted in the corresponding protein solution, blots were treated with luminol and exposed to Kodak XAR-5 film for 10 s (*upper panel*) or 1 h (*lower panel*). The signal was more rapidly detected when albumin was used as the blocking protein, but the background was much cleaner using milk as a blocking reagent.

5. Krajewski, S., Zapata, J. M., and Reed, J. C. (1996) Detection of multiple antigens on Western blots. *Analyt. Biochem.* **236,** 221–228.
6. Gingrich, J. C., Davis, D. R., and Nguyen, Q. (2000) Multiplex detection and quantitation of proteins on Western blots using fluorescent probes. *Biotechniques* **29,** 636–642.
7. Oda, T., Heaney, C., Hagopian, J. R., Okuda, K., Griffin, J. D., and Druker, B. J. (1994) Crkl is the major tyrosine–phosphorylated protein in neurophils from patients with chronic myelogenous leukemia. *J. Biol. Chem.* **269,** 22,925–22,928.
8. Viera, L., Ye, Y. Z., Estevez, A. G., and Beckman, J. S. (1999) Immunohistochemical methods to detect nitrotyrosine. *Meth. Enzymol.* **301,** 373–381.
9. Guther, M. L. S., de Almeida, M. L. C., Rosenberry, T.L., and Ferguson, M. A. J. (1994) The detection of phospholipase-resistant and -sensitive glycosyl phosphatidylinositol membrane anchors by Western blotting. *Analyt. Biochem.* **219,** 249–255.

10. Laycock, C. A., Phelan, M. J. I., Bucknall, R. C., and Coleman, J. W. (1994) A Western blot approach to detection of human plasma protein conjugates derived from D-penicillamine. *Ann. Rheum. Dis.* **53,** 256–260.
11. Hashida, S., Hashinaka, K., Ishikawa, S., and Ishikawa, E. (1997) More reliable diagnosis of infection with human immunodeficiency virus type 1 (HIV-1) by detection of antibody IgGs to Pol and Gag proteins of HIV-1 and p24 antigen of HIV-1 in urine, saliva, and/or serum with highly sensitive and specific enzyme immunoassay (immune complex transfer enzyme immunoassay): a review. *J. Clin. Lab. Analyt.* **11,** 267–286.
12. Steere, A. C. (1997) Diagnosis and treatment of Lyme arthritis. *Med. Clin. North Am.* **81,** 179–194.
13. Sambrook, J., Fritsch, E. F., and Maniatis, T. (1989) in *Molecular Cloning: A Laboratory Manual* Cold Spring Harbor Laboratory Press, Cold Spring Harbor, New York.
14. Legocki, R. P. and Verma, D. P. S. (1981) Multiple immunoreplica technique: screening for specific proteins with a series of different antibodies using one polyacrylamide gel. *Analyt. Biochem.* **111,** 385–392.
15. Erickson, P. F., Minier, L. N., and Lasher, R. S. (1982) Quantitative electrophoretic transfer of polypeptides from SDS polyacrylamide gels to nitrocellulose sheets: a method for their re-use in immunoautoradiographic detection of antigens. *J. Immunol. Meth.* **51,** 241–249.
16. Kaufmann, S. H., Ewing, C. M., and Shaper, J. H. (1987) The erasable Western blot. *Analyt. Biochem.* **161,** 89–95.
17. Parekh, B. S., Mehta, H. B., West, M. D., and Montelaro, R. C. (1985) Preparative elution of proteins from nitrocellulose membranes after separation by sodium dodecylsulfate-polyacrylamide gel electrophoresis. *Analyt. Biochem.* **148,** 87–92.
18. Salinovich, O. and Montelaro, R. C. (1986) Reversible staining and peptide mapping of proteins transferred to nitrocellulose after separation by sodium dodceylsulfate-polyacrylamide gel electrophoresis. *Analyt. Biochem.* **156,** 341–347.
19. Gallagher, S. R. (1999) One-dimensional SDS gel electrophoresis of proteins, in *Current Protocols in Molecular Biology,* Vol. 2 , John Wiley & Sons, New York, pp. 10.2A.1–10.2A.34.
20. Laskey, R. A. and Mills, A. D. (1977) Enhanced autoradiographic detection of ^{32}P and ^{125}I using intensifying screens and hypersensitized film. *FEBS Lett.* **82,** 314–316.
21. Lin, F., Fan, W., and Wise, G. E. (1991) Eosin Y staining of proteins in polyacrylamide gels. *Analyt. Biochem.* **196,** 279–283.
22. Lin, W. and Kasamatsu, H. (1983) On the electrotransfer of polypeptides from gels to nitrocellulose membranes. *Analyt. Biochem.* **128,** 302–311.
23. Domin, B. A., Serabjit-Singh, C. J., and Philpot, R. M. (1984) Quantitation of rabbit cytochrome P-450, form 2, in microsomal preparations bound directly to nitrocellulose paper using a modified peroxidase-immunostaining procedure. *Analyt. Biochem.* **136,** 390–396.
24. Chu, N. M., Janckila, A. J., Wallace, J. H., and Yam, L. T. (1989) Assessment of a method for immunochemical detection of antigen on nitrocellulose membranes. *J. Histochem. Cytochem.* **37,** 257–263.
25. Kaufmann, S. H. (1989) Additional members of the rat liver lamin polypeptide family: structural and immunological characterization. *J. Biol. Chem.* **264,** 13,946–13,955.
26. Kaufmann, S. H. (2001) Reutilization of immunoblots after chemiluminiscent detection. *Analyt. Biochem.* **296,** 283–286.

Part IV

Chemical Modification of Proteins and Peptide Production and Purification

59

Carboxymethylation of Cysteine Using Iodoacetamide/ Iodoacetic Acid

Alastair Aitken and Michèle Learmonth

1. Introduction

If cysteine or cystine is identified in a protein it requires modification in order to be quantified. Thiol groups may be blocked by a variety of reagents including iodoacetic acid and iodoacetamide. Iodoacetate produces the *S*-carboxymethyl derivative of cysteine, effectively introducing new negative charges into the protein. Where such a charge difference is undesirable, iodoacetamide may be used to derivatize cysteine to *S*-carboxyamidomethylcysteine (on acid hydrolysis, as for amino acid analysis, this yields *S*-carboxymethylcysteine). The charge difference between these two derivatives has been utilized in a method to quantify the number of cysteine residues in a protein (*[1]*, *see* Chapter 89).

Carboxymethylation may be carried out without prior reduction to modify only those cysteine residues that are not involved in disulfide bridges.

If the protein is to be analyzed using gas-phase protein microsequencing, the derivatizing agent of choice is commonly vinylpyridine as this produces a well separated phenylthiohydantoin (PTH) derivative (PTH-*S*-pyridylethylcysteine), *see* Chapter 62.

2. Materials

1. Denaturing buffer: 6 *M* guanidinium hydrochloride (or 8 *M* deionized urea, *see* **Note 1**) in 0.6 *M* Tris-HCl, pH 8.6.
2. 4 m*M* Dithiothreitol (DTT), freshly prepared in distilled water.
3. β-Mercaptoethanol.
4. Sodium dodecyl sulfate (SDS).
5. Oxygen-free nitrogen source.
6. Iodoacetic acid: 500 m*M* in distilled water, pH adjusted to 8.5 with NaOH. This is light sensitive and although the solution may be stored in the dark at –20°C, it is better to use it freshly prepared.
7. Iodoacetamide: 500 m*M*, freshly prepared in distilled water.
8. Ammonium bicarbonate: 5 m*M* and 50 m*M* in distilled water.
9. 0.1% (v/v) Trifluoracetic acid (TFA), high-performance liquid chromatography (HPLC) grade.
10. Acetonitrile, far UV HPLC grade (e.g. Romil, Cambridge, UK).

From: *The Protein Protocols Handbook, 2nd Edition*
Edited by: J. M. Walker © Humana Press Inc., Totowa, NJ

11. Microdialysis kit (such as that supplied by supplied by Pierce-Warriner).
12. HPLC apparatus.

3. Method

3.1. Large-scale Reaction

1. Dissolve protein (up to 30 mg) in 3 mL of 6 *M* guanidium hydrochoride (or 8 *M* deionized urea), 0.6 *M* Tris-HCl, pH 8.6.
2. Add 30 µL of β-mercaptoethanol (or 5 µL 4 *M* DTT) and incubate under N_2 for 3 h at room temperature. Some proteins require the presence of 1% SDS with an incubation time of up to 18 h at 30–40°C.
3. Add 0.3 mL of colorless iodoacetate (*see* **Note 2**) or iodoacetamide, with stirring, and incubate in the dark for 30 min at 37°C.

3.2 Small-scale Reaction

1. Dissolve the protein (1–10 µg) in 50 µL of denaturing buffer in an Eppendorf tube and flush with N_2.
2. Add an equal volume of 4 m*M* DTT solution to give a final concentration of 2 m*M*.
3. Wrap the tube in aluminium foil and then add 40 mL of iodoacetic acid (or iodoacetamide) dropwise with stirring. Keep under N_2, and incubate in the dark for 30 min at 37°C (*see* **Note 3**). Proceed to **Subheading 3.3.**

3.3. Purification

The removal of excess reagents may be carried out in a variety of ways.

1. Desalting on a 20 × 1 cm Sephadex G10 or Biogel P10 column, eluting with 5 m*M* ammonium bicarbonate, pH 7.8.
2. Dialysis or microdialysis using 50 m*M* ammonium bicarbonate, pH 7.8, as dialysis buffer.
3. By HPLC using a C_4 or C_8 matrix such as an Aquapore RP-300 column, equilibrated with 0.1% (v/v) TFA and eluting with an acetonitrile –0.1% TFA gradient (*see* **Note 4**).

4. Notes

1. Deionization of urea: Urea should be deionized immediately before use to remove cyanates, which react with amino and thiol groups. The method is to filter the urea solution through a mixed bed Dowex or Amberlite resin in a filter flask.
2. The iodoacetic acid used must be colorless. A yellow color indicates the presence of iodine, which will rapidly oxidize thiol groups, preventing alkylation and may also modify tyrosine residues. It is possible to recrystallize from hexane.
3. Reductive alkylation may also be carried out using iodo–[^{14}C]acetic acid. The radiolabeled material should be diluted to the desired specific activity before use with carrier iodoacetic acid to ensure an excess over total thiol groups.
4. If the HPLC separation is combined with mass spectrometric characterization, this level of TFA (which is required to produce sharp peaks and good resolution of peptide) results in almost or complete suppression of signal. In this case it is recommended to use the new "low TFA," 218MS54, reverse-phase HPLC columns from Vydac (300 Å pore size). They are available in C_4 and two forms of C_{18} chemistries and in 1 mm diameter columns that can be used with as little as 0.005% TFA without major loss of resolution and minimal signal loss (*see* Chapters 78 and 81).

Reference

1. Creighton, T. E. (1980) Counting integral numbers of amino acid residues per polypeptide chain. *Nature* **284,** 487–488.

60

Performic Acid Oxidation

Alastair Aitken and Michèle Learmonth

1. Introduction

Where it is necessary to carry out quantitative amino acid analysis of cysteine or cystine residues, carboxymethylation (Chapter 59) or pyridethylation (Chapter 62) is not the method of choice as it may be difficult to assess the completeness of the reactions. It is often preferable to oxidize the thiol or disulfide groups with performic acid to give cysteic acid. This also has the effect of converting methionine residues to methionine sulfone. Tyrosine residues may be protected from reaction by the use of phenol in the reaction mixture.

Performic acid oxidation is also used in the determination of disulfide linkages by diagonal electrophoresis (*see* Chapter 87).

2. Materials

1. 30% (v/v) Hydrogen peroxide. **Note**: Strong oxidizing agent.
2. 88% (v/v) Formic acid.
3. Phenol (CARE). **Note:** Toxic and corrosive.
4. 1 m*M* Mercaptoethanol.

3. Method

1. Add 0.5 mL of 30% (v/v) H_2O_2 to 4.5 mL of 88% formic acid containing 25 mg of phenol.
2. After at least 30 min at room temperature, cool to 0°C.
3. Add this reagent (performic acid, HCOOOH) to the protein sample at 0°C to result in a final protein concentration of 1%.
4. After 16–18 h, dilute the solution with an equal volume of water and dialyze against two changes of 100 volumes of water at 4°C, and finally dialyze against 100 volumes of 1 m*M* mercaptoethanol at 4°C.
5. The derivatized protein can then be lyophilized or dried in a hydrolysis tube ready for acid cleavage and subsequent amino acid analysis.

Elution positions for separations of derivatized amino acids by ion-exchange chromatography are shown in **Table 1**.

From: *The Protein Protocols Handbook, 2nd Edition*
Edited by: J. M. Walker © Humana Press Inc., Totowa, NJ

Table 1
Elution Order of Amino Acids and Derivatives from Ion-Exchange Amino Acid Analyzers

1. O-Phosphoserine	24. S-Ethylcysteine
2. O-Phosphothreonine	25. Glucosamine
3. Cysteic acid	26. Mannosamine
4. Urea	27. Galactosamine
5. Glucosaminic acid	28. Valine
6. Methionine sulphoxides	29. Cysteine
7. Hydroxyproline	30. Methionine
8. CM-Cysteine	31. α-Methylmethionine
9. Aspartic acid	32. Isoleucine
10. Methionine sulphone	33. Leucine
11. α-Methyl aspartic acid	34. Norleucine
12. Threonine	35. Tyrosine
13. Serine	36. Phenylalanine
14. Asparagine	37. Ammonia
15. Glutamine	38. Hydroxylysine
16. α-Methyl serine	39. Lysine
17. Homoserine	40. 1-Methylhistidine
18. Glutamic acid	41. Histidine
19. α-Methyl glutamic acid	42. 3-Methylhistidine
20. Proline	43. Tryptophan
21. S-methylcysteine	44. Pyridylethyl-cysteine
22. Glycine	45. Homocysteine thiolactone
23. Alanine	46. Arginine

The precise elution order may vary depending on the exact instrument and the precise temperature, molarity, and pH of buffers.

61

Succinylation of Proteins

Alastair Aitken and Michèle Learmonth

1. Introduction

Modification of lysine with dicarboxylic anhydrides such as succinic anhydride prevents the subsequent cleavage of lysyl peptide bonds with trypsin. In addition, modified proteins act as better substrates for proteases. Succinic anhydride reacts with the ε-amino group of lysine and the amino-N-terminal α-amino group of proteins, in their non-protonated forms, converting them from basic to acidic groups *(1)*. Thus one effect of succinylation is to alter the net charge of the protein by up to two charge units. This is an effect that has been exploited for the counting of integral numbers of lysine residues within a protein *(2)*.

Succinic anhydride has been reported to react with sulfydryl groups. It is therefore advisable to modify any cysteines within the protein (*see* Chapters 59 and 62) prior to succinylation.

The reaction occurs between pH 7.0 and 9.0. The reaction is carried out using an approx 50-fold excess of the anhydride over native or carboxymethylated protein.

2. Materials

1. 8 *M* deionized urea (*see* **Note 1**).
2. Distilled water containing up to 0.1 *M* NaCl or 0.2 *M* sodium borate buffer, pH 8.5.
3. Succinic anhydride.

3. Method

1. The protein (5 mg) should be dissolved in 5 mL of buffer (*see* **Note 2**).
2. A pH electrode should be placed within the solution to allow monitoring of the pH. The solution should be continuously stirred.
3. The solid succinic anhydride should be added in 0.5 mg portions over a period of 15 min to 1 h to give a 50-fold excess. The pH should be adjusted back to 7 with sodium hydroxide after each addition.
4. The reaction should be allowed to proceed for at least 30 min after the last addition.
5. The modified protein may be separated from the side products of the reaction by dialysis against 50 m*M* ammonium carbonate buffer or by gel filtration (e.g., Sephadex G25) (*see* **Note 3**).

From: *The Protein Protocols Handbook, 2nd Edition*
Edited by: J. M. Walker © Humana Press Inc., Totowa, NJ

4. Notes

1. Deionization of urea: Urea should be deionized immediately before use to remove cyanates, which react with amino and thiol groups. The method is to filter the urea solution through a mixed-bed Dowex or Amberlite resin in a filter flask.
2. The choice of buffer is dependent on the solubility of the protein. Distilled water may be used or it may be necessary to use 0.1 *M* NaCl or 0.2 *M* sodium borate buffer in the presence of 8 *M* deionized urea.
3. Other dicarboxylic anhydrides may be used. Particularly common is citraconic anhydride *(3)*. This reagent has the advantage of reversibility, being readily removed with acid. When used in conjunction with trypsin, the proteolytic reaction may be stopped by reducing the pH, which will also remove the lysine blocking groups.

References

1. Klapper, M. H. and Klotz, I. M. (1972) *Meth. Enzymol.* **25,** 531–536.
2. Hollecker, M. and Creighton, T. E. (1980) *Counting integral numbers of amino groups per polypeptide chain: FEBS Lett.* **119,** 187–189.
3. Yarwood, A. (1989) Manual methods of protein sequencing, in *Protein Sequencing-A Practical Approach,* (Findlay J. B. C. and Geisow, M. J., eds.) IRL Press, Oxford, pp. 119–145.

62

Pyridylethylation of Cysteine Residues

Malcolm Ward

1. Introduction

To help maintain their three-dimensional structure, many proteins contain disulfide bridges between cysteine residues. Cysteine residues can cause problems during Edman sequence analysis, and quantification of cysteine and cystine by amino acid analysis is difficult since these residues are unstable during acid hydrolysis.

Chemical modification of cysteine residues can enhance the solubility of the protein, enable more effective enzymatic digestion with proteases, such as trypsin, and facilitate quantification by amino acid analysis.

Oxidation with performic acid can be used to convert cysteine and cystine to cysteic acid *(1)* (and *see* Chapter 60). This, however, can lead to other nondesirable side reactions, such as oxidation of methionine residues and destruction of tryptophan residues. Alkylation with iodoacetic acid has been used extensively, since this enables the addition of negative charges to the protein. The use of iodoacetic acid containing ^{14}C provides a means of incorporating a radioactive label into the polypeptide chain (*see* Chapter 59) *(2)*.

The method described in this chapter is an effective alternative, where reduction and alkylation can be achieved in one step. 4-Vinylpyridine is used to convert cysteine residues to *S*-pyridylethyl derivatives. The *S*-pyridylethyl group is a strong chromophore at λ254 nm, which facilitates the detection of cysteine containing peptides as well as aiding the identification of cysteine residues during Edman sequencing (*see* **Note 1**).

2. Materials (*see* Note 2)

1. Denaturing buffer: 0.1 *M* Tris-HCl, pH 8.5, 6 *M* guanidine hydrochloride.
2. 4-Vinylpyridine: Store at –20°C.
3. 2-Mercaptoethanol.

3. Method

1. Dissolve 10–50 μg protein/peptide in the denaturing buffer (1 mL).
2. Add 2-mercaptoethanol (5 μL) and 4-vinyl-pyridine (2 μL) to the sample tube and shake. Since both reagents are extremely volatile and toxic, all experimental work should be carried out in a fumehood wearing appropriate safety clothing.

From: *The Protein Protocols Handbook, 2nd Edition*
Edited by: J. M. Walker © Humana Press Inc., Totowa, NJ

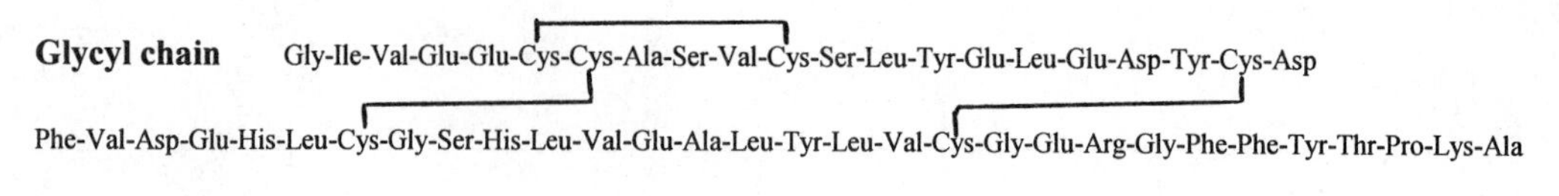

Fig. 1. Amino acid sequence of insulin showing position of disulfide bonds.

3. Blow nitrogen gas over the reaction mixture to expel any oxygen.
4. Seal the tubes with a screw cap.
5. Allow the reaction to proceed at 37°C for 30 min.
6. After this time, an aliquot of the reaction mixture may be taken and analyzed by mass spectrometry. Bovine insulin is shown as an example. This protein contains two interchain bridges and one intrachain bridge (**Fig. 1**). The matrix-assisted laser desorption ionization (MALDI) mass spectrum of reduced and alkylated insulin shows molecular ions at m/z 2763 and m/z 3612 corresponding to the two fully alkylated peptide chains (**Fig. 2**) (*see* **Note 3**). There are no other signals present, indicating that no side reactions have occurred.
7. The sample can now be loaded onto a hydrophobic column for either N-terminal sequence analysis, using the HPG1000A protein sequencer, or *in situ* enzymatic or chemical digestion.

4. Notes

1. The PTH-pyridylethylcysteine derivative can be readily assigned during Edman sequence analysis. The relative elution position on the HPLC system of the Applied Biosystems gas-phase sequencer is between PTH-valine and diphenylthiourea. On the Hewlett Packard G1000A system, PTH-pyridylethylcysteine elutes at 14.2 min just before PTH-methionine.
2. All reagents should be of the highest quality available.
3. The pyridylethylation reaction as described allows for fast, effective derivatization of cysteine residues. The reaction is efficient, giving a high yield of fully alkylated product with no side products.
4. The concept of monitoring chemical reactions by mass spectrometry is not new, yet the sensitivity and speed of MALDI provide a means of examining reaction products to enable the controlled use of reagents that have previously proven troublesome. A recent publication by Vestling et al. *(3)* describes the controlled use of BNPS Skatole, a reagent that to date has been seldom used owing to side reactions. The reactivity of such a reagent may be more widely used in the future now that the progress of the reaction can be easily monitored. *See* Chapter 91 for further uses of MALDI.

References

1. Glazer, A. N., Delange, R. J., and Sigman, D. S. (1975) Chemical characterization of proteins and their derivatives, in *Chemical Modifications of Proteins* (Work, T. S and Work. E., eds.), North Holland, Amsterdam, 21–24.
2. Allen, G. (1989) Sequencing of proteins and peptides, in *Laboratory Techniques in Biochemistry and Molecular Biology* (Burdon, R. H. and Van Knippenberg, P. H., eds.), Elsevier, Amsterdam.
3. Vestling, M. M., Kelly, M. A., and Fenselau, C. (1994) Optimisation by mass spectrometry of a tryptophan-specific protein cleavage reaction. *Rapid Commun. Mass Spectrom.* **8,** 786–790.

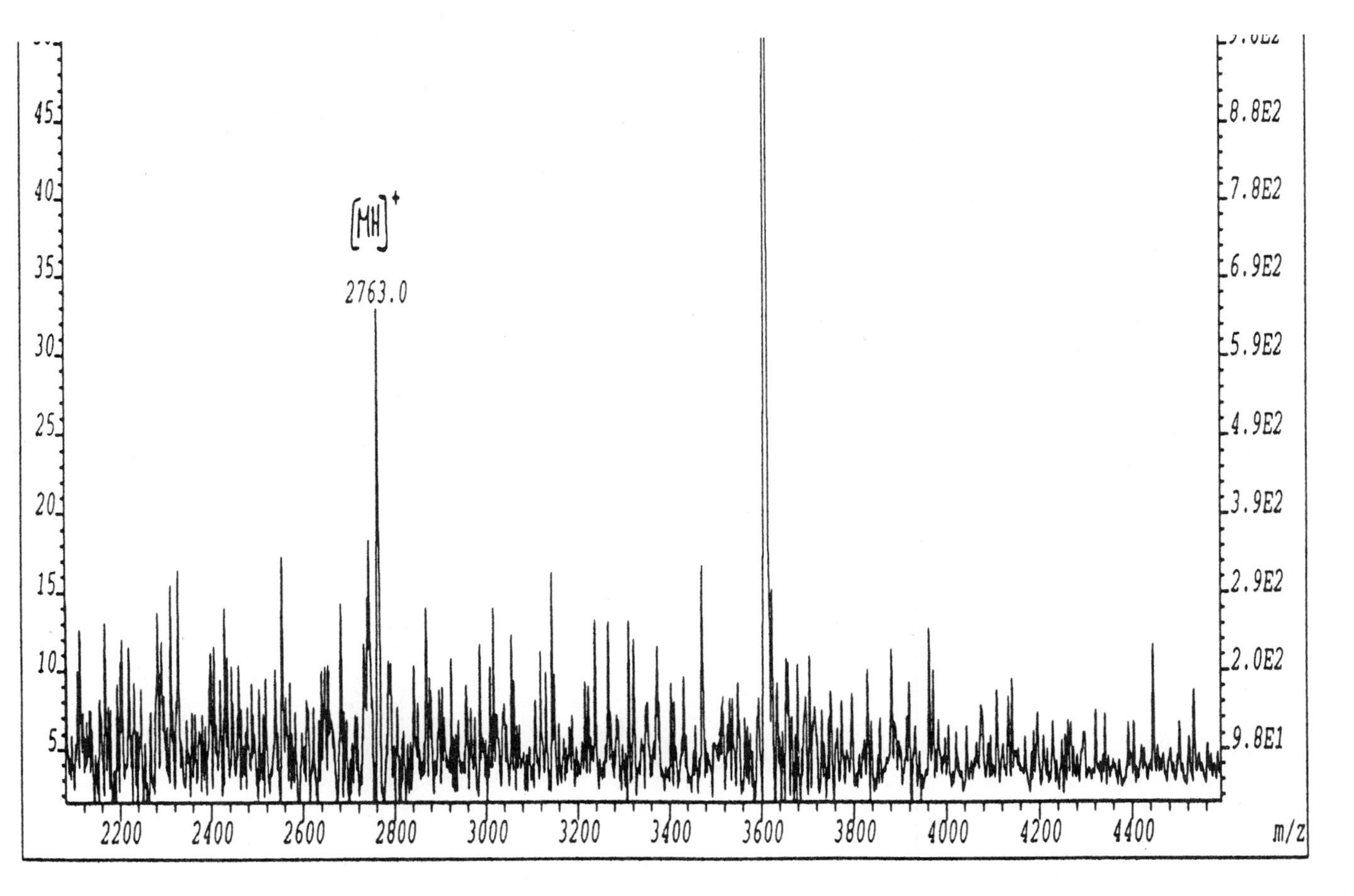

Fig. 2. The MALDI mass spectrum of reduced and alkylated insulin.

63

Side Chain Selective Chemical Modifications of Proteins

Dan S. Tawfik

1. Introduction

Chemical modifications of proteins may be performed simply and rapidly to provide preliminary data regarding the role of particular amino acids in a given protein. Many reviews and books cover these aspects of protein chemistry; only a few are cited here *(1–4)*. In particular, the book by Means and Feeney *(1)*, although about 30 yr old, is an excellent introduction and a practical guide to this field. Notably, even in the era of molecular biology, when site-directed mutagenesis has become widely accessible (including with non-natural amino acids), selective chemical modifications are still applied regularly. In most cases, chemical modifications are used, often together with site-directed mutagenesis, to either identify or confirm the role of active site residues (for recent examples *see* **ref. *5–9***). But chemical modifications are also applied for the generation of improved and modified proteins for a variety of applications *(10–12)*.

The modifications discussed in the following chapters are side chain selective, that is, *under appropriate conditions*, the reagents mentioned in this chapter (and additional reagents mentioned in **refs. *1–4***) react specifically with a single type of amino acid side chain. Hence, loss of activity (enzymatic, binding, or other biological activity) following treatment of the protein with such a modifying reagent is considered to be an indication for the role of that side chain in the active site of the protein.

Data obtained by side chain modifications must be analyzed with caution (as is the case for data obtained by genetic site-directed mutagenesis). Loss of activity on treatment with a reagent might be the result of conformational changes or other changes that occur far from the active site. Some of the reagents, in particular when applied in large excess or under inappropriate conditions, may react with more than one type of side chain or may even disrupt the overall fold of the protein. In general, the type of modifications that alter the size of a particular residue, but not its charge, are preferred (*see* Chapters 67 and 68). In addition, the reactivity of a certain type of side chain in a protein varies by several orders of magnitude owing to interactions with neighboring groups that affect the accessibility and reactivity. For example, the pK_a of the carboxylate side chain of aspartic acid is generally approx 4.5; however, interactions with other side chains may increase the pK_a by more than three units. This change will have a major effect on the reactivity of such a carboxylate group toward the modifying

From: *The Protein Protocols Handbook, 2nd Edition*
Edited by: J. M. Walker © Humana Press Inc., Totowa, NJ

reagent. Dramatic pK_a changes are often found in active sites; thus, certain residues might be particularly difficult to modify, thereby forcing conditions that result in a nonspecific loss of activity.

A number of quite simple experiments may strongly support the results obtained by chemical modifications:

1) A simple control that allows the modification to be ascribed to the active site of a protein is to demonstrate protection (i.e. lack of modification) in the presence of a specific ligand to that site, for example, a hapten for an antibody, a substrate or an inhibitor for an enzyme (for an example *see* Chapter 64).
2) The extent of modification is determined primarily by the molar excess of the modifying reagent, but also by other conditions such as pH, temperature, and reaction time. Reliable and reproducible results are generally obtained only after a wide range of reagent concentrations are applied under different reaction conditions. Following these modifications, one should determine not only the remaining biological activity but also, when possible, the number of modified side chains (details for each reaction are provided in the following chapters). These data may allow one to assess to what extent the modification is indeed site specific (e.g., loss of activity is the result of modification of one or two amino acids of the type modified) or whether loss of activity is due to a complete disruption of the protein due to, for example, the modification of a large number of amino acids.
3) Some of the modifications are reversible; for example, histidine side chains modified by diethyl pyrocarbonate can be recovered by a short treatment with hydroxylamine. Recovery of the activity of the modified protein following this treatment may demonstrate the specificity of the modification (*see* Chapter 65). Additional reversible modifications are described in Chapters 68 and 70.
4) Demonstrating pseudo-first-order kinetics for the inactivation of the protein may indicate that the modification proceeds like an ordinary bimolecular reaction and not via the formation of a binding complex as with affinity labelers or suicide inhibitors.

In the following chapters I have provided basic protocols for the specific modifications of different side chains; these, or very similar, protocols can be applied with other reagents as well (*see* **refs.** *1–4*). The first protocol for the nitration of tyrosine side chains with tetranitromethane (Chapter 64) is written for a specific protein (an anti-dinitrophenyl [DNP] antibody) and provides as many experimental details as possible. Hence it is recommended to read Chapter 64 (including the Notes) before applying any of the other modifications described in Chapters 65 to 70.

1.1. General Notes

1. To avoid misleading results it is important to be familiar with the *chemistry* of each of the reagents (*see* **refs.** *1–4* and references therein). A detailed mechanistic discussion is beyond the scope of this book, although examples for typical problems or side reactions are given in each chapter.
2. Many of the reagents and solvents described are harmful and should be used only in a well-ventilated hood and while applying other precautions such as wearing suitable gloves.
3. Examine the buffer you intend to use with your protein in light of the modification reaction. For example, while amidating carboxylate groups (*see* Chapter 67), the use of an acetate buffer, or of any other buffer that contains carboxylate groups or other nucleophiles (e.g., Tris buffer), should obviously be avoided. Likewise, the use of certain organic solvents should be avoided (e.g., acetone with 2-hydroxy-5-nitrobenzylbromide, *see* Chapter 69).

4. Most of the reagents described in the following chapters are reactive with water; however, in some cases, quenching the excess of unreacted reagent is required (for an example *see* Chapter 64).
5. Chemically modified proteins are often unstable. Likewise, some of the modifications are removed even under mild conditions. This must be taken into consideration while the protein is purified and its activity is being determined. Hence, when possible, it is best to determine the results of the modification reaction (i.e., the remaining biological activity and the number of modified residues) immediately after the reaction and minimize further manipulations of the protein (e.g., dialysis or gel filtration).

References

1. Means, G. E. and Feeney, R. E. (1971) *Chemical Modifications of Proteins.* Holden-Day, San Francisco.
2. Hirs, C. H. N. and Timasheff, S. N. (eds.) (1972) Enzyme structure B. *Meth. Enzymol.* **25**.
3. Lundblad, R. L. and Noyes, C. M. (1984) *Chemical Reagents for Protein Modifications* Vols. 1 and 2. CRC Press, Boca Raton, FL.
4. Feeney, R. E. (1987) Chemical modification of proteins: comments and perspectives. *Int. J. Pept. Protein Res.* **27,** 145–161.
5. Stoughton, D. M., Zapata, G., Picone, R., and Vann, W. F. (1999) Identification of Arg-12 in the active site of *Escherichia coli* K1CMP-sialic acid synthetase. *Biochem. J.* **343,** 397–402.
6. Sheflyan, G. Y., Duewel, H. S., Chen, G., and Woodard, R. W. (1999) Identification of essential histidine residues in 3-deoxy-D- manno-octulosonic acid 8-phosphate synthase: analysis by chemical modification with diethyl *pyrocarbonate and site-directed mutagenesis. Biochemistry* **38,** 14,320–14,329.
7. Nichols, D. J., Keeling, P. L., Spalding, M., and Guan, H. P. (2000) Involvement of conserved aspartate and glutamate residues in the catalysis and substrate binding of maize starch synthase. *Biochemistry* **39,** 7820–7825.
8. Jiang, W. J., Locke, G., Harpel, M. R., Copeland, R. A., and Marcinkeviciene, J. (2000) Role of Lys100 in human dihydroorotate dehydrogenase: mutagenesis studies and chemical rescue by external amines. *Biochemistry* **39,** 7990–7997.
9. Suzuki, K., Asao, E., Nakamura, Y., Nakamura, M., Ohnishi, K., and Fukuda, S. (2000) Overexpression of salicylate hydroxylase and the crucial role of Lys(163) as its NADH binding site. *J. Biochem.* **128,** 293–299.
10. Davis, B. G., Khumtaveeporn, K., Bott, R. R., and Jones, J. B. (1999) Altering the specificity of subtilisin Bacillus lentus through the introduction of positive charge at single amino acid sites. *Bioorg. Med. Chem.* **7,** 2303–2311.
11. Lundblad, R. L. and Bradshaw, R. A. (1997) Applications of site-specific chemical modification in the manufacture of biopharmaceuticals 1. An overview. *Biotechnol. Appl. Biochem.* **26,** 143–151.
12. Altamirano, M. M., Garcia, C., Possani, L. D., and Fersht, A. R. (1999) Oxidative refolding chromatography: folding of the scorpion toxin Cn5 [see comments]. *Nat. Biotechnol.* **17,** 187–191.

64

Nitration of Tyrosines

Dan S. Tawfik

1. Introduction

Tetranitromethane (TNM) reacts with the phenolic side chain of tyrosine under relatively mild conditions to give 3-nitrotyrosine *(1)*. The protocol described in this chapter was developed for an anti-DNP antibody, but can be used with any other antibody or protein. The major side reaction is oxidation of thiols although under more extreme conditions tryptophans and methionines might be oxidized as well. As the reactive species is the phenolate ion, the main factor controlling the reactivity of protein tyrosyl side chains toward nitration by TNM is pH. Increasing the pH will usually enhance the rate of modification.

The number of nitrated tyrosines can be determined spectrophotometrically during the reaction. In addition, the stability of nitrotyrosine allows the specific site of modification to be determined by amino acid analysis of cleaved fragments of the protein. Finally, 3-nitrotyrosine has a much lower pK_a than tyrosine. Thus, examining the pH activity profile of the nitrated protein, for example, the binding of the protein to its ligand can be readily exploited not only to demonstrate the specificity of the modification but also to engineer the binding properties of the protein (*see* **Subheading 3.2.** and **refs.** ***2–5***).

2. Materials

1. Tris-buffered saline (TBS) 8.0: 0.05 *M* Tris-HCl, pH 8.0, 0.15 *M* NaCl.
2. 1–2 mg/mL monoclonal or polyclonal antibody preparation (purified by protein A affinity chromatography).
3. Acetonitrile.
4. TNM. Store in aliquots at –20°C.
   Note: TNM should be handled with care. Preparation of aliquots and of stock solutions should be done in a ventilated hood and with suitable gloves!
5. 2-Mercaptoethanol.
6. Phosphate buffered saline (PBS): 0.01 *M* phosphate buffer, pH 7.4, 0.15 *M* NaCl.
7. Dialysis tubes: 10,000 molecular weight cutoff.

From: *The Protein Protocols Handbook, 2nd Edition*
Edited by: J. M. Walker © Humana Press Inc., Totowa, NJ

3. Method

3.1. Nitration of Tyrosines with Tetranitromethane

1. Dialyze the antibody against TBS 8.0 (4 h at 4°C). Determine the protein concentration in the sample by measuring the absorbance at 280 nm (for IgG, $\varepsilon = 1.45\ cm^{-1} \cdot mg^{-1} \cdot mL$), and adjust it with TBS 8.0 to optical density (OD) 1.09 or 0.75 mg/mL (= 5 μ*M* antibody = 10 m*M* sites).
2. Prepare a set of TNM solutions in acetonitrile: 0, 2.1, 10.5, 42, 105, and 420 m*M* (corresponding to 21 times the final reagent concentration or to 0, 10, 50, 200, 500 and 2000 molar ratios of TNM per antibody sites) (*see* **Note 1**).
3. Add 5 μL of each of the TNM solutions to 100 μL aliquots of the cold antibody solution in Eppendorf tubes immersed in an ice bath. Incubate, with occasional stirring, for 1.5 h at 4°C and then for 30 min at room temperature.
4. Quench the reaction by adding 2 μL of 2-mercaptoethanol, to the samples containing 0–200 fold excess of TNM, and 10 μL to the samples containing 500–2000 fold excess of TNM. Incubate for 15 min at room temperature.
5. Dilute the samples with TBS to a total volume of 500 μL and dialyze them against TBS or PBS (at least twice; each round for 4 h at 4°C).
6. Determine the activity of the various antibody samples by enzyme-linked immunosorbent assay (ELISA) (or any other immunoassay) at increasing antibody dilutions (e.g., 1:50 up to 1:50,000 in PBS).
7. Determine the number of 3-nitrotyrosines per antibody molecule by measuring the optical density at 428 nm ($\varepsilon = 4100\ M^{-1}cm^{-1}$ at pH ≥ 8.5; *see* **Note 2**).
8. For modification in the presence of the hapten–dinitrophenol (DNPOH): Incubate the antibody with 1 m*M* DNPOH for 30–60 min at 4°C.
9. Proceed with the addition of teranitromethane, quenching and dialysis as described in the preceding (*see* **Note 3**).
10. Determine the number of 3-nitrotyrosines by measuring the optical density at 428 nm (and compare it to the number of tyrosines modified under the same conditions in the absence of the hapten) (*see* **Note 4**).

3.2. pH Dependency of Binding of Nitrated Antibodies

Nitration of the phenolic group of tyrosine induces a dramatic shift in the pK_a of this residue. The pK_a of tyrosine is normally around 10; thus the hydroxyl group is mostly protonated under pH ≤ 9.5. In contrast, the pK_a of 3-nitrotyrosine is around 7.0 ***(1)***; hence, loss of activity of the nitrated protein is often the result of deprotonation of the hydroxyl group at pH above 7.0. In such cases, activity could be recovered at pH < 6.0, at which the hydroxyl of the nitrated tyrosine regains its proton. This was originally demonstrated with several antibodies (including an anti-DNP antibody) in which a pH dependency of binding was observed after site-specific modification of tyrosine with TNM ***(2)***. More recently, this approach was applied to other antibodies ***(3)***, and also with avidin, leading to pH-dependent, reversible biotin binding ***(4)***. Recovery and loss of binding of these antibodies to the corresponding haptens (at pH < 6.0 and at pH > 8.0, respectively) were ascribed to the protonation and deprotonation of the hydroxyl group of a 3-nitrotyrosine side chain at their binding sites.

This approach can be utilized to determine the role of the modified tyrosine residue at the binding site; it may also find use in a variety of applications in which controlled modulation of binding under mild conditions is required —for example, affinity chromatography ***(4)***, cell sorting or immunosensors.

1. Nitrate the antibody as described in **Subheading 3.1., steps 1–5**).
2. Perform a series of dilutions of the nitrated antibody (1:50 – 1:50,000; *see* **Note 5**) in 50 m*M* 2-morpholinoethane sulfonic acid (MES) saline buffer pH 5.8, and in TBS, pH 9.0.
3. Determine the binding activity of the diluted antibody at pH 5.8 and 9.0 by ELISA on microtiter plates coated with DNP–bovine serum albumin (BSA) (*see* **Note 6**).

4. Notes

1. In the first modification experiment of a protein a wide range of TNM concentrations of should be applied, for example, 0–10,000 molar excess. If loss of activity is not observed it is recommended to try again at higher pH (e.g., at pH 9.0) and with longer incubations at room temperature.
2. Although proteins hardly absorb at this 428 nm, a sample of the same concentration of the unmodified protein should be used as blank. Relatively high quantities of a protein are required for the determination of low modification ratios; for example, a single 3-nitrotyrosine per site (i.e., two per antibody molecule) would give an OD_{428nm} of approx 0.11 at antibody concentration of 2 mg/mL (13.3 m*M*).
3. Demonstrating specificity by modifying the protein in the presence of an active site specific ligand should be done under the mildest conditions that cause full loss of activity; these conditions (e.g., excess of TNM and pH) should be determined in a preliminary experiment. In some cases dialysis with 6 *M* urea (or a similar reagent) is required to release the ligand from the protein to allow the determination of its activity. In any case, control samples (without the addition of TNM) containing the protein alone and the protein incubated with the ligand should be included for comparison of the remaining activity after modification.
4. Nitrated proteins (and other chemically modified proteins as well) are often unstable. Therefore, measure the residual activity soon after modification and avoid freezing and defrosting of the samples. In those cases in which the activity assay can be performed at low protein concentrations, for example, measuring the binding activity of an antibody by ELISA, dialysis that follows the quenching can be avoided. Protein stability can be improved by adding an equal volume of a 10 mg/mL BSA solution after the addition of the 2-mercaptoethanol.
5. To be able to observe pH-dependent binding the nitration should be performed under mild modification conditions (e.g., 200 molar excess of TNM at pH 8.3); under more extreme conditions, e.g., 1000 molar excess of TNM, the antibody is irreversibly inactivated, that is, hapten-binding is not recovered at pH 5.8 *(2)*.
6. Conditions of the binding assay (e.g., ELISA) should be optimized to clearly identify the pH dependency of binding; in particular, the concentration of the immobilized ligand (DNP in this example) must be low enough so that the differences in binding affinities at pH 5.8 vs pH 9.0 may be observed. Thus, antigen carrying low ratios of DNP (3–10 DNPs per molecule of BSA) should be used and its concentration for the coating of the ELISA microtiter plates should be determined.

References

1. Riordan, J. F., Sokolovsky, M., and Valee, B. L. (1966) Tetranitromethane: a reagent for the nitration of tyrosine and tyrosyl residues in proteins. *J. Am. Chem. Soc.* **88,** 4104–4105.
2. Tawfik, D. S., Chap, R., Eshhar, Z., and Green, B. S. (1994) pH 'on-off' switching of antibody-hapten binding obtained by site-specific chemical modification of tyrosine. *Protein Eng.* **7,** 431–434.
3. Resmini, M., Vigna, R., Simms, C., Barber, N. J., HagiPavli, E. P., Watts, A. B., et al. (1997) Characterisation of the hydrolytic activity of a polyclonal catalytic antibody

preparation by pH-dependence and chemical modification studies: evidence for the involvement of Tyr and Arg side chains as hydrogen-bond donors. *Biochem. J.* **326,** 279–287.

4. Morag, E., Bayer, E. A., and Wilchek, M. (1996) Reversibility of biotin-binding by selective modification of tyrosine in avidin. *Biochem. J.* **316,** 193–199.
5. Walton, D. J. and Heptinstall, J. (2000) Electrochemical modification of proteins. A review. *Prep. Biochem. Biotech.* **30,** 1–14.

65

Ethoxyformylation of Histidine

Dan S. Tawfik

1. Introduction

Diethylpyrocarbonate (DEP) reacts with various nucleophiles (amines, alcohols, thiols, imidazoles, or guanido groups) to yield the respective ethoxyformyl derivatives. At low pH (generally < 6.0) the reaction is quite selective for histidine, as the main side reaction with the ε-amino group of lysine proceeds very slowly (*see* e.g., **ref. *1***). Still, side reactions even with hydroxyl groups (e.g., of serine or tyrosine) were observed ***(2)***. The fact that the ethoxyformyl group can be removed from the imidazole side chain by mild treatment with hydroxylamine can be exploited to ascribe the modification to a histidine residue. In addition, ethoxyformylation of histidines is characterized by an increase in absorbance at 242 nm, which is also used to determine the number of modified histidines ***(1–3)***.

2. Materials

1. (DEP) (*see* **Note 1**).
2. Acetonitrile.
3. Protein for modification (approx 5 μ*M*) diluted in 0.1 *M* sodium acetate buffer, pH 5.0.
4. 1 *M* Hydroxylamine, pH 7.0.

3. Method

1. Prepare a series of *fresh* DEP solutions in acetonitrile (1–30 m*M*) (*see* **Note 2**).
2. Add 5-μL aliquots of each of the DEP solutions to 95-μL aliquots of the protein solution (*see* **Note 3**).
3. Incubate for 15–60 min.
4. Determine the activity of the modified protein.
5. To assay the recovery of the ethoxyformylated histidine residues (*see also* **Notes 4** and **5**):
   a. Add to a solution of the modified protein 1/10 of a volume of 1 *M* hydroxylamine, pH 7.0, and incubate for 10 min.
   b. Dilute with acetate buffer and dialyze extensively against a buffer suitable for the protein being studied.
   c. Determine the activity of the protein.

From: *The Protein Protocols Handbook, 2nd Edition*
Edited by: J. M. Walker © Humana Press Inc., Totowa, NJ

4. Notes

1. The concentration of commercial DEP is often lower than indicated owing to hydrolysis. The concentration of the sample after dilution with an organic solvent can be readily determined by adding an aliquot of an imidazole solution (1–10 m*M* in phosphate, pH 7.0) and measuring the increase in absorbance at 230 nm after 5 min ($\varepsilon = 3000\ M^{-1}\ cm^{-1}$).
2. The final acetonitrile concentration in the protein reaction mixture should be 5< %, *see* Chapter 64, **Subheading 3.1.** for typical dilutions and reaction volumes.
3. A molar excess of 10–300 of DEP is usually sufficient; however, with some proteins higher concentrations of the reagent might be needed. Likewise, if no modification is observed at pH 5.0, the reaction can be performed at higher pH. In such cases it is recommended that one ensure that loss of activity is indeed due to the ethxyformylation of histidine (*see* **Note 4**).
4. Measuring the differential UV spectra during modification with DEP is useful not only to determine the number of modified histidines but also to eliminate the possibility that residues other than histidines were modified. The ethoxyformylation of histidine side chains by DEP should result in an increase in the absorbance of the protein at 242 nm ($\varepsilon = 3200\ M^{-1}\ cm^{-1}$). Likewise, restoration of the activity of the modified protein after treatment with hydroxylamine should be accompanied by a parallel decrease in the absorbance at 242 nm. Ethoxyformylation of the hydroxyl of tyrosine would increase the absorbance at approx 280 nm whereas similar modification of serine or threonine does not cause a significant changes of absorbance at this range.
5. The ethoxyformyl product is quite unstable. Because the reagent is rapidly hydrolyzed (to give ethanol and carbonate) there is hardly a need to purify the protein after modification. In any case, dialysis even in neutral buffers may result in a significant removal of the ethoxyformyl group. Hence, purification by exclusion chromatography (e.g., on Sephadex G-25) is preferred.

References

1. Dominicini, P., Tancini, B., and Voltattorni, C. B. (1985) Chemical modification of pig kidney 3, 4-dihydroxy-phenylalanine decarboxylase with diethyl pyrocarbonate. *J. Biol. Chem.* **260,** 10,583–10,589.
2. Melchior, W. B., Jr. and Fahrney, D. (1970) Ethoxyformylation of proteins. Reaction of ethoxyformic anhydride with a-chymotrypsin, pepsin and pancreatic ribonuclease at pH 4. *Biochemistry* **9,** 251–258.
3. Sheflyan, G. Y., Duewel, H. S., Chen, G., and Woodard, R. W. (1999) Identification of essential histidine residues in 3-deoxy-D- manno-octulosonic acid 8-phosphate synthase: analysis by chemical modification with diethyl pyrocarbonate and site-directed mutagenesis. *Biochemistry* **38,** 14,320–14,329.

66

Modification of Arginine Side Chains with *p*-Hydroxyphenylglyoxal

Dan S. Tawfik

1. Introduction

A variety of dicarbonyl compounds including phenylglyoxal, 2,3-butanendione, and 1,2-cyclohexanedione selectively modify the guanidine group of arginine ***(1–3)***. The main advantage of *p*-hydroxyphenylglyoxal is in the ability to determine the number of modified arginines spectrophotometrically. This reagent is also reactive at mildly alkaline pH (usually 8.0–9.0) and yields a single product that is relatively stable ***(1)***.

2. Materials

1. *p*-Hydroxyphenylglyoxal.
2. 1 *M* NaOH.
3. Protein for modification (approx 10 µ*M*) diluted in 0.1 *M* sodium pyrophosphate buffer, pH 9.0.
4. Sephadex G-25 column.

3. Method

1. Prepare a 100 m*M* solution of *p*-hydroxyphenylglyoxal in water and adjust the pH of the solution with 1 *M* NaOH to 9.0.
2. Prepare a series of dilutions (5–50 m*M*; *see* **Note 2**) of the solution in **Step 1** in 0.1 *M* sodium pyrophosphate buffer, pH 9.0 (*see* **Note 1**).
3. Add 10-µL aliquots of the p-hydroxyphenylglyoxal solutions to 90-µL aliquots of the protein solution. Check the pH and if necessary adjust it back to pH 9.0.
4. Incubate for 60–180 min *in the dark.*
5. Pass the sample through a Sephadex G–25 column. Elute with deionized water or with an appropriate buffer (*see* **Note 4**).
6. Determine the activity of the modified protein.
7. Determine the number of modified arginines by measuring the absorbance of the *purified* protein (*see* **Notes 3** and **4**) at 340 nm (at pH 9.0, $\varepsilon = 18{,}300\ M^{-1}\ cm^{-1}$).

4. Notes

1. An optimal rate and selectivity of modification is generally obtained at pH 8.0–9.0; however, in some proteins a higher pH might be required to modify a particular arginine residue.

From: *The Protein Protocols Handbook, 2nd Edition*
Edited by: J. M. Walker © Humana Press Inc., Totowa, NJ

2. A molar excess of *p*-hydroxyphenylglyoxal in the range of 50–500 is usually sufficient for a first trial.
3. The absorbance of *p*-hydroxyphenylglyoxal modified arginines changes with the pH. Maximal absorbance is observed at 340 nm at pH ≥ 9.0 (ε = 18,300 $M^{-1}cm^{-1}$) *(1)*.
4. Prolonged dialysis in neutral or mildly alkaline buffers may cause a significant release of the modifying group. Purification of the protein to determine the extent of modification should therefore be performed using exclusion chromatography *(1)*.

References

1. Yamasaki, R. B., Vega, A., and Feeney, R. E. (1980) Modification of available arginine residues in proteins by *p*-hydroxyphenylglyoxal. *Analyt. Biochem.* **109,** 32–40.
2. Rogers, T. B., Børresen, T., and Feeney, R. E. (1978) Chemical modification of the arginines in transferrins. *Biochemistry* **17,** 1105–1109.
3. Stoughton, D. M., Zapata, G., Picone, R., and Vann, W. F. (1999) Identification of Arg-12 in the active site of *Escherichia coli* K1CMP-sialic acid synthetase. *Biochem. J.* **343,** 397–402.

67

Amidation of Carboxyl Groups

Dan S. Tawfik

1. Introduction

Several reactions have been described for the modification of the carboxylic side chains of aspartic and glutamic acids *(1,2)*; of these, amidation, using an amine and a water-soluble coupling carbodiimide reagent, is most often applied to proteins. The advantages of this approach are the stability of the modification (an amide bond) and the ability to achieve some site specificity (namely, to selectively modify a particular carboxylic side chain) by using different carbodiimide reagents and by variations in the structure (e.g., size, charge, hydrophobicity) of the amine *(3–4)*.

The methyl or ethyl esters of glycine are commonly used as the amine nucleophile. Hydrolysis of these groups by a short treatment with 0.1 *M* hydroxyl amine (pH 8.0, 5–30 min) or a base (0.1 *M* carbonate pH 10.8, 2–6 h) affords a free carboxyl group (e.g., protein–COOH is converted into protein–CO-$NHCH_2COOH$). Hence, a mild modification that affects only the size but not the charge of the aspartyl or glutaryl side chains of the protein is achieved. Determination of the number of modified carboxylate group can be performed only by labeling the amine group, for example, by using a radiolabeled glycine ethyl ester (which is commercially available). In principle, different amines can be used to achieve selectivity or to assist the identification of the modified residues. It is important, however, to ensure that the reaction with these amines is rapid enough and yields a single product *(3)*.

Several side reactions (e.g., with tyrosines and cysteines) may occur mainly at neutral or mildly basic pH; most of these can be ruled out by demonstrating that the activity of the modified protein is not regained after treatment with hydroxylamine.

2. Materials

1. 1-Ethyl-3-(3-dimethylaminopropyl)carbodiimide (EDC) (*see* **Note 1**).
2. Glycine ethyl ester.
3. Protein for modification (approx 5 µg/mL) diluted in 0.1 *M* 2-morpholinoethanesulfonic acid (MES) buffer, pH 5.5.
4. 0.1 *M* Acetate buffer, pH 5.0.

From: *The Protein Protocols Handbook, 2nd Edition*
Edited by: J. M. Walker © Humana Press Inc., Totowa, NJ

3. Method

1. Add the glycine ethyl ester to the protein solution to give a final concentration of up to 50 m*M* (*see* **Note 2**); check the pH and if necessary adjust it back to pH 5.5 (*see* **Note 3**).
2. Add EDC to a final concentration of 0.5–10 m*M* and incubate for 1–6 h (*see* **Note 4**).
3. Add one volume of acetate buffer to quench the reaction.
4. Dialyze against an appropriate buffer or pass the sample through a Sephadex G-25 column.
5. Determine the activity of the modified protein.

4. Notes

1. 1-Ethyl-3-(3-dimethylaminopropyl) carbodiimide (EDC) is most commonly used. Several other water-soluble carbodiimide reagents are available (e.g., 1-cyclohexyl-3-[2-morpholinoethyl]carbodiimide) and can be used for this reaction; these reagents, however, are usually derived from more bulky side chains than EDC and may therefore have more limited accessibility to certain carboxylic residues of the protein.
2. Crosslinking of the protein in the presence of the coupling reagent is avoided by using relatively dilute protein solutions (≤ 0.5 mg/mL) and a large excess of the amine nucleophile (e.g., glycine ethyl ester).
3. The reaction is usually performed at acidic pH (4.5–5.5) to minimize side reactions. However, the modification of particular carboxylate side chains may require higher pH (*see* Chapter 63).
4. The reaction is usually run at ambient temperature; nevertheless, lowering the temperature to 4°C may eliminate the appearance of certain side products.

References

1. Means, G. E. and Feeney, R. E. (1971) *Chemical Modifications of Proteins.* Holden-Day, San Francisco.
2. Lundblad, R. L. and Noyes, C. M. (1984) *Chemical Reagents for Protein Modifications* Vols. 1 and 2. CRC Press, Boca Raton, FL.
3. Hoare, D. G. and Koshland, D. E., Jr. (1967) A method for quantitative modification and estimation of carboxylic acid groups in proteins. *J. Biol. Chem.* **242,** 2447–2453.
4. Nichols, D. J., Keeling, P. L., Spalding, M., and Guan, H. P. (2000) Involvement of conserved aspartate and glutamate residues in the catalysis and substrate binding of maize starch synthase. *Biochemistry* **39,** 7820–7825.

68

Amidination of Lysine Side Chains

Dan S. Tawfik

1. Introduction

Perhaps the largest variety of modifications available is that for ε-amino group of lysine *(1–4)*. The amino side chain can be acylated (using e.g., acetic anhydride) or alkylated by trinitrobenzenesulfonic acid (TNBS); these reactions alter both the size and the charge of the amino group. Other modifications, using anhydrides of dicarboxylic acids (e.g., succinic anhydride), replace the positively charged amino group with a negatively charged carboxyl group. Amidinations *(5,6)* and reductive alkylations (*see* **ref.** *7*) offer an opportunity to modify the structure of the ε-amino group of lysines, while maintaining the positive charge. Modifications that usually do not disrupt the overall structure of the protein are preferred, particularly in those cases when one wishes to identify the specific role of lysine in the active site of the protein being studied.

Amidination is performed by reacting the protein with imidoesters such as methyl or ethyl acetimidate at basic pH. The reaction proceeds solely with amino groups to give mainly the positively charged acetimidine derivative, which is stable under acidic and mildly basic pH. Side products can be avoided by maintaining the pH above 9.5 throughout the reaction (*see* **Note 1** and **2**, **ref.** *6*). The modification can be removed at a higher pH (≥ 11.0) and in the presence of amine nucleophiles (e.g., ammonia) (*5,6*, and *see* **Note 2**).

A major drawback of this modification is that the number of amidinated lysines cannot be readily determined. However, it is possible to take advantage of the fact that the amidine group is not reactive with amine modifying reagents such as TNBS and thereby to indirectly determine the number of the remaining unmodified lysine residues after the reaction *(8)*.

2. Materials

1. Methyl acetimidate hydrochloride.
2. 0.1 *M* and 1 *M* NaOH.
3. Protein for modification.
4. 0.1 *M* borate buffer, pH 9.5.

From: *The Protein Protocols Handbook, 2nd Edition*
Edited by: J. M. Walker © Humana Press Inc., Totowa, NJ

3. Method

1. Dissolve the protein (1–2 mg/mL) in 0.1 *M* borate buffer pH 9.5; check the pH and if necessary adjust it back to 9.5 using 0.1 *M* NaOH .
2. Dissolve 110 mg of methyl acetimidate hydrochloride in approx 1.1 mL of 1 *M* NaOH (approx 0.9 *M*); check the pH and if necessary adjust it to approx 10 with 1 *M* NaOH (*see* **Note 1**).
3. Add an aliquot of the methyl acetimidate solution to the protein solution and check the pH again (*see* **Note 1**).
4. Incubate for 40 min.
5. Dialyze the sample against an appropriate buffer (pH < 8.5) or filter on a Sephadex G-25 column.
6. Determine the activity of the protein.

4. Notes

1. The amidination reaction proceeds with almost no side products only at pH > 9.5; at lower pH, the side reactions proceed very rapidly. Hence it is important to add the methyl acetimidine solution to the buffered protein solution without causing a change of pH *(6)*. The methyl acetimidine is purchased as the hydrochloride salt, which is neutralized by dissolving it in 1 *M* NaOH (*see* **Subheading 3., step 2**). The pH of the resulting solution should be approx 10; if necessary the pH may be adjusted before the addition to the protein solution with 1 *M* NaOH or 1 *M* HCl. As acetimidates are rapidly hydrolyzed at basic pH, the entire process should be performed very rapidly. It is therefore recommended, in a preliminary experiment, to dissolve the methyl acetimidine hydrochloride and determine the exact amount of 1 *M* NaOH that yields a solution of pH 10. The same process is then repeated with a freshly prepared methyl acetimidine solution which is rapidly added to the protein.
2. The acetimidyl group may be removed by treatment of the modified protein with an ammonium acetate buffer prepared by adding concentrated ammonium hydroxide solution to acetic acid to a pH of 11.3 **(Caution! Preparation of this buffer must be done carefully and in a well-ventilated chemical hood)**.
3. Amidination is obviously unsuitable for the modification of proteins that are sensitive to basic pH. Reductive alkylation, using formaldehyde and sodium cyanoborohydride (*see* **ref.** *7*), can be performed at neutral pH and is recommended for the modification of such proteins.

References

1. Means, G. E. and Feeney, R. E. (1971) *Chemical Modifications of Proteins.* Holden-Day, San Francisco.
2. Hirs, C. H. N. and Timasheff, S. N. (eds.) (1972) Enzyme structure B. *Meth. Enzymol.* **25**.
3. Lundblad, R. L. and Noyes, C. M. (1984) *Chemical Reagents for Protein Modifications,* Vols. 1 and 2. CRC Press, Boca Raton, FL.
4. Feeney, R. E. (1987) Chemical modification of proteins: comments and perspectives. *Int. J. Pept. Protein Res.* **27,** 145–161.
5. Hunter, M. J. and Ludwig, M. L. (1962) The reaction of imidoesters with small proteins and related small molecules. *J. Amer. Chem. Soc.* **84,** 3491–3504.
6. Wallace, C. J. A. and Harris, D. E. (1984) The preparation of fully N-e-acetimidylated cytochrome *c*. *Biochem. J.* **217,** 589–594.
7. Jentoft, N. and Dearborn, D. G. (1979) Labeling of proteins by reductive methylation using sodium cyanoborohydride. *J. Biol. Chem.* **254,** 4359–4365.
8. Fields, R. (1972) The rapid determination of amino groups with TNBS. *Meth. Enzymol.* **25,** 464–468.

69

Modification of Tryptophan with 2-Hydroxy-5-Nitrobenzylbromide

Dan S. Tawfik

1. Introduction

2-Hydroxy-5-nitrobenzylbromide, Koshland's reagent *(1)*, reacts rapidly and under mild conditions with tryptophan residues. At low pH (< 7.5) this reagent exhibits a marked selectivity for tryptophan; under more basic pH or at higher reagent concentrations, cysteine, tyrosine, and even lysine residues can be modified as well (*see* **Note 1**). The reaction is extremely rapid either with the protein or with water. The reagent is relatively insoluble in water; it is therefore necessary to first dissolve it in an organic solvent (e.g., dioxane) and then to add it to the protein solution in buffer. Unlike that of most other modifying reagents, the final organic solvent concentration in the reaction mixture is relatively high (5–15%). Determination of the number of modified tryptophans is done spectrophotometrically.

2. Materials

1. 2-Hydroxy-5-nitrobenzylbromide.
2. Dioxane (water free) (*see* **Note 2**).
3. Protein for modification: 1 mg/mL in 0.1 *M* phosphate buffer, pH 7.0.
4. 1 *M* NaOH.
5. A Sephadex G-25 column.

3. Method

1. Prepare a *fresh* solution of 200 m*M* 2-hydroxy-5-nitrobenzylbromide in dioxane (keep the solution *in the dark*).
2. Dilute this solution in dioxane to 10× the final reagent concentration (*see* **Note 3**).
3. Add 10 µL of the 2-hydroxy-5-nitrobenzylbromide solution to a 90-µL aliquot of the protein solution and shake for 2 min (e.g., on a Vortex).
4. Centrifuge (for few minutes at 10,000 rpm) if a precipitate forms (*see* **Note 4**).
5. Filter on a Sephadex G-25 and then dialyze against an appropriate buffer (*see* **Note 4**).
6. Determine the number of modified tryptophans by measuring the absorbance of the purified protein at 410 nm at pH ≥ 10 ($\varepsilon = 18{,}450\ M^{-1}\ \text{cm}^{-1}$).
7. Determine the activity of the protein.

From: *The Protein Protocols Handbook, 2nd Edition*
Edited by: J. M. Walker © Humana Press Inc., Totowa, NJ

4. Notes

1. At neutral and slightly acidic pH the reaction is generally selective. At higher pH, the major side reaction is the benzylation of sulfhydryl groups; to prevent the modification of free cysteine residues the protein can be carboxymethylated prior to the modification with 2-hydroxy-5-nitrobenzylbromide.
2. Avoid the use of acetone, methanol, or ethanol as organic co-solvents for this reagent. Water-free dioxane is commercially available or can be prepared by drying dioxane over sodium hydroxide pellets.
3. At high reagent concentrations a precipitate of 2-hydroxy-5-nitrobenzyl alcohol is sometimes observed and can be removed by centrifugation. Purification of the modified protein from *all* the remaining alcohol product is sometimes difficult and may require extensive dialysis in addition to gel filtration.
4. The reaction of hydroxy-5-nitrobenzylbromide is accompanied by the release of hydrobromic acid. The pH of the reaction should be maintained if necessary by the subsequent addition of small aliquots of 1 *M* NaOH solution.

References

1. Horton, H. R. and Koshland, D. E., Jr. (1972) Modification of proteins with active benzyl halides. *Meth. Enzymol.* **25,** 468–482.

70

Modification of Sulfhydryl Groups with DTNB

Dan S. Tawfik

1. Introduction

5-5'-Dithio-*bis* (2-nitrobenzoic acid) (DTNB or Ellman's reagent; **ref.** ***1***) reacts with the free sulfhydryl side chain of cysteine to form an S–S bond between the protein and a thionitrobenzoic acid (TNB) residue. The modification is generally rapid and selective. The main advantage of DTNB over alternative reagents (e.g., *N*-ethylmaleimide or iodoacetamide) is in the selectivity of this reagent and in the ability to follow the course of the reaction spectrophotometrically. The reaction is usually performed at pH 7.0–8.0 and the modification is stable under oxidative conditions. The TNB group can be released from modified protein by treatment with reagents that are routinely used to reduce S–S bonds, for example, mercaptoethanol, or by potassium cyanide ***(2)*** (*see* **Note 1**). In addition, the often highly pronounced differences in reactivity of different cysteine side chains in the same protein or even active site, and the availability of a variety of thiol-modifying reagents can be exploited to selectively modify cysteine side chains in proteins in the presence of other, more reactive cysteine residues ***(3)***.

2. Materials

1. DTNB.
2. 0.1 *M* Tris-HCl, pH 8.0.
3. Protein for modification: approx 5 μ*M* in 0.1 *M* Tris-HCl, pH 8.0.

3. Method

1. Prepare *fresh* solutions of DTNB (0.5–5 m*M*; *see* **Note 2**) in 0.1 *M* Tris-HCl, pH 8.0.
2. Add 20-μL aliquots of the DTNB solution to 180 μL of the protein solution and incubate for 30 min.
3. Determine the number of modified cysteines by measuring the absorbance of the released TNB anion at 412 nm ($\varepsilon = 14{,}150\ M^{-1}\ cm^{-1}$) (*see* **Note 3**).
4. Dialyze against an appropriate buffer (*see* **Note 4**).
5. Determine the activity of the protein.

From: *The Protein Protocols Handbook, 2nd Edition*
Edited by: J. M. Walker © Humana Press Inc., Totowa, NJ

4. Notes

1. Release of the TNB modification can be achieved by treatment of the modified protein (preferably after dialysis) with thiols (e.g., mercaptoethanol) or by potassium cyanide (20 m*M* final concentration; 10–60 min). The release of the TNB anion can be followed spectrophotometrically at 412 nm *(2)*.
2. A molar excess of 10–100 of DTNB and an incubation time of 30 min is usually sufficient with most proteins. The modification of a particular cysteine residue may require a higher DTNB concentration, a longer reaction time, or a higher pH.
3. The extent of modification is determined by measuring the absorbance of the *released* TNB anion and hence can be followed during the reaction and in the presence of an excess of unreacted DTNB.
4. Dialysis of the reaction mixture is not always necessary; in many cases the activity of the protein (enzymatic or binding) can be determined directly after the reaction (for an example *see* **ref.** *2*).

References

1. Ellman, G. L. (1959) Tissue sulfhydryl groups. *Arch. Biochem. Biophys.* **82,** 70–77.
2. Fujioka, M., Takata, Y., Konishi, K., and Ogawa, H. (1987) Function and reactivity of sulfhydryl groups of rat liver glycine methyltransferase. *Biochemistry* **26,** 5696–5702.
3. Altamirano, M. M., Garcia, C., Possani, L. D., and Fersht, A. R. (1999) Oxidative refolding chromatography: folding of the scorpion toxin Cn5 [see comments]. *Nat. Biotechnol.* **17,** 187–191.

71

Chemical Cleavage of Proteins at Methionyl-X Peptide Bonds

Bryan John Smith

1. Introduction

One of the most commonly used methods for proteolysis uses cyanogen bromide to cleave the bond to the carboxy-(C)-terminal side of methionyl residues. The reaction is highly specific, with few side reactions and a typical yield of 90–100%. It is also relatively simple and adaptable to large or small scale. Because methionine is one of the least abundant amino acids, cleavage at that residue tends to generate a relatively small number of peptides of large size—up to 10,000–20,000 Da. For this reason the technique is usually less useful than some other methods (such as cleavage by trypsin) for identification of proteins by mass mapping, which is better done with a larger number of peptides. Cleavage at Met-X can be useful for other purposes, however:

1. Generation of internal sequence data, from the large peptides produced *(1)*.
2. Peptide mapping.
3. Mapping of the binding sites of antibodies *(2)* or ligands *(3)*.
4. Generation of large, functionally distinct domains (e.g., from hirudin, by Wallace et al. *[4]*) or proteins of interest from fusion proteins *(5)*.
5. Confirmation of estimates of methionine content by amino acid analysis, which has a tendency to be somewhat inaccurate for this residue *(6)*. This is by determination of the number of peptides produced by cleavage at an assumed 100% efficiency.

2. Materials

1. Ammonium bicarbonate (0.4 *M*) solution in distilled water (high-performance liquid chromatography [HPLC] grade). Stable for weeks in refrigerated stoppered bottle.
2. 2-Mercaptoethanol. Stable for months in dark, stoppered, refrigerated bottle.
3. Trifluoroacetic acid (TFA), Aristar grade. Make to 70% v/v by addition of distilled water (HPLC grade), and use fresh. *See* **Notes 1** and **2**.
4. CNBr. Stable for months in dry, dark, refrigerated storage. Warm to room temperature before opening. Use only white crystals, not yellow ones. **Beware of the toxic nature of this reagent: hydrogen cyanide is a breakdown product. Use in a fume cupboard.** *See* **Note 3**.
5. Sodium hypochlorite solution (domestic bleach).

From: *The Protein Protocols Handbook, 2nd Edition*
Edited by: J. M. Walker © Humana Press Inc., Totowa, NJ

6. Equipment includes a nitrogen supply, fume hood, and suitably sized and capped tubes (e.g., Eppendorf microcentrifuge tubes).

3. Methods

3.1. Reduction

1. Dissolve the polypeptide in water to between 1 and 5 mg/mL, in a suitable tube. Add one volume of ammonium bicarbonate solution, and add 2-mercaptoethanol to between 1% and 5% (v/v).
2. Blow nitrogen over the solution to displace oxygen, seal the tube, and incubate at room temperature for approx 18 h.

3.2. Cleavage

1. Dry down the sample under vacuum, warming if necessary to help drive off all of the bicarbonate. Any remaining ammonium bicarbonate will form a nonvolatile salt on subsequent reaction with the TFA that follows. If a white salty deposit remains, redissolve in water and dry down again.
2. Redissolve the dried sample in 70% v/v TFA, to a concentration of 1–5 mg/mL.
3. Add excess white crystalline CNBr to the sample solution, to 10-fold or more molar excess over methionyl residues. Practically, this amounts to approx equal weights of protein and CNBr. To very small amounts of protein, add one small crystal of reagent. Carry out this stage in the fume hood (*see* **Note 3** and **4**).
4. Seal the tube and incubate at room temperature for 24 h.
5. Terminate the reaction by drying down under vacuum. Store samples at –10°C or use immediately (*see* **Note 5**).
6. Immediately after use decontaminate equipment that has contacted CNBr by immersion in hypochlorite solution (bleach) until effervescence stops (a few minutes). Wash decontaminated equipment thoroughly.

4. Notes

1. The mechanism of the action of cyanogen on methionine-containing peptides is shown in **Fig. 1**. For further details, see the review by Fontana and Gross *(7)*. The methioninyl residue is converted to homoseryl or homoseryl lactone. The relative amounts of these two depend on the acid used, but when 70% TFA is the solvent, homoserine lactone is the major derivative. Peptides generated are suitable for peptide sequencing by Edman chemistry. Methionine sulfoxide does not take part in this reaction and the first step in the method is intended to convert any methionyl sulfoxide to methionyl residues, and so maximize cleavage efficiency. If the reduction is not carried out, the efficiency of cleavage may not be greatly diminished. If virtually complete cleavage is not necessary, partial cleavage products are desired (*see* **Note 6**), the sample is small and difficult to handle without loss, or speed is critical, the reduction step may be omitted.

   An acid environment is required to protonate basic groups and so prevent reaction there and maintain a high degree of specificity. Met-Ser and Met-Thr bonds may give significantly less than 100% yields of cleavage and simultaneous conversion to methionyl to homoseryl residues within the uncleaved polypeptide. This is because of the involvement of the β-hydroxyl groups of seryl and threonyl residues in alternative reactions, which do not result in cleavage *(7)*. Morrison et al. *(8)* however, have found that use of 70% v/v TFA gives a better yield of cleavage of a Met-Ser bond in apolipoprotein A1 than does use of 70% formic acid (*see* **Note 2**). Using model peptides, Kaiser and Metzka *(9)* analyzed

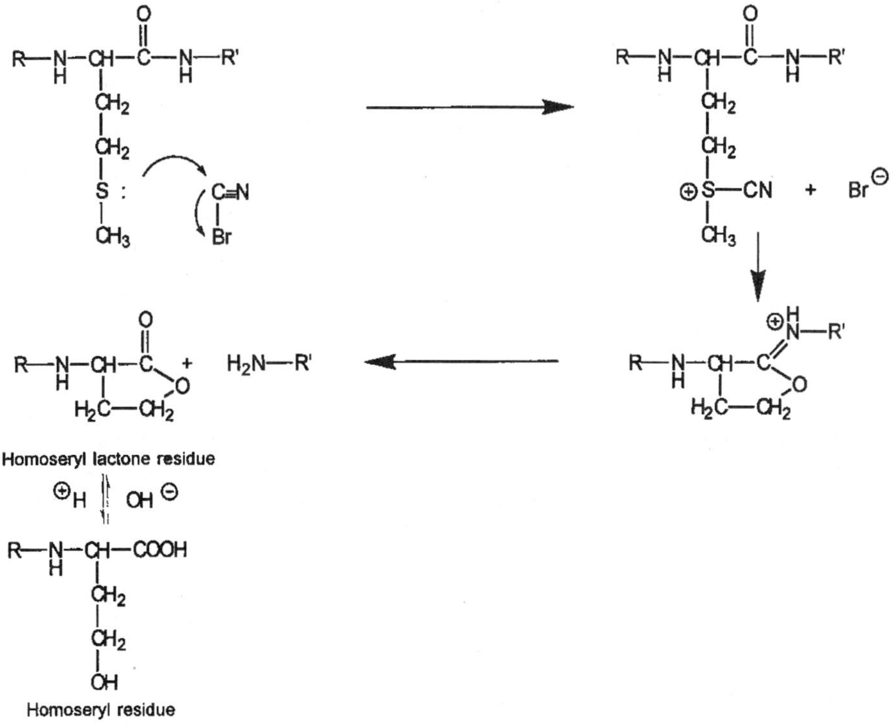

Fig. 1. Mechanism of cleavage of Met-X bonds by CNBr.

the cleavage reaction at Met-Ser and Met-Thr and concluded that cleavage that efficiency is improved by increasing the amount of water present, and for practical purposes 0.1 *M* HCl is a good acid to use, giving about 50% cleavage of these difficult bonds. Remaining uncleaved molecules contained either homoserine or methionyl sulfoxide instead of the original methionyl. Cleavage efficiency improved with increasing strength of acid, but there was an accompanying risk of degradation in the stronger acids. Utilization of C-terminal homoseryl lactone for linkage to solid phase is discussed in **Note 7**.

2. Acid conditions are required for the reaction to occur. In the past, 70% v/v formic acid (pH 1) was commonly used because it is a good protein solvent and denaturant, and also volatile. However, it may damage tryptophan and tyrosine residues ***(8)*** and also cause formation of seryl and threonyl side chains (showing up during analysis by mass spectroscopy as an increase of 28 amu per modification ***[9,10]***). Use of other acids avoids this problem. TFA (also volatile) may be used in concentrations in the range 50%–100% (v/v). The pH of such solutions is approx pH 0.5 or less. The rate of cleavage in 50% TFA may be somewhat slower than in 70% formic acid, but similar reaction times of hours, up to 24 h will provide satisfactory results. Caprioli et al. ***(11)*** and Andrews et al. ***(12)*** have illustrated the use of 60% and 70% TFA (respectively) for cyanogen bromide cleavage of proteins. Acetic acid (50%–100% v/v) may be used as an alternative but reaction is somewhat slower than in TFA. Alternatively, 0.1 *M* HCl has been used ***(9,10)***. To increase solubilization of proteins, urea or guanidine·HCl may be added to the solution. Thus, in 0.1 *M* HCl, 7 *M* urea, for 12 h at ambient temperature, a Met-Ala bond was cleaved with 83% efficiency, and the more problematical Met-Ser and Met-Thr bonds with 56% and 38% efficiency, respectively ***(9)***.

3. Although the specificity of this reaction is excellent, some side reactions may occur. This is particularly so if colored (yellow or orange) CNBr crystals are used, when there may be destruction of tyrosyl and tryptophanyl residues and bromination may also be detected by mass spectroscopy.
   Treatment of samples in other formats is discussed in **Notes 8** and **10**.
4. The above protocol describes addition of solid CNBr to the acidic protein solution, to give a molar excess of CNBr over methionyl residues. This has the advantage that pure white crystals may be selected in favor of pale yellow ones showing signs of degradation (*see* **Note 3**). It does not allow accurate estimation of the quantity of reagent used, however. The work of Kaiser and Metzka ***(9)*** suggests that more than a 10-fold molar excess of CNBr over methionyl residues does not increase the extent of cleavage. If in doubt as to the concentration of methionyl residues, however, err on the side of higher cyanogen concentration.
   If accurate quantification of CNBr is required, solid cyanogen bromide may be weighed out and dissolved to a given concentration by addition of the appropriate volume of 70% v/v TFA, and the appropriate volume of that solution added to the sample. The CNBr will start to degrade once in aqueous acid, so use when fresh. An alternative is to dissolve the CNBr in acetonitrile, in which it is more stable. CNBr in acetonitrile solution is available commercially, for instance, at a concentration of 5 *M* (Aldrich). While such a solution may be seen to be degrading by its darkening color, this is not so obvious as it is with CNBr in solid form. For use, sufficient acetonitrile solution is added to the acidic protein solution to give the desired excess of cyanogen bromide over protein (e.g., 1/20 dilution of a 5 *M* CNBr solution to give a final 250 m*M* solution). The data of Kaiser and Metzka ***(9)*** indicate that high concentrations (70–100%) of acetonitrile can interfere with the cleavage reaction by decreasing the amount of water present, but below a concentration of 30% (in 0.1 *M* HCl) the effect is noticeable in causing a small decrease of Met-Ser and Met-Thr bond cleavage, but negligible for the Met-Ala bond.
5. The reagents used are removed by lyophilization, unless salt has formed following failure to remove all the ammonium bicarbonate. The products of cleavage may be fractionated by the various forms of electrophoresis and chromatography currently available. If analyzed by reverse phase HPLC, the reaction mixture may be applied to the column directly without lyophilization. Since methionyl residues are among the less common residues, peptides resulting from cleavage at Met-X may be large and therefore in HPLC, use of wide-pore column materials may be advisable (e.g., 30-µm pore size reverse-phase column, using gradients of acetonitrile in 0.1% v/v TFA in water). Beware that some large peptides that are generated by this technique may prove to be insoluble (fore instance, if the solution is neutralized after the cleavage reaction) and therefore form aggregates and precipitates.
6. Incomplete cleavage, generating combinations of (otherwise) potentially cleaved peptides, may be advantageous, for ordering peptides within a protein sequence. Mass spectrometric methods are suitable for this type of analysis ***(10)***. Such partial cleavage may be achieved by reducing the duration of reaction, even to less than 1 h ***(10)***.
   The acid conditions employed for the reaction may lead to small degrees of deamidation of glutamine and asparagine side chains (which occurs below pH 3) and cleavage of acid-labile bonds, for example, Asp-Pro. A small amount of oxidation of cysteine to cysteic acid may occur, if these residues have not previously been reduced and carboxymethylated. Occasional cleavage of Trp-X bonds may be seen, but this does not occur with good efficiency, as it does when the reduction step of this technique is replaced by an oxidation step (for a description of this approach to cleavage of Trp-X bonds). Rosa et al. ***(13)***

cleaved both Met-X and Trp-X bonds simultaneously by treatment of protein with 12 m*M* CNBr in 70% TFA solution, plus 240 μM potassium bromide.

7. As described in **Note 1**, the peptide to the N-terminal side of the point of cleavage, has at its C-terminus a homoserine or homoserine lactone residue. The lactone derivative of methionine can be coupled selectively and in good yield ***(17)*** to solid supports of the amino type, for example, 3-amino propyl glass. This is a useful technique for sequencing peptides on solid supports. The peptide from the C-terminus of the cleaved protein will, of course, not end in homoserine lactone (unless the C-terminal residue was methionine!) and so cannot be so readily coupled. Similarly, the C-terminal peptide carboxyl can react (if not amidated) with acidic methanol, to become a methyl ester (with a corresponding mass increment of 14 amu). Homoserine lactone, present as the C-terminal residue on other peptides in a CNBr digest, will react with acidic methanol and show a mass increase of 32 amu. With account made for side chain carboxyl residues, this is a means to identify C-terminal peptides by mass spectroscopy ***(18)***.
8. Frequently, the protein of interest is impure, in a preparation containing other proteins. Polyacrylamide gel electrophoresis (PAGE) is a popular means by which to resolve such mixtures. Proteins in gel slices may be subjected to treatment with CNBr ***(14)***, as follows: The piece of gel containing the band of interest is cut out, lyophilized, and then exposed to vapor from a solution of CNBr in TFA for 24 h, at room temperature in the dark. The vapor is generated from a solution of 20 mg CNBr in 1 mL of 50% v/v TFA by causing it to boil by placing it under reduced pressure, in a sealed container together with the sample. The gel piece is then lyophilized again, and the peptides in it analyzed by PAGE.
9. Proteins that have been transferred from polyacrylamide gel to polyvinylidene difluoride (PVDF) membrane may be cleaved *in situ*, as described by Stone et al. ***(15)***. The protein band (of a few micrograms) is first cut from the membrane on the minimum size of PVDF (as excess membrane reduces final yield). The dried membrane is then wetted with about 50 μL of CNBr solution in acid solution—Stone et al. ***(15)*** report the use of CNBr applied in the ratio of about 70 μg per 1 g of protein. Note that although PVDF does not wet directly in water, it *does* do so in 70% formic acid, or in the alternative of 50% or 70% v/v TFA. Cleavage is achieved by incubation at room temperature, in the dark for 24 h. Oxidation of methionine during electrophoresis and blotting was not found to be a significant problem in causing reduction in cleavage yield, being about 100% in the case of myoglobin ***(15)***. The peptides generated by cleavage may be extracted for further analysis, first in the solution of CNBr in the acid solution, second in 100 μL of acetonitrile (40% v/v, 37°C, 3 h), and thirdly in 100 μL of TFA (0.05% v/v in 40% acetonitrile, 50°C. All extracts are pooled and dried under vacuum before any subsequent analysis.
10. Analysis of protein samples in automated peptide sequencing may sometimes yield no result. The alternative causes (lack of sample or amino-(N)-terminal blockage) may be tested by cleavage at methionyl residues by CNBr. Generation of new sequence(s) indicates blockage of the original N-terminus. The method is similar if the sample has been applied to a glass fiber disk, or to a piece of PVDF membrane in the sequencing cartridge. The cartridge containing the filter and/or membrane is removed from the sequencer, and the filter and/or membrane saturated with a fresh solution of CNBr in acid solution. The cartridge is wrapped in sealing film to prevent drying out, and then incubated in the dark at room temperature for 24 h. The sample is then dried under vacuum, replaced in the sequencer and sequence started again. Yields tend to be poorer than in the standard method described in the preceding. If the sample contains more than one methionine, more than one new N-terminus is generated, leading to a complex of sequences. This may be simplified by subsequent reaction with orthophthaladehyde which blocks all N-termini except

those bearing a prolyl residue ***(16)***. For success with this approach, prior knowledge of the location of proline in the sequence is required, the reaction with orthophthaladehyde being conducted at the appropriate cycle of sequencing.

11. The reagents used are removed by lyophilization, unless salt has formed following failure to remove all of the ammonium bicarbonate. The products of cleavage may be fractionated by the various forms of electrophoresis and chromatography currently available. If analyzed by reverse-phase HPLC, the reaction mixture may be applied to the column directly without lyophilization. As methionyl residues are among the less common residues, peptides resulting from cleavage at Met-X may be large, and so in HPLC use of wide-pore column materials may be advisable (e.g., 30-μm pore size reverse-phase columns, using gradients of acetonitrile in 0.1% v/v TFA in water). Beware that some large peptides that are generated by this technique may prove to be insoluble (e.g., if the solution is neutralized after the cleavage reaction) and so form aggregates and precipitates.

References

1. Yuan, G, Bin, J. C., McKay, D. J., and Snyder, F. F. (1999) Cloning and characterization of human guanine deaminase. Purification and partial amino acid sequence of the mouse protein. *J. Biol. Chem.* **274,** 8175–8180.
2. Malouf, N. N., McMahon, D., Oakeley, A. E., and Anderson, P. A. W. (1992). A cardiac troponin T epitope conserved across phyla. *J. Biol. Chem.* **267,** 9269–9274.
3. Dong, M., Ding, X.-Q., Pinon, D. I., Hadac, E. M., Oda, R. P., Landers, J. P., and Miller, L. J. (1999) Structurally related peptide agonist, partial agonist, and antagonist occupy a similar binding pocket within the cholecystokinin receptor. *J. Biol. Chem.* **274,** 4778–4785.
4. Wallace, D. S., Hofsteenge, J., and Store, S. R. (1990). Use of fragments of hirudin to investigate thrombin-hirudin interaction. *Eur. J. Biochem.* **188,** 61–66.
5. Callaway, J. E., Lai, J., Haselbeck, B., Baltaian, M., Bonnesen, S. P., Weickman, J., et al. (1993). *Antimicrob. Agents Chemother.* **17,** 1614–1619.
6. Strydom, D. J., Tarr, G. E., Pan, Y.-C, E., and Paxton, R. J. (1992). Collaborative trial analyses of ABRF-91AAA, in *Techniques in Protein Chemistry III* (Angeletti, R. H., ed.) Academic Press, San Diego, New York, Boston, London, Sydney, Tokyo, Toronto, pp. 261–274.
7. Fontana, A. and Gross, E. (1986) Fragmentation of polypeptides by chemical methods in Practical Protein Chemistry A Handbook (Darbre, A., ed. John Wiley and Sons, Chichester, pp. 67–120.
8. Morrison, J. R., Fidge, N. H., and Greo, B. (1990) studies on the formation, separation, and characterisation of cyanogen bromide fragments of human A1 apolipoprotein. *Analyt. Biochem.* **186,** 145–152.
9. Kaiser, R. and Metzka, L. (1999) Enhancement of cyanogen bromide cleavage yields for methionyl-serine and methionyl-threonine peptide bonds. *Analyt. Biochem.* **266,** 1–8.
10. Beavis, R. C. and Chait, B. T. (1990) Rapid, sensitive analysis of protein mixtures by mass spectrometry. *Proc. Natl. Acad. Sci. USA* **87,** 6873–6877.
11. Caprioli, R. M., Whaley, B., Mock, K. K., and Cottrell, J. S. (1991). Sequence-ordered peptide mapping by time-course analysis of protease digests using laser description mass spectrometry in *Techniques in Protein Chemistry II* (Angeletti, R. M., ed.) Academic Press, San Diego, pp. 497–510.
12. Andrews, P. C., Allen, M. M., Vestal, M. L., and Nelson, R. W. (1992) Large scale protein mapping using infrequent cleavage reagents, LD T*OF MS, and ES MS, in* Techniques in Protein Chemistry II (Angeletti, R. M., ed.) Academic Press, San Diego pp. 515–523.

13. Rosa, J. C., de Oliveira, P. S. L., Garrat, R., Beltramini, L., Roque-Barreira, M.-C., and Greene, L. J. (1999) KM+, a mannose-binding lectin from Artocarpus integrifolia: amino acid sequence, predicted tertiary structure, carbohydrate recognition, and analysis of the beta-prism fold. *Protein Science* **8,** 13–24.
14. Wang, M. B., Boulter, D., and Gatehouse, J. A. (1994) Characterisation and sequencing of cDNA clone encoding the phloem protein pp2 of Cucurbita pepo Plant *Mol. Biol.* **24,** 159–170.
15. Stone, K. L., McNulty, D. E., LoPresti, M. L., Crawford, J. M., DeAngelis, R., and Williams K. R. (1992). Elution and internal amino acid sequencing of PVDF blotted proteins, in *Techniques in Protein Chemistry III* (Angeletti, R. M., ed.) Academic Press, San Diego, pp. 23–34.
16. Wadsworth, C. L., Knowth, M. W., Burrus, L. W., Olivi, B. B., and Niece, R. L. (1992) Reusing PVDF electroblotted protein samples after N-terminal sequencing to obtain unique internal amino acid sequence, in *Techniques in Protein Chemistry III* (Angeletti, R. M., ed.) Academic Press, San Diego, pp. 61–68.
17. Horn, M. and Laursen, R. A. (1973) Solid-phase Edman degradation. Attachment of carboxyl-terminal homoserine peptides to an insoluble resin. *FEBS Lett.* **36,** 285–288.
18. Murphy, C. M. and Fenselau, C. (1995) Recognition of the carboxy-terminal peptide in cyanogen bromide digests of proteins. *Analyt. Chem.* **67,** 1644–1645.

72

Chemical Cleavage of Proteins at Tryptophanyl-X Peptide Bonds

Bryan John Smith

1. Introduction

Tryptophan is represented in the genetic code by a single codon and has proven useful in cloning exercises in providing an unambiguous oligonucleotide sequence as part of a probe or primer. It is also one of the less abundant amino acids found in polypeptides, and cleavage of bonds involving tryptophan generates large peptides. This may be convenient for generation of internal sequence information (the usual purpose to which this technique is put), but is less useful for identification of proteins by mass mapping, where a larger number of smaller peptides makes for more successful database searching.

While there is no protease that shows specificity for tryptophanyl residues, various chemical methods have been devised for cleavage of the bond to the carboxy-(C)-terminal side of tryptophan. These are summarized in **Table 1**. Some show relatively poor yields of cleavage, and/or result in modification of other residues (such as irreversible oxidation of methionine to its sulfone), or cleavage of other bonds (such as those to the C-terminal side of tyrosine or histidine). The method described in this chapter is one of the better ones, involving the use of cyanogen bromide (CNBr). Cleavage of bonds to the C-terminal side of methionyl residues (*see* Chapter 71) is prevented by prior reversible oxidation to methionine sulphoxide. Cleavage by use of *N*-bromosuccinimide or *N*-chlorosuccinimide remains a popular method, however, despite some chance of alternative reactions (*see* **Table 1**).

Apart from its use in peptide mapping and generation of peptides for peptide sequencing, cleavage at tryptophan residues has also been used to generate peptides used to map the binding site of an antibody *(1)*, or the sites of phosphorylation *(2)*. Another application has been generation of a recombinant protein from a fusion protein *(3)*: tryptophan was engineered at the end of a β-galactosidase leader peptide, adjacent to the N-terminus of phospholipase A_2. Reaction with *N*-chlorosuccinimide allowed subsequent purification of the enzyme without leader peptide.

From: *The Protein Protocols Handbook, 2nd Edition*
Edited by: J. M. Walker © Humana Press Inc., Totowa, NJ

Table 1
Summary of Methods for Cleavage of Trp-X Bonds in Polypeptides

Brief method details[a]	Comments	Example ref.
1. Incubation in molar excess of cyanogen bromide: a. Incubation in glacial acetic acid: 9 *M* HCl (2/1 [v/v] with DMSO, room temperature, 30 min. b. Neutralization c. Incubation with CNBr in acid (e.g., 60% formic acid) 4°C, 30 h in the dark.	*See text* for further details. Yields and specificity excellent; Met oxidized to sulfoxide by step a.	*4*
2. Incubation in very large (up to 10,000-fold) molar excess of CNBr over Met, in heptafluorobutyric acid–88%–HCOOH (1/1 [v/v]), room temperature, 24 h in the dark.	To inhibit Met-X cleavage, Met is photooxidized irreversibly to Met sulfone; yield poor.	*8*
3. a. Incubate protein for 30 min, room temperature, in: 21.6 mg/mL of phenol in glacial acetic acid 12 *M* HCl–DMSO:24:12:1,	Cys and Met oxidized; some deamidation and cleavage around Asp may occur; fresh, colorless HBr required.	*9*
b. Add 0.1 vol of 48% HBr, 0.03 by volume of DMSO, and incubate for 30 min, room temperature.	Trp converted to dioxindolylalanyl lactone. Protein containing dioxindolylalanine derivatives(s) but remaining uncleaved may be cleaved to improve yield to approx 80% as follows: *Incubate in* 10% acetic acid, 60°C 10–15 h.	
4. Incubate in BNPS-skatole[b] 100-fold molar excess over Trp) in 50% acetic acid (v/v) room temperature, 48 h, in the dark.	Reagent is unstable—fresh reagent required to minimize side reactions; some reactions with Tyr may occur, but addition of free Tyr to reaction mixture minimizes this; Met and Xys may be oxidized, yields up to 60%.	*10*

	Alternative conditions for rapid reaction: neat acetic acid 47°C 15–60 min; adaptable for cleavage of proteins bound to glass fiber or PVDF.	*11*
5. Incubation in *N*-bromosuccinimide (NBS, 3-fold molar excess over His, Trp, and Tyr), pH 3.0–4.0 (e.g., pyridine–acetic acid pH 3.3) 1 h, 100°C.	Also less rapid cleavage at His-X and Tyr; yields moderate to poor; Trp converted to lactone derivative.	*12*
6. Incubation in *N*-chlorosuccinimide (NCS,10-fold molar excess) 13 over (protein) in 27.5% acetic acid, 4.68 *M* urea; room temperature, 30 min; stopped by addition of *N*-acetyl-L-methionine.	Yield approx 50%; oxidation of methionine (to sulfone) and (to cysteic acid), especially in higher concentrations; *N*-chloro–succinate; Tyr and His not modified or cleaved (cf. *N*-bromo–succinimide; method 5); adaptable for cleavage of proteins in gels.	*13* *14*
7. Incubate in 80% acetic acid containing 4 *M* guanidine; HCI, 13 mg/mL iodosobenzoic acid; 20 mL *p*-cresol, for 20 h, room temperature, in the dark.	Specificity and yields good; the *p*-cresol is used to prevent cleavage at Tyr; Trp converted to lactone derivative.	*15*

[a]All methods cleave bond to the C-terminal side of residue.
[b]BNPS-skatole = 3-bromo-3-methyl-2-(2'-nitrophenylsulfenyl) indolenine.
[c]PVDF = polyvinylidene difluoride.

2. Materials

1. Oxidizing solution: Mix together 30 vol of glacial acetic acid, 15 vol of 9 *M* HCl, and 4 vol of dimethyl sulfoxide (DMSO). Use best grade reagents. Although each of the constituents is stable separately, mix and use the oxidizing solution when fresh.
2. Ammonium hydroxide (15 *M*). (*See* **Note 4**.)
3. CNBr solution in formic acid (60% v/v): Bring 6 mL of formic acid (minimum assay 98%, Aristar grade) to 10 mL with distilled water. Add white crystalline cyanogen bromide to a concentration of 0.3 g/mL. Use when fresh. (*See* **Note 5**.)
   Store CNBr refrigerated in the dry and dark, where it is stable for months. Use only white crystals. **Beware of the toxic nature of this reagent. Use in a fume hood.**
4. Sodium hypocholorite solution (domestic bleach).
5. Equipment includes a fume hood and suitably sized capped tubes (e.g., Eppendorf microcentrifuge tubes).

3. Methods

1. Oxidation: Dissolve the sample to approx 0.5 nmol/μL in oxidizing solution (e.g., 2–3 nmol in 4.9 μL of oxidizing solution). Incubate at 4°C for 2 h. (*See* **Notes 6** and **7**).
2. Partial neutralization: To the cold sample, add 0.9 volume of ice-cold NH_4OH (e.g., 4.4 μL of NH_4OH to 4.9 μL of oxidized sample solution). Make this addition carefully so as to maintain a low temperature. (*See* **Note 7**.)
3. Cleavage: Add 8 vol of CNBr solution. Incubate at 4°C for 30 h in the dark. Carry out this step in a fume hood.
4. To terminate the reaction, lyophilize the sample (all reagents are volatile). (*See* **Note 8**).
5. Decontaminate equipment such as spatulas that have contacted CNBr, by immersion in bleach until the effervescence stops (a few minutes) and thorough washing.

4. Notes

1. The method described is that of Huang et al. *(4)*. Although full details of the mechanism of this reaction are not clear, it is apparent that tryptophanyl residues are converted to oxindolylalanyl residues in the oxidation step, and the bond to the C-terminal side at each of these is readily cleaved in excellent yield (approaching 100% in **ref. *4***) by the subsequent CNBr treatment. The result is seemingly unaffected by the nature of the residues surrounding the cleavage site.
   During the oxidation step, methionyl residues become protected by conversion to sulfoxides, bonds at these residues not being cleaved by the cyanogen bromide treatment. Cysteinyl residues will also suffer oxidation if they have not been reduced and alkylated beforehand. Rosa et al. *(5)* cleaved both Trp-X and Met-X bonds simultaneously by omission of the oxidation step and inclusion of 240 μ*M* potassium iodide in the reaction of protein with 12 m*M* CNBr in 70% TFA solution.
   The peptide to the C-terminal side of the cleavage point has a free amino-(N)-terminus and so is suitable for amino acid sequencing by Edman chemistry.
2. Methionyl sulfoxide residues in the peptides produced may be converted back to the methionyl residues by incubation in aqueous solution with thiols (e.g., dithiothreitol, as described in **ref. *3***, or see use of 2-mercaptoethanol above).
3. The acid conditions used for oxidation and cleavage reactions seem to cause little deamidation *(4)*, but one side reaction that can occur is hydrolysis of acid-labile bonds. The use of low temperature minimizes this problem. If a greater degree of such acid hydrolysis is not unacceptable, speedier and warmer alternatives to the reaction conditions described above can be used as follows:

   a. Oxidation at room temperature for 30 min, but cool to 4°C before neutralization.
   b. Cleavage at room temperature for 12–15 h.
4. As alternatives to the volatile base NH_4OH, other bases may be used (e.g., the nonvolatile potassium hydroxide or Tris base).
5. Formic acid is a good protein denaturant and solvent, and is volatile and so relatively easy to remove. However, it has been noted that use of formic acid can cause formulation of seryl and threonyl residues *(6)* in the polypeptide (seen as an 28 amu increase in molecular mass) and damage to tryptophan and tyrosine (evidenced by spectral changes *[7]*). As an alternative to formic acid, 5 *M* acetic acid may be used, or as in the use of CNBr in cleaving methionyl-X bonds, 70% (v/v) trifluoroacetic acid may prove an acceptable alternative *(5)*.
6. Samples eluted from sodium dodecylsulfate (SDS) gels may be treated as described, but for good yields of cleavage, Huang et al. (4) recommend that the sample solutions are acidified to pH 1.5 before lyophilization in preparation for dissolution in the oxidizing solution. Any SDS present may help to solubilize the substrate and, in small amounts at least, does not interfere with the reaction. However, nonionic detergents that are phenolic or contain unsaturated hydrocarbon chains (e.g., Triton, Nonidet P-40) and reducing agents are to be avoided.
7. The method is suitable for large-scale protein cleavage, requiring simple scaling up. Huang et al. *(4)* made two points, however:
   a. The neutralization reaction generates heat. As this might lead to protein or peptide aggregation, cooling is important at this stage. Ensure that the reagents are cold and are mixed together slowly and with cooling. A transient precipitate is seen at this stage. If the precipitate is insoluble, addition of SDS may solubilize it (but will not interfere with the subsequent treatment).
   b. The neutralization reaction generates gases. Allow for this when choosing a reaction vessel.
8. At the end of the reaction, all reagents may be removed by lyophilization and the peptide mixture analyzed, for example, by polyacrylamide gel electrophoresis or by reverse-phase high-performance liquid chromatography (HPLC). Peptides generated may tend to be large, ranging up to a size of the order of 10,000 Da or more. Some of these large peptides may not be soluble, for example, if the solution is neutralized following the cleavage reaction, and consequently they aggregate and precipitate.
9. Note that all reactions are performed in one reaction vial, eliminating transfer of sample between vessels, and so minimizing peptide losses that can occur in such exercises.
10. Alternative methods for cleavage of tryptophanyl-X bonds are outlined in **Table 1**. The method (vi) that employs *N*-chlorosuccinimide is the most specific but shows only about 50% yield. BNPS-skatole is a popular Trp-X-cleaving reagent whose reaction and products have been studied in some detail (e.g., *see* **refs.** ***16*** and ***17***)
11. Both BNPS-skatole and N-chlorosuccinimide methods (**Table 1**, iv and vi, respectively) have been adapted to cleave small amounts (micrograms or less) of proteins on solid supports or in gels. Thus, proteins bound to glass fiber (as used in automated peptide sequences) may be cleaved by wetting the glass fiber with 1 μg/mL BNPS-skatole in 70%, (v/v) acetic acid, followed by incubation in the dark for 1 h at 47°C. After drying, sequencing may proceed as normal. Alternatively protein blotted to polyvinylidene difluoride membrane may be similarly treated and resulting peptides eluted for further analysis *(11)*. Alternatively, protein in slices of polyacrylamide gel (following polyacrylamide gel electrophoresis [SDS–PAGE]) may be cleaved by soaking for 30 min 0.015 *M* in N-chlorosuccinimide, in 0.5 g/mL of urea in 50%, (v/v) acetic acid. Following washing, peptides may be electrophoresed to generate peptide maps *(14)*.

References

1. Kilic, F. and Ball, E. H. (1991). Partial cleavage mapping of the cytoskeletal protein vinculin. Antibody and talin binding sites. *J. Biol. Chem.* **266,** 8734–8740.
2. Litchfield, D. W., Lozeman, F. J., Cicirelli, M. F., Harrylock, M., Ericsson, L. H., Piening, C. J., and Krebs. E. G. (1991) Phosphorylation of the beta subunit of casein kinase II in human A431 cells. Identification of the autophosphorylation site and a site phosphorylated by p34cdc2. *J. Biol. Chem.* **266,** 20,380–20,389.
3. Tseng, A., Buchta, R., Goodman, A. E., Loughman, M., Cairns D., Seilhammer, J., et al. (1991) A strategy for obtaining active mammalian enzyme from a fusion protein expressed in bacteria using phospholipase A2 as a model. *Pro. Express. and Purificat.* **2,** 127–135.
4. Huang, H. V., Bond, M. W., Hunkapillar, M. W., and Hood, L. E. (1983) Cleavage at tryptophanyl residues with dimethyl sulfoxide-hydrochloric acid and cyanogen bromide. *Meth. Enzymol.* **91,** 318–324.
5. Rosa, J. C., de Oliveira, P. S. L., Garrat, R., Beltramini, L., Roque-Barreira, M.-C. and Greene, L. J. (1999) KM+, a mannose-binding lectin from *Artocarpus integrifolia*: amino acid sequence, predicted tertiary structure, carbohydrate recognition, and analysis of the beta-prism fold. *Prot. Sci.* **8,** 13–24.
6. Beavis, R. C. and Chait. B. T. (1990). Rapid, sensitive analysis of protein mixtures by mass spectrometry. *Proc. Natl. Acad. Sci. USA* **87,** 6873–6877.
7. Morrison, J. R., Fidge, N. H., and Grego, B. (1990) Studies on the formation, separation, and characterisation of cyanogen bromide fragments of human A1 apolipoprotein. *Analyt. Biochem.* **186,** 145–152.
8. Ozols, J. and Gerard, C. (1977). Covalent structure of the membranous segment of horse cytochrome b5. Chemical cleavage of the native hemprotein. *J. Biol. Chem.* **252,** 8549–8553.
9. Savige, W. E. and Fontana, A. (1977) Cleavage of the tryptophanyl peptide bond by dimethyl sulfoxide-hydrobromic acid. *Meth. Enzymol.* **47,** 459–469.
10. Fontana, A. (1972) Modification of tryptophan with BNPS skatole (2-(2-nitro-phenylsulfenyl)-3-methyl-3'-bromoindolenine). *Meth. Enzymol.* **25,** 419–423.
11. Crimmins, D. L., McCourt, D. W., Thoma, R. S., Scott, M. G., Macke, K. and Schwartz, B. D. (1990) In situ cleavage of proteins immobilised to glass-fibre and polyvinylidene difluoride membranes: cleavage at tryptophan residues with 2-(2'-nitropheylsulfenyl)-3-methyl-3'-bromoindolenine to obtain internal amino acid sequence. *Analyt. Biochem.* **187,** 27–38.
12. Ramachandran, L. K. and Witkop, B. (1976). -Bromosuccinimide cleavage of peptides. *Meth. Enzymol.* **11,** 283–299.
13. Lischwe, M. A. and Sung, M. T. (1977). Use of -chlorosuccinimide/urea for the selective cleavage of tryptophanyl peptide bonds in proteins. *J. Biol. Chem.* **252,** 4976–4980.
14. Lischwe, M. A. and Ochs, D. (1982). A new method for partial peptide mapping using chlorosuccinimide/urea and peptide silver staining in sodium dodecyl sulphate-polyacrylamide gels. *Analyt. Biochem.* **127,** 453–457.
15. Fontana, A., Dalzoppo, D., Grandi, C., and Zambonin, M. (1983) Cleavage at tryptophan with iodosobenzoic acid. *Meth. Enzymol.* **91,** 311–318.
16. Vestling, M. M., Kelly, M. A., and Fenselau, C. (1994) Optimization by mass spectrometry of a tryptophan-specific protein cleavage reaction. *Rapid Commun. Mass Spectr.* **8,** 786–790.
17. Rahali, V. and Gueguen, J. (1999) Chemical cleavage of bovine (β-lactoglobulin by BNPS-skatole for preparative purposes: comparative study of hydrolytic procedures and peptide characterization. *J. Prot. Chem.* **18,** 1–12.

73

Chemical Cleavage of Proteins at Aspartyl-X Peptide Bonds

Bryan John Smith

1. Introduction

Some methods for chemically cleaving proteins, such as those described in Chapters 71 and 72, are fairly specific for a particular residue, show good yields, and generate usefully large peptides (as reaction occurs at relatively rare amino acid residues). There may be cases, however, when such peptides (or indeed, proteins) lacking these rarer residues need to be further fragmented, and in such instances cleavage at the more common aspartyl residue may prove useful. The method described in this chapter (and in **ref.** ***1***) for cleavage to the carboxy-(C)-terminal side of aspartyl residues is best limited to smaller polypeptides rather than larger proteins because yields are <100% and somewhat variable according to sequence. Partial cleavage of Asp-X bonds in a larger protein leads to a very complex set of peptides that may be difficult to analyze. Partial hydrolysis of smaller peptides yields correspondingly simpler mixtures. This may even be preferable to complete fragmentation for some purposes, such as peptide sequencing or mass spectrometry whereby the series of overlapping peptides may be used to order the peptides in the protein sequence (as in **ref.** ***2*** in which partial cleavage at methionyl-X bonds by CNBr was used).

2. Materials

1. Dilute HCl (approx 0.013 *M*) pH 2 ± 0.04: Dilute 220 μL of constant boiling (6 *M*) HCl to 100 mL with distilled water.
2. Pyrex glass hydrolysis tubes.
3. Equipment includes a blowtorch suitable for sealing the hydrolysis tubes, a vacuum line, and an oven for incubation of samples at 108°C.

3. Method

1. Dissolve the protein or peptide in the dilute acid to a concentration of 1–2 mg/mL in a hydrolysis tube.
2. Seal the hydrolysis tube under vacuum; that is, with the hydrolysis (sample) tube connected to a vacuum line, using a suitably hot flame, draw out and finally seal the neck of the tube.

From: *The Protein Protocols Handbook, 2nd Edition*
Edited by: J. M. Walker © Humana Press Inc., Totowa, NJ

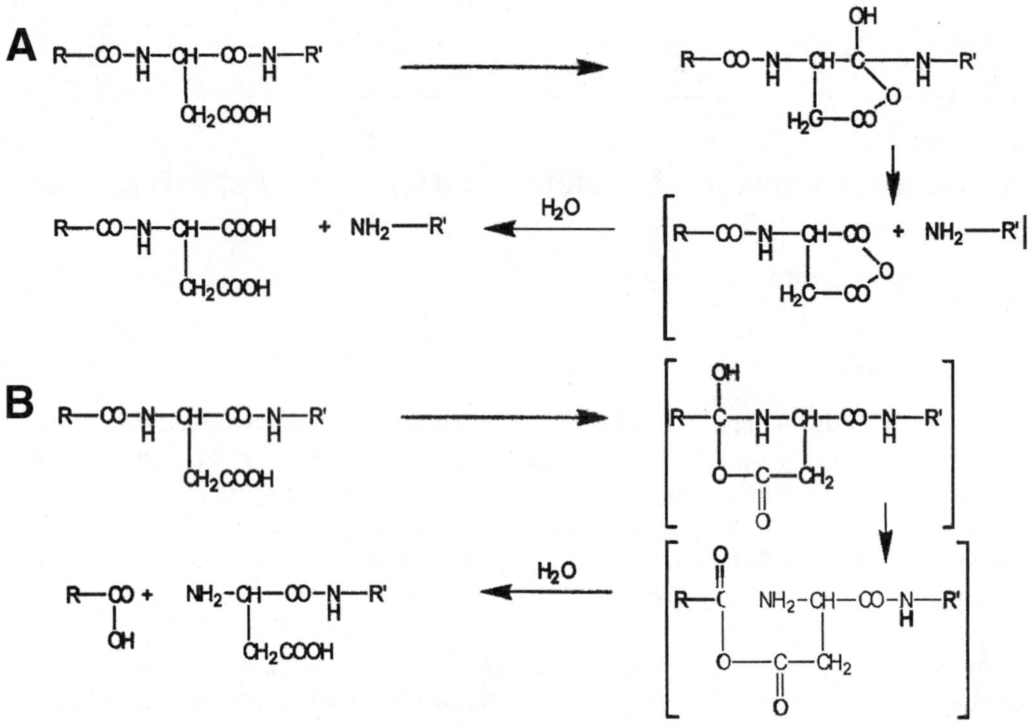

Fig. 1 Mechanisms of the cleavage of bonds to the COOH site (**scheme A**) and to the NH_2 site (**scheme B**) of aspartyl residues in dilute acid.

3. Incubate at 108°C for 2 h. (*See* **Note 5**.)
4. To terminate the reaction, cool and open the hydrolysis tube, dilute the sample with water, and lyophilize.

4. Notes

1. The bond most readily cleaved in dilute acid is the Asp-X bond, by the mechanism outlined in **Fig. 1(A)**. The bond X-Asp may also be cleaved, in lesser yields (*see* **Fig. 1[B]**). Thus, either of the peptides resulting from any one cleavage may keep the aspartyl residue at the point of cleavage, or neither might, if free aspartic acid is generated by a double cleavage event. Any of these peptides is suitable for sequencing.
2. The amino acid sequence of the protein can affect the lability of the affected bond. Thus, the aspartyl-prolyl bond is particularly labile in acid conditions (*see* **Note 3**). Ionic interaction between the aspartic acid side chains and basic residue side chains elsewhere in the molecule can adversely affect the rate of cleavage at the labile bond. Such problems as these make prediction of cleavage points somewhat difficult, particularly if the protein is folded up (e.g., a native protein). The method may well prove suitable for use in cleaving small proteins or peptides, in which such intramolecular interactions are less likely. Nevertheless, yields are <100%—up to about 70% have been reported *(1)*.
3. As noted in the preceding the aspartyl-prolyl bond is particularly acid labile and the following conditions have been proposed to promote cleavage of this particular bond *(3)*: dissolution of the sample in guanidine·HCl (7 *M*) in dilute acid (e.g., acetic acid, 10% v/v, adjusted to pH 2.5 with pyridine); incubation at moderate temperature (e.g., 37°C) for

prolonged periods (e.g., 24 h); terminate by lyophilization. Inclusion of guanidine·HCl, intended to denature the protein, may still fail to render all aspartyl-prolyl bonds sensitive to cleavage.

4. The conditions of low pH can be expected to cause a number of side reactions: cleavage at glutamyl residues; deamidation of (and possibly some subsequent cleavage at) glutaminyl and asparaginyl residues; partial destruction of tryptophan; cyclization of amino-(N)-terminal glutaminyl residues to residues of pyrrolidone carboxylic acid; α-β shift at aspartyl residues. The last two changes create a blockage to Edman degradation. The short reaction time of 2 h is intended to minimize these side reactions. A small degree of loss of formyl or acetyl groups from N-termini *(1)* is another possible side reaction but is not recognized as a significant problem, generally.
5. A polypeptide substrate that is insoluble in cold dilute HCl may dissolve during the incubation at 108°C. Formic acid is a good protein denaturant and solvent and may be used instead of HCl as follows: dissolve the sample in formic acid (minimum assay 98%, Aristar grade), then dilute 50–fold to pH 2; proceed as in method for HCl. Note, however, that incubation of protein in formic acid may result in formulation (increased molecular mass *[2]*) and damage to tryptophan and tyrosine residues (altered spectral properties *[4]*).
6. The comments concerning the effect of the amino acid sequence and of the environment around potentially labile bonds, and the various side reactions that can occur, indicate that the consequences of incubation of a protein in dilute acid are difficult to predict—they are best investigated empirically by monitoring production of peptides by electrophoresis or high-performance liquid chromatography (HPLC).
7. The method described has the benefit of simplicity. It is carried out in a single reaction vessel, with reagents being removed by lyophilization at the end of reaction. Thus, sample handling and losses incurred during this are minimized. This makes it suitable for subnanomolar quantities of protein, although the method may be scaled up for larger amounts also.
8. Note that bonds involving aspartyl residues may also be cleaved by commercially available enzymes. Endoproteinase Asp-N hydrolyzes the bond to the N-terminal side of an aspartyl residue, but also of a cysteinyl residue. Glu-C cleaves the bond to the C-terminal side of glutamyl and aspartyl residues.
9. The literature has various examples of unwanted cleavage of Asp-X bonds as the result of incubation of protein in acid conditions. The conditions encountered during N-terminal sequencing by Edman chemistry cause cleavage at Asp-X, generally somewhat inefficiently but sufficient to cause gradually increasing background. Cleavage may occur during protein isolation, for instance, of guanylin by 1 *M* acetic acid, which causes artificial cleavage from its corresponding prohormone *(5)*. Cleavage of labile bonds may also occur during heating of samples in sodium dodecyl sulfate (SDS) solution in preparation for SDS-polyacrylamide gel elctrophoresis (SDS-PAGE) *(6)*. Cleavage of Asp-X bonds has also been noted to occur in the process of matrix-assisted mass spectrometry *(7)*.

References

1. Ingris, A. S. (1983) Cleavage at aspartic acid. *Meth. Enzymol.* **91,** 324–332.
2. Beavis, R. C., and Chait, B. T. (1990) Rapid, sensitive analysis of protein mixtures by mass spectrometry. *Proc. Natl. Acad. Sci. USA* **87,** 6873–6877.
3. Landon, M. (1977) Cleavage at aspartyl-prolyl bonds. *Meth. Enzymol.* **47,** 132–145.
4. Morrison, J. R., Fiolge, N. H., and Grego, B. (1990) Studies on the formation, separation and characterisation of CNBr fragments of human A1 apolipoprotein. *Analyt. Biochem.* **186,** 145–152.

5. Schulz, A., Marx, U. C., Hidaka, Y., Shimonishi, Y., Rosch, P., Forssmann, W.-G., and Adermann, K. (1999) Role of the prosequence of guanylin. *Protein Science,* **8,** 1850–1859.
6. Correia, J. J., Lipscomb, L. D., and Lobert, S. (1993) Nondisulfide crosslinking and chemical cleavage of tubulin subunits: pH and temperature dependence. *Arch. Biochem. Biophys.* **300,** 105–114.
7. Yu, W., Vath, J. E., Huberty, M. C., and Martin, S. A. (1993) Identification of the facile gas-phase cleavage of the Asp-Pro and Asp-Xxx peptide bonds in matrix-assisted laser desorption time-of-flight mass spectrometry. *Analyt. Chem.*, **65,** 3015–3023.

74

Chemical Cleavage of Proteins at Cysteinyl-X Peptide Bonds

Bryan John Smith

1. Introduction

Cysteine is a significant amino acid residue in that it can form a disulfide bridge with another cysteine (to form cystine). Such disulfide bridges are important determinants of protein structure. No known endoproteinase shows specificity solely for cysteinyl, or cystinyl, residues, although endoproteinase Asp-N is able to hydrolyze bonds to the amino-(N)-terminal side of aspartyl or cysteinyl residues. Modification of aspartyl residues *(1)* can generate specificity for cysteinyl residues. Again, as discussed by Aitken *(2)*, modification of cysteinyl to 2-aminoethylcysteinyl residues makes the bond to the carboxy-(C)-terminal side susceptible to cleavage by trypsin, or the bond to the N-terminal side sensitive to Lys-N. However, specific cleavage of bonds to the N-terminal side of cysteinyl residues may be achieved in good yield by chemical means [*(3)*; *see* **Note 1**]. The cleavage generates a peptide blocked at its N-terminus as the cysteinyl residue is converted to an iminothiazolidinyl residue, but peptide sequencing can be carried out after conversion of this residue to an alanyl residue *(4)*.

Cysteine is one of the less frequent amino acids in proteins, and cleavage at cysteinyl residues tends to generate relatively large peptides. Thus, the method described in this chapter has been used to generate separate domains of troponin C for the purpose of studying Ca and peptide binding *(5)*, and peptides from vinculin to study talin- and anti-vinculin antibody binding *(6)*. It has also been adapted for the purposes of "footprinting": Cys residues that remain protected in a native protein remain non-cyanylated on reaction with 2-nitro-5-thiocyanobenzoate, and remain uncleaved upon alteration of the pH to 9.0. Thus the protected Cys residues can be mapped within the proteins sequence *(7)*.

2. Materials

1. Modification buffer: 0.2 *M* Tris-acetate, pH 8.0, 6 *M* guanidine·HCl, 10 m*M* dithiothreitol. Use Analar grade reagents and high-performance liquid chromatography (HPLC)-grade water. (*See* **Note 4.**)
2. 2-Nitro-5-thiocyanobenzoate (NTCB): Commercially available (Sigma) as a yellowish powder. Contact with skin, eyes, etc. may cause short-term irritation. Long-term effects are unknown, so handle with care (wear protective clothing). Sweep up spillages. Store at 0–5°C.

From: *The Protein Protocols Handbook, 2nd Edition*
Edited by: J. M. Walker © Humana Press Inc., Totowa, NJ

3. NaOH solution, sufficiently concentrated to allow convenient alteration of reaction pH. For example: 2 *M* in HPLC-grade water.
4. Deblocking buffer: 50 m*M* Tris-HCl, pH 7.0.
5. Raney nickel activated catalyst: Commercially available (e.g., from Sigma as 50% slurry in water, pH >9). Wash in deblocking buffer prior to use. A supply of N_2 gas is also required for use with the Raney nickel. (*See* **Note 7**.)

3. Methods (*see* Note 3)

1. Dissolve the polypeptide to a suitable concentration (say, 2 mg/mL) in the modification buffer (pH 8.0). To reduce disulfides in the dithiothreitol, incubate at 37°C for 1–2 h. (*See* **Note 4**.)
2. Add NTCB to 10-fold excess over sulfhydryl groups in polypeptide and buffer. Incubate at 37°C for 20 min.
3. To cleave the modified polypeptide, adjust to pH 9 by addition of NaOH solution. Incubate at 37°C for 16 h or longer (*see* **Note 2**).
4. Dialyze against water. Alternatively, submit to gel filtration or reverse-phase HPLC to separate salts and peptides. Lyophilize peptides.
5. If it is necessary to convert the newly formed iminothiazolidinyl N-terminal residue to an alanyl group, proceed as follows: Dissolve the sample to, say, 0.5 mg/mL in deblocking buffer, (pH 7); add to Raney nickel (10-fold excess, [w/w] over polypeptide); and incubate at 50°C for 7 h under an atmosphere of nitrogen. Cool and centrifuge briefly to pellet the Raney nickel. Store supernatant at –20°C, or further analyze as required. (*See* **Notes 5–8**.)

4. Notes

1. The reactions described in **Subheading 3.** are illustrated in **Fig. 1**. The method described is basically that used by Swenson and Fredrickson *(5)*, an adaptation of that of Jacobson et al. (*8*; *also see* **ref.** *3*). The principal difference is that the earlier method *(3,8)* describes desalting (by gel filtration or dialysis) at the end of the modification step (*see* **Subheading 3.**), followed by lyophilization and redissolution in a pH 9 buffer to achieve cleavage. Simple adjustment of pH as described in **Subheading 3., step 3** has the advantages of speed and avoiding the danger of sample loss upon desalting.
2. Swenson and Fredrickson *(5)* describe cleavage (**Subheading 3., step 3**) at 37°C for 6 h, but report yields of 60–80%. Other references recommend longer incubations of 12 h or 16 h at 37°C to obtain better yields *(3,8,9)*.
3. A slightly modified procedure is described in *(9)*:
   a. 1 mg/mL sample in 20 m*M* borate buffer, pH 8.0, 6 *M* urea, mixed with NTCB, added as a 0.1 *M* solution in 33% (v/v) dimethylformamide, at the rate of 40 µL/mL sample solution. Incubation was at 25°C for 1 h.
   b. Cleavage was by adjusting to pH 9.0 with NaOH and incubation at 55°C for 3 h.
   c. Reaction was quenched by addition of 2-mercaptoethanol to an 80-fold excess over NCTB.
4. If the sample contains no intramolecular or intermolecular disulfide bonds, the DDT content of the modification buffer may be made less, at 1 m*M*. Note that nominally nonbonded cysteinyl residues may be involved in mixed disulfides with such molecules as glutathione or free cysteine.
5. If blockage of the N-terminal residue of the newly generated peptide(s) to the C-terminal side of the cleavage point(s) is not a problem (i.e., if sequencing is not required, e.g., **ref.** *5*), **step 5** in **Subheading 3.**, may be omitted.

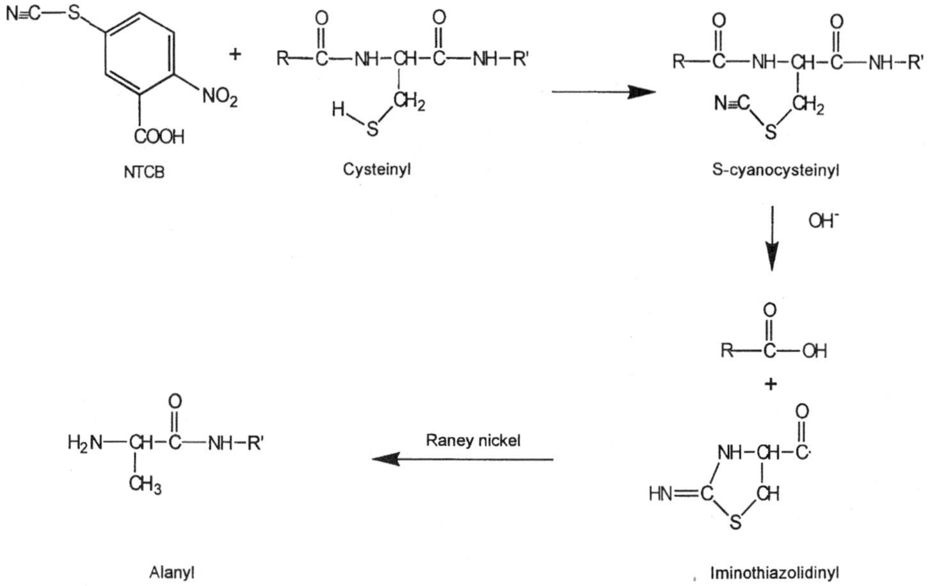

Fig. 1. Reactions in modification of, and cleavage at, cysteinyl residues by NTCB, and subsequent generation of alanyl N-terminal residue.

6. Reaction with Raney nickel (**Subheading 3., step 5**) converts methionyl residues to (β-aminobutyryl residues.
7. Although Raney nickel is available commercially, Otieno *(4)* has reported that a more efficient catalyst may be obtained by the method he described, starting from Raney nickel–aluminum alloy. This is reacted with NaOH, washed, deionized, and washed again (under H_2 gas).
8. Treatment of protein with Raney nickel *without* prior treatment with NTCB causes desulfurization of methioninyl residues (to give (β-aminobutyryl residues) and of cysteinyl and cystinyl residues (to alanyl residues). Otieno *(4)* has suggested that this modification might be used to study dependence of protein function on Met and Cys content.

References

1. Wilson, K. J., Fischer, S., and Yuau, P. M. (1989). Specific enzymatic cleavage at cystine/cysteine residues. The use of Asp-N endoproteinase, in *Methods in Protein Sequence Analysis* (Wittman-Liebold, B., ed.), *Springer-Verlag, Berlin*, pp. 310–314.
2. Aitken, A. (1994) Analysis of cysteine residues and disulfide bonds, in *Methods in Molecular Biology*, Vol. 32: *Basic Protein and Peptide Protocols* (Walker, J. M., ed.), *Humana Press*, Totowa, NJ, pp. 351–360.
3. Stark, G. R. (1977) Cleavage at cysteine after cyanylation. *Meth. Enzymol.* **47,** 129–132.
4. Otieno, S. (1978) Generation of a free α-amino group by Raney nickel after 2-nitrothiocyanobenzoic acid cleavage at cysteine residues: applications to automated sequencing. *Biochemistry* **17,** 5468–5474.
5. Swenson, C. A. and Fredrickson, R. S. (1992) Interaction of troponin C and troponin C fragments with troponin I and the troponin I inhibitory peptide. *Biochemistry* **31,** 3420–3427.
6. Kilic, F. and Ball, E. H. (1991) Partial cleavage mapping of the cytoskeletal protein vinculin. Antibody and talin binding sites. *J. Biol. Chem.* **266,** 8734–8740.

7. Tu, B. P. and Wang, J. C. (1999) Protein footprinting at cysteines: probing ATP-modulated contacts in cysteine-substitution mutants of yeast DNA topoisomerase II. *Proc. Natl. Acad. Sci. USA* **96,** 4862–4867.
8. Jacobson, G. R., Schaffer, M. H., Stark, G., and Vanaman, T. C. (1973) Specific chemical cleavage in high yield at the amino peptide bonds of cysteine and cystine residues. *J. Biol. Chem.* **248,** 6583–6591.
9. Peyser, Y. M., Muhlrod, A., and Werber, M. M. (1990). Tryptophan-130 is the most reactive tryptophan residue in rabbit skeletal myosin subfragment-1. *FEBS Lett.* **259,** 346–348.

75

Chemical Cleavage of Proteins at Asparaginyl-Glycyl Peptide Bonds

Bryan John Smith

1. Introduction

Reaction with hydroxylamine has been used to cleave DNA and to deacylate proteins (at neutral pH; *see* **Note 6**). At alkaline pH, however, hydroxylamine may be used to cleave the asparaginyl-glycyl bond (*see* **Note 1**). This cleavage tends to generate large peptides, as this pairing of relatively common residues is relatively uncommon, representing about 0.25% of amino acid pairs, according to Bornstein and Balian ***(1)***. It is therefore generally not useful for identification of proteins by mass mapping, which is better served by larger numbers of peptides. It may be a useful method for further cleavage of large peptides.

Cleavage at Asn-Gly bonds has been used for:

1. Generation of peptides for sequencing purposes (e.g., **ref.** ***2***).
2. Generation of peptides for use in mapping ligand binding sites (e.g. **ref.** ***3***) and phosphorylation sites (e.g. **ref.** ***4***).
3. Identification of sites of succinimide (cyclic amide) sites in proteins ***(5)***.
4. Cleavage of a fusion protein at the point of fusion of the constituent polypeptides (although it was noted that formation of hydroxamates may occur ***[6]***, as may minor cleavage reactions ***[3]***).

2. Materials

1. Cleavage buffer: 2 *M* hydroxylamine·HCl,
2 *M* guanidine·HCl,
0.2 *M* KCO_3, pH 9.0

   Use Analar grade reagents and HPLC grade water (*see* **Note 2**).

   Beware the mutagenic, toxic and irritant properties of hydroxylamine. Wear protective clothing. Clear wet spillages with absorbent material or clear dry spillages with a shovel, and store material in containers prior to disposal.
2. Stopping solution: Trifluoroacetic acid, 2% (v/v) in water (high-performance liquid chromatography [HPLC] grade).

From: *The Protein Protocols Handbook, 2nd Edition*
Edited by: J. M. Walker © Humana Press Inc., Totowa, NJ

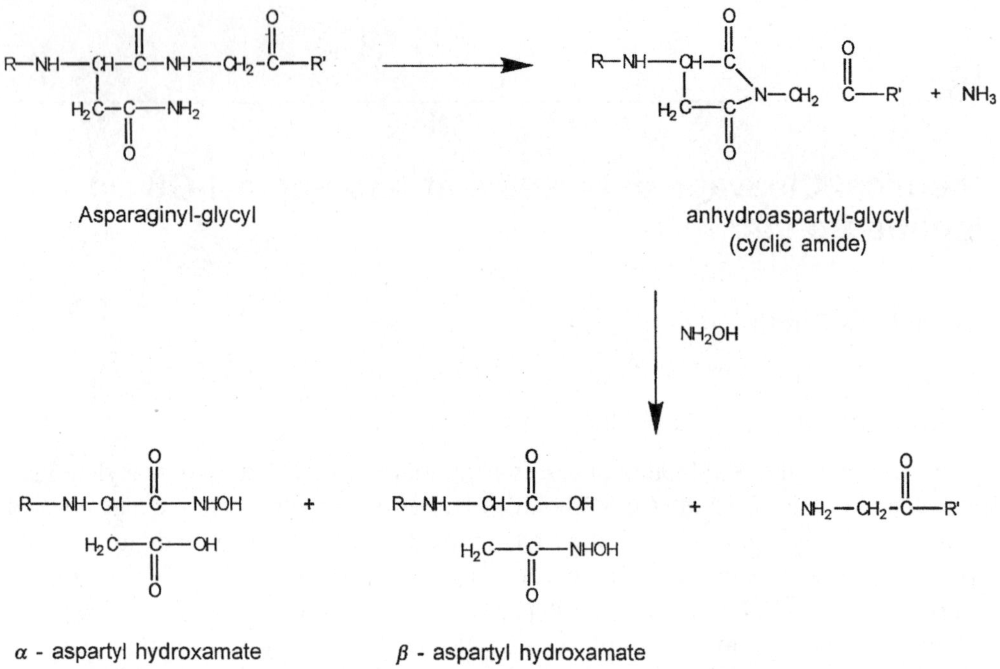

Asparaginyl-glycyl

anhydroaspartyl-glycyl (cyclic amide)

NH_2OH

α - aspartyl hydroxamate

β - aspartyl hydroxamate

Fig. 1. Illustration of reactions leading to cleavage of Asn-Gly bonds by hydroxylamine.

3. Method

1. Dissolve the protein sample directly in the cleavage buffer, to give a concentration in the range 0.1–5 mg/mL. Alternatively, if the protein is in aqueous solution already, add 10 vol of the cleavage buffer (i.e., sufficient buffer to maintain pH 9.0 and high concentration of guanidine·HCl and hydroxylamine). Use a stoppered container (Eppendorf tube or similar) with small headspace, so that the sample does not dry out during the following incubation (*see* **Notes 3** and **4**).
2. Incubate the sample (in stoppered vial) at 45°C for 4 h.
3. To stop reaction, cool and acidify by addition of three volumes of stopping solution. Store frozen (–20°C) or analyze immediately (*see* **Note 5**).

4. Notes

1. The reaction involved in this cleavage is illustrated in **Fig. 1** and is described in more detail in **ref. *1***, with the proposed role of the succinimide being confirmed by Blodgett et al. *(7)*. Note that the reaction of hydroxylamine is actually with the cyclic imide that derives from the Asn-Gly pair. The Asn-Gly bond itself is resistant to cleavage by hydroxylamine. Kwong and Harris *(5)* have reported cleavage at a presumed succinimide at an Asp-Gly bond. The succinimide occurs as an intermediate in the isomerization of Aspartyl to *iso*-Aspartyl, a reaction that involves the Gly residue. This can occur at sites that are in regions flexible enough to accommodate the three structures, so it is influenced by the neighboring sequence. This therefore affects cleavage efficiency. Bornstein and Balian *(1)* have reported an Asn-Gly cleavage yield of about 80% but lower efficiency may be experienced (as reported in **ref. 5**, for various conditions). Cleavage is to the carboxy-(C)-terminal side of the succinimide. The peptide generated to the C-terminal

side is available for amino-(N)-terminal sequence analysis, for generation of internal sequence or to identify the site of the succinimide.

The succinimide is stable enough to be found in proteins *(5,8)*. Isomerization of the Asp to iso-Asp can affect immunogenicity and function (for instance, *see [8]*). The *iso*-Asp may be detected by other assays (see *[9]* and references therein), or by termination of N-terminal sequencing, for both it and the succinimidyl residue are refractory to Edman sequencing chemistry.

The succinimidyl version of a polypeptide is slightly more basic (by one net negative charge) than the aspartate version that forms after incubation in neutral pH *(5)*.

In addition to cleavage at Asn-Gly, treatment with hydroxylamine may generate other, lower yielding, cleavages. Thus, Bornstein and Balian *(1)* mention cleavage of Asn-Leu, Asn-Met, and Asn-Ala, while Hiller et al *(3)* report cleavage of Asn-Gln, Asp-Lys, Gln-Pro, and Asn-Asp. Prolonged reaction times tend to generate more of such cleavages. Treatment of protein with hydroxylamine may generate hydroxamates of asparagine and glutamine, these modifications producing more acidic variants of the protein *(6)*.

2. Inclusion of guanidine·HCl as a denaturant seems to be a factor in improving yields. Kwong and Harris *(5)* reported that omission of guanidine·HCl eliminated cleavage at an Asn-Gly site while allowing cleavage at an Asp-Gly site. The literature has other examples of the use of buffers lacking guanidine·HCl, for instance, **refs.** *3* and *5*. Both of these examples report use of a Tris-HCl buffer of approx pH 9.0, with **ref.** *5* including 1 m*M* EDTA and ethanol (10% v/v). Yet other examples *(1,2,10)* describe the use of more concentrated guanidine·HCl, at 6 *M*, the cleavage buffer being prepared by titrating a solution of guanidine·HCl (6 *M*, final) and hydroxylamine·HCl (2 *M*, final) to pH9.0 by addition of a solution of lithium hydroxide (4.5 *M*). Note that preparation of this lithium hydroxide solution may generate insoluble carbonates that can be removed by filtration. Other reaction conditions were as described in **Subheading 3.**
3. As when making peptides by other cleavage methods, it may be advisable, prior to the above operations, to reduce disulfide bonds and alkylate cysteinyl residues (*see* Chapter 59). This denatures the substrate and prevents formation of interpeptide disulfide bonds. Niles and Christen *(10)* describe alkylation and subsequent cleavage by hydroxylamine on scale of a few microliters.
4. The hydroxylamine cleavage method has been adapted to cleave proteins in polyacrylamide gel pieces *(11)*. The cleavage buffer was 2 *M* hydroxylamine·HCl, 6 *M* guanidine·HCl, in 15 m*M* Tris titrated to pH 9.3 by addition of 4.5 *M* lithium hydroxide solution. Pieces of gel that had been washed in 5% methanol to remove sodium dodecyl sulfate (SDS), and then dried *in vacuo* were submerged in the cleavage solution (50–200 μL per 3 μL of gel) and incubated at 45°C for 3 h. Analysis by electrophoresis on a second SDS gel then followed. Peptides of about 10,000 Da or less tended to be lost during washing steps, and about 10% of sample remained bound to the treated gel piece. Recoveries in the second (analytical) SDS gel were reported to be approx 60% and cleavage yield was about 25%.
5. After the cleavage reaction has been stopped by acidification, the sample may be loaded directly onto reverse-phase HPLC or gel filtration for analysis/peptide preparation. Alternatives are dialysis and polyacrylamide gel electrophoresis (PAGE). The reaction itself may be stopped not by acidification, but by mixing with SDS-PAGE sample solvent and immediate electrophoresis *(3)*.
6. In approximately neutral pH conditions, reaction of protein with hydroxylamine may cause esterolysis, and so may be a useful method in studying posttranslational modification of proteins. Thus, incubation in 1 *M* hydroxylamine, pH 7.0, 37°C for up to 4 h cleaved carboxylate ester-type ADP–ribose–protein bonds (on histones H2A and H2B) and argin-

ine–ADP–ribose bonds (in histones H3 and H4) ***(12)***. Again, Weimbs and Stoffel ***(13)*** identified sites of fatty acid-acylated cysteine residues by reaction with 0.4 *M* hydroxylamine at pH 7.4, such that the fatty acids were released as hydroxamates. Omary and Trowbridge ***(14)*** adapted the method to release [^{3}H] palmitate from transferrin receptor in polyacrylamide gel pieces, soaking these for 2 h in 1 *M* hydroxylamine·HCl titrated to pH 6.6 by addition of sodium hydroxide.

References

1. Bornstein, P., and Balian, G. (1977) Cleavage at Asn-Gly bonds with hydroxlyamine. *Meth. Enzymol.* **47,** 132–145.
2. Arselin, G., Gandar, J. G., Guérin, B., and Velours, J. (1991) Isolation and complete amino acid sequence of the mitochondrial ATP synthase (-subunit of the yeast Saccharomyces cerevisiae. *J. Biol. Chem.* **266,** 723–727.
3. Hiller, Y., Bayer, E. A., and Wilchek, M. (1991) Studies on the biotin-binding site of avidin. Minimised fragments that bind biotin. *Biochem. J.* **278,** 573–585.
4. Hoeck, W. and Groner, B. (1990) Hormone-dependent phosphorylation of the glucocorticoid receptor occurs mainly in the amino-terminal transactivation domain. *J. Biol. Chem.* **265,** 5403–5408.
5. Kwong, M. Y. and Harris, R. J. (1994) Identification of succinimide sites in proteins by N-terminal sequence analysis after alkaline hydroxylamine cleavage. *Protein Sci.* **3,** 147–149.
6. Canova-Davis, E., Eng, M., Mukku, V., Reifsnyder, D. H., Olson, C. V., and Ling, V. T. (1992) Chemical heterogeneity as a result of hydroxylamine cleavage of a fusion protein of human insulin-like growth factor I. *Biochem. J.* **278,** 207–213.
7. Blodgett, J. K., Londin, G. M., and Collins, K. D. (1985) Specific cleavage of peptides containing an aspartic acid (beta-hydroxamic) residue. *J. Am. Chem. Soc.* **107,** 4305–4313.
8. Cacia, J., Keck, R., Presta, L. G., and Frenz, J. (1996) Isomerization of an aspartic acid residue in the complementarity-determining regions of a recombinant antibody to human IgE: identification and effect on binding affinity. *Biochemistry* **35,**1897–1903.
9. Schurter, B. T., and Aswad, D. A. (2000) Analysis of isoaspartate in peptides and proteins without the use of radioisotopes. *Analyt. Biochem.* **282,** 227–231.
10. Niles, E. G. and Christen, L. (1993) Identification of the vaccinia virus mRNA guanyltransferase active site lysine. *J. Biol. Chem.* **268,** 24,986–24,989.
11. Saris, C. J. M., van Eenbergen, J., Jenks, B. G., and Bloemers, H. P. J. (1983) Hydroxylamine cleavage of proteins in polyacrylamide gels. *Analyt. Biochem.* **132,** 54–67.
12. Golderer, G., and Gröbner, P. (1991) ADP-ribosylation of core histones and their acetylated subspecies. *Biochem. J.* **277,** 607–610.
13. Wiembs, T. and Stoffel, W. (1992) Proteolipid protein (PLP) of CNS myelin: Positions of free, disulfide-bonded and fatty acid thioester-linked cysteine residues and implications for the membrane topology of PLP. *Biochemistry* **31,** 12,289–12,296.
14. Omary, M. B. and Trowbridge I. S. (1981) Covalent binding of fatty acid to the transferrin receptor in cultured human cells. *J. Biol. Chem.* **256,** 4715–4718.

76

Enzymatic Digestion of Proteins in Solution and in SDS Polyacrylamide Gels

Kathryn L. Stone and Kenneth R. Williams

1. Introduction

Although most prokaryotic proteins have free NH_2-termini and therefore can be directly sequenced, most eukaryotic proteins have blocked NH_2-termini which precludes Edman degradation *(1,2)*. In these instances, one of the most direct approaches to obtaining partial amino acid sequences is via enzymatic or chemical cleavage followed by peptide fractionation and sequencing. Although several different approaches may be taken to cleave proteins, one of the most common is to digest the protein enzymatically with a relatively specific protease such as trypsin or lysyl endopeptidase. Since final purification is often dependent on SDS-PAGE, cleavage procedures that can either be carried out in the polyacrylamide gel matrix *(3,4)* or that may be used on samples that have been blotted from SDS polyacrylamide gels onto PVDF *(5)* or nitrocellulose *(5,6)* membranes are extremely useful. In both instances, the proteins are usually stained with Coomassie blue or Ponceau S prior to excision, and proteolytic digestion and the resulting peptides are separated by reverse-phase HPLC. Relatively straightforward solution and in-gel digestion procedures that have been used extensively in the Keck Foundation Biotechnology Resource Laboratory at Yale University will be described in this chapter, whereas a procedure suitable for in situ digestion of SDS-PAGE blotted proteins is described in Chapter 77.

2. Materials

2.1. Enzymatic Digestion of Proteins

1. Enzymatic digestion of proteins is usually accomplished using either sequencing grade, modified trypsin (from Promega [Madison, WI] or Boehringer Mannheim [Indianapolis, IN]) or lysyl endopeptidase (Achromobacter Protease I from Achromobacter lyticus) from Wako Pure Chemical Industries, Ltd. (Osaka, Japan). Occasionally, chymotrypsin or endoproteinase Glu-C from Boehringer Mannheim may also be used. All enzyme stocks are divided into 100 µL aliquots and stored at –20°C.
2. Sequencing-grade chymotrypsin and modified trypsin (Boehringer Mannheim): prepare by dissolving the dried 100-µg aliquots (obtained from the manufacturer) in 1 mL 1 m*M* HCl to make a 0.1 mg/mL stock solution that appears to be stable for at least 6 mo at –20°C.

From: *The Protein Protocols Handbook, 2nd Edition*
Edited by: J. M. Walker © Humana Press Inc., Totowa, NJ

3. Sequencing grade, modified trypsin (Promega): dissolve the 20-µg aliquot (obtained from the manufacturer) in 200 µL 1 m*M* HCl to make a 0.1 mg/mL stock solution that appears to be stable for at least 6 mo at –20°C.
4. Endoproteinase Glu-C: dissolve the 50-µg aliquot from the manufacturer in 500 µL 50 m*M* NH_4HCO_3. According to the manufacturer, the dissolved enzyme is stable for 1 mo at –20°C.
5. Lysyl endopeptidase: dissolve 2.2 mg as purchased in 2.2 mL of 2 m*M* Tris-HCl, pH 8.0, to make a 1 mg/mL stock, which according to the manufacturer is stable for at least 2 yr when stored at –20°C. More dilute solutions are made by adding 10 µL of this 1 mg/mL stock solution to 90 µL 2 m*M* Tris-HCl, pH 8.0, for a 0.1 mg/mL stock.
6. Pepsin (Sigma Chemical Co., St. Louis, MO): dissolve in 5% formic acid (Baker) at a concentration of 0.1 mg/mL.
7. Digestion buffer for in-solution digestion: 8 *M* urea, 0.4 *M* NH_4HCO_3. Prepare by dissolving 4.8 g Pierce Sequanal-Grade urea and 0.316 g Baker ammonium bicarbonate in H_2O to make a final volume of 10 mL.
8. 50% CH_3CN/0.1 *M* Tris-HCl, pH 8.0: prepare by diluting 5 mL 1.0 *M* Tris-HCl, pH 8.0 (12.1 g Tris-HCl dissolved in 100 mL H_2O with the pH adjusted to 8.0 with HCl) with 20 mL H_2O and 25 mL 100% acetonitrile.
9. In gel digestion buffer: 0.1 *M* Tris-HCl, pH 8.0/0.1% Tween 20. Prepared by adding 1 mL 1.0 *M* Tris-HCl, pH 8.0, and 10 µL polyoxyethylene-sorbitan monolaurate (Tween 20 from Sigma Chemical Co.) to 9.0 mL H_2O.
10. Cysteine modification buffer: 0.1 *M* Tris-HCl, pH 8.0/60% CH_3CN is prepared by diluting 5.0 mL 1.0 *M* Tris-HCl, pH 8.0 with 15 mL H_2O plus 30 mL 100% CH_3CN.
11. 45 m*M* DTT (Pierce Chemical Co.) solution for protein reduction: dissolve 69 mg DTT in 10 mL H_2O.
12. 100 m*M* iodoacetic acid (IAA) (Pierce Chemical Co.) solution for alkylation: dissolve 185.9 mg IAA in 10 mL H_2O.
13. 200 m*M* methyl 4-nitrobenzene sulfonate (Aldrich Chemical Co.) solution for cysteine modification: is made by dissolving 0.0434 g methyl 4-nitrobenzene sulfonate in 100% CH_3CN. This solution is made immediately before use and is not stored.
14. 0.1% TFA/60% CH_3CN solution for peptide extraction from the gel slices: add 50 µL 100% trifluoracetic acid (TFA) to 20 mL H_2O and 30 mL 100% CH_3CN.

3. Methods

3.1. Enzymatic Digestion of Proteins

3.1.1. Digestion of Proteins in Solution

3.1.1.1. Sample Preparation for in Solution Digestion

Proper sample preparation is critical both in avoiding sample loss and in ensuring successful digestion (*see* **Note 1**). If the sample contains a sufficiently low level of salt and glycerol (such that the concentration of salt and glycerol in the final digest will be less than the equivalent of 1 *M* NaCl and 15% glycerol, respectively), it may simply be reduced to dryness in a SpeedVac prior to carrying out the digest in the tube in which it was dried. Samples containing higher levels of salts, glycerol, and/or detergents, such as SDS, may be precipitated using either trichloroacetic acid (TCA) or acetone in order to remove the salts, glycerol, and detergents. To TCA-precipitate the protein, 1/9th vol of 100% TCA is added to the sample prior to incubating on ice for 30 min. The sample is then centrifuged (10,000g/15 min) and the supernatant is removed. Residual TCA is then removed by suspending the pellet in 50 μL cold acetone. After vortexing and centrifuging, the supernatant is removed with a pipet. The air-dried pellet is then digested as described below. To acetone precipitate the protein, it is often necessary first to reduce the salt concentration by dialysis vs 0.05% SDS, 5 m*M* NH_4HCO_3. At this point, the volume is reduced to <50 μL in a SpeedVac, and 9 vol cold acetone are added followed by a 1 h incubation at –20°C. The sample is then centrifuged and the pellet washed (as described in **Subheading 3.1.**) with 50 μL cold acetone. When large amounts of SDS are present, the acetone wash should be repeated once or twice more to remove excess detergent. The recommended minimum protein concentration during either precipitation is 100 μg/mL, and for TCA precipitation, the glycerol concentration should be below 15%.

3.1.1.2. Digestion of Proteins in Solution with Trypsin, Lysyl Endopeptidase, Chymotrypsin, and Endoproteinase Glu-C

After the salts and detergents have been removed/minimized, the sample is ready for digestion. Typically, trypsin is the enzyme of choice because of its relatively high cleavage specificity (the COOH-terminal side of lysine and arginine) and its ability to digest insoluble substrates. That is, proteins that are only partially soluble in the digest buffer will often cleave with trypsin as evidenced by rapid clearing of the solution. Cleavage occurs more slowly when there is an acidic residue following the lysine or arginine and not at all when a lysine-proline or arginine-proline linkage is present. Since lysyl endopeptidase only cleaves on the COOH-terminal side of lysine, it provides longer peptides than trypsin. Based on the average occurrence of lysine (5.7%) and arginine (5.4%) in proteins in the Protein Identification Resource Data Base, the average length of a tryptic and lysyl endopeptidase peptide is about 9 and 18 residues, respectively. However, since lysyl endopeptidase does not generally cleave insoluble substrates and since we have occasionally encountered proteins that will digest with trypsin, but not with lysyl endopeptidase, we generally use trypsin unless there is sufficient protein to carry out two digests (if needed), in which case we would initially use lysyl endopeptidase.

Since chymotrypsin readily cleaves on the COOH-terminal side of tryptophan, tyrosine, and phenylalanine, and generally gives partial cleavage after leucine, methionine, and several other amino acids, its specificity is too broad to be of general use. However, it is occasionally used to redigest larger peptides or proteins that fail to digest with trypsin. Although endoproteinase Glu-C is relatively specific in that it cleaves after glutamic acid in either ammonium bicarbonate (pH 8.0) or ammonium acetate (pH 4.0) buffers and after both aspartic acid and glutamic acid in phosphate buffers (pH 7.8), it also does not generally cleave insoluble substrates, and the HPLC profiles we have obtained with this enzyme usually suggest relatively incomplete cleavage and the resulting generation of overlapping peptides in lower yield than might often be obtained with trypsin.

Although this brief survey of proteolytic enzymes is far from complete, it does cover most of the enzymes that are frequently used. The enzymatic digestion protocol outlined below may be used with any of these four enzymes, providing the appropriate buffer changes are made in the case of Glu-C.

1. Dissolve the dried or precipitated protein in 20 μL 8 *M* urea, 0.4 *M* NH_4HCO_3 and then remove a 10–15% aliquot for acid hydrolysis/ion-exchange amino acid analysis. If the

analysis indicates there is sufficient protein to digest (*see* **Note 2**), proceed with **step 2**; otherwise additional protein should be prepared to pool with the sample.

2. Check the pH of the sample by spotting 1–2 µL on pH paper. If necessary, adjust the pH to between 7.5 and 8.5.
3. Add 5 µL 45 m*M* DTT and incubate at 50°C for 15 min to reduce the protein. (*See* **Note 3**).
4. After cooling to room temperature, alkylate the protein by adding 5 µL 100 m*M* IAA and incubating at room temperature for 15 min. (*See* **Note 3**).
5. Dilute the digestion buffer with H_2O so the final digestion will be carried out in 2 *M* urea, 0.1 *M* NH_4HCO_3.
6. Add the enzyme in a 1/25, enzyme/protein (wt/wt) ratio. (*See* **Note 4**).
7. Incubate at 37°C for 24 h.
8. Stop the digest by freezing, acidifying the sample with TFA, or by injecting onto a reverse-phase HPLC system.

3.1.1.3. Digestion of Proteins in Solution with Pepsin

Although the very broad specificity of pepsin hinders its routine use for comparative peptide mapping, its low pH optimum enables it to cleave proteins that might otherwise be intransigent. It is also applicable for digesting relatively small peptides and, particularly, for studies directed at identifying disulfide bonds (*see* **Note 5**). Although pepsin cleaves preferentially between adjacent aromatic or leucine residues, it has been shown to cleave at either the NH_2- or COOH-terminal side of any amino acid, except proline. A typical digestion procedure follows:

1. Dissolve the dried protein in 100 µL 5% formic acid.
2. Add pepsin at a 1:50, enzyme:protein, (wt:wt) ratio.
3. Incubate the sample at room temperature for 1–24 h with the time of incubation being dependent on the desired extent of digestion.
4. Dry the digest in a SpeedVac prior to dissolving in 0.05% TFA and immediately injecting onto a reverse-phase HPLC system.

3.1.2. Digestion of Proteins in SDS Polyacrylamide Gels

3.1.2.1. Sample Preparation for In Gel Digestion

As in the case of samples destined for in-solution digests, care must be exercised in preparing samples for SDS-PAGE so that sample losses are minimized (*see* **Note 6**) and so that the final ratio of protein/gel matrix is as high as possible (*see* **Note 7**). Although prior carboxymethylation does not appear to be essential with in-gel digests, the presence of greater than 10–20% carbohydrate (by weight) often appears to hinder cleavage significantly (*see* **Note 3**). Samples that have been purified by SDS-PAGE can be digested directly in the gel matrix, thereby eliminating the need for electroelution or electroblotting of the intact protein from the gel. SDS PAGE-separated proteins destined for in-gel digestion should be stained with 0.1% Coomassie blue in 50% methanol, 10% acetic acid for 1 h prior to destaining with 50% methanol, 10% acetic acid for a minimum of 2 h (*see* **Note 8**). Alternatively, when sufficient sample is available and the band of interest is well separated, a guide lane in the gel can be stained and the protein of interest excised from the gel using this guide. However, in this instance, the gel still must be exposed to 50% methanol, 10% acetic acid for 3 h (to ensure adequate removal of SDS) prior to excising the protein band of interest.

The % polyacrylamide gel that is used is determined by the size of the protein. Proteins >100 kDa are typically electrophoresed in 7–10% polyacrylamide gels, whereas smaller proteins are electrophoresed in 10–17.5% polyacrylamide.

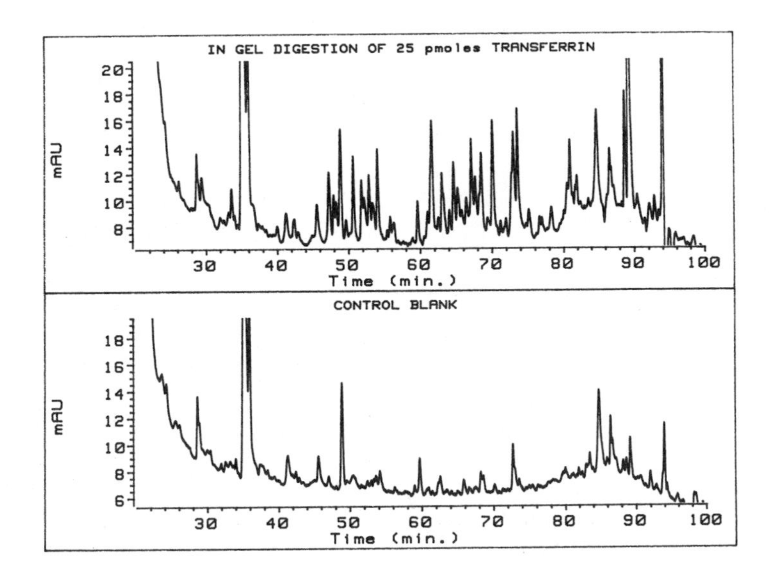

Fig. 1. In-gel lysyl endopeptidase digestion of transferrin. Following SDS-PAGE of 25 pmol of transferrin in a 12.5% polyacrylamide gel, the gel was stained and destained as described in **Subheading 3.** The protein band was then excised, along with a control blank, and digested in the gel with lysyl endopeptidase as described in **Subheading 3.** Peptides were chromatographed on a Vydac C-18, 2.1 × 250 mm reverse-phase column that was eluted at a flow rate of 150 µL/min as described in Chapter 72. A comparison of the transferrin digest (*top profile*) with the control digest (*bottom profile*), which was carried out on a blank section of gel, indicates that the digest proceeded well.

Although the following procedure, which is a modification of the Rosenfeld et al. *(4)* procedure as reported in Williams and Stone *(3)*, is capable of succeeding with as little as 25 pmol protein (e.g., *see* **Fig. 1**), the minimum amount of protein recommended for in gel digestion is 50–100 pmol at a minimum protein to gel density (after staining/destaining) of 0.05 µg/mm^3 (*see* **Notes 7** and **9**). The protein of interest, along with a blank section of gel that serves as a control, are excised using a razor blade and tweezers. To prevent going ahead with insufficient protein, we recommend that a 10–15% section of the gel band of interest be subjected to hydrolysis/ion exchange amino acid analysis to quantitate the amount of protein remaining (*see* **Note 10**).

3.1.2.2. In-Gel Digestion of Proteins with Trypsin and Lysyl Endopeptidase

1. Determine the approximate volume of gel to be digested (length × width × thickness).
2. Cut the gel band(s) containing the protein of interest into approx 1 × 2 mm pieces, and place in an Eppendorf tube. Repeat for the "blank" section of gel.
3. Add 150 µL (or more, if necessary to cover the gel pieces) 50% CH_3CN/0.1 *M* Tris-HCl, pH 8.0, to the gel pieces.
4. Wash for 15 min at room temperature on a rocker table.
5. Remove wash.
6. Semidry the washed gel pieces in a SpeedVac to approx 25–50% of original volume.
7. Make up the enzyme solution by diluting 5 µL 0.1 mg/mL enzyme stock solution with 10 µL 0.1 *M* Tris-HCl, pH 8.0, 0.1% Tween 20 for every 15 mm^3 gel that is to be digested.
8. Rehydrate gel pieces with the enzyme solution from **step 7**, which should be equal in volume to that of the gel pieces and should provide a final enzyme ratio of about 0.5 µg/15-mm^3 gel volume.

9. If the gel pieces are not totally immersed in the enzyme solution, add an additional volume of 0.1 *M* Tris-HCl, pH 8.0, 0.1% Tween 20 to submerge the gel pieces totally.
10. Incubate at 37°C for 24 h.
11. After incubation, add a volume of 0.1 *M* Tris-HCl, pH 8.0/60% CH_3CN equal to the total volume of buffer added in **steps 8** and **9**.
12. Estimate the total volume of the sample plus gel.
13. Calculate the volume of 45 m*M* DTT needed to give a final (DTT) of about 1 m*M* in the sample.
14. Add the above volume of 45 m*M* DTT and then incubate at 50°C for 20 min.
15. Remove samples from the incubator; cool to room temperature and add an equal volume of 100 m*M* IAA or 200 m*M* methyl 4-nitrobenzene sulfonate (MNS) (*see* **Note 11**) as the volume of DTT added in **step 14**.
16. For IAA alkylation, incubate at room temperature in the dark for 20 min; for MNS treatment, incubate at 37°C for 40 min.
17. Extract peptides by adding at least 100 µL 0.1% TFA, 60% CH_3CN (or, if it is greater, a volume equal to the gel volume estimated in **step 2**), and shake on a rocker table at room temperature for at least 40 min.
18. Sonicate for 5 min in a water bath sonicator.
19. Remove and save the supernatant, which contains the released peptides, and repeat **steps 17** and **18**.
20. SpeedVac dry the combined washes from **step 19**.
21. Redissolve the dried samples in 80 µL H_2O and bring to a final volume of 110 µL.
22. Filter through a Millipore Ultrafree-MC 0.22-µm filter unit.
23. The sample is now ready for reverse-phase HPLC peptide separation as described in Chapter 71.

4. Notes

1. In many instances, large losses occur during the final purification steps when the protein concentrations are invariably lower. Hence, although ultrafiltration or dialysis of a 5 mg/mL crude solution of a partially purified enzyme may lead to nearly 100% recovery of activity, similar treatment of a 25 µg/mL solution of the purified protein might well lead to significant, if not total loss of activity owing to nonspecific adsorption. Similarly, the effectiveness of organic and acid-precipitation procedures often decreases substantially as the final protein concentration is decreased below about 100 µg/mL. Whenever possible, therefore, the final purification step should be arranged such that the resulting protein solution is as concentrated as possible and, ideally, can simply be dried in a SpeedVac prior to enzymatic digestion. In this regard, it should be noted that a final NaCl concentration of 1 *M* does not significantly affect the extent of trypsin digestion *(7)*. When it is necessary to carry out an organic or acid precipitation to remove salts or detergents, the protein should first be dried in a SpeedVac (in the 1.5-mL tube in which it will ultimately be digested) prior to redissolving or suspending in a minimum volume of water and then adding the acetone or TCA. In this way, the protein concentration will be as high as possible during the precipitation and losses will be minimized. Two common contaminants that are extremely deleterious to enzymatic cleavage are detergents (as little as 0.005% SDS will noticeably decrease the rate of tryptic digests carried out in the presence of 2 *M* urea *[7]*) and ampholines. Since detergent removal is often associated with protein precipitation, and since many detergents (such as SDS) form large micelles, which cannot be effectively dialyzed, it is usually preferable to extract the detergent from the protein (that has been dried in the tube in which it will be digested) rather than to dialyze it away from the protein.

In the case of ampholines, our experience is that even prolonged dialysis extending over several days with a 15,000-Dalton cutoff membrane is not sufficient to decrease the ampholine concentration to a level that permits efficient trypsin digestion. Rather, the only effective methods we have found for complete removal of ampholines are TCA precipitation, hydrophobic or reverse-phase chromatography, or SDS-PAGE followed by staining and destaining.

2. One of the most common causes of "failed" digests is that the amount of protein being subjected to digestion has been overestimated. Often this is because of the inaccuracy of dye binding and colorimetric assays. For this reason, we recommend that an aliquot of the sample be taken for hydrolysis and amino acid analysis prior to digestion. The aliquot for amino acid analysis should be taken either immediately prior to drying the sample in the tube in which it will be digested or after redissolving the sample in 8 *M* urea, 0.4 *M* NH_4HCO_3. Although up to 10 µL of 8 *M* urea is compatible with acid hydrolysis/ion-exchange amino acid analysis, this amount of urea may not be well tolerated by PTC amino acid analysis. Hence, in the latter case, the amino acid analysis could be carried out prior to drying and redissolving the sample in urea. Although it is possible to succeed with less material, to ensure a high probability of success, we recommend that a minimum of 50–100 pmol protein be digested. Typically, 10–15% of this sample would be taken for amino acid analysis. In the case of a 25-kDa protein, the latter would correspond to only 0.125–0.188 µg protein being analyzed. When such small amounts of protein are being analyzed, it is important to control for the ever-present background of free amino acids that are in buffers, dialysis tubing, plastic tubes and tips, and so forth. If sufficient protein is available, aliquots should be analyzed both before (to determine the free amino acid concentration) and after hydrolysis. Alternatively, an equal volume of sample buffer should be hydrolyzed and analyzed and this concentration of amino acids should then be subtracted from the sample analysis.
3. Since many native proteins are resistant to enzymatic cleavage, it is usually best to denature the protein prior to digestion. Although some proteins may be irreversibly denatured by heating in 8 *M* urea (as described in the above protocol), this treatment is not sufficient to denature transferrin. In this instance, prior carboxymethylation, which irreversibly modifies cysteine residues, brings about a marked improvement in the resulting tryptic peptide map *(7)*. Another advantage of carboxymethylating the protein is that this procedure enables cysteine residues to be identified during amino acid sequencing. Cysteine residues have to be modified in some manner prior to sequencing to enable their unambiguous identification. Under the conditions that are described in **Subheading 3.**, the excess dithiothreitol and iodoacetic acid do not interfere with subsequent digestion. Although carboxymethylated proteins are usually relatively insoluble, the 2 *M* urea that is present throughout the digest is frequently sufficient to maintain their solubility. However, even in those instances where the carboxymethylated protein precipitates following dilution of the 8 *M* urea to 2 *M*, trypsin and chymotrypsin will usually still provide complete digestion. Often, the latter is evidenced by clearing of the solution within a few minutes of adding the enzyme.

   If carboxymethylation is insufficient to bring about complete denaturation of the substrate, an alternative approach is to cleave the substrate with cyanogen bromide (1000-fold molar excess over methionine, 24 h at room temperature in 70% formic acid). The resulting peptides can then either be separated by SDS-PAGE (since they usually do not separate well by reverse-phase HPLC) or, preferably, they can be enzymatically digested with trypsin or lysyl endopeptidase and then separated by reverse phase HPLC. If this approach fails, the protein may be digested with pepsin, which, as described above, is carried out under very acidic conditions or can be subjected to partial acid cleavage *(8)*. However, the disadvantage of these later two approaches is that they produce an extremely complex mixture of overlapping peptides. Finally, extensive glycosylation (i.e., typically >10–20% by weight) can also hinder enzymatic cleavage. In these instances, it is usually best to remove the carbohydrate prior to beginning the digest. In the case of in gel digests, this may often be best carried out immediately prior to SDS-PAGE, which thus prevents loss (owing to insolubility) of the deglycosylated protein and effectively removes the added glycosidases.

4. Every effort should be made to use as a high substrate and enzyme concentration as possible to maximize the extent of cleavage. Although the traditional 1:25, weight:weight ratio of enzyme to substrate provides excellent results with milligram amounts of protein, it will often fail to provide complete digestion with low microgram amounts of protein. For instance, using the procedures outlined above, this weight:weight ratio is insufficient to provide complete digestion when the substrate concentration falls below about 20 µg/mL *(7)*. The only reasonable alternative to purifying additional protein is either to decrease the final digestion volume below the 80-µL value used above or to compensate for the low substrate concentration by increasing the enzyme concentration. The only danger in doing this, of course, is the increasing risk that some peptides may be isolated that are autolysis products of the enzyme. Assuming that only enzymes, such as trypsin, chymotrypsin, lysyl endopeptidase, and Protease V8 are used, whose sequences are known, it is usually better to risk sequencing a peptide obtained from the enzyme (which can be quickly identified via a data base search) than it is to risk incomplete digestion of the substrate. Often, protease autolysis products can be identified by comparative HPLC peptide mapping of an enzyme (i.e., no substrate) control that has been incubated in the same manner as the sample and by subjecting candidate HPLC peptide peaks to matrix-assisted laser desorption mass spectrometry prior to sequencing. The latter can be extremely beneficial both in identifying (via their mass) expected protease autolysis products and in ascertaining the purity of candidate peptide peaks prior to sequencing. To promote more extensive digestion, we have sometimes used enzyme:substrate mole ratios that approach unity. If there is any doubt concerning the appropriate enzyme concentration to use with a particular substrate concentration, it is usually well worth the effort to carry out a control study (using a similar concentration and size standard protein) where the extent of digestion (as judged by the resulting HPLC profile) is determined as a function of enzyme concentration.
5. One approach to identifying disulfide-linked peptides is to comparatively HPLC peptide map a digest that has been reduced/carboxymethylated vs one that has only been carboxymethylated (thus leaving disulfide-linked peptides intact). In this instance, pepsin offers an advantage in that the digest can be carried out under acidic conditions where disulfide interchange is less likely to occur.
6. As in the case of in-solution digests, care must be exercised to guard against sample loss during final purification. Whenever possible, SDS (0.05%) should simply be added to the sample prior to drying in a SpeedVac and subjecting to SDS-PAGE. Oftentimes, however, if the latter procedure is followed, the final salt concentration in the sample will be too high (i.e., >1 *M*) to enable it to be directly subjected to SDS-PAGE. In this instance, the sample may either be concentrated in a SpeedVac and then precipitated with TCA (as described above) or it may first be dialyzed to lower the salt concentration. If dialysis is required, the dialysis tubing should be rinsed with 0.05% SDS prior to adding the sample, which should also be made 0.05% in SDS. After dialysis versus a few micromolar NH_4HCO_3 containing 0.05% SDS, the sample may be concentrated in a SpeedVac and then subjected to SDS PAGE (note that samples destined for SDS PAGE may contain several % SDS).

Another approach that works extremely well is to use an SDS polyacrylamide gel containing a funnel-shaped well that allows samples to be loaded in volumes as large as 300 µL *(9)*.

7. In general, the sample should be run in as few SDS-PAGE lanes as possible to maximize the substrate concentration and to minimize the total gel volume present during the digest. Whenever possible, a 0.5–0.75-mm thick gel should be used and at least 0.5–0.75 µg of the protein of interest should be run in each gel lane so the density of the protein band is at least 0.05 µg/mm^3. As shown in **Table 1**, the in-gel procedure has an average success rate of close to 98%. The least amount of an unknown protein that we have successfully digested and sequenced was 5 pmol of a protein that was submitted at a comparatively high density of 0.1 µg/mm$_3$. We believe the latter factor contributed to the success of this sample.

   Although several enzymes (i.e., trypsin, chymotrypsin, lysyl endopeptidase, and endoproteinase GluC) may be used with the in gel procedure, nearly all our experience has been with trypsin. In general, we recommend using 0.5 µg enzyme/15 mm^3 of gel with the only caveat being that we use a corresponding lower amount of enzyme if the mole ratio of protease/substrate protein would exceed unity.
8. Since high concentrations of Coomassie blue interfere with digestion, it is best to use the lowest Coomassie blue concentration possible and to stain for the minimum time necessary to visualize the bands of interest. In addition, the gel should be well destained so that the background is close to clear.
9. As shown in **Table 1**, the in-gel digestion procedure outlined in this chapter appears to have a success rate of nearly 98% with unknown proteins. Surprisingly, this success rate does not appear to vary significantly over the range of protein extending from an average of about 37–323 pmol. Obviously, however, the quality of the resulting sequencing data is improved by going to the higher levels, and this is probably reflected by the increased number of positively called residues/peptide sequenced that was observed at the >200 pmol level. One critical fact that has so far not been noted is that 71% of the proteins on which the data in **Table 1** are based were identified via data base searching of the first peptide sequence obtained. Hence, all internal peptide sequences obtained should be immediately searched against all available databases to determine if the protein that has been digested is unique.
10. Although most estimates of protein amounts are based on relative staining intensities, our data suggest there is a 5–10-fold range in the relative staining intensity of different proteins. Obviously, when working in the 50–100 pmol level, such a 5–10-fold range could well mean the difference between success and failure. Hence, we routinely subject an aliquot of the SDS-PAGE gel (usually 10–15% based on the length of the band) to hydrolysis and ion-exchange amino acid analysis prior to proceeding with the digest. As these analyses will often contain <0.5 µg protein, it is important that a "blank" section of gel, which is about the same size as that containing the sample, be hydrolyzed and analyzed as a control to correct for the background level of free amino acids that are usually present in polyacrylamide gels. Based on samples taken from about 20 different gels submitted by users of the W. M. Keck Foundation Biotechnology Resource Laboratory, the background ranges up to about 0.2 µg and averages about 0.09 µg in these 10–15% aliquots. Although amino acid analyses on gel slices are complicated by this background level of free amino acids and by the fact that some amino acids (i.e., glycine, histidine, methionine, and arginine) usually cannot be quantitated following hydrolysis of gel slices, these estimates are still considerably more accurate than are estimates based on relative staining intensities. In those instances where amino acid analysis indicates <50 pmol protein, the stained band may be stored frozen while additional material is purified.

Table 1
Summary of 88 In-Gel Digests Carried Out in the W. M. Keck Foundation Biotechnology Resource Laboratory at Yale University

Parameter	<50 pmol	51–100 pmol	101–200 pmol	>200 pmol	Total
Number of proteins digested	7	19	34	28	88
Average mass of protein, kDa	67	53	56	63	58
Average amount of protein digested, pmol[a]	37	77	143	323	177
Average density of protein band, μg/mm^3 gel volume	0.117	0.178	0.255	0.489	0.302
Average amount of protease added, μg	0.943	1.11	1.48	3.30	
Total number of peptides sequenced	14	31	74	53	172
Peptides sequenced that provided >6 positively called residues, %	80	77	85	87	84
Average initial yield, %[b]	21.1	8.1	9.0	12.3	10.8
Median initial yield, %[b]	10.2	3.2	6.1	10.0	7.1
Average number of positively called residues/peptide sequenced	12.5	11.0	12.6	14.5	12.9
Overall average background, μg[c]	—	—	—	—	0.09
Overall digest success rate[d]	100	94.7	100	96.4	97.7

[a]Data on the amount of protein digested and on the density of the protein band are based on hydrolysis/amino acid analysis of 10–15% of the stained band.

[b]Calculated as the yield of Pth-amino acid sequencing at the first or second cycle compared to the amount of protein that was digested. Since initial yields of purified peptides directly applied onto the sequencer are usually about 50%, the actual % recovery of peptides from the above digests is probably twice the % initial yields that are given.

[c]Based on hydrolysis/amino acid analysis of a 10–15% aliquot of a piece of SDS polyacrylamide gel that is equal in volume to that of the sample band and that is from an area of the sample gel that should not contain protein.

[d]As judged by the ability to sequence a sufficient number of internal residues to either identify the protein via database searches or to allow oligonucleotide probes/primers to be synthesized to enable cDNA cloning studies to proceed. The total number of positively identified residues for those proteins that were scored as a success and that were not identified via database searches varied (at the request of the submitting investigator) from 9 to 56 residues.

11. Modification of cysteine residues with MNS provides the advantage that the resulting S-methyl cysteine phenylthiohydantoin derivative elutes in a favorable position (i.e., between PTH-Tyr and PTH-Pro) using an Applied Biosystems/Perkin Elmer PTH C-18 column. Either the Applied Biosystems Premix or sodium acetate buffer system can be used, with a linear gradient extending from 10 to 38% buffer B over 27 min.

References

1. Brown, J. L. and Roberts, W. K. (1976) Evidence that approximately eighty percent of the soluble proteins from Ehrlich Ascites cells are amino terminally acetylated. *J. Biol. Chem.* **251,** 1009–1014.
2. Driessen, H. P., DeJong, W. W., Tesser, G. I., and Bloemendal, H. (1985) The mechanism of N-terminal acetylation of proteins in *Critical Reviews in Biochemistry* vol. 18 (Fasman, G. D., ed.), CRC, Boca Raton, FL, pp. 281–325.
3. Williams. K. R. and Stone, K. L. (1995) In gel digestion of SDS PAGE-separated protein: observations from internal sequencing of 25 proteins in *Techniques in Protein Chemistry VI* (Crabb, J. W., ed.), Academic, San Diego, pp. 143–152.
4. Rosenfeld, J., Capdevielle, J., Guillemot, J., and Ferrara, P. (1992) In gel digestions of proteins for internal sequence analysis after 1 or 2 dimensional gel electrophoresis. *Analyt. Biochem.* **203,** 173–179.
5. Fernandez, J., DeMott, M., Atherton, D., and Mische, S. M. (1992) Internal protein sequence analysis: enzymatic digestion for less than 10 μg of protein bound to polyvinylidene difluoride or nitrocellulose membranes. *Analyt. Biochem.* **201,** 255–264.
6. Aebersold, R. H., Leavitt, J., Saavedra, R. A., Hood, L. E., and Kent, S. B. (1987) Internal amino acid sequence analysis of proteins separated by one- or two-dimensional gel electrophoresis after in situ protease digestion on nitrocellulose. *Proc. Natl. Acad. Sci. USA* **84,** 6970–6974.
7. Stone, K. L., LoPresti, M. B., and Williams, K. R. (1990) Enzymatic digestion of proteins and HPLC peptide isolation in the subnanomole range in *Laboratory Methodology in Biochemistry: Amino Acid Analysis and Protein Sequencing* (Fini, C., Floridi, A., Finelli, V., and Wittman-Liebold, B., eds.), CRC, Boca Raton, FL, pp. 181–205.
8. Sun, Y., Zhou, Z., and Smith, D. (1989) Location of disulfide bonds in proteins by partial acid hydrolysis and mass spectrometry in *Techniques in Protein Chemistry* (Hugli, T., ed.), Academic, San Diego, pp. 176–185.
9. Lombard-Platet, G. and Jalinot, P. (1993) Funnel-well SDS-PAGE: a rapid technique for obtaining sufficient quantities of low-abundance proteins for internal sequence analysis. *BioTechniques* **15,** 669–672.

77

Enzymatic Digestion of Membrane-Bound Proteins for Peptide Mapping and Internal Sequence Analysis

Joseph Fernandez and Sheenah M. Mische

1. Introduction

Enzymatic digestion of membrane-bound proteins is a sensitive procedure for obtaining internal sequence data of proteins that either have a blocked amino terminus or require two or more stretches of sequence data for DNA cloning or confirmation of protein identification. Since the final step of protein purification is usually SDS-PAGE, electroblotting to either PVDF or nitrocellulose is the simplest and most common procedure for recovering protein free of contaminants (SDS, acrylamide, and so forth) with a high yield. The first report for enzymatic digestion of a nitrocellulose-bound protein for internal sequence analysis was by Aebersold et al. in 1987, with a more detailed procedure later reported by Tempst et al. in 1990 *(1,2)*. Basically, these procedures first treated the nitrocellulose-bound protein with PVP-40 (polyvinyl pyrrolidone, M_r 40,000) to prevent enzyme adsorption to any remaining nonspecific protein binding sites on the membrane, washed extensively to remove excess PVP-40, and the sample was enzymatically digested at 37°C overnight. Attempts with PVDF-bound protein using the above procedures *(3,4)* give poor results and generally require >25 µg of protein. PVDF is preferred over nitrocellulose because it can be used for a variety of other structural analysis procedures, such as amino-terminal sequence analysis and amino acid analysis. In addition, peptide recovery from PVDF-bound protein is higher, particularly from higher retention PVDF (ProBlott, Westran, Immobilon P^{sq}). Finally, PVDF-bound protein can be stored dry as opposed to nitrocellulose, which must remain wet during storage and work up to prevent losses during digestion.

Enzymatic digestion of both PVDF- and nitrocellulose-bound protein in the presence of 1% hydrogenated Triton X-100 (RTX-100) buffers as listed in **Table 1** was first performed after treating the protein band with PVP-40 *(5)*. Unfortunately, the RTX-100 buffer also removes PVP-40 from the membrane, which can interfere with subsequent reverse-phase HPLC. Further studies *(6,7)* demonstrate that treatment with PVP-40 is unnecessary when RTX-100 is used in the digestion buffer. It appears that RTX-100 acts as both a blocking reagent and a strong elution reagent.

PVDF-bound proteins are visualized by staining and subsequently excised from the blot. Protein bands are immersed in hydrogenated Triton X-100 (RTX-100), which

From: *The Protein Protocols Handbook, 2nd Edition*
Edited by: J. M. Walker © Humana Press Inc., Totowa, NJ

Table 1
Digestion Buffers Recipes for Various Enzymes

Enzyme	Digestion buffer	Recipe (using RTX-100)[a]	Recipe (using OGP)[c]	Comments
Trypsin or Lys-C	1% Detergent/ 10% acetonitrile/ 100 m*M* Tris, pH 8.0	100 µL 10% RTX-100 stock, 100 µL acetonitrile, 300 µL HPLC-grade water, and 500 µL 200 m*M* Tris stock[b]	10 mg OGP, 100 µL acetonitrile, 400 µL HPLC-grade water, and 500 µL 200 m*M* Tris stock[b]	Detergent prevents enzyme adsorption to membrane and increases recovery of peptides
Glu-C	1% Detergent/ 100 m*M* Tris, pH 8.0	100 µL 10% RTX-100 stock, 400 µL HPLC-grade water, and 500 µL 200 m*M* Tris stock[b]	10 mg OGP, 500 µL HPLC-grade water, and 500 µL 200 m*M* Tris stock[b]	Acetonitrile decreases digestion efficiency of Glu-C
Clostripain	1% Detergent/ 10% acetonitrile/ 2 m*M* DTT/ 1 m*M* $CaCl_2$/ 100 m*M* Tris, pH 8.0	100 µL 10% RTX-100 stock, 100 µL acetonitrile, 45 µL 45 m*M* DTT, 10 µL 100 m*M* $CaCl_2$, 245 µL HPLC-grade water, and 500 µL 200 m*M* Tris stock[b]	10 mg OGP, 100 µL acetonitrile, 45 µL 45 m*M* DTT, 10 µL 100 m*M* $CaCl_2$, 345 µL HPLC-grade water, and 500 µL 200 m*M* Tris stock[b]	DTT and $CaCl_2$ are necessary for Clostripain activity

[a]Hydrogenated Triton X-100 (RTX-100) as described in Tiller et al. ***(13)***.
[b]200 m*M* Tris stock (pH 8.0) is made up as follows: 157.6 mg Tris-HCl and 121.1 mg trizma base to a final volume of 10 mL with HPLC-grade water.
[c]Octyl glucopyranoside can be substituted for RTX-100.

acts as both a reagent for peptide extraction and a blocking reagent for preventing enzyme adsorption to the membrane during digestion. Remaining cystine bonds are reduced with dithiotreitol (DTT) and carboxyamidomethylated with iodoacetamide. After incubation with the enzyme of choice, the peptides are recovered in the digestion buffer. Further washes of the membrane remove the remaining peptides, which can be analyzed by microbore HPLC. Purified peptides can then be subjected to automated sequence analysis.

Additional studies have been reported that have enhanced this procedure. Best et al. *(8)* have reported that a second aliquot of enzyme several hours later improves the yield of peptides. Reduction and alkylation of cysteine is possible directly in the digestion buffer, allowing identification of cysteine during sequence analysis *(9)*. Finally, octyl- or decylglucopyranoside can be substituted for RTX-100 in order to obtain cleaner mass spectrometric analysis of the digestion mixture *(10)*.

2. Materials

The key to success with this procedure is cleanliness. Use of clean buffers, tubes, and staining/destaining solutions, as well as using only hydrogenated Triton X-100 as opposed to the nonhydrogenated form, greatly reduces contaminant peaks obscured during reverse-phase HPLC. A corresponding blank piece of membrane must always be analyzed at the same time as a sample is digested, as a negative control. Contaminants can occur from may sources and particularly protein contaminants are a cocnern. Human Kertain may contaminate samples if gloves are not worn. Proteins from Western blotting can contaminate the PVDF membrane if previously used dishes are used for staining. UV absorbing contaminants can arise from dirty tubes and sub quality detergents. All solutions should be prepared with either HPLC-grade water or double-glass-distilled water that has been filtered through an activated charcoal filter, and passed through a 0.22-μm filter *(11)*.

2.1. Preparation of the Membrane-Bound Sample

Protein should be analyzed by SDS-PAGE or 2D-IEF using standard laboratory techniques. Electrophoretic transfer of proteins to the membrane should be performed in a full immersion tank rather than a semidry transfer system to avoid sample loss and obtain efficient transfer *(12)*. PVDF membranes with higher protein binding capacity such as Immobilon P^{sq} (Millipore, Bedford, MA), Problott (Applied Biosystems, Foster City, CA), and Westran (Bio-Rad, Hercules, CA), are preferred owing to greater protein recovery on the blot, although all types of PVDF and nitrocellulose can be used with this procedure. The following stains are compatible with the technique: Ponceau S, Amido black, india ink, and chromatographically pure Coomassie brilliant blue with a dye content >90%. **A blank region of the membrane should be excised to serve as a negative control.**

2.2. Enzymatic Digestion Buffers

Digestion buffer should be made as described in Table 1. Make up 1 mL of buffer at a time and store at –20°C for up to 1 wk. Hydrogenated Triton X-100 (RTX-100) (protein grade, cat. # 648464, Calbiochem, LaJolla, CA) is purchased as a 10% solution, which should be stored at –20°C. Note: Only hydrogenated Triton X-100 should be used

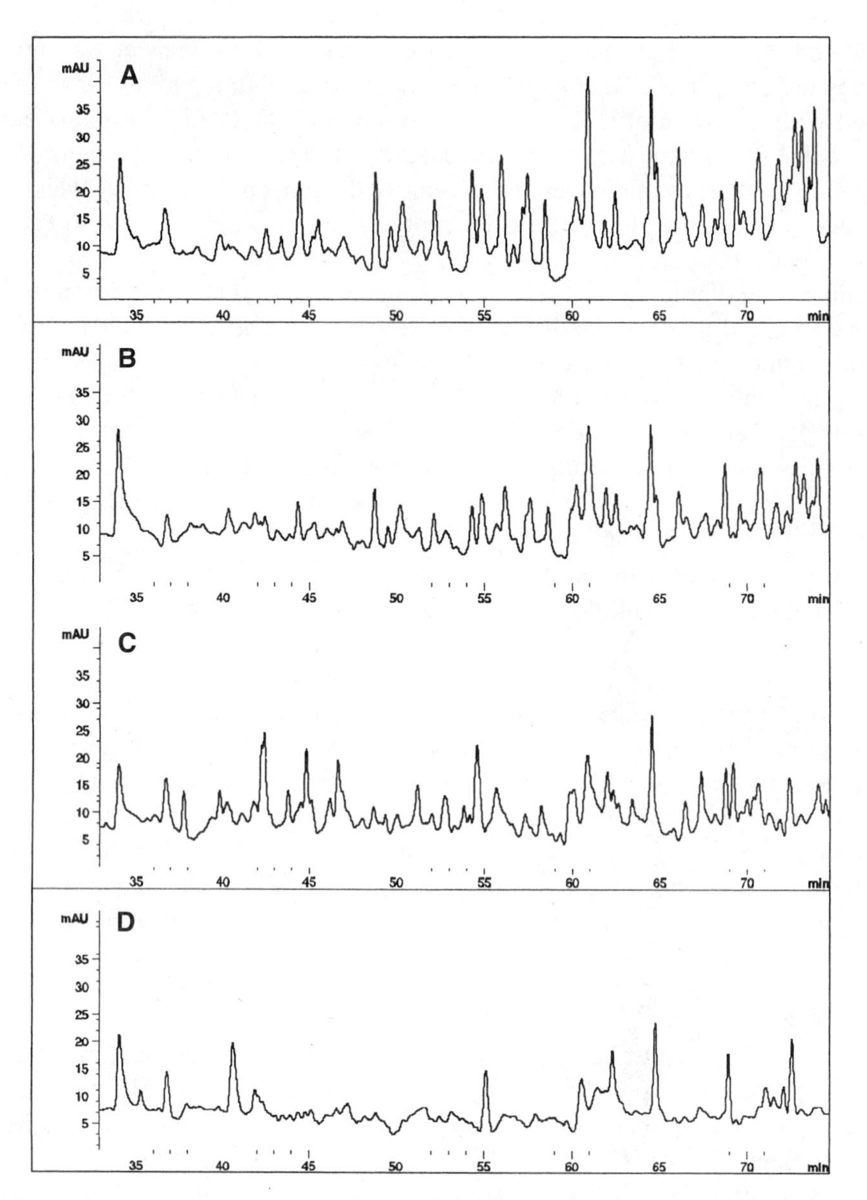

Fig. 1. Peptide maps of trypsin digestion of beta galactosidase bound to PVDF. Panels A–D represent varying amounts of proteins, 40 pmol, 20 pmol, 10 pmol, and 5 pmol respectively. Proteins were analyzed by SDS-PAGE, transferred to PVDF, and stained with Coomassie Brilliant Blue G-250.

since UV-absorbing contaminants are present in ordinary Triton X-100, making identification of peptides on subsequent HPLC impossible (*see* **Fig. 1**). Alternately, octyl glucopyranoside (OGP) (Ultrol-grade, Calbiochem) or decyl glucopyranoside (DGP) (Ultrol-grade, Calbiochem) can be substituted for RTX-100 as described in **Table 1**.

2.3. Reduction and Carboxyamidomethylation

1. 45 m*M* DTT: bring 3.5 mg DTT (Ultrol-grade, Calbiochem) up in 500 µL HPLC-grade water. This can be stored at –20°C for up to 3 mo.

2. 100 m*M* iodoacetamide solution: bring 9.25 mg iodoacetamide (reagent-grade) up in 500 µL HPLC-grade water. This solution must be made fresh just prior to use. Dry DTT and iodoacetamide should be stored at 4°C or –20°C.

2.4. Enzyme Solutions and Inhibitors

Enzymes should be stored as small aliquots at –20°C, and made up as 0.1 µg/µL solutions immediately before use. These aliquots can be stored for at least 1 mo at –20°C without significant loss of enzymatic activity.

1. Trypsin (25 µg, sequencing-grade, Boehringer Mannheim, Indianapolis, IN): Solubilize trypsin in 25 µL of 0.01% trifluoroacetic acid (TFA), and let stand ~10 min. Aliquot 5 µL (5 µg) quantities to clean microcentrifuge tubes, dry in a SpeedVac, and store at –20°C. Reconstitute the dry enzyme in 50 µL 0.01% TFA for a 0.1 µg/µL working solution, which is good for 1 d. **Trypsin cleaves at arginine and lysine residues.**
2. Endoproteinase Lys-C (3.57 mg, Wako Pure Chemicals, Osaka, Japan): Solubilize enzyme in 1000 µL HPLC-grade water, and let stand ~10 min. Make nine 100-µL quantities to clean tubes, and store at –20°C. Disperse remaining 100 µL into 20 × 5 µL aliquots (17.85 µg each), and store at –20°C. When needed, take one 5-µL aliquot and add 173 µL of HPLC-grade water to establish a 0.1 µg/µL working solution, which can be used for up to 1 wk if stored at –20°C between uses. When 5-µL aliquots are used up, disperse another 100-µL aliquot. **Endoptoteinase Lys-C cleaves only at lysine residues.**
3. Endoproteinase Glu-C (50 µg, sequencing-grade, Boehringer Mannheim) : Solubilize in 100 µL of HPLC-grade water, and let stand ~10 min. Aliquot 10-µL (5 µg) quantities to clean tubes, dry in a SpeedVac, and store at –20°C. Reconstitute the enzyme in 50 µL HPLC-grade water for a 0.1 µg/µL solution, which is good for only 1 d. **Under these conditions, endoproteinase Glu-C cleaves predominantly at glutamic acid residues, but can sometimes cleave at aspartic acid residues.**
4. Clostripain (20 µg, sequencing-grade, Promega, Madison, WI): Solubilize enzyme in 200 µL of manufacturer's supplied buffer for a concentration of 0.1 µg/µL. Enzyme solution can be stored at –20°C for 1 mo. **Clostripain cleaves at arginine residues only.**
5. 1% Diisopropyl fluorophosphate (DFP) solution: DFP is a dangerous neurotoxin that must be handled with double gloves in a chemical hood. Please follow all precautions listed with this chemical. Add 10 µL of DFP to 990 µL absolute ethanol in a capped microcentrifuge tube. Store at –20°C.

3. Method

3.1. Preparation of the Membrane-Bound Sample

Protein should be electrophoresed in one or two dimensions, followed by electrophoretic transfer to the membrane and visualization of the bands according to the following suggestions.

1. Electroblotting of proteins will be most efficient with a tank transfer system, rather than a semidry system. Concentrate as much protein into a lane as possible; however, if protein resolution is a concern, up to 5 cm^2 of membrane can be combined for digestion (*see* **Notes 3** and **5**).
2. Stain PVDF membrane with either Ponceau S, Amido black, india ink, or chromatographically pure Coomassie brilliant blue with a dye content of 90%. Destain the blot until the background is clean enough to visualize the stained protein. Complete destaining of the blot is unnecessary, but at least three washes (~5 min) with distilled water should be done to remove excess acetic acid, which is used during destaining (*see* **Note 5**).

3. Excise protein band(s), and place into a clean 1.5-mL microcentrifuge tube. In addition, excise a blank region of the membrane approximately the same size as the protein blot to serve as a negative control (*see* **Note 4**).
4. Air dry PVDF-bound protein dry at room temperature and store at –20°C or 4°C.

3.2. Digestion of the Membrane-Bound Protein

NOTE: Gloves should be worn during all steps to avoid contamination of sample with skin keratin.

1. Place ~100 µL of HPLC-grade water onto a clean glass plate, and submerge the membrane-bound protein into the water. Transfer the wet membrane to a dry region of the plate and with a clean razor blade, cut the membrane first lengthwise into 1-mm wide strips, and then perpendicular so that the membrane pieces are 1 × 1 mm. Treat the negative control under the same conditions as the sample. Keeping the membrane wet will simplify manipulation of the sample as well as minimize static charge, which could cause PVDF to "jump" off the plate. The 1 × 1 mm pieces of membrane will settle to the bottom of the tube and require less digestion buffer to immerse the membrane completely (*see* **Note 7**).
2. Slide the cut membrane onto the forceps with the razor blade, and return it to a clean 1.5-mL microcentrifuge tube. Use the cleanest tubes possible to minimize contamination during peptide mapping. Surprisingly, many UV-absorbing contaminants can be found in microcentrifuge tubes, and this appears to vary with supplier and lot number. Tubes can be cleaned by rinsing with 1 mL of HPLC-grade methanol followed by 2 rinses of 1 mL HPLC-grade water prior to adding the protein band.
3. Add 50 µL of the appropriate digestion buffer (**Table 1**), and vortex thoroughly for 10–20 s. Optionally, add 50 µL digestion buffer to an empty microcentrifuge tube to serve as a digestion blank for HPLC analysis. The amount of digestion buffer can be increased or decreased depending on the amount of membrane; however, the best results will be obtained with a minimum amount of digestion buffer. PVDF membrane will float in the solution at first, but will submerge after a short while, depending on the type and amount of PVDF .
4. Add 5 µL of 45 mM DTT, vortex thoroughly for 10–20 s, seal the tube cap with parafilm, and incubate at 55°C for 30 min. DTT will reduce any remaining cystine bonds. DTT should be of the highest grade to reduce uv absorbing contaminants that might interfere with subsequent peptide mapping.
5. Allow the sample to cool to room temperature. Add 5 µL of 100 mM iodoacetamide, vortex thoroughly for 10–20 s, and incubate at room temperature for 30 min in the dark. Iodoacetamide alkylates cysteine residues to generate carboxyamidomethyl cysteine, allowing identification of cysteine during sequence analysis. Allowing the sample to cool prior to adding iodoacetamide and incubating at room temperature are necessary to avoid side reaction to other amino acids.
6. Add enough of the required enzyme solution to obtain an estimated enzyme to substrate ratio of 1:10 (w/w) and vortex thoroughly for 10–20 s. Incubate the sample (including digestion buffer blank) at 37°C for 22–24 h. The amount of protein (substrate) can be estimated by comparison of staining intensity to that of known quantities of stained standard proteins. The 1:10 ratio is a general guideline. Ratios of 1:2 through 1:50 can be used without loss of enzyme efficiency or peptide recovery. An enzyme should be selected that would likely produce peptides of >10 amino acids long. Amino acid analysis of the protein would be informative for estimating the number of cleavage sites. A second aliquot of enzyme can be added after 4–6 h (*see* **Note 6**).

3.3. Extraction of the Peptides

1. After digestion, vortex the sample for 5–10 s, sonicate for 5 min by holding in a sonicating water bath, spin in a centrifuge (~1800g) for 2 min, and transfer the supernatant to a separate vial that will be used directly for HPLC analysis.
2. Add a fresh 50 µL of digestion buffer to the sample, repeat **step 11**, and pool the supernatant with the original buffer supernatant.
3. Add 100 µL 0.1 % TFA to the sample, and repeat **step 1**. The total volume for injection onto the HPLC is 200 µL. Most of the peptides (~80%) are recovered in the original digestion buffer; however, these additional washes will ensure maximum recovery of peptides from the membrane.
4. Terminate the enzymatic reaction by either analyzing immediately by HPLC or adding 2 µL of the DFP solution.

   CAUTION: DFP is a dangerous neurotoxin and must be handled with double gloves under a chemical fume hood. Please follow all precautions listed for this chemical.

3.4. Analysis of Samples by Reverse-Phase HPLC and Storage of Peptide Fractions

1. Prior to reverse-phase HPLC, inspect the pooled supernatants for small pieces of membrane or particles that could clog the HPLC tubing. If membrane or particles are observed, either remove the membrane with a clean probe (such as thin tweezers, a thin wire, thin pipet tip, and so forth), or spin in a centrifuge for 2 min and transfer the sample to a clean vial. A precolumn filter will help increase the life of HPLC columns, which frequently have problems with clogged frits.
2. Sample is ready to be fractionated by HPLC (*see* Chapter 78). Fractions can be collected in capless 1.5-mL plastic tubes, capped, and stored at –20°C until sequenced (*see* **Note 9**). A typical fractionation is shown in **Fig. 1**.

4. Notes

1. This procedure is generally applicable to proteins that need to have their primary structure determined and offers a simple method for obtaining internal sequence data in addition to amino terminal sequence analysis data. The procedure is highly reproducible and is suitable to peptide mapping by reverse-phase HPLC. Proteins 12–300 kDa have been successfully digested with this procedure with the average size around 100 kDa. Types of proteins analyzed by this technique include DNA binding, cystolic, peripheral, and integral-membrane proteins, including glycosylated and phosphorylated species. The limits of the procedure appear to be dependent on the sensitivity of both the HPLC used for peptide isolation and the protein sequencer.
2. There are several clear advantages of this procedure over existing methods. First, it is applicable to PVDF (especially high-retention PVDF membranes), which is the preferred membrane owing to higher recovery of peptides after digestion, as well as being applicable to other structural analysis. The earlier procedures *(1,2)* have not been successful with PVDF. Second, because of the RTX-100 buffer, recovery of peptides from nitrocellulose is higher than earlier nitrocellulose procedures *(5)*. Third, the procedure is a one-step procedure and does not require pretreatment of the protein band with PVP-40. Fourth, since the procedure does not require all the washes that the PVP-40 procedures do, there is less chance of protein washout. Fifth, the time required is considerably less than with the other procedures. Overall, the protocol described here is the simplest and quickest method to obtain quantitative recovery of peptides.

3. The largest source of sample loss is generally not the digestion itself, but rather electroblotting of the sample. Protein electroblotting should be performed with the following considerations. Use PVDF (preferably a higher binding type, i.e., ProBlott, Immobilon P[sq]) rather than nitrocellulose, since peptide recovery after digestion is usually higher with PVDF *(5,8)*. If nitrocellulose must be used, e.g., protein is already bound to nitrocellulse before digestion is required, never allow the membrane to dry out since this will decrease yields. Always electroblot protein using a transfer tank system, since yields from semidry systems are not as high *(12)*. Using stains such as Ponceau S, Amido black, or chromatographically pure Coomassie brilliant blue, with a dye content >90% will increase detection of peptide fragments during reverse-phase HPLC.

   Note: Most commercial sources of Coomassie brilliant blue are extremely dirty and should be avoided. Only chromatographically pure Coomassie brilliant blue with a dye content of 90% appears suitable for this procedure.
4. The greatest source of failure in obtaining internal sequence data is not enough protein on the blot, which results in the failure to detect peptide during HPLC analysis. An indication of insufficient protein is that either the intensity of the stain is weak, i.e., cannot be seen with Amido black even though observable with india ink (about 10-fold more sensitive), or possibly detectable by radioactivity or immunostaining, but not by protein stain. Amino acid analysis, amino-terminal sequence analysis, or at the very least, comparison with stained standard proteins on the blot should be performed to help determine if enough material is present. When <10 μg of protein is present, the most problematic item is misidentification of peptides on reverse-phase HPLC owing to artifact peaks and contaminants. Although elimination of every contaminant is usually impossible, there are several strategic points and steps that can be taken to help alleviate these contaminants. A negative control of a blank region of the membrane blot (preferably from a blank lane) that is approximately the same size as the protein band will help to identify contaminants present that are associated with the membrane and digestion buffer (*see* **Fig. 1D**). The blank membrane should have gone through the same preparation steps as the sample, including electroblotting and staining, and should be analyzed by HPLC immediately before or after the sample. A positive control (membrane-bound standard protein) is generally unnecessary, but should be performed if the activity of the enzyme is in question or a new lot number of enzyme is to be used.
5. Major sources of contaminants are stains used to visualize the protein, the microcentrifuge tubes used for digestion, reagents used during digestion and extraction of peptides, and the HPLC itself. Stains are the greatest source of contaminants, and Coomassie brilliant blue in particular is a problem (*see* **Fig. 1A**). Amido black and Ponceau S are generally the cleanest, whereas chromatographically pure Coomassie brilliant blue with a dye content <90% does appear to generate less contaminants than most other commercially available Coomassie brilliant blue stains. Surprisingly, microcentrifuge tubes can produce significant artifact peaks, which seem to vary with supplier and lot number. A digestion blank of just the microcentrifuge tube should be done, since some contaminants only appear after incubation in the RTX-100 buffer. The major concern with the digestion buffer is the hydrogenated Triton X-100 (*see* **Fig. 1C**). Additional late-eluting peaks may be observed with certain lots of RTX-100, whereas other lots are completely free of UV-absorbing contaminants. HPLC-grade water or water prepared as described by Atherton *(11)* should be used for all solution preparation. An HPLC blank (gradient run with no injection) should always be performed to determine what peaks are related to the HPLC (*see* **Fig. 1E**).
6. The key to success of the procedure and quantitative recovery of peptides from both PVDF and nitrocellulose membranes is the use of hydrogenated Triton X-100 (RTX-100) in the

buffer. This should be purchased from Calbiochem as a 10% stock solution, protein-grade (cat. no. 648464). **Figure 1C** demonstrates why hydrogenated Triton X-100 should be used, since nonhydrogenated Triton X-100 has several strong UV-absorbing contaminants ***(13)***. RTX-100 acts as a block to prevent enzyme adsorption to the membrane as well as a strong elution reagent of peptides ***(6,7)***. In addition, RTX-100 does not inhibit enzyme activity or interfere with peak resolution during HPLC, as do ionic detergents, such as SDS ***(5)***. The concentration of RTX-100 can be decreased to 0.1% ***(7)***. However, with a large amount of membrane, there could be a decrease in peptide recovery. Optionally, octyl- or decylglucopyranoside can be substituted for RTX-100 with no loss in peptide recovery ***(10)***.

7. The membrane should be cut into 1 × 1 mm pieces while keeping it wet to avoid static charge buildup. The 1 × 1 mm pieces allow using the minimum volume of digestion buffer to cover the membrane. The volume of the digestion buffer should be enough to cover the membrane (~ 50 µL), but can be increased or decreased depending on the amount of membrane present. The enzyme solution should be selected based on additional knowledge of the protein, such as amino acid composition or whether the protein is basic or acidic. If the protein is a complete unknown, endoproteinase Lys-C or Glu-C would be a good choice. The enzyme-to-substrate ratio should be about 1:10; however, if the exact amount of protein is unknown, ratios of 1:2 through 1:50 will not affect the quantitative recovery of peptides. After digestion, most of the peptides are recovered in the original buffer (about 80%), and the additional washes are performed to ensure maximum peptide recovery. Microbore reverse-phase HPLC is the best isolation procedure for peptides.
8. As mentioned earlier, previous procedures ***(1–3,5)*** require pretreatment with PVP-40 to prevent enzyme adsorption to the membrane. RTX-100 is essential for quantitative recovery of peptides from the membrane; however, RTX-100 also strips PVP-40 from the membrane, resulting in a broad, large UV-absorbing contaminant that can interfere with peptide identification. The PVP-40 contaminant is not dependent on the age or lot number of PVP-40, and making fresh solutions did not help as previously suggested ***(4)***. This appears to be more of a problem with nitrocellulose and higher binding PVDF (ProBlott and Immobilon P^{sq}) than lower binding PVDF (Immobilon P), and also is dependent on the amount of membrane. The PVP-40 contaminant also appears to elute earlier in the chromatogram as the HPLC column ages, becoming more of a nuisance in visualizing peptides. Therefore, using PVP-40 to prevent enzyme adsorbtion to the membrane should be avoided.
9. There are a few considerations that should be addressed regarding peptide mapping by HPLC using the protocol described here. A precolumn filter (Upchurch Scientific) must be used to prevent small membrane particles from reaching the HPLC column, thus decreasing its life. Inspection of the pooled supernatants for visible pieces of PVDF can prevent clogs in the microbore tubing, and can be removed either with a probe, or by spinning in a centrifuge and transferring the sample to a clean vial.

   Peptide mapping by reverse-phase HPLC after digestion of the membrane-bound protein should result in several peaks on the HPLC. Representative peptide maps from trypsin digestions of human transferrin bound to PVDF and stained with Coomassie blue or amido black (Immobilon P^{sq}) is shown in Figs. 1A and B. Peptide maps should be reproducible under identical digestion and HPLC conditions. In addition, the peptide maps from proteins digested on membranes are comparable if not identical to those digested in solution, indicating that the same number of peptides are recovered from the membrane as from in solution. The average peptide recovery is generally 40–70 % based on the amount analyzed by SDS-PAGE, and 70–100 % based on the amount bound to PVDF as determined by amino acid analysis or radioactivity counting. The recovery of peptides from the membrane appears to be quantitative, and the greatest loss of sample tends to be in the electroblotting.

10. The entire procedure can be done in approx 24 h plus the time required for peptide mapping by reverse-phase HPLC. Cutting the membrane takes about 10 minutes, reduction with DTT takes 30 min, carboxyamidomethylation take another 30 min, digestion at 37°C takes 22–24 h, and extraction of the peptides requires about 20 min.

References

1. Aebersold, R. H., Leavitt, J., Saavedra, R. A., Hood, L. E., and Kent, S. B. (1987) Internal amino acid sequence analysis of proteins separated by one- or two-dimensional gel electrophoresis after *in situ* protease digestion on nitrocellulose. *Proc. Natl. Acad. Sci. USA* **84,** 6970–6974.
2. Tempst, P., Link, A. J., Riviere, L. R., Fleming, M., and Elicone, C., (1990) Internal sequence analysis of proteins separated on polyacrylamide gels at the submicrogram level: improved methods, applications and gene cloning strategies. *Electrophoresis* **11,** 537–553.
3. Bauw, G., Van Damme, J., Puype, M., Vandekerckhove, J., Gesser, B., Ratz, G. P., Lauridsen, J. B., and Celis, J. E. (1989) Protein-electroblotting and -microsequencing strategies in generating protein data bases from two-dimensional gels. *Proc. Natl. Acad. Sci. USA* **86,** 7701–7705.
4. Aebersold, R. (1993) Internal amino acid sequence analysis of proteins after in situ protease digestion on nitrocellulose, in *A Practical Guide to Protein and Peptide Purification for Microsequencing,* 2nd ed. (Matsudaira, P., ed.), Academic, New York, pp. 105–154.
5. Fernandez, J., DeMott, M., Atherton, D., and Mische, S. M. (1992) Internal protein sequence analysis: enzymatic digestion for less than 10 µg of protein bound to polyvinylidene difluoride or nitrocellulose membranes. *Analyt. Biochem.* **201,** 255–264.
6. Fernandez, J., Andrews, L., and Mische, S. M. (1994) An improved procedure for enzymatic digestion of polyvinylidene difluoride-bound proteins for internal sequence analysis. *Analyt. Biochem.* **218,** 112–117.
7. Fernandez, J., Andrews, L., and Mische, S. M. (1994) A one-step enzymatic digestion procedure for PVDF-bound proteins that does not require PVP-40, in *Techniques in Protein Chemistry V* (Crabb, J., ed.), Academic Press, San Diego, pp. 215–222.
8. Best, S., Reim, D. F., Mozdzanowski, J., and Speicher, D. W., (1994) High sensitivity sequence analysis using *in-situ* proteolysis on high retention PVDF membranes and a biphasic reaction column sequencer, in *Techniques in Protein Chemistry V* (Crabb, J., ed.), Academic, New York, pp. 205–213.
9. Atherton, D., Fernandez, J., and Mische, S. M. (1993) Identification of cysteine residues at the 10 pmol level by carboxamidomethylation of protein bound to sequencer membrane supports. *Analyt. Biochem.* **212,** 98–105.
10. Kirchner, M., Fernandez, J., Agashakey, A., Gharahdaghi, F., and Mische, S. M. (1996) in *Techniques in Protein Chemistry VII* (Marshak, O., ed.), Academic, New York (in press).
11. Atherton, D., (1989) Successful PTC amino acid analysis at the picomole level, in *Techniques in Protein Chemistry* (Hugli, T., ed.) Academic, New York, pp. 273–283.
12. Mozdzanowski, J. and Speicher, D. W. (1990) Quantitative electrotransfer of proteins from polyacrylamide gels onto PVDF membranes, in *Current Research in Protein Chemistry: Techniques, Structure, and Function* (Villafranca, J., ed.), Academic, New York, pp. 87–94.
13. Tiller, G. E., Mueller, T. J., Dockter, M. E., and Struve, W. G. (1984) Hydrogenation of Triton X-100 Eliminates Its Fluorescence and Ultraviolet Light Absorbance while Preserving Its Detergent Properties. *Analyt. Biochem.* **141,** 262–266.

78

Reverse-Phase HPLC Separation of Enzymatic Digests of Proteins

Kathryn L. Stone and Kenneth R. Williams

1. Introduction

The ability of reverse-phase HPLC to resolve complex mixtures of peptides within a few hours' time in a volatile solvent makes it the current method of choice for fractionating enzymatic digests of proteins. In general, we find that peptides that are less than about 30 residues in length usually separate based on their content of hydrophobic amino acids and that their relative elution positions can be reasonably accurately predicted from published retention coefficients *(1,2)*. Since proteins often retain some degree of folding under the conditions used for reverse-phase HPLC, the more relevant parameter in this instance is probably surface rather than total hydrophobicity. Although larger peptides and proteins may be separated on HPLC, sometimes their tight binding, slow kinetics of release, propensity to aggregate, and relative insolubility in the usual acetonitrile/0.05% trifluoroacetic acid mobile phase results in broad peaks and/or carryover to successive chromatograms. In our experience, these problems are seldom seen with peptides that are less than about 30 residues in length, which thus makes reverse-phase HPLC an ideal method for fractionating tryptic and lysyl endopeptidase digests of proteins. Although it is sometimes possible to improve a particular separation by lessening the gradient slope in that region of the chromatogram, generally, enzymatic digests from a wide variety of proteins can be reasonably well fractionated using a single gradient that might extend over 1–2 h. Another advantage of reverse-phase HPLC is its excellent reproducibility which greatly facilitates using comparative HPLC peptide mapping to detect subtle alterations between otherwise identical proteins. Applications of this approach might include identifying point mutations as well as sites of chemical and posttranslational modification and demonstrating precursor/product relationships. Finally, since peptides are isolated from reverse-phase HPLC in aqueous mixtures of acetonitrile and 0.05% TFA, they are ideally suited for subsequent analysis by matrix-assisted laser desorption mass spectrometry (MALDI-MS, *8*) and automated Edman sequencing.

From: *The Protein Protocols Handbook, 2nd Edition*
Edited by: J. M. Walker © Humana Press Inc., Totowa, NJ

2. Materials

1. HPLC system: Digests of 25 pmol to 10 nmol amounts of proteins may be fractionated on a Hewlett Packard 1090M or comparable HPLC system (*see* **Note 1** for suggestions on evaluating HPLC systems and for a general discussion of important parameters that affect HPLC reproducibility, resolution, and sensitivity) capable of generating reproducible gradients in the flowrate range extending at least from about 75 to 500 µL/min. The HP 1090M HPLC used in this study was equipped with an optical bench upgrade (HP #79891A), a 1.7-µL high-pressure microflow cell with a 0.6-cm path length, a diode array detector, a Waters Chromatography static mixer, and a 250-µL injection loop. The detector outlet was connected to an Isco Foxy fraction collector and Isco Model 2150 Peak Separator to permit collection by peaks into 1.5-mL Sarstedt capless Eppendorf tubes that were positioned in the tops of 13 × 100 mm test tubes. We have found that by applying a varying resistance in parallel to the input detector signal that enters the Model 2150 Peak Separator, it is possible to improve peak detection significantly in the <500 pmol range. This can be easily accomplished via a small "black box" that is equipped with a four-position switch and that contains four different resistors labeled and configured as follows: 50 pmol/2,000 Ω, 100 pmol/1000 Ω, 250 pmol/470 Ω, 500 pmol/220 Ω. With this arrangement, best peak detection is obtained with the following settings on the 2150 Peak Separator: input = 10 mV, peak duration = 1 min, slope sensitivity = high. To minimize the "transit time" during which a given peak is traveling from the flow cell to the fraction collector, these two components are connected with 91 cm of 75-µ id fused silica capillary tubing (*see* **Note 2**). With this configuration, the dead volume is equal to about 7.5 µL, which corresponds to a peak delay of about 6 s at a flowrate of 75 µL/min, and the drop size is sufficiently small that even extremely small peak volumes can be accurately fractionated.
2. A 5-µ particle size, 2–2.1 × 250 mm C18 column is recommended for fractionating 25–250 pmol amounts of digests, whereas 250 pmol to 10 nmol amounts may be fractionated on 3.9–4.6 mm id columns. Although many commercially available columns would undoubtedly be satisfactory (*see* **Note 3**), the two columns that are currently being used in our laboratory are the 300-Å Vydac (cat. # 218TP52 for the 2.1-mm column, Separations Group, Hesperia, CA) and the 120 Å YMC column (cat. #MCQ-112 for the 2.0-mm column, YMC, Inc., Wilmington, NC).
3. pH 2.0 Buffer system: *see* **Notes 4** and **5**.
   Buffer A: 0.06% trifluoroacetic acid (TFA) (3 mL 20% TFA/H_2O/L final volume HPLC-grade H_2O).
   Buffer B: 0.052% TFA/80% acetonitrile (2.7 mL 20% TFA/H_2O, 800 mL CH_3CN [HPLC-grade], HPLC-grade H_2O to a final volume of 1000 mL).
4. pH 6.0 Buffer system: *see* **Note 5**.
   Buffer A: 5 m*M* potassium phosphate, pH 6.0 (10 mL 0.5*M* KH_2PO_4 in a total volume of 1000 mL HPLC-grade H_2O).
   Buffer B: 80% (v/v) CH_3CN (200 mL HPLC-grade H_2O and 800 mL acetonitrile [HPLC-grade]).
5. Peptide dilution buffer: 2 *M* urea, 0.1 *M* NH_4HCO_3. Dissolve 1.2 g Pierce Sequanal Grade urea and 79 mg Baker NH_4HCO_3 in a final volume of 10 mL HPLC-grade H_2O. This solution should be made up at least weekly and stored at –20°C.
6. 0.02% (v/v) Tween 20 solution for peptide dilution: Add 2 µL polyoxyethylene-sorbitan monolaurate (Tween 20 from Sigma Chemical Co., St. Louis, MO) to 10 mL HPLC-grade H_2O.

3. Methods

3.1. HPLC Separation of Peptides

The TFA acetonitrile buffer system described in **Materials**, **item 3** is an almost universal reverse-phase solvent system owing to its low-UV absorbance, high resolution, and excellent peptide solubilizing properties. The gradient we generally use is:

Time	% B
0–60 min	2–37.5%
60–90 min	37.5–75%
90–105 min	75–98%

In the case of extremely complex digests (i.e., tryptic digests of proteins that are above about 100 kDa), the gradient times may be doubled. In general, we recommend using the lowest flowrates consistent with near-optimum resolution for the column diameter that is being used *(3,4)*. Hence, we recommend a flowrate of 75 μL/min for 2.0–2.1 mm columns and a flowrate of 0.5 mL/min for 3.9–4.6 mm columns. Immediately following their collection, all fractions are tightly capped (to prevent evaporation of the acetonitrile—*see* **Note 6**) and are then stored in plastic boxes (USA/Scientific Plastics, Part #2350-5000) at 5°C. With the reduced flowrate of 75 μL/min, the average peak detected fraction volume is about 50 μL, which is sufficiently small that, if necessary, the entire fraction can be directly spotted onto support disks used for automated Edman degradation. To prevent adsorptive peptide losses onto the plastic tubes, fractions should not be concentrated prior to further analysis and, after spotting the peptide sample, the empty tube should be rinsed with 50 μL 100% TFA, which is then overlaid on top of the sample. If <100% of the sample is to be sequenced, we recommend that the fraction that is to be saved be transferred to a second tube, so that the tube in which the sample was collected can be rinsed in the same manner as described with 100% TFA. Because one of the important applications of reverse-phase HPLC is comparative peptide mapping, we have included below (*see* **Note 7**) a brief discussion of the use of this approach to identifying subtle structural modifications between otherwise identical proteins.

3.2. HPLC Repurification of Peptides

Peptides whose absorbance profile and/or MALDI-MS spectrum indicate they are insufficiently pure for amino acid sequencing may be further purified by chromatography on a second (different) C18 column developed with the same mobile phase and gradient as was used for the initial separation. Because of their differing selectivity (*see* **Note 8**), we recommend that (when necessary) peptides that are initially separated on a Vydac C-18 column be further purified by injection onto a YMC C-18 column that is eluted with the same mobile phase and gradient as was used for the initial separation. Peptides destined for repurification are mixed with 20 μL 0.02% Tween 20 and then a volume of 2 *M* urea, 0.1 *M* NH_4HCO_3 such that the volume of Tween 20 and 2 *M* urea, 0.1 *M* NH_4HCO_3 is equal to or greater than the volume of 0.05% TFA, acetonitrile in which the fraction was originally isolated (*see* **Note 9**). In this way, the acetonitrile concentration is diluted by at least 50%, which, in our experience, is sufficient to permit peptide binding to the second C-18 column.

In those few instances where the sequential use of Vydac and YMC C-18 columns fails to bring about sufficient purification, the sample may be further purified by chromatography at pH 6.0 on either of these columns (*see* **Note 10**). Again, the same gradient is used with the only difference being the change in mobile phase.

4. Notes

1. Although general suggestions for selecting suitable HPLC systems may be found in Stone et al. *(5)*, three factors that critically impact on peptide HPLC and that will be briefly discussed are reproducibility, resolution, and sensitivity.
   Although reproducibility will not have a significant impact on the success of a single analytical HPLC separation, comparative peptide mapping requires that successive chromatograms of digests of the same protein be sufficiently similar that they can be overlaid onto one another with little or no detectable differences. In general, the latter requires that average peak retention times not vary by more than about ± 0.20% *(5)*. Assuming the digests were carried out under identical conditions, problems with regard to reproducibility often relate to the inability of the HPLC pumps to deliver accurate flowrates at the extremes of the gradient range. That is, to accurately deliver a 99% buffer A/1% buffer B composition at an overall flowrate of 75 µL/min requires that pump B be able to accurately pump at a flowrate of only 0.75 µL/min. The latter is well beyond the capabilities of many conventional HPLC systems. Although reproducibility can be improved somewhat by restricting the gradient range to 2–98%, as opposed to 0–100% buffer B, the reproducibility of each HPLC system will be inherently limited in this regard by the ability of its pumps to deliver low flowrates accurately. Obviously, some HPLC systems that provide reproducible chromatograms at an overall flowrate of 0.5 mL/min might be unable to do so at 75 µL/min *(5)*. Similarly, minor check valve, piston seal, and injection valve leaks that go unnoticed at 0.5 mL/min might well account for reproducibility problems at 75 µL/min.
   The ability of HPLC to discriminate between chemically similar peptides and to resolve adequately a reasonable number of peptides from a high-mol-wt protein, which might well produce 100 or more tryptic peptides, is critically dependent on resolution, which, in turn, depends on a large number of parameters, including the flowrate, gradient time, column packing, and dimensions as well as the mobile phase *(3–5)*. Studies with tryptic digests of transferrin suggest that, within reasonable limits, gradient time is a more important determinant of resolution than is gradient volume. In general, a total gradient time of ~100 min seems to represent a reasonable compromise between optimizing resolution and maintaining reasonable gradient times *(3)*. As mentioned in **Subheading 3.1.**, optimal flowrates depend on the inner diameter of the column being used, which is dictated primarily by the amount of protein that has been digested. In general, we find that amounts of protein digests in the 25–250 pmol range are best chromatographed at ~75 µL/min on 2.0–2.1 mm id columns, whereas larger amounts are best chromatographed at ~0.5 mL/min on 3.9–4.6 mm id columns. Amounts of digests that are below ~25 pmol can probably be best chromatographed on 1-mm id columns developed at flowrates near 25 µL/min. Unless precautions are taken to minimize dead volumes, significant problems may, however, be encountered in terms of automated peak detection/collection and postcolumn mixing as flowrates are lowered much below 0.15 mL/min *(5)*. Typically, the use of flowrates in the 25–75 µL/min range require that fused silica tubing be used between the detector and the fraction collector, and that a low volume flow cell (i.e., 1–2 µL) be substituted for the standard flow cell in the UV detector. Several commercially available C-18, reverse-phase supports provide high resolution, including (but certainly

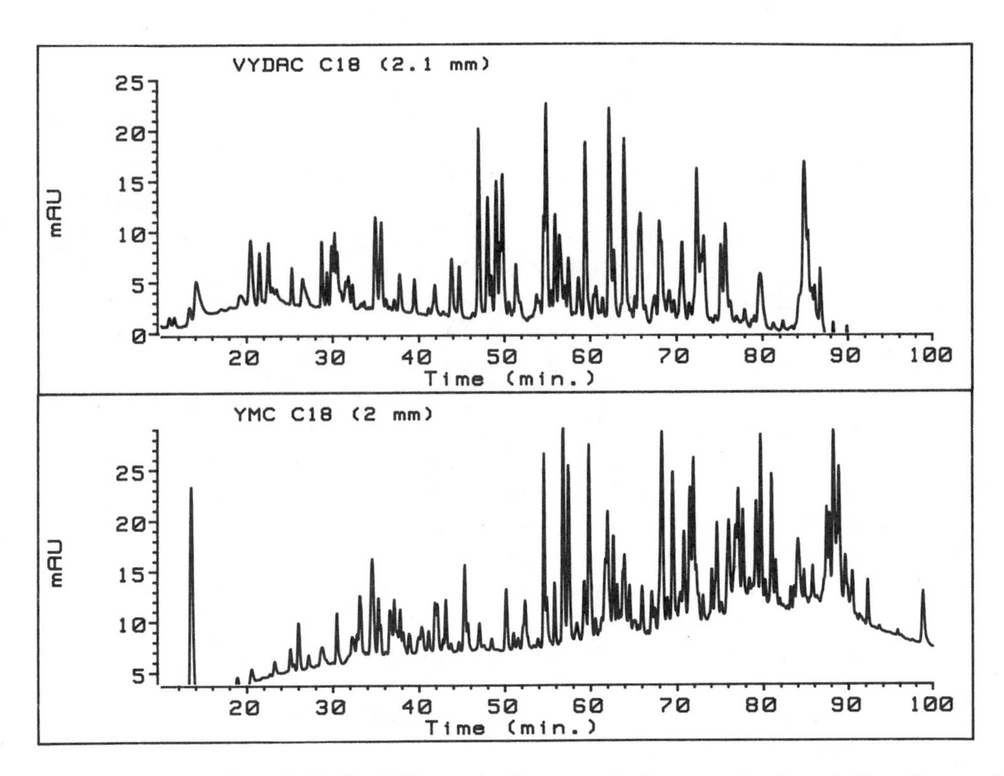

Fig. 1. Reverse-phase HPLC of 50-pmol aliquots of a large scale, in solution digest of carboxamidomethylated transferrin *(5)* that was eluted with the pH 2.0 mobile phase at a flow rate of 75 μL/min as described in **Methods 3**. The top chromatogram was obtained on a 2.1 mm × 25 cm Vydac-18 column (5-μ particle size, 300-Å pore size), whereas the bottom chromatogram was obtained on a YMC C-18 column (5-μ particle size, 120-Å pore size). In both instances, detection was at 210 nm.

not limited to) Alltech Macrosphere, Reliasil, Vydac (**Fig. 1**), Waters' Delta Pak *(3–5)* and YMC (**Fig. 1**). Since peptide resolution appears to be directly related to column length *(3–5)*, whenever possible, the 25-cm versions of these columns should be used. One caveat with regard to the latter is that under the conditions tested, we found that a 15-cm Delta-Pak C-18 column provided similar resolution as that obtained on a 25-cm Vydac C-18 column *(3)*. Although the low-UV absorbance, high resolution, and excellent solubilizing properties of the 0.05% TFA/acetonitrile, pH 2.0, buffer system have made it the almost universal mobile phase for reverse-phase HPLC, there are occasions when a different mobile phase might be advantageous. Hence, the differing selectivity of the 5 m*M*, pH 6.0, phosphate system *(5)* makes this a valuable mobile phase for detecting posttranslational modifications (such as deamidation) that may be more difficult to detect at the lower pH of the TFA system (where ionization of side-chain carboxyl groups would be suppressed). In addition, as noted in **Subheading 3.2.**, changing the mobile phase provides another approach for further purifying peptides that were originally isolated in the TFA system. In our experience, however, the pH 2.0 mobile phase provides somewhat better resolution than the higher pH mobile phase *(5)*. Hence, we recommend using the pH 2.0 system for the initial separation.

The sensitivity of detection of HPLC is dictated primarily by the volume in which each peak is eluted. Although sensitivity can be increased by simply decreasing the flowrate (while maintaining a constant gradient time program) eventually, the linear flow velocity on the column will be reduced to such an extent that optimal resolution will be lost *(3–5)*. At this point, the column diameter needs to be decreased so that a more optimal linear flow velocity can be maintained at a lower flowrate. In general, the sensitivity of detection is increased as the wavelength is decreased with the practical limit in 0.05% TFA being about 210 nm. Finally, an important determinant of sensitivity (that is often overlooked) is the path length of the flow cell. For instance, an HP1090 equipped with a 0.6-cm path length cell provides (at the same flowrate) a threefold increase in sensitivity over that afforded by a Michrom UMA System equipped with a 0.2-cm path length cell.

2. If the transit time (i.e., peak delay) between the detector and the fraction collector is too long, closely eluting peaks will be pooled together. The reason is that if a second peak is detected by the Isco Model 2150 Peak Separator while it is "counting down" the peak delay for the first peak, the two peaks will be pooled together. Although our experience is this phenomenon seldom occurs with a peak delay of 6 s (at a flowrate of 75 µL/min), which corresponds to a "dead volume" of about 7.5 µL, it often occurs if the peak delay exceeds about 15 s. To our knowledge, no commercial peak detector is currently available that can simultaneously track more than one peak.
3. The procedure we use to evaluate C-18 reverse-phase columns is to determine the relative number of peaks that are detected at a given slope sensitivity during the fractionation of an aliquot of a large-scale tryptic digest of transferrin *(3–5)*.
4. For high-sensitivity work, the baseline may be "balanced" (after running the first blank run) by adding a small volume of 20% TFA (i.e., typically 10 to 100 µL) to either buffer A or B as needed *(5)*.
5. Because filtering HPLC solvents may result in their contamination *(6)*, we recommend they be made with HPLC-grade water and acetonitrile, and that they not be filtered prior to use.
6. Provided the fractions are tightly capped within a few hours of collection (to prevent loss of acetonitrile owing to evaporation), the acetonitrile is extremely effective at preventing microbial growth and peptide loss owing to adsorption. Under these conditions, we have often successfully sequenced peptide fractions that have been stored for longer than a year.
7. Provided that samples of both the modified and unmodified protein are available, comparative HPLC peptide mapping provides an extremely facile means of rapidly identifying peptides that contain posttranslational modifications. In the case of proteins that have been expressed in *E. coli*, the latter can often serve as the unmodified control, since relatively few posttranslational modifications occur in this organism. Certainly, the first attempt at comparative HPLC peptide mapping should be with enzymes, such as trypsin or lysyl endopeptidase, that have high specificity, and the digests should be separated using acetonitrile gradients in 0.05% TFA. Although elution position (as detected by absorbance at 210 nm) provides a sensitive criterion to detect subtle alterations in structure, the value of comparative HPLC peptide mapping can be further enhanced by multiwavelength monitoring and, especially, by on-line or off-line mass spectrometry of the resulting peptide -fractions (*see* Chapter 8, which details the off-line use of MALDI-MS for analyzing peptides). If comparative peptide mapping fails to reveal any significant changes, it is often worthwhile running the same digest in the pH 6.0 phosphate-buffered system. At this higher pH, some changes, such as deamidation of asparagine and glutamine, produce a larger effect on elution position than at pH 2.0, where ionization of the side-chain carboxyl groups would be suppressed. Another possible reason for failing to detect differences on comparative HPLC is that the peptide(s) containing the

modifications are either too hydrophilic to bind or too hydrophobic to elute from reverse-phase supports. Hence, in addition to trying a different HPLC solvent system, another approach that may be taken to expand the capabilities of comparative HPLC peptide mapping is to try a different proteolytic enzyme, such as chymotrypsin or Protease V8. Finally, the failure to observe any difference on comparative HPLC peptide mapping may result from loss of the posttranslational modification during either the cleavage or the subsequent HPLC. Assignment of disulfide bonds is one example where this can be a problem in that disulfide interchange may occur during enzymatic cleavage, which is typically carried out at pH 8.0. This problem can be addressed by either going to shorter digestion times *(7)* or by carrying out the cleavage under acidic conditions, where disulfide interchange is less likely to occur. For this reason, pepsin (which is active in 5% formic acid) digests are often used for isolating disulfide bonded peptides. Providing that the control sample is reduced, comparative HPLC peptide mapping can be used to identify disulfide-linked peptides rapidly. If the sequence of the protein of interest is known, then comparison of the MALDI-MS spectra obtained before and after reduction of the disulfide-linked peptide can be used to identify the two peptides that are disulfide-bonded.

8. The differing selectivity of the YMC and Vydac C-18 columns that is evident in **Fig. 1** explains why peptides that are initially separated on a Vydac C-18 column usually may be purified further by chromatography on a YMC C-18 column that is eluted with the same mobile phase and gradient as was used for the initial separation.
9. The purpose of the 2 *M* urea and Tween 20 is to minimize adsorptive losses on diluting the peptide fraction (which is accompanied by a 50% decrease in the acetonitrile concentration). We have found that this amount of Tween 20 has no effect on the subsequent HPLC separation and that provided the urea is made up (at least weekly) in NH_4HCO, no detectable NH_2-terminal blocking occurs as the result of cyanate formation.
10. The advantage of increasing the pH from 2.0, which is the approximate pH of the usual 0.05% TFA mobile phase, to pH 6.0 is that (as mentioned above) this change accentuates the separation of peptides based on their content of acidic amino acid residues. That is, since the P*Ka* of the acidic side chains of aspartic and glutamic acid is about 4, increasing the pH of the mobile phase from 2.0 to 6.0 results in ionization of their COOH side-chains. The increased charge that accompanies ionization greatly decreases retention of peptides on reverse-phase supports *(1)*.

References

1. Guo, D., Mant, C. T., Taneja, A. K., Parker, J. M. R., and Hodges, R. S. (1986) Prediction of peptide retention times in reversed-phase high-performance liquid chromatography. I. Determination of retention coefficients of amino acid residues of model synthetic peptides. *J. Chromatog.* **359,** 499–517.
2. Guo, D., Mant, C. T., Taneja, A. K., and Hodges, R. S. (1986) Prediction of peptide retention times in reversed-phase high-performance liquid chromatography. II. Correlation of observed and predicted peptide retention times and factors influencing the retention times of peptides. *J. Chromatog.* **359,** 519–532.
3. Stone, K. L., LoPresti, M. B., Crawford, J. M., DeAngelis, R., and Williams, K. R. (1991) Reversed-phase HPLC separation of sub-nanomole amounts of peptides obtained from enzymatic digests, in *High-Performance Liquid Chromatography of Peptides and Proteins: Separation, Analysis, and Conformation* (Mant, C. T. and Hodges, R. S., eds.), CRC, Boca Raton, FL, pp. 669–677.
4. Stone, K. L., LoPresti, M. B., and Williams, K. R. (1990) Enzymatic digestion of proteins and HPLC peptide isolation in the subnanomole range, in *Laboratory Methodology in*

Biochemistry (Fini, C., Floridi, A., Finelli, V. N., and Wittman-Liebold, B., eds.), CRC, Boca Raton, FL, pp. 181–205.
5. Stone, K. L., Elliott, J. I., Peterson, G., McMurray, W., and Williams, K. R. (1990) Reversed-phase high-performance chromatography for fractionation of enzymatic digests and chemical cleavage products of proteins, in *Methods in Enzymology*, vol. 193, Academic, New York, pp. 389–412.
6. Stone, K. L. and Williams, K. R. (1986) High-performance liquid chromatographic peptide mapping and amino acid analysis in the sub-nanomole range. *J. Chromatog.* **359,** 203–212.
7. Glocker, M. O., Arbogast, B., Schreurs, J., and Deinzer, M. L. (1993) Assignment of the inter- and intramolecular disulfide linkages in recombinant human macrophage colony stimulating factor using fast atom bombardment mass spectrometry. *Biochemistry* **32,** 482–488.
8. Williams, K. R., Samandar, S. M., Stone, K. L., Saylor, M., and Rush, J. (1996) Matrix assisted-laser desorption ionization mass spectrometry as a complement to internal protein sequencing in the *The Protein Protocols Handbook* (Walker, J. M., ed.) Humana Press, Totowa, NJ, pp. 541–555.

Part V

Protein/Peptide Characterization

79

Peptide Mapping by Two-Dimensional Thin-Layer Electrophoresis–Thin-Layer Chromatography

Ralph C. Judd

1. Introduction

The principle behind peptide mapping is straightforward: If two proteins have the same primary structures, then cleavage of each protein with a specific protease or chemical cleavage reagent will yield identical peptide fragments. However, if the proteins have different primary structures, and then the cleavage will generate unrelated peptides. The similarity or dissimilarity of the proteins' primary structure is reflected in the similarity or dissimilarity of the peptide fragments. Separation of peptides by 2-D thin-layer electrophoresis–thin-layer chromatography (2-D TLE-TLC) results in very high resolution of the peptides, making subtle comparisons possible. There are four phases to the 2-D TLE-TLC peptide mapping process:

1. Identification and purification of the proteins to be compared;
2. Radiolabeling of the proteins, and thus the peptide fragments, to minimize the quantity of protein required;
3. Cleavage of the proteins with specific endopeptidic reagents, either chemical or enzymatic; and
4. Separation and visualization of the peptide fragments for comparison.

Each step can be accomplished in different ways depending on the amount of protein available, the technologies available, and the needs of the researcher. Basic procedures that have proven reliable are presented for each phase. Because peptide mapping is empirical by nature, reaction times, reagent concentrations, and amounts of proteins and peptides may need to be altered to accommodate different research requirements.

2. Materials

2.1. Sodium Dodecyl Sulfate-Polyacrylamide Gel Electrophoresis (SDS-PAGE)

See Chapter 11 for reagents and procedures for SDS-PAGE.

2.2. Electroblotting

1. Blotting chamber with cooling coil (e.g., Transblot chamber, Bio-Rad, Inc., Hercules, CA, or equivalent).

From: *The Protein Protocols Handbook, 2nd Edition*
Edited by: J. M. Walker © Humana Press Inc., Totowa, NJ

2. Power pack (e.g., EC 420, EC Apparatus Inc. [St. Petersburg, FL], or equivalent).
3. Nitrocellulose paper (NCP).
4. Polyvinylidene difluoride (PVDF) nylon membrane.
5. Ponceau S: 1–2 mL/100 mL H_2O.
6. Naphthol blue black (NBB): 1% in H_2O.
7. India ink (Pelikan, Hannover, Germany).
8. 0.05% Tween-20 in phosphate buffered saline (PBS), pH 7.4.
9. 20 m*M* phosphate buffer, pH 8.0: 89 mL 0.2 *M* Na_2HPO_4 stock, 11 mL 0.2*M* stock in 900 mL H_2O.

2.3. Radiolabeling

1. γ-Radiation detector.
2. Speed-Vac concentrator (Savant Inst. Inc., Farmingdale, NY) or any other drying system, such as heat lamps, warm air, and so forth, will suffice.
3. Carrier-free ^{125}I (*see* **Notes 1** and **2**).
4. 1,3,4,6-tetrachloro-3α,6α-glycouril (Iodogen). Iodogen-tubes are prepared by placing 10 µL of chloroform containing 1 mg/mL Iodogen in the bottom of 1.5-mL polypropylene microfuge tubes and allowing to air-dry. Iodogen tubes can be stored at –20°C for up to 6 mo.
5. PBS, pH 7.4 (any dilute, neutral buffer should work).
6. Dowex 1-X-8, 20–50 mesh, anion-exchange resin (#451421, Bio-Rad).
7. Twenty-four-well disposable microtiter plate.
8. Sephadex G-25 or G-50 (Pharmacia, Piscataway, NJ).
9. 15% Methanol in H_2O.
10. XAR-5 film (Kodak, Rochester, NY, or equivalent).
11. Lightening Plus intensifying screens (DuPont, Wilmington, DE, or equivalent).

2.4. Protein Cleavage

1. Enzymatic and chemical cleavage reagents and appropriate buffers as described in Chapter 71.
2. 88 or 70% Formic acid.
3. 50 m*M* NH_4HCO_3 adjusted to the appropriate pH with sodium hydroxide.
4. 50% Glacial acetic acid in H_2O.
5. Glacial acetic acid added to H_2O to bring pH to 3.0.

2.5. Peptide Separation

2.5.1. 2-D TLE-TLC

1. Forma 2095 refrigerated cooling bath (Forma Scientific, Marietta, OH) or equivalent.
2. Immersion TLE chamber (e.g., Savant TLE 20 electrophoresis chamber, or equivalent).
3. 1200-V Power pack.
4. Chromatography chambers.
5. "Varsol" (EC123, Savant, or equivalent).
6. 0.1-mm Mylar-backed cellulose sheets (E. Merck, MCB Reagents, Gibbstown, NJ, or equivalent).
7. TLE buffer: 2 L H_2O, 100 mL glacial acetic acid, 10 mL pyridine.
8. TLC buffer: 260 mL *n*-butanol, 200 mL pyridine, 160 mL H_2O, 40 mL glacial acetic acid.
9. H_2O containing Tyr, Ile, and Asp (1 mg/mL). These are 2-D TLC-TLE amino acid markers.
10. 1% Methyl green in H_2O (w/v).
11. Laboratory sprayer.

12. 0.25% Ninhydrin in acetone.
13. XAR-5 film (Kodak or equivalent).
14. Lightening Plus intensifying screens (DuPont or equivalent).

3. Methods

3.1. Protein Punfication

3.1.1. SDS-PAGE

Any protein purification procedure that results in 95–100% purity is suitable for peptide mapping. For analytical purposes, the discontinuous buffer, SDS-PAGE procedure is the best choice ***(1,2)***. SDS-PAGE provides apparent molecular-mass information, and the ability to probe SDS-PAGE-separated proteins by immunoblotting, helps ensure that the proper proteins are being studied.

If the proteins to be compared are abundant (>100 μg), peptide fragments can be visualized following 2-D TLE-TLC by ninhydrin staining. Much smaller amounts <0.05 μg) can be visualized by autoradiography if the proteins are extrinsically labeled with ^{125}I. Resolution of peptides increases as the amount of each peptide decreases. For these reasons, it is highly recommended that radiolabeled proteins be used ***(1,3–5)***. Proteins separated in SDS-PAGE gels can be labeled and cleaved directly in gel slices (***3–7*** and *see* Chapter 76), but labeling and cleavage are much more efficient if the proteins are first electroblotted to NCP ***(4)***. Proteins can also be intrinsically or extrinsically labeled before SDS-PAGE separation ***(5)***. It is strongly recommended that even highly pure proteins be separated in SDS-PAGE gels and transferred to NCP because of the ease of labeling and cleavage using this system. Blotted proteins can be readily located by staining with Ponceau S in water (preferred), NBB in water, or India ink-0.05% Tween-20-PBS ***(8)***. Proteins of interest can then be excised, labeled using ^{125}I, and cleaved directly on the NCP ***(4)***. The peptides are released into the supernatant and can then be separated using 2-D TLE-TLC ***(4)***.

A single SDS-PAGE separation is often adequate to purify proteins for peptide mapping. Occasionally, a second separation may be required. Alternately, 2-D isoelectric focusing–SDS-PAGE (Chapters 20–22) can be used to purify proteins. If ^{125}I-labeling is used, a single protein band from a single lane of a 24-tooth comb is ample material for numerous separations of peptide fragments. Again, labeling and cleavage are greatly facilitated by electroblotting the protein to NCP.

3.1.1.1. Single SDS-PAGE Separation

1. Samples to be compared can be separated in individual lanes of an SDS-PAGE gel; "preparative" gels, where each sample is loaded over the entire stacking gel ***(4)***, may also be used.
2. After electrophoresis, fixation, Coomassie brilliant blue (CBB) staining, and destaining (*see* Chapter 11), excise the protein bands of interest for use in the "gel slice" methods described below. The preferred method is to electroblot the protein to NCP, at 20 V constant current, 0.6 A for 16 h in degassed 20 m*M* phosphate buffer, pH 8.0 ***(9)*** (*see* Chapters 37–39).
3. To stain the proteins on NCP, shake the NCP in Ponceau S for 15 min, then destain with H_2O, or shake in 0.1% NBB in H_2O for 1 h, and then destain with H_2O. If the proteins cannot be located by using these stains, place the NCP in 100 mL of 0.05% Tween-20 PBS, and mix for 1 h. Then add three drops of India ink and mix for another hour. Protein bands will be black and the background white (*see* **Note 3**).

4. Excise the protein band from the NCP (a 1 × 5-mm band is more than ample), and place the excised strip in a 1.5-mL microfuge tube. Wash with H_2O until no stain is released into the supernatant. The protein is now ready for labeling and cleavage (*see* **Note 4**).

3.1.1.2. Double SDS-PAGE Separation

1. Separate the samples in individual lanes of an SDS-PAGE gel or in "preparative" SDS-PAGE gels. Fix, stain with CBB, and then destain (*see* Chapter 11).
2. Excise the protein bands of interest. Soak the bands in 50% ethanol–50% stacking buffer (1 *M* Tris-HCl, pH 6.8) for 30 min to shrink the gel strip to facilitate loading onto a second SDS-PAGE gel.
3. Push the excised band into contact with the stacking gel of a second SDS-PAGE gel of a different acrylamide concentration (generally use high concentration in the first gel and lower concentration in the second gel).
4. Separate proteins in a second gel (CBB runs just behind the dye front). Stain or electroblot the proteins as in **Subheading 3.1.1.** The protein is now ready for labeling and cleavage.

3.2. Protein Labeling

Proteins can be intrinsically labeled by growing organisms in the presence of a uniform mixture of ^{14}C-amino acids *(5)*, but this is quite expensive. Intrinsic labeling with individual amino acids, such as ^{35}S-Met or ^{35}S-Cys, will not work, since many peptide fragments will not be labeled. Iodination with ^{125}I is inexpensive and reproducible. Iodinated peptides are readily visualized by autoradiography *(1,3,4)*. Comparative cleavages of a 40,000-Dalton protein intrinsically labeled with ^{14}C-amino acids extrinsically labeled with ^{125}I showed that 61 of 66 α-chymotryptic peptides were labeled with ^{125}I, whereas all 22 *Staphylococcus aureus* V8 protease-generated peptides were labeled with ^{125}I, demonstrating the effectiveness of radioiodination *(10)*. This demonstrates that tyrosine (Tyr) is not the only amino acid labeled using this procedure.

Iodination mediated by chloramine-T (CT) *(11)* produces extremely high specific activities, but the procedure requires an extra step to remove the CT and can cleave some proteins at tryptophan residues *(12)*. This can be beneficial since it is specific and increases the number of peptides, thus increasing the sensitivity of the procedure (*see* **ref.** *13*) for peptide maps of CT- vs Iodogen-labeled proteins). Unfortunately, small peptides generated by CT cleavage, followed by a second enzymatic or chemical cleavage, can be lost during the removal of the CT and unbound ^{125}I.

The 1,3,4,6-tetrachloro-3α,6α-glycouril (Iodogen) *(13)* procedure, where the oxidizing agent is bound to the reaction vessel, does not damage the protein and produces high specific activities. Aspiration of the reaction mixture stops the iodination and separates the protein from the oxidant in a single step. For these reasons, Iodogen-mediated labeling is the preferred method for radioiodination. (Radioemission of ^{125}I will be expressed as counts per minute [cpm]. This assumes a detector efficiency of 70%. If detector efficiency varies, multiply the cpm presented here by 1.43 to determine decays per minute (dpm), and then multiply the dpm by the efficiency of your detector.)

3.2.1. NCP Strip (Preferred Method)

1. Put the protein-containing NCP strip in an Iodogen-coated (10 μg) microfuge tube.
2. Add 50–100 μL PBS, pH 7.4 (any dilute, neutral buffer should work) and 50–100 μCi ^{125}I (as NaI, carrier-free, 25 μCi/μL) (*see* **Notes 1** and **2**).

3. Incubate at room temperature for 1 h. Aspirate the supernatant. **(Caution: supernatant is radioactive).**
4. Place the NCP strip in a fresh microfuge tube and wash three to five times with 1.5 mL H_2O (radioactivity released should stabilize at <10,000 cpm/wash).
5. The protein on the NCP strip is now ready for cleavage (*see* **Note 5**).

3.2.2. Gel Slice

1. Dry the gel slice containing protein using a Speed-Vac concentrator or other drying system, such as heat lamps, warm air, and so forth.
2. Put the slice in an Iodogen-coated (10-µg) microfuge tube.
3. Add 100 µL of PBS, pH 7.4 (any dilute, neutral buffer should work) plus 50–100 µCi ^{125}I (as NaI, carrier-free, 25 µCi/µL) (*see* **Notes 1** and **2**).
4. Incubate at room temperature for 1 h. Aspirate the supernatant. **(Caution: supernatant is radioactive).**
5. Remove the gel slice and soak for 0.5 to 1 h in 1.5 mL of H_2O. Repeat three times.
6. Place 0.5 g Dowex 1-X-8, 20–50 mesh, anion-exchange resin and 1.5 mL of 15% methanol in H_2O in the wells of a 24-well microtiter plate.
7. Add the iodinated gel slice to a well with anion-exchange resin and incubate at room temperature for 16 h. The resin binds unreacted iodine, becoming **extremely** radioactive.
8. Remove the gel slice from the resin, and soak it in 1.5 mL of H_2O. Repeat several times, and dry the gel slice. The protein is now ready for cleavage (*see* **Note 6**).

3.2.3. Lyophilized/Soluble Protein

1. Suspend up to 1 mg/mL of protein in 100–200 µL of PBS, pH 7.4 (any dilute, neutral buffer should work), in a 1.5-mL microfuge tube containing 10 µg Iodogen.
2. Add 100–200 µCi ^{125}I (as NaI, carrier-free, 25 µCi/µL) (*see* **Notes 1** and **2**).
3. Incubate on ice for 1 h.
4. Remove the protein-containing supernatant and separate the protein from salts and unbound iodine by the following methods:
   a. (Preferred method) Separate on a Sephadex G-25 or G-50 desalting column using H_2O as the eluant and lyophilize.
   b. (Relatively easy) Solubilize the sample in 2X sample buffer (10–20 µg/lane), and separate in an SDS-PAGE gel. Stop the electrophoresis before the ion front reaches the bottom of the gel and cut the gel just above the dye front. Unbound iodine will be in this portion of the gel. Either fix, stain, and destain the gel to locate the protein band, or electroblot onto NCP and locate the protein by Ponceau S. NBB, or India ink staining (*see* **Subheading 3.1.1., step 3**). Excise the protein band from the gel or NCP.
   c. (Excellent if available) Separate the protein using reverse-phase or molecular-exclusion HPLC columns, and then dialyze and lyophilize.
   d. (Least preferred) Dialysis, followed by lyophilization, can be used, but it produces excessive radioactive liquid waste.
5. The protein is now ready for cleavage.

3.3. Protein Cleavage

The use of cleavage reagents (e.g., α-chymotrypsin) or combinations of cleavage reagents, which generate many fragments, tends to accentuate differences in primary structure, whereas cleavage reagents that produce small numbers of fragments (e.g., V8 protease, thermolysin, CNBr, BNPS-skatole) emphasize similarities in primary structure. Enzymatic reagents are often easiest to use, safest, and most reliable, but they can interfere with results, since they are themselves proteins. Chemical reagents are also easy to use and reliable, but can be toxic, requiring careful handling (*see* **Note 7**). Several practical cleavage reagents are presented in **Table 1**.

Volatile buffers must be used when using 2-D TLE-TLC peptide separation, since this system is negatively affected by salts. For formate and CNBr cleavages, the acid is diluted in H_2O to 88 or 70%, respectively. BNPS-skatole works well in 50% glacial acetic acid–50% H_2O. Ammonium bicarbonate (50 m*M*), adjusted to the appropriate pH with sodium hydroxide, is excellent for enzymes requiring weak base environments (trypsin, α-chymotrypsin, thermolysin, V8 protease). The acid peptidase, Pepsin A, is active in H_2O adjusted to pH 3.0 with glacial acetic acid (*see also* Chapter 76).

3.3.1. Protein on NCP Strip

1 Put the NCP strip containing the radiolabeled protein in a 1.5-mL microfuge tube, and measure the radioemission using a γ-radiation detector.
2. Add 90 µL of the appropriate buffer and 10 µL of chemical or enzymatic cleavage reagent in buffer (1 mg/mL) to the NCP strip.
3. Incubate with shaking at 37°C for 4 h (for enzymes) or at room temperature for 24–48 h in dark under nitrogen (for chemical reagents).
4. Aspirate the peptide-containing supernatant, and count the NCP strip and supernatant. Enzymes should release 60–70% of counts in the slice into the supernatant; CNBr should release >80%, BNPS-skatole rarely releases more than 50% (*see* **Notes 8** and **9**).
5. Completely dry the supernatant in a Speed-Vac, and wash the sample at least four times by adding 50 µL of H_2O, vortexing, and redrying in a Speed-Vac. Alternate drying systems will work.
6. The sample is now ready for peptide separation (*see* **Note 10**).

3.3.2. Protein in Gel Slice

1. Put the dry gel slice containing the radiolabeled protein in a 1.5-mL microfuge tube and measure the radioemission using a γ-radiation detector.
2. Add 10 µL of cleavage reagent in buffer (1 mg/mL) directly to the dry gel slice. Allow slice to absorb cleavage reagent, and then add 90 µL of appropriate buffer.
3. Continue as from **step 3, Subheading 3.3.1.** Release of peptides into the supernatant will be less efficient than with the NCP strip.
4. The sample is now ready for peptide separation (*see* **Note 10**).

3.3.3. Lyophilized/SolubleProteins

1. Rehydrate the lyophilized radiolabeled proteins in the appropriate buffer at 1 mg/mL (less concentrated samples can be used successfully).
2. Add up to 25 µL of the appropriate cleavage reagent (1 mg/mL) to 25 µL of suspended protein.
3. Continue as from **step 3, Subheading 3.3.1.**, except there is no strip to count.
4. The sample is now ready for peptide separation. Be aware that the sample will usually contain uncleared protein along with the peptide fragments (*see* **Note 10**).

3.4. 2-D TLE-TLC Peptide Separation

It is strongly recommended that iodinated samples be used in the 2-D TLE-TLC system. The technique described is precise enough that peptide maps can be overlaid to facilitate comparisons. Flat-bed electrophoresis can be used, but systems that cool by immersion of the thin-layer sheet in an inert coolant, such as "varsol" (such as the Savant TLE 20 electrophoresis chamber or equivalent), yield superior results. Cooling

Table 1
Cleavage Reagents

Reagent	Site of cleavage	Buffer
Chemical		
Cyanogen bromide[a,d]	Carboxy side of Met	70% formate in H_2O
BNPS-skatole[a] (-bromo-methyl-2-(nitrophenylmercapto)-3H-indole)	Carboxy side of Trp	50% H_2O-50% glacial acetic acid
Formic acid	Between Asp and Pro	88% in H_2O
Chloramine T[b]	Carboxy side of Trp	H_2O
Enzymatic[c]		
α-chymotrypsin	Carboxy side of Tyr, Trp, Phe, Leu	50 m*M* NH_4HCO_3, pH 8.5
Pepsin A	Amino side of Phe>Leu	Acetate-H_2O, pH 3.0
Thermolysin	Carboxy side of Leu>Phe	50 m*M* NH_4HCO_3, pH 7.85
Trypsin	Carboxy side of Arg, Lys	50 m*M* NH_4HCO_3, pH 8.5
V8 protease	Carboxy side of Glu, Asp, or carboxy side of Glu	50 m*M* NH_4HCO_3, pH 8.5 pH 6–7 (H_2O)

[a]Cyanogen bromide and BNPS-skatole are used at 1 mg/mL. Room temperature incubation should proceed for 24–48 h under nitrogen in the dark.
[b]Chloramine T is used at 10 mg/mL in H_2O.
[c]All enzymes are used at 1 mg/mL in the appropriate buffer.
[d]**Caution:** CNBr is extremely toxic—handle with care in chemical hood.

should be supplied by as large a refrigerated bath as possible, such as the Forma 2095 refrigerated cooling bath. Extra cooling coils, made by bending 1/4 in. aluminum tubing, are helpful. Peptides migrate based on charge, which is a function of pH, in an electric field. The buffer pH is a function of temperature. Therefore, maintenance of the running buffer temperature is crucial. Inconsistent cooling results in inconsistent peptide migration. For best results, use only 0.1 mm Mylar backed cellulose sheets of E. Merck. Run two or three peptide maps per 20 × 20-cm sheet. If necessary, increased resolution can be obtained by running one sample/sheet and increasing running times (*see* **Note 11**).

1. Set the cooling bath at 8.5°C to keep the electrophoresis tank at 10–13.5°C. The temperature of the cooling tank should not increase more than 1.5°C during a run.
2. Rehydrate the peptide sample to 10^5 cpm/μL in H_2O containing Tyr, Ile, and Asp (1 mg/mL) as amino acid markers.
3. For two samples/run: draw a line down the center of the back of the sheet with a laboratory marker parallel to the machine lines (they can be subtle, but always electrophorese parallel to these lines). Mark two spots 2.5 cm from the end of the sheet and 1.5 cm from center line on the back of the sheet to indicate where to load samples. When the sheet is turned with the cellulose facing up, the marks will show through. For three samples/run: draw two lines of the back of the sheet (parallel to machine lines) 6.7 and 13.4 cm from left edge of sheet and mark three spots, each 8 cm from the end of the plate and 1 cm to the right of the left edge and each line (*see* **Note 12**).

4. Use a graduated, 1–5 μL capillary pipette to spot 2 μL (~2×10^5 cpm) if two samples are used, or 1.5 μL (~1.5×10^5 cpm) if three samples are used, 0.5 μL at a time (dry spot with hair-dryer each time) to one mark on the sheet. Repeat for each sample on the other mark(s). More sample can be run, but resolution will decrease. To verify proper electrophoresis, spot 1 μL of 1% methyl green on the center line. The methyl green should migrate rapidly toward the cathode in a straight line. Veering indicates a problem.
5. Spray the plate with TLE buffer using a laboratory sprayer. Do not over wet. Remove any standing buffer with one paper towel. Always blot TLE plates in exactly the same manner.
6. Place in the electrophoresis chamber with the samples toward the anode. Run the electrophoresis at 1200 V (about 20 W and 20 mA) for 45 min (two samples/run) or 31 min (three samples/run).
7. Remove the sheet from the chamber, and immediately dry with a hair-dryer. The "varsol" will dry first, and then the buffer. Cut along the lines on the back of the sheet. Score the cellulose 0.5 cm down from top edge of each piece (bottom is the edge closest to the sample) to form a moat.
8. Place the sheets in the chromatography chamber so that chromatography can proceed perpendicularly to the electrophoresis. The TLC buffer should be about 0.5-cm deep. Chromatograph until the buffer reaches the moat. Remove and dry with a hair-dryer (best done in hood).
9. Spray the sheet with 0.25% ninhydrin in acetone (do not saturate) and dry with a hair-dryer to locate amino acid markers. Ninhydrin can also be used to locate peptides if larger amounts of sample are separated (10–100 μg). Be sure to run enzyme controls to distinguish sample from enzyme. Markers should migrate identically in all separations.
10. Overlay the sheets with X-ray film, place Lightening Plus intensifying screen over film, and place in cassette. Expose for 16–24 h at –70°C or expose film without a screen for about 4 d at room temperature. Develop the film (*see* **Note 13**).

Figure 1 is presented to demonstrate the separation of peptides generated by cleavage with trypin by 2-D TLE-TLC. These peptide maps indicate that the porin protein (POR) of *Neisseria gonorrhoeae* is structurally unrelated to the 44-kDa proteins, whereas the 44-kDa protein from the sarkosyl insoluble (membrane) extract (44-kDa Mem.) is structurally indistinguishable from the 44-kDa protein from the periplasmic extract (44-kDa Peri.). Note the high resolution of the peptide fragments using this technique.

4. Notes

1. Regardless of the labeling procedure, never use ^{125}I that is over one half-life (60 d) old. Do not increase the amount of older ^{125}I to bring up the activity; it does not work.
2. Never use carrier-free ^{125}I in acid buffers. The iodine becomes volatile and could be inhaled.
3. Ponceau S can be completely removed with H_2O, but NBB or India ink cannot.
4. Do not compare proteins stained with Ponceau S with those stained with NBB or India ink. Use the same staining procedure for all proteins to be compared.
5. It is common to have between 3×10^6 and 6×10^6 cpm for a strip 1×5 mm. This provides enough material to run 15–30 peptide maps/strip.
6. It is common to have between 2×10^6 and 4×10^6 cpm in a gel slice (1×5 mm). This provides enough material to run 10–20 peptide maps/slice.
7. Chemical cleavage reagents are preferred when peptides are not radiolabeled since enzymes cleave themselves, resulting in confusing data. **CNBr is extremely toxic—handle with great care in a chemical hood.**

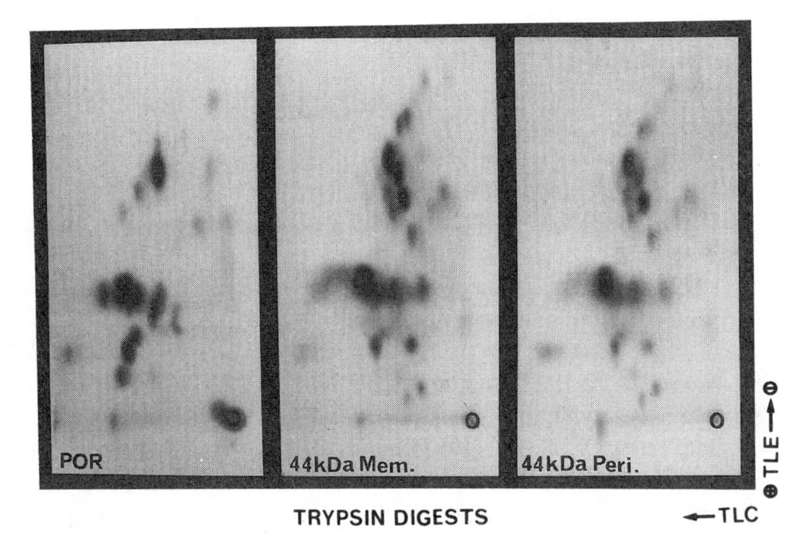

Fig. 1. Example of peptides separated by 2-D TLE-TLC. Proteins were radiolabeled on NCP strips as described in **Subheading 3.2.1.** and cleaved with trypsin as described in **Subheading 3.3.1.** The peptides (1×10^5 cpm) were spotted on a thin-layer cellulose sheet and subjected to 2-D TLE-TLC as described in **Subheading 3.4.** The origin (O) is at the lower right of each map. TLE-direction of thin-layer electrophoresis; TLC—direction of thin-layer chromatography. The ^{125}I-labeled peptides were visualized by autoradiography.

8. Repeated digestions will release about the same percentage of counts. Generally, one or two digestions are adequate. Only 1×10^5 cpm are necessary to produce a peptide map.
9. If >95% of counts are released in the first enzyme digestion, there may be excess, unbound iodine left in sample. This could cause serious problems, since the enzyme may become labeled. The resultant peptide maps will all be identical maps of the enzyme and not your sample.
10. Peptides prepared in this manner can also be separated by high-performance liquid chromatography (*see* Chapter 75).
11. CNBr should not be used to generate peptides for use in 2-D TLE-TLC, since it produces very hydrophobic peptides, which tend to compress at the top of the chromatogram. CNBr is excellent for generating peptides to be separated by SDS-PAGE or HPLC.
12. Always spot sample to be compared the same distance from the anode (positive) terminal, since peptides migrate more rapidly close to the anode and more slowly far from the anode.
13. Migration of peptides should be consistent enough to overlay peptide maps directly for comparisons. Coordinates of amino acid markers and peptides can be determined and used to compare migration. The labeling procedures described here are precise enough to use emission intensities as a criterion for comparison.

Acknowledgments

The author thanks Pam Gannon for her assistance and the Public Health Service, NIH, NIAID (grant RO1 AI21236) and UM Research Program for their continued support.

References

1. Judd, R. C. (1986) Evidence for N-terminal exposure of the PIA subclass of protein I of *Neisseria gonorrhoeae*. *Infect. Immunol.* **54,** 408–414.

2. Laemmli, U. K. (1970) Cleavage of structural proteins during the assembly of the head of bacteriophage T4. *Nature* **227,** 680–695.
3. Judd, R. C. (1982) I^{125}-peptide mapping of protein III isolated from four strains of *Neisseria gonorrhoeae. Infect. Immunol.* **37,** 622–631.
4. Judd, R. C. (1987) Radioiodination and ^{125}I-labeled peptide mapping on nitrocellulose membranes. *Analyt. Biochem.* **160,** 306–315.
5. Caldwell, H. D. and Judd, R. C. (1982) Structural analysis of chlamydial proteins. *Infect. Immunol.* **38,** 960–968.
6. Swanson, J. (1979) Studies on gonococcus infection. XVIII. ^{125}I-labeled peptide mapping of the major protein of the gonococcal cell wall outer membrane. *Infect. Immunol.* **23,** 799–810.
7. Elder, J. H., Pickett, R. A. III, Hampton, J., and Lerner, R. A. (1977) Radioiodination of proteins in single polyacrylamide gel slices. *J. Biol. Chem.* **252,** 6510–6515.
8. Hancock, K. and Tsang, V. C. W. (1983) India ink staining of protein on nitrocellulose paper. *Analyt. Biochem.* **133,** 157–162.
9. Batteiger, B., Newhall, W. J., V., and Jones, R. B. (1982) The use of Tween-20 as a blocking agent in the immunological detection of proteins transferred to nitrocellulose membranes. *J. Immunol. Meth.* **55,** 297–307.
10. Judd, R. C. and Caldwell, H. D. (1985) Comparison of ^{125}I- and ^{14}C-radiolabeled peptides of the major outer membrane protein of *Chlamydia trachomatis* strain L2/434 separated by high-performance liquid chromatography. *J. Liq. Chromatogr.* **8,** 1109–1120.
11. Greenwood, F. C., Hunter, W. M., and Glover, J. S. (1963) The preparation of ^{131}H-labeled human growth hormone of high specific radioactivity. *Biochem. J.* **89,** 114–123.
12. Alexander, N. M. (1973) Oxidation and oxidative cleavage of tryptophanyl peptide bonds during iodination. *Biochem. Biophys. Res. Commun.* **54,** 614–621.
13. Markwell, M. A. K. and Fox, C. F. (1978) Surface-specific iodination of membrane proteins of viruses and eukaryotic cells using 1,2,3,6-tetrachloro-3α, 6α-diphenylglycoluril. *Biochemistry* **112,** 278–281.

80

Peptide Mapping by Sodium Dodecyl Sulfate-Polyacrylamide Gel Electrophoresis

Ralph C. Judd

1. Introduction

The comparison of the primary structure of proteins is an important facet in the characterization of families of proteins from the same organism, similar proteins from different organisms, and cloned gene products. There are many methods available to establish the sequence similarities of proteins. A relatively uncomplicated approach is to compare the peptide fragments of proteins generated by enzymatic or chemical cleavage, i.e., peptide mapping. The similarity or dissimilarity of the resultant peptides reflects the similarity or dissimilarity of the parent proteins.

One reliable method of peptide mapping is to separate peptide fragments by sodium dodecyl sulfate-polyacrylamide gel electrophoresis (SDS-PAGE). Comparison of the separation patterns reveals the structural relationship of the proteins. Moderate separation of peptides can be accomplished using this procedure. The technique, first described by Cleveland et al. *(1)*, is simple, inexpensive, requires no special equipment, and can be combined with Western blotting to locate epitopes (i.e., epitope mapping *[1–3]*) or blotting to nylon membranes for microsequencing *(4)*. Microgram amounts of peptide fragments can be visualized by in-gel staining, making this system fairly sensitive. Sensitivity can be greatly enhanced by radiolabeling (*see* Chapter 78).

2. Materials

2.1. SDS-PAGE

1. SDS-PAGE gel apparatus and power pack (e.g., EC 500, EC Apparatus, Inc., St. Petersburg, FL, or equivalent).
2. SDS-PAGE Solubilization buffer: 2 mL 10% SDS (w/v) in H_2O, 1.0 mL glycerol, 0.625 mL 1 *M* Tris-HCl, pH 6.8, 6 mL H_2O, bromophenol blue to color.
3. Enzyme buffer for Cleveland et al. *(1)* "in-gel" digestion: 1% SDS, 1 m*M* EDTA, 1% glycerol, 0.1 *M* Tris-HCl, pH 6.8.
4. All buffers and acrylamide solutions necessary for running SDS-PAGE (*see* Chapter 11).
5. Ethanol: 1 *M* Tris-HCl, pH 6.8 (50:50; [v/v]).
6. Fixer/destainer: 7% acetic acid, 25% isopropanol in H_2O (v/v/v).
7. Coomassie brilliant blue: 1% in fixer/destainer (w/v).
8. Mol-wt markers, e.g., low-mol-wt kit (Bio-Rad, Hercules, CA, or equivalent) or peptide mol-wt markers (Pharmacia, Piscataway, NJ, or equivalent).

From: *The Protein Protocols Handbook, 2nd Edition*
Edited by: J. M. Walker © Humana Press Inc., Totowa, NJ

2.2. Electroblotting (for Epitope or Sequence Analyses)

1. Blotting chamber with cooling coil (e.g., Transblot chamber, Bio-Rad, Inc. or equivalent).
2. Power pack (e.g., EC 420, EC Apparatus, Inc. or equivalent).
3. Nitrocellulose paper (NCP).
4. Polyvinylidene difluoride (PVDF) nylon membrane.
5. Ponceau S: 1–2 mL/100 mL H_2O.
6. 1% Naphthol blue black (NBB) in H_2O (w/v).
7. 0.05% Tween-20 in phosphate-buffered saline (PBS), pH 7.4.
8. India ink, three drops in 0.05% Tween-20 in PBS, pH 7.4.
9. 20 m*M* phosphate buffer, pH 8.0: 89 mL 0.2 *M* Na_2HPO_4 stock, 11 mL 0.2 *M* NaH_2PO_4 stock in 900 mL H_2O.

3. Methods

3.1. Protein Purification

Purification of the proteins to be compared is the first step in peptide mapping. Any protein purification procedure that results in 95–100% purity is suitable for peptide mapping. For analytical purposes, the discontinuous buffer, SDS-PAGE procedure is often the best choice (Chapter 11 and **refs.** ***2*** and ***5***). The advantages of SDS-PAGE are: Its resolving power generally can bring proteins to adequate purity in one separation, whereas a second SDS-PAGE separation almost always provides the required purity for even the most difficult proteins; the simple reliability of the procedure; both soluble and insoluble proteins can be purified; apparent molecular-mass information; and the ability to probe SDS-PAGE-separated proteins by immunoblotting help ensure that the proper proteins are being studied. If even greater separation is required, 2D isoelectric focusing-SDS-PAGE can be used (*see* Chapters 22–24).

Proteins separated in SDS-PAGE gels can be cleaved directly in gel slices ***(6)***. However, cleavage is more efficient if the proteins are first electroblotted to NCP ***(3)***. If required, proteins can also be intrinsically or extrinsically labeled before SDS-PAGE separation (Chapters 21 and 78 and **ref.** ***6***).

3.2. Protein Cleavage

Cleavages of purified proteins can be accomplished in the stacking gel with the resultant peptides separated directly in the separating gel, or they can be performed prior to loading the gel (preferred). Several lanes, with increasing incubation time or increasing concentration of enzyme in each lane, should be run to determine optimal proteolysis conditions. Standard Laemmli SDS-PAGE ***(5)*** system is able to resolve peptide fragments >3000 Dalton. Smaller peptides are best separated in the tricine gel system of Schagger and von Jagow ***(7)*** (*see also* Chapter 13). The methods described in Chapter 74, **Subheading 3.3.** can be used to generate peptides. Cleavage technique varies slightly depending on the form of the purified proteins to be compared.

3.2.1. In-Gel Cleavage/Separation

3.2.1.1. Lyophilized/Soluble Protein

1. Boil the purified protein (~1 mg/mL) for 5 min in enzyme buffer for Cleveland “in-gel” digestion.
2. Load 10–30 µL (10–30 µg) of each protein to be compared into three separate wells of an SDS-PAGE gel (*see* **Note 1**).

3. Overlay samples of each protein with 0.005, 0.05, and 0.5 µg enzyme in 10–20 µL of Cleveland "in-gel" digestion buffer to separate wells. V8 protease (endoproteinase Glu-C) works very well in this system (*see* **Note 2**). SDS hinders the activity of trypsin, α-chymotrypsin, and themolysin, so cleavage with these enzymes may be very slow. In-gel cleavage with chemical reagents is not generally recommended, since they can be inefficient in neutral, oxygenated environments. Gently fill wells and top chamber with running buffer.
4. Subject the samples to electrophoresis until the dye reaches the bottom of the stacking gel. Turn off the power, and incubate for 2 h at 37°C to allow the enzyme to digest the protein partially. Following incubation, continue electrophoresis until the dye reaches the bottom of the gel. Fix, stain, and destain, or electroblot onto NCP for immunoanalysis (*see* **Note 3**).

3.2.1.2. Protein In-Gel-Slice

Peptides of proteins purified by SDS-PAGE usually retain adequate SDS to migrate into a second gel without further treatment.

1. Run SDS-PAGE gels, fix, stain with Coomassie brilliant blue, and destain.
2. Excise protein bands to be compared with a razor blade.
3. Soak excised gel slices containing proteins to be compared in ethanol–1 *M* Tris-HCl, pH 6.8, for 30 min to shrink the gel, making loading on the second gel easier. Place the gel slices into the wells of a second gel.
4. Continue from **step 3, Subheading 3.2.1.1.**

3.2.1.3. Protein on NCP Strip

1. Run SDS-PAGE gels, and electroblot to NCP.
2. Stain blot with Ponceau S, NBB, or block with 0.05% Tween-20 in PBS for 30 min, and then add three drops India ink (*see* **Note 4**).
3. Excise protein bands to be compared with a razor blade.
4. Push NCP strips to bottom of the wells of a second SDS-PAGE gel.
5. Continue from **step 3, Subheading 3.2.1.1.**

Once the separation conditions, protein concentrations, and enzyme concentrations have been established, a single digestion lane for each sample can be used for comparative purposes.

3.2.2. Protein Cleavage Followed by SDS-PAGE

It is often preferable to cleave the proteins to be compared prior to loading onto an SDS-PAGE gel for separation. This generally gives more complete, reproducible cleavage and allows for the use of chemical cleavage reagents not suitable for in-gel cleavages.

3.2.2.1. Lyophilized/Soluble Protein

1. Rehydrate the lyophilized proteins in the appropriate buffer at 1 mg/mL (less concentrated samples can be used successfully). Soluble proteins may need to be dialyzed against the proper buffer.
2. Place 10–30 µL (10–30 µg) of each protein to be compared in 1.5-mL microfuge tubes.
3. Add up to 25 µL of the appropriate chemical cleavage reagent (1 mg/mL) to suspended protein. For enzymes, use only as much enzyme as needed to achieve complete digestion (1:50 enzyme to sample maximum) (*see* **Note 5**).
4. Incubate with shaking at 37°C for 4 h (for enzymes) or at room temperature for 24–48 h in dark under nitrogen (for chemical reagents).
5. Add equal volume of 2X SDS-PAGE solubilization buffer and boil for 5 min (*see* **Note 6**).
6. Load entire sample on SDS-PAGE gel and proceed with electrophoresis.

3.2.2.2. Protein In-Gel Slice

1. Run SDS-PAGE gels, fix, stain with Coomassie brilliant blue, and destain.
2. Excise protein bands to be compared with a razor blade.
3. Dry gel slices containing proteins using a Speed-Vac concentrator or other drying system, such as heat lamps, warm air, and so forth.
4. Put the dry gel slice containing the protein in a 1.5-mL microfuge tube.
5. Add up to 10 µL of the appropriate chemical cleavage reagent (1 mg/mL) directly to the gel slice protein, and then add 90 µL of appropriate buffer. For enzymes, use only as much enzyme as needed to achieve complete digestion (1:50 enzyme to sample maximum) (*see* **Note 5**).
6. Incubate with shaking at 37°C for 4 h (for enzymes) or at room temperature for 24–48 h in dark under nitrogen (for chemical reagents).
7. Aspirate peptide-containing supernatant.
8. Completely dry-down the supernatant in a Speed-Vac, and wash the sample several times by adding 50 µL of H_2O, vortexing, and redrying in a Speed-Vac. Alternate drying systems will work.
9. Add 10–20 µL SDS-PAGE solubilizing solution to samples and boil for 5 min (*see* **Note 6**).
10. Load samples onto SDS-PAGE gel and proceed with electrophoresis.

3.2.2.3. Protein on NCP Strip

1. Run SDS-PAGE gels, and electroblot to NCP.
2. Stain blot with Ponceau S, NBB, or block with 0.05% Tween-20 in PBS for 30 min, and then add three drops India ink (*see* **Note 3**).
3. Excise protein bands to be compared with a razor blade.
4. Put the NCP strip containing the protein in a 1.5-mL microfuge tube.
5. Continue from **step 5, Subheading 3.2.2.1.**

Figure 1 is presented to demonstrate the separation of peptides generated by cleavage with BNPS-skatole in an SDS-PAGE gel. These peptide maps indicate that the porin protein (POR) was structurally unrelated to the 44-kDa proteins, whereas the 44-kDa proteins from a sarkosyl insoluble (membrane) extract (44-kDa Mem.) and a periplasmic extract (44-kDa Peri.) appeared to have similar primary structures.

4. Notes

1. Between 5×10^4 cpm and 10^5 cpm should be loaded in each lane if radioiodinated samples are to be used. Autoradiography should be performed on unfixed gels. Fixation and staining may wash out small peptides. Place gel in a plastic bag and overlay with XAR-5 film, place in a cassette with a Lightening Plus intensifying screen, and expose for 16 h at –70°C.
2. Control lanes containing only enzyme must be run to distinguish enzyme bands from sample bands.
3. PVDF nylon membrane is preferable to NCP when blotting small peptides.
4. Do not compare proteins stained with Ponceau S with those stained with NBB or India ink. Use the same staining procedure for all proteins to be compared.
5. To ensure complete digestion, it is advisable to incubate the samples for increasing periods of time or to digest with increasing amounts of enzyme. Once optimal conditions are established, a single incubation time and enzyme concentration can be used.
6. It is often advisable to solubilize protein in SDS prior to cleavage, since some peptides do not bind SDS well.

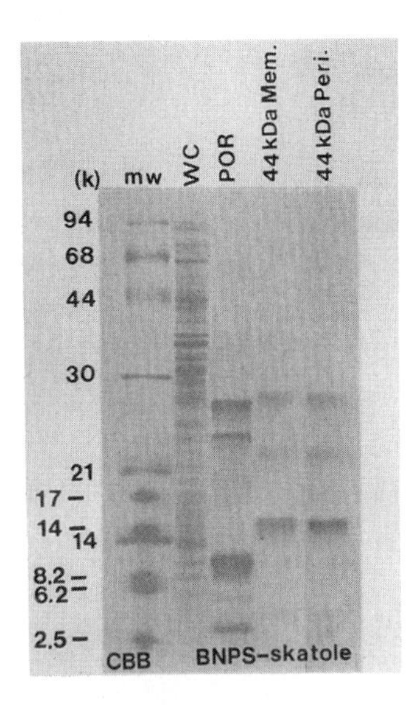

Fig. 1. Example of peptides separated in an SDS-PAGE gel. Whole cells (WC), a sarkosyl insoluble pellet, and a periplasmic extract of *Neisseria gonorrhoeae* were separated in "preparative" 15% SDS-PAGE gels and blotted to NCP as described in Chapter 79, **Subheading 3.1.1.** The 37,000-Dalton major outer membrane protein (POR) and two 44,000-Dalton (44-kDa) proteins, one isolated from a sarkosyl insoluble (membrane) extract (44-kDa Mem.), and the other isolated from a periplasmic extract (44-kDa Peri.), were located on the NCP by Ponceau S staining, excised, and cleaved with BNPS-skatole as described in Chapter 73, **Subheading 3.3.1.** Approximately 30 μg of peptides of each protein were solubilized and separated in an SDS-PAGE gel along with whole-cells (WC), Bio-Rad low-mol-wt markers, and Pharmacia peptide mol-wt markers (mw) (expressed in thousands of Dalton [k]). The gel was stained with Coomassie brilliant blue (CBB) to visualize peptides.

Acknowledgments

The author thanks Pam Gannon for her assistance and the Public Health Service, NIH, NIAID (grant RO1 AI21236), and UM Research Program for their continued support.

References

1. Cleveland, D. W., Fischer, S. G., Kirschner, M. W., and Laemmli, U. K. (1977) Peptide mapping by limited proteolysis in sodium dodecyl sulfate and analysis by gel electrophoresis. *J. Biol. Chem.* **252,** 1102–1106.
2. Judd, R. C. (1986) Evidence for *N*-terminal exposure of the PIA subclass of protein I of *Neisseria gonorrhoeae. Infect. Immunol.* **54,** 408–414.
3. Judd, R. C. (1987) Radioiodination and ^{125}I-labeled peptide mapping on nitrocellulose membranes. *Analyt. Biochem.* **160,** 306–315.
4. Moos, M., Jr. and Nguyen, N. Y. (1988) Reproducible high-yield sequencing of proteins electrophoretically separated and transferred to an inert support. *J. Biol. Chem.* **263,** 6005–6008.
5. Laemmli, U. K. (1970) Cleavage of structural proteins during the assembly of the head of bacteriophage T4. *Nature* **227,** 680–695.
6. Caldwell, H. D. and Judd, R. C. (1982) Structural analysis of chlamydial proteins. *Infect. Immunol.* **38,** 960–968.
7. Schagger, H. and von Jagow, G. (1987) Tricine-sodium dodecyl sulfate-polyacrylamide gel electrophoresis for the separation of proteins in the range from 1 to 100 kDa. *Analyt. Biochem.* **166,** 368–397.

81

Peptide Mapping by High-Performance Liquid Chromatography

Ralph C. Judd

1. Introduction

Peptide mapping is a convenient method for comparing the primary structures of proteins in the absence of sequence data. There are many techniques for specifically cleaving peptides with enzymes or chemical cleavage reagents (*see* Chapters 71–78).

Peptides generated by specific endopeptidic cleavage must be separated and visualized if comparisons are to be made. A convenient separation system is reverse-phase high-performance liquid chromatography (HPLC). The precision of this technique allows for rigorous comparison of primary structure with the added benefit that peptides can be recovered for further analysis. The availability of extremely sensitive in-line UV and γ-detectors makes it possible to visualize extremely small amounts of material.

2. Materials

2.1. Protein Purification, Radiolabeling, and Cleavage

See Chapter 79, **Subheading 2.** for materials needed to purify, radiolabel, and cleave proteins to be compared (*see* **Note 1**).

2.2. HPLC

1. HPLC capable of generating binary gradients.
2. *Preferred:* In-line UV detector and in-line γ-radiation detector (e.g., Model 170, Beckman [Fullerton, CA], or equivalent). *Alternate:* Manual UV and γ-radiation detectors.
3. Fraction collector.
4. Computing integrator or strip chart recorder.
5. Reverse-phase C_{18} column (P/N 27324 S/N, Millipore, Bedford, MA).
6. 0.005% Trifluoroacetic acid (TFA).
7. HPLC-grade methanol.
8. H_2O containing Phe, Trp, and Tyr (1 mg/mL). These are HPLC amino acid markers.

3. Method

HPLC separation of peptides can be used for structural comparisons, but its main advantage is the ability to recover peptides for further studies ***(1–4)***. If radioiodinated

From: *The Protein Protocols Handbook, 2nd Edition*
Edited by: J. M. Walker © Humana Press Inc., Totowa, NJ

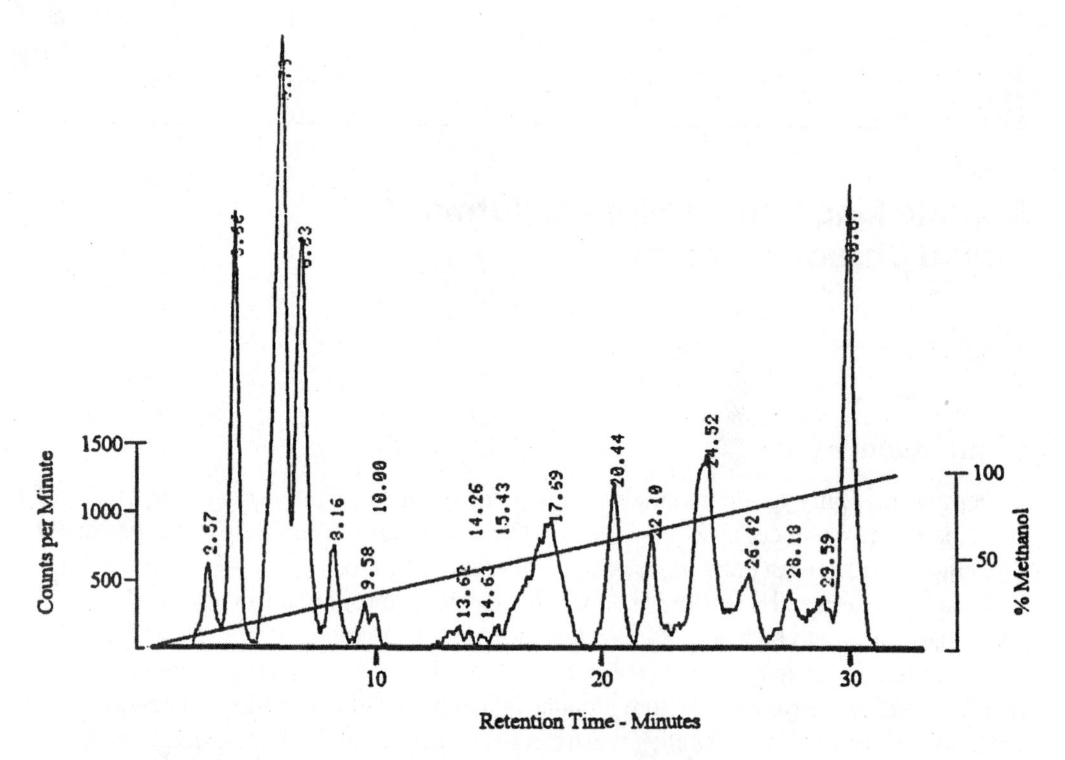

Fig. 1. Example of peptides separated by HPLC. A 37,000-Dalton membrane protein of *Neisseria gonorrhoeae* was radiolabeled and cleaved with thermolysin on an NCP strip as described in Chapter 79, **Subheading 3.3.1.** The peptides (2×10^5 cpm) were injected into a mobile phase of 0.05% TFA-H_2O (flow rate of 1 mL/min) and separated over a 35-min linear gradient to 100% methanol as described in **Subheading 3.** The ^{125}I-labeled peptides were visualized using an in-line γ-radiation detector linked to a computing integrator.

peptides are to be separated, the iodogen method of radiolabeling should be used (*see* Chapter 79, **Subheadings 2.3.** and **3.2.**), since chloramine-T-mediated labeling results in considerable "noise" in HPLC chromatograms. Most peptides can be separated by reverse-phase chromatography using a C_{18} column and a linear gradient of H_2O-0.005% TFA to 100% methanol *(2)*. Different gradients using these solvents or other solvents, such as acetonitrile–0.005% TFA ***(1,2,4)***, isopropanol, and so forth, may be needed to achieve adequate separation. An in-line γ-radiation detector (e.g., Beckman Model 170) is helpful, but fractions can be collected and counted. Ten to 100 µg of sample should be used if UV absorbance (280_{nm}) is to be used to detect peptides. The sensitivity of UV detection roughly parallels that of Coomassie brilliant blue (CBB) staining in gels. Precision of repeated separation should be ± 0.005 min retention time for all peaks, allowing for direct comparisons of elution profiles of different samples (*see* **Note 1**).

1. Rehydrate the peptides in about 0.1 mL of H_2O containing 1 mg/mL of Phe, Trp, and Tyr (internal amino acid markers to verify consistent separations).
2. Inject between 1.5×10^5 and 1×10^6 cpm/separation into H_2O–0.005% TFA mobile phase running at 1 mL/min. A linear gradient (0.05% TFA to 100% methanol) over 0.5–1 h

should yield adequate preliminary separation of peptides (but *see* **Note 2**). Time and gradient profile will vary with the number of peptides to be separated and the nature of the peptides.

3. Monitor amino acid marker elution, or peptide elution if using UV absorbance, using an in-line UV detector at 280_{nm}. Monitor radiolabeled peptide elution using an in-line γ-radiation detector. Alternatively, fractions can be collected, and the marker and peptide elution times monitored manually. Peaks can be collected, washed, dried, and reseparated by SDS-PAGE or 2D TLE-TLC.

4. Notes

1. CNBr and BNPS-skatole are excellent cleavage reagents to generate peptides to be separated by HPLC (*see* Chapter 71, 72, and 79 for details).
2. **Figure 1** demonstrates an HPLC separation of peptides generated by cleavage of a 37,000-Dalton membrane protein of *Neisseria gonorrhoeae* with themolysin. Note that there appears to be several peptides eluting in a single diffuse peak in the 17-min region of the chromatogram. Compression of peptides in this manner is relatively common. The gradient must be modified in this region to resolve these peptides adequately. Common modifications include changing the gradient slope, addition of a third solvent, such as acetonitrile, alteration of flow rate, or extending the time of separation.

Acknowledgments

The author thanks Pam Gannon for her assistance and the Public Health Service, NIH, NIAID (grant RO1 AI21236) and UM Research Program for their continued support.

References

1. Judd, R. C. (1983) ^{125}I-labeled peptide mapping and high-performance liquid chromatography ^{125}I-peptide separation of protein I of four strains of *Neisseria gonorrhoeae*. *J. Liq. Chromatogr.* **6,** 1421–1439.
2. Judd, R. C. (1987) Radioiodination and ^{125}I-labeled peptide mapping on nitrocellulose membranes. *Analyt. Biochem.* **160,** 306–315.
3. Judd, R. C. and Caldwell, H. D. (1985) Comparison of ^{125}I- and ^{14}C-radio-labeled peptides of the major outer membrane protein of *Chlamydia trachomatis* strain L2/434 separated by high-performance liquid chromatography. *J. Liq. Chromatogr.* **8,** 1109–1120.
4. Judd, R. C. and Caldwell, H. D. (1985) Identification and isolation of surface-exposed portions of the major outer membrane protein of *Chlamydia trachomatis* by 2D peptide mapping and high-performance liquid chromatography. *J. Liq. Chromatogr.* **8,** 1559–1571.

82

Production of Protein Hydrolysates Using Enzymes

John M. Walker and Patricia J. Sweeney

1. Introduction

Traditionally, protein hydrolysates for amino acid analysis are produced by hydrolysis in 6 *N* HCl. However, this method has the disadvantage that tryptophan is totally destroyed, serine and threonine partially (5–10%) destroyed, and most importantly, asparagine and glutamine are hydrolyzed to the corresponding acids. Digestion of the protein/peptide with enzymes to produce protein hydrolysate overcomes these problems, and is particularly useful when the concentration of asparagine and glutamine is required. For peptides less than about 35 residues in size, complete digestion can be achieved by digestion with aminopeptidase M and prolidase. For larger polypeptides and proteins, an initial digestion with the nonspecific protease Pronase is required, followed by treatment with aminopeptidase M and prolidase. Since it is important that all enzymes have maximum activity, the following sections will discuss the general characteristics of these enzymes.

1.1. Pronase

Pronase (EC 3.4.24.4) is the name given to a group of proteolytic enzymes that are produced in the culture supernatant of *Streptomyces griseus* K-1 ***(1–3)***. Pronase is known to contain at least ten proteolytic components: five serine-type proteases, two Zn^{2+} endopeptidases, two Zn^{2+}-leucine aminopeptidases, and one Zn^{2+} carboxypeptidase ***(4,5)***. Pronase therefore has very broad specificity, hence its use in cases where extensive or complete degradation of protein is required. The enzyme has optimal activity at pH 7.0–8.0. However, individual components are reported to retain activity over a much wider pH range ***(6–9)***. The neutral components are stable in the pH range 5.0–9.0 in the presence of calcium, and have optimal activity at pH 7.0–8.0. The alkaline components are stable in the pH range 3.0–9.0 in the presence of calcium, and have optimal activity at pH 9.0–10.0 ***(4)***. The aminopeptidase and carboxypeptidase components are stable at pH 5.0–8.0 in the presence of calcium ***(9)***. Calcium ion dependence for the stability of some of the components (mainly exopeptidases) was one of the earliest observations made of Pronase ***(2)***. Pronase is therefore normally used in the presence of 5–20 m*M* calcium. The addition of excess EDTA results in the irreversible loss of 70% of proteolytic activity ***(10)***. Two peptidase components are inactivated by EDTA, but activity is restored by the addition of Co^{2+} or Ca^{2+}. One of these components, the

From: *The Protein Protocols Handbook, 2nd Edition*
Edited by: J. M. Walker © Humana Press Inc., Totowa, NJ

leucine aminopeptidase, is heat stable up to 70°C. All other components of Pronase lose 90% of their activity at this temperature *(5)*. The leucine aminopeptidase is not inactivated by 9 *M* urea, but is labile on dialysis against distilled water *(2)*. Some of the other components of Pronase are also reported to be stable in 8 *M* urea *(2)*, and one of the serine proteases retains activity in 6 *M* guanidinium chloride *(11)*. Pronase retains activity in 1% SDS (w/v) and 1% Triton (w/v) *(12)*.

Among the alkaline proteases, there are at least three that are inhibited by diisopropyl phosphofluoridate (DFP) *(10)*. In general, the neutral proteinases are inhibited by EDTA, and the alkaline proteinases are inhibited by DFP *(4)*. No single enzyme inhibitor will inhibit all the proteolytic activity in a Pronase sample.

1.2. Aminopeptidase M

Aminopeptidase M (EC 3.4.11.2), a zinc-containing metalloprotease, from swine kidney microsomes *(13–16)* removes amino acids sequentially from the N-terminus of peptides and proteins. The enzyme cleaves N-terminal residues from all peptides having a free α-amino or α-imino group. However, in peptides containing an X-Pro sequence, where X is a bulky hydrophobic residue (Leu, Tyr, Trp, Met sulfone), or in the case of an N-blocked amino acid, cleavage does not occur. It is for this reason that prolidase is used in conjunction with aminopeptidase M to produce total hydrolysis of peptides. The enzyme is stable at pH 7.0 at temperatures up to 65°C, and is stable between pH 3.5 and 11.0 at room temperature for at least 3 h *(15)*. It is not affected by sulfhydryl reagents, has no requirements for divalent metal ions, is stable in the presence of trypsin, and is active in 6 *M* urea. It is not inhibited by PMSF, DFP, or PCMB. It is, however, irreversibly denatured by alcohols and acetone, and 0.5 *M* guanidinium chloride, but cannot be precipitated by trichloroacetic acid *(15)*. It is inhibited by 1,10-phenanthroline (10M) *(16)*.

Alternative names for the enzyme are amino acid arylamidase, microsomal alanyl aminopeptidase, and α-aminoacyl peptide hydrolase.

1.3. Prolidase

Prolidase (EC 3.4.13.9) is highly specific, and cleaves dipeptides with a prolyl or hydroxyprolyl residue in the carboxyl-terminal position *(17,18)*. It has no activity with tripeptides *(19)*. The rate of release is inversely proportional to the size of the amino-terminal residue *(19)*. The enzyme's activity depends on the nature of the amino acids bound to the imino acid. For optimal activity, amino acid side chains must be as small as possible and apolar to avoid steric competition with the enzyme receptor site. The enzyme has the best affinity for alanyl proline and glycyl proline. The enzyme has optimal activity at pH 6.0–8.0, but it is normally used at pH 7.8–8.0 *(20)*. Manganous ions are essential for optimal catalytic activity. The enzyme is inhibited by 4-chloromercuribenzoic acid, iodoacetamide, EDTA, fluoride, and citrate. However, if Mn^{2+} is added before iodoacetamide, no inhibition is observed *(21)*.

Alternative names for the enzyme are imidodipeptidase, proline dipeptidase, amino acyl L-proline hydrolyase, and peptidase D.

2. Materials

1. Buffer: 0.05 *M* ammonium bicarbonate, pH 8.0 (no pH adjustment needed) or 0.2 *M* sodium phosphate, pH 7.0 (*see* **Note 1**).

2. Pronase: The enzyme is stable at 4°C for at least 6 mo and is usually stored as a stock solution of 5–20 mg/mL in water at –20°C.
3. Aminopeptidase M: The lyophilized enzyme is stable for several years at –20°C. A working solution can be prepared by dissolving about 0.25 mg of protein in 1 mL of deionized water to give a solution of approx 6 U of activity/mL. This solution can be aliquoted and stored frozen for several months at –20°C.
4. Prolidase: The lyophilized enzyme is stable for many months when stored at –20°C and is stable for several weeks at 4°C if stored in the presence of 2 m*M* $MnCl_2$ and 2 m*M* β-mercaptoethanol *(18)*.

3. Methods

3.1. Digestion of Proteins (22)

1. Dissolve 0.2-μmol of protein in 0.2 mL of 0.05*M* ammonium bicarbonate buffer, pH 8.0, or 0.2 *M* sodium phosphate, pH 7.0 (*see* **Note 1**).
2. Add Pronase to 1% (w/w), and incubate at 37°C for 24 h.
3. Add aminopeptidase M at 4% (w/w), and incubate at 37°C for a further 18 h.
4. Since in many cases the X-Pro- bond is not completely cleaved by these enzymes, to ensure complete cleavage of proline-containing polypeptides, the aminopeptidase M digest should be finally treated with 1 μg of prolidase for 2 h at 37°C.
5. The sample can now be lyophilized and is ready for amino acid analysis (*see* **Note 2**).

3.2. Digestion of Peptides (22)

This procedure is appropriate for polypeptides less than about 35 residues in length. For larger polypeptides, use the procedure described in **Subheading 3.1.**

1. Dissolve the polypeptides (1 nmol) in 24 μL of 0.2 *M* sodium phosphate buffer, pH 7.0, or 0.05 *M* ammonium bicarbonate buffer, pH 8.0 (*see* **Note 1**).
2. Add 1 μg of aminopeptidase M (1 μL), and incubate at 37°C.
3. For peptides containing 2–10 residues, 8 h are sufficient for complete digestion. For larger peptides (11–35 residues), a further addition of enzyme after 8 h is needed, followed by a further 16-h incubation.
4. To ensure complete cleavage at proline residues, finally treat the digest with 1 μg of prolidase for 2 h at 37°C.
5. The sample can now be lyophilized and is ready for amino acid analysis (*see* **Note 2**).

4. Notes

1. Sodium phosphate buffer should be used if ammonia interferes with the amino acid analysis.
2. When using two enzymes or more, there is often an increase in the background amino acids owing to hydrolysis of each enzyme. It is therefore important to carry out a digestion blank to correct for these background amino acids.

References

1. Hiramatsu, A. and Ouchi, T. (1963) On the proteolytic enzymes from the commercial protease preparation of *Streptomyces griseus* (Pronase P). *J. Biochem. (Tokyo)* **54(4),** 462–464.
2. Narahashi, Y. and Yanagita, M. (1967) Studies on proteolytic enzymes (Pronase) of *Streptomyces griseus* K-1. Nature and properties of the proteolytic enzyme system. *J. Biochem. (Tokyo)* **62(6),** 633–641.
3. Wählby, S. and Engström, L. (1968) Studies on *Streptomyces griseus* protease. Amino acid sequence around the reactive serine residue of DFP-sensitive components with esterase activity. *Biochim. Biophys. Acta* **151,** 402–408.

4. Narahashi, Y., Shibuya, K., and Yanagita, M. (1968) Studies on proteolytic enzymes (Pronase) of *Streptomyces griseus* K-1. Separation of exo- and endopeptidases of Pronase. *J. Biochem.* **64(4),** 427–437.
5. Yamskov, I. A., Tichonova, T. V., and Davankov, V. A. (1986) Pronase-catalysed hydrolysis of amino acid amides. *Enzyme Microb. Technol.* **8,** 241–244.
6. Gertler, A. and Trop, M. (1971) The elastase-like enzymes from *Streptomyces griseus* (Pronase). Isolation and partial characterization. *Eur. J. Biochem.* **19,** 90–96.
7. Wählby, S. (1969) Studies on *Streptomyces griseus* protease. Purification of two DFP-reactin enzymes. *Biochim. Biophys. Acta* **185,** 178–185.
8. Yoshida, N., Tsuruyama, S., Nagata, K., Hirayama, K., Noda, K., and Makisumi, S. (1988) Purification and characterisation of an acidic amino acid specific endopeptidase of *Streptomyces griseus* obtained from a commercial preparation (Pronase). *J. Biochem.* **104,** 451–456.
9. Narahashi, Y. (1970) Pronase. *Meth. Enzymol.* **19,** 651–664.
10. Awad, W. M., Soto, A. R., Siegei, S., Skiba, W. E., Bernstrom, G. G., and Ochoa, M. S. (1972) The proteolytic enzymes of the K-1 strain of *Streptomyces griseus* obtained from a commercial preparation (Pronase). Purification of four serine endopeptidases. *J. Biol. Chem.* **257,** 4144–4154.
11. Siegel, S., Brady, A. H., and Awad, W. M. (1972) Proteolytic enzymes of the K-1 strain of *Streptomyces griseus* obtained from a commercial preparation (Pronase). Activity of a serine enzyme in 6*M* guanidinium chloride. *J. Biol. Chem.* **247,** 4155–4159.
12. Chang, C. N., Model, P., and Blobel, G. (1979) Membrane biogenesis: cotranslational integration of the bacteriophage F1 coat protein into an *Escherichia coli* membrane fraction. *Proc. Natl. Acad. Sci. USA* **76(3),** 1251–1255.
13. Pfleiderer, G., Celliers, P. G., Stanulovic, M., Wachsmuth, E. D., Determann, H., and Braunitzer, G. (1964) Eizenschafter und analytische anwendung der aminopeptidase aus nierenpartikceln. *Biochem. Z.* **340,** 552–564.
14. Pfleiderer, G. and Celliers, P. G. (1963) Isolation of an aminopeptidase in kidney tissue. *Biochem. Z.* **339,** 186–189.
15. Wachsmuth, E. D., Fritze, I., and Pfleiderer, G. (1966) An aminopeptidase occuring in pig kidney. An improved method of preparation. Physical and enzymic properties. *Biochemistry* **5(1),** 169–174.
16. Pfleiderer, G. (1970) Particle found aminaopeptidase from pig kidney. *Meth. Enzymol.* **19,** 514–521.
17. Sjöstrom, H., Noren, O., and Jossefsson, L. (1973) Purification and specificity of pig intestinal prolidase. *Biochim. Biophys. Acta* **327,** 457–470.
18. Manao, G., Nassi, P., Cappugi, G., Camici, G., and Ramponi, G. (1972) Swine kidney prolidase: Assay isolation procedure and molecular properties. *Physiol. Chem. Phys.* **4,** 75–87.
19. Endo, F., Tanoue, A., Nakai, H., Hata, A., Indo, Y., Titani, K., and Matsuela, I. (1989) Primary structure and gene localization of human prolidase. *J. Biol. Chem.* **264(8),** 4476–4481.
20. Myara, I., Charpentier, C., and Lemonnier, A. (1982) Optimal conditions for prolidase assay by proline colourimetric determination: application to iminodipeptiduria. *Clin. Chim. Acta* **125,** 193–205.
21. Myara, I., Charpentier, C., and Lemonnier, A. (1984) Minireview: Prolidase and prolidase deficiency. *Life Sci.* **34,** 1985–1998.
22. Jones, B. N. (1986) Amino acid analysis by *o*-phthaldialdehyde precolumn derivitization and reverse-phase HPLC, in *Methods of Protein Microcharacterization* (Shively, J. E., ed.), Humana Press, Totowa, NJ, pp. 127–145.

83

Amino Acid Analysis by Precolumn Derivatization with 1-Fluoro-2,4-Dinitrophenyl-5-L-Alanine Amide (Marfey's Reagent)

Sunil Kochhar, Barbara Mouratou, and Philipp Christen

1. Introduction

Precolumn modification of amino acids and the subsequent resolution of their derivatives by reverse-phase high-performance liquid chromatography (RP-HPLC) is now the preferred method for quantitative amino acid analysis. The derivatization step introduces covalently bound chromophores necessary not only for interactions with the apolar stationary phase for high resolution but also for photometric or fluorometric detection.

Marfey's reagent, 1-fluoro-2,4-dinitrophenyl-5-L-alanine amide or (*S*)-2-[(5-fluoro-2,4-dinitrophenyl)-amino]propanamide, can be used to separate and to determine enantiomeric amino acids *(1)*. The reagent reacts stoichiometrically with the amino group of enantiomeric amino acids to produce stable diastereomeric derivatives, which can readily be separated by reverse-phase HPLC (**Fig. 1**). The dinitrophenyl alanine amide moiety strongly absorbs at 340 nm ($\varepsilon = 30{,}000\ M^{-1} \cdot cm^{-1}$), allowing detection in the subnanomolar range.

Precolumn derivatization with the reagent is also used to quantify the 19 commonly analyzed L-amino acids *(2)*. Major advantages of Marfey's reagent over other precolumn derivatizations are: (1) possibility to carry out chromatography on any multipurpose HPLC instrument without column heating; (2) detection at 340 nm is insensitive to most solvent impurities; (3) simultaneous detection of proline in a single chromatographic run; and (4) stable amino acid derivatives. For the occasional user, the simple methodology provides an attractive and inexpensive alternative to the dedicated amino acid analyzer. Further development to on-line derivatization and microbore chromatography has been reported *(3)*. The precolumn derivatization with Marfey's reagent has found applications in many diverse areas of biochemical research *(4–15)* including determination of substrates and products in enzymic reactions of amino acids (*see* **Note 1**).

From: *The Protein Protocols Handbook, 2nd Edition*
Edited by: J. M. Walker © Humana Press Inc., Totowa, NJ

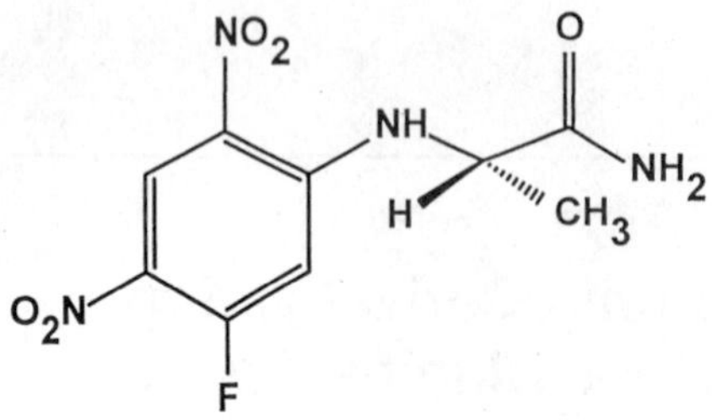

Fig.1. Marfey's reagent (I) and L,L-diastereomer derivative from L-amino acid and the reagent (II).

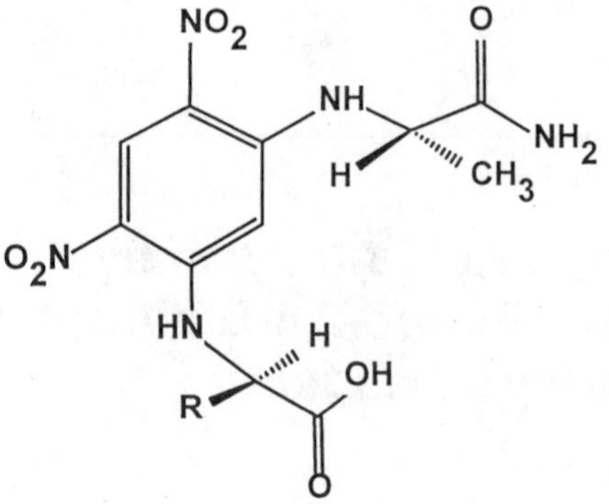

2. Materials

2.1. Vapor-Phase Protein Hydrolysis

1. Pyrex glass vials (25–50 × 5 mm) from Corning.
2. Screw-capped glass vials.
3. 6 *N* HCl from Pierce.

2.2. Derivatization Reaction

1. Amino acid standard solution H from Pierce.
2. L-Amino acid standard kit LAA21 from Sigma.
3. Marfey's reagent from Pierce. **Caution: Marfey's reagent is a derivative of 1-fluoro-2,4-dinitrobenzene, a suspected carcinogen.** Recommended precautions should be followed in its handling ***(16)***.
4. HPLC/spectroscopic grade triethylamine, methanol, acetone, and dimethyl sulfoxide (DMSO) from Fluka.

2.3. Chromatographic Analysis

1. Solvent delivery system: Any typical HPLC system available from a number of manufacturers can be used for resolution of derivatized amino acids. We have recently used HPLC systems from Bio-Rad (HRLC 800 equipped with an autoinjector AST 100), Hewlett-Packard 1050 system and Waters system.
2. Column: A silica-based C_8 reverse-phase column, for example, LiChrospher 100 RP 8 (250 × 4.6 mm; 5 µm from Macherey-Nagel), Aquapore RP 300 (220 × 4.6 mm; 7 µm from Perkin-Elmer), Nucleosil 100-C_8 (250 × 4.6 mm; 5 µm from Macherey-Nagel), or Vydac C_8 (250 × 4.6 mm; 5 µm from The Sep/a/ra/tions group).
3. Detector: A UV/VIS HPLC detector equipped with a flow cell of lightpath 0.5–1 cm and a total volume of 3–10 µL (e.g., Bio-Rad 1790 UV/VIS monitor, Hewlett-Packard Photodiode Array 1050, Waters Photodiode Array 996). The derivatized amino acids are detected at 340 nm.

4. Peak integration: Standard PC-based HPLC software with data analysis program can be employed to integrate and quantify the amino acid peaks (e.g., ValueChrom from Bio-Rad, Chemstation from Hewlett-Packard or Millennium from Waters).
5. Solvents: The solvents should be prepared with HPLC grade water and degassed.
   a. Solvent A: 13 m*M* trifluoroacetic acid plus 4% (v/v) tetrahydrofuran in water.
   b. Solvent B: Acetonitrile (50% v/v) in solvent A.

3. Methods

3.1. Vapor-Phase Protein Hydrolysis

1. Transfer 50–100 pmol of protein sample or 20 μL of amino acid standard solution H containing 2.5 μmol/mL each of 17 amino acids into glass vials and dry under reduced pressure in a SpeedVac concentrator.
2. Place the sample vials in a screw cap glass vial containing 200 μL of 6 *N* HCl.
3. Flush the vial with argon for 5–15 min and cap it airtight.
4. Incubate the glass vial at 110°C for 24 h or 150°C for 2 h in a dry-block heater.
5. After hydrolysis, remove the glass vials from the heater, cool to room temperature, and open them slowly.
6. Remove the insert vials, wipe their outside clean with a soft tissue paper, and dry under reduced pressure.

3.2. Derivatization Reaction

1. Add 50 μL of triethylamine/methanol/water (1:1:2) to the dried sample vials, mix vigorously by vortexing, and dry them under reduced pressure (*see* **Note 2**).
2. Prepare derivatization reagent solution (18.4 m*M*) by dissolving 5 mg of Marfey's reagent in 1 mL of acetone.
3. Dissolve the dried amino acid mixture or the hydrolyzed protein sample in 100 μL of 25% (v/v) triethylamine and add 100 μL of the Marfey's reagent solution and mix by vortexing (*see* **Note 3**).
4. Incubate the reaction vial at 40°C for 60 min with gentle shaking protected from light (*see* **Note 4**).
5. After incubation, stop the reaction by adding 20 μL of 2 *N* HCl (*see* **Note 5**). Dry the reaction mixture under reduced pressure.
6. Store the dried samples at –20°C in the dark until used (*see* **Note 6**).
7. Dissolve the dry sample in 1 mL of 50% (v/v) DMSO.

3.3. Chromatographic Analysis

1. Prime the HPLC system according to the manufacturer's instructions with solvent A and solvent B.
2. Equilibrate the column and detector with 90% solvent A and 10% solvent B.
3. Bring samples for analysis to room temperature; dilute, if necessary, with solvent A, and inject 20 μL (50–1000 pmol) onto the column (*see* **Note 7**).
4. Elute with gradient as described in **Table 1** (*see* **Note 8**). Resolution of 2,4-dinitrophenyl-5-L-alanine amide derivatives of 18 commonly occurring L-amino acids and of cysteic acid is achieved within 120 min (**Fig. 2**) (*see* **Note 9**).
5. Determine the response factor for each amino acid from the average peak area of standard amino acid chromatograms at 100, 250, 500, and 1000 pmol amounts.
6. When HPLC is completed, wash the column and fill the pumps with 20% (v/v) degassed methanol for storage of the system. Before reuse, purge with 100% methanol.

Table 1
HPLC Program for Resolution of Amino Acids Derivatized with Marfey's Reagent

Time (min)	% Solvent A	% Solvent B	Input
0	90	10	
15	90	10	Detector auto zero
15.1	90	10	Inject
115	50	50	
165	0	100	
170	0	100	

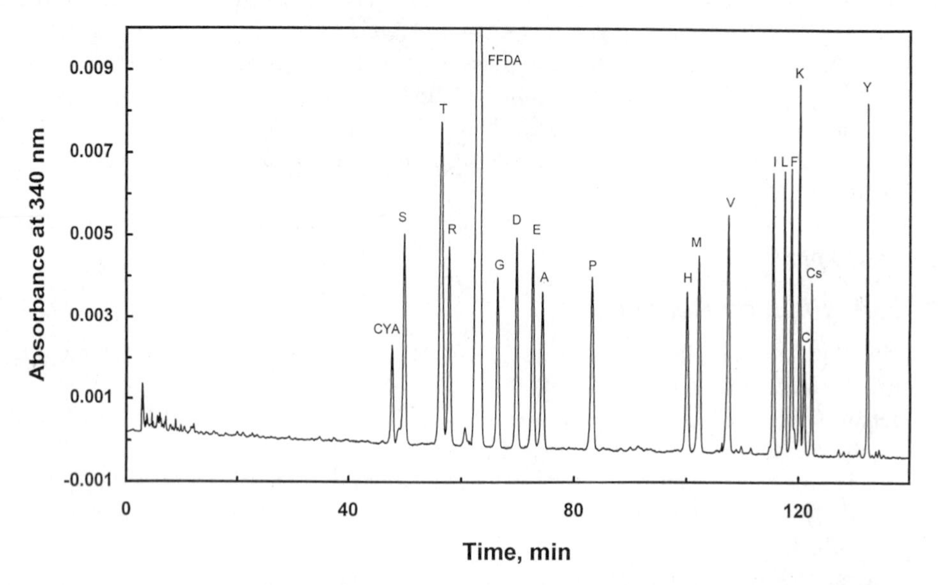

Fig. 2. Separation of 2,4-dinitrophenyl-5-L-alanine amide derivatives of L-amino acids by HPLC. A 20-µL aliquot from the amino acid standard mixture (standard H from Pierce Chemical Company) was derivatized with Marfey's reagent. Chromatographic conditions: Column, LiChrospher 100 RP 8 (250 × 4.6 mm; 5 µm from Macherey-Nagel); sample, 100 pmol of diastereomeric derivatives of 18 L-amino acids; solvent A, 13 m*M* trifluoroacetic acid plus 4% (v/v) tetrahydrofuran in water; solvent B, 50% (v/v) acetonitrile in solvent A; flow rate, 1 mL/min; detection at 340 nm; elution with a linear gradient of solvent B in solvent A (**Table 1**). The amino acid derivatives are denoted by single-letter code, cysteic acid as CYA and cystine as Cs. FFDA indicates the reagent peak; the peaks without denotations are reagent-related as shown in an independent chromatographic run of the reagent alone.

4. Notes

1. For analysis of amino acids as substrates or products in an enzymic reaction, deproteinization is required. Add 4 *M* perchloric acid to a final concentration of 1 *M* and incubate on ice for at least 15 min. Excess perchloric acid is precipitated as $KClO_4$ by adding an equal volume of ice-cold 2 *M* KOH. After centrifugation for 15 min at 4°C, the supernatant is collected, dried, and used for derivatization.
2. Complete removal of HCl is absolutely essential for quantitative reaction between amino acids and Marfey's reagent.

3. The molar ratio of the Marfey's reagent to the total amino acids should not be more than 3:1.
4. During derivatization, the color of the reaction mixture turns from yellow to orange-red.
5. On acidification, the color of the reaction mixture turns to yellow.
6. The dried amino acids are stable for over one month when stored at –20°C in the dark. In 50% (v/v) DMSO, derivatives are stable for 72 h at 4°C and for > 6 wk at –20°C.
7. To obtain good reproducibility, an HPLC autoinjector is highly recommended.
8. Silica-based C_8 columns from different commercial sources produce baseline resolution of the 19 amino acid derivatives; nevertheless, the gradient slope of solvent B should be optimized for each column. *S*-carboxymethyl-L-cysteine and tryptophan, if included, are also separated in a single chromatographic run (*see* **ref. 2**).
9. The identity of each peak is established by adding a threefold molar excess of the amino acid in question. A reagent blank should be run with each batch of the Marfey's reagent to identify reagent-related peaks. Excess reagent interferes with baseline separation of the arginine and glycine peaks. Lysine and tyrosine are separated as disubstituted derivatives.

References

1. Marfey, P. (1984) Determination of D-amino acids. II. Use of a bifunctional reagent, 1,5-difluoro-2,4-dinitrobenzene. *Carlsberg Res. Commun.* **49,** 591–596.
2. Kochhar, S. and Christen, P. (1989) Amino acid analysis by high-performance liquid chromatography after derivatization with 1-fluoro-2,4-dinitrophenyl-5-L-alanine amide. *Analyt. Biochem.* **178,** 17–21.
3. Scaloni, A., Simmaco, M., and Bossa, F. (1995) D-L Amino acid analysis using automated precolumn derivatization with 1-fluoro-2,4-dinitrophenyl-5-alanine amide. *Amino Acids* **8,** 305–313.
4. Kochhar, S. and Christen, P. (1988) The enantiomeric error frequency of aspartate aminotransferase. *Eur. J. Biochem.* **175,** 433–438.
5. Martínez del Pozo, A., Merola, M., Ueno, H., Manning, J. M., Tanizawa, K., Nishimura, K., et al. (1989) Stereospecificity of reactions catalyzed by bacterial D-amino acid transaminase. *J. Biol. Chem.* **264,** 17,784–17,789.
6. Szókán, G., Mezö, G., Hudecz, F., Majer, Z., Schön, I., Nyéki, O., et al. (1989) Racemization analyses of peptides and amino acid derivatives by chromatography with pre-column derivatization. *J. Liq. Chromatogr.* **12,** 2855–2875.
7. Adamson, J. G., Hoang, T., Crivici, A., and Lajoie, G. A. (1992) Use of Marfey's reagent to quantitate racemization upon anchoring of amino acids to solid supports for peptide synthesis. *Analyt. Biochem.* **202,** 210–214.
8. Goodlett, D. R., Abuaf, P. A., Savage, P. A., Kowalski, K. A., Mukherjee, T. K., Tolan, J. W., et al. (1995) Peptide chiral purity determination: hydrolysis in deuterated acid, derivatization with Marfey's reagent and analysis using high-performance liquid chromatography-electrospray ionization-mass spectrometry. *J. Chromatogr. A* **707,** 233–244.
9. Moormann, M., Zähringer, U., Moll, H., Kaufmann, R., Schmid, R., and Altendorf, K. (1997) New glycosylated lipopeptide incorporated into the cell wall of smooth variant of *Gordona hydrophobica. J. Biol. Chem.* **272,** 10,729–10,738.
10. Gramatikova, S. and Christen, P. (1997) Monoclonal antibodies against N^{a}-(5'-phosphopyridoxyl)-L-lysine. Screening and spectrum of pyridoxal 5'-phosphate-dependent activities toward amino acids. *J. Biol. Chem.* **272,** 9779–9784.
11. Vacca, R. A., Giannattasio, S., Graber, R., Sandmeier, E., Marra, E., and Christen, P. (1997) Active-site Arg→Lys substitutions alter reaction and substrate specificity of aspartate aminotransferase. *J. Biol. Chem.* **272,** 21,932–21,937.

12. Wu, G. and Furlanut, M. (1997) Separation of DL-dopa by means of micellar electrokinetic capillary chromatography after derivatization with Marfey's reagent. *Pharmacol. Res.* **35,** 553–556.
13. Goodnough, D. B., Lutz, M. P., and Wood, P. L. (1995) Separation and quantification of D- and L-phosphoserine in rat brain using *N*-alpha-(2,4-dinitro-5-fluorophenyl)-L-alaninamide (Marfey's reagent) by high-performance liquid chromatography with ultraviolet detection. *J. Chromatogr. B. Biomed. Appl.* **672,** 290–294.
14. Yoshino, K., Takao, T., Suhara, M., Kitai, T., Hori, H., Nomura, K., et al. (1991) Identification of a novel amino acid, *o*-bromo-L-phenylalanine, in egg-associated peptides that activate spermatozoa. *Biochemistry* **30,** 6203–6209.
15. Tran, A. D., Blanc, T., and Leopold, E. J. (1990) Free solution capillary electrophoresis and micellar electrokinetic resolution of amino acid enantiomers and peptide isomers with L- and D-Marfey's reagents. *J. Chromatogr.* **516,** 241–249.
16. Thompson, J. S. and Edmonds, O. P. (1980) Safety aspects of handling the potent allergen FDNB. *Ann. Occup. Hyg.* **23,** 27–33.

84

Molecular Weight Estimation for Native Proteins Using High-Performance Size-Exclusion Chromatography

G. Brent Irvine

1. Introduction

The chromatographic separation of proteins from small molecules on the basis of size was first described by Porath and Flodin, who called the process "gel filtration" *(1)*. Moore applied a similar principle to the separation of polymers on crosslinked polystyrene gels in organic solvents, but named this "gel permeation chromatography" *(2)*. Both terms came to be used by manufacturers of such supports for the separation of proteins, leading to some confusion. The term size-exclusion chromatography is more descriptive of the principle on which separation is based and has largely replaced the older names, although the expression "gel filtration" is still commonly used by biochemists to describe separation of proteins in aqueous mobile phases. Hagel gives a comprehensive review of the subject with emphasis on proteins *(3)* and there is also much useful information in a technical booklet published by a leading manufacturer *(4)*.

Conventional liquid chromatography is carried out on soft gels, with flow controlled by peristaltic pumps and run times on the order of several hours. These gels are too compressible for use in high-performance liquid chromatography (HPLC), but improvements in technology have led to the introduction of new supports with decreased particle size (5–13 µm) and improved rigidity. This technique is usually called high-performance size-exclusion chromatography (HPSEC), but is sometimes referred to as size-exclusion high-performance liquid chromatography (SE-HPLC). Columns (about 1 × 30 cm) are sold prepacked with the support, and typically have many thousands of theoretical plates. They can be operated at flow rates of about 1 mL/min, giving run times of about 12 min, 10–100 times faster than conventional chromatography on soft gels. A medium pressure high-performance system called fast protein liquid chromatography (FPLC) that avoids the use of stainless steel components has been developed by one major supplier (Amersham Pharmacia Biotech). Prepacked columns of Superose and Superdex developed for this system can also be operated on ordinary HPLC systems.

From: *The Protein Protocols Handbook, 2nd Edition*
Edited by: J. M. Walker © Humana Press Inc., Totowa, NJ

As well as being a standard chromatographic mode for desalting and purifying proteins, size-exclusion chromatography can be used for estimation of molecular weights. This is true only for ideal size-exclusion chromatography, in which the support does not interact with solute molecules (*see* **Note 1**). Because the physical basis for discrimination in size-exclusion chromatography is size, one would expect that there would be some parameter related to the dimensions and shape of the solute molecule that would determine its K_d value (*see* **Note 2**). All molecules with the same value for this parameter should have identical K_d values on a size-exclusion chromatography column. Dimensional parameters that have been suggested for such a "universal calibration" include two terms related to the hydrodynamic properties of proteins, namely Stokes radius (R_S) and viscosity radius (R_h). Current evidence appears to indicate that R_h is more closely correlated to size-exclusion chromatography behavior than is R_S ***(5)***. However, even this parameter is not suitable for a universal calibration that includes globular proteins, random coils, and rods ***(6)***. Hence calibration curves prepared with globular proteins as standards cannot be used for the assignment of molecular weights to proteins with different shapes, such as the rodlike protein myosin. Several mathematical models that relate size-exclusion chromatography to the geometries of protein solutes and support pores have been proposed (*see* **Note 3**). However, in practical terms, for polymers of the same shape, plots of log molecular weight against K_d have been found to give straight lines within the range $0.1 < K_d < 0.9$ (*see* **Note 4**). It has been found that the most reliable measurements of molecular weight by HPSEC are obtained under denaturing conditions, when all proteins have the same random coil structure. Disulfide bonds must be reduced, usually with dithiothreitol, in a buffer that destroys secondary and tertiary structure. Buffers containing guanidine hydrochloride ***(7,8)*** or sodium dodecyl sulfate (SDS) have been used for this purpose. However, the use of denaturants has many drawbacks, which are described in **Note 5**. In any case, polyacrylamide gel electrophoresis in SDS-containing buffers is widely used for determining the molecular weights of protein subunits. This technique can also accommodate multiple samples in the same run, although each run takes longer than for HPSEC.

The method described in this chapter is for the estimation of molecular weights of native proteins. The abilities to measure bioactivity, and to recover native protein in high yield, make this an important method, even though a number of very basic, acidic, or hydrophobic proteins will undergo non-ideal size exclusion under these conditions.

2. Materials

2.1. Apparatus

An HPLC system for isocratic elution is required. This comprises a pump, an injector, a size-exclusion column and a guard column (*see* **Note 6**), a UV-detector, and a data recorder. There are many size-exclusion columns based on surface-modified silica on the market. The results described here were obtained using a Zorbax Bioseries GF-250 column (0.94 × 25 cm), of particle size 4 µm, sold by Agilent Technologies. This column can withstand very high back pressures (up to 380 Bar or 5500 psi) and can be run at flow rates up to 2 mL/min with little loss of resolution ***(10)***. It has a molecular weight exclusion limit for globular proteins of several hundred thousand.

Table 1
Protein Standards

Protein	Molecular weight
Thyroglobulin	669,000
Apoferritin	443,000
β-Amylase	200,000
Immunoglobulin G	160,000
Alcohol dehydrogenase	150,000
Bovine serum albumin	66,000
Ovalbumin	42,700
β-Lactoglobulin	36,800
Carbonic anhydrase	29,000
Trypsinogen	24,000
Soybean trypsin inhibitor	20,100
Myoglobin	16,900
Ribonuclease A	13,690
Insulin	5900
Glucagon	3550

All proteins listed were obtained from Sigma, Poole, England.

The exclusion limit for the related GF-450 column is about a million. Other suitable columns include TSK-GEL SW columns (Tosoh Biosep), and Superdex and Superose columns (Amersham Pharmacia Biotech).

2.2. Chemicals

1. 0.2 *M* Disodium hydrogen orthophosphate (Na_2HPO_4).
2. 0.2 *M* Sodium dihydrogen orthophosphate (NaH_2PO_4).
3. 0.2 *M* Sodium phosphate buffer, pH 7.0: Mix 610 mL of 0.2 *M* Na_2HPO_4 with 390 mL of 0.2 *M* NaH_2PO_4. Filter through a 0.22-µm filter (Millipore type GV).
4. Solutions of standard proteins: Dissolve each protein in 0.2 *M* sodium phosphate buffer, pH 7.0 at a concentration of about 0.5 mg/mL. Filter the samples through a 0.45-µm filter (Millipore type HV). Proteins suitable for use as standards are listed in **Table 1** (*see* **Note 7**).
5. Blue dextran, average molecular weight 2,000,000 (Sigma), 1 mg/mL in 0.2 *M* sodium phosphate buffer, pH 7.0. Filter through a 0.45-µm filter (Millipore type HV).
6. Glycine, 10 mg/mL in 0.2 *M* sodium phosphate buffer, pH 7.0. Filter through a 0.45-µm filter (Millipore type HV).

3. Method

1. Allow the column to equilibrate in the 0.2 *M* sodium phosphate buffer, pH 7.0, at a flow rate of 1 mL/min (*see* **Note 8**) until the absorbance at 214 nm is constant.
2. Inject a solution (20 µL) of a very large molecule, such as blue dextran (*see* **Note 9**) to determine V_0. Repeat with 20 µL of water (negative absorbance peak) or a solution of a small molecule, such as glycine, to determine V_t.
3. Inject a solution (20 µL) of one of the standard proteins and determine its elution volume, V_e, from the time at which the absorbance peak is at a maximum. Repeat this procedure until all the standards have been injected. A chromatogram showing the separation of seven solutes during a single run on a Zorbax Bio-series GF-250 column is shown in **Fig. 1**.

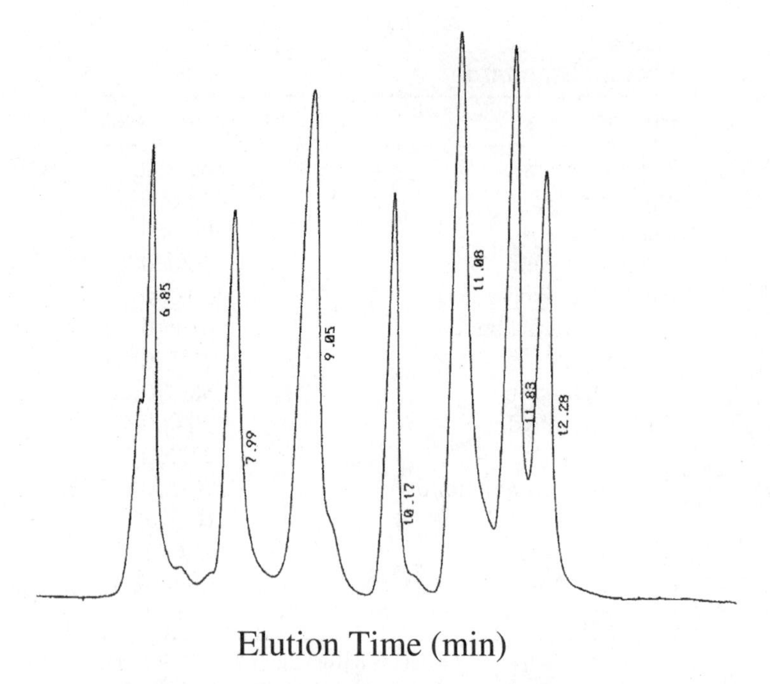

Fig. 1. Separation of a mixture of seven solutes on a Zorbax Bio-series GF-250 column. Twenty microliters of a mixture containing about 1.5 µg of each protein was injected. The solutes were, in order of elution, thyroglobulin, alcohol dehydrogenase, ovalbumin, myoglobin, insulin, glucagon, and sodium azide. The number beside each peak is the elution time in minutes. The absorbance of the highest peak, insulin, was 0.105. The equipment was a Model 501 Pump, a 441 Absorbance Detector operating at 214 nm, a 746 Data Module, all from Waters (Millipore, Milford, MA, USA) and a Rheodyne model 7125 injector (Rheodyne, Cotati, CA, USA) with 20-µL loop. The column was a Zorbax Bio-Series GF-250 with guard column (Aligent Technologies). The flow rate was 1 mL/min and the chart speed was 1 cm/min. The attenuation setting on the Data Module was 128.

Calculate K_d (*see* **Note 2**) for each protein and plot K_d against log molecular weight. A typical plot is shown in **Fig. 2**.

4. Inject a solution (20 µL) of the protein of unknown molecular weight and measure its elution volume, V_e, from the absorbance profile. If the sample contains more than one protein, and the peaks cannot be assigned with certainty, collect fractions and assay each fraction for the relevant activity.
5. Calculate K_d for the unknown protein and use the calibration plot to obtain an estimate of its molecular weight.

4. Notes

1. Silica to which a hydrophilic phase such as a diol has been bonded still contains underivatized silanol groups. Above pH 3.0 these are largely anionic and will interact with ionic solutes, leading to non-ideal size-exclusion chromatography. Depending on the value of its isoelectric point, a protein can be cationic or anionic at pH 7.0. Proteins that are positively charged will undergo ion exchange, causing them to be retarded. Conversely, anionic proteins will experience electrostatic repulsion from the pores, referred to as "ion exclusion," and will be eluted earlier than expected on the basis of size alone. When size-

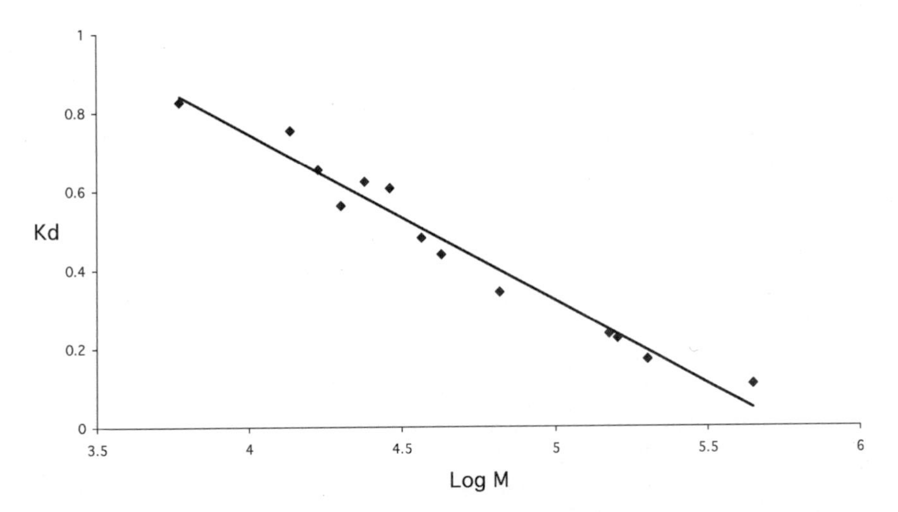

Fig. 2. Plot of K_d vs log molecular weight for those proteins listed in **Table 1** with K_d values between 0.1 and 0.9 (i.e., all except glucagon and thyroglobulin). Chromatography was carried out as described in the caption to **Fig. 1**. V_0 was determined to be 6.76 mL from the elution peak of blue dextran. *Vt* was determined from the elution peak of glycine and from the negative peak given by injecting water, both of which gave a value of 11.99 mL. The regression line $y = -0.4234x + 2.4378$, $r^2 = 0.9657$ was computed using all the points shown.

exclusion chromatography is carried out at a low pH, the opposite behavior is found, with highly cationic proteins being eluted early and anionic ones being retarded. To explain this behavior, it has been suggested that at pH 2.0 the column may have a net positive charge ***(11)***. To reduce ionic interactions it is necessary to use a mobile phase of high ionic strength. On the other hand, as ionic strength increases, this promotes the formation of hydrophobic interactions. To minimize both ionic and hydrophobic interactions, the mobile phase should have an ionic strength between 0.2 and 0.5 *M* ***(12)***.

2. The support used in size-exclusion chromatography consists of particles containing pores. The molecular size of a solute molecule determines the degree to which it can penetrate these pores. Molecules that are wholly excluded from the packing emerge from the column first, at the void volume, V_0. This represents the volume in the interstitial space (outside the support particles) and is determined by chromatography of very large molecules, such as blue dextran or DNA. Molecules that can enter the pores freely have full access to an additional space, the internal pore volume, V_i. Such molecules emerge at V_t, the total volume available to the mobile phase, which can be determined from the elution volume of small molecules. Hence $V_t = V_0 + V_i$. A solute molecule that is partially restricted from the pores will emerge with elution volume, *Ve*, between the two extremes, V_0 and V_t. The distribution coefficient, K_d, for such a molecule represents the fraction of V_i available to it for diffusion. Hence

$$V_e = V_0 + K_d V_i$$
$$\text{and } K_d = (V_e - V_0)/V_i = (V_e - V_0)/(V_t - V_0)$$

3. These models give rise to various plots that should result in a linear relationship between a function of solute radius, usually R_S, and some size-exclusion chromatography parameter, usually K_d ***(13)***. However, none has proven totally satisfactory. Recently it has been suggested that size-exclusion chromatography does not require pores at all, but rather that

K_d can be calculated from a thermodynamic model for the free energy of mixing of the solute and the gel phase ***(14)***.

4. Manufacturers' catalogs often show plots of K_d against log molecular weight, M. For most globular proteins, this plot is a sigmoidal curve that is approximately linear in the middle section, where

$$K_d = a - b \log M$$

where a is the intercept on the ordinal axis, and b is the slope.
From such plots one can estimate fractionation range (the working range lies in the linear portion, between about $0.9 > K_d > 0.1$) and the selectivity. This latter parameter depends on the slope of the plot and is a function of the pore size distribution. A support with average pore size distribution in a narrow band gives high selectivity (large value of b) but a restricted separation range (the larger the value of b, the lower the range of values for M). If no estimate is available for the size of the protein under investigation it is better to use a support with a broad fractionation range. A support of higher selectivity in a more restricted fractionation range can then be used later to give a more precise value for solute size.
5. Problems arising when size-exclusion chromatography is carried out under denaturing conditions include:
   a. For a particular column, the molecular weight range in which separation occurs is reduced. This is because the radius of gyration, and hence the hydrodynamic size, of a molecule increases when it changes from a sphere to a random coil. For example, the separation range of a TSK G3000SW column operating with denatured proteins is 2000–70,000, compared to 10,000–500,000 for native proteins ***(8,15)***. Of course this may actually be an advantage when working with small proteins or peptides.
   b. Proteins are broken down into their constituent subunits and polypeptide chains, so that the molecular weight of the intact protein is not obtained.
   c. Bioactivity is usually destroyed or reduced and it is not usually possible to monitor enzyme activity. This can be a serious disadvantage when trying to identify a protein in an impure preparation.
   d. The denaturants usually absorb light in the far UV range, so that monitoring the absorbance in the most sensitive region for proteins (200–220 nm) is no longer possible.
   e. Manufacturers often advise that once a column has been exposed to a mobile phase containing denaturants it should be dedicated to applications using that mobile phase, as the properties of the column may be irreversibly altered. In addition, the denaturant, especially if it is SDS, may be difficult to remove completely.
   f. Because these mobile phases have high viscosities, flow rates may have to be reduced to avoid high back pressures.
   g. High concentrations of salts, especially those containing halide ions, can adversely affect pumps and stainless steel.
6. Most manufacturers sell guard columns appropriate for use with their size-exclusion columns. To protect the expensive size-exclusion column it is strongly recommended that a guard column be used.
7. Even in the presence of high ionic strength buffers several proteins show non-ideal behavior and are thus unsuitable as standards. For example, the basic proteins cytochrome c (pI ≅ 10) and lysozyme (pI ≅ 11) have K_d values > 1.0, under the conditions described for **Fig. 1**, because ion-exchange interactions are not totally suppressed. On the other hand, the very acidic protein pepsin (pI ≅ 1) emerges earlier than expected on the basis of size, because of ion exclusion. One should be aware that such behavior might also occur when interpreting results for proteins of unknown pI.
8. Columns are often stored in ethanol–water mixtures or in 0.02% sodium azide to prevent bacterial growth. Caution: ***Sodium azide is believed to be a mutagen. It should be handled***

with care (see suppliers' safety advice) and measures taken to avoid contact with solutions. It can also lead to explosions when disposed of via lead pipes. Solutions should be collected in waste bottles.

When changing mobile phase some manufacturers recommend that the flow rate should not be greater than half the maximum flow rate.

9. Although V_0 is most commonly measured using blue dextran, Himmel and Squire suggested that it is not a suitable marker for the TSK G3000SW column, owing to tailing under nondenaturing conditions, and measured V_0 using glutamic dehydrogenase from bovine liver (Sigma Type II; mol. wt 998,000) ***(16)***. Calf thymus DNA is also a commonly used marker for V_0.

References

1. Porath, J. and Flodin, P. (1959) Gel filtration: a method for desalting and group separation. *Nature* **183,** 1657–1659.
2. Moore, J. C. (1964) Gel permeation chromatography. I. A new method for molecular weight distribution of high polymers. *J. Polymer Sci.* **A2,** 835–843.
3. Hagel, L. (1989) Gel filtration, in *Protein Purification: Principles, High Resolution Methods and Applications* (Janson, J. C., and Ryden, L., eds.), pp. 63–106. VCH, New York, pp. 63–106.
4. Amersham Pharmacia Biotech. 1998. *Gel Filtration Principles and Methods*, 8th edit., Lund, Sweden.
5. Potschka, M. (1987) Universal calibration of gel permeation chromatography and determination of molecular shape in solution. *Analyt. Biochem.* **162,** 47–64.
6. Dubin, P. L. and Principi, J. M. (1989) Failure of universal calibration for size exclusion chromatography of rodlike macromolecules versus random coils and globular proteins. *Macromolecules* **22,** 1891–1896.
7. Ui, N. (1979) Rapid estimation of molecular weights of protein polypeptide chains using high-pressure liquid chromatography in 6 *M* guanidine hydrochloride. *Analyt. Biochem.* **97,** 65–71.
8. Kato, Y., Komiya, K., Sasaki, H., and Hashimoto, T. (1980) High-speed gel filtration of proteins in 6 *M* guanidine hydrochloride on TSK-GEL SW columns. *J. Chromatogr.* **193,** 458–463.
9. Josic, D., Baumann, H., and Reutter, W. (1984) Size-exclusion high-performance liquid chromatography and sodium dodecyl sulphate-polyacrylamide gel electrophoresis of proteins: a comparison. *Analyt. Biochem.* **142,** 473–479.
10. Anspach, B., Gierlich, H. U., and Unger, K. K. (1988) Comparative study of Zorbax Bio Series GF 250 and GF 450 and TSK-GEL 3000 SW and SWXL columns in size-exclusion chromatography of proteins. *J. Chromatogr.* **443,** 45–54.
11. Irvine, G. B. (1987) High-performance size-exclusion chromatography of polypeptides on a TSK G2000SW column in acidic mobile phases. *J. Chromatogr.* **404,** 215–222.
12. Regnier, F. E. (1983) High performance liquid chromatography of proteins. *Meth. Enzymol.* **91,** 137–192.
13. Ackers, G. K. (1970) Analytical gel chromatography of proteins. *Adv. Protein Chem.* **24,** 343–446.
14. Brooks, D. E., Haynes, C. A., Hritcu, D., Steels, B. M., and Muller, W. (2000) Size exclusion chromatography does not require pores. *Proc. Natl. Acad. Sci. USA* **97,** 7064–7067.
15. Kato, Y., Komiya, K., Sasaki, H., and Hashimoto, T. (1980) Separation range and separation efficiency in high-speed gel filtration on TSK-GEL SW columns. *J. Chromatogr.* **190,** 297–303.
16. Himmel, M. E. and Squire, P. G. (1981) High pressure gel permeation chromatography of native proteins on TSK-SW columns. *Int. J. Peptide Protein Res.* **17,** 365–373.

85

Detection of Disulfide-Linked Peptides by HPLC

Alastair Aitken and Michèle Learmonth

Introduction

Classical techniques for determining disulfide bond patterns usually require the fragmentation of proteins into peptides under low pH conditions to prevent disulfide exchange. Pepsin or cyanogen bromide are particularly useful (*see* Chapters 76 and 71 respectively).

Diagonal techniques to identify disulphide-linked peptides were developed by Brown and Hartley (*see* Chapter 87). A modern micromethod employing reverse-phase high-performance liquid chromatography (HPLC) is described here.

2. Materials

1. 1 *M* Dithiothreitol (DTT, good quality, e.g., Calbiochem).
2. 100 m*M* Tris-HCl, pH 8.5.
3. 4-Vinylpyridine.
4. 95% Ethanol.
5. Isopropanol.
6. 1 *M* triethylamine-acetic acid, pH 10.0.
7. Tri-n-butyl-phosphine (1% in isopropanol).
8. HPLC system.
9. Vydac C_4, C_8, or C_{18} reverse phase HPLC columns (*see* **Note 1**).

3. Method (*see* refs. *1* and *2*)

1. Alkylate the protein (1–10 mg in 20–50 mL of buffer) without reduction to prevent possible disulfide exchange by dissolving in 100 m*M* Tris-HCl, pH 8.5, and adding 1 µL of 4-vinylpyridine (*see* **Note 2**).
2. Incubate for 1 h at room temperature and desalt by HPLC or precipitate with 95% ice-cold ethanol followed by bench centrifugation.
3. The pellet obtained after the latter treatment may be difficult to redissolve and may require addition of 10-fold concentrated acid (HCl, formic, or acetic acid) before digestion at low pH. It may be sufficient to resuspend the pellet with acid using a sonic bath if necessary, then commence the digestion. Vortex-mix the suspension during the initial period until the solution clarifies.
4. Fragment the protein under conditions of low pH (*see* **Note 3**) and subject the peptides from half the digest to reverse-phase HPLC. Vydac C_4, C_8, or C_{18} columns give particu-

From: *The Protein Protocols Handbook, 2nd Edition*
Edited by: J. M. Walker © Humana Press Inc., Totowa, NJ

larly good resolution depending on the size range of fragments produced. Typical separation conditions are;- column equilibrated with 0.1% (v/v) aqueous trifluoro acetic acid (TFA) , elution with an acetonitrile–0.1% TFA gradient. A combination of different cleavages, both chemical and enzymatic, may be required if peptide fragments of interest remain large after one digestion method.

5 To the other half of the digest (dried and resuspended in 10 µL of isopropanol) add 5 µL of 1 *M* triethylamine-acetic acid pH 10.0; 5 µL of tri-*n*-butyl-phosphine (1% in isopropanol) and 5 µL of 4-vinylpyridine. Incubate for 30 min at 37°C, and dry *in vacuo,* resuspending in 30 µL of isopropanol twice. This procedure cleaves the disulfides and modifies the resultant –SH groups.
6 Run the reduced and alkylated sample on the same column, under identical conditions on reverse-phase HPLC. Cysteine-linked peptides are identified by the differences between elution of peaks from reduced and unreduced samples.
7 Collection of the alkylated peptides (which can be identified by rechromatography on reverse-phase HPLC with detection at 254 nm) and a combination of sequence analysis and mass spectrometry *(see* **Note 1** and Chapter 86) will allow disulfide assignments to be made.

4. Notes

1. If the HPLC separation is combined with mass spectrometric characterization, the level of TFA required to produce sharp peaks and good resolution of peptide (approx 0.1% v/v) results in almost or complete suppression of signal. This does not permit true on-line HPLC-MS as the concentration of TFA in the eluted peptide must first be drastically reduced. However, the new "low TFA," 218MS54, reverse-phase HPLC columns from Vydac (300 Å pore size) are available in C_4 and two forms of C_{18} chemistries. They are also supplied in 1 mm diameter columns that are ideal for low levels of sample eluted in minimal volume. We have used as little as 0.005% TFA without major loss of resolution and have observed minimal signal loss. There may be a difference in selectivity compared to "classical" reverse-phase columns; for example, we have observed phosphopeptides eluting approx. 1% acetonitrile later than their unphosphorylated counterparts (the opposite to that conventionally seen). This is not a problem, but it is something of which one should be aware, and could be turned to advantage.
2. The iodoacetic acid used must be colorless. A yellow color indicates the presence of iodine; this will rapidly oxidize thiol groups, preventing alkylation and may also modify tyrosine residues. It is possible to recrystallize from hexane. Reductive alkylation may also be carried out using iodo-[^{14}C]-acetic acid or iodoacetamide (*see* Chapter 59). The radiolabelled material should be diluted to the desired specific activity before use with carrier iodoacetic acid or iodoacetamide to ensure an excess of this reagent over total thiol groups.
3. Fragmentation of proteins into peptides under low pH conditions to prevent disulfide exchange is important. Pepsin, Glu-C, or cyanogen bromide are particularly useful (*see* Chapters 76 and 71 respectively). Typical conditions for pepsin are 25°C for 1–2 h at pH 2.0–3.0 (10 m*M* HCl, 5% acetic or formic acid) with an enzyme:substrate ratio of about 1:50. Endoproteinase Glu-C has a pH optimum at 4 as well as an optimum at pH 8.0. Digestion at the acid pH (typical conditions are 37°C overnight in ammonium acetate at pH 4.0 with an enzyme/substrate ratio of about 1:50) will also help minimize disulfide exchange. CNBr digestion in guanidinium 6 *M* HCl/ 0.1–0.2 *M* HCl may be more suitable acid medium due to the inherent redox potential of formic acid which is the most commonly used protein solvent. When analyzing proteins that contain multiple disulfide

bonds it may be appropriate to carry out an initial chemical cleavage (CNBr is particularly useful), followed by a suitable proteolytic digestion. The initial acid chemical treatment will cause sufficient denaturation and unfolding as well as peptide bond cleavage to assist the complete digestion by the protease. If a protein has two adjacent cysteine residues this peptide bond will not be readily cleaved by specific endopeptidases. For example, this problem was overcome during mass spectrometric analysis of the disulfide bonds in insulin by using a combination of an acid proteinase (pepsin) and carboxypeptidase A as well as Edman degradation *(3)*.

References

1. Friedman, M., Zahnley, J. C., and Wagner, J.R. (1980) Estimation of the disulfide content of trypsin inhibitors as *S*-b-(2-pyridylethyl)-L-cysteine. *Analyt. Biochem.* **106,** 27–34.
2. Amons, R. (1987) Vapor-phase modification of sulfhydryl groups in proteins. *FEBS Lett.* **212,** 68–72.
3. Toren, P., Smith, D., Chance, R., and Hoffman, J. (1988). Determination of Interchain Crosslinkages in Insulin B-Chain Dimers by Fast Atom Bombardment Mass Spectrometry. *Analyt. Biochem.* **169,** 287–299.

86

Detection of Disulfide-Linked Peptides by Mass Spectrometry

Alastair Aitken and Michèle Learmonth

1. Introduction

Mass spectrometry is playing a rapidly increasing role in protein chemistry and sequencing (*see* Chapters 91 and 97–100) and is particularly useful in determining sites of co- and posttranslational modification ***(1,2)***, and application in locating disulfide bonds is no exception. This technique can of course readily analyze peptide mixtures; therefore it is not always necessary to isolate the constituent peptides. However, a cleanup step to remove interfering compounds such as salt and detergent may be necessary. Thus can be achieved using matrices such as 10-μm porous resins slurry-packed into columns 0.25 mm diameter. Polypeptides can be separated on stepwise gradients of 5–75% acetonitrile in 0.1% formic or acetic acid ***(3)***. On-line electrospray mass spectrometry (ES-MS) coupled to capillary electrophoresis or high-performance liquid chromatography (HPLC) has proved particularly valuable in the identification of modified peptides. If HPLC separation on conventional columns is attempted on-line with mass spectrometry, the level of trifluoroacetic acid (TFA) (0.1%) required to produce sharp peaks and good resolution of peptides results in almost or complete suppression of signal. In this case it is recommended to use the new "low TFA," 218MS54, reverse-phase HPLC columns from Vydac (300 Å pore size) which can be used with as little as 0.005% TFA without major loss of resolution and minimal signal loss (*see* further details in Chapter 85).

Sequence information is readily obtained using triple quadrupole tandem mass spectrometry after collision-induced disassocation ***(4)***. Ion trap mass spectrometry technology (called LCQ) is now well established which also permits sequence information to be readily obtained. Not only can MS-MS analysis be carried out, but owing to the high efficiency of each stage, further fragmentation of selected ions may be carried out to MS." The charge state of peptide ions is readily determined by a "zoom-scan" technique that resolves the isotopic envelopes of multiply charged peptide ions. The instrument still allows accurate molecular mass determination to 100,000 Da at 0.01% mass accuracy. The recent development of Fourier transform ion cyclotron resonance mass spectrometry ***(5)***, in which the ions can be generated by a wide variety of techniques, has very high resolution and sensitivity.

From: *The Protein Protocols Handbook, 2nd Edition*
Edited by: J. M. Walker © Humana Press Inc., Totowa, NJ

Selective detection of modified peptides is possible on ES-MS. For example, phosphopeptides can be identified from the production of phosphate-specific fragment ions of 63 Da (PO) and 79 Da (PO) by collision-induced dissociation during negative ion HPLC–ES-MS. This technique of selective detection of posttranslational modifications through collision-induced formation of low-mass fragment ions that serve as characteristic and selective markers for the modification of interest has been extended to identify other groups such as glycosylation, sulpfation, and acylation *(6)*.

2. Materials

Materials for proteolytic and chemical cleavage of proteins are described in Chapters 71–76.

3. Method

3.1. Detection of Disulfide-Linked Peptides by Mass Spectrometry

1. Peptides generated by any suitable proteolytic or chemical method that minimizes disulfide exchange (i.e., acid pH, *see* **Note 1**). Partial acid hydrolysis, although nonspecific, has been successfully used in a number of instances. The peptides are then analyzed by a variety of mass spectrometry techniques *(7)*. The use of thiol and related compounds should be avoided for obvious reasons. Despite this, it is possible that disulfide bonds will be partially reduced during the analysis and peaks corresponding to the individual components of the disulfide-linked peptides will be observed. Control samples with the above reagents are essential to avoid misleading results (*see* **Note 2**).
2. The peptide mixture is incubated with reducing agents, such as mercaptoethanol and dithiothreitol (DTT), and reanalyzed as before. Peptides that were disulfide linked disappear from the spectrum and reappear at the appropriate positions for the individual components. For example, in the positive ion mode the mass (M) of disulfide-linked peptides (of individual masses A and B) will be detected as the pseudomolecular ion at $(M+H)^+$, and after reduction this will be replaced by two additional peaks for each disulfide bond in the polypeptide at masses $(A+H)^+$ and $(B+H)^+$. Remember that $A + B = M + 2$, as reduction of the disulfide bond will be accompanied by a consistent increase in mass due to the conversion of cystine to two cysteine residues, that is, -S-S $\rightarrow$ -SH + HS- and peptides containing an intramolecular disulfide bond will appear at 2 amu higher. Such peptides, if they are in the reduced state can normally be readily reoxidized to form intramolecular disulfide bonds by bubbling a stream of air through a solution of the peptide for a few minutes (*see* **Note 3**).
3. Computer programs are readily available on the Internet that are supplied with the mass spectrometer software package and will predict the cleavage position of any particular proteinase or chemical reagent. Simple knowledge of the mass of the fragment will, in most instances, give unequivocal answers as to which segments of the polypeptide chain are disulfide linked. If necessary, one cycle of Edman degradation can be carried out on the peptide mixture and the FAB-MS analysis repeated. The shift in mass(es) that correlates with loss of specific residues will confirm the assignment. Development in techniques such as ladder sequencing are proving extremely useful (*see* **Note 4**).

4. Notes

1. Fragmentation of proteins into peptides under low pH conditions is important (e.g., with pepsin, Glu-C, or cyanogen bromide; *see* Chapter 76 and 71 respectively) to prevent disulfide exchange. Analysis can also be carried out by electrospray mass spectrometers,

(ES-MS) which will give an accurate molecular mass up to 80–100,000 Da (and in favorable cases up to 150,000 Da). The increased mass of 2 Da for each disulfide bond will be too small to obtain an accurate estimate for polypeptide of mass Ca 10,000 (accuracy obtainable is >0.01%). There has been a recent marked increase in resolution obtained with both electrospray mass spectrometers and laser desorption time-of-flight mass spectrometers that could now permit a meaningful analysis. On the other hand, oxidation with performic acid will cause a mass increase of 48 Da for each cysteine and 49 Da for each half-cystine residue. Note that Met and Trp will also be oxidized. Met sulfoxide, the result of incomplete oxidation and that is often found after gel electrophoresis, has a mass identical to that of Phe. Met sulfone is identical in mass to Tyr. Fourier transform mass spectrometers have ppm accuracy that is well within the necessary range of resolution even for large proteins.

2. Mass analysis by ES-MS ***(3,8)*** and matrix-assisted laser desorption mass spectrometry with time-of-flight detection (MALDI-TOF) ***(9)*** is affected, seriously in some cases, by the presence of particular salts, buffers, and detergents. In some cases, using nonionic saccharide detergents such as *n*-dodecyl-β-D-glucopyranoside, improvements in signal-to-noise ratios of peptides and proteins were observed ***(9)***. The effect on ES-MS sensitivity of different buffer salts, detergents, and tolerance to acid type may vary widely with the instrument and particularly with the ionization source. Critical micelle concentration is not a good predictor of how well a surfactant will perform ***(8)***.
3. The difference of 2 Da may allow satisfactory estimation of the number of intramolecular disulfide bonds by mass spectrometry. If necessary a larger mass difference may be generated by oxidation with performic acid (***10***, and *see* Chapter 60). This will cause a mass increase of 48 Da for each cysteine and 49 Da for each half-cystine residue. (Remember that Met and Trp will also be oxidized.)
4. Ladder sequencing has particular application in MALDI-TOF MS, which has high sensitivity and greater ability to analyze mixtures. The technique involves the generation of a set of nested fragments of a polypeptide chain followed by analysis of the mass of each component. Each component in the ragged polypeptide mixture differs from the next by loss of a mass that is characteristic of the residue weight (which may involve a modified side chain). In this manner, the sequence of the polypeptide can be read from the masses obtained in MS. The ladder of degraded peptides can be generated by Edman chemistry ***(11)*** or by exopeptidase digestion from the N- and C-termini. This essentially a subtractive technique (one looks at the mass of the remaining fragment after each cycle). For example, when a phosphoserine residue is encountered, a loss of 167.1 Da is observed in place of 87.1 for a serine residue. This technique therefore avoids one of the major problems of analyzing posttranslational modifications. Although the majority of modifications are stable during the Edman chemistry, *O*- or *S*-linked esters, for example (which are very numerous), may be lost by β-elimination (e.g., *O*-phosphate) during the cleavage step to form the anilinothioazolidone or undergo *O*- (or *S*- in the case of palmitoylated cysteine) to *N*-acyl shifts which block further Edman degradation. Exopeptidase digestion may be difficult and the rate of release of amino acid may vary greatly. The use of modified Edman chemistry has great possibilities ***(12)***. The modification consists of carrying out the coupling step with PITC in the presence of a small amount of phenylisocyanate which acts as a chain-terminating agent. A development of this technique involves the addition of volatile trifluorethylisothiocyanate (TFEITC) to the reaction tube to which a fresh aliquot of peptide is added after each cycle. This avoids steps to remove excess reagent and byproducts. This may be combined with subsequent modification of the terminal NH_2 group with quaternary ammonium alkyl NHS esters, which allows increased sensitivity in MALDI-TOF mass spectrometry.

Acknowledgment

We thank Steve Howell (NIMR) for the mass spectrometry analysis.

References

1. Costello, C. E. (1999) Bioanalytic applications of mass spectrometry. *Curr. Opin. Biotechnol.* **10,** 22–28.
2. Burlingame, A. L., Boyd, R. K., and Gaskell, S. J. (1998) Mass spectrometry. *Analyt. Chem.* **70,** 647R–716R.
3. Aitken, A., Howell, S., Jones, D., Madrazo, J., and Patel, Y. (1995) 14-3-3 α and δ are the phosphorylated forms of Raf-activating 14-3-3 β and ζ. *In vivo* stoichiometric phosphorylation in brain at a Ser-Pro-Glu-Lys motif. *J. Biol. Chem.* **270,** 5706–5709.
4. Hunt, D. F., Yates, J. R., Shabanowitz, J., Winston, S., and Hauer, C. R. (1986) Protein sequencing by tandem mass spectrometry. *Proc. Natl. Acad. Sci. USA* **83,** 6233–6237.
5. Hendrickson, C. L. and Emmett, M. R. (1999) Electrospray ionization Fourier transform ion cyclotron resonance mass spectrometry. *Annu. Rev. Phys. Chem.* **50,** 517–536.
6. Bean, M. F., Annan, R. S., Hemling, M. E., Mentzer, M., Huddleston, M. J., and Carr, S. A. (1994) LC-MS methods for selective detection of posttranslational modifications in proteins, in *Techniques in Protein Chemistry* VI (Crabb, J. W., ed.), Academic Press, San Diego, pp. 107–116.
7. Morris, R. H. and Pucci, P. (1985). A new method for rapid assignment of S—S bridges in proteins. *Biochem. Biophys. Res. Commun.* **126,** 1122–1128.
8. Loo, R., Dales, N., and Andrews, P. C. (1994) Surfactant effects on protein structure examined by electrospray ionisation mass spectrometry. *Protein Sci.* **3,** 1975–1983.
9. O., Vorm, Chait, B. T., and Roepstorff, (1993) Mass spectrometry of protein samples containing detergents, in *Proceedings of the 41st ASMS Conference on Mass Spectrometry and Allied Topics*, pp. 621–622.
10. Sun, Y. and Smith, D. L. (1988) Identification of disulfide-containing peptides by performic acid oxidation and mass spectrometry. *Analyt. Biochem.* **172,** 130–138.
11. Chait, B. T., Wang, R., Beavis, R. C., and Kent, S. B. H. (1993) Protein ladder sequencing. *Science* **262,** 89–92.
12. Bartlet-Jones, M., Jeffery, W. A., Hansen, H. F., and Pappin, D. J. C. (1994) Peptide ladder sequencing by mass spectrometry using a novel, volatile degradation reagent. *Rapid Commun. Mass Spectrosc.* **8,** 737–742.

87

Diagonal Electrophoresis for Detecting Disulfide Bridges

Alastair Aitken and Michèle Learmonth

1. Introduction

Methods for identifying disulfide bridges have routinely employed "diagonal" procedures using two-dimensional paper or thin-layer electrophoresis. This essentially utilizes the difference in electrophoretic mobility of peptides containing either cysteine or cystine in a disulfide link, before and after oxidation with performic acid. It was first described by Brown and Hartley *(1)*. Peptides unaltered by the performic acid oxidation have the same mobility in both dimensions and, therefore, lie on a diagonal. After oxidation, peptides that contain cysteine or were previously covalently linked produce one or two spots off the diagonal, respectively. This method has also been adapted for HPLC methodology and is discussed in Chapter 85.

First the protein has to be fragmented into suitable peptides containing individual cysteine residues. It is preferable to carry out cleavages at low pH to prevent possible disulfide bond exchange. In this respect, pepsin (active at pH 3.0) and cyanogen bromide (CNBr) are particularly useful reagents. Proteases with active-site thiols should be avoided (e.g., papain, bromelain).

Before the advent of HPLC, paper electrophoresis was the most commonly used method for peptide separation *(2)*. Laboratories with a history of involvement with protein characterization are likely to have retained the equipment, but it is no longer commercially available. Although it is possible to make a simple electrophoresis tank in house *(3)*, thin-layer electrophoresis equipment is still commercially available, and it is advisable that this be used, owing to the safety implications.

For visualization of the peptides, ninhydrin is the classical amino group stain. However, if amino acid analysis or sequencing is to be carried out, fluorescamine is the reagent of choice.

2. Materials

2.1. Equipment

1. Electrophoresis tank.
2. Flat bed electrophoresis apparatus (e.g., Hunter thin layer peptide mapping system, Orme, Manchester, UK).

From: *The Protein Protocols Handbook, 2nd Edition*
Edited by: J. M. Walker © Humana Press Inc., Totowa, NJ

3. Whatman (Maidstone, UK) 3MM and Whatman No. 1 paper.
4. Cellulose TLC plates (Machery Camlab [Cambridge, UK] Nagel or Merck [Poole, UK]).

2.2. Reagents

1. Electrophoresis buffers (*see* **Note 1**): Commonly used volatile buffers are:
   pH 2.1 acetic acid/formic acid/water 15/5/80, v/v/v
   pH 3.5 pyridine/acetic acid/water 5/50/945 v/v/v
   pH 6.5 pyridine/acetic acid/water 25/1/225, v/v/v
2. Nonmiscible organic solvents: toluene, for use with pH 6.5 buffer: white spirit for use with pH 2.1 and 3.5 buffers.
3. Formic acid. **Care!**
4. 30% w/v Hydrogen peroxide. **Care!**
5. Fluorescamine: 1 mg/100 mL in acetone.
6. Marker dyes: 2% orange G, 1% acid fuschin, and 1% xylene cyanol dissolved in appropriate electrophoresis buffer.
7. 1% (v/v) Triethylamine in acetone.

3. Methods

3.1. First-Dimension Electrophoresis

3.1.1. Paper Electrophoresis

1. Dissolve the peptide digest from 0.1–0.3 μmol protein in 20–25 μL electrophoresis buffer.
2. Place the electrophoresis sheet on a clean glass sheet (use Whatman No. 1 for analytical work, Whatman 3MM for preparative work). Support the origin at which the sample is to be applied on glass rods. Where paper is used, multiple samples can be run side by side. Individual strips can then be cut out for running in the second dimension.
3. Apply the sample slowly without allowing to dry (covering an area of about 2 × 1 cm) perpendicular to the direction intended for electrophoresis. NB: for pH 6.5 buffer, apply near the center of the sheet, for acidic buffers, apply nearer the anode end.
4. Apply a small volume of marker dyes (2% orange G, 1% acid fuschin, 1% xylene cyanol, in electrophoresis buffer) on the origin and additionally in a position that will not overlap the peptides after the second dimension.
5. Once the sample is applied, wet the sheet with electrophoresis buffer slowly and uniformly on either side of the origin so that the sample concentrates in a thin line. Remove excess buffer from the rest of the sheet with blotting paper.
6. Place the wet sheet in the electrophoresis tank previously set up with electrophoresis buffer covering bottom electrode. An immiscible organic solvent (toluene where pH 6.5 buffer is used, white spirit for acidic buffers) used to fill the tank to the top. Upper trough filled with electrophoresis buffer. **Care!**
7. Electrophorese at 3 kV for about 1 h, with cooling if necessary. The progress of the electrophoresis can be monitored by the movement of the marker dyes (*see* **Note 1**).
8. Dry sheet in a well-ventilated place overnight, at room temperature, secured from a glass rod with Bulldog clips.

3.1.2. Thin-Layer Electrophoresis

1. Dissolve the peptide digest from 0.1–0.3 μmol protein in at least 10 μL electrophoresis buffer.
2. Mark the sample and dye origins on the cellulose side of a TLC plate with a cross using an extrasoft blunt-ended pencil, or on the reverse side with a permanent marker.

3. Spot the sample on the origin. This can be done using a micropipet fitted with a disposable capillary tip. To keep the spot small, apply 0.5–1 µL spots and dry between applications.
4. Apply 0.5 µL marker dye to the dye origin.
5. Set up electrophoresis apparatus with electrophoresis buffer in both buffer tanks. Prepare electrophoresis wicks from Whatman 3MM paper, wet with buffer, and place in buffer tanks.
6. Prepare a blotter from a double sheet of Whatman 3MM, with holes cut at the positions of the sample and marker origins. Wet with buffer and place over TLC plate. Ensure concentration at the origins by pressing lightly around the holes.
7. Place TLC plate in apparatus.
8. Electrophorese at 1.5 kV for about 30–60 min.
9. Dry plate in a well-ventilated place at room temperature, overnight.

3.2. Performic Acid Oxidation (see Note 2)

1. Prepare performic acid by mixing 19 mL formic acid with 1 mL 30% (w/v) hydrogen peroxide. The reaction is spontaneous. **Care!**
2. Place the dry electrophoresis sheet/plate in a container where it can be supported without touching the sides.
3. Place the performic acid in a shallow dish inside the container. Close the container and leave to oxidize for 2–3 h. (Note the marker dyes change from blue to green.)
4. Dry sheet thoroughly at room temperature overnight.

3.3. Second Dimension Electrophoresis

3.3.1. Paper Electrophoresis

1. To prepare for the second dimension, individual strips from the first dimension can be machine zigzag stitched onto a second sheet. The overlap of the second sheet should be carefully excised with a razor blade/scalpel.
2. Wet the sheet with electrophoresis buffer, applying the buffer along both sides of the sample, thus concentrating the peptides in a straight line.
3. Repeat electrophoresis, at right angles to the original direction.
4. Thoroughly dry sheet, as before.

3.3.2. Thin-Layer Electrophoresis

1. Wet TLC plate with electrophoresis buffer using two sheets of prewetted Whatman 3 MM paper on either side of sample line.
2. Repeat electrophoresis at right angles to the original direction.
3. Thoroughly dry plate as before.

3.4. Visualization

Peptide spots can be seen after reaction with fluorescamine.

1. The reaction should be carried out under alkaline conditions. The sheet or plate should be dipped in a solution of triethylamine (1% [v/v]) in acetone. This should be carried out at least twice if the electrophoresis buffer employed was acidic. Dry the sheet well.
2. Dip the sheet in a solution of fluorescamine in acetone (1 mg/100 mL).
3. Allow most of the acetone to evaporate.
4. View the map under a UV lamp at 300–365 nm (**Care:** Goggles must be worn). Peptides and amino-containing compounds fluoresce. Encircle all fluorescent spots with a soft pencil.

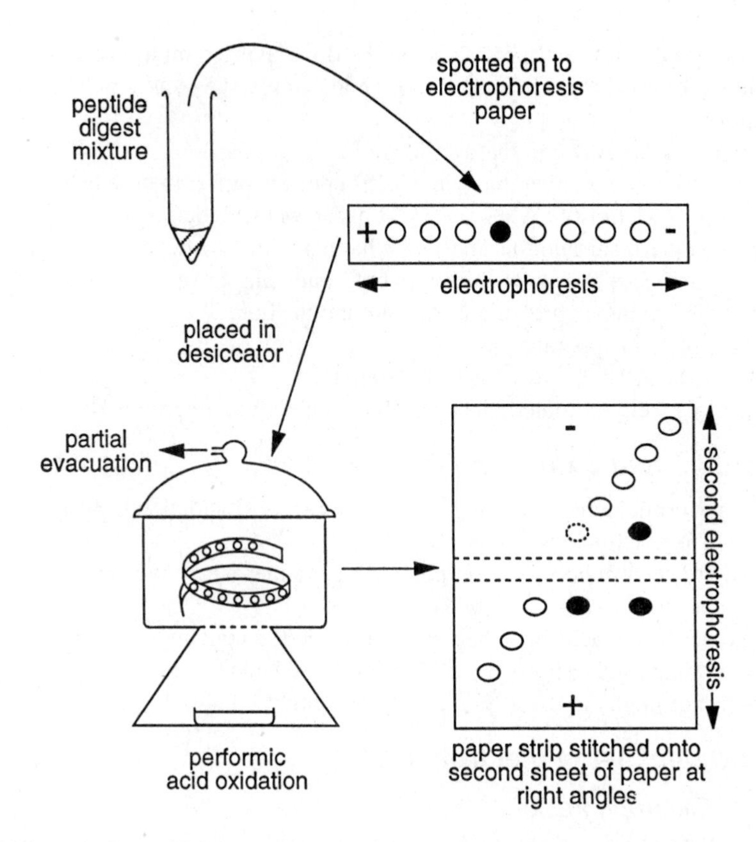

Fig. 1. Diagonal electrophoresis for identification and purification of peptides containing cysteine or disulfide bonds. This figure shows that peptides unaltered by the performic acid treatment (open circles) have the same mobility in both dimensions and therefore lie on a diagonal. Peptides that contain cysteine or were previously covalently linked (closed circles) produce one or two spots respectively, that lie off the diagonal after oxidation.

5. Interpretation of diagonal maps: **Figure 1** shows that peptides unaltered by the performic acid treatment have the same mobility in both dimensions and therefore lie on a diagonal. Peptides that contain cysteine or were previously covalently linked produce one or two spots, respectively, that lie off the diagonal after oxidation.

3.5. Elution

Peptides may be eluted from the paper using, for example, 0.1 M NH_3 or 25% acetic acid. Peptides may be extracted from TLC plates by scraping off the spot into an Eppendorf tube containing elution buffer. This should be vortexed for 5 min and then centrifuged for 5 min. The cellulose can then be re-extracted once or twice with the same buffer to ensure optimal recovery of peptide.

4. Notes

1. The buffer of choice for the initial analysis is pH 6.5. However, if the cysteine residues have already been blocked with iodoacetate (*see* Chapter 59), the pH 3.5 buffer is very useful, since peptides containing these residues lie slightly off the diagonal, being slightly more acidic in the second dimension after the performic acid oxidation.

2. The movement of the marker dyes will enable progress of the electrophoresis to be followed to ensure that the samples do not run off the end of the paper.
3. It is important to exclude halide ions rigorously, since these are readily oxidized to halogens, which will react with histidine, tyrosine, and phenylalanine residues in the protein.

References

1. Brown, J. R., and Hartley, B. S. (1966) Location of disulphide bridges by diagonal paper electrophoresis *Biochem. J.* **101,** 214–228.
2. Michl, H. (1951) Paper electrophoresis at potential differences of 50 volts per centimetre. *Monatschr. Chem.* **82,** 489–493.
3. Creighton, T. E. (1983) Disulphide bonds between cysteine residues, in *Protein Structure—a Practical Approach* (Rickwood, D. and Hames B. D., eds.), IRL, Oxford, UK, pp. 155–167.

88

Estimation of Disulfide Bonds Using Ellman's Reagent

Alastair Aitken and Michèle Learmonth

1. Introduction

Ellman's reagent 5,5'-dithiobis(2-nitrobenzoic acid) (DTNB) was first introduced in 1959 for the estimation of free thiol groups *(1)*. The procedure is based on the reaction of the thiol with DTNB to give the mixed disulfide and 2-nitro-5-thiobenzoic acid (TNB) which is quantified by the absorbance of the anion (TNB^{2-}) at 412 nm.

The reagent has been widely used for the quantitation of thiols in peptides and proteins. It has also been used to assay disulfides present after blocking any free thiols (e.g., by carboxymethylation) and reducing the disulfides prior to reaction with the reagent *(2,3)*. It is also commonly used to check the efficiency of conjugation of sulfhydryl-containing peptides to carrier proteins in the production of antibodies.

2. Materials

1. Reaction buffer: 0.1 *M* phosphate buffer, pH 8.0.
2. Denaturing buffer: 6 *M* guanidinium chloride, 0.1 *M* Na_2HPO_4, pH 8.0 (*see* **Note 1**).
3. Ellman's solution: 10 m*M* (4 mg/mL) DTNB (Pierce, Chester, UK) in 0.1 *M* phosphate buffer, pH 8.0 (*see* **Note 2**).
4. Dithiothreitol (DTT) (Boerhinger or Calbiochem) solution: 200 m*M* in distilled water.

3. Methods

3.1. Analysis of Free Thiols

1. It may be necessary to expose thiol groups, which may be buried in the interior of the protein. The sample may therefore be dissolved in reaction buffer or denaturing buffer. A solution of known concentration should be prepared with a reference mixture without protein. Sufficient protein should be used to ensure at least one thiol per protein molecule can be detected; in practice, at least 2 nmol of protein (in 100 µL) are usually required.
2. Sample and reference cuvets containing 3 mL of the reaction buffer or denaturing buffer should be prepared and should be read at 412 nm. The absorbance should be adjusted to zero (A_{buffer}).
3. Add 100 µL of buffer to the reference cuvet.
4. Add 100 µL of Ellman's solution to the sample cuvet. Record the absorbance (A_{DTNB}).
5. Add 100 µL of protein solution to the reference cuvet.

From: *The Protein Protocols Handbook, 2nd Edition*
Edited by: J. M. Walker © Humana Press Inc., Totowa, NJ

6. Finally, add 100 µL protein solution to the sample cuvet, and after thorough mixing, record the absorbance until there is no further increase. This may take a few minutes. Record the final reading (A_{final}).
7. The concentration of thiols present may be calculated from the molar absorbance of the TNB anion. (*See* **Note 3**.)

$$\Delta A_{412} = E_{412}TNB^{2-}[RSH] \quad (1)$$

Where $\Delta A_{412} = A_{final} - (3.1/3.2)\,(A_{DTNB} - A_{buffer})$
and $E_{412}\ TNB^{2-} = 1.415 \times 10^4\ cm^{-1}M^{-1}$.

If using denaturing buffer, use the value $E_{412}\ TNB^{2-} = 1.415 \times 10^4\ cm^{-1}\ M^{-1}$.

3.2. Analysis of Disulfide Thiols

1. Sample should be carboxymethylated (*see* Chapter 59) or pyridethylated (*see* Chapter 62) without prior reduction. This will derivatize any free thiols in the sample, but will leave intact any disulfide bonds.
2. The sample (at least 2 nmol of protein in 100 µL, is usually required) should be dissolved in 6 *M* guanidinium HCl, 0.1 *M* Tris-HCl, pH 8.0 or denaturing buffer, under a nitrogen atmosphere.
3. Add freshly prepared DTT solution to give a final concentration of 10–100 m*M*. Carry out reduction for 1–2 h at room temperature.
4. Remove sample from excess DTT by dialysis for a few hours each time, with two changes of a few hundred mL of the reaction buffer or denaturing buffer (*see* **Subheading 3.1.**). Alternatively, gel filtration into the same buffer may be carried out.
5. Analysis of newly exposed disulfide thiols can thus be carried out as described in **Subheading 3.1.**

4. Notes

1. It is not recommended to use urea in place of guanidinium HCl, since this can readily degrade to form cyanates, which will react with thiol groups.
2. Unless newly purchased, it is usually recommended to recrystallize DTNB from aqueous ethanol.
3. Standard protocols for use of Ellman's reagent often give $E_{412}\ TNB^{2-} = 1.36 \times 10^4\ cm^{-1}M^{-1}$. A more recent examination of the chemistry of the reagent indicates that these are more suitable values ***(4)***, and these have been used in this chapter.

References

1. Ellman, G. L. (1959) Tissue sulfhydryl groups. *Arch. Biochem. Biophys.* **82,** 70–77.
2. Zahler, W. L. and Cleland, W. W. (1968) A specific and sensitive assay for disulfides. *J. Biol. Chem.* **243,** 716–719.
3. Anderson, W. L. and Wetlaufer, D. B. (1975) A new method for disulfide analysis of peptides. *Analyt. Biochem.* **67,** 493–502.
4. Riddles P. W., Blakeley, R. L., and Zerner, B. (1983) Reassessment of Ellman's reagent. *Meth. Enzymol.* **91,** 49–60.

89

Quantitation of Cysteine Residues and Disulfide Bonds by Electrophoresis

Alastair Aitken and Michèle Learmonth

1. Introduction

Amino acid analysis quantifies the molar ratios of amino acids per mole of protein. This generally gives a nonintegral result, yet clearly there are integral numbers of the amino acids in each protein. A method was developed by Creighton *(1)* to count integral numbers of amino acid residues, and it is particularly useful for the determination of cysteine residues. Sulfhydryl and disulfide groups are of great structural, functional, and biological importance in protein molecules. For example, the Cys sulfhydryl is essential for the catalytic activity of some enzymes (e.g., thiol proteases) and the interconversion of Cys SH to Cystine S—S is directly involved in the activity of protein disulfide isomerase *(2)*. The conformation of many proteins is stabilized by the presence of disulfide bonds *(3)*, and the formation of disulfide bonds is an important posttranslational modification of secretory proteins *(4)*.

Creighton's method exploits the charge differences introduced by specific chemical modifications of cysteine. A similar method was first used in the study of immunoglobulins by Feinstein in 1966 *(5)*. Cys residues may be reacted with iodoacetic acid, which introduces acidic carboxymethyl (—O_2CCH_2S—) groups, or with iodoacetamide, which introduces the neutral carboxyamidomethyl (H_2NCOCH_2S—) groups. The reaction with either reagent is essentially irreversible, thereby producing a stable product for analysis. Using a varying ratio of iodoacetamide/iodoacetate, these acidic and neutral agents will compete for the available cysteines, and a spectrum of fully modified protein molecules having 0,1,2, ... *n* acidic carboxymethyl residues per molecule is produced (where *n* is the number of cysteine residues in the protein). These species will have, correspondingly, n, $n - 1$, $n - 2$, ... 0 neutral carboxyamidomethyl groups. These species may then be separated by electrophoresis, isoelectric focusing, or by a combination of both *(1,6,7)*. The examples of the analysis of the cysteine residues in bovine pancreatic trypsin inhibitor and ovotransferrin are shown in **Fig. 1**.

Creighton used a low-pH discontinuous system *(1)*. Takahasi and Hirose recommend a high-pH system *(6)*, whereas Stan-Lotter and Bragg used the Laemmli electrophoresis system followed by isoelectric focusing *(7)*. It may therefore be necessary to carry out preliminary experiments to find the best separation conditions for the protein under analysis. The commonly used methods are given below.

From: *The Protein Protocols Handbook, 2nd Edition*
Edited by: J. M. Walker © Humana Press Inc., Totowa, NJ

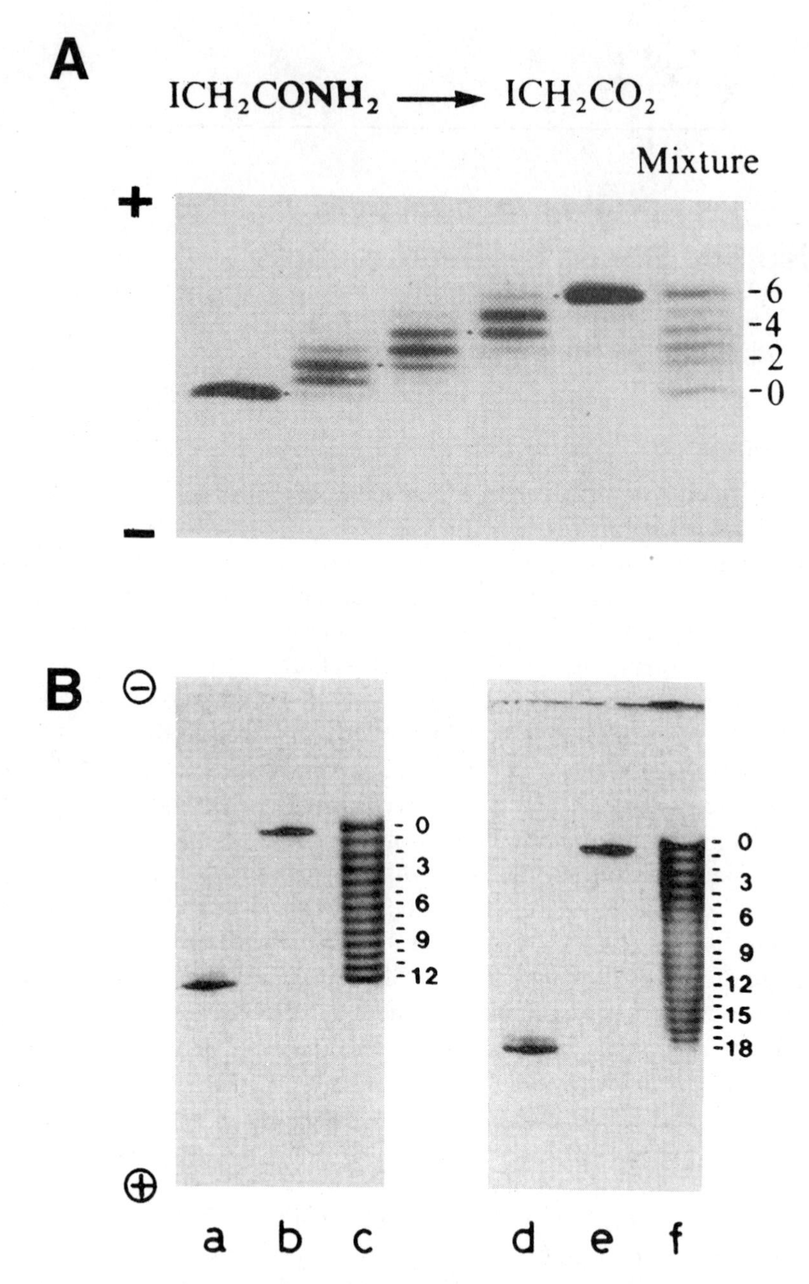

Fig. 1. (**A**) Electrophoretic analysis of the cysteine residues in bovine pancreatic trypsin inhibitor *(1)* with six cysteine residues, run with the low-pH system. Lanes 1–5 contain (respectively) samples reacted with neutral iodoacetamide; 1:1; 1:3; 1:9 ratios of neutral to acidic reagent; acidic iodoacetate. Lane 6 contains a mixture of equal portions of the samples in lanes 1–5. (**B**) Electrophoretic analysis of the cysteine residues in the N-terminal (lanes a–c) and the C-terminal (lanes d–f) domains of ovotransferrin *(6)* run with the high pH system. The subunits contain 13 and 19 cysteines, respectively. Lanes a and d contain samples alkylated with iodoacetic acid. Lanes b and e contain samples alkylated with iodoacetamide. Lanes c and e contain mixtures of the samples alkylated with different ratios of neutral to acidic reagent.

In order to ensure that all thiol groups are chemically equivalent, the reactions must be carried out in denaturing (in the presence of urea) and reducing (in the presence of dithiothreitol, DTT) conditions. The electrophoretic separation must also be carried out with the unfolded protein (i.e., in the presence of urea) in order that the modification has the same effect irrespective of where it is in the polypeptide chain.

The original method has been modified into a two-stage process to allow for the quantification of both sulfhydryl groups and disulfide bonds (*see* **Notes 1** and **2**) *(6,8)*. The principle of the method has also been adapted to counting the numbers of lysine residues after progressive modification of the ε-amino groups with succinic anhydride, which converts this basic group to a carboxylic acid-containing moiety.

2. Materials

2.1. Reaction Solutions

1. 1 *M* Tris-HCl, pH 8.0.
2. 0.1 *M* EDTA, pH 7.0.
3. 1 *M* DTT (good-quality, e.g., Calbiochem, Nottingham, UK).
4. 8 *M* Urea (BDH [Poole, UK], Aristar-grade, *see* Note 2).
5. Solution A: 0.25 *M* iodoacetamide, 0.25 *M* Tris-HCl, pH 8.0.
6. Solution B: 0.25 *M* iodoacetic acid, prepared in 0.25 *M* Tris-HCl, pH readjusted to 8.0 with 1 *M* KOH.

2.2. Solutions for Electrophoretic Analysis in the Low-pH System (pH 4.0) (9)

1. 30% Acrylamide solution containing 30 g acrylamide, 0.8 g *bis*-acrylamide **(extreme caution: work in fume hood)**, made up to 100 mL with distilled water.
2. 10% Acrylamide solution containing 10 g acrylamide **(extreme caution: work in fume hood)** and 0.25 g *bis*-acrylamide made up to 100 mL with distilled water.
3. Low-pH buffer (eight times concentrated stock) for separating gel; 12.8 mL glacial acetic acid, 1 mL *N,N,N'N'*-tetramethylethylenediamine (TEMED), 1 *M* KOH (approx 35 mL) to pH 4.0, made up to 100 mL with distilled water.
4. Low pH buffer (8 times concentrated) for stacking gel; 4.3 mL glacial acetic acid, 0.46 mL TEMED, 1 *M* KOH to pH 5.0, to 100 mL with distilled water.
5. 4 mg riboflavin/100 mL water.
6. Low-pH buffer for electrode buffer; dissolve 14.2 g β-alanine in ~800 mL water then adjust to pH 4.0 with acetic acid. Make up to a final volume of 1 L with distilled water.
7. Tracking dye solution (five times concentrated); 20 mg methyl green, 5 mL water, and 5 g glycerol.

2.3. Gel Solution Recipes for Low-pH Electrophoresis (pH 4.0) (see Note 3)

1. 30 mL Separating gel (10% acrylamide, photopolymerized with riboflavin) is made up as follows: 10 mL 30% acrylamide stock, 4 mL pH 4.0 buffer stock, 3 mL riboflavin stock, 14.7 g urea, water (approx 2.5 mL) to 30 mL . Degas on a water vacuum pump (to remove oxygen which inhibits polymerization).
2. 8 mL Stacking gel (2.5% acrylamide, photopolymerized with riboflavin) is made up with: 2 mL 10% acrylamide stock, 1 mL pH 5.0 buffer stock, 1 mL riboflavin stock, 3.9 g urea, water (approx 1.2 mL) to 8 mL. Degas.

2.4. Electrophoresis Buffers for High-pH Separation (pH 8.9)

1. 30% Acrylamide solution containing 30 g acrylamide, 0.8 g *bis*-acrylamide **(extreme caution: work in fume hood)**, made up to 100 mL with distilled water.
2. 10% Acrylamide solution containing 10 g acrylamide **(extreme caution: work in fume hood)** and 0.25 g *bis*-acrylamide made up to 100 mL with distilled water.
3. High-pH buffer (four times concentrated stock) for separating gel; 18.2 g Tris base (in ~40 mL water), 0.23 mL TEMED, 1 *M* HCl to pH 8.9, made up to 100 mL with distilled water.
4. High-pH buffer (four times concentrated) for stacking gel; 5.7 g Tris base (in ~40 mL water), 0.46 mL TEMED, 1 *M* H_3PO_4 to pH 6.9, made up to 100 mL with distilled water.
5. 4 mg Riboflavin in 100 mL water
6. 10% Ammonium persulfate solution (consisting of 0.1 g ammonium persulfate in 1 mL water).
7. High-pH buffer for electrode buffer; 3 g Tris base, 14.4 g glycine, distilled water to 1 L.
8. Tracking dye solution (five times concentrated): 1 mL 0.1% Bromophenol blue, 4 mL water, and 5 g glycerol.

2.5. Gel Solution Recipes for High-pH Electrophoresis (see Note 3)

1. 30 mL Separating gel (7.5% acrylamide, polymerized with ammonium persulfate) is made up with: 7.5 mL 30% acrylamide stock, 7.5 mL pH 8.9 buffer stock, 0.2 mL 10% ammonium persulfate (add immediately before casting), 14.7 g urea, water (approx 4.5 mL) to 30 mL. Degas.
2. 8 mL Stacking gel (2.5% acrylamide, photopolymerized with riboflavin) is made up with: 2 mL 10% acrylamide stock, 1 mL pH 6.9 buffer stock, 1 mL riboflavin stock solution, 3.9 g urea, water (approx 1.2 mL) to 8 mL. Degas.

3. Methods

3.1. Reduction and Denaturation

1. To a 0.2-mg aliquot of lyophilized protein add 10 µL of each of the solutions containing 1 *M* Tris-HCl, pH 8.0, 0.1 *M* EDTA, and 1 *M* DTT (*see* **Note 4**).
2. Add 1 mL of the 8 *M* urea solution (*see* **Note 2**).
3. Mix and incubate at 37°C for at least 30 min.

3.2. Reaction

1. Freshly prepare the following solutions using solutions A and B listed in **Subheading 2.**
   a. Mix 50 µL of solution A with 50 µL solution B (to give solution C).
   b. Mix 50 µL of solution A with 150 µL solution B (to give solution D).
   c. Mix 50 µL of solution A with 450 µL of solution B (to give solution E).
2. Label six Eppendorf tubes 1–6.
3. Add 10 µL of solutions A, B, C, D, and E to tubes 1–5. Reserve tube 6.
4. Add 40 µL of denatured, reduced protein solution prepared as in **Subheading 3.1.** to each of tubes 1–5.
5. Gently mix each tube and leave at room temperature for 15 min. Thereafter, store on ice.
6. After the 15 min incubation period, place 10 µL aliquots from each of tubes 1–5 into tube 6. Mix.

The samples are now ready for analysis (*see* **Note 5**).

3.3. Electrophoretic Analysis

1. 50 µL aliquots of each sample, Labeled 1–6, mixed with 12 µL of appropriate tracking dye solution, are loaded onto successive lanes of a polymerized high- or low-pH gel, set up in a suitable slab gel electrophoresis apparatus.
2. Low-pH buffer system: Electrophoresis is carried out toward the negative electrode, using a current of 5–20 mA for each gel, overnight at 8°C.
3. High-pH buffer system *(10)*; Electrophoresis is carried out toward the positive electrode at 10–20 mA per gel (or 100–180 V) for 3–4 h.
4. Electrophoresis is stopped when the tracking dye reaches bottom of the gel.
5. Proteins are visualized using conventional stains e.g., Coomassie blue (Pierce, Chester, UK), silver staining (*see* Chapters 33).

4. Notes

1. The method of Takahashi and Hirose ***(6)*** can be used to categorize the half-cystines in a native protein as:
   a. Disulfide bonded;
   b. Reactive sulfhydryls; and
   c. Nonreactive sulfhydryls.

   In the first step, the protein sulfhydryls are alkylated with iodoacetic acid in the presence and absence of 8 *M* urea. In the second step, the disulfide bonded sulfhydryls are fully reduced and reacted with iodoacetamide. The method described above is then used to give a ladder of half-cystines so that the number of introduced carboxymethyl groups can be quantified.
2. Urea is an unstable compound; it degrades to give cyanates which may react with protein amino and thiol groups. For this reason, the highest grade of urea should always be used, and solutions should be prepared immediately before use.
3. Electrophoresis in gels containing higher or lower percent acrylamide may have to be employed depending on the molecular weight of the particular protein being studied.
4. Where protein is already in solution, it is important to note that the pH should be adjusted to around 8.0, and the DTT and urea concentrations should be made at least 10 m*M* and 8 *M,* respectively.
5. Other ratios of iodoacetic acid to iodoacetamide may need to be used if more than about eight cysteine residues are expected, since a sufficiently intense band corresponding to every component in the complete range of charged species may not be visible. A greater ratio of iodoacetic acid should be used if the more acidic species are too faint (and vice versa).

References

1. Creighton, T. E. (1980) Counting integral numbers of amino acid residues per polypeptide chain. *Nature* **284,** 487,488.
2. Freedman R. B., Hirst, T. R., and Tuite, M. F. (1994) Protein disulphide isomerase: building bridges in protein folding. *Trends Biochem. Sci.* **19,** 331–336.
3. Creighton, T. E. (1989) Disulphide bonds between cysteine residues, in *Protein Structure—a Practical Approach* (Creighton, T. E., ed.), IRL, Oxford, pp. 155–167.
4. Freedman, R. B. (1984) Native disulphide bond formation in protein biosynthesis; evidence for the role of protein disulphide isomerase. *Trends Biochem. Sci.* **9,** 438–441.
5. Feinstein, A. (1966) Use of charged thiol reagents in interpreting the electrophoretic patterns of immune globulin chains and fragments. *Nature* **210,** 135–137.

6. Takahashi, N. and Hirose, M. (1990) Determination of sulfhydryl groups and disulfide bonds in a protein by polyacrylamide gel electrophoresis. *Analyt. Biochem.* **188,** 359–365.
7. Stan-Lotter, H. and Bragg, P. (1985) Electrophoretic determination of sulfhydryl groups and its application to complex protein samples, *in vitro* protein synthesis mixtures and cross-linked proteins. *Biochem. Cell Biol.* **64,** 154–160.
8. Hirose, M., Takahashi, N., Oe, H., and Doi, E. (1988) Analyses of intramolecular disulfide bonds in proteins by polyacrylamide gel electrophoresis following two-step alkylation. *Analyt. Biochem.* **168,** 193–201.
9. Reisfield, R. A., Lewis, U. J., and Williams, D. E. (1962) Disk electrophoresis of basic proteins and peptides on polyacrylamide gels. *Nature* **195,** 281–283.
10. Davis, B. J. (1964) Disk electrophoresis II method and application to human serum proteins. *Ann. N.Y. Acad. Sci.* **21,** 404–427.

90

Analyzing Protein Phosphorylation

John Colyer

1. Introduction

Protein phosphorylation is a ubiquitous modification used by eukaryotic cells to alter the function of enzymes, ion channels, and other proteins in response to extracellular stimuli, or mechanical or metabolic change within the cell. In many instances, phosphorylation results in a change in the catalytic activity of the phosphoprotein, which influences one particular aspect of cellular physiology, thereby allowing the cell to respond to the initiating stimulus. A number of different residues within a protein can be modified by phosphorylation. Serine, threonine, and tyrosine residues can be phosphorylated on the side chain hydroxyl group (*o*-phosphoamino acids), whereas others become phosphorylated on nitrogen atoms (*N*-phosphoamino acids, lysine, histidine, and arginine). The former group are involved in dynamic "regulatory" functions and have been studied extensively ***(1)***, whereas the latter group may perform both structural/catalytic roles and signaling functions, the study of which has occurred more recently ***(2)***. The disparity in our understanding of the role of *o*- and *N*-phosphoamino acids is in part a consequence of the acid lability of *N*-phosphoamino acids, which leads to their destruction during the analysis of many phosphorylation experiments.

In terms of the process of studying an individual phosphoprotein, a number of key issues can be identified. First, one must demonstrate that phosphorylation of the protein takes place; then define the number of sites within the primary sequence that can be phosphorylated and by which protein kinase; identify the individual residue(s) phosphorylated; the functional implication of phosphorylation of each site; and describe the use of each site of phosphorylation in vivo. This chapter aims to describe the conduct of an experiment performed to identify a protein as a phosphoprotein. In the case of oligomeric enzymes, it will identify the subunit(s) phosphorylated by a particular kinase. The determination of the stoichiometry of phosphorylation is also described, which provides the first information concerning the number of phosphorylation sites within a polypeptide. These procedures are most straightforward if one has access to the purified protein kinase of interest and the protein substrate. In the case of the kinase, this can be served in many instances by a number of commercial sources, but the approach may be limited by the availability of sufficient pure protein substrate. If this is the case, phosphorylation of a particular target as part of a complex mixture of proteins (e.g., whole-cell extract) can be performed. Under these conditions, identification

From: *The Protein Protocols Handbook, 2nd Edition*
Edited by: J. M. Walker © Humana Press Inc., Totowa, NJ

of the protein of interest will require exploitation of a unique electrophoretic property of the protein *(3)* or require purification of the protein by immunoprecipitation or other comparable affinity-interaction means prior to electrophoresis. The identification of the protein as a phosphoprotein can thereby be achieved, although analysis of the stoichiometry of phosphorylation in this way is inadvisable. In each case, the experimental procedure has a common design: an in vitro phosphorylation reaction is followed by separation of the phosphoproteins by SDS-PAGE and subsequent identification by autoradiography. The incorporation of labeled phosphate can be determined by excising the phosphoprotein band from the dried gel, scintillation counting this gel piece, and converting ^{32}P cpm into molar terms from the knowledge of the specific activity of the initial ATP stock, and the amount of protein substrate analyzed.

2. Materials

1. Purified and partially purified multifunctional protein kinases can be obtained from several commercial sources. The availability of a number of enzymes is illustrated in **Table 1**. The list is not exhaustive, and inclusion in the table does not constitute endorsement of the product:
2. Phosphorylation buffer for the catalytic subunit of protein kinase A (c-PKA): 50 m*M* Histidine-KOH, pH 7.0, 5 m*M* $MgSO_4$, 5 m*M* NaF, 100 n*M* c-pKA, 100 μ*M* ATP.
   a. Histidine-KOH, pH 7.0 is prepared as a concentrated stock, 200 m*M* stored at 4°C for 1 mo, or –20°C for >12 mo. (Warm to 30°C for 30 min to dissolve histidine following storage at –20°C.)
   b. $MgSO_4$, EGTA, NaF: all prepared as 100-m*M* stock, stable at 4°C >12 mo.
   c. ATP, nonradioactive: 20- or 100-m*M* stock (pH corrected to 7.0 with KOH) stable at –20°C for >12 mo. Aliquot to avoid repeat freeze-thaw cycles.
   d. c-pKA, M_r 39,000 *(4)*, sources **Table 1**: stable at –70°C for ~12 mo. Avoid dilute solutions—enzyme tends to aggregate and inactivate.
3. γ-^{32}P-ATP, ICN Pharmaceuticals: dispense into small aliquots and store –20°C. Avoid freeze to thaw cycles. $T_{1/2}$ ~14 d; discard 1 mo after reference date.
4. SDS-PAGE sample buffer, (double-strength): 125 m*M* Tris-HCl, pH 6.8, 20% glycerol, 2% SDS, 10% 2-mercaptoethanol, 0.01% bromophenol blue, stable at –20°C >12 mo. Aliquot and avoid freeze-thaw cycles.
5. Filter paper: Whatman 3MM.
6. X-ray film, X-ray cassettes, intensifying screens, developer and fixative: X-Ograph Ltd.; developer and fixative are stored at room temperature, are reusable, and are stable for approx 2 wk.
7. Scintillation fluid: Emulsifier-safe, Packard.

3. Method

3.1. Phosphorylation Reaction Using Purified Protein Substrates

1. In a designated radioactive area with appropriate acrylic screening prepare a stock of radioactive ATP. The addition of 50 μCi [γ-^{32}P]-ATP to a 0.5-mL solution of 1 m*M* ATP will produce a suitable experimental ATP stock of 220 cpm/pmol (*see* **Note 1**). Warm to 37°C.
2. Incubate the purified protein of interest (0.1–1.0 mg/mL) in the phosphorylation assay medium lacking ATP. Allow the sample to warm to 37°C for 2 min (*see* **Note 2**). A number of control samples should be set up in parallel. One control should contain target protein, but no exogenous kinase, and another, kinase but no target protein.

Table 1
Commercial Source of Multifunctional Protein Kinases

Kinase	Source
Protein kinase A	*a,c,d,e*
Protein kinase C	*a,b,c,d*
Calmodulin kinase II	*c*
Casein kinase II	*a,b,d*
CDC2 kinase	*d*
src kinase	*d*

[a]Boehringer Mannheim.
[b]Calbiochem-Novabiochem.
[c]Sigma Chemical Co.
[d]TCS Biologicals Ltd.

3. Start the phosphorylation reaction by the addition of ATP, containing [γ-^{32}P]-ATP (as defined in **step 1, Subheading 3.**). Cap the tube and vortex briefly. Follow the phosphorylation as a function of time by removing aliquots of the reaction at specific points in time, every 20 s for the first minute, and then at 60-s intervals for the next 4 min.
4. Terminate the reaction by mixing the sample with an equal volume of double-strength SDS-PAGE sample buffer at room temperature. Dispense the phosphorylation sample into an Eppendorf tube containing an equal volume of double-strength sample buffer, cap the tube, and vortex briefly. Store these samples at room temperature, behind appropriate screens, until all samples have been collected.
5. Incubate all samples for 30 min at 37°C. Perform SDS-PAGE in a gel of suitable acrylamide composition, loading a minimum of 5 µg pure target protein/lane (details of electrophoresis in Chapter 11). Allow the dye front to migrate off the bottom of the gel (depositing most of the radioactivity into the electrode buffer), stain with Coomassie brilliant blue, and destain the gel (*see* Chapter 11).
6. Mount the gel on filter paper and cover with clingfilm. The filter paper should be wet with unused destain solution prior to contact with the SDS-PAGE gel. Lower the gel onto the wet filter paper slowly, and flatten to remove air bubbles. Cover with clingfilm, and dry using a vacuum-assisted gel drier for 2 h at 90°C.
7. Once dry, an X-ray film should be placed in contact with the gel for a protracted period to image the location of phosphoproteins. **This procedure must be performed in the dark,** although light emitted by dark room safety lamps is permitted. An X-ray film is first exposed to a conditioning flash of light from a flash gun. Hold a single piece of X-ray film and the flash gun 75 cm apart. Set the flash gun to the minimum power output, and discharge a single flash directly onto the film (*see* **Note 3**). Place the film on top of a clean intensifying screen within an X-ray cassette. Take the dried SDS-PAGE gel, still sandwiched between filter paper and clingfilm, and place it gel side down onto the X-ray film. Do not allow the gel to move once in contact with the film, and use adhesive tape to secure the contact. With a permanent marker pen, draw distinctive markings from the filter paper backing of the gel onto the X-ray film to facilitate orientation of film and gel once autoradiography is complete (*see* **Note 4**). Close the X-ray cassette, label the cassette with experimental details, including the current date and time, and store at –70°C.

8. After 16 h of exposure, the autoradiograph can be developed. Remove the cassette from the –70°C freezer, and allow at least 30 min for it to thaw. Once in the dark room, with safety light illumination only, the cassette should be opened, and SDS-PAGE gel removed from the X-ray film. The film should be placed in 2 L developer and agitated for 4 min at room temperature, in the dark. Using plastic forceps, the film is removed from developer, rinsed in water, and then agitated for a further 90 s in 2 L of fixative (room temperature, dark). At the end of 90 s, the autoradiograph is no longer light-sensitive, and normal lighting can be resumed. Wash the autoradiograph extensively, for 10 min in a constant flow of water, and allow to air-dry.
9. Identification of phosphorylated polypeptides can be performed by superimposition of autoradiograph on the SDS-PAGE gel (*see* **Note 5**). A uniform, almost transparent background should be achieved, with phosphorylated proteins identified as black bands on the autoradiograph of variable intensities (depending on the level of phosphorylation), but regular width and shape (*see* **Note 6**). Exposure times can be altered in the light of results obtained, and repeated autoradiographs of various durations performed on a single gel (*see* **Note 7**).

3.2. Phosphorylation of Components in Complex Protein Mixtures

1. The phosphorylation experiment is performed largely as detailed in **Subheading 3.1.** with the following modifications. A protein concentration of 1–5 mg/mL is recommended supplemented with 100 n*M* purified c-pKA and 10 μ*M* adenosine cyclic 3,5-monophosphate, and with 1 μ*M* microcystin-LR for additional Ser/Thr phosphatase control.
2. Perform a phosphorylation time-course experiment and process as described in **Subheading 3.1.** If the identity of a particular protein cannot be gauged from a peculiar electrophoretic feature (e.g., dissociation of oligomer to monomer on boiling; *3*), then an affinity purification step must be introduced prior to electrophoresis.
3. In this case, phosphorylation will be terminated by placing the sample on ice. Immunoprecipitation of the protein of interest should be performed as detailed in Chapter 57 taking care to solubilize membrane proteins effectively if they are of interest. Immunoprecipitates should be processed as described in **Subheading 3.1., step 5** onward (*see* **Note 8**).

3.3. Determination of Phosphorylation Stoichiometry

1. The specific activity of the ATP (cpm/pmol) needs to be determined empirically. At the time of the phosphorylation experiment, dilute a sample of the experimental ATP (1 m*M* containing 100 μCi/mL [γ-^{32}P]-ATP, as described in **Subheading 3.1., step 1**) to 1 μ*M* by serial dilution in water. Dispense triplicate 10-μL aliquots of the 1-μ*M* ATP (10 pmol) into separate scintillation vials, and add 4.6 mL scintillation fluid to each. Cap and label 10 pmol ATP (*see* **Note 9**).
2. Perform **steps 1–9** of **Subheading 3.1.** To excise the phosphorylated protein bands from the dry SDS-PAGE gel, identify the location by superimposition of gel and autoradiograph. With a marker pen, outline the autoradiographic limits of the phosphoprotein on the gel. Overlay again to confirm the accuracy of demarkation. Mark similar-sized areas of gel that do not contain phosphoproteins to determine background [^{32}P]. Excise these gel pieces with scissors, remove the clingfilm, and place the acrylamide piece and filter paper support in a scintillation vial. Add 4.6 mL scintillation fluid, and cap the vial.
3. Scintillation count each vial for 5 min or longer, using a program defined for ^{32}P radionucleotides. Minimal quenching of ^{32}P occurs under these conditions.

4. To calculate the pmol phosphate incorporated/µg protein, subtract the background radioactivity (cpm) from experimental data (cpm) to obtain phosphate incorporation into the protein sample (in units of cpm). Convert this to pmol incorporation/µg using the formula:

$$\text{Phosphorylation (pmol/µg protein)} = [\text{protein phosphorylation (cpm)}/\text{SA of ATP (cpm/pmol)} \times \text{µg protein}] \quad (1)$$

   With a knowledge of the molecular weight of the polypeptide, a molar phosphorylation stoichiometry can be calculated from these data using the formula:

$$\text{Phosphorylation (mol/mol protein)} = [\text{phosphorylation (pmol/µg protein)} \times \text{molecular weight}/10^6] \quad (2)$$

5. Experimental conditions that result in maximal protein phosphorylation will have to be optimized. Parameters worth considering include alteration of the pH of the reaction, extension of the time-course of phosphorylation (up to several hours), addition of extra protein kinase during the phosphorylation process, and addition of extra ATP throughout the time-course.

4. Notes

1. The specific activity of ATP must be tailored to the experiment intended. Phosphorylation and autoradiography of proteins require ≥200 cpm/pmol, studies that require analysis beyond this point (e.g., phosphoamino acid identification) require ~2000 cpm/pmol, while phosphorylation of peptides requires ~20–50 cpm/pmol.
2. The stability of proteins at low concentration is sometimes an issue; inclusion of an irrelevant protein, but not a phosphoprotein (e.g., bovine serum albumin) at 1 mg/mL (final) is recommended. The phosphorylation example used to illustrate this method (c-pKA) displays catalytic activity in the absence of signaling molecules. In other instances this is not so. Therefore, relevant activators should be included as dictated by the kinase (e.g., Ca^{2+}, calmodulin, acidic phospholipids, and so forth).
3. Film developed at this stage will exhibit very slight discoloration compared to unexposed film.
4. Phosphorescent labels (Sigma-Techware) can be used to label an SDS-PAGE gel prior to autoradiography. It will also facilitate superimposition of gel and autoradiograph. These can also highlight the position of mol-wt markers, an image of which will be captured on the X-ray film.
5. Protein kinases invariably autophosphorylate. This can be identified clearly in control samples lacking phosphorylation target (**Subheading 3.1., step 2**).
6. Autoradiographs sometimes have a high background signal. Uniform black coloration over the whole film, extending beyond the area exposed to the gel is indicative of illumination of the X-ray film. Discoloration of part of the film is indicative of light entering the X-ray cassette. Examine the cassette carefully, particularly the corners that are prone to damage by rough handling.
7. Phosphoimage technology represents an alternative to autoradiography, it has the advantage of collecting the image quickly. Exposure times are reduced by an order of magnitude. However, in my experience, this benefit is at the expense of the quality of the image, which is granular.
8. The time required for immunoprecipitation or similar procedure should be kept to a minimum to limit the dephosphorylation of proteins by endogenous phosphatase enzymes. A cocktail of phosphatase inhibitors should be included for the same reason.
9. The hydrolysis of ATP to ADP and Pi occurs at a low rate in the absence of any enzyme. The extent of hydrolysis of [γ-^{32}P]-ATP can affect the determination of phosphorylation

stoichiometry, since it will result in the overestimation of the specific activity of ATP if a correction is not made. Quantification of the purity of ATP is quoted in the product specification from suppliers. Only fresh [γ-^{32}P]-ATP should be used in these procedures or the degree of hydrolysis confirmed by thin-layer chromatography *(6)*.

References

1. Krebs, E. G. (1994) The growth of research on protein phosphorylation. *Trends Biochem. Sci.* **19,** 439.
2. Swanson, R. V., Alex, L. A., and Simon, M. I. (1994) Histidine and aspartate phosphorylation: two component systems and the limits of homology. *Trends Biochem. Sci.* **19,** 485–490.
3. Drago, G. A. and Colyer, J. (1994) Discrimination between two sites of phosphorylation on adjacent amino acids by phosphorylation site-specific antibodies to phospholamban. *J. Biol. Chem.* **269,** 25,073–25,077.
4. Peters, K. A., Demaille, J. G., and Fischer, E. H. (1977) Adenosine 3':5'-monophosphate dependent protein kinase from bovine heart. Characterisation of the catalytic subunit. *Biochemistry* **16,** 5691–5697.
5. Otto, J. J. and Lee, S. W. (1993) Immunoprecipitation methods. *Meth. Cell Biol.* **37,** 119–127.
6. Bochner, B. R. and Ames, B. N. (1982) Complete analysis of cellular nucleotides by two dimensional thin layer chromatography. *J. Biol. Chem.* **257,** 9759–9769.

91

Mass Spectrometric Analysis of Protein Phosphorylation

Débora Bonenfant, Thierry Mini, and Paul Jenö

1. Introduction

Phosphorylation is one the most frequently occurring posttranslational modifications in proteins, playing an essential role in transferring signals from the outside to the inside of a cell and in regulating many diverse cellular processes such as growth, metabolism, proliferation, motility, and differentiation. It is estimated that up to one third of all proteins in a typical mammalian cell are phosphorylated *(1)*. Phosphorylation is carried out by a vast group of protein kinases which are thought to constitute 3% of the entire eukaryotic genome *(1–3)*. To decipher the recognition signal of protein kinases and protein phosphatases acting on a given molecular target, and to understand how the activity of the target protein is regulated by phosphorylation, it is important to define the sites and the extent of phosphorylation at each specific site.

The most commonly employed technique of phosphoprotein analysis involves in vivo or in vitro labeling with [^{32}P]phosphate *(4–8)*. The radiolabeled protein is subsequently digested with a suitable protease and radiolabeled peptides are separated either by high-performance liquid chromatography (HPLC) *(7,8)* or by two-dimensional phosphopeptide mapping *(4–8)*. The site of phosphorylation is then determined by solid-phase Edman sequencing *(6)*.

Identification of phosphorylation sites by Edman degradation of [^{32}P]-labeled proteins remains a valuable approach when working with limited quantities of protein. In the last couple of years, however, electrospray ionization-mass spectrometry (ESI-MS) and matrix-assisted laser desorption ionization-mass spectrometry (MALDI-MS) have been used quite successfully for phosphoprotein analysis *(9–12)*. Mass spectrometry is particularly useful when radioactive protein labeling cannot be performed, when phosphorylation sites are clustered within a short peptide sequence, or when the sites of phosphorylation are located more than 10–15 residues from the amino-(N)-terminus of a peptide. The basic elements of mass spectrometric phosphoprotein analysis consist of proteolytic digestion of the protein of interest followed by measuring the masses of the resulting peptides. When the protein sequence is known, peptides are identified by comparing observed masses to those predicted based on the specificity of the protease. Phosphopeptides have masses 80 Da greater (due to the presence of an HPO_3 group) than predicted from the peptide sequence; hence masses increasing in multiples of 80 Da

From: *The Protein Protocols Handbook, 2nd Edition*
Edited by: J. M. Walker © Humana Press Inc., Totowa, NJ

indicate the presence of several phosphorylated residues. The site of phosphorylation in a given phosphopeptide can be identified by subjecting it to fragmentation in the collision cell of the mass spectrometer. From the mass differences between the fragment ions, the sequence can be "read" and the site of phosphorylation can be identified when the mass difference is either 167, 181, or 243 Da, corresponding to the residue masses of P-Ser, P-Thr, or P-Tyr. Because the activity of many proteins is regulated by the extent to which they become phosphorylated by a particular stimulus, it is important to be able to follow the extent of phosphorylation of a given site. Stoichiometries can be quantified by comparison of the peak intensities of a given phosphopeptide and its unphosphorylated counterpart.

Owing to its speed and sensitivity, it is evident that mass spectrometric tracing of phosphorylation sites is a viable alternative to radioactive phosphopeptide mapping. The main problem, however, lies in the complexity of the data. Proteins with large molecular weights generate some 50 peptides on enzymatic digestion. In addition, electrospray ionization produces multiple signals for each peptide due to multiple charging. Therefore, data interpretation of phosphoprotein digests can be quite cumbersome. Furthermore, phosphopeptides usually make up only a small portion of all peptides in a digest. To reduce data complexity, procedures have been developed to selectively detect phosphopeptides either by appropriate scanning procedures ***(13)***, or by exploiting the phosphate group for selective phosphopeptide purification prior to mass spectrometric analysis ***(14,15)***. This enables the researcher to find phosphopeptides in crude protein digests with high speed and sensitivity.

In this chapter, we describe the basic steps of mass spectrometric phosphopeptide mapping, namely enzymatic digestion of phosphoproteins, reverse-phase high-performance liquid chromatography of peptides with mass spectral analysis (LC/MS), and determination of the site of phosphorylation by tandem mass spectrometry (MS/MS).

2. Materials

2.1. Equipment

1. HPLC system: Two micro-gradient pumps are currently used in our laboratory: a Hewlett-Packard 1090M (Palo Alto, CA), whose outlet is connected to a stainless steel T-piece to reduce the flow to 1–2 µL/min and a splitless microgradient system consisting of an Evolution 200 system (Prolab, Reinach, Switzerland) operated at flow rates of 1 µL/min.
2. Model P-2000 quartz micropipet puller (Sutter Instrument Company, Novato, CA).
3. Capillary reverse-phase columns prepared from 100 µm inner diameter (i.d.) × 280 µm outer diameter (o.d.) fused silica capillaries (LC Packings, Amsterdam, The Netherlands).
4. *xyz*-Positioner, model M-PRC-3 (Newport, Irvine, CA).
5. Stereo microscope, model Leica MZ12 (Wild Heerbrugg, Switzerland) and a CLS100 lamp (Wild Heerbrugg, Switzerland) equipped with two fiber-optic light guides installed on the mass spectrometer.
6. Mass spectrometer: A TSQ7000 triple quadrupole instrument (Finnigan, San José, CA) equipped with a homemade micro-electrospray ion source is used.

2.2. Materials and Reagents

1. Fused silica capillaries (100 µm i.d., 280 µm o.d.) and PEEK sleeves (300 µm i.d.) were purchased from LC Packings (Amsterdam, The Netherlands). F120 PEEK Fingertight fittings were from Upchurch Scientific (Oak Harbor, WA).

2. C_{18} reverse-phase packing material: For preparing the capillary columns, Vydac C_{18} material (5 μm particle size), removed from old Vydac 218TP51 columns (Vydac, Hesperia, CA), was used.
3. Chemicals: Trifluoroacetic acid (TFA) and acetonitrile were from Pierce (Rockford, IL) and from J. T. Baker (Phillipsburg, NJ), respectively. $Ba(OH)_2$ was purchased from Merck (Darmstadt, Germany). Dithiothreitol (DTT) and 10× times concentrated dephosphorylation buffer (0.5 *M* Tris-HCl, pH 8.5; 1 m*M* EDTA) were from Roche Diagnostics (Mannheim, Germany). Iodoacetamide was from Fluka Chemie AG (Buchs, Switzerland).
4. Enzymes: Enzymes were obtained from the following suppliers: modified trypsin: Promega (Madison, WI), endoproteinase LysC (*Achromobacter* protease): Wako Pure Chemical Industries (Osaka, Japan), endoproteinase GluC, sequencing grade, and calf intestinal alkaline phosphatase (1000 U/mL): Roche Diagnostics (Mannheim, Germany).

3. Methods

3.1. Enzymatic Digestions of Phosphoproteins

To efficiently trace sites of phosphorylation in proteins, it is important to obtain an essentially complete proteolysis of the protein of interest (*see* **Note 1**). This is best achieved if the protein is fully denatured by reduction and alkylation with DTT and iodoacetamide. For reduction, the protein is dissolved in 10 μL of 100 m*M* Tris-HCl, pH 8.0 and 8 *M* urea (freshly prepared); 1.5 μL of 75 m*M* DTT (dissolved in water) is added and the protein is reduced for 1–2 h at 37°C. Alkylation is achieved by adding 1 μL of 625 m*M* iodoacetamide and incubating for 15 min at room temperature.

At this point the protein can be digested with the endoproteinase LysC. The residual iodoacetamide and urea (approx 6 *M*) do not appreciably inhibit the enzymatic activity. The enzyme-to-substrate ratio is kept between 1:50 and 1:20 (w/w). Incubation with endoproteinase LysC is carried out at 37°C for 1 h. The reaction is stopped by the addition of 10% TFA to a final concentration of 0.5% TFA.

For trypsin digestion, the urea concentration has to be lowered to 2 *M* by dilution with 100 m*M* Tris-HCl, pH 8.0, whereas endoproteinase GluC was found to be severely inhibited even by 2 *M* urea. When using this enzyme, it is best to remove urea completely. This can be achieved by reverse-phase chromatography or, if the protein fails to chromatograph with acceptable yields, by acetone precipitation. Digestion with trypsin or endoproteinase GluC is carried out at an enzyme-to-substrate ratio of 1:50 to 1:20 (w/w) for 2 h at 37°C.

3.2. Enzymatic Dephosphorylation with Alkaline Phosphatase

Dephosphorylation is achieved by dissolving the phosphoprotein or the enzymatic digest in 10 μL of 100 m*M* Tris-HCl, pH 8.0. To this, 1 μL of 10× concentrated dephosphorylation buffer (0.5 *M* Tris-HCl, pH 8.5, 1 m*M* EDTA) is added, followed by 1 μL of calf intestinal alkaline phosphatase (CIP), and dephosphorylation is allowed to proceed at 37°C for 15 min. The solution is acidified with 3 μL of 1% TFA and analyzed by liquid chromatography.

3.3. β-Elimination with $Ba(OH)_2$

A dilute alkali solution is prepared from 150 m*M* Tris-HCl, pH 8.0, to which solid $Ba(OH)_2$ is added in excess of saturation (0.16 *M* at 20°C) ***(19)***. Peptides resulting from proteolytic digests are dissolved in 10 μL of $Ba(OH)_2$ solution and incubated at 37°C for 1 h. The reaction is stopped by the addition of 2 μL of 10% TFA and immediately injected onto the reverse-phase column for LC/MS.

3.4. Construction of Capillary Columns

Glass capillary columns are constructed from polyimide fused silica capillaries (100 μm i.d. × 280 μm o.d.) according to **ref. *16***. The capillaries are cut into pieces of approx 20 cm length and placed into a quartz micropipet puller. Tips with an aperture of 0.5–1 μm are produced with the following puller settings: heat: 205, velocity: 40, delay: 128, line: 1. A reservoir made from stainless steel tubing (0.5–1.0 mm i.d.) is connected to a blunt-ended Luer-lock needle with the aid of a stainless steel union (0.75- mm bore) and a slurry of reverse-phase resin (20% suspension in 50% acetonitrile) is drawn into the stainless steel capillary. The syringe is removed, connected to an HPLC pump, and the capillary is connected to the other end of the reservoir. The flow of the HPLC pump is set to 200 μL/min. During the pressurization phase of the pump the resin is forced into the capillary. The resin is allowed to settle at a pressure of approx 100 bars for 10 min. After packing, the column is removed from the reservoir and cut at the top so as to be filled completely with resin. The column is then washed repeatedly with 0.1% TFA and 80% acetonitrile–0.1% TFA.

3.5. Liquid Chromatography–Mass Spectrometry

The capillary column is connected to a stainless steel union with a bore of 0.2 mm with the aid of PEEK sleeves and F120 PEEK Fingertight fittings (*see* **Note 2**). The outlet of the injector is connected to the other end of the union. The column is placed into a *xyz*-positioner to allow for precise alignment of the column outlet with the mass spectrometer inlet system. With the aid of a stereo microscope the tip of the column is placed into the center of the heated capillary of the mass spectrometer and retracted 2–3 mm from the heated capillary to avoid arcing. A spray voltage of 1000–1300 V is applied onto the stainless steel union. Care should be taken that all the liquid leaving the column tip is nebulized to a fine spray (visible as a fine whitish spray leaving the tip of the column). Once a stable spray has been developed, the gradient for peptide elution is started and data collection is initiated. Positive ionization is almost exclusively used, as it yields greater sensitivity in most mass spectrometers. The column effluent is scanned over the 200–2000–Da mass-to-charge ratio (*m/z*) in 3 s at unit resolution.

Once all the peptides have been eluted from the column, the observed masses of the peptides are compared to those predicted from the sequence of the protein based on the specificity of the protease. To find all sites of phosphorylation, it is important to cover as much of the protein sequence as possible (*see* **Note 3**). Once all observed masses have been assigned to peptides corresponding to completely (or incompletely) cleaved peptides, the spectra are analyzed for the presence of phosphopeptides that can be recognized by peptides having masses 80 Da heavier than the mass predicted by the peptide sequence (or multiples thereof in the case of several phosphorylation sites) (*see* **Note 4**). **Fig. 1** shows the elution of a tryptic peptide derived from the nuclear phosphoprotein lamin isolated from *Drosophila melanogaster* embryos ***(17)***. The peptide can be observed as the triply and doubly charged ion with *m/z* values of 890.9 Da and 1336.7 Da, respectively. Also apparent in the spectrum is the singly phosphorylated

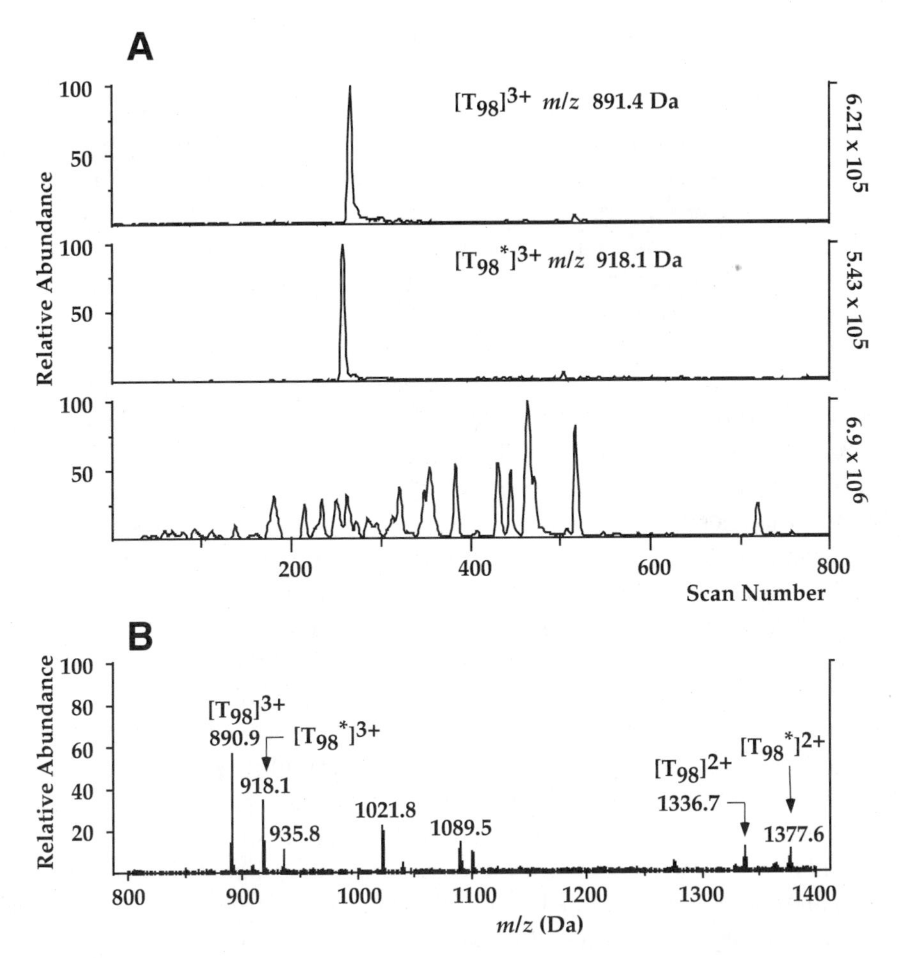

Fig. 1. **(A)** Ion chromatograms of the tryptic peptide T_{98} (SVTAVDGNEQLYHQQ-GDPQQSNEK, comprising residues 595–618) derived from the C-terminal tail domain of *Drosophila melanogaster* interphase lamin. The LC/MS data were scanned for ions with *m/z* predicted for the triply charged unphosphorylated (*top*) and the triply charged phosphorylated peptide (the calculated mass of the singly charged peptide is 2672.2 Da). The *bottom panel* shows the reconstructed ion current of the sum of all ions present during the LC/MS run. **(B)** Spectrum of the triply and doubly charged T_{98} and its phosphorylated form (labeled with an *asterisk*). The spectrum was generated by summing all scans acquired during the elution of T_{98} and phospho-T_{98}. (Modified from **ref. *17***).

peptide with a signal approx 27 Da higher for the triply charged (80 Da divided by 3), and 40 Da (80 Da divided by 2) higher for the doubly charged ion. Note that in the corresponding ion chromatograms, the phosphopeptide elutes slightly earlier than the non phosphorylated peptide, an effect that is frequently observed due to the higher polarity of the phosphate group.

3.6. Stoichiometry of Phosphorylation

Stoichiometries of phosphorylation can be quantified by dividing the sum of the peak areas of all intensities of a given phosphopeptide (all charge states must be included) by the sum of all intensities of the phosphorylated and its non-phosphorylated forms (*see* **Note 5**). For example, comparison of the integrated peak areas of the various charge forms for the two phosphopeptides K_2 and K_{26} of the *Drosophila* lamin indicates an approximate extent of phosphorylation of 23–29%, and 13–19%, respectively (**Fig. 2A,C**). This assumes that the ionization efficiencies of the phospho- and the nonphosphopeptide are equal. Our observations indicate that the relative areas between phosphorylated and unphosphorylated peptide varies with the charge state: the higher charge forms tend to overrepresent the phosphoform, whereas the lower charge forms tend to underrepresent the phosphoform. Accordingly, the stoichiometry of phosphorylation can be determined only as an average from a number of charge states with a certain variation. Nevertheless, large changes in phosphorylation that occur at a given site can still be followed that peak areas from spectra of the corresponding phospho-nonphospho- forms are compared.

For example, the extent of phosphorylation of the K_2 and K_{26} fragment is changed considerably when a soluble isoform of the protein (the so-called mitotic form) is isolated from young embryos. The extent of phosphorylation of the K_2 fragment increases from 13–19% to 66–71%, whereas phosphorylation of the K_{26} fragment decreases to levels too low to be quantified with certainty (**Fig. 2B,D**).

These examples clearly demonstrate that quantitation of the extent of phosphorylation of a given phosphopeptide is possible within certain limits. However, it must be stressed that this becomes more difficult when multiple phosphorylation is clustered within a short stretch of a protein. In our studies to determine the insulin-induced phosphorylation sites of ribosomal protein S6 (dS6A) in the fruit fly *Drosophila melanogaster*, the sites were found to be located in a short stretch at the carboxy-(C)-terminal end of the protein ***(18)***. Two-dimensional polyacrylamide gel electrophoresis of ribosomal proteins indicated that up to five phosphates are incorporated into dS6A upon insulin/cycloheximide stimulation of Kc167 cells. Although LC/MS analysis of an endoproteinase LysC digest of dS6A demonstrated the presence of singly, doubly, and triply phosphorylated forms of the C-terminal fragment, the quadruply, and fivefold phosphorylated forms could not be detected. Because polyacrylamide gel electrophoresis clearly demonstrated that the quadruply, or fivefold phosphorylated form of the protein was predominant, it was concluded that ionization efficiency decreases with increasing degree of phosphorylation. Therefore, an endoproteinase LysC digest of dS6A from insulin- and cycloheximide-stimulated Kc167 cells was treated with $Ba(OH)_2$ to induce β-elimination of the phosphate groups ***(10,19)***. Subsequent capillary LC/MS analysis revealed strong signals with mass-to-charge ratios of 692.1, 698.4, and 704.5 Da, corresponding to the triply charged ions of the RRRSASIRESKSSVSSDKK ($K_{35–37}$) peptide containing three, four, and five dehydroalanines (**Fig. 3B**). The intensity of the corresponding ions closely resembled the staining intensity of the individual phosphoderivatives of dS6A, with the tetradehydroalanine derivative being the most abundant species (**Fig. 3B**). Furthermore, when dS6A was isolated from cells pretreated with okadaic acid, a type 2A phosphatase inhibitor, prior to stimulation, substantially higher amounts of the penta-dehydroalanine derivative could be observed in the spectrum of the $K_{35–37}$ peptide (**Fig. 3C**).

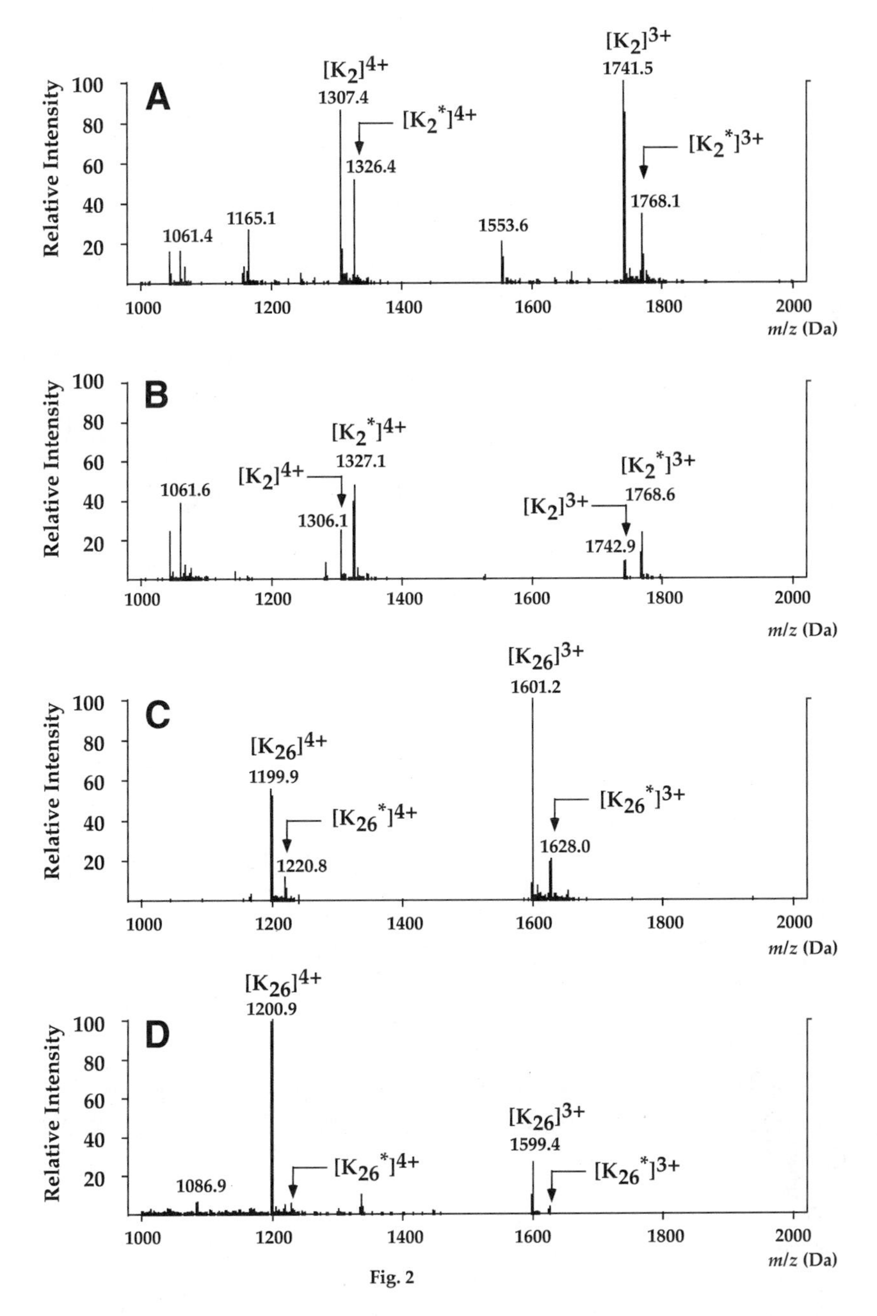

Fig. 2. Mass spectrum of the endoproteinase LysC peptides K_2 and K_{26} from the head and the central domain of interphase (**A,C**) and mitotic (**B,D**) lamin. Because of the large size, each peptide produces a complex spectrum of multiply charged ions. Therefore, only the triply and quadruply charged peptides (labeled with 3^+ and 4^+) in the mass range between 1000 and 2000 Da are shown. The phosphorylated peptides are labeled with an *asterisk*. Because it was not possible to separate all of the lamin peptides into single components during the LC/MS run, the spectra in (**A**) and (**B**) contain signals from coeluting peaks other than K_2. (From **ref.** ***17***).

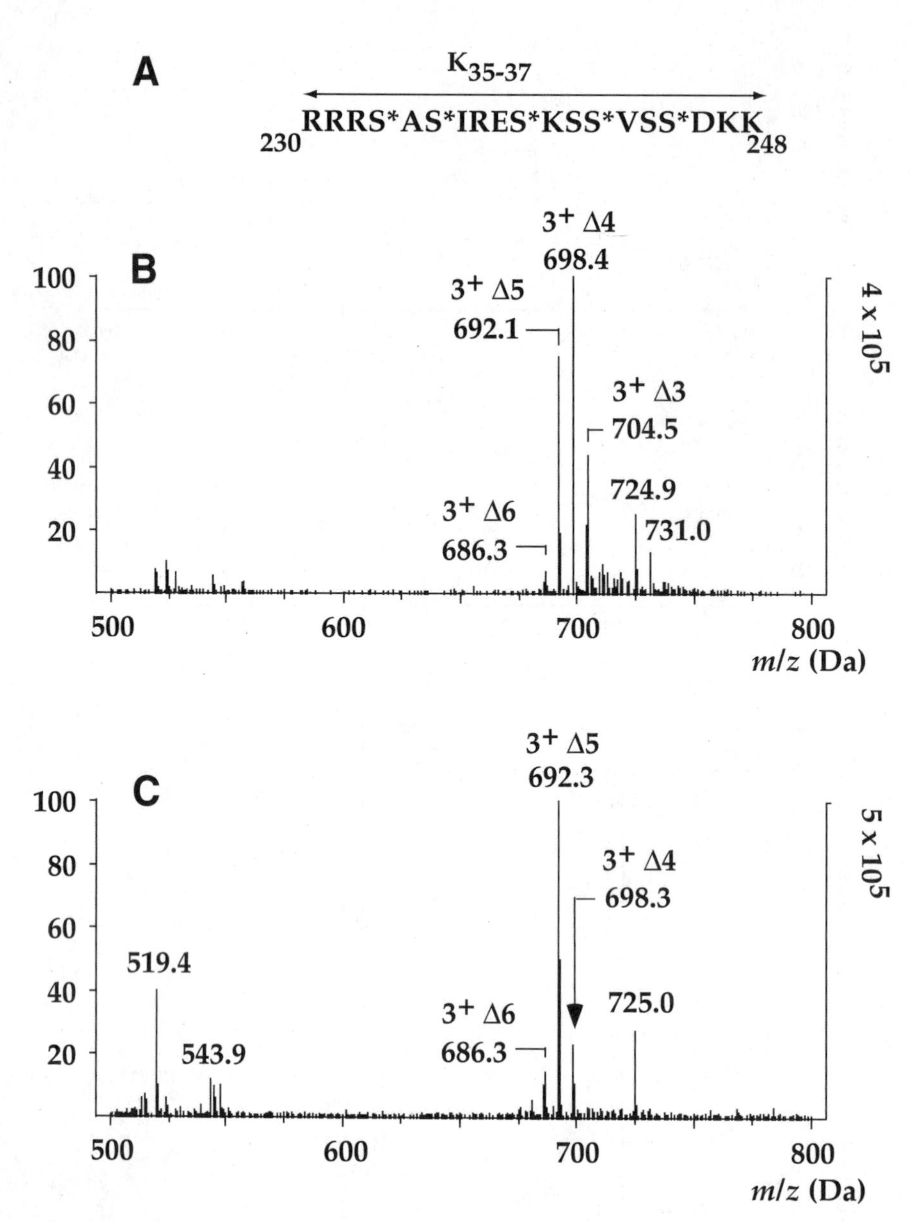

Fig. 3. **(A)** Amino acid sequence of the K_{35-37} peptide obtained by endoproteinase LysC cleavage of *Drosophila melanogaster* S6A (corresponding to residues 230–248 of ribosomal protein S6). The residues labeled with an asterisk indicate the amino acids which become phosphorylated upon insulin/cycloheximide treatment. **(B)** Mass spectrum of $Ba(OH)_2$-treated RRRSASIRESKSSVSSDKK (K_{35-37}) from cells stimulated with insulin and cycloheximide. Δ6, Δ5, ΔD4, and Δ3 indicate the number of dehydroalanines arising from $Ba(OH)_2$ treatment of K_{35-37} carrying 6, 5, 4, and 3 phosphates. The spectrum was obtained by summing up all scans acquired during the elution of Δ3–Δ5 K_{35-37} from the capillary reverse-phase column. **(C)** Spectrum of $Ba(OH)_2$-treated K_{35-37} from cells that had been treated with okadaic acid prior to insulin/cycloheximide treatment to suppress phosphatase activity. For labeling of the spectrum, *see* **(B)**. (From **ref.** ***18***).

3.7. Localization of Sites of Phosphorylation by LC/MS/MS

After identifying a candidate ion from the initial LC/MS analysis, the sample is applied to LC and the ion is selected for on-line collision-induced dissociation (CID) to determine the site of phosphorylation. Peptides usually fragment at peptide bonds to produce a series of daughter ions containing the N-terminal or C-terminal ends of the molecule ("b" ions or "y" ions, respectively) allowing the sequence to be "read" from the mass differences between the fragment ions. For example, **Fig. 4** shows a comparison of the MS/MS spectra of unphosphorylated and phosphorylated CP_4 peptide derived from the head domain of the interphase form of *Drosophila melanogaster* lamin *(17)*. The peptide with the sequence PPSAGPQPPPPSTHSQTASSPLSPTR has nine potential phosphorylation sites. An identical set of unphosphorylated fragment ions, whose masses correspond to those predicted for y_{19}-y_6 (**Fig. 4A,B**) was obtained. This places the site of phosphorylation beyond y_{19}. As there is only one candidate serine beyond y_{19}, it can be concluded that the serine at position three (which corresponds to Ser_{25} in the lamin protein) is the site of phosphorylation. Positive identification is seen in a set of ions, corresponding to b_2–b_7 where b_3–b_7 is shifted by 80 Da in the phospho-CP_4 MS/MS spectrum compared with the nonphosphorylated CP_4 peptide. Also note the loss of phosphoric acid, giving rise to an intense signal 27 Da lower than the triply charged precursor ion (**Fig. 4B**).

Figure 2 demonstrates that the phosphorylation of K_2 isolated from M-phase lamin was increased when compared with the same fragment isolated from interphase lamin. This could mean either that the previously identified serine at position 3 (corresponding to Ser_{25} in the lamin protein) becomes phosphorylated to a greater extent than in interphase lamin, or that a site other than Ser_{25} is phosphorylated to higher stoichiometry. These two possibilities can be distinguished by subjecting the same peptide isolated from interphase and M-phase lamin to fragmentation. 80 Da shifts due to the presence of a phosphorylated residue can be observed for y_{10}–y_{18} in the spectrum of phospho-CP_4 derived from M-phase lamin, as well as fragments corresponding to y_6–y_{13} after neutral loss (**Fig. 5A**). This indicates that the site of phosphorylation of the head region in M-phase lamin has changed to a residue located C-terminally of Ser_{25}. Identification of the phosphorylated residue in the M-phase CP_4 fragment becomes evident upon inspection of the low mass range of the spectrum. In both phosphopeptides (interphase and M-phase forms) the y_3 ions are identical (compare **Figs. 4** and **5**). This eliminates the Thr_{47} as the phosphorylated site. As neutral loss of phosphate occurs from y_6 on in the M-phase CP_4 peptide, and because there is only one serine residue between y_3 and y_6, it can be concluded that Ser_{45} is the residue that becomes phosphorylated in M-phase lamin (**Fig. 5**). Therefore, during embryonal development of the fruit fly, phosphorylation of the head domain of lamin is regulated by changing the site of phosphorylation from Ser_{45} in early embryos to Ser_{25} in older embryos. This leads to profound changes in the ability of lamin to self polymerize: whereas lamin phosphorylated on Ser_{45} is soluble, lamin phosphorylated on Ser_{25} polymerizes to form the underlying structure of the nuclear envelope. This example demonstrates the use of mass spectrometry to trace protein phosphorylation in organisms that are not amenable to classical phosphoprotein mapping by metabolic ^{32}P-labeling.

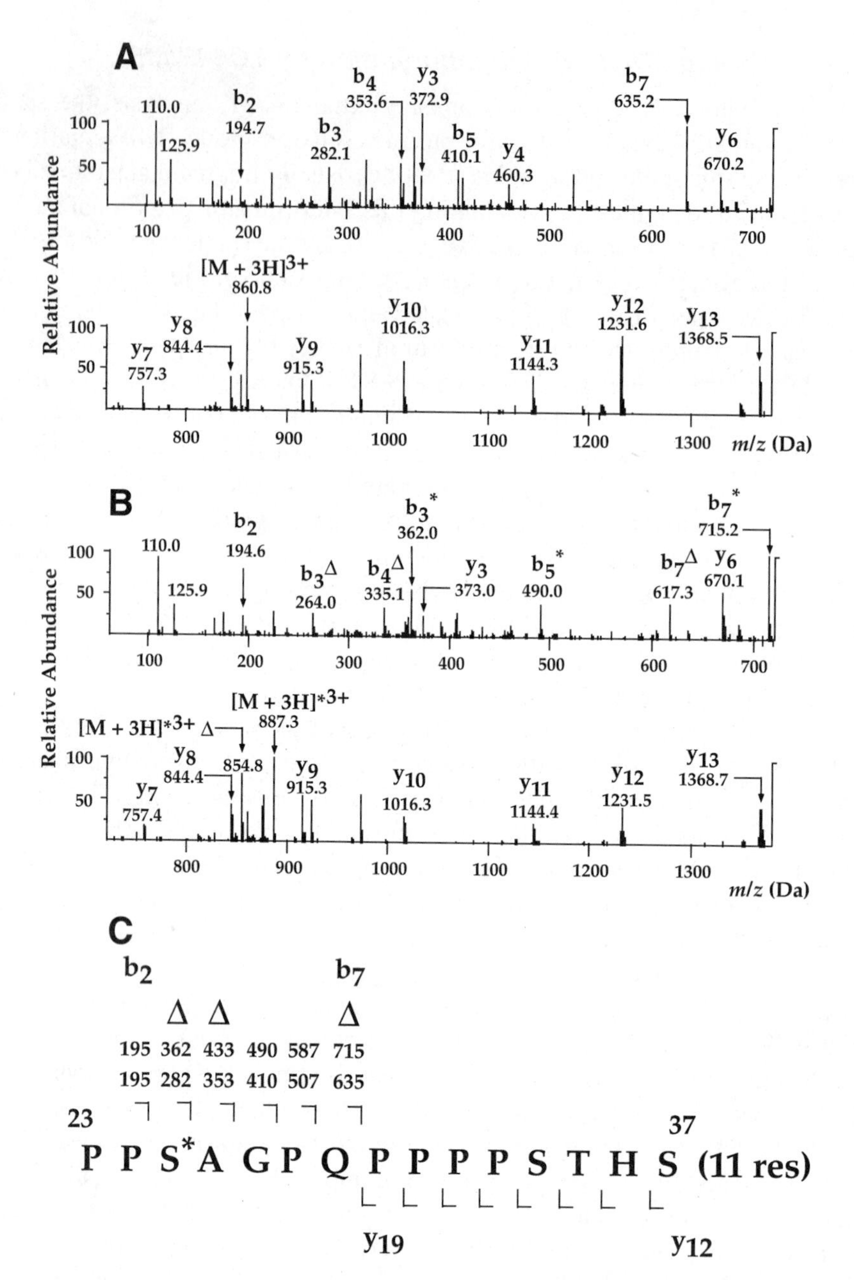

Fig. 4. Comparison of the MS/MS spectra of **(A)** the unphosphorylated and **(B)** the phosphorylated CP_4 peptide (comprising residues 23–48) from interphase lamin. In both experiments, the triply charged ion was selected for collision-induced dissociation. Phosphorylated fragment ions are indicated by an *asterisk*, ions that have undergone neutral loss of phosphoric acid are indicated by Δ. **(C)** Summary of the observed b- and y-ions in the MS/MS spectrum of CP_4. The predicted b-ions are listed above the sequence for unphosphorylated CP_4 (*lower row*) and for phospho- CP_4 (*upper row*) by assuming phosphorylation on the serine indicated with an *asterisk* (Ser_{25} in the amino acid sequence of *Drosophila* lamin). The symbol Δ marks those ions that have undergone neutral loss of phosphoric acid during the collision process. The amino sequence of CP_4 is shown only for residues 23–37, the C-terminal part is abbreviated as (11 res). (From **ref.** ***17***).

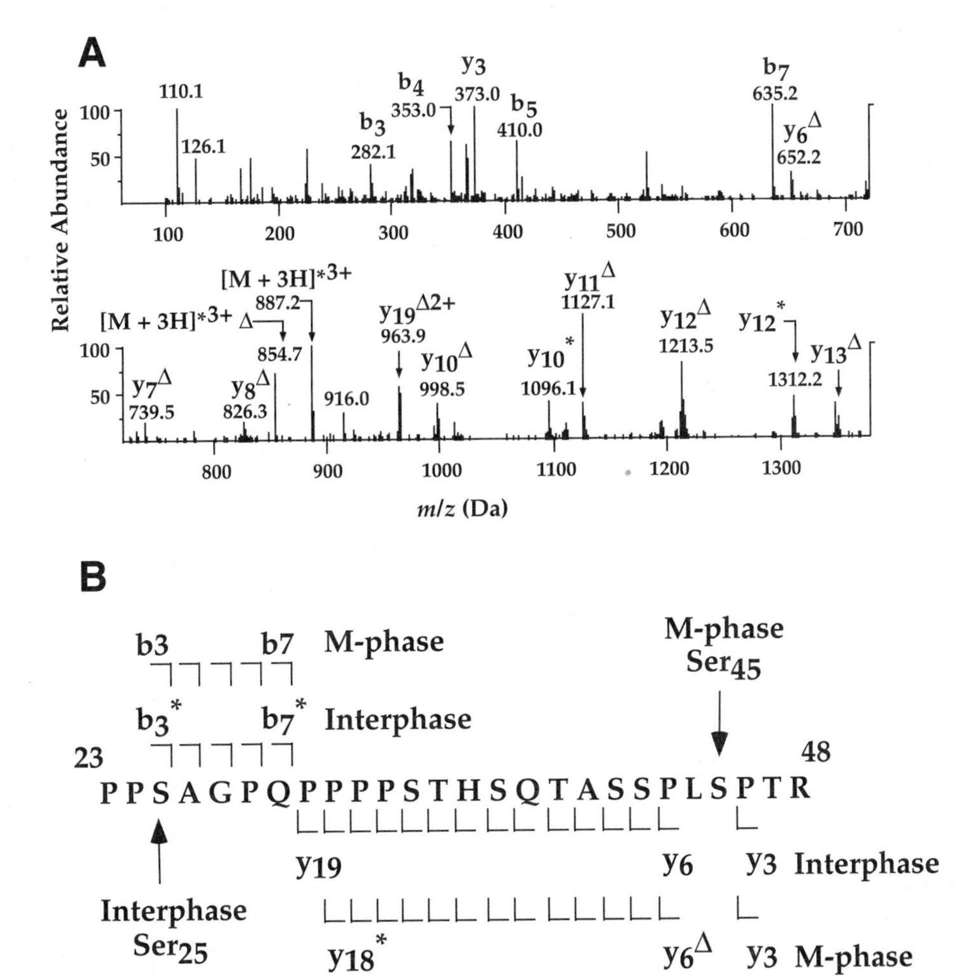

Fig. 5. **(A)** MS/MS spectrum of the phosphorylated clostripain peptide CP_4 from M-phase lamin. The triply charged ion was selected for fragmentation. The corresponding MS/MS spectrum for the phosphorylated CP_4 peptide from interphase lamin is shown in **Fig. 4B**. For labeling of ions *see* **Fig. 4**. **(B)** Summary of the observed b- and y-ions of the CP_4 fragment from interphase and M-phase lamin. The *arrows* indicate the site of phosphorylation that occurs on Ser_{25} during interphase and on Ser_{45} during M-phase. (From **ref.** ***17***).

4. Notes

1. The choice of the protease should be guided by the protein sequence such that as many peptides as possible between 5 and 20 amino acids are generated. Smaller peptides are usually lost in the flow through of the column during LC/MS. Large peptides are sometimes recovered with low yields from the column and are particularly difficult to fragment in the collision cell of the mass spectrometer. Trypsin is the preferred protease because it generates peptides with at least two positive charges on each peptide for efficient ionization. Furthermore, the C-terminal lysine or arginine favors fragmentation in the collision cell, which leads to a high sequence coverage, a prerequisite for unambiguous localization of the site of phosphorylation. For phosphoproteins soluble only in high concentrations of urea, endoproteinase LysC should be used, as the protease tolerates urea concentrations

up to 6 *M*. However, due to the cleavage specificity of the enzyme, peptides containing internal arginine residues are generated, which in some cases suppresses extensive fragmentation, therefore making the assignment of the site of phosphorylation more difficult.

2. For interfacing peptide separation with an ESI source, the flow rates applied to the separation system have to meet the requirements of the inlet system of the mass spectrometer. HPLC can be carried out on commercially packed reverse-phase columns provided the solvent delivery rate is within the limits of nebulization for the particular inlet system. Otherwise, either post- or precolumn splitting ***(20)*** is required to reduce the flow rate. For the capillary column system described in this chapter, a practical limitation in solvent delivery is the ability to generate gradients at low flow rates (1–20 μL/min). Whereas dual-syringe pumps capable of delivering gradients at 10 μL/min are appropriate for capillary columns with 500 μm i.d., 100 μm i.d. columns require flow rates of 1 μL/min and less. Newer models (e.g., Evolution 200, Prolab) have been introduced that are able to deliver flows of 1 ml/min without the use of flow splitters.
3. Peptides can be lost during the chromatographic separation process. Therefore, it is mandatory to analyze the complete LC/MS data set with respect to sequence coverage of the protein. Particular attention should be paid to missing peptides containing phosphorylatable residues (Ser, Thr, and Tyr). The most common reason for missing peptides is that they are either too small (therefore often too hydrophilic) to bind to the reverse-phase support, or too large (too hydrophobic) to be eluted from the column. These peptides can often be recovered by selecting a different digestion strategy. The data set should also be examined for incompletely cleaved peptides, as phosphorylation located close to a cleavage site often reduces digestion efficiency. Large phosphopeptides can also be subdigested with a different protease to yield smaller peptides with better elution and fragmentation behavior.
4. The presence of a phosphorylation site can be observed when the mass of a peptide is 80 Da larger than predicted from the peptide sequence. Such candidates, however, have to be checked rigorously for the presence of a phosphorylation site. Often, calculating the masses of incompletely cleaved peptides with various combinations of missed cleavage sites yield peptide masses identical to the phosphopeptide candidate. A simple test is to dephosphorylate the digest with alkaline phosphatase and to repeat the analysis. If the candidate peptide is phosphorylated, dephosphorylation leads to a reduction of its mass by 80 Da (or multiples thereof in the case of multiple phosphorylation). Enzymatic dephosphorylation with (CIP) can be directly carried out in the digestion mix, as CIP is completely resistant to proteolysis by enzymes such as trypsin, endoproteinase LysC, or endoproteinase GluC, although urea tends to inactivate the enzyme. For protein digests in the 1–500 picomol range, 1 U of CIP was found to be sufficient to lead to complete dephosphorylation of phosphopeptides. Alternatively, peptides phosphorylated on serine and threonine tend to undergo neutral loss of H_3PO_4 on fragmentation, while phosphotyrosine shows no neutral loss. H_3PO_4 loss can be exploited to selectively screen protein digests for the presence of phosphopeptides, by scanning for decreases in *m/z* of all ions between two mass analyzers of a tandem instrument ***(21)***. Separate experiments have to be performed to scan for mass decreases corresponding to each charged form (e.g., 98 Da for the singly, 48 Da for the doubly, 33 Da for the triply charged ion). Also, by raising the orifice potential during LC/MS, and scanning in negative ion mode for the appearance of the PO_3^- (79 Da) and PO_2^- (63 Da) fragment ions derived from phosphoserine, phosphothreonine, or phosphotyrosine, selective phosphopeptide detection is possible ***(11)***. However, the latter two methods can be as much as 10-fold less sensitive than simple mass detection, and many peptides do not show the required fragmentation loss.

5. On fragmentation of a phosphopeptide in the collision cell of the mass spectrometer, the phosphorylated amino acid can usually be identified. Peptides fragment at peptide bonds, producing a series of daughter ions containing the N-terminal or C-terminal ends of the molecule ("b" ions or "y" ions, respectively). When a phosphorylated residue is encountered, the fragment ion series beyond the phosphorylated residue is shifted by 80 Da. Therefore, comparison of the fragment spectrum of the phosphorylated and nonphosphorylated peptide allows one to pinpoint the residue from which the mass shift occurs. However, precise localization of the phosphorylated residue is often obscured owing to chemical instability of ions containing phosphoserine or phosphothreonine. As mentioned earlier, on collision-activated dissociation, the phosphate group can be lost as phosphoric acid, reducing the mass of the corresponding fragment ion by 98 Da. Also, partial dehydration of serine and threonine residues is frequently observed during the collision process, generating signals 18 Da lower than those of the nondehydrated ion. Therefore, the spectrum can be misinterpreted by assuming complete loss of phosphoric acid from a phosphorylated residue, rather than loss of water from a nonphosphorylated Ser or Thr. Such ambiguities can be solved by comparing the fragmentation patterns of the phosphopeptide with those of the unphosphorylated peptide. Fragment ions that undergo dehydration by loss of water will be apparent in the spectrum of the unphosphorylated peptide, giving rise to a mixture of nondehydrated and dehydrated ions. Fragment ions from the phosphorylated peptide that undergo neutral loss produce a signal at the same m/z value as the dehydrated ion, but the signal of the nondehydrated ion is missing.
6. When phosphorylation stoichiometries are quantified by comparing the integrated peak intensities of a phosphopeptide and its nonphosphorylated counterpart, it is assumed that both forms of the peptide ionize with equal efficiencies. This estimation of stoichiometries is in most cases valid for singly and doubly phosphorylated peptides. However, this should be proven by showing that the peak intensities of a peptide derived from the unphosphorylated protein equals the sum of intensities of phosphorylated and unphosphorylated peptide derived from the phosphoprotein (provided there is partial phosphorylation at this given site). We have found that with increasing clustering of phosphorylation sites extensively phosphorylated peptides tend to become poorly ionized. In such cases, the extent of phosphorylation can be estimated more reliably by β-elimination of the phosphate group. In the case of the *Drosophila* ribosomal protein S6 ***(18)*** this was found to be extremely useful for phosphoserines. Although phosphothreonine has been reported to undergo β-elimination ***(19)***, we have not yet tested the procedure for threonine phosphorylation. Furthermore, the stoichiometries of phosphorylation observed by mass spectrometry and in vitro or in vivo ^{32}P-labeling often differ. The difference might arise from the fact that preexisting phosphate resides on the protein at the time of ^{32}P-labeling, preventing the incorporation of labeled phosphate.

References

1. Hubbard, M. J., and Cohen, P. (1993) On target with a new mechanism for the regulation of protein phosphorylation. *Trends Biochem. Sci.* **18,** 172–177.
2. Hunter, T. (1991) Protein kinase classification. *Meth. Enzymol.* **200,** 3–37.
3. Cohen, P. (1992) Signal integration at the level of protein kinases, protein phosphatases, and their substrates. *Trends Biochem. Sci.* **17,** 408–413.
4. Boyle, W. J., van der Geer, P., and Hunter, T. (1991) Phosphopeptide mapping and phosphoamino acid analysis by two-dimensional separation on thin-layer cellulose plates. *Meth. Enzymol.* **201,** 110–152.

5. Luo, K., Hurley, T. R., and Sefton, B. M. (1991) Cyanogen bromide cleavage and proteolytic peptide mapping of proteins immobilized to membranes. *Meth. Enzymol.* **201,** 149–152.
6. Wettenhall, R. E. H., Aebersold, R. H., and Hood, L. E. (1991) Solid-phase sequencing of 32P-labeled phosphopeptides at picomole and subpicomole levels. *Meth. Enzymol.* **201,** 186–199.
7. Kuiper, G. G. J. M. and Brinkmann, A. O. (1995) Phosphotryptic peptide analysis of the human androgen receptor: detection of a hormone-induced phosphopeptide. *Biochemistry* **34,** 1851–1857.
8. Winz, R., Hess, D., Aebersold, R., and Brownsey, R. W. (1994) Unique structural features and differential phosphorylation of the 280-kDa component (isozyme) of rat liver acetyl-CoA carboxylase. *J. Biol. Chem.* **269,** 14,438–14,445.
9. Payne, M. D., Rossomando, A. J., Martino, P., Erickson, A., K., Her, J.-H., and Shabanowitz, J. (1991) Identification of the regulatory phosphorylation sites in pp42/mitogen-activated protein kinase (MAP kinase). *EMBO J.* **10,** 885–892.
10. Resing, K. A., Johnson, R. S. & Walsh, K. A. (1995) Mass spectrometric analysis of 21 phosphorylation sites in the internal repeat of rat profilaggrin, precursor of an intermediate filament associated protein. *Biochemistry* **34,** 9477–9487.
11. Verma, R., Annan, R. S., Huddleston, M. J., Carr, S. A., Reynard, G., and Deshaies, R. J. (1997) Phosphorylation of Siclp by G1 Cdk required for its degradation and entry in S phase. *Science* **278,** 455–460.
12. Kalo, M. S. and Pasquale, E. B. (1999) Multiple in vivo tyrosine phosphorylation sites in EphB receptors. *Biochemistry* **38,** 14,396–14,408.
13. Carr, S. A., Huddleston, M. J., and Annan, R. S. (1996) Selective detection and sequencing of phosphopeptides at the femtomole level by mass spectrometry. *Analyt. Biochem.* **239,** 180–192.
14. Nuwaysir, L. M. and Stults, J. T. (1993) Electropsray ionization mass spectrometry of phosphopeptides isolated by on-line imobilized metal-ion affinity chromatography. *J. Am. Soc. Mass Spectrom.* **4,** 662–669.
15. Watts, J. D., Affolter, M., Krebs, D. L., Wange, R. L., Samelson, L. E., and Aebersold, R. (1994) Identification by electrospray ionization mass spectrometry of the sites of tyrosine phosphorylation induced in activated Jurkat T cells on the protein tyrosine kinase ZAP-70. *J. Biol. Chem.* **269,** 29,520–29,529.
16. Davis, M. T. and Lee, T. D. (1992) Analysis of peptide mixture by capillary high performance liquid chromatography: a practical guide to small-scale separations. *Protein Sci.* **1,** 935–944.
17. Schneider, U., Mini, T., Jenö, P., Fisher, P. A., and Stuurman, N. (1999) Phosphorylation of the major Drosophila lamin in vivo: site identification during both M-phase (meiosis) and interphase by electrospray ionization tandem mass spectrometry. *Biochemistry* **38,** 4620–4632.
18. Radimerski, T., Mini, T., Schneider, U., Wettenhall, R. E. H., Thomas, G., and Jenö, P. (2000) Identification of insulin-induced sites of ribosomal protein S6 phosphorylation in Drosophila melanogaster. *Biochemistry* **39,** 5766–5774.
19. Byford, M. F. (1991) *Biochem. J.* **280,** 261–265.
20. Covey, T. R. (1995) in *Methods in Molecular Biology,* Vol. 61, (Chapman, J. R., ed.), Humana Press, Totowa, NJ, pp. 83–99.
21. Covey, T., Shushan, B., Bonner, R., Schröder, W., and Hucho, F., (1991) in *Methods in Protein Sequence Analysis* (Jörnvall, H., Hoog, J. O., and Gustavsson, A. M. eds.), Birkhäuser, Basel, pp. 249–256.

92

Identification of Proteins Modified by Protein (D-Aspartyl/L-Isoaspartyl) Carboxyl Methyltransferase

Darin J. Weber and Philip N. McFadden

1. Introduction

The several classes of *S*-adenosylmethionine-dependent protein methyltransferases are distinguishable by the type of amino acid they modify in a substrate protein. The protein carboxyl methyltransferases constitute the subclass of enzymes that incorporate a methyl group into a methyl ester linkage with the carboxyl groups of proteins. Of these, protein (D-aspartyl/L-isoaspartyl) carboxyl methyltransferase, EC 2.1.1.77 (PCM) specifically methyl esterifies aspartyl residues that through age-dependent alterations are in either the D-aspartyl or the L-isoaspartyl configuration ***(1,2)***. There are two major reasons for wishing to know the identity of protein substrates for PCM. First, the proteins that are methylated by PCM in the living cell, most of which have not yet been identified, are facets in the age-dependent metabolism of cells. Second, the fact that PCM can methylate many proteins in vitro, including products of overexpression systems, can be taken as evidence of spontaneous damage that has occurred in these proteins since the time of their translation.

The biggest hurdle in identification of substrates for PCM arises from the extreme base-lability of the incorporated methyl esters, which typically hydrolyze in a few hours or less at neutral pH. Thus, many standard biochemical techniques for separating and characterizing proteins are not usefully applied to the identification of these methylated proteins. In particular, the electrophoresis of proteins by the most commonly employed techniques of sodium dodecyl sulfate polyacrylamide gel electrophoresis (SDS-PAGE) results in a complete loss of methyl esters incorporated by PCM, owing to the alkaline pH of the buffers employed. Consequently, a series of systems employing polyacrylamide gel electrophoresis at acidic pH have been utilized in efforts to identify the substrates of PCM. A pH 2.4 SDS system ***(3)*** using a continuous sodium phosphate buffering system has received the most attention ***(1,3–14)***. The main drawback of this system is that it produces broad electrophoretic bands. Acidic discontinuous gel systems using cationic detergents, ***(15)***, have proven useful in certain situations ***(16–21)*** and can be recommended if the cationic detergent is compatible with other procedures that might be utilized by the investigator (e.g., immunoblotting, protein sequencing). Recently, we have developed an electrophoresis system that employs SDS

From: *The Protein Protocols Handbook, 2nd Edition*
Edited by: J. M. Walker © Humana Press Inc., Totowa, NJ

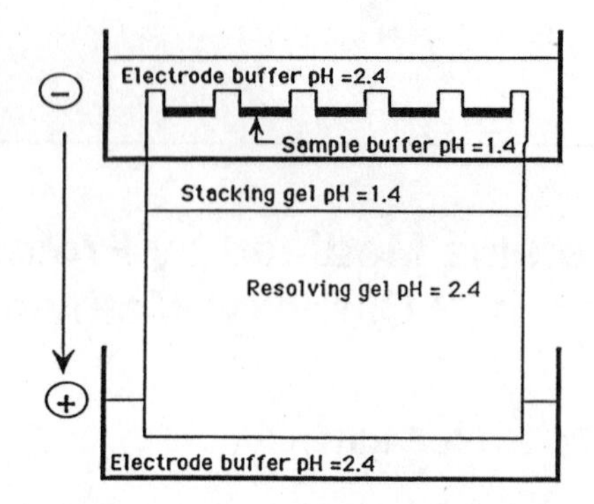

Fig. 1. Schematic of acidic discontinuous gel system. The system employs a pH 1.4 stacking gel on top of a pH 2.4 resolving gel with chloride as the leading ion and acetate as the trailing ion to stack the proteins tightly. The presence of the anionic detergent SDS allows the separation of proteins on the basis of molecular weight. The low pH preserves labile protein methyl esters, and so allows the identification of age-altered substrates of PCM.

and an acidic discontinuous buffering system (**Fig. 1**). This procedure results in sharp electrophoretic bands and would be a good choice for investigators wishing to adhere to SDS as the anionic detergent. This system is described below, and examples of its ability to resolve proteins are provided.

2. Materials

2.1. Equipment

1. Slab gel electrophoresis unit: We have used the mini-gel electrophoresis units from Idea Scientific (Minneapolis, MN) (10 × 10 × 0.1 cm) with great success, as well as the Sturdier large-format gel system from Hoeffer Scientific Instruments (San Francisco, CA) (16 × 18 × 0.15 cm).
2. Electrophoresis power supply, constant current.
3. X-ray film and photo darkroom.
4. Scintillation counter.

2.2. Reagents

1. 40% (w/v) Acrylamide stock solution containing 37:1 ratio of acrylamide to *N,N'*-methylene-bis-acrylamide (Bio-Rad, Richmond, CA).
2. Resolving gel buffer: 0.1 *M* NaH_2PO_4 (Sigma, St. Louis, MO), 2.0 % SDS (United States Biochemical, Cleveland, OH [USB], ultrapure), 6 *M* urea [USB], ultrapure) pH 2.4, with HCl.
3. Modified Clark and Lubs buffer (C & L buffer): Add 25.0 mL of 0.2 *M* NaCl to 26 mL of 0.2 *M* HCl, and bring to a final volume of 0.1 L. The buffer pH should be ~1.4 *(22)* (*see* **Note 1**).
4. Stacking gel buffer (2X): 2.0% (w/v) SDS, 6.0 *M* urea and C & L buffer such that the C & L buffer makes up 66% (v/v) of the total volume with the remaining 34% consisting of water and other buffer components. Buffer will be 0.033 *M* in NaCl. Readjust pH to 1.4 with HCl.

5. Sample solubilization buffer (2X): 2.0% (w/v) SDS, 6.0 *M* urea, 10% glycerol ([USB], ultrapure), 0.01% pyronin Y dye (Sigma), and C & L 33% (v/v) of the total volume, with the remaining 67% consisting of water and other buffer components. Buffer will be 0.0165 *M* in NaCl. Readjust pH to 1.4 with HCl.
6. Electrode buffer (1X): 0.03 *M* NaH_2PO_4, 0.1% SDS, 0.2 *M* acetate, pH 2.4 with HCl.
7. Gel polymerization catalysts: 0.06% $FeSO_4$, 1.0% H_2O_2, 1.0% ascorbic acid, prepared fresh in separate containers.
8. Colloidal Coomassie G-250 protein stain stock solution: 125 g ammonium sulfate, 25 mL 86% phosphoric acid, 1.25 g Coomassie brilliant blue G-250 (Sigma), deionized water to 1.0 L. The dye will precipitate, so shake well immediately before use. Stable indefinitely at room temperature ***(11)***.
9. Destain solution: 10% (v/v) acetic acid.
10. Fluorography solution: 1.0 *M* sodium salicylate brought to pH 6.0 with acetic acid ***(23)***.
11. X-ray film: Kodak X-Omat AR or equivalent.

2.3. Gel Recipes

2.3.1. 12% Acrylamide Resolving Gel, pH 2.4

The following volumes are sufficient to prepare one 7.5 × 10.5 × 0.15 cm slab gel:

2X Resolving gel buffer	3.75 mL
40% Acrylamide (37:1)	2.25 mL
0.06% $FeSO_4$	0.06 mL
1.0% Ascorbic acid	0.06 mL
0.3% H_2O_2	0.06 mL

2.3.2. 4.0% Stacking Gel, pH 1.4

The following volumes are sufficient to prepare one 3.0 × 10.5 × 0.15 cm stacking gel:

2X Stacking gel buffer	3.75 mL
40% Acrylamide (37:1)	0.75 mL
0.06% $FeSO_4$	0.06 mL
1.0% Ascorbic acid	0.06 mL
0.3% H_2O_2	0.06 mL

3. Methods

3.1. Sample Solubilization

1. An equal volume of 2X solubilization buffer is added and the samples are heated in a 95°C heating block for no more than 30 s (*see* **Note 2**).
2. To separate samples under reducing conditions, it is critical to add any reducing agent before addition of the solubilization buffer. Up to 10 µg/gel protein band can be resolved with this system; total sample loaded in a single well should not exceed about 100 µg.

3.2. Gel Preparation

3.2.1. Resolving Gel

1. Add all the components listed under **Subheading 2.3.1.**, except the 0.3% H_2O_2, together in a small Erlenmeyer side-arm flask.
2. Degas the solution for at least 5.0 min using an in-house vacuum.
3. Assemble together gel plates and spacers that have been scrupulously cleaned and dried.

4. Using a pen, make a mark 3.0 cm from the top of the gel plates to denote the space left for the stacking gel.
5. To the degassed gel solution, add the H_2O_2 catalyst. Gently mix solution by pipeting the solution in and out several times. Avoid introducing air bubbles into the solution.
6. Quickly pipet the acrylamide solution between the glass gel plates to the mark denoting 3.0 cm from the top of the gel plates.
7. Carefully overlay the acrylamide solution with about a 2.0-mm layer of water-saturated butanol using a Pasteur pipet, so that the interface will be flat on polymerization.
8. Allow the gel to polymerize at room temperature until a distinct gel–butanol interface is visible.
9. After polymerization is complete, pour off the overlay, and gently rinse the top of the gel with deionized water. Invert the gel on paper towels to blot away any remaining water between the gel plates.

3.2.2. Stacking Gel

1. Add all the components listed under **Subheading 2.3.2.**, except the 0.3% H_2O_2, together in a small Erlenmeyer sidearm flask.
2. Degas the solution for at least 5.0 min using in-house vacuum.
3. Soak the well-forming combs in the H_2O_2 catalyst solution while preparing the stacking gel.
4. To the degassed gel solution, add the H_2O_2 catalyst. Gently mix solution by pipeting the solution in and out several times. Avoid introducing air bubbles into the solution.
5. Quickly pipet the acrylamide solution between the glass gel plates to the very top of the gel plates.
6. Remove the combs from the H_2O_2 catalyst solution, shake off some of the excess solution, and insert the comb in between the gel plates by angling the comb with one hand and guiding the comb between the plates with the other hand.
7. Ensure the comb is level relative to the top of the resolving gel and that no bubbles are trapped under the comb.
8. After the stacking gel has completely polymerized, carefully remove the comb. Remove any unpolymerized acrylamide from each well by rinsing with deionized water (*see* **Note 3**).

3.3. Electrophoresis

1. Assemble the gel in the electrophoresis unit.
2. Add sufficient electrode buffer to cover the electrodes in both the upper and lower reservoirs.
3. Remove any air bubbles trapped under the bottom edge of the plates with a bent 25-gage needle and syringe containing electrode buffer.
4. Rinse each well with electrode buffer immediately before adding samples.
5. Load all wells with 40–60 μL of protein samples; load 1X solubilization buffer into any empty wells.
6. Run the gels at 15 mA, constant current, at room temperature for 4–6 h, or until proteins of interest have been adequately resolved (*see* **Note 4**).
7. Fix the gel in 12% (v/v) TCA with gentle shaking for 30 min.
8. Pour off fixative, and mix 1 part methanol with 4 parts of colloidal Coomassie G-250 stock solution. Slowly shake gel with staining solution for ~12 h (*see* **Note 5**).
9. Pour off staining solution. Any nonspecific background staining of the gel can be removed by soaking the gel in 10% (v/v) acetic acid. Little destaining of protein bands occurs even after prolonged times in 10% acetic acid.

3.4. Detection of Radioactively Methylated Proteins

Several protocols exist for radioactively methylating proteins with PCM *(2,16)*. Following electrophoresis, gel bands containing radioactively labeled proteins can be detected by fluorography or scintillation counting of gel slices.

3.4.1. Fluorography of Gels

1. If gels have been stained, they are destained using 10% methanol/7% acetic acid to decolorize the gel bands.
2. Expose the gel to fluorography solution for 30 min at room temperature with gentle shaking.
3. The gels are then placed on a piece of filter paper and dried under vacuum without heat for 3 h.
4. In a dark room, X-ray film (Kodak X-Omat AR) is preflashed twice at a distance of 15 cm with a camera flash unit fitted with white filter paper (3M) to act as a diffuser.
5. The gel is placed in direct contact with the film and taped in place. For future alignment, puncture holes in an asymmetric pattern in a noncrucial area of the gel-film sandwich. A 25-gage needle is useful for this purpose.
6. After sealing in a film cassette, the cassette is wrapped with aluminum foil, and exposure takes place at –70°C for several weeks.
7. After exposure, remove the cassette from –70°C, and allow to warm to room temperature. Develop film in darkroom.
8. An example of this technique is shown in **Fig. 2**.

3.4.2. Scintillation Counting Radioactive Methanol Evolved by Base Hydrolysis of Protein Methyl Esters in Gel Slices

1. Stained gels are soaked in 10% acetic acid containing 3% (v/v) glycerol for 1.0 h, placed on a piece of filter paper, and dried under vacuum without heat for 3.0 h. The glycerol keeps the gel from cracking and keeps it pliable for the steps described below.
2. Using a ruler and fine-tip marking pen, a grid is drawn directly on the surface of the dried gel.
3. A sharp scalpel is then used to cut out uniform slices precisely from each gel lane. Alternatively, selected bands can be individually excised from the gel.
4. 4.0 mL of scintillation fluid are added to each 20-mL scintillation vial. A glass 1-dram vial is then placed inside the scintillation vial, carefully avoiding spilling scintillation fluid into the 1-dram vial.
5. Each dried gel slice is then placed in the inner dram vial of a scintillation vial.
6. After all the gel slices have been placed in a separate inner 1-dram vial, 0.3 mL of 0.2 *M* sodium hydroxide is added to each inner vial. The scintillation vial is then immediately tightly capped and allowed to sit undisturbed for at least 3 h. Several hours are required for any volatile methanol that has formed by methyl ester hydrolysis to equilibrate and partition into the organic scintillation fluid, where it can then be detected by the scintillation counter.
7. Controls for measuring the efficiency of equilibration are performed by using a ^{14}C methanol standard, which is added to an inner 1-dram vial containing a nonradioactive gel slice and base hydrolysis solution. This is then placed inside a scintillation vial and allowed to equilibrate along with the other samples. An equal aliquot of the ^{14}C methanol standard is mixed directly with the scintillation fluid.
8. **Figures 3** and **4** show examples of gels that have been sliced and counted in a scintillation counter under the conditions just described.

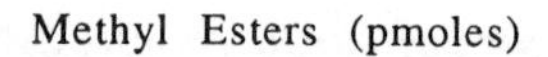

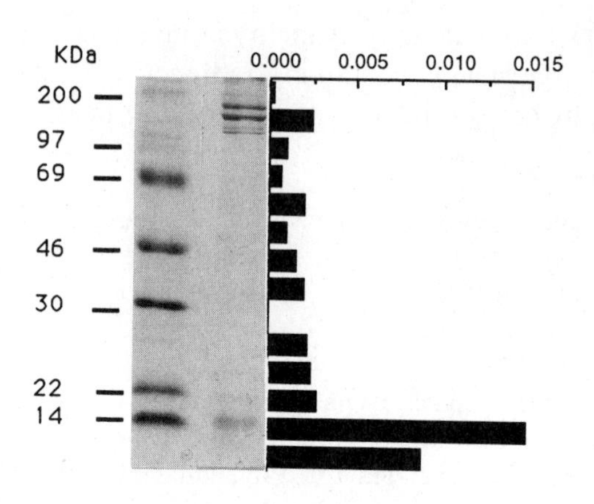

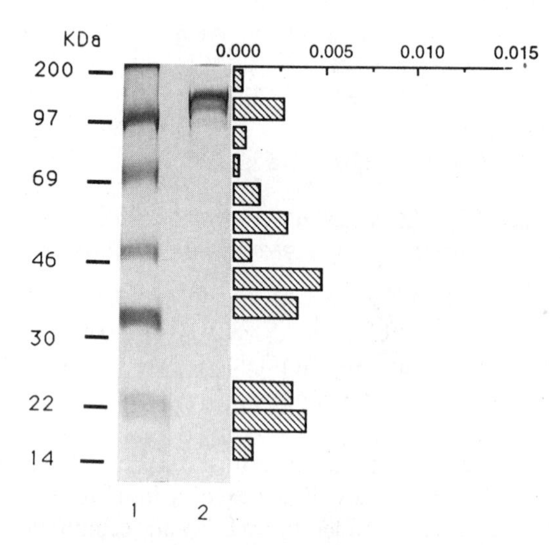

Fig. 2. Comparison of discontinuous acid gel with continuous acid gel. The following experiment tested for the presence of age-altered proteins in a commercial preparation of collagenase. The collagenase preparation (Sigma, type IV) was methylated with purified rabbit erythrocyte PCM and ^{3}H-AdoMet. Aliquots from the same methylation reaction were then resolved on (top) 12% discontinuous acid gel, described in text, or (bottom) 12% continuous acid gel system prepared according to the method of Fairbanks and Auruch *(3)*. *Lane 1*: Rainbow mol-wt markers (Amersham); *Lane 2*: 18 μg of methylated collagenase were loaded on each gel. Following electrophoresis and staining, 0.5-cm gel slices were treated with base to detect radioactivity as described under **Subheading 3.** Both gel systems are capable of preventing the loss of methyl esters from protein samples, but the discontinuous system provides much higher resolution of individual polypeptide bands.

4. Notes

1. The buffering system employed in the stacking gel and sample solubilization buffer is based on a modification of the C & L buffering system. NaCl is employed rather than KCl of the original system, because K^+ ions cause SDS to precipitate out of solution.

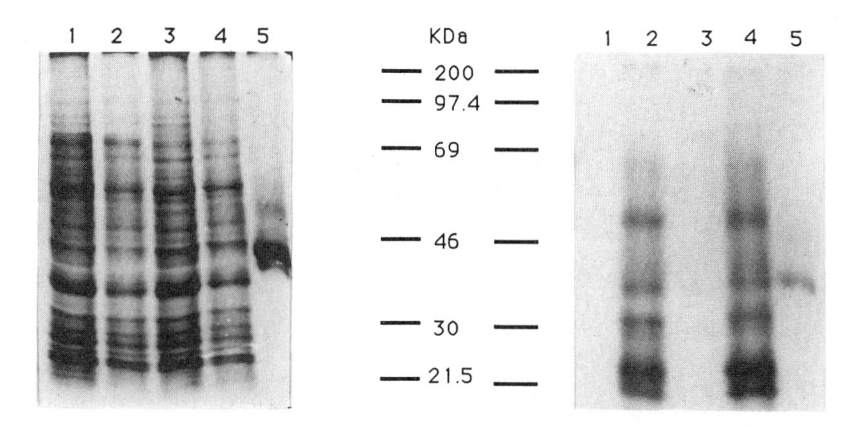

Fig. 3. Coomassie staining and autoradiography of complex protein mixtures by acidic discontinuous SDS gel electrophoresis. The following experiment was performed to measure the varieties of age-altered proteins in a cell cytoplasm. (**Left Panel**) Coomasie G-250 stained acidic discontinuous gel. (**Right Panel**) Autoradiogram of same gel. *Lane 1*: Cytoplasmic proteins, following incubation cells with ^{3}H-*S*-adenosyl-methione (AdoMet). *Lane 2*: Cytoplasmic proteins, following incubation of PC12 cytoplasm with ^{3}H-AdoMet. *Lane 3*: Cytoplasmic proteins, following incubation of intact PC12 cells with ^{3}H-AdoMet and purified rabbit PCM. *Lane 4*: Cytoplasmic proteins, following incubation of lysed PC12 cells with ^{3}H-AdoMet and purified rabbit PCM. *Lane 5*: Positive control; 20 µg of ovalbumin methylated with ^{3}H-AdoMet and purified rabbit PCM. Cells, lysates, and subfraction incubated with IU PCM, 300 pmol Adomet in final volume of 50 µL 0.2 *M* citrate, pH 6.0, 20 min at 37°C.

2. On occasion, the solubilization buffer will contain precipitates. These can be brought back into solution by briefly heating the buffer at 37°C . The solubilization buffer is stable for at least 2 wk at room temperature; by storing the buffer in small aliqouts at –20°C, it is stable indefinitely.
3. Gels can be stored for up to two weeks at 4°C by wrapping them in damp paper towels and sealing tightly with plastic wrap.
4. Since the dye front is a poor indicator of protein migration, use of prestained mol-wt markers, such as the colored Rainbow markers from Amersham, allows the progress of protein separation to be monitored by simply identifying the colored bands, which are coded according to molecular weight.
5. It is essential that all fixing and staining steps occur at acidic pH. The colloidal Coomassie G-250 procedure described under **Subheading 3.** has several advantages: it is acidic, simple to perform, and has higher sensitivity than other dye-based staining methods, including those using Coomassie R-250. Additionally, nonspecific background staining is very low, so only minimal destaining is necessary to visualize protein bands.
6. Avoid using higher glycerol concentrations or prolonged incubation of the gel in this solution. Otherwise, the gel will be sticky after drying and contract sharply away from the paper backing on cutting.

References

1. Aswad, D. W. and Deight, E. A. (1983) Endogenous substrates for protein carboxyl methyltransferase in cytosolic fractions of bovine brain. *J. Neurochem.* **31,** 1702–1709.

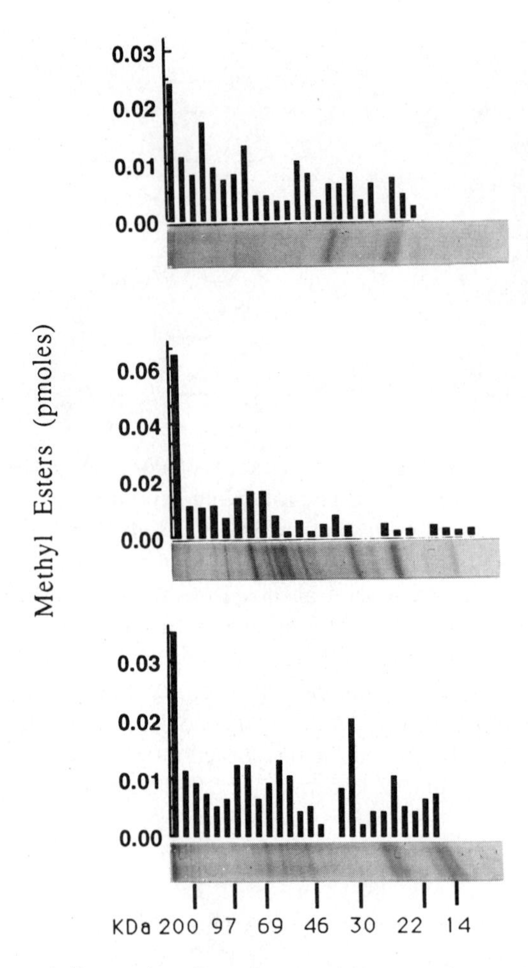

Fig. 4. Coomassie staining and radioactive methyl ester determination in gel slices of electrophoresed proteins from diseased human brain tissue. Extracts prepared from homogenates of Alzheimer's diseased brain (obtained from the Department of Pathology, Oregon Health Sciences University) were methylated in vitro with purified rabbit erythrocyte PCM and ^{3}H-AdoMet. Methylated proteins were then separated on 12% discontinuous, acidic gels, and radioactivity in each gel slice was quantified with scintillation counting as described under **Subheading 3.** *Top*: Distribution of methyl acceptor proteins in tissue protein that was insoluble in the nonionic detergent Triton X-100. *Middle*: Distribution of methyl acceptor proteins in tissue proteins that was soluble in an aqueous homogenization buffer. *Bottom*: Distribution of methyl acceptor proteins in crude homogenates of Alzheimer's diseased brains. Incubation conditions are similar to those described in **Fig. 3**.

2. Lou, L. L. and Clarke, S. (1987) Enzymatic methylation of band 3 anion transporter in intact human erythrocytes. *Biochemistry* **26,** 52–59.
3. Fairbanks, G. and Avruch J. (1973) Four gel systems for electrophoretic fractionation of membrane proteins using ionic detergents. *J. Supramol. Struct.* **1,** 66–75.
4. Barber, J. R. and Clarke, S. (1984) Inhibition of protein carboxyl methylation by *S*-adenosyl-L-homocysteine in intact erythrocytes. *J. Biol. Chem.* **259(11),** 7115–7122.
5. Bower, V. E. and Bates, R. G. (1955) pH Values of the Clark and Lubs buffer solutions at 25°C. *J. Res. Natl. Bureau Stand.* **55(4),** 197–200.

6. Gingras, D., Menard, P., and Beliveau, R. (1991) Protein carboxyl methylation in kidney brush-border membranes. *Biochim. Biophys. Acta.* **1066,** 261–267.
7. Johnson, B. A., Najbauer, J., and Aswad, D. W. (1993) Accumulation of substrates for protein L-isoaspartyl methyltransferase in adenosine dialdehyde-treated PC12 cells. *J. Biol. Chem.* **268(9),** 6174–6181.
8. Johnson, B. A., Freitag, N. E., and Aswad, D. W. (1985) Protein carboxyl methyltransferase selectively modifies an atypical form of calmodulin. *J. Biol. Chem.* **260(20),** 10,913–10,916.
9. Lowenson, J. D. and Clarke, S. (1995) Recognition of isomerized and racemized aspartyl residues in peptides by the protein L-isoaspartate (D-aspartate) *O*-methyltransferase, in *Deamidation and Isoaspartate Formation in Peptides and Proteins.* (Aswad, D. W., ed.), CRC, Boca Raton, pp. 47–64.
10. McFadden, P. N., Horwitz, J., and Clarke, S. (1983) Protein carboxyl methytransferase from cow eye lens. *Biochem. Biophys. Res. Comm.* **113(2),** 418–424.
11. Neuhoff, V, Stamm, R., Pardowitz, I., Arold, N., Ehrhardt, W., and Taube, D. (1988) Essential problems in quantification of proteins following colloidal staining with Coomassie brilliant blue dyes in polyacrylamide gels, and their solutions. *Electrophoresis* **9,** 255–262.
12. O'Conner, C. M. and Clarke, S. (1985) Analysis of erythrocyte protein methyl esters by two-dimensional gel electrophoresis under acidic separating conditions. *Analyt. Biochem.* **148,** 79–86.
13. O'Conner, C. M. and Clarke, S. (1984) Carboxyl methylation of cytosolic proteins in intact human erythrocytes. *J. Biol. Chem.* **259(4),** 2570–2578.
14. Sellinger, O. Z. and Wolfson, M. F. (1991) Carboxyl methylation affects the proteolysis of myelin basic protein by staphylococcus aureus V8 proteinase. *Biochim. Biophys. Acta.* **1080,** 110–118.
15. MacFarlane, D. E. (1984) Inhibitors of cyclic nucleotides phosphodiesterases inhibit protein carboxyl methylation in intact blood platelets. *J. Biol. Chem.* **259(2),** 1357–1362.
16. Aswad, D. W. (1995) Methods for analysis of deamidation and isoaspartate formation in peptides, in *Deamidation and Isoaspartate Formation in Peptides and Proteins* (Aswad, D. W., ed.), CRC, Boca Raton, pp. 7–30.
17. Freitag, C. and Clarke, S. (1981) Reversible methylation of cytoskeletal and membrane proteins in intact human erythrocytes. *J. Biol. Chem.* **256(12),** 6102–6108.
18. Gingras, D., Boivin, D., and Beliveau, R. (1994) Asymmetrical distribution of L-isoaspartyl protein carboxyl methyltransferases in the plasma membranes of rat kidney cortex. *Biochem. J.* **297,** 145–150.
19. O'Conner, C. M., Aswad, D. W., and Clarke, S. (1984) Mammalian brain and erythrocyte carboxyl methyltranserases are similar enzymes that recognize both D-aspartyl and L-isoaspartyl residues in structurally altered protein substrates. *Proc. Natl. Acad. Sci. USA* **81,** 7757–7761.
20. O'Conner, C. M. and Clarke, S. (1983) Methylation of erythrocyte membrane proteins at extracellular and intracellular D-aspartyl sites in vitro. *J. Biol. Chem.* **258(13),** 8485–8492.
21. Ohta, K., Seo, N., Yoshida, T., Hiraga, K., and Tuboi, S. (1987) Tubulin and high molecular weight microtubule-associated proteins as endogenous substrates for protein carboxyl methyltransferase in brain. *Biochemie* **69,** 1227–1234.
22. Barber, J. R. and Clarke, S. (1983) Membrane protein carboxyl methylation increase with human erythrocyte age. *J. Biol. Chem.* **258(2),** 1189–1196.
23. Chamberlain, J. P. (1979) Fluorographic detection of radioactivity in polyacrylamide gels with the water soluble fluor, sodium salicylate. *Analyt. Biochem.* **98,** 132.

93

Analysis of Protein Palmitoylation

Morag A. Grassie and Graeme Milligan

1. Introduction

The incorporation of many membrane proteins into the lipid environment is based on sequences of largely hydrophobic amino acids that can form membrane-spanning domains. However, a number of other proteins are membrane-associated, but do not display such hydrophobic elements within their primary sequence. Membrane association in these cases is often provided by covalent attachment, either cotranslationally or posttranslationally, of lipid groups to the polypeptide chain. Acylation of proteins by either addition of C14:0 myristic acid to an N-terminal glycine residue or addition of C16:0 palmitic acid by thioester linkage to cysteine residues, in a variety of positions within the primary sequence, has been recorded for a wide range of proteins. Palmitoylation of proteins is not restricted to thioester linkage and may occur also through oxyester linkages to serine and threonine residues. Furthermore, thioester linkage of fatty acyl groups to proteins is not restricted to palmitate. Longer chain fatty acids, such as stearic acid (C18:0) and arachidonic acid (C20:4), have also been detected. Artificial peptide studies have provided evidence to support the concept that attachment of palmitate to a protein can provide sufficient binding energy to anchor a protein to a lipid bilayer, but that attachment of myristate is insufficient, in isolation, to achieve this.

Mammalian proteins that have been demonstrated to be palmitoylated include a range of G protein-coupled receptors and G protein α subunits, members of the Src family of nonreceptor tyrosine kinases, growth cone-associated protein GAP 43, endothelial nitric oxide synthase, spectrin, and glutamic acid decarboxylase. Since many of these proteins play central roles in information transfer across the plasma membrane of cells, there has been considerable interest in examining both the steady-state palmitoylation status of these proteins and, because the thioester linkage is labile, the possibility that it may be a dynamic, regulated process *(1–7)*. Palmitoylation thus provides a means to provide membrane anchorage for many proteins and, as such, can allow effective concentration of an enzyme or other regulatory protein at the two-dimensional surface of the membrane. Turnover of the protein-associated palmitate may regulate membrane association of polypeptides and, thus, their functions.

There has been considerable pharmaceutical interest in the development of small-mol-wt inhibitors of the enzyme farnesyl transferase, since attachment of the farnesyl

From: *The Protein Protocols Handbook, 2nd Edition*
Edited by: J. M. Walker © Humana Press Inc., Totowa, NJ

group to the protooncogene p21[ras] is integral both for its membrane association and transforming activities. Whether there will be similar interest in molecules able to interfere with protein palmitoylation is more difficult to ascertain as the wide range of proteins modified by palmitate is likely to limit specificity of such effects. It is true, however, that compounds able to interfere with protein palmitoylation are available *(8,9)*. These may prove to be useful experimental reagents in a wide range of studies designed to explore further the role of protein palmitoylation.

This chapter will describe methodology to determine if a protein expressed in a cell maintained in tissue culture is in fact modified by the addition of palmitate. Specific conditions are taken from our own experiences in analysis of the palmitoylation of G protein α subunits *(10–12)*, but should have universal relevance to studies of protein palmitoylation.

2. Materials

1. (9,10-^{3}H [*N*])palmitic acid (Dupont/NEN or Amersham International) (*see* **Note 1**) and Trans[^{35}*S*]-label (ICN Biomedicals, Inc.).
2. Growth medium for fibroblast derived cell lines: DMEM containing 5% newborn calf serum (NCS), 20 m*M* glutamine, 100 U/mL penicillin, and 100 mg/mL streptomycin (Life Technologies).
3. [^{3}H]palmitate labeling medium: as for growth medium with NCS replaced with 5% dialyzed NCS (*see* **item 5** below), 5 m*M* Na pyruvate, and 150 μCi/mL (9,10-^{3}H [*N*]) palmitic acid (*see* **Note 2**). (9,10-^{3}H [*N*])palmitic acid is usually supplied at 1 μCi/mL in ethanol (*see* **Note 3**). Dry under N_2 in a glass tube to remove ethanol, and then redissolve in labeling medium to give a final concentration of 150 μCi/mL of medium.
4. [^{35}S] methionine/cysteine labeling medium: 1 part growth medium, 3 parts DMEM lacking methionine and cysteine (Life Technologies) supplemented with 50 μCi/mL Trans[^{35}S]-label (ICN Biomedicals, Inc.) (*see* **Note 4**).
5. Dialyzed NCS: Prepare in dialysis tubing that has been boiled twice in 10 m*M* EDTA for 10 min (**Note 5**); 50 mL of NCS are dialyzed against 2 L of Earle's salts (6.8 g NaCl, 0.1 g KCl, 0.2 g $MgSO_4 \cdot 7H_2O$, 0.14 g NaH_2PO_4, 1.0 g glucose) over a period of 12–36 h with three changes of buffer. Remove serum from dialysis tubing and filter-sterilize before storing at –20°C in 2 mL aliquots until required.
6. Phosphate-buffered saline (PBS): 0.2 g KCl, 0.2 g KH_2PO_4, 8 g NaCl, 1.14 g $NaHPO_4$ (anhydrous) to 1000 mL with H_2O (pH should be in range of 7.0–7.4).
7. 1 and 1.33% (w/v) SDS.
8. TE buffer: 10 m*M* Tris-HCl, 0.1 m*M* EDTA, pH 7.5.
9. Solubilization buffer: 1% Triton X-100, 10 m*M* EDTA, 100 m*M* NaH_2PO_4, 10 m*M* NaF, 100 μ*M* Na_3VO_4, 50 m*M* HEPES, pH 7.2.
10. Immunoprecipitation wash buffer: 1% Triton X-100, 0.5% SDS, 100 m*M* NaCl, 100 m*M* NaF, 50 m*M* NaH_2PO_4, 50 m*M* HEPES, pH 7.2.
11. Gel solutions for 10% gels:
    a. Acrylamide: 30 g acrylamide, 0.8 g *bis*-acrylamide to 100 mL with H_2O.
    b. Buffer 1: 18.17 g Tris, 4 mL 10% SDS (pH 8.8) to 100 mL with H_2O.
    c. Buffer 2: 6 g Tris, 4 mL 10% SDS (pH 6.8) to 100 mL with H_2O.
    d. 50% (v/v) Glycerol.
    e. 10% (w/v) Ammonium persulfate (APS).
    f. TEMED.
    g. 0.1% (w/v) SDS.

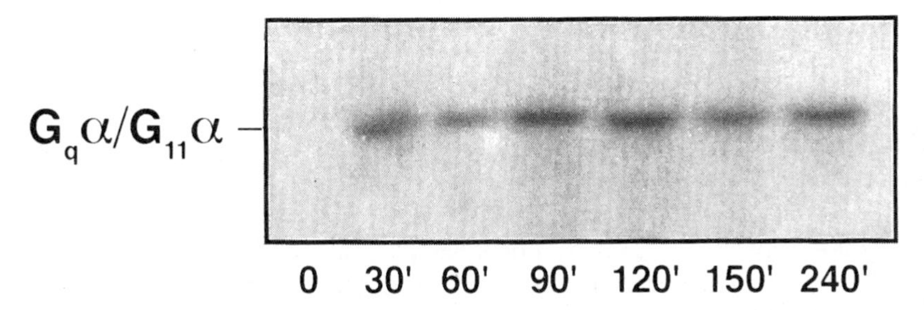

Fig. 1. Incorporation of [^{3}H] palmitate into the α subunits of the G proteins G_q and G_{11}. Rat 1 fibroblasts were metabolically labeled with (9,10-^{3}H [*N*])palmitic acid (150 μCi/mL) for the times indicated (in minutes) as described in **Subheading 3.1.** The cells were harvested and the α subunits of the G proteins G_q and G_{11} were immunoprecipitated as in **Subheading 3.3.** using an antipeptide antiserum directed against the C-teminal decapeptide, which is conserved between these two polypeptides ***(15)***. Following SDS-PAGE, the gel was treated as in **Subheading 3.5.** and exposed to X-ray film for 6 wk.

   h. Laemmli sample buffer: 3 g urea, 0.5 g SDS, 0.6 g DTT (*see* **Note 6**), 0.5 mL 1 *M* Tris-HCl, pH 8.0, to 10 mL with H_2O.
   i. Running buffer: 28.9 g glycine, 6 g Tris, 2 g SDS to 2 L with H_2O.
12. Fixing solution: 25% propan-2-ol, 65% H_2O, 10% acetic acid.
13. Amplify (Amersham International plc, UK) (*see* **Note 7**).
14. 1 *M* hydroxylamine adjusted to pH 8.0 with KOH.

3. Methods

3.1. Cell Culture and Metabolic Labeling

1. Seed equal numbers of cells to be analyzed into six-well tissue-culture dishes using 1.5–2 mL growth medium/well (or in 100 mm tissue-culture dishes with 10 mL of growth medium if cell fractionation is to be carried out). Incubate at 37°C in an atmosphere of 5% CO_2 until cells are 90% confluent. Remove growth medium and replace with 1 mL of (9,10-^{3}H [*N*])palmitic acid-labeling medium for 2 h at 37°C (*see* **Fig. 1**).
2. Parallel control experiments using Trans[^{35}*S*]-label (50 μCi/mL) are performed. However, addition of [^{35}*S*] methionine/cysteine labeling medium should occur when cell are 60–80% confluent (*see* **Note 8**), and the cells labeled over a period of 18 h (*see* **Note 9**).

3.2. Cell Harvesting and Sample Solubilization

1. At the end of the labeling period, remove the labeling medium, and add 200 μL of 1% (w/v) SDS/well. Scrape the monolayer of cells into the SDS solution, and transfer to a 2 mL screw-top plastic tube.
2. Heat to 80°C for 20 min (*see* **Note 10**) in a heating block to denature proteins. If the samples have a stringy consistency after this stage, pass through a 20–25 gage needle and reboil for 10 min.
3. Remove the samples from the heating block, and allow to cool for 2 min at room temperature. Pulse each tube briefly at high speed in a microfuge to bring the contents to the bottom of the tube. The samples can either now be frozen at –20°C until required or processed immediately.

3.3. Immunoprecipitation

1. Add 800 μL of ice-cold solubilization buffer to each tube, and mix by inverting. Pulse the samples in a microfuge to collect contents at the bottom of the tubes.
2. Retain fractions of these samples to allow analysis of total incorporation into cellular protein of [^{3}H]palmitate and the [^{35}S]labeled amino acids.
3. Before immunoprecipitating the protein of interest, preclear the sample by adding 100 μL of protein A-Sepharose (Sigma, St. Louis, MO) or 100 μL of Pansorbin (a cheaper alternative of bacterial membranes containing protein A, [Calbiochem]) (*see* **Note 11**), and mix at 4°C for 1–2 h on a rotating wheel. (Ensure caps are firmly closed before rotating.)
4. Centrifuge samples for 2 min at maximum speed in a microfuge to pellet the protein A or Pansorbin. Transfer the precleared (*see* **Note 12**) supernatants to fresh screw-top tubes.
5. To the precleared samples, add 5–15 μL of protein-specific antibody (volume will vary depending on the antibody used) and 100 μL of protein A-Sepharose. Ensure caps are firmly closed and rotate at 4°C as before for 2–5 h (*see* **Note 13**).
6. Spin samples for 2 min at maximum speed in a microfuge to pellet the immunocomplex. Remove the supernatant and resuspend the pellet in 1 mL of immunoprecipitation wash buffer. Invert the tube 10 times (do not vortex). (The supernatant can be retained to analyze the efficiency of immunoprecipitation if required.)
7. Repeat **step 6**.
8. Centrifuge samples for 2 min at maximum speed in a microfuge and discard supernatant. Resuspend agarose-immunocomplex pellet in 40 μL of Laemmli sample buffer (*see* **Note 6**).
9. Heat samples to 80°C (*see* **Note 10**) for 5 min and then centrifuge for 2 min at maximum speed in a microfuge. Analyze the samples by SDS-PAGE by loading an equal volume of each sample (e.g., 55 μL) on a 10% (w/v) acrylamide gel (*see* **Subheading 3.4.**).

3.4. SDS-PAGE Analysis

Recipes given below are for one 16 × 18 cm 10% acrylamide gel. Resolving gel (lower): 8.2 mL H_2O, 6 mL buffer 1.8 mL acrylamide, 1.6 mL 50% glycerol, 90 μL APS, 8 μL TEMED.

1. Add all reagents in the order given and mix thoroughly. Carefully pour into prepared gel plates.
2. Very carefully overlay gel mixture with approx 1 mL of 0.1% (w/v) SDS, and leave gel to polymerize.
3. Once gel has polymerized, pour off SDS.
4. Prepare stacker gel mixture as indicated below and mix thoroughly. Stacker gel mixture: 9.75 mL H_2O, 3.75 mL buffer 2, 1.5 mL acrylamide, 150 mL APS, 8 mL TEMED.
5. Pour stacker gel on top of resolving gel, and place well-forming comb in top of gel, ensuring no air bubbles are trapped under the comb. Leave to polymerize.
6. Once gel has polymerized remove the comb, place the gel in the gel tank containing enough running buffer in the base to cover the bottom edge of the gel, and add the remaining running buffer to the top.
7. Load the prepared samples in the preformed wells using a Hamilton syringe.
8. Run the gel overnight (approx 16 h) at 12 mA and 60 V until the dye front reaches the bottom of the gel plates.

3.5. Enhancement of [^{3}H] Fatty Acid Signal from Gel

In order the increase the effectiveness of detection of the weak β-particle signal emitted by [^{3}H]palmitic acid, the gel is treated with Amplify (Amersham International) (*see* **Note 7**) according to the manufacturer's instructions.

1. Fix proteins in gel using fixing solution for 30 min.
2. Pour off fixing solution (**caution—this may contain radioactivity**), and soak gel in Amplify with agitation for 15–30 min. Wash.
3. Remove gel from solution, and dry under vacuum at 60–80°C.
4. Expose the gel to X-ray film (Hyperfilm-MP, Amersham International, or equivalent) at –70°C for an appropriate time (*see* **Note 14**) before developing.

3.6. The Nature of the Linkage Between [^{3}H] Palmitate and Protein

Hydrolysis of the thioester bond between palmitate and cysteine can be achieved by treatment with near neutral hydroxylamine ***(9,13)***. Such an approach is amenable to samples following SDS-PAGE resolution and provides clear information on the chemical nature of the linkage. Hydrolysis by hydroxylamine of *O*-esters requires strongly alkaline (pH >10.0) conditions, and amide linkages are stable to treatment with this agent.

1. Following immunoprecipitation, resolution, and fixing of the protein in SDS-PAGE, lanes of the gels are exposed to 1 *M* hydroxylamine pH 7.4, or 1 *M* Tris-HCl 7.4, for 1 h at 25°C (*see* **Note 15**).
2. The gels are then washed with H_2O (2 15-min washes) and prepared for fluorography as described in **Subheading 3.5.**

3.7. Metabolic Interconversion Between [^{3}H] Fatty Acids

Metabolic interconversion between fatty acids is a well-appreciated and established phenomenon. As such, it is vital in experiments designed to demonstrate that a polypeptide is truly a target for palmitoylation that the [^{3}H] fatty acid incorporated following labeling of cells with [^{3}H] palmitate is shown actually to be palmitate. An excellent and detailed description of the strategies that may be used to examine the identity of protein-linked radiolabeled fatty acids, and the limitations of such analyses, has recently appeared ***(14)*** and readers are referred to this for details.

4. Notes

1. [^{3}H]-labeled palmitic acid available from commercial sources is not at a sufficiently high concentration that it can be directly added to cells for labeling experiments. This can be concentrated by evaporation of the solvent and dissolution in either ethanol or dimethylsulfoxide.
2. Since the cellular pool of palmitate is large, radiolabeling experiments have to use relatively high amounts of [^{3}H]palmitate (0.1–1.0 µCi/mL) to obtain sufficient incorporation into a protein of interest, such that detection of the [^{3}H]radiolabeled polypeptide can be achieved in a reasonable time frame. Although it would theoretically be possible to use [^{14}C]palmititc acid, which is also available commercially, a combination of low specific activity and concerns about degradation by β-oxidation removing the radiolabel (particularly if labeled in the 1 position) restricts its usefulness.
3. When preparing [^{3}H] palmitic acid (a) ensure radiolabel is dried in a glass tube to minimize loss of material by adsorption, and (b) when resuspending [^{3}H] palmitic acid, take great care to ensure all material is recovered from the sides of the glass tube and fully redissolved in the labeling medium. This can be confirmed by counting the amount of radioactivity present in a small proportion of the labeling medium and comparing it to the amount of radioactivity added originally. Generally, addition of [^{3}H]palmitate in metabolic labeling studies is regulated such that the addition of vehicle is limited to 1% (v/v) or less.
4. **Caution: [^{35}S] is volatile and therefore stocks should be opened in a fume hood.**
5. Dialysis tubing can be stored at this point in 20% ethanol until required.

6. When preparing samples for SDS-PAGE, care must be taken with addition of nucleophilic reducing agents, such as dithiothreitol (DTT) and 2-mercaptoethanol, since these can cause the cleavage of thioester linkages. We limit the concentration of DTT in the sample buffer to 20 m*M*. This may not be a universal problem, occurring only with certain proteins.
7. Other commercially available solutions for fluorography, or indeed methods based on salicylate or 2,5 diphenyloxazole (PPO) may be substituted.
8. The confluency of cells required for [^{35}S]metabolic labeling will varying depending on the speed of growth of the cell line used. For rapidly growing cells, such as fibroblasts, add radiolabel at 60% confluency; for slow-growing cells, e.g., of neuronal derivation, add radiolabel when cells are approaching 80–85% confluency.
9. Subsequent immunoprecipitation (*see* **Subheading 3.3.**) of [^{35}S] labeled protein can thus provide controls for immunoprecipitation efficiency and confirm that the lack of immunoprecipitation of a [^{3}H]palmitate containing polypeptide was not owing to lack of immunoprecipitation of the relevant polypeptide by the antiserum.
10. It is recommended that heating is not prolonged and does not exceed 80°C.
11. We have found Pansorbin to be a good and cheap alternative to protein A-Sepharose, especially for preclearing; however, the use of protein A-Sepharose is recommended for the immunoprecipitation reaction itself, since the use of Pansorbin has been found to give increased nonspecific background for some antibodies.
12. Preclearing samples before immunoprecipitation removes material that binds nonspecifically to protein A, thus reducing the background levels in the final sample. This is especially useful for [^{35}S] radiolabeled material where it may be advisable to preclear for the maximum 2 h.
13 The length of time of immunoprecipitation should be determined empirically for each antibody used, in order to minimize the amount of nonspecific material present in the final sample. For most antibodies with high titer, the shorter the incubation, the less nonspecific material immunoprecipitated.
14. Time will obviously depend on the levels of expression of the protein and the amount of [^{3}H]palmitate used. For cell lines that have not been transfected to express high levels of a particular protein, 30–40 d of exposure is not an unusual period of time.
15. Stability of incorporation of the [^{3}H]radiolabel to treatment with Tris-HCl, but removal by hydroxylamine under these conditions can be taken to reflect linkage via a thioester bond.

References

1. Milligan, G., Parenti, M., and Magee, A. I. (1995) The dynamic role of palmitoylation in signal transduction. *Trends Biochem. Sci.* **20,** 181–186.
2. Wedegaertner, P. B. and Bourne, H. R. (1994) Activation and depalmitoylation of $G_{s\alpha}$. *Cell* **77,** 1063–1070.
3. Degtyarev, M. Y., Spiegel, A. M., and Jones, T. L. Z. (1993) Increased palmitoylation of the G_s protein α subunit after activation by the β-adrenergic receptor or cholera toxin. *J. Biol. Chem.* **268,** 23,769–23,772.
4. Mouillac, B., Caron, M., Bonin, H., Dennis, M., and Bouvier, M. (1992) Agonist-modulated palmitoylation of β_2-adrenergic receptor in Sf9 cells. *J. Biol. Chem.* **267,** 21,733–21,737.
5. Robinson, L. J., Busconi, L., and Michel, T. (1995) Agonist-modulated palmitoylation of endothelial nitric oxide synthase. *J. Biol. Chem.* **270,** 995–998.
6. Stoffel. R. H., Randall, R. R., Premont, R. T., Lefkowitz, R. J., and Inglese, J. (1994) Palmitoylation of G protein-coupled receptor kinase, GRK6. Lipid modification diversity in the GRK family. *J. Biol. Chem.* **269,** 27,791–27,794.

7. Mumby, S. M., Kleuss, C., and Gilman, A. G. (1994) Receptor regulation of G-protein palmitoylation. *Proc. Natl. Acad. Sci. USA* **91,** 2800–2804.
8. Hess, D. T., Patterson, S. I., Smith, D. S., and Skene, J. H. (1993) Neuronal growth cone collapse and inhibition of protein fatty acylation by nitric oxide. *Nature* **366,** 562–565.
9. Patterson, S. I. and Skene, J. H. P. (1995) Inhibition of dynamic protein palmitoylation in intact cells with tunicamycin. *Meth. Enzymol.* **250,** 284–300.
10. Parenti, M., Vigano, M. A., Newman, C. M. H., Milligan, G., and Magee, A. I. (1993) A novel N-terminal motif for palmitoylation of G-protein α subunits. *Biochem. J.* **291,** 349–353.
11. Grassie, M. A., McCallum, J. F., Guzzi, F., Magee, A. I., Milligan, G., and Parenti, M. (1994) The palmitoylation status of the G-protein $G_01\alpha$ regulates its avidity of interaction with the plasma membrane. *Biochem. J.* **302,** 913–920.
12. McCallum, J. F., Wise, A., Grassie, M. A., Magee, A. I., Guzzi, F., Parenti, M., and Milligan, G. (1995) The role of palmitoylation of the guanine nucleotide binding protein $G_{11}\alpha$ in defining interaction with the plasma membrane. *Biochem. J.* **310,** 1021–1027.
13. Magee, A. I., Wooton, J., and De Bony, J. (1995) Detecting radiolabeled lipid-modified proteins in polyacrylamide gels. *Meth. Enzymol.* **250,** 330–336.
14. Linder, M. A., Kleuss, C., and Mumby, S. M. (1995) Palmitoylation of G-protein α subunits. *Meth. Enzymol.* **250,** 314–330.
15. Mitchell, F. M., Mullaney, I., Godfrey, P. P., Arkinstall, S. J., Wakelam, M. J. O., and Milligan, G. (1991) *FEBS Lett.* **287,** 171–174.

94

Incorporation of Radiolabeled Prenyl Alcohols and Their Analogs into Mammalian Cell Proteins

A Useful Tool for Studying Protein Prenylation

Alberto Corsini, Christopher C. Farnsworth, Paul McGeady, Michael H. Gelb, and John A. Glomset

1. Introduction

Prenylated proteins comprise a diverse family of proteins that are posttranslationally modified by either a farnesyl group or one or more geranylgeranyl groups *(1–3)*. Recent studies suggest that members of this family are involved in a number of cellular processes, including cell signaling *(4–6)*, differentiation *(7–9)*, proliferation *(10–12)*, cytoskeletal dynamics *(13–15)*, and endocytic and exocytic transport *(4,16,17)*. The authors' studies have focused on the role of prenylated proteins in the cell cycle *(18)*. Exposure of cultured cells to competitive inhibitors (statins) of 3-hydroxy-3-methylglutaryl Coenzyme A (HMG-CoA) reductase not only blocks the biosynthesis of mevalonic acid (MVA), the biosynthetic precursor of both farnesyl and geranylgeranyl groups, but pleiotropically inhibits DNA replication and cell-cycle progression *(10,18–20)*. Both phenomena can be prevented by the addition of exogenous MVA *(10,18,19)*. The authors have observed that all-*trans*-geranylgeraniol (GGOH) and, in a few cases, all-*trans*-farnesol (FOH) can prevent the statin-induced inhibition of DNA synthesis *(21)*. In an effort to understand the biochemical basis of these effects, the authors have developed methods for the labeling and two-dimensional gel analysis of prenylated proteins that should be widely applicable.

Because of the relative diversity of prenylated proteins, it is important to use analytical methods that differentiate between them. A useful approach discussed here is to selectively label farnesylated or geranylgeranylated proteins using [^{3}H] labeled FOH or GGOH *(22–24)*, followed by one-dimensional SDS-PAGE or high-resolution two-dimensional gel electrophoresis (2DE) of the labeled proteins. Since the enzymes that transfer prenyl groups to proteins utilize the corresponding prenyl alcohol pyrophosphate (FPP or GGPP) as substrate, these prenols are thought to undergo two phosphorylation steps prior to their subsequent utilization *(2,3)*. The discovery of a GGOH

From: *The Protein Protocols Handbook, 2nd Edition*
Edited by: J. M. Walker © Humana Press Inc., Totowa, NJ

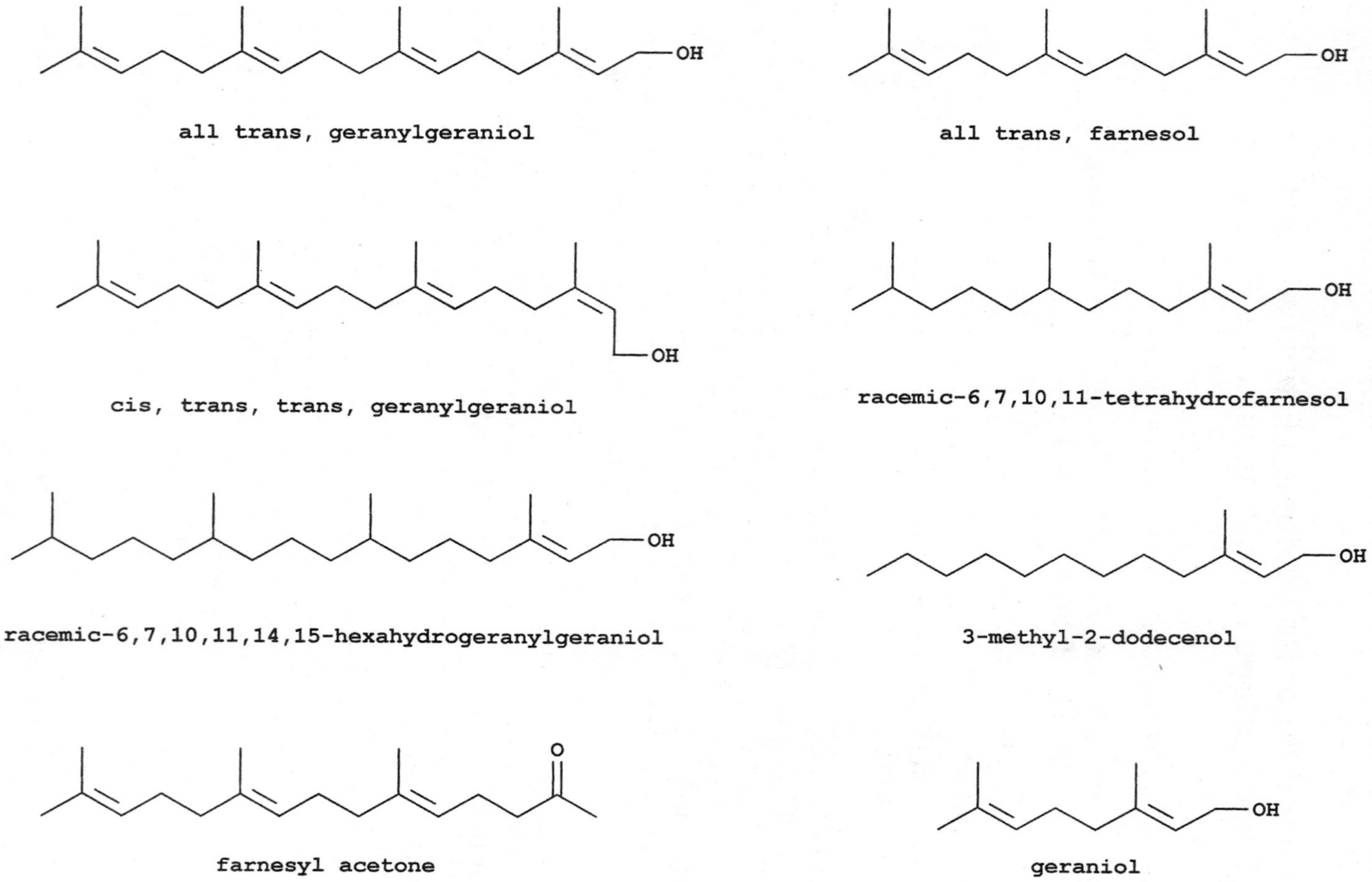

all trans, geranylgeraniol

all trans, farnesol

cis, trans, trans, geranylgeraniol

racemic-6,7,10,11-tetrahydrofarnesol

racemic-6,7,10,11,14,15-hexahydrogeranylgeraniol

3-methyl-2-dodecenol

farnesyl acetone

geraniol

Fig. 1. Structure of the natural and synthetic prenol analogs used in these labeling studies.

kinase and a geranylgeranyl phosphate kinase in eubacteria *(25)*, together with the ability of rat liver microsomal and peroxisomal fractions to form FPP, provide additional evidence that these prenol pools serve as a source of lipid precursor for protein prenylation. When proteins are labeled in this way, and are subsequently analyzed by one-dimensional SDS-PAGE, it is possible to distinguish several major bands of radioactivity that correspond to two apparently distinct subsets of proteins: those that incorporate [^{3}H]-FOH and those that incorporate [^{3}H]-GGOH. But when the labeled proteins are analyzed by high-resolution 2DE, the number of radioactive proteins that are observed is at least fivefold greater, and three subsets of prenylated proteins can be identified: one subset of proteins that incorporates only farnesol, a second that incorporates only geranylgeraniol, and a third that can incorporate either prenol.

In this chapter, the advantages of using labeled prenols to dissect the differential effects of FOH and GGOH on cellular function will be presented in the context of studies of the role of prenylated proteins in cell-cycle progression. The ability of mammalian cells to incorporate natural and synthetic prenol analogs (**Fig. 1**) into specific proteins also will be discussed. In **Subheading 4.** some of the advantages and limitations of the methods will be discussed.

2. Materials

2.1. Prenols and Analogs: Synthesis and Labeling

1. All-*trans*-FOH, d^{20} 0.89 g/mL; mevalonic acid lactone; geraniol, d^{20} 0.89 g/mL (Sigma, St. Louis, MO).
2. Tetrahydrofarnesol was a gift from Hoffman la Roche (Basel, Switzerland). A racemic form can be made from farnesyl acetone following the procedure for hexahydro-GGOH (*see* **Subheading 3.1.1.**).
3. All-*trans*-GGOH, d^{20} 0.89 g/mL; all-*trans*-GGOH [1-^{3}H], 50–60 Ci/mmol; all-*trans*-FOH [1-^{3}H], 15–20 Ci/mmol; geraniol [1-^{3}H], 15–20 Ci/mmol (American Radiolabeled Chemicals, St. Louis, MO).
4. Mevalonolactone RS-[5-^{3}H(N)], 35.00 Ci/mmol (NEN, Boston, MA).
5. Charcoal; $LiAlH_4$ powder; MnO_2; $NaBH_4$ powder; triethyl phosphonoacetate; phosphonoacetone; sodium ethoxide (Aldrich, Milwaukee, WI).
6. [^{3}H]-$NaBH_4$ solid, 70 Ci/mmol (Amersham, Arlington Heights, IL).
7. Silica gel 60 (F254) (Merck, Gibbstown, NJ).
8. Aquamix liquid scintillation solution (ICN Radiochemicals, Irvine, CA).

2.2. Cell Culture Reagents

1. Dulbecco's modified Eagle media (DMEM) (glucose 4.5 g/L); penicillin (10,000 U/mL)–streptomycin (10 mg/mL); trypsin (0.25% w/v)–1 m*M* ethylenediaminetetraacetate (EDTA); nonessential amino acids (NEAA) solution 10 m*M*; phosphate-buffered saline (PBS) without Ca^{2+} and Mg^{2+}; fetal calf serum (FCS) (Gibco, Grand Island, NY).
2. Disposable culture Petri dishes (100 × 10 mm and 35 × 10 mm) (Corning Glass Works, Corning, NY).
3. Filters, 0.22 μm (Millipore, Bedford, MA).
4. Plasma-derived, bovine serum (PDS) (Irvine Scientific, Santa Ana, CA).
5. Aquasol scintillation cocktail; thymidine [methyl-^{3}H], 2 Ci/mmol] (NEN).
6. Trichloroacetic acid (TCA) (J. T. Baker, Phillipsburg, NJ).
7. Phenylmethylsulfonyl fluoride (PMSF), aprotinin, leupeptin, and pepstatin A (Sigma).

8. Simvastatin in its lactone form (gift from Merck, Sharp and Dohme; Rahway, NJ) is dissolved in 0.1 *M* NaOH (60°C, 3 h), to give the active form. Adjust the pH to 7.4 and the concentration to 50 m*M*, then sterilize by filtration.
9. Swiss 3T3-albino mouse cell line (3T3) were from American Type Culture Collection (ATCC), Rockville, MD.
10. Human skin fibroblasts (HSF) are grown from explants of skin biopsies obtained from healthy individuals. The cells are used between the fifth and fifteenth passages.

2.3. One-Dimensional and Two-Dimensional Gel Electrophoresis

1. SDS, TEMED, ammonium persulfate, mol wt protein standards, glycine, Bradford protein assay kit, heavyweight filter paper, and Bio Gel P6-DG (Bio-Rad, Hercules, CA).
2. Duracryl™ preblended acrylamide solution (30% T, 0.65% C, used for both one-dimensional and 2DE gels) and all other 2DE gel reagents, were obtained from ESA, Chelmsford, MA.
3. Ethylmaleimide, N-[ethyl-1,2-^{3}H] ([^{3}H]-NEM), 57 Ci/mmol (NEN).
4. N-ethylmaleimide (Aldrich).
5. Soybean trypsin inhibitor and SDS-7 mol wt standards used with 2DE gels (Sigma).
6. Amplify and Hyperfilm (Amersham).
7. Reacti-Vials™ (Pierce, Rockford, IL).
8. DNase1 and RNaseA (Worthington Biochem, Freehold, NJ).

3. Methods

3.1. Synthesis and Labeling of Prenols and Analogs

3.1.1. Synthesis of Racemic 6,7,10,11,14,15 Hexahydrogeranylgeraniol

1. This synthesis is outlined in **Fig. 2**. Weigh 10.4 g of farnesyl acetone and 500 mg 10% Pd on charcoal, dissolve in 100 mL methanol and hydrogenate at 50 psi in a Parr hydrogenator for 2 d. Analysis by TLC (7:3 hexanes/diethyl ether) and GC-MS should indicate that the material is fully converted to the hexahydrofarnesyl acetone.
2. Filter the reaction mixture over filter paper, wash with ethanol, rotary-evaporate to dryness, and take up in a small volume of ethanol, then filter over Fluorosil to remove the last traces of catalyst. Concentrate the resultant oil to dryness with a rotary evaporator (yield 8.64 g, 81% recovery).
3. Dissolve 1.56 g hexahydrofarnesyl acetone in 10 g dry dimethylformamide in a round-bottom flask submerged in ice water and equipped with a dropping funnel containing 1.37 mL triethyl phosphonoacetone. Flush the entire apparatus with argon (Ar). Add triethyl phosphonoacetone over the course of 45 min, followed by 2.2 mL 21% (w/w) sodium ethoxide in ethanol, which is added over the course of 1 h.
4. Remove the mixture from the ice water bath and continue to stir for 48 h under Ar. Analysis by TLC (7:3 hexanes/diethyl ether) should show that the majority of the starting material (R_f 0.5) has reacted (final product R_f 0.6).
5. Transfer the reaction mixture to a separatory funnel with hexane, and wash 2× with a NaCl-saturated solution. Dry the organic layer with $MgS0_4$, pass through filter paper, and evaporate the solvent to dryness by rotary evaporation (yield 1.41 g, 72% recovery).
6. Dissolve 0.75 g of the product from the previous step in 10 mL anhydrous diethyl ether in a round-bottom flask equipped with a stir bar. Cool the apparatus by stirring in an ice-water bath.
7. Add 267 mg of $LiAlH_4$, and stir the reaction mixture overnight under Ar. The following day, add 25 mL saturated NH_4Cl and stir the reaction mixture overnight under Ar.

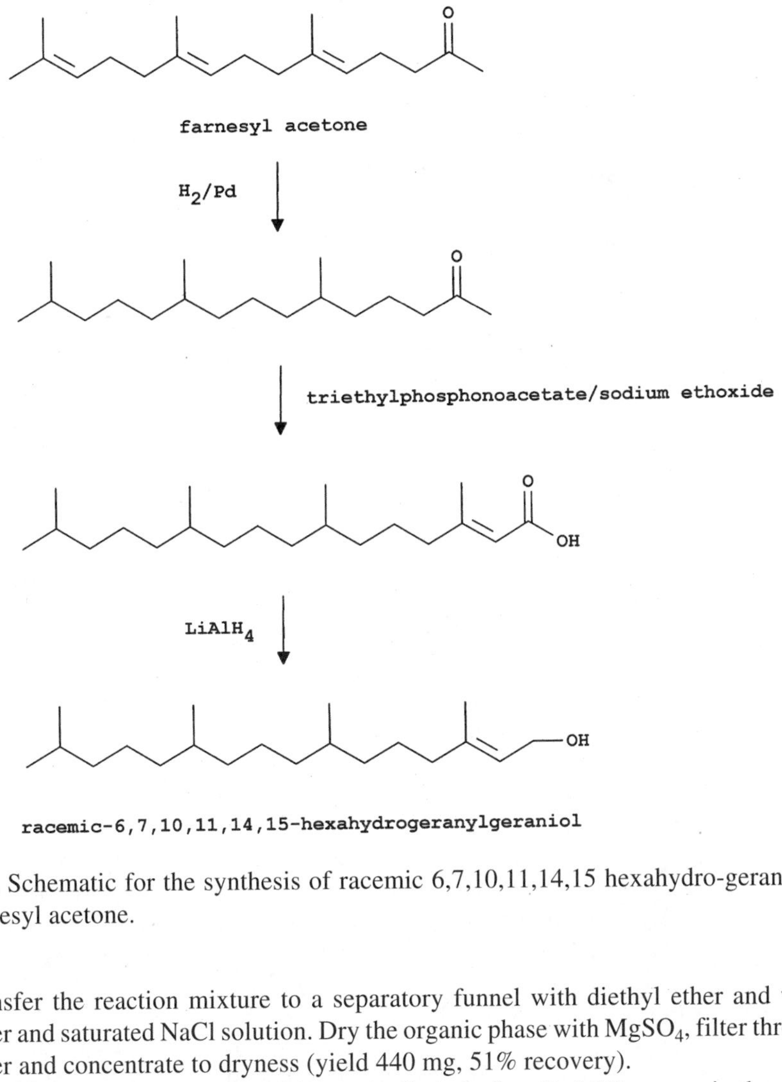

farnesyl acetone

H_2/Pd

triethylphosphonoacetate/sodium ethoxide

OH

$LiAlH_4$

OH

racemic-6,7,10,11,14,15-hexahydrogeranylgeraniol

Fig. 2. Schematic for the synthesis of racemic 6,7,10,11,14,15 hexahydro-geranylgeraniol from farnesyl acetone.

8. Transfer the reaction mixture to a separatory funnel with diethyl ether and wash with water and saturated NaCl solution. Dry the organic phase with $MgSO_4$, filter through filter paper and concentrate to dryness (yield 440 mg, 51% recovery).
9. Purify the material by TLC (7:3 hexanes–diethyl ether, R_f 0.25), or use in the crude state if proceeding on to make radiolabeled material.

3.1.2. Synthesis of cis-Isomers

1. Form the *cis*-isomers of the allylic alcohols from the respective *trans*-alcohols after oxidation to the aldehyde (*see* **Subheading 3.1.3.**). Allow the mixture of *cis/trans* isomers to come to equilibrium (approx 36 *cis*:65 *trans*) by standing at room temperature for several days.
2. Purify the *cis*-isomer by TLC (7:3 hexanes/diethyl ether).

3.1.3. Tritium Labeling of Prenols

1. The method for radiolabeling the prenols is outlined in **Fig. 3**. First, oxidize the alcohol to the aldehyde with an excess of MnO_2. Dissolve 44.7 mg of hexahydrogeranylgeraniol in 2 mL benzene, and add 517 mg of MnO_2. Mix the reaction by tissue culture rotator overnight.

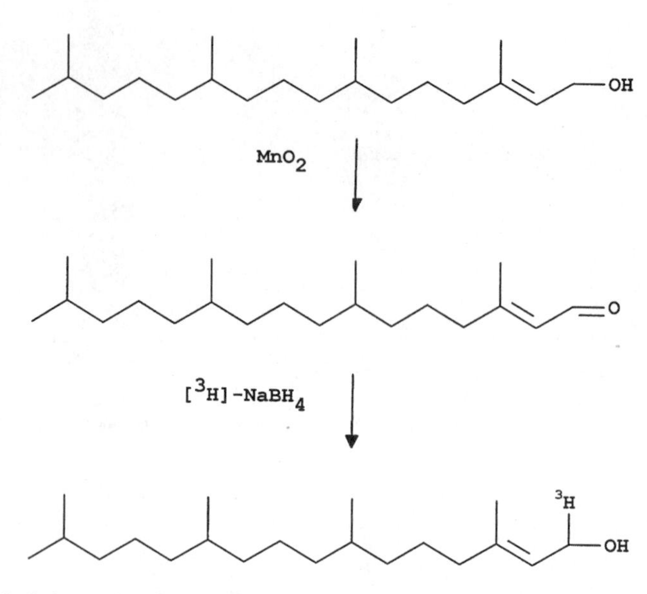

Fig. 3. Schematic for the tritium labeling of naturally occurring prenols.

2. The next day, allow the reaction mixture to settle, remove the liquid phase, and filter over glass wool to remove any remaining MnO_2. Add diethyl ether to the original reaction vessel, and repeat the process. Combine the two solutions, and concentrate to dryness under a stream of nitrogen.
3. Chromatograph the material by TLC (7:3 hexanes/diethyl ether, R_f 0.35). Two overlapping bands are present corresponding to the *cis*- and *trans*-double bond isomer, the upper band being the *cis*-isomer and the lower band being the *trans*.
4. Remove the lower one-third of the overlapping bands, elute with diethyl ether, dry with $MgSO_4$, and concentrate to dryness (yield 20%). If not used immediately, this material should be stored at –70°C to prevent isomerization of the *trans*-aldehyde to the equilibrium mixture (~30% *cis*).
5. Elute the remaining material, and rechromatograph if desired.
6. Reduce the aldehyde to the alcohol using [^{3}H]$NaBH_4$. Dissolve 2 mg of the aldehyde in 1 mL absolute ethanol, and add 5 µL 14 *N* NH_4OH. Dissolve the [^{3}H]$NaBH_4$ in ethanol at a concentration of 100 mCi/mL. Add 200 µL of the [^{3}H]$NaBH_4$ to the aldehyde solution, shake the solution, and then leave it vented in the hood for at least 4 h.
7. Concentrate the material to dryness in a gentle stream of Ar, take up in diethyl ether, and place over a column of silica gel overlaid with $MgSO_4$. Purify the material by TLC (7:3 hexanes/diethyl ether) to remove unreacted starting material and the small amount of the *cis*-isomer that is generated during the reaction. Elute the labeled alcohols from the silica using diethyl ether. Dry using a stream of Ar.
8. Dissolve the product in ethanol, and bring to a final concentration of 2.1 m*M*, 2.6 m*M*, 6.5 m*M*, or 0.4 m*M* for labeled 2 *cis*-GGOH, hexahydrogeranylgeraniol, geraniol, and tetrahydrofarnesol, respectively.

3.2. Cell Culture Experiments

1. Grow HSF and 3T3 cells in monolayers, and maintain in 100-mm Petri dishes at 37°C in a humidified atmosphere of 95% air, 5% CO_2 in DMEM, pH 7.4, supplemented with 10% FCS v/v, 1% (v/v) NEAA, penicillin (100 U/mL), and streptomycin (0.1 mg/mL).

2. Dissociate confluent stock cultures with 0.05% trypsin-0.02% EDTA, and seed HSF cells (3×10^5 cells/35-mm Petri dish) or 3T3 cells (2×10^5 cells/35-mm Petri dish) in a medium containing 0.4% FCS or 1% PDS, respectively, to stop cell replication.
3. HSF cells become quiescent within 3 d and the experiments can begin on d 4. For 3T3 cells, change the medium on d 2 and 4; cells become quiescent within 5 d, and the experiments can begin on d 6.
4. At this time, stimulate the cells by replacing the medium with one containing 10% FCS, in the presence or absence of the tested compounds, and continue the incubation as needed at 37°C. Simvastatin is used at a final concentration of 40 μ*M*, when required.
5. Dissolve unlabeled prenols in absolute ethanol and prepare stock solutions as follows:
   a. 2 m*M* all-*trans*-GGOH (the density for this and all other prenols used in these studies is d^{20} 0.89 g/mL). Add 30 μL of the prenol to 49 mL ethanol and store in 1-mL aliquots at –20°C. On the day of the experiment, dilute an aliquot with ethanol (e.g., 300 μL of GGOH to 900 μL of ethanol) to obtain a working solution of 0.5 m*M*.
   b. 4 m*M* all-*trans*-FOH. Add 20 μL of the prenol to 20 mL of ethanol and store in 1-mL aliquots at –20°C. On the day of the experiment, dilute this 1:3 with ethanol to make 1 m*M* working stock solution.
   c. 4 m*M* geraniol. Add 20 μL of the prenol to 28.83 mL ethanol and store in 1-mL aliquots at –20°C. On the day of the experiment, dilute this 1:3 with ethanol to make 1 m*M* working stock solution.
6. Prenols are light-sensitive, so experiments should be performed under dim light, and incubation should be done with Petri dishes covered with foil. Be careful not to exceed 1% (v/v) ethanol in the culture medium.
7. For estimation of DNA synthesis after mitogenic stimulation, incubate cells for 22–24 h at 37°C then change the medium to one containing 10% FCS and 2 μCi/mL [^{3}H]-thymidine and incubate the cells for another 2 h at 37°C ***(19)***. Remove the medium and wash the monolayers once with PBS at room temperature, then add 2 mL of fresh, ice-cold 5% TCA (w/v), and keep the cells at 4°C on ice for at least 10 min. Remove TCA and wash once with 5% TCA, and dissolve the monolayers in 1 mL of 1 *N* NaOH for 15 min. Transfer 500 μL of the cell lysates to a liquid scintillation vial, add 150 μL glacial acetic acid and 5 mL Aquasol, and measure incorporated radioactivity in a liquid scintillation counter (**Table 1**; **Fig. 4**).

3.3. Cell Labeling and Prenylated Protein Analysis

3.3.1. Labeling of Proteins with [^{3}H] Farnesol or [^{3}H] Geranylgeraniol and One-Dimensional SDS-PAGE Analysis

1. Dry 9 mCi [^{3}H]-MVA, 500 μCi [^{3}H]-GGOH, or 1 mCi [^{3}H]-FOH, and resuspend in 50 μL ethanol, to bring to the concentration of 4.2 m*M*, 166 μ*M*, or 1.1 m*M*, respectively.
2. Incubate the cells (3 × 35-mm Petri dishes/sample) (*see* **Notes 2–6**) for 20 h at 37°C, with the appropriate labeled isoprenoids. Incubation periods as short as 5 h are sufficient to label most proteins, making it possible to perform time course experiments.
3. To harvest the cells, place the Petri dishes on ice, remove the medium, and wash 3× with ice-cold PBS containing 1 m*M* PMSF. Scrape the cells into 1.5 mL PBS/PMSF per Petri dish, collect the resuspended cells, and centrifuge for 5 min at 200*g*.
4. Aspirate the PBS, add 150 μL PBS/PMSF to the cell pellet, and sonicate in a bath sonicator to disrupt the cell pellet. Transfer the resuspended pellet to a 1.5-mL Eppendorf tube.
5. Add 1.3 mL of cold (–20°C) acetone, mix, sonicate, and allow to stand on ice for 15 min. Centrifuge for 5 min (13,000 g at 4°C) to sediment the delipidated proteins. Re-extract the protein pellet twice with 1 mL cold acetone.

Table 1
Mevalonate, Prenols, and Synthetic Analogs Vary in Their Ability to Prevent the Inhibitory Effect of Simvastatin on DNA Synthesisin Cultured Cellsas Measured by Incorporation of [^{3}H]-Thymidine

Cell type	HSF	3T3
Treatment Conditions:	(Reported as % control)	
Simvastatin (S) 40 μ*M*	30	30
S + mevalonate 100 μ*M*	100	90
S+ all-*trans*-GGOH 5 μ*M*	85	96
S+ 2-*cis*-GGOH 5 μ*M*	29	30
S+ all-*trans*-FOH 10 μ*M*	37	33
S+ geraniol 5 m*M*	29	30
S+ 6,7,10,11,14,15-hexahydro-GGOH 5 μ*M*	34	46
S+ tetrahydro FOH 5 μ*M*	N.A.	36

Experimental conditions: *see* **Subheading 3.2.**
N.A., not assayed.
The mean value of control (100%) without inhibitor was $104 \times 10^3 \pm 3 \times 10^3$ dpm/plate and $187 \times 10^3 \pm 10 \times 10^3$ dpm/plate for HSF and 3T3 cells, respectively.

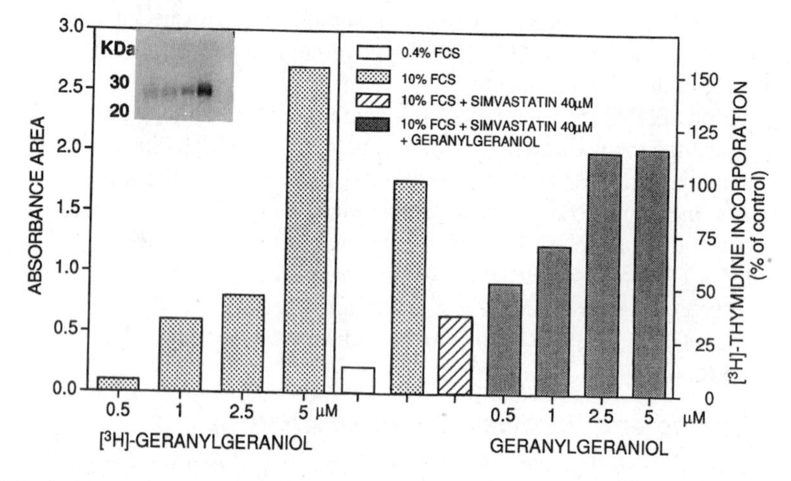

Fig. 4. The ability of simvastatin-treated HSF cells to synthesize DNA was used as a measure of their ability to traverse the cell cycle. HSF cells were seeded at a density of 3×10^5/35-mm dish in medium supplemented with 0.4% FCS, and the cultures were incubated for 72 h. Quiescent cells were incubated for 24 h in fresh medium containing 10% FCS, 40 μ*M* simvastatin, and

6. Add 1 mL chloroform:methanol (2:1), mix, sonicate, and incubate for 30 min at 37°C. Centrifuge for 5 min (13,000 g, 4°C), and collect the lipid extracts.
7. Remove the residual organic solvent by evaporation, and solubilize the delipidated proteins overnight at room temperature in 100 μL 3% SDS, 62.5 mM Tris-HCl, pH 6.8.
8. Determine the protein content in 10 μL of sample, according to Lowry *(27)*.
9. Transfer 10 μL of the sample to a liquid scintillation vial, add 5 mL Aquamix, and measure the incorporated radioactivity.
10. Add 80 μL of sample application buffer to the remainder, and analyze an aliquot (15–80 μg of protein) by one-dimensional SDS-PAGE, according to Laemmli *(28)*, using a 12% gel.

11. After electrophoresis, wash the gel with destain (methanol–MilliQ water–acetic acid, 25:65:10) for 10 min, then stain the gel (0.1% Coomassie R250 in methanol–MilliQ–acetic acid, 40:50:10) for 20 min, and destain overnight.
12. Treat the gel with Amplify for 30 min in preparation for fluorography. Wash the gel twice with water, dry, and expose to Kodak XOMAT-AR film at –70°C (**Figs. 5** and **6**).

3.3.2. High-Resolution Two-Dimensional Gel Electrophoresis of Labeled Proteins

1. In 2DE analysis, proteins are first separated on the basis of pI, using isoelectric focusing (IEF), followed by separation on the basis of mol mass. Therefore, differentiation of closely related proteins with similar mol masses is facile ***(29,30)***. Two different IEF methods have been developed for use in high-resolution 2DE. The one described here uses carrier ampholytes ***(31–33)*** to obtain the appropriate pH gradient in which sample proteins are focused. This system is available from Genomics Solutions, Chelmsford, MA. The second IEF method uses precast gel strips containing an immobilized pH gradient made by crosslinking ampholytes into a gel matrix ***(29,34)***. This system is available from Amersham Pharmacia, Piscataway, NJ (*see* **Notes 7–10** for additional comments).
2. Prior to analysis by 2DE, cells are labeled as described in **Subheading 3.3.1.**
3. On ice, wash each 100-mm dish once with 10 mL PBS, then twice with a solution containing 10 m*M* Tris buffer, pH 7.5, 0.1 m*M* PMSF, and 1.0 µg/mL each of aprotinin, leupeptin, and pepstatin A. Drain for 45 s, and remove residual buffer.
4. Add 240 µL of boiling sample buffer (28 m*M* Tris-HCl, and 22 m*M* Tris base, 0.3% SDS, 200 m*M* DTT, pH 8.0), and collect cells with a cell scraper. Transfer lysate to a 1.5-mL microcentrifuge tube, boil for 5 min, and cool on ice. This and subsequent steps are a modification of methods previously described ***(32,33)***.
5. Add 30 µL (or one-tenth vol) of protein precipitation buffer (24 m*M* Tris base, 476 m*M* Tris-HCl, 50 m*M* $MgCl_2$, 1 mg/mL DNaseI, 0.25 mg/mL RNaseA, pH 8.0) to each cell lysate sample, and incubate on ice for 8 min. Subsequent steps are performed at room temperature, unless otherwise indicated.
6. Add ~6000 cpm of the internal standard, [^{3}H]NEM-labeled soybean trypsin inhibitor (preparation described in **Subheading 3.3.3.**), to each sample (*see* **Note 10**).
7. Add 4 vol of acetone at room temperature to each lysate, and let stand for 20 min. Centrifuge the protein precipitate for 10 min at 16,000 g.
8. Remove the acetone phase, resuspend the precipitate in 800 µL fresh acetone, using a bath sonicator, and centrifuge as above for 5 min. Repeat this wash step once.

---

the indicated concentrations of unlabeled GGOH or [^{3}H]-GGOH (15 Ci/mmol). Control cells received only 10% FCS. **(Left panel)** the labeled cell lysates from each treatment group were analyzed by SDS-PAGE (*see* **Subheading 3.3., step 1**), the gel was fluorographed, and the resulting films analyzed by densitometry. The optical density reported for each treatment group represents the sum of all protein bands between 20 and 30 kDa observed in the corresponding gel lane (*see insert*). An equal amount of cell lysate (15 µg cell protein/lane) was applied to each lane. In the experiment shown, 9×10^3, 12×10^3, 42×10^3, and 142×10^3 cpm/lane were analyzed for the 0.5, 1, 2.5, and 5 µ*M* [^{3}H]-GGOH samples, for lanes left to right respectively. **(Right panel)** In a parallel experiment, DNA synthesis was determined for each treatment group. Cells were incubated as above, but with a corresponding concentration of unlabeled GGOH. After 22 h, [^{3}H]-thymidine (2 µCi/mL) was added to treated and control samples, and the incubation was continued for another 2 h. Nuclear DNA was analyzed for incorporation of [^{3}H]-thymidine (*see* **Subheading 3.2., step 7**). The incorporation of [^{3}H]-thymidine for each treatment group is reported as a percentage of the control. The mean value of control (100%) was $114 \times 10^3 \pm 6 \times 10^3$ dpm/plate.

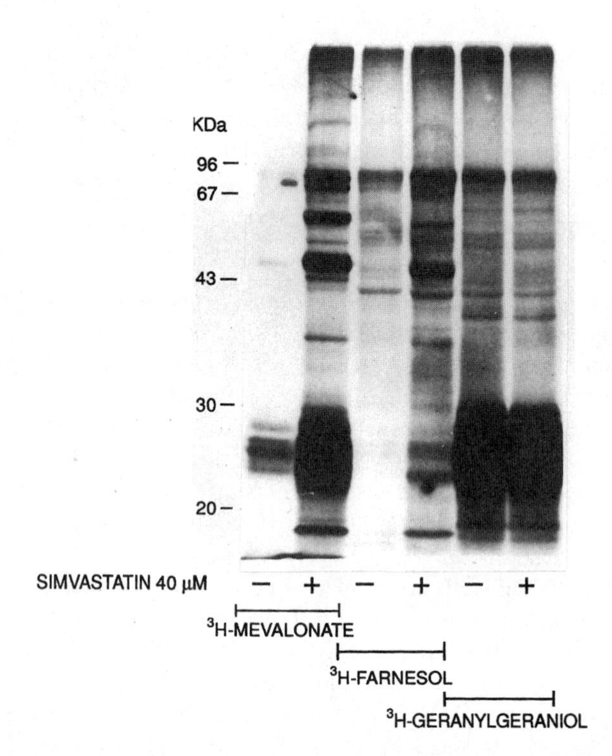

Fig. 5. Metabolic labeling of prenylated proteins in Swiss 3T3 cells after various treatments. Cells were seeded at a density of 2×10^5/dish in medium containing 1% PDS and the cultures incubated for 5 d. Quiescent cells were incubated for 20 h in fresh medium containing 10% FCS, 50 μ*M* [^{3}H]-MVA (35 Ci/mmol), 2.5 μ*M* [^{3}H]-FOH (15 Ci/mmol), or 1 μ*M* [^{3}H] GGOH (60 Ci/mmol), in the presence or absence of 40 μ*M* simvastatin. Cell pellets were delipidated, and equal amounts of cell extracts (80 μg cell protein/lane) were separated by 12% SDS-PAGE and fluorographed. In the experiment shown, 8×10^8, 2.7×10^7, 6×10^7 cpm/lane were analyzed for cells labeled with [^{3}H]-MVA, [^{3}H]-FOH or [^{3}H]-GGOH, respectively.

9. Let the pellet air dry for 3 min (avoid overdrying), before adding 30 μL of a 1:4 mixture of the SDS buffer (from **Subheading 3.2.2., step 4**) and urea solution (9.9 *M* urea; 4% Triton X-100; 2.2% ampholytes, pH 3–10; 100 m*M* DTT). Warm the sample to 37°C, and bath-sonicate to solubilize the protein (avoid overheating the sample). Centrifuge for 5 min at 16,000 g. Set aside 1 μL for scintillation counting and 1 μL for Bradford protein assay (*see* **Subheading 2.3.1.**).
10. Apply the sample (26 μL) onto a 1 × 180 mm-tube gel (4.1% T, 0.35% C; ampholyte pH range of 3.0–10.0). Focus for 17.5 h at 1000 V, then for 0.5 h at 2000 V.
11. Extrude the gel from the tube into equilibration buffer (300 m*M* Tris base, 75 m*M* Tris-HCl, 3% SDS, 50 m*M* DTT, and 0.01% bromophenol blue), and incubate it for 2 min before overlaying onto the second dimension SDS-PAGE gel (12.5% T, 0.27% C).
12. Perform SDS-PAGE on large format gels (220 × 240 × 1 mm) for 6 h, using the constant power mode (16 W/gel) in a prechilled tank equipped with Peltier cooling (Genomics Solutions). When five gels are used in this mode, the running buffer temperature will increase 14°C over the course of 6 h when started at 0°C.
13. Fix gels for 1 h in methanol:Milli Q:acetic acid (50:40:10), then stain overnight with a solution containing 0.001% Coomassie R250 in methanol:Milli Q:acetic acid (25:65:10).

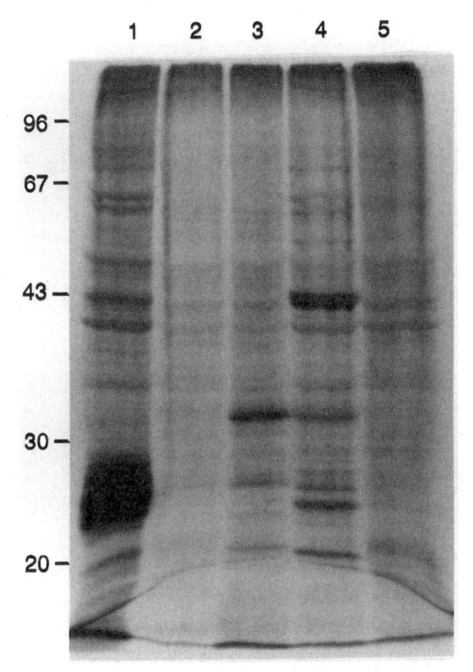

Fig. 6. Metabolic labeling of prenylated proteins in Swiss 3T3 cells after various treatments. Experimental conditions as in **Fig. 5**. Quiescent cells were incubated for 20 h in medium containing 10% FCS, 40 μ*M* simvastatin, and one of the following: lane 1, 5 μ*M* [^{3}H]-all *trans*-GGOH (20 Ci/mmol); lane 2, 5 μ*M* [^{3}H]-*cis, trans*-GGOH (17.5 Ci/mmol); lane 3, 5 μ*M* [^{3}H]-6,7,10,11,14,15 hexahydrogeranylgeraniol (17.5 Ci/mmol); lane 4, 5 μ*M* [^{3}H]-tetrahydrofarnesol (17.5 Ci/mmol); and lane 5, 5 μ*M* [^{3}H]-geraniol (17.5 Ci/mmol). Cell pellets were delipidated, and equal amounts of cell extract (60 μg cell protein/lane) were analyzed by SDS-PAGE on 12% gels, and gels were fluorographed. In the experiment shown, 334×10^3, 197×10^3, 247×10^3, 271×10^3, 285×10^3 cpm/lane were analyzed for lanes 1–5, respectively.

14. Immerse each stained gel in 200 mL of Amplify, and incubate for 20 min, with rocking. Without rinsing, dry the treated gel onto heavyweight blotting paper and expose to preflashed Hyperfilm MP for 7–14 d at –70°C. Exposures of >10 wk may be needed to visualize low abundance labeled proteins (**Fig. 7A,B**).

3.3.3. Preparation of the Two-Dimensional Gel Internal Standard, [^{3}H]NEM-Labeled Soybean Trypsin Inhibitor

1. Transfer 100 μL of a freshly prepared solution of 22 m*M* N-ethylmaleimide (NEM) in acetonitrile to a 0.5-mL Reacti-Vial, and add 250 μL [^{3}H]NEM (1 μCi/ μL in pentane; 56 Ci/mmol). Mix, and let stand for 2 h uncapped in a fume hood to evaporate the pentane. The final volume will be ~110 μL, and the new specific activity will be 95.4 mCi/mmol.
2. Prepare 1 mg/mL solution of soybean trypsin inhibitor (STI) in 50 m*M* Tris, pH 8.5, and reduce the protein by adding DTT (2 m*M*, final concentration). Incubate overnight at 4°C.
3. Remove excess DTT by applying 300 μL of the reduced STI (in six 50-μL aliquots) to six SW rotor equibbed, spin columns, and centrifuge for 4 min at 1000 g in a preequilibrated countertop centrifuge. Prepare spin columns by placing 1 mL bed-volume of Bio Gel P6-DG (previously rehydrated in 50 m*M* Tris HCl, pH 7.5) into a 1-mL plastic tuberculin-syringe equipped with a plug of glass wool, followed by centrifugation for 2 min at 1000 g.

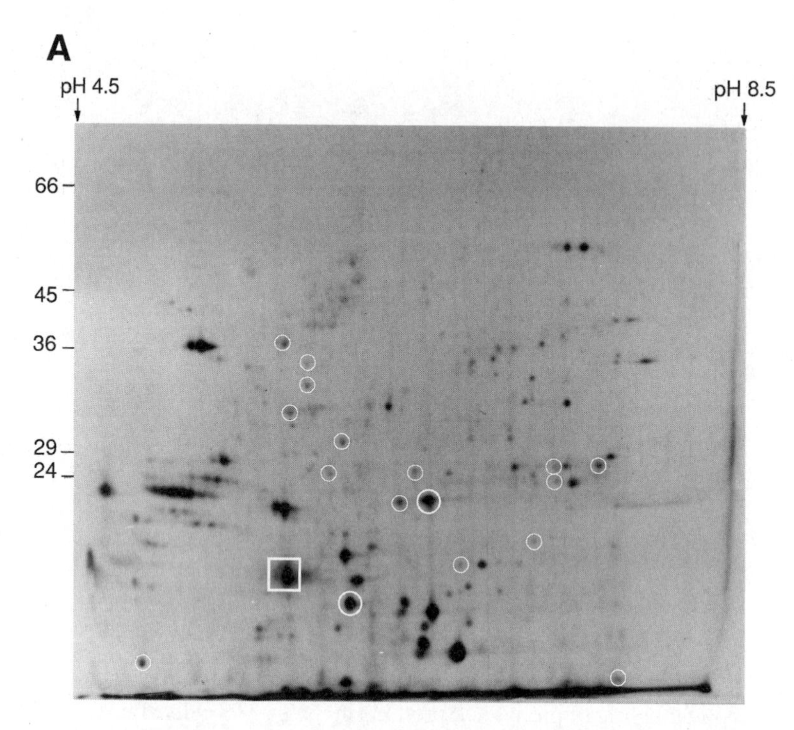

Fig. 7. Two-dimensional gel fluorographs of [^{3}H]-prenol-labeled cell-lysates. HSF cells were labeled with either 5 µm [^{3}H]-FOH or 5 µm [^{3}H]-GGOH (*see* **Subheading 3.3.1.**), lysed, delipidated and analyzed by 2DE and fluorography (*see* **Subheading 3.3.2.**). **(A)** Whole-cell lysate from [^{3}H]-FOH labeled cells that contained 385 µg protein and 170 × 10^3 cpm. Film was exposed for 23 wk (*continued*).

(In addition to removing DTT, this step also permits buffer exchange to provide the proper pH required for the NEM reaction step which follows.)

4. Pool the six column eluates (250–300 µL total volume), and add 10–12 µL of the [^{3}H]NEM solution (~24 µCi; *see* **Subheading 3.3.3., step 1**). Incubate for 6 h on ice.
5. Separate the [^{3}H]NEM-labeled STI from the unreacted [^{3}H]NEM using five or six fresh, preequibbelated spin columns (*see* **Subheading 3.3.3., step 3**). Pool the column eluates containing the purified [^{3}H]NEM-labeled STI (~1 µCi/nmol protein), divide into 30-µL aliquots, and store at –70°C. To make up a working stock solution, dilute the eluate 1:39 with buffer (~6000 cpm/5 µL), and use as an internal standard for 2DE gel fluorography (*see* **Subheading 3.3.2., step 6**; *see also* **Note 10**).

4. Notes

1. Since either all-*trans*-GGOH or MVA, but not all-*trans*-FOH, can completely prevent the inhibitory effect of simvastatin on DNA synthesis in cultured HSF and 3T3 cells (**Table 1**), it may be worthwhile to determine whether this effect is related to the cells' ability to incorporate labeled isoprenoid derivatives into specific proteins, or into other isoprenoid metabolites.
2. To maximize the incorporation of [^{3}H]-MVA into cellular proteins, it is necessary to block endogenous MVA synthesis with a statin such as simvastatin *(35)*. One-dimensional SDS-PAGE reveals that whole-cell homogenates incorporated [^{3}H]-MVA into at least 10 major bands, with molecular masses ranging from 21 to 72 kDa (**Fig. 5**). Intense bands of

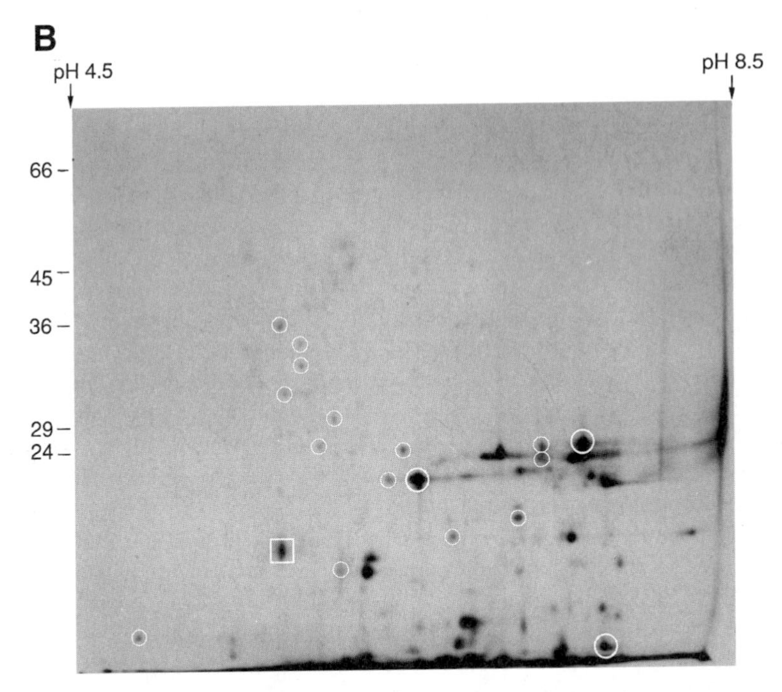

Fig. 7. *(continued)* **(B)** Whole-cell lysate from [^{3}H]-GGOH-labeled cells that contained 475 μg protein and 390 × 10^3 cpm. Film was exposed for 10 wk. The major proteins that incorporated both FOH and GGOH are circled. At least nine additional clusters of minor proteins display a similar ability to incorporate either prenol, but have not been circled (*see also* **Note 7**). The position of the internal standard, [^{3}H]-NEM-labeled soybean trypsin inhibitor (6250 cpm per gel; *see* **Subheading 3.3.3.**), is indicated by a square.

radioactivity are seen in the 21–28 kDa range, corresponding to Ras and Ras-related small GTPases *(2,3)*. In contrast, only a few protein bands are visible when simvastatin is omitted from the incubation.

3. It also is necessary to block endogenous MVA synthesis with simvastatin, for efficient incorporation of [^{3}H]-FOH into specific cellular proteins. However, this is not the case for [^{3}H]-GGOH labeling, which proceeds equally well in the presence or absence of simvastatin (**Fig. 5**).
4. In investigations of the cellular uptake of MVA and its derivatives, the authors typically found uptake of less than 0.1% of the added [^{3}H]-MVA, 2% of the added [^{3}H]-FOH, and more than 10–15% of the added [^{3}H]-GGOH. The efficient uptake of GGOH by cells permits rapid labeling of geranylgeranylated proteins. Protein bands from one-dimensional SDS-PAGE are detectable after labeling the cells for only 1–2 h, and subsequent gels require only a few days of film exposure.
5. Both [^{3}H]-GGOH labeling of proteins and the GGOH-mediated prevention of simvastatin-induced inhibition of DNA synthesis show parallel dose-response sensitivities to a similar range of GGOH concentrations (up to 2.5 μ*M*; **Fig. 4**).
6. The labeling approach described here can also be utilized for investigating natural or synthetic isoprenoid analogs (**Fig. 6**). When cells were treated with [^{3}H]-2-*cis*-GGOH (precursor of dolichols), proteins are only weakly labeled, compared to all *trans*-GGOH. Under parallel conditions, the addition of unlabeled 2-*cis*-GGOH failed to prevent the inhibitory effect of simvastatin on DNA synthesis (**Table 1**).

7. When prenylated proteins are selectively labeled with [^{3}H]-FOH or [^{3}H]-GGOH, and analyzed by one-dimensional SDS-PAGE, two subsets of proteins can be identified: a subset of 17 protein bands that incorporate [^{3}H]-FOH, and a separate subset of 12 protein bands that incorporate [^{3}H]-GGOH (*see* **Fig. 5**). However, when similarly labeled proteins are analyzed by 2DE, which allows much greater resolution, a third subset of proteins is identified. Using this approach, 82 proteins are visualized that incorporate only [^{3}H]-FOH, 34 proteins that incorporate only [^{3}H]-GGOH, and 25 proteins that can incorporate either prenol (compare **Fig. 7A,B**), although with varying efficiencies. The presence of unlabeled GGOH in the media during [^{3}H]-FOH labeling experiments or unlabeled FOH during [^{3}H]-GGOH labeling experiments, does not affect the protein composition of these subsets.
8. As mentioned above (*see* **Subheading 3.3.2., step 1**), there are primarily two different methods of IEF available for use in 2DE: one that uses soluble carrier ampholytes ***(31,32)*** and one that uses immobilized ampholytes in Immobiline™ gel-strips, discussed in **Note 9** below ***(29,30,34)***. The analyses presented here were performed using the first method, which was, until recently, the only large format system commercially available. This system is especially convenient for small analytical samples, but can also be used reproducibly for large-sample analyses ***(33)***. When 2DE is coupled with immunoanalysis, this system permits the detection of some proteins that cannot be visualized on blots from Immobiline gels (M. Aepfelbacker, personal communication; C. Farnsworth, unpublished results).
9. When large samples (>500 μg protein; e.g., cell lysates) are analyzed, the Immobiline system is currently preferable, because of advances in gel technology and sample-handling techniques. These gels are now commercially available in an extended format (0.5 × 3 × 180 mm) and come as dehydrated gel strips bonded to a rigid plastic support. This permits the gel to be rehydrated directly with dilute sample lysates. Additionally, samples containing up to 5 mg protein can be analyzed in this way ***(36)*** without the need for concentration by techniques such as protein precipitation. The gels can accommodate a relatively large sample volume (upper limit of 350 μL vs 26 μL for the carrier ampholyte system). Another advantage of immobilized ampholyte pH gradients is that they allow the use of very narrow pH gradients (e.g., pH 5.5–6.5). This permits the separation of charge-isoforms that differ by only pH 0.05 units ***(29)***. Lastly, when 2DE is coupled with immunoanalysis and overlay assays, detailed maps of multimember protein families can be constructed ***(37)***.
10. In control experiments using lysates of unlabeled cells, recovery of the labeled internal standard, [^{3}H]-NEM STI (*see* **Subheading 3.3.3.**), following complete sample workup (*see* **Subheading 3.3.2., steps 6–9**) was 91 ± 4%, $n = 3$.

Acknowledgment

Alberto Corsini (1993–1999) was visiting scientist under the terms of the US (National Heart, Lung, and Blood Institute)–Italy bilateral agreement in the cardiovascular area.

References

1. Zhang, F. L. and Casey, P. J. (1996) Protein Prenylation: Molecular mechanisms and functional consequences. *Annu. Rev. Biochem.* **65,** 241–269.
2. Glomset, J. A., Gelb, M. H., and Farnsworth, C. C. (1990) Prenyl proteins in eukariotic cells: a new type of membrane anchor. *Trends Biochem. Sci.* **15,** 139–142.
3. Maltese, W. A. (1990) Posttranslational modification of proteins by isoprenoids in mammalian cells. *FASEB J.* **4,** 3319–3328.

4. Glomset, J. A. and Farnsworth, C. C. (1994) Role of protein modification reactions in programming interactions between ras-related GTPases and cell membranes. *Annu. Rev. Cell Biol.* **10,** 181–205.
5. Casey, P. J., Moomaw, J. F., Zhang, F. L., Higgins, J. B., and Thissen, J. A. (1994) Prenylation and G protein signaling, in *Recent Progress in Hormone Research*, Academic, New York.
6. Inglese, J., Koch, W. J., Touhara, K., and Lefkowitz, R. J. (1995) $G_{\beta\gamma}$ interactions with pH domains and Ras-MPK signaling pathways. *Trends Biochem. Sci.* **20,** 151–156.
7. Marshall, M. S. (1995) Ras target proteins in eukaryotic cells. *FASEB J.* **9,** 1311–1318.
8. Kato, K., Cox, A. D., Hisaka, M. M., Graham, S. M., Buss, J. E., and Der, C. J. (1992) Isoprenoid addition to Ras protein is the critical modification for its membrane association and transforming activity. *Proc. Natl. Acad. Sci. USA* **89,** 6403–6407.
9. Boguski, M. S. and McCormick, F. (1993) Proteins regulating Ras and its relatives. *Nature* **365,** 643–654.
10. Jakobisiak, M., Bruno, S., Skiersky, J. S., and Darzynkiewicz, Z. (1991) Cell cycle-specific effects of lovastatin. *Proc. Natl. Acad. Sci. USA* **88,** 3628–3632.
11. Taylor, S. J. and Shalloway, D. (1996) Cell cycle-dependent activation of Ras. *Curr. Biol.* **6,** 1621–1627.
12. Olson, M. F., Ashworth, A., and Hall, A. (1995) An essential role for Rho, Rac, and Cdc42 GTPases in cell cycle progression through G_1. *Science* **269,** 1270–1272.
13. Fenton, R. G., Kung, H., Longo, D. L., and Smith, M. R. (1992) Regulation of intracellular actin polymerization by prenylated cellular proteins. *J. Cell Biol.* **117,** 347–356.
14. Pittler, S. J., Fliester, S. J., Fisher, P. L., Keller, R. K., and Rapp, L. M. In vivo requirement of protein prenylation for maintenance of retinal cytoarchitecture and photoreceptor structure. *J. Cell Biol.* **130,** 431–439.
15. Tapon, N. and Hall, A. (1997) Rho, Rac, and Cdc42 GTPases regulate the organization of the actin cytoskeleton. *Curr. Opin. Cell Biol.* **9,** 86–92.
16. Zerial, M. and Stenmark, H. (1993) Rab GTPases in vesicular transport. *Curr. Opin. Cell Biol.* **5,** 613–620.
17. Novick, P. and Brennwald, P. (1993) Friends and family: the role of the Rab GTPases in vesicular traffic. *Cell* **75,** 597–601.
18. Raiteri, M., Arnaboldi, L., McGeady, P., Gelb, M. H., Verri, D., Tagliabue, C., Quarato, P., et al. (1997) Pharmacological control of the mevalonate pathway: effect on smooth muscle cell proliferation. *J. Pharmacol. Exp. Ther.* **281,** 1144–1153.
19. Habenicht, A. J. R., Glomset, J. A., and Ross, R. (1980) Relation of cholesterol and mevalonic acid to the cell cycle in smooth muscle and Swiss 3T3 cells stimulated to divide by platelet-derived growth factor. *J. Biol. Chem.* **255,** 5134–5140.
20. Goldstein, J. L. and Brown, M. S. (1990) Regulation of the mevalonate pathway. *Nature* **343,** 425–430.
21. Corsini, A., Mazzotti, M., Raiteri, M., Soma, M. R., Gabbiani, G., Fumagalli, R., and Paoletti, R. (1993) Relationship between mevalonate pathway and arterial myocyte proliferation: *in vitro* studies with inhibitor of HMG-CoA reductase. *Atherosclerosis* **101,** 117–125.
22. Crick, D. C., Waechter, C. J., and Andres, D. A. (1994) Utilization of geranylgeraniol for protein isoprenylation in C6 glial cells. *Biochem. Biophys. Res. Commun.* **205,** 955–961.
23. Crick, D. C., Andres, D. A., and Waechter, C. J. (1995) Farnesol is utilized for protein isoprenylation and the biosynthesis of cholesterol in mammalian cells. *Biochem. Biophys. Res. Commun.* **211,** 590–599.

24. Danesi, R., McLellan, C. A., and Myers, C. E. (1995) Specific labeling of isoprenylated proteins: application to study inhibitors of the posttranslational farnesylation and geranylgeranylation. *Biochem. Biophys. Res. Commun.* **206,** 637–643.
25. Shin-ichi, O., Watanabe, M., and Nishino,T. (1996) Identification and characterization of geranylgeraniol kinase and geranylgeranyl phosphate kinase from the *Archebacterium Sulfolobus acidocaldarius. J. Biochem.* **119,** 541–547.
26. Westfall, D., Aboushadi, N., Shackelford, J. E., and Krisans, S. (1997) Metabolism of farnesol: phosphorylation of farnesol by rat liver microsomial and peroxisomal fractions. *Biochem. Biophys. Res. Commun.* **230,** 562–568.
27. Lowry, O. H., Rosebrough, N. J., Farr, A. L., and Randall, R. J. (1951) Protein measurement with the folin phenol reagent. *J. Biol. Chem.* **193,** 265–275.
28. Laemmli, U. K. (1970) Cleavage of structural proteins during the assembly of the head of bacteriophage T4. *Nature* **227,** 680–685.
29. Görg, A., Postel, W., and Günther, S. (1988) The current state of two-dimensional electrophoresis with immobilized pH gradients. *Electrophoresis* **9,** 531–546.
30. Jungblutt, P., Thiede, B., Simny-Arndt, U., Müller, E.-C., Wittmann-Liebold, B., and Otto, A. (1997) Resolution power of two-dimensional electrophoresis and identification of proteins from gels. *Electrophoresis* **17,** 839–847.
31. O'Farrell, P. (1975) High resolution two-dimensional electrophoresis of proteins. *J. Biol. Chem.* **250,** 4007–4021.
32. Patton, W. F., Pluskal, M. G., Skea, W. M., Buecker, J. L., Lopez, M. F., Zimmermann, R., Lelanger, L. M., and Hatch, P. D. (1990) Development of a dedicated two-dimensional gel electrophoresis system that provides optimal pattern reproducibility and polypeptide resolution. *BioTech.* **8,** 518–529.
33. Lopez, M. F. and Patton, W. F. (1990) Reproducibility of polypeptide spot positions in two-dimensional gels run using carrier ampholytes in the isolelectric focusing dimension. *Electrophoresis* **18,** 338–343.
34. Görg, A., Günther, B., Obermaier, C., Posch, A., and Weiss, W. (1995) Two-dimensional polyacrylamide gel electrophoresis with immobilized pH gradients in the first dimension (IPG-Dalt): the state of the art and the controversy of vertical versus horizontal systems. *Electrophoresis* **16,** 1079–1086.
35. Schmidt, R. A., Schneider, C. J., and Glomset, J. A. (1984) Evidence for post-translational incorporation of a product of mavalonic acid into Swiss 3T3 cell proteins. *J. Biol. Chem.* **259,** 10,175–10,180.
36. Sanchez, J.-C., Rouge, V., Pisteur, M., Ravier, F., Tonella, L., Moosmayer, M., Wilkins, M. R., and Hochstrasser, D. F. (1997) Improved and simplified in-gel sample application using reswelling of dry immobilized pH gradients. *Electrophoresis* **18,** 324–327.
37. Huber, L. A., Ullrich, O., Takai, Y., Lütcke, A., Dupree, P., Olkkonen, et al. (1994) Mapping of Ras-related GTP-binding proteins by GTP overlay following two-dimensional gel electrophoresis. *Proc. Natl. Acad. Sci. USA* **91,** 7874–7878.

95

The Metabolic Labeling and Analysis of Isoprenylated Proteins

Douglas A. Andres, Dean C. Crick, Brian S. Finlin, and Charles J. Waechter

1. Introduction

1.1. Background

The posttranslational modification of proteins by the covalent attachment of farnesyl and geranylgeranyl groups to cysteine residues at or near the carboxyl-(C)-terminus via a thioether bond is now well established in mammalian cells ***(1–6)***. Most isoprenylated proteins are thought to serve as regulators of cell signaling and membrane trafficking. Farnesylation and geranylgeranylation of the cysteinyl residues has been shown to promote both protein–protein and protein–membrane interactions ***(6–8)***. Isoprenylation, and in some cases the subsequent palmitoylation, provide a mechanism for the membrane association of polypeptides that lack a transmembrane domain, and appear to be prerequisite for their in vivo activity ***(6,9,10)***.

Three distinct protein prenyltransferases catalyzing these modifications have been identified ***(1–5)***. Two geranylgeranyltransferases (GGTases) have been characterized and are known to modify distinct protein substrates. The CaaX GGTase (also known as GGTase-1) geranylgeranylates proteins that end in a CaaL(F) sequence, where C is cysteine, A is usually an aliphatic amino acid, and the C-terminal amino acyl group is leucine (L) or phenylalanine (F). Rab GGTase (also known as GGTase-2) catalyzes the attachment of two geranylgeranyl groups to paired C-terminal cysteines in most members of the Rab family of GTP-binding proteins ***(11)***. These proteins terminate in a Cys-Cys, Cys-X-Cys, or Cys-Cys-X-X motif where X is a small hydrophobic amino acid. Another set of regulatory proteins is modified by protein farnesyltransferase (FTase). All known farnesylated proteins terminate in a tetrapeptide CaaX box, wherein C is cysteine, A is an aliphatic amino acid, and X has been shown to be a COOH-terminal methionine, serine, glutamine, cysteine, or alanine.

Isoprenylated proteins are commonly studied by metabolic labeling of cultured cells by incubation with [^{3}H]mevalonate which is enzymatically converted to [^{3}H]farnesyl pyrophosphate (F-P-P) and [^{3}H]geranylgeranyl pyrophosphate (GG-P-P) prior to being incorporated into protein (**ref. *12*, Fig. 1**). Cellular proteins can then be analyzed by gel

From: *The Protein Protocols Handbook, 2nd Edition*
Edited by: J. M. Walker © Humana Press Inc., Totowa, NJ

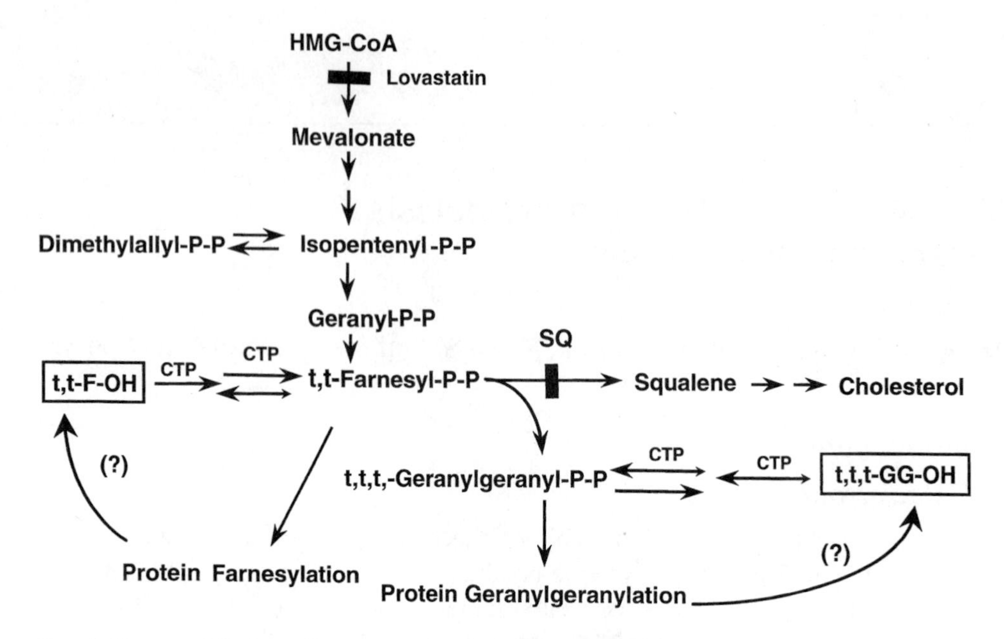

Fig. 1. Proposed "salvage" pathway for the utilization of F-OH and GG-OH for isoprenoid biosynthesis. Evidence for the presence of microsomal enzymes catalyzing the conversion of F-P-P and GG-P-P to the free isoprenols has been reported by Bansal and Vaidya *(30)*.

electrophoresis and autoradiography. However, [^{3}H]mevalonate labeling does not distinguish between farnesylated and geranylgeranylated proteins directly. Other methods for the identification of the isoprenyl group attached to protein that make use of [^{3}H]mevalonolactone labeling have been described. These methods generally require the isolation of large amounts of labeled protein, extensive proteolysis, a number of column chromatographic steps, and cleavage of the thioether bond with Raney nickel or methyl iodide, followed by gas chromatography or HPLC and mass spectrometry to identify the volatile cleavage products (*see* **ref. *1*** and papers cited therein).

This chapter describes a novel approach for the use of free [^{3}H]farnesol (F-OH) and [^{3}H]geranylgeraniol (GG-OH) in the selective metabolic labeling of farnesylated and geranylgeranylated proteins, respectively, in cultured mammalian cells. In addition, the methods use materials and equipment readily available in most laboratories.

We have recently shown that mammalian cells can utilize the free isoprenols, GG-OH and F-OH, for the isoprenylation of cellular proteins *(13,14)*. When C6 glioma cells were incubated with [^{3}H]F-OH, radioactivity was also incorporated into cholesterol *(14)*. The observation that the incorporation of label into sterol was blocked by squalestatin 1 (SQ), a potent inhibitor of squalene synthetase *(15–19)*, suggested that F-OH, and probably GG-OH, are utilized for isoprenoid biosynthesis after being converted to the corresponding activated allylic pyrophosphates, F-P-P and GG-P-P (**Fig. 1**). Preliminary studies have suggested the presence of enzyme systems in mammals and lower organisms that are capable of phosphorylating F-OH and GG-OH *(20–22)*. More recently microsomal fractions from *N. tabacum* have been shown to contain CTP-mediated kinases that catalyze the conversion of F-OH and GG-OH to the respective allylic pyrophosphate intermediates by two successive monophosphorylation reactions *(23)*. Further work is certainly warranted on the isolation and characterization of these enzymes. The early developments in understanding the mechanism and physiological significance of the salvage pathway for the utilization of F-OH and GG-OH have been reviewed *(24)*.

1.2. Experimental Strategy

Utilizing free F-OH or GG-OH as the isotopic precursors has several experimental advantages over metabolic labeling of isoprenylated proteins with mevalonate. First, F-OH and GG-OH are more hydrophobic and rapidly enter cultured mammalian cells. They are efficiently utilized in a range of mammalian cell lines, and obviate the need to include HMG-CoA reductase inhibitors to lower endogenous pools of mevalonate. The experimental strategy is illustrated in **Fig. 1**. A key advantage of the strategy is that F-OH and GG-OH are selectively incorporated into distinct subsets of isoprenylated proteins, providing a simple and convenient approach to specifically label farnesylated or geranylgeranylated proteins (**Fig. 2**).

Following metabolic labeling using [^{3}H]F-OH, [^{3}H]GG-OH, or [^{3}H]mevalonolactone, the metabolically labeled proteins are exhaustively digested with Pronase E to liberate the specific isoprenyl-cysteine residues (**Fig. 3**). The identity of the isoprenylated cysteine residue can then be readily identified by normal or reverse-phase thin layer chromatography (TLC). **Figs. 4** and **5** show representative TLC analysis of isoprenyl-cysteines released from metabolically labelled cellular proteins and from recombinant proteins that were isoprenylated in vitro.

The isoprenol labeling and Pronase E methods may also be applied to the analysis of individual metabolically labeled proteins. These methods provide a simple and convenient approach for the identification of the isoprenyl group found on a specific protein. Two experimental approaches are available. In the first, separate cell cultures are incubated with [^{3}H]mevalonate, [^{3}H]F-OH, or [^{3}H]GG-OH the metabolically labeled protein of interest is then purified and subjected to sodium dodecyl sulfate-polyacrylamide gel electrophoresis (SDS-PAGE) analysis. The specificity of the F-OH and GG-OH incorporation allows the identity of the prenyl group to be directly assessed (**Fig. 6**). In the second, the cells are metabolically labeled with [^{3}H]mevalonolactone and the isolated protein of interest is subjected to Pronase E treatment followed by TLC analysis of the butanol-soluble products. In this case analysis of the chromatogram reveals the nature of the isoprenyl group (**Fig. 7**). The experimental methods for these analytical procedures are presented in detail in this chapter.

2. Materials

2.1. Metabolic Radiolabeling of Mammalian Cells in Culture with [^{3}H]Farnesol, [^{3}H]Geranylgeraniol, and [^{3}H]Mevalonolactone

1. ω,*t*,*t*-[1-^{3}H]Farnesol (20 Ci/mmol, American Radiolabeled Chemicals, Inc., St. Louis, MO).
2. ω,*t*,*t*,*t*-[^{3}H]Geranylgeraniol (60 Ci/mmol, American Radiolabeled Chemicals, Inc., St. Louis, MO).
3. [^{3}H]Mevalonolactone (60 Ci/mmol, American Radiolabeled Chemicals, Inc., St. Louis, MO).
4. Appropriate cell culture media, plastic ware, and cell incubator.

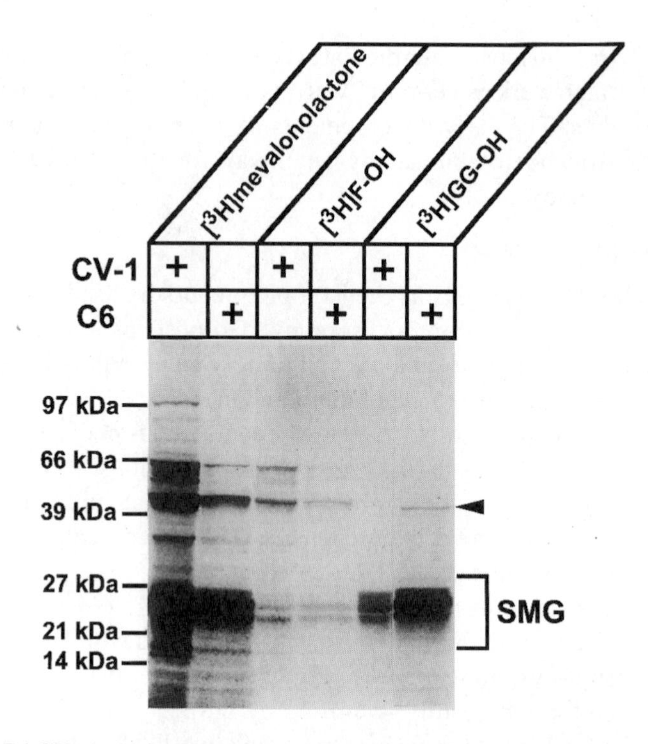

Fig. 2. SDS-PAGE analysis of proteins labeled by incubating C6 glioma cells and CV-1 cells with [^{3}H]mevalonate, [^{3}H]F-OH, or [^{3}H]GG-OH. The details of the metabolic labeling procedure and SDS-PAGE analysis are described in **Subheadings 3.1.–3.3.** For these analyses proteins were metabolically labeled by incubating the indicated cultured cells with [^{3}H]F-OH in the presence of lovastatin (5 μg/mL) because it increased the amount of radioactivity incorporated into protein during long-term incubations, presumably by reducing the size of the endogenous pool of F-P-P. The gel patterns reveal that distinctly different sets of proteins are labeled by each precursor in C6 and CV-1 cells. Consistent with selective labeling by [^{3}H]F-OH and [^{3}H]GG-OH, the labeling pattern for [^{3}H]mevalonate, which can be converted to ^{3}H-labeled F-P-P and GG-P-P, appears to be a composite of the patterns seen with the individual [^{3}H]isoprenols. SMG = proteins in the size range (19–27 kDa) of small GTP-binding proteins. (Figure reprinted with permission) (*see* **ref. 14**).

5. 95% Ethanol.
6. Serum Supreme (SS), an inexpensive fetal bovine serum (FBS) substitute obtained from BioWhittaker, has been successfully used with C6 glioma, Chinese hamster ovary (CHO) clone UT-2 and green monkey kidney (CV-1) cells in this laboratory.
7. Bath sonicator.

2.2. Delipidation of Labeled Proteins

1. Phosphate-buffered saline (PBS).
2. PBS containing 2 m*M* EDTA.
3. Methanol.
4. Chloroform–methanol (2:1, v/v).
5. 12-mL Disposable conical screw-capped glass centrifuge tubes.
6. Benchtop centrifuge.
7. Probe sonicator.

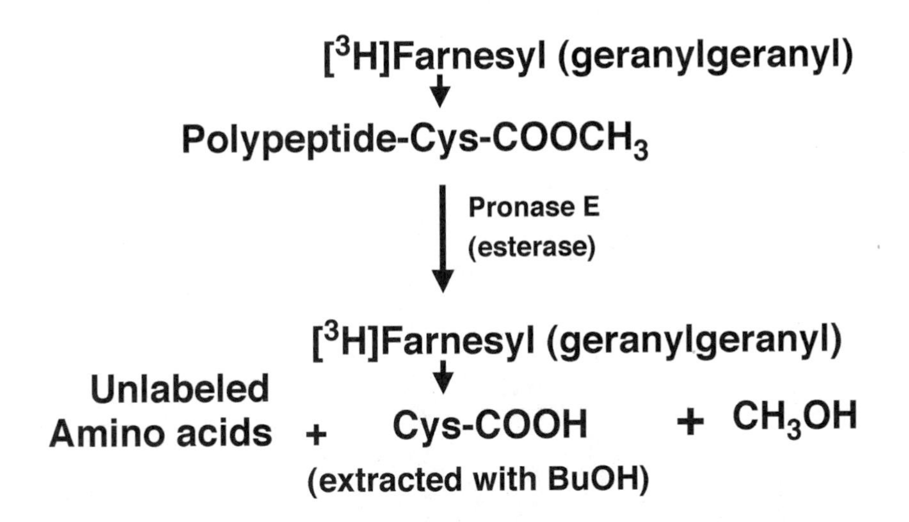

Fig. 3. Experimental scheme for the rapid analysis of metabolically labeled isoprenylated cysteine residues labeled by various isotopic precursors. The experimental details of this procedure are described in **Subheading 3.4.**

2.3. SDS-PAGE Analysis of Metabolically Labeled Proteins

1. SDS-PAGE apparatus.
2. Western blot transfer apparatus.
3. Nitrocellulose membrane (Schleicher and Schuell, Protran BA83).
4. 2% SDS, 5 m*M* 2-mercaptoethanol.
5. Ponceau S: 0.2% Ponceau S, 3% trichloracetic acid (TCA), 3% sulfosalicylic acid (Sigma, S 3147).
6. Fluorographic reagent (Amplify, Amersham Corp.).
7. Sheets of stiff plastic (such as previously exposed X-ray film).

2.4. Pronase E Digestion and Chromatographic Analysis of Radiolabeled Proteins

1. *N*-[2-hydroxyethyl]piperazine-*N'*-[2'-ethanesulfonic acid] Hepes, pH 7.4.
2. Calcium acetate.
3. Bath sonicator.
4. Pronase E (Sigma, St. Louis, MO).
5. 37°C water bath.
6. *n*-Butanol saturated with water.
7. Bench centrifuge.
8. Oxygen-free nitrogen gas.
9. Chloroform–methanol–H_2O (10:10:3, by vol).
10. Farnesyl-cysteine (F-Cys) was synthesized as described by Kamiya et al. *(25)*.
11. Geranylgeranyl-cysteine (GG-Cys) was synthesized as described by Kamiya et al. *(25)* except that isopropanol is substituted for methanol in the reaction solvent to improve the yield of the synthetic reaction.
12. Silica Gel G 60 TLC plates (Sigma, St. Louis, MO).
13. Chloroform–methanol–7 *N* ammonia hydroxide (45:50:5, by vol).
14. Si*C_{18} reverse-phase plates (J. T. Baker Inc., Phillipsburg, NJ).
15. Acetonitrile–H_2O–acetic acid (75:25:1, by vol).

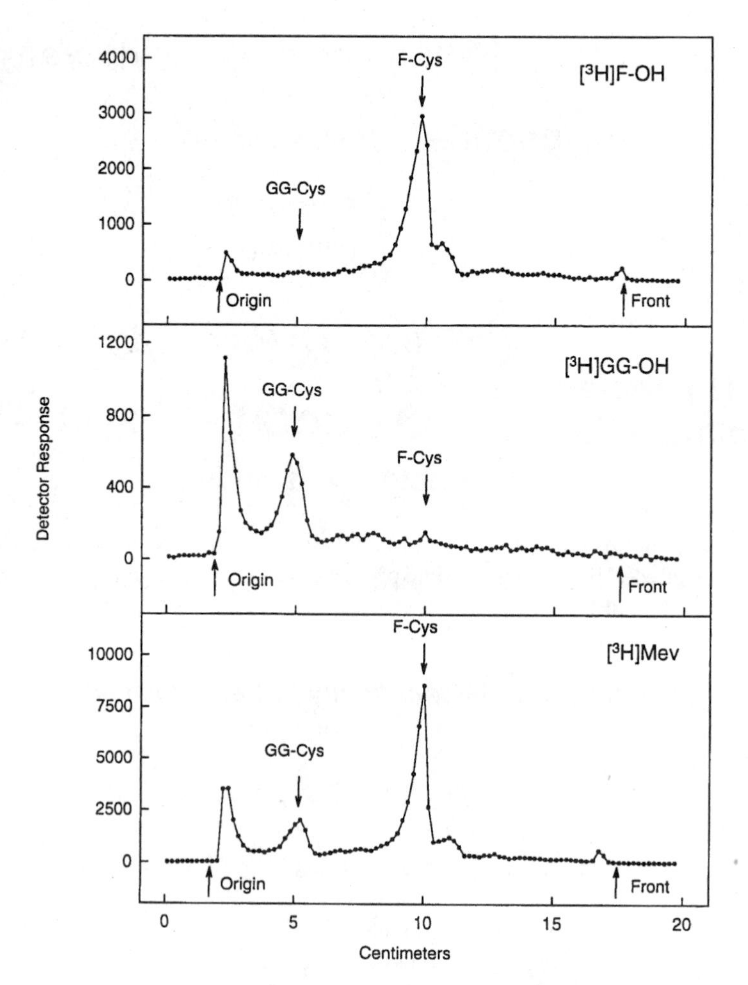

Fig. 4. Chromatographic analysis of isoprenyl-cysteine residues metabolically labeled by incubating C6 glioma cells with, [^{3}H]F-OH (*upper panel*), [^{3}H]GG-OH (*middle panel*) or [^{3}H]mevalonate (*lower panel*). From the traces illustrated, in it can be seen that F-Cys and GG-Cys are both metabolically labeled when C6 cells are incubated with [^{3}H]mevalonate, but only F-Cys or GG-Cys are labeled when C6 cells are incubated with [^{3}H]F-OH or [^{3}H]GG-OH, respectively. Virtually identical results were obtained by chromatographic analysis of the Pronase E digests of CV-1 proteins metabolically labeled by each isotopic precursor. Radiolabeled products are also observed at the origin of the TLC which could be incompletely digested isoprenylated peptides, or in the case of [^{3}H]mevalonate and [^{3}H]GG-OH, possibly mono- or digeranylgeranylated Cys-Cys or Cys-X-Cys sequences reprinted with permission (**ref.** ***11*****, Fig. 5**).

16. Conical glass tubes.
17. Anisaldehyde spray reagent *(26)* (*see* **Note 1**).
18. Ninhydrin spray reagent (*see* **Note 2**).

2.5. Immuniprecipation of Specific Radiolabeled Protein

1. Cell lysis buffer: 20 m*M* Tris-HCl, pH 7.5, 150 m*M* NaCl, 1% Nonidet P-40 (NP-40), 1 m*M* phenylmethyl+sulfonyl fluoride (PMSF) (add fresh PMSF solution and lysis is carried out at 4°C.

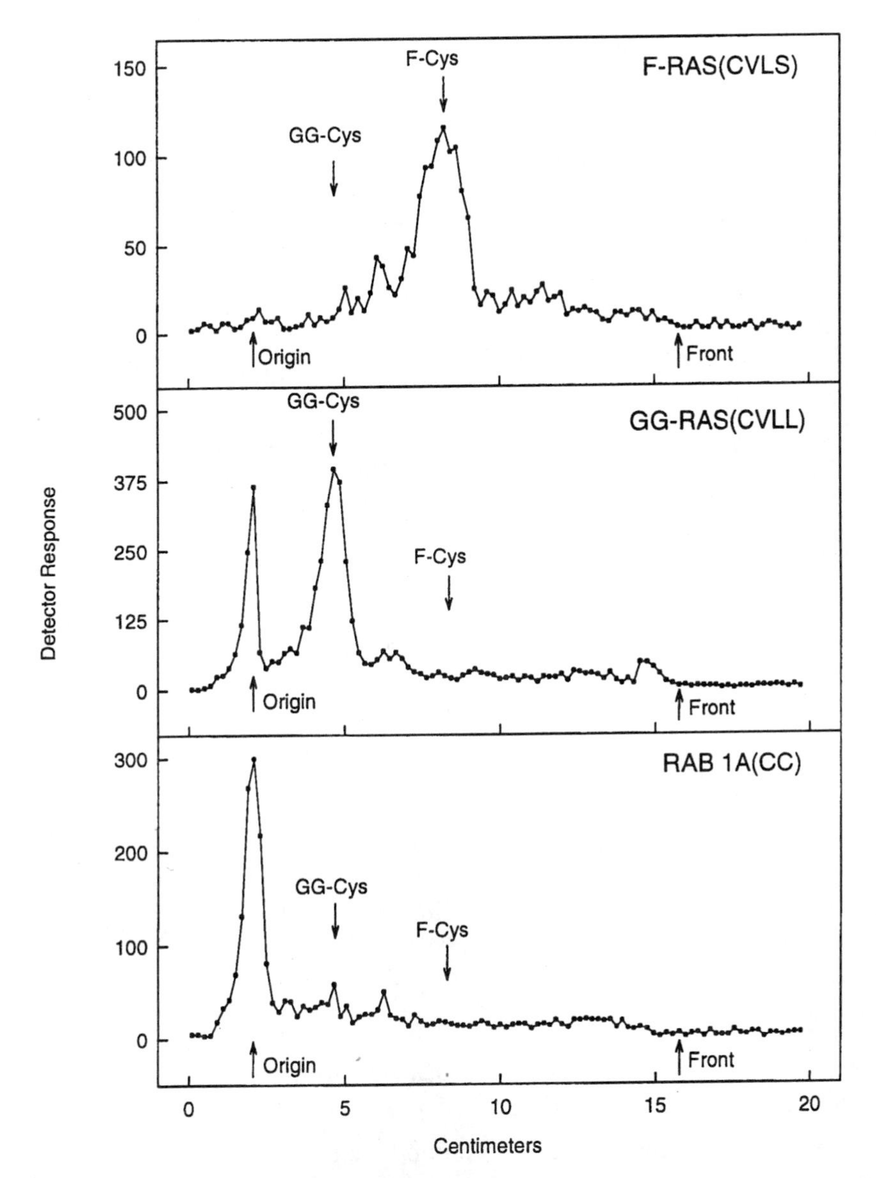

Fig. 5. Chromatographic analyses of isoprenyl-cysteine residues liberated by Pronase E digestion of recombinant protein substrates enzymatically labeled in vitro by [^{3}H]F-P-P or [^{3}H]GG-P-P. The recombinant proteins were labeled by incubation with recombinant FTase (*upper panel*), GGTase I (*upper panel*) or GGTase II (*lower panel*), essentially under the conditions described previously ***(19)***.

2. Protein-specific immunoprecipitating antibody.
3. Protein A-Sepharose (Pharmacia).
4. Wash buffer A: 20 m*M* Tris-HCl, pH 7.5, 150 m*M* NaCl, 0.2% NP-40.
5. Wash buffer B: 20 m*M* Tris-HCl, pH 7.5, 500 m*M* NaCl, 0.2% NP-40.
6. Wash buffer C: 20 m*M* Tris-HCl, pH 7.5, 150 m*M* NaCl.
7. *n*-Butanol saturated with water.
8. Tabletop ultracentrifuge (Beckman).

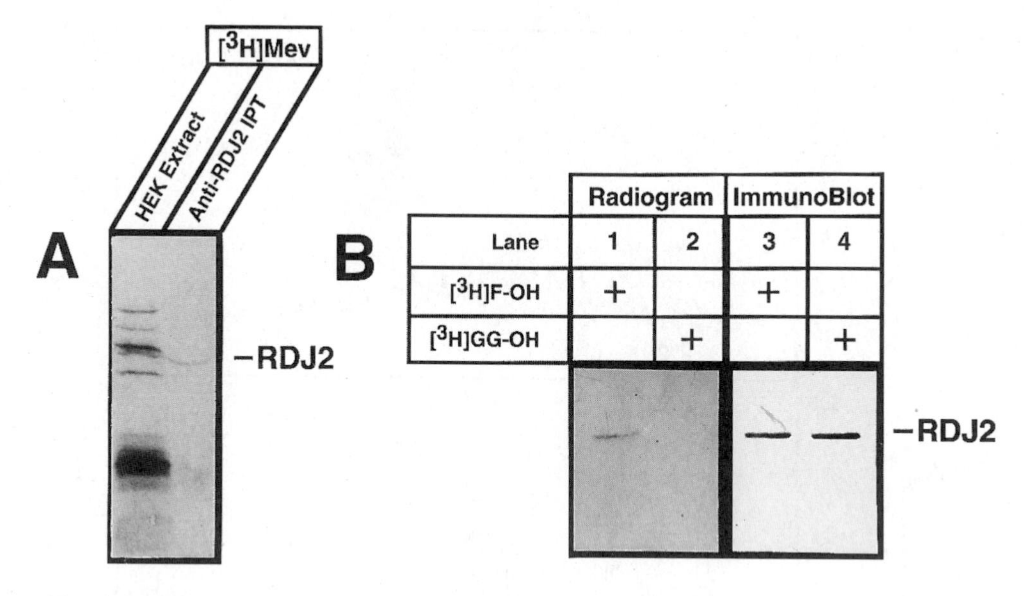

Fig. 6. Specific metabolic labeling of RDJ2-transfected HEK cells. **(A)** Monolayers of human embryonic kidney HEK cells transfected with pRDJ2 (an expression plasmid that contained the RDJ2 cDNA under the control of a CMV promoter) were radiolabeled with [^{3}H]mevalonate and the expressed RDJ2 immunoprecipitated from detergent-solubilized cell extracts using a RDJ2-specific antibody. A portion of the resulting immunoprecipitate as well as a portion of the cell extract were subjected to SDS-PAGE. The gel was treated with Amplify, dried, and exposed to film for 14 d. **(B)** Immunoprecipitated RDJ2 after metabolic labeling of transfected HEK cells with [^{3}H]F-OH (*lane 1*) or [^{3}H]GG-OH (*lane 2*) was subjected to SDS-PAGE analysis and fluorography. The remaining immunoprecipitated protein fractions isolated from HEK cells after metabolic labeling with [^{3}H]F-OH (*lane 3*) or [^{3}H]GG-OH (*lane 4*) were immunoblotted using anti-RDJ2 IgG and subjected to chemiluminescence detection.

3. Methods

3.1. Procedure for the Selective Metabolic Labeling of Farnesylated and Geranylgeranylated Proteins

Procedures are described for the metabolic labeling of mammalian cells grown to near-confluence in Falcon 3001 tissue culture dishes. These protocols can be scaled up or down as appropriate. The incorporation of [^{3}H]F-OH or [^{3}H]GG-OH into protein was linear with respect to time and the concentration of [^{3}H]isoprenol in tissue culture dishes ranging from 10 to 35 mm in diameter under the conditions described here *(13,14)* (*see* **Note 3**).

1. For metabolic labeling experiments with either [^{3}H]F-OH or [^{3}H]GG-OH, a disposable conical glass centrifuge tube (screw capped) and Teflon-lined cap are flame sterilized. The labeled isoprenol dissolved in ethanol is added to the tube and the ethanol is evaporated under a sterile stream of air.
2. An appropriate volume of sterile SS is added to yield a final concentration of 0.5–1 mCi/mL. The labeled isoprenols are dispersed in the SS by sonication in a Branson bath sonicator for 10 min. After sonication an aliquot is taken for liquid scintillation counting to verify that the ^{3}H-labeled isoprenol has been quantitatively dispersed in SS.

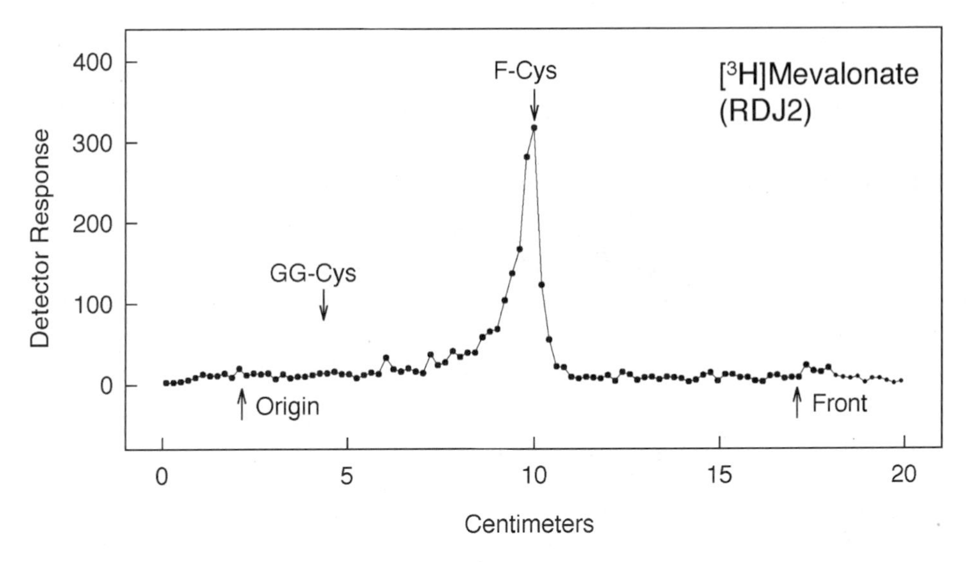

Fig. 7. Chromatographic analysis of butanol-soluble products released by Pronase E digestion of RDJ2 protein metabolically labeled by incubation with [^{3}H]mevalonolactone. Immunoprecipitated RDJ2, isolated from [^{3}H]mevalonolactone-labeled HEK cells, was subjected to Pronase E digestion. The labeled products were extracted with 1-butanol, and analyzed by reverse-phase chromatography using C18 reverse-phase TLC plates, and developed in acetonitrile–water–acetic acid (75:25:1). Radioactive zones were located with a Bioscan Imaging System 200-IBM. The *arrows* indicate the position of authentic F-Cys and GG-Cys.

3. The growth medium is removed from the cultured cells by aspiration and 500 μL of labeling medium, consisting of an appropriate medium and SS (final concentration = 3–5%) containing [^{3}H]GG-OH or [^{3}H]F-OH, is added.
4. Cell cultures are typically incubated at 37°C under 5% CO_2 for 6–24 h. Actual culture media and incubation conditions will vary depending on the specific cell type being studied.

3.2. Recovery of Metabolically Labeled Proteins from Adherent Cell Lines

1. The labeling medium is removed by gentle aspiration.
2. The metabolically labeled cells are washed with 1–2 mL of ice-cold PBS, to remove unincorporated isotopic precursor.
3. One milliliter of PBS containing 2 m*M* EDTA is added, and cells are incubated for 5 min at room temperature. The washed cells are gently scraped from the dish with a disposable cell scraper, and transferred to a 12-mL conical glass centrifuge tube.
4. The metabolically labeled cells are sedimented by centrifugation (500*g*, 5 min), and the PBS–EDTA is removed by aspiration (avoid disrupting the cell pellet).
5. The cells are resuspended in PBS (1–2 mL) and the PBS is removed by aspiration after the cells are sedimented by centrifugation.
6. Two milliliters of Methanol (CH_3OH) is added to the cell pellet, and the pellet is disrupted by sonication using a probe sonicator.
7. The suspension is sedimented by centrifugation (1500*g* for 5 min).
8. The CH_3OH extract is carefully removed, to avoid disturbing the partially delipidated pellet, and transferred to a glass conical tube (*see* **Note 4**).
9. The protein pellet is reextraced twice with 2 mL of $CHCl_3$–CH_3OH (2:1), and the extracts are pooled with the CH_3OH extract (*see* **Note 4**).

10. Residual organic solvent is removed from the delipidated protein pellet by evaporation under a stream of nitrogen. The delipidated protein fractions are then subjected to Pronase E digestion for analysis of isoprenyl-cysteine analysis by TLC (*see* **Subheading 3.4.**), or dissolved in 2% SDS, 5 m*M* 2-mercaptoethanol for SDS-PAGE analysis.

3.3. SDS-PAGE Analysis of Metabolically Labeled Proteins

To examine the molecular weight and number of proteins metabolically labeled by incubation with [^{3}H]F-OH or [^{3}H]GG-OH, the delipidated protein fractions can be analyzed by SDS-PAGE. Because these experiments rely on the detection of low-energy ^{3}H-labeled compounds, two procedures are described for the use of fluorography to increase the sensitivity of detection. (*See* **Note 5** before proceeding.)

1. The delipidated protein fractions are solubilized in 2% SDS, 5 m*M* β-mercaptoethanol. An aliquot is used to determine the amount of labeled precursor incorporated into protein.
2. The radiolabeled polypeptides (20–60 μg of protein) were analyzed by SDS-PAGE using an appropriate percentage polyacrylamide resolving gel (4–20%) for the proteins of interest.
3. Following SDS-PAGE, gels can be analyzed using two distinct methods. In the first, the gel is directly soaked in the fluorographic reagent, Amplify (Amersham), according to the manufacturer's protocol, dried, and exposed to X-ray film as described in **step 5**. In a second approach, proteins were electrophoretically transferred to nitrocellulose filters and stained with Ponceau S to determine the efficiency of transfer. The nitrocellulose filters are destained by brief washing in distilled water and allowed to air-dry. (*See* **Note 5** for a discussion of the merits of each method before continuing.)
4. The filters were then dipped briefly in the fluorographic reagent, Amplify (Amersham), placed on a sheet of plastic backing, and dried for 1 h at 50°C. It is important that a thin and even film of Amplify reagent remain on the filter and that it be placed protein side up to dry.
5. Fluorograms were produced by exposing preflashed X-ray film to the nitrocellulose filter, or dried SDS-PAGE gel, for 5–30 d at –80°C.

3.4. Methods for the Identification of Cysteine-Linked Isoprenyl Group

These simple methods are inexpensive, rapid, and allow the identification of the isoprenyl-cysteine residue(s) from isoprenylated protein(s). Examples of this method for the identification of isoprenyl-cysteine groups from metabolically labeled cells and recombinant proteins labeled in vitro are shown in **Figs. 4** and **5**. As expected, [^{3}H]F-Cys and [^{3}H]GG-Cys were liberated from RAS(CVLS) and RAS(CVLL), respectively. A radioactive peak is also seen at the origin in the analysis of the Pronase digest of radiolabeled RAS(CVLL) (**Fig. 5**, *middle panel*). This radiolabeled product(s) is probably incompletely digested [^{3}H]geranylgeranylated peptides. Rab 1A terminates in two cysteine residues, both of which are isoprenylated ***(11)***. **Fig. 5** (*lower panel*) indicates that Pronase E is incapable of cleaving between the two cysteine residues.

1. To liberate the labeled isoprenyl-cysteine residues for analysis, the delipidated protein fractions (50–100 μg) are incubated with 2 mg of Pronase E; 50 m*M* HEPES, pH 7.4; and 2 m*M* calcium acetate in a total volume of 0.1 mL at 37°C for approx 16 h. The experimental scheme for this analysis is illustrated in **Fig. 3** (*see* **Note 6**).
2. Proteolysis is terminated by the addition of 1 mL of *n*-butanol saturated with H_2O and mixing vigorously.

3. Centrifuge the mixture for 5 min in a benchtop centrifuge at 1500*g*. Two phases will form, and the upper phase should be clear (*see* **Note 7**).
4. Transfer the upper phase, containing *n*-butanol, to a separate conical glass centrifuge tube.
5. Add 1 mL of H_2O to the butanol extract and mix vigorously. Centrifuge at 1500*g* for 5 min, to effect a phase separation. Remove the lower aqueous phase with a Pasteur pipet (*see* **Note 8**). Evaporate the *n*-butanol under a stream of nitrogen at 30–40°C (*see* **Note 9**).
6. The labeled isoprenyl-cysteines are dissolved in 250 µL of $CHCl_3$–CH_3OH–H_2O (10:10:3, by vol) by mixing vigorously, and an aliquot (10 µL) is taken to determine the amount of radioactivity.
7. The radiolabeled products are analyzed chromatographically on a normal-phase system, using silica gel G 60 TLC plates developed in $CHCl_3$–CH_3OH–7 *N* NH_4OH (45:50:5, by vol) or by reverse-phase chromatography using silica gel Si*C18 reverse-phase plates developed with acetonitrile–H_20–acetic acid (75:25:1, by vol) (*see* **Note 10**).
8. The desired developing solvent mixture is added to the chromatography tank to a depth of 0.5–1.0 cm and allowed to equilibrate.
9. Dry entire sample under nitrogen stream. Redissolve in chloroform–methanol–water (10:10:3, by vol).
10. Apply an aliquot (approx 10,000 dpm containing 10–12 µg of authentic F-Cys and GG-Cys) to the origin on the TLC plate, using a fine glass capillary or a Hamilton syringe. The addition of the unlabeled standards improves the resolution of the metabolically labeled isoprenylated cysteines.
11. Allow the sample to dry (this can be facilitated with a stream of warm air). Complete application of the labeled isoprenyl-cysteine extract to the plate in 5-µL aliquots, allowing time for the spot to dry between applications.
12. Place the plate(s) in the preequilibrated chromatography tank, and after the solvent has reached the top of each plate (1–2 h) remove and allow to air-dry in a fume hood.
13. When the plates are thoroughly dried, the radioactive zones are located with a Bioscan Imaging Scanner System 200-IBM or autoradiography (*see* **Note 11**).
14. Standard compounds are localized by exposure of the plate to the anisaldehyde spray reagent *(26)* or a ninhydrin spray reagent (*see* **Note 12**).

3.5. Application of These Methods to the Analysis of Individual Proteins

To illustrate the utility of this approach for individual isoprenylated proteins, these methods are applied to the analysis of a recently isolated farnesylated protein, RDJ2 (rat DnaJ homologue 2). The cDNA clone of this DnaJ-related protein was recently identified *(27)*. The predicted amino acid sequence is found to terminate with the tetrapeptide Cys-Ala-His-Gln, which conforms to the consensus sequence for recognition by protein farnesyltransferase, and was shown to undergo farnesylation in vivo.

To perform this analysis, a means of specifically identifying the protein of interest must be available. In this example, a protein-specific immunoprecipitating antibody was used to isolate the protein from isotopically radiolabeled mammalian cells. However, other experimental approaches are available. (*See* **Note 13** for a discussion of these stategies.)

1. Mammalian cells are grown to near confluence and metabolically labeled with either [^{3}H]mevalonate, [^{3}H]F-OH, or [^{3}H]GG-OH as described in **Subheading 3.1.** If a recombinant protein is to be analyzed, the mammalian cells should be transfected either stably or transiently with the expression vector prior to labeling *(28)*.

2. The metabolically labeled cells are washed with 1–2 mL of ice-cold PBS to remove unincorporated isotopic precursor.
3. One milliliter of PBS containing 2 m*M* EDTA is added, and the cells are incubated for 5 min at room temperature. The washed cells are gently scraped from the dish with a disposable cell scraper and transferred to a 12-mL conical glass centrifuge tube.
4. The metabolically labeled cells are sedimented by centrifugation (500*g*, 5 min), and the PBS–EDTA is removed by aspiration (avoid disrupting the cell pellet).
5. The cells are resuspended in lysis buffer (2 mL), disrupted by passage through a 20-gauge needle (3–4×), and centrifuged for 15 min at 100,000*g* in a Beckman table top ultracentrifuge.
6. For immunoprecipitation of recombinant RDJ2, 20 μg of rabbit anti–RDJ2 antibody was added to the cleared supernatant and the mixture was incubated for 12 h at 4°C with gentle rocking (*see* **Note 14**).
7. Immune complexes were then precipitated by addition of 100 μL of a 50% slurry of protein A-Sepharose for 2 h at 4°C with gentle rocking.
8. Protein A beads were collected by centrifugation (1 min in a tabletop microfuge at 10,000 rpm). The pellet was washed 3× by resuspending in 1 mL of wash buffer A, 1 mL of wash buffer B, and 1 mL of wash buffer C, sequentially.
9. The protein is dissolved in 2% SDS, 5 m*M* β-mercaptoethanol for SDS-PAGE analysis (*see* **Subheading 3.3.** and **Fig. 6**). Alternatively, the protein A beads are subjected to pronase E digestion for analysis of isoprenyl-cysteine analysis by TLC (*see* **Subheading 3.4., Fig. 7**, and **Note 15**).

4. Notes

1. Anisaldehyde spray reagent contains 10 mL of anisaldehyde, 180 mL of 95% ethanol, and 10 mL of conc. sulfuric acid, added in that order.
2. Ninhydrin spray reagent contains 0.2% ninhydrin in 95% ethanol.
3. All mammalian cells tested (CHO, C6 glioma cells and green monkey kidney [CV-1] cells) utilized F-OH and GG-OH for protein isoprenylation except murine B cells before or after activation by lipopolysaccharide (LPS). Thus, it is possible that the "salvage" pathway for F-OH and GG-OH (**Fig. 1**) may not be ubiquitous in mammalian cells.
4. The pooled organic extracts can be used for analysis of lipid products metabolically labeled by $[^3H]$F-OH (*see* **ref. 14**). The organic solvent is evaporated under a stream of nitrogen, and the lipid residue redissolved in $CHCl_3$–CH_3OH (2:1) containing 20 μg each of authentic cholesterol and squalene. An aliquot is taken to determine the amount of radioactivity incorporated into the lipid extracts. The lipid products are analyzed on Merck silica gel G 60 TLC plates (Sigma) by developing with hexane–diethyl ether–acetic acid (65:35:1) or chloroform. Radioactive zones were located with a Bioscan Imaging Scanner System 200-IBM. Standard compounds are located with iodine vapor or anisaldehyde spray reagent ***(26)***.
5. Western transfer is preferred because transferring labeled proteins to nitrocellulose membrane appears to give a gain of 2–10-fold in sensitivity. One suspects that the polyacrylamide gel acts to quench the signal from radiolabeled protein. The transfer step serves to collect proteins in a single plane, and eliminates this problem. However, care should be taken with the intrepretation of these experiments. It is possible that some radiolabeled proteins, particularly those of either small (<10 kDa) or very large (>100 kDa) molecular mass, may be inefficiently transferred. The properties of the protein, percentage of acrylamide, transfer buffer components, and transfer time will each influence the transfer efficiency. Direct analysis of the gel will be less sensitive, requiring more labeled protein and longer exposure times, but material will not be lost during transfer.

6. Pronase E will completely dissolve over a period of 30 min at 37°C, and these protease preparations contain sufficient esterase activity to hydrolyze the carboxymethyl esters at the C-termini of isoprenylated proteins ***(29)***.
7. The lower aqueous phase remains cloudy, and contains precipitated proteins and peptides.
8. The addition of H_2O will significantly reduce the volume of the n-butanol phase at this step (reducing the time required for the following steps)
9. The addition of an equal volume of *n*-hexane to the *n*-butanol will speed this evaporation, by forming an azeotrope. The addition of the hexane will cause the solution to become cloudy and biphasic. The upper (butanol–hexane phase) evaporates quite rapidly, and the subsequent addition of 1 mL of 100% ethanol speeds the evaporation of the lower aqueous phase.
10. The resolution of GG-Cys and F-Cys is better in the reverse–phase chromatography system. If the normal phase system is used, the plates should be activated by heating in a 100°C oven for at least 1 h.
11. At least 5000 dpm are required for good detection using a Bioscan Imaging System, but analysis will be better (and faster) with 10,000 dpm or more. The entire sample from the digestion of 100 μg of labeled protein can usually be loaded on the TLC plates without any significant loss of chromatographic resolution.
12. Spray until plate is moist, and heat in 100°C oven until spots appear. The plates are scanned prior to spraying to avoid the reagent vapors, and because the spray reagents quench the detection of radioactivity.
13. A consideration before beginning these experiments is the abundance of the protein of interest. The analysis of a very low abundance protein will require a large number of radiolabeled tissue culture cells. Therefore, the overproduction of the protein using a mammalian expression vector may present a distinct experimental advantage. This approach also allows the cDNA to be modified to contain a unique epitope or affinity sequence at the N-terminus. In this way, proteins for which specific antibodies are not available may be studied.
14. Optimal immunoprecipitation conditions must be established for each protein-antibody complex.
15. Pronase E digestion can be carried out directly on the protein A beads without further processing. Follow directions given in **Subheading 3.4.**, scaling up the volume of the proteolysis reaction mixture to provide sufficient liquid to amply cover the protein A beads. Follow **steps 2–15** as directed. Initially, proteins should be labeled with mevalonate and both free isoprenols, to ensure correct interpretation of the results. Although a variety of proteins have been tested (*see* **Figs. 2,4**, and **7**) using [^{3}H]isoprenol labeling, the list of individual proteins is quite limited. It will be necessary to analyze a wide variety of defined isoprenylated proteins to further establish the reliability and limitations of this method.

Acknowledgments

The methods described in this chapter were developed with support from NIH grants EY11231 (D.A.A.) and GM36065 (C.J.W.).

References

1. Maltese, W. A. (1990) Posttranslational modification of proteins by isoprenoids in mammalian cells. *FASEB J.* **4,** 3319–3328.
2. Glomset, J. A., Gelb, M. H., and Farnsworth, C. C. (1990) Prenyl proteins in eukaryotic cells: a new type of membrane anchor. *TIBS* **15,** 139–142.

3. Clarke, S. (1992) Protein isoprenylation and methylation at carboxyl-terminal cysteine residues. *Annu. Rev. Biochem.* **61,** 355–386.
4. Schafer, W. R. and Rine, J. (1992) Protein prenylation: genes, enzymes, targets, and functions. *Annu. Rev. Genet.* **30,** 209–237.
5. Zhang, F. L. and Casey, P. J. (1996) Protein prenylation: molecular mechanisms and functional consequences. *Annu. Rev. Biochem.* **65,** 241–269.
6. Hancock, J. F., Magee, A. I., Childs, J. E., and Marshall, C. (1989) All ras proteins are polyisoprenylated but only some are palmitoylated. *Cell* **57,** 1167–1177.
7. Hancock, J. F., Paterson, H., and Marshall, C. J. (1990) A polybasic domain or palmitoylation is required in addition to the CAAX motif to localize $p21^{ras}$ to the plasma membrane. *Cell* **63,** 133–139.
8. Hancock, J. F., Cadwallader, K., and Marshall, C. J. (1991) Methylation and proteolysis are essential for efficient membrane binding of prenylated $p21^{K\text{-}ras(B)}$. *EMBO J.* **10,** 641–646.
9. Schafer, W. R., Kim, R., Sterne, R., Thorner, J., Kim, S.-H., and Rine, J. (1989) Genetic and pharmacological suppression of oncogenic mutations in RAS genes of yeast and humans. *Science* **245,** 379–385.
10. Kato, K., Cox, A. D., Hisaka, M. M., Graham, S. M., Buss, J. E., and Der, C. J. (1992) Isoprenoid addition to Ras protein is the critical modification for its membrane association and transforming activity. *Proc. Natl. Acad. Sci. USA* **89,** 6403–6407.
11. Seabra, M. C., Goldstein, J. L., Sudhof, and Brown, M. S. (1992) Rab geranylgeranyl transferase: a multisubunit enzyme that prenylates GTP-binding proteins terminating in cys-x-cys or cys-cys. *J. Biol. Chem.* **267,** 14,497–14,503.
12. Grunler, J., Ericsson, J., and Dallner, G. (1994) Branch-point reactions in the biosynthesis of cholesterol, dolichol, ubiquinone and prenylated proteins. *Biochim Biophys. Acta* **1212,** 259–277.
13. Crick, D. C, Waechter, C. J., and Andres, D. A. (1994) Utilization of geranylgeraniol for protein isoprenylation in C6 glial cells. Biochem. *Biophys. Res. Commun.* **205,** 955–961.
14. Crick, D. C., Andres, D. A., and Waechter, C. J. (1995) Farnesol is utilized for protein isoprenylation and the biosynthesis of cholesterol in mammalian cells. *Biochem. Biophys. Res. Commun.* **211,** 590–599.
15. Baxter, A., Fitzgerald, B. J., Hutson, J. L., McCarthy, A. D., Motteram, J. M., Ross, B. C., et al. (1992) Squalestatin 1, a potent inhibitor of squalene synthase, which lowers serum cholesterol in vivo. *J. Biol. Chem.* **267,** 11,705–11,708.
16. Bergstrom, J. D., Kurtz, M. M., Rew, D. J., Amend, A. M., Karkas, J. D., Bostedor, R. G., et al. (1993) Zaragozic acids: a family of fungal metabolites that are picomolar competitive inhibitors of squalene synthase. *Proc. Natl. Acad. Sci. USA* **90,** 80–84.
17. Hasumi, K., Tachikawa, K., Sakai, K., Murakawa, S., Yoshikawa, N., Kumizawa, S., and Endo, A. (1993) Competitive inhibition of squalene synthetase by squalestatin 1. *J. Antibiot.* (Tokyo) **46,** 689–691.
18. Thelin. A., Peterson, E., Hutson, J. L., McCarthy, A. D., Ericcson, J., and Dallner, G. (1994) Effect of squalestatin1 on the biosynthesis of the mevalonate pathway lipids. *Biochim. Biophys. Acta* **1215,** 245–249.
19. Crick, D. A., Suders, J., Kluthe, C. M., Andres, D. A., and Waechter, C. J. (1995) Selective inhibition of cholesterol biosynthesis in brain cells by squalestatin 1. *J. Neurochem.* **65,** 1365–1373.
20. Inoue, H., Korenaga, T., Sagami, H., Koyama, T., and Ogura, K. (1994) Phosphorylation of farnesol by a cell-free system from Botryococcus brauni. *Biochem. Biophys. Res. Commun.* **200,** 1036–1041.

21. Ohnuma, S.-I., Watanabe, M., and Nishino, T. (1996) Identification and characterization of geranylgeraniol kinase and geranylgeranyl phosphate kinase from the archebacterium *Sulfolobus acidocaldarius. J. Biochem.* (Tokyo) **119,** 541–547.
22. Westfall, D., Aboushadi, N., Shackelford, J. E., and Krisans, S. K. (1997) Metabolism of farnesol: phosphorylation of farnesol by rat liver microsomal and peroxisomal fractions. *Biochem. Biophys. Res. Commun.* **230,** 562–568.
23. Thai, L., Rush, J. S., Maul, J. E., Devarenne, T., Rodgers, D. L., Chappell, J., and Waechter, C. J. (1999) Farnesol is utilized for isoprenoid biosynthesis in plant cells via farnesyl pyrophosphate formed by successive monophosphorylation reactions. *Proc. Natl. Acad. Sci USA* **96,** 13,080–13,085.
24. Crick, D. C., Andres, D. A., and Waechter, C. J. (1997) Novel salvage pathway utilizing farnesol and geranylgeraniol for protein isoprenylation. *Biochem. Biophys. Res. Commun.* **237,** 483–487.
25. Kamiya, Y., Sakurai, A., Tamura, S., Takahashi, N., Tsuchiya, E., Abe, K., and Fukui, S. (1979) Structure of rhodotorucine A, a peptidyl factor, inducing mating tube formation in *Rhodosporidium toruloides. Agric. Biol. Chem.* **43,** 363–369.
26. Dunphy P. J., Kerr, J. D., Pennock, J. F., Whittle, K. J., and Feeney, J. (1967) The plurality of long chain isoprenoid (polyprenols) alcohols from natural sources. *Biochim. Biophys. Acta* **13,** 136–147.
27. Andres, D. A., Shao, H., Crick, D. C., and Finlin, B. S. (1997) Expression cloning of a novel farnesylated protein, RDJ2, encoding a DnaJ protein homologue. *Arch. Biochem. Biophys.* **346,** 113–124.
28. Sambrook, J., Fritsch, E. F., and Maniatis, T. (1989) *Molecular Cloning: A Laboratory Manual*, Cold Spring Harbor Laboratory Press, Cold Spring Harbor, NY.
29. Stimmel, J. B., Deschenes, R. J., Volker, C., Stock, J., and Clarke, S. (1990) Evidence for an S-farnesylcysteine methyl ester at the carboxyl terminus of the Saccharomyces cerevisiae RAS2 protein. *Biochemistry* **29,** 9651–9659.
30. Bansal, V. S. and Vaidya, S. (1994) Characterization of two distinct allyl pyrophosphatase activities from rat liver microsomes. *Arch. Biochem. Biophys.* **315,** 393–399.

96

2-D Phosphopeptide Mapping

Hikaru Nagahara, Robert R. Latek, Sergei A. Ezhevsky, and Steven F. Dowdy

1. Introduction

A major mechanism that cells use to regulate protein function is by phosphorylation and/or dephosphorylation of serine, threonine, and tyrosine residues. Phosphopeptide mapping of these phosphorylated residues allows investigation into the positive and negative regulatory roles these sites may play in vivo. In addition, phosphopeptide mapping can uncover the specific phosphorylated residue and, hence, kinase recognition sites, thus helping in the identification of the relevant kinase(s) and/or phosphatase(s).

Two-dimensional (2-D) phosphopeptide mapping can utilize in vivo and in vitro ^{32}P-labeled proteins ***(1–6)***. Briefly, ^{32}P-labeled proteins are purified by sodium docyl sulfate-polyacrylamide gel electrophoresis (SDS-PAGE), transferred to a nitrocellulose filter, and digested by proteases or chemicals. The phosphopeptides are then separated by electrophoresis on thin-layer cellulose (TLC) plate in the first dimension followed by thin-layer chromatography in an organic buffer in the second dimension. The TLC plate is then exposed to autoradiographic (ARG) film or phosphor-imager screen, and the positions of the ^{32}P-containing peptides are thus identified. Specific phosphopeptides can then be excised from the TLC plate and analyzed further by amino acid hydrolysis to identify the specific phosphorylated residue(s) and/or by manual amino-terminal sequencing to obtain the position of the phosphorylated residue(s) relative to the protease cleavage site ***(3)***. In addition, mixing in vivo with in vitro ^{32}P-labeled proteins can yield confirmation of the specific phosphorylated residue and the relevant kinase.

2. Materials

2.1. Equipment

1. Multiphor II horizontal electrophoresis apparatus (Pharmacia).
2. Power pack capable of 1000-V constant.
3. Refrigerated circulating water bath.
4. Thin-layer chromatography (TLC) chamber, ~30 cm L × 10 cm W × 28 cm H, and internal standard.

From: *The Protein Protocols Handbook, 2nd Edition*
Edited by: J. M. Walker © Humana Press Inc., Totowa, NJ

5. Speed Vac or lyophilizier.
6. Shaking water bath.
7. SDS-PAGE apparatus.
8. Semidry blotting apparatus (Owl Scientific).
9. Small fan.
10. Rotating wheel or apparatus.

2.2. Reagents

1. Phosphate-free tissue-culture media.
2. Phosphate-free dialyzed fetal bovine serum (FBS). Alternatively, dialyze 100 mL FBS against 4 L dialysis buffer for 12 h, and repeat two more times using 10,000 MWCO dialysis tubing. Dialysis buffer: 32 g NaCl + 0.8 g KCl + 12 g Tris in 4 L, pH to 7.4, with HCl.
3. $^{32}PO_4$-Orthophosphate: 3–5 mCi/tissue-culture dish.
4. Protein extraction buffer (ELB): 20 m*M* HEPES (pH 7.2), 250 m*M* NaCl, 1 m*M* EDTA, 1 m*M* DTT, 0.1–0.5% NP40 or Triton X-100, 1 µg/mL leupeptin, 50 µg/mL PMSF, 1 µg/mL aprotinin, and containing the following phosphatase inhibitors: 0.5 m*M* NaP_2O_7, 0.1 m*M* $NaVO_4$, 5.0 m*M* NaF.
5. Rabbit antimouse IgG.
6. Killed *Staphylococcus aureus* cells (Zyzorbin).
7. Protein A agarose.
8. 2X Sample buffer: 100 m*M* Tris-HCl (pH 6.8), 200 m*M* DTT, 4% SDS, 0.2% bromophenol blue, 20% glycerol.
9. Protein transfer buffer: 20% methanol, 0.037% SDS, 50 m*M* Tris, 40 m*M* glycine.
10. 50 m*M* NH_4HCO_3, made fresh each usage from powder.
11. 0.5% (w/v) Polyvinyl pyrolidone-360 in 100 m*M* acetic acid.
12. Sequencing grade trypsin (Boehringer Mannheim 1047 841): Resuspend 100 µg in 1.5 mL fresh 50 m*M* NH_4HCO_3, and store 10 µg/150-µL aliquots at –20°C.
13. Performic acid: Mix 1 vol hydrogen peroxide with 9 vol formic acid. Incubate for 1.5 h on ice.
14. Scintillation fluid.
15. TLC cellulose plastic-backed plates, 20 × 20 cm (Baker-flex/VWR).
16. pH 1.9 Electrophorese running buffer: 50 mL 88% formic acid, 56 mL acetic acid, and 1894 mL H_2O. Do not adjust pH.
17. Electrophoresis color marker: 5 mg/mL DNP-lysine, 1 mg/mL xylene cyanol FF in 50 µL *n*-butanol, 25 µL pyridine, 25 µL acetic acid, 1.9 mL H_2O.
18. TLC chamber buffer: 75 mL *n*-butanol, 50 mL pyridine, 15 mL glacial acetic acid, 60 mL H_2O.

3. Methods

3.1. ^{32}P-Orthophosphate Labeling

1. For in vivo ^{32}P-orthophosphate labeling of cellular proteins, preplate approx 1×10^6 cells in a 10-cm dish. Rinse adherent cells three times with 5 mL phosphate-free media. Suspension cells can be rinsed and collected by centrifugation at ~1800 rpm for 5 min at room temperature or 30°C, aspirate the media, and repeat as above. Add 3–5 mCi of ^{32}P-orthophosphate in 3.5 mL of phosphate-free media containing 10% dialyzed serum to the 10-cm dish and incubate cells at 37°C for 4–6 h (*see* **Note 1**).
2. Aspirate the ^{32}P-containing media with a plastic pipet, and transfer supernatant waste into a 50-mL conical disposable tube. Rinse the cells twice with 10 mL PBS(–) and combine with ^{32}P-media waste in a 50-mL tube, and dispose of properly.

3. Add 1 mL ice-cold extraction buffer (ELB; **ref.** *7*) and place dish on a flat bed of ice behind a shield. Tilt dish slightly every 30 s for 3–5 min to cover the cells continually. Collect cellular lysate and debris by tilting dish approx 30° using a P-1000 pipetman tip, transfer to a 1.5-mL Eppendorf tube, and mix. Alternatively, adherent cells can be released by trypsin/EDTA addition, collected, and washed twice in media containing serum to inactivate the trypsin. After the final centrifugation, add 1 mL ELB, mix by using a P-1000 tip, and transfer to an eppendorf tube. Place tube on ice for 20–25 min with occasional mild inverting (*see* **Note 2**).
4. Spin out insoluble particular matter from the cellular lysate in microfuge at 12,000, 4°C, for 10 min. Transfer supernatant to new Eppendorf tube, and preclear lysate by the addition of 50 µL killed *S. aureus* cells, cap tube, and place on rotating wheel at 4°C for 30–60 min.
5. To remove *S. aureus* cells, centrifuge lysate at 12,000, 4°C for 10 min. Transfer lysate to fresh Eppendorf tube, making sure to leave the last 50 µL or so of lysate behind with the pellet. The presence of contaminating *S. aureus* cells in this portion of the sample will reduce the amount of immunocomplexes recovered if included.

3.2. Immunoprecipitation and Transfer of ^{32}P-Labeled Protein

1. Add primary antibody to the precleared lysate supernatant, usually 100–200 µL in the case of hybridoma supernatants and 3–5 µL of commercially purified antibodies. Add 30 µL protein A agarose beads, cap the tube, and place on a wheel at 4°C for 2–4 h to overnight. If primary mouse antibody isotype is IgG1 or unknown, add 1 µL rabbit antimouse IgG to allow indirect binding of the primary antibodies to the protein A agarose *(7)*.
2. After incubation with primary antibody, perform a "1*g* spin" by placing the tube on ice for ~15 min, and aspirate the supernatant to just above the protein A agarose bed level. Stop aspirating at the top of the agarose beads to avoid drying the beads. This supernatant waste is still highly radioactive. Wash the agarose beads by addition of 1 mL ice-cold ELB, cap, invert tube several times, and centrifuge at 12,000, 4°C, 30 s. Aspirate supernatant, and repeat two to three more times. Again, avoid drying the beads.
3. After the final 30-s spin, aspirate supernatant off of the protein A beads until just dry, and add 30 µL of 2X sample buffer. Boil sample for ~5 min, centrifuge 12K for 10 s, cool tube on ice, and load immunocomplexes onto an SDS-PAGE *(8)*.
4. After running the SDS-PAGE, separate the glass plates, and trim down the gel with a razor blade by removing the stacking gel and any excess on the sides or bottom. Measure the trimmed gel size, and cut six sheets of Whatman 3MM filter paper and one sheet of nitrocellulose (NC) filter to the same size.
5. Soak two sheets of cut 3MM filter paper in transfer buffer, and place on semidry transfer unit. This can be done by filling a Tupperware-like container with transfer buffer and dipping the 3MM paper into it. Place one soaked 3MM sheet on the back of gel and rub slightly to adhere to gel. Then invert the glass plate with the 3MM gel still stuck to it and peel 3MM gel away from the glass with the use of a razor blade. Place it 3MM side down onto the soaked 3MM sheets on the semidry unit. Soak the cut NC filter with transfer buffer, and place it on top of gel followed by three presoaked 3MM sheets on top of filter. Gently squeegee out bubbles and excess buffer with the back of your little finger or by rolling a small pipet over the stack. It is important to squeegee out the bubbles, but avoid excess squeegeeing that would result in drying the stack. Mop up the excess buffer on the sides of the stack present on the transfer unit. Place the top on the transfer unit. In this configuration, the bottom plate is the cathode (negative) and the top the anode (positive).
6. Transfer the ^{32}P-labeled proteins to the NC filter at 10 V constant for 1.5–2 h. The starting current will vary from ~180 to 300 mA, depending on the surface area of the gel and will drop to around 80–140 mAmp by the end of the transfer.

3.3. Trypsinization of Protein on Nitrocellulose Filter

1. Following transfer of ^{32}P-labeled proteins to NC filter, open the transfer unit. Using a pair of filter forceps, place the NC filter, protein side down onto Saran wrap, and cover. Expose the Saran wrap-covered NC filter to ARG film for approx 1–2 h with protein side toward the film. The length of the exposure will vary with the abundance of ^{32}P-content and protein, and with the efficiencies of recovery from immunoprecipitation and transfer. Radioactive or luminescent markers are needed on the filter to determine the orientation, position, and alignment of the NC filter with respect to the ARG film. Using the markers, line up your filter on top of its ARG. This is best done on a light box. Use a razor blade to excise the slice of NC filter that corresponds to the ^{32}P-labeled protein band.
2. Place the NC filter slice into an Eppendorf tube, add ~200 µL of 0.5% PVP-360 in 100 m*M* acetic acid, cap tube, and incubate in a shaking water bath at 37°C for 30 min (*see* **Note 3**).
3. Wash the filter slice five times with 1 mL H_2O and then twice with 1 mL fresh 50 m*M* NH_4HCO_3.
4. Add 10 µg of trypsin in 150 µL of 50 m*M* NH_4HCO_3 to the NC filter slice. Incubate in a 37°C shaking water bath overnight. Following this incubation, spike the digestion of the NC filter with another 10 µg of trypsin in 150 µL of 50 m*M* NH_4HCO_3, and incubate at 37°C for an additional 4 h.
5. Vortex the tube containing the NC filter/trypsin sample for 1 min, and centrifuge at 12,000 and 4°C for 30 s. Transfer the supernatant to a new Eppendorf tube. Wash the NC filter slice by addition of 300 µL H_2O, vortex for 1 min, centrifuge at 12,000 for 30 s, and combine the supernatants.
6. Freeze the trypsinized ^{32}P-labeled peptide sample on dry ice and then completely dry in a Speed Vac (without heat). This generally takes about 4 h to complete. Prepare the performic acid for the oxidation step (**step 8**).
7. After the sample has been completely dried, add 50 µL of ice-cold performic acid, and place on ice for 1 h. Stop the reaction by addition of 400 µL H_2O to the sample, followed by freezing on dry ice and then drying in a Speed Vac.
8. Resuspend the sample in 8–10 µL of H_2O. Determine the level of radioactivity by counting 0.5 µL of the sample on a scintillation counter. Usually a total of ≥2000 cpm is sufficient for 2-D phosphopeptide mapping.

3.4. Phosphopeptide Separation: First and Second Dimensions

1. Mark the origin on the TLC plate by lightly touching a pencil to the TLC plate at the position indicated in **Fig. 1**. Apply multiple 0.5-µL aliquots of the trypsinized ^{32}P-labeled peptides to the origin to achieve ≥2000 cpm. Dry the TLC plate thoroughly between each aliquot application by use of a gentle fan. Pay particular attention to adding each subsequent aliquot to the same small area at the origin. Add 0.5 µL of the color marker 3 cm from the top edge of the TLC plate (**Fig. 1**) and then dry the TLC plate under a fan for an additional 30–60 min.
2. Prepare the Multiphor II apparatus for electrophoresis. Place the Multiphor II in a cold room, connect the cooling plate to the cooling circulator bath hoses, and precool to 5°C. Prepare and chill the pH 1.9 running buffer, and add to Multiphor II buffer tanks. Insert the electrode paddles into innermost chambers, and attach the wire connections. We have used the IEF electrodes in direct contact with the cellulose plate; however, using the paddles provided and wicking buffer onto the plate yield the best results (*see* **Figs. 1** and **2**). Place the cooling plate into the Multiphor apparatus. Add 1 L of prechilled pH 1.9 running buffer to each chamber of the Multiphor II. (These instructions are provided with the Multiphor II unit.)

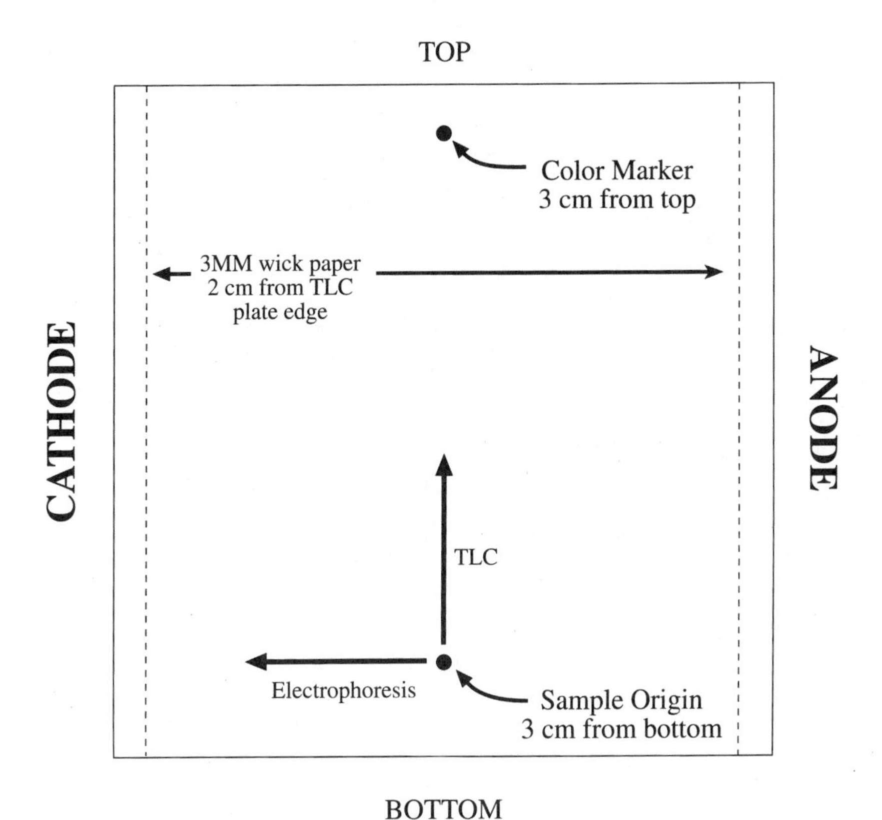

Fig. 1. The location of the origin, anode, cathode, and color dye marker dye relative to each other on the 20 × 20 cm TLC plate are depicted (*see* **Subheading 3.4., steps 1** and **2**).

3. Place the loaded TLC plate on top of the cooling plate. To dampen the TLC plate with buffer, first cut a 21 × 21-cm piece of Whattman 3MM paper, and make an approx 1-cm hole at the origin by puncturing the 3MM paper with a pencil. Soak the cut 3MM paper in pH 1.9 running buffer, blotting it between two sheets of dry 3MM paper, and then placing it over the loaded TLC plate sitting on top of the cooling plate. Slowly pipet running buffer onto the 3MM paper until the entire cellulose plate is damp beneath, avoiding excessive puddling. Remove the paper, and wick a single piece of 13 × 21 cm buffer-soaked 3MM filter paper from the buffer chamber onto the 2-cm outer edge on both sides of the plate (*see* **Fig. 2**). Be sure to fold the paper neatly over the edge of the cooling plate, and make sure that it is evenly contacting the TLC plate. Place the glass cover over the TLC plate touching/resting on the 2-cm overhang of the 3MM paper wicks. Attach the Multiphor II cover.
4. Electrophorese the peptides on the loaded TLC plate at 1000-V constant for 28–30 min. The run time may be increased up to 38 min. If further separation of ^{32}P-labeled peptides in this dimension is required, the run time may be increased up to 38 min.
5. Following the first-dimension separation by electrophoresis, remove the TLC plate from the Multiphor II apparatus, and dry for 1 h with a fan.
6. Place the TLC plate on a stand in a thin-layer chromatography chamber pre-equilibrated for 48 h in chromatography buffer. The TLC buffer should cover approx 1 cm of the bottom of the TLC plate when placed on the stand. Leave the TLC plate in the chamber

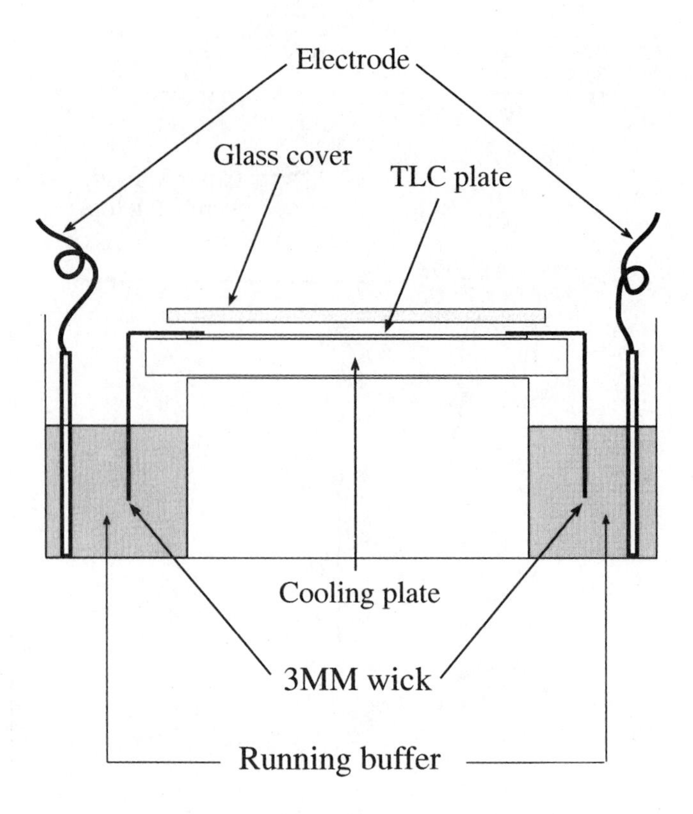

Fig. 2. A cross-sectional view of the Multiphor II apparatus. Loaded TLC plate, Whatman 3MM filter paper wicks, and glass cover plate are depicted in the running position (*see* **Subheading 3.4., steps 2** and **3**).

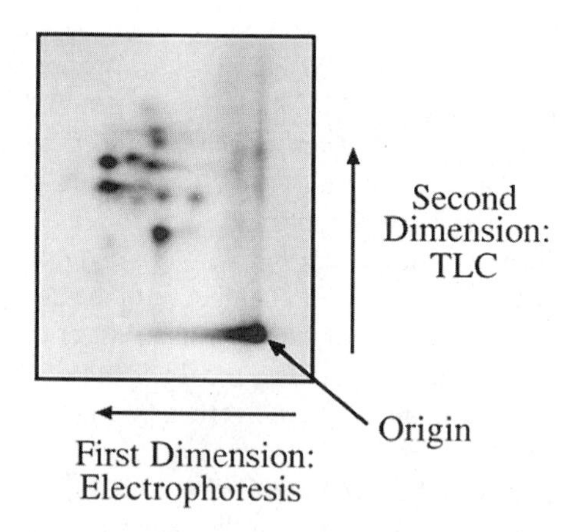

Fig. 3. Two-dimensional phosphopeptide map of the retinoblastoma tumor-suppressor gene product (pRb) labeled with ^{32}P-phosphate in vivo. pRb contains 13 cyclin-dependent kinase (cdk) phosphorylation sites, hence the complexity of the phosphopeptide map. Note the presence of several levels of ^{32}P intensity associated with specific peptides. This can arise by a number of mechanisms, including in vivo site preferences and/or accessibility of the nitrocellulose immobilized ^{32}P-labeled protein to trypsin. The origin, first-, and second-dimension runs are as indicated (*see* **Subheading 3.4., step 7**).

until the solvent line diffuses to the dye position at the top of the TLC plate, ~3 cm below the top of the plate. This usually takes 7–8 h (*see* **Notes 4** and **5**).

7. Remove the TLC plate from the chamber and dry for >1 h. Expose the TLC plate to ARG film or a phosphor-imager screen overnight, and develop. If the signal is too weak, expose for 5–7 d. The use of a phosphor-imager greatly diminishes the length of time required to obtain a 2-D phosphopeptide map (*see* **Fig. 3**).

4. Notes

1. We routinely label ~1×10^6 cells; however, depending on the abundance of the specific protein of interest and on the number of phosphorylation sites, this number may vary from 1×10^5 to 1×10^7 cells. In addition, adherent and nonadherent cells may both be labeled in suspension. The cells can be labeled with ^{32}P-orthophosphate in a 15- or 50-mL disposable conical tube or T-flask. Please note that the activity of kinase(s)/phosphatase(s) present in adherent cells may be altered when labeling in suspension. Dish size is not important as long as enough media are added to cover the bottom of the dish/flask. Attempt to achieve a final ^{32}P-orthophosphate concentration of ~1.0–1.3 mCi/mL of media.
2. We use ELB to lyse cells and generate cellular extracts; however, any extraction buffer containing Triton X-100, SDS, DOC, NP40, or similar detergent that will lyse cells will suffice. If a strong background is observed following the SDS-PAGE, transferring the immobilized protein A agarose immunocomplexes to a new microcentrifuge tube prior to the final wash and centrifugation step can result in a reduced background with minimal loss of specific signal.
3. When running multiple lanes of the same ^{32}P-labeled protein immunocomplexes on SDS-PAGE, treat each lane/NC filter slice separately. The trypsinized peptides from as many as five NC filter slices can be combined together, then frozen on dry ice, and dried.
4. Seal the top of the TLC chamber with vacuum grease, and minimize the amount of time the lid is off of the chamber. Pre-equilibrate chamber for >48 h prior to use. We routinely change the TLC chamber buffer every 8 wk. Poor separation in the second dimension is usually indicative of buffer alterations owing to evaporation and/or hydration.
5. The procedures described in this chapter can be stopped at the following steps:
   a. When the immunocomplexes are in 2X sample buffer following the immunoprecipitation.
   b. After lyophilization following trypsinization of the NC filter slices.
   c. After drying the TLC plate following the first-dimension electrophoresis.

Acknowledgments

We thank Fung Chen for protocols and Jeff Settleman for the "1g spin." S. F. D. is an Assistant Investigator of the Howard Hughes Medical Institute.

References

1. Kamp, M. P. and Sefton, B. M. (1989) Acid and base hydrolysis of phosphoproteins bound to immmobilon facilitates analysis of phosphoamino acid in gel-fractionated proteins. *Analyt. Biochem.* **176,** 22–27.
2. Lees, J. A., Buchkovich, K. J., Marshak, D. R., Anderson, C. W., and Harlow, E. (1991) The retinoblastoma protein is phosphorylated on multiple sites by human cdc2. *EMBO* **13,** 4279–4290.
3. Luo, K., Hurley, T. R, and Sefton, B. M. (1991) Cynogen bromide cleage and proteolytic peptide mapping of proteins immobilized to membranes. *Meth. Enzymol.* **201,** 149–152.
4. Desai, D., Gu, Y., and Morgan, D. O. (1992) Activation of human cyclin-dependent kinase in vitro. *Mol. Biol. Cell.* **3,** 571–582.

5. Hardie, D. G. (1993) *Protein Phosphorylation: A Practical Approach.* IRL, Oxford, UK.
6. van der Geer, P., Luo, K., Sefton, B. M., and Hunter, T. (1993) Phosphopeptide mapping and phosphoamino acid analysis on cellulose thin-layer plates, in *Protein phosphorylation—A Practical Approach.* Oxford University Press, New York.
7. Harlow, E. and Lane, D. (1988) *Antibodies: A Laboratory Manual.* Cold Spring Harbor Laboratory, Cold Spring Harbor, NY.
8. Judd, R. C. (1994) Electrophoresis of peptides. *Meth. Mol. Biol.* **32,** 49–57.

97

Detection and Characterization of Protein Mutations by Mass Spectrometry

Yoshinao Wada

1. Introduction

A significant proportion of genetic disorders are caused by point mutations in proteins. Structural studies of these mutated proteins can be carried out by genetic methods, whereas mass spectrometry (MS) is a more rapid method of mutation analysis if sufficient amounts of the gene product are obtained. It is also useful for verifying the structure of recombinant proteins.

Mutation-based variants are analyzed by comparing them with normal reference proteins, which is in contrast to more traditional methods of elucidating the primary amino acid sequence of structurally unknown proteins. In 1981, MS was first used for the analysis of a complex mixture of peptides from hemoglobin variants *(1)*. A few years later, fast-atom bombardment (FAB) replaced field desorption as an ionization method for the same purpose *(2)*. The strategy was the peptide mass mapping, aiming at both detection and characterization of mutations *(3)*. In the early 1990s, MS has joined the pool of detection methods dealing with whole protein molecules *(4)*. This was mainly brought about by the advent of electrospray ionization (ESI), which allows proteins with molecular masses of many kilodaltons that were previously considered to be well beyond the mass range of MS, to be analyzed by using multiply charged species that bring the mass-to-charge ration (*m/z*) down to within the limits of the available mass analyzers. Currently, the matrix-assisted laser desorption/ionization (MALDI) combined with a time-of-flight (TOF) mass analyzer enables the measurement of singly charged molecular ions of intact proteins. Although these detection methods are not absolutely perfect, MS seems to be effective for the proteins of molecular mass <20,000–30,000 Da.

For the characterization of protein mutations, a chemical approach consists of two steps: peptide mapping and peptide sequencing. As described, these successive steps may be replaced by a mass spectrometric analysis of a complex mixture of peptides, followed by collision-induced dissociation (CID) of the species of interest.

They are summarized as follows:

1. Procedures prior to MS (if necessary): Purification. alkylation, and desalting by reversed-phase liquid chromatography (RP-LC).

From: *The Protein Protocols Handbook, 2nd Edition*
Edited by: J. M. Walker © Humana Press Inc., Totowa, NJ

2. Detection of variants through mass measurement of the intact protein molecule with ESI- or MALDI-MS.
3. Chemical or enzymatic cleavage.
4. Identification of mutated peptides through "peptide mass mapping" of peptide mixtures using ESI-, MALDI-, or FAB-MS.
5. Characterization of the mutation.

2. Materials

2.1. Sample Preparation

1. LC column (C4, 300 Å; diameter 4.6 mm; length 150 mm) (Vydac, Hestperia, CA), or small devises packed with RP media such as ZipTip™ (Millipore, Bedford, MA) or Poros™ (PerSeptive Biosystems, Foster City, CA).
2. Trifluoroacetic acid (TFA) (Sigma).
3. Acetonitrile (high-performance liquid chromatography [HPLC]-grade) (Sigma).

2.2. Reduction and Alkylation (Carboxymethylation)

1. Stock solution A: 1 *M* Tris-HCl containing 4 m*M* EDTA, adjusted to pH 8.5 with HCl.
2. Stock solution B: 8 *M* Guanidium chloride.
3. Mix stock solutions A and B at a 1:3 ratio to make a solution of 6 *M* Guanidium chloride, 0.25 *M* Tris-HCl, and 1 m*M* EDTA prior to use.
4. Dithiothreitol (DTT, Sigma).
5. Iodoacetic acid, free acid (Sigma).

2.3. Cleavage

1. Cyanogen bromide (CNBr) (Sigma) **(toxic, handle with gloves in hood)**.
2. Trypsin (*N*-tosyl-L-phenylalanine chloromethyl ketone [TPCK]-treated) (Sigma type XIII).
3. Lysylendopeptidase (Wako, Osaka, Japan) or endoproteinase Lys-C (Boehringer).
4. Endoproteinase Glu-C (Boehringer).
5. Endoproteinase Asp-N (Boehringer).
6. 50 m*M* Ammonium hydrogen-carbonate, pH 7.8 (pH adjustment is not required).
7. 50 m*M* Ammonium carbonate, pH 8.3 (pH adjustment is not required).
8. 50 m*M* Tris-HCl, pH 7.4.

2.4. Mass Spectrometry

1. Sinapinic acid (10 mg/mL) in acetonitrile–water–TFA (30:69.9:0.1, by vol) for MALDI matrix.
2. Acetonitrile–water–acetic acid (49:49:2, by vol) or methanol–water–acetic acid (75:24.8:0.2, by vol) for ESI solvent.
3. Glycerol (Sigma)–trichloroacetic acid (Sigma) (95:5 [w/w]) for FAB matrix.

3. Methods

3.1. Procedures Prior to MS

3.1.1. Purification and Desalting by RP-LC

In general, MS requires no special procedures for sample preparation. However, removing salts (desalting) is preferred because the formation of adduct ions with alkali metals such as sodium and potassium will reduce the abundance of the protonated molecular ion species, which are the informative ions for the mass measurement of proteins and peptides. Adducts also unnecessarily complicate mass spectra. To prevent this, RP-LC is recommended as the final step of sample preparation prior to MS, and volatile buffers are preferably chosen for the enzymatic cleavage prior to peptide mass mapping. RP-LC is also a convenient method for estimating the amount of the sample prepared.

1. Separation conditions: Flow rate 1 mL/min; temperature ambient.
2. Buffer system and gradient condition: A solution of 0.1% TFA and 30% acetonitrile to a solution of 0.1% TFA in 50% acetonitrile using a linear gradient for 45 min.
3. Detection: UV absorbance at 220 nm.
4. Collection in Eppendorf tubes and recovery by lyophilization.
5. Sample loading: <1 mg for a 4.6-mm diameter column.
6. For the purposes of desalting, a pipet tip packed with RP media, ZipTip™ (Millipore) is convenient for a small amount (less than a few micrograms) of samples. (see instruction from the manufacturer.)

3.1.2. Reduction and Carboxymethylation

Cysteine residues are often subject to auto-oxidation during preparation procedures, leading to a variety of products and random formation of disulfide bridges during preparation procedures. To avoid these unexpected modifications, conversion to stable derivatives by alkylation with iodoacetic acid often performed after reduction. Such a modification destroys the higher order structure of the protein by cleaving intra- or inter-molecular bridges, thus facilitating the action of proteolytic enzymes by allowing them to access cleavage sites.

1. Dissolve the protein at 1–20 µg/mL (<500 µL in volume) in a solution containing 6 *M* guanidium chloride, 0.25 *M* Tris-HCl, and 1 m*M* EDTA, pH 8.5.
2. Add a 100-fold molar excess, over cysteine residues, of DTT directly and incubate under nitrogen gas at room temperature for 2 h.
3. Add the same amount (in moles) of iodoacetic acid (*X* mg) directly to the solution, quickly add *X* µL of 10 *N* NaOH to maintain the pH, and then incubate at room temperature for 30 min in the dark.
4. Desalt by RP-LC.

3.2. Detection of Variants Through Mass Measurement of the Intact Protein Molecule with MS

Among the naturally occurring amino acids, Leu and Ile have the same elemental composition, and thus exactly the same mass. Lys and Gln are different in composition, but have the same nominal mass. All other amino acids differ from each other in their molecular mass by ≤129 Da. Variants having substitutions of these amino acids are different in molecular mass from the corresponding normal proteins and are thus, theoretically, detectable by measuring the molecular mass of the intact protein by MS.

For ESI-MS, the sample at a concentration of 1–10 µ*M* in the ESI solvent is introduced into the ion source. Any sample remaining in the delivery tube after analysis is recovered and lyophilized for further analysis, if necessary. For MALDI TOF MS, the sample solution is mixed with the same volume of a matrix solution on the MALDI sample plate and then dried. **Figure 1** shows the MALDI-TOF mass spectrum of the globins from a patient with a mutant hemoglobin. Two different β-globin species are

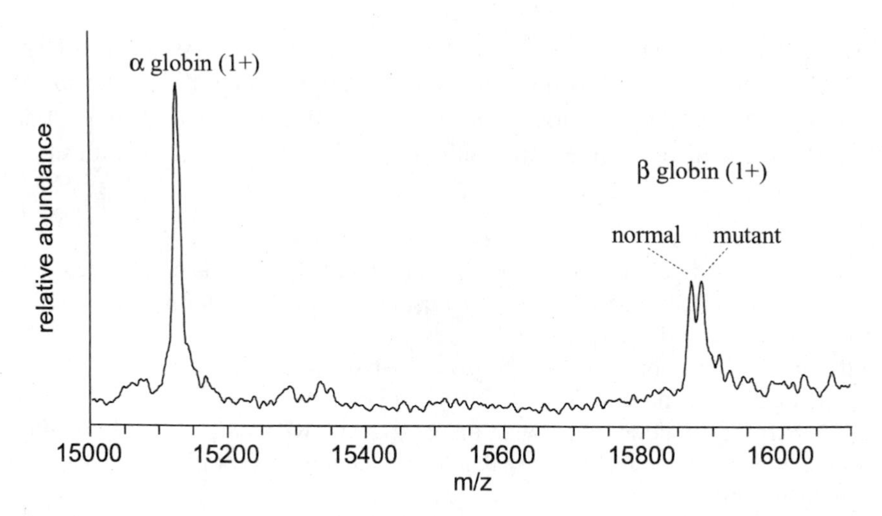

Fig. 1. Detection of a variant protein by MALDI-TOF-MS. The mass separation is determined to be 14 Da.

clearly differentiated in the mass spectrum taken at a resolution of 3000, and the mass separation was determined from this analysis to be 14 Da (*see* **Note 1**).

The smallest difference in molecular masses detectable by MS taken at a modest resolution of 2000–5000 increases with the increasing size of the proteins. A theoretical analysis can be considered using two model proteins: β-globin (M_r 15,867) and transferrin (M_r 79,556) (**Fig. 2**). A separation of 14 Da in transferrin is not clear (C in **Fig. 2B**), whereas the shapes of the clusters for 20 and 31 Da differences, corresponding to D and E, respectively, in **Fig. 2B**, are similar to those for 9 and 14 Da differences, corresponding to C and D, respectively, of the β-globin in **Fig. 2A**. The minimum difference in molecular mass detectable as pattern D in **Fig. 2B** for an equimolar mixture with normal proteins of $M_r > 100,000$ is about 20 Da. This means that about half of the possible amino acid substitutions in a protein of this size will escape detection. In those cases, a strategy to detect these substitutions preferably starts with the analysis of CNBr-cleaved peptides (*see* **Subheading 3.3.**).

3.3. Chemical or Enzymatic Cleavage

3.3.1. Cleavage with CNBr Followed by MS

CNBr cleaves only at a relatively rare amino acid, methionine (Met), in proteins. Therefore, it will generate large peptide fragments compared with enzymatic cleavage, but their molecular size will be <20,000–30,000 Da, which is well within the range that can be measured to the nearest integer mass. Most of the resulting C-terminal homoserine residues of the peptides from this type of cleavage are in lactone form. Cleavage by CNBr helps to confirm a suspected site of mutation to a small part of the protein molecule and for determining the size of the molecular shift precisely. The choice of ionization method in this instance is ESI or MALDI.

1. Dissolve the protein at 1 mg/mL in 70% (v/v) TFA (*see* **Note 2**).
2. Add a 100-fold molar excess, over Met residues, of CNBr dissolved in 70% TFA solution, and incubate in the dark under nitrogen at room temperature for 2 h.

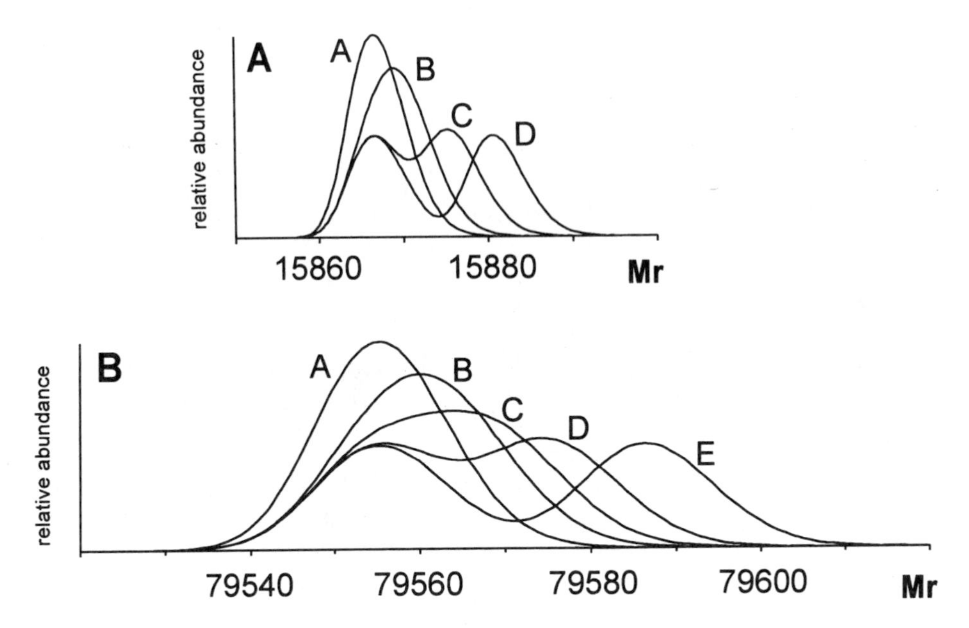

Fig. 2. Theoretical distributions of isotopic clusters corresponding to an equimolar mixture of **(A)** normal and variant β-globins. A, Pure normal protein; B, normal and +4 Da mutant; C, normal and +9 mutant; D, normal and +14 mutant. **(B)** Normal and variant transferrin. A, Pure normal protein; B, normal and +9 Da mutant; C, normal and +14 mutant; D, normal and +20 mutant; E, normal and +31 mutant. The curves have been drawn by connecting the peak top of each isotopic ion from a display at sufficient resolution to achieve complete separation of ^{13}C isotopic peaks.

3. Add 10 vol of water, rotary evaporate to remove excess CNBr, freeze, and lyophilize. Alternatively, the cleavage mixture is directly subjected to RP-LC similar to the desalting procedure described in **Subheading 3.1.1.**
4. ESI- or MALDI-MS, as described in **Subheading 3.2.**

In the example shown in **Fig. 3**, a mutation causing a difference of 14 Da is identified in peptide CB2 (M_r 9821.4 for the normal species) of β-globin. The separation of the normal and the variant species is obviously better than that of the intact proteins shown in **Fig. 1**.

3.3.2. Enzymatic Digestion and MS of Peptide Mixtures

The following enzymes are amino acid specific in their cleavage, and are employed for peptide mass mapping: trypsin (TPCK-trypsin), endoproteinase Lys-C (lysylendopeptidase), endoproteinase Glu-C, and endoproteinase Asp-N.

1. Dissolve protein in the digestion buffer at about 1 mg/mL.
2. Optimized buffers: For trypsin and endoproteinase Glu-C, 50 m*M* ammonium hydrogencarbonate, pH 7.8; for endoproteinase Lys-C, 50 m*M* ammonium carbonate, pH 8.3; for endoproteinase Asp-N, 50 m*M* Tris-HCl, pH 7.4 (*see* **Note 3**).
3. Add protease at 1% (w/w) ratio and digest at 37°C for 6 h.
4. Lyophilize.
5. ESI- or MALDI-MS, as described in **Subheading 3.2.** The matrix for the MALDI of peptides is α-cyano-4-hydroxycinnamic acid (10 mg/mL) dissolved in acetonitrile–water–

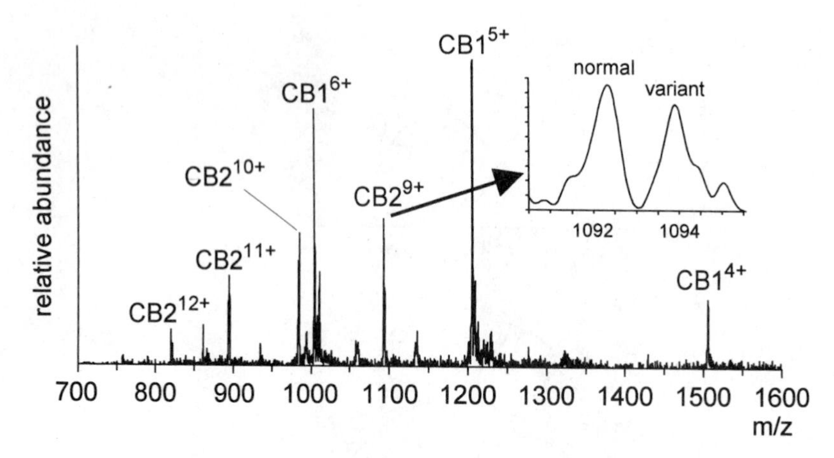

Fig. 3. ESI mass spectrum of CNBr peptides of the β-globin isolated from the same patient shown in **Fig. 1**. The region of the nine-charged ion for peptide CB2 is depicted in an *inset*. A Cys-containing peptide CB2 is detected at the expected mass without alkylation. The resolution is 2000.

TFA (50:49.9:0.1, by vol). For, FAB-MS, approx 100 μg of lyophilized digest are mixed with 50 μL of the matrix, and 2 μL of the resultant mixture are placed on the FAB probe.

3.4. Peptide Mapping by MS

The peptide mass maps of a tryptic digest of β-globin by ESI-, MALDI-TOF-, and FAB-MS are shown in **Fig. 4**. Most of the sequence of this protein is covered by the mass spectrum. Low-mass peptides comprised of a few amino acid residues are difficult to identify, but this may be overcome by using multiple analyses with different enzymes. ESI-MS of peptides can be directly forwarded to the succeeding CID for sequence analysis, although the mass spectrum is complicated due to a number of multiply charged ions from the component peptides (**Fig. 4A**). LC-MS can be used with ESI, facilitating detection of component peptides with good signal-to-noise ratio. MALDI-TOF produces a lucid mass spectrum that is comprised of singly charged ions, but the relative intensity of ions is not correlative with the abundance of the corresponding peptides (**Fig. 4B**). FAB ionization presents a simple mass spectrum as MALDI but is less sensitive than other ionizations (**Fig. 4C**).

3.5. Characterization of the Mutation

Categorized strategies are described here using examples of variant hemoglobins (*see* **Note 4**).

3.5.1. Simple Cases: Mutation at the Cleavage Site of a Specific Enzyme

1. New cleavage site: The example shown in **Fig. 5** is a hemoglobin variant with a mutation in α-globin. Tryptic peptide 6 (T6), which should be detected at *m/z* 1833.9, is missing, and a new peptide appears at *m/z* 1634.8 in the mass spectrum. The substitution of Arg for Gln at position 54 is the only possible substitution that fits the molecular mass change. The substitution can be verified by a digestion with endoproteinase Lys-C, which does not cleave arginyl peptide bonds; the peptide ion is shifted from the normal *m/z* 1833.9 to *m/z* 1862.0, with the increase of 28 Da confirming the Gln→Arg mutation.

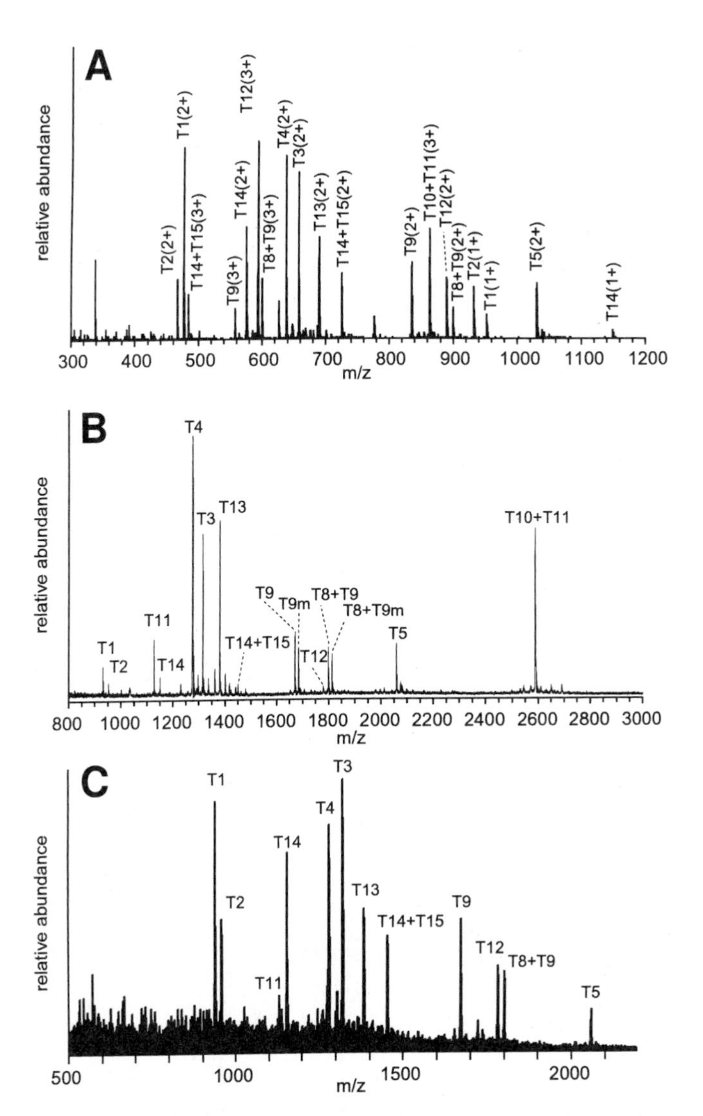

Fig. 4. Mass spectra of the tryptic digest of carboxymethylated β-globin acquired by different ionizations: **(A)** ESI, **(B)** MALDI, and **(C)** FAB. The number above each peak is that of the predicted tryptic peptide starting from the N-terminus. In the MALDI-TOF mass spectrum, mutated peptides, T9m and T8+T9m, are detected beside the corresponding normal peptides.

2. Missing cleavage site: In this example, a β-globin variant (**Fig. 6**), tryptic peptides 13 (T13) and 14 (T14) are missing in the digest, indicating a mutation at the Lys residue at position 132. The new peptide is observed at *m/z* 2495.3, indicating the replacement of Lys by Asn.
3. Unique replacement: In the case shown in **Fig. 7**, tryptic peptide 11 (T11) at *m/z* 1126.5 is missing and a new peptide appears at *m/z* 1152.5 in the digest of purified β-globin of the variant hemoglobin. This increase of 26 Da suggests five possible mutations; His→Tyr, Ala→Pro, Cys→Glu, Ser→Leu, and Ser→Ile. Because the normal peptide has His, but not Ala, Cys, or Ser residues, and has a single His residue at position 97 in the sequence, it follows that the latter has been replaced by Tyr.

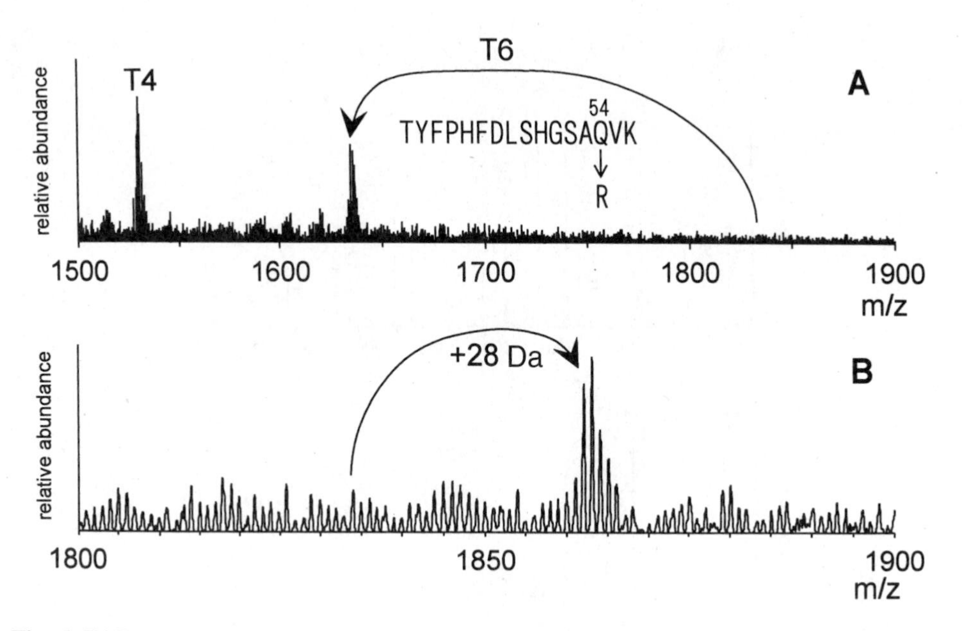

Fig. 5. FAB mass spectra of **(A)** tryptic and **(B)** endoproteinase Lys-C digests of an α-globin variant.

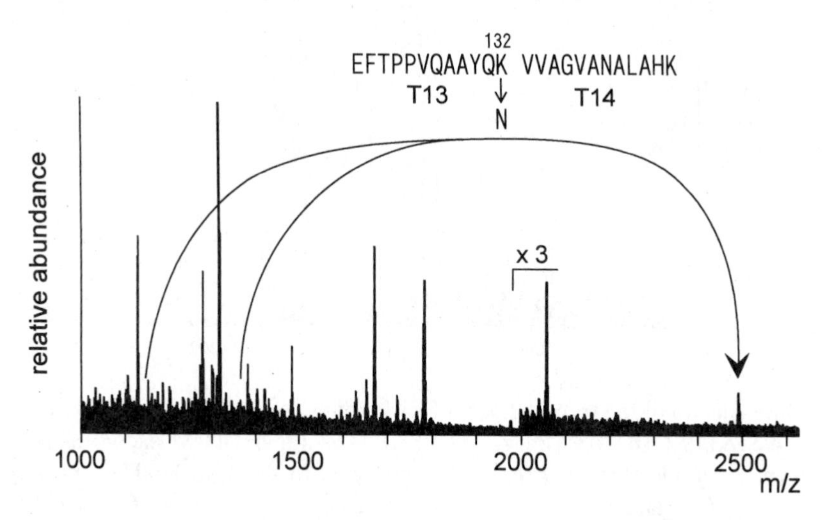

Fig. 6. FAB mass spectrum of the tryptic digest of a β-globin variant.

3.5.2. Complex Cases

1. Characterization based on additional information: In cases in which the molecular weight of the mutated peptide does not permit the assignment of a unique type and/or position of substitution, chemical data, such as the charge carried on the variant, may provide an answer. In the mass spectrum of a tryptic digest of an α-globin shown in **Fig. 8**, a combined peptide T10+T11, which is normally present, at *m/z* 1087.6 owing to incomplete cleavage, is missing and a new peptide is observed at *m/z* 1059.6. The decrease of 28 Da suggests six possible mutations: Arg→Lys, Arg→Gln, Val→Ala, Asp→Ser, Glu→Thr, and Met→Cys. The first four substitutions are possible on the basis of the normal sequence

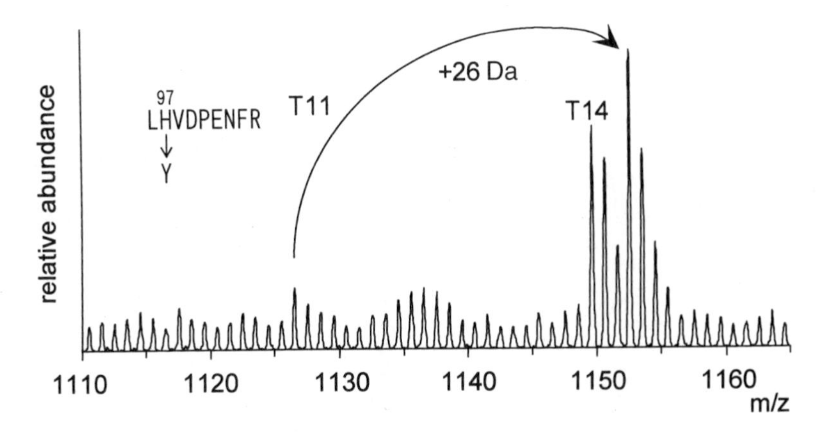

Fig. 7. FAB mass spectrum of the tryptic digest of a β-globin variant. The peak at *m/z* 1149.6 is the normal peptide T14.

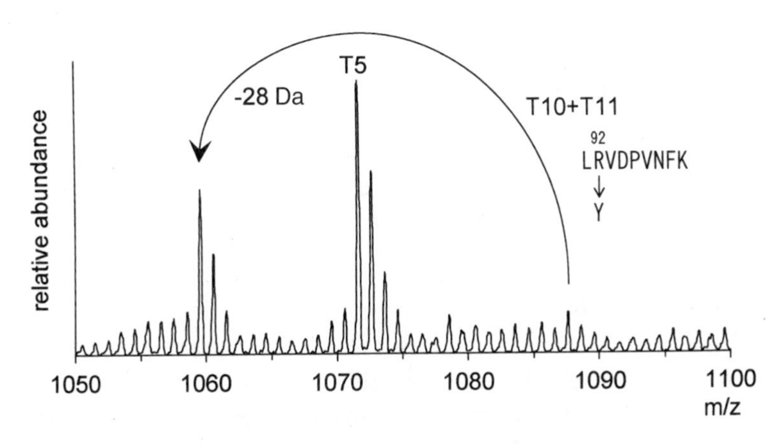

Fig. 8. FAB mass spectrum of the tryptic digest of an α-globin variant. Peptide T10+T11 is normally identified at *m/z* 1087.6 owing to incomplete cleavage of arginyl bond at position 92. In this case, the peptide is missing and a new peak appears at *m/z* 1059.6. The peak at *m/z* 1071.6 is the normal peptide T5.

of the peptide. The variant was determined to be acidic, relative to the normal globin, on electrophoresis under denaturing conditions, and therefore only the Arg→Gln mutation is plausible (*see* **Note 5**).

2. Combination of different enzymes: In the case shown in **Fig. 9**, the structure could not be determined directly from the mass spectrum of the peptide mixture. The mass spectrum of a tryptic digest of β-globin from the patient discloses new peaks near the normal peptide 9 (T9), suggesting that the mutation must be in T9 (**Fig. 9**). The difference of 14 Da suggests seven possible substitutions: Gly→Ala, Ser→Thr, Val→Leu, Val→lle, Asn→Gln, Asp→Glu, Asn→Lys, and Thr→Asp. However, even on the basis of both the normal sequence and the electrophoretic data, six different substitutions at seven possible positions are left as candidates. In this instance, the substitution of Glu for Asp at position 79 is determined, when the tryptic digest is then hydrolyzed with endoproteinase Glu-C. Otherwise, the mutation can be characterized by CID.

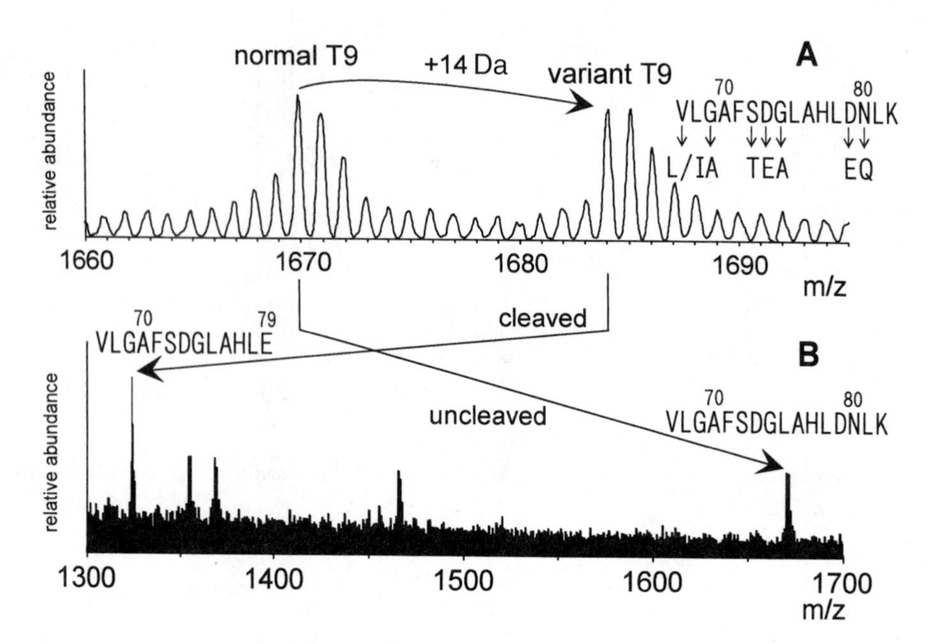

Fig. 9. FAB mass spectra of **(A)** tryptic and **(B)** endoproteinase Glu-C digests of a β-globin variant. The protonated molecular ions at *m/z* 1669.9 and at *m/z* 1683.9 are derived from the normal and variant proteins, respectively. The increase of 14 Da for this peptide gives several possible types of mutations even with supporting chemical data. The substitution Asp79→Glu is determined by endoproteinase Glu-C digestion.

3.5.3. CID or Tandem Mass Spectrometry

The CID technique is described in Chapter 98. It is easy to read or interpret the CID mass spectrum for characterization of mutations, because the product ions from the mutated peptide can be compared with those from the normal one (**Fig. 10**). It can be applied to a peptide mixture, in which the ion for the mutated peptide is selected as a precursor.

The CID does not always produce any key fragments in good quantity sufficient to define the mutation. These cases are due either to the mutation present in large peptides or to inefficient dissociation at the mutated residues. The use of different kinds of enzymes possibly solves the former problem, by cutting the mutated peptide into an appropriate size for CID analysis. Alternatively, high-energy collision may produce key fragments *(5)*.

The CID analysis of intact proteins without digestion is an issue of challenge, while the information is still confined to a part of the molecule *(6)*.

4. Notes

1. The theoretical isotopic distributions of human β-globin (146 residues, M_r 15,867) at resolutions of 15,000 **(A)** and 2,000 **(B)** are shown in **Fig. 11**. In actual measurements of proteins of molecular mass over several kilodaltons, it is usually impossible to obtain the high ion current necessary for unit resolution as **(A)**, nor is it possible to identify the ^{12}C species corresponding to the exact mass in the molecular ion cluster. Consequently, the average mass of the unresolved isotopic cluster or "chemical mass" is usually used.

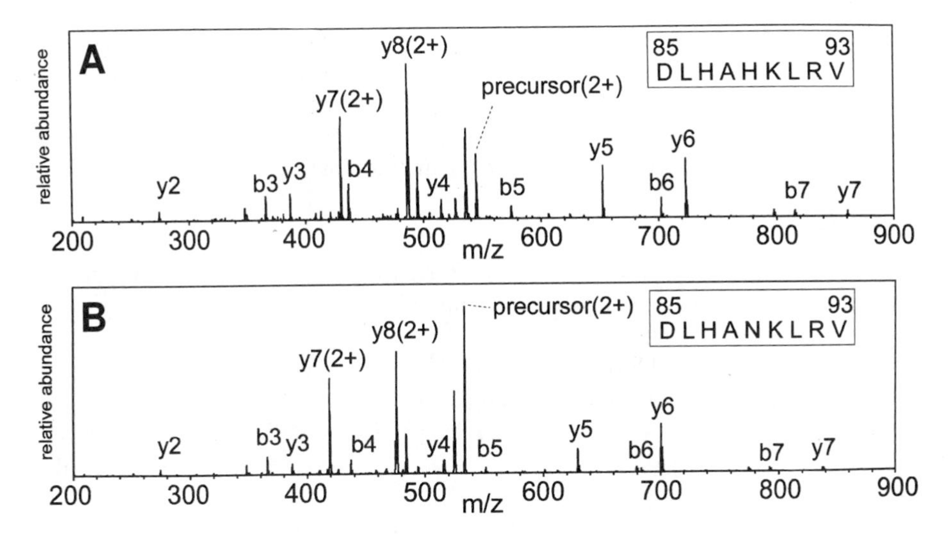

Fig. 10. CID mass spectra taken for the structural characterization of a mutated peptide by ESI ion trap MS. **(A)** As a reference, the $[M+2H]^{2+}$ ion for the normal peptide at *m/z* 544.8 was selected as the precursor. **(B)** The $[M+2H]^{2+}$ ion for the mutated peptide at *m/z* 533.3 was selected as the precursor. Note the difference in the *m/z* values for the b5 or the y5 product ion between the two spectra, indicating the mutation at position 89.

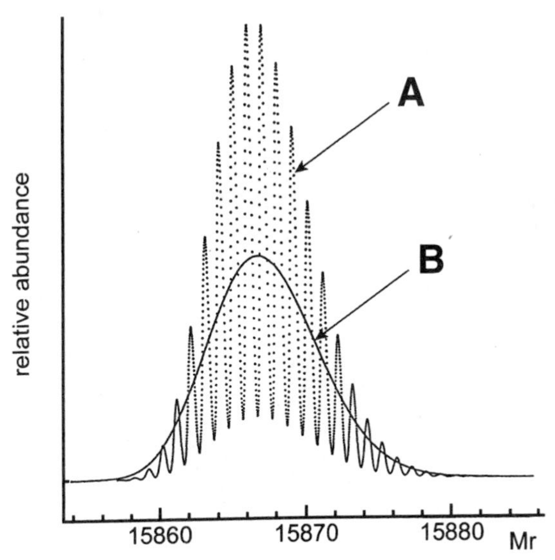

Fig. 11. Theoretical isotopic distributions of β-globin at resolutions of 15,000 **(A)** and 2000 **(B)** (10% valley definition).

2. Cleavage in formic acid is not recommended, because it has a tendency to formylate amino and hydroxyl groups, which creates a mixture consisting of products differing in molecular mass by multiples of 28 Da.
3. Endoproteinase Lys-C is active in urea solution at a concentration up to 4 *M*. The digestion is often carried out in a solution of 4 *M* urea, 0.1 *M* Tris-HCl, pH 9.0, in which substrates are denatured and become susceptible to the cleavage.

4. Protein variants are heterozygous in most cases; namely both normal and mutated proteins are present in the patients as shown in **Fig. 1**. In this chapter, variant proteins, isolated from their normal counterparts, are presented, except for the cases shown in **Figs. 1**, **3**, and **9**.
5. When mutations are suggested by MS of intact proteins, the sequence analysis of the corresponding genes is another choice of the method for characterization *(7)*.

References

1. Wada, Y., Hayashi, A., Fujita, T., Matsuo, T., Katakuse, I., and Matsuda, H. (1981) Structural analysis of human hemoglobin variants with field desorption mass spectrometry. *Biochim. Biophys. Acta* **667,** 233–241.
2. Wada, Y., Matsuo, T., and Sakurai, T. (1989) Structure elucidation of hemoglobin variants and other proteins by digit-printing method. *Mass Spectrom. Rev.* **8,** 379–434.
3. Wada, Y. and Matsuo, T. (1994) Structure determination of aberrant proteins, in *Biological Mass Spectrometry* (Matsuo, T., Caprioli, R. M., Gross, M. L., and Seyama. Y., eds.), John Wiley & Sons, Chichester. UK, pp. 369–399.
4. Green, B. N., Oliver, R. W. A., Falick, A. M., Shackleton, C. H. L., Roitmann, E., and Witkowska, H. E. (1990) Electrospray MS, LSIMS, and MS/MS for the rapid detection and characterization of variant hemoglobins, in *Biological Mass Spectrometry* (Burlingame, A. L. and McCleskey, J. A., eds.) Elsevier, Amsterdam, pp. 129–146.
5. Wada, Y., Matsuo, T., Papayannopoulos, I. A., Costello, C., and Biemann, K. (1992) Fast atom bombardment and tandem mass spectrometry for the characterization of hemoglobin variants including a new variant. *Int. J. Mass Spectrom. Ion Processes* **122,** 219–229.
6. Light-Wahl, K. J., Loo, J. A., Edmonds, C. G., Smith, R. D., Witkowska, H. E., Shackleton, C. H., and Wu, C. S. (1993) Collisionally activated dissociation and tandem mass spectrometry of intact hemoglobin β-chain variant proteins with electrospray ionization. *Biol. Mass Spectrom.* **22,** 112–120.
7. Kaneko, R., Wada, Y., Hisada, M., Naoki, H., and Matsuo, T. (1999) Establishment of a combined strategy of genetic and mass spectrometric analyses for characterizing hemoglobin mutations. An example of Hb Hoshida (β43Glu→Gln). *J. Chromatogr. B Biomed. Sci. Appl.* **731,** 125–130.

98

Peptide Sequencing by Nanoelectrospray Tandem Mass Spectrometry

Ole Nørregaard Jensen and Matthias Wilm

1. Introduction

Electrospray ionization mass spectrometry (ESI-MS) *(1)* has had a profound influence on biological research over the last decade. With this technique it is possible to generate and characterize gas-phase analyte ions from aqueous solutions of proteins, peptides, and other classes of biomolecules. ESI is performed at atmospheric pressure, which simplifies sample preparation and handling and allows on-line coupling of chromatography, such as capillary high-performance liquid chromatography (HPLC), to mass spectrometers (LC-MS). The use of ESI in combination with tandem mass spectrometry (MS/MS) provides the capability for amino acid sequencing of peptides. The optimization and miniaturization of peptide sample preparation methods for ESI as well as the development of highly sensitive tandem mass spectrometers, such as triple quadrupoles, ion traps, and quadrupole-time-of-flight (Q-TOF) hybrid instruments, makes it possible and almost routine to obtain amino acid sequences from subpicomole levels of protein in many laboratories. Peptide sequencing is typically performed by nanoelectrospray MS/MS analysis of crude, concentrated peptide mixtures or by hyphenated techniques, such as capillary HPLC coupled to micro/nanoelectrospray-MS/MS. The sets of peptide tandem mass spectra generated in such experiments are used to query biological sequence databases with the aim to identify all protein components present in the sample. In this chapter we describe practical aspects of nanoelectrospray mass spectrometry aimed at amino acid sequencing of peptides at subpicomole levels. We do not consider LC-MS/MS, although many of the features of the two analytical approaches are very similar.

1.1. The Nanoelectrospray Ion Source

The main characteristics of the nanoelectrospray ion source are low flow rate, high ionization efficiency, and extended measurement time with concomitant improvements in absolute sensitivity *(2,3)*. These features are crucial for peptide sequencing by tandem mass spectrometry. Nanoelectrospray tandem mass spectrometry is a reliable and robust technique for identification or sequencing of gel-isolated proteins available in sub-picomole amounts *(4,5)*. The very low flow rate of 10–25 nL/min of the

From: *The Protein Protocols Handbook, 2nd Edition*
Edited by: J. M. Walker © Humana Press Inc., Totowa, NJ

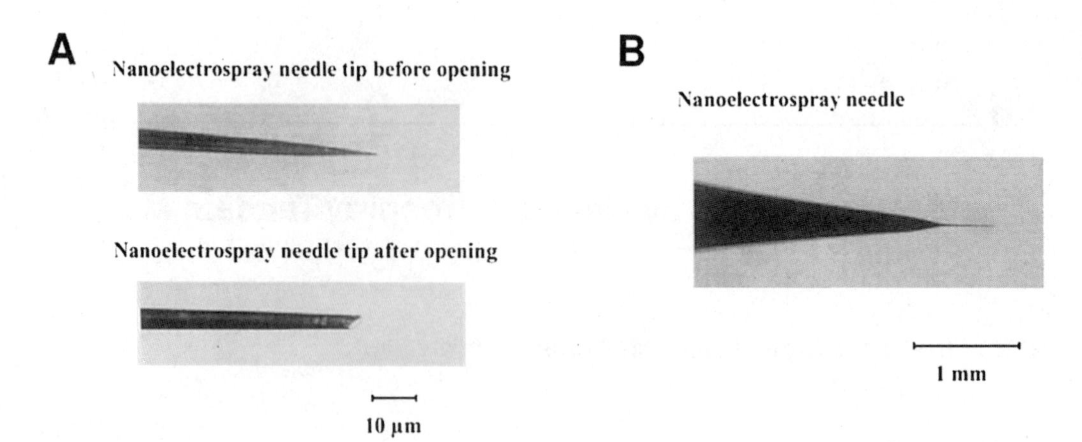

Fig. 1. Nanoelectrospray needles. Pulled glass capillary needle tip before **(A)** and after **(B)** opening. The needle tip should be 200–500 μm long with an opening of only 1–2 μm. **(C)** A two-stage heat/pull cycle on a capillary puller produces two identical glass capillary needles with a thin, tapered needle tip.

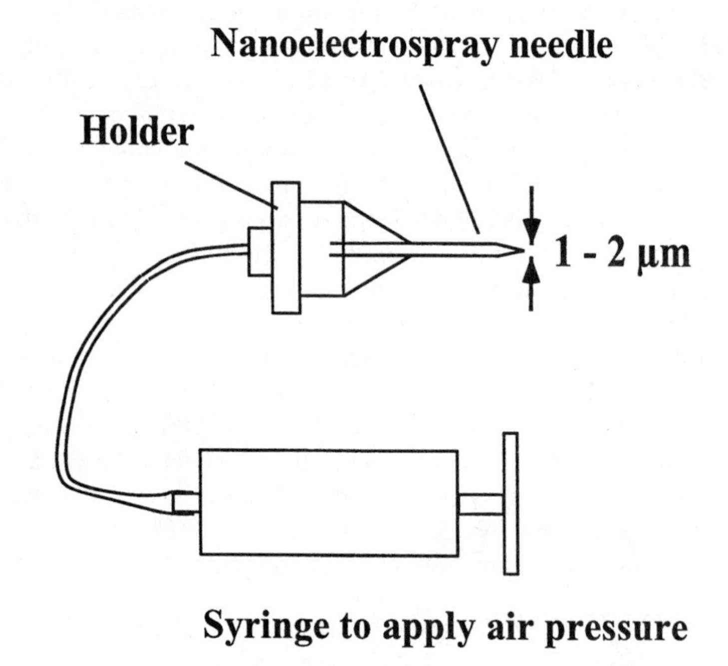

Fig. 2. Nanoelectrospray ion source. The nanoelectrospray needle is mounted in a metal holder that is connected to the ion source power supply and to a syringe that provides the air back pressure. The needle assembly is mounted onto a *x-y-z* manipulator and positioned on axis and 1.5–2 mm from the orifice of the mass spectrometer.

nanoelectrospray source has required a change of sample preparation method from on-line HPLC separation to sequencing directly from desalted/concentrated peptide mixtures as outlined.

For nanoelectrospray tandem mass spectrometry and other mass spectrometric methods it is important to choose the most suitable protease for generation of peptides from a protein. Trypsin has several advantages for the generation of peptides for tandem mass spectrometric sequencing. It is an aggressive and extremely specific protease that cleaves the carboxy-(C)-terminal amide bond of lysine and arginine residues. Tryptic peptides typically have a mass between 800 Da and 2500 Da and both the amino-(N)-terminal amino group and the C-terminal residue (Lys or Arg) are basic so they generate mainly doubly charged peptide ions in electrospray. Such ions predominantly fragment at the peptide amide bond to generate singly charged N-terminal (b-type) or C-terminal (y-type) ions *(6,7)*. Tryptic peptides rarely contain internal arginines that generate complicated fragmentation patterns. The in-gel digestion procedure *(5,8)* is compatible with matrix-assisted laser desorption/ionization mass spectrometry (MALDI) MS as well as nanoelectrospray mass spectrometry. In the latter case a sample desalting/concentration step is necessary (**Protocol 2** or **3**)

1.2. Analytical Characteristics of the Nanoelectrospray Ion Source

The flow rate of a nanoelectrospray source is about 25 nL/min. At this flow rate an analyte concentration of 1 pmol/µL results in one analyte molecule per droplet on the average *(2)*. Furthermore, 1 µL of sample is consumed in 40 min extending the time available for optimization of experiments. The overall sensitivity is limited by the signal-to-noise level and it is therefore a function of the ionization efficiency, desolvation efficiency, ion transmission, the level of chemical background ions, and detector characteristics.

The electrospray generated with the nanoelectrospray ion source is very stable. This allows purely aqueous solutions to be sprayed even in negative ion mode without nebulizer assistance. The source can be operated with solutions containing up to 1 *M* NaCl *(3)*, although this is not recommended. The high stability allows optimization of experimental conditions based on analyte characteristics rather than electrospray requirements, that is when choosing buffer composition. For example, preservation of noncovalent complexes often does not allow addition of organic solvent to the sample to facilitate spraying. Because the ion source exclusively produces very small droplets, relatively soft desolvation conditions in the interface region of the mass spectrometer can be chosen.

The nanoelectrospray source consists of a metal-coated glass capillary whose tip is pulled into a needle shape with an outer diameter of about 2 µm and an orifice diameter of 1 µm (**Fig. 1**). The glass capillary is mounted in a gas-tight holder that can be pressurized by air up to about 1 bar (**Fig. 2**). The needle assembly is connected to the ion source power supply. Holder and needle are electrically connected, for example, by a small droplet of conductive carbon cement (Neubauer Chemikalien, Münster, Germany). The metal coating of the glass capillary needle ensures that the electrical potential is transferred to the liquid sample at the needle tip.

The nanoelectrospray needles are made from borosilicate glass capillaries (GC 120 F–10, Clark Electromedical Instruments, Pangbourne, UK). The desired needle shape is obtained by using a two-stage pulling cycle on a microcapillary puller (Model P-97, Sutter Instruments Co., USA; parameter set [heat, pull, velocity, time]: 1—520, 100,

10, 200; 2–490, 160, 12, 165. The first heating/pulling stage reduces the diameter of the capillary to about 0.5 mm, while the second stage pulls the glass capillary apart, producing two nanoelectrospray needles. These needles should have an opening of 1–2 µm. However, after pulling the opening diameter can be less than 100 nm and has to be widened (*see* **Methods**). Nanolectrospray needles and ion sources are commercially available from New Objective (Boston, MA) and MDS Proteomics (Odense, Denmark) as well as from several MS instrument manufacturers.

Liquid injected into the needle is drawn to the tip by capillary force. To reduce the flow resistance for a stable flow rate in the 10–25 nL/min range the narrow part of the tip should not be longer than 500 µm (**Fig. 1**). Needles with very short constrictions (50–100 µm) can be operated easily but with a higher risk of losing sample due to a higher flow rate. Short needles are preferred for rapid mass measurements when abundant sample is available, for example, recombinant proteins, synthetic peptides, or oligonucleotides. Longer tips (200–500 µm) are used for tandem mass spectrometry experiments when the longest possible operation time is desirable and when the sample load volume will not exceed 1 µL. A major advantage of these types of nanelectrospray needles is that they do not easily block due to the relatively short length of the needle tip.

Metal coating of the glass capillaries is achieved by metal (gold) vapor deposition in a sputter chamber (Polaron SC 7610 sputter coater, Fisons Instruments, East Sussex, UK). The needles are used only once so it is not a problem that the coating is not tightly fixed to the glass and can be rubbed off. Methods to produce a more stable metal coating include pretreatment with (3-mercaptopropyl)trimethoxysilane *(9)* or protecting the metal layer by a second layer of SiO_x *(10)*. A stable gold coating is necessary when a glass needle is used for several samples over a prolonged time. As the needle tip is fragile it should be handled carefully when loading the sample and when mounting it in the holder.

2. Methods

2.1. Protocol 1. Operation of the Nanoelectrospray Ion Source

Nanoelectrospray ion sources are now optional on a number of electrospray mass spectrometers or they can be custom made for individual instruments. Note that electrospray ion sources are operated at high voltages and care should be taken to avoid electrical shock.

1. *The geometrical location of the ion source.*
   The nanoelectrospray source is mounted directly in front of the orifice of the mass spectrometer, usually at a distance of 1.5–2 mm from the orifice. Electrospray at a low flow rate generates very small droplets with a diameter of 200 nm or less. Desolvation is therefore achieved in a very short time and distance.
2. *The voltage applied to the source.*
   To initiate the electrospray a minimal electrical field strength at the surface of the liquid has to be reached *(11)*. Conventional electrospray ion sources are operated with a 3 –5 kV potential difference between the needle and counterelectrode (i.e., the orifice plate of the mass spectrometer). The very small tip diameter of the nanoelectrospray needle allows a spray cone to be established at a much lower electrical potential, typically 500–900 V.
3. *The desolvation conditions in the interface region.*
   Because the charged droplets generated by the nanoelectrospray ion source are very small, softer desolvation conditions in the interface (skimmer) region of the mass spectrometer

are used. The electrical gradient and the countercurrent gas flow can be reduced. This appears to lead to the generation of colder molecular ions as compared to conventional electrospray sources ,facilitating, for example, studies of noncovalent molecular interactions.

The main objective when operating the nanoelectrospray source is to achieve a low and stable flow rate despite sample-to-sample variations. This is achieved by applying air pressure to the needle which helps to adjust the flow rate and thereby compensates for differences in needle orifice size and sample viscosity.

A step-by-step guide for installing the nanoelectrospray source is given in the following list. The order of events may not be compatible with all nanoelectrospray ion sources:

1. Mount the pulled and metal coated nanoelectrospray needle in the holder. Connect the holder to the ion source power supply.
2. Make electrical contact between the needle and the holder, for example, by applying a droplet of graphite paste. (Alternatively, the power supply can be connected directly to the needle with a clamp.)
3. Mount the ion source on an x-y-z manipulator in front of the mass spectrometer.
4. Inject the sample. Dissolve proteins and peptides in 5% formic acid in 20–50% methanol and inject 0.5–2 µL into the nanoelectrospray needle using a micropipet with a gel loader tip.
5. Connect the needle holder to a 20-mL syringe or pressurized gas which provides the backing air pressure for the nanoelectrospray ion source.
6. Position the nanoelectrospray needle 1.5–2 mm from the orifice of the mass spectrometer. Monitor the position of the needle tip with a microscope or a video camera.
7. Gently pressurize the needle by air using the 20 mL syringe or an adjustable gas valve.
8. If no liquid appears at the needle tip then briefly and gently touch it against the interface plate of the mass spectrometer under microscopic control. The needle is not visibly shortened but a small sample droplet appears after the contact ,indicating the opening of the needle tip (*see* **Fig. 1**).
   Note: Needle and plate should be at the same electric potential. A tiny droplet will appear on the metal plate—it spreads out as a faint shadow in a few seconds.
9. Reposition the needle in front of the orifice of the mass spectrometer.
10. Apply the voltage to needle/holder and mass spectrometer interface and start scanning the mass spectrometer. If there are ions in the spectrum reduce the air pressure in the needle to the lowest value that still keeps the flow stable.

2.2. Troubleshooting

The nanoelectrospray ion source is a very robust device. However, problems may occur if the sample contains high concentrations of buffers, polymers or salts or if the shape of the needle tip is not within the dimensions described in the preceding. In this subheading we provide a few troubleshooting tips.

If there are no ions or no noise in the spectrum then the needle is not spraying. Repeat **step 8** in **Protocol 1** to open the needle tip.

If the spray becomes instable then jitter and spikes will appear in the spectrum or the spectrum will contain an unusually high level of chemical background. Increasing the air pressure of the needle helps stabilize the flow. If this simple measure is insufficient then the spraying voltage may be too low (increase it by 100–150 V) or the opening of the needle tip is still too small (repeat **step 8** in **Protocol 1**).

Be aware that the nanoelectrospray needle has a very small diameter. Small changes in the applied potential (increases of 100 V) change the field density at the tip considerably. Electrical discharge can be initiated and ionize atmospheric gas which generates a mass spectrum. These ions are usually small (<400 Da), and therefore chemical background ions in the higher mass region are missing. Atmospheric gas ionization can be visible as a blue corona around the needle tip (when light sources are switched off) and may lead to oxidations of methionine containing peptides ***(12)***.

If opening of a needle is not successful by the means described in the preceding then apply voltage to the needle, pressurize it, and briefly touch it against the mass spectrometer interface plate (the potential difference between needle and interface plate should be about 500 V). The combined mechanical and electrical stress opens almost every needle. It is not advisable to routinely open needles by this approach because it tends to damage the metal coating at the tip. High electrical current drawn from very thin tips can heat the glass to a degree that the glass melts. The damaged piece of the tip can often be broken off and analysis can proceed with a larger opening.

Two effects can prevent or stop spraying from an opened needle: The high surface tension of the liquid or precipitation of salts, polymers, or other nonvolatile sample constituents when their concentration is high. In the former case the needle can easily be reopened by slightly touching it to the interface plate, thereby destroying the surface tension by physical contact. In the latter case, the harsher procedure employing mechanical and electrical stress for needle opening (described earlier) can be attempted.

2.3. Sample Preparation for Nanoelectrospray Mass Spectrometry

The nanoelectrospray ion source unfolds its full potential when the available sample is concentrated to 1 μL or less. Microcolumns packed in glass capillaries or in gelloader pipette tips can be used to desalt and concentrate protein and peptide samples to microliter volumes ***(3,13,14)***. A pulled glass capillary or a partially constricted gelloader tip is packed with a few hundred nanoliters of Poros resin (Applied Biosystems, Boston, MA). Working in the perfusion mode, Poros material generates only a small flow resistance when packed into a capillary. Peptide solutions are normally desalted/concentrated on Poros R2 resin, protein solutions on Poros R1 resin, and hydrophilic samples (such as small peptides and phosphopeptides) on R3 material (anion-exchange resin). Purification of phosphopeptides or highly negatively charged peptides can be accomplished by immobilized metal affinity chromatography (IMAC) using Fe(III) or Ga(III) ions ***(15,16)***.

2.4. Protocol 2. Sample Desalting and Concentration Employing Glass Capillary Microcolumns

For peptide analysis we routinely use the Poros R2 material (PerSeptive Biosystems) as chromatographic resin. Remove the smallest particles by sedimenting the resin 3–5× in methanol. In a 1.5-mL microcentrifuge tube make a slurry of 10–20 μL resin in 1.2 mL of methanol.

2.4.1. Materials

Chromatographic resin (e.g., Poros R1, R2, or Oligo R3); pulled glass capillaries for microcolumns and nanoelectrospray needle; custom-made holder for centrifugal load-

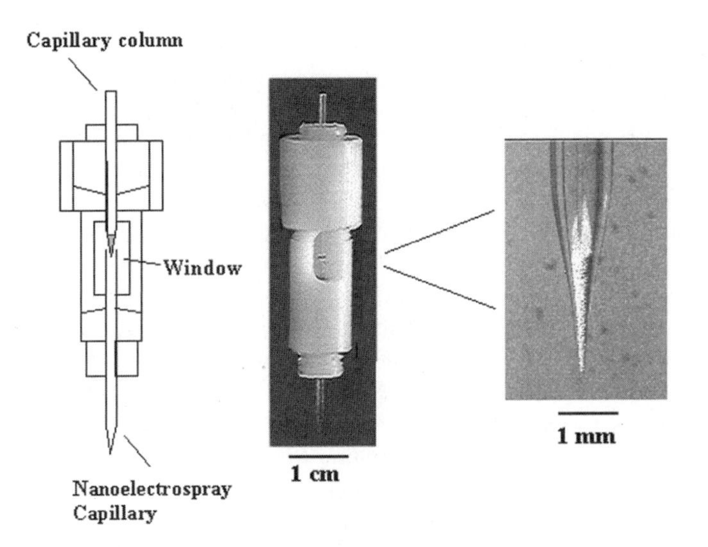

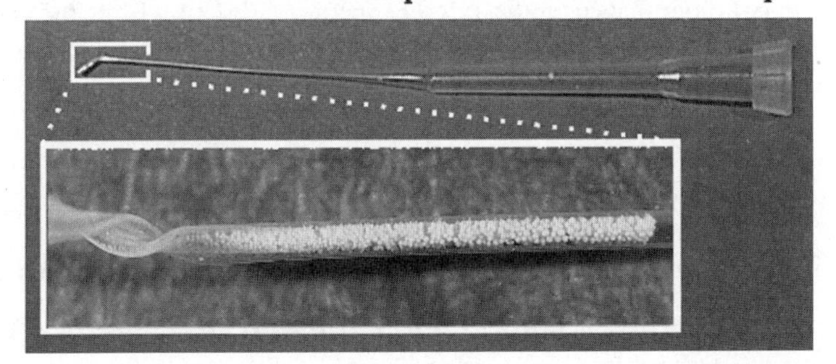

Fig. 3. Microcolumns for peptide and protein desalting/concentration prior to nanoelectrospray mass spectrometry. **(A)** Capillary column made from a pulled glass capillary and filled with a small volume of Poros resin. The sample is rinsed and then eluted from the column capillary directly into the nanoelectrospray needle by centrifugal force using a tabletop centrifuge. **(B)** Microcolumn made from a GELoader tip. The sample is loaded and eluted by air pressure from a syringe adapted to the GELoader tip.

ing and transfer of sample from capillary column to nanoelectrospray needle (available from MDS Proteomics, Odense, Denmark).

1. Mount a pulled glass capillary into a custom-made capillary holder (**Fig. 3A**) or into a pierced lid of a 1.5-mL microcentrifuge tube. Use a micropipet with a gel loader tip to transfer 5 μL of resin slurry into the capillary.
2. Load the chromatographic resin into the tip of the glass capillary by centrifugal force using a manually operated tabletop minicentrifuge (e.g. PicoFuge, Stratagene, Palo Alto, CA) at low speed (500–2000 rpm). The pulled glass capillaries used for the columns are

the same as used for nanoelectrospray emitters but they are not sputter-coated with metal. The chromatographic material is visible against a dark background but nearly invisible in front of white paper.

3. When sufficient chromatographic resin has been loaded into the capillary (1–2 mm resin height) the glass tip is widened/broken by gently touching it against the tabletop. The opening should allow liquid but not column resin to exit the capillary during centrifugation. Do not centrifuge the resin too fast; otherwise it becomes compressed and may block the flow of liquid. The centrifugal capillary column is only used once to avoid sample-to-sample contamination.
4. Rinse the capillary column by injecting 5 µL of 50% MeOH followed by gentle centrifugation.
5. Equilibrate the column resin by injecting 5 µL of 5% formic acid into the capillary followed by centrifugation until all liquid has passed through the column.
6. Dissolve the sample in 10–20 µL of 5% formic acid and inject it onto the capillary column in aliquots of 5 µL followed by centrifugation. For best recovery, dissolve the dried protein or peptide sample in 80% formic acid and then immediately dilute it to 5% by addition of water.
7. Wash the column resin twice by centrifugation with 5 µl of 5% formic acid solution. This desalting step is very efficient as the column is washed with 50–100× its resin volume. Before beginning the elution step the washing solution must be completely removed by gentle centrifugation.
8. Elution of sample into a nanoelectrospray needle.
   Mount the capillary column in-line with a premade nanoelectrospray needle in a custom-made capillary holder that fits into a microcentrifuge (**Fig. 3A**). Elute the peptide mixture into the nanoelectrospray capillary by centrifugating twice with 0.5 µL of 60% methanol–5% formic acid. Elute proteins with 60–70 % acetonitrile–5 % formic acid. This procedure allows handling of elution volumes between 10 µL and 0.2 µL. Elution, however, should be performed twice because the first elution does not completely deplete the column. Keep in mind that signal intensity in an electrospray spectrum is concentration dependent, so keep the elution volume as small as possible.
9. Mount the loaded nanoelectrospray needle onto the ion source and begin the experiment.

2.5. Protocol 3. Sample Desalting and Concentration Employing Microcolumns Packed in GELoader Tips

This sample desalting/concentration method is very simple and can be used prior to MALDI-MS and nanoelctrospray MS. The resin is held in place by making a constriction at the end of a GELoader pipettor tip. Sample loading, washing and elution is performed by loading liquid on top of the resin and applying air pressure to generate a low flow through the column. No frits are necessary.

2.5.1. Materials

1-mL plastic syringe; GELoader tips (Eppendorf brand). The syringe is adapted to the GELoader tips by using the top part of a "yellow" plastic tip.

1 Prepare a slurry of 100–200 µL of chromatographic resin, for example, Poros R1, R2, or OligoR3, in 1 mL of methanol.
2 Make a partially constricted GELoader pipet tip by gently squeezing or twisting the end of the tip (**Fig. 3B**). This allows liquid to flow through the tip while retaining the chromatographic resin.
3 Use another GELoader tip to load 5 µL of slurry of resin into the GELoader tip from the top and pack it by applying air pressure with the 1-µL syringe adapted to fit the GELoader tip microcolumn. The column height should be 2–4 mm.

4 Equilibrate resin by flushing 10–20 mL of 5% formic acid through the GELoader tip by air pressure (syringe).
5 Redissolve the peptide or protein sample in 20–40 mL of 5 % formic acid.
6 Load 5–20 μL of sample onto the microcolumn and gently press it through the column by air pressure using the syringe
7 Wash resin by flushing 10–20 μL of 5% formic acid through the packed GELoader tip by air pressure (syringe)
8 Elute the sample using a small volume of 5% formic acid–50% methanol. The eluate can be collected in a microcentrifuge tube, deposited directly onto the MALDI probe tip, or eluted directly into a nanoelectrospray needle. In the latter two cases the elution volume should be 1–2 μL only.

2.5.2. Notes and Options

Peptide separation may be improved by eluting the sample from the microcolumn by a step gradient using a series of mobile phases containing 5% formic acid in 15% methanol, 30% methanol, and 50% methanol, respectively.

For MALDI-MS analysis, elute the sample directly onto the MALDI probe using matrix solution, for example, HCCA, SA, or dhydroxybutyrate (DHB) dissolved in 30–50% methanol or acetonitrile. Deposit the eluate in a series of tiny droplets rather than one large drop ***(13)***. Trifluoroacetic acid or acetic acid can be used instead of formic acid in the mobile phases, whereas acetonitrile can substitute for methanol. Formic acid and methanol are recommended for nanoelectrospray mass spectrometry.

2.6. Nanoelectrospray Tandem Mass Spectrometry of Unseparated Peptide Mixtures

Peptide sequencing with tandem mass spectrometry ***(17,18)*** consists of three steps: (1) measuring the *m/z* values of peptides in a sample and determination of the charge state z; (2) acquiring the tandem mass spectra after collision-induced dissociation (i.e., fragmentation) of selected peptides; and (3) interpreting the tandem mass spectrometry data. With the nanoelectrospray source the first two steps are performed in one experiment with the unseparated peptide mixture.

The *m/z* values of analyte peptides are detected by comparing a "single MS" mass spectrum to a representative spectrum of the autolytic peptides of the enzyme used (i.e., trypsin) or a representative control from a particular experiment. For electrophoretically isolated proteins it is often advantageous to process an empty gel piece excised near the protein band of interest as control. The charge states of peptide ions are determined by considering the isotopically resolved ion signal: An isotope spacing of 0.5 Da corresponds to a doubly charged species whereas an isotope spacing of 0.33 Da reflects a triply charged ion.

At subpicomole protein amounts on the gel it is possible to employ the precursor ion scan capability of the triple-quadrupole tandem mass spectrometer to detect peptide ion signals that were below the chemical noise in the normal Q_1 spectrum (**Fig. 4**) ***(19)***. Precursor ion scans of the abundant *m/z* 86 immonium ion of isoleucine/leucine detect peptides that contain these common amino acids. For the parent ion scan the mass spectrometer parameters are adjusted to obtain optimum detection efficiency at reduced mass resolution. The parent ion scan technique can also be used to selectively detect phosphopeptides by monitoring *m/z* 79 in the negative ion mode ***(19–21)***, to selectively detect glycopeptides by monitoring *m/z* 162 or 204 in the positive ion mode ***(19,22)***,

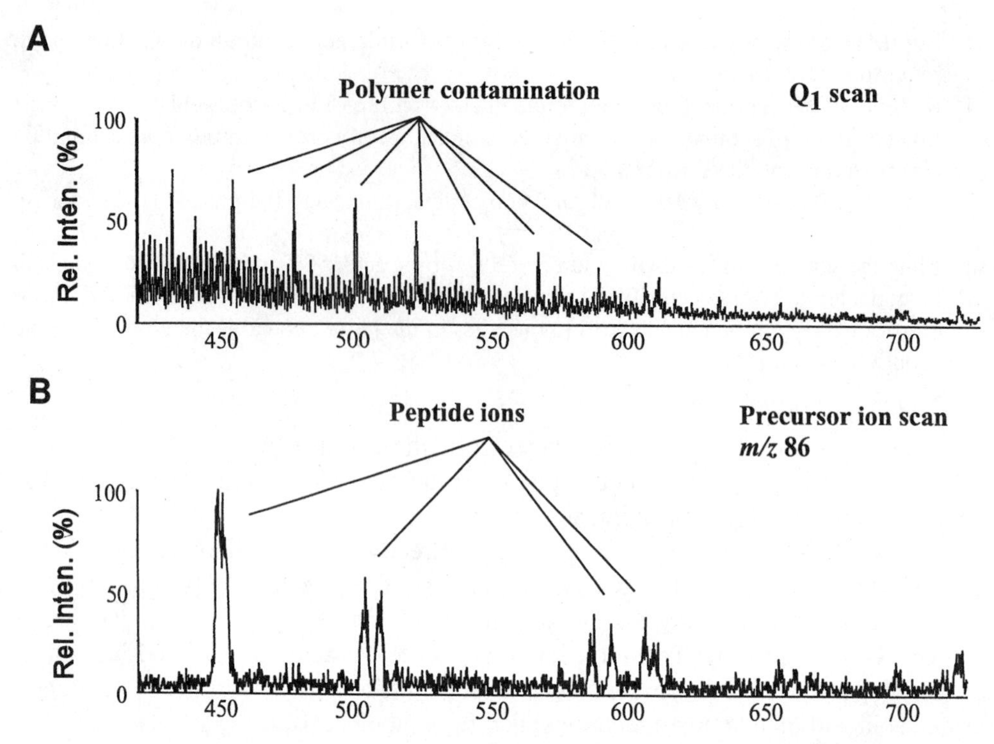

Fig. 4. Parent ion scan for the leucine/isoleucine immonium ion (*m/z* 86) detects peptide ions below the chemical noise level. **(A)** The nanoelectrospray quadrupole mass spectrum (Q_1 scan) of a tryptic peptide mixture displays only polymer ion signals and chemical background which suppress peptide ions. **(B)** The nanoelectrospray triple-quadrupole MS/MS precursor ion (*m/z* 86) analysis of the same sample reveals several peptide ion signals that subsequently can be selected for sequencing by product ion analysis by tandem mass spectrometry.

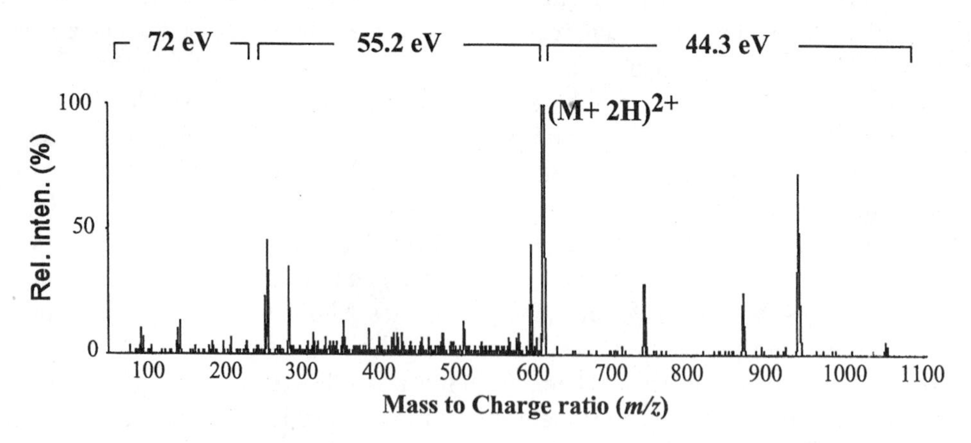

Fig. 5. Peptide tandem mass spectrum acquired in separate segments in a triple-quadrupole instrument. The *m/z* range above the precursor ion, $[M+2H]^{2+}$, is acquired at low collision energy and at relatively low mass resolution to efficiently generate and detect large peptide fragment ions. The *m/z* range below the precursor ion is acquired at higher resolution and with stepped collision energies, that is, intermediate collision energy to generate low *m/z* sequence ions and high collision energy to generate immonium ions and the a_2 and b_2 fragments.

and to detect intact proteins or oligonucleotides in contaminated samples *(23)*. A similar method has recently been implemented on Q-TOF type instruments *(24)*.

Once a set of peptide *m/z* values has been determined by either "single MS" scans or by precursor ion scans, high-resolution scans can be performed for selected peptide ion signals in the "multiple ion monitoring" mode to determine the exact peptide mass and the peptide charge state based on the isotope spacing. The reduction of sensitivity when measuring at high resolution is compensated by adding many scans, for example, 50 or more, to one spectrum. The latter feature further demonstrates the utility of long measurement times that the nanoelectrospray source provides. It is advantageous to select doubly charged tryptic peptide ions for tandem mass spectrometry experiments because they generate relatively simple fragment ion spectra. Triply charged tryptic peptides can also be fragmented and often allow determination of long stretches of amino acids sequence, that is, 15–25 consecutive residues, via doubly charged fragment ion series.

2.7. Fragmenting Peptides by Collision-Induced Dissociation

Once a set of peptide *m/z* values has been accurately determined each peptide is fragmented in turn. For peptide sequencing by tandem mass spectrometry using a triple quadrupole or a Q-TOF instrument two main instrumental parameters are adjusted to obtain high quality amino acid sequence information. First, the resolution setting of the first quadrupole (Q_1) is adjusted according to the abundance of the peptide ion signal, that is, the lower the ion intensity the higher the resolution setting in order to reduce the chemical background noise in the lower half of the tandem mass spectrum for a better signal-to-noise ratio. Second, the collision energy can be adjusted according to the peptide mass and varied depending on the mass range scanned (**Fig. 5**). The collision gas pressure is kept constant throughout the MS/MS experiment.

It may be advantageous to acquire a tandem mass spectrum in two or three segments. The high *m/z* segment is acquired with a wide parent ion selection window (low resolution) and a low collision energy to generate and detect relatively large peptide ion fragments. The low *m/z* region is acquired at higher resolution and at higher collision energies to generate and detect low *m/z* fragments and immonium ions. The nanoelectrospray allow this and other types of optimization due to the stability and long duration of the spray.

When investigating a peptide mixture by tandem mass spectrometry as many peptides as possible should be fragmented. This motivated the development of semiautomatic software routines to assist in data acquisition. The list of peptide *m/z* values is stored by customized software that calculates the optimum hardware settings for subsequent sequencing of each individual peptide. However, for *de novo* sequencing of long stretches of amino acid sequence it is not yet advisable to use automated software routines for data acquisition. Careful adjustments of collision energy and mass resolution is required to obtain high-quality data for unambiguous sequence assignments.

2.8. Generation of Peptide Sequence Tags from Tandem Mass Spectra of Peptides

Complete interpretation of tandem mass spectra of peptides can be complicated and requires some experience. However, it is often relatively straightforward to generate

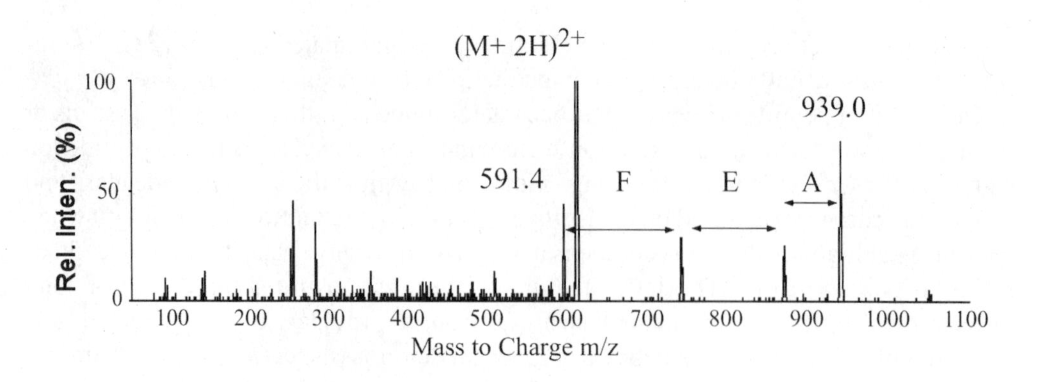

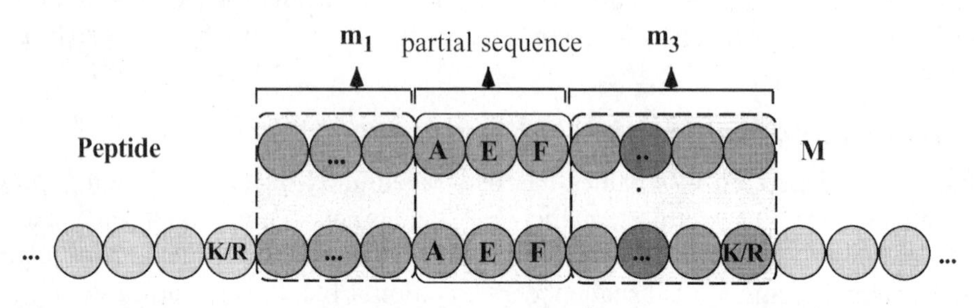

Fig.6. Peptide sequence tag generated from a tandem mass spectrum. A short search string, (591.4)FEA(939.0), is readily identified in the tandem mass spectrum of this tryptic peptide with mass M. It is subsequently converted to a peptide sequence tag by the PeptideSearch or PepSea software and used to query a database. The modular composition of a peptide sequence tag permits error tolerant searches where one or two of the modules are allowed to contain an error.

short consecutive sequences of two to five amino acid residues from a tandem mass spectrum. This information is valuable for sequence database searches as follows. A "peptide sequence tag" is assembled from the peptide mass, a short internal sequence of consecutive amino acid residues, and the distance in mass to the N- and C-terminus of the peptide *(25)* (**Fig. 6**). The search specificity of this construct is very high because the amino acid sequence is "locked" in place by the masses of the "unknown" parts of the peptide. The modular composition of a peptide sequence tag makes it tolerant to errors in any one of the modules. As only a fraction of the information content of the tandem mass spectrum is used to generate a peptide sequence tag used to query the sequence database, the remaining information confirm a retrieved peptide sequence: Every significant fragment ion signal should correlate to the peptide sequence. A peptide sequence tag consisting of three residues typically retrieve only one protein from a database containing more than 500,000 sequences when using high mass accuracy data obtained on a Q-TOF tandem mass spectrometer *(26)*. If longer stretches of sequence can be read out of a tandem mass spectrum, that is, six or more residues, it is advantageous to search by amino acid sequence instead of by peptide sequence tags. Searching by amino acid sequence is more flexible and allows sequence homology searches.

As mentioned previously, tryptic peptides have the desirable feature that they contain an N-terminal amino group and a C-terminal Lys or Arg residue, localizing protons at both the N-terminus and the C-terminus of the peptide. Tandem mass spectra of tryptic peptides very often contain a continuous y-ion series that can be readily assigned in the *m/z* range above the doubly charged parent ion signal. The spectrum interpretation strategy builds on this characteristic. It is guided by the demand to identify a protein in sequence databases or to reliably sequence peptides for cloning of the cognate protein.

Several algorithms can identify proteins based on sequence database searches with uninterpreted peptide tandem mass spectra *(27,28)*. Such software tools are very useful for a first screening of peptide tandem mass spectra because they immediately identify peptides originating from known proteins. The Mascot search engine and a number of other useful Internet-based services can be found via the URL http://www.protein.sdu.dk.

2.9. Guidelines for Interpreting Tandem Mass Spectra of Tryptic Peptides

The following list summarizes a few basic empirical rules that we use in interpreting tandem mass spectra of tryptic peptides. Because peptides differ in their fragmentation behavior in a sequence-dependent manner it is possible to find exceptions to these rules.

1. The goal of the interpretation is to find a series of peaks that belong to one ion series—for tryptic peptides mostly y-ions (C-terminal fragments).
2. Initial peak selection: The high *m/z* region of a tandem mass spectrum is often straightforward to interpret. Choose a large ion signal in this region as the "starting peak".
3. Assembly of a partial amino acid sequence: Try to find ion signals that are precisely one amino acid residue mass away from the starting peak (up or down in mass). We use software which marks all the possibilities in the spectrum. This provides a good overview whether there is more than one possibility for sequence assignments. If there is a repeating pattern of fragment ion peaks with satellite peaks representing an H_2O loss (–18 Da) or an NH_3 loss (–17 Da) a fragment ion series has been identified (for tryptic peptides a y–ion series is more likely).
4. By repeating step 3 a peptide sequence tag consisting of two to four amino acids is assembled that is subsequently used to identify a protein in the sequence database. As default for tryptic peptides, the database is searched under the assumption that a y–ion series was determined. However, even for tryptic peptides the tandem mass spectrum can be dominated by b-ions if a peptide contained an internal basic residue or when the C-terminal peptide of the protein had been sequenced.

2.10. Confirming Protein Identifications Made via Peptide Tandem Mass Spectra

If a protein sequence is retrieved by a database search with a peptide sequence tag or by similar methods then the amino acid sequence of the retrieved peptide should fit the tandem mass spectrum in order to be called as a positive match. Two or more peptides from a sample should independently identify the same protein in a database. To verify a match, the peptide fragment masses must be correct within the expected error of the mass measurement. For tryptic peptides the y-ion series should be nearly complete, except when a peptide contains an internal proline residue (*see below*). The N-terminal

b_2 and a_2 fragment ions, generated at relatively high collision energy, should be present in the low *m/z* region. Odd fragmentation patterns should reflect the amino acid sequence as discussed in the following paragraphs.

Peptides that contain internal basic residues (lysine or arginine) do not fragment in the vicinity of these residues because a charge is localized at the sidechain of the basic residue and therefore not available for amide backbone cleavage. If the triply charged precursor ion was fragmented then doubly charged y ions are present in the spectrum.

Internal proline residues deserve special attention. Cleavage of the C-terminal bond of a proline is observed to a low degree. The N-terminal bond of a proline is labile giving rise to an intense y–ion fragment. Internal fragmentation of peptides containing a proline often confirm a sequence: The y–ion generated by fragmentation at the N-terminal side of Pro will dissociate a second time to produce a y–ion series which is superimposed on the original y–ion series. However, the b–ions generated from this Pro containing fragment serve to confirm the C-terminal part of a peptide sequence up to and including the internal proline residue.

Isoleucine and leucine cannot be differentiated by amide bond cleavage alone because they have the same elemental composition and therefore identical molecular weight. Pairs of amino acid with identical nominal mass, Lys/Gln (128 Da) and oxidized Met/Phe (147 Da), can often be distinguished. Lys and Gln differ in their basicity and trypsin cleaves C-terminal to Lys and not at Gln so internal Lys is rarely found in tryptic peptides. However, if the latter is the case then the tryptic peptide usually acquire an additional proton for a total of three charges and the tandem mass spectrum will often contain a doubly charged y-ion series. If an internal Lys residue is suspected then the peptide mixture can be inspected for the presence of the limit peptide produced by tryptic cleavage at this Lys residue. Oxidized methionine (147.02 Da) and phenylalanine (147.07 Da) residues are differentiated relatively easily as follows. Tandem mass spectra of peptides which contain an oxidized methionine residue (i.e., methionine sulfoxide) display satellite peaks that appear 64 Da below each methionine sulfoxide-containing y-ion fragment due to loss of CH_3SOH from methionine sulfoxide ***(29,30)***. The oxidation reaction is often not complete so inspection of the peptide mass spectrum (Q_1 spectrum) may reveal a peptide 16 Da below the one containing oxidized methionine. The high resolution and mass accuracy of Q-TOF type instruments enable differentiation of Gln/Lys and OxMet/Phe by accurate mass determination.

If no proteins are retrieved with a simple database search then redo the search under the assumption that some of the amino acids in the peptide are modified, containing for example, an oxidized methionine or a *S*-acrylamidocysteine. An error tolerant search can be launched in which only partial correspondence between the peptide sequence tag and a database entry is required ***(25)***. Additional information about the protein can be used to select possible candidates if more than one protein sequence is retrieved (such as protein size, organism, or function).

If a protein cannot be identified by any of its peptide sequence tags we conclude that it is unknown. For proteins from human, mouse, or other model organisms, the peptide sequence tags are then screened against a database of expressed sequence tags (ESTs) ***(31)***. ESTs are short stretches of cDNA, that is, single-stranded DNA generated from expressed mRNA by reverse transcriptase, and thus represent the set of expressed genes in a given cell line. If a database search retrieves a cDNA sequence then library screen-

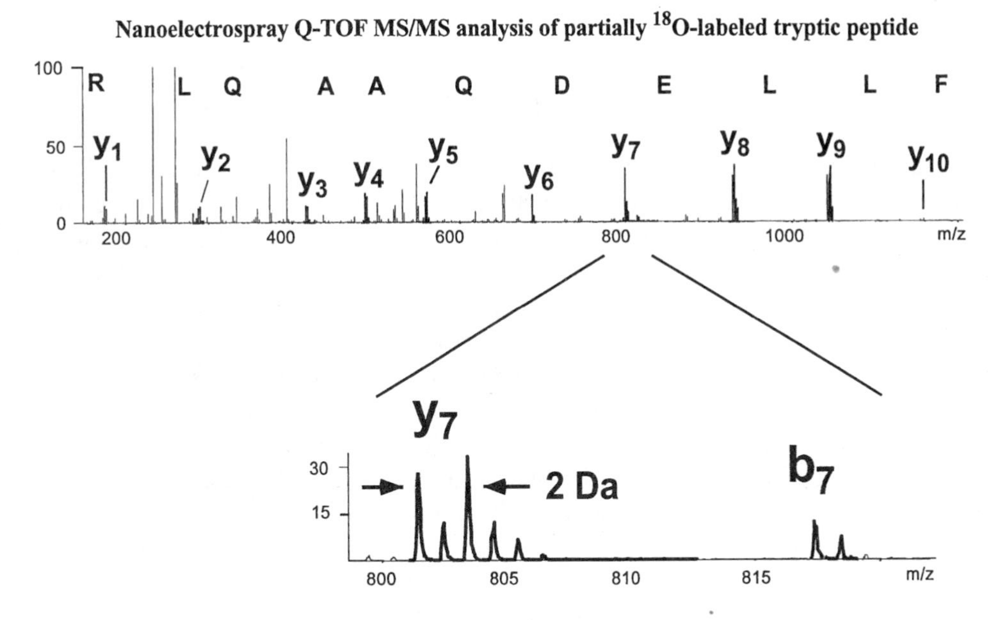

Fig. 7. Incorporation of the O stable isotope in peptides facilitates sequence interpretation in MS/MS. Digestion of protein by trypsin a solution of $H_2{}^{16}O$/ $H_2{}^{18}O$ water (1:1) results in partial labeling of the C-terminal carboxyl group in each peptide, except the C-terminal peptide of the protein. Identification of y-ion series is now straightforward due to appearance of y-ion signal peak splitting in a 1:1 ratio and a mass spacing of 2 Da.

ing and cloning is straightforward. If the EST database search produces no hit then *de novo* peptide sequencing has to be pursued to obtain longer stretches of amino acid sequences for homology searches. It has been demonstrated that peptide MS/MS data is useful to identify genes in raw genomic sequences ***(32)***.

2.11. De Novo *Peptide Sequencing by Nanoelectrospray Tandem Mass Spectrometry*

To sequence unambiguously an unknown protein for homology searching and cloning, it is necessary to be able to confidently call extended series of b-type and y-type fragment ion signals.

A useful method to recognize y-ion series employes ^{18}O-labeled water (*see*, e.g., **refs. *26,33,34***). By performing the trypsin digestion in a 1:1 mixture of normal water and ^{18}O labeled water all the tryptic peptides will incorporate ^{18}O at the C-terminus with a yield of approx 50%. Each peptide ion appears in a mass spectrum as a doublet separated by 2 Da. Selecting both isotope species together for fragmentation (low resolution setting in Q_1 precursor ion selection) produce tandem mass spectra that display y–ions as a series of split peaks, that is, separated by 2 Da, whereas b–ions are single peaks. This isotope-splitting is easily resolved in modern tandem mass spectrometers, particularly Q-TOF instruments ***(26)*** and aids in the interpretation of a peptide tandem mass spectrum because y-ion series are confidently assigned (**Fig. 7**). Note that the ^{18}O-labeled water has to be very pure to avoid chemical background noise. Redistillation of commercially available ^{18}O-labeled water is recommended.

Another method uses derivatization of free carboxyl groups, including the C-terminal carboxyl group of peptides. The tryptic peptide mixture is split in two portions. The first portion of the mixture is analyzed by nanoelectrospray tandem mass spectrometry and long peptide sequences are generated through complete interpretation of tandem mass spectra using the guidelines described in the preceding. The other portion of the peptide mixture is *O*-methyl-esterified ***(17,35,36)*** and then analyzed. Every free carboxyl-group including the C-terminus of peptides is esterified and therefore increase in mass by 14 Da. The number of methyl-esters in a peptide can be determined by the mass shift of peptides which is predictable from the previously interpreted tandem mass spectra of the native peptides. Because all y–ion fragments produced from an esterified peptide contain the C-terminus they are all shifted up in mass. Comparison of a set of tandem mass spectra obtained from a peptide and the corresponding esterified peptide serve to confirm the amino acid sequence because the y–ion series can unambiguosly be assigned. In addition, internal acidic residues, Asp and Glu, are methylated as well and can easily be differentiated from their corresponding amide residues, Asn and Gln, which otherwise differ in mass by only 1 Da.

2.12. Perspectives

Novel peptide sample preparation methods, mass analyzer configurations, and peptide dissociation techniques are continuously developed to increase sensitivity, mass accuracy, mass resolution, or sample throughput for peptide mass analysis and sequencing by mass spectrometry. The combination of the MALDI source with a Q-TOF hybrid instrument ***(37)*** and the development of electron capture dissociation (ECD) for MS/MS sequencing of large peptides and small intact proteins ***(38)*** are just two recent examples. Posttranslational modification of proteins leads to a change in the molecular mass of the affected residues and mass spectrometry is, therefore, a versatile analytical tool for structural characterization of modified peptides and proteins ***(39,40)***. MS-based approaches to peptide and protein quantitation using stable isotope labeling are also being pursued ***(41)*** and the use of multidimensional chromatography methods combined with ESI-MS/MS is an alternative or complement to two-dimensional gel electrophoresis for analysis of very complex protein mixtures ***(42–44)***. There is no doubt that applications of mass spectrometry in biological research will expand substantially in the future as the structure and function of all the gene products encoded in genomes of model organisms, including humans, have to be characterized in molecular details.

References

1. Fenn, J. B., Mann, M., Meng, C. K., Wong, S. F., and Whitehouse, C. M. (1989) Electrospray ionization for the mass spectrometry of large biomolecules. *Science* **246,** 64 –71.
2. Wilm, M. S. and Mann, M. (1994) Electrospray and Taylor - cone theory, Dole's beam of macromolecules at last ? *Int. J. Mass Spectrom. Ion Proces.* **136,** 167–180.
3. Wilm, M. and Mann, M. (1996) Analytical properties of the nano electrospray ion source. *Analyt. Chem.* **66,** 1–8.
4. Wilm, M. and Mann, M. (1996) Femtomole sequencing of proteins from polyacrylamide gels by nano electrospray mass spectrometry. *Nature* **379,** 466–469.
5. Shevchenko, A., Wilm, M., Vorm, O., and Mann, M. Mass spectrometric sequencing of proteins from silver stained polyacrylamide gels. *Analyt. Chem.* **68,** 850–858.
6. Roepstorff, P. and Fohlmann, J. (1984) Proposal for a common nomenclature for sequence ions in mass spectra of peptides. *Biomed. Mass Spectrom.* **11,** 601.

7. Biemann, K. (1988) Contributions of mass spectrometry to peptide and protein structure. *Biomed. Environm. Mass Spectrom.* **16,** 99–111.
8. Jensen, O. N., Wilm, M., Shevchenko, A., and Mann, M. (1999) Sample preparation methods for mass spectrometric peptide mapping directly from 2-DE gels. *Meth. Mol. Biol.* **112,** 513–530.
9. Kriger, M. S., Cook, K. D., and Ramsey, R. S. (1995) Durable gold coated fused silica capillaries for use in electrospray mass spectrometry. *Analyt. Chem.* **67,** 385–389.
10. Valaskovic, G. A. and McLafferty, F. W. (1996) Long-lived metallized tips for nanoliter electrospray mass spectrometry. *J. Am. Soc. Mass Spectrom.* **7,** 1270–1272.
11. Taylor, G. (1964) Taylor cone theory. *Proc. R. Soc. London Ser. A.* **280,** 383 (1964).
12. Morand, K., Talbo, G., and Mann, M. (1993) Oxidation of peptides during electrospray ionization. *Rapid Commun. Mass Spectrom.* **7,** 738–743.
13. Gobom, J., Nordhoff, E., Mirgorodskaya, E., Ekman, R., and Roepstorff, P. (1999) Sample purification and preparation technique based on nano-scale reversed-phase columns for the sensitive analysis of complex peptide mixtures by matrix-assisted laser desorption/ionization mass spectrometry. *J. Mass Spectrom.* **34,** 105–116.
14. Erdjument-Bromage, H., et al. (1998) Examination of micro-tip reversed-phase liquid chromatographic extraction of peptide pools for mass spectrometric analysis. *J. Chromatogr. A.* **826,** 167–181.
15. Posewitz, M., and Tempst, P. (1999) Immobilized gallium(III) affinity chromatography of phosphopeptides. *Analyt. Chem.* **71,** 2883–2892.
16. Stensballe, A., Andersen, S., and Jensen, O. N. (2001) Characterization of phosphoproteins from electrophoretic gels by nanoscale Fe(III) affinity chromatography with off-line mass spectrometry analysis. *Proteomics* **1,** 207–222.
17. Hunt, D. F., Yates, J. R., Shabanowitz, J., Winston, S., and Hauer, C. R. (1986) Peptide Sequencing by Tandem Mass Spectrometry. *Proc. Natl. Acad. Sci. USA* **83,** 6233–6237.
18. Biemann, K., and Scoble, H. A. (1987) Characterization by tandem mass spectrometry of structural modifications in proteins. *Science* **237,** 992.
19. Wilm, M., Neubauer, G., and Mann, M. (1996) Parent ion scans of unseparated peptide mixtures. *Analyt. Chem.* **68,** 527–533.
20. Carr, S. A., Huddleston, M. J., and Annan, R. S. (1996) Selective detection and sequencing of phosphopeptides at the femtomole level by mass spectrometry. *Analyt. Biochem.* **239,** 180–192.
21. Huddleston, M. J., Annan, R. S., Bean, M. T., and Carr, S. A. (1993) Selective detection of 5phosphopeptides in complex mixtures by electrospray liquid chromatography Mass Spectrom. *J. Am. Soc. Mass Spectrom.* **4,** 710–717.
22. Carr, S. A., Huddleston, M. J., and Bean, M. F. (1993) Selective identification and differentiation of N- and O-linked oligosaccharides in glycoproteins by liquid chromatography-mass spectrometry. *Prot. Sci.* **2,** 183–196.
23. Neubauer, G. and Mann, M. (1997) Parent ion scans of large molecules. *J. Mass Spectrom.* **32,** 94–98.
24. Steen, H. and Mann, M. (2001) Precursor scan on QTOF. *Analyt. Chem.* (in press).
25. Mann, M. and Wilm, M. S. (1994) Error tolerant identification of peptides in sequence databases by peptide sequence tags. *Analyt. Chem.* **66,** 4390–4399.
26. Shevchenko, A., Chemushevich, I., and Ens, W. (1997) Rapid de novo pept5–ide sequencing by a combination of nanoelectrospray, isotope labeling and a quadrupole/time-of-flight mass spectrometer. *Rapid Commun. Mass Spectrom.* **11,** 1015–1024.
27. Perkins, D. N., Pappin, D. J., Creasy, D. M., and Cottrell, J. S. (1999) Probability-based protein identification by searching sequence databases using mass spectrometry data. *Electrophoresis* **20,** 3551–3567.

28. Yates, J. R., 3rd, McCormack, A. L., and Eng, J. (1996) Mining genomes with MS. *Analyt. Chem.* **68,** 534A–540A.
29. Jiang, X., Smith, J. B., and Abraham, E. C. Identification of a MSMS diagnostic for methionine sulfoxide. *J. Mass Spectrom.* **31,** 1309–1310.
30. Lagerwerf, F. M., van der Weert, M., Heerma, W., and Haverkamp, J. (1996) Identification of oxidized methionine in peptides. *Rapid Commun. Mass Spectrom.* **10,** 1905–1910.
31. Mann, M. (1996) A shortcut to interesting human genes: Peptide Sequence Tags, ESTs and Computers. *Trends Biol. Sci.* **21,** 494–495.
32. Mann, M. and Pandey, A. (2001) Use of mass spectrometry-derived data to annotate nucleotide and protein sequence databases. *Trends Biochem. Sci.* **26,** 54–61.
33. Takao, T., et al. (1991) Facile assignment of sequence ions of a peptide labelledwith ^{18}O at the carboxyl terminus. *Rapid Commun. Mass Spectrom.* **5,** 312–315.
34. Schnölzer, M., Jedrzejewski, P., and Lehmann, W. D. (1996) Protease catalyzed incorporation of 18–O into peptide fragments and its application for protein sequencing by electrospray and MALD ionization mass spectrometry. *Electrophoresis* **17,** 945–953.
35. Shevchenko, A., Wilm, M., and Mann, M. (1997) Peptide sequencing by mass spectrometry for homology searches and cloning of genes. *J. Protein Chem.* **16,** 481–190.
36. Jensen, O. N., Wilm, M., Shevchenko, A., and Mann, M. (1999) Peptide sequencing of 2-DE gel-isolated proteins by nanoelectrospray tandem mass spectrometry. *Meth. Mol. Biol.* **112,** 571–588.
37. Loboda, A. V., Krutchinsky, A. N., Bromirski, M., Ens, W., and Standing, K. G. (2000) A tandem quadrupole/time-of-flight mass spectrometer with a matrix-assisted laser desorption/ionization source: design and performance. *Rapid Commun. Mass Spectrom.* **14,** 1047–1057.
38. Zubarev, R. A. (2000) Electron capture dissociation for structural characterization of multiply charged protein cations. *Analyt. Chem.* **72,** 563–573.
39. Kuster, B. and Mann, M. (1998) Identifying proteins and post-translational modifications by mass spectrometry. *Curr. Opin. Struct. Biol.* **8,** 393–400.
40. Jensen, O. N. (Elsevier Science, 2000) Modification-specific proteomics, in *Proteomics: A Trends Guide* (Blackstock, W., ed.) (Elsevier, Amsterdam, pp. 36–42.)
41. Gygi, S. P., et al. (1999) Quantitative analysis of complex protein mixtures using isotope-coded affinity tags. *Nat. Biotechnol.* **17,** 994–999.
42. Link, A. J., et al. (1999) Direct analysis of protein complexes using mass spectrometry. *Nat. Biotechnol.* **17,** 676–682.
43. Yates, J. R., 3rd., (2000) Mass spectrometry. From genomics to proteomics. *Trends Genet.* **16,** 5–8.
44. Tong, W., Link, A., Eng, J. K., and Yates, J. R., 3rd. (1999) Identification of proteins in complexes by solid-phase microextraction/multistep elution/capillary electrophoresis/tandem mass spectrometry. *Analyt. Chem.* **71,** 22,770–22,788.

99

Matrix-Assisted Laser Desorption/Ionization Mass Spectrometry for Protein Identification Using Peptide and Fragment Ion Masses

Paul L. Courchesne and Scott D. Patterson

1. Introduction

Methods for the identification of proteins advanced dramatically in the past decade through the introduction of mass spectrometric techniques and instrumentation sensitive enough to be applicable to biological systems *(1,2)*. The two mass spectrometric techniques that have provided these advantages are electrospray-ionization mass spectrometry (ESI-MS) and matrix-assisted laser desorption/ionization mass spectrometry (MALDI-MS).

This chapter deals with the identification of gel-separated proteins, whether or not they have been electroblotted to a polyvinylidene difluoride (PVDF)-based membrane. Therefore, we detail methods used to generate peptide fragments from in-gel digests, or on-membrane digests from Immobilon-P or Immobilon-CD, using essentially the methods of Moritz et al. *(3)*, Pappin et al. *(4,5)*, and Patterson *(6)* (as modified from Zhang et al. *[7]*), respectively (*see* **Fig. 1**). The microcolumn sample clean-up approach is a more detailed description of that which we have published previously *(8)*. We will not reiterate the electrophoretic methods required to separate and electroblot the proteins to a membrane or visualize the proteins (*see* Chapters 11, 12, 22, 39–41). MALDI-MS methods for accurate mass determination of the released peptide fragments together with approaches that utilize this peptide-mass data for identification will then be covered.

The principle behind identification of proteins via accurate mass measurements of enzymatically derived peptides relies upon the frequency of specific cleavage sites within a protein yielding a set of potential peptide masses that are unique to that sequence entry when compared with all of the others in the database. Studies have shown that only four to six peptide masses are required to identify proteins in searches of a database of more than 100,000 entries *(9)*. Masses sufficiently accurate for this purpose are easily obtained using MALDI-MS, and with the availability of instruments fitted with delayed extraction ion sources mass accuracy to 5 ppm is achievable *(10,11)*. The approach requires that the protein to be identified (or a very close sequence

From: *The Protein Protocols Handbook, 2nd Edition*
Edited by: J. M. Walker © Humana Press Inc., Totowa, NJ

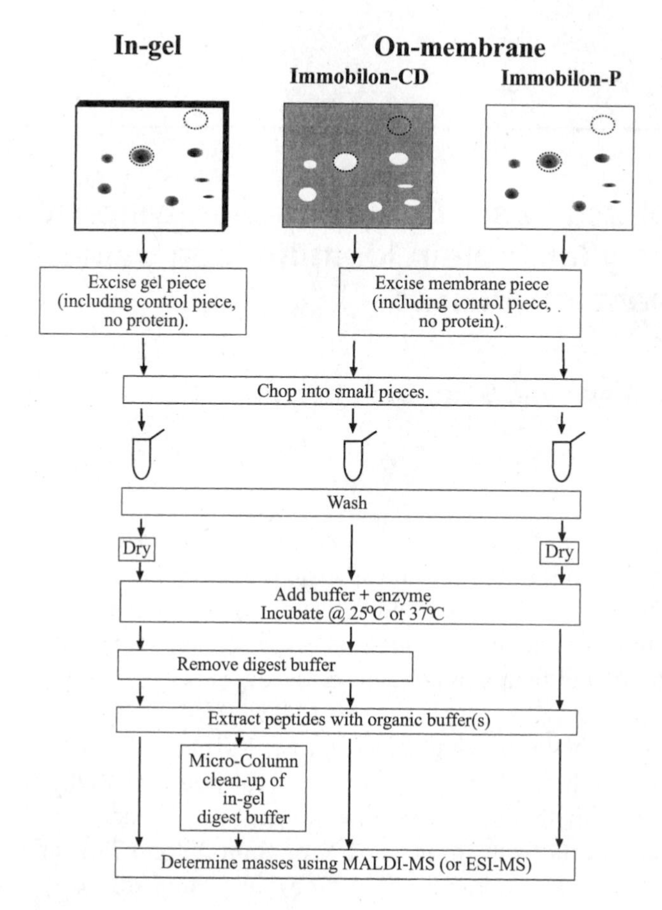

Fig. 1. Flow chart outlining methods for obtaining peptides from gel-separated proteins by either digestion in-gel or on-membrane for subsequent MS analysis. In all cases, the spot (or band) of interest, together with a control piece containing no protein, is excised and subjected to digestion and extraction of the peptides. For the Immobilon-CD protocol, the washing and chopping into small pieces steps can be reversed.

homologue) exists in the database. The caveats to this approach and considerations for interpretation of the results are described.

Finally, we cover generation and analysis of sequence-specific fragment ions using post-source decay MALDI-MS. Either partially interpreted or uninterpreted fragment ion data can be used in search programs to identify proteins with high accuracy in sequence databases, and the use of this approach is also described.

2. Materials

2.1. Equipment

1. Water bath (that can be set at 25°C or 37°C).
2. Sonication bath.
3. Vacuum concentrator (e.g., Speed Vac, Savant).

4. Polypropylene microcentrifuge tubes (500 µL).
5. MALDI-MS instrument (linear and/or reflector capability).
6. MALDI-MS instrument fitted with a reflector capable of variable voltage or a curved-field reflector (the example employs the Kompact MALDI IV—a curved-field reflector MALDI-MS ([Kratos Analytical]).

2.2. General Reagents

1. 50% v/v methanol (MeOH) containing 0.1% v/v aqueous trifluoroacetic acid (TFA).
2. Deionized water (dH_2O).

2.3. Enzymatic Digestion In-Gel

1. 50 m*M* Citric acid.
2. 20 m*M* NH_4HCO_3 containing 50% v/v acetonitrile (MeCN).
3. Stock enzyme: Endoproteinase LysC (5-µg vial, sequencing grade) (Boehringer Manheim) (LysC).
4. Digest buffer: 20 m*M* NH_4HCO_3, 1.0 m*M* $CaCl_2$.
5. Laboratory rotating mixer.
6. 1% TFA.
7. 60% v/v MeCN containing 0.1% v/v aqueous TFA.

2.4. Enzymatic Digestion on Immobilon-CD

1. 20 m*M* Tris-HCl, pH 9.0, containing 50% v/v MeOH.
2. 25 m*M* Tris-HCl, 1 m*M* EDTA, pH 8.0.
3. Stock enzyme (LysC): As for **Subheading 2.3., item 3**.
4. 30% v/v MeCN containing 2.5% v/v aqueous TFA.
5. 60% v/v MeCN containing 2.5% v/v aqueous TFA.

2.5. Enzymatic Digestion on Immobilon-P

1. 70% v/v Aqueous MeCN.
2. Digest buffer: 25 m*M* NH_4CO_3, 1% octyl-β-glucoside, 10% v/v MeOH.
3. Stock enzyme (LysC): As for **Subheading 2.3., item 3**.
4. 50% v/v Ethanol–50% v/v formic acid (98% HCOOH) (prepared immediately prior to use).
5. 10% v/v Aqueous MeCN.

2.6. Microcolumn Chromatography

1. µ-Guard™ column, 300 µm internal diameter (i.d.) × 1 mm, C8 packing, 300 Å wide-pore (LC Packings) (*see* **Note 1**).
2. Lite-Touch® ferrules for 1/16-in-outer diameter (o.d.) tubing.
3. Inlet tubing, 2 cm, 300 µm i.d., 1/16-in. o.d. Teflon® tubing (PE-Applied Biosystems Division) or similar.
4. Outlet tubing, 2 cm, 0.005-in. i.d., 1/16-in. o.d. PEEK (Red)(Upchurch Scientific).
5. 10-µL Hamilton syringe, with a fixed or removeable beveled needle and Chaney adaptor (Baxter Healthcare).
6. 1% v/v Aqueous 98% HCOOH (make fresh weekly).
7. 10% v/v Increments, or as desired, of MeCN to 90% in 1% v/v Aqueous 98% HCOOH (make fresh weekly).
8. MeOH.
9. Parafilm™.

2.7. MALDI-MS Reagents

1. 10 g/L of α-cyano-4-hydroxycinnamic acid (97% 4-HCCA, Aldrich Chemical) (50 mg/5 mL) in 70% v/v MeCN containing 30% 0.1% v/v aqueous TFA.
2. 33 m*M* 4-HCCA in MeCN–MeOH–dH_2O (5:3:2, by vol) (Hewlett-Packard).
3. 29 m*M* α 4-HCCA in HCOOH–dH_2O–2-propanol (1:3:2, by vol) (also referred to as formic acid–water–isopropanol [FWI]; *see* next item).
4. HCOOH–dH_2O–2-propanol (1:3:2, by vol) (FWI).
5. 10 μ*M* Bovine insulin B-chain, oxidized (Sigma Chemical Co.).
6. Ice-cold 0.1% aqueous TFA.
7. Synthetic peptide $Pro_{14}Arg$.
8. 99%+ Thionyl chloride (Aldrich Chemical). **Note: Use this reagent only in a well-ventilated fume hood, as it reacts violently with water to yield HCl vapor.**
9. 99%+ Anhydrous methanol, (Aldrich Chemical).
10. Heating block capable of maintaining 50°C.

3. Methods

3.1. Generation of Peptide Fragments

3.1.1. In-Gel Digestion Protocol

This protocol is for proteins visualized in a sodium dodecyl sulfate-polyacrylamide gel electrophoresis (SDS-PAGE) gel using the "reverse-staining" protocol of Ortiz et al. ***(12)***. In brief, this method involves immersion of the gel in 0.2 *M* Imidazole for at least 15 min, after which time the solution is changed to 50 m*M* $ZnCl_2$ until the background becomes opaque. The gel is then rinsed and stored in dH_2O at 4°C. The stain is sensitive to approx 100 fmol (loaded on the gel) level except for heavily glycosylated and sialylated proteins which are not readily observed ***(13)***. The following protocol was originally described for use with Coomassie Blue stained proteins ***(3)***. We have not included a reduction and alkylation protocol, but this can be used if required (*see* Chapter 76, and **refs. *14,15***). The only difference between that method and what is listed below is the use of citric acid for "destaining" or mobilizing the proteins.

1. Clean work area, microfuge tubes, and all utensils that will be used with 50% v/v MeOH–0.1% v/v TFA solution and let dry (*see* **Note 2**).
2. Identify the band or spot of interest on the gel and carefully excise (*see* **Note 3**, and **Fig. 1** for schematic). Do not touch the gel except with forceps. Remove the gel piece to a "cleaned" area where the piece (depending on its size) can be chopped into approx 1 mm^2 or smaller cubes for digestion. Excise a blank region (containing no protein) from the same gel that is equivalent in size to the band or spot of interest to serve as a control (*see* **Note 4**).
3. Place the chopped pieces into a microfuge tube and wash with a 50 m*M* citric acid solution (usually 200 μL, but depending on band size, more may be used without effect) for 20 min on a rotating mixer. This mobilizes the proteins in the reverse-stained gel. Decant the citric acid solution when completed.
4. Add 500 μL of 20 m*M* NH_4HCO_3–50% v/v MeCN to the tube containing the gel pieces and again place on the rotating mixer. After 30 min, replace the wash solution with a fresh buffer and continue to wash for another 30 min. Decant the wash solution when completed.
5. Place the tubes in a vacuum concentrator (e.g., Speed Vac, Savant) and dry the gel pieces completely (minimum of 30 min) (*see* **Note 5**).

6. To the dry pieces in the bottom of the tube, add 1 µL of stock enzyme (LysC at 0.1 µg/µL in dH_2O) and 15 µL of 20 m*M* NH_4HCO_3, 1.0 m*M* $CaCl_2$. Leave at room temperature for 15 min. Repeat this step until the gel pieces are totally rehydrated. Add 3× more digest buffer than what was used for the rehydration (i.e., if two additions of enzyme and digest buffer were added, in this step add an additional 90 µL of digestion buffer) (*see* **Note 6**). Gently mix the tube and place in a 37°C water bath for overnight digestion.
7. After brief centrifugation, remove the digest buffer supernatant and add it to a "clean" microfuge tube and reduce the volume of the supernatant to ~5 µL by vacuum concentration in a Speed Vac. Make every effort to not let the samples dry completely to avoid additional loss of peptides to the tube. Store these tubes for later microcolumn clean-up (*see* **Subheading 3.2.**). To the gel pieces now add 200 µL of 1% v/v TFA and sonicate in a warm (~37°C) sonication bath for 30 min (*see* **Note 7**).
8. Again, after a brief centrifugation, remove the bufffer and place it in a "clean" microfuge tube and reduce the volume to 1–2 µL without drying (if possible). These samples are now ready for MALDI-MS analysis (*see* **Notes 8** and **9**). Next add 200 µL of 60% v/v MeCN/ 0.1% v/v TFA to the microfuge tube containing the gel pieces and again place in a warm (~37°C) sonication bath for 30 min (*see* **Note 8**).
9. After a brief centrifugation, remove the buffer and place it in a "clean" microfuge tube and reduce the volume to 1–2 µL without drying completely (if possible). These samples are now ready for MALDI-MS analysis (*see* **Note 10**).

3.1.2. Immobilon-CD Digestion Protocol

Proteins are generally visualized on Immobilon-CD using a commercial negative stain kit (Immobilon-CD Stain Kit, Millipore) that yields a purple background with white areas indicating the presence of protein. The sensitivity of the stain is about 0.5 pmol of protein loaded on the gel ***(13)***.

1. Clean work area, microfuge tubes, and all utensils that will be used with a 50% v/v MeOH–0.1% v/v TFA solution and let dry.
2. Identify the band or spot of interest on the wet membrane and carefully excise (*see* **Note 11**, and **Fig. 1** for schematic). Do not touch the membrane. Place the membrane pieces into separate microfuge tubes and wash in 200 µL of 20 m*M* Tris, pH 9.0; 50% MeOH 4× prior to any further manipulation to remove any residual SDS from the membrane.
3. Remove the membrane piece to a "cleaned" area where the piece (depending on its size) can be cut into approx 1 mm^2 or smaller squares for digestion. Excise a piece of equivalent size of blank region (containing no protein) from the same membrane as the band or spot of interest to serve as a control (*see* **Note 4**). Keep the membrane pieces wet while cutting them with a drop of dH_2O if necessary.
4. Add the diced membrane pieces to a new 500-µL microfuge tube and add 2 µL (for a membrane piece of approx 6 mm^2) or 10 µL (for a membrane piece of approx 20 mm^2) of 25 m*M* Tris, 1 m*M* EDTA, pH 8.0. Then add either 0.4 or 2 µL of Endo Lys-C (5-µg vial reconstituted in 50 µL of deionized water) for the 6- or 20-mm^2 membrane pieces, respectively. Incubate at 25°C for at least 20 h (*see* **Note 12**).
5. Remove the digest solution containing any passively eluted peptides and store in a microfuge tube. Extract the membrane pieces with 3 or 10 µL (6-or 20-mm^2 membrane pieces, respectively) of 30% MeCN–2.5% TFA. After vortex-mixing for at least 30 s, centrifuge the samples, remove the extract, and store separately in another microfuge tube. Repeat the procedure using the same volume of 60% MeCN–2.5% TFA with sonication

for 3 min. The three aliquots are now ready for MALDI-MS analysis, preferably on the same day (*see* **Note 8**).

3.1.3. Immobilon-P Digestion Protocol

The proteins can be visualized on Immobilon-P using Coomassie Blue or Sulforhodamine B as described *(5)*. The sensitivity of Coomassie is about the same as that on Immobilon-CD, whereas Sulforhodamine B can visualize proteins to the approx 100 fmol level.

1. Clean work area, microfuge tubes, and all utensils that will be used with a 50% v/v MeOH–0.1% v/v TFA solution and let dry.
2. Identify the band or spot of interest on the dry membrane and carefully excise (*see* **Note 10**, and **Fig. 1** for schematic). Do not touch the membrane. Remove the sample to a "cleaned" area where the piece (depending on its size) can be cut into approx 1-mm^2 or smaller squares for digestion. Excise a piece of blank membrane (containing no protein) the same size as the band or spot of interest to serve as a control (*see* **Note 4**). Rehydration of the membrane pieces while cutting them with a drop of 50% v/v MeOH–dH_2O is sometimes helpful.
3. The Coomassie Blue stained membrane pieces are then put into a microfuge tube and washed/destained using 200 µL of 70% v/v MeCN and vortex-mixing. When the bands are totally destained (approx 30 s), pipet off the solvent and dry the pieces in a vacuum concentrator for 10 min or until completely dry.
4. To the dry membrane pieces, add 1 µL of your stock enzyme (LysC at 0.1 µg/µL in dH_2O) and just enough digest buffer (NH_4HCO_3–octyl-β-glucoside–MeOH) to cover all of the membrane pieces, usually 3–10 µL depending on the size of the protein band. It is sometimes helpful at this point to use a clean gel loading pipet tip or needle of a syringe to keep the dry membrane pieces in the digest buffer while they rehydrate (they tend to float on the surface of the liquid in the tube). Incubate the sample at 26–27°C overnight.
5. Remove and store separately the digest buffer from the tube (*see* **Note 13**). If necessary this can be cleaned up later using the microcolumn (*see* **Subheading 3.2.**, and **Note 14**). Add 10 µL of freshly prepared 50% v/v ethanol–50% v/v HCOOH, and place in a sonicator for 30 min. After sonication, remove the extract and dry completely in a vacuum concentrator. Rehydrate in 1–5 µL of 10% v/v MeCN–1% v/v HCOOH and perform MALDI-MS analysis.

3.2. Microcolumn Chromatography for Simple Clean-Up

This technique has been developed as a manual chromatographic clean-up step to remove salts and contaminants from samples (digest supernatants) that would be below the detectable limit for normal microbore chromatography (i.e., a few picomoles or less loaded on the gel). This quick, easy clean-up step has not only helped the overall recovery of peptides from techniques of this nature, but has also dramatically lowered the effective limit of in-gel and on-membrane digests to low picomolar to subpicomolar levels *(8,13,16)*. It should also be noted that Millipore has a commercial product, ZipTips, that is designed for this purpose and uses a similar approach.

1. Assemble the microcolumn by attaching the outlet and inlet tubing to the µ-Guard™ column using the Lite-Touch® ferrules and tighten by hand. The outlet tubing (Red PEEK) is shaved to yield a conical end allowing 3-µL droplets to form. This will not occur with a blunt end.

2. Equilibrate the microcolumn using 30 µL (3 × 10 µL) of MeOH using a 10-µL Hamilton syringe to introduce the solution into the inlet tubing (*see* **Note 15**).
3. Clean work area, microfuge tubes, and all utensils that will be used with a 50% v/v MeOH–0.1% v/v TFA solution and let dry.
4. Prepare the digest supernatant for microcolumn clean-up by reducing its volume to 3–10 µL.
5. Equilibrate the column with 30 µL of 1% v/v HCOOH at a flow rate of approx 1–2 µL/s.
6. Using a-10 µL Hamilton syringe, load your sample into the syringe and attach to the column inlet. Using the same approximate flow rate as with the equilibration, pass the sample over the column, collecting the flow-through as the "void/wash" into a microfuge tube. Reload 10 µL of 1% v/v HCOOH back into the same syringe and pass over column. Continue to collect this "wash" into the "void/wash" tube (*see* **Note 16**).
7. Rinse the syringe quickly (2–3×) with 30% v/v MeCN–1% v/v HCOOH (*see* **Note 17**). Load 3 µL of 30% v/v MeCN–1% v/v HCOOH into the syringe and attach to the column inlet. Again at the same approximate flow rate, pass the solvent over the column, collecting the eluant in a microfuge tube labeled 30% MeCN.
8. Repeat **step 7** using 3 µL of 70% v/v MeCN–1% v/v HCOOH and elute into an appropriately labeled tube.
9. Repeat **step 7** using 6 µL of 90% v/v MeCN–1% v/v HCOOH and elute into an appropriately labeled tube. The larger elution volume is used to attempt to ensure complete recovery of peptides from the column.
10. Wash the column with 30 µL of 90% v/v MeCN/1% v/v HCOOH.
11. Repeat the series of steps on the next sample.
12. When all samples are finished, clean the column in MeOH (at least 3 × 10 µL) and wrap the column in Parafilm™, being sure the inlet and outlet are sealed to avoid drying out.
13. Proceed with MALDI-MS analysis on all fractions collected (*see* **Note 18**).

3.3. MALDI-MS

3.3.1. Matrix Preparation and Selection

1. Matrix solutions are prepared in small volumes (5 mL or less) so that they will not be stored for excessive periods of time. Add the appropriate quantity of powdered 4-HCCA to the organic solution, and then add the remaining solution(s). The matrix solution is stored in a container (e.g., glass vial or microfuge tube) protected from the light and may need to be centrifuged briefly to pellet any undissolved chemical prior to use (*see* **Note 19**).
2. The matrix 4-HCCA has been listed in **Subheading 2.7.** dissolved in different solvents. In general we use the commercial matrix preparation, but this is for convenience only and we have not found an appreciable difference in spectra obtained with this and the MeCN–TFA solvent mixture, although one should remember that the commercial matrix preparation also has a limited life. However, the FWI mixture can dramatically change the peptides observed in complex mixtures *(17)*, and this can be particularly useful when both the "standard" and FWI–based matrices are used in parallel as more peptides can be observed than with either matrix alone.
3. 4-HCCA is the most widely used matrix for peptide mixtures (which this chapter is concentrating on), and can also be employed for proteins up to the size of at least serum albumin. However, sinapinic acid is often used to analyze proteins (*see* **Note 20**).

3.3.2. Preparation and Loading of the MALDI-MS Target

1. Rinse the MALDI-MS target with MeOH and wipe dry with a lint-free tissue, or follow the procedure described by the MALDI-MS manufacturer.

2. The simplest MALDI-MS sample preparation method is to add a small aliquot of the sample, approx 0.3 µL (in the low to subpicomolar range), to the target well on the sample slide followed by an equal amount of matrix. This solution is mixed in the pipet tip, and then allowed to dry at room temperature. When small volumes are used, drying takes only a few minutes.
3. Other sample preparation methods include mixing an aliquot of the sample and the matrix in a separate microfuge tube prior to loading the mixture on the target well. The sample–matrix mixture can also be subjected to vacuum to assist in even crystallization.
4. After the sample has dried, and hopefully an even crystalline surface is visible, the sample slide is ready for loading into the MALDI-MS instrument and data analysis.

3.3.3. Calibration

It is important to ensure that masses measured with the MALDI-MS are as accurate as possible. This can be achieved through "external" calibration, in which the calibrant is applied to a target well separate from the sample, or by "internal" calibration, in which the calibrant is mixed with the sample or when using trypsin and its autocleavage products. In either case the aim is to use ions of known mass that bracket the sample to be measured. We recommend incorporating a calibrant with every sample set to provide the opportunity for external calibration with each experiment, or at least to confirm that the instrument calibration is stable.

1. Calibration routines are instrument (software) dependent, but in their simplest form they employ a two-point calibration using a matrix ion and a large peptide of known molecular mass.
2. Calibration for peptide mixtures can be performed using 0.3 µL of a 10 µ*M* solution of the oxidized B-chain of bovine insulin (MH^+ 3496.9) dissolved in dH_2O into the target well, followed by 0.3 µL of matrix. A spectrum averaged from at least 50 shots is obtained at a laser fluence intensity just above threshold to ionize the calibrant, in the same manner that sample data is acquired. The signal from the calibrant (oxidized B-chain of bovine insulin), together with either the dehydrated matrix ion of 4-HCCA at MH^+ 172.2 or the matrix dimer at MH^+ 379.2, allows us to bracket many of the peptides generated in a LysC or tryptic digest.
3. The specific software calibration routine is then followed to fit a straight line to these two ions (matrix and calibrant).
4. "Internal'' calibration can be performed after the sample spectrum has been obtained by redissolving the sample in a small volume (approx 0.2 µL) of calibrant mixed with matrix (e.g., a ratio of calibrant/matrix of 4:1). The amount of calibrant applied needs to yield an intensity equivalent to that of the sample. This can be judged from the intensity of the calibrant in the "external" calibrant, although ionization suppression can sometimes occur (*see* **Note 21**).

3.3.4. On Probe Sample Clean-Up

Occassionally, despite the procedures outlined in the preceding, some samples may still not yield signals by MALDI-MS. There are a number of on-probe sample clean-up procedures, but it is not the aim of this chapter to detail them all. The following procedure first described by Beavis and Chait ***(18)*** is one of the simplest, and is often very effective. It relies on the difference in solubility between the peptide/protein-matrix crystals, and salts and other low molecular weight contaminants. The salts are soluble in ice-cold acidic solution whereas the peptide/protein-matrix crystals are not.

1. Add 2–3 µL of ice-cold 0.1% TFA to the dried sample in the target well for approx 5 s and then remove this with the same pipet (or blowing it off with forced air).
2. **Step 1** can be repeated at least twice (*see* **Note 22**).

3.4. Peptide-Mass Searches

After the accurate peptide masses are obtained from the protein band or spot, they can be used in search programs to determine whether the protein exists in the current full-length protein sequence databases, or translations of nucleotide sequence databases. There are a number of publicly accessible sites on the Internet (World Wide Web) that can be used to search these databases. The following is a list of servers currently available together with their affiliations and URL addresses. All supply full instructions on-line as to their use.

1. MassSearch from CBRG, ETHZ, Zurich: http://cbrg.inf.ethz.ch/subsection3_1_3.html
2. Mascot search program from Matrix Science Ltd., London:http:// www.matrix-science.com/cgi/index.pl?page=/home.html
3. MOWSE search program from the UK Human Genome Mapping Project Resource Centre (HGMP-RC): http://www.hgmp.mrc.ac.uk/Bioinformatics/Webapp/mowse/
4. MS-Fit (and MS-Tag) from the University of California, San Francisco: http://prospector.ucsf.edu/
5. PeptIdent and TagIdent, part of the ExPasy site from the Swiss Institute for Bioinformatics: http://www.expasy.ch/tools/#proteome
6. PeptideSearch from EMBL, Heidelberg: http://www.narrador.embl-heidelberg.de/GroupPages/PageLink/peptidesearchpage.html
7. Prowl search program from Rockefeller University, New York: http://prowl.rockefeller.edu/

It is essential to obtain the most accurate masses possible. Therefore, we recommend internal calibration for such analyses as described in the preceding (i.e, we obtain externally calibrated spectra prior to adding a small amount of calibrant to the sample).

The various peptide-mass search programs each have their own idiosyncrasies but all require a set of peptide masses (together with a stated tolerance or mass accuracy), the enzyme or chemical reagent used to generate the peptides, whether the cysteine residues have been modified (i.e., carboxylmethylated, etc.), and whether missed cleavage sites should be considered. Additional input to the program can include modifications to other specified amino acid residues (e.g., methionine sulfoxide), peptide masses following deuteration (an amino acid composition dependent mass increase), a selection of which database(s) to search, whether the search should be restricted to a subset of proteins whose intact mass falls within a specified range around the mass of the unknown protein of interest, and the species from which the sample was derived (for reviews *see* **refs.** ***1,9,19,20***). The MOWSE program can use further information on individual peptides such as partial composition and partial sequence, as can TagIdent from the ExPasy server. It should be noted that electrophoresis can induce artifactual modifications, for example, acrylamide adducted to free cysteines and the amino-N-terminus, and methionine oxidation (*see* **Note 23**).

Most programs will yield some result given a set of input peptide masses even if only a few masses are found to match. Rarely do all of the input peptide masses match with the top ranked candidate. Therefore, it is critical to attempt to determine how these peptide masses arose. The following is a list of possibilities for peptide masses that are not matched with the top-ranked candidate:

1. The correct protein was identified and the nonmatching peptides are due to either post-translational modification (including artifactual) or processing (*see* **Note 24**).
2. The correct protein was matched but some peptides were derived following either unspecific cleavage of the protease or specific cleavage from a contaminating protease (*see* **Note 25**).
3. The correct protein was identified but was part of a mixture of two or more proteins (*see* **Note 26**).
4. A homologue (or processing/splice variant) from the same or a different species was identified (*see* **Note 27**).
5. The result is a FALSE positive! (*see* **Note 28**). The possibility should not be overlooked that the real protein does not exist as an entry in the database being searched and may be truly novel.

When possible, it is advisable to attempt to gain further information on the peptides by either chemical treatments or chemical or mass spectrometric sequencing (*see* **Subheading 3.6.**). Of course this is dependent on the quantity of peptide available for further analysis.

3.5. Post-Source Decay MALDI-MS

To improve the confidence in the results of a peptide-mass search one needs to determine additional primary structural information on the peptides observed *(1,2)*. One means of achieving this aim is to isolate the peptide in the gas phase of the mass spectrometer and induce fragmentation of the peptide and measure the mass of the fragment (or daughter) ions. There are a number of mass spectrometric instrument designs with various ionization sources that allow gas-phase isolated peptides to be fragmented, for example, a triple-quadrupole MS, an ion-trap MS, and a time-of-flight (MALDI-)MS when the instrument is fitted with either a variable voltage reflector or curved-field reflector. We will describe the use of the latter type of MALDI-MS instrument, that is, a curved-field reflector instrument. However, it should be noted that the same principles apply to the use of a MALDI-MS fitted with a variable voltage reflector. Fragment ion spectra observed in a MALDI-MS instrument are generated by "post-source decay" or metastable fragmentation of the ions during the ionization process (*see* **Note 29**). This is distinct from "in-source" or "prompt" fragmentation that occur very early during the ionization (and therefore acceleration) process, and so the fragment ions are resolved in the linear dimension because of their different velocities (e.g., peptides linked by a single disulfide bond have been oberved to undergo "prompt" fragmentation *[21,22]*). Fragmentation is sequence specific, and generates the same ion series observed during other types of gas phase fragmentation including both single and multiple bond cleavages (*see* **Note 30**). It should be noted that if the peptide carries a posttranslational modification, the bond linking it to the peptide may be weaker than the peptide bonds themselves and so the fragment ions observed sometimes result from loss of the modifying group and little if no peptide bond fragmentation (e.g., *O*-linked carbohydrate, and sometimes phosphate on serine or threonine residues *[23]*).

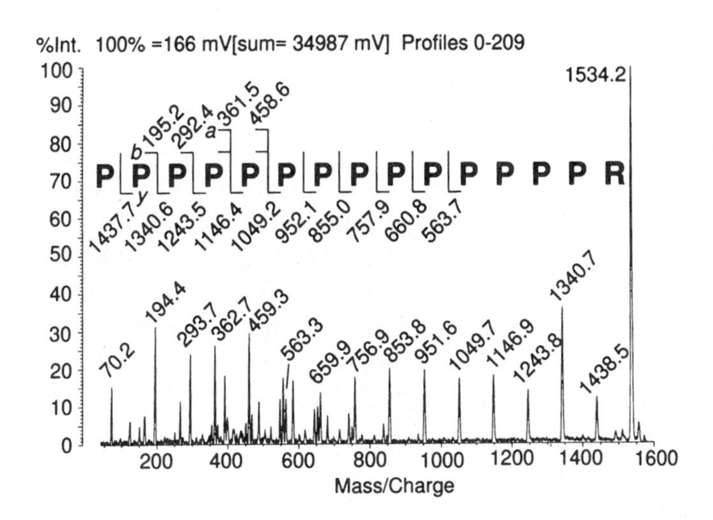

Fig. 2. PSD-MALDI-mass spectrum of $Pro_{14}Arg$ ($M_{ave}H^+$ = 1534,8) using a curved-field reflector MALDI-MS. The observed fragment ion masses following calibration are listed over the peaks used in the calibration routine. The expected fragment ion masses for the observed *a*, *b*, and *y* series ions are listed above and below the peptide sequence. The immonium of Pro has a mass of 70.1 Da. The spectrum was smoothed over four bins, and a baseline subtraction of 40 was used.

3.5.1. Matrix Selection, Sample Loading and Calibration

1. The most commonly used matrix for PSD-MALDI-MS is 4-HCCA, as this is considered a relatively "hot" matrix, that is, there is often considerable metastable fragmentation. The two other most common matrices, sinapinic acid (3,5-dimethoxy cinnamic acid) and gentisic acid (or dihydroxybenzoic acid), yield less metastable decay products. Matrix preparation and sample loading are as described in **Subheadings 3.3.1.** and **3.3.2.**
2. PSD-MALDI mass spectra can be calibrated only externally as the sample is a single molecular species. The manufacturer will have a specific protocol to follow but we use the calibrant first described by Cordero et al. *(24)*, which is a synthetic peptide of 14 prolines with an arginine carboxy-*(C)*-terminus ($Pro_{14}Arg$)($M_{ave}H^+$ = 1534.8). It should also be noted that Ala_nArg (when $n > 10$) is also a good substitute. However, other peptides such as angiotensin II, the fragment ion spectrum of which is well known, can also be employed for fragment ion calibration. The $Pro_{14}Arg$ peptide readily generates strong *y* series ions with a few *b* series ions as well as the proline immonium ion, thereby providing an excellent range of fragment ions for calibration (*see* **Fig. 2**).
3. Using the Kompact MALDI IV, the reflector is first calibrated with intact molecular ions in the same manner as in linear mode (*see* **Subheading 3.2.3.**), and then the calibrant peptide is fragmented and the masses of the fragment ions observed without any correction is entered into the calibration program. The expected fragment ion mass for each selected ion is then entered and the software uses a curve-fitting program to construct a calibration curve for use with other spectra.
4. Although we routinely check the calibration, in our experience the calibration is very stable. This is expected, as it is an instrument-dependent parameter. The same would be expected for MALDI-MS instruments where the reflector is scanned with different set voltages, so long as these are stable between analyses.

3.5.2. Methyl Esterification as a Means to Assist Fragment Ion Interpretation

Pappin et al. *(4)* described a simple method for methyl esterification of peptides that can be used both for peptide mass mapping in linear MALDI-MS and PSD-MALDI-MS (*see* **Note 31**). This method, in most instances, quantitatively esterifies all carboxylic acid groups on peptides to the corresponding methyl esters, thereby increasing the mass by 14 Da for each group. A free C-terminus will result in a mass increase of 14 Da and any acidic residues (Asp or Glu) will result in additional 14 Da mass increases for the peptide allowing the number of acidic residues to be calculated. This is of assistance to both peptide mass and fragment ion searches as well as for *de novo* interpretation of the fragment ion sepctrum. A simple example using a four-residue peptide of sequence FGSR is shown in **Fig. 3A**, and following methyl esterification in **Fig. 3B**.

1. Aliquot a portion, 1–1.5 μL, of the sample (either digest or individual peptide fraction) and dry by vacuum concentration in a 500 μL microfuge tube (e.g., in a Speed Vac) (*see* **Note 32**).
2. Make a 1% v/v thionyl chloride solution in anhydrous methanol (*see* **Note 33**).
3. Add 3–6 μL of the thionyl chloride solution (3× the volume of the original sample). Cap the tube and heat at 50°C for 30 min in a heat block (*see* **Note 34**).
4. Dry the reaction mixture by vacuum concentration and resuspend with 3 μL of 30% v/v MeCN–1% v/v HCOOH (**Subheading 2.6., item 7**).
5. An aliquot of this solution can be applied to the MALDI sample well, and allowed to dry prior to adding matrix (*see* **Subheading 3.3.2.**).

3.5.3. Interpretation of PSD-MALDI-MS Spectra

We do not usually manually interpret PSD-MALDI-MS spectra, instead we use the uninterpreted fragment ions in a computer search (*see* **Subheading 3.6.**). If all of the ions used in the search have not been matched by the program we then use the matched sequence to see if any unmatched ions can be explained by fragmentations not included in the search program (e.g., some of the search programs do not include internal fragmentations). However, it is certainly possible to manually interpret some PSD-MALDI-MS spectra. The simple spectrum in **Fig. 3** is a good example. The following is a description of how the spectrum could be interpreted, in a manner similar to that described by Kaufmann et al. *(25)*. The calculated fragment ions for this peptide are shown in **Fig. 3C**. Basically, the strategy is to look for immonium ions, and then to determine whether an ion series can be identified by mass differences (some of which could correspond to the identified immonium ions). Once a sequence has been formulated, one attempts to corrrelate all of the observed ions in the spectrum with the sequence.

1. The immonium ion region can often provide information on the amino acid content of the peptide (*see* **Table 1** for immonium ion masses). In **Fig. 3A** the following ions are observed in this region: 43, 61, 71, 88, 113, 117, 121, 131, 145, and 158. Given the expected mass tolerance (±1 Da) these ions could represent immonium ions from Arg (expected 43, 70, 87, 112, and 129), Pro (expected 70), Ser (expected 60), Leu/Ile (expected 86), Phe (expected 120), and Trp (expected 159). However, given the ion intensities — Pro and Leu/Ile too weak, and Trp too strong — these are not likely candidiates. The 117 and 145 ions are not matched.
2. The peptide was derived from a tryptic digest and so it would be expected that the C-terminus would be Lys or Arg (although of course there are exceptions, e.g., the C-terminal peptide or a nonspecific cleavage). A strong ion at 175.4 is observed in

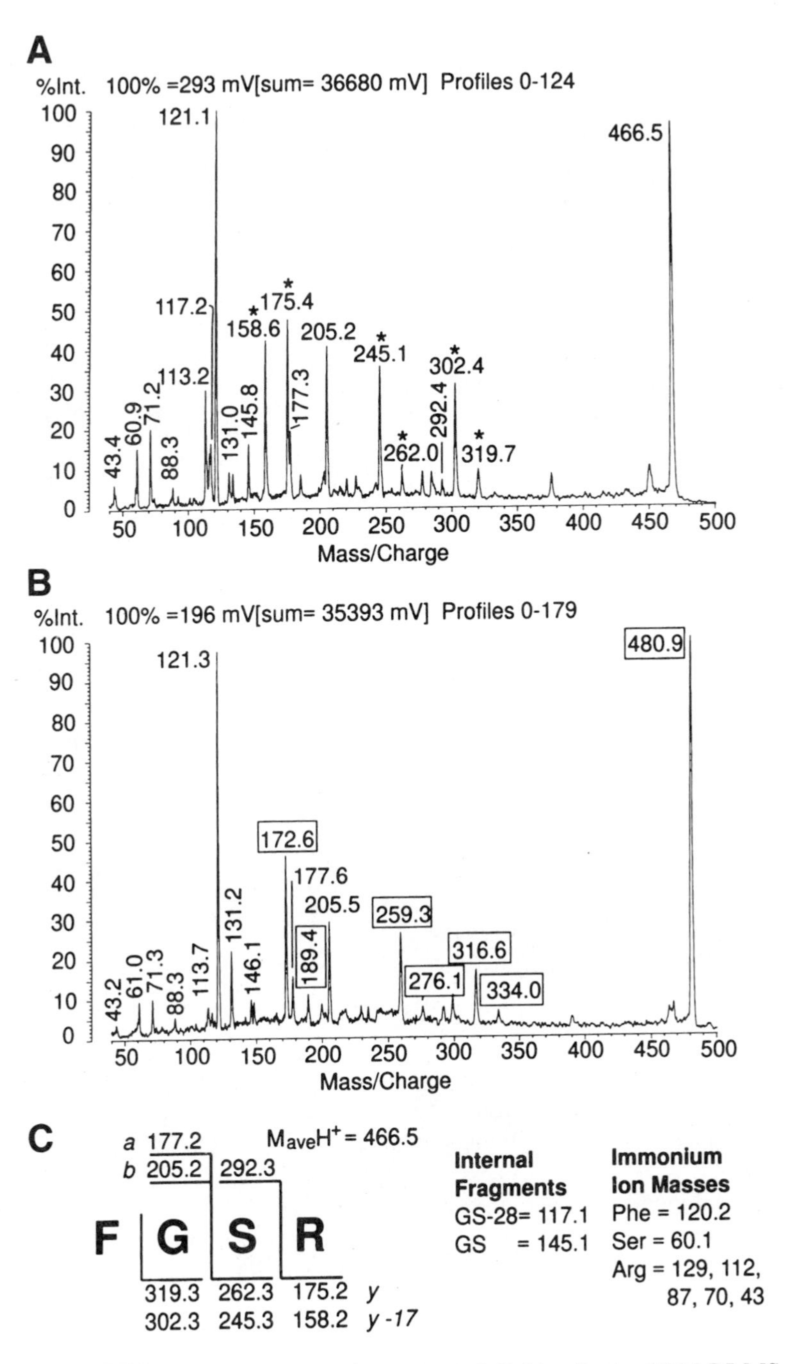

Fig. 3. PSD-MALDI-mass spectrum, using a curved-field reflector MALDI-MS, of a peptide of sequence FGSR with and without methyl esterification. (**A**) The fragment-ion spectrum without methyl esterification and (**B**) with methyl esterification; (**C**) shows the theoretical fragment ion masses for the peptide FGSR (corresponding to those labeled in **A** and **B**). The *y* and *y*-17 (loss of ammonia) series ions are labeled with an * in **A**. All of the ions (*y* and *y*-17 series ions, and the intact molecule) that increase in mass by 14 Da following methyl esterification have their observed masses boxed in panel **B**. Both spectra were smoothed over two bins with no baseline subtraction.

Table 1
Residue Masses of Amino Acids Together with Their Corresponding Immonium Ion Masses[a]

Amino acid	Abbreviations: three letter and (single letter)	Residue mass[b]	Immonium ion mass[b]
Glycine	Gly (G)	57	30
Alanine	Ala (A)	71	44
Serine	Ser (S)	87	60
Proline	Pro (P)	97	70
Valine	Val (V)	99	72
Threonine	Thr (T)	101	74
Cysteine	Cys (C)	103	76
Isoleucine	Ile (I)	113	86
Leucine	Leu (L)	113	86
Asparagine	Asn (N)	114	87
Aspartate	Asp (D)	115	88
Glutamine	Gln (Q)	128	101
Lysine	Lys (K)	128	129, 101, 84[c]
Glutamate	Glu (E)	129	102
Methionine	Met (M)	131	104
Histidine	His (H)	137	110
Phenylalanine	Phe (F)	147	120
Arginine	Arg (R)	156	129, 112, 100, 87, 70, 43[c]
Tyrosine	Tyr (Y)	163	136
Tryptophan	Trp (W)	186	159

[a]The values were obtained from Jardine ***(33)***, and Spengler et al. ***(34)***.
[b]All masses are given as average integer values.
[c]Arginine and lysine both exhibit multiple immonium ions, and these are listed. (They are not of equal intensity.)

Fig. 3A, which is the expected y_1 ion for an Arg C-terminus (expected 175.2). The neutral loss ion at 158.6, which is expected for an Arg, is also present (given the intensity, a more plausible explanation than a Trp immonium ion). If Lys were at the C-terminus, the y_1 ion would be 147.

3. We can now look for ions with masses between 232 (175 + 57, Gly) and 361 (175 + 186, Trp), because the next series ion has to have a mass between that of an additional Gly or Trp residue (if no residues are modified). A number of ions are present in this mass range. As there is an Arg at the C-terminus the *y*-series ions are expected to exhibit strong neutral losses. The ions at 245.1 and 262.0 could represent a *y* and *y-17* pair, as could 302.4 and 319.7. In fact, the mass difference between 262.0 and 175.4 is 86.6 (which could correspond to Ser, observed in the immonium ions). The difference between 319.7 and 175.4 (144.3) does not match an amino acid residue; however, the difference between 319.7 and 262.0 of 57.7 could correspond to Gly. This would make the sequence RSG, reading from the C-terminus.
4. The difference in mass from the molecular ion (466.5) and the last assigned *y* series ion (319.7) is 146.8, which matches with Phe (expected mass of 147.2). Thus, the complete *y* series has been observed and given that the immonium ion masses indicated — Arg, Ser, and Phe — the sequence can be assigned with some confidence as FGSR.

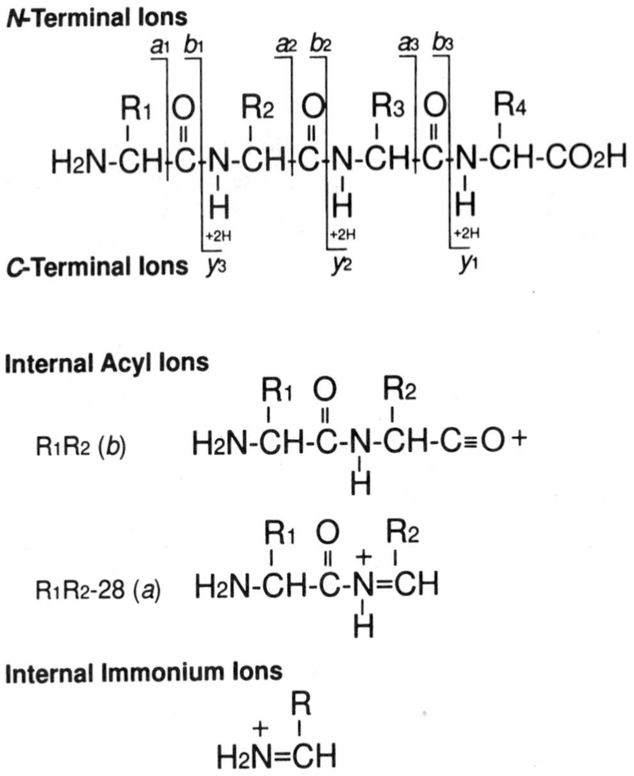

N-Terminal Ions

a_1 b_1 a_2 b_2 a_3 b_3

$H_2N\text{-}CH(R_1)\text{-}C(=O)\text{-}NH\text{-}CH(R_2)\text{-}C(=O)\text{-}NH\text{-}CH(R_3)\text{-}C(=O)\text{-}NH\text{-}CH(R_4)\text{-}CO_2H$

+2H +2H +2H

C-Terminal Ions y_3 y_2 y_1

Internal Acyl Ions

R_1R_2 (*b*) $H_2N\text{-}CH(R_1)\text{-}C(=O)\text{-}NH\text{-}CH(R_2)\text{-}C{\equiv}O^+$

R_1R_2-28 (*a*) $H_2N\text{-}CH(R_1)\text{-}C(=O)\text{-}N^+H{=}CH(R_2)$

Internal Immonium Ions

$H_2N^+{=}CH(R)$

Fig. 4. Fragmentation nomenclature for the most common positive ions observed by PSD-MALDI-MS (after ***30, 32***). Fragmentation is usually only unimolecular, that is, only one break occurs in the peptide backbone, and the charge is retained on either the N-terminus (*a* and *b* series ions) or the C-terminus (*y* series ions). However, internal acyl ions and immonium ions can also be formed from multiple fragmentation events. The internal acyl ions are referred to by their amino acid sequence (e.g., R_1R_2)(in a form similar to *b* series ions), or their sequence – 28 (for those similar to *a* series ions). There are also neutral losses form the internal acyl ions where part of an amino acid side chain is lost (.e.g, ammonia [17 Da] can be lost from Q, K, and R side chains, and water [18 Da] can be lost from S and T side chains). Similar losses can also occur from the molecular ion. (A very good list can be found at the MS-Tag WWW site, http:// prospector.ucsf.edu/ucsfhtml3.4/instruct/tagman.htm#Fragment Ion types). Amino acid side chains are referred to as R_1, R_2, and so forth in the formulae.

5. One can also examine the spectrum for ion pairs with a mass difference of 28 Da that could represent a and b series ions or internal fragments (which can also exit as a pair of ions separated by 28 Da [*see* **Fig. 4**]). There are two sets of ions that fit this criteria: 117.2–145.8, and 177.3–205.2. Given our interpretation of the sequence from the y series ions, these ion pairs correspond to the internal fragment GS and GS-28, and a_2 and b_2, respectively. The b_3 ion is also observed at 292.4. The mass difference between the molecular ion and the b_3 ion corresponds to 156.1 + 18 which matches Arg + water. We could also have used dipeptide tables (such as those listed at the WWW site of Burlingame's group at http://prospector.ucsf.edu/ucsfhtml3.4/misc/dipep.htm), which in this case reveal that GS is the only pair with a mass close to that observed, whereas the 205 could be from CT, M(Ox)G, or the correct pair, FG. Therefore, we have been able to assign ions consistent with the putative sequence FGSR using the strategy of observing

the immonium ion masses followed by looking for mass differences that correspond to amino acid residues (and pairs of ions separated by either a neutral loss of ammonia [17 Da] or water [18 Da] or an a–b ion series pair [28 Da], in this case starting with a putative y_1 ion.

6. The methyl esterification experiment shown in **Fig. 3B** provides additional confirmation of the sequence interpreted in the preceding. The mass of the parent ion only increased only from 466.5 to 480.9 corresponding to one methyl group that would be expected to be added to the C-terminal carboxyl residue. Therefore, all of the *y* series ions would be expected to shift by 14 Da and no other ions should shift compared to unmodified spectrum in **Fig. 3A**.
7. As expected all of the putative *y*-series fragment ions have increased in mass by 14 Da (boxed in **Fig. 3B**), confirming their identity. None of the other fragment ions have shifted. When methyl esterification is possible, even if acidic residues are present in the peptide, and therefore increase the parent ion mass by multiples of 14 Da, careful examination of the spectrum can often allow almost complete interpretation of spectrum. An example of when this becomes difficult is when there is one acidic residue and it is at the N-terminus. Then all fragment ions (except internal fragment ions) are shifted by 14 Da and the parent ion mass is 28 Da higher than the unmodified ion. The immonium ion masses of Glu and Asp are also increased by 14 Da following methyl esterification.

3.6. Fragment Ion Searches

Search programs, in addition to the peptide-mass search programs (*see* **Subheading 3.4.**), which use either partially or uninterpreted fragment ion spectra to search protein or translated nucleotide sequence databases are publicly available on the Internet (World Wide Web). The following is a list of servers currently available together with their affiliations and URL addresses. All supply full instructions on-line as to their use.

1. Mascot search program from Matrix Science Ltd., London: http:// www.matrix-science.com/cgi/index.pl?page=/home.html
2. MOWSE search program from the UK Human Genome Mapping Project Resource Centre (HGMP-RC): http://www.hgmp.mrc.ac.uk/Bioinformatics/Webapp/mowse/
3. MS-Tag from the University of California, San Francisco: http://prospector.ucsf.edu/
4. TagIdent, part of the ExPasy site from the Swiss Institute for Bioinformatics: http:// www.expasy.ch/tools/#proteome
5. PeptideSearch from EMBL, Heidelberg: http://www.narrador.embl-heidelberg.de/ GroupPages/PageLinked/PeptideSearchpages.html./PeptideSearch/PeptideSearchIntro.html
6. Prowl search program from Rockefeller University, New York: http://prowl.rockefeller.edu/

The uninterpreted fragment ion programs from San Francisco (MS-Tag) and New York (Prowl) require input of the fragment ion masses together with the ion series from which they may have been derived (e.g., *a*, *b*, *y* and neutral losses, etc). The MOWSE and TagIdent programs are not really fragment ion search programs but they do allow information to be added to individual peptides in a peptide-mass search, such as how many acidic residues are present (from a methyl esterification experiment), or what amino acid residues are present (which can be derived from immonium ion mass information). The peptide sequence tag program, which is part of PeptideSearch from Heidelberg, requires at least partial interpretation of the spectrum and assignment of the ions as being either *b* or *y* series. This can sometimes be difficult for PSD-MALDI-MS spectra. However, in the spectrum in **Fig. 3A** a *y* series tag of (262.0)G(319.7) from a parent of 466.5 could be entered. It should be noted that one would normally use data from a peptide of at least eight or nine residues for database searches. A peptide that is too short will not be as useful in searches of large databases, as too many proteins will have the same (or similar) sequence. With the Heidelberg program the search can be conducted using both ion series independently, and without constraints on potential modifications on either side of the assigned tag residue(s). The program has detailed on-line instructions.

When a result is obtained from any of these search programs one should attempt to assign all of the ions in the spectrum to the matched sequence. This will allow evaluation of the match and determination of whether any fragment ions are present that are not normally included in the expected ion lists generated by the search software. As for the peptide-mass search programs, confidence in the search results can also be gained from evaluation of the scores associated with the top ranked matches (for those programs that have scores), for example, if the second ranked score is considerably less than the top ranked score this may be indicative of a good match. However, each program also describes how to evaluate the output. The greatest confidence (and highest score) is achieved when there are a large number of fragment ions observed that match expected ion series.

Of course, if other peptides have been observed from the same protein, even if it has not been possible to obtain fragment ion spectra from them, they can still be used to evaluate the result of the search, for example, by determining if these other peptides could be derived from the matched protein. One should always attempt to obtain fragment ion spectra from as many peptides in the mixture as possible to increase one's confidence in the match, and to rule out any possibility of there being more than one protein in a particular band.

4. Notes

1. Although we use a column with C8 packing, columns packed with media of differing selectivity may be more appropriate for other applications.
2. When staining the gel, DO NOT use a container that has been previously used for immunoblotting protocols, for example, for blocking with milk, etc. as even in rinsed containers milk proteins can be adsorbed by either a blot or gel.
3. Trim the protein band carefully so that the gel piece contains only protein.
4. A control gel/membrane piece should always be included in any analysis to allow autodigestion products from the enzyme and any gel/membrane derived ionizing species to be accounted for.
5. This step is crucial; complete dryness must be achieved for adequate protein digestion, particularly at low picomolar levels.
6. Do not add extra enzyme, as this could lead to increased autolysis and potentially spurious cleavages.
7. Add warm tap water to the sonication bath and check the temperature prior to sonication.
8. Analysis should be performed as soon as possible, as peptides can be lost due to adsorption to the microfuge tube, although this seems to be alleviated somewhat with the Immobilon-CD protocol possibly due to the TFA concentration.
9. Just prior to MALDI analysis, a small amount (1–5 µL) of 30% v/v MeCN–1% v/v HCOOH can be added to extract any peptides from the walls of the tube.
10. On lower level samples (<2.0 pmol) the majority of recovered peptides can usually (but not always) be found in the digest buffer as opposed to the acidic and organic extracts. Therefore, to increase the possibility of detecting low level peptides in the respective extracts, these are often pooled and analyzed as one sample on the MALDI-MS. Likewise,

it is advantageous to perform the MALDI-MS analysis as soon as possible to again avoid the possibility of peptides adsorbing to the walls of the tubes.

11. When excising the band try to only include membrane that contains protein by cutting close to the edge of the band.
12. Incubation is conducted at 25°C to avoid the membrane pieces drying out due to condensation of the buffer solution at the top of the tube. However, this is not a problem if the samples are incubated in a thermostated oven.
13. Separation of the digest buffer is not essential, and is not in the original protocol *(5)*.
14. MALDI-mass spectra can be obtained from this sample directly applied to the sample well with matrix, but we have observed stronger signals, presumably due to partial fractionation of the peptide mixture, if we first use the microcolumn to clean up the sample of salts and contaminants.
15. In addition to equilibrating the column, we suggest that a solution enzymatic digest of a known protein be used to "condition" the column prior to use. That is, load approx 5 pmol of an enzymatic digest onto to the microcolumn as described in **Subheading 3.2., steps 7–13**, and step elute the peptides in 30%, 70%, and 90% v/v MeCN–1% v/v HCOOH and analyze by MALDI-MS. This will allow the user to become familiar with the use of the microcolumn and potentially block any nonspecific peptide binding sites in the microcolumn. We use the equivalent of a load of 5 pmol, as that is the amount we found to be retained by the C8 microcolumn *(8)*. Therefore, this method is recommended for sample loads of 5 pmol or less.
16. The volume of the wash is sample dependent; with very dirty/salty samples, wash with an additional 20–30 μL of 1% v/v HCOOH prior to elution, or even use 5% v/v MeCN–1% v/v HCOOH.
17. The step elutions can take place at whatever MeCN concentration you desire, for this example we use 30%, 70%, and 90% v/v MeCN–1% v/v HCOOH and a step volume of 3 μL.
18. In most cases the "void/wash" sample will give no MALDI data due to high salt and contaminant concentration. If peptides are observed, this may be an indication of an overloaded microcolumn, a very old microcolumn, or peptides that do not bind to the column type chosen for the analysis. Also, if additional sensitivity is desired, the 3-μL fractions can be successfully reduced in volume with a gentle N_2 stream down to 1.0–0.5 μL.
19. Plasticizers and other contaminants can leech from the tubes over time. Be sure to run matrix only on the target well at various times to determine the mass of any contaminants. In addition, always centrifuge the matrix if there is any particulate matter (i.e., undissolved or precipitated matrix); otherwise the intact matrix crystals can act as crystallization seeds, causing inhomogeneous crystal formation *(26)*.
20. Even some small proteins will not ionize well with 4-HCCA but will with sinapinic acid, for example, the phosphoprotein β-casein (approx 24 kDa).
21. With this approach it may be necessary to ablate several layers of the calibrant to yield signal from both calibrant and sample.
22. One can also redissolve the crystals in a small amount of additional matrix, or solvent only, if the signal has not improved.
23. The artifactual modifications characterized to date include *(1)*: cysteine- acrylamide (+71 u), oxidized acrylamide (+86 u), β-mercaptoethanol (+76 u); N-terminus-acrylamide (+71 u); methionine oxidation (to sulfoxide) (+16 u).
24. It may be possible to rationalize the unassignable masses by taking into account common posttranslational modifications; however, these should always be considered tentative unless confirmatory experimental evidence is obtained.
25. Some confidence in this theory can be gained by determining whether the observed masses can be derived without assuming cleavage specificity of the enzyme (i.e., determining whether a set of contiguous residues sum.to the masses observed).

26. Depending on how many masses were obtained, the masses corresponding to the matched protein can be removed and the database searched with this remaining subset of masses. This may even result in identification of the extra protein.
27. This will not occur if the database to be searched is restricted to the species of interest, but can be of assistance if the protein has not been sequenced in your species of interest. This is particularly true when working in genetically poorly characterized species.
28. This is the worst possible outcome but one that can be interpreted (sometimes) from the difference in the scores between the first and second ranked candidates (i.e., if there is little difference in all of the scores this may be a false positive).
29. Although the ionization by MALDI is said to be relatively soft, it was observed that the intact molecular ions formed undergo significant metastable fragmentation referred to as "post-source decay" (PSD) ***(27,28)***. This term refers to the fact that the fragmentation is thought to result from multiple early collisions between the analyte (sample) and matrix ions during plume expansion and ion acceleration (i.e., after the source), as well as from collision events in the field-free drift region of the time-of-flight analyzer. Because the metastable fragments, both neutral and charged, have the same velocity as their parent ions, they all reach the linear detector at approximately the same time. The metastable fragments are observed by decelerating ions of discrete energies as a function of their *m/z* ratios with an ion mirror, and then accelerating them back through the field-free flight tube to a second detector. Fragment ions have a lower kinetic energy than, although the same velocity as, their unfragmented parent ion due to their smaller mass. These ions are resolved by lowering the potential of the ion mirror (reflector) while maintaining a constant accelerating potential. With a dual stage reflector this operation of decreasing the voltage (or scanning of the reflector) is performed between 7 and 14 times to generate a series of spectra that can be concatenated with appropriate software to generate a full fragment ion spectrum. With the curved-field reflector design of Cornish and Cotter ***(29)***, there is no need to step down the voltage as all of the fragment ions are focused at once, making the process simpler and more rapid. Both types of MALDI-MS instruments have the ability to observe fragment ion spectra from specific ions in a mixture by allowing only ions of a particular flight time into the field-free drift tube. The resolution of this gating procedure is only about ±2.5% of the parent ion mass.
30. The predominant ion series observed using PSD-MALDI-MS, after the nomenclature of Biemann ***(30)*** and Roepstorff and Fohlmann ***(31)*** are: N-terminal derived fragments (unimolecular cleavage with charge retained on the N-terminus): *a* and *b*; C-terminal derived fragments (unimolecular cleavage with charge retained on the *C*-terminus): *y*; internal (acyl) fragments (two peptide bond cleavages) of the *a* and *b* type: referred to by the sequence of the component residues (with a –28 suffix for the *a* series type); neutral losses of ammonia (17 Da, particularly strong if Arg is in the fragment of any series, and may be even more intense than the intact fragment ion) or water (18 Da) from any of the previously listed ions depending on the residues contained in the sequence (particularly if Ser or Thr are in the fragment): designated as the ion series with the sufix –17 or –18; and immonium ions:- designated as their single-letter abbreviation (*see* **Fig. 4** and **Table 1**). If Pro is present in the peptide, this often results in a strong internal ion series starting at the Pro and extending C-terminal from this site. For example, for peptides of sequence ILPEFTEAR, a series of PE, PEF, PEFT, and so forth and PE-28, PEF-28, PEFT-28, and so forth may be observed. Other internal fragment ions are not unidirectional, but tend to cluster around basic residues present in the peptide. Whether the *a* and *b* or *y* series ions will predominate the spectrum will depend on which terminus has the strongest basic charge, for example, if Arg is at the C-terminus the *y* series ions will probably predominate the spectrum. The strongest immonium ion we have observed is generated by His at 110. Even in its methylated form, as in actin ***(32)***, it still yields a strong signal, in this case at 124.

31. More recently Pappin has updated the procedure and a combination of both procedures that we have tried is presented here *(5)*.
32. Pappin *(5)* recommends using 100-µL glass tapered vials that have been rinsed briefly in 6 *M* HCl, rinsed thoroughly with dH_2O, and dried at 110°C before storing in the presence of desiccant. These vials can be crimp capped.
33. To ensure no water comes in contact with the solutions a dry glass syringe stored with desiccant can be used. Any residual water will result in a strong reaction with the thionyl chloride so it is recommended that all vials and measuring implements be absolutely dry.
34. Although we have generally used only 3× the volume of the original sample, Pappin *(5)* recommends using 10–15 µL of the thionyl chloride solution.

Acknowledgments

We would like to thank Dr. Tony Polverino for critical reading of this chapter, and Dr's Hsieng Lu and Mike Rohde for support of this work.

References

1. Patterson, S. D. and Aebersold, R. (1995) Mass spectrometric approaches for the identification of gel-separated proteins. *Electrophoresis* **16,** 1791–1814.
2. Aebersold, R. and Patterson, S. D. (1998) Current problems and technical solutions in protein biochemistry, in *PROTEINS: Analysis & Design* (Angeletti, R. H., ed., Academic Press, Inc., San Diego, pp. 3–120.
3. Moritz, R. L., Eddes, J., Ji, H., Reid, G. E., and Simpson, R. J. (1995) Rapid separation of proteins and peptides using conventional silica-based supports: identification of 2-D gel proteins following in-gel proteolysis, in *Techniques in Protein Chemistry VI* (Crabb, J. W., ed., Academic Press, San Diego, pp. 31–319.
4. Pappin, D. J. C., Rahman, D., Hansen, H. F., Bartlet-Jones, M., Jeffery, W., and Bleasby, A. J. (1995) Chemistry, mass spectrometry and peptide-mass databases: evolution of methods for the rapid identification and mapping of cellular proteins, in *Mass Spectrometry in the Biological Sciences* (Burlingame, A. L. and Carr, S. A., eds.,) Humana Press, Totowa, NJ, pp. 135–150.
5. Pappin, D. J. C. (1997) Peptide mass fingerprinting using MALDI-TOF mass spectrometry, Vol. 64: in *Methods in Molecular Biology, Protein Sequencing Protocols* (Smith, B. J., ed.), Humana Press, Totowa, NJ, pp. 165–173.
6. Patterson, S. D. (1995) Matrix-assisted laser-desorption/ionization mass spectrometric approaches for the identification of gel-separated proteins in the 5–50 pmol range. *Electrophoresis* **16,** 1104–1114.
7. Zhang, W., Czernik, A. J., Yungwirth, T., Aebersold, R., and Chait, B. T. (1994) Matrix-assisted laser desorption mass spectrometric peptide mapping of proteins separated by two-dimensional gel electrophoresis: determination of phosphorylation in synapsin I. *Protein Sci.* **3,** 677–686.
8. Courchesne, P. L. and Patterson, S. D. (1997) Manual microcolumn chromatography for sample clean-up before mass spectrometry. *BioTechniques* **22,** 244–250.
9. Fenyo, D., Qin, J., and Chait, B. T. (1998) Protein identification using mass spectrometric information. *Electrophoresis* **19,** 998–1005.
10. Jensen, O. N., Podtelejnikov, A., and Mann, M. (1996) Delayed extraction improves specificity in database searches by matrix-assisted laser desorption/ionization peptide maps. *Rapid Commun. Mass Spectrom.* **10,** 1371–1378.

11. Vestal, M. L., Juhasz, P., and Martin, S. A. (1995) Delayed extraction matrix-assisted laser desorption time-of-flight mass spectrometry. *Rapid Commun. Mass Spectrom.* **9,** 1044–1050.
12. Ortiz, M. L., Calero, M., Fernandez Patron, C., Castellanos, L., and Mendez, E. (1992) Imidazole-SDS-Zn reverse staining of proteins in gels containing or not SDS and microsequence of individual unmodified electroblotted proteins. *FEBS Lett.* **296,** 300–304.
13. Courchesne, P. L., Luethy, R., and Patterson, S. D. (1997) Comparison of in-gel and on-membrane digestion methods at low to sub pmol level for subsequent peptide and fragment-ion mass analysis using matrix-assisted laser-desorption/ionization mass spectrometry. *Electrophoresis* **18,** 369–381.
14. Moritz, R. L., Eddes, J. S., Reid, G. E., and Simpson, R. J. (1996) *S*-Pyridylethylation of intact polyacrylamide gels and *in-situ* digestion of electrophoretically separated proteins—a rapid mass-spectrometric method for identifying cysteine-containing peptides. *Electrophoresis* **17,** 907–917.
15. Shevchenko, A., Wilm, M., Vorm, O., and Mann, M. (1996) Mass-spectrometric sequencing of proteins from silver-stained polyacrylamide gels. *Analyt. Chem.* **68,** 850–858.
16. Erdjument-Bromage, H., Lui, M., Lacomis, L., Grewal, A., Annan, R. S., McNulty, D. E., et al. (1998) Examination of micro-tip reversed-phase liquid chromatographic extraction of peptide pools for mass spectrometric analysis. *J. Chromatogr. A* **826,** 167–181.
17. Cohen, S. L. and Chait, B. T. (1996) Influence of matrix solution conditions on the MALDI-MS analysis of peptides and proteins. *Analyt. Chem.* **68,** 31–37.
18. Beavis, R. C. and Chait, B. T. (1990) Rapid, sensitive analysis of protein mixtures by mass spectrometry. *Proc. Natl. Acad. Sci. USA* **87,** 6873–6877.
19. Patterson, S. D. (1994) From electrophoretically separated protein to identification: strategies for sequence and mass analysis. *Analyt. Biochem.* **221,** 1–15.
20. Cottrell, J. S. (1994) Protein identification by peptide mass fingerprinting. *Peptide Res.* **7,** 115–124.
21. Patterson, S. D. and Katta, V. (1994) Prompt fragmentation of disulfide-linked peptides during marix-assisted laser desorption ionization mass spectrometry. *Analyt. Chem.* **66,** 3727–3732.
22. Crimmins, D. L., Saylor, M., Rush, J., and Thoma, R. S. (1995) Facile, in-situ matrix-assisted laser-desorption ionization mass-spectrometry analysis and assignment of disulfide pairings in heteropeptide molecules. *Analyt. Biochem.* **226,** 355–361.
23. Annan, R. S. and Carr, S. A. (1996) Phosphopeptide analysis by matrix-assisted laser desorption time-of-flight mass spectrometry. *Analyt. Chem.* **68,** 3413–3421.
24. Cordero, M. M., Cornish, T. J., Cotter, R. J., and Lys, I. A. (1995) Sequencing peptides without scanning the reflectron: post-source decay with a curved-field reflectron time-of flight mass spectrometer. *Rapid Commun. Mass Spectrom.* **9,** 1356–1361.
25. Kaufmann, R., Kirsch, D., Tourmann, J. L., Machold, J., Hucho, F., Utkin, Y., and Tsetlin, V. (1995) Matrix-assisted laser-desorption ionization (MALDI) and post-source decay (PSD) product ion mass analysis localize a photolabel cross-linked to the delta-subunit of NACHR protein by neurotoxin-II. *Eur. Mass Spectrom.* **1,** 313–325.
26. Beavis, R. C. and Chait, B. T. (1996) Matrix-assisted laser desorption ionization mass-spectrometry of proteins. *Meth. Enzymol.* **270,** 519–551.
27. Spengler, B., Kirsch, D., Kaufmann, R., and Jaeger, E. (1992) Peptide sequencing by matrix-assisted laser-desorption mass spectrometry. *Rapid Commun. Mass Spectrom.* **6,** 105–108.
28. Spengler, B., Kirsch, D., and Kaufmann, R. (1992) Fundamental aspects of postsource decay in matrix-assisted laser desorption mass spectrometry. 1. Residual gas effects. *J. Phys. Chem.* **96,** 9678–9684.
29. Cornish, T. J. and Cotter, R. J. (1994) A curved field reflectron time-of-flight mass-spectrometer for the simultaneous focusing of metastable product ions. *Rapid Commun. Mass Spectrom.* **8,** 781–785.

30. Biemann, K. (1990) Appendix 5. Nomenclature for peptide fragment ions (positive ions). *Meth. Enzymol.* **193,** 886–887.
31. Roepstorff, P. and Fohlman, J. (1984) Proposal for a common nomenclature for sequence ions in mass spectra of peptides [letter]. *Biomed. Mass. Spectrom.* **11,** 601–612.
32. Patterson, S. D., Thomas, D., and Bradshaw, R. A. (1996) Application of combined mass spectrometry and partial amino acid sequence to the identification of gel-separated proteins. *Electrophoresis* **17,** 877–891.
33. Jardine, I. (1990) Molecular weight analysis of proteins. *Meth. Enzymol.* **193,** 441–455.
34. Spengler, B., Lutzenkirchen, F., and Kaufmann, R. (1993) On-target deuteration for peptide sequencing by laser mass spectrometry. *Org. Mass Spectrom.* **28,** 1482–1490.

100

Protein Ladder Sequencing

Rong Wang and Brian T. Chait

1. Introduction

In protein ladder sequencing, the amino acid sequence of a peptide is determined by measuring the molecular mass differences between members of a family of fragments produced from the peptide. Adjacent members of this family of sequence-defining peptide fragments differ from one another by one amino acid residue. The molecular masses of the family of peptide fragments are measured by mass spectrometry in a single operation to read-out of the amino acid sequence. The identity of the amino acid residues is assigned by the mass differences between adjacent peaks and the amino acid sequence is determined by the order of their occurrence in the data set.

The ladder peptides are generated by stepwise chemical degradation in a controlled manner. In our first experiment, we obtained ladders by a manual method using stepwise Edman degradation in the presence of a small amount of terminating agent *(1)*. Subsequently, two additional chemical approaches have been developed for generating N-terminal peptide ladders. The first of these employed short coupling reaction times (incomplete reaction) in an automatic gas-phase amino acid sequencer *(2)*, while the second used volatile coupling reagents in a set of manual reactions *(3)*. Here, we will describe our early manual approach, using a terminating agent to produce the ladder peptides. A small quantity of 5% phenylisocyanate (PIC) is used as the terminating agent in the coupling step. The resulting phenylcarbamyl (PC) peptide derivatives are stable to the trifluoroacetic acid (TFA) used in the subsequent cleavage step. A small fraction of N-terminally blocked peptide is generated at each cycle. A predetermined number of cycles is performed without intermediate separation or analysis of the released amino acid derivatives. The resulting mixture of peptides (the ladder) is measured by matrix-assisted laser desorption/ionization mass spectrometry *(4)*. The mass spectrum contains peaks corresponding to each terminated peptide species present. The amino acid residues are identified from the mass differences between consecutive peaks. The order of these mass differences in the data set gives the amino acid sequence of the original peptide (**Fig. 1**).

Protein ladder sequencing has several potential advantages compared with conventional amino acid sequencing.

From: *The Protein Protocols Handbook, 2nd Edition*
Edited by: J. M. Walker © Humana Press Inc., Totowa, NJ

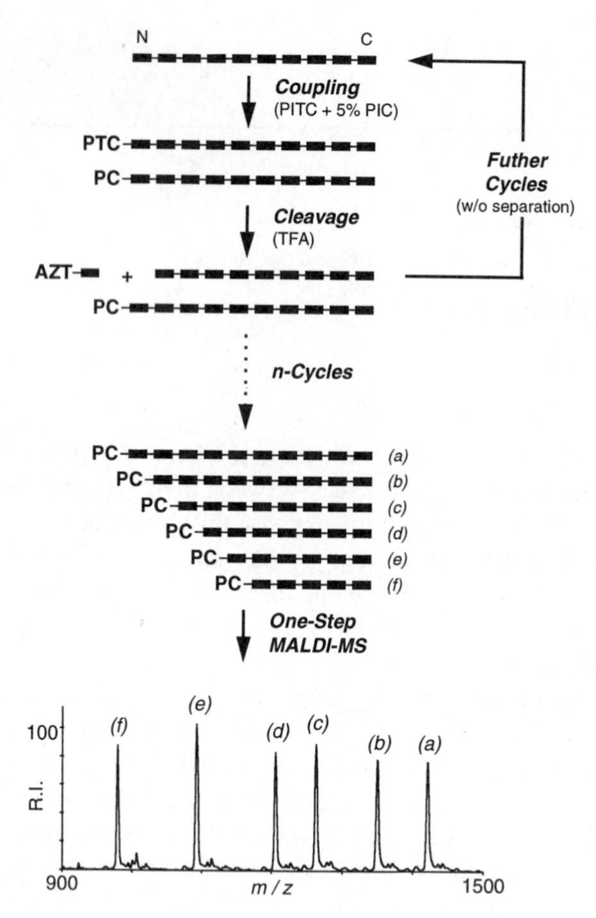

Fig. 1. The principle of N-terminal protein ladder sequencing. A stepwise Edman degradation is carried out in the presence of small amount of terminating reagent (phenylisocyanate, PIC) in the coupling step. The resulting phenylcarbomyl (PC) peptides are stable to trifluoroacetic acid (TFA) used to cleave the terminal amino acid from the phenylthiocarbomyl (PTC) peptides in the cleavage step. A ladder peptide mixture is formed as the result of PC-peptide accumulation in successive Edman cycles. Each black block represents an amino acid residue.

1. The amino acid sequence is read-out from one spectrum. Thus, optimally, all members of the sequence-defining fragments of the peptide, each differing by one amino acid residue, are simultaneously examined. The sequence of the peptide is deduced from the set of fragmentation products. Such a data set contains mutually interdependent information that determines the identity and the order of each amino acid residue in the parent peptide. Carry-over resulting from incomplete sequencing reactions cause no ambiguities in the present method.
2. The method can be used to obtain sequence information from peptide mixtures.
3. Direct detection of posttranslational modifications can be made with ladder sequencing *(5)*.
4. The ladder generation chemistry does not require complicated apparatus and can be done in any chemistry laboratory. However, the current procedure requires at least 100 pmol of peptide, since sample loss generally occurs in the liquid-phase extraction.

Protein ladder sequencing has the potential to be used in the following biological applications.

1. In conjunction with peptide fragment mass mapping, the method can provide additional sequence information and facilitate protein sequence database searching for protein identification.
2. The method can be used to study the nature and to determine the sites of chemical modifications and posttranslational modifications (e.g., phosphorylation, and glycosylation).
3. In protein processing and metabolism pathway studies, the method can be used to identify and confirm the termini of protein processing products.

2. Materials

2.1. Chemicals and Reagents (see Note 1)

1. Phenylisothiocyanate (PITC) (Pierce [Rockford, IL] or Sigma [St. Louis, MO].
2. Phenylisocyanate (PIC), >98% (Aldrich, Milwaukee, WI).
3. Pyridine (Pierce).
4. Hexafluoroisopropanol (HFIP), 99.8+% (Aldrich).
5. Heptane (ABI [Foster City, CA] or Pierce).
6. Ethyl acetate (ABI, Pierce, or Sigma).
7. Trifluoroacetic acid (TFA), anhydrous (Pierce).
8. 12.5% Trimethylamine (TMA) (ABI).
9. α-Cyano-4-hydroxycinnamic acid, 97% (Aldrich).
10. Acetonitrile (ABI).
11. Nitrogen, 99.99% (Matheson, Montgomeryville, PA).
12. Distilled and deionized water (prepared by using Milli-Q UV Plus water purification system).

2.2. Laboratory Equipment

1. Eppendorf Micro-Centrifuge (Brinkmann Instruments, Westbury, NY).
2. Savant SC110 Speed-Vac (Savant Instruments, Farmingdale, NY).
3. Multi-Block Heater (Lab-Line Instruments, Melrose Park, IL).
4. Fisher Vortex (Fisher Scientifics, Pittsburg, PA).
5. In-house laboratory vacuum.
6. Polypropylene microcentrifuge tube (PGC Scientifics, Gaithersburg, MD).

2.3. Mass Spectrometry

Matrix-assisted laser desorption/ionization time-of-flight mass spectrometer (*see* **Note 2**).

2.4. Preparing Reagents (see Note 3)

1. Coupling buffer: pyridine/water (1:1 v/v, pH 10.1). 250 µL water, 250 µL pyridine.
2. Coupling reagent: PITC/PIC/pyridine/HFIP (24:1:72:3 v/v). (Prepared under a blanket of dry nitrogen). 120 µL PITC, 5 µL PIC, 360 µL pyridine, 15 µL HFIP.
3. Cleavage reagent: 500 µL TFA.
4. Conversion buffer: 12.5% TMA/pyridine (1:1 v/v). 100 µL 12.5% TMA, 100 µL pyridine.
5. Conversion reagent: PIC/pyridine/HFIP (1:80:2.5 v/v). 2 µL of PIC, 160 µL pyridine, 5 µL HFIP.
6. Washing solvent A: Heptane/ethyl acetate (10:1 v/v). 10 mL heptane, 1 mL ethyl acetate.
7. Washing solvent B: Heptane/ethyl acetate (2:1 v/v). 8 mL heptane, 4 mL ethyl acetate.

8. Mass spectrometric matrix solution. 2 mg α-cyano-4-hydroxycinnamic acid, 200 µL 0.1% TFA in water, 100 µL acetonitrile. Vortex vial thoroughly to suspend the solid into liquid phase to form turbid mixture and then centrifuge the turbid mixture at 16,000*g* for 2 min. The saturated matrix solution will be used in the sample preparation for mass spectrometric analysis.

3. Methods

3.1. Ladder Generating Reaction by PITC/PIC—Wet Chemistry

This is a manual method to produce ladder peptides based on a manual Edman chemistry *(7)* (*see* **Note 4**). All of the reactions are carried out in a 0.6-mL polypropylene microcentrifuge tube. Since PITC is very sensitive to oxidation, some of the operations are perform under a blanket of dry nitrogen. Operations requiring a nitrogen blanket will be indicated.

3.1.1. Coupling Reaction

1. Dissolve peptide sample in 20 µL coupling buffer.
2. Add 20 µL coupling reagent (under a blanket of dry nitrogen).
3. Incubate the reaction vial at 50°C for 3 min with a block heater.

3.1.2. Removal of the Coupling/Terminating Reagents and Byproducts by Two-Phase Liquid Extraction (see **Note 5**)

1. Remove the reaction vial from the block heater.
2. Wipe the outside of the reaction vial with a prewet paper towel to condense the reaction liquid.
3. Centrifuge the reaction vial briefly at 16,000*g*.
4. Add 200 µL of washing solvent A to the reaction vial, vortex gently, and centrifuge at 14,000 rpm for 1 min to clear the phase.
5. Aspirate the upper phase by vacuum suction. Extreme care should be taken to avoid loss of the lower phase liquid.
6. Repeat steps 4 and 5 one more times with washing solvent A and twice with washing solvent B. It is advisable to use a fine bore pipet tip for this task.
7. After these four extractions, dry the remaining solution in a Speed-vac centrifuge for about 5–8 min.

3.1.3. Cleavage Reaction

After the reaction mixture is dried, add 20 µL of anhydrous TFA to the reaction vial (under a blanket of dry nitrogen). Incubate the reaction vial at 50°C for 2 min with a block heater.

3.1.4. Removal of Cleavage Reagent

Remove the cleavage reagent, TFA, by drying the reaction vial in a Speed-vac centrifuge for about 5 min.

3.1.5. Conversion of Free N-Termini to Phenylcarbamyl Derivatives

After the last cycle of coupling-washing-cleavage steps, subject the total peptide mixture to an additional treatment with PIC to convert any remaining unblocked peptides to their phenylcarbamyl derivatives.

1. Add 20 µL of conversion buffer and 20 µL of conversion reagent to the reaction vial.
2. Carry out the reaction at 50°C for 5 min.

3. Extract the reagents as described in **Subheading 3.1.2.**
4. Repeat **steps 1–3** one more time (*see* **Note 6**).

*3.1.6. Acidify the Peptide Mixture (see **Note 7**)*

1. Add 5 µL of TFA to each reaction vial and vortex briefly.
2. Remove TFA by a Speed-vac centrifuge for about 5 min.

3.2. Mass Spectrometric Analysis

Add 3 µL of matrix solution to the reaction vial and vortex briefly. Apply 1 µL of this peptide/matrix solution on the sample probe tip and dry at ambient temperature. Acquire mass spectra in the positive ion mode using a laser desorption/ionization time-of-flight mass spectrometer.

3.3. Sequencing Data Interpretation

A successful measurement will result in a mass spectrum (*see* **Note 8**) that contains a group of peaks as illustrated in **Fig. 2**. The mass spectrum is plotted using relative intensity (y-axis) vs mass-to-charge ratio (x-axis). The measured value is the ratio of mass/charge (m/z) for each ion detected by the mass spectrometer. Since the peaks observed in the spectra usually represent singly protonated peptide ions (unity charge, $[M+1]^+$) in matrix-assisted laser desorption/ionization mass spectrometry, the amino acid sequence is determined by calculating the m/z or mass differences between two peaks sequentially from high mass to low mass in the mass spectrum. In the example shown in **Fig. 2**, peak *(a)* has an m/z value of 1419.6 Dalton and peak *(b)* has an m/z value of 1348.5 Dalton. The mass difference is calculated as 1419.6 – 1348.5 = 71.1 Dalton. The identity of this amino acid residue (Ala) is determined by comparing the calculated mass difference with the amino acid residue masses listed in Appendix I. The identities of other amino acid residues are determined in the same fashion as Ser (86.8), Gly (57.2), Ile/Leu (113.0), and Ile/Leu (113.2) (*see* **Note 9**). The amino acid sequence (amino-terminal to carboxy-terminal) is determined from the order of the residues identified from high to low mass of the sequencing ladder (i.e., Ala-Ser-Gly-Ile/Leu-Ile/Leu).

Chemically modified amino acid residues can be identified by the same procedure using the mass differences obtained from the spectrum *(5)*. The amino acid residue masses resulting from common posttranslational modifications and from some of the chemical modifications are listed in Appendix I (*see* **Note 10**). These residue masses can be used to determine the identity of the modifications.

4. Notes

1. All chemicals and solvents used in the ladder-generating chemistry should be protein sequencing grade or the highest grade available.
2. The laser desorption/ionization time-of-flight mass spectrometer was constructed at The Rockefeller University and described in detail elsewhere *(6)*. This instrument is equipped with a 2-m-long flight tube and uses a Nd-YAG laser source (HY 400, Lumonics, Kanata, Ontario, Canada) that generates pulsed laser light (wavelength 355 nm) with a duration of 10 nsec. Commercial matrix-assisted laser desorption/ionization time-of-flight mass spectrometers should also be suitable for reading out the masses of the ladder peptides.
3. Amount of sequencing reagents prepared here are enough for sequencing four peptide samples and five cycles per sample. Fresh PITC and TFA (both in ampule) are used for each experiment.

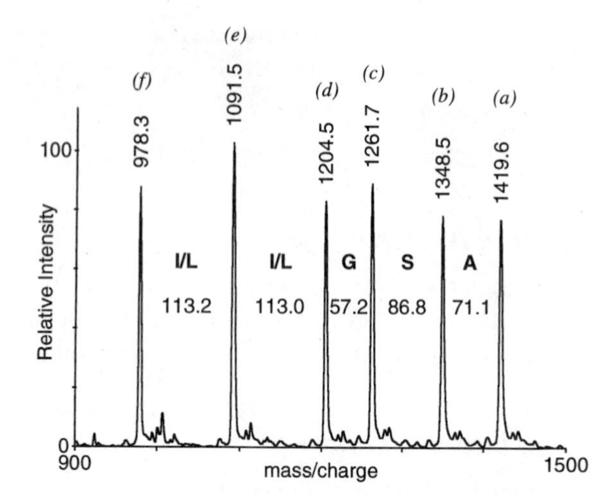

Fig. 2. Amino acid sequence read-out from a mass spectrum of ladder peptides. The vertical labels are measured m/z values for the peaks in the spectrum. The horizontal numbers are calculated mass differences between adjacent peaks. The identified amino acid residues are indicated. The amino-terminal sequence is read from right to left.

4. A detailed discussion on manual Edman chemistry is given in **ref. 7**.
5. Special care should be taken to remove the coupling reagents and byproducts of the coupling reaction. Otherwise, nonvolatile salt will form and affect subsequent reactions and mass spectrometric analysis.
6. It is necessary to perform the conversion twice, since the concentration of PIC is limited by its solubility and may be too low to yield 100% conversion.
7. The matrix-assisted laser desorption/ionization mass spectrometric analysis of protein and peptide requires acidic conditions. Unless the pH is lowered, the base used in the coupling reaction will significantly reduce the sensitivity of the mass spectrometric measurement.
8. It is important to note that the mass spectrometric response of a peptide is dependent on its amino acid composition. The sensitivity, in general, is enhanced when the peptide contains basic amino acid residues. On the other hand, the sensitivity is reduced when no basic amino acid residues are present. The modification on ε-amino groups of lysine residues by phenylisocyanate may decrease the mass spectrometric response, especially for peptides that do not contain arginine residues.
9. Leucine and isoleucine have identical residue masses (113 Dalton). These isobaric amino acid residues cannot be distinguished by the present ladder sequencing method. Lysine and glutamine have the same nominal residue masses (128 Dalton). However, they can be distinguished from each other, since the ε-amino group of lysyl residue is modified by PIC and results in a mass shift of 119 (residue mass of PC-lysine is 247.3 Dalton).
10. To detect a posttranslational modification using ladder sequencing, the modification should not be labile to the ladder sequencing chemistry. In certain cases, ambiguous results may result from posttranslationally modified amino acid residues that have similar masses to natural amino acid residues.

References

1. Chait, B. T., Wang, R., Beavis, R. C., and Kent, S. B. H. (1993) Protein ladder sequencing. *Science* **262,** 89–92.

2. Wang, R., Chait, B. T., and Kent, S. B. H. (1994) Protein ladder sequencing: towards automation, in *Techniques in Protein Chemistry*, vol. 5 (Crabb, J. W., ed.), Academic, San Diego, CA, pp. 19–26.
3. Bartlet-Jones, M., Jeffery, W. A., Hansen, H. F., and Pappin, D. J. C. (1994) Peptide ladder sequencing by mass spectrometry using a novel, volatile degradation reagent. *Rapid Commun. Mass Spectrom.* **8,** 737–742.
4. Hillenkamp, F., Karas, M., Beavis, R. C., and Chait, B. T. (1991) Matrix-assisted laser desorption/ionization mass spectrometry of biopolymers. *Anal. Chem.* **63,** 1193A–1203A.
5. Wang, R. and Chait, B. T. (1996) Post-translational modifications analyzed by automated protein ladder sequencing, in *Protein and Peptide Analysis by Mass Spectrometry* (Chapman, J. R., ed.), Humana, Totowa, NJ.
6. Beavis, R. C. and Chait, B. T. (1989) Matrix-assisted laser-desorption mass spectrometry using 355 nm radiation. *Rapid Commun. Mass Spectrom.* **3,** 436–439.
7. Tarr, G. E. (1977) Improved manual sequencing methods. *Meth. Enzymol.* **47,** 335–357.

101

Sequence Analysis with WinGene/WinPep

Lars Hennig

1. Introduction

Retrieving information from sequence data is becoming increasingly important. Currently, a multitude of public services is available via the Internet. However, many scientists are annoyed by slow network connections and busy servers. Moreover, identifying appropriate web pages while struggling with outdated links is often extremely time consuming. For these reasons, PC-based stand-alone applications constitute a useful alternative to Internet services. As available commercial software is usually quite expensive, WinGene/WinPep were developed that may be used free of charge by researchers in academic institutions *(1)*

WinGene/WinPep merge diverse analysis possibilities with maximal comfort for the user. Thus, the programs combine data analysis and data presentation. Simple, general principles of Windows® programs allow immediate usage without the need for lengthy training. Specialized analysis tasks for which Internet capabilities are extremely advantageous, for example, structure prediction and databank searches, were not included in the software. The first version of WinPep was released in 1998 *(1)* and became very popular. The latest versions, WinGene 2.2 and WinPep 2.3, released in fall 2000, contain several improvements and new functions. **Tables 1** and **2** summarize the abilities of WinGene 2.0 and WinPep 2.2, respectively. The following subheadings describe some features in more detail.

2. Materials

2.1. Hardware

The computer should be at least a PC/486 equipped with 8MB RAM. WinGene/WinPep requires one of the following operating systems: Windows®95, Windows®98, Windows®NT, or Windows®2000. At least 3 MB hard disk space must be available. Connection to the Internet is needed only for downloading and updating the software.

2.2. Software

WinGene/WinPep can be downloaded free of charge from http://www.ipw.agrl.ethz.ch/~lhennig/winpep.html. WinGene performs restriction analysis of nucleotide

From: *The Protein Protocols Handbook, 2nd Edition*
Edited by: J. M. Walker © Humana Press Inc., Totowa, NJ

Table 1
Features of WinGene 2.0

	Ref.
Reverse-complement of sequence	—
Restriction analysis	*2*
Calculation of composition and melting temperature for oligonucleotides	*3*
Translation and identification of open reading frames	—

Table 2
Features of WinPep 2.2

	Ref.
Determination of amino acid composition	—
Estimation of isoelectric point	*6*
Calculation of molecular weight	*6*
Calculation of molar absorption coefficients	*4,5*
Batch analysis of multiple sequences	—
Display of helical wheels	*6*
Display of hydropathy plots	*7,8*
Display of amphipathy plots	*1*
Search for sequence motifs	—
Simulation of site-specific cleavage	—
Identification of posttranslational modifications	—
Display of domain structure	—

sequences (*see* **Subheading 3.2.2.**) requiring an enzyme database. Recent versions can be obtained from REBASE (ftp.neb.com, http://rebase.neb.com) *(2)*.

3. Methods

3.1. General Features

The programs come with extensive online help that explains many procedures and contains main references. Most parameters of WinGene/WinPep are saved in INI files and can be modified from within the program. Alternatively, direct editing of the INI files is possible with any text processor.

Sequence data can be entered by different means: Direct typing, "Copy & Paste" via the clipboard, and import of text files or files in FASTA format are possible (*see* **Note 1**). Moreover, WinGene allows one to translate nucleotide sequence data and export them directly into WinPep. All results can be easily printed or copied to the clipboard. After pasting in other applications, further editing of text and graphics is possible. Moreover, graphical presentations can be saved to a disk as an Enhanced Windows Metafile (EMF). Sequences can be exported in FASTA format.

3.2. WinGene

3.2.1. Miscellaneous

With WinGene, nucleotide sequences can be easily reverse-complemented by clicking on the "Complement the Sequence" button. After selecting part of the sequence the "Primer check" function calculates base composition and melting temperature of the selected oligonucleotides *(3)*. If there was no selection the dialog will be opened for *de novo* entry of the oligo-sequence. The resulting report can be printed or pasted into any text processor software.

3.2.2. Restriction Analysis

Starting with version 2.0, WinGene performs restriction analysis of a given sequence. The source file for restriction enzymes is gcg.009 from the REBASE website *(2)* (*see* **Note 2**). In the "Restriction Analysis Dialog," the options for the current task can be set. The user can choose one of the following enzyme subgroups: (1) all enzymes, (2) only enzymes with recognition sites equal to or greater than a defined value, (3) a prechosen, saved subset of enzymes (YFE, your favorite enzymes), or (4) a chosen subset out of the list of all enzymes. To generate a YFE list: Select all desired enzymes into the right list box. Then, click on "Make YFE." The file YFE01.txt will be created in the program directory. This file contains the ID numbers of the selected favorite enzymes for fast and convenient access with subsequent analysis (*see* **Note 3**). In addition, the selected subset can be restricted further by activating "Only commercially available enzymes." The provided sequence will be treated as linear unless the "Circular DNA" item is checked.

Output may contain the sequence with indicated restriction sites ("Map") and/or a table of all occurrences of the recognition sequences in the input sequence ("Table"). Recognition sites can be marked automatically by bold letters if desired.

3.2.3. Translation and Identification of Open Reading Frames

Translation of the nucleotide sequence is achieved by clicking on the "Translate the Sequence" button. The translate window displays the sequence with its translation in three frames below. Basic text formatting (bold, underlined, italic) is available. WinGene will indicate which sequence (original or complementary) is used, that is, corresponds to the upper strand. The individual strands and frames can be deselected. This feature is useful if display of only the upper strand with one translated frame is desired. A list of all ORFs longer than the preset minimum (*see* **Note 4**) is available with the position of the first occurring Met given. If only one frame is displayed the longest ORF in this frame can be exported directly into WinPep ("Export into WinPep" in the file menu).

3.3. WinPep

3.3.1. Physicochemical Properties

A basic set of calculations determines some physicochemical properties of the given peptide, including amino acid composition, molecular weight, molar absorption coefficients for denatured *(4)* and native *(5)* states, and isoelectric point *(6)* (*see* **Note 5**). Furthermore, the sequence can be searched for the occurrence of sequence motifs.

In the age of genomics, there is an increasing need for processing large groups of sequences instead of just individual ones. WinPep 3.0 serves such demands by accepting a file containing a basically unlimited number of polypeptide sequences in FASTA format. Physicochemical properties of these polypeptides will be calculated. Results are saved in a tab delimited text file that can easily be read by MS-Excel or other spreadsheet software. Furthermore, some statistical parameters can be included in the output. The "Batch Options Dialog" allows specifying the traits to be analyzed. The only field always present in the output file is the sequence name. By default, the settings of the user's last analysis are activated when next opening the dialog. It is possible to determine length, molecular weight, molar absorption coefficient (native or denatured), relative and absolute amino acid composition, isoelectric point, grand average of hydropathy (GRAVY) *(7)*, and number of predicted transmembrane helices. The following statistical parameters can be selected for each molecular trait: mean, median, maximum, minimum, standard deviation, and standard error.

If desired, a log file of the analysis is generated. This file includes date, name of input file, name of output file, number of recognized sequences, used hydropathy scale for GRAVY, and used threshold value for putative transmembrane helices. By default, analysis of a file "name.txt" yields the files "name_results.txt" and "name_results_log.txt."

3.3.2. Sequence-Specific Cleavage and Identification of Posttranslational Modifications

For the simulation of sequence-specific cleavage, proteases can be selected out of a provided list or can be defined individually. Use the "Select" or "Unselect" buttons to obtain a subgroup of agents to be used (*see* **Note 6**). The "Specificity" field displays the sequence specificity of selected agents; user entry is not possible. Subsequently, either all possible fragments or only those resulting from a complete digest are displayed (*see* **Note 7**).

The identification of potential posttranslational modifications is a novel function of WinPep 2.11 and later. If mass spectroscopy yields peptides with an unexpected molecular weight, an algorithm tests whether this value might be caused by a posttranslational modification. First, a simulation of a sequence-specific cleavage must be performed. The generated list of fragments is the basis for subsequent analysis. WinPep displays a list of potential modifications. Furthermore, the presence of residues, which are required for this modification, is indicated.

3.3.3. Helical Wheel

Part of a sequence can be displayed in a circular way with an offset of 100 degree, representing the typical α-helix. The view menu can be used to change the size of the image and to switch between color and monochrome display. In color mode positive charged amino acids are red, negative charged blue, uncharged but hydrophilic ones green, and hydrophobic residues grey (*see* **Note 8**).

Clicking with the right mouse button invokes a floating menu where the displayed part of the sequence may be modified by opening the helical wheel properties dialog or the amphipathy moment may be calculated. This quantitative value was defined to facilitate the identification of potential amphiphilic helices: Considering the hydropathy values of all amino acids within the given sequence segment but not their distribution along the axis, the gradient across the helix is calculated. Accordingly, an amphipathy moment close to zero results from a relatively even distribution of hydrophilic and hydrophobic side chains, while large values indicate clustering of hydrophilic and hydrophobic residues on opposite sides. The amphipathy moment can be plotted along the entire sequence to further facilitate the identification of amphiphilic helices. Thus, the amphipathy moment provides indications for further experimental validation.

3.3.4. Hydropathy Plots

Selected subsequences can be displayed as a hydropathy plot (*see* **Note 9**). For most scales, parts that average below zero are mainly hydrophilic and therefore likely to be surface exposed, whereas parts that average above zero are mainly hydrophobic and therefore likely to be buried or membrane spanning. Mean hydropathy (GRAVY score) and scale in use are indicated. Kyte and Doolittle *(7)* showed for their scale that none of 84 analyzed soluble proteins had a GRAVY score > 0.5. Nevertheless, although a GRAVY score > 0.5 is a good indication that a protein is membrane bound, several membrane-embedded proteins have GRAVY scores even below zero. Therefore, the amino acid composition alone does not sufficient to distinguish soluble from membrane-bound proteins *(7)*. If the scale by Kyte and Doolittle is applied putative transmembrane stretches are marked by a red line; the used threshold value is displayed next to the size of the sliding window. Kyte and Doolittle showed in a statistical analysis of several proteins that the mean hydropathy of the most hydrophobic 19-residue-stretch averages at 1.08 ± 0.22 for soluble and at 1.86 ± 0.38 for membrane-bound proteins *(7)*. Therefore, they stated that if a given 19-residue-stretch displays a mean hydropathy of at least 1.6 there is a high probability for membrane spanning. By default WinPep uses a threshold of 1.6 to predict putative transmembrane segments. Nevertheless, it was also shown that the occurrence of multiple membrane spanning segments often correlates with a lower hydrophobicity of the individual segments *(7)*. Therefore, WinPep was designed to enable modifications of the default threshold by the user.

Furthermore, individual residues can be marked with an arrow. Line color and font may be changed for each mark by double-clicking on it with the "Select-Tool." Using the "Insert-text-Tool" user defined text items can be inserted. The position, font, and content may be changed by double-clicking on them with the "Select-Tool." Some text items are inserted by default (e.g., the currently used scale or mean hydropathy). They can be modified, repositioned, or deleted by the user. However, once edited, the automatic update of the content ceases.

Clicking the right mouse button invokes a floating menu where the displayed part of the sequence may be modified by opening the "Hydropathy Plot Properties Dialog" or the display may be unzoomed to 100%. In addition, it can be decided whether selections with the left mouse button zoom into the hydropathy plot or display a helical wheel.

3.3.5. Domain Drawing

A sketch of the domain composition of the sequence is easily generated. Every display contains at least one domain: the entire sequence. Further domains can be added. Domain color, name, font, and text color need to be specified. If domains overlap, the last defined (last in the list) will appear on top of the others. Therefore, the smallest domain should be defined last. By changing the default scale, drawings of proteins of different length may be produced that can be aligned in any graphic processing software. Individual residues can be marked. Line color and font are adjustable by double clicking on the mark. Furthermore, text items can be inserted and modified easily.

4. Notes

1. When sequences are pasted into WinGene or WinPep, any numbers, spaces, or special characters will be omitted. Previous manual deletion of sequence numbering, and so forth is therefore not necessary.
2. Make sure to use the file in "gcg" format.
3. Installation of an updated "gcg.*" file may corrupt the YFE file by introducing a new numbering.
4. The minimal ORF length can be changed in the options menu.
5. The default pK_a values on which the calculations are based were taken from **ref. *6***. However, these values can easily be changed from the options menu.
6. Available proteases can be edited and new ones added from the options menu.
7. Using the view menu the list of fragments can be sorted by weight, start position, or length.
8. The default settings for color coding amino acid residues may be changed for each helical wheel with the "Change Colors" option in the options menu.
9. The program comes with two predefined scales *(7,8)*. However, additional scales can be supplied from the options menu.

Acknowledgments

The author wants to thank everyone who tested the program, reported bugs, or contributed suggestions for additional functional features, especially C. Köhler, S. Frye, and J. Swire.

References

1. Hennig, L. (1999) WinGene/WinPep: user-friendly software for the analysis of aminoacid sequences. *BioTechniques* **26,** 1170–1172.
2. Roberts, R. J. and Macelis, D. (2000) REBASE - restriction enzymes and methylases. *Nucleic Acids Res.* **28,** 306–307.
3. Baldino, F., Chesselet, M.-F., and Lewis, M. E. (1983) High-resolution *in situ* hybridization histochemistry. *Methods Enzymol.* **168,** 761–777.
4. Gill, S. C. and von Hippel, P. H. (1989) Calculation of protein extinction coefficients from amino acid sequence data. *Analyt. Biochem.* **182,** 319–326.
5. Pace, C. N., Vajdos, F., Fee, L., Grimsley, G., and Gray, T. (1995) How to measure and predict the molar absorption coefficient of a protein. *Prot. Sci.* **4,** 2411–2423.
6. Creighton, T. E. (1993) *Proteins: Structures and Molecular Properties.* Freeman, New York.
7. Kyte, J. and Doolittle, R. F. (1982) A simple method for displaying the hydropathic character of a protein. *J. Mol. Biol.* **157,** 105–132.
8. Sweet, R. M. and Eisenberg, D. (1983) Correlation of sequence hydrophobicities measures similarity in three-dimensional protein structure. *J. Mol. Biol.* **171,** 479–488.

102

Isolation of Proteins Cross-linked to DNA by Cisplatin

Virginia A. Spencer and James R. Davie

1. Introduction

One way of identifying and further characterizing transcription factors is to study their association with DNA *in situ*. Many studies have performed this task using agents such as formaldehyde that crosslink proteins to DNA. However, the treatment of cells with agents such as formaldehyde results in the cross-linking of protein to DNA, and protein to protein. Thus, proteins cross-linked to DNA binding proteins may be misinterpreted as DNA binding proteins. To overcome this obstacle, researchers have focussed their attention on cisplatin (*cis*-DDP; *cis*-platinum (II)diamminedichloride), a cross-linking agent shown to crosslink protein to DNA and not to protein *(1)*. Recent studies have shown that the majority of proteins cross-linked to DNA by cisplatin *in situ* are nuclear matrix proteins *(2–4)*. We have also shown that cisplatin cross-links nuclear matrix-associated transcription factors and cofactors to DNA in the MCF-7 human breast cancer cell line *(5)*. Thus, cisplatin appears to be an effective cross-linking agent for studying the role of transcription factors and nuclear matrix proteins in transcription. In support of this, we have discovered cisplatin DNA-cross-linked nuclear matrix (NM) proteins whose levels vary between well and poorly differentiated human breast cancer cell lines *(6)*. Such changes in protein levels indicate that breast cancer development most likely involves changes in DNA organization, and, likely, changes in transcriptional events. Currently used nuclear matrix protein extraction protocols have been effective in identifying diagnostic NM protein markers for bladder cancer detection *(7)*. Thus, the isolation of cisplatin DNA-cross-linked proteins is a complementary approach to these nuclear matrix extraction protocols, which may also be useful in the detection of additional nuclear matrix proteins for cancer diagnosis.

This chapter provides a detailed description of the isolation of proteins cross-linked to DNA by cisplatin. The method is similar to that previously published by Ferraro and colleagues *(8)*. In addition, this method has been successfully performed on human breast cancer cells and avian erythrocytes. (*See* **Fig. 1**).

2. Materials

All solutions are prepared fresh from analytical grade reagents dissolved in double-distilled water. 1 m*M* Phenylmethylsulfonyl flouride (PMSF) was added to all solu-

From: *The Protein Protocols Handbook, 2nd Edition*
Edited by: J. M. Walker © Humana Press Inc., Totowa, NJ

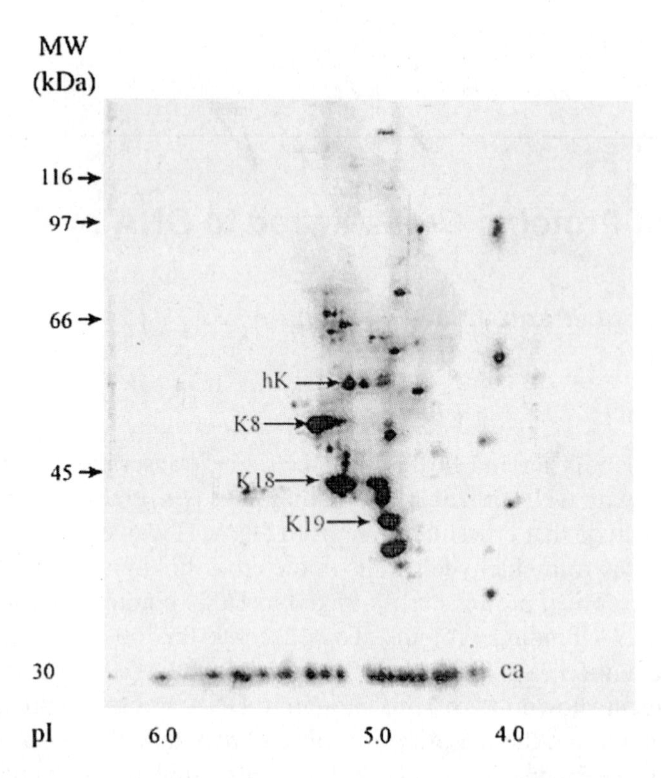

Fig. 1. Two-dimension gel electrophoresis profile of cisplatin DNA-cross-linked proteins. Thirty micrograms of proteins cross-linked to DNA in situ by cisplatin in MCF-7 human breast cancer cells were electrophoretically resolved on a two-dimensional gel, and the gel was stained with silver. K8, K18, and K19 represent cytokeratins 8, 18, and 19, respectively. The transcription factor hnRNPK is designated as hK. The position of the carbamylated forms of carbonic anhydrase is indicated by ca. The position of the molecular weight standards (in thousands) is shown to the left of the gel pattern.

tions immediately before use. With the exception of the cisplatin solution, all solutions are cooled on ice before use.

1. Hanks buffer: 0.2 g KCl (5.4 m*M*), 0.025 g Na_2HPO_4 (0.3 m*M*), 0.03 g KH_2PO_4 (0.4 m*M*), 0.175 g $NaHCO_3$ (4.2 m*M*), 0.07 g $CaCl_2$ (Ann) (1.3 m*M*), 62.5 µL 4 *M* $MgCl_2$ (0.5 m*M*), 0.075 g $MgSO_4$ (0.6 m*M*), 4 g NaCl (137 m*M*), and 0.5 g D-glucose (5.6 m*M*) in 500 mL of double distilled water.
2. Hanks buffer with Na acetate instead of NaCl: Refer to recipe for Hanks buffer except replace the NaCl with 9.3 g sodium acetate (137 m*M*).
3. 1 m*M* Cisplatin Cross-linking Solution: Add 0.003 g of cisplatin (*cis*-platinum (II)-diammine dichloride; Sigma) to 10 mL of Hanks buffer containing sodium acetate instead of NaCl. Cover the solution with foil to protect the cisplatin from the light and stir over gentle heat (approx 40°C) to dissolve the cisplatin. Once the cisplatin is dissolved, keep the solution at room temperature in a foil-covered or amber bottle (*see* **Notes 1–3**).
4. Lysis buffer: Add 150 g urea (5 *M*), 95.5 g guanidine hydrochloride (2 *M*), and 58.5 g NaCl (2 *M*) to 16 mL of 1 *M* KH_2PO_4 and 84 mL of 1 *M* K_2HPO_4 (200 m*M* potassium phosphate buffer, pH 7.5). Stir while heating solution to approx. 50°C to speed up solubi-

lization, make up to 500 mL with double distilled water, and filter solution with 1 *M* Whatman filter (*see* **Note 4**).

5. Hydroxyapatite preparation: Weigh out 1 g of hydroxyapatite Bio-Gel® HTP Gel (Bio-Rad, CA) for every 4 mg of total cellular DNA in the cell lysate, as determined by A_{260} measurements. Place hydroxyapatite into a 30 mL polypropylene tube and pre-equilibrate the hydroxyapatite by suspending the resin in 6 vol of lysis buffer. Gently invert the resin in the lysis buffer to mix, let the resin settle for approx 20 min on ice, then decant the lysis buffer off the hydroxyapatite (*see* **Notes 5,6**).
6. Reverse lysis buffer: Add 3.8 g Thiourea (1 *M*), 9.55 g guanidine hydrochloride (2 *M*), and 5.85 g NaCl (2 *M*), to 1.6 mL 1 *M* KH_2PO_4 and 8.4 mL 1 *M* K_2HPO_4 (200 m*M* potassium phosphate buffer, pH 7.5). Stir while heating solution to approx. 50°C, make up to 50 mL with double distilled water.

3. Methods

3.1. Isolation of Cisplatin DNA-Cross-linked Proteins

1. Rinse 1×10^7 cells in 30 mL of cold Hank's buffer.
2. Centrifuge at 50*g* for 5 min at room temperature.
3. Repeat the rinse two more times.
4. Decant the Hank's buffer from the cell pellet and add 10 mL of 1 m*M* cisplatin solution to the pellet.
5. Incubate at 37°C for 2 h with shaking.
6. Centrifuge at 50*g* for 5 min at room temperature.
7. Resuspend the cell pellet in 10 mL of cold lysis buffer, and store on ice.
8. Measure the A_{260} of 10 µL of the cell lysate and use this value to determine the grams of hydroxyapatite required for DNA isolation (*see* **Notes 6**, **8**, and **9**).
9. Preequilibrate the hydroxyapatite in a 30-mL tube.
10. Transfer the cell lysate into the tube containing the preequilibrated hydroxyapatite, and mix by gentle inversion until all the hydroxyapatite is resuspended.
11. Incubate 1 h at 4°C on an orbitron.
12. Centrifuge at 5000*g* for 5 min at 4°C.
13. Remove the supernatant that contains proteins not cross-linked to DNA.
14. Wash the hydroxyapatite resin with 20 mL of ice-cold lysis buffer by gentle inversion until the resin is completely resuspended.
15. Centrifuge the washed resin at 5000*g* for 5 min at 4°C.
16. Repeat this wash two more times.
17. Add 10 mL of cold reverse lysis buffer to the hydroxyapatite resin.
18. Incubate at 4°C for 2 h on the orbitron to reverse the crosslink between the protein and resin-bound DNA.
19. Centrifuge the hydroxyapatite resin at 5000*g* for 5 min at 4°C.
20. Carefully remove the supernatant and place it in dialysis tubing that has been soaked in distilled water for at least 30 min.
21. Dialyze the protein sample at 4°C against four 2-L changes of double-distilled water over a 24-h period. Include 0.5 m*M* PMSF in the first change of double-distilled water (*see* **Notes 10** and **11**).
22. Transfer the dialyzed solution from the dialysis tubing to a 13-mL centrifuge tube, and freeze at –80°C until the solution is completely frozen.
23. Lyophilize the solution to a dry powder.
24. Resuspend the dry powder in 100 µL of 8 *M* urea and store at –20°C (**Fig. 1**).

4. Notes

1. The conditions for cisplatin cross-linking and protein isolation may vary for other cell types.
2. NaCl is excluded from the cross-linking solution because chloride ions impair the efficiency of the cross-linking reaction by competing with cisplatin for cellular proteins *(9)*.
3. Human breast cancer cells treated with 1 μ*M* cisplatin display a drastic decrease in cell number after 2 h. Thus, a prolonged incubation time (i.e., > 2 h) in the presence of cisplatin may result in the activation of pro-apoptotic proteins involved in protein degradation *(5)*.
4. Filtering the lysis solution with 1M Whatman filter paper will remove particulates that may interfere with A_{260} measurements.
5. Hydroxyapatite is a calcium phosphate resin that binds to the phosphate backbone of DNA.
6. One gram of hydroxyapatite is used for every 4 mg of genomic DNA in the cell lysate, as determined by A_{260} measurements, as this ratio has been shown previously in our laboratory to bind all cellular DNA with approx 100% efficiency (data not shown).
7. Gentle inversion of the hydroxyapatite resin is important to avoid damaging the integrity of the resin.
8. For measuring the A_{260} of the cell lysate, transfer 10 μL of cell lysate into a tube containing 990 μL of lysis buffer.
9. For determining an approximate amount of total cellular DNA within the lysate, the following equation is used.
10. The porosity of the dialysis tubing will depend on the size of the protein of interest.
11. Dialysis tubing should be soaked in distilled water for at least 30 min before use.

$$A_{260} \times 50\ \mu g/mL \times 100 \times 10\ mL/1000\ \mu g/mg \text{ of DNA}$$

One A_{260} unit represents 50 mg of DNA/mL of cell lysate; thus the absorbance reading is first multiplied by 50 and then by the dilution factor (i.e., 100) to determine the micrograms of DNA in 1 mL of cell lysate. The resulting value is multiplied by the total volume of cell lysate (i.e., 10 mL) to determine the total micrograms of DNA in the cell lysate and then divided by 1000 to convert this value into milligrams of DNA. The determined amount of cellular DNA is considered only an approximation, as the cell lysate contains some proteins with a peak absorption at 260 nm.

Acknowledgments

Our research was supported by grants from the Medical Research Council of Canada (MT-9186, RO-15183), CancerCare Manitoba, U.S. Army Medical and Materiel Command Breast Cancer Research Program (DAMD17-001-10319), and the National Cancer Institute of Canada (NCIC) with funds from the Canadian Cancer Society. A Medical Research Council of Canada Senior Scientist to J. R. D. and a NCIC Studentship to V. A. S. are gratefully acknowledged.

References

1. Foka, M. and Paoletti, J. (1986) Interaction of *cis*-diamminedichloroplatinum (II) to chromatin. *Biochem. Pharmacol.* **35,** 3283–3291.
2. Davie, J. R., Samuel, S. K., Spencer, V. A., Bajno, L., Sun, J. M., Chen, H. Y., and Holth, L. T. (1998) Nuclear matrix: application to diagnosis of cancer and role in transcription and modulation of chromatin structure. *Gene Ther. Mol. Biol.* **1,** 509–528.

3. Ferraro, A., Eufemi, M., Cervoni, L., Altieri, F., and Turano, C. (1995) DNA–nuclear matrix interactions analyzed by crosslinking reactions in intact nuclei from avian liver. *Acta Biochim. Pol.* **42,** 145–151.
4. Ferraro, A., Cervoni, L., Eufemi, M., Altieri, F., and Turano, C. (1996) A comparison of DNA–protein interactions in intact nuclei from avian liver and erythrocytes: a crosslinking study. *J. Cell. Biochem.* **62,** 495–505.
5. Samuel, S. K., Spencer, V. A., Bajno, L., Sun, J. M., Holth, L. T., Oesterreich, S., and Davie, J. R. (1998) *In situ* crosslinking by cisplatin of nuclear matrix-bound transcription factors to nuclear DNA of human breast cancer cells. *Cancer Res.* **58,** 3004–3008.
6. Spencer, V. A., Samuel, S. K., and Davie, J. R. (2000) Nuclear matrix proteins associated with DNA *in situ* in hormone-dependent and hormone-independent human breast cancer cell lines. *Cancer Res.* **60,** 288–292.
7. Konety, B. R., Nguyen, T. S., Brenes, G., Sholder, A., Lewis, N., Bastacky, S., et al. (2000) Clinical usefulness of the novel marker BLCA-4 for the detection of bladder cancer. *J. Urol.* **164,** 634–639.
8. Ferraro, A., Grandi, P., Eufemi, M., Altieri, F., Cervoni, L., and Turano, C. (1991) The presence of *N*-glycosylated proteins in cell nuclei. *Biochem. Biophys. Res. Commun.* **178,** 1365–1370.
9. Lippard, S. J. (1982) New chemistry of an old molecule: *cis*-$[Pt(NH_3)_2Cl_2]$. *Science* **218,** 1075–1082.

103

Isolation of Proteins Cross-linked to DNA by Formaldehyde

Virginia A. Spencer and James R. Davie

1. Introduction

Formaldehyde is a reversible cross-linker that will cross-link protein to DNA, RNA, or protein *(1)*. Because of its high-resolution (2 Å) cross-linking, formaldehyde is a useful agent to cross-link a DNA binding protein of interest to DNA. For example, formaldehyde has been used to cross-link proteins to DNA in studies fine-mapping the distribution of particular DNA binding proteins along specific DNA sequences *(1,2)*.

When applied to a cell, formaldehyde will initially begin to cross-link protein to DNA. As the time of exposure to formaldehyde increases, proteins become cross-linked to one another. Soluble cellular components become more insoluble as they become cross-linked to one another and to the insoluble cellular material. Sonication is most commonly used to release cross-linked DNA–protein complexes from the insoluble nuclear material. Excessively cross-linking a cell with formaldehyde will cause nuclear DNA cross-linked to protein to become trapped within the insoluble nuclear material. Such an event will protect this cross-linked DNA from breakage by sonication. Moreover, the efficiency of formaldehyde DNA–protein cross-linking varies with cell type *(3)*. Therefore, two parameters must be considered when using formaldehyde as an agent for cross-linking DNA to a protein of interest: the release of DNA from the insoluble nuclear material after cross-linking and the extent of sonication of cross-linked cells. This chapter describes an approach for determining the optimal formaldehyde cross-linking conditions of a cell and for isolating proteins cross-linked to DNA by formaldehyde. (*See* **Fig. 1**).

2. Materials

All solutions are prepared from analytical grade reagents dissolved in double-distilled water. 1 m*M* phenylmethylsulfonyl fluoride (PMSF) was added to all solutions immediately before use. All solutions were cooled on ice before use.

1. RSB buffer: Add 10 mL of 1 *M* Tris-HCl, pH 7.5 (10 m*M*), and 2.5 mL of 4 *M* NaCl (10 m*M*) to approx 800 mL of double-distilled water. Adjust the pH to 7.5 with NaOH then add 0.75 mL of 4 *M* $MgCl_2$ (3 m*M*). Readjust the pH if necessary, then make volume up to 1 L with double-distilled water.

From: *The Protein Protocols Handbook, 2nd Edition*
Edited by: J. M. Walker © Humana Press Inc., Totowa, NJ

2. Hepes buffer: Add 2.38 g of Hepes (10 m*M*), and 2.5 mL of 4 *M* NaCl (10 m*M*) to 800 mL of double-distilled water. Adjust the pH to 7.5 with NaOH then add 0.75 mL of 4 *M* $MgCl_2$ (3 m*M*). Re-adjust the pH if necessary, then make volume up to 1 L with double-distilled water.
3. Lysis buffer: Add 150 g urea (5 *M*), 95.5 g guanidine hydrochloride (2 *M*), and 58.5 g NaCl (2 *M*) to 16 mL of 1 *M* KH_2PO_4 and 84 mL of 1 *M* K_2HPO_4 (200 m*M* potassium phosphate buffer, pH 7.5). Stir while heating solution to approx. 50°C to speed up solubilization, make up to 500 mL with double distilled water, and filter solution with 1 *M* Whatman filter.
4. Dounce homogenizer (for 20-mL sample volume).
5. Hydroxyapatite preparation: Weigh out 1 g of hydroxyapatite Bio-Gel® HTP Gel (Bio-Rad, CA) for every 4 mg of total cellular DNA in the cell lysate, as determined by A_{260} measurements. Place hydroxyapatite into a 30-mL polypropylene tube and preequilibrate the hydroxyapatite by suspending the resin in six volumes of lysis buffer. Gently invert the resin in the lysis buffer to mix, let the resin settle for approx 20 min on ice, and then decant the lysis buffer off the hydroxyapatite.

3. Methods

3.1. Formaldehyde Cross-linking of Immature Chicken Erythrocyte Nuclei

This procedure was performed on immature chicken erythrocytes isolated from adult white Leghorn chickens treated with phenylhydrazine.

1. 2 mL of packed erythrocytes are resuspended in 15 mL of RSB buffer containing 0.25% Nonidet P-40 (NP-40).
2. The cells are homogenized 5× in a Dounce homogenizer and centrifuged at 1500*g* for 10 min at 4°C.
3. **Steps 1** and **2** are performed two more times, leaving a pellet of nuclei.
4. The nuclei are resuspended in HEPES buffer to an A_{260} of 20 U/mL (*see* **Note 1**).
5. Formaldehyde is then added to the suspension of nuclei to a final concentration of 1% (v/v).
6. The nuclei are mixed gently by inversion, and incubated for up to 15 min at room temperature.
7. After 0, 5, 10, and 15 min of formaldehyde cross-linking, 4-mL aliquots of nuclei are collected and made up to 125 m*M* glycine on ice to stop the cross-linking reaction.
8. The aliquots of nuclei are then centrifuged at 1500*g* for 10 min at 4°C (*see* **Note 2**).
9. The nuclear pellets are then washed in 10 mL of RSB buffer and centrifuged at 1500*g* for 10 min at 4°C.
10. The nuclear pellets are resuspended in 10 mL of ice-cold lysis buffer.

3.2. Sonication of Formaldehyde-Cross-linked Cells

For each formaldehyde-cross-linked sample from **Subheading 3.1.**:

1. The 4 mL of the lysed nuclei are transferred to a 50-mL Falcon tube and sonicated under 170 W for ten 30-s pulses with a Braun-sonic 1510 Sonicator. The sample is cooled on ice for 1-min waiting intervals between each pulse (*see* **Notes 3–5**).
2. A 100-μL aliquot of nuclei is dialyzed against double-distilled water (without PMSF) overnight at 4°C to remove excess urea and salt that may interfere with proteinase K digestion. The dialyzed sample is made to 0.5% sodium dodecyl sulfate (SDS) and 0.4 mg/mL of proteinase K and incubated for 2 h at 55°C to digest the protein.
3. The sample is then incubated at 65°C for 6 h to reverse the formaldehyde cross-links between the DNA and protein.

4. The digested mixture is extracted 3× with an equal volume of solution composed of phenol–chloroform–isoamyl alcohol in a 25:24:1 ratio, respectively.
5. To precipitate the DNA from the sample, 1/10th the sample volume of 3 *M* sodium acetate, pH 5.5, along with three volumes of absolute ethanol is added to the sample and the sample is incubated at –80°C for 20 min.
6. The sample is centrifuged at 12,000*g* for 10 min at 4°C to pellet the DNA.
7. The DNA pellet is washed with 1 mL of ice-cold 70% ethanol, and centrifuged at 12,000*g* for 10 min at 4°C.
8. The resulting DNA pellet is resuspended in 30 µL of double-distilled water, and 4 µL of this pellet is electrophoresed on a 0.8% agarose gel to identify high molecular weight DNA bands indicative of extensive cross-linking.

3.3. Efficiency of Solubilization of Formaldehyde-Cross-linked Cells

For each formaldehyde-cross-linked sample from **Subheading 3.1.**:

1. Transfer the 4 mL of cross-linked, lysed, and sonicated nuclei to a 15-mL tube.
2. Determine the A_{260} of 10 µL of the total nuclear lysate.
3. Centrifuge the sample at 9000*g* for 10 min at 4°C.
4. The supernatant contains solubilized DNA and protein. Transfer the supernatant to a clean 15-mL tube.
5. Determine the A_{260} of 10 µL of the supernatant in 990 µL of lysis buffer.
6. Divide the A_{260} of the supernatant by the A_{260} of the total nuclear lysate and multiply this value by 100 to determine the percent of DNA released from the nuclei following formaldehyde cross-link and sonication.

3.4. Isolation of Proteins Cross-linked to DNA by Formaldehyde

For each formaldehyde-cross-linked sample from **Subheading 3.1.**:

1. Determine the approximate amount of DNA present in the 4 mL of lysed nuclei suspension from **Subheading 3.3.**
2. Add the lysed nuclei suspension to preequilibrated hydroxyapatite (*see* **Notes 6–8**) and mix by gentle inversion.
3. Incubate at 4°C for 1 h on an orbitron.
4. Centrifuge the hydroxyapatite at 5000*g* for 5 min at 4°C.
5. Wash the hydroxyapatite with 10 mL of lysis buffer, mix by gentle inversion, and centrifuge at 5000*g* for 5 min at 4°C.
6. Repeat **step 4** an additional two times.
7. Add 2 mL of lysis buffer to the washed hydroxyapatite and mix by gentle inversion.
8. Incubate this suspension at 68°C for 6 h to reverse the formaldehyde cross-links between the DNA and protein.
9. Centrifuge the sample at 4°C or room temperature for 5 min at 5000*g*.
10. Place the supernatant containing protein that was cross-linked to DNA into presoaked dialysis tubing (*see* **Notes 10** and **11**).
11. Dialyze the sample overnight at 4°C against 2–3-L changes of double-distilled water (include 0.5 m*M* PMSF in the first change).
12. Lyophilize the sample to a powder form.
13. Resuspend the powder in double distilled water and store at –20°C (*see* **Fig. 1**).

4. Notes

1. Sonication conditions will vary for different cell types.
2. Perform sonication on ice to avoid protein denaturation.

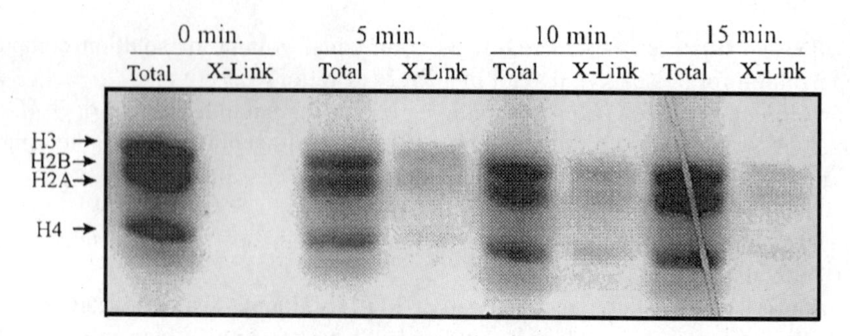

Fig. 1. Sodium dodecyl sulfate-polyacrylamide gel electrophoresis (SDS-PAGE) of total cellular proteins and formaldehyde-DNA-crosslinked proteins. Formaldehyde-crosslinked nuclei were separated into two fractions of equal volume. One fraction was crosslinked with formaldehyde for 0, 5, 10, and 15 min, and the DNA-crosslinked proteins were isolated by hydroxyapatite chromatography, dialyzed against double-distilled water, lyophilized to a powder form, and made up to a final volume of 200 μL with double-distilled water. The other fraction was lysed, dialyzed, lyophilized to a powder form, and made up to a final volume of 1 mL with double-distilled water. Equal volumes of total nuclear protein were loaded onto a 15% SDS-PAGE gel for each treatment. In addition, equal volumes of DNA-crosslinked proteins were loaded onto the same gel for each treatment. The gel was electrophoresed at 170 V for 70 min at room temperature, stained with Coomassie Blue overnight, and then destained. H3, H2B, H2A, and H4 represent histones H3, H2B, H2A, and H4, respectively.

3. The extent of sonication depends on the length of the target DNA sequence. For example, if the association of a protein with a specific 1000-basepair region needs to be determined, then the DNA should be sonicated to 500 basepairs to avoid the immunoprecipitation of DNA sequences surrounding the target region (*see* **ref. *4*** for further explanation). However, if one is simply trying to determine if a protein of interest is associated with DNA, then the formaldehyde-cross-linked cells need only be sonicated to an extent that allows the efficient release of nuclear DNA from the insoluble nuclear material (*see* **Note 5**).
4. Formaldehyde reacts with amine groups of proteins. Thus, to ensure a high efficiency of cross-linking, it is important to resuspend the nuclei in a HEPES buffer before treatment with formaldehyde.
5. The duration of formaldehyde cross-linking will vary according to cell type, cell treatment, and the degree to which the DNA associated with the nuclear material can be solubilized. For example, the sonication and subsequent centrifugation of a nuclear lysate may result in the solubilization of only 80% of total nuclear DNA. The acceptability of this percentage of DNA release from the nucleus depends on the location of the protein of interest. If the target protein is tightly associated with DNA that is associated with the nuclear matrix, the fraction of solubilized DNA–protein complexes may be somewhat depleted of the target protein even though as much as 80% of the nuclear DNA is released from the insoluble material.
6. Hydroxyapatite is a calcium phosphate resin that binds to the phosphate backbone of DNA.
7. A ratio of 1 g of hydroxyapatite for every 4 mg of genomic DNA has been shown in our lab to bind all cellular DNA with 100% efficiency (data not shown).
8. The following equation can be used to determine the approximate amount of DNA within the cell lysate:

$$A_{260} \times 50\ \mu g/mL \times 100 \times \text{(volume of cell lysate)}/1000\ \mu g/mg \text{ of DNA}$$

One A_{260} unit represents 50 mg of DNA/mL of cell lysate. To determine the micrograms of DNA in 1 mL of cell lysate, multiply the absorbance reading by 50 and then by the dilution factor (i.e., 100). Multiply the resulting concentration by the total volume of cell lysate to determine the total mg of DNA in the cell lysate. Divide the total micrograms by 1000 to convert this value into milligrams of DNA. The resulting amount of cellular DNA is only an approximation, as some proteins within the cell lysate will have a peak absorption at 260 nm.

9. Gently inverting the hydroxyapatite resin when mixing avoids damaging the integrity of the resin.
10. The porosity of the dialysis tubing will depend on the size of the protein of interest.
11. Dialysis tubing should be soaked in distilled water for at least 30 min before use.

Acknowledgments

Our research was supported by grants from the Medical Research Council of Canada (MT-9186, RO-15183), CancerCare Manitoba, U.S. Army Medical and Materiel Command Breast Cancer Research Program (DAMD17-001-10319), and the National Cancer Institute of Canada (NCIC) with funds from the Canadian Cancer Society. A Medical Research Council of Canada Senior Scientist to J. R. D. and a NCIC Studentship to V. A. S. are gratefully acknowledged.

References

1. Orlando, V. (2000) Mapping chromosomal proteins by in vivo formaldehyde-crosslinked-chromatin immunoprecipitation. *Trends Biochem. Sci.* **25,** 99–104.
2. Dedon, P. C., Soults, J. A., Allis, C. D., and Gorovsky, M. A. (1991) A simplified formaldehyde fixation and immunoprecipitation technique for studying protein-DNA interactions. *Analyt. Biochem.* **197,** 83–90.
3. Orlando, V., Strutt H., and Paro, R. (1997) Analysis of chromatin structure by in vivo formaldehyde crosslinking. *Methods* **11,** 205–214.
4. Kadosh, D. and Struhl, K. (1998) Targeted recruitment of the Sin3-Rpd3 histone deacetylase complex generates a highly localized domain of repressed chromatin in vivo. *Mol. Cell. Biol.* **18,** 5121–5127.

Part VI

Glycoproteins

104

Detection of Glycoproteins in Gels and Blots

Nicolle H. Packer, Malcolm S. Ball, Peter L. Devine, and Wayne F. Patton

1. Introduction

As we become more aware of the significance of posttranslational modifications, such as glycosylation, in the production of recombinant proteins and in the proteomic studies of development and disease, techniques for the identification and characterization of the oligosaccharides attached to proteins need to be established.

After separation of the proteins by either one-diemensional-(1D)- or two-dimensional-(2D)-polyacrylamide gel electrophoresis (PAGE), the initial step is to identify which proteins are glycoproteins so that further characterization can proceed. Various methods have been developed since the early detection of glycoproteins on gels *(1)* and blots, with color, fluorescence, and lectin detection now carried out at the analytical level. The actual level of detection of course depends on the extent of glycosylation of the protein, as the reagents react only with the carbohydrate moiety. We have chosen to describe here the stains that we have found to be the most useful for visualizing glycoproteins, both on gels and blots, after separation by electrophoresis. It should be noted that all the staining procedures currently in use destroy the structure of the carbohydrate and thus prevent further analysis of the glycan component once the glycoprotein is visualized.

As an initial characterization step, once the glycoprotein of interest has been located, a protocol for analyzing the monosaccharide composition on replicate blots of these proteins is also described.

1. Periodic acid-Schiff staining is a generally useful technique for locating glycoproteins on gels and nitrocellulose blots, although the sensitivity may not be sufficient for many applications. Realistically, only 1–10 µg of a highly glycosylated protein can be detected and the stain is most useful for mucins and proteoglycans. Periodic acid oxidizes vicinal diols of glycosyl residues to dialdehydes. The aldehydes are then allowed to react with fuchsin (Schiff's reagent) to form a Schiff base. Glycoproteins stain pink with fuchsin on a clear background.
2. Digoxigenin (DIG)–anti-digoxigenin, alkaline phosphate (AP) labeling is an extension of the periodic acid–Schiff method above although the sensitivity is much greater (realistically, depending on the degree of glycosylation, about 0.1 µg of glycoprotein). Glycoproteins can be detected on dot blots or after Western transfer to membranes such as nitrocellulose or polyvinylidene difluoride (PVDF). Vicinal (adjacent) hydroxyl groups

From: *The Protein Protocols Handbook, 2nd Edition*
Edited by: J. M. Walker © Humana Press Inc., Totowa, NJ

in sugars of glycoconjugates are oxidized to aldehyde groups by mild periodate treatment. The spacer-linked steroid hapten digoxigenin (DIG) is then covalently attached to these aldehydes via a hydrazide group. DIG-labeled glycoconjugates are subsequently detected in an enzyme immunoassay using a DIG-specific antibody conjugated with AP. DIG glycan detection is known to label almost all known *N*- and *O*-linked glycans including GPI anchors.

3. Lectins are carbohydrate binding proteins which are particularly useful in glycoprotein and carbohydrate analysis as they can be conjugated to a variety of enzymes or haptens for use in sensitive detection systems. Their specificity can be used to probe for specific structures in the glycoconjugates. Lectins are usually classified on the basis of the monosaccharides with which they interact best, but it is important to note that complex glycoconjugates are generally found to be much better ligands. In addition, the position of a particular monosaccharide in a glycan chain (i.e., to what it is attached) will affect lectin binding, so results obtained in lectin binding studies should be treated with caution. For example, (a) the wheat germ agglutin (WGA) is inhibited most strongly by dimeric GlcNAc, but in glycoproteins this lectin also reacts very strongly with sialic acid and peptide-linked GalNAc; (b) the peanut agglutin binds to Gα1,3GalNAc but does not react when this structure is sialylated. The labeled lectin–carbohydrate conjugate can then be visualized by enzyme immunoassay in the same way as in (b).
4. Recently a new kit that uses a fluorescent hydrazide to react with the periodate-oxidized carbohydrate groups on glycoproteins has been released commercially. The fluorescent tag (Pro-Q™ Emerald 300) is excited by ultraviolet light and emits at a visible light (green) wavelength. The fluorescent signal allows an increased level of detection of the glycoproteins (down to 1 ng of protein) on both gels and PVDF blots, while allowing the subsequent visualization of the total proteins with another fluorescent stain emitting at a different wavelength.
5. The monosaccharide composition of a glycoprotein is a useful start to full characterization. This is obtained by hydrolysis of the separated glycoprotein spots that have been electroblotted to PVDF, followed by monosaccharide analysis using high pressure anion exchange chromatography with pulsed amperometric detection (HPAEC-PAD) *(2)*.

In all methods, it is useful to include a glycosylated protein such as transferrin or ovalbumin in the marker lane as a control.

2. Materials

2.1. Periodic Acid–Schiff Staining

1. Solution A: 1.0% (v/v) periodic acid in 3% acetic acid. Periodic acid is corrosive and volatile—**handle with caution**. Be aware of the concentration of periodic acid in the solution that is being diluted, as periodic acid is only about 50% out of the reagent bottle.
2. Solution B: 0.1% (w/v) Sodium metabisulfite in 10 m*M* HCl.
3. Schiff's reagent: A commercial reagent from Sigma Chemical Co. (St. Louis, MO, USA) may be used or better staining can often be achieved by making fresh reagent:
   a. Dissolve 1 g of basic fuchsin in 200 mL of boiling distilled water, stir for 5 min and cool to 50°C.
   b. Filter and add 20 mL of 1 *M* HCl to filtrate.
   c. Cool to 25°C, add 1 g of potassium metabisulfite, and leave to stand in the dark for 24 h.
   d. Add 2 g of activated charcoal, shake for 1 min, and filter. Store at room temperature in the dark.

Schiff's reagent is corrosive and slightly toxic and a very dilute solution will stain anything with oxidized carbohydrates a pink-purple color **Note:** wear gloves and protective clothing when using this solution and washing it out of the gel/blot.
4. Solution C: 50% (v/v) Ethanol.
5. Solution D: 0.5% (w/v) Sodium metabisulfite in 10 m*M* HCl.
6. Solution E: 7.5% (v/v) Acetic acid–5% (v/v) methanol in distilled water.

Solutions A, B, and D should be made up freshly.

2.2. DIG–Anti-DIG AP Labeling

1. Buffer A (TBS): 50 m*M* Tris-HCl, pH 7.5, 150 m*M* NaCl.
2. Buffer B: 100 m*M* Sodium acetate, pH 5.5.
3. Buffer C: 100 m*M* Tris-HCl, 50 m*M* $MgCl_2$, 100 m*M* NaCl.
4. Buffer D: 250 m*M* Tris-HCl, pH 6.8, 8% (w/v) sodium dodecyl sulfate (SDS), 40% (v/v) glycerol, 20% (v/v) 2-mercaptoethanol, bromophenol blue as tracking dye.
5. Buffer E: 50 m*M* potassium phosphate, 150 m*M* NaCl, pH 6.5 (PBS).
6. Blocking reagent: A fraction of milk proteins that are low in glycoproteins. 0.5g sample is dissolved in 100 mL of buffer A. The solution should be heated to 60°C with stirring; the solution will remain turbid. Allow the solution to cool before immersing the membrane.
7. DIG Glycan Detection Kit (Roche Diagnostics, Basel, Switzerland) containing:
   a. Solution 1: 10 m*M* sodium metaperiodate in buffer B.
   b. Solution 2: 3.3 mg/mL of sodium metabisulfite.
   c. DIG-succinyl-ε-amidocaproic acid hydrazide.
   d. Anti-DIG-AP: Polyclonal sheep anti-DIG Fab fragments, conjugated with AP (750 U/mL).
   e. Solution 3: 75 mg/mL 4-nitroblue tetrazolium chloride dissolved in 70% (v/v) dimethylformamide.
   f. Solution 4: 50 mg/mL of 5-bromo-4-chloro-3-indolyl-phosphate dissolved in dimethylformamide.

   Make sure that solutions 3 and 4 are still good, as after a few weeks the solutions may begin to precipitate thus reducing the staining efficiency. Store these solutions (3 and 4) in the dark.

2.3. DIG-Labeled Lectin Staining

1. 1% KOH.
2. Blocking reagent (Roche Diagnostics): A fraction of milk proteins that are low in glycoproteins. A 0.5 g sample is dissolved in 100 mL of TBS. The solution should be heated to 60°C with stirring; the solution will remain turbid. Allow the solution to cool before immersing the membrane. Other blockers such as skim milk powder, gelatin, or bovine serum albumin (BSA) may lead to high background.
3. TBS: 50 m*M* Tris-HCl, pH 7.5, 150 m*M* NaCl.
4. TBS-Tween: TBS + 0.05% Tween 20.
5. Divalent cation stock solution: 0.1 *M* $CaCl_2$, 0.1 *M* $MgCl_2$, 0.1 *M* $MnCl_2$.
6. DIG-labeled lectins (Roche Diagnostics): Sambucus sieboldiana agglutinin (SNA), Maackia amurensis agglutinin (MAA), Arachis hypogaea (peanut) agglutinin (PNA), and Datura stramonium agglutinin (DSA).
7. Part of DIG Glycan Detection Kit (Roche Diagnostics) comprising:
   a. Anti-DIG-AP: Polyclonal sheep anti-DIG Fab fragments, conjugated with AP (750 U/mL)
   b. Buffer C: Tris-100 m*M* HCl, 50 m*M* $MgCl_2$, 100 m*M* NaCl.

c. Solution 3: 75 mg/mL of 4-nitrobluetetrazolium chloride dissolved in 70% (v/v) dimethylformamide.
d. Solution 4: 50 mg/mL of 5-bromo-4-chloro-3-indolyl-phosphate dissolved in dimethylformamide.

Store these solutions (3 and 4) in the dark.

2.4 Pro-Q Emerald 300 Dye Staining

1. Pro-Q Emerald 300 glycoprotein gel stain kit (Molecular Probes, OR, USA) containing:
   a. Pro-Q Emerald 300 reagent, 50× concentrate in dimethyl formamide (DMF) (component A), 5 mL.
   b. Pro-Q Emerald 300 dilution buffer (component B), 250 mL.
   c. Periodic acid (component C), 2.5 g.
   d. SYPRO Ruby protein gel stain (component D), 500 µL.
   e. SDS, component E), 500 µL of a 10% solution.
   f. CandyCane glycoprotein molecular weight standards (component F), 40 µL, sufficient volume for approx 20 gel lanes. Each protein is present at approx 0.5 mg/mL. The standards contain a mixture of glycosylated and nonglycosylated proteins, which, when separated by electrophoresis, provide alternating positive and negative controls.
2. Fix solution: Prepare a solution of 50% methanol and 50% dH_2O. One 8 × 10 gel will require approx 100 mL of fix solution.
3. Wash solution. Prepare a solution of 3% glacial acetic acid in dH_2O. One 8 × 10 gel will require about approx 250 mL of wash solution.
4. Oxidizing solution. Add 250 mL of 3% acetic acid to the bottle containing the periodic acid (component C) and mix until completely dissolved.
5. CandyCane molecular weight standards diluted in sample buffer. For a standard lane on a 8 cm × 10 cm gel, dilute 0.5 µL of the standards with 7.5 µL of sample buffer and vortex-mix. This will result in approx 250 ng of each protein per lane, a sufficient amount for detection of the glycoproteins by the Pro-Q Emerald 300 stain. For large 16 cm × 18 cm gels, double the amount of standard and buffer used.

2.5. HPAEC Analysis of Monosaccharide Composition

1. Methanol.
2. 0.1 *M* TFA.
3. 2 *M* TFA.
4. Standard sugars:
   a. 0.1 mmole/mL of lactobionic acid.
   b. 0.1 mmole/mL of 2-deoxyglucose.
   c. 0.1 mmole/mL of *N*-acetyl neuraminic acid and *N*-glycolyl neuraminic acid.
   d. 0.1 mmole/mL of mixture of fucose, 2-deoxyglucose, galactosamine, glucosamine, galactose, glucose and mannose.

   Sugars should be dried thoroughly over phosphorus pentoxide in a desiccator before weighing.
5. Metal-free HPLC System with DIONEX CarboPac PA1 PA10 column, 4 mm × 25 cm and pulsed amperometric detector (HPAEC-PAD).

3. Methods

3.1. Periodic Acid–Schiff Staining

This method is essentially as described in **ref. *3*** for glycoproteins transferred to nitrocellulose membranes. A modification of the method for PVDF membranes is described in **ref. *4***.

*3.1.1. Gel Staining (see **Note 1**)*

1. Soak the gel in solution C for 30 min (*see* **Note 2**).
2. Wash in distilled water for 10 min. All of the ethanol must be removed from the gel, so make sure that the gel is immersed in the water properly. If necessary wash a second time to ensure the removal of the ethanol.
3. Incubate in solution A for 30 min. **Beware of the fumes from the acid.** From this point onwards the gel should be placed in the fume hood.
4. Wash in distilled water for at least 6 × 5 min or 5 × 5 min and 1× overnight.
5. Wash in solution B for 2 × 10 min. At this stage make up 100 mL of solution B and perform 2 × 30 mL washes; save the final 40 mL for **step 7**.
6. Incubate in Schiff's reagent for 1 h in the dark. It is essential after adding the Schiff's reagent that the gel is kept in the dark, as any light will stop the color from developing.
7. Incubate in solution B for 1 h in the dark.
8. Wash several times in solution D for a total of at least 2 h and leave as long as overnight to ensure good color detection (*see* **Notes 3** and **4**).
9. Store the gel in solution E.

*3.1.2. Membrane Staining (see **Note 5**)*

1. Wash the membrane in distilled water for 5 min (*see* **Note 6**).
2. Incubate in solution A for 30 min.
3. Wash in distilled water for 2 × 5 min.
4. Wash in solution B for 2 × 5 min.
5. Incubate for 15 min in Schiff's reagent (*see* **Note 3**).
6. Wash in solution B for 2 × 5 min.
7. Air dry the membrane.

3.2. DIG–Anti-DIG AP Labeling (See Note7)

This staining procedure can only be used on membranes; however, the proteins can be prelabeled in solution before electrophoresis, or labeled on the membrane after blotting (*see* **Notes 8** and **9**). In both cases the color development is the same. Nitrocellulose membranes can be used but some background staining can occur with postlabeling. In preference, the proteins should be blotted onto PVDF.

The reagents for this method are provided in the Roche Diagnostics DIG Glycan Detection Kit and the methods described are essentially taken from that kit.

A similar kit based on biotin/streptavidin binding instead of the DIG–Anti-DIG interaction is marketed by Bio-Rad as the Immun-Blot® Glycoprotein Detection Kit.

*3.2.1. Prelabeling (see **Note 8**)*

1. Dilute protein solution 1:1 to 20 µL with buffer B (*see* **Note 7**).
2. Add 10 µL of solution 1 and incubate for 20 min in the dark at room temperature.
3. Add 10 µL of solution 2 and leave for 5 min. The addition of the sodium bisulfite destroys the excess periodate.
4. Add 5 µL of DIG-succinyl-e-amidocaproic acid hydrazide, mix, and incubate at room temperature for 1 h. Sensitivity may be increased by increasing the incubation time to several hours.
5. Add 15 µL buffer D and heat the mixture to 100°C for 5 min to stop the labeling.
6. Separate the labeled glycoproteins by SDS-PAGE and blot to the membrane (*see* **Note 10**) which is now ready for the staining reaction (*see* **Subheading 3.2.3.**).

*3.2.2. Postlabeling (see **Note 9**)*

1. Wash the membrane for 10 min in 50 mL buffer E (PBS). (*see* **Note 11**).

2. Incubate the membrane in 20 mL of solution 1 for 20 min at room temperature. For low amounts of oligosaccharide it may be necessary to increase the amount of sodium metaperiodate in the solution. Increasing the concentration up to 200 m*M* increases the final staining.
3. Wash in buffer E, 3 × 10 min
4. Incubate the membrane in 5 mL of buffer B containing 1 mL of DIG-succinyl-ε-amidocaproic acid hydrazide for 1 h at room temperature. For low amounts of glycoproteins greater sensitivity can be obtained by increasing the concentration of DIG-succinyl-ε-amidocaproic acid hydrazide; however no further benefit is gained by raising the concentration greater than 3 mL in 5 mL (*see* **Note 12**).
5. Wash for 3 × 10 min in buffer A. TBS may now be used to wash the membrane as the DIG labeling has taken place.

3.2.3. DIG Staining Reaction

1. Incubate the membrane for at least 30 min in the blocking reagent (*see* **Note 10**). The membrane can be stored for several days at 4°C at this stage, and in fact a lower background staining can be achieved by allowing the filter membrane to wash in the solution at 4°C overnight (shaking is not necessary) and then for 30 min at room temperature with shaking.
2. Wash for 3 × 10 min in buffer A.
3. Incubate the membrane with 10 mL of buffer A containing 10 µL of anti-digoxigenin-AP at room temperature for 1 h. Sensitivity can be increased by increasing the amount of anti-DIG in the solution by a factor of 2 (although any more has no appreciable effect), or by increasing the incubation time to several hours.
4. Wash for 3 × 10 min with buffer A.
5. Immerse the membrane without shaking into 10 mL of buffer C containing 37.5 µL solution 4 and 50 µL solution 3 (mix just before use). If there is a large amount of sugar present in the bands the color reaction can take only a few minutes; however, if there is little material the reaction could take several hours or overnight. The reaction is best done in the dark, as light can cause nonspecific staining of the membrane. If solutions 3 and 4 are not fresh then the reaction can take several hours and cause a large amount of background staining. It is possible to speed up the reaction by doubling the amounts of solutions 3 and 4 added to buffer C.
6. Wash the membrane several times with Milli-Q water and allow to air dry. The membrane is best stored in foil to reduce fading of the bands once the reaction is stopped.

3.3. DIG - Labeled Lectin Staining

A wide range of lectins are commercially available as free lectin or conjugates of peroxidase, biotin, DIG, fluorescein isothiocyanate (FITC), alcohol dehydrogenase, colloidal gold, or on solid supports such as agarose. A list of commonly used commercially available lectins is shown in **Table 1**. Peroxidase or AP-labeled lectins can be detected directly. Alternatively, lectins can be detected using anti-DIG peroxidase (if DIG labeled) or streptavidin peroxidase (if biotin labeled), followed by an insoluble substrate. Sensitivity is generally increased with these indirect methods. The detection can be carried out with blots on nitrocellulose or PVDF (*see* **Note 13**).

1. Fix membrane for 5 min with 1% KOH. (*See* **Note 12**).
2. Rinse for 1 min with distilled water.
3. Block unbound sites with a 1-h incubation at room temperature with blocking reagent.

Table 1
Some Commonly Used Lectins[a]

Taxonomic name	Common name	Specificity[b,c]
Aleuria aurantia	AAA	α-(1,6)Fuc
Amanranthus caudatus	Amaranthin (ACA)	Galb-(1,3)GalNAc
Canavalia ensiformis	Jackbean concancavalin A	α-Man > α-Glc (Con A)
Datura stramonium	Jimson weed (DSA)	β-(1,4)GlcNAc Terminal GlcNAc
Maackia amurensis	MAA	α-(2,3)Neu-N-Ac
Phaseolis vulgaris	Red kidney bean (PHA-L)	complex *N*-linked
Arachis hypogaea	Peanut (PNA)	Gal·(1,3)GalNAc
Ricinus communis	Castor bean	Terminal Gal (RCA120)
Sambucus sieboldiana	Elderberry (SNA)	α-(2,6)Neu-N-Ac
Triticum vulgaris	Wheat germ (WGA)	$(GlcNAc)_2$ > GlcNAc
Helix pomatia	Snail, edible (HPA)	Terminal GalNAc
Lens culinaris	Lentil	α-Man, α-Glc
Glycine max	Soybean (SBA)	GalNAc
Erythrina cristagalli	Coral tree (ECA)	Gal (1,4)GlcNAc

[a]Many lectins are available commercially as free (unlabeled); digoxigenin (DIG) labeled; biotinylated; peroxidase labelled; or conjugated to agarose beads from companies such as Roche Diagnostics, Sigma Chemical Company, and Amersham-Pharmacia.

[b]GalNAc, *N*-acetyl galactosamine; GlcNAc, *N*-acetylglucosamine; Gal, galactose; Glc, glucose; Man, mannose; Fuc, fucose; NeuNAc, *N*-acetylneuraminic acid (sialic acid).

[c]D-sugars are the preferred sugars.

4. Rinse away blocking reagent with 3 × 1 min washes with TBS-Tween.
5. Add DIG-labeled lectins diluted in TBS-Tween containing 0.1 m*M* $CaCl_2$, 0.1 m*M* $MgCl_2$, 0.1 m*M* $MnCl_2$. Leave overnight at 4°C (*see* **Notes 14** and **15**).
6. Remove unbound lectin by washing with 6 × 5 min in TBS-Tween.
7. Incubate nitrocellulose with 10 µL of anti-DIG AP diluted in 10 µL of TBS for 1 h at room temperature.
8. Repeat washing **step 6**.
9. Immerse the membrane without shaking into 10 µL of buffer C containing 37.5 µL of solution 4 and 50 µL of solution 3 (mix just before use).
10. Stop reaction when gray to black spots are seen (few minutes to overnight) by rinsing with water.

3.4. Pro-Q Emerald 300 Dye Staining

Pro-Q™ Emerald 300 Glycoprotein Gel Stain Kit (Molecular Probes, OR) provides a method for differentially staining glycosylated and nonglycosylated proteins in the same gel. The technique combines the green fluorescent Pro-Q™ Emerald 300 glycoprotein stain with the red fluorescent SYPRO Ruby total protein gel stain. A related product, Pro-Q™ Emerald 300 Glycoprotein Blot Stain Kit, allows similar capabilities for proteins electroblotted to PVDF membranes.

Using this stain allows detection of <1 ng of glycoprotein/band, depending on the nature and the degree of glycosylation, making it 500-fold more sensitive than the standard periodic acid–Schiff base method using acidic fuchsin dye. The green fluorescent signal from Pro-Q Emerald 300 stain can be visualized with 300 nm UV illumination.

1. Separate proteins by standard SDS-PAGE. Typically, the sample is diluted to about 10–100 µg/mL with sample buffer and 5–10 µL of diluted sample is added per lane for 8 cm × 10 cm gels. Large, 16 cm × 18 cm gels require twice as much material.
2. After electrophoresis, fix the gel by immersing it in 75–100 mL of fix solution and incubating at room temperature with gentle agitation (e.g., on an orbital shaker at 50 rpm) for 45 min.
3. Wash the gel by incubating it in 50 mL of wash solution with gentle agitation for 10 min. Repeat this step once.
4. Incubate the gel in 25 mL of oxidizing solution with gentle agitation for 30 min.
5. Wash the gel in 50 mL of wash solution with gentle agitation for 5–10 min. Repeat this step twice more.
6. Prepare fresh Pro-Q Emerald 300 Staining Solution by diluting the Pro-Q Emerald 300 reagent (component A) 50-fold into Pro-Q Emerald 300 dilution buffer (component B). For example, dilute 500 µL of Pro-Q Emerald 300 reagent into 25 mL of dilution buffer to make enough staining solution for one 8 × 10 cm gel.
7. Incubate the gel in the dark in 25 mL of Pro-Q Emerald 300 Staining Solution while gently agitating for 90–120 min. The signal can be seen after about 20 min and maximum sensitivity is reached at about 120 min. Staining overnight is not recommended.
8. Wash the gel with 50 mL of wash solution at room temperature for 15 min. Repeat this wash once for a total of two washes. Do not leave the gel in wash solution for more than 2 h, as the staining signal will start to decrease.
9. Visualize the stain using a standard UV transilluminator (*see* **Notes 16–20**).
10. To counter-stain nonglycosylated proteins in the sample, completely thaw the 10% SDS solution (component F), vortex-mixing to completely dissolve the SDS.
11. Dilute the 10% SDS solution 2000-fold into the SYPRO Ruby protein gel stain to make a final concentration of 0.005% SDS.
12. Proceed with staining as outlined in Chapter 34 in this volume (Patton) describing the SYPRO Ruby protein gel staining procedure (*see* **Note 21**).

3.5. HPAEC Analysis of Monosaccharide Composition (See Note 22)

This technique requires about 100 pmol (5 µg) of glycosylated protein.

1. Excise spots that have been visualized by coomassie blue or amido black on PVDF membranes and place into screw-capped Eppendorf tubes (*see* **Note 23**).
2. Wet membrane with methanol. If the membrane is not properly wetted it will float and the hydrolysis will not be complete.

3.5.1. Analysis of Sialic Acids

1. Add 100 µL 0.1 *M* TFA to the wetted membrane. Mix and close the tube.
2. Incubate in heating block at 80°C for 40 min.
3. Remove membrane from tube and wash with 50 µL of water.
4. Dry combined solutions in a Speed Vac concentrator.

5. Resuspend in 25 μL and add 0.5 nmol lactobionic acid as internal standard for quantitation. Analyze by HPAEC-PAD using a linear gradient of 0–200 m*M* NaAcetate in 250 m*M* NaOH for 30 min (*see* **Notes 24–26**).
6. Compare retention time with standard of 1 nmol *N*-acetyl and *N*-glycolylneuraminic acids.

3.5.2. Analysis of Neutral Monosaccharides

1. Add 100 μL of 2 *M* TFA to the desialylated membrane spots (*see* **Note 27**). Mix and close tube (*see* **Note 24**). Re-wet membrane with methanol before addition of acid if dried out.
2. Incubate in a boiling water bath for 4 h.
3. Remove membrane from tube and wash with 50 μL of water. Keep membrane for subsequent amino acid analysis if required.
4. Dry combined solutions in a Speed Vac concentrator.
5. Resuspend in 25 μl of water and add 0.25 nmol of 2-deoxyglucose as internal standard for quantitation. Analyze by HPAEC-PAD eluted isocratically with 12 m*M* NaOH for 30 min (*see* **Notes 24–26**).
6. Compare with a standard mixture containing fucose, 2-deoxyglucose, galactosamine, glucosamine, galactose, glucose, and mannose (5 μl of 1 m*M* solution) (*see* **Note 28**). This is the order of elution of the monosaccharides from the CarboPac column.

4. Notes

4.1. Periodic Acid/Schiff Staining

1. This method can be used on PAGE gels, agarose, or polyacrylamide–agarose composite gels. The procedure should be carried out on an orbital shaker or rocker (an orbital shaker may cause background swirls on some gels). High background staining can occur when staining some batches of agarose or composite gels.
2. All steps in this procedure should be performed with shaking and should be carried out in a fume hood.
3. A negative result often means that insufficient protein is present or that the protein has little glycosylation
4. Be aware that highly glycosylated proteins do not transfer well to nitrocellulose membranes, and failure to detect may be due to inefficient transfer to the membrane. It is sometimes useful to carry out the stain on the gel after transfer, to test for remaining glycoprotein.
5. PAS staining after transfer of glycoproteins to membranes eliminates the need for extensive fixation steps, as well as shortening the time needed for washing steps , without loss of staining intensity. Results are easier to visualize and are easier to store. Before Western transfer, it is suggested to gently rock the gel for 30–60 min in transfer buffer so as to remove SDS, which can lead to higher backgrounds during membrane staining.
6. Periodic acid–Schiff staining on nitrocellulose membranes increases sensitivity slightly and reduces washing times considerably compared with gel staining.

4.2. DIG–Anti-DIG AP Labeling

7. All steps in the procedure except for color development should be done on a shaker or rocker. Swirling background staining may occur when using a orbital shaker.
8. Prelabeling gives a higher sensitivity than postlabeling, requires less DIG (1 μl per labeling), and can be used with <0.25% SDS, Nonidet P-40 (NP-40), Triton X-100, but not with octylglucoside. Prelabeling results in broader bands than post-labeling and may result in a change of p*I* of the proteins in the first dimension. Pre-labeling is negatively influenced by mercaptoethanol, dithiothreitol (DTT), glycerol and Tris.

9. The postlabeling procedure is used for proteins that have already been separated by gel electrophoresis and immobilized on blots. It is important to use PBS (buffer E) instead of TBS (buffer A) in the initial stages as Tris inhibits the DIG labeling process.
10. Nitrocellulose and PVDF membranes can be used. Nylon membranes result in high background.
11. For mucins and other heavily glycosylated proteins it may be necessary to fix the bands to the membrane before labeling by washing in 1% KOH for 5 min. If this procedure is done, increase the initial washes to 3 × 10 min in buffer B.
12. Postlabeling results in sharper bands than prelabeling but has a lower sensitivity and requires more DIG (5 μL per gel).

4.3. DIG-Labeled Lectin Staining

13. The binding of some lectins will be increased by desialylation while the binding of others will be decreased.
14. Suggested dilutions of the DIG labeled lectins: SNA—1:1000, MAA—1:500, DSA—1:1000, PNA—1:100.
15. The divalent ions are necessary for optimal lectin reactivity. The stock solution should be diluted to give a final concentration of 1 m*M* of each ion.

4.4. Pro-Q Emerald 300 Dye Staining

16. The Pro-Q Emerald 300 stain has an excitation maximum at approx 280 nm and an emission maximum near 530 nm. Stained glycoproteins can be visualized using a 300 nm UV transilluminator. *The use of a photographic camera or charge coupled device (CCD) camera and the appropriate filters is essential to obtain the greatest sensitivity.*
17. It is important to clean the surface of the transilluminator after each use with deionized water and a soft cloth (such as cheesecloth). Otherwise, fluorescent dyes can accumulate on the glass surface and cause a high background fluorescence.
18. Use a 300 nm transilluminator with six 15W watt bulbs. Excitation with different light sources may not give the same sensitivity.
19. Using a Polaroid® camera and Polaroid 667 black-and-white print film, the highest sensitivity is achieved with a 490 nm longpass filter, such as the SYPRO protein gel stain photographic filter (S-6656), available from Molecular Probes. Gels are typically photographed using an f-stop of 4.5 for 2–4 s, using multiple 1-s exposures.
20. Using a CCD camera, images are best obtained by digitizing at about 1024 × 1024 pixels resolution with 12–, 14– or 16– bit gray scale levels per pixel. A 520 nm long pass filter is suitable for visualizing the stain. A CCD camera-based image analysis system can gather quantitative information that will allow comparison of fluorescence intensities between different bands or spots. Using such a system, the Pro-Q Emerald stain has a linear dynamic range over three orders of magnitude. The polyester backing on some premade gels is highly fluorescent. For maximum sensitivity using a UV transilluminator, these gels should be placed polyacrylamide side down and an emission filter used to screen out the blue fluorescence of the plastic.
21. The green-fluorescent Pro-Q Emerald 300 staining should be viewed and documented before staining total proteins with SYPRO Ruby protein gel stain. Pro-Q Emerald dye signal will fade somewhat after SYPRO Ruby dye staining.

4.5. HPAEC Analysis of Monosaccharide Composition

22. Data on the composition of both the acidic sialic acids and the hexoses and amino sugars can be obtained sequentially from a single spot. After the 4 *M* TFA hydrolysis,

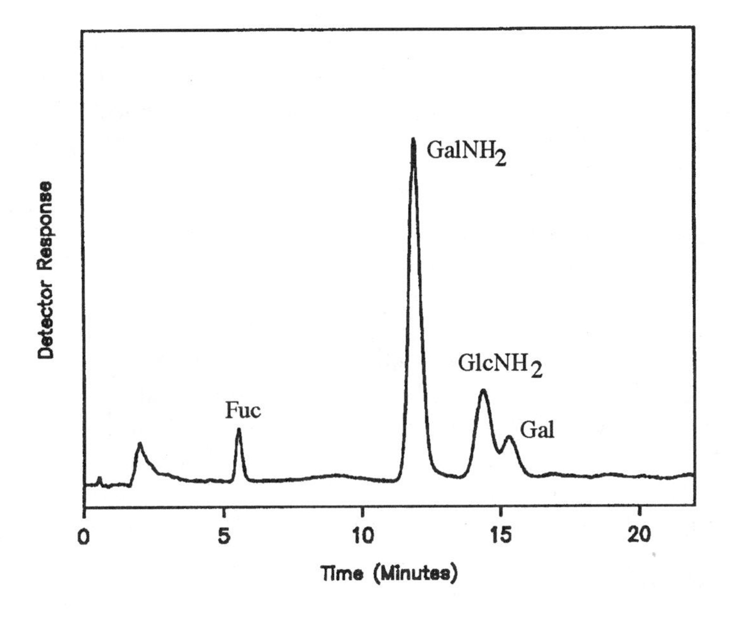

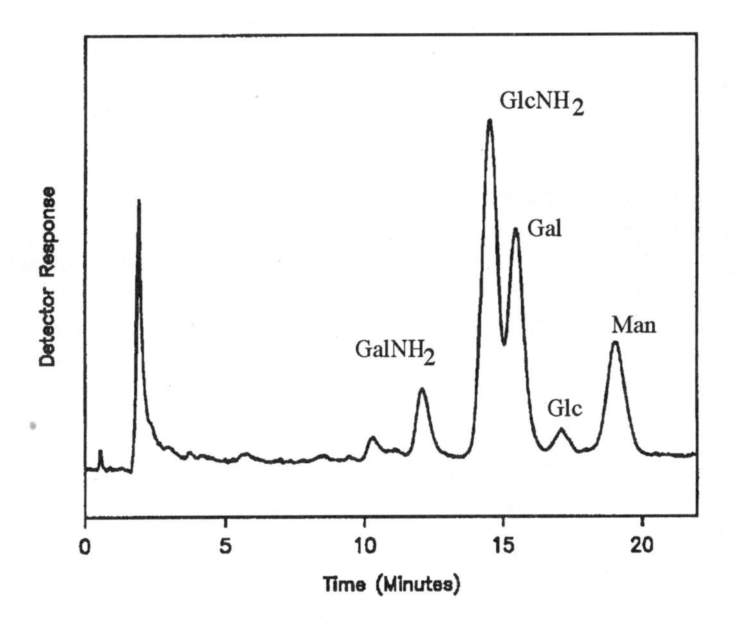

Fig. 1. Monosaccharide composition of glycoprotein hydrolysates. (*Upper*), Bovine submaxillary mucin; (*lower*), fetuin. Separation by HPAEC-PAD.

amino acid analysis (6 *M* HCl, 1 h, 155°C) can be performed subsequently on the same spot for identification of the protein.

23. It is important for monosaccharide analysis that the tubes be kept clean in a cellulose-free environment (i.e., no paper, cotton wool, or dust!) because of the common glucose contamination from the hydrolysis of cellulose and other ubiquitous polysaccharides. For the same reason a blank of all reagents should be run to establish background contamination levels.

24. The HPLC system must be totally metal free because of the caustic reagents. All solvents must be made free of carbon dioxide by boiling of the MilliQ water used. The water is cooled under argon and the solutions kept under helium to reduce the precipitation of sodium carbonate from the sodium hydroxide solutions and the resultant effect on the anion-exchange chromatography.
25. The DIONEX CarboPac PA10 column gives better separation of the monosaccharides than the CarboPac PA1 but the latter may still be used effectively and can be used for both monosaccharide and oligosaccharide analysis.
26. The columns must be washed with strong alkali (0.4 *M* NaOH) for 10 min and re-equilibrated between injections.
27. If more accurate quantitation of the amino sugars is required, stronger hydrolysis using 4 *M* HCl for 4 h at 100°C can be carried out instead of the milder 4 *M* TFA hydrolysis. Subsequent amino acid analysis cannot be performed in this case.
28. Glucose will almost always be present as a contaminant and so is exceptionally difficult to quantitate if it is a constituent. Examples of the monosaccharide composition of typical *N*-linked sugars (fetuin) and *O*-linked sugars (mucin) are shown in **Fig. 1**.

Acknowledgments

I thank Margaret Lawson for her invaluable technical assistance. I also thank the NHMRC (Grant No. 950475) for their support for this work.

References

1. Gander J. E. (1984) Gel protein stains: glycoproteins. *Meth. Enzymol.* **104,** 447–451.
2. Packer, N. H., Wilkins, M. R., Golaz, O., Lawson, M. A., Gooley, A. A., Hochstrasser, D. F., et al. (1996) Characterization of human plasma glycoproteins separated by two-dimensional gel electrophoresis. *Bio. Technol.* **14,** 66–70
3. Thornton, D. J., Carlstedt I., and Sheehan, J. K. (1996) Identification of Glycoproteins on Nitrocellulose Membranes and Gels. *Mol. Biotechnol.* **5,** 171–176.
4. Devine, P. L. and Warren, J. A. (1990) Glycoprotein detection on Immobilon PVDF transfer membrane using the periodic acid/Schiff reagent. *BioTechniques* **8,** 492–495.

105

Staining of Glycoproteins/Proteoglycans in SDS-Gels

Holger J. Møller and Jørgen H. Poulsen

1. Introduction

Sodium dodecyl sulfate-polyacrylamide gel electrophoresis (SDS-PAGE) is a commonly employed technique for separation of proteins according to size. Among other applications, it is used for identification and characterization of proteins on the basis of molecular weight determinations. In this chapter, staining methods are described that permit detection of highly glycosylated proteins on SDS gels at levels of a few nanograms.

Proteins with limited glycosylation are most often stained with Coomassie brilliant blue or, if high sensitivity is needed, with a silver stain (*[1]*, and *see* Chapter 33). These stains, however, are far less sensitive when used for detection of highly glycosylated proteoglycans (protein glycosaminoglycans) or glycoproteins (protein oligosaccharides), leading to weak staining or even failure of detection. This is presumably the result of steric interference by the carbohydrates with the binding of silver ions.

Proteoglycans are traditionally stained with cationic dyes, such as Alcian blue or Toluidine blue *(2)*, that bind to the negatively charged glycosaminoglycan side chains, whereas more neutral glycoproteins can be detected by some variation of the Schiff base reaction involving initial oxidation of carbohydrates by periodic acid and subsequent staining with Schiff's reagent (PAS) *(3)*, alcian blue *(4)*, or a hydrazine derivate *(5)*. A protocol for PAS staining of small SDS gels is described in **ref. *6***. However, these methods, although useful in many instances, are characterized by low sensitivity, generally requiring microgram amounts of protein for detection. At the same time, they are carbohydrate specific, which means that nonglycosylated proteins are not stained.

The two methods described here are based on silver enhancement of traditional staining methods for proteoglycan and glycoprotein *(7–12)*, which result in a twofold increase in sensitivity as compared with alcian blue or PAS alone. In both methods, Alcian blue is used as the primary staining agent, binding either directly to the proteoglycans (**Subheading 3.1.**) or to oxidized glycoproteins (**Subheading 3.2.**), subsequently enhanced by a neutral silver staining protocol.

The methods also stain highly negatively charged phosphoproteins (e.g., osteopontin and bone sialoprotein, rich in phosphate and acidic amino acids) that stain weakly with ordinary methods *(13)*. Furthermore, both glycosylated and nonglycosylated proteins

From: *The Protein Protocols Handbook, 2nd Edition*
Edited by: J. M. Walker © Humana Press Inc., Totowa, NJ

are stained. Nonglycolylated proteins, however, stain weaker, and it is possible to exclude the staining of nonglycosylated proteins (*see* **Note 1**).

A protocol for staining by the cationic dye "Stains-All", instead of Alcian blue, combined with silver is described in **ref.** ***13***. This method seems to more uniformly stain both nonglycosylated and negatively charged proteins

The high sensitivity of the methods makes them very suitable for detection of proteoglycan/glycoprotein in dilute mixed samples, and for characterization of small amounts of purified materials, for example, by determination of molecular weight before and after deglycosylation ***(14,15)***. In this connection, it is important to note that, owing to heterogeneity of carbohydrate substitution, highly glycosylated proteins usually move as diffuse bands or broader smears, so that unambiguous determination of molecular weight is not always possible.

For more specific purposes, other detection systems are usually employed. Sensitive detection of glycoproteins can be achieved with lectins or specific antibodies, most often after blotting onto a membrane ***(16,17)***. Proteoglycans can similarly be identified by antibodies directed toward the core protein or glycosaminoglycan structures ***(18)***.

The methods described here are optimized for the small, supported gels used in the PhastSystem (Pharmacia, Uppsala, Sweden), which can be programmed for automatic staining resulting in fast and reproducible results. If other systems (larger gels) are used, generally a longer time is needed in each step (*see* **Note 2**).

2. Materials

2.1. Staining of Proteoglycans

All chemicals used should be of analytical grade, and water should be of high purity (e.g., Maxima from Elga [High Wycombe, UK] or Milli-Q from Millipore [Bedford, MA]).

1. Washing solution I: 25% ethanol (v/v), 10% acetic acid (v/v) in water.
2. Washing solution II: 10% ethanol (v/v), 5% acetic acid (v/v) in water.
3. Staining solution: 0.125% alcian blue (w/v) (e.g., Bio-Rad [Hercules, CA] no. 161-0401) in washing solution I. Stir extensively and filter before use.
4. Stopping solution: 10% acetic acid (v/v), 10% glycerol (v/v) in water.
5. Developer stock: 2.5% (w/v) sodium carbonate in water.
6. Sensitizing solution: 5% (v/v) glutardialdehyde in water. (Prepare fresh when required.)
7. Silvering solution: 0.4% (w/v) silver nitrate in water. (Prepare fresh when required.)
8. Developer: Add formaldehyde to the developer stock to a final concentration of 0.013% (v/v). For example, add 35 μL of a 37% formaldehyde solution to 100 mL of developer stock, and stir for a few seconds. (Prepare fresh just before use.)

2.2. Staining of Glycoproteins

1. Prepare all solutions described in **Subheading 2.1.**
2. Fixing solution: 10% (v/v) trichloroacetic acid in water.
3. Washing solution III: 5% (v/v) acetic acid in water.
4. Oxidizing solution: 1% (w/v) periodic acid in water.
5. Reducing solution: 0.5% (w/v) potassium metabisulfite in water.

3. Methods

3.1. Alcian Blue/Silver Staining of Proteoglycans (See Note 1)

This method stains proteoglycans, glycosaminoglycans, phosphoproteins, non-glycosylated proteins, and some glycoproteins with a high content of negatively charged groups. The method can be varied for specific staining of proteoglycan/glycosaminoglycan (*see* **Note 1**).

1. Immediately after electrophoresis, transfer the gel to the development chamber/staining tray (*see* **Note 3**), and wash the gel 3× in washing solution I for 5, 10, and 15 min at 50°C (*see* **Note 4**).
2. Add the staining solution and stain the gel for 15 min at 50°C.
3. Wash the gel 3× in washing solution I for 1, 4, and 5 min and subsequently twice in washing solution II for 2 and 4 min at 50°C (*see* **Note 5**).
4. Add the sensitizing solution for 6 min at 50°C (*see* **Note 6**).
5. Wash the gel twice in washing solution II for 3 and 5 min, and subsequently in water twice for 2 min at 50°C.
6. Submerge the gel in the silvering solution for 6.5 min at 40°C.
7. Wash the gel twice in water for 30 s at 30°C (*see* **Note 7**).
8. Develop the gel with the freshly prepared developer at room temperature for an initial period of 30 s and subsequently for 4–8 min, depending on the desired sensitivity, background staining, and specificity (*see* **Note 1**).
9. Stop the development and preserve the gel by adding the stopping solution for 5 min (*see* **Note 8**).

3.2. Periodic Acid Oxidation and Subsequent Alcian Blue/Silver Staining of Glycoproteins (See Note 1)

Some glycoproteins stain weakly with alcian blue/silver alone, probably owing to the low content of negatively charged groups for binding of Alcian blue. This is overcome by initial oxidation of carbohydrates by periodic acid.

1. After electrophoresis, transfer the gel to the development chamber/Petri dish, and add the fixing solution for 10 min at 30°C (*see* **Note 9**).
2. Wash the gel twice for 2 min in washing solution III and submerge the gel in the oxidizing solution for 20 min at 30°C.
3. Wash the gel twice for 2 min in washing solution III and subsequently in water twice for 2 min before adding the reducing solution for 12 min at 30°C.
4. Wash the gel with water twice for 2 min and subsequently with washing solution I twice for 2 min at 50°C.
5. Continue with **steps 2–9** of **Subheading 3.1.** exactly as described.

4. Notes

1. The development time can be varied according to the desired sensitivity and background staining. Even if an automatic staining chamber is used, these steps can be performed advantageously in staining trays/Petri dishes. In the first development step, a dark precipitate is created and the gel is transferred to a fresh developing solution. After 1–3 min proteoglycans will appear, whereas nonglycosylated proteins appear after 4–8 min. The

gel can be photographed sequentially on a light box during the development step for identification of proteoglycans. As a control, a gel can be stained for proteins with ordinary silver (**Subheading 3.1., steps 3–9**), in which proteoglycans will not stain.

2. The methods are optimized for the PhastSystem, which uses small, supported gels that are stained in an automatic development chamber. If larger gels are stained in staining trays at room temperature, longer incubation times are needed. For unsupported 1-mm thick gels, good results are obtained by increasing the washing with washing solution I in **steps 1** and **3** (**Subheading 3.1.**) to a total of 2.5 and 1.5 h, respectively, with four changes of the solution, and furthermore doubling the incubation times in all other steps.
3. Handle gels with gloves or forceps, as fingerprints stain. The staining methods are highly sensitive, and it is essential that the equipment (development chamber/staining trays) is scrupulously clean and that high-quality water is used. Use a separate tray for the staining solution. Gels should be agitated during all staining and washing procedures.
4. This rather extensive washing procedure is necessary to remove the SDS from the gel, which otherwise precipitates alcian blue, resulting in excessively high background. If gels are run in native PAGE, one short washing step is sufficient.
5. Alcian blue is irreversibly fixed in the gel by the subsequent silver staining and results in a greenish-black background if the stain is not washed out by dilute acetic acid. Only a weak bluish nuance in the background should remain.
6. Glutaraldehyde is injurious to health. If the procedure is not carried out in a closed chamber, a fume cupboard should be used.
7. This step is important for washing out excess silver ions without losing silver bound to Alcian blue/proteoglycan for autocatalytic reduction in the development step. Too little washing leads to formation of metallic silver in the background, whereas too intense washing leads to decreased sensitivity.
8. The image is stable, but over time, increased background staining will develop, especially from light exposure.
9. In this staining procedure, Acetic acid/ethanol solutions are not always efficient for fixation of the proteins in the gel, whereas Trichloroacetic acid works well.

Acknowledgments

This work was supported by the Danish Rheumatism Association and the Danish Medical Research Council.

References

1. Switzer, R. C., Merril, C. R., and Shifrin, S. (1979) A highly sensitive silver stain for detecting proteins and peptides in polyacrylamide gels. *Analyt. Biochem.* **98,** 231–237.
2. Heinegård, D. and Sommarin, Y. (1987) Isolation and characterization of proteoglycans, in *Methods in Enzymology*, Vol. 144: *Structural and Contractile Proteins* (Cunningham, L. W., ed.), Academic, Press, New York, pp. 319–372.
3. Zacharius, R. M., Zell, T. E., Morrison, J. H., and Woodlock, J. J. (1969) Glycoprotein staining following electrophoresis on acrylamide gels. *Analyt. Biochem.* **30,** 148–152.
4. Wardi, A. H. and Michos, G. A. (1972) Alcian blue staining of glycoproteins in acrylamide disc electrophoresis. *Analyt. Biochem.* **49,** 607–609.
5. Eckhardt, A. E., Hayes, C. E., and Goldstein, I. J. (1976) A sensitive fluorescent method for the detection of glycoproteins in polyacrylamide gels. *Analyt. Biochem.* **73,** 192–197.
6. Van-Seuningen, I. and Davril, M. (1992) A rapid periodic acid-Schiff staining procedure for the detection of glycoproteins using the PhastSystem. *Electrophoresis* **13,** 97–99.

7. Min, H. and Cowman, M. K. (1986) Combined Alcian blue and silver staining of glycosaminoglycans in polyacrylamide gels: application to electrophoretic analysis of molecular weight distribution. *Analyt. Biochem.* **155,** 275–285.
8. Krueger, R. C. and Schwartz, N. B. (1987) An improved method of sequential Alcian blue and Ammoniacal silver staining of chondroitin sulfate proteoglycan in polyacrylamide gels. *Analyt. Biochem.* **167,** 295–300.
9. Lyon, M. and Gallagher, J. T. (1990) A general method for the detection and mapping of submicrogram quantities of glycosaminoglycan oligosaccharides on polyacrylamide gels by sequential staining with Azure A and Ammoniacal silver. *Analyt. Biochem.* **185,** 63–70.
10. Jay, G. D., Culp, D. J., and Jahnke, M. R. (1990) Silver staining of extensively glycosylated proteins on sodium dodecyl sulfate-polyacrylamide gels: enhancement by carbohydrate-binding dyes. *Analyt. Biochem.* **185,** 324–330.
11. Møller, H. J., Heinegård, D., and Poulsen, J. H. (1993) Combined Alcian blue and silver staining of subnanogram quantities of proteoglycans and glycosaminoglycans in sodium dodecyl sulfate-polyacrylamide gels. *Analyt. Biochem.* **209,** 169–175.
12. Møller, H. J. and Poulsen, J. H. (1995) Improved method for silver staining of glycoproteins in thin sodium dodecyl sulfate polyacrylamide gels. *Analyt. Biochem.* **226,** 371–374.
13. Goldberg, H. A. and Warner, K. J. (1997) The staining of acidic proteins on polyacrylamide gels: enhanced sensitivity and stability of "Stains-All" staining in combination with silver nitrate. *Analyt. Biochem.* **251,** 227–233.
14. Raulo, E., Chernousov, M. A., Carey, D. J., Nolo, R., and Rauvala, H. (1994) Isolation of a neuronal cell surface receptor of heparin binding growth-associated molecule (HB-GAM). *J. Biol. Chem.* **269,** 12,999–13,004.
15. Halfter, W. and Schurer, B. (1994) A new heparan sulfate proteoglycan in the extracellular matrix of the developing chick embryo. *Exp. Cell. Res.* **214,** 285–296.
16. Furlan, M., Perret, B. A., and Beck, E. A. (1979) Staining of glycoproteins in polyacrylamide and agarose gels with fluorescent lectins. *Analyt. Biochem.* **96,** 208–214.
17. Moroi, M. and Jung, S. M. (1984) Selective staining of human platelet glycoproteins using nitrocellulose transfer of electrophoresed proteins and peroxidase-conjugated lectins. *Biochem. Biophys. Acta* **798,** 295–301.
18. Heimer, R. (1989) Proteoglycan profiles obtained by electrophoresis and triple immunoblotting. *Analyt. Biochem.* **180,** 211–215.

106

Identification of Glycoproteins on Nitrocellulose Membranes Using Lectin Blotting

Patricia Gravel

1. Introduction

Glycoproteins result from the covalent association of carbohydrate moieties (glycans) with proteins. The enzymatic glycosylation of proteins is a common and complex form of posttranslational modification. The precise roles played by the carbohydrate moieties of glycoproteins are beginning to be understood *(1–3)*. It has been established that glycans perform important biological roles including: stabilization of the protein structure; protection from degradation; and control of protein solubility, protein transport in cells, and protein half-life in blood. They also mediate the interactions with other macromolecules and the recognition and association with viruses, enzymes, and lectins *(4–6)*.

Carbohydrate moieties are known to play a role in several pathological processes. Alterations in protein glycosylation have been observed, for example, in the membrane glycoproteins of cancer cells; in the plasma glycoproteins of alcoholic patients and patients with liver disease; and in glycoproteins in brains from patients with Alzheimer disease, inflammation, and infection. These changes provide the basis for more sensitive and discriminative clinical tests *(2,7–11)*.

Lectins are carbohydrate binding proteins and can be used to discriminate and analyze the glycan structures of glycoproteins. The lectin blotting technique detects glycoproteins separated by electrophoresis (sodium dodecyl sulfate-polyacrylamide gel electrophoresis [SDS-PAGE] or two-dimensional polyacrylamide gel electrophoresis [2-D PAGE]) and transferred to membranes. This chapter describes the protocol for the detection of glycoproteins on nitrocellulose membrane using biotinylated lectins and avidin conjugated with horseradish peroxidase (HRP) or with alkaline phosphatase (AP) (*see* **Note 1**). These complexes are subsequently revealed either by a chemiluminescent or a colorimetric reaction. To illustrate this method, human plasma glycoprotein patterns, obtained after the separation of proteins by 2-D PAGE and different lectin incubations, are presented in **Figs. 1–4** (*see* **Note 2**). The principal merits of this separation technique are the high resolution and reproducibility of the separation and the high loading capacity. When the lectin blotting technique is used after 2-D PAGE separation. One can observe multiple is oforms caused by modification of the change (e.g.,

From: *The Protein Protocols Handbook, 2nd Edition*
Edited by: J. M. Walker © Humana Press Inc., Totowa, NJ

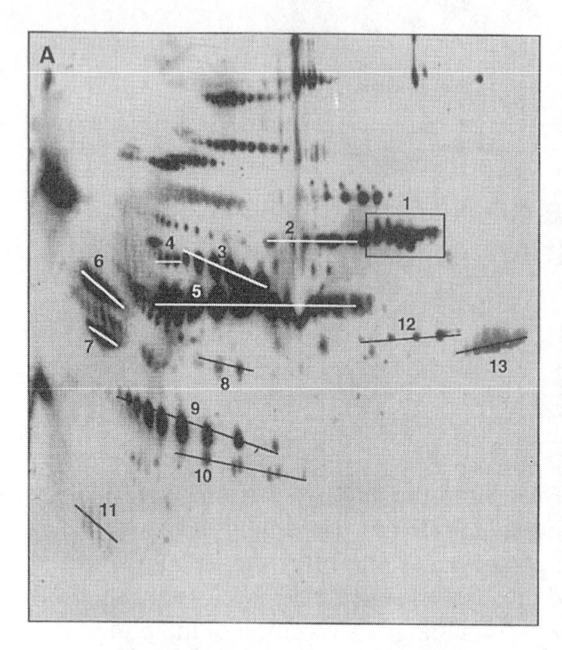

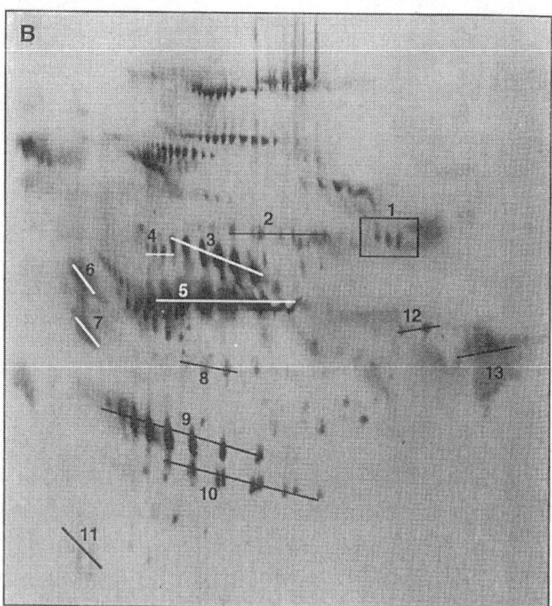

Fig. 1. Glycoprotein blot pattern of 2-D PAGE separation of plasma proteins (120 µg) probed with WGA (specific for *N*-acetylglucosamine and neuraminic acid) and **(A)** revealed with chemiluminesence (15 s film exposure) and **(B)** with NBT/BCIP (20 min for the development of the color reaction). Since most of the glycoproteins in plasma contain one or more *N*-linked glycans with at least two *N*-acetylglucosamine residues, the use of WGA allows a general staining of *N*-linked glycoproteins. *1*, Transferrin; *2*, IgM µ-chain; *3*, hemopexin; *4*, α_1-β-glycoprotein; *5*, IgA α-chain; *6*, α_1-antichymotrypsin; *7*, α_2-HS-glycoprotein; *8*, fibrinogen γ-chain; *9*, haptoglobin β-chain; *10*, haptoglobin cleaved β-chain; *11*, apolipoprotein D; *12*, fibrinogen β-chain; *13*, IgG γ-chain.

alteration of the sugar sialic acid) and/or the molecular weight of glycoprotein. Different protein alterations associated with alcohol-related disease have been identified and confirmed using this lectin blotting technique coupled to 2-D PAGE ***(12)***.

1.1. Classification of Glycoproteins

A useful classification of glycoproteins is based on the type of glycosidic linkage involved in the attachment of the carbohydrate to the peptide backbone. Three types of glycan–protein linkage have been described:

1. *N*-Glycosidic linkage of *N*-acetylglucosamine to the amide group of asparagine: GlcNAc (β-1, *N*) Asn.
2. *O*-Glycosidic linkage of *N*-acetylgalactosamine to the hydroxyl group of either serine or threonine: GalNAc (α-1,3) Ser; GalNAc (α-1,3) Thr.
3. *O*-Glycosidic linkage of galactose to 5-hydroxylysine: Gal (β-1,5) OH-Lys.

Most plasma glycoproteins bear exclusively asparagine-linked oligosaccharides ***(3,13)***, and these *N*-glycosidic linkages are by far the most diverse. They can be subdivided into three groups according to structure and common oligosaccharide sequences

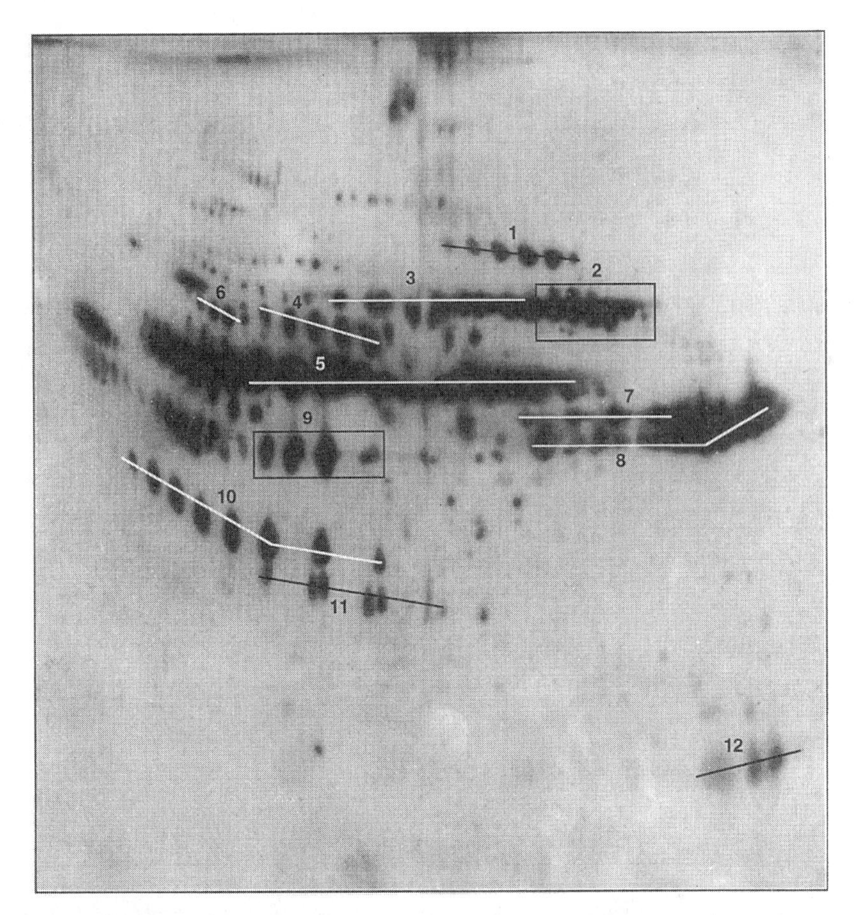

Fig. 2. Glycoprotein blot pattern of 2-D PAGE separation of plasma proteins (120 µg) probed with RCA (specific for galactose) and revealed using chemiluminescence (5 s film exposure). *1*, Complement 3 α-chain; *2*, transferrin; *3*, IgM µ-chain; *4*, hemopexin; *5*, IgA α-chain; *6*, α_1-β-glycoprotein; *7*, fibrinogen β-chain; *8*, IgG γ-chain; *9*, fibrinogen γ-chain; *10*, haptoglobin β-chain; *11*, haptoglobin cleaved β-chain; *12*, Ig light chain.

(**Table 1**). They all have in common the inner core structure presented in **Fig. 5**. The presence of this common core structure reflects the fact that all these asparagine-linked oligosaccharides originate from the same precursor.

O-glycans are found frequently in mucins, but rarely in plasma glycoproteins. **Table 2** illustrates the three different groups of *O*-glycosyl protein glycans.

1.2. Lectins as a Tool for Glycoprotein Detection

Lectins are a class of carbohydrate binding proteins, commonly detected by their ability to precipitate glycoconjugates or to agglutinate cells. (Some lectins react selectively with erythrocytes of different blood types.) Lectins are present in plants, animals, and microorganisms *(14)*.

Each lectin binds specifically and noncovalently to a certain sugar sequence in oligosaccharides and glycoconjugates. Lectins are traditionally classified into specificity

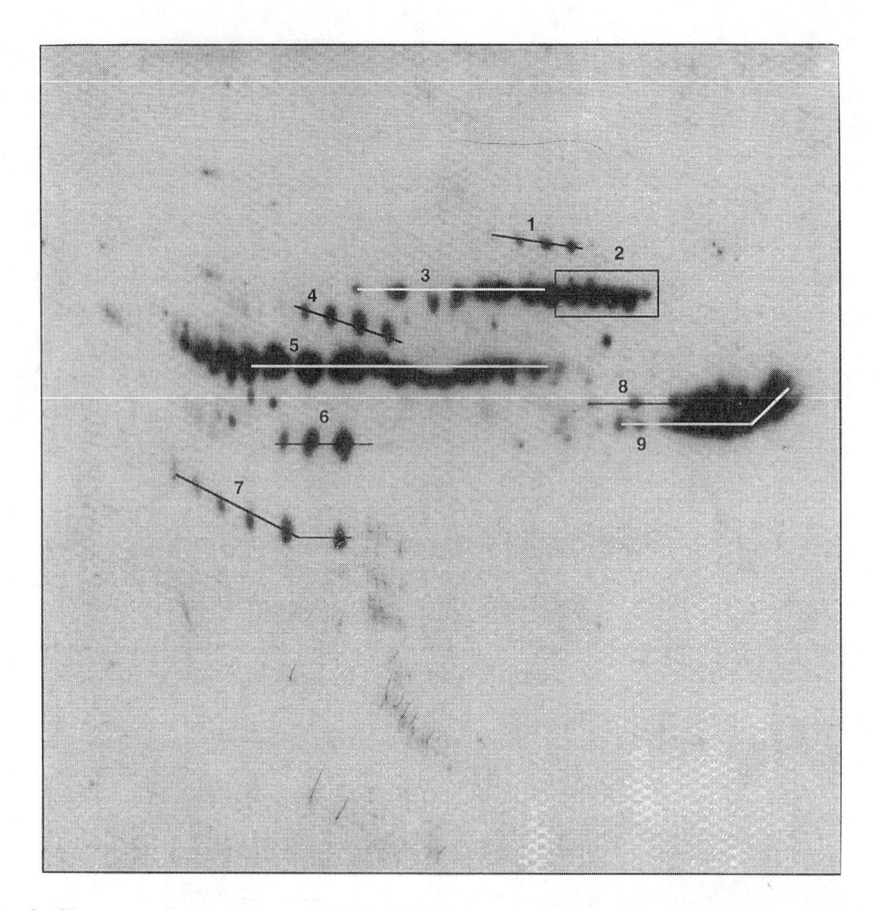

Fig. 3. Glycoprotein blot pattern of 2-D PAGE separation of plasma proteins (120 μg) probed with AAA (specific for fucose) and revealed using chemiluminescence (5 s film exposure). *1*, Complement 3 α-chain; *2*, transferrin; *3*, IgM μ-chain; *4*, hemopexin; *5*, IgA α-chain, *6*, fibrinogen γ-chain; *7*, haptoglobin β-chain; *8*, fibrinogen β-chain; *9*, IgG γ-chain.

groups (mannose, galactose, *N*-acetylglucosamine, *N*-acetylgalactosamine, fucose, neuraminic acid) according to the monosaccharide that is the most effective inhibitor of the agglutination of erythrocytes or precipitation of polysaccharides or glycoproteins by the lectin (fixation-site saturation method). Another method to determine the carbohydrate specificity of a given lectin consists in the determination of the association constant, by equilibrium analysis.

In most cases, lectins bind more strongly to oligosaccharides (di-, tri-, and tetrasaccharides) than to monosaccharides ***(15,16)***. Therefore, the concept of "lectin monosaccharide specificity" should advantageously be replaced by that of "lectin oligosaccharide specificity." **Table 3** summarizes the carbohydrate specificity of lectins commonly used in biochemical/biological research, both in terms of the best monosaccharide inhibitor of the precipitation of polysaccharides or glycoproteins and in terms of the structure of oligosaccharides recognized by immobilized lectins. The interactions of lectins with glycans are complex and not fully understood. Many lectins recognize terminal nonreducing saccharides, while others also recognize internal sugar sequences. Moreover, lectins within each group may differ markedly in their affinity

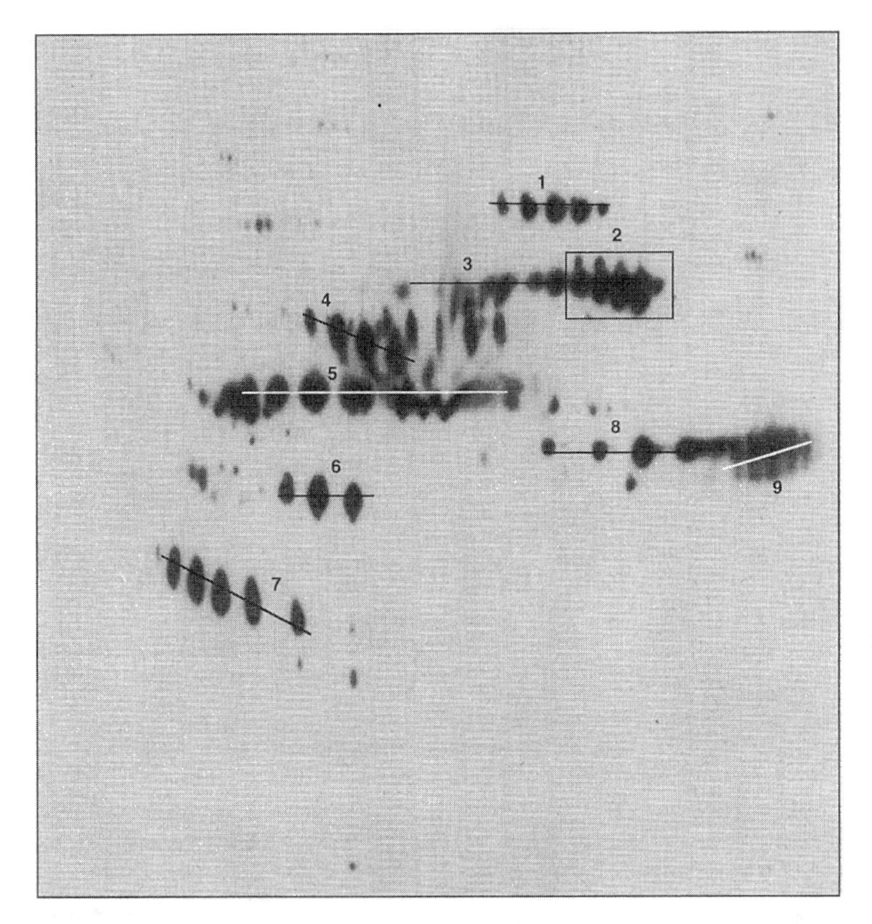

Fig. 4. Glycoprotein blot pattern of 2-D PAGE separation of plasma proteins (120 μg) probed with SNA (specific for neuraminic acid linked α-2,6 to galactose) and revealed using chemiluminescence (30 s film exposure). *1*, Complement 3 α-chain; *2*, transferrin; *3*, IgM μ-chain; *4*, hemopexin; *5*, IgA α-chain; *6*, fibrinogen γ-chain; *7*, haptoglobin β-chain; *8*, fibrinogen β-chain; *9*, IgG γ-chain.

for the monosaccharides or their derivatives. They do not have an absolute specificity and therefore can bind with different affinities to a number of similar carbohydrate groups. Because lectin binding can also be affected by structural changes unrelated to the primary binding site, the results obtained with lectin-based methods must be interpreted with caution ***(2)***.

Despite these limitations, lectin probes do provide some information as to the nature and composition of oligosaccharide substituents on glycoproteins. Their use together with the blotting technique provides a convenient method of screening complex protein samples for abnormalities in the glycosylation of the component proteins. Lectin blotting requires low amounts of proteins and is easy to perform; it is therefore particularly indicated for analysis of biological samples.

When the lectin blotting method described in **Subheading 3.** is combined with the high resolution and reproducibility of 2-D PAGE and with the sensitivity of enhanced chemiluminescence, it is possible to identify rapidly the glycoproteins of

Table 1
Glycans *N*-Glycosidically Conjugated to Proteins

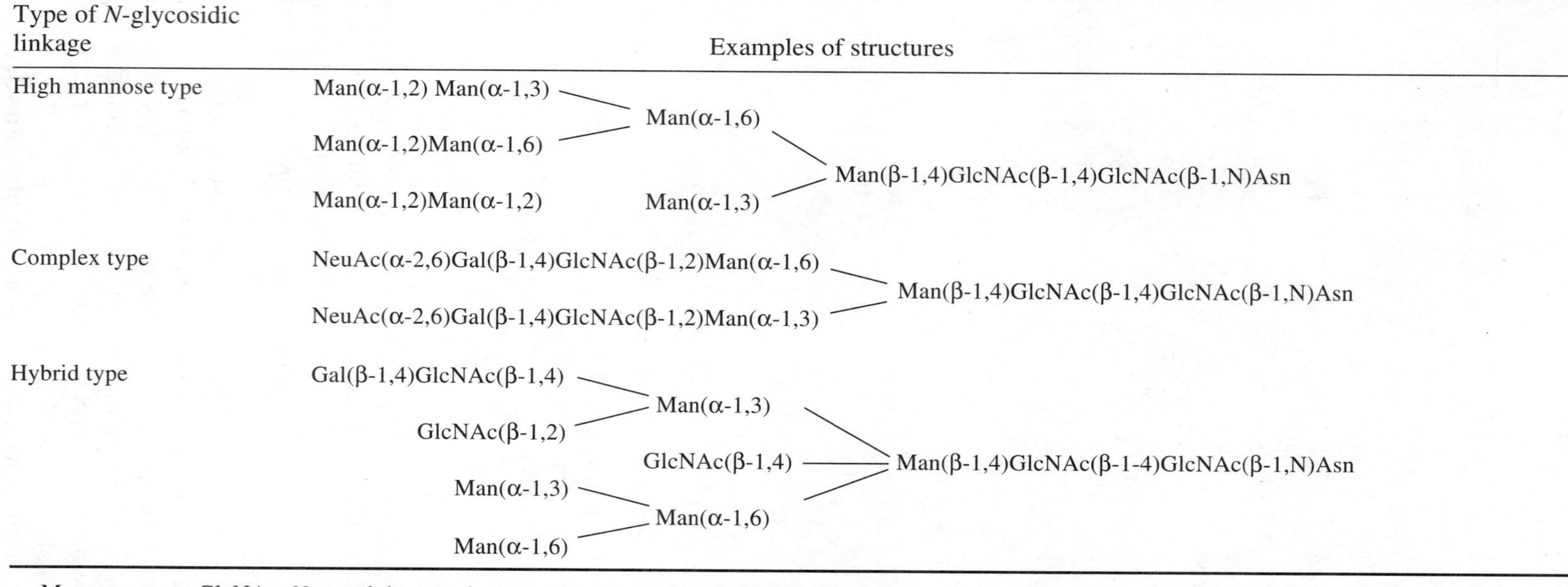

Type of *N*-glycosidic linkage	Examples of structures
High mannose type	Man(α-1,2) Man(α-1,3) — Man(α-1,6) Man(α-1,2)Man(α-1,6) — Man(α-1,6) Man(α-1,6) — Man(β-1,4)GlcNAc(β-1,4)GlcNAc(β-1,N)Asn Man(α-1,2)Man(α-1,2) Man(α-1,3) — Man(β-1,4)GlcNAc(β-1,4)GlcNAc(β-1,N)Asn
Complex type	NeuAc(α-2,6)Gal(β-1,4)GlcNAc(β-1,2)Man(α-1,6) — Man(β-1,4)GlcNAc(β-1,4)GlcNAc(β-1,N)Asn NeuAc(α-2,6)Gal(β-1,4)GlcNAc(β-1,2)Man(α-1,3) — Man(β-1,4)GlcNAc(β-1,4)GlcNAc(β-1,N)Asn
Hybrid type	Gal(β-1,4)GlcNAc(β-1,4) — Man(α-1,3) GlcNAc(β-1,2) — Man(α-1,3) Man(α-1,3) — Man(β-1,4)GlcNAc(β-1-4)GlcNAc(β-1,N)Asn GlcNAc(β-1,4) — Man(β-1,4)GlcNAc(β-1-4)GlcNAc(β-1,N)Asn Man(α-1,3) — Man(α-1,6) Man(α-1,6) — Man(α-1,6) Man(α-1,6) — Man(β-1,4)GlcNAc(β-1-4)GlcNAc(β-1,N)Asn

Man, mannose; GlcNAc, *N*-acetylglucosamine; NeuAc, neuraminic acid; Gal, galactose.

NeuAc located at the terminal position of oligosaccharide side chains are *N*-acetylated (NeuAcNAc) and are often referred to as sialic acid. Sialic acid is in fact the generic term given to a family of acetylated derivatives of neuraminic acid.

Table 2
Glycans *O*-Glycosidically Conjugated to Proteins

Type of *O*-glycosidic linkage	Examples of structures
Mucin type	NeuAc(α-2,6)GalNAc(α-1,3) Ser (or) Thr Gal(β-1,3)GalNAc(α-1,3) Ser (or) Thr The presence of such *O*-glycosidically linked moieties has been established for a number of plasma proteins (IgA, IgD, hemopexin, plasminogen, apoE)
Proteoglycan type	[GalNAc(β-1,4)GlcUA(β-1,3)]*n*GalNAc(β-1,4)GlcUA(β-1,3) Gal(β-1,3)Gal(β-1,4) Xyl(β-1,3)Ser They are generally linear polymers made up from repeating disaccharide units
Collagen type	Glc(β-1,2)Gal(β-1,5)OH-Lys

NeuAc, neuraminic acid; GalNAc, *N*-acetylgalactosamine; Gal, galactose; GlcUA, Glucuronic acid; Glc, glucose; Xyl, xylose.

Man (α–1,6)
Man(β–1,4) GlcNAc(β–1,4) GlcNAc (β–1,N) Asn
Man (α–1,3)

Fig. 5. Oligosaccharide inner core common to all *N*-glycosylproteins.

interest by comparison with a reference 2-D PAGE protein map and to obtain reliable and reproducible results ***(30–32)***.

2. Materials

1. The transfer buffer for protein blotting by the semidry method (*see* Chapter 40) or by the tank method (*see* Chapter 39) is the Towbin buffer diluted 1:2 with distilled water: 12.5 m*M* Tris, 96 m*M* glycine, and 10% (v/v) methanol ***(33)***.
2. Polyvinylidene difluoride (PVDF) membranes (0.2 mm, 200 × 200 mm) are supplied by Bio-Rad, the filter papers (chromatography paper grade 3 mm CHr) by Whatman, and the nitrocellulose membranes (0.45 mm) by Schleicher & Shuell.
3. Blocking solution: 0.5% (w/v) Tween 20 in phosphate-buffered saline (PBS) ***(34)***. All further incubations and washing steps are carried out in the same blocking solution.
4. PBS: 137 m*M* NaCl, 27 m*M* KCl, 10 m*M* Na_2HPO_4, 1.8 m*M* KH_2PO_4, pH 7.4. A 10-fold concentrated PBS solution is prepared, sterilized by autoclaving, and stored at room temperature for many weeks (10X PBS: 80 g of NaCl, 2 g of KCl, 14.4 g of Na_2HPO_4, 2.4 g of KH_2PO_4. Add H_2O to 1 L).
5. Biotinylated lectins are obtained from Boehringer-Mannheim. HRP-labeled Extravidin (a modified form of affinity-purified avidin) and AP-labeled Extravidin were from Sigma.
6. Two different detection methods are described: the chemiluminescent detection of HRP activity using the luminol reagent (ECL Kit, Amersham International) and the conventional colorimetric reaction of AP revealed by nitroblue tetrazolium/bromochloroindolyl phosphate (NBT/BCIP).
   a. Stock solutions of NBT and BCIP (Fluka) are prepared by solubilizing 50 mg of NBT in 1 mL of 70% (v/v) dimethylformamide and 50 mg of BCIP disodium salt in 1 mL of

Table 3
Specificity of Lectins for Glycan Linked to Asparagine (*N*-Linked Oligosaccharide Chains)

Lectin	Monosaccharide specificity	Oligosaccharide specificity[a]
Wheat germ (WGA)	GlcNAc NeuAc	NeuAc(α-2,6)Gal(β-1,4)GlcNAc(β-1,2)Man(α-1,6) \ GlcNac(β-1,4) — Man(β-1,4)GlcNAc(β-1,4)GlcNAc(β-1,N)Asn NeuAc(α-2,6)Gal(β-1,4)GlcNAc(β-1,2)Man(α-1,3) / Note: The GlcNAc (β-1,4) Man (β-1,4) GlcNAc (β-1,4) GlcNAc (β-1,N) Asn structure is important for tight binding of glycopeptides to a WGA agarose column ***(17)***. Neuraminic acids (NeuAc) are implicated as important factors in WGA interactions but the inhibitory effect of NeuAc is weaker than that of GlcNAc. The presence of clustering sialyl residues may be necessary for the stronginteraction of sialoglycoconjugates with WGA ***(18)***.
Ricinus communis (RCA1, RCA120)	Gal	Gal(β-1,4)GlcNAc(β-1,6) \ Man(α-1,6) \ Gal(β-1,4)GlcNAc(β-1,2) / Man(β-1-4)GlcNAc(β-1,4)GlcNAc(β-1,N)Asn Gal(β-1,4)GlcNAc(β-1,4) \ Man(α-1,3) / Gal(β-1,4)GlcNAc(β-1,2) / Note: RCA is specific for terminal β-galactosyl residues ***(19)***. It binds primarily to the terminal Gal (β-1,4) GlcNAc sugar sequence and much more weakly to the Gal (β-1,3) GalNAc sugar sequence ***(20)***. As galactose is the subterminal sugar in fully formed *N*-linked oligosaccharides, RCA provides a means of identifying the asialo forms of glycoproteins.
Concanavalin A (Con A)	Man	GlcNAc(β-1,2)Man(α-1,6) \ Man(β-1,4)GlcNAc(β-1,4)GlcNAc(β-1,N)Asn GlcNAc(β-1,2)Man(α-1,3) / Note: Con A interacts with glycoconjugates that have at least two nonsubstituted or 2-*O*-substituted α-mannosyl residues. It detects the core portion of *N*-linked oligosaccharide chains. The most potent hapten is illustrated here ***(17)***.

Lens culinaris (LCA)	Man	GlcNAc(β-1,2)Man(α-1,6) \ Man(β-1,4)GlcNAc(β-1,4)[Fuc(α-1,6)]GlcNAc(β-1,N)Asn GlcNAc(β-1,2)Man(α-1,3) / <u>Note</u>: LCA is specific of mannose (α-1,3) or (α-1,6). Fucose (α-1,6) linked to GlcNAc is required for the reaction ***(3)***.
Galanthus nivalis (GNA)	Terminal Man	Man(α-1,2)(α-1,3) or (α-1,6)Man <u>Note</u>: GNA is particularly useful for identifying high mannose or hybrid type oligosaccharide structure because it does not react (unlike ConA) with biantennary complex type chains ***(21,22)***.
Datura stramonium (DSA)	GlcNAc	[Gal(β-1,4)GlcNAc(β-1,3)]nGal(β-1,4)GlcNAc(β-1,2) \ Man(α-1,6) [Gal(β-1,4)GlcNAc(β-1,3)]$_n$Gal(β-1,4)GlcNAc(β-1,6) / Man(β-1,m4)GlcNAc(β-1,4)GlcNAc(β-1,N)Asn [Gal(β-1,4)GlcNAc(β-1,3)]$_n$Gal(β-1,4)GlcNAc(β-1,2) Man(α-1,3) / <u>Note</u>: Datura lectin is specific for Gal-GlcNAc termini of complex oligosaccharides ***(22)***.
Peanut agglutinin (PNA)	Gal	Gal(β-1,3)GalNAc **refs. *22* and *23***
Phaseolus vulgaris (E-PHA) erythroagglutinating phytohemagglutinin	GlcNAc	Gal(β-1,4)GlcNAc(β-1,2)Man(α-1,6) \ GlcNAc(β-1,4) — Man(β-1,4)GlcNAc(β-1,4)GlcNAc(β-1,N)Asn Gal(β-1,4)GlcNAc(β-1,2)Man(α-1,3) / <u>Note</u>: A bisecting GlcNAc residue that links (β-1,4) to the β-linked mannose residue in the core is an essential and specific determinant for high-affinity binding to E-PHA-agarose column. Without this bisecting GlcNAc residue, a complex-type glycopeptide is not retained by E-PHA-agarose ***(24)***.

(continued)

Table 3
Specificity of Lectins for Glycan Linked to Asparagine (*N*-Linked Oligosaccharide Chains) *(continued)*

Lectin	Monosaccharide specificity	Oligosaccharide specificity[a]
Phaseolus vulgaris (L-PHA) leukoagglutinating phytohemagglutinin	GlcNAc	Gal(β-1,4)GlcNAc(B-1,2) and Gal(β-1,4)GlcNAc(β-1,6) → Man (α-1,6); Gal(β-1,4)GlcNAc(β-1,4) and Gal(β-1,4)GlcNAc(β-1,2) → Man(α-1,3); Man (α-1,6) and Man(α-1,3) → Man(β-1,4)GlcNAc(β-1,4)GlcNAc(β-1,N)Asn Note: The Gal (β-1,4) GlcNAc Man sugar sequence is essential for the binding of L-PHA (and E-PHA), as the interaction is completely abolished by the removal of the β-galactosyl residue in the peripheral portion ***(24)***.
Ulex europeus 1 (UEA1)	Fuc	Fuc(α-1,6) – GlcNAc(β-1,N)Asn **ref. *3***
Lotus tetragonolobus (LTA)	Fuc	Fuc(α-1,6) – GlcNAc(β-1,N)Asn; Fuc(α-1,3) – Gal(β-1,4)GlcNAc; Fuc(α-1,2) – Gal(β-1,4)GlcNAc **refs. *3*** and ***25***
Aleuria aurantia (AAA)	Fuc	Fuc(α-1,6) – GlcNAc(β-1,N)Asn; Fuc(α-1,2) – Gal(β-1,4)GlcNAc; Fuc(α-1,3) – Gal(β-1,4)GlcNAc Note: The AAA lectin is particularly useful for identifying fucose bound (α-1,6) to GlcNAc. Fucose bound (α-1,6) to GlcNAc in the core region is necessary for strong binding of AAA. Fucose (α-1,3) linked to GlcNAc and fucose (α-1,2) linked to Gal (β-1,4) GlcNAc in the outer chain react only very weakly with AAA. However, fucose (α-1,3) or (α-1,2), in addition to a fucose-linked (α-1,6) to the proximal core GlcNAc enhances the binding ***(17,26)***.

Sambucus nigra (SNA)	NeuAc	NeuAc(α-2,6)Gal Note: SNA binds specifically to glycoconjugates containing the (α-2,6)-linked NeuAc while isomeric structures containing terminal NeuAc in (α-2,3)- linkage are very weakly bound *(27)*.
Maackia amurensis (MAA)	NeuAc	NeuAc(α-2,3)Gal Note: Immobilized MAA interacts with high affinity with complex-type tri- and tetraantennary Asn-linked oligosaccharides containing outer NeuAc linked (α-2,3) to penultimate Gal residues. Glycopeptide containing NeuAc linked only (α-2,6) to Gal do not interact with MAA *(28)*.
Limulus polyphemus (LPA)	NeuAc	Terminal NeuAc **ref.** ***29***

Man, mannose; GlcNAc, *N*-acetylglucosamine; NeuAc, neuraminic acid; Gal, galactose; GalNAc: *N*-acetylgalactosamine; Fuc, fucose.
*a*Examples of carbohydrate structures recognized by the lectins commonly used in biochemical/biological research are presented in this table.

dimethylformamide. These two solutions are stable when stored in closed containers at room temperature.

b. AP buffer, composed of 100 m*M* NaCl, 5 m*M* $MgCl_2$, and 100 m*M* Tris-HCl, pH 9.5 is prepared just before the colorimetric detection.

7. For the chemiluminescent detection of HRP activity, X-ray films (X-OMAT S, 18 × 24 cm) and a cassette (X-OMATIC cassette, regular screens) are available from Kodak.
8. For the development of X-ray film, an automatic developer machine is used (Kodak RP X-OMAT Processor, Model M6B). Manual development is also possible using the developer for autoradiography films (ref. no. P-7042, Sigma) and the fixer (ref. no. P-7167, Sigma).

3. Method

The use of gloves is strongly recommended to prevent blot contamination.

1. After electrophoresis and protein blotting procedures, the membrane is first washed with distilled water (3 × 5 min) and then treated for 1 h at room temperature and under gentle agitation with 100 mL of blocking solution (PBS containing 0.5% [w/v] Tween 20) (*see* **Note 3**).
2. The blot is then incubated for 2 h in biotinylated lectin, at a concentration of 1 µg/mL in the blocking solution, under agitation, in a glass dish and at room temperature. Twenty-five milliliters of lectin solution is used for the incubation of a membrane of 16.5 cm × 21 cm (*see* **Note 4**).
3. A washing step is then performed for 1 h with six changes of 200 mL of PBS–Tween 20.
4. Extravidin-HRP or Extravidin-AP diluted 1:2000 in the blocking solution is added for 1 h at room temperature under agitation. The membrane is then washed for 1 h with six changes of PBS–Tween 20.
5. The colorimetric reaction with AP is carried out under gentle agitation by incubating the blot in the following solution: 156 µL BCIP stock solution, 312 µL of NBT stock solution in 50 mL of AP buffer. The colorimetric reaction is normally completed within 10–20 min. The blotted proteins are colored in blue.
6. The chemiluminescence detection of peroxidase activity is performed according to the manufacturer's instructions (Amersham). The enhanced chemiluminescent assay involves reacting peroxidase with a mixture of luminol, peroxide, and an enhancer such as phenol. (*See also* Chapter 57.) Five milliliters of detection solution 1 are mixed with 5 mL of detection solution 2 (supplied with the ECL Kit). The washed blot is placed in a glass plate and the 10 mL of chemiluminescent reagents are added directly to the blot with a 10-mL pipet, to cover all the surface carrying the proteins. The blot is incubated for 1 min at room temperature without agitation.
7. The excess chemiluminescent solution is drained off by holding the blot vertically. The blot is then wrapped in plastic sheet (Saran Wrap™), without introducing air bubbles and exposed (protein side up) to X-ray film in a dark room, using red safelights. The exposure time of the film depends on the amount of target proteins on the blot (*see* **Note 5**).
8. The development of the X-ray films can be done with an automatic developer or manually with the following protocol:
   a. The developer and fixer solutions are prepared according to the manufacturer's instructions (1:4 dilution with distilled water).
   b. In a dark room, the film is attached to the film hanger and immersed in the developer solution until the bands (or spots) appear (the maximum incubation time is 4 min). The film should not be agitated during development.
   c. The hanger is immersed into water and the film is rinsed for 30 s to 1 min.

d. The hanger is then placed in the fixer solution for 4–6 min. Intermittent agitation should be used throughout the fixing procedure.
e. The film is washed in clean running water for 5–30 min.
f. Finally, the film is dried at room temperature. All previous incubations are also undertaken at room temperature.

4. Notes

1. Instead of nitrocellulose, PVDF membrane can be used for lectin blotting.
2. **Fig. 1A** shows the plasma glycoprotein signals detected with wheat germ agglutinin (WGA) and generated on a film after chemiluminescence detection. **Figure 1B** shows an identical blot stained with NBT/BCIP. As previously reported *(32)*, the same pattern of glycoproteins or glycoprotein subunits are revealed by both methods but the chemiluminescent detection system shows higher sensitivity (about 10-fold) than NBT/BCIP staining. The spots in the former case are more intense and the detection with enhanced chemiluminescence is more reliable and easier to control than the colorimetric reaction.

   Albumin that does not contain any carbohydrate moiety represents a negative protein control in all blots.

   Lectin blotting of plasma proteins with *Ricinus communis* agglutin (RCA), *Aleuria aurantia* agglutinin (AAA), and *Sambucus nigra* agglutin (SNA) are presented in **Figs. 2**, **3**, and **4**, respectively.
3. Generally, bovine serum albumin (BSA) or nonfat dry milk is used to block membranes. For lectin blotting, we have tested BSA, which has produced a very high background, probably owing to glycoprotein contamination in the commercial preparations of this protein. When low-fat dried milk was used, no glycoprotein signals were obtained on the blot after the chemiluminescent reaction. This could be attributed to the presence of biotin, which competed with biotinylated lectin and prevented its binding to the sugar moieties of glycoproteins *(35)*. The PBS–Tween 20 blocking solution gave a very low background both for colorimetric and chemiluminescence detections.
4. The blot should be immersed in a sufficient volume of solution to allow a good exchange of fluid over its entire surface. The use of plastic bags to incubate protein blots is not recommended because it often leads to uneven background and may cause areas without signal.
5. Most of the proteins in plasma are glycosylated. Therefore, a very short exposure time of the blot to X-ray film is needed. In general, when 120 μg of plasma proteins are loaded onto 2-D PAGE, the blot is exposed for a period of 3–30 s. The difference in the exposure time depends on the lectin used. For example, WGA lectin detects most of the *N*-linked glycoproteins and an exposure of only 3 s is sufficient *(32)*. In this case, a longer exposure time leads to a very high background with no additional spot signal.

   For an unknown sample, the blot can be exposed for 1 and 5 min as a first attempt. The enhanced chemiluminescence using luminol (ECL Kit, Amersham) leads to a "flash" of light owing to the addition of enhancers. The light emission on membrane peaks at 5–10 min after the addition of substrate and lasts for 2–3 h *(36)*. Therefore longer exposure times (10 min–3 h) may allow the detection of weak signals.

References

1. Varki, A. (1993) Biological roles of oligosaccharides: all of the theories are correct. *Glycobiology* **3,** 97–130.
2. Turner, G. A. (1992) *N*-Glycosylation of serum proteins in disease and its investigation using lectins. *Clin. Chim. Acta* **208,** 149–171.

3. Montreuil, J., Bouquelet, S., Debray, H., Fournet, B., Spik, G., and Strecker, G. (1986) Glycoproteins, in *Carbohydrate Analysis: A Practical Approach* (Chaplin, M. F. and Kennedy, J. F., eds.), Academic Press, Oxford, pp. 143–204.
4. Baenziger, J. U. (1984) The oligosaccharides of plasma glycoproteins: synthesis, structure and function, in *The Plasma Proteins*, Vol. 4 (Putnam, F. W., ed.), Academic Press, NewYork, pp. 272–315.
5. Rademacher, T. W., Parekh, R. B., and Dwek, R. A. (1988) Glycobiology. *Annu. Rev. Biochem.* **57,** 785–838.
6. Berger, E. G., Buddecke, E., Kamerling, J. P., Kobata, A., Paulson, J. C., and Vliegenthart, J. F. G. (1982) Structure, biosynthesis and functions of glycoprotein glycans. *Experientia* **38,** 1129–1158.
7. Lundy, F. T. and Wisdom, G. B. (1992) The determination of asialoglycoforms of serum glycoproteins by lectin blotting with *Ricinus communis* agglutinin. *Clin. Chim. Acta* **205,** 187–195.
8. Thompson, S. and Turner, G. A. (1987) Elevated levels of abnormally-fucosylated haptoglobins in cancer sera. *Br. J. Cancer* **56,** 605–610.
9. Stibler, H. and Borg, S. (1981) Evidence of a reduced sialic acid content in serum transferrin in male alcoholics. *Alcohol. Clin. Exp. Res.* **5,** 545–549.
10. Takahashi, M., Tsujioka, Y., Yamada, T., Tsuboi, Y., Okada, H., Yamamoto, T., and Liposits, Z. (1999) Glycosylation of microtubule-associated protein tau in Alzheimer's disease brain. *Acta Neuropathol.* **97,** 635–641.
11. Guevara, J., Espinosa, B., Zenteno, E., Vazquez, L., Luna, J., Perry, G., and Mena, R. (1998) Altered glycosylation pattern of proteins in Alzheimer disease. *J. Neuropathol. Exp. Neurol.* **57,** 905–914.
12. Gravel, P., Walzer, C., Aubry, C., Balant, L. P., Yersin, B., Hochstrasser, D. F., and Guimon, J. (1996) New alterations of serum glycoproteins in alcoholic and cirrhotic patients revealed by high resolution two-dimensional gel electrophoresis. *Biochem. Biophys. Res. Commun.* **220,** 78–85.
13. Clamp, J. R. (1984) The oligosaccharides of plasma protein, in *The Plasma Proteins*, Vol. 2 (Putnam, F. W., ed.), Academic Press, NewYork, pp. 163–211.
14. Lis, H. and Sharon, N. (1986) Lectins as molecules and as tools. *Annu. Rev. Biochem.* **55,** 35–67.
15. Goldstein, I. J. and Hayes, C. E. (1978) The lectins: carbohydrate-binding proteins of plants and animals. *Adv. Carbohydr. Chem. Biochem.* **35,** 127–340.
16. Goldstein, I. J. and Poretz, R. D. (1986) Isolation, physicochemical characterization, and carbohydrate-binding specificity of lectins, in *The Lectins: Properties, Functions and Applications in Biology and Medicine* (Liener, I. E., Sharon, N., and Goldstein, I. J., eds.), Academic Press, Orlando, pp. 35–247.
17. Osawa, T. and Tsuji, T. (1987) Fractionation and structural assessment of oligosaccharides and glycopeptides by use of immobilized lectins. *Annu. Rev. Biochem.* **56,** 21–42.
18. Bhavanandan, V. P. and Katlic, A. W. (1979) The interaction of wheat germ agglutinin with sialoglycoproteins. The role of sialic acid. *J. Biol. Chem.* **254,** 4000–4008.
19. Debray, H., Decout, D., Strecker, G., Spik, G., and Montreuil, J. (1981) Specificity of twelve lectins towards oligosaccharides and glycopeptides related to *N*-glycosylproteins. *Eur. J. Biochem.* **117,** 41–55.
20. Kaifu, R. and Osawa, T. (1976) Synthesis of *O*-β-D-galactopyranosyl-(1-4)-*O*-(2-acetamido-2-deoxy-β-D-glucopyranosyl)-(1-2)-*n*-mannose and its interaction with various lectins. *Carbohydr. Res.* **52,** 179–185.

21. Animashaun, T. and Hughes, R. C. (1989) *Bowringia milbraedii* agglutinin. Specificity of binding to early processing intermediates of asparagine-linked oligosaccharide and use as a marker of endoplasmic reticulum glycoproteins. *J. Biol. Chem.* **264,** 4657–4663.
22. Haselbeck, A., Schickaneder, E., Von der Eltz, H., and Hosel, W. (1990) Structural characterization of glycoprotein carbohydrate chains by using digoxigenin-labeled lectins on blots. *Analyt. Biochem.* **191,** 25–30.
23. Sueyoshi, S., Tsuji, T., and Osawa, T. (1988) Carbohydrate-binding specificities of five lectins that bind to *O*-glycosyl-linked carbohydrate chains. Quantitative analysis by frontal-affinity chromatography. *Carbohydr. Res.* **178,** 213–224.
24. Cummings, R. D. and Kornfeld, S. (1982) Characterization of the structural determinants required for the high affinity interaction of asparagine-linked oligosaccharides with immobilized phaseolus vulgaris leukoagglutinating and erythroagglutinating lectins. *J. Biol. Chem.* **257,** 11,230–11,234.
25. Pereira, M. E. A. and Kabat, E. A. (1974) Blood group specificity of the lectin from lotus tetragonolobus. *Ann. NY Acad. Sci.* **334,** 301–305.
26. Debray, H. and Montreuil, J. (1989) *Aleuria aurantia* agglutinin. A new isolation procedure and further study of its specificity towards various glycopeptides and oligosaccharides. *Carbohydr. Res.* **185,** 15–26.
27. Shibuya, N., Goldstein, I. J., Broekaert, W. F., Nsimba-Lubaki, M., Peeters, B., and Peumans, W. J. (1987) The elderberry (sambucus nigra l.) bark lectin recognizes the Neu5Ac (α2-6) Gal/GalNAc sequence. *J. Biol. Chem.* **262,** 1596–1601.
28. Wang, W. C. and Cummings, R. D. (1988) The immobilized leukoagglutinin from the seeds of *Maackia amurensis* binds with high affinity to complex-type Asn-linked oligosaccharides containing terminal sialic acid-linked α-2,3 to penultimate galactose residues. *J. Biol. Chem.* **263,** 4576–4585.
29. Cohen, E., Roberts, S. C., Nordling, S., and Uhlenbruck, G. (1972) Specificity of *Limulus polyphemus* agglutinins for erythrocyte receptor sites common to M and N antigenic determinants. *Vox Sang.* **23,** 300–307.
30. Appel, R. D., Sanchez, J. C., Bairoch, A., Golaz, O., Miu, M., Vargas, J. R., and Hochstrasser, D. F. (1993) Swiss-2D PAGE: a database of two-dimensional gel electrophoresis images. *Electrophoresis* **14,** 1232–1238.
31. Jadach, J. and Turner, G. A. (1993) An ultrasensitive technique for the analysis of glycoproteins using lecting blotting enhanced chemiluminescence. *Analyt. Biochem.* **212,** 293–295.
32. Gravel, P., Golaz, O., Walzer, C., Hochstrasser, D. F., Turler, H., and Balant, L. P. (1994) Analysis of glycoproteins separated by two-dimensional gel electrophoresis using lectin blotting revealed by chemiluminescence. *Analyt. Biochem.* **221,** 66–71.
33. Sanchez, J. C., Ravier, F., Pasquali, C., Frutiger, S., Bjellqvist, B., Hochstrasser, D. F., and Hughes, G. J. (1992) Improving the detection of proteins after transfer to polyvinylidene difluoride membranes. *Electrophoresis* **13,** 715–717.
34. Becker, B., Salzburg, M., and Melkonian, M. (1993) Blot analysis of glycoconjugates using digoxigenin-labeled lectins: an optimized procedure. *BioTechniques* **15,** 232–235.
35. Garfin, D. E. and Bers, G. (1982) Basic aspects of protein blotting, in *Protein Blotting* (Baldo, B. A. and Tovey, E. R., eds.), Karger, Basel, pp. 5–42.
36. Durrant, I. (1990) Light-based detection of biomolecules. *Nature* **346,** 297–298.

107

A Lectin-Binding Assay for the Rapid Characterization of the Glycosylation of Purified Glycoproteins

Mohammad T. Goodarzi, Angeliki Fotinopoulou, and Graham A. Turner

Introduction

Most proteins have carbohydrate chains (glycosylation) attached covalently to various sites on their polypeptide backbone. These posttranslational modifications, which are carried out by cytoplasmic enzymes, confer subtle changes in the structure and behavior of a molecule, and their composition is very sensitive to many environmental influences *(1–3)*. There is increasing interest in determining the glycosylation of a molecule because of the importance of glycosylation in affecting its reactivity *(1,4,5)*. This is particularly true in the production of therapeutic glycoproteins by recombinant methods, in which glycosylation can be determined by the type of host cell used or the production process employed *(2)*. Glycosylation is also important in disease situations in which changes in the carbohydrate structure can be involved in the pathological processes *(3,6,7,8)*. Unfortunately, the glycosylation of proteins is very complex; there are variations in the site of glycosylation, the type of amino acid–carbohydrate bond, the composition of the chains, and the particular carbohydrate sequences and linkages in each chain *(3,4)*. In addition, within any population of molecules there is considerable heterogeneity in the carbohydrate structures (glycoforms) that are synthesized at any one time *(1)*. This is typified by some molecules showing increased branching, reduced chain length, and further addition of single carbohydrate moieties to the internal chain.

To unravel completely the complexities of the glycosylation of a molecule is a substantial task, which requires considerable effort and resources. However, in many situations this is unnecessary, because only a limited amount of information, on a single or a group of structural features, is needed. Lectins can be useful for this purpose. These substances are carbohydrate binding proteins with particular specificity *(9)*. Although their specificity is not absolute, there is usually one carbohydrate or group of carbohydrates to which the lectin binds with a higher affinity than the rest of the group. The major carbohydrate specificity of a number of commonly used lectins is shown in **Table 1**.

Lectins have previously been used for investigating glycoprotein glycosylation by incorporating them into existing technologies such as affinity chromatography, blotting, and electrophoresis. Although these modifications give workable methods, they

From: *The Protein Protocols Handbook, 2nd Edition*
Edited by: J. M. Walker © Humana Press Inc., Totowa, NJ

Table 1
Specificity of Different Lectins

Lectin	Abbreviation	Specificity
Concanavalin A	ConA	Mannose α1–3 or mannnose α1–6
Datura stramonium agglutinin	DSA	Gallactose β1–4 *N*-acetylglucosamine
Aleuria aurantia agglutinin	AAA	Fucose α1–6 *N*-acetylglucosamine
Lens culinaris agglutinin	LCA	Mannose α1–3 or mannose α1–6 (Fucose α1–6 GlcNAc is also required)
Lotus tetragonolobus agglutinin	LTA	Fucose α1–2 galactose β1–4 [Fucose α1–3] *N*-acetylglucosamine
Sambucus nigra agglutinin	SNA	*N*-acetylneuraminic acid α2–6 galactose
Maackia amurensis agglutinin	MAA	*N*-acetylneuraminic acid α2–3 galactose
Peanut agglutinin	PNA	Galactose β1–3 *N*-acetylgalactosamine on O-linked chains

frequently use large amounts of lectin, which is expensive, requires considerable technical skill, cannot handle large numbers of specimens, and provides only semiquantitative results. For a more detailed discussion of these methods the reader should refer to a previous review ***(3)***.

Another approach is to use lectins in the familiar sandwich enzyme-linked immunosorbent assay (ELISA) technology in multiwell plates. Procedures of this type have been described for the measurement of particular carbohydrate structures in α-fetoprotein ***(10)***, glycodelin ***(11)***, haptoglobin (HP) ***(12)***, human chorionic gonadotrophin (hCG) ***(13)***, α1-acid glycoprotein (AGP) ***(14)***, fibronectin ***(15)***, c-*erb*-B2 ***(16),*** immunoglobulins (IgG) ***(17)***, mucins ***(18),*** plasminogen (Pg) activator ***(19)***, and transferrin (TF) ***(20)***. In this method, the lectin is used either to capture the molecule of interest or to identify it. An antibody is used as the other partner in the sandwich and the degree of binding is measured by the presence of an enzyme label on the identifier. Both configurations suffer from the disadvantage that the lectin may bind to carbohydrate determinants on the immunoglobulin used as the antibody. Furthermore, using the lectin as the capture molecule, the immobilized lectin may bind to glycans of other glycoproteins in the sample and these will compete with the molecule of interest for the available binding sites. When immobilized antibody is used as the capture molecule, care must be taken to ensure that the antibody is not binding to the same determinant as that reacting with lectin.

The lectin binding assay (LBA) described herein ***(21)*** overcomes many of the disadvantages with the previous lectin immunoassays. It was developed from a previously reported procedure ***(22)***. A purified glycoprotein is absorbed onto the plastic surface of a well in a microtiter plate. After the unbound protein is removed by washing, uncoated sites on the plate are blocked using a non-ionic detergent. A lectin labeled with digoxigenin (DIG) or biotin is added and allowed to interact with the carbohydrate on the absorbed glycoprotein. Unbound lectin is removed by further washing and the amount of bound lectin is measured by adding an anti-DIG antibody or streptavidin conjugated to an enzyme. Streptavidin has a very high affinity for biotin. Following further washing, the bound enzyme is used to develop a color reaction by the addition

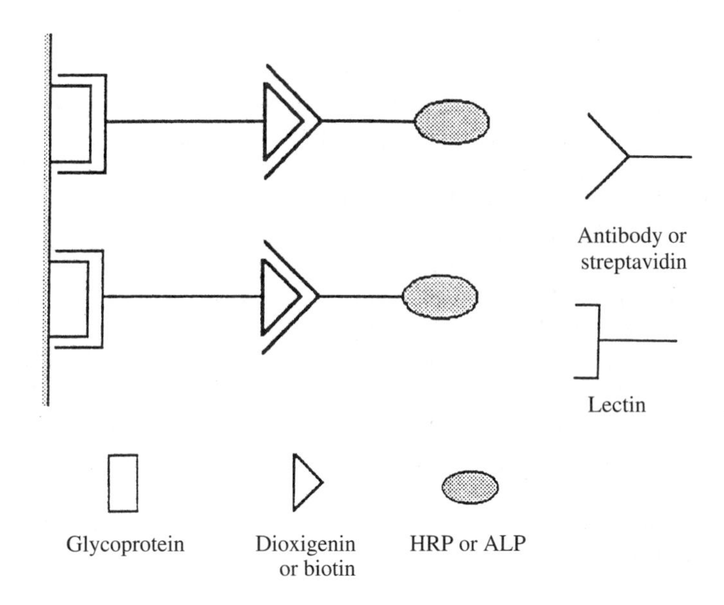

Fig. 1. Schematic diagram of the interactions in the LBA.

of the appropriate substrates. The principle of the procedure is summarized schematically in **Fig. 1**.

Using this method it is possible to rapidly screen multiple specimens, with high sensitivity and excellent precision. In addition, very small amounts of lectin are used, background absorbances are low, and the procedure does not require a high degree of technical skill other than some experience with micropipets and ELISA. Because such small amounts of glycoprotein are needed, a glycoprotein can be rapidly purified by a batch affinity chromatography method ***(23,24)***. The LBA has successfully been applied to the investigation of the glycosylation of purified alpha-1-proteinase inhibitor (API) ***(24)*** and HP (25) using concanavalin A (ConA), *Maackia amurensis* agglutinin (MAA), and *Sambucus nigra* agglutinin (SNA). Other studies of API and HP have been carried out using *Lens culinaris* agglutinin (LCA) and *Lotus tetragonolobus* agglutinin (LTA) (unpublished observations). If a panel of lectins is used, an overall picture of the carbohydrate structure of a glycoprotein can be built up very quickly and cheaply. Furthermore, subtle differences in the glycosylation of the same glycoprotein in different situations can be identified, for example, different diseases ***(24,25)*** or between different batches of a recombinant protein ***(26)***.

Materials

2.1. Reagents

1. Tris-buffered saline (TBS): 125 m*M* Tris-HCl, pH 7.5, containing 100 m*M* NaCl pH 7.5.
2. Control glycoprotein (*see* **Notes 1** and **2**).
3. Sample glycoprotein (*see* **Notes 2** and **3**).
4. Tween–TBS (TTBS): 125 m*M* Tris-HCl, pH 7.5, containing 100 m*M* NaCl and 0.1% (v/v) Tween 20.
5. Tris cations (TC) : 1 m*M* Tris-HCl, pH 7.5, containing 1 m*M* $CaCl_2$, 1 m*M* $MgCl_2$, 1 m*M* $MnCl_2$, and 0.1% (v/v) Tween 20.

6. DIG-labeled lectins (Roche Molecular Biochemicals): Lectin-DIG dissolved in TC (*see* **Note 3**).
7. Biotin-labeled lectins (Sigma): LCA-biotin and LTA-biotin dissolved in TC (*see* **Note 3**).
8. Anti-DIG antibody conjugated with horseradish peroxidase (HRP) (Roche Molecular Biochemicals): 50 mU/mL (Fab fragment) in TTBS (*see* **Note 4**).
9. Streptavidin-alkaline phosphatase (SALP) (Sigma): 1 ng/μL (1.3 mU/μL) in TTBS.
10. Citrate buffer: 34.8 m*M* citric acid, 67.4 m*M* Na_2HPO_4, pH 5.0.
11. Diethanolamine buffer: 100 m*M* diethanolamine, pH 9.8, containing 1m*M* $MgCl_2$. Store in the dark at 4°C and prepare monthly.
12. *O*-Phenylenediamine (OPD): 37 m*M* solution which is stored in 1 mL aliquots at –20°C.
13. *P*-nitrophenylphosphate (PNPP) (BDH, Atherstone, UK).
14. Hydrogen peroxide: 3% (v/v) solution prepared by diluting concentrated (30%) H_2O_2 1:10 with deionized water. Prepared monthly and stored at 4°C.
15. Color reagent:
    a. HRP : 9 mL citrate buffer + 1 mL of OPD + 50 mL of 10% H_2O_2. Make up fresh 10 min before use. H_2O_2 is added just prior to use.
    b. SALP : 20 mL of diethanolamine buffer + 20 mg of PNPP. Prepare fresh.
16. 1.25 *M* Sulfuric acid (H_2SO_4).
17. 1 *M* Sodium hydroxide (NaOH).

All other reagents were of analytical grade or better (BDH or Sigma), and prepared as required. All solutions were prepared with double-distilled deionized water (Millipore, Bedford, MA).

2.2. Equipment

1. Multiwell plastic plate: 96-Well plate (Immunolon 4, Dynex, Chantilly, VA).
2. Plate reader: Multiskan MCC/340 (Titerteck [McLean, VA]).
3. Multichannel micropipet: Finnpipet, Labsystem, Finland.

3. Methods

1. For each lectin add 100 μL of sample glycoprotein in triplicate to a multiwell plate, and for each batch of samples, add 100 μL of control glycoprotein or TBS in triplicate.
2. Incubate the plate for 2 h at 37°C. Pour off liquid by turning upside down, and tap on a pad of tissue paper to dry completely.
3. Add 200 μL of TTBS to each well, pour off liquid, and dry again. Repeat this operation 3×.
4. Add 200 μL of TTBS to each well and incubate for 1 h at 37°C and then overnight at 4°C.
5. Wash each well 3× with 200 mL of TC, removing liquid as described in **step 2**.
6. Add 200 mL of DIG-lectin or biotin-lectin to each well and incubate for 1 h at 37°C .
7. Wash each well 4× with 200 mL TTBS, removing liquid as described in **step 2**.
8. Add 150 mL anti-DIG/HRP or SALP to each well and incubate the DIG/HRP for 1 h at room temperature and the SALP for 1 h at 37°C.
9. Wash each well 4× with TTBS, removing liquid as described in **step 2**.
10. Add 100 μL of color reagent to each well as appropriate.
11. Leave plate for 10–30 min (*see* **Note 5**) at room temperature (in the dark for HRP).
12. Stop reaction:
    a. For HRP, add 100 μL of H_2SO_4 to each well.
    b. For SALP, add 100 μL of NaOH to each well.
13. Read the absorbance for each well with plate reader:
    a. For HRP, 492 nm.
    b. For SALP, 405 nm.

Table 2
Optimum Protein Coating Concentration for DifferentLectins and Different Serum Glycoproteins

	Optimum protein concentration, µg/mL			
Lectin	API	HP	TF	AGP
ConA	0.15	0.05	0.15	2.50
SNA	0.15	0.05	0.07	0.50
MAA	1.00	0.50	3.00	0.50
LCA	1.50	0.50	NR	ND
LTA	1.50	1.00	NR	ND
DSA	0.05	0.01	0.35	0.06

NR, no reactivity; ND, not determined.

14. Calculate the mean of triplicate measurements and subtract the absorbance value (OD) of the blank from the values of all samples and controls.

4. Notes

1. A pool of purified glycoprotein that reacts with the lectin is recommended as a positive control in the assay and also for preliminary experiments to develop assay conditions. Various serum glycoproteins can be obtained commercially in a purified form for this purpose, but with other glycoproteins this may not be possible, and suitable material may have to be prepared in the laboratory.
2. Sample and control glycoproteins can be conveniently purified for assay by extracting with an antibody coupled to Sepharose. We have described a batch method that is very rapid, can handle many specimens, produces very pure protein, and works for a number of different glycoproteins ***(23,24)***. Although the yield from this procedure can be low (5–20%), this does not matter because only low amounts of glycoprotein are needed in the assay. A similar method could be equally well adapted for the extraction of other types of glycoprotein.
3. For each combination of lectin and protein, preliminary experiments are carried out to determine the optimum concentration of protein required to coat the well. This is done by coating the well with different amounts of protein (1–1000 ng) and probing with a lectin solution of 1 µg/mL ***(22)***. The protein concentration is chosen that gives an OD value of approx 1. If this value cannot be established, the lectin concentration is increased (e.g., 1.5 µg/mL) or decreased (e.g., 0.5 µg/mL) as required, and the protein is reassayed. **Table 2** shows the optimum glycoprotein concentrations for various lectins we have previously studied. The concentrations of Con A, LCA, LTA, MAA, and SNA used in these experiments were 1, 0.5, 1, 1.5, and 1 µg/mL, respectively.
4. The activity of anti-DIG HRP varies from batch to batch and should be routinely checked. This reagent is usually used at a dilution of 1:3000 to 1:6000.
5. The time required to develop the color reaction depends on the reactivity of lectin with the particular protein being investigated. For some lectins (e.g., Con A and SNA) it is between 10 and 30 min whereas for other lectins (e.g., LTA, MAA) it is as long as 120 min.
6. To minimize background, TBS and TTBS must be filtered through a 0.2-µm membrane (Tuffryn membrane). The absorbances (without glycoprotein but with all other reagents) for Con A, SNA, and MAA were (mean ± SD, no. of observations) 0.28 ± 0.02, 9, 0.16 ± 0.03, 9; 0.14 ± 0.02,9 respectively.

Table 3
Analysis of Purified API and Hp in the LBA using ConA , MAA , and SNA

		Lectin[b] ($OD_{492\ nm}$)		
Glycoprotein	Source	ConA	SNA	MAA
Hp	Healthy	1.40 ± 0.35	1.70 ± 0.22	0.18 ± 0.07
	Cancer[a]	0.86 ± 0.15	1.69 ± 0.29	0.45 ± 0.10
API	Healthy	1.29 ± 0.28	1.31 ± 0.19	0.47 ± 0.08
	Cancer[a]	1.87 ± 0.14	1.95 ± 0.09	0.23 ± 0.07

[a]Cancer specimen from patients with stage III/IV ovarian carcinoma.
[b]Mean ± SD calculated from the values of 8 healthy individuals or 12 cancer patients.

7. The method gives very good reproducibility. For example, in the case of API, the interassay precision using ConA, SNA, and MAA was 2.2%, 4.5%, and 6.4%, respectively, and the intraassay precision for this glycoprotein with the same lectins was 1.5, 3 and 5%, respectively.
8. The specificity of each lectin can be checked by assaying the glycoprotein in the presence of a competitive sugar. For example, in the presence of 100 m*M* α-D-methyl mannoside, a competitive inhibitor for ConA, the absorbance obtained for API is reduced from 2 to 0.09. Alternatively, glycoproteins can be used that are known to lack the carbohydrate grouping under investigation, for example, carboxypeptidase Y-MAA and TF-LTA.
9. The method is only semiquantitative, and if high absorbance values are obtained for a glycoprotein this suggests that it has high amounts of a particular carbohydrate grouping, and vice versa. Positive and negative controls must always be run in the assay, because it does not use standards. The method is at its most useful when comparing different samples in the same assay. This is illustrated in **Table 3**, which shows the lectin binding characteristics for Hp from healthy individuals and cancer patients. The results are interpreted according to the known properties of each lectin *(27)*. Therefore, in cancer, the branching of Hp is increased, there is more α2–3 *N*-acetyneuraminic acid (Neu5A), and the α2–6 Neu5A content is unchanged. On the other hand, with API in cancer, the branching decreases, the α2–6 Neu5A content increases, and the α2–3 Neu5A content decreases. All these changes are consistent with the reported monosaccharide composition of these specimens *(23,24)*, and the known carbohydrate structures present in *N*-glycans *(4)*. It is important to emphasize that the method does not give an indication of the glycosylation of individual glycoforms in a population, but represents the overall glycosylation.
10. More information about carbohydrate structure can be obtained from the LBA by treating the immobilized glycoprotein with a glycosidase, for example, neuraminidase (NANase), galactosidase. **Table 4** shows the effect of NANase treatment (the glycosidase that removes terminal Neu5Ac from oligosaccharide chains] on the binding of ConA, MAA, PNA, and SNA to IgG, fetuin, and plasminogen (Pg). The presence of terminal α2–3 and α2–6 Neu5Ac can be clearly shown using lectins and NANase treatment. Furthermore, the presence of a cryptic grouping (galactose β1–3 *N*-acetylgalactosamine) on *O*-Linked chains can be demonstrated on fetuin and Pg with PNA after the removal of Neu5Ac. *O*-Linked chains were not detected on IgG, which appears also to lack α2–3 Neu5Ac.
11. Because OPD is carcinogenic, the noncarcinogenic substance, 3,3',5, 5'-tetramethyl benzidine (dihydrochloride) can be used as a substrate for HRP *(28)*.

Table 4
Effect of Neuraminidase Treatment of Immobilized Glycoproteins on Their Lectin Binding Activity

	Lectin binding ($OD_{492\ nm}$)							
	Before NANase				After NANase			
Glycoprotein	ConA	MAA	PNA	SNA	ConA	MAA	PNA	SNA
Fetuin	0.81	1.15	0.04	2.78	1.08	0.09	1.23	0.08
IgG	1.61	0.06	—	1.62	1.29	—	0.11	0.34
Pg	1.45	1.37	0.53	1.13	0.99	0.32	1.24	0.25

Immobilized proteins were treated with 9 mU of neuraminidase (*Vibrio cholerae*, Roche Molecular Biochemicals) for 16 h at 37°C. After treatment the well was washed 4× with TC prior to measurement of lectin binding. Fetuin and IgG are from Sigma and Pg is from Biopool, Sweden.

References

1. Rademacher, T. W., Parekh, R. B., and Dwek, R. A. (1988) Glycobiology. *Ann. Rev. Biochem.* **57,** 785–838.
2. Goochee, C. F., Gramer, M. J., Andersen, D. C., Bahr, J. B., and Rasmussen. J. R. (1991) The oligosaccharide of glycoproteins: bioprocess factors affecting oligosaccharide structure and their effect on glycoprotein properties. *BioTechnology* **9,** 1347–1355.
3. Turner, G. A. (1992) N-glycosylation of serum proteins in disease and its investigation using lectins. *Clin. Chim. Acta.* **208,** 149–171.
4. Kobata, A. (1992) Structure and function of the sugar chains of glycoproteins. *J. Biochem.* **209,** 483–501.
5. Varki, A. (1993) Biological role of oligosaccharides: all of the theories are correct. *Glycobiology* **3,** 97–130.
6. Van den Steen, P., Rudd, P. M., Dwek, R. A., Van Damme, J., and Opdenakker, G. (1998) Cytokine and protease glycosylation as a regulatory mechanism in inflammation and autoimmunity. *Glycoimmunology 2* **435,** 133–143.
7. Brockhausen, I., Schutzbach, J., and Kuhns, W. (1998) Glycoproteins and their relationship to human disease. *Acta Anatomica* **161,** 36–78.
8. Durand, G. and Seta, N. (2000) Protein glycosylation and diseases: blood and urinary oligosaccharides as markers for diagnosis and therapeutic monitoring. *Clin. Chem,* **46:6,** 795–805.
9. Singh, R. S., Tiwary, A. K., and Kennedy, J. F. (1999) Lectins: Sources, activities, and applications. *Crit. Rev. Biotechnol.* **19,** 145–178.
10. Suzuki, Y., Aoyagi, Y., Muramatsu, M., et. al. (1990). A lectin-based monoclonal enzyme immunoassay to distinguish fucosylated and non-fucosylated α-fetoprotein molecular variants. *Ann. Clin. Biochem.* **27,** 121–128.
11. Van den Nieuwenholf, I. M., Koistinen, H., Easton, R. L., Koistinen, R., Kamarainen, M., Morris, H. R., Van Die, I., Seppala, M., Dell, A., and Van den Eijnden, D. H. (2000) Recombinant glycodelin carrying the same type of glycan structures as contraceptive glycodelin-A can be produced in human kidney 293 cells but not in Chinese hamster ovary cells. *Eur. J. Biochem.* **267,** 4753–4762
12. Thompson, S. Stappenbeck, R., and Turner, G. A. (1989) A multiwell lectin-binding assay using Lotus tetragonolobus for measuring different glycosylated forms of haptoglobin. *Clin. Chim. Acta.* **180,** 277–284.

13. Abushoufa, R. A., Talbot, J. A., Brownbill, K., Rafferty, B., Kane, J. W., and Robertson, W. R. (2000) The development of a sialic acid specific lectin-immunoassay for the measurement of human chorionic gonadotrophin glycoforms in serum and its application in normal and Down's syndrome pregnancies. *Clin. Endocrinol.* **52,** 499–508.
14. Ryden, I., Lundblad, A., and Pahlsson, P. (1999) Lectin ELISA for analysis of alpha 1-acid glycoprotein fucosylation in the acute phase response. *Clin. Chem.* **45,** 2010–2012.
15. Hampel, D. J., Kottgen, B., Dudenhausen, J. W., and Kottgen, E (1999) Fetal fibronectin as a marker for an imminent (preterm) delivery. A new technique using the glycoprotein lectin immunosorbent assay. *J. Immunol. Meth.* **224(1–2),** 31–42.
16. Cook, D. B., Bustamam, A. A., Brotherick, I., Shenton, B. K., and Self, C. H. (1999) Lectin ELISA for the c-*erb*-B2 tumor marker protein p185 in patients with breast cancer and controls. *Clin. Chem.* **4,** 292–295.
17. Kinoshita, N., Ohno, M., Nishiura, T., Fujii, S., Nishikowa, A., Kawakami, Y., Uozumi N., and Taniguchi, N. (1991) Glycosylation of the Fab portion of myeloma immunoglobulin G and increased fucosylated biantennary chains: structural analysis by high-performance liquid chromatography and antibody-lectin immunoassay using Lens culinaris agglutinin. *Can. Res.* **70,** 5888–5892.
18. Parker, N., Makin, C. A., Ching, C. K., et. al. (1992) A new enzyme-linked lectin/mucin antibody sandwich assay (CAM 17.1/WGA) assessed in combination with CA 19-9 and peanut lectin binding assay for the diagnosis of pancreatic cancer. *Cancer* **70,** 1062–1068.
19. Hayashi, S. and Yamada, K. (1992) Quantitative assay of the carbohydrate in urokinase-type plasminogen activator by lectin-enzyme immunoassay. *Blood Coagul. Fibrin.* **3,** 423–428.
20. Pekelharing, J. M., Vissers, P., and Leijnse, B. (1987) Lectin-Enzyme immunoassay of transferrin sialovariants using immobilized antitransferrin and enzyme-labeled galactose-binding lectin from Ricinus communis. *Analyt. Biochem.* **165,** 320–326.
21. Goodarzi, M. T., Rafiq, M., and Turner, G. A. (1995) An improved multiwell immunoassay using digoxigenin-labelled lectins to study the glycosylation of purified glycoproteins. *Biochem. Soc. Trans.* **23,** 168S.
22. Katnik, I., Jadach, J., Krotkiewski, H., and Gerber, J. (1994) Investigation the glycosylation of normal and ovarian cancer haptoglobins using digoxigenin-labelled lectins. *Glycosyl. Dis.* **1,** 97–104.
23. Thompson, S., Dargan, E., and Turner, G. A. (1992). Increased fucosylation and other carbohydrate changes in haptoglobin in ovarian cancer. *Can. lett.* **66,** 43–48.
24. Goodarzi, M. T. and Turner, G. A. (1995) Decreased branching, increased fucosylation and changed sialylation of alpha-1-proteinase inhibitor in breast and ovarian cancer. *Clin. Chim. Acta* **236,** 161–171.
25. Turner, G. A., Goodarzi, M. T., and Thompson, S. (1995) Glycosylation of alpha-1-proteinase inhibitor and haptoglobin in ovarian cancer: evidence for two different mechanisms. *Glycoconju. J.* **12,** 211–218.
26. Fotinopoulou, A., Cook, A., and Turner, G. A. (1999) Rapid methods for investigating the carbohydrate profile of therapeutic recombinant glycoproteins. *Glycoconju. J.* **16,** S174.
27. Wu, A. M., Sugii, S., and Herp, A. (1988) A table of lectin carbohydrate specificities, in *Lectins- Biology, Biochemistry, Clinical Biochemistry*, (Bog-Hansen, T. C. and Freed, D. L. J. eds.) **vol. 6,** Sigma Chemical Company, St. Louis, Missouri, pp. 723–740.
28. Bos, E. S., Van der Doelen, A. A., Van Rooy, N., and Schuurs, A. H. W. M. (1981) 3,3'5,5'-Tetramethylbenzidine as an Ames test negative chromogen for horse-radish peroxidase in enzyme immunoassay. *J. Immuno.* **2(384)** 187–204.

108

Chemical Methods of Analysis of Glycoproteins

Elizabeth F. Hounsell, Michael J. Davies, and Kevin D. Smith

1. Introduction

The first analysis of glycoconjugates that often needs to be carried out is to see if they indeed contain sugar. For glycoproteins in gels or oligosaccharides in solution, this can be readily achieved by periodate oxidation at two concentrations, the first to detect sialic acids, and the second, any monosaccharide that has two free vicinal hydroxyl groups *(1)*. Periodate cleaves between the hydroxyl groups to yield reactive aldehydes, which can be detected by reduction with NaB^3H_4 or coupled to high sensitivity probes available in commercial kits, e.g., from Boeringher Mannheim (Mannheim, Germany) or Oxford GlycoSciences (Abingdon, UK). In solution, a quick spot assay can be carried out for the presence of any monosaccharide or oligosaccharide having a C-2 hydroxyl group by visualization with charring by phenol/sulfuric acid reagent *(2)*. These methods are relatively specific for mono/oligosaccharides *(3)*.

2. Materials

2.1. Periodate Oxidation

1. 0.1 *M* Acetate buffer, pH 5.5, containing 8 or 15 m*M* sodium periodate.
2. Ethylene glycol.
3. 0.1 *M* Sodium hydroxide.
4. Reducing agent: sodium borohydride, tritiated sodium borohydride, or sodium borodeuteride.
5. Glacial acetic acid.
6. Methanol.

2.2. Phenol Sulfuric Acid Assay

1. H_2O (HPLC-grade).
2. 4% Aqueous phenol.
3. Concentrated H_2SO_4.
4. 1 mg/mL galactose.
5. 1 mg/mL mannose.

3. Methods

3.1. Periodate Oxidation

1. Dissolve 0.1–1.0 mg glycoprotein in 2 mL of acetate buffer containing sodium periodate (15 m*M* for all monosaccharides, 8 m*M* for alditols, and 1 m*M* specifically for oxidation of sialic acids).

From: *The Protein Protocols Handbook, 2nd Edition*
Edited by: J. M. Walker © Humana Press Inc., Totowa, NJ

2. Carry out the periodate oxidation in the dark at room temperature for 1 h for oligosaccharides, or at 4°C for 48 h for alditols, or 0°C for 1 h for sialic acids (*see* **Note 1**).
3. Decompose excess periodate by the addition of 25 μL of ethylene glycol, and leave the sample at 4°C overnight.
4. Add 0.1 *M* sodium hydroxide (about 1.5 mL) until pH 7.0 is reached.
5. Reduce the oxidized compound with 25 mg of $NaB[^3H]_4$ at 4°C overnight (*see* **Note 2**).
6. Add acetic acid to pH 4.0, and concentrate the sample to dryness.
7. Remove boric acid by evaporations with 3 × 100 μL methanol (*see* **Note 3**).

3.2. Phenol/Sulfuric Acid Assay

1. Aliquot a solution of the unknown sample containing a range of approx 1 μg/10 μL into a microtiter plate along with a range of concentrations of a hexose standard (galactose or mannose, usually 1–10 μg).
2. Add 25 μL of 4% aquenous phenol to each well, mix thoroughly, and leave for 5 min (*see* **Note 4**).
3. Add 200 μL of H_2SO_4 to each well, and mix prior to reading on a microtiter plate reader at 492 nm (*see* **Note 5**).

4. Notes

1. It is important that the periodate oxidation is carried out in the dark to avoid unspecific oxidation. The periodate reagent has to be prepared fresh, since it is degraded when exposed to light.
2. The reactive aldehydes can also be detected by coupling to an amine-containing compound, such as digoxigenin *(1)*.
3. Addition of methanol in an acidic environment leads to the formation of volatile methyl borate.
4. Do not overfill wells during the hexose assay, since the conc. H_2SO_4 will severely damage the microtiter plate reader if spilled.
5. Exercise care when adding the conc. H_2SO_4 to the phenol/alditol mature, since it is likely to "spit," particularly in the presence of salt.

References

1. Haselbeck, A. and Hösel, W. (1993) Immunological detection of glycoproteins on blots based on labeling with digoxigenin, in *Methods in Molecular Biology, vol. 14: Glycoprotein Analysis in Biomedicine* (Hounsell, E. F., ed.), Humana, Totowa, NJ, pp. 161–173.
2. Smith, K., Harbin, A. M., Carruthers, R. A., Lawson, A. M., and Hounsell, E. F. (1990) Enzyme degradation, high performance liquid chromatography and liquid secondary ion mass spectrometry in the analysis of glycoproteins. *Biomed. Chromatogr.* **4,** 261–266.
3. Hounsell, E. F. (1993) A general strategy for glycoprotein oligosaccharide analysis, in *Methods in Molecular Biology, vol. 14: Glycoprotein Analysis in Biomedicine* (Hounsell, E. F., ed.), Humana, Totowa, NJ, pp. 1–15.

109

Monosaccharide Analysis by HPAEC

Elizabeth F. Hounsell, Michael J. Davies, and Kevin D. Smith

1. Introduction

Once the presence of monosaccharides has been established by chemical methods (*see* Chapter 108), the next stage of any glycoconjugate or polysaccharide analysis is to find out the amount of sugar and monosaccharide composition. The latter can give an idea as to the type of oligosaccharides present and, hence, indicate further strategies for analysis *(1)*. Analysis by high-pH anion-exchange chromatography (HPAEC) with pulsed electrochemical detection, that is described here, is the most sensitive and easiest technique *(2)*. If the laboratory does not have a biocompatible HPLC available that will withstand high salt concentrations, a sensitive labeling technique can be used with gel electrophoresis *(3)*, e.g., that marketed by Glyko Inc. (Navato, CA) or Oxford GlycoSciences (Abingdon, UK), to include release of oligosaccharides/monosaccharides and labeling via reductive amination with a fluorescence label *(4)*.

2. Materials

1. Dionex DX500 (Dionex Camberley, Surrey UK) or other salt/biocompatible gradient HPLC system (titanium or PEEK lined), e.g., 2 Gilson 302 pumps with 10-mL titanium pump heads, 802Ti manometric module, 811B titanium dynamic mixer, Rheodyne 7125 titanium injection valve with Tefzel rotor seal (Gilson Medical Electronics, Villiers-le-Bel, France).
2. CarboPac PA1 separator (4 × 250 mm) and PA1 guard column (Dionex).
3. Pulsed amperometric detector with Au working electrode (Dionex), set up with the following parameters:
   a. Time = 0 s   E = + 0.1 V
   b. Time = 0.5 s   E = + 0.1 V
   c. Time = 0.51 s   E = + 0.6 V
   d. Time = 0.61 s   E = + 0.6 V
   e. Time = 0.62 s   E = – 0.8 V
   f. Time = 0.72 s   E = – 0.8 V
4. Reagent reservoir and postcolumn pneumatic controller (Dionex).
5. High-purity helium.
6. 12.5 *M* NaOH (BDH, Poole, UK).
7. Reagent grade sodium acetate (Aldrich, Poole, UK).
8. HPLC-grade H_2O.

From: *The Protein Protocols Handbook, 2nd Edition*
Edited by: J. M. Walker © Humana Press Inc., Totowa, NJ

9. 2 *M* Trifluoroacetic acid HPLC-grade.
10. 2 *M* HCl.
11. Dowex 50 W × 12 H^+ of cation-exchange resin.
12. 3.5-mL screw-cap septum vials (Pierce, Chester, UK) cleaned with chromic acid (2 L H_2SO_4/350 mL H_2O/100 g Cr_2O_3) (**Use care! extremely corrosive,** *see* **Note 1**), and coated with Repelcote (BDH, Poole, UK).
13. Teflon-backed silicone septa for 3.5-mL vials (Aldrich).

3. Method

1. Dry down the glycoprotein (10 µg) or oligosaccharide (1 µg) in a clean screw-top vial with Teflon-backed silicone lid insert (*see* **Note 2**).
2. Hydrolyze in an inert N_2 atmosphere for 4 h at 100°C with 2 *M* HCl.
3. Dry the hydrolyzate and re-evaporate three times with HPLC-grade H_2O.
4. Purify on a 1-mL Dowex 50 W × 12 H^+ column eluted in water.
5. Dry down the monosaccharides ready for injection onto the HPLC system.
6. Prepare the following eluants:
   Eluant A = 500 mL HPLC-grade H_2O.
   Eluant B = 500 mL of 50 m*M* NaOH, 1.5 m*M* sodium acetate.
   Eluant C = 100 mL of 100 m*M* NaOH.
7. Degas the eluants by bubbling through helium.
8. Place the postcolumn reagent in a pressurized reagent reservoir (300 mm NaOH) and use the pneumatic controller to adjust helium pressure to give a flowrate of 1 mL/min (approx 10 psi).
9. Equilibrate the column with 98% eluant A and 2% eluant B at a flowrate of 1 mL/min.
10. Add the postcolumn reagent between column and detector cell at a flowrate of 1 mL/min via a mixing tee.
11. Inject approx 100 pmol of monosaccharide and elute isocratically as follows: eluant A = 98%; eluant B = 2%; flowrate = 1 mL/min; for 30 min.
12. Calculate monosaccharide amounts by comparison with a range of known monosaccharide standards run on the same day with deoxyglucose as an internal standard. From this, it is possible to infer the type and amount of glycosylation of the glycoprotein.
13. Regenerate the column in eluant C for 10 min at 1 mL/min (*see* **Note 3**).
14. Re-equilibrate the column with 98%A/2%B before the next injection.
15. At the end of the analysis, regenerate the column in eluant C, and flush pumps with H_2O (*see* **Note 4**).

4. Notes

1. If required, an equivalent detergent-based cleaner may be used.
2. Use polypropylene reagent vessels as far as possible for HPAEC-PAD because of the corrosive nature of the NaOH, and to minimize leaching of contaminants from the reservoirs.
3. Some drift in retention times may be observed during the monosaccharide analysis. This can be minimized by thorough regeneration of the column and use of a column jacket to maintain a stable column temperature.
4. Failure to wash out the eluants from the pumps at the end of an analysis may result in crystallization and serious damage to the pump heads.

References

1. Hounsell, E. F. (1993) A general strategy for glycoprotein oligosaccharide analysis, in *Methods in Molecular Biology, vol. 14: Glycoprotein Analysis in Biomedicine* (Hounsell, E. F., ed.), Humana, Totowa, NJ, pp. 1–15.

2. Townsend, R. R. (1995) Analysis of glycoconjugates using high-pH anion-exchange chromatography. *J. Chromatog. Library* **58,** 181–209.
3. Linhardt, R. J., Gu, D., Loganathan, D., and Carter, S. R. (1989) Analysis of glycosaminoglycan-derived oligosaccharides using reversed-phase ion-pairing and ion-exchange chromatography with suppressed conductivity detection. *Anal. Biochem.* **181,** 288–296.
4. Davies, M. J. and Hounsell, E. F. (1996) Comparison of separation modes for high performance liquid chromatography for the analysis of glycoprotein- and proteoglycan-derived oligosaccharides. *J. Chromatogr.* **720,** 227–234.

110

Monosaccharide Analysis by Gas Chromatography (GC)

Elizabeth F. Hounsell, Michael J. Davies, and Kevin D. Smith

1. Introduction

Although the methods given in Chapter 109 can give an approximate idea of oligosaccharide amount or composition, they would not be able to distinguish the multiple monosaccharides and substituents present in nature. For this, the high resolution of gas chromatography (GC) is required ***(1,2)***. The most unambiguous results are provided by analysis of trimethylsilyl ethers (TMS) of methyl glycosides with on line mass selective detection (MS).

2. Materials

1. 0.5 *M* Methanolic HCl (Supelco, Bellefont, PA).
2. Screw-top PTFE septum vials.
3. Phosphorus pentoxide.
4. Silver carbonate.
5. Acetic anhydride.
6. Trimethylsilylating (TMS) reagent (Sylon HTP kit, Supelco: pyridine hexamethyldisilazane, trimethylchlorosilane; **use care; corrosive**).
7. Toluene stored over 3-Å molecular sieve.
8. GC apparatus fitted with flame ionization or MS detector and column, e.g., for TMS ethers 25 m × 0.22 mm id BP10 (SGE, Austin, TX) 30 m × 0.2 mm id ultra-2 (Hewlett Packard, Bracknell, Berkshire, UK).

3. Method

1. Concentrate glycoproteins or oligosaccharides containing 1–50 μg carbohydrate and 10 μg internal standard (e.g., inositol or perseitol) in screw-top septum vials. Dry under vacuum in a desiccator over phosphorus pentoxide.
2. Place the sample under a gentle stream of nitrogen, and add 200 μL methanolic HCl (*see* **Note 1**).
3. Cap immediately, and heat at 80°C for 18 h (*see* **Note 2**).
4. Cool the vial, open, and add approx 50 mg Silver carbonate.
5. Mix the contents, and test for neutrality (*see* **Note 3**).
6. Add 50 μL acetic anhydride, and stand at room temperature for 4 h in the dark (*see* **Note 4**).
7. Spin down the solid residue (*see* **Note 5**), and remove the supernatant to a clean vial.
8. Add 100 μL methanol, and repeat **step 7**, adding the supernatants together.

From: *The Protein Protocols Handbook, 2nd Edition*
Edited by: J. M. Walker © Humana Press Inc., Totowa, NJ

9. Repeat **step 8** and evaporate the combined supernatants under a stream of nitrogen.
10. Dry over phosphorus pentoxide before adding 20 µL TMS reagent.
11. Heat at 60°C for 5 min, evaporate remaining solvent under a stream of nitrogen, and add 20 µL dry toluene.
12. Inject onto a capillary GC column with 14 psi He head pressure and a temperature program from 130 to 230°C over 20 min and held at 230°C for 20 min.
13. Calculate the total peak area of each monosaccharide by adding individual peaks and dividing by the peak area ratio of the internal standard. Compare to standard curves for molar calculation determination.

4. Notes

1. The use of methanolic HCl for cleavage of glycosidic bonds and oligosaccharide-peptide cleavage yields methyl glycosides and carboxyl group methyl esters, which gives acid stability to the released monosaccharides, and thus, monosaccharides of different chemical lability can be measured in one run. If required as free reducing monosaccharides (e.g., for HPLC), the methyl glycoside can be removed by hydrolysis. The reagent can be obtained from commercial sources or made in laboratory by bubbling HCl gas through methanol until the desired pH is reached or by adding a molar equivalent of acetyl chloride to methanol.
2. An equilibrium of the α and β methyl glycosides of monosaccharide furanose (*f*) and pyranose (*p*) rings is achieved after 18 h so that a characteristic ratio of the four possible (fα, fβ, pα, pβ) molecules is formed to aid in unambiguous monosaccharide assignment.
3. Solid-silver carbonate has a pink hue in an acidic environment, and, therefore, neutrality can be assumed when green coloration is achieved.
4. The acidic conditions remove *N*-acetyl groups, which are replaced by acetic anhydride. This means that the original status of *N*-acetylation of hexosamines and sialic acids is not determined in the analysis procedure. If overacetylation occurs, the time can be reduced.
5. Direct re-*N*-acetylation by the addition of pyridine-acetic anhydride 1:1 in the absence of silver carbonate can be achieved, but this gives more variable results.

References

1. Hounsell, E. F. (1993) A general strategy for glycoprotein oligosaccharide analysis, in *Methods in Molecular Biology, vol. 14: Glycoprotein Analysis in Biomedicine* (Hounsell, E. F., ed.), Humana, Totowa, NJ, pp. 1–15.
2. Hounsell, E. F. (1994) Physicochemical analyses of oligosaccharide determinants of glycoproteins. *Adv. Carbohyd. Chem. Biochem.* **50,** 311–350.

111

Determination of Monosaccharide Linkage and Substitution Patterns by GC-MS Methylation Analysis

Elizabeth F. Hounsell, Michael J. Davies, and Kevin D. Smith

1. Introduction

The GC or GC-MS method discussed in Chapter 110 can distinguish substituted monosaccharides, but to characterize the position of acyl groups together with the linkages between the monosaccharides, a strategy has been developed to "capture" the substitution pattern by methylation of all free hydroxyl groups. The constituent monosaccharides are then analyzed after hydrolysis, reduction, and acetylation as partially methylated alditol acetates in a procedure known as methylation analysis *(1–3)*.

2. Materials

1. DMSO.
2. Methyl iodide.
3. Sodium hydroxide (anhydrous).
4. Chloroform.
5. Acetonitrile.
6. Pyridine.
7. Ethanol.
8. Water.
9. V-bottomed reacti-vials (Pierce [Rockford, IL]) with Teflon-backed silicone lid septa.
10. Sodium borodeuteride.
11. Ammonium hydroxide.
12. Trifluoroacetic acid.
13. Acetic acid.
14. GC-MS (e.g., Hewlett Packard 5890/5972A with ultra-2 or HP-5MS capillary column).

3. Method

1. Dry 20 nmol pure, desalted oligosaccharide into "V"-bottomed reacti-vials in a dessicator containing P_2O_5.
2. Resuspend samples into 150 μL of a suspension of powdered NaOH/anhydrous DMSO (approx 60 mg/mL) under an inert atmosphere (*see* **Note 1**).

From: *The Protein Protocols Handbook, 2nd Edition*
Edited by: J. M. Walker © Humana Press Inc., Totowa, NJ

3. Add 75 μL of methyl iodide (care) and sonicate for 15 min (*see* **Note 2**).
4. Extract the permethylated glycans with 1 mL $CHCl_3$ 3 mL H_2O, washing the aqueous phase with 3 × 1 mL $CHCl_3$. Wash the combined $CHCl_3$ washes with 3 × 5 mL H_2O (*see* **Note 3**).
5. Dry the $CHCl_3$ phase under N_2 after taking an aliqout for liquid secondary ion mass spectrometry (LSIMS) (*see* **Note 4**).
6. Hydrolyze for 1 h in 2 *M* TFA at 100°C.
7. Reduce samples with 50 m*M* $NaBD_4$/50 m*M* NH_4OH (*see* **Note 5**) at 4°C or 4 h at room temperature.
8. Evaporate once from AcOH and three times from 1:10 AcOH/MeOH.
9. Re *N*-acetylate samples with 50:50 acetic anhydride/pyridine 100°C, 90 min.
10. Analyze samples by GC-MS on low-bleed 5% capillary column (e.g., HP ultra-2 or HP5-MS or equivalent), with either on-column or splitless injection and a temperature gradient from to 50–265°C over 21 min, held for a further 10 min, and a constant gas flow of 1 mL/min into the MS. Ionization is in EI mode and a mass range of 45–400 (*see* **Note 6**).

4. Notes

1. The NaOH can be powdered either in a mortar and pestle or glass homogenizer and thoroughly vortexed with the DMSO prior to addition to the glycan.
2. If at the end of the methylation a yellow color is present, the reaction can be stopped by adding a crystal of sodium thiosulfate with aqueous extraction.
3. The permethylated oligosaccharides may also be purified on a Sep-Pak C_{18} column (Waters, Wafford, UK) by elution of permethylated oligosaccharides with acetonitrile or acetonitrile–water mixture.
4. The reaction with the NaOH/DMSO suspension deprotonates all the free hydroxyl groups and NH of acetamido groups forming an unstable carbanion. The addition of methyl iodide then rapidly reacts with the carbanions to form O-Me groups, and thus, permethylate the oligosaccharide. When subjected to LSIMS, these permethylated oligosaccharides will fragment about their glycosidic bonds particularly at hexosacetamido residues. This means each oligosaccharide generates a unique fragmentation pattern allowing the determination of the oligosaccharide sequence. For example, the oligosaccharide Hex-HexNAc-HexNAc-Hex will generate the following fragments: Hex-HexNAc and Hex-HexNAc-HexNAc, where Hex denotes a hexose residue and HexNAc an *N*-acetylhexosamine.
5. The reduction can also be carried out with 50 m*M* NaOH, but this is not volatile and is harder to remove prior to further derivitization.
6. The hydrolysis step generates monosaccharides with the hydroxyl groups involved in the glycosidic linkages still retaining their protons. Reduction of these monosaccharides with $NaBD_4$ will break the ring structure to form monosaccharide alditols with the anomeric (C_1) carbon being monodeuterated. Acetylation of the free hydroxyls to *O*-acetyl groups completes the derivitization. The retention times of the PMAAs on the GC allow the assignment of the monosaccharide type (galactose, *N*-acetylgalactosamine, and so on). On-line mass spectrometric detection identifies fragment ions formed by the cleavage of C—C bonds of the monosaccharide alditols with the preference: methoxy-methoxy > methoxy-acetoxy > acetoxy-acetoxy. The resulting spectra are diagnostic for the substitution pattern, and hence, the previous position of linkage, e.g., a 2-linked hexose will produce a different set of ions to a 3-linked hexose, and a 2,3-linked hexose being different again. Selected ions from the spectra of all commonly occurring linkages can be used to analyze across the chromatogram (selected ion monitoring).

References

1. Hansson, G. C. and Karlsson, H. (1993) Gas chromatography and gas chromatography-mass spectrometry of glycoprotein oligosaccharides, in *Methods in Molecular Biology, vol. 14: Glycoprotein Analysis in Biomedicine* (Hounsell, E. F., ed.), Humana, Totowa, NJ, pp. 47–54.
2. Güther, M. L. and Ferguson, M. A. J. (1993) The microanalysis of glycosyl phosphatidylinositol glycans, in *Methods in Molecular Biology, vol. 14: Glycoprotein Analysis in Biomedicine* (Hounsell, E. F., ed.), Humana, Totowa, NJ, pp. 99–117.
3. Hounsell, E. F. (1994) Physicochemical analysis of oligosaccharide determinants of glycoproteins. *Adv. Carbohyd. Chem. Biochem.* **50,** 311–350.

112

Sialic Acid Analysis by HPAEC-PAD

Elizabeth F. Hounsell, Michael J. Davies, and Kevin D. Smith

1. Introduction

The most labile monosaccharides are the family of sialic acids, which are usually chain-terminating substituents. These are therefore usually released first by either mild acid hydrolysis or enzyme digestion, and can be analyzed with great sensitivity by high pH anion-exchange chromatography with pulsed amperometric detection (HPAEC-PAD *[1,2]*). The remaining oligosaccharide is analyzed as discussed in Chapter 111 to identify the position of linkage of the sialic acid.

2. Materials

1. HCl (0.01, 0.1, and 0.5 *M*).
2. 100 m*M* NaOH.
3. 1 *M* NaOAc.
4. HPLC and PA1 columns as described in Chapter 109.

3. Method

1. Dry the glycoprotein into a clean screw-top vial with a Teflon-backed silicone lid insert (*see* **Note 1**).
2. Release the sialic acids by hydrolysis with 0.01 or 0.1 *M* HCl for 60 min at 70°C in an inert N_2 atmosphere (*see* **Note 2**).
3. Dry down the hydrolysate, and wash three times with HPLC-grade H_2O.
4. Prepare 500 mL of 100 m*M* NaOH, 1.0 *M* sodium acetate (eluant A).
5. Prepare 500 mL of 100 m*M* NaOH (eluant B).
6. Degas eluants by bubbling helium through them in their reservoirs (*see* **Note 3**).
7. Regenerate the HPLC column in 100% eluant B for 30 min at a flow rate of 1 mL/min.
8. Equilibrate the column in 95% eluant B for 30 min at a flow rate of 1 mL/min (*see* **Note 4**).
9. Inject approx 0.2 nmol of sialic acid onto the column, and elute using the following gradient at a flow rate of 1 mL/min:
   a. Time = 0 min; 95% eluant B.
   b. Time = 4 min; 95% eluant B.
   c. Time = 29 min; 70% eluant B.
   d. Time = 34 min; 70% eluant B.
   e. Time = 35 min; 100% eluant B.
   f. Time = 44 min; 100% eluant B.
   g. Time = 45 min; 95% eluant B; 1 mL/min.

From: *The Protein Protocols Handbook, 2nd Edition*
Edited by: J. M. Walker © Humana Press Inc., Totowa, NJ

10. Quantitate the sialic acids by comparison with known standards run on the same day.
11. When the baseline has stabilized, the system is ready for the next injection.
12. When the analyses have been completed, regenerate the column in eluant B, and flush pumps with HPLC-grade H_2O (*see* **Note 5**).

4. Notes

1. If problems with contaminants are encountered, it may be necessary to wash the vials with chromic acid overnight, wash them thoroughly with distilled water, and then treat with a hydrophobic coating, such as repelcoat (*see* Chapter 109).
2. 0.01 *M* HCl will release sialic acids with intact *N*- or *O*-acyl groups, but without quantitative release of the sialic acids. These can also be detected by HPAEC-PAD *(3)*. At 0.1 *M* HCl, quantitive release is achieved, but with some loss of *O*- and *N*-acylation. At this concentration, some fucose residues may also be labile. Alternatively, the sialic acids can be released by neuraminidase treatment, which can be specific for α2-6 or α2-3 linkage, e.g., with α-sialidase of *Arthrobacter ureafaciens* for α2-6 and α-sialidase of Newcastle disease virus for α2-3 using the manufacturer's instructions.
3. Use polypropylene reagent vessels as far as possible for HPAEC-PAD because of the corrosive nature of the NaOH, and to minimize leaching of contaminants from the reservoirs.
4. For maximum efficiency of detection, always ensure that the PAD reference electrode is accurately calibrated, the working electrode is clean, and the solvents are thoroughly degassed.
5. Failure to wash out the eluants from the pumps at the end of an analysis may result in crystallization and serious damage to the pump heads.

References

1. Townsend, R. R. (1995) Analysis of glycoconjugates using high-pH anion-exchange chromatography. *J. Chromatog. Library* **58,** 181–209.
2. Manzi, A. E., Diaz, S., and Varki, A. (1990) HPLC of sialic acids on a pellicular resin anion exchange column with pulsed amperometry. *Anal. Biochem.* **188,** 20–32.

113

Chemical Release of *O*-Linked Oligosaccharide Chains

Elizabeth F. Hounsell, Michael J. Davies, and Kevin D. Smith

1. Introduction

O-linked oligosaccharides having the core sequences shown below can be released specifically from protein via a β-elimination reaction catalyzed by alkali. The reaction is usually carried out with concomitant reduction to prevent peeling, a reaction caused by further β-elimination around the ring of 3-substituted monosaccharides *(1)*. The reduced oligosaccharides can be specifically bound by solid sorbent extraction on phenylboronic acid (PBA) columns *(2)*.

O-linked protein glycosylation core structures linked to Ser/Thr:

Galβ1-3GalNAcα1⁻
GlcNAcβ1-3GalNAcα1⁻

```
GlcNAcβ1-6
          \
           GalNAcα1-
          /
   Galβ1-3
GlcNAcβ1-6
          \
           GalNAcα1-
          /
GlcNAcβ1-3
```

GalNAcα1-3GalNAcα1⁻
GlcNAcβ1-6GalNAcα1⁻
GalNAcα1-6GalNAcα1⁻

2. Materials

1. 1 *M* $NaBH_4$ (Sigma, Poole, UK) in 50 m*M* NaOH. This is made up fresh each time from 50% (w/v) NaOH and HPLC-grade H_2O.
2. Methanol (HPLC-grade containing 1% acetic acid).
3. Acetic acid.
4. 1 mL Dowex H^+ (50 W × 12) strong cation ion-exchange column (Sigma, St. Louis, MO).
5. Bond elute phenyl boronic acid column (Jones Chromatography, Hengoed, UK; activated with MeOH).
6. 0.1 *M* HCl.
7. 0.2 *M* NH_4OH.
8. 0.1 *M* Acetic acid.
9. Methanol.

From: *The Protein Protocols Handbook, 2nd Edition*
Edited by: J. M. Walker © Humana Press Inc., Totowa, NJ

3. Method

1. Dry the glycoprotein (100 µg to 1 mg) in a screw-topped vial and resuspend in 100 µL 50 m*M* NaOH containing 1 *M* $NaBH_4$ (*see* **Notes 1** and **2**).
2. Incubate at 55°C for 18 h.
3. Quench the reduction by the addition of ice-cold acetic acid until no further effervescence is seen.
4. Dry the reaction mixture down, and then wash and dry three times with a 1% acetic acid, 99% methanol solution to remove methyl borate.
5. Resuspend the alditols in H_2O, and pass down a 1-mL H^+ cation exchange resin. The alditols will not be retained and will elute by washing the column with water.
6. Dry the alditols and resuspend them in 100 µL of 0.2 *M* NH_4OH.
7. Activate a phenyl boronic acid (PBA) column with 2 × 1 mL MeOH.
8. Equilibrate the PBA column with 2 × 1 mL 0.1 *M* HCl, 2 × 1 mL H_2O, and 2 × 1 mL 0.2 *M* NH_4OH.
9. Add the sample in 100 µL 0.2 *M* NH_4OH and elute with 2 × 100 µL 0.2 *M* NH_4OH, 2 × 100 µL H_2O, and 6 × 100 µL 0.1 *M* Acetic acid. Collect these fractions and test for monosaccharide (Chapter 108) or combine and analyze according to Chapter 114.
10. Regenerate the PBA column with 0.1 *M* HCl and 2 × 1 mL H_2O before storing and reactivation in 2 × 1 mL MeOH.

4. Notes

1. The $NaBH_4$/NaOH solution is made up <6 h before it is required.
2. Reduction with NaB^3H_4 allows the incorporation of a radioactive label into the alditol to enable a higher degree of sensitivity to be achieved while profiling.
3. Protein degradation can be minimized by the omission of the $NaBH_4$, although this results in the degradation of sugar chains having a 3-substituted GalNAc-Ser/Thr, i.e., most types. The addition of 6 m*M* Cadmium acetate, 6 m*M* Na_2EDTA to the $NaBH_4$/NaOH solution reduces protein degradation without the loss of oligosaccharide alditol.

References

1. Hounsell, E. F. (1994) Physicochemical analyses of oligosaccharides determinants of glycoproteins. *Adv. Carbohyd. Chem. Biochem.* **30,** 311–350.
2. Stoll, M. S. and Hounsell, E. F. (1988) Selective purification of reduced oligosaccharides using a phenylboronic acid bond elut column: potential application in HPLC, mass spectrometry, reductive amination procedures and antigenic/serum analysis. *Biomed. Chromatogr.* **2,** 249–253.

114

O-Linked Oligosaccharide Profiling by HPLC

Elizabeth F. Hounsell, Michael J. Davies, and Kevin D. Smith

1. Introduction

The several different core regions of *O*-linked chains (Chapter 113) can be further extended by Gal and GlcNAc containing backbones or by addition of blood group antigen-type glycosylation and the presence of sialic acid or sulfate. We sought a universal column that can be applied with high resolution to the various oligosaccharide alditols released from glycoproteins by β-elimination (*see* **Note 1**) and have pioneered *(1,2)* the use of porous graphitized carbon (PGC). This is an alternative to C_{18} reversed-phase HPLC and normal-phase amino-bonded columns, which can be used together in the presence and absence of high-salt buffers *(3).*

2. Materials

1. Gradient HPLC system: e.g., 2 × 302 pumps, 802C manometric module, 811 dynamic mixer, 116 UV detector, 201 fraction collector, 715 chromatography system control software (all Gilson Medical Electronics, Villiers-le-Bel, France).
2. IBM PS-2 personal computer (or compatible model) with Microsoft Windows.
3. Hypercarb S HPLC Column (100 × 4.6 mm) (Shandon Scientific, Runcorn, Cheshire, England) or Glycosep H (100 × 3 mm, Oxford Glycosystems, Abingdon, UK).
4. HPLC-grade H_2O.
5. HPLC-grade acetonitrile.
6. HPLC-grade trifluoroacetic acid (Pierce and Warriner, Chester, UK).

3. Methods

1. Prepare eluant A: 500 mL of 0.05% TFA.
2. Prepare eluant B: 250 mL of acetonitrile containing 0.05% TFA.
3. Degas eluants by sparging with helium.
4. Equilibrate the column in 100% eluant A and 0% eluant B for 30 min at 0.75 mL/min prior to injection samples.
5. Elute 10 nmol oligosaccharide alditol (obtained as in Chapter 113) with the following gradient at a flow rate of 0.75 mL/min and UV detection at 206 nm/0.08 AUFS.
   a. Time = 0 min A = 100%; B = 0%
   b. Time = 5 min A = 100%; B = 0%
   c. Time = 40 min A = 60%; B = 40%
   d. Time = 45 min A = 60%; B = 40%
   e. Time = 50 min A = 100%; B = 0%

From: *The Protein Protocols Handbook, 2nd Edition*
Edited by: J. M. Walker © Humana Press Inc., Totowa, NJ

6. The resulting oligosaccharide containing fractions are then derivatized for LSIMS and GC-MS or analyzed by NMR.

4. Note

1. The GalNAc residue linked to Ser/Thr in *O*-linked chains is normally substituted at least at C-3 and therefore the alkali catalyzed β-elimination reaction will also result in "peeling" of the released oligosaccharide. This is obviated by concomitant reduction to give oligosaccharide alditols ending in GalNAcol. Endo-α-*N*-acetylgalactosaminidase digestion at hydrazinolysis under mild conditions can be used to release intact reducing sugars, which will have longer retention times on HPLC *(4)*.

References

1. Davies, M. J., Smith, K. D., Harbin, A.-M., and Hounsell, E. F. (1992) High performance liquid chromatography of oligosaccharide alditols and glycopeptides on graphitized carbon column. *J. Chromatogr.* **609,** 125–131.
2. Davies, M. J., Smith, K. D., Carruthers, R. A., Chai, W., Lawson, A. M., and Hounsell, E. F. (1993) The use of a porous graphitized carbon (PGC) column for the HPLC of oligosaccharides, alditols and glycopeptides with subsequent mass spectrometry analysis. *J. Chromatogr.* **646,** 317–326.
3. Davies, M. J. and Hounsell, E. F. (1996) Comparison of separation modes for high performance liquid chromatography of glycoprotein- and proteoglycan-derived oligosaccharides. *J. Chromatogr.* **720,** 227–234.
4. Hounsel, E. F., ed. (1993) *Methods in Molecular Biology, vol 14: Glycoprotein Analysis in Biomedicine,* Humana, Totowa, NJ.

115

O-Linked Oligosaccharide Profiling by HPAEC-PAD

Elizabeth F. Hounsell, Michael J. Davies, and Kevin D. Smith

1. Introduction

Although it can often be an advantage to be able to chromatograph neutral and sialylated oligosaccharides/alditols in one run (*see* Chapter 114), the added resolution of HPAEC and sensitivity with PAD detection means that this is an additional desirable technique for analysis. Neutral oligosaccharide alditols are poorly retained on CarboPac PA1, but can be resolved on two consecutive columns ***(1)***. The carboPac PA1 column is ideal for sialylated oligosaccharides ***(2)*** and can also be used for sialylated alditols.

2. Materials

1. Dionex D500 (Dionex, Camberley Surrey, UK) or other salt/biocompatible gradient HPLC system (titanium or PEEK) lined, e.g., 2 Gilson 302 pumps with 10-mL titanium pump heads, 802Ti manometric module, 811B titanium dynamic mixer, Rheodyne 7125 titanium injection valve with Tefzel rotor seal, and Gilson 712 chromatography system control software (Gilson Medical Electronics, Villiers-le-Bel, France).
2. CarboPac PA1 separator (4 × 250 mm) and PA1 Guard column (Dionex).
3. Pulsed amperometric detector with Au working electrode (Dionex), set up with the following parameters:
   a. Time = 0 s E = +0.1 V
   b. Time = 0.5 s E = +0.1 V
   c. Time = 0.51 s E = +0.6 V
   d. Time = 0.61 s E = +0.6 V
   e. Time = 0.62 s E = −0.8 V
   f. Time = 0.72 s E = −0.8 V
4. Anion micromembrane suppressor 2 (AMMS2) (Dionex).
5. Autoregen unit with anion regenerant cartridge (Dionex).
6. High-purity helium.
7. NaOH 50% w/v.
8. Reagent-grade sodium acetate (Aldrich, Poole, UK).
9. HPLC-grade H_2O.
10. 500 mL of 50 m*M* H_2SO_4 (reagent-grade).

From: *The Protein Protocols Handbook, 2nd Edition*
Edited by: J. M. Walker © Humana Press Inc., Totowa, NJ

3. Methods

1. Prepare eluant A: 100 m*M* NaOH, 500 m*M* Sodium acetate.
2. Prepare eluant B: 100 m*M* NaOH.
3. Prepare the column by elution with 50% A/50% B for 30 min at a flow rate of 1 mL/min.
4. Equilibrate column is 5% eluant A/95% eluant B at a flow of 1 mL/min.
5. Connect the AMMS2 to eluant out line and autoregen unit containing 500 mL of 50 m*M* reagent-grade H_2SO_4, and pump regenerant at a flow of 10 mL/min (*see* **Note 1**).
6. Inject 200 pmol of each oligosaccharide or sialylated oligosaccharide alditol (more if required for NMR or LSIMS), and elute with the following gradient at a flow of 1 mL/min:
   a. Time = 0 min A = 5%; B = 95%
   b. Time = 15 min A = 5%; B = 95%
   c. Time = 50 min A = 40%; B = 60%
   d. Time = 55 min A = 40%; B = 60%
   e. Time = 58 min A = 0%; B = 100%
7. Equilibrate the column in 5%A/95%B prior to the next injection.
8. At the end of the analyses, regenerate the column in 100%, and flush at the pumps with H_2O.
9. Desalt oligosaccharide-containing fractions by AMMS, and derivatize for LSIMS and GC-MS.

4. Notes

1. A better desalting profile may be achieved with an AMMS membrane (rather than AMMS2) if a flow rate of <1 mL/min can be used. In addition, it is important that the membranes of the suppressor remain fully hydrated and that the regenerant solution is replaced about once a week.

References

1. Campbell, B. J., Davies, M. J., Rhodes, J. M., and Hounsell, E. F. (1993) Separation of neutral oligosacharide alditols from human meconium using high-pH anion-exchange chromatography. *J. Chromatogr.* **622,** 137–146.
2. Lloyd, K. O. and Savage, A. (1991) High performance anion exchange chromatography of reduced oligosaccharides from sialomucins. *Glycoconjugate J.* **8,** 493–498.

116

Release of *N*-Linked Oligosaccharide Chains by Hydrazinolysis

Tsuguo Mizuochi and Elizabeth F. Hounsell

1. Introduction

Hydrazinolysis is the most efficient method of releasing all classes of *N*-linked oligosaccharide chains from glycoproteins. Disadvantages compared to enzymic release (*see* Chapter 117) is the use of hazardous chemicals, break up of the protein backbone, and destruction of some chains. The first of these can be obviated by use of a commercial machine for hydrazinolysis—the Glyco Prep (Oxford GlycoSystems, Abingdon, UK), but this is expensive, and as long as caution is exercised, the following method is excellent *(1)*.

2. Materials

1. Anhydrous hydrazine (*see* **Note 1**).
2. Toluene.
3. Saturated sodium bicarbonate solution prepared at room temperature.
4. Acetic anhydride.
5. 1-Octanol.
6. Lactose.
7. 1 *N* Acetic acid.
8. 1 *N* NaOH.
9. Methanol.
10. NaOH (0.05 *N*) freshly prepared from 1 *N* NaOH just before use.
11. Sodium borotritide (NaB^3H_4, approx 22 GBq/mmol) and approx 40 m*M* in dimethylformamide (silylation grade # 20672, Pierce Chemical Co., Rockford, IL).
12. 1-Butanol:ethanol:water (4:1:1, v/v).
13. Ethyl acetate:pyridine:acetic acid:water (5:5:1:3 v/v).
14. Dowex 50W-X12 (H^+ form, 50–100 mesh).
15. Whatman 3MM chromatography paper.
16. Air-tight screw-cap tube with a Teflon disk seal.
17. Dry heat block capable of maintaining 100°C.
18. Vacuum desiccator.
19. High-vacuum oil pump.
20. Descending paper chromatography tank.

From: *The Protein Protocols Handbook, 2nd Edition*
Edited by: J. M. Walker © Humana Press Inc., Totowa, NJ

3. Method (*see* ref. 1)

1. Add the glycoprotein (0.1–100 mg) to an air-tight screw-cap tube with a Teflon disk seal, and dry *in vacuo* overnight in a desiccator over P_2O_5 and NaOH.
2. Add anhydrous hydrazine (0.2–1.0 mL) with a glass pipet. The pipet must be dried to avoid introducing moisture into the anhydrous hydrazine (*see* **Note 2**).
3. Heat at 100°C for 10 h using a dry heat block. Glycoprotein is readily dissolved at 100°C.
4. Remove hydrazine by evaporation *in vacuo* in a desiccator. To protect the vacuum oil pump from hydrazine, connect traps between the desiccator and the pump in the following order: cold trap with dry ice and methanol, concentration, H_2SO_4-trap, and NaOH-trap on the desiccator side. Remove the last trace of hydrazine by coevaporation with several drops of toluene.
5. To re-*N*-acetylate, dissolve the residue in ice-cold saturated $NaHCO_3$ solution (1 mL/mg of protein). Add 10 μL of acetic anhydride, mix, and incubate for 10 min at room temperature. Re-*N*-acetylation is continued at room temperature by further addition of 10 μL (three times) and then 20 μL (three times) of acetic anhydride at 10-min intervals. The total volume of acetic anhydride is 100 μL/1 mL of saturated $NaHCO_3$ solution. Keep the solution on ice until the addition of acetic anhydride to avoid epimerization of the reducing terminal sugar.
6. To desalt, pass the reaction mixture through a column (1 mL for 1 mL of the $NaHCO_3$ solution) of Dowex 50W-X12, and wash with five column bed volumes of distilled water. Evaporate the effluent to dryness under reduced pressure at a temperature below 30°C. Addition of a drop of 1-octanol is effective in preventing bubbling over.
7. Dissolve the residue in a small amount of distilled water, and spot on a sheet of Whatman 3MM paper. Perform paper chromatography overnight using 1-butanol:ethanol:water (4:1:1, v/v) as developing solvent (*see* **Note 3**).
8. Cut the area 0–4 cm from the origin, recover the oligosaccharides by elution with distilled water, and then evaporate to dryness under reduced pressure. On this chromatogram, lactose migrates <4 cm from the origin.
9. To label oligosaccharides with tritium, dissolve the oligosaccharide fraction thus obtained in 100 μL of ice-cold 0.05 *N* NaOH (freshly prepared from 1 *N* NaOH). Verify that the pH of the oligosaccharide solution is above 11 with pH test paper using a <1-μL aliquot. If not, adjust the pH paper as 0.05 *N* NaOH. After addition of NaOH solution, keep the oligosaccharide solution on ice; otherwise, part of the reducing terminal *N*-acetylglucosamine may be converted to *N*-acetylmannosamine by epimerization (*see* **Note 4**).
10. Add a 20 *M* excess of NaB^3H_4 solution to the oligosaccharide solution, mix, and incubate at 30°C for 4 h to reduce the oligosaccharides. Then, add an equal weight of $NaBH_4$ (20 mg/mL of 0.05 *M* NaOH, freshly prepared) as the original glycoprotein, and continue the incubation for an additional 1 h to reduce the oligosaccharides completely. Stop the reaction by acidifying the mixture with 1 *N* acetic acid. During the reduction, and addition of acetic acid, keep the reaction mixture in a draft chamber, since tritium gas is generated (*see* **Note 5**).
11. To desalt, apply the reaction mixture to a small Dowex 50W-X12 column, and wash with five column bed volumes of distilled water, and then evaporate the effluent under reduced pressure below 30°C. The volume of the column should be calculated based on the amount of NaOH and $NaBH_4$, and the capacity of the resin. Then, remove the boric acid by repeated (three to five times) evaporation with methanol under reduced pressure. Dimethyl formamide used to dissolve NaB^3H_4 is usually coevaporated during the repeated evaporation.
12. Dissolve the residue in a small amount of distilled water, spot on a sheet of Whatman 3MM paper, and perform paper chromatography overnight using ethyl acetate: pyridine:acetic

acid:water (5:5:1:3, v/v) as developing solvent. This procedure is effective in removing the radioactive components originating from NaB^3H_4, which migrate a significant distance on the chromatogram.

13. Recover radioactive oligosaccharides, which migrate slower than lactitol (about 20 cm from origin), from the paper by elution with distilled water, and evaporate to dryness under reduced pressure.
14. Finally, subject the radioactive *N*-linked oligosaccharides thus obtained to high-voltage paper electrophoresis at pH 5.4 or HPLC to separate oligosaccharides by charge.

4. Notes

1. Anhydrous hydrazine is prepared by mixing 80% hydrazine hydrate (50 g), toluene (500 g), and CaO (500 g) and allowing to stand overnight. The mixture is refluxed for 3 h using a cold condenser and an NaOH tube. The mixture is then subjected to azeotropic distillation with toluene at 93–94°C under anhydrous conditions. Anhydrous hydrazine is collected from the bottom layer and stored in an air-tight screw-cap tube with a Teflon disk seal under dry conditions at 4°C in the dark. Commercially available anhydrous hydrazine (such as that from Aldrich Chemical Co., Inc., Milwaukee, WI) can also be used. It is important to check the quality with a glycoprotein of which the oligosaccharide structure has already been established before using for analysis, because contamination by trace amounts of water in some lots could modify the reducing terminal *N*-acetylglucosamine of *N*-linked oligosaccharides.

   Caution: Anhydrous hydrazine is a strong reducing agent, highly toxic, corrosive, suspected to be carcinogenic, and flammable. Therefore, great caution should be exercised during handling.
2. When stored in a small air-tight tube with a Teflon disk seal at –18°C, this NaB^3H_4 solution is stable for at least 1 yr. Dimethylformamide should be stored with molecular sieves in a small screw-cap bottle with a Teflon disk seal under dry conditions.
3. This procedure is indispensable for the next tritium-labeling step of the liberated oligosaccharides. This is because oligosaccharides larger than trisaccharides remain very close to the origin, whereas the degradation products derived from the peptide moiety, which react with NaB^3H_4, move a significant distance on the paper.
4. For quantitative liberation of intact *N*-linked oligosaccharides from glycoproteins by hydrazinolysis, great care should be taken to maintain anhydrous condition until the re-*N*-acetylation step. Introduction of moisture into glycoprotein samples or anhydrous hydrazine results in diverse modifications of reducing terminal *N*-acetylglucosamine residues, especially when unsubstituted with an Fucα 1-6 group, and causes the release of *O*-linked oligosaccharides accompanied with various degradations of the reducing end.
5. When oligosaccharides are reduced by NaB^3H_4 with high specific activity (e.g., 555 GBq/mmol), the sensitivity of detection of oligosaccharides increases about 20-fold. To label oligosaccharides with tritium at high efficiency, it is recommended to keep the concentration of NaB^3H_4 high in the incubation mixture by reducing the volume of 0.05*N* NaOH (e.g., to the same volume as the NaB^3H_4 solution). A 20 *M* excess of NaB^3H_4 solution is required for complete reduction of *N*-linked oligosaccharides, whereas a 5 *M* excess of NaB^3H_4 solution required for complete reduction of *N*-linked oligosaccharides derived from glycoprotein samples is roughly estimated from data on the carbohydrate content or amino acid sequence. If generation of tritium gas is to be avoided, continue the incubation for an additional 1 h with large amounts of glucose to absorb excess NaB^3H_4 before acidifying the mixture.

Reference

1. Mizuochi, T. (1993) Microscale sequencing of *N*-linked oligosaccharides of glycoproteins using hydrazinalysis, Bio-Gel P-4 and sequential exoglycosidase digestion, in *Methods in Molecular Biology, vol. 14: Glycoprotein Analysis in Biomedicine* (Hounsell, E. F., ed.) Humana, Totowa, NJ, pp. 55–68.

117

Enzymatic Release of *O*- and *N*-Linked Oligosaccharide Chains

Elizabeth F. Hounsell, Michael J. Davies, and Kevin D. Smith

1. Introduction

Enzymes involved in both the synthesis and degradation of glycoconjugates are highly specific for monosaccharide, linkage position, and anomeric configuration factors further away in the oligosaccharide sequence or protein. Not withstanding this, endo- and exoglycosidases are extremely useful tools in structural analysis. The RAAM technique has automated the use of exoglycosidase digestion (Oxford Glycosystems, Abingdon, UK). Here we discuss the release of intact oligosaccharide chains from proteins that can be further analyzed for separate functions *(1,2)*.

2. Materials

2.1. Desalting

1. 1 mL Spectra/Chrom desalting cartridge (Orme, Manchester, UK) or Biogel P_{-2} minicolumn.
2. HPLC-grade H_2O.

2.2. Glycosidases

1. Endoglycosidase H (EC 3.2.1.96) (e.g., *E. coli,* Boehringer Mannheim, Lewes, UK). Digestion buffer: 250 m*M* sodium citrate buffer adjusted to pH 5.5 with 1 *M* HCl.
2. Test-neuraminidase (EC 3.2.1.18) (e.g., *Vibrio cholerae,* Behring Ag, Marburg, Germany). Made up as 1 U/mL enzyme in digestion buffer and stored at 4°C. Digestion buffer: 50 m*M* sodium acetate, 134 m*M* NaCl, 9 m*M* $CaCl_2$.
3. Peptide-*N*-glycosidase F (EC 3.2.2.18) (e.g., *Flavobacterium meningosepticum,* Boehringer Mannheim). Digestion buffer: 40 m*M* potassium dihydrogen orthophosphate (KH_2PO_4), 10 m*M* EDTA adjusted to pH 6.2 with 1.0 *M* NaOH.
4. *O*-glycosidase (EC 3.2.1.97) (e.g., *Diplococcus pneunomiae,* Boehringer Mannheim). Digestion buffer: 40 m*M* KH_2PO_4/10 m*M* EDTA adjusted to pH 6.0 with 1.0 *M* NaOH.
5. Ice-cold ethanol.
6. Toluene.

3. Methods

3.1. Desalting

1. Wash the cartridge with 5 mL of HPLC-grade H_2O.
2. Load the sample onto the cartridge in a volume between 50 and 200 µL H_2O.
3. Elute the column with 200 µL of H_2O (including sample load).

From: *The Protein Protocols Handbook, 2nd Edition*
Edited by: J. M. Walker © Humana Press Inc., Totowa, NJ

4. Elute the glycoprotein in 350 µL of H_2O.
5. Elute the salt with an additional 1 mL of H_2O.

3.2. Glycosidase Digestions

1. Dissolve 1 nmol of glycoprotein in 100 µL of H_2O, and boil for 30 min to denature. Remove a 10% aliquot for sodium dodecyl sulfate-polyacrylamide gel electrophoresis (SDS-PAGE), as a control for detection of enzyme digestion. Dry the remainder by lyophilization (*see* **Note 1**).
2. Resuspend the glycoprotein in 500 µL of Endo H digestion buffer and 5 µL of toluene. Add 1 mU of Endo H/l nmol glycoprotein, and incubate at 37°C for 72 h (*see* **Notes 2** and **3**).
3. Precipitate the protein with a twofold excess of ice-cold ethanol and centrifuge at 15,000*g* for 20 min. Wash the pellet three more times with ice-cold ethanol. Dry the protein pellet, take up in water, and aliquot 10% (relative to original amount) for SDS-PAGE.
4. Dry the remaining pellet, and resuspend in neuraminidase/neuraminidase digestion buffer at a concentration of 2 nmol of glycoprotein/10 µL of buffer. Incubate for 18 h at 37°C, and then ethanol-precipitate and aliquot as in **step 3**.
5. Resuspend the remaining glycoprotein in 500 µL of PNGase F digestion buffer, 5 µL toluene, and 1 U PNGase F/10 nmol glycoprotein. Incubate at 37°C for 72 h before precipitation and aliquoting as in **step 3** (*see* **Note 4**).
6. Digest the final pellet with *O*-glycosidase under the same conditions as for the PNGase F digestion. Precipitate the pellet from ethanol washing and dry.
7. Apply all the pellet to SDS-PAGE.
8. The supernatants containing *N*- and *O*-linked oligosaccharides can be analyzed as discussed in Chapters 111, 118, and 119.

4. Notes

1. The described procedure assumes approx 10% glycosylation of the glycoprotein. The amount of glycoprotein treated may have to be increased to obtain oligosaccharides for further analysis with less highly glycosylated glycoproteins.
2. PNGase F is stored at –20°C, and all other enzymes at 4°C. Endo H removes high-mannose oligosaccharide chains, but not complex type.
3. The toluene is added to prevent bacterial growth.
4. The PNGase F digestion can be performed directly on the boiled glycoprotein if all *N*-linked glycoprotein chains (both high mannose and complex) are required to be removed. Digests may also be performed in 0.2 *M* sodium phosphate buffer, pH 8.4, but this will result in the release of sialic acid residues as monosaccharides.

References

1. Yamamoto, K., Tsuji, T., and Osawa, T. (1993) Analysis of asparagine-linked oligosaccharides by sequential lectin-affinity chromatography, in *Methods in Molecular Biology, vol. 14: Glycoprotein Analysis in Biomedicine* (Hounsell, E. F., ed.), Humana, Totowa, NJ, pp. 17–34.
2. Davies, M. J., Smith, K. D., and Hounsell, E. F. (1994) The release of oligosaccharides from glycoproteins, in *Methods in Molecular Biology, vol. 32: Basic Protein and Peptide Protocols* (Walker, J. M., ed.), Humana, Totowa, NJ, pp. 129–141.

118

N-Linked Oligosaccharide Profiling by HPLC on Porous Graphitized Carbon (PGC)

Elizabeth F. Hounsell, Michael J. Davies, and Kevin D. Smith

1. Introduction

The vast array of possible *N*-linked oligosaccharides demands high-resolution HPLC columns for their purification *(1,2)*. Reverse-phase (C_{18}) and normal-phase (NH_2) columns have been used for the separation (singly or in concert) of many *N*-linked oligosaccharides. The porous graphitized carbon (PGC) column described in Chapter 114 for *O*-linked alditol separation will give improved *N*-linked oligosaccharide resolution over C_{18} columns, and has the advantage of using salt-free buffers for preparative work *(3,4)*.

2. Materials

1. Biocompatible gradient HPLC system, e.g., 2 × 302 pumps, 802C manometric module, 811 dynamic mixer, 116 UV detector, 201 fraction collector, 715 chromatography system control software (all Gilson Medical Electronics, France).
2. IBM PS-2 personal computer (or compatible model) with Microsoft Windows.
3. Hypercarb S HPEC Column (100 × 4.6 mm) (Shandon Scientific, Runcorn, Cheshire, England) or Glyco H (OGS, Abingdon, Oxon, UK).
4. HPLC-grade H_2O.
5. HPLC-grade acetonitrile.
6. HPLC-grade trifluoroacetic acid (Pierce and Warriner, Chester, UK).

3. Methods

1. Prepare eluant A: 500 mL of 0.05% TFA.
2. Prepare eluant B: 250 mL of acetonitrile containing 0.05% TFA.
3. Degas eluants by sparging with helium.
4. Equilibrate the column in 100% eluant A, and 0% eluant B for 30 min at 0.75 mL/min prior to injection of samples.
5. Elute 10 nmol oligosaccharides with the following gradient at a flow rate of 0.75 mL/min and UV detection at 206 nm/0.08 AUFS (*see* **Note 1**).
   a. Time = 0 min A = 100%; B = 0%
   b. Time = 5 min A = 100%; B = 0%
   c. Time = 40 min A = 60%; B = 40%
   d. Time = 45 min A = 60%; B = 40%
   e. Time = 50 min A = 100%; B = 0%

From: *The Protein Protocols Handbook, 2nd Edition*
Edited by: J. M. Walker © Humana Press Inc., Totowa, NJ

6. The resulting oligosaccharide containing fractions are then derivatized for LSIMS and GC-MS or analyzed by NMR.

4. Notes

1. Reverse-phase or normal-phase HPLC may also be required for the complete separation of some oligosaccharide isomers.
2. To prevent anomerization, the oligosaccharides can be reduced to their alditols (either after PNGase F digestion or hydrazinolysis). The inclusion of a ^{3}H-label on reduction will give increased sensitivity over UV. Alternatively, the oligosaccharides can be fluorescently labeled at their reducing terminus (with 2-amino-benzamide or 2-amino-pyridine) by reductive amination to give a fluorescent chromophore and increased sensitivity. Sensitivity of detection may also be increased by the postcolumn addition of 300 m*M* NaOH and pulsed amperometric detection as described for HPAEC-PAD.

References

1. Hase, S. (1993) Analysis of sugar chains by pyridylamination, in *Methods in Molecular Biology, vol. 14: Glycoprotein Analysis in Biomedicine* (Hounsell, E. F., ed.), Humana, Totowa, NJ, pp. 69–80.
2. Kakehi, K. and Honda, S. (1993) Analysis of carbohydrates in glycoproteins by high performance liquid chromatography and high performance capillary electrophoresis, in *Methods in Molecular Biology, vol. 14: Glycoprotein Analysis in Biomedicine* (Hounsell, E. F., ed.), Humana, Totowa, NJ, pp. 81–98.
3. Smith, D. D., Davies, M. J., Hounsell, E. F. (1994) Structural profing of oligosaccharides of glycoproteins, in *Methods in Molecular Biology, vol. 32: Basic Protein and Peptide Protocols* (Walker, J. M., ed.), Humana, Totowa, NJ, pp. 143–155.
4. Davies, M. J. and Hounsell, E. F. (1995) Comparison of separation modes for high performance liquid chromatography of glycoprotein- and proteoglycan-derived oligosaccharides. *J. Chromatogr.* **720,** 227–234.

119

N-Linked Oligosaccharide Profiling by HPAEC-PAD

Elizabeth F. Hounsell, Michael J. Davies, and Kevin D. Smith

1. Introduction

High-pH anion-exchange chromatography with pulsed amperometric detection (HPAEC-PAD) is a very powerful tool for the profiling of *N*-linked oligosaccharides *(1,2)*, only being limited by the concentrated sodium hydroxide and sodium acetate required to achieve the separation and sensitive detection. Oligosaccharides are separated on the basis of charge (i.e., the number of sialic acid residues) and the linkage isomers present. Neutral oligosaccharide alditols are most weakly retained, retention increasing with increasing sialylation, and NeuGc- and sulfate-bearing oligosaccharides being most strongly retained. Fucosylation results in a shorter retention. If the oligosaccharides can be effectively desalted after HPAEC-PAD, this remains the method of choice for oligosaccharide purification, and is a very powerful analytical tool.

2. Materials

1. Dionex DX300 or salt/biocompatible gradient HPLC system (titanium or PEEK lined), e.g., 2 Gilson 302 pumps with 10-mL titanium pump heads, 802Ti manometric module, 811B titanium dynamic mixer, Rheodyne 7125 titanium injection valve with Tefzel rotor seal, and Gilson 715 chromatography system control software (all Gilson Medical Electronics, Villiers-le-Bel, France).
2. IBM PS-2 personal computer (or compatible model) with Microsoft Windows.
3. CarboPac PA100 separator (4 × 250 mm) and PA100 guard column (Dionex, Camberley, UK).
4. Pulsed amperometric detector with Au working electrode (Dionex), set up with the following parameters:
   a. Time = 0 s E = +0.1 V
   b. Time = 0.5 s E = +0.1 V
   c. Time = 0.51 s E = +0.6 V
   d. Time = 0.61 s E = +0.6 V
   e. Time = 0.62 s E = –0.8 V
   f. Time = 0.72 s E = –0.8 V
5. High-purity helium.
6. NaOH 50% (w/v).
7. ACS-grade sodium acetate (Aldrich, UK).
8. HPLC-grade H_2O.

From: *The Protein Protocols Handbook, 2nd Edition*
Edited by: J. M. Walker © Humana Press Inc., Totowa, NJ

3. Methods

1. Prepare eluant A: 100 m*M* NaOH, 500 m*M* sodium acetate.
2. Prepare eluant B: 100 m*M* NaOH.
3. Prepare the column by elution with 50% A/50% B for 30 min at a flow rate of 1 mL/min.
4. Equilibrate column in 5% eluant A/95% eluant B at a flow of 1 mL/min.
5. Inject 200 pmol of each oligosaccharide or sialylated oligosaccharide alditol (more if required for NMR or LSIMS) and elute with the following gradient at a flow of 1 mL/min:
   a. Time = 0 min A = 5%; B = 95%
   b. Time = 15 min A = 5%; B = 95%
   c. Time = 50 min A = 40%; B = 60%
   d. Time = 55 min A = 40%; B = 60%
   e. Time = 58 min A = 0%; B = 100%
6. Equilibrate the column in 5%A/95%B prior to the next injection.
7. At the end of the analyzes, regenerate the column in 100%B, and flush the pumps with H_2O.
8. Desalt oligosaccharide-containing fractions and derivatize for LSIMS, GC-MS, or NMR (*see* **Note 1**).

4. Notes

1. Desalting can either be achieved by means of a ion-suppression system (e.g., Dionex AMMS or SRS system, *see* Chapter 107) or by off-line desalting on Biogel P2 mini-columns or H^+ cation-exchange resins. Fractions containing multiply sialylated oligosaccharides will probably contain too much salt to be totally desalted by an ion-suppressor, and column methods will be required. This can lead to losses of minor oligosaccharides.
2. All previous notes for HPAEC-PAD methods in this volume also apply.

References

1. Townsend, R. R. (1995) Analysis of glycoconjugates using high pH anion-exchange chromatography. *J. Chromatogr. Library* **58,** 181–209.
2. Smith, K. D., Davies, M. J., and Hounsell, E. F. (1994) Structural profiling of oligosaccharides of glycoproteins, in *Methods in Molecular Biology, vol. 32: Basic Protein and Peptide Protocols* (Walker, J. M., ed.), Humana, Totowa, NJ, pp. 143–155.

120

HPAEC-PAD Analysis of Monosaccharides Released by Exoglycosidase Digestion Using the CarboPac MA1 Column

Michael Weitzhandler, Jeffrey Rohrer,
James R. Thayer, and Nebojsa Avdalovic

1. Introduction

Exoglycosidases are useful reagents for the structural determination of glycoconjugates. Their anomeric, residue, and linkage specificity for terminal monosaccharides have been used to assess monosaccharide sequence and structure in a variety of glycoconjugates *(1)*. Their usefulness depends on the absence of contaminating exoglycosidases and an understanding of their specificity. Digestions of oligosaccharides with exoglycosidases give two classes of products: monosaccharides and the shortened oligosaccharides. Most assays of such reactions have monitored the reaction by following oligosaccharides that are labeled at their reducing ends. In these assays, after exoglycosidase digestion the shortened oligosaccharide retains the label at its reducing end. The other digestion product, the released monosaccharide, does not carry a label and thus cannot be quantified. Additionally, identification of any other monosaccharide that could be the result of a contaminating exoglycosidase activity would not be possible. Quantitative measurement of all products (all released monosaccharide[s] as well as the shortened oligosaccharide product) would be useful because it would enable the determination of any contaminating exoglycosidase activities by determining the extent of release of other monosaccharides. High pH anion exchange chromatography with pulsed amperometric detection (HPAEC-PAD) detects the appearance of monosaccharide product(s), the shortened oligosaccharide product(s) as well as the disappearance of the oligosaccharide substrate(s) in a single chromatographic analysis without labeling. Thus, HPAEC-PAD has been used extensively to monitor the activities of several different exoglycosidases on glycoconjugates, usually using the CarboPac PA1 column to separate the digestion products (*see* **refs. *1–8***).

A problem encountered when using HPAEC-PAD to monitor exoglycosidase digestions is that *N*-acetylglucosamine (GlcNAc) is not baseline resolved from other glycoconjugate monosaccharides on the CarboPac PA1 column using the isocratic conditions that successively baseline resolve glycoprotein hydrolysis products: fucose, galactosamine, glucosamine, galactose, glucose, and mannose *(9)*. We recently discovered that the CarboPac MA1 column, developed for the separation of neutral sugar

From: *The Protein Protocols Handbook, 2nd Edition*
Edited by: J. M. Walker © Humana Press Inc., Totowa, NJ

alditols *(10,11)*, gives an isocratic baseline separation of GlcNAc, GalNAc, fucose, mannose, glucose, and galactose, and simultaneously resolves many neutral oligosaccharides. This separation extends the usefulness of the CarboPac MA1 column to the assay of reducing monosaccharides released by exoglycosidases. In the following, an assay of exposed GlcNAc after β-*N*-acetylhexosaminidase treatment of a variety of glycoconjugates is shown.

To determine the suitability of HPAEC-PAD and the MA1 column for analyzing both released monosaccharide and oligosaccharide products in a single analysis, we subjected an asialo agalacto biantennary oligosaccharide standard (**Table 1**, structure 3; **Fig. 1A**, peak 3) and an asialo agalacto tetraantennary oligosaccharide standard (**Table 1**, structure 5; **Fig. 1B**, peak 5) to Jack bean β-*N*-acetylhexosaminidase digestion. In addition to differences in numbers of antennae (2 vs 4) and retention times (22.3 vs 25.2 min), these two oligosaccharides differ in that the tetraantennary oligosaccharide has terminal GlcNAc linked to mannose β(1→4) and β(1→6) in addition to the β(1→2) linkages present in the biantennary oligosaccharide standard.

The complete disappearance of the asialo agalacto biantennary substrate is indicated by the disappearance of peak 3 (**Fig. 1A**; compare dashed vs solid line in bottom tracing). The complete disappearance of the asialo agalacto tetraantennary substrate is indicated by the disappearance of peak 5 (**Fig. 1B**; compare dashed vs solid line in bottom tracing). The expected digestion products of both structures 3 and 5 would be the released monosaccharide GlcNAc, (*see* **Fig. 1A,B**; peak at 15.6 min) and the released, shortened oligosaccharide product, $Man_3GlcNAc_2$ (**Table 1**, structure 1; *see* **Fig. 1A,B**, peak at 17.5 min).

In addition to being useful for monitoring the β-*N*-acetylhexosaminidase release of terminal GlcNAc from isolated oligosaccharides, HPAEC-PAD and the MA1 column can be used to directly monitor the presence of terminal GlcNAc on glycoproteins. Such an assay could be useful for monitoring terminal carbohydrate modifications in therapeutic glycoproteins; these modifications have been shown to affect the stability and efficacy of therapeutic glycoproteins *(12)*. The β-*N*-acetylhexosaminidase release of terminal GlcNAc from a monoclonal IgG is shown in **Fig. 2A** (bottom tracing; compare solid vs dashed line). Additionally, the absence of contaminating exoglycosidases is apparent by the absence of release of any other monosaccharides.

To assess whether the released GlcNAc was derived from a GlcNAc-terminated *N*-linked oligosaccharide, the monoclonal IgG was treated with peptide-*N*-glycosidase (PNGase F), an amidase that nonspecifically releases *N*-linked oligosaccharides from glycoproteins. The *N*-linked oligosaccharide map of the monoclonal IgG is shown in **Fig. 2B** (bottom tracing; solid line). The major PNGase F product peak eluted between 18.5 and 19 min with a retention time similar to a core fucosylated asialo agalacto biantennary oligosaccharide standard (18.7 min, *see* **Table 1**, structure 4; *see also* **Fig. 2B**, peak 4). The second PNGase F released peak had a retention time of 21 min and could represent a monogalactosylated, biantennary oligosaccharide with core fucosylation, as has been reported in other monoclonal IgGs *(7)*. To assess whether the PNGase F released oligosaccharides were terminated with GlcNAc, the PNGase F released oligosaccharides were treated with β-*N*-acetylhexosaminidase.

Table 1
Oligosaccharide Substrates for Exoglycosidase Analyses

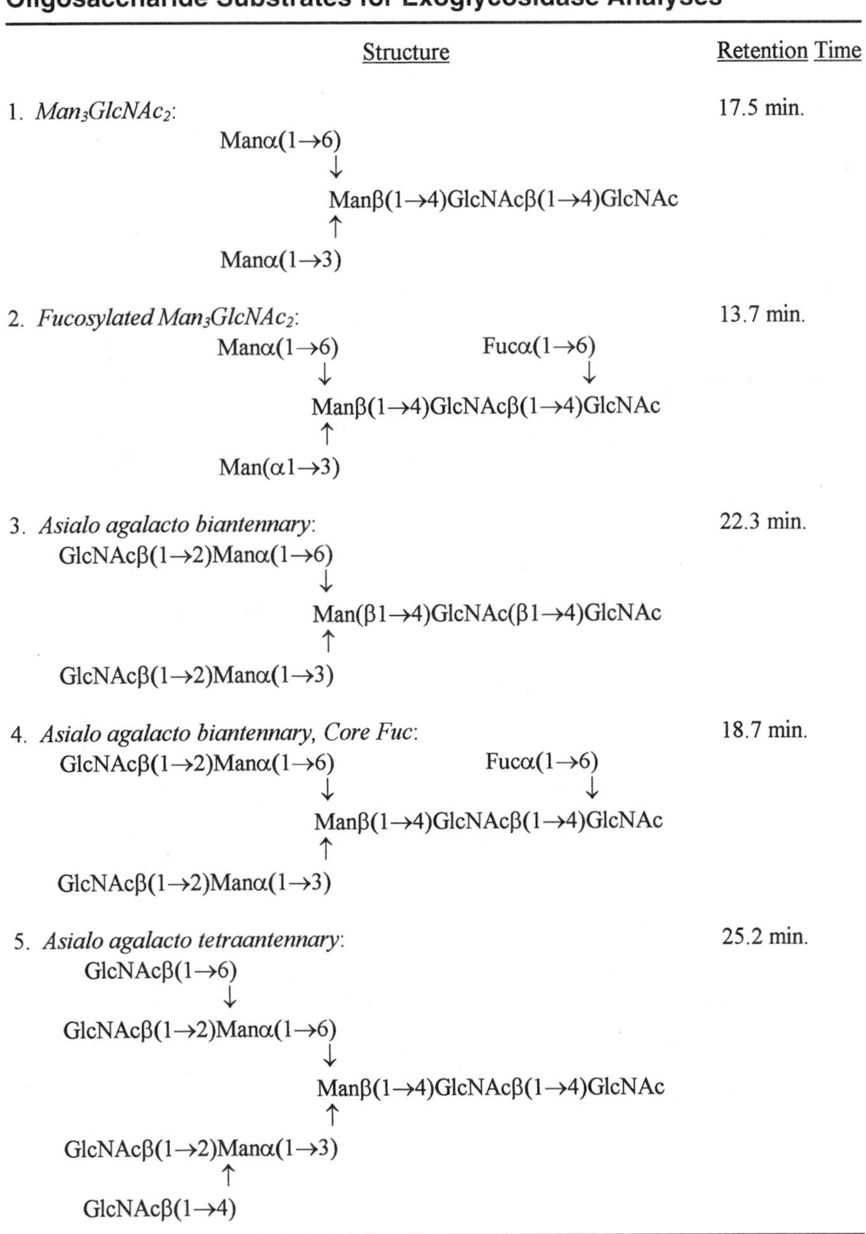

Structure	Retention Time
1. *Man₃GlcNAc₂*: Manα(1→6) ↓ Manβ(1→4)GlcNAcβ(1→4)GlcNAc ↑ Manα(1→3)	17.5 min.
2. *Fucosylated Man₃GlcNAc₂*: Manα(1→6) Fucα(1→6) ↓ ↓ Manβ(1→4)GlcNAcβ(1→4)GlcNAc ↑ Man(α1→3)	13.7 min.
3. *Asialo agalacto biantennary*: GlcNAcβ(1→2)Manα(1→6) ↓ Man(β1→4)GlcNAc(β1→4)GlcNAc ↑ GlcNAcβ(1→2)Manα(1→3)	22.3 min.
4. *Asialo agalacto biantennary, Core Fuc*: GlcNAcβ(1→2)Manα(1→6) Fucα(1→6) ↓ ↓ Manβ(1→4)GlcNAcβ(1→4)GlcNAc ↑ GlcNAcβ(1→2)Manα(1→3)	18.7 min.
5. *Asialo agalacto tetraantennary*: GlcNAcβ(1→6) ↓ GlcNAcβ(1→2)Manα(1→6) ↓ Manβ(1→4)GlcNAcβ(1→4)GlcNAc ↑ GlcNAcβ(1→2)Manα(1→3) ↑ GlcNAcβ(1→4)	25.2 min.

The results of β-*N*-acetylhexosaminidase treatment of the PNGase F released oligosaccharides are depicted in **Fig. 2B** (second from bottom tracing). Both peaks corresponding to PNGase F released oligosaccharides disappeared and two new peaks appeared. The two new peaks had retention times corresponding to released GlcNAc (15.6 min) and to a fucosylated $Man_3GlcNAc_2$ standard (17.5 min, *see* **Table 1**, struc-

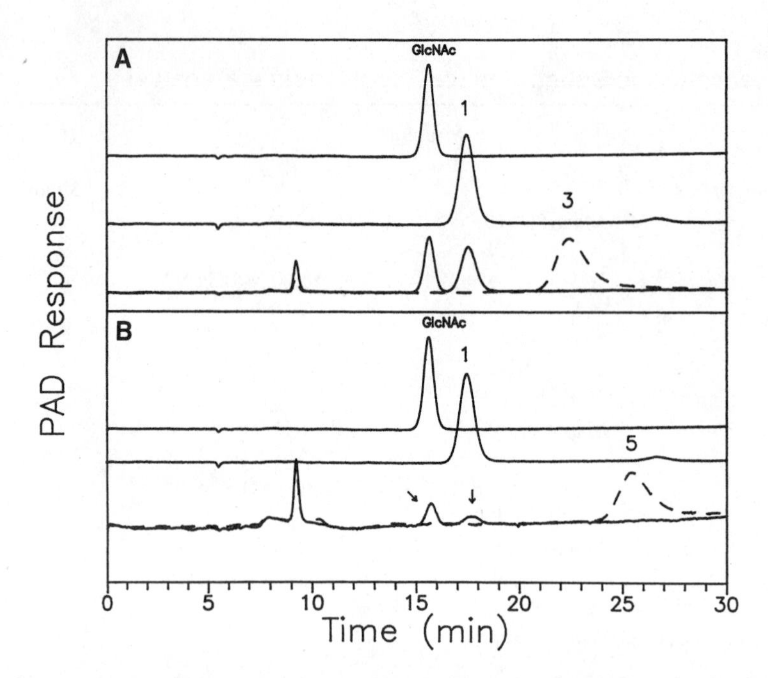

Fig. 1. HPAEC-PAD of β-*N*-acetylhexosaminidase digests of oligosaccharide standards. **(A)** Digestion of an asialo agalacto biantennary oligosaccharide standard. **(B)** Digestion of an asialo agalacto tetraantennary oligosaccharide standard. Peaks 1, 3, and 5 refer to structures 1, 3, and 5 identified in **Table 1**. (Bottom tracings in **A** and **B**: dashed line, chromatography of substrate; solid line, chromatography of digestion products.)

ture 2; *see also* **Fig. 2B**, peak 2). Thus, the two *N*-linked oligosaccharide peaks released from the monoclonal IgG are fucosylated and have terminal GlcNAc. To obtain further structural information regarding GlcNAc linkages, one could use the *Streptococcus pneumoniae* β-*N*-acetylhexosaminidase, an exoglycosidase which shows much more efficient cleavage of the GlcNAcβ(1→2)Man if the Man residue is not substituted with GlcNAc at the C-6 *(13)*. This method is also useful for monitoring the β-*N*-acetylhexosaminidase preparation for contaminating exoglycosidase activities as other monosaccharide digestion products would be readily discernible.

2. Materials

1. HPLC grade deionized water (*see* **Note 1**).
2. 50% NaOH solution (w/w) (Fisher Scientific, Pittsburgh, PA) (*see* **Note 2**).
3. Reference-grade monosaccharides (Pfanstiehl Laboratories, Waukegan, IL).
4. Oligosaccharide standards $Man_3GlcNAc_2$ (**Table 1**, structure 1) fucosylated $Man_3GlcNAc_2$ (**Table 1**, structure 2), asialo agalacto biantennary (**Table 1**, structure 3), asialo agalacto biantennary, core fucose (**Table 1**, structure 4), and asialo agalacto tetraantennary oligosaccharide (**Table 1**, structure 5) (Oxford GlycoSciences, Abingdon, UK).
5. β-*N*-acetylhexosaminidase, Jack bean (Oxford GlycoSciences) (*see* **Note 3**).
6. PNGase F (New England BioLabs, Beverly, MA) (*see* **Note 4**).
7. 25-mL Plastic pipets (Fisher Scientific).
8. 1.5-mL Polypropylene microcentrifuge tubes, caps, and O rings (Sarstedt, Newton, NC).

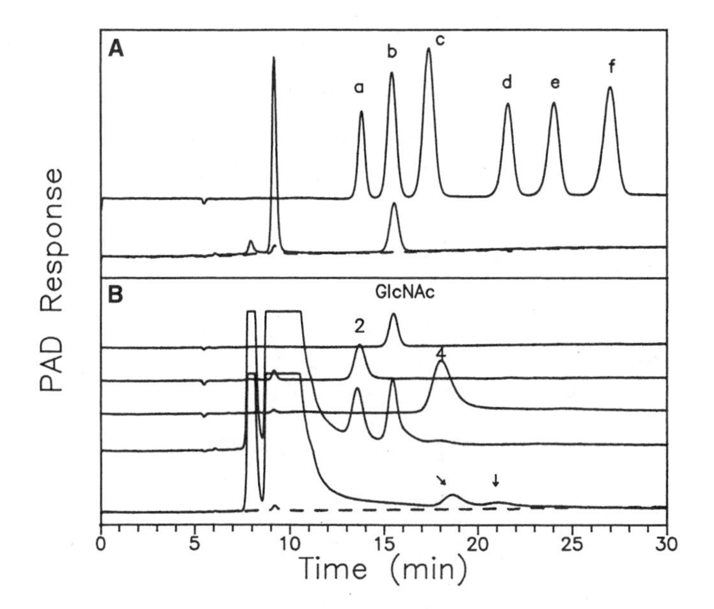

Fig. 2. HPAEC-PAD of β-*N*-acetylhexosaminidase digests of a monoclonal IgG **(A)** Bottom tracing is β-*N*-acetylhexosaminidase digests of a monoclonal IgG (dashed line, chromatography of IgG substrate; solid line, chromatography of digestion products). Top tracing is chromatography of monosaccharide standards; Peak a, fucose; b, GlcNAc; c, GalNAc; d, mannose; e, glucose; f, galactose. **(B)** Bottom tracing is PNGase F digest of a monoclonal IgG (dashed line, chromatography of IgG substrate; solid line, chromatography of PNGase F-released *N*-linked oligosaccharides from the monoclonal IgG; arrows indicate oligosaccharide peaks). β-*N*-acetylhexosaminidase treatment of PNGase F-released *N*-linked oligosaccharides is shown in the second to bottom tracing. Note appearance of product peaks corresponding to GlcNAc and fucosylated $Man_3GlcNAc_2$. Peaks 2 and 4 represent chromatography of oligosaccharide standards corresponding to structures 2 and 4 in **Table 1**.

9. Autosampler vials: 12- × 32-mm disposable, limited-volume sample vials, Teflon/silicone septa, and caps (Sun Brokers, Wilmington, NC).
10. Nylon filters (Gelman Sciences, Ann Arbor, MI).
11. The chromatograph (Dionex, Sunnyvale, CA) consists of a gradient pump, a PAD II or PED, and an eluent degas module (EDM). The EDM is used to sparge and pressurize the eluents with helium. The system was controlled and data were collected using Dionex AI450 software. Sample injection was accomplished with a Spectra Physics SP8880 autosampler (Fremont, CA) equipped with a 200-µL sample loop. The Rheodyne (Cotati, CA) injection valve is fitted with a Tefzel rotor seal to withstand the alkalinity of the eluents. We also used a DX 500 BioLC system (Dionex) configured for carbohydrate analysis with PeakNet software (Dionex).

3. Methods

3.1. β-N-Acetylhexosaminidase Digestion

1. Reconstitute approx 2 µg each of the neutral agalacto biantennary and agalacto tetraantennary oligosaccharides (**Table 1**, structures 3 and 4, respectively) in 10 µL of 25 m*M* sodium citrate-phosphate buffer, pH 5.0.

2. Add 0.1 U of Jack bean β-*N*-acetylhexosaminidase in 2 μL of 25 m*M* sodium citrate-phosphate buffer, pH 5.0.
3. Incubate the digest for 20 h at 37°C.
4. Inject 10 μL of each digest directly onto the column.

3.2. PNGase F Digestion

1. Reconstitute approx 100 μg of a monoclonal IgG in 10 μL of 25 m*M* sodium citrate-phosphate buffer, pH 5.0.
2. Add 2 μL of the PNGase F preparation.
3. To an identifical PNGase F digest of the same monoclonal IgG, add 0.1 U of Jack bean β-*N*-acetylhexosaminidase in 2 μL of 25 m*M* sodium citrate-phosphate buffer, pH 5.0.
4. Separately inject 10 μL of each digest directly onto the column.

3.3. Chromatography and Detection of Carbohydrates

Separations of monosaccharides and oligosaccharides can be achieved using a Dionex BioLC system equipped with a CarboPac MA1 column (4 × 250 mm) and a CarboPac MA1 guard column working at an isocratic concentration of 480 m*M* NaOH and a flow rate of 0.4 mL/min at ambient temperature over 35 min. Separated mono- and oligosaccharides are detected by PAD with a gold electrode and triple-pulse amperometry (E_1 = 0.05 V, t_1 = 420 ms; E_2 = 0.80 V, t_2 = 360 ms; E_3 = –0.15 V, t_3 = 540 ms), measuring at 1000 nA full scale. Alternatively, a more recently described quadruple potential waveform can be used (Dionex Technical Note 21; Waveform A). (E1 = +0.1 V, t1 = 400 ms. The first 200 ms is t del and the second 200 ms is t det; E2 = –2.0 V, t2 = 10 ms; E3 = +0.6 V, t3 = 40 ms; E4 = –0.1 V, t4 = 60 ms).

4. Notes

1. It is essential to use high quality water of high resistivity (18 MeΩ) and to have as little dissolved carbon dioxide in the water as possible. Biological contamination should be absent. The use of fresh Pyrex glass-distilled water is recommended. The still should be fed with high-resistivity (18 MeΩ) water. The use of plastic tubing in the system should be avoided, as plastic tubing often supports microbial growth. Degas appropriately.
2. It is extremely important to minimize contamination with carbonate. Carbonate, a divalent anion at pH ≥ 12, binds strongly to the columns and interferes with carbohydrate chromatography. Thus carbonate is known to affect column selectivity and produce a loss of resolution and efficiency. Commercially available NaOH pellets are covered with a thin layer of sodium carbonate and should NOT be used. Fifty percent (w/w) sodium hydroxide solution is much lower in carbonate and is the preferred source for NaOH. Diluting 104 mL of a 50% NaOH solution into 2 L of water yields a 1.0 *M* NaOH solution. Degas appropriately.
3. This enzyme has a broad specificity, cleaving nonreducing terminal β-*N*-acetylglucosamine residues (GlcNAc) and β-*N*-acetylgalactosamine (GalNAc) with 1–2,3,4, and 6 linkages.
4. PNGase F is an amidase from *Flavobacterium meningosepticum*. The enzyme cleaves between the innermost GlcNAc and asparagine residues of high mannose, hybrid, and complex oligosaccharides from *N*-linked glycoproteins.

Acknowledgment

We thank Sylvia Morris for her work on the manuscript.

References

1. Jacob, J. S. and Scudder, P. (1994) Glycosidases in structural analysis. *Meth. Enzymol.* **230,** 280–299.
2. Scudder, P., Neville, D. C. A., Butters, T. D., Fleet, G. W. J., Dwek, R. A., Rademacher, T. W., and Jacob, G. S. (1990) The isolation by ligand affinity chromatography of a novel form of α-L-fucosidase from almond. *J. Biol. Chem.* **265,** 16,472–16,477.
3. Butters, T. D., Scudder, P., Willenbrock, F. W., Rotsaert, J. M. V., Rademacher, T., Dwek, R. A., and Jacob, G. S. (1989) A serial affinity chromatographic method for the purification of *charonia lampas* α-L-fucosidase, in *Proceedings of the 10th International Symposium on Glycoconjugates*, Sept. 10–15 (Sharon, N., Lis, H., Duksin, D., and Kahane, I., eds.), Magnes, Jerusalem, Israel, pp. 312,313.
4. Davidson, D. J. and Castellino, F. J. (1991) Structures of the asparagine-289-linked oligosaccharides assembled on recombinant human plasminogen expressed in a *Mamestra brassicae* cell line (IZD-MBO503). *Biochemistry* **30,** 6689–6696.
5. Grollman, E. F., Saji, M., Shimura, Y., Lau, J. T., and Ashwell, G. (1993) Thyrotropin regulation of sialic acid expression in rat thyroid cells. *J. Biol. Chem.* **268,** 3604–3609.
6. Willenbrock, F. W., Neville, D. C. A., Jacob, G. S., and Scudder P. (1991) The use of HPLC-pulsed amperometry for the characterization and assay of glycosidases and glycosyltransferases. *Glycobiology* **1,** 223–227.
7. Weitzhandler, M., Hardy, M., Co, M. S., and Avdalovic, N. (1994) Analysis of carbohydrates on IgG preparations. *J. Pharm. Sci.* **83,** 1670–1675.
8. Lin, A. I., Philipsberg, G. A., and Haltiwanger, R. S. (1994) Core fucosylation of high-mannose-type oligosaccharides in GlcNAc transferase I-deficient (Lec 1) CHO cells. *Glycobiology* **4,** 895–901.
9. Hardy, M. R., Townsend, R. R., and Lee, Y. C. (1988) Monosaccharide analysis of glycoconjugates by anion exchange chromatography with pulsed amperometric detection. *Analyt. Biochem.* **170,** 54–62.
10. Chou, T. Y., Dang, C. V., and Hart, G. W. (1995) Glycosylation of the c-Myc transactivation domain. *Proc. Natl. Acad. Sci. USA* **92,** 4417–4421.
11. Lin, A. I., Polk, C., Xiang, W. K., Philipsberg, G. A., and Haltiwanger, R. S. (1993) Novel fucosylation pathways in parental and GlcNAc transferase I deficient (Lec 1) CHO cells. *Glycobiology* **3,** 524 (Abstract).
12. Varki, A. (1993) Biological roles of oligosaccharides: all of the theories are correct. *Glycobiology* **3,** 97–130.
13. Yamashita, K., Ohkura, T., Yoshima, H. and Kobata, A. (1981) Substrate specificity of Diplococcal β-*N*-acetylhexosaminidase, a useful enzyme for the structural studies of complex type asparagine-linked sugar chains. *Biochem. Biophys. Res. Commun.* **100,** 226–232.

121

Microassay Analyses of Protein Glycosylation

Nicky K. C. Wong, Nnennaya Kanu, Natasha Thandrayen, Geert Jan Rademaker, Christopher I. Baldwin, David V. Renouf, and Elizabeth F. Hounsell

1. Introduction

The majority of secreted and cell surface proteins are glycosylated. To characterize accurately the carbohydrate moieties of oligosaccharide chains in glycosylated proteins, it is necessary to distinguish exactly which types of oligosaccharide are present and at which sites in the protein. The goal of the microassays presented here is to determine oligosaccharide structure and occupancy using lectin overlay assays. Lectins are multivalent binding proteins with the ability to bind specifically to certain sugar sequences in oligosaccharides and glycopeptides *(1)*. Lectin overlay assays take advantage of the ability of lectins to distinguish between different types of glycoproteins by virtue of their ability to recognize terminal sugars, thus allowing the chain type and peripheral antigenic components to be determined. In addition, the assay is important as it provides structural information of protein glycosylation prior to conformational analysis and identification of antigenically and biologically active oligosaccharides. The lectin overlay, in the form of a microassay, can be developed to give a high level of sensitivity for glycan detection. The specificities of the lectins in such assays appears to vary from previous studies using lectin affinity chromatography and hence must be characterized further.

Three microassays are standardized here: The first is the analysis of non-protease-treated intact glycoproteins blotted onto nitrocellulose membranes (NCMs). Second, glycopeptides are released by prior digestion of the glycoprotein with a suitable protease that will cleave it into glycopeptides of ideally 5–15 amino acids and then separated by high performance liquid chromatography (HPLC) prior to lectin assay. Glycopeptides do not usually adhere to gels or membranes and therefore the latter need to be modified. Third, oligosaccharide chain structure has been characterized by the release of sugars from glycoproteins by endoglycosidase digestion or hydrazinolysis and coupling them to a multivalent support followed by lectin overlay analysis. Coupling of oligosaccharides is achieved by reductive amination. This reaction takes advantage of the availability of reducing sugars in the open chain aldehydo form for reductive amination with primary amino groups, for example, εNH_2 of lysine in pro-

From: *The Protein Protocols Handbook, 2nd Edition*
Edited by: J. M. Walker © Humana Press Inc., Totowa, NJ

teins. The base is stabilized by Schiff base formation via reduction in sodium cyanoborohydride. We have used this method, for example, to characterize the structure and antigenicity of a series of reducing *O*-linked oligosaccharides released from pigeon mucin *(2)*.

2. Materials

1. Glycorelease *N*- and *O*-glycan recovery kit (Oxford GlycoSystems; Abingdon, UK).
2. Porous graphitized carbon (PGC) column (Hypersil Hypercarb, 100 mm × 4.6 mm, 7 μm, Hypersil; Runcorn, UK).
3. Reagents used in reverse-phase (RP)-HPLC (*see* **Subheading 3.2.**): acetonitrile and trifloroacetic acid (TFA) (Merck & Co.; Hoddesdon, UK).
4. Rat Thy-1 protein obtained by immunoaffinity purification and the lipid moiety removed by phospholipase C *(3)*.
5. Pigeon intestinal mucin obtained from the intestines of freshly killed pigeons by CsCl density gradient *(4)*.
6. Bovine fetuin, asialylated bovine fetuin, bovine serum albumin, chicken ovalbumin, ExtraAvidin peroxidase conjugate, 4-chloro-1 naphthol, poly-L-lysine, ammonium carbonate, and all reagents used in the modification and activation of NCM (*see* **Subheading 3.4.**): divinyl sulfone, dimethylformamide, sodium hydrogen carbonate, disodium carbonate, ethylenediamine, glutaraldehyde, and methanol (Sigma; Poole, UK).
7. Bovine pancreatic RNase B (Oxford GlycoSystems; Abingdon, UK).
8. Immobilon-P nitrocellulose membranes (0.2-μm and 0.45-μm pore size) (Millipore Corp.; Bedford, MA, USA).
9. Digoxigenin-labeled lectins of the glycan differentiation kit (Boehringer Mannheim; Mannheim, Germany).
10. Biotinylated lectins (Vector Labs, Peterborough, UK).
11. Benchmark microplate reader (Bio-Rad Lab. Ltd.; Hemel Hempstead, UK).

3. Methods

All procedures are carried out at room temperature unless otherwise stated.

3.1. Hydrozinolysis for Release of N- *and* O-*Linked Oligosaccharides (e.g.,* O-*Linked Glycoproteins) Using the Glycorelease Kit*

1. Directly prior to hydrazinolysis, lyophilize glycoproteins overnight in a screw-capped) vial and store in a desiccator in the presence of phosphorus pentoxide for 24 h.
2. Add 5 mL of anhydrous hydrazine to the vial, seal, and heat in a heating block at 60°C for 4 h for *O*-linked glycans (95°C for 5 h for *N*-linked glycans).
3. Cool the reaction vessel to room temperature and remove excess hydrazine by centrifugal evaporation under reduced pressure (*see* **Note 1**).
4. Acetylate the released oligosaccharides by addition of 0.5 mL of ice-cold acetylation buffer and 50 μL of acetylating reagent (supplied with the kit). Incubate mixture for 20 min at 4°C and then 30 min at room temperature.
5. Desalt oligosaccharides on a cation-exchange resin and lyophilize using a centrifugal evaporator.
6. Dissolve dried samples in a minimal amount of pure water. Purify the released glycans from peptide material by ascending paper chromatography using high-purity butanol–ethanol–water (16:4:1 by vol). Elute oligosaccharides at and a few centimeters in front

of the origin using water. Filter through a 0.2-μm PTFE filter. Evaporate to dryness and finally dissolve in 100 μL of water and store at –80°C.

3.2. Glycopeptide Preparation and RP-HPLC on Porous Graphitized Carbon Column

3.2.1. N-Linked Sugars

1. Dissolve 1 mg of trypsin in 500 μL of digestion buffer (500 m*M* NH_4CO_3, pH 8.5) immediately before it is required.
2. Dissolve either native or reduced carboxymethylated glycoprotein in enzyme–buffer solution at 20 μg of enzyme/1 mg of glycoprotein with 5 μL of toulene to prevent bacterial growth. Incubate for up to 72 h at 37°C, with an addition of enzyme (10 μg of enzyme/1 mg of glycoprotein) after 24 h.
3. Wash the digest 3× with 0.1 mL of water.
4. Lyophilize the digest by evaporation prior to chromatography.
5. Redissolve dried sample in 0.1% aqueous TFA and inject onto an HPLC column of PGC *(5)*.
6. Run a starting eluent for 2 min at 2% buffer A (0.1% TFA in acetonitrile) in buffer B (0.1% aqueous TFA) followed by a linear gradient to 82% buffer A for 78 min at a flow rate of 1 mL/min.
7. Collect 2-mL fractions up to 70 min and then lyophilize each fraction. Spot samples onto NCM modified to bind peptides and glycopeptides as described in **Subheading 1.4.1.** (adapted from the method of **ref.** ***6***) (*see* **Notes 2** and **3**).

3.2.2. O-Linked Sugars

1. Lyophilize *O*-linked sugars released by hydrazinolysis (*see* **Subheading 3.1.**).
2. Redissolve dried sample in 100 μL of 0.1% TFA and inject onto an HPLC column of PGC.
3. Run a gradient of 0.1% TFA in water (buffer A) and 0.1% TFA in acetonitrile (buffer B) at a flow rate of 1 mL/min with 0–60% buffer B for 60 min, 60% buffer B for 10 min, and 60–0% buffer B for 2 min. Collect 70 fractions of 1 mL and then lyophilize each faction.
4. To assay the amount of hexose present, redissolve each lyophilized fraction by addition of 60 μL of water.
5. Pipet 10 μL of each fraction into a microtiter plate along with a range of concentrations of a hexose standard (Gal or Man, usually 1–10 μg). To each well, add 50 μL of 2% aqueous phenol, mix thoroughly, and leave for 5 min. Add 200 μL of concentrated sulfuric acid to each well and mix prior reading on a benchmark microplate reader at 490 nm (*see* **Note 4**).

3.3. Coupling of Oligosaccharides to PolyL-Lysine (PLL) to Reintroduce Multivalency (see Note 5)

1. To each oligosaccharide fraction recovered from RP-HPLC on PGC (assuming 20 μg of oligosaccharide in each tube), add 50 μL of 0.092 *M* sodium phosphate, pH 7.4, containing 32 μg/mL of PLL (200 pmol in 50 μL) and incubate at 37°C for 2 h.
2. After incubation, add to each fraction 60 nmol of $NaCNBH_3$ (4μg in 5 μL of water) and incubate for 16 h.
3. Prior to lectin overlay analysis, dilute samples 1:2 in phosphate-buffered saline (PBS), and spot 2 × 1 μL of each sample, 1 cm apart, onto a nonmodified NCM (0.2-μm pore size) (*see* **Note 6**).

3.4. Modification and Activation of NCM

3.4.1. Modification of NCM

1. To modify the NCM (0.2-μm pore size) for adherence of peptides and glycopeptides, incubate the membrane with 1 : 2 divinyl sulfone in dimethylformamide dissolved for 1 h in 0.5 *M* $NaHCO_3$–Na_2CO_3, pH 10.0.
2. Wash the membrane with distilled water and then incubate in 1% aqueous ethylenediamine for 30 min, followed by a wash with water.
3. Incubate the membrane in 1% glutaraldehyde in 0.5 *M* $NaHCO_3$–Na_2CO_3 buffer, pH 10.0, for 15 min.
4. Wash the membrane with water and air dry. The membrane is ready for use for post-HPLC tryptic-digest fractions.

3.4.2. Activation of NCM for Glycoprotein Adhesion

1. To hydrophilically activate NCM without modification (0.45-μm pore size), incubate the membrane in 100% methanol for 5 min.
2. Wash the membrane in distilled water for 1 min and 50 m*M* Tris–0.15 *M* NaCl (TBS) buffer for an additional 1 min and then air dry. The membrane is now ready for loading intact glycoproteins samples.

3.5. Spotting of Samples on the NCM

1. Spot (1 μL) intact non-protease-treated glycoproteins (total protein concentration ranging from 2 μg to 10 ng) or HPLC fractions containing peptides and glycopeptides from the tryptic digests 1 cm apart either onto the activated or modified NCM, respectively (*see* **Notes 7** and **8**).
2. Allow samples to air dry for at least 10 min before identification of glycans (*see* **Note 6**).

3.6. Dot-Blot Lectin Overlay Analysis

3.6.1. Digoxigenin-Labeled Lectins

1. Block the surface of the activated membrane by incubating in the blocking solution (supplied with the kit) for 30 min with gentle shaking (*see* **Notes 9–11**).
2. Wash the membrane twice (5 min each) with at least 50 mL of TBS, pH 7.5.
3. Incubate with the respective lectin solutions: *Datura stramonium* agglutinin (DSA), *Galanthus nivalis* agglutinin (GNA), *Maackia amurenisis* agglutinin (MAA), and *Sambucus nigra* agglutinin (SNA) (an aliquot of the digoxigenin-labeled lectin in 50 m*M* Tris containing 1 m*M* $MgCl_2$ and 1 m*M* $CaCl_2$) for 1 h (*see* **Note 12**).
4. After incubation, wash the membrane 3× with TBS (10 min each).
5. Incubate with anti-digoxigenin (polyclonal sheep anti-digoxigenin Fab fragments conjugated with alkaline phosphatase) (supplied with the kit) in TBS for 1 h.
6. After washing extensively in TBS, develop the membranes using a freshly prepared staining solution of 4-nitroblue tetrazolium chloride–5-bromo-4-chloro-3-indolyl phosphate (NBT/X-phosphate) (supplied with the kit) in buffer (0.1 *M* Tris-HCl, 0.05 *M* $MgCl_2$, 0.1 *M* NaCl, pH 9.5) (*see* **Notes 13** and **14**).
7. Stop the reaction by washing the membranes extensively (at least 20 changes) in distilled water and leave to air dry (*see* **Note 15**).
8. Scan the membrane using a Hewlett-Packard flat bed scanner (*see* **Notes 16** and **17**).

3.6.2. Biotinylated Lectins

1. Wash the activated or modified membrane twice (5 min each wash), in PBS containing 0.1% Tween 20 (PBS-T) (*see* **Notes 9** and **10**).

2. After washing, block the membrane in PBS-T containing 5% Bovine serum albumin (PBS–T-BSA) for 1 h and wash for a further 4× (5 min each) with at least 50 mL of PBS-T.
3. Dilute in 5 µg/mL of PBS–T-BSA the following biotinylated lectins: *Aleuria aurantia* (AAL), *Lotus tetragononolobus* (LTL), *Lycopersicon esculentum* (LEL), *Maackia amurensis* (MAL-1), *Maclura pomifera* (MPL), soybean agglutinin (SBA), *Sophora japonica* (SJA), *Ricinus communis* agglutinin (RCA-1), *Ulex europeus* II (UEA-II), and wheat germ agglutinin (WGA). Apply respective lectin solutions (50 mL to membranes and incubate for 2 h (*see* **Note 9**).
4. Wash the membranes 4× (5 min each) with at least 50 mL of PBS-T.
5. Incubate in ExtraAvidin peroxidase conjugate, diluted 1:500 in PBS–T-BSA, for 45 min.
6. Wash the membranes 3× in PBS-T and twice in PBS and then incubate in developing solution (60 mg of 4-chloro-1 naphthol dissolved in 20 mL of cold methanol, 80 mL of cold PBS, and 60 mL of 6% hydrogen peroxide) until spots appear on the membrane.
7. Stop the reaction by washing the membranes extensively (at least 20 changes) in distilled water and leave to air dry (*see* **Note 15**).
8. Scan the membrane using a Hewlett-Packard flat bed scanner (*see* **Notes 17** and **18**).

4. Notes

1. For safety, the evaporator is vented to a fume cupboard and kept on overnight after the sample had been dried.
2. To identify and study the site occupancy of individual glycans of glycoproteins, the protein is first cleaved with proteases. The relative hydrophobicities *(7,8)* of the trypsin digest are calculated to predict the elution of the HPLC column as shown for disulfide-bonded Thy-1 (**Table 1**). We have shown earlier that in HPLC of peptides and glycopeptides using a PGC column *(5)*, the main determinator of retention is peptide and not oligosaccharide; thus, glycopeptides will elute in the same relative order as their peptides. The predicted hydrophobicities of the peptides containing each *N*-glycosylation site had large differences in value (**Table 1**) as mirrored in the time of elution between the glycopeptide HPLC fractions.
3. As an example for Thy-1, of the 35 fractions collected from the HPLC purification, the lectin-DSA (specific for Galβ1–4GlcNAc linkages present in complex and hybrid chains) bound well to fractions 13 and 14 and not as well to fractions 2 and 3. Staining with the lectin-GNA (specific for nonreducing terminal Man present in high mannose and hybrid chains) showed binding to fraction 10 and fractions 8, 9, and 11. This suggested that the least hydrophobic glycopeptide containing site $_{98}$NKT (**Table 1**) had complex chains (factions 2 and 3), the glycopeptide containing $_{23}$NKT has high mannose chains (fractions 8–10), and the most hydrophobic glycopeptide containing $_{74}$NFT had complex chains (fractions 13 and 14). There was some evidence (*not shown*) for staining of fractions 12–14 with GNA as well as DSA, suggesting that hybrid chains could also be present at this site. The GPI anchor glycan attached to the C-terminal tryptic peptideµs) was not expected to interact well with either of the lectins, but may possibly have added to the binding of GNA in the fractions 8–10 due to the Manα1–2Man shown near the protein.
4. The hexose assay is used to quantitate the amount of material after coupling to PLL (*see* **Subheading 3.3.**) an dialysis during optimization of the reaction. Do not overfill wells during the hexose assay, as concentrated sulfuric acid will severely damage the microtiter plate reader if spilled.
5. To screen for lectin and antigenic activity to multiple oligosaccharides released from mucin glycoproteins by hydrazinolysis, the oligosaccharides are coupled to PLL, which provides a multivalent molecule to which dot blots adhere throughout washing procedures

Table 1
Theoretical Tryptic Peptides of Thy-1 in Order of Increasing Hydrophobic Character

HPLC[a]	Residues	Sequences[b]	
–15.4	79–99	K <DEGDYMCELRVSGQNPTSSNK> T	Fx 2–3
–12.9	89–99	R <VSGQNPTSSNK> T	complex
–7.9	38–49	R <EK> K	chain on
–4.7	89–105	R <VSGQNPTSSNKTINVIR> D	N^{98}KT
–4.2	38–39	R <EK> K	
–2.6	111–111	K <C>	
–1.1	57–58	R <SR> V	
–0.8	40–41	K <KK> H	
–0.0	106–107	R <DK> L	
–2.7	1–2	<QR> V	
2.9	40–40	K <K> K	
2.9	41–41	K <K> H	
4.1	79–85	K <DEGDYMCELR> V	
8.2	100–107	K <TINVIRDK> L	
10.9	17–20	R <LDCR> H	
14.8	100–105	K <TINVIR> D	
17.2	108–111	K <LVKC>	Fx 8–9
19.8	106–110	R <DKLVK> C	Possible GPI
24.8	21–39	R <HENNTNLPIQHEFSLTREK> K	Anchor at
25.3	57–65	R <SRVNLFSDR> F	VKC- or high
26.4	108–110	K <LVK> C	Man at N^{23} NT
28.7	66–68	R <FIK> V	
33.0	59–65	R <VNLFSDR> F	
35.6	21–37	R <HENNTNLPIQHEFSLTR> E	Fx 10
37.1	42–58	K <HVLSGTLGFPEHTYRSR> V	High Man
39.9	17–37	R <LDCRHENNTNLPIQHEFSLTR> E	at N^{23} NT
41.1	41–56	K <KHVLSGTLGVPEHTYR> S	
44.8	42–56	K <HVLSGTLGVPEHTYR> S	
55.1	59–68	R <VNLFSDRFIK> V	
55.8	1–16	<QRVISLTACLVNQNLR> L	
59.7	3–16	R <VISLTACLVNQNLR> L	
64.0	3–26	R <VISLTACLVNQNLRLDCR> H	
67.1	69–88	K <VLTLANFTTKDEGDYMCELR> V	Fx 13–14
69.6	69–78	K <VLTLANFTTK> D	Complex and hybrid at N^{74} FT
91.7	66–78	R <FIKVLTLANFTTK> D	

[a]Relative hydrophobicity index on C18 HPLC.
[b]< > denotes the possible alternative tryptic cleavage sites.

Table 2
The Sensitivity Levels of Commercial Lectins on Glycoproteins by Lectin Overlay Analysis

Lectin	Fetuin	Asialylated fetuin	Bovine serum albumin	Chicken ovalbumin	RNase B
MAA	10 ng	1 μg	NB	NB	NB
SNA	0.1 μg	NB	NB	NB	NB
DSA	0.1 μg	0.1 μg	NB	NB	NB
GNA	1 μg	1 μg	2 μg	1 μg	NB

NB, No binding of lectin detected to sample.

required for lectin overlay. This is a convenient "one-pot" reaction before purification and detailed quantitative studies of active oligosaccharides can be done.

6. Take care to allow the samples to air dry completely before the second application on the NCM.
7. To avoid unnecessary cross-contamination between samples, a 96-well dot-blotter (Bio-Rad) was used instead of spotting the samples directly onto the membrane. Prior to activation or modification of the NCM, cut the membrane to size for placing onto the dot blotter cassette. Once the membrane is in a fixed position, apply respective samples as indicated in **Subheading 3.5.** After loading samples remove membrane from dot blotter and leave to air dry. Once dried, mark the membrane with a pencil to indicate the order of the samples. Also, draw a grid on paper to indicate the location of samples.
8. When applying respective samples onto the membrane, always use fresh Gilson tips for each sample to avoid contamination.
9. All incubations in the lectin overlay analysis are carried out on a rocking table at room temperature.
10. During the lectin overlay analysis, use gloves and avoid direct contact of the membrane with skin. Handle the membrane carefully with a flat-ended twiser to avoid damage.
11. For a clearer background, the blocking solution can be incubated with the membrane for overnight at 4°C. After, incubate the membrane with the blocking solution for a further 30 min at room temperature prior to washing.
12. For convenience, use large Petri dishes to incubate each membrane with respective lectin solutions. Make sure that the lectin solution always covers the membrane surface during incubation.
13. In contrast to the manufacturer's instructions the staining reaction can be extended until significant staining could be observed (1–18 h).
14. Standard glycopeptides should also spotted onto the membrane to monitor the progress of staining.
15. To protect the membrane from dust and damage, place a clear plastic film over the membrane and store in the dark.
16. **Table 2** presents a summary of results of a series of whole non-protease-treated glycoproteins (at various concentrations) analyzed for lectin specificity and sensitivity by dot-blot lectin overlay. The results from this assay clearly show that three lectins (MAA, SNA, and DSA) recognize Bovine fetuin (BF) at high sensitivity protein levels ranging from 10 ng to 0.1 μg. Asialylated BF (ABF) were detected by lectins MAA and DSA but not SNA. As expected, all three lectins showed no binding to Bovine serum albumin (BSA), which is not glycosylated. According to the lectin specificity as published by the manufacturer,

Table 3
Binding Preferences of Digoxigenin-Labeled and Biotinylated Lectins

Lectin	Specificity	References
Digoxigenin-labeled Lectins		
Datura stramonium agglutinin	Galβ1–4GlcNAc in complex and hybrid *N*-glycans, in *O*-glycans and GlcNAc in *O*-glycans	*9*
Galanthus nivalis agglutinin	Terminal Man, α1–3, α1–6, or α1–2 linked to Man; suitable for identifying "high Man" *N*-glycan chains or *O*-glycosidically linked Man in yeast glycoproteins	*10*
Maackia amurensis agglutinin	Sialic linked α2–3 to galactose; suitable for indentifying complex, sialylated carbohydrate chains and type of sialic acid linkage	*11*
Sambucus nigra agglutinin	Sialic acid linked α2–6 to galactose; suitable for identifying complex, sialylated *N*-glycan chains in combination with lectin MAA	*12*
Biotinylated lectins		
Aleuria aurantia lectin	Fucose linked α1–6 to GlcNAc or to fucose linked α1–3 to *N*-acetyllactosamine	*13*
Lotus tetragonolobus lectin	α-Linked L-fucose containing oligosaccharides	*14*
Lycopersicon esculentum (tomato) lectin	Trimers and tetramers of GlcNAc	*15*
Maclura pomifera lectin	α-Linked GalNAc	*16*
Ricinus communis agglutinin I	Terminal galactose and GalNAc	*17*
Sophora japonica agglutinin	Terminal GalNAc and galactose residues, with preferential binding to β-anomers	*18*
Soybean agglutinin	Terminal α- or β-linked GalNAc, and to a lesser extent, galactose residues	*19*
Ulex europaeus agglutinin II	2′Flucosyllactose (fucosyl α1–2 galactosyl β1–4 glucose)	*14,20*
Wheat germ agglutinin	Terminal GlcNAc or chitobiose and sialic acid	*21*

GNA should stain positive for RNase B and ovalbumin but not the other glycoproteins/proteins; however, negative results with RNase B were shown even after repeated tests such as using increased concentrations of RNase B (up to 20 μg). The sugar of RNase B was analyzed by hexose assay and shown to contain the expected 3%; thus the specificity of GNA is not as published, and the lectin can distinguish between ovalbumin and RNase B glycosylation.

17. **Table 3** shows the binding preferences of the lectins used. From the results obtained, the sensitivity of lectins can detect protein concentrations down to 10 ng. The choice between the commercial digoxigenin-labeled and biotinylated lectins used relate to their recognition of a range of defined and limited sugar structures depending on the samples tested. One drawback to the procedure is that because a number of structurally distinct oligosaccharides interact identically with certain lectins, this suggests that the structural heterogeneity frequently encountered with glycoproteins may not always be reflected in the interactions of these moieties with lectins.
18. For pigeon intestinal mucin *(2)* 8 of the 10 lectins (AAL, LEL, LTL, MAL-1, MPL, RCA-1, SBA, SJA, UEA-II, and WGA) reacted specifically with a different spectrum of PLL-oligosaccharide conjugates. LTL stained the entire blot, suggesting that this lectin interacted with the nitrocellulose. SJA reacted with all of the fractions and controls, indicative of nonspecific interaction with PLL itself. For the other lectins there was a wide range of activities with the various HPLC fractions. Difference in staining patterns between the lectins demonstrated that oligosaccharides of varying structure have been released from pigeon mucin by hydrazinolysis and have successfully been coupled to PLL, thus allowing binding to the NCM. Optimization of conditions at the macro-level included both neutral and sialylated oligosaccharides, with the chosen conditions giving maximum yield of both. The principle of this assay is derived essentially from enzyme-linked immunosorbent assay (ELISA) and commercial kits are available, where lectins are already linked to antibodies or other methods for detection.

Acknowledgments

The authors wish to thank Erminia Barboni for collaboration in a larger study that furnished the analytical amount of Thy-1 used in this study and to the following for support: E. B. PRIN97 program grant, University for Scientific and Technological Research, Rome, Italy; N. K. C. W., EU-BIOTECH research grant PL9765055; G. J. R., EU-HCM grant; C. I. B., Wellcome Trust grants 042462 and 052676; and E. F. H. and D. V. R., the UK Medical Research Council.

References

1. Sharon, N. (1998) Lectins: from obscurity into the limelight. *Protein Sci.* **7,** 2042–2048.
2. Baldwin, C. I., Calvert, J. E., Renouf, D. V., Kwok, C., and Hounsell, E. F. (1999) Analysis of pigeon intestinal mucin allergens using a novel dot blot assay. *Carbohydr. Res.*, in press.
3. Barboni, E., Pliego Rivero, B., George, A. J. T., Martin, S. R., Renouf, D. V., Hounsell, E. F., et al. (1995) The glycophosphatidylinositol anchor affects the conformation of the Thy-1 protein. *J. Cell Sci.* **108,** 487–497.
4. Baldwin, C. I., Todd, A., Bourke, S. J., Allen, A., and Clavert, J. E. (1998) IgG subclass responses to pigeon intestinal mucin are related to development of pigeon fanciers' lung. *Clin. Exp. Allergy* **28,** 349–357.
5. Davies, M., Smith, K. D., Harbin, A.-M., and Hounsell, E. F. (1992) High performance liquid chromatography of oligosaccharide alditols and glycopeptides on a graphitised carbon column. *J. Chromatogr.* **609,** 125–131.

6. Lauritzen, E., Masson, M., Rubin, I., and Holm, A. (1990) Dot immunobinding and immunoblotting of picogram and nanogram quantities of small peptides on activated nitrocellulose. *J. Immunol. Meth.* **131,** 257–267.
7. Browne, C. A., Bennett, H. P. I., and Solomon, S. (1982) The isolation of peptides by high performance liquid chromatography using predicted elution positions. *Analyt. Chem.* **124,** 201–208.
8. Davies, M. J. and Hounsell, E. F. (1998) HPLC and HPAEC of oligosaccharides and glycopeptides, in *Glycoanalysis Protocols* (Hounsell, E. F., ed.), Humana Press, Totowa, NJ, pp. 79–107.
9. Green, E. D., Brodbeck, R. M., and Baenziger, J. U. (1987) Lectin affinity high-performance liquid chromatography: interactions of *N*-glycanase-released oligosaccharides with leukoagglutinating phytohemagglutinin, concanavalin A, *Datura stramonium* agglutinin, and *Vicia villosa* agglutinin. *Analyt. Biochem.* **167,** 62–75.
10. Shibuya, N., Goldstein, I. J., Van Damme, E. J., and Peumans, W. J. (1988) Binding properties of a mannose-specific lectin from the snowdrop (*Galanthus nivalis*) bulb. *J. Biol. Chem.* **263,** 728–734.
11. Wang, W. C. and Cummings, R. D. (1988) The immobilized leukoagglutinin from the seeds of *Maackia amurensis* binds with high affinity to complex-type Asn-linked oligosaccharides containing terminal sialic acid-linked alpha-2,3 to penultimate galactose residues. *J. Biol. Chem.* **263,** 4576–4585.
12. Shibuya, N., Goldstein, J., Broeknaert, W. F., Nsimba-Luvaki, M., Peeters, B., and Peumanns, W. J. (1987) The elderberry (*Sambucus nigra L.*) bark lectin recognizes the Neu5Ac(alpha 2-6)Gal/GalNAc sequence. *J. Biol. Chem.* **262,** 1596–1601.
13. Yazawa, S., Furukawa, K., and Kochibe, N. (1984) Isolation of fucosyl glycoproteins from human erythrocyte membranes by affinity chromatography using *Aleuria aurantia* lectin. *J. Biochem. (Tokyo)* **96,** 1737–1742.
14. Allen, H. J., Johnson, E. A., and Matta, K. L. (1977) A comparison of the binding specificities of lectins from *Ulex europaeus* and *Lotus tetragonolobus*. *Immunol. Commun.* **6,** 585–602.
15. Zhu, B. C. and Laine, R. A. (1989) Purification of acetyllactosamine-specific tomato lectin by erythroglycan-sepharose affinity chromatography. *Prep. Biochem.* **19,** 341–350.
16. Lee, X., Thompson, A., Zhang, Z., Ton-that, H., Biesterfeldt, J., Ogata, C., et al. (1998) Structure of the complex of *Maclura pomifera* agglutinin and the T-antigen disaccharide, Galbeta1,3GalNAc. *J. Biol. Chem.* **273,** 6312–6318.
17. Wu, J. H., Herp, A., and Wu, A. M. (1993) Defining carbohydrate specificity of *Ricinus communis* agglutinin as Galβ1–4GlcNAc>Galβ1–3GlcNAc>Galα1–3Gal>Galβ1–3GalNAc. *Mol. Immunol.* **30,** 333–339.
18. Fournet, B., Leroy, Y., Wieruszeski, J. M. , Montreuil, J., Poretz, R. D., and Goldberg, R. (1987) Primary structure of an *N*-glycosidic carbohydrate unit derived from *Sophora japonica* lectin. *Eur. J. Biochem.* **166,** 321–324.
19. Dessen, A., Gupta, D., Sabesan, S., Brewer, C. F., and Sacchettini, J. C. (1995) X-ray crystal structure of the soybean agglutinin cross-linked with a biantennary analog of the blood group I carbohydrate antigen. *Biochemistry* **34,** 4933–4942.
20. Akiyama, M., Hayakawa, K., Watanabe, Y., and Nishikawa, T. (1990) Lectin-binding sites in clear cell acanthoma. *J. Cutan. Pathol.* **17,** 197–201.
21. Renkonen, O., Penttila, L., Niemela, R., Vainio, A., Leppanen, A., Helin, J., et al. (1991) *N*-Acetyllactosaminooligosaccharides that contain the βD-Glc*p*Nac-(1→6)-D-GlcpNAc-(1→6)-D-GalNAc sequences reveal reduction-sensitive affinities for wheat germ agglutinin. *Carbohydr. Res.* **213,** 169–183.

122

Polyacrylamide Gel Electrophoresis of Fluorophore-Labeled Carbohydrates from Glycoproteins

Brian K. Brandley, John C. Klock and Christopher M. Starr

1. Introduction

Prior to 1980, most methods for analysis of glycoprotein carbohydrates utilized column, thin-layer, and paper chromatography, gas chromatography, mass spectroscopy, and rarely nuclear magnetic resonance spectroscopy. These methods required relatively large amounts of materials (micromoles), specialized training and experience, and in some cases, significant capital equipment outlays. Because of these restrictions, convenient carbohydrate analysis on small samples was not available to most biologists. Recently, improvements in chromatographic methods, labeling methods for carbohydrates, carbohydrate-specific enzymes, and higher resolution electrophoresis methods have allowed carbohydrate analysis to be done on nanomolar amounts of material. Because of these improvements, today's biologist now has an improved ability to evaluate the role of carbohydrates in their research and development work.

Polyacrylamide gel electrophoresis (PAGE) of fluorophore-labeled carbohydrates has also been referred to as fluorophore-assisted carbohydrate electrophoresis, or FACE®. The technique was first developed in England by Williams and Jackson ***(1–4)*** and utilizes reductive amination of carbohydrates by low molecular weight, negatively charged or neutral fluorophores and electrophoresis on 20–40% polyacrylamide slab gels. This method permits separation of charged or uncharged sugars or oligosaccharides with high resolution and can detect single hydroxyl anomeric differences between mono- and oligosaccharides of sugars with otherwise identical molecular weight, charge, and sequence. The separation of sugars by electrophoresis is largely empirical, and it is not always possible to predict relative mobilities of structures. For the most part the success of this approach has been based on experimental observations and the use of highly specific reagents, enzymes, and standards.

What follows in this chapter are descriptions of the materials and methods required to perform two of the most common manipulations of oligosaccharides used in biologic research today. The method with some modification is useful for analysis of a variety of carbohydrates including reducing and nonreducing sugars, substituted and

From: *The Protein Protocols Handbook, 2nd Edition*
Edited by: J. M. Walker © Humana Press Inc., Totowa, NJ

unsubstituted monosaccharides, and oligosaccharides from glycoproteins, proteoglycans, glycolipids, and polysaccharides. In this chapter only profiling of *N*-linked and *O*-linked glycoprotein oligosaccharides is discussed. The method is also useful for a variety of other manipulations including structure–function studies, preparative work, and synthesis of oligosaccharides that are not discussed in detail here. Other reviews of this technique have been published previously ***(5,6)*** and a number of research studies have been performed using this method ***(7–31)***.

2. Materials

2.1 N-linked Oligosaccharide Analysis

For profiling of Asn-linked oligosaccharides released by peptide *N*-glycosidase F.

1. *N*-linked Gels: 10 × 10 cm low fluorescence glass plates with 0.5 mm spacers and 8-well combs filled with 20% T acrylamide:bis in 0.448 *M* Tris-acetate buffer pH 7.0 and containing a 5-mm stack of 10% polyacrylamide in the same buffer (gels should be made fresh or obtained precast from a commercial source).
2. Releasing enzyme: peptide *N*-glycosidase F (from commercial sources).
3. Running buffer: 50 m*M* Tris tricine buffer (pH 8.2).
4. Enzyme buffer: 100 m*M* sodium phosphate buffer (pH 7.5).
5. Labeling dye: 1 *M* 8-aminonaphthalene-1,3,6-trisulfonic acid (ANTS) in 15% acetic acid (reagent is stable for 2 wk at –70°C in the dark).
6. Reducing agent: 1 *M* sodium cyanoborohydride ($NaBH_3CN$) in dimethyl sulfoxide (DMSO) (reagent is stable for 2 wk at –70°C).
7. Sample loading solution: 25% glycerol with thorin 1 dye (store at 4°C).
8. Tracking dye: Mixture of Thorin 1, Bromphenol blue, Direct red 75, and Xylene cyanole in water.
9. Electrophoresis gel box (with 2-sided cooling).
10. Deionized or distilled water.
11. Assorted pipeting devices including a 0–10 μL positive displacement pipet (e.g., Hamilton syringe).
12. Centrifugal vacuum evaporator.
13. Oven or water bath at 45°C and 37°C.
14. Sodium dodecyl sulfate (SDS), 5%.
15. β-Mercaptoethanol (βME).
16. 7.5% nonidet P-40 (NP-40).
17. 100% Cold ethanol (undenatured).
18. Microcentrifuge tubes (1.5 mL).
19. Microcentrifuge.
20. Chicken trypsin inhibitor control (optional).
21. Maltotetraose or partially hydrolyzed starch standard (optional).

2.2. O-linked Glycoprotein Oligosaccharide Analysis

For profiling of Ser/Thr-linked oligosaccharides released by hydrazine.

1. *O*-linked gels: 10 × 10 cm low fluorescence glass plates with 0.5-mm spacers and eight-well combs filled with 35% T acrylamide:bis in 0.448 *M* Tris-acetate buffer and containing a 5-mm stack of 16% polyacrylamide in the same buffer (gels should be made fresh or obtained as precast gels from a commercial source).

2. Gel running buffer: 50 m*M* Tris-glycine (pH 8.2).
3. *O*-linked cleavage reagent: anhydrous hydrazine, 1 mL amp. **Hydrazine is toxic and flammable; discard ampoule and residual contents after using once; dispose of safely according to your institution's regulations.**
4. Re-*N*-acetylation reagent: acetic anhydride.
5. Re-*N*-acetylation buffer: 0.2 *M* ammonium carbonate, pH 9.4.
6. Desalting resin: Dowex AG50X8.
7. Tracking dye: Mixture of Thorin 1, Bromphenol blue, Direct red 75, and Xylene cyanole in water.
8. Sample loading solution: 25% glycerol with Direct red 75 (store at 4°C).
9. Labeling dye: 1 *M* 8-aminonaphthalene-1,3,6-trisulfonic acid (ANTS) in 15% acetic acid (reagent is stable for 2 wk if stored in the dark at –70°C).
10. Reducing agent: 1 *M* sodium cyanoborohydride ($NaBH_3CN$) in DMSO (reagent is stable for 2 wk at –70°C).
11. Distilled-deionized water.
12. Microcentrifuge tubes (2-mL, glass-lined).
13. Assorted pipeting devices including a 0–100 µL capillary positive displacement pipet.
14. Centrifugal vacuum evaporator.
15. Oven or water bath set at 37°C.
16. Sand filled heat block set at 60°C.
17. Vacuum dessicator.
18. Microfuge.
19. Phosphorus pentoxide (P_2O_5).
20. Bovine submaxillary mucin control (optional).
21. Maltotetraose or partially hydrolyzed starch standard (optional).

3. Methods

3.1. Principle

Using fluorescent PAGE, individual oligosaccharides can be quantified to obtain molar ratios, to obtain degree of glycosylation, and to detect changes in the extent or nature of glycosylation. Oligosaccharide profiling involves four steps:

1. Release of the oligosaccharides from the glycoprotein enzymatically or chemically;
2. Labeling of the mixture of released oligosaccharides with a fluorescent tag;
3. Separation of the fluorophore-labeled oligosaccharides by PAGE; and
4. Imaging of the gel either on a UV lightbox to obtain qualitative band information or using a commercial imaging system to determine the amount of oligosaccharide present in each band and the relative mobility of the bands.

Once separated on the gel, individual oligosaccharide bands can also be purified for further study.

3.2. Preparation of Glycoprotein

1. Isolate the glycoprotein according to your usual procedures. The sample should be relatively salt-free and contain no extraneous carbohydrates (e.g., sephadex-purified material contains large amounts of glucose) (*see* **Note 1**).
2. If the volume of the glycoprotein solution required is >100 µL, dry the glycoprotein in a 1.5-mL microcentrifuge tube. Generally 50–200 µg of glycoprotein is required for

N-linked analysis (*see* **Note 2**). For analysis of *O*-linked oligosaccharides 100–500 µg of glycoprotein may be required.

3. For the *N*-linked oligosaccharide control (*see* **Note 3**) remove 100 µg of the chicken trypsin inhibitor in a 45 mL aliquot, and place it in a 1.5 mL microfuge tube. Proceed with the enzymatic digestion as described below. For the *O*-linked oligosaccharide control remove 50 µL (100 µg) of Bovine submaxillary mucin, and place it in a reaction vial. Proceed with lyophilization, P_2O_5 drying and process along with the sample glycoprotein (*see* **Subheading 3.4.2.**). Store remaining glycoproteins at 4°C for future use.

3.3. Release of Asn-Linked Oligosaccharides with Peptide N-Glycosidase F

1. Add an equal volume of enzyme buffer to the glycoprotein in solution or dissolve the dried glycoprotein in 22.5 µL of water and add 22.5 µL enzyme buffer.
2. SDS is often required to completely denature the glycoprotein prior to enzymatic digestion. To denature the protein add SDS to 0.1% (1.0 µL of 5% SDS to 45 µL reaction) and β-ME to 50 m*M* (1.5 µL of a 1:10 dilution of 14.4 *M* stock β-ME to 45 µL reaction). Boil for 5 min (*see* **Note 4**), cool to room temperature and add NP-40 to 0.75% (5 µL of 7.5% NP-40 into 45 µL). Mix with finger flicks.
3. Add 2.0 µL (5 U of peptide-*N*-glycosidase F or as specified by the manufacturer) of enzyme to the glycoprotein sample. Mix with finger flicks and centrifuge for 5 s. Store remaining enzyme at 4°C.
4. Incubate sample for 2 h at 37°C.
5. Precipitate protein by adding 3 vol of cold 100% ethanol.
   Keep samples on ice for 10 min. Spin samples in microcentrifuge for 5 min to pellet protein.
6. Remove the supernatant and transfer to a clean 1.5-mL microcentrifuge tube.
   IMPORTANT! Do not discard the supernatant. It contains the released carbohydrates!
7. If a large amount of protein was digested (>250 µg) 5–10% of the released oligosaccharides may remain in the pellet. The recovery of these oligosaccharides can be accomplished by drying the pellet completely in a centrifugal vacuum evaporator or lyophilizer. Add 50 µL H_2O to resuspend, then 150 µL 100% cold ethanol and precipitate on ice. Centrifuge and combine the supernatants.
8. Dry the supernatants in a centrifugal vacuum evaporator or lyophilize to a translucent pellet. At this point samples may be stored at –20°C or proceed with the fluorophore labeling procedure described in **Subheading 3.4.2.**

3.4. O-Linked Oligosaccharide Release Using Hydrazine

3.4.1. Isolation of Glycoprotein

1. Isolate glycoprotein according to your usual procedures. The purified glycoprotein should be prepared in a non-Tris buffer containing a minimum amount of salt. The presence of nonvolatile salts may cause the breakdown of the oligosaccharides during hydrazinolysis. If the glycoprotein is in a buffer containing salt it is recommended that the sample be dialyzed against distilled water to remove salts prior to chemical digestion.

3.4.2. Hydrazinolysis

1. Dry 100–500 µg of glycoprotein in a glass-lined reaction vial, using a centrifugal vacuum evaporator or lyophilizer. The actual amount of glycoprotein required will depend on the size of the protein and the extent of glycosylation (*see* **Note 2**).
2. The sample must be completely dry before hydrazinolysis. Following lyophilization, dry the sample overnight under vacuum in the presence of P_2O_5 to remove all traces of H_2O.

Place samples in a dessicator flask with a beaker containing a small amount of P_2O_5. Attach the dessicator directly to the pump without a cold-trap—any water remaining in the sample will be trapped by the P_2O_5.

3. Open a fresh ampoule of anhydrous hydrazine *O*-linked cleavage reagent. Add 50 µL of hydrazine to the dried sample using a glass transfer pipet, or a positive displacement capillary pipet (metal or plastic should not be used). Resuspend the dried sample. Overlay the sample with dry nitrogen and cap tightly. **Hydrazine is very hygroscopic. Discard unused hydrazine according to your hazardous waste regulations. Do not reuse.**
4. Incubate samples for 3 h in a sand bath or dry heat block set at 60°C (do not use a water bath) to release *O*-linked oligosaccharides (higher temperatures may result in the degradation of *O*-linked sugars or in the release of non-*O*-linked sugar chains, such as *N*-linked sugars, from the sample if they are present).
5. Dry samples in vacuum evaporator on low heat setting.

3.4.3. Re-N-Acetylation Procedure

1. Add 30 µL of Re-*N*-acetylation buffer to the dried pellet from **step 5** in **Subheading 3.4.2.**
2. Resuspend by vortexing. Spin 2 s in a microfuge.
3. Add 2 µL of re-*N*-acetylation reagent to the solution. Mix well. Spin 2 s in a microfuge.
4. Incubate tubes on ice for 15 min.
5. Following the 15 min incubation, stop the reaction by adding 60 µL of the desalting resin. Desalting resin is prepared by adding 0.5 g of Dowex AG50X8 to 0.7 mL water. Invert or vortex the resin immediately prior to removing the 60 µL for each tube. Incubate the resin with the sample at room temperature for 5 min mixing by placing the tube on a shaker or by continuously inverting the tube to keep the resin suspended.
6. Briefly centrifuge to pellet the resin, remove the supernatant (save supernatant) and wash the resin 2× with 120 µL of water for 2 min each (save supernatant).
7. Combine the resin supernatants in a 1.5-mL microcentrifuge tube.
8. Dry the supernatants in a vacuum evaporator on low heat setting.

3.5. Labeling Oligosaccharides

3.5.1. Preparation of Samples and Standards

If quantitation of the oligosaccharide bands in the samples is required, then one must compare the intensity of an internal standard (e.g., maltotetraose) band with the intensity of sample bands and it is, therefore, essential that the standard is present on each gel used (*see* **Note 5** for preparation and use of this material).

3.5.2. ANTS Labeling

1. Prepare the labeling dye as 1 *M* ANTS in 15% acetic acid (dye solution can be stored in the dark at –70°C for up to 2 wk).
2. Prepare 1 *M* solution of $NaBH_3CN$ in DMSO and mix well by vortexing until crystals are completely dissolved (this reducing agent can be stored for 2 wk at –70°C).
3. Add 5 µL of labeling dye to each dried oligosaccharide pellet. Mix well until the oligosaccharide pellet is dissolved.
4. Add 5 µL of reducing agent. Mix well by vortexing. Centrifuge 5 s in microcentrifuge.
5. Incubate samples at 45°C for 3 h (temperatures higher than 45°C or times longer than 3 h can destroy or modify carbohydrates, e.g., sialic acids). Greater than 90% of the oligosaccharides are labeled under these conditions. As a convenient alternative samples can be labeled at 37°C (not 45°C) overnight (or approx 16 h). These latter conditions result in labeling of >90% of the oligosaccharides (*see* **Note 6**).

6. After labeling, dry the samples in centrifugal vacuum evaporator for approx 15 min or until the sample reaches a viscous gel stage.

3.6. Electrophoresis

3.6.1. Preparation of a Sample for Electrophoresis

1. Resuspend the dried fluorophore "labeled" oligosaccharide in 5–20 μL H_2O. The actual volume of H_2O used to resuspend the sample will depend on the amount of oligosaccharide present in the sample (start with 10 μL, this will enable the sample to be diluted further if necessary).
2. Remove an aliquot of the sample (generally 1–2 μL) and dilute it with an equal volume of sample loading solution. Load the entire aliquot into one lane of a gel. Best results are obtained by loading 4 μL/lane on a 10 × 10 cm gel with eight lanes.

3.6.2. Electrophoresis

1. For *N*-linked analysis chill the running buffer to 4–6°C prior to use. For *N*-linked analysis perform electrophoresis at a buffer temperature of 5–8°C. All *O*-linked gels are run at 15–20°C.
2. For *N*-linked analysis set up electrophoresis with a recirculating chiller and place the electrophoresis tank containing a stir bar on a mechanical stirrer. Connect the gel box cooling chamber to a refrigerating circulator. Turn on the circulator and stirrer and set the coolant temperature to 5°C.
3. For *N*-linked analysis pour the precooled running buffer into the electrophoresis tank up to the appropriate level. The temperature of the buffer should be monitored during the run using a thermometer inserted through the hole in the lid or other method. The temperature will probably increase a few degrees during electrophoresis, but should not exceed 10°C. For *O*-linked gels the temperature should not exceed 23°C.
4. Determine the number of gels required for the samples prepared. Each gel should contain eight lanes. The outside lanes should be used for the tracking dye and glucose polymer standard leaving the six inner lanes for samples and quantitation standard (maltotetraose).
5. Gently remove the comb(s) from the gel(s). To avoid distorting the wells, gently wiggle each comb to free the teeth from the gel, then lift up slowly until the comb is released.
6. Place the gel cassette(s), one on each side of the center core unit of the gel apparatus with the short glass plate against the gasket. Be sure the cassette is centered and that the cassette is resting on the "feet" at the bottom of the apparatus. If only one gel is being run place the buffer dam on the other side.
7. It is essential that the wells of the gel are thoroughly rinsed out with the running buffer from the upper buffer reservoir prior to sample loading. This is best accomplished by using a syringe with a blunt needle (a Pasteur pipet is not recommended because of the possibility of breakage into the wells).
8. With the core unit containing the gels placed securely on the bench, load samples into the wells by underlaying the upper buffer. Use flat sequencing pipet tips to load by delivering the sample to the bottom of each well. Optimal resolution will be achieved by using 4 μL of sample per lane.
   Note: For the most reliable quantitation of oligosaccharide bands the use of a positive displacement pipet (e.g., Hamilton syringe) is recommended.
9. Load 4 μL of the standard in a lane when prepared as described in **Note 5**.
10. Load 2 μL of tracking dye in a lane directly from the vial.
11. Load 4 μL of each labeled oligosaccharide sample in a lane. Samples should be diluted 1:1 in the sample loading solution (*see* **Note 7**).

12. To prevent possible lane distortions as a result of different loading volumes it is recommended that 4 µL of Sample Loading Solution be loaded in any unused lanes. Best results are obtained when the same volume of sample is added to each lane.
13. Place the core unit containing the loaded gels into the electrophoresis tank and connect the power cords to the electrophoresis tank then connect the power supply.
14. Place the thermometer into the lower buffer chamber through the hole in the lid. For *N*-linked analysis the initial temperature of the lower buffer must be 5–8°C; for *O*-linked analysis 15–20°C is optimal.
15. Turn on the power supply and select the proper current. Gels should be run at a constant current of 15 mA/gel (30 mA for 2 gels; 15 mA for 1 gel and 1 buffer dam). Limits on the power supply should be set for 1000V and 60W. These run conditions will result in voltages of 100–400V at the beginning of the run and may approach 800–1000V at the end of the run. If the initial voltage is significantly different check to be sure that the leads are connected properly and that the buffers are at the recommended levels (*see* **Notes 8** and **9**).
16. Most *N*-linked oligosaccharides fall in the Glucose_4–Glucose_{12} range (also referred to as G4–G12 or DP4–DP12), so the time of electrophoresis should be adjusted to optimize the separation of this region of the gel. Most *O*-linked oligosaccharides run in the G1–G6 range (DP1–DP6).
17. Monitor the electrophoresis by following the migration of the fast moving thorin dye (orange band). Generally, electrophoresis of *N*-linked oligosaccharides is complete when the orange dye just exits the bottom of the gel in approx 1 to $1^1/_4$ h. For the *O*-linked oligosaccharides the gel run is complete when the orange dye is 1 cm above the bottom of the gel. In a darkened room, the migration of the labeled oligosaccharides can be monitored directly during electrophoresis by turning off the power supply, removing the leads and the gel box cover and holding a hand-held UV light over the gels. The run can be continued by repositioning the gel in the electrophoresis box and reconnecting the power supply as described in **step 15**. The amount of time the current is off should be as short as possible (<5 min) to minimize diffusion of the oligosaccharides in the gel.
18. When the electrophoresis is complete, turn off the power supply. Disconnect the power cords from the power supply and the electrophoresis tank. Turn off the refrigerated cooler and discard the buffer (*see* **Note 10**).

3.7. Gel Imaging

CAUTION: UV protective eyeware or faceshield should be worn. Avoid prolonged exposure to UV light.

1. Allow UV lightbox to "warm-up" for at least 2 min in order to get maximum intensity output. The lightbox must be long-wave UV and have a peak output at approx 360 nm; this is *not* the type of box typically used for ethidium bromide-stained DNA gels.
2. Remove the tape from the gel cassette, which may be fluorescent, and clean the surfaces of the cassette if it is required to image the gel within the cassette. The cassette glass must be of special low-fluorescence type to obtain an image of a gel within the cassette. If the glass is not of a low fluorescence type and if the type of gel cassette being used permits disassembly, remove the gel completely from the cassette and place it on the UV lightbox (*see* **Notes 11** and **12**).
3. An electronic image of the *N*-linked oligosaccharides from several glycoproteins is shown in **Fig. 1**; an electronic image of the *O*-linked oligosaccharides from several different glycoproteins is shown in **Fig. 2**. In each image you can also see the position of maltotetraose and a ladder of maltooligosaccharides from partially hydrolyzed wheat starch (*see* **Notes 9**, **10**, **13–16** for problems with bands).

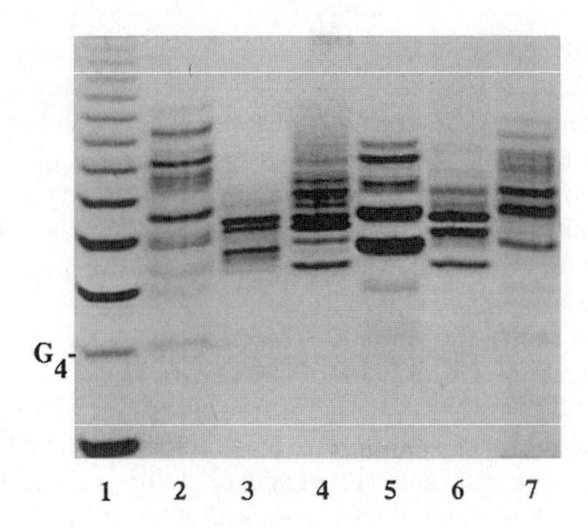

Fig. 1. Profiles of *N*-linked oligosaccharides from several glycoproteins. ANTS-labeled oligosaccharides released by peptide *N*-glycosidase (PNGase F) from six different glycoproteins are shown. *Lane 1* contains an oligosaccharide ladder standard of partially hydrolyzed wheat starch with G4 representing glucotetraose; *lane 2*, chicken trypsin inhibitor; *lane 3*, bovine fetuin; *lane 4*, human α-acid glycoprotein; *lane 5*, bovine ribonuclease B; *lane 6*, human chorionic gonadotropin (hCG); and *lane 7*, chicken ovalbumin. The profiles show a wide variety of different glycosylation patterns, indicating the minimum number (some oligosaccharides will comigrate) of different types of oligosaccharide present and their relative quantities. The images presented in this review were obtained by imaging the gels following electrophoresis using a CCD (charge-coupled-device)-based imaging system (Glyko).

3.8. Gel Handling

1. If the gel is no longer needed it should be properly discarded.
2. Following imaging of the oligosaccharide gels, the glass plates can be separated and the gels dried on a flat bed gel drier between sheets of Teflon™ membrane at 80°C for 1 h. After the gel is dry, carefully peel the Teflon sheets away from the gel. Gels dried in this manner can be stored indefinitely in a dark dry location and can reimaged at any time with minimal bleaching.
3. Following imaging, the gel cassette can be placed back in the electrophoresis apparatus and the run continued in order to improve the resolution of the oligosaccharide bands. In this case, the *upper* buffer should be saved and reused until the run is finally terminated. Note that diffusion of carbohydrate bands and subsequent poor resolution will occur if the time between electrophoresis and imaging exceeds 15 min.

4. Notes

1. The glycoprotein sample should ideally first be dialyzed against distilled water and stored lyophilized in a 1.5 mL microfuge tube. If the sample needs to be in a buffered solution, one can place the sample in 50 m*M* sodium phosphate buffer, pH 7.5, at a final concentration of at least 100 µg/50 µL or 2 mg/mL. Best results are obtained if the total salt concentration of the solution is <100 m*M*. The use of a Tris-based buffer is not recommended. If a detergent is required, the sample may be suspended in up to 0.1% SDS, 0.75% NP-40, or 1% n-octyl β-D-glucopyranoside. If desired, the sample may also contain 0.05% sodium azide.
2. We recommend that you use at least 250 µg of glycoprotein for analysis.

   The actual amount of glycoprotein required for profiling will depend on the size of the

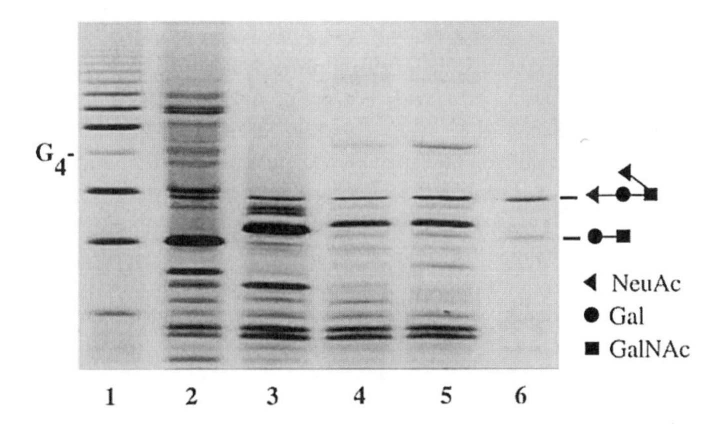

Fig. 2. Profiles of *O*-linked oligosaccharides from several glycoproteins. ANTS-labeled Ser/Thr-lined oligosaccharides released using hydrazine at 60°C from four different glycoproteins are shown. *Lane 1* contains an oligosaccharide ladder standard of partially hydrolyzed wheat starch with G4 representing glucotetraose; *lane 2*, porcine stomach mucin, Type II; *lane 3*, bovine submaxillary mucin, Type I; *lane 4*, bovine fetuin; and *lane 5*, human chorionic gonadotropin (hCG). *Lane 6* contains the NeuAc(α2-3)Gal(β1-3)[NeuAc(α2-6)]GalNAc and Gal(β1-3)GalNAc standards. The images presented in this review were obtained by imaging the gels following electrophoresis using a CCD-based imaging system (Glyko).

protein, the amount of glycosylation, and the degree of oligosaccharide heterogeneity. In general, the amount of glycoprotein required increases with the size of the protein or the degree of heterogeneity and decreases with the percent of glycosylation. As a general guideline, one would start with approx 50–100 µg to profile the *N*-linked oligosaccharides of a 60 kDa glycoprotein that contains 10–20% carbohydrate by weight. For *O*-linked oligosaccharide analysis we suggest 100–500 µg of starting glycoprotein. This amount would normally provide sufficient material for several electrophoretic runs. For isolation of individual oligosaccharides, and carrying out sequencing, additional material will be required.

3. The control for *N*-linked profiling consists of trypsin inhibitor that is used as a control for enzyme digestion and fluorophore labeling. This control should be included in the analysis for the following reasons:
   a. If this control is used for the first time it will help the user to become familiar with the procedures;
   b. If the profile obtained looks appropriate then this assures the user that things are working properly; and
   c. In an unknown sample that may not contain *N*-linked oligosaccharides, the user can be certain that the reagents are good and that the release and labeling procedures were performed properly.

   Similarly a control for *O*-linked profiling such as Bovine submaxillary mucin should be used.
4. Some proteins will precipitate when boiled, i.e., immunoglobulins. The following procedure should be used if your protein precipitates: Add SDS/β-ME at the recommended concentration and incubate for 5 min at room temperature, add NP-40 according to directions, and then add PNGase F and incubate overnight.
5. The quantitation control consists of maltotetraose (Glucose $_4$). This control should be prepared so that once prepared 5 µL will contain 200 pmol of maltotetraose (standards prelabeled with ANTS are also available commercially). Accurate quantitation will be

achieved when using an electronic imaging system. If using a commercial imaging system, refer to the manual for a detailed description of the quantitation procedures.

6. The stoichiometry of labeling is such that only one molecule of fluorophore is attached to each molecule of oligosaccharide. When labeling 20 nmol or less of total sugar using the reagents and labeling conditions described, the fluorophore labeling efficiency is >95% *(5)*. Labeling more than 20 nmol in each reaction will result in reduced labeling efficiency. When labeling >20 nmol it is recommended that an internal labeling control is included.
7. Sample handling and storage:
   a. Always avoid exposing labeled samples and dyes to light or excess heat;
   b. Labeled samples are stable when stored for 3 mo at –70°C in the dark;
   c. Unused solutions of the dye and reducing agent can be stored for as long as 2 wk at –70°C. Thaw immediately before use.
8. Band distortion in gels caused by vertical streaking or smearing may result if the sample is overloaded. Use a maximum of 1/5 of the volume of the labeling reaction for each lane. The sample may have a high concentration of salt. Remove salts by dialysis, desalting column, and so on, prior to enzymatic digestion. Distorted sample wells in gel may be caused by tearing of wells when the comb was removed. Remove the comb slowly using a gentle back and forth rocking motion and lift vertically. Alternatively, gels may have been in contact with upper buffer too long prior to sample loading. Samples should be loaded within 5 min of placing the gel in the upper buffer tank.
9. Voltage and/or current leaks can result when high voltages are used. If at the beginning of the run the voltage is >400V or readings are unstable, turn the power off before checking the following: possible electrical leak, check for cracks in glass plates. Remove inner core assembly and check for buffer leak between gaskets and cassette plates. If leaks are evident check that the plates are clean and not cracked or chipped, and that they are installed properly.
10. Buffers should not be reused as they have fluorophore contamination after use. Reuse of buffers may result in no bands being visible on the gel owing to "washout" of the fluorophore-labeled oligosaccharides.
11. Accurate quantification is essential for detailed carbohydrate analysis. Although oligosaccharide patterns on PAGE gels can be viewed and photographed on a standard laboratory UV lightbox, it is not reliable for accurate quantification. Images of gels can be recorded using a Polaroid camera. The proper choice of light source, filters, and film must be made. A filter must be fitted to the camera lens that completely covers the glass of the lens (stray UV contacting the lens will cause it to fluoresce and subsequently lower the sensitivity of the film). A suitable filter will have no inherent fluorescence, peak transmission at approx 500 nm and bandwidth of 80 nm FWHM. A medium speed, medium resolution, Polaroid film is recommended. Use Polaroid 53 film for cameras which use single 4 × 5" sheet film; use Polaroid 553 film for cameras that use 8 sheet film cartridges.

    To visualize the carbohydrate banding patterns, the low fluorescent glass cassette containing the gel (or the gel removed from the casette) is placed on a longwave UV lightbox with a peak excitation output at approx 360 nm.

    Photograph the gel using the lowest practical f-stop setting on the lens with the gel filling as much of the frame as possible. E.g., exposures at f5.6 using Polaroid 53 film have ranged from 5–40 s using the equipment specified above. **Keep UV exposure of the gel to a minimum to prevent bleaching.**

    Develop the film according to the manufacturer's instructions.
12. For electronic archiving and quantitation, several types of imaging systems are available. To give best results these systems must have an illumination source with an excitation

wavelength of 365 nm and a 520 nm emission filter placed in the light path between the gel and the image capturing device. The use of an internal standard in the gels is also required for quantitation.

Following electrophoresis, the gel is inserted into the imager under long-wave UV excitation, and an electronic image of the fluorescent carbohydrate banding pattern of the gel is acquired by the imager's CCD as a digital image. The gel image is displayed on a computer screen using the imaging software. The imaging system should allow for detection and quantification of individual carbohydrate bands into the low picomole range of 1.6–300 pmol. In practice, the most useful and accurate range of the imager for band quantification is between 5 and 500 pmol of carbohydrate and this range was used for the experiments described in this chapter.

13. You may have "smile effect" gel distortions at both sides of the gel. This can happen if the gel is not being cooled uniformly. Check that the cooling system is on and working properly. Check the buffer temperature. Check that the power supply is set for the proper current level.
14. You may have band distortions or "fuzzy bands." This can be caused by wells that may have not been rinsed thoroughly with electrophoresis upper buffer prior to loading samples, or the current may have not been set properly, i.e., the current was too high.
15. Incomplete re-*N*-acetylation may result in little or no labeling of the released oligosaccharides, presumably by the hydrazide interfering with the reductive amination using ANTS. To check the re-*N*-acetylation reagents you can use glucosamine that will migrate at a DP of approx 2.5 on the gel when *N*-acetylated and below DP1 when unacetylated. You can also use *N*-acetyl-glucosamine as an internal control at the beginning of the experiment and take it through hydrazinolysis and re-*N*-acetylation expecting the same migrations as stated.
16. Oligosaccharides are small molecules that can diffuse rapidly in the gel matrix. Band diffusion and resulting broad or "fuzzy" bands can occur if:
    a. Electrophoresis is run too slowly;
    b. Electrophoresis is run at higher than optimal temperatures;
    c. Electrophoresis is stopped and started repeatedly; and
    d. The gel is removed for visualization for longer than 10 min and then re-electrophoresed.

References

1. Williams, G. R. and Jackson, P. (1992) Analysis of carbohydrates. *U.S. Patent* **5,** 508.
2. Jackson, P. (1990) The use of polyacrylamide-gel electrophoresis for the high resolution separation of reducing sugars labeled with the fluorophore 8-aminonaphthalene-1,3, 6-trisulfonic acid. *Biochem. J.* **270,** 705–713.
3. Jackson, P. and Williams, G. R. (1991) Polyacrylamide gel electrophoresis of reducing saccharides labeled with the fluorophore 8-aminonaphthalene-1,3,6- trisulfonic acid: application to the enzymatic and structural analysis of oligosaccharides. *Electrophoresis* **12,** 94–96.
4. Jackson, P. (1994) The analysis of fluorophore-labeled glycans by high-resolution polyacrylamide gel electrophoresis. *Analyt. Biochem.* **216,** 243–252.
5. Starr, C., Masada, R. I., Hague, C., Skop, E., and Klock, J. (1996) Fluorophore-assisted-carbohydrate- electrophoresis, FACE® in the separation, analysis, and sequencing of carbohydrates. *J. Chromatogr. A* **720,** 295–321.
6. Masada, R. I., Hague, C., Seid, R., Ho, S., McAlister, S., Pigiet, V., and Starr, C. (1995) Fluorophore-assisted-carbohydrate-electrophoresis, FACE®, for determining the nature and consistency of recombinant protein glycosylation. *Trends Glycosci. Glycotech.* **7,** 133–147.

7. Hu, G. (1995) Fluorophore assisted carbohydrate electrophoresis technology and applications. *J. Chromatogr.* **705,** 89–103
8. Higgins, E. and Friedman, Y. (1995) A method for monitoring the glycosylation of recombinant glycoproteins from conditioned medium, using fluorophore assisted carbohydrate electrophoresis. *Analyt. Biochem.* **228,** 221–225.
9. Basu, S. S., Dastgheib-Hosseini, S., Hoover, G., Li, Z., and Basu, S. (1994) Analysis of glycosphingolipids by fluorophore-assisted carbohydrate electrophoresis using ceramide glycanase from *Mercenaria mercenaria. Analyt. Biochem.* **222,** 270–274.
10. Flesher, A. R., Marzowski, J, Wang, W., and Raff, H. V. (1995) Fluorophore-labeled carbohydrate analysis of immunoglobulin fusion proteins: correlation of oligosaccharide content with in vivo clearance profile. *Biotech. Bioeng.* **46,** 399–407.
11. Lee, K. B., Al-Hakim, A., Loganathan, D., and Linhard, R. J. (1991) A new method for sequencing carbohydrates using charged and fluorescent conjugates. *Carbohydr. Res.* **214,** 155–162.
12. Stack, R. J. and Sullivan, M. T. (1992) Electrophoretic resolution and fluorescence detection of *N*-linked glycoprotein oligosaccharides after reductive amination with 8-aminonaphthalene-1,3,6-trisulfonic acid. *Glycobiology* **2,** 85–92.
13. Roy, S. N., Kudryk, B., and Redman, C. M. (1995) Secretion of biologically active recombinant fibrinogen by yeast. *J. Biol. Chem.* **270,** 23,761–23,767.
14. Denny, P. C., Denny, P. A., and Hong-Le, N. H. (1995) Asparagine-linked oligosaccharides in mouse mucin. *Glycobiology* **5,** 589–597.
15. Qu, Z., Sharkey, R. M., Hansen, H. J., Goldenberg, D. M., and Leung, S.-O. (1997) Structure determination of N-linked oligosaccharides engineered at the CH1 domain of humanized LL2. *Glycobiology* **7,** 803–809.
16. Prieto, P. A., Mukerji, P., Kelder, B., Erney, R., Gonzalez, D., Yun, J. S., et al. (1995) Remodeling of mouse milk glycoconjugates by transgenic expression of a human glycosyltransferase. *J. Biol. Chem.* **270,** 29,515–29,519.
17. Pirie-Shepherd, S., Jett, E. A., Andon, N. L., and Pizzo, S. V. (1995) Sialic acid content of plasminogen 2 glycoforms as a regulator of fibrinolytic activity. *J. Biol. Chem.* **270,** 5877–5881.
18. Goss, P. E., Baptiste, J., Fernandes, B., Baker, M., and Dennis, J. W. (1994) A phase I study of swainsonine in patients with advanced malignancies. *Cancer Res.* **54,** 1450–1457.
19. Sango, K., McDonald, M. P., Crawley, J. N., Mack, M. L., Tifft, C. J., Skop, E., et al. (1996) Mice lacking both subunits of lysosomal beta-hexosaminidase display gangliosidosis and mucopolysaccharidosis. *Nature Genet.* **14,** 348–352.
20. Masada, R. I., Skop, E., and Starr, C. M. (1996) Fluorophore-assisted-carbohydrate-electrophoresis for quality control of recombinant protein glycosylation. *Biotechnol. Appl. Biochem.* **24,** 195–205.
21. Kumar, H. P. M., Hague, C., Haley, T., Starr, C. M., Besman, M. J., Lundblad, R., and Baker, D. (1996) Elucidation of *N*-linked oligosaccharide structures of recombinant factor VII using fluorophore assisted carbohydrate electrophoresis. *Biotechnol. Appl. Biochem.* **24,** 207–216.
22. Tsuchida, F., Tanaka, T., Matuo, Y., Moriyama, N., Okada, K., Jimbo, H., et al. (1997) Oligosaccharide Analysis Of Renal Cell Carcinoma by Fluorophore-Assisted Carbohydrate Electrophoresis (FACE). *Jpn J. Electroph.* **41,** 39–41, 81–83.
23. Hague, C., Masada, R. I., and Starr, C. (1998) Structural determination of oligosaccharides from recombinant iduronidase released with peptide N-glycase F, using Fluorophore-Assisted Carbohydrate Electrophoresis. *Electrophoresis* **19,** 2612–2620

24. Morimoto, K., Maeda, N., Foad, A., Toyoshima, S., and Hayakawa, T. (1999) Structural Characterization of Recombinant Human Erythropoietins by Fluorophore-assisted Carbohydrate Electrophoresis. *Biol. Pharm. Bull.* **22,** 5–10.
25. Allen, A. C., Bailey, E. M., Barratt, J., Buck, K. S., and Feehally, J. (1999) Analysis of IgA1 O-glycans in IgA nephropathy by fluorophore-assisted carbohydrate electrophoresis. *J. Am. Soc. Nephrol.* **10,** 1763–1771.
26. Bardor, M., Cabanes-Macheteau, M., Faye, L., and Lerouge, P. (2000) Monitoring the N-glycosylation of plant glycoproteins by fluorophore-assisted carbohydrate electrophoresis. *Electrophoresis.* **21,** 2250–2256.
27. Calabro, A., Benavides, M., Tammi, M., Hascall, V. C., and Midura, R. J., (2000) Microanalysis of enzyme digests hyaluronan and chondroitin/dermatan sulfate by Fluorophore-Assisted Carbohydrate Electrophoresis (FACE). *Glycobiology* **10,** 273–281.
28. Calabro, A. C., Hascall, V. J., and Midura, R., (2000) Structural characterization of oligosaccharides in recombinant soluble human interferon receptor 2 using Fluorophore-Assisted Carbohydrate Electrophoresis. *Electrophoresis* **21,** 2296–2308.
29. Frado, L. Y. and Strickler, J. E. (2000) Relative abundance of oligosaccharides in candida species as determined by Fluorophore-Assisted Carbohydrate Electrophoresis. *J. Clin. Microbiol.* **38,** 2862–2869.
31. Lemoine, J., Cabanes-Macheteau, M., Bardor, M., Michalaski, J. C., Faye, L., and Lerouge, P. (2000) Analysis of 8-aminoapthalene-1,3,6-trisulfonic acid labelled N-glycans by matrix assisted laser desorption/ionisation time-of-flight mass spectrometry. *Rapid Commun Mass Spectrom,* **14,** 100–104.

123

HPLC Analysis of Fluorescently Labeled Glycans

Tony Merry

1. Introduction

The study of the glycan chains (oligosaccharides) of glycoproteins presents a number of analytical problems that generally make their analysis more difficult than that of the peptides to which they are attached. A number of different analytical techniques have been applied to the study of glycans. Definitive characterization has traditionally been performed using nuclear magnetic resonance spectroscopy (NMR) and this still remains the only way to unequivocally assign the structure to novel glycans. But the techniques do require relatively large amounts of material (in the milligram range). In addition, access to the sophisticated equipment and the expertise required for the interpretation of data mean that the technique is available only to specialized dedicated laboratories. Another technique, which has also been applied extensively to studies of glycan structure, is that of mass spectrometry in various forms. A number of structures have been solved by means of fast atom bombardment (FAB-MS) ***(1,2)*** although generally further information from linkage analysis by gas chromatography-MS of permethylated alditol acetates (PMAA ***[3–5]***) or by use of specific exoglycosidase enzymes ***(6)*** is required. Other techniques for mass spectrometric analysis are becoming more widely used as a result of recent developments in instrumentation ***(7–9)***. Both matrix-assisted laser desorption mass spectrometry with time-of-flight detection ***(8)*** (MALDI-TOF) and electrospray coupled to post source decay ***(9)*** to produce fragmentation ions are now becoming routinely used. The instrumentation is still expensive, however, and requires skilled operation and interpretation. There are also still problems with optimization and contamination of samples. Another technique that is also used is capillary electrophoresis, which can offer rapid characterization but that can also be difficult to optimize ***(10)***. Polyacrylamide gel electrophoresis of fluorescently labeled glycans, a technique that requires no expensive instrumentation or specialized expertise, may also be used for screening or profiling. Although this is very useful for routine screening of a large number of samples, detailed characterization or sequencing of complex mixtures is problematic.

High-performance liquid chromatography (HPLC) techniques offer a compromise in that they give the capability of full characterization of complex glycan mixtures relatively quickly, and although they do require good instrumentation this is considera-

From: *The Protein Protocols Handbook, 2nd Edition*
Edited by: J. M. Walker © Humana Press Inc., Totowa, NJ

blly less expensive than mass spectrometry or NMR. HPLC is therefore the method of choice for routine analysis of glycosylation of protein, which have glycans of a type that have been previously characterized. Even when the glycan type is novel it is a technique that is compatible with others that can give further structural information. The development of HPLC has presented several problems, which are unique to glycan analysis. In particular they generally do not possess a strong chromophore or fluorophore for detection. In addition they may not be charged and different glycans may have very similar composition and physicochemical properties. Therefore, their study has required the development of techniques for their derivatization and also for their separation, which differ from those used for peptides.

When the glycans are released with a free reducing terminus, by either hydrazinolysis or endo-glycosidase, or endo-glycopeptidase enzymes, they may be derivatized by the relatively simple reaction of reductive animation *(11)*. It is desirable that the derivatization is nonselective to achieve quantitative analysis and it should not cause structural changes such as desialylation. The incorporation of tritium into the C1 position of the reducing terminal monosaccharide fulfills these requirements most effectively, and many studies have been performed by this technique *(12–14)*. This technique does, however, require the use of relatively large amounts or radioactivity and the use of scintillation counting for high sensitivity work. Fluorescent labels may also be introduced into the C1 position by similar reductive amidation reactions and a number of these have now been described including 2-amino pyridine *(15–17)*, 2-aminoacridone *(18)*, 3-(acetylamino)-6-aminoacridine (AA-Ac) *(19)*, 2-aminoanthanilic acid *(20,21)*, and 2-aminobenzoic acid (2-AB) *(20)*.

A number of chromatographic systems have been developed for separation including gas–liquid chromatography *(22)* size-exclusion chromatography on polyacylamide-based beads, notably the BioGel P4 series *(24,24)*, ion-exchange chromatography on standard matrices *(25)*, and the development of specialized matrices for anion-exchange (high-performance anion-exchange chromatography [(HPAEC]) *(26–29)*. They have been used in a large number of studies on protein glycosylation but these techniques have certain drawbacks. Low-pressure size-exclusion chromatography requires great care in column packing and the run times are long, HPAEC requires the use of high-pH and high-salt buffers that have to be removed before further analysis of separated glycans, such as exoglycosidase sequencing. The use of glycans labeled with a fluorescent tag such as 2-AB *(20)* and separated on suitable HPLC matrices *(30,31)* provides a convenient and sensitive means of profiling and sequencing of glycoprotein glycans *(30,32,33)* that can now be performed in many laboratories.

The analysis of glycans by this technique proceeds in a number of stages:

1. Release of glycans from the glycoprotein.
2. 2-AB labeling of the glycans.
3. Initial HPLC profiling.
4. Enzymatic sequencing of glycans.
5. Conformation of structures by mass spectrometry.

This approach is summarized in **Fig. 1**. The glycan release techniques may be adapted according to the source and purity of the proteins under study.

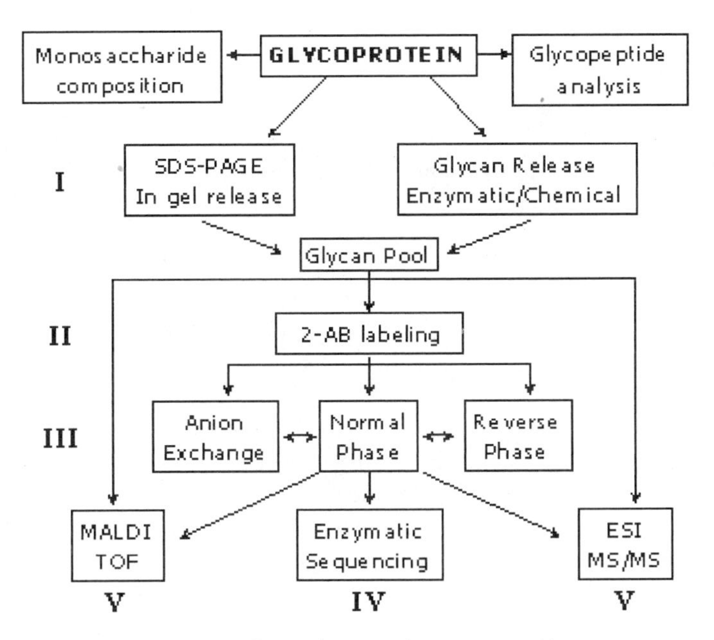

Fig. 1. The steps involved in complete analysis are shown diagrammatically. The Roman numerals relate to the stages given in the Introduction.

2. Materials

2.1. Equipment

1. HPLC system: High-pressure mixing two-solvent system capable of delivering 0.1–1.0 mL/min with shallow gradient (*see* **Note 1**).
2. Solvent degasser: Additional solvent degasser aids reproducibility.
3. Detector fluorescent detector capable of excitation at λ_{330nm} and emission at λ_{420nm} (*see* **Note 2**).
4. PC computer system for data analysis.
5. Data acquisition software, for example, Waters Millenium™.
6. Curve-fitting software, Microsoft Excel™ or Peak time (*see* **Note 3**).

2.2. Reagents

1. Anhydrous hydrazine was prepared by distillation from reagent grade hydrazine (Pierce Aldrich 21515-1) by mixture with calcium oxide and toluene as previously described ***(34)*** (*see* **Note 4**). This is available commercially from Glyko.
2. Acetic anhydride (ACS reagent, Sigma).
3. Anion-exchange resin: Bio-Rad AG50 X12.
4. Acetonitrile: HPLC grade with low background fluorescence such as E Chromosolv® (Reiedel-de-Haën, from Sigma).
5. Formic acid aristar grade (BDH).
6. 26% Extra pure ammonia solution (Reiedel-de-Haëen, from Sigma).
7. Water MilliQ or equivalent, subboiling point double distilled water (for mass spectrometric analysis).

8. Whatman no. 3 chromatography paper.
9. 2-Aminobenzamide labeling kit (*see* **Note 5**).
10. Pro-Mem Filter, 0.45 μm cellulose nitrate (R. B. Radley and Co. Ltd., Shire Hill, Saffron Walden, Essex, UK).
11. 3-([3-Cholamidopropyl)dimethylammonio)-1-propanesulfonate (CHAPS) detergent buffer: 50 m*M* Ammonium formate, pH 8.6; 1.29. w/v CHAPS; 0.1 *M* EDTA.

2.3. Enzymes

1. Sequencing grade enzymes were obtained from Glyko (Upper Heyford, UK) unless specified and were used with the buffers supplied as shown in **Table 1**. The following enzymes were used for exoglycosidase sequencing and given the abbreviations shown below along with their specificities.
   a. ABS α2–6 + α2,3 Specific sialidase (*Arthrobacter ureafaciens*).
   b. NDVS α2–3 Specific sialidase (Newcastle disease virus).
   c. BTG β1–3 (>β1,4 or 6) Specific galactosidase (bovine testes).
   d. SPG β1–4 Specific galactosidase (*Streptococcus pneumoniae*).
   e. AMF α1–3 Specific fucosidase (almond meal).
   f. BKF α1–6 (>α1,2 or 3 or 4) Specific fucosidase (bovine kidney).
   g. JBH β1–2,3, 4, or 6 Specific hexosaminidase (GlcNAc or GalNAc, Jack Bean).
   h. SPH β1,3 or 6 to Gal β1,2,3 or 6 to Man *N*-acetylglucosaminidase (*Streptococcus pneumoniae*).
   i. JBM α1–2, 1–3, 1–6 Specific mannosidase (Jack Bean).
   j. HPM β1–4 Specific mannosidase (*Helix pomatia*).

2.4. Standards

1. Partially hydrolyzed dextran (glucose oligomer standard) (Glyko, Upper Heyford, UK).
2. Arabinose oligomers (monomer to octamer are available from Dextra Labs, Reading, UK).
3. 2-AB labeled glycan standards, all available from Glyko (Upper Heyford, UK):
   a. Biantennary galactosylated.
   b. Biantennary galactosylated (with core 1,6-linked fucose).
   c. Biantennary galactosylated (with bisecting GlcNAc).
   d. Biantennary sialylated.
   e. Triantennary.
   f. Tetraantennary.
   g. Oligomannose (Man 5,6,7,8,9).
   h. Type 2 core *O*-glycan.

3. Methods

3.1. Glycan Release

3.1.1. Enzymatic Release with PNGaseF from Glycoproteins in Solution (Method from **ref. 35**)

1. PNGaseF solution at 1000 U/mL (Boehringer Mannheim).
2. Isolate the glycoprotein according to your usual procedures.
3. The sample should be relatively salt free and contain no extraneous carbohydrates (e.g., Sepharose-purified material contains large amounts of free carbohydrate that should be removed by dialysis using a 15–18,000 mol wt cutoff membrane)
4. If the volume of the glycoprotein solution required is >100 μL, dry the glycoprotein in a 1.5-mL microcentrifuge tube. Generally 50–200 μg of glycoprotein is required.

5. The incubation of a control glycoprotein with known glycosylation alongside experimental samples is recommended (*see* **Note 6**).
6. Proceed with the enzymatic digestion as described in **Subheading 3.1.2.**
7. Store remaining glycoprotein at 4°C for future use.
8. Dissolve sample in 50 μL of 50 m*M* ammonium formate, pH 8.6, 0.4% sodium dodecyl sulfate (SDS).
9. Incubate for 3 min at 100°C.
10. Cool and add 50 μL of CHAPS detergent buffer
11. Add 1 U of PNGaseF (1 μL).
12. Incubate for 24 h at 37°C (add 5 μL of toluene to prevent bacterial growth).
13. Remove 5 μL and analyze the reaction mixture by SDS-polyacrylamide gel electrophoresis (SDS-PAGE).
14. If sample is completely deglycosylated proceed with step otherwise continue with incubation (*see* **Note 8**).
15. Filter samples through protein binding membrane or perform gel filtration.
16. Dry sample in vacuum centrifuge.

3.1.2. Enzymatic Release with PNGaseF from Glycoprotein Bands in SDS-PAGE Gels

Suitable for analysis of low microgram amounts of protein or for unpurified proteins separated by SDS-PAGE or 2-dimensional electrophoresis. Method of Kuster et al. ***(36)***.

1. Destain Coomassie Blue stained gels in 30% methanol–7.5% acetic acid.
2. Cut out gel pieces with band of interest using a washed scalpel blade, keeping pieces as small as possible.
3. Put into 1.5-mL tubes and wash with 1 mL of 20 m*M* $NaHCO_3$, pH 7.0, twice using a rotating mixer.
4. Add 300 μL of $NaHCO_3$, pH 7.0.
5. Add 20 μL 45 m*M* dithiothreitol (DTT).
6. Incubate at 60°C for 30 min.
7. Cool to room temperature and add 20 μL of 100 m*M* iodoacetamide.
8. Incubate for 30 min at room temperature in the dark.
9. Add 5 mL of 1:1 acetonitrile–20 m*M* $NaHCO_3$, pH 7.0.
10. Incubate for 60 min to wash out reducing agents and SDS.
11. Cut gel into pieces of 1 mm^2.
12. Place in a vacuum centrifuge to dry.
13. Prepare PNGaseF solution at 1000 U/mL (Boehringer Mannheim): Add 30 U 30 μL of PNGaseF in 20 m*M* $NaHCO_3$, pH 7.0.
14. Allow gel to swell and then add a further 100-μL aliquot of buffer.
15. Incubate at 37°C for 12–16 h.

3.1.3. Release by Hydrazinolysis

Suitable for analysis of *N*- or *O*-linked glycans where the amount of protein is limited, where steric hindrance to enzymatic release is known, or where selective release of glycans by enzymatic means is suspected.

3.1.3.1. Preparation of Samples for Hydrazinolysis

1. Desalt and dry the samples completely as follows (*see* **Note 8**).
2. Dissolve the sample in 0.1% trifluorocetic acid (TFA) in as small a volume as possible.
3. Set up dialysis at 4°C.

4. Dialyze for a minimum of 48 h in a microdialysis apparatus fitted with a 6000–8000-Dalton cutoff membrane.
5. Recover sample from dialysis membrane. Wash membrane with 0.1% TFA to ensure recovery.
6. Transfer to a suitable tube for hydrazinolysis (*see* **Note 9**).
7. Lyophilize sample for 48 h.
8. For *O*-glycan analysis further drying is recommended (*see* **Note 10**).
9. Remove sample from lyophilizer immediately prior to addition of hydrazine.

3.1.3.2. Manual Hydrazinolysis Procedure

Suitable for analysis of *N*- and *O*-linked glycans when expertise and equipment for the procedure are available ***(34)***.

1. Remove tubes from drying immediately prior to hydrazine addition.
2. Flush tube with argon, taking care not to dislodge lyophilized protein.
3. Rinse dried syringe fitted with stainless steel needle with anhydrous hydrazine.
4. Take up fresh hydrazine and dispense onto sample. A 0.1-mL volume of hydrazine is sufficient to dissolve up to 2 mg of glycoprotein. For larger amounts add more hydrazine.
5. Seal the tube.
6. Gently shake tube—the protein should dissolve.
7. Place in an incubator (do not use water bath).
8. For release of *N*-linked glycans incubate at 95°C for 5 h; for *O*-glycan release incubate for 60°C for 6 h.
9. Allow to cool and remove hydrazine by evaporation.
10. Add 250 μL of toluene and evaporate. Repeat 5×.
11. Place tube on ice and add 100 μL of saturated sodium bicarbonate solution.
12. Add 20 μL of acetic anhydride.
13. Mix gently and leave at 4°C for 10 min.
14. Add a further 20 μL acetic anhydride.
15. Incubate at room temperature for 50 min.
16. Pass solution through a column of Dowex AG50 X12 (H^+ form)—0.5 mL bed volume.
17. Wash tube with 4 × 0.5 mL of water and pass through a Dowex column.
18. Evaporate solution to dryness. This should be done in stages by redissolving in decreasing volumes of water.
19. Prepare 50 × 2.5 cm strips of Whatman no. 1 chromatography paper (prewashed in water by descending chromatography for 2 d).
20. Spot sample on strip.
21. Perform descending paper chromatography for 2 d in 4:4:1 by vol butanol–ethanol–water for *N*-glycans and 8:2:1 1 by vol butanol–ethanol–water for *O*-glycans.
22. Remove strip from tank and allow all traces of solvent to evaporate.
23. Cut out region of strip from –2 cm to +5 cm from application.
24. Roll up cut chromatography paper and place in a 2.5-mL nonlubricated syringe.
25. Fit a 0.45 μm PTFE filter to syringe.
26. Add 0.5 mL of water and allow to soak into paper for 15 min.
27. Fit syringe plunger and force solution through filter.
28. Wash filter with water 4×, and pass through filter.
29. Evaporate sample to dryness, dissolve in 50 μL of water, transfer to microcentrifuge tubes, and store at –20°C until required.

3.2. Glycan Labeling

Fluorescent labeling with 2-aminobenzamide was performed as described by Bigge et al. *(20)* using the kit provided by Glyko (Upper Heyford, UK). The procedure is as follows:

1. Take an aqueous solution of glycan that should contain a minimum of 0.1 pmol and a maximum of 50 pmol of glycans. The volume should be no more than 50 µL.
2. Place in a 0.5-mL microcentrifuge tube and dry down in centrifugal evaporator.
3. Prepare labeling solvent by addition of 150 µL of glacial acetic acid to 100 µL of dimethyl sulfoxide (DMSO).
4. Add 100 µL of the acidified dimethyl sulfoxide solvent to 2-aminobenzamide to make a 0.25 *M* solution. Mix well to dissolve the 2-AB (may require gentle warming).
5. Add all this solution to sodium cyanoborohydride to make a 1.0 *M* solution.
6. Mix well for 5 min to dissolve the reductant.
7. Add 5 µL of the labeling reagent to the dried sample.
8. Mix well and centrifuge briefly.
9. Incubate at 65°C for 1 h and mix well. Incubate for a further 2 h.
10. Cool the labeling mixture on ice.
11. Remove free label by either technique given in **Subheadings 3.2.1.** and **3.2.2.**

3.2.1. Removal of Free 2-AB Label by Ascending Paper Chromatography

1. Cool the 2-AB reaction mixture in freezer.
2. Apply all 5 µL of sample to a point in center of strip 1 cm from the bottom.
3. Allow to dry for 2 h.
4. Perform ascending chromatography for 1 h in acetonitrile.
5. Examine strip under UV to see if the spot of free dye has migrated to the top of the strip.
6. If it is not at the top continue chromatography until it is.
7. Dry the paper completely.
8. Cut out the origin on the strip.
9. Place in a 2.5-mL syringe fitted with a 0.45-µm PTFE filter.
10. Apply 0.5 mL of water and leave for 15 min.
11. Push water through the filter.
12. Wash twice with a further 0.5 mL of water.
13. Dry down labeled sample in a vacuum centrifuge.

3.2.2. Removal of 2-AB Label on Filter Disks

1. Before use wash Whatman no. 1 filter paper in 500 mL of MilliQ water. Place filter paper in a beaker and add water. Leave for 15 min at room temperature. Decant water. Repeat this 4 times.
2. Dry paper in a 65°C over for 2 h. The paper may be stored at room temperature after this step.
3. Cut 6-mm discs from washed paper.
4. Place discs in a Bio-Rad disposable column. For a sample of < 10 µg of glycoprotein use two discs; for more use five discs).
5. Add 2 mL of water and leave for 5 min with columns capped.
6. Uncap columns and wash with another 4 × 2 mL of water (*see* **Note 11**).
7. Wash with a further 5 × 2 mL of 30% acetic in water (*see* **Note 11**).
8. Cap column and add 2 mL of acetonitrile.

9. Leave for 5 min and uncap the column.
10. Wash with a further 2 mL of acetonitrile just before applying sample; cap column.
11. Remove incubation vial and put in freezer for 5 min to cool down.
12. Centrifuge the tube and spot all the sample onto the disc.
13. Leave for 15 min and rinse tube with 100 μL of acetonitrile and add to disc. Leave for 5 min.
14. Add 2 mL of acetonitrile—uncap column and discard wash.
15. Wash with a further 4 × 2 mL of acetonitrile.
16. Place syringes (2.5-mL Fortuna) fitted with a 0.45-μm filter and stoppers, under the filter.
17. Elute glycans with 4 × 0.25 mL of water (*see* **Note 11**).
18. Dry sample down for analysis.

3.3. Normal Phase Chromatography

The following conditions are recommended:

1. Buffers:
   Ammonium formate: 50 m*M* formic acid adjusted to pH 4.4 with ammonia solution.
   Acetonitrile: HPLC grade; *see* **Subheading 2.2.**
2. Column: GlycoSep N™ column or TokoHas TSK-amide 80.
3. Gradient:

a. Startup method run time, 30 min

Time (min)	Flow (mL/min)	A (%)	B (%)
0	0	20	80
4	1	20	80
8	1	95	5
13	1	95	5
16	1	20	80
25	1	20	80
26	0.4	20	80
60	0.4	20	80
61	0	20	80

b. Separating method run time, 180 min

Time (min)	Flow (mL/min)	A (%)	B (%)
0	0.4	20	80
152	0.4	58	42
155	0.4	100	0
157	1	100	0
162	1	100	0
163	1	20	80
177	1	20	80
178	0.4	20	80
260	0	20	80

4. Sample loading:
   a. The sample should be loaded in 80% acetonitrile–20% water (v/v). In practice take 20 μL of the sample in water and add 80 μL of acetonitrile.
   b. Inject 95 μL of the sample.
   c. A standard dextran should be included with all sample runs.
   d. **Figure 2** shows a typical separation of human serum IgG with dextran ladder calibration.
5. Data analysis:
   a. The elution times of all peaks in the dextran ladder should be recorded.
   b. To assign glucose unit values a polynomial fit should be applied to the data to generate a standard curve. A third-order polynomial will generally give a good fit (*see* **Fig. 2**).
   c. The glucose unit values of sample peaks may then be calculated (*see also* **Note 3**).
   d. Published values of glucose unit values for a wide series of glycans may be used to give an indication of possible structures present. The most recent values are published in *Current Protocols on Protein Science Supp* 22 ***(37)***.

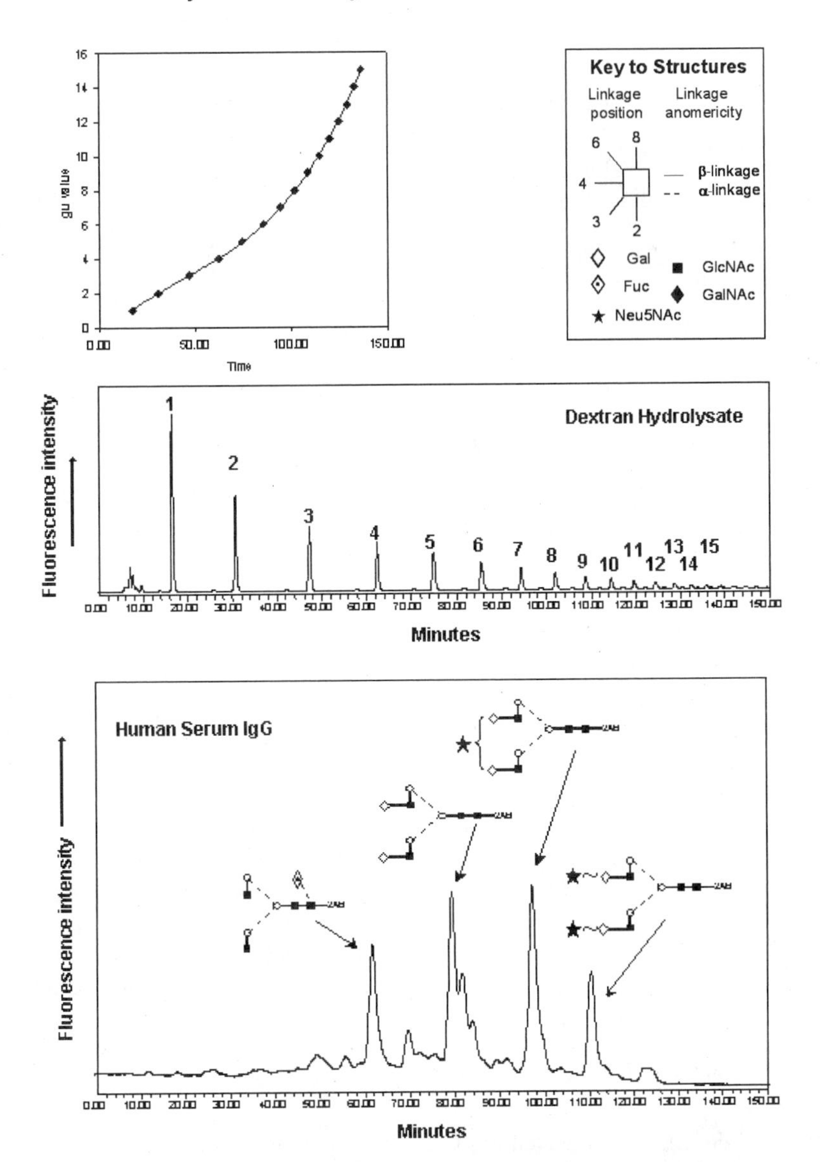

Fig. 2. Analysis of IgG N-glycans realeased by hydrazinolysis on normal phase HPLC. The *top panel* shows the calibration curve for conversion of retention time into glucose units based on a third order polynomial. The *middle panel* shows the separations of glucose oligomers from the dextran ladder. The *lower panel* shows the separation of the fluorescently labeled glycans from IgG with structures corresponding to the major peaks indicated.

e. Software for this purpose is currently being developed at the Oxford Glycobiology Institute (*see* **Note 3**).

3.4. Reverse-Phase Chromatography

1. Buffers:
   a. ammonium formate–triethylamine 50 m*M* formic acid adjusted to pH 5.0 with triethylamine.
   b. Acetonitrile: HPLC grade; *see* **Subheading 2.2.**

2. Column: Reverse-phase column GlycoSep™ R—Glyko (Upper Heyford, UK) or equivalent.

a. Startup method run time, 30 min

Time (min)	Flow (mL/min)	A (%)	B (%)
0	0.05	5	95
5	1	5	95
10	1	5	95
20	1	5	95
21	1	95	5
28	1	95	5
29	0.5	95	5
90	0.5	95	5
91	0	95	5

b. Separating method run time, 180 min

Time (min)	Flow (mL/min)	A (%)	B (%)
0	0.5	95	5
30	0.5	95	5
160	0.5	85	15
165	0.5	76	24
166	1.5	5	95
172	1.5	5	95
173	1.5	95	5
178	1.5	95	5
179	0.5	95	5
220	0.5	95	5
221	0	95	5

3. Sample loading:
   a. The sample should be loaded as a aqueous solution. A volume of 95 µL of sample should be loaded.
   b. A standard of an arabinose oligomers (available from Dextra Labs, Reading) should be included with all sample runs.
4. Data analysis:
   a. The data may be analyzed in a similar way to that of normal phase but using the values for the arabinose ladder.
   b. Typical values for the AU units for a number of structures are available *(37)*.
   c. It can often be helpful to compare normal and reverse-phase profiles for the same glycans as shown in **Fig. 3**. The different principles for separation are illustrated by the differences in separation on the two systems. By combining both types of analysis all four structures can readily be identified, for example, the separation of A2G2F and A2G2FB.

3.5. Weak Anion-Exchange Chromatography (WAX)

1. Buffers:
   a. 500 m*M* Formic acid adjusted to pH 9.0 with ammonia solution.
   b. Methanol water: 10:90 (v/v).
2. Gradient

a. Startup method run time, 30 min

Time (min)	Flow (mL/min)	A (%)	B (%)
0	0	0	0
5	1	5	95
20	1	5	95
21	1	95	5
28	1	95	5
29	0.5	95	5
90	0.5	95	5
91	0	95	5

b. Separating method run time, 180 min

Time (min)	Flow (mL/min)	A (%)	B (%)
0	1	0	100
12	1	5	95
50	1	80	20
55	1	100	0
65	1	100	0
66	2	0	100
77	2	0	100
78	1	0	100

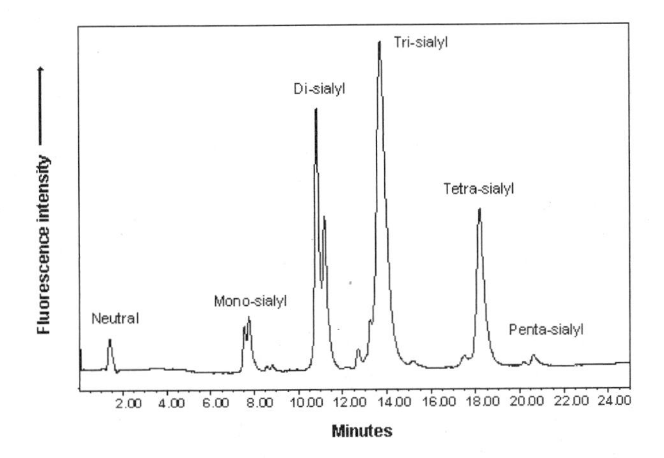

Fig. 3. Analysis of fetuin on Weak Anion Exchange Chromatography. The separation of glycans possessing from 1 to 5 sialic acids is shown.

3. Sample loading:
   a. The sample should be loaded as an aqueous solution. A volume of 95 μL of sample should be loaded.
   b. No suitable standards are currently available although it is useful to run a well characterized glycoprotein such as fetuin.
   c. The glycans generally elute on the basis of charge although the size of the glycan also contributes to the position of elution.
   d. A typical separation of charged *N*-glycans from bovine serum fetuin is shown in **Fig. 3**.

3.6. Exoglycosidase Digestion of N-Glycans

Digestions are generally performed on the pool of glycans by the application of exoglycosidases under the incubation conditions shown in **Table 1** in a series of enzyme arrays of increasing complexity as shown in **Table 2**.

1. The amount required for each digestion can be judged from the previous profiling run, but in general 100 fmol of glycan is detectable. In practice more may be used with the enzyme concentrations given below if sufficient material is available.
2. Pipet no more than 50 μL of solution into a 0.5-mL microcentrifuge tube.
3. Evaporate glycan solutions to dryness in a vacuum centrifuge.
4. Add 2–5 μL of enzyme solution and 2 μL of incubation buffer.
5. Make the total volume up to 10 μL with water.
6. Incubate for 16–24 h at 37°C.
7. Cool and load the digestion mixture onto a protein binding filter (Microspin 45).
8. Leave for 15 min at room temperature.
9. Centrifuge for 15 min at 5000*g*.
10. Wash tube with 10 μL of freshly prepared 5% acetonitrile in water and load onto filter.
11. Leave for 15 min at room temperature.
12. Centrifuge for 15 min at 5000*g*.
13. Repeat **steps 10–12**.
14. Twenty microliters of sample may be directly analyzed by HPLC if sufficient material is available; if not, then dry down sample and redissolve in 20 μL of water.
15. An example of digestion of *N*-glycans from IgG is shown in **Fig. 4**.

Table 1
Incubation Conditions for Exoglycosidase Enzymes

Enzyme Code	Specificity	Reconstitution	Percent of final volume	Final enzyme concentration	Final buffer concentration	PH optimum
ABS	Sialic acid α2–6, 2–3 >2.8	0.2 U + 20 µL of water	10	1 mU/µL	100 m*M* Sodium acetate	5
NDVS	Sialic acid α2–3	20 mU + 10 µL of water	10	200 µU/µL	50 m*M* Sodium acetate	5.5
BTG	Gal β1–3, 1,4 > 1–6	As supplied	20	1 mU/µL	100 m*M* Sodium citrate/phosphate	5
SPG	Gal β1–4	40 mU + 100 µL of water	20	80 µU/µL	100 m*M* Sodium acetate	6
AMF	Fuc α1–3, 1–4	20 µU + 2 µL of water	10	1 µ/µL	50 m*M* Sodium acetate	5
BKF	Fuc α1–6 (>1–2, 1–3, 1–4)	0.1 U + 10 µL of water	10	1 mU/µL	100 m*M* Sodium citrate	6
JBH	GlcNAc, GalNAc β1–2, 1–3, 1–4, 1–6	As supplied	20	10 mU/µL	100 m*M* Sodium citrate/phosphate	5
SPH	GlcNAc β1–2 Man>>GlcNAcβ1–3Gal>GlcNAc1–6Gal	30 mU + 100 µL of water	40	120 mU/µL	100 m*M* Sodium citrate/phosphate	5
JBM (High)	Man α1–2, 1–3, 1–6	2 U in 30 µL of buffer	100	67 mU/µL	100 m*M* Sodium acetate, 2 m*M* zinc	5
JBM (Low)	Man α1–6	2 U in 30 µL of buffer	8	5.4 mU/µL	100 m*M* sodium acetate, 2 m*M* ainc	5
HPM	Man β1–4 GlcNAc	0.2 U in 10 µL of water	10	2 mU/µL	100 m*M* Sodium citrate/phosphate	4

Table 2
Typical Arrays for *N*-Glycan Sequencing

Array	Vol enzyme	Vol buffer	Water
ABS	1	2	7
ABS BTG	1 2	2	5
ABS BTG BKF	1 2 1	2	4
ABS BTG BKF AMF	1 2 1 1	2	3
ABS BTG BKF AMF JBH	1 2 1 1 2	2	1
JBM (high)	10		
JBM (low)	0.8		9.2

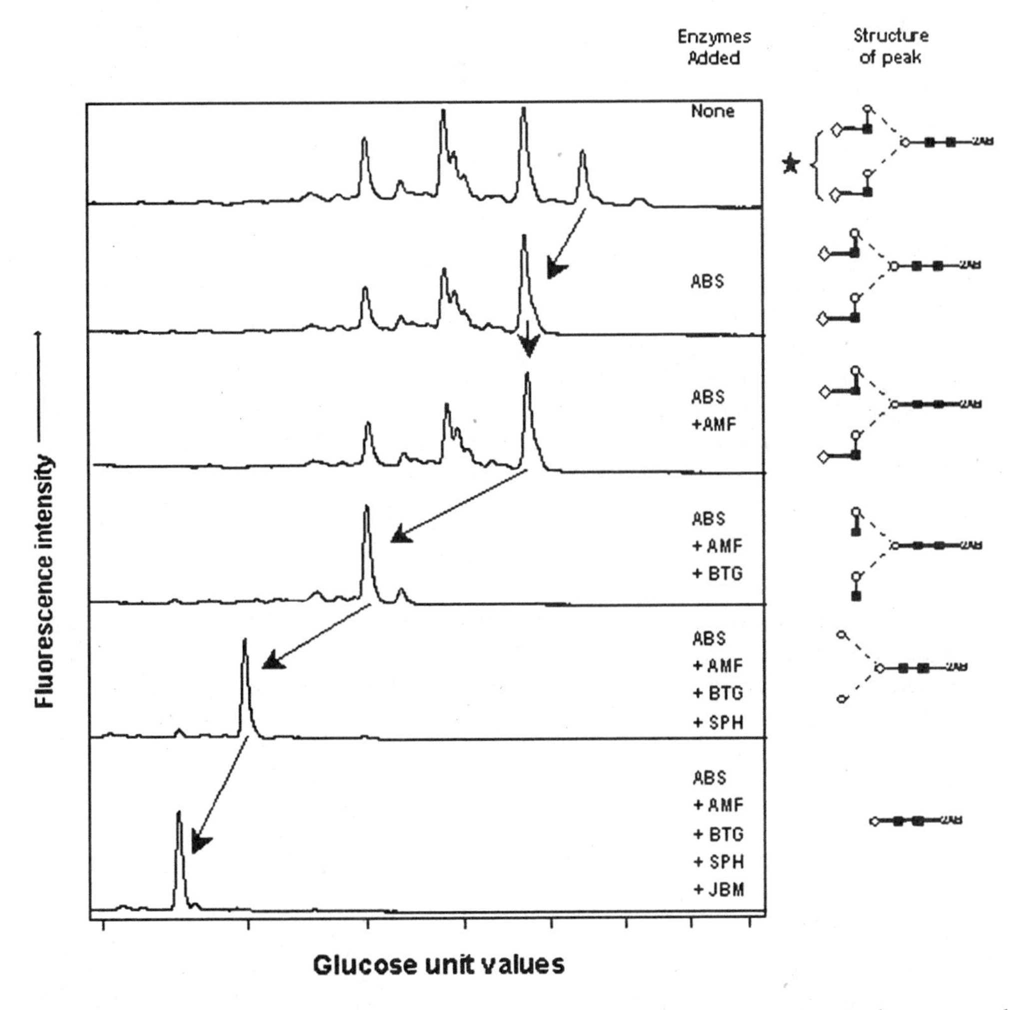

Fig. 4. Exoglycosidase digestion of IgG glycans monitored on normal phase column normal phase HPLC. Exoglycosidase abbreviations ABS - α2,6 + α2,3 specific sialidase (*Arthrobacter Ureafaciens*), BTG β1,2 (+β 1–4) specific galactosidase (Bovine Testes), AMF-α2,3 specific fucosidase (Almond Meal) SPH-β1,3 (4,6) specific hexosaminidase (*Streptococcus Pneumonae*) JBM-α-mannosidase (Jack Bean).

3.7. Exoglycosidase Digestion of O-Glycans

Digestions are generally performed on the pool of glycans; however, the strategy differs from that for *N*-glycans. Digestions should be performed by the following mixtures of glycans where the choice of glycosidases is dictated by the outcome of previous digestions. Since several *O*-glycans may coelute on any matrix the analysis on both normal and reverse phase is recommended. A detailed description of the procedure and reference values may be found in Royle et al. (*Submitted to Analyt. Biochem.*).

1. The samples are prepared for digestion as described in **steps 1–3** (*see* **Subheading 3.6.**).
2. First perform ABS digestion to detect presence of sialic acid (in any linkage).
3. Perform ABS + BTG digestion to show the presence of galactose.
4. Perform ABS + BKF digestion to show the presence of fucose.
5. Perform ABS + JBH digestion to show presence of *N*-acetylglucosamine or *N*-acetylgalactosamine.
6. In each case digestion will be apparent from a shift in peaks. If the digests are analyzed by normal phase and reverse phase then possible identities of structures may be found from the GU and AU values.
7. If digestion with ABS occurs then also digest pool with NDVS to determine if sialic acid linkage is 2,3.
8. If digestion with ABS + BTG occurs then digest with ABS + SPG to determine if linkage of galactose is 1,4.
9. If digestion occurs with ABS + BKF then digest with ABS + AMF to determine if fucose linkage is 1,3 or 1,4.
10. If digestion with ABS + JBH occurs then digest with ABS + SPH to determine if *N*-acetylglucosamine or *N*-acetylgalactosamine is present.
11. Note that the complete characterization may require repeated digestions as there is no common core sequence as found for *N*-glycans.
12. An example of *O*-glycan sequencing is shown **Fig. 5**.

3.7.1. Data Analysis

1. The movement of a peak in an exoglycosidase incubation indicates that it contains the monosaccharides in the linkages removed by that enzyme.
2. A shift in the position of peaks will be seen with decreasing complexity of the chromatogram with increase in the number of enzymes applied.
3. Reference to the last chromatogram in a series will allow identification of the basic core *N*-glycan structure as shown in **Fig. 4**.
4. The last digest in a series will indicate the core glycans present.
5. Working back through the digests, the structures from which a glycan has been removed are identified.
   a. In the example the presence of terminal *N*-acetylglucosamine in the structure is shown comparing digest 5 to 4.
   b. The presence of terminal galactose is shown by comparing digest 3 to 4.
   c. The presence of sialic acid is shown by comparing digest 3 to 1.
   d. In this way the structure shown can be assigned to peak A.
6. The presence of other peaks indicates that monosaccharides not covered by the array are present (such as oligomannose) or that residues are substituted by groups such as sulfate or phosphate.
7. An examples of *O*-glycan sequencing is shown in **Fig. 5** with sequential application of glycosidases.

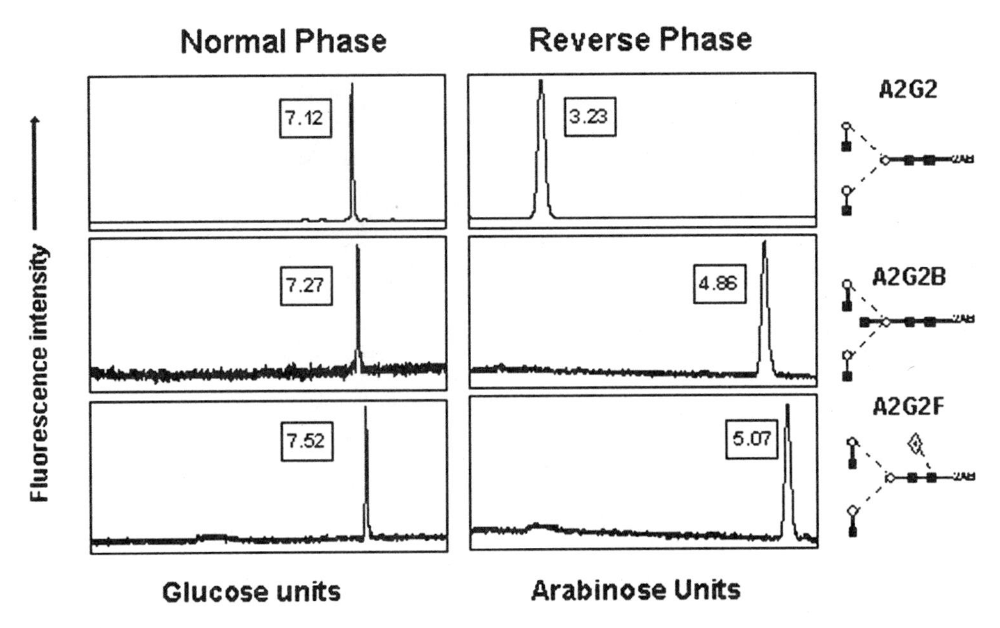

Fig. 5. Analysis of bi-anennary, fucosylated, bisected glycans by reversse phase HPLC – comparison with normal phase HPLC showing complimentarily of separation techniques.

3.8. Mass Spectrometry

Samples may be further analyzed by mass spectrometry (MALDI-TOF or ESI/MS/MS) for confirmation of structures but generally require further desalting.

1. It is often useful to perform the MS analysis on the glycan pool alongside the HPLC analysis but approx 10× more material is generally required.
2. As all solvents used are volatile, fractions containing peaks from the HPLC run may be collected and dried in a vacuum centrifuge but cleanup may be required (*see* **Note 12**).

4. Notes

1. Waters System is recommended. Hewlett-Packard is also suitable. System with similar specification from other manufacturers may be suitable but must be capable of high-pressure mixing of acetonitrile and aqueous solvents.
2. Jasco or Waters is recommended but other detectors of similar specificity may be suitable.
3. PeakTime is a program under development by Dr. Ed Hart at the Oxford Glycobiology Institute, which computes the GU or AU values and then compares them to those of standards. It will also analyze the products of enzymatic digestions for sequence analysis.
4. Great care should be taking in the handling of hydrazine. Dry hydrazine of suitable quality may be obtained from Glyko or from Ludger Ltd. (Oxford, UK) .
5. Availible from Glyko Inc.
6. Suitable glycoproteins include ribonuclease B, bovine serum fetuin, and human serum haptoglobulin.
7. The change in position of a band on SDS-PAGE indicates *N*-glycosylation. In some cases incomplete digestion may lead to the production of a series of bands corresponding to partially deglycosylated glycoprotein. Complete release is indicated by conversion to a single band at the lowest molecular weight.

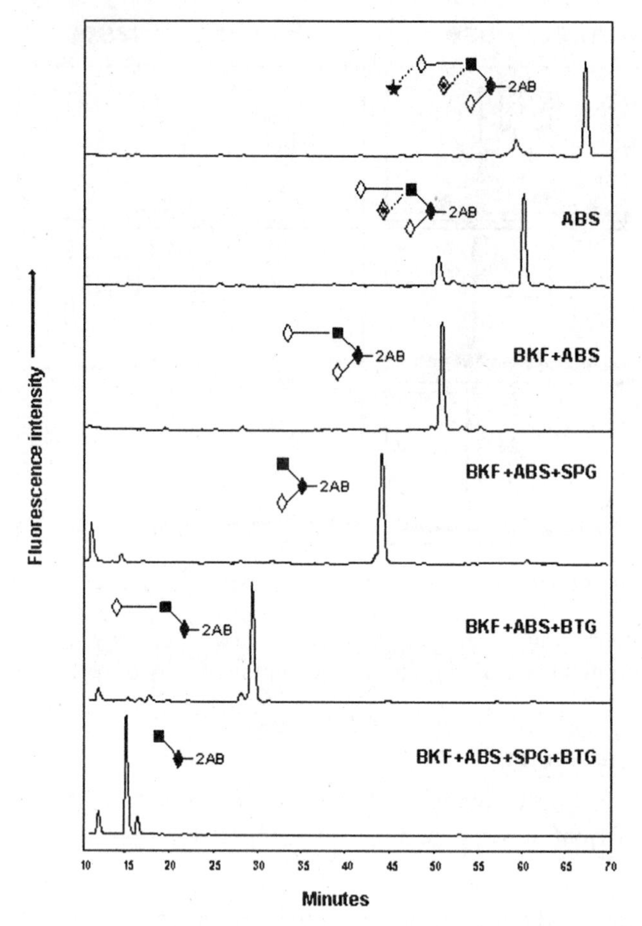

Fig. 6. Example of exoglycosidase sequencing an *O*-glycan on normal phase HPLC. SPG - β1,4 specific galactosidase (*Streptococcus pneumoniae*), other exoglycosidse abbreviations as in **Fig. 4**.

8. The presence of most salts, dyes, or detergents will interfere with hydrazinolysis. The sample must also be completely anhydrous.
9. All glassware used should be soaked in 4 *M* nitric acid for 4 h and thoroughly washed in water before use.
10. The use of cryogenic drying by placing under high vacuum connected to a trap of charcoal submerged in liquid nitrogen for a period of 72 h as described by Ashford et al. ***(34)*** is recommended.
11. It may be necessary to apply slight pressure to the top of the column or to use a suitable vacuum apparatus.
12. Double-distilled (subboiling point) water should be used for mass spectrometric analysis. Suitable techniques for post-HPLC cleanup and for mass spectrometry by electrospray ***(8)*** and by MALDI ***(9)*** have been described recently.

Acknowledgments

The advice and help of the members of the Glycoimmunology group at the Oxford Glycobiology Institute, Director Professor R. A. Dwek and Group Leader Dr. Pauline Rudd, is gratefully acknowledged. In particular, I would like to thank Dr. Louise Royle for her valued help and comments. Thanks also to Max Crispin for data in **Fig. 4** and for his helpful comments on the manuscript.

The techniques described here are the result of many years of development by current and previous group members, and Dr. Geoffrey Guile and Dr. Taj Mattu were instrumental in the development of the protocols described in this chapter.

References

1. Dell, A., Thomas Oates, J. E., Rogers, M. E., and Tiller, P. R. (1988) Novel fast atom bombardment mass spectrometric procedures for glycoprotein analysis. *Biochimie* **70,** 1435–1444.
2. Carr, S. A. and Roberts, G. D. (1986) Carbohydrate mapping by mass spectrometry: a novel method for identifying attachment sites of Asn-linked sugars in glycoproteins. *Analyt. Biochem.* **157,** 396–406.
3. Karlsson, H., Karlsson, N., and Hansson, G. C. (1994) High-temperature gas chromatography and gas chromatography-mass spectrometry of glycoprotein and glycosphingolipid oligosaccharides. *Mol. Biotechnol.* **1,** 165–180.
4. Anumula, K. R. and Taylor, P. B. (1992) A comprehensive procedure for preparation of partially methylated alditol acetates from glycoprotein carbohydrates. *Analyt. Biochem.* **203,** 101–108.
5. Cumming, D. A., Hellerqvist, C. G., Harris Brandts, M., Michnick, S. W., Carver, J. P., and Bendiak, B. (1989) Structures of asparagine-linked oligosaccharides of the glycoprotein fetuin having sialic acid linked to *N*-acetylglucosamine. *Biochemistry* **28,** 6500–6512.
6. Edge, C. J., Rademacher, T. W., Wormald, M. R., Parekh, R. B., Butters, T. D., Wing, D. R., and Dwek, R. A. (1992) Fast sequencing of oligosaccharides: the reagent-array analysis method. *Proc. Natl. Acad. Sci. USA* **89,** 6338–6342.
7. Dwek, R. A., Edge, C. J., Harvey, D. J., Wormald, M. R., and Parekh, R. B. (1993) Analysis of glycoprotein-associated oligosaccharides. *Annu. Rev. Biochem.* **62,** 65–100.
8. Harvey, D. J. (1999) Matrix-assisted laser desorption/ionization mass spectrometry of carbohydrates. *Mass Spectrom. Rev.* **18,** 349–450.
9. Harvey, D. J. (2000) Electrospray mass spectrometry and fragmentation of *N*-linked carbohydrates derivatized at the reducing terminus. *J. Am. Soc. Mass Spectrom.* **11,** 900–915.
10. Taverna, M., Tran, N. T., Merry, T., Horvath, E., and Ferrier, D. (1998) Electrophoretic methods for process monitoring and the quality assessment of recombinant glycoproteins. *Electrophoresis* **19,** 2572–2594.
11. Hase, S., Ikenaka, T., and Matsushima, Y. (1979) Analyses of oligosaccharides by tagging the reducing end with a fluorescent compound. I. Application to glycoproteins. *J. Biochem. Tokyo* **85,** 989–994.
12. Takasaki, S. and Kobata, A. (1978) Microdetermination of sugar composition by radioisotope labeling. *Meth. Enzymol.* **50,** 50–54.
13. Takasaki, S., Mizuochi, T., and Kobata, A. (1982) Hydrazinolysis of asparagine-linked sugar chains to produce free oligosaccharides. *Meth. Enzymol.* **83,** 263–268.
14. Endo, T., Amano, J., Berger, E. G., and Kobata, A. (1986) Structure identification of the complex-type, asparagine-linked sugar chains of beta D-galactosyl-transferase purified from human milk. *Carbohydr. Res.* **150,** 241–263.
15. Takahashi, N., Hitotsuya, H., Hanzawa, H., Arata, Y., and Kurihara, Y. (1990) Structural study of asparagine-linked oligosaccharide moiety of taste-modifying protein, miraculin. *J. Biol. Chem.* **265,** 7793–7798.
16. Nakagawa, H., Kawamura, Y., Kato, K., Shimada, I., Arata, Y., and Takahashi, N. (1995) Identification of neutral and sialyl *N*-linked oligosaccharide structures. *Analyt. Biochem.* **226,** 130–138.

17. Rice, K. G., Takahashi, N., Namiki, Y., Tran, A. D., Lisi, P. J., and Lee, Y. C. (1992) Quantitative mapping of the *N*-linked sialyloligosaccharides of recombinant erythropoietin: combination of direct high-performance anion-exchange chromatography and 2-aminopyridine derivatization. *Analyt. Biochem.* **206,** 278–287.
18. Camilleri, P., Harland, G. B., and Okafo, G. (1995) High resolution and rapid analysis of branched oligosaccharides by capillary electrophoresis. *Analyt. Biochem.* **230,** 115–122.
19. Charlwood, J., Birrell, H., Gribble, A., Burdes, V., Tolson, D., and Camilleri, P. (2000) A probe for the versatile analysis and characterization of *N*-linked oligosaccharides. *Analyt. Chem.* **72,** 1453–1461.
20. Bigge, J. C., Patel, T. P., Bruce, J. A., Goulding, P. N., Charles, S. M., and Parekh, R. B. (1995) Nonselective and efficient fluorescent labeling of glycans using 2-amino benzamide and anthranilic acid. *Analyt. Biochem.* **230,** 229–238.
21. Anumula, K. R. and Dhume, S. T. (1998) High resolution and high sensitivity methods for oligosaccharide mapping. *Glycobiology* **8,** 685–694.
22. Endo, Y., Yamashita, K., Han, Y. N., Iwanaga, S., and Kobata, A. (1977) The carbohydrate structure of a glycopeptide release by the action of plasma kallikrein on bovine plasma high-molecular-weight kininogen. *J. Biochem. Tokyo* **82,** 545–550.
23. Kobata, A., Yamashita, K., and Tachibana, Y. (1978) Oligosaccharides from human milk. *Meth. Enzymol.* **50,** 216–220.
24. Kobata, A., Yamashita, K., and Takasaki, S. (1987) BioGel P-4 column chromatography of oligosaccharides: effective size of oligosaccharides expressed in glucose units. *Meth. Enzymol.* **138,** 84–94.
25. Takamoto, M., Endo, T., Isemura, M., Kochibe, N., and Kobata, A. (1989) Structures of asparagine-linked oligosaccharides of human placental fibronectin. *J. Biochem. Tokyo* **105,** 742–750.
26. Field, M., Papac, D., and Jones, A. (1996) The use of high-performance anion-exchange chromatography and matrix-assisted laser desorption/ionization time-of-flight mass spectrometry to monitor and identify oligosaccharide degradation. *Analyt. Biochem.* **239,** 92–98.
27. Townsend, R. R. and Hardy, M. R. (1991) Analysis of glycoprotein oligosaccharides using high-pH anion exchange chromatography. *Glycobiology* **1,** 139–147.
28. Martens, D. A. and Frankenberger, W. T., Jr. (1991) Determination of saccharides in biological materials by high-performance anion-exchange chromatography with pulsed amperometric detection. *J. Chromatogr.* **546,** 297–309.
29. Rohrer, J. S. and Avdalovic, N. (1996) Separation of human serum transferring isoforms by high-performance pellicular anion-exchange chromatography. *Protein Exp. Purif.* **7,** 39–44.
30. Guile, G. R., Rudd, P. M., Wing, D. R., Prime, S. B., and Dwek, R. A. (1996) A rapid high-resolution high-performance liquid chromatographic method for separating glycan mixtures and analyzing oligosaccharide profiles. *Analyt. Biochem.* **240,** 210–226.
31. Guile, G. R., Harvey, D. J., O'Donnell, N., Powell, A. K., Hunter, A. P., Zamze, S., et al. (1998) Identification of highly fucosylated *N*-linked oligosaccharides from the human parotid gland. *Eur. J. Biochem.* **258,** 623–656.
32. Rudd, P. M., Mattu, T. S., Zitzmann, N., Mehta, A., Colominas, C., Hart, E., et al. (1999) Glycoproteins: rapid sequencing technology for *N*-linked and GPI anchor glycans. *Biotechnol. Genet. Eng. Rev.* **16,** 1–21.
33. Rudd, P. M., Guile, G. R., Kuster, B., Harvey, D. J., Opdenakker, G., and Dwek, R. A. (1997) Oligosaccharide sequencing technology. *Nature* **388,** 205–7.
34. Ashford, D., Dwek, R. A., Welply, J. K., Amatayakul, S., Homans, S. W., Lis, H., et al. (1987) The beta 1→2-D-xylose and alpha 1→3-L-fucose substituted *N*-linked oligosaccha-

rides from *Erythrina cristagalli* lectin. Isolation, characterisation and comparison with other legume lectins. *Eur. J. Biochem.* **166,** 311–320.

35. Goodarzi, M. T. and Turner, G. A. (1998) Reproducible and sensitive determination of charged oligosaccharides from haptoglobin by PNGase F digestion and HPAEC/PAD analysis: glycan composition varies with disease. *Glycoconj. J.* **15,** 469–475.
36. Kuster, B., Hunter, A. P., Wheeler, S. F., Dwek, R. A., and Harvey, D. J. (1998) Structural determination of *N*-linked carbohydrates by matrix-assisted laser desorption/ionization-mass spectrometry following enzymatic release within sodium dodecyl sulphate-polyacrylamide electrophoresis gels: application to species-specific glycosylation of alpha1-acid glycoprotein. *Electrophoresis* **19,** 1950–1959.
37. Rudd, P. M. and Dwek, R. A. (2001) *Determining the Structures of Oligosaccharides* N- *and* O-*Linked to Glycoproteins*, Vol. 2 Suppl. 22, John Wiley & Sons, New York.

124

Glycoprofiling Purified Glycoproteins Using Surface Plasmon Resonance

Angeliki Fotinopoulou and Graham A. Turner

1. Introduction

1.1. Glycosylation and Its Investigation

The carbohydrate part of glycoproteins can define several of their biological properties, including the clearance rate, immunogenicity, thermal stability, solubility, the specific activity, and conformation *(1)*. Often, small differences in the composition of the sugar side chains of a glycoprotein can affect the biological properties *(2)*. Characterization of oligosaccharide structures on glycoproteins is essential for glycoprotein therapeutic products, because inflammatory and systemic responses may arise if glycosylation of the product is different from that of the native substance.

Determination of the precise oligosaccharide profile of a glycoprotein, however, cannot yet be considered as a routine laboratory task. Methods for glycosylation analysis can be generally divided in two major categories, direct methods and methods using lectins. This chapter describes a recently developed method that measures the binding of lectins to glycoproteins using surface plasmon resonance (SPR). Lectins are a class of proteins that bind to carbohydrates in a noncovalent reversible way that does not chemically modify the sugar molecule *(3)*. The lectin specificity is not absolute, but usually there is one carbohydrate grouping to which a lectin binds with much higher affinity than to other carbohydrate structures. This property makes them ideal for recognizing oligosaccharide structures. Although lectin methods are indirect, they can be used when comparisons are needed. They provide many advantages compared to other methods for investigating carbohydrate structures. There are a large number of lectins available with different and distinct specificities. Lectin methods are also quick and simple *(4)*.

1.2. Surface Plasmon Resonance and Lectins

SPR is an optical sensing phenomenon that allows one to monitor biomolecule interactions in real time (**Fig. 1**). The sensor device is composed of a sensor chip consisting of three layers (glass, a thin gold film, and a carboxymethylated dextran matrix); a prism placed on the glass surface of the chip; and a microfluidic cartridge, which controls the delivery of liquid to the sensor chip surface *(5)*. When light illuminates the thin gold

From: *The Protein Protocols Handbook, 2nd Edition*
Edited by: J. M. Walker © Humana Press Inc., Totowa, NJ

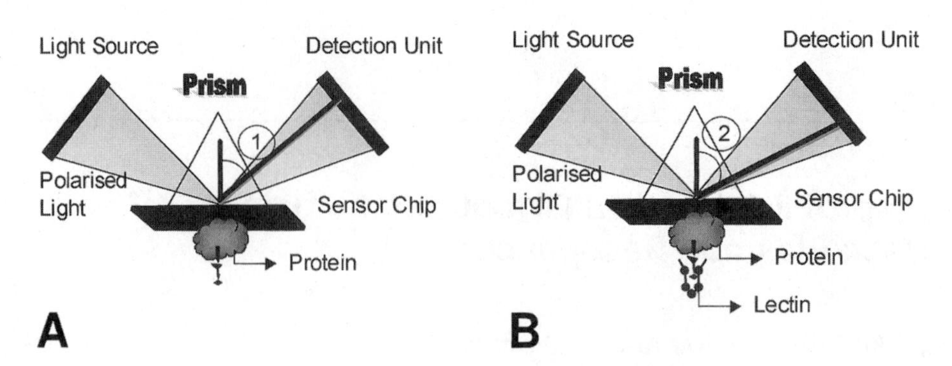

Fig. 1. SPR detects changes in the angle of nonreflectance. This angle changes if the refractive index of the chip surface changes, for example, when a protein is immobilized on the chip. **(A)** A glycoprotein is immobilized resulting in *angle 1*; **(B)** a lectin is bound to the immobilized glycoprotein resulting in *angle 2*.

film, energy is transferred to the electrons in the metal surface causing the reflected light to have reduced intensity at a specific incident angle. This angle of nonreflectance changes as the refractive index in the vicinity of the metal surface changes. The refractive index depends upon the mass on the surface of the gold film. A response of 1000 resonance units (RUs) corresponds to a shift of 0.1° in the resonance angle and represents a change in the surface protein concentration of about 1 ng/mm^2 ***(6)***.

Because all proteins, independent of sequence, contribute the same refractive index, SPR can be used as a mass detector. A glycoprotein, therefore, can be immobilized onto a sensor chip surface and then be probed by a panel of lectins. The binding or nonbinding of the lectins provides information about the oligosaccharide structures found on the carbohydrate chains. For example, if a glycoprotein has a trimannose core on its *N*-linked chains, it will bind to the lectin concanavalin A (Con A), or if it has terminal α2–3 or α2-6 *N*-acetylneuraminic acid (Neu5NAc) on its oligosaccharide chains it will bind to the Neu5NAc specific lectins *Maackia amurensis* agglutinin (MAA) or *Sambucus nigra* agglutinin (SNA), respectively.

Lectin binding can also be performed after treatment with glycosidases to gain more information about the oligosaccharide structures present. By sequentially treating the immobilized glycoprotein with glycosidases and measuring the binding of a panel of lectins after every treatment it is possible to gain information on the oligosaccharide sequence.

Some lectins will bind to the immobilized glycoprotein only if they are pretreated with sialidase to remove the terminal Neu5NAc. An example of this is peanut agglutinin (PNA), which binds to Galactose β1–3 *N*-acetylgalactosamine groupings on *O*-linked chains after the removal of Neu5NAc.

SPR has been used for investigating the kinetics of the interaction between oligosaccharides and lectins ***(7–9)***, for characterization of fetuin glycosylation ***(10)***, for the determination of agalactoIgG in rheumatoid arthritis patients ***(11)***, and recently, we have developed a method for investigating the glycosylation of recombinant glycoproteins ***(12)***.

2. Materials

2.1. Reagents

1. 0.05 *M* *N*-Hydroxysuccinimide (NHS).
2. 0.2 *M* *N*-ethyl-*N*'(dimethylaminopropyl)-carbodiimide (EDC).
3. 1 *M* Ethanolamine hydrochloride, pH 8.5.
4. 0.1 *M* HCl.
5. Running buffer (HBS): 10 m*M* 2-[4-(2-hydroxyethyl)-1-piperazinyl] ethanesulfonic acid (HEPES), pH 7.4, 150 m*M* NaCl, 3.4 m*M* ethylenediaminetetraacetic acid (EDTA), 0.0005% Surfactant P20.
6. 10 m*M* Sodium acetate, pH 4.0.
7. 10 m*M* Sodium acetate, pH 4.5.
8. 10 m*M* Sodium acetate, pH 5.0.
9. 10 m*M* Sodium acetate, pH 5.5.
10. Lectin buffer: 10 m*M* sodium acetate, pH 5.0 with cations 2 m*M* $MgCl_2$, $MnCl_2$, $ZnCl_2$, $CaCl_2$.
11. Lectins used: *MAA, SNA, Datura stramonium Agglutinin (DSA), ConA, PNA, aleuria aurantia Agglutinin (AAA).*

2.2. Equipment

BIAcore 1000 apparatus (Biacore AB, Uppsala) is fully automated and controlled by the manufacturer's software. During the operation of the equipment the output from the chip is displayed on VDU as resonance units (RUs) vs time.

3. Method

The procedure used for lectin/SPR can be divided into four steps. The first three are essential for any SPR experiments and the last one is optional.

1. "Preconcentration" step.
2. Ligand immobilization step.
3. Lectin binding.
4. Enzyme treatment.

3.1. Preconcentration

Before immobilizing a ligand to the chip surface a procedure called "preconcentration" needs to be performed. The latter step is important for efficient chemical immobilization of the ligand. It is accomplished by passing the ligand over the chip surface and utilizing the electrostatic attraction between the negative charges on the surface matrix (carboxymethyl dextran on a CM chip) and positive charges on the ligand at pH values below the ligand pI. Preconcentration is performed at 25°C. The ligand is injected at different pHs and the pH that provides the steepest curve is used (*see* **Note 1**). The detailed sequence of steps is as follows:

1. A continuous flow of HBS buffer is applied at flow rate of 10 mL/min.
2. A 2-min pulse of ligand in 10 m*M* sodium acetate, pH 4.0, is applied.
3. A 2-min pulse of ligand in 10 m*M* sodium acetate, pH 4.5, follows.

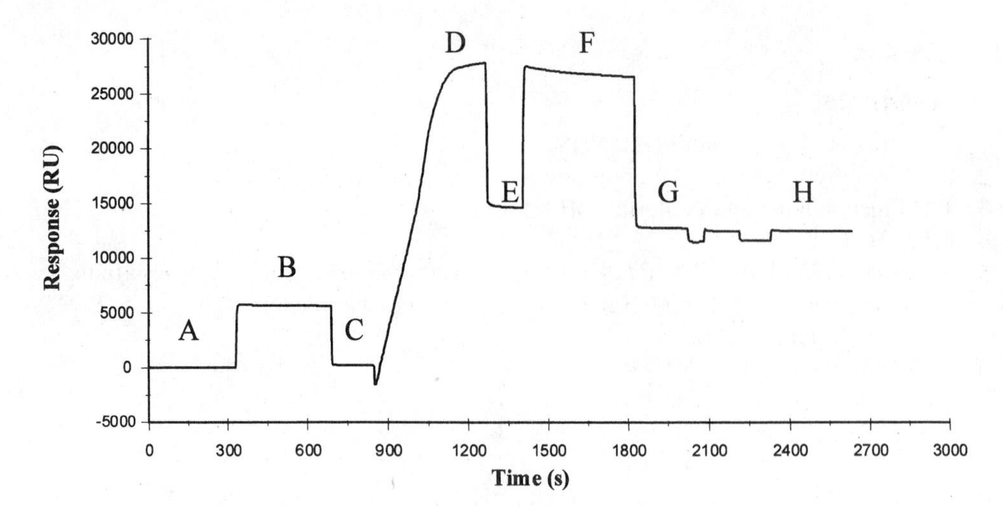

Fig. 2. **(A)** Baseline during continuous buffer flow; **(B)** injection of NHS/EDC to activate the surface; **(C)** baseline after activation; **(D)** injection of ligand; **(E)** immobilized ligand before deactivation; **(F)** deactivation of chip surface using ethanolamine; **(G)** immobilized ligand after deactivation; **(H)** final immobilization level after two HCl washes.

4. A 2-min pulse of ligand in 10 m*M* sodium acetate, pH 5.0, follows.
5. A 2-min pulse of ligand in 10 m*M* sodium acetate, pH 5.5, follows.
6. Two 2-min washes with 0.1 *M* HCl follow for regeneration.

3.2. Glycoprotein Immobilization

Choosing the correct matrix for immobilization depends on properties of the glycoprotein to be used and the functional groups on the chip surface. There are various types of chips available, which utilize different immobilization chemistries. Amine coupling is used for neutral and basic proteins ***(13)***. This introduces *N*-hydroxysuccinimide esters into the surface matrix by modifying the carboxymethyl groups of the matrix with a mixture of NHS and EDC. These esters react with the amines and other nucleophilic groups on the ligand to form covalent links.

Optimization of the amount of immobilized ligand is an important factor in SPR measurements and different buffers, ligand concentrations and injection times must be investigated. This procedure is performed at 25°C (**Fig. 2**).

1. A continuous flow of HBS buffer is applied at flow rate of 5 μL/min.
2. The sensor chip surface is activated with a 7-min pulse of NHS and EDC.
3. Immobilization of ligand is performed by a 7-min pulse of the ligand solution (*see* **Notes 2** and **3**).
4. Deactivation of excess reactive groups on the surface and removal of noncovalently bound material is performed by a 7-min pulse of ethanolamine.
5. Two 4-min pulses with HCl perform regeneration (complete removal of nonbound material).

3.3. Lectin Binding

Lectin binding is performed at 37°C. A continuous liquid flow is applied to the chip and the lectin is injected as a short pulse. If the lectin recognizes a carbohydrate group-

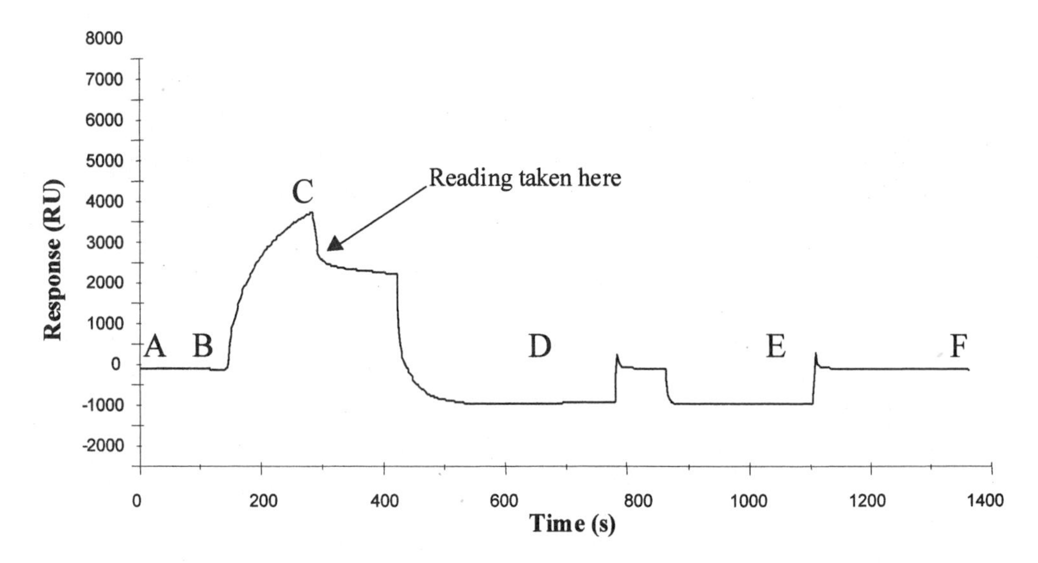

Fig. 3. Typical sensogram of ConA binding. AB, baseline; B, sample injection; BC, association; CD, dissociation; D, E, HCl injections; F, baseline.

ing on the glycoprotein then it will bind resulting in an increasing RU value. A RU reading is taken approx 15 s after dissociation starts. Bound lectin is removed by passing HCl over the chip and then the binding of another lectin is investigated (**Fig. 3**).

1. A continuous flow of HBS buffer is applied at flow rate of 5 µL/min.
2. Lectin is injected by a 7-min pulse (*see* **Note 4**).
3. A reading is taken 15 s after dissociation starts (*see* **Note 5**).
4. Lectin is removed by two 4-min pulses with HCl (*see* **Note 6**).
5. **Steps 1–4** are repeated for each lectin and buffer solution.

3.4. Enzyme Treatment

The specificity of lectin binding measured by SPR can be confirmed by enzyme treatment at 37°C with glycosidases. These are usually very specific for one carbohydrate structure. After enzyme treatment lectin binding should not be detected if the interaction was specific. **Figure 4** shows the binding of MAA before and after treatment with neuraminidase. Immobilized glycoproteins can be treated with other enzymes such as galactosidase and PGNase F (*see* **Note 7**).

1. Neuraminidase treatment is performed by a 7-min injection of the enzyme followed by a stop in the flow for 6 h.
2. The flow rate restored back to 5 µL/mL.
3. Regeneration (removal of neuraminidase) is performed by two 4-min pulses with HCl.
4. Lectin injection, reading, and regeneration as in **steps 1–5** of **Subheading 3.3.**
5. The same procedure as described in **steps 1–4** can be repeated for other enzymes until all residues on a sugar chain are chopped off.

4. Notes

1. A preliminary choice of pH for immobilization can be made if the pI of the ligand is known according to the following rule of thumb: for pI > 7, use pH 6, for pI 5.5–7 use

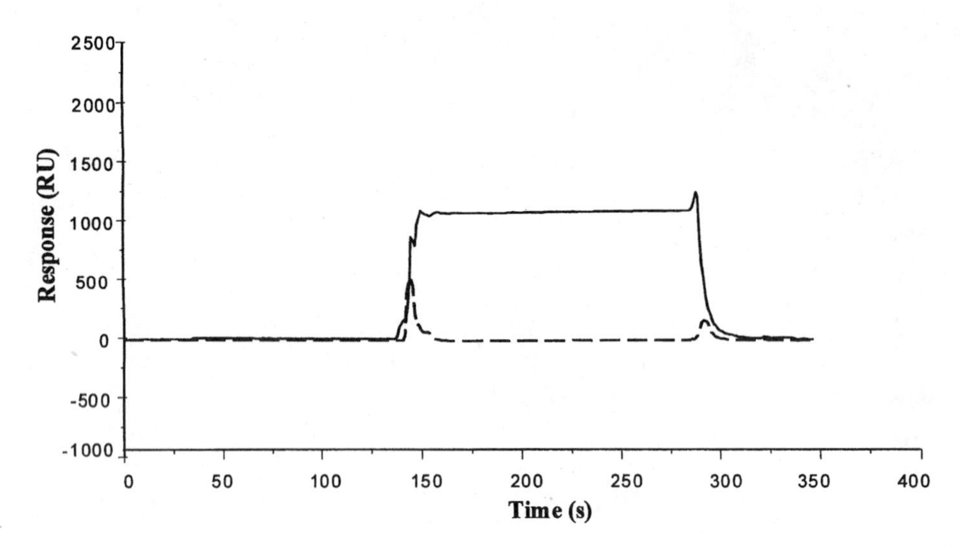

Fig. 4. MAA binding before (*solid line*) and after (*dotted line*) neuraminidase treatment (buffer plot subtracted from both sensograms).

Table 1
Typical Results for Different Glycoproteins with Different Lectins

	ConA	SNA[b]	SNA[a]	MAA[b]	MAA[a]	PNA[b]	PNA[a]	DSA	AAA
IgG	1680	344	50	25	26	24	23	33	71
Fetuin	776	1230	78	624	5	5	80	230	32
Rec1	2768	22	20	10	25	26	30	11	22
Rec2	1203	10	25	325	22	4	297	42	15

[b,a]Before and after neuraminidase treatment. IgG and fetuin were commercial products, while rec1 and rec2 are recombinant glycoproteins. Buffer readings gave values of 20–35 RU. *See* **Note 8**.

1 unit pH below pI, for pI 3.5–5.5, use 0.5 pH units below pI. For pH range of 4–5.5 10 m*M* acetate buffer is recommended.

2. The time of activation may be modified to regulate the amount of ligand immobilized.
3. For a given pH the ideal ligand concentration is the lowest value that gives maximum preconcentration. Usually, a suitable concentration will be in the range 10–200 µg/mL. A target response is a level of immobilization that gives about 0.07 pmol ligand/mm^2. That corresponds to mol wt/15 RU.
4. Analyte concentration can be chosen between 5 µg/mL and 500 µg/mL.
5. A reading value above 30–40 RU is considered as real binding.
6. After regeneration the baseline should be at the same level as before lectin binding. A drift in the baseline may appear after repeated regeneration, which may be due to loss of Neu5Ac.
7. Different enzyme incubation times may apply depending on the enzyme properties.
8. **Table 1** gives typical results for different glycoproteins with different lectins. The following conclusions can be drawn from these results: IgG has *N*-linked chains (ConA), α2–6 Neu5NAc (SNA) and fucose (AAA) and it does not have any *O*-linked chains (PNA) or any α2–3 Neu5NAc (MAA); Fetuin has *N*- and *O*-linked chains, α2–6 Neu5NAc and

α2–3 Neu5NAc, galactose β1–4 *N*-acetylglucosamine (DSA) and no fucose; Rec1 has *N*-linked chains and no fucose, no Neu5NAc, no *O*-linked chains; Rec2 has *N*- and *O*-linked chains, α2–6 Neu5NAc, no α2–3 Neu5NAc and no fucose.

Acknowledgments

This work was supported by funds from Biomed Laboratories, Newcastle upon Tyne, UK, British Biotech Ltd., Oxford, UK and Cambridge Antibody Technology, Royston, UK.

References

1. Imperiali, B. and O'Connor, S. E. (1999) Effect of N-linked glycosylation on glycopeptide and glycoprotein structure. *Curr. Opin. Chem. Biol.* **3,** 643–649.
2. Varki, A. (1993) Biological roles of oligosaccharides: all of the theories are correct. *Glycobiology* **3,** 97–130.
3. Singh, R. S., Tiwary, A. K., and Kennedy, J. F. (1999) Lectins: sources, activities, and applications. *Crit. Rev. Biotechnol.* **19,** 145–178.
4. Turner, G. A. (1992) *N*-Glycosylation of serum proteins in disease and its investigation using lectins. *Clin. Chim. Acta* **208,** 149–171.
5. Fivash, M., Towler, E. M., and Fisher, R. J. (1998) BIAcore for macromolecular interaction. *Curr. Opin. Biotechnol.* **9,** 97–101.
6. (1994) The SPR signal, in *BIAtechnology Handbook*, Pharmacia Biosensor AB, Uppsala, Sweden, pp. 4–3.
7. Haseley, S. R., Talaga, P., Kamerling, J. P., and Vliegenthart, J. F. G. (1999) Characterization of the carbohydrate binding specificity and kinetic parameters of lectins by using surface plasmon resonance. *Analyt. Biochem.* **274,** 203–210.
8. Shinohara, Y., Kimi, F., Shimizu, M., Goto, M., Tosu, M., and Hasegawa, Y. (1994) Kinetic measurement of the interaction between an oligosaccharide and lectins by a biosensor based on surface plasmon resonance. *Eur. J. Biochem.* **223,** 189–194.
9. Satoh, A. and Matsumoto, I. (1999) Analysis of interaction between lectin and carbohydrate by surface plasmon resonance. *Analyt. Biochem.* **275,** 268–270.
10. Hutchinson, M. (1994) Characterization of glycoprotein oligosaccharides using surface plasmon resonance. *Analyt. Biochem.* **220,** 303–307.
11. Liljeblad, M., Lundblad, A., and Pahlsson, P. (2001) Analysis of agalacto-IgG in rheumatoid arthritis using surface plasmon resonance. *Glycoconj. J.* **17,** 323–329.
12. Fotinopoulou, A., Cooke, A., and Turner, G. A. (2000) Does the 'glyco' part of recombinant proteins affect biological activity. *Immunol. Lett.* **73,** 105.
13. (1994) Ligand immobilization chemistry, in *BIA applications Handbook*, Pharmacia Biosensor AB, Uppsala, Sweden, pp. 4–1.

125

Sequencing Heparan Sulfate Saccharides

Jeremy E. Turnbull

1. Introduction

The functions of the heparan sulfates (HSs) are determined by specific saccharide motifs within HS chains. These sequences confer selective protein binding properties and the ability to modulate protein activities *(1,2)*. HS chains consist of an alternating disaccharide repeat of glucosamine (GlcN; *N*-acetylated or *N*-sulfated) and uronic acid (glucuronic [GlcA] or iduronic acid [IdoA]). The initial biosynthetic product containing N-acetylglucosamine (GlcNAc) and GlcA is modified by *N*-sulfation of the GlcN, ester (*O*)-sulfation (at positions 3 and 6 on the GlcN and position 2 on the uronic acids) and by epimerization of GlcA to IdoA. The extent of these modifications is incomplete and their degree and distribution varies in HS between different cell types. In HS chains *N*- and *O*-sulfated sugars are predominantly clustered in sequences of up to eight disaccharide units separated by *N*-acetyl-rich regions with relatively low sulfate content *(3)*.

Sequence analysis of HS saccharides is a difficult analytical problem and until recently sequence information had been obtained for only relatively short saccharides from HS and heparin. Gel chromatography and high-performance liquid chromatography (HPLC) methods have been used to obtain information on disaccharide composition *(3,4)*. Other methods such as nuclear magnetic resonance (NMR) spectroscopy and mass spectroscopy *(5–9)* have provided direct sequence information, but are difficult for even moderately sized oligosaccharides and in the case of NMR requires large amounts of material (micromoles). This situation has changed rapidly in the last few years with the availability of recombinant exolytic lysosomal enzymes. These exoglycosidases and exosulfatases remove specific sulfate groups or monosaccharide residues from the nonreducing end (NRE) of saccharides *(10)*. They can be employed in combination with polyacrylamide gel electrophoresis (PAGE) separations to derive direct information (based on band shifts) on the structures present at the nonreducing end of GAG saccharides (*11*; *see* **Fig. 1** for an example).

Integral glycan sequencing (IGS), a PAGE-based method using the exoenzymes, was recently developed as the first strategy for rapid and direct sequencing of HS and heparin saccharides *(11)*. Its introduction has been quickly followed by a variety of similar approaches using other separation methods including HPLC and matrix-assisted laser desorption (MALDI) mass spectrometry *(12–14)*. An outline of the IGS sequenc-

From: *The Protein Protocols Handbook, 2nd Edition*
Edited by: J. M. Walker © Humana Press Inc., Totowa, NJ

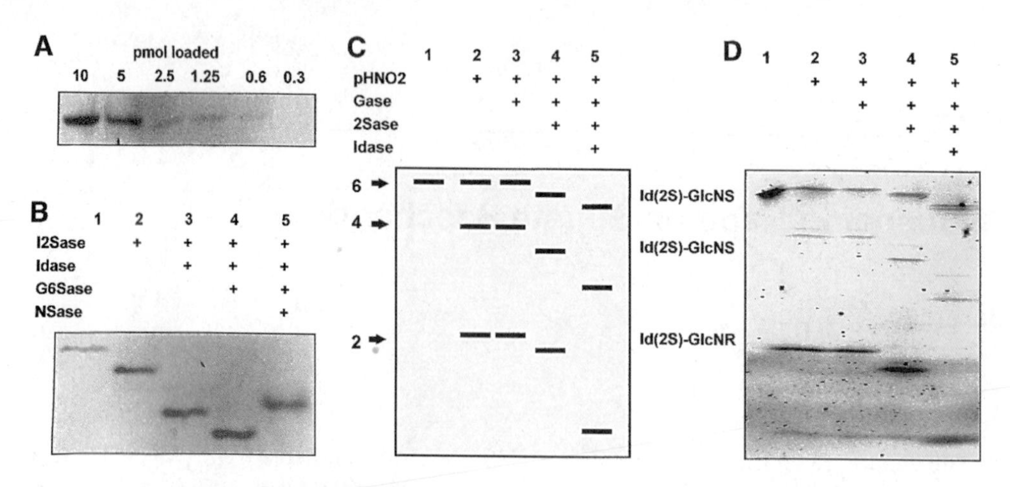

Fig. 1. Basic principles of integral glycan sequencing. **(A)** Fluorescence detection of different amounts of a 2-AA-tagged heparin tetrasaccharide run on a 33% minigel. **(B)** Exosequencing of a 2-AA-tagged heparin tetrasaccharide with lysosomal enzymes and separation of the products on a 33% minigel (15 pmol per track). Band shifts following the exoenzyme treatments shown reveal the structure of the nonreducing end disaccharide unit (*track 1*, untreated). I2Sase, iduronate-2-sulfatase; Idase, iduronidase; G6Sase, glucosamine-6-sulfatase; Nsase, sulfamidase. **(C)** Schematic representation of IGS of a hexasaccharide ($pHNO_2$, partial nitrous acid treatment). **(D)** Actual example of IGS performed on a purified heparin hexasaccharide, corresponding to the scheme in **C**, using the combinations of $pHNO_2$ and exoenzyme treatments indicated (*track 1*, untreated, 25 pmol; other tracks correspond to approx 200 pmol per track of starting sample for $pHNO_2$ digest). The hexasaccharide (purified from bovine lung heparin) has the putative structure IdoA(2*S*)-GlcNSO$_3$(6*S*)-IdoA(2*S*)-GlcNSO$_3$(6*S*)-IdoA(2*S*)-AMannR(6*S*). Electrophoresis was performed on a 16-cm 35% gel (from **ref. *11***, Copyright 1999 National Academy of Sciences, USA).

ing strategy is given in **Fig. 2**. An HS (or heparin) saccharide (previously obtained from the polysaccharide by partial chemical or enzymatic degradation and purification) is labeled at its reducing terminus with a fluorescent tag. It is then subjected to partial nitrous acid treatment to give a ladder of evenly numbered oligosaccharides (di-, tetra-, hexa-, etc.) each having a fluorescent tag at their reducing end. Portions of this material are then treated with a variety of highly specific exolytic lysosomal enzymes (exosulfatases and exoglycosidases) that act at the nonreducing end of each saccharide (if it is a suitable substrate). The various digests are then separated on a high-density polyacrylamide gel and the positions of the fragments detected by excitation of the fluorescent tag with a UV transilluminator. Band shifts resulting from the different treatments permit the sequence to be read directly from the banding pattern (*see* **Fig. 1** for an example). This novel strategy allows direct read-out sequencing of a saccharide in a single set of adjacent gel tracks in a manner analogous to DNA sequencing. IGS provides a rapid approach for sequencing HS saccharides, and has proved very useful in recent structure–function studies ***(15)***. It should be noted that this methodology is designed for sequencing purified saccharides, not whole HS preparations. An important factor in all sequencing methods is the availability of sufficiently pure saccharide starting material. HS and heparin saccharides can be prepared following selective scission

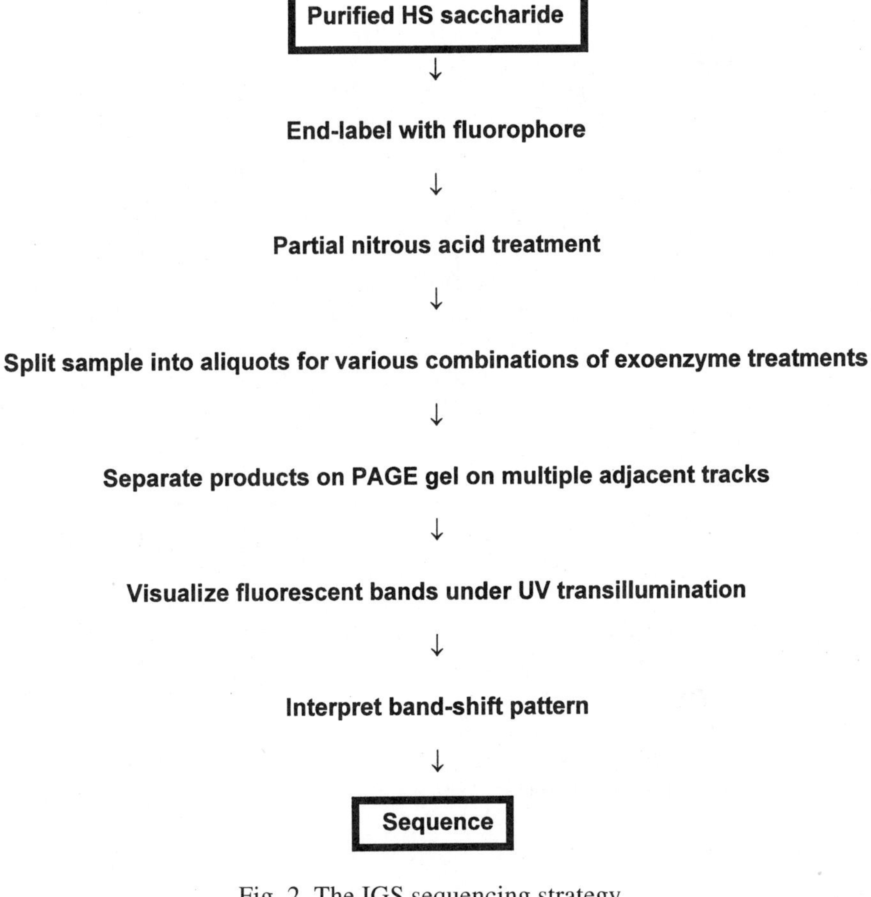

Fig. 2. The IGS sequencing strategy.

by enzymic (or chemical) reagents and isolation by methods such as affinity chromatography ***(4)***. Final purification usually requires the use of strong anion-exchange HPLC ***(11,15)***.

2. Materials

1. 2-Aminobenzoic acid (2-AA; Fluka Chemicals).
2. 7-Aminonaphthalene-1,3-disulfonic acid monopotassium salt (ANDSA; Fluka Chemicals).
3. Formamide.
4. Sodium cyanoborohydride (>98% purity).
5. Sodium triacetoxyborohydride (Aldrich).
6. Distilled water.
7. Oven or heating block at 37°C.
8. Desalting column (Sephadex G-25; e.g., HiTrap™ desalting columns, Pharmacia).
9. Centrifugal evaporator.
10. 200 m*M* HCl.
11. 20 m*M* Sodium nitrite (1.38 mg/mL in distilled water; prepare fresh).
12. 200 m*M* Sodium acetate, pH 6.0 (27.2 g/L of sodium acetate trihydrate; adjust pH to 6.0 using acetic acid).

13. Enzyme buffer (0.2 *M* Na acetate, pH 4.5). Make 0.2 *M* sodium acetate (27.2 g/L of sodium acetate trihydrate) and 0.2 *M* acetic acid (11.6 mL/L) and mix in a ratio of 45 mL to 55 mL, respectively.
14. Enzyme stock solutions (typically at concentrations of 500 mU/mL, where 1 U = 1 μmol substrate hydrolyzed per minute). Available from Glyko, Novato, CA.
15. Vortex tube mixer.
16. Microcentrifuge.
17. Acrylamide stock solution (T50%–C5%). **Caution: Acrylamide is neurotoxic.** Wear gloves (and a face mask when handling powdered forms). It is convenient to use premixed *bis*-acrylamide such as Sigma A-2917. Add 43 mL of distilled water to the 100-mL bottle containing the premixed chemicals and dissolve using a small stirrer bar (approx 2 h). The final volume should be approx 80 mL. Store the stock solution at 4°C. Note that it is usually necessary to warm gently to redissolve the acrylamide after storage.
18. Resolving gel buffer stock solutions: 2 *M* Tris-HCl, pH 8.8 (242.2 g/L of Tris base; adjust pH to 8.8 with HCl).
19. Stacking gel buffer stock solution: 1 *M* Tris-HCl, pH 6.8 (121.1 g/L of Tris base; adjust pH to 6.8 with HCl).
20. Electrophoresis buffer (25 m*M* Tris, 192 m*M* glycine, pH 8.3): 3 g/L of Tris base, 14.4 g/L of glycine; adjust pH to 8.3 if necessary with HCl.
21. 10% Ammonium persulfate in water (made fresh or stored at –20°C in aliquots).
22. *N,N,N',N'*-Tetramethylethylenediamine (TEMED).
23. Vertical slab gel electrophoresis system (minigel or standard size).
24. D.C. Power supply unit (to supply up to 500–1000 V and 200 mA).
25. UV transilluminator (312 nm maximum emission wavelength).
26. Glass UV bandpass filter larger than gel size (type UG-11, or M-UG2).
27. Charge coupled device (CCD) imaging camera fitted with a 450-nm (blue) bandpass filter.

3. Methods

3.1. Tagging Saccharides with a Fluorophore

HS (and heparin) saccharides can be endlabeled by reaction of their reducing aldehyde functional group with a primary amino group of a fluorophore (reductive amination). For sulfated saccharides anthranilic acid (2-AA; ***11***) has been found to be effective for the IGS methodology. 2-AA conjugates display an excitation maxima in the range 300–320 nm, which is ideal for visualization with a commonly available 312-nm UV source (e.g., transilluminators used for visualizing ethidium bromide stained DNA). Emission maxima are typically in the range 410–420 nm (bright violet fluorescence). Recently it has been found that approx 10-fold more sensitive detection is possible using an alternative fluorophore ANDSA ***(16)***. ANDSA has an excitation maxima of 350 nm and emission maxima of 450 nm. Both approaches described in **Subheadings 3.1.1.** and **3.1.2.** allow rapid labeling and purification of tagged saccharide from free tagging reagent, and give quantitative recoveries and products free of salts that might interfere with subsequent enzymic conditions. For saccharides in the size range hexa- to dodecasaccharides, approx 2–3 nmol (approx 2–10 μg) of purified starting material is the minimum required using the 2-AA label and approx 5- to 10-fold less for ANDSA labeling.

3.1.1. Labeling Saccharides with 2-AA

1. Dry down the purified saccharide (typically 2–10 nmol) in a microcentrifuge tube by centrifugal evaporation.

2. Dissolve directly in 10–25 µL of formamide containing freshly prepared 400 m*M* 2-AA (54.8 mg/mL) and 200 m*M* reductant (sodium cyanoborohydride; 12.6 mg/mL) and incubate at 37°C for 16–24 h in a heating block or oven. **(Caution: The reductant is toxic and should be handled with care.)** The volume used should be sufficient to provide a 500–1000-fold molar excess of 2-AA over saccharide (*see* **Note 1**).
3. Remove free 2-AA, reductant, and formamide from the labeled saccharides by gel filtration chromatography (Sephadex G-25 Superfine). Dilute the sample (maximum 250 µL of reaction mixture) to a total of 1 mL with distilled water (*see* **Note 2**).
4. Load sample onto two 5-mL HiTrap™ Desalting columns (Pharmacia Ltd.) connected in series. Alternatively it is possible to use self-packed columns of other dimensions.
5. Elute with distilled water at a flow rate of 1 mL/min and collect fractions of 0.5 mL. Saccharides consisting of four or more monosaccharide units typically elute in the void volume (approx fractions 7–12). Note that the HiTrap™ columns can be eluted by hand with a syringe without need for a pump.
6. Pool and concentrate these fractions by centrifugal evaporation or freeze drying.

3.1.2. Labeling Saccharides with ANDSA

1. Dry down the purified saccharide (typically 2–10 nmol) in a microcentrifuge tube by centrifugal evaporation.
2. Dissolve directly in 10 µL of formamide.
3. Mix with 15 µL of formamide containing ANDSA at a concentration of 80 mg/mL (approaching saturation) and incubate at 25°C for 16 h.
4. Mix with 10 µL of formamide containing 1 mg of the reductant sodium triacetoxyborohydride (*see* **Note 1**) and incubate for 2 h at 25°C.
5. Remove free ANDSA by gel filtration as described in **Subheading 3.1.1.** for 2-AA.

3.2. Nitrous Acid Treatment of Saccharides

Low pH nitrous acid cleaves HS only at linkages between *N*-sulfated glucosamine and adjacent hexuronic acid residues ***(17,18)***. Under mild controlled conditions nitrous acid cleavage creates a ladder of bands corresponding to the positions of internal *N*-sulfated glucosamine residues in the original intact saccharide ***(11)***. A series of different reaction stop points are pooled, resulting in a partial digest with a range of different fragment sizes.

1. Dry down 1–2 nmol of labeled saccharide by centrifugal evaporation.
2. Redissolve in 80 µL of distilled water and chill on ice.
3. Add 10 µL of 200 m*M* HCl and 10 µL of 20 m*M* sodium nitrite (both prechilled on ice) and incubate on ice.
4. At a series of individual time points (typically 15, 30, 60, 120, and 180 min), remove an aliquot and stop the reaction by raising the pH to approx 5.0 by the addition of 1/5 volume of 200 m*M* sodium acetate buffer, pH 6.0 (*see* **Note 3**).
5. Pool the set of aliquots and either use directly for enzyme digests or desalt as described in **Subheading 3.1.**

3.3. Exoenzyme Treatment of Saccharides

The approach for treatment of HS samples with exoenzymes is described below. Details of the specificities of the exoenzymes is given in **Table 1**. These enzymes have differing optimal pH and buffer conditions, but in general they can be used under the single set of conditions given here, which simplifies the multiple enzyme treatments required (*see* **Note 4**).

Table 1
Exoenzymes for Sequencing Heparan Sulfate and Heparin

Enzyme[a]	Substrate specificity[b]
Sulfatases	
Iduronate-2-sulfatase	IdoA(2*S*)
Glucosamine-6-sulfatase	GlcNAc(6*S*), $GlcNSO_3(6S)$
Sulphamidase (glucosamine *N*-sulfatase)	$GlcNSO_3$
Glucuronate-2-sulfatase	GlcA(2*S*)
Glucosamine-3-sulfatase	$GlcNSO_3(3S)$
Glycosidases	
Iduronidase	IdoA
Glucuronidase	GlcA
α-*N*-Acetylglucosaminidase	GlcNAc
Bacterial exoenzymes	
Δ4,5-Glycuronate-2-sulfatase	ΔUA(2*S*)
Δ4,5-Glycuronidase	ΔUA

[a]Enzyme availability: Glucuronidase is widely available commercially as purified or recombinant enzyme. Recombinant iduronate-2-sulfatase, iduronidase, glucosamine-6-sulfatase, sulfamidase, and α-*N*-acetylglucosaminidase are available from Glyko (Novato, CA; www.glyko.com). Glucuronate-2-sulfatase and glucosamine-3-sulfatase have only been purified from cell and tissue sources to date. The bacterial exoenzymes are available from Grampian Enzymes, Nisthouse, Harray, Orkney, Scotland; e-mail, grampenz@aol.com.

[b]The specificities are shown as the nonreducing terminal group recognized by the enzymes. Sulfatases remove only the sulfate group, whereas the glycosidases cleave the whole nonsulfated monosaccharide.

1. Dissolve the sample (typically 10–200 pmol of saccharide) in 10 μL of H_2O in a microcentrifuge tube.
2. Add 5 μL of exoenzyme buffer (100 m*M* sodium acetate buffer, pH 4.5), 1 μL of 0.5 mg/mL bovine serum albumin, 2 μL of appropriate exoenzyme [0.2–0.5 mU], and distilled water to bring the final volume to 20 μL.
3. Mix the contents well on a vortex mixer, and centrifuge briefly to ensure that the reactants are at the tip of the tube.
4. Incubate the samples at 37°C for 16 h in a heating block or oven.

3.4. Separation of Saccharides by PAGE

PAGE is a high-resolution technique for the separation of HS and heparin saccharides of variable sulfate content and disposition. Its resolution is generally superior to gel filtration or anion-exchange HPLC ***(19,20)***. Improved resolution can be obtained using gradient gels, although these are more difficult to prepare and use routinely. In most cases sufficient resolution can be obtained with isocratic gels (*see* **Note 5**). PAGE provides a simple but powerful approach for separating the saccharide products generated in the sequencing process.

3.4.1. Preparing the PAGE Gel

1. Assemble the gel unit (consisting of glass plates and spacers, etc).
2. Prepare and degas the resolving gel acrylamide solution without ammonium persulfate or TEMED. To make a 30% acrylamide gel solution for a 16 cm × 12 cm × 0.75 mm gel,

16 mL is required. Mix 9.6 mL of T50%–C5% acrylamide stock with 3 mL of 2 *M* Tris, pH 8.8, and 3.4 mL of distilled water.

3. Add 10% ammonium persulfate (30 µL) and TEMED (10 µL) to the gel solution, mix well, and immediately pour into the gel unit.
4. Overlay the unpolymerized gel with resolving gel buffer (375 m*M* Tris-HCl, pH 8.8, diluted from the 2 *M* stock solution) or water-saturated butanol. Polymerization should occur within approx 30–60 min. The gel can then be used immediately or stored at 4°C for 1–2 wk.

3.4.2. Electrophoresis

1. Immediately before electrophoresis, rinse the resolving gel surface with stacking gel buffer (0.125 *M* Tris-HC1 buffer, pH 6.8, diluted from the 1 *M* stock solution).
2. Prepare and degas the stacking gel solution (for 5 mL, mix 0.5 mL of T50%–C5% acrylamide stock with 0.6 mL of 1 *M* Tris, pH 6.8, and 3.9 mL of distilled water).
3. Add 10% ammonium persulfate (10 µL) and TEMED (5 µL). Immediately pour onto the top of the resolving gel and insert the well-forming comb.
3. After polymerization (approx 15 min) remove the comb and rinse the wells thoroughly with electrophoresis buffer.
4. Place the gel unit into the electrophoresis tank and fill the buffer chambers with electrophoresis buffer.
5. Load the oligosaccharide samples (5–20 µL dependent on well capacity, containing approx 10% [v/v] glycerol or sucrose in 125 m*M* Tris-HCl, pH 6.8) carefully into the wells with a microsyringe. Marker samples containing bromophenol blue and phenol red should also be loaded into separate tracks.
6. Run the samples into the stacking gel at 150–200 V (typically 20–30 mA) for 30–60 min, followed by electrophoresis at 300–400 V (typically 20–30 mA and decreasing during run) for approx 5–8 h (for a 16-cm gel). Heat generated during the run should be dissipated using a heat exchanger with circulating tap water, or by running the gel in a cold room or in a refrigerator.
7. Electrophoresis should be terminated before the phenol red marker dye is about 5 cm from the bottom of the gel. (At this point, disaccharides should be 3–4 cm from the bottom of the gel.)

3.5. Gel Imaging

The most effective approach for gel imaging requires a CCD camera that can detect faint fluorescent banding patterns by capturing multiple frames. Systems commonly used for detection of ethidium bromide stained DNA can usually be adapted with appropriate filters as described below (*see* **Note 6**).

1. Place a UV filter (UG-1, UG-11, or MUG-2) onto the transilluminator, and fit a 450-nm blue filter onto the camera lens.
2. Remove the gel carefully from the glass plates after completion of the run and place on the UV transilluminator surface wetted with electrophoresis buffer. Wet the upper surface of the gel to reduce gel drying and curling.
3. Switch on the transilluminator and capture the image using the CCD camera. Exposure times are typically 1–5 s depending on the amount of labeled saccharide (*see* **Note 7**).

3.6. Data Interpretation

The sequence of saccharides can be read directly from the banding pattern by interpreting the band shifts due to removal of specific sulfate or sugar moieties. **Figure 1**

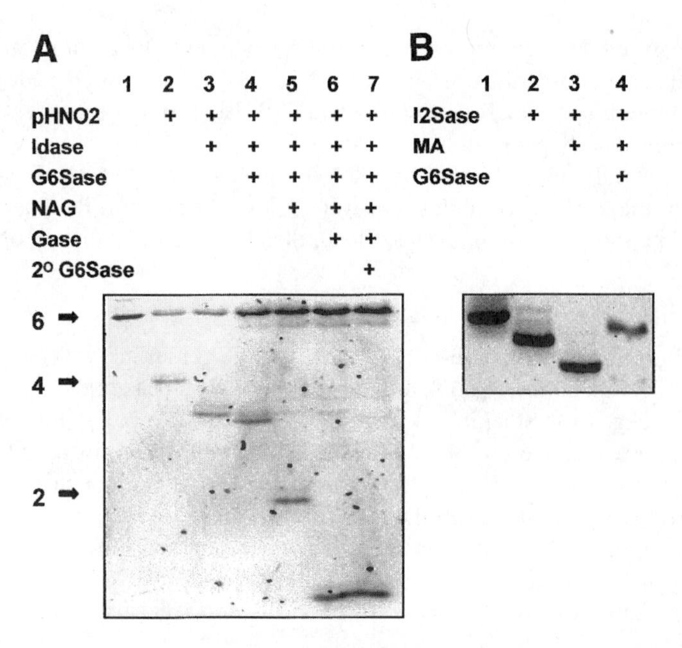

Fig. 3. IGS of a heparin hexasaccharide. A heparin hexasaccharide with the structure ΔHexA(2*S*)-GlcNSO$_3$(6*S*)-IdoA-GlcNAc(6*S*)-GlcA-GlcNSO$_3$(6*S*) was 2-AA-tagged and subjected to sequencing on a 16-cm 33% gel. **(A)** IGS of hexasaccharide using the combinations of pHNO$_2$ and exoenzyme treatments indicated (*track 1*, untreated, 20 pmol; other tracks correspond to approx 90 pmol per track of starting sample for pHNO$_2$ digest). NAG, *N*-acetylglucosaminidase. **(B)** Determining the sequence of the nonreducing disaccharide unit of the hexasaccharide using the I2Sase, G6Sase, and mercuric acetate (MA) treatments shown (approx 20 pmol per track; *track 1*, untreated). (From **ref.** ***11***, Copyright 1999 National Academy of Sciences, USA.)

shows an actual example and a schematic representation. First, bands generated by the partial nitrous acid treatment indicate the positions of *N*-sulfated glucosamine residues in the original saccharide (**Fig. 1C**, *track 2*). A "missing" band in the ladder at a particular position indicates the presence of an *N*-acetylated glucosamine residue in the original saccharide at that position (an example of this is shown in **Fig. 3**). Such saccharides can be sequenced by the additional use of the exoenzyme *N*-acetylglucosaminidase, which removes this residue and allows further sequencing of an otherwise "blocked" fragment. Following the nitrous acid treatment, the "ladder" of bands is then subjected to various exoenzyme digestions. The presence of specific sulfate or sugar residues can be deduced from the band shifts that occur (**Fig. 1C**, *tracks 3–5*). **Figure 4** shows an example of a decasaccharide from HS that has been purified by SAX-HPLC and sequenced using IGS *(11)*.

Although the band shifts are usually downwards (because of to the lower molecular mass and thus higher mobility of the product) it should be noted that occasionally upward shifts occur, probably due to subtle differences in charge/mass ratio (for examples, *see* **Figs. 1B**, **3B**, and **4C**). Note also that minor "ghost" bands sometimes appear after

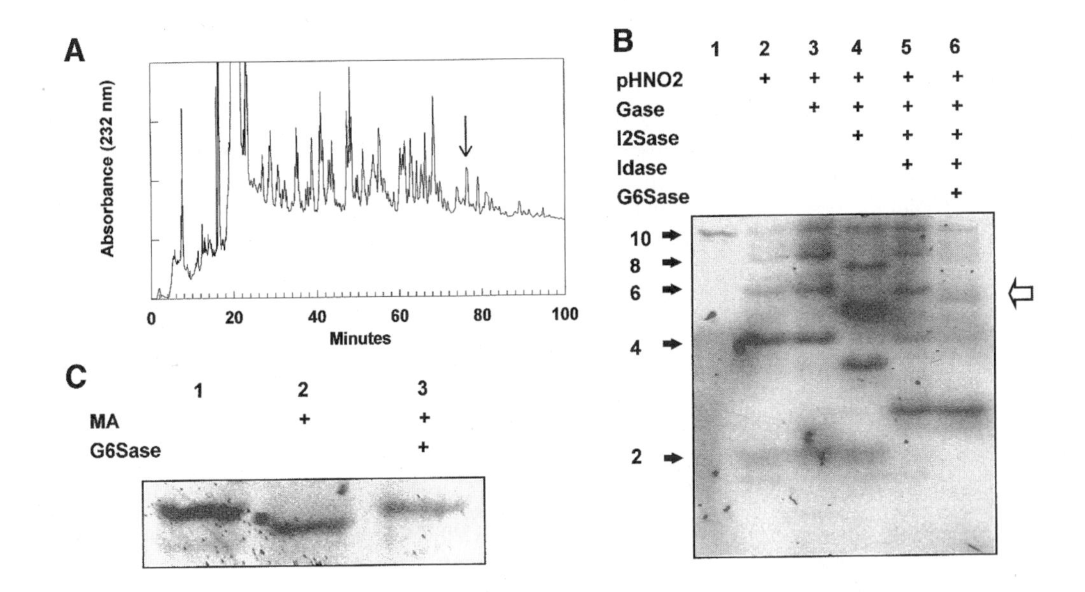

Fig. 4. HPLC purification and IGS of a HS decasaccharide. **(A)** SAX-HPLC of a pool of HS decasaccharides derived by heparitinase treatment of porcine mucosal HS (for details *see* **ref. *11***). The *arrowed peak* was selected for sequencing. **(B)** IGS of the purified HS decasaccharide on a 16-cm 33% gel using the combinations of $pHNO_2$ and exoenzyme treatments indicated (*track 1*, untreated, 20 pmol; other tracks correspond to approx 400 pmol per track of starting sample for $pHNO_2$ digest). **(C)** Determining the sequence of the nonreducing disaccharide unit of the HS decasaccharide using the mercuric acetate (MA) and G6Sase treatments shown (approx 40 pmol per track; *track 1*, untreated). (From **ref. *11***, Copyright 1999 National Academy of Sciences, USA.)

the nitrous acid treatment. They are probably due to loss of an *N*-sulfate group, and normally these do not affect interpretation of the shifts in the major bands ***(11)***.

If the saccharide being sequenced was derived by bacterial lyase treatment, it will have a Δ4,5-unsaturated uronate residue at its nonreducing terminus. If this residue has a 2-*O*-sulfate attached, this can be detected by susceptibility to I2Sase (*see* **Fig. 3B**), but the sugar residue itself is resistant to both Idase and Gase. Its removal is required to confirm whether there is a 6-*O*-sulfate on the adjacent nonreducing end glucosamine (*see* **Figs. 3B** and **4C** for examples). Bacterial enzymes that specifically remove the Δ4,5-unsaturated uronate residues (and the 2-*O*-sulfate groups that may be present on them) are now available commercially (*see* **Table 1**). Alternatively, they can be removed chemically with mercuric acetate (***21***; *see* **Figs. 3B** and **4C**).

In addition to the basic sequencing experiment, it is wise to confirm agreement of the data with an independent analysis of the disaccharide composition of the saccharide ***(11)***. It can sometimes be difficult to sequence the reducing terminal monosaccharide owing to it being a poor substrate for the exoenzymes. In these cases it has proved more effective to analyze the terminal 2AA-labeled disaccharide unit in comparison to 2AA-labeled disaccharide standards ***(11)***.

4. Notes

1. Using large excesses of reagent as described, saccharides derived from HS and heparin by bacterial lyase scission generally couple with 2-AA with efficiencies in the range of 60–70%. Note that saccharides derived from HS and heparin by low pH nitrous acid scissioning (i.e., having an anhydromannose residue at their reducing ends) label more efficiently (approx 70–80% coupling efficiency). Labeling with ANDSA achieves similar coupling efficiencies, and the alternative reducing agent, sodium triacetoxyborohydride, is less toxic than sodium cyanoborohydride.
2. Unwanted reactants and solvent can also be removed from labeled saccharides by methods such as dialysis but the rapid gel filtration chromatography step described above using the HiTrap desalting columns is convenient and usually allows good recoveries of loaded sample (typically 70–80%).
3. It is best to perform some trial incubations to test for optimal time points needed to generate a balanced mix of all fragments in the partial nitrous acid digestion. With longer saccharides (octasaccharides and larger) it is observed that the largest products are generated quickly and thus a bias toward shorter incubation times is required as saccharide length increases ***(16)***.
4. The enzyme conditions described should provide for complete digestion of all susceptible residues. This is important to the sequencing process, as incomplete digestion would create a more complex banding pattern and would give a false indication of sequence heterogeneity. It is useful to run parallel controls with standard saccharides to enable monitoring of reaction conditions. Where combinations of exoenzymes are required, these can be incubated simultaneously with the sample. If required, the activity of one enzyme can be destroyed prior to a secondary digestion with a different enzyme by heating the sample at 100°C for 2–5 min.
5. Adequate separations, particularly over limited size ranges of saccharides, can be obtained using single concentration gels, typically in the range 25–35% acrylamide. Improvements in resolution can be made by using longer gel sizes. Different voltage conditions (usually in the range 200–600 V) and running times are required for different gel formats, and should be established by trial and error with the particular samples being analyzed. Gels up to 24 cm in length can usually be run in 5–8 h using high voltages, whereas for longer gels it is often convenient to use lower voltage conditions and overnight runs. Minigels can also be used effectively for separation of small HS–heparin saccharides up to octasaccharides in size (*see* **Fig. 1**). Note that it is also possible to run Tris-acetate gels with a Tris-MES electrophoresis buffer (*see* **Fig. 1**; ***11***).
6. Because the emission wavelength of 2-AA tagged saccharides is 410–420 nm, there is a need to filter out background visible wavelength light from the UV lamps. This can be done effectively with special glass filters that permit transmission of UV light but do not allow light of wavelengths >400 nm to pass. A blue bandpass filter on the camera also improves sensitivity. Suitable filters are available from HV Skan (Stratford Road, Solihull, UK; Tel: 0121 733 3003) or UVItec Ltd. (St. John's Innovation Centre, Cowley Road, Cambridge, UK; www.uvitec.demon.co.uk).
7. Required exposure times are strongly dependent on sample loading and the level of detection required. Over-long exposures will result in excessive background signal. Note that negative images are usually better for band identification (*see* figures). Under the conditions described the limit of sensitivity is approx 10–20 pmol per band of original starting material for 2-AA (*see* **Fig. 1**) and 2–5 pmol per band for ANDSA.

References

1. Turnbull, J. E., Powell, A., and Guimond, S. E. (2001) Heparan sulphate: decoding a dynamic multifunctional cellular regulator. *Trends Cell Biol.* **11,** 75–82.
2. Bernfield, M., Gotte, M., Park, P. W., et al. (1999) Functions of cell surface heparan sulfate proteoglycans. *Annu. Rev. Biochem.* **68,** 729–777.
3. Turnbull, J. E. and Gallagher, J. T. (1991) Distribution of iduronate-2-sulfate residues in HS: evidence for an ordered polymeric structure. *Biochem. J.* **273,** 553–559.
4. Turnbull, J. E., Fernig, D., Ke, Y., Wilkinson, M. C., and Gallagher, J. T. (1992) Identification of the basic FGF binding sequence in fibroblast HS. *J. Biol. Chem.* **267,** 10,337–10,341.
5. Pervin, A., Gallo, C., Jandik, K., Han, X., and Linhardt, R. (1995) Preparation and structural characterisation of heparin-derived oligosaccharides. *Glycobiology* **5,** 83–95.
6. Yamada, S., Yamane, Y., Tsude, H., Yoshida, K., and Sugahara, K. (1998) A major common trisulfated hexasaccharide isolated from the low sulfated irregular region of porcine intestinal heparin. *J. Biol. Chem.* **273,** 1863–1871.
7. Yamada, S., Yoshida, K., Sugiura, M., Sugahara, K., Khoo, K., Morris, H., and Dell, A. (1993) Structural studies on the bacterial lyase-resistant tetrasaccharides derived from the antithrombin binding site of porcine mucosal intestinal heparin. *J. Biol. Chem.* **268,** 4780–4787.
8. Mallis, L., Wang, H., Loganathan, D., and Linhardt, R. (1989) Sequence analysis of highly sulfated heparin-derived oligosaccharides using FAB-MS. *Analyt. Chem.* **61,** 1453–1458.
9. Rhomberg, A. J., Ernst, S., Sasisekharan, R., and Bieman, K. (1998) Mass spectrometric and capillary electrophoretic investigation of the enzymatic degradation of heparin-like glycosaminoglycans. *Proc. Natl. Acad. Sci. USA* **95,** 4176–4181.
10. Hopwood, J. (1989) Enzymes that degrade heparin and heparan sulfate, in *Heparin* (Lane and Lindahl, eds.), Edward Arnold Press, pp. 191–227.
11. Turnbull, J. E., Hopwood, J. J., and Gallagher, J. T. (1999) A strategy for rapid sequencing of heparan sulfate/heparin saccharides. *Proc. Natl. Acad. Sci. USA* **96,** 2698–2703.
12. Merry, C. L. R., Lyon, M., Deakin, J. A., Hopwood, J. J., and Gallagher, J. T. (1999) Highly sensitive sequencing of the sulfated domains of heparan sulfate. *J. Biol. Chem.* **274,** 18,455–18,462.
13. Vives, R. R., Pye, D. A., Samivirta, M., Hopwood, J. J., Lindahl, U., and Gallagher, J. T. (1999) Sequence analysis of heparan sulphate and heparin oligosaccharides. *Biochem. J.* **339,** 767–773.
14. Venkataraman, G., Shriver, Z., Ramar, R., and Sasisekharan, R. (1999) Sequencing complex polysaccharides. *Science* **286,** 537–542.
15. Guimond, S. E. and Turnbull, J. E. (1999) Fibroblast growth factor receptor signalling is dictated by specific heparan sulfate saccharides. *Curr. Biol.* **9,** 1343–1346.
16. Drummond, K. J., Yates, E. A., and Turnbull, J. (2001) Electrophoretic sequencing of heparin/heparan sulfate oligosaccharides using a highly sensitive fluorescent end label. *Proteomics* **1,** 304–310.
17. Shively, J. and Conrad, H. (1976) Formation of anhydrosugars in the chemical depolymerisation of heparin. *Biochemistry* **15,** 3932–3942.
18. Bienkowski and Conrad, H. (1985) Structural characterisation of the oligosaccharides formed by depolymerisation of heparin with nitrous acid. *J. Biol. Chem.* **260,** 356–365.
19. Turnbull, J. E. and Gallagher, J. T. (1988) Oligosaccharide mapping of heparan sulfate by polyacrylamide-gradient-gel electrophoresis and electrotransfer to nylon membrane. *Biochem. J.* **251,** 597–608.

20. Rice, K., Rottink, M., and Linhardt, R. (1987) Fractionation of heparin-derived oligosaccharides by gradient PAGE. *Biochem. J.* **244,** 515–522.
21. Ludwigs, U., Elgavish, A., Esko, J., and Roden, L. (1987) Reaction of unsaturated uronic acid residues with mercuric salts. *Biochem. J.* **245,** 795–804.

126

Analysis of Glycoprotein Heterogeneity by Capillary Electrophoresis and Mass Spectrometry

Andrew D. Hooker and David C. James

1. Introduction

The drive toward protein-based therapeutic agents requires both product quality and consistency to be maintained throughout the development and implementation of a production process. Differences in host cell type, the physiological status of the cell, and protein structural constraints are known to result in variations in posttranslational modifications that can affect the bioactivity, receptor binding, susceptibility to proteolysis, immunogenicity, and clearance rate of a therapeutic recombinant protein in vivo *(1)*. Glycosylation is the most extensive source of protein heterogeneity, and many recent developments in analytical biotechnology have enhanced our ability to monitor and structurally define changes in oligosaccharides associated with recombinant proteins.

Variable occupancy of potential glycosylation sites may result in extensive glycosylation macroheterogeneity in addition to the considerable diversity of carbohydrate structures that can occur at individual glycosylation sites, often referred to as glycosylation microheterogeneity. Variation within a heterogeneous population of glycoforms may lead to functional consequences for the glycoprotein product. Therefore, regulatory authorities such as the US Food and Drug Administration (FDA) and the Committee for Proprietary Medical Productions demand increasingly sophisticated analysis for biologics produced by the biotechnology and pharmaceutical industries *(2)*. The FDA has described a "well-characterized biologic" as "a chemical entity whose identity, purity, impurities, potency and quantity can be determined and controlled."

The glycosylation of a recombinant protein product can be examined by:

1. Analysis of glycans released by chemical or enzymatic means.
2. Site-specific analysis of glycans associated with glycopeptide fragments following proteolysis of the intact glycoprotein.
3. Direct analysis of the whole glycoprotein.

A number of techniques are currently available to provide rapid and detailed analysis of glycan heterogeneity:

1. High-pH anion-exchange chromatography with pulsed amperometric detection (HPAEC-PAD).
2. Enzymatic analysis methods such as the reagent array analysis method (RAAM; ***3,4***).

From: *The Protein Protocols Handbook, 2nd Edition*
Edited by: J. M. Walker © Humana Press Inc., Totowa, NJ

3. High-performance capillary electrophoresis (HPCE).
4. Matrix-assisted laser desorption/ionization mass spectrometry (MALDI-MS).
5. Electrospray ionization mass spectrometry (ESI-MS).

In particular, novel mass spectrometric strategies continue to rapidly advance the frontiers of biomolecular analysis, with technical innovations and methodologies yielding improvements in sensitivity, mass accuracy, and resolution.

1.1. High-Performance Capillary Electrophoresis

Capillary electrophoresis has been employed in various modes in the high-resolution separation and detection of glycoprotein glycoforms, glycoconjugates, glycopeptides, and oligosaccharides, even though carbohydrate molecules do not absorb or fluorese and are not readily ionized ***(5–8)***. A number of approaches have been employed to render carbohydrates more amenable to analysis which include *in situ* complex formation with ions such as borate and metal cations ***(9)*** and the addition of ultraviolet (UV)-absorbing or fluorescent tags to functional groups ***(10)***.

1.2. MALDI-MS

MALDI-MS has been extensively used to determine the mass of proteins and polypeptides, confirm protein primary structure, and to characterize posttranslational modifications. MALDI-MS generally employs simple time-of-flight analysis of biopolymers that are co-crystallized with a molar excess of a low molecular weight, strongly UV absorbing matrix, such as 2,5-dihydroxybenzoic acid, on a metal sample disk. Both the biopolymer and matrix ions are desorbed by pulses of a UV laser. Following a linear flight path the molecular ions are detected, the time between the initial laser pulse and ion detection being directly proportional to the square root of the molecular ion mass/charge (*m/z*) ratio. For maximum mass accuracy, internal and external protein or peptide calibrants of known molecular mass are required. In addition to this "linear" mode, many instruments offer a "reflectron" mode that effectively lengthens the flight path by redirecting the ions toward an additional ion detector that may enhance resolution, at the expense of decreased sensitivity. MALDI-MS is tolerant to low (micromolar) salt concentrations, can determine the molecular weight of biomolecules in excess of 200 kDa with a mass accuracy of ±0.1%, and is capable of analyzing heterogeneous samples with picomole to femtomole sensitivity. These properties combined with its rapid analysis time and ease of use for the nonspecialist have made it an attractive technique for the analysis of glycoproteins, glycopeptides, and oligosaccharides.

1.3. Electrospray Ionization Mass Spectrometry

ESI-MS is another mild ionization method, where the covalent bonding of the biopolymer is maintained and is typically used in combination with a single or triple quadrupole. This technique is capable of determining the molecular weight of biopolymers up to 100 kDa with a greater mass accuracy (±0.01%) and resolution (±2000) than MALDI-MS. Multiply charged molecular ions are generated by the ionization of biopolymers in volatile solvents, the resulting spectrum being convoluted to produce noncharged peaks.

ESI-MS has been extensively used for the direct mass analysis of glycopeptides and glycoproteins and is often interfaced with liquid chromatography ***(11–13)***, but has found limited application for the direct analysis of oligosaccharides ***(14)***. ESI-MS is better suited to the analysis of whole glycoprotein populations than MALDI-MS, its superior resolution permitting the identification of individual glycoforms ***(15,16)***.

2. Materials

1. P/ACE 2100 HPCE System (Beckman Instruments Ltd., High Wycombe, UK).
2. Phosphoric acid (Sigma Chemical Co., Poole, UK).
3. Boric acid (Sigma).
4. Trypsin, sequencing grade (Boehringer Mannheim, UK, Lewes, UK).
5. Waters 626 Millenium HPLC System (Millipore Ltd., Watford, UK).
6. Vydac 218TP52 reverse-phase column: C18, 2.1 × 250 mm (Hichrom Ltd., Reading, UK).
7. HPLC grade water–acetonitrile (Fischer Scientific, Loughborough, UK).
8. α-Cyano-4-hydroxy cinnaminic acid (Aldrich Chemical Co., Gillingham, UK).
9. VG Tof Spec Mass Spectrometer (Fisons Instruments, Manchester, UK).
10. Vasoactive intestinal peptide—fragment 1–12 (Sigma).
11. Peptide-*N*-glycosidase F and glycosidases (Glyko Inc., Upper Heyford, UK).
12. 2,5-Dihydroxybenzoic acid (Aldrich).
13. 2,4,6-Trihydroxyacetophenone (Aldrich).
14. Ammonium citrate (Sigma).
15. VG Quattro II triple quadrupole mass spectrometer (VG Organic, Altrincham, UK).
16. Horse heart myoglobin (Sigma).

3. Methods

This chapter describes some of the recent technological advances in the analysis of posttranslational modifications made to recombinant proteins and focuses on the application of HPCE, MALDI-MS, and ESI-MS to the monitoring of glycosylation heterogeneity. These techniques are illustrated by describing their application to the analysis of recombinant human γ-interferon (IFN-γ), a well-characterized model glycoprotein that has *N*-linked glycans at Asn_{25} and at the variably occupied site, Asn_{97} ***(17)***.

3.1. Glycosylation Analysis by HPCE

Micellar electrokinetic capillary chromatography (MECC) can be used to rapidly "fingerprint" glycoforms of recombinant human IFN-γ produced by Chinese Hamster Ovary (CHO) cells ***(8)*** and to quantitate variable-site occupancy (macroheterogeneity; *see* **Fig. 1**). This approach allows glycoforms to be rapidly resolved and quantified without the need for oligosaccharide release, derivatization or labeling.

1. Separations are performed with a P/ACE 2100 capillary electrophoresis system using a capillary cartridge containing a 50 mm internal diameter (i.d.) × 57 cm length of underivatized fused silica capillary.
2. Buffer solutions are prepared from phosphoric and boric acids using NaOH to adjust the pH.
3. Capillaries are prepared for use by rinsing with 0.1 *M* NaOH for 10 min, water for 5 min, 0.1 *M* borate, pH 8.5, for 1 h, then 0.1 *M* NaOH and water for 10 min, respectively. Prior to use, capillaries are equilibrated with electrophoresis buffer (400 m*M* borate + 100 m*M* sodium dodecyl sulfate [SDS], pH 8.5) for 1 h.

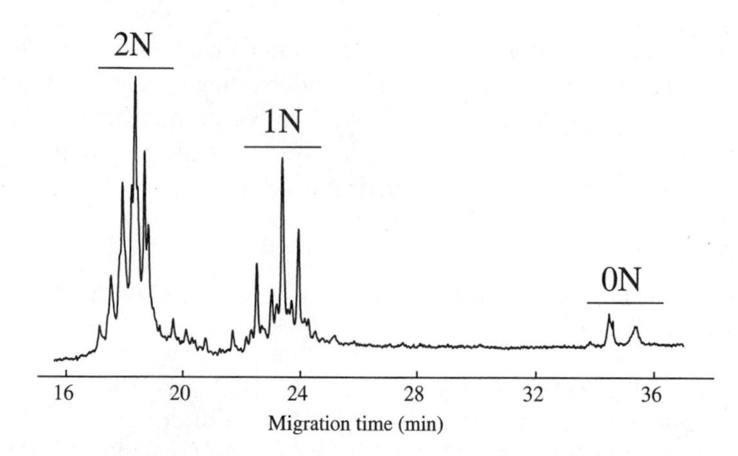

Fig. 1. Whole recombinant human IFN-γ analyzed by capillary electrophoresis. Recombinant human IFN-γ glycoforms were "fingerprinted" by micellar electrokinetic capillary chromatography. Peak groups represent IFN-γ variants with both Asn sites occupied (2N), one site occupied (1N), or no sites occupied (0N).

4. Voltages are applied over a 0.2-min linear ramping period at a detection wavelength of 200 nm and operating temperature of 25°C. Recombinant human IFN-γ (1 mg/mL in 50 m*M* borate, 50 m*M* SDS, pH 8.5) is injected for 5 s prior to electrophoresis at 22 kV. Between each separation, the capillary is rinsed with 0.1 *M* NaOH, water, and electrophoresis buffer for 5 min, respectively (*see* **Note 1**).

3.2. Glycosylation Analysis by MALDI-MS

There are only a few reports of the analysis of whole glycoproteins due to the limited resolution of this technique ***(18)***. As a result, analysis of intact glycoproteins is generally limited to those proteins that contain one glycosylation site and are <15–20 kDa ***(15)***.

MALDI-MS has proved more useful for the identification and characterization of glycopeptides following their separation and purification by reverse-phase HPLC ***(19)***. The advantage of this approach over other methods is that site-specific glycosylation data can be obtained (***20,21***; *see* **Note 2**). This approach has been successfully used to determine the differences in *N*-linked glycosylation for the Asn_{25} and Asn_{97} sites of recombinant human IFN-γ when produced in different expression systems ***(22)***, to monitor changes during batch culture ***(23)***, and to monitor intracellular populations (***24***; *see* **Fig. 2**).

1. Purified IFN-γ samples are digested with trypsin (1.5 μγ; 50 μg) for 24 h at 30°C.
2. The glycopeptides containing the *N*-glycan populations are isolated following their separation by reverse-phase HPLC. Samples are applied in 0.06% (v/v) trifluoroacetic acid (TFA) and the peptides separated with a linear gradient (0–70%) of 80% (v/v) aqueous CH3CN with 0.052% TFA over 100 min at a flow rate of 0.1 ml/min. Peptide peaks are detected at a wavelength of 210 nm and collected individually.
3. The glycopeptides are reduced to the aqueous phase in a Speed Vac concentrator, lyophilized overnight, and stored at –20°C.
4. A 0.5-μL aliquot of the digest samples is mixed with 0.5 μL of a saturated solution of α-cyano-4-hydroxy cinnaminic acid in 60% (v/v) aqueous CH_3CN and allowed to co-crystallize on stainless steel sample discs.

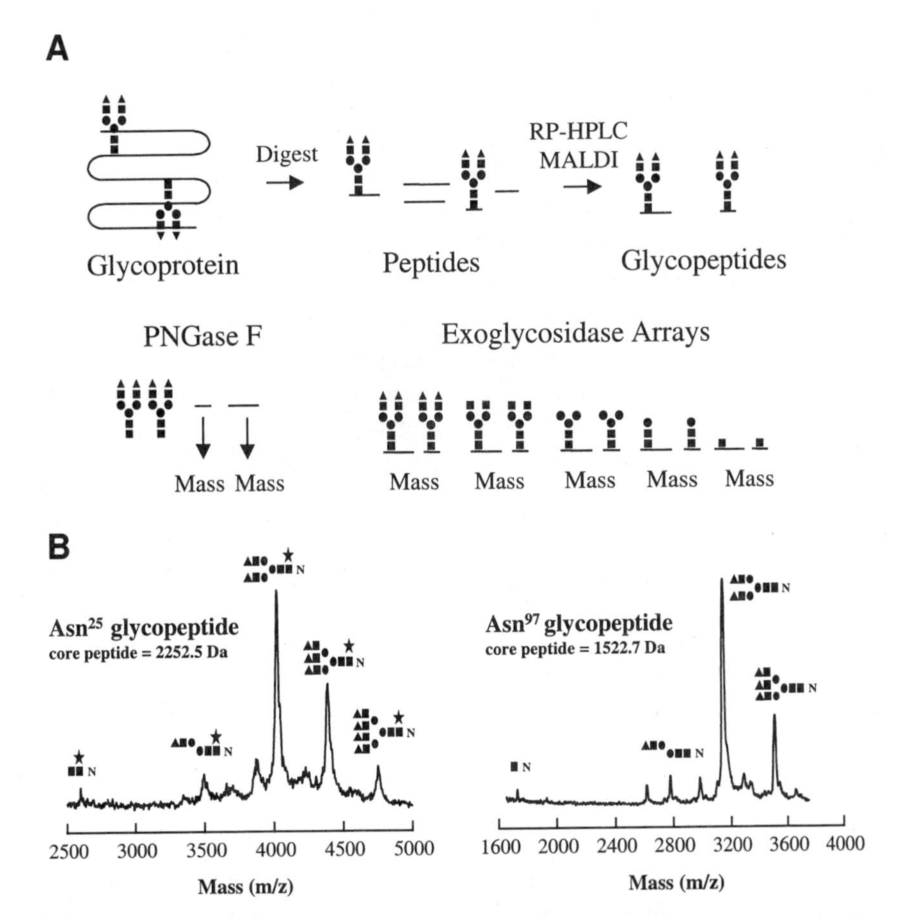

Fig. 2. Site-specific *N*-glycosylation of recombinant human IFN-γ examined by MALDI-MS analyses of glycopeptides. **(A)** The complete analytical protocol. The masses of individual *N*-glycans at a single glycosylation site were calculated by subtracting the known mass of the core peptide moiety from each component in a glycopeptide spectrum. **(B)** An *N*-glycan structure was then tentatively assigned, based on mass criteria alone. As ionization is entirely dependent on the core peptide moiety after desialylation, individual *N*-glycan structures can be quantified. Monosaccharide structures are schematically represented as: (▲), galactose (162.14); (■), *N*-acetylglucosamine (203.20 Da); (●), mannose (162.14 Da); and (★), fucose (146.14 Da).

5. MALDI-MS is performed with a N_2 laser at 337 nm. Desorbed positive ions are detected after a linear flight path by a microchannel plate detector and the digitalized output signal adjusted to obtain an optimum output signal-to-noise ratio from 20 averaged laser pulses. Mass spectra are calibrated with an external standard, vasoactive intestinal peptide with an average molecular mass of 1425.5.
6. Digestion of glycopeptides with peptide-*N*-glycosidase F (PNGaseF) prior to analysis by mass spectrometry confirms the mass of the core peptide. For this determination, 0.5 μL of sample and 0.5 μL of PNGaseF are incubated at 30°C for 24 h and 0.5-μL aliquots are removed for MALDI-MS analysis.
7. Simultaneous digestion of purified glycopeptides with linkage-specific exoglycosidase arrays for the sequential removal of oligosaccharides permit the sequencing of *N*-glycans

at individual glycosylation sites by MALDI-MS *(21–24)*. Sample (0.5 µL) and glycosidase (0.5 µL), or a combination of glycosidases, are incubated at 30°C for 24 h and 0.5-µL aliquots are removed for MALDI-MS analysis (*see* **Note 3**).

3.3. Glycosylation Analysis by ESI-MS

ESI-MS has been used to aid the analysis of glycosylation macro- and micro-heterogeneity and proteolytic cleavage of the C-terminal in conjunction with information obtained by HPLC and MALDI-MS of released oligosaccharides (*25*; *see* **Note 4**).

1. Spectra are obtained with a VG Quattro II triple quadrupole mass spectrometer having a mass range for singly charged ions of 4000 Da (**Fig. 3**).
2. Lyophilized proteins are dissolved in 50% aqueous acetonitrile, 0.2% formic acid to a concentration of 0.1 µg/µL and introduced into the electrospray source at 4 µL/min. The mass-to-charge (*m/z*) range of 600–1800 Da are scanned at 10 s/scan and data are summed for 3–10 min, depending on the intensity and complexity of the spectra. During each scan, the sample orifice-to-skimmer potential (cone voltage) are scanned from 30 V at *m/z* 600 to 75 V at *m/z* 1800. The capillary voltage is set to 3.5 kV.
3. Mass scale calibration employ the multiply charged ion series from a separate introduction of horse heart myoglobin (average molecular mass of 16,951.49). Molecular weights are based on the following atomic weights of the elements: C = 12.011, H = 1.00794, N = 14.00674, O = 15.9994, and S = 32.066 (*see* **Note 5**).
4. Background subtracted *m/z* data are processed by software employing a maximum-entropy (MaxEnt) based analysis to produce zero-charge protein molecular weight information with optimum signal-to-noise ratio, resolution, and mass accuracy.

4. Notes

1. Attempts to separate IFN-γ with borate alone, as used by Landers et al. *(26)* for the separation of ovalbumin glycoforms *(26)*, were unsuccessful, because SDS is required to disrupt the hydrogen-bonded dimers. Application of this technique to ribonuclease B and fetuin met with variable success. The glycoprotein microheterogeneity of a monoclonal antibody with a single glycosylation site has been mapped using a borate buffer at high pH; the glycans were enzymatically or chemically cleaved and the resulting profile used for testing batch-to-batch consistency in conjunction with MALDI-MS analysis *(27)*.
2. MALDI-MS of free *N*-linked oligosaccharides, following chemical release with hydrazinolysis or enzymatic release with an endoglycosidase such as PNGaseF, is also popular as it requires no prior structural knowledge of the glycoprotein of interest and is ideal for the analysis of underivatized populations of oligosaccharides. However, there is a loss of glycosylation site-specific data. Enzymatic release of oligosaccharides is preferred where an intact deglycosylated protein product is required, as *N*-glycan release by hydrazinolysis may result in peptide bond cleavage and the oligosaccharide product requires reacetylation. A drawback to the enzymatic release of oligosaccharides is that the presence of SDS is often required to denature the glycoprotein and has to be removed prior to MALDI-MS analysis. MALDI-MS has also been used for the analysis of IFN-γ glycoforms separated by SDS-polyacrylamide gel electrophoresis (PAGE) *(28)*.
3. Until recently, only desialylated oligosaccharides could be analyzed successfully by MALDI-MS using 2,5-dihydroxybenzoic acid as matrix, as negatively charged sialic acids interfere with the efficiency of ionization using this procedure *(29,30)*. However, advances in matrix mixtures and sample preparation schemes now promise to further improve analytical protocols. For example, sialylated oligosaccharides have recently been

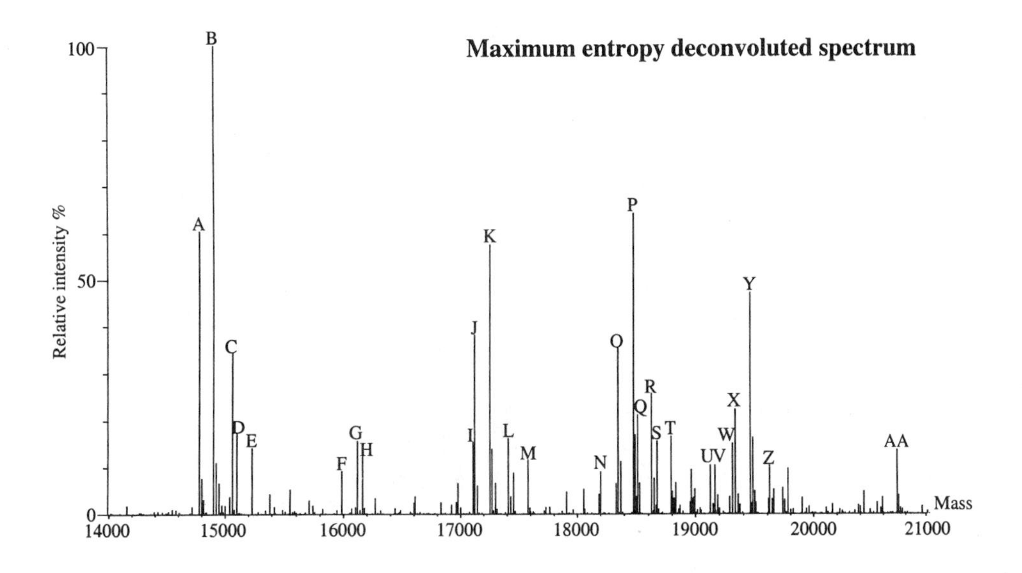

Component	Observed mass (Da)	Correlating mass (Da)	Polypeptide length	Asn sites occupied	Monosaccharide composition
A	14772.5	14771.8	1-127	0	
B	14900.4	14900.0	1-128	0	
C	15056.8	15056.1	1-129	0	
D	15096.9	NA			
E	15224.3	NA			
F	15988.3	15988.9	1-127	1	H5 N2
G	16116.5	16117.1	1-128	1	H5 N2
H	16160.4	16160.5	1-138	0	
I	17107.0	17106.0	1-128	1	H5 N4 S2
J	17124.4	17123.9	1-127	1	H5 N4 S2 D1
K	17253.0	17252.1	1-128	1	H5 N4 S2 D1
L	17409.1	17408.3	1-129	1	H5 N4 S2 D1
M	17576.0	NA			
N	18195.6	18194.9	1-127	2	H10 N6 S2
O	18341.0	18341.0	1-127	2	H10 N6 S2 D1
P	18470.0	18469.2	1-128	2	H10 N6 S2 D1
Q	18511.9	18512.6	1-138	1	H5 N4 S2 D1
R	18627.2	18625.4	1-129	2	H10 N6 S2 D1
S	18668.8	18668.8	1-139	1	H5 N4 S2 D1
T	18793.8	NA			
U	19124.3	19125.0	1-133	2	H10 N6 S2 D1
V	19165.5	19166.9	1-128	2	H10 N8 S3 D1
W	19312.8	19312.0	1-128	2	H10 N8 S4
		19313.0*	1-128	2	H10 N8 S3 D2
X	19329.2	19329.9	1-127	2	H10 N8 S4 D1
Y	19458.6	19458.1	1-128	2	H10 N8 S4 D1
Z	19614.7	19614.3	1-129	2	H10 N8 S4 D1
AA	20719.0	20718.6	1-138	2	H10 N8 S4 D1

Fig. 3. Heterogeneous glycoprotein populations directly analyzed by ESI-MS. This technique provides highly resolved mass analyses with a mass accuracy of 0.01% (1 Dalton/10 kDa). In this example, individual transgenic mouse derived recombinant human IFN-γ components were assigned a C-terminal polypeptide cleavage site and an overall monosaccharide composition.

shown to ionize effectively, with picomole to femtomole sensitivity, as deprotonated molecular ions using 2,4,6-trihydroxyacetophenone as matrix in the presence of ammonium citrate ***(31)***. Further improvements in sensitivity and resolution may be obtained by derivatization of oligosaccharides with fluorophores such as 2-aminoacridome (AMAC) or 2-aminobenzamidine (AB; **Fig. 4**). Integration of MALDI-MS peak areas obtained on

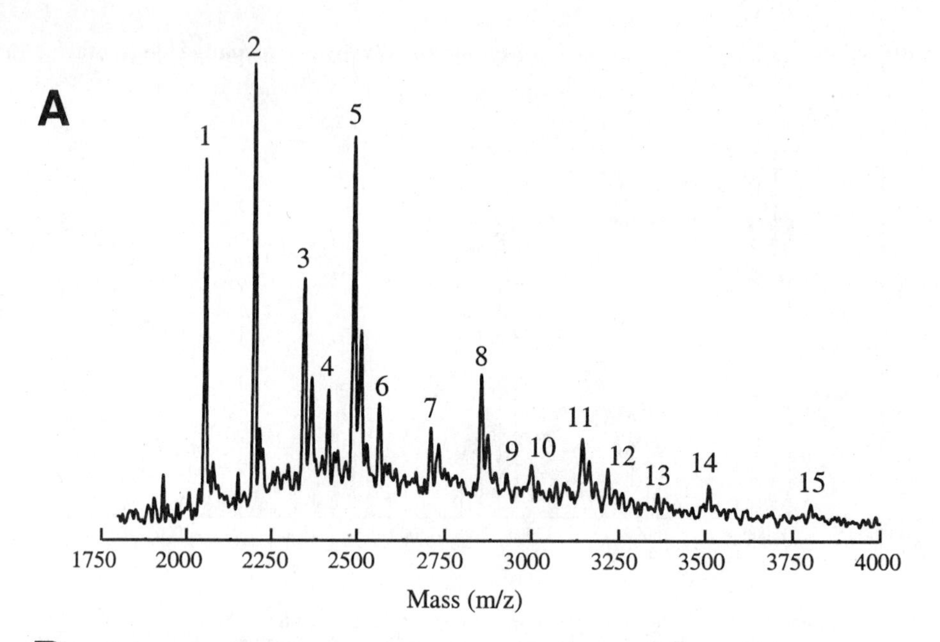

B

Component	Observed Mass (Da)	Correlating Mass (Da)	Monosaccharide Composition
1	2052	2052.7	H5N4D0S1
2	2198	2198.8	H5N4D1S1
3	2343	2344.0	H5N4D0S2
4	2416	2418.1	H6N5D0S1
5	2490	2490.1	H5N4D1S2
6	2563	2564.2	H6N5D1S1
7	2709	2709.3	H6N5D0S2
8	2855	2855.5	H6N5D1S2
9	2928	2929.5	H7N6D1S1
10	2998	3000.6	H6N5D0S3
11	3146	3146.7	H6N5D1S3
12	3219	3220.8	H7N6D1S2
13	3364	3366.0	H7N6D0S3
14	3511	3512.0	H7N6D1S3
15	3801	3803.3	H7N6D1S4

Fig. 4. Sialylated *N*-glycans associated with recombinant human IFN-γ released with PNGaseF, labeled with 2-aminobenzamide, and analyzed by MALDI-MS using 2,4,6-trihydroxyacetophenone as matrix (**A**). The masses of individual *N*-glycans are used to assign an overall monosaccharide composition, including degree of sialylation (**B**). H, hexose; N, *N*-acetylhexosamine; D, deoxyhexose; and S, *N*-acetylneuraminic acid.

analysis of sialylated glycans from IFN-γ provided quantitative information that compared favorably with analysis of the derivatized sialylated glycans by ion-exchange HPLC.

4. Glycopeptides may be directly analyzed by ESI-MS ***(32,33)***, and the oligosaccharides sequenced following digestion with combinations of exoglycosidases ***(34)*** or glycoforms separated by liquid chromatography prior to analysis ***(35)***. Possibly the most powerful application of this technique has resulted from its interfacing with liquid chromatography which permits the separation and on-line identification of glycoproteins from protein

digests *(36)*. Although ESI-MS may be interfaced with capillary electrophoresis, the separation of glycoproteins and glycopeptides under acidic conditions makes the analysis of sialylated oligosaccharides unsuitable.

5. ESI-MS is not ideally suited for the analysis of neutral and anionic oligosaccharides that have been chemically or enzymatically released from a glycoprotein of interest as they do not readily form multiply charged ions. However, the characterization of methylated derivatives of oligosaccharides from recombinant erythropoietin by ESI-MS has been reported *(37)*.

References

1. Jenkins, N. and Curling, E. M. (1994) Glycosylation of recombinant proteins: problems and prospects. *Enzyme Microb. Technol.* **16,** 354–364.
2. Liu, D. T. Y (1992) Glycoprotein pharmaceuticals—scientific and regulatory considerations and the United States Orphan Drug Act. *Trends Biotechnol.* **10,** 114–120.
3. Dwek, R. A., Edge, C. J., Harvey, D. J., and Wormald, M. R. (1993) Analysis of glycoprotein associated oligosaccharides. *Ann. Rev. Biochem.* **62,** 65–100.
4. Huberty, M. C., Vath, J. E., Yu, W., and Martin, S. A. (1993) Site-specific carbohydrate identification in recombinant proteins using MALDI-TOF MS. *Analyt. Chem.* **65,** 2791–2800.
5. Frenz, J. and Hancock, W. S. (1991) High performance capillary electrophoresis. *Trends Biotechnol.* **9,** 243–250.
6. Novotny, M. V. and Sudor, J. (1993) High performance capillary electrophoresis of glycoconjugates. *Electrophoresis* **14,** 372–389.
7. Rush, R., Derby, P., Strickland, T., and Rohde, M. (1993) Peptide mapping and evaluation of glycopeptide microheterogeneity derived from endoproteinase digestion of erythropoietin by affinity high-performance capillary electrophoresis. *Analyt. Chem.* **65,** 1834–1842.
8. James, D. C., Freedman,, R. B. Hoare, M., and Jenkins, N. (1994) High resolution separation of recombinant human interferon-γ glycoforms by micellar electrokinetic capillary chromatography. *Analyt. Biochem.* **222,** 315–322.
9. Rudd, P. M., Scragg, I. G., Coghill, E., and Dwek, R. A. (1992) Separation and analysis of the glycoform populations of ribonuclease b using capillary electrophoresis. *Glycoconjugate J.* **9,** 86–91.
10. El-Rassi, Z. and Mechref, Y. (1996) Recent advances in capillary electrophoresis of carbohydrates. *Electrophoresis* **17,** 275–301.
11. Müller, D., Domon, B., Karas, M., van Oostrum, J., and Richter, W. J. (1994) Characterization and direct glycoform profiling of a hybrid plasminogen-activator by matrix-assisted laser-desorption and electrospray mass spectrometry—correlation with high performance liquid-chromatographic and nuclear-magnetic-resonance analyses of the released glycans. *Biol. Mass Spectrom.* **23,** 330–338.
12. Hunter, A. P. and Games, D. E. (1995) Evaluation of glycosylation site heterogeneity and selective identification of glycopeptides in proteolytic digests of bovine α1-acid glycoprotein by mass spectrometry. *Rapid Commun. Mass Spectrom.* **9,** 42–56.
13. Mann, M. and Wilm, M. (1995) Electrospray mass spectrometry for protein characterization. *Trends Biochem. Sci.* **20,** 219–224.
14. Gu, J., Hiraga, T., and Wada, Y. (1994) Electrospray ionization mass spectrometry of pyridylaminated oligosaccharide derivatives: sensitivity and in-source fragmentation. *Biol. Mass Spectrom.* **23,** 212–217.
15. Tsarbopoulos, A., Pramanik, B. N., Nagabhushan, T. L., and Covey, T. R. (1995) Structural analysis of the CHO-derived interleukin-4 by liquid chromatography electrospray ionization mass spectrometry. *J. Mass. Spectrom.* **30,** 1752–1763.

16. Ashton, D. S., Beddell, C. R., Cooper, D. J., Craig, S. J., Lines, A. C., Oliver, R. W. A., and Smith, M. A. (1995) Mass spectrometry of the humanized monoclonal antibody CAMPATH 1H. *Analyt. Chem.* **67,** 835–842.
17. Hooker, A. D. and James, D. C. (1998) The glycosylation heterogeneity of recombinant human IFN-γ. *J. Interf. Cytok. Res.* **18,** 287–295.
18. Bihoreau, N., Veillon, J. F., Ramon, C., Scohyers, J. M., and Schmitter, J. M. (1995) Rapid Commun. Characterization of a recombinant antihaemophilia-A factor (factor-VIII-delta-II) by matrix-assisted laser desorption ionization mass spectrometry. *Rapid Commun. Mass. Spectrom.* **9,** 1584–1588.
19. Stone, K. L., LoPresti, M. B., Crawford, J. M., DeAngelis, R., and Williams, K. R. (1989) *A Practical Guide to Protein and Peptide Purification for Microsequencing,* Academic Press, San Diego, CA, p. 36.
20. Treuheit, M. J., Costello, C. E., and Halsall, H. B. (1992) Analysis of the five glycosylation sites of human α1-acid glycoprotein. *Biochem. J.* **283,** 105–112.
21. Sutton, C. W., O'Neill, J., and Cottrell, J. S. (1994) Site specific characterization of glycoprotein carbohydrates by exoglycosidase digestion and laser desorption mass spectrometry. *Analyt. Biochem.* **218,** 34–46.
22. James, D. C., Freedman, R. B., Hoare, M., Ogonah, O. W., Rooney, B. C., Larionov, O. A., et al. (1995) *N*-Glycosylation of recombinant human interferon-gamma produced in different animal expression systems. *BioTechnology* **13,** 592–596.
23. Hooker, A. D., Goldman, M. H., Markham, N. H., James, D. C., Ison, A. P., Bull, A. T., et al. (1995) *N*-Glycans of recombinant human interferon-γ change during batch culture of Chinese hamster ovary cells. *Biotechnol. Bioeng.* **48,** 639–648.
24. Hooker, A. D., Green, N. H., Baines, A. J., Bull, A. T., Jenkins, N., Strange, P. G., and James, D. C. (1999) Constraints on the transport and glycosylation of recombinant IFN-γ in Chinese hamster ovary and insect cells. *Biotechnol. Bioeng.* **63,** 559–572.
25. James, D. C., Goldman, M. H., Hoare, M., Jenkins, N., Oliver, R. W. A., Green, B. N., and Freedman, R. B. (1996) Post-translational processing of recombinant human interferon-gamma in animal expression systems. *Protein Sci.* **5,** 331–340.
26. Landers, J. P., Oda, R. P., Madden, B. J., and Spelsberg, T. C. (1992) High-performance capillary electrophoresis of glycoproteins: the use of modifiers of electroosmotic flow for analysis of microheterogeneity. *Analyt. Biochem.* **205,** 115–124.
27. Hoffstetter-Kuhn, S. H., Alt, G., and Kuhn, R. (1996) Profiling of oligosaccharide-mediated microheterogeneity of a monoclonal antibody by capillary electrophoresis. *Electrophoresis* **17,** 418–422.
28. Mortz, E., Sareneva, T., Haebel, S., Julkunen, I., and Roepstorff, P. (1996) Mass spectrometric characterization of glycosylated interferon-gamma variants by gel electrophoresis. *Electrophoresis* **17,** 925–931.
29. Stahl, B., Steup, M., Karas, M., and Hillenkamp, F. (1991) Analysis of neutral oligosaccharides by matrix-assisted laser desorption/ionization mass spectrometry. *Analyt. Chem.* **63,** 1463–1466.
30. Tsarbopoulos, A., Karas, M., Strupat, K., Pramanik, B., Nagabhushan, T., and Hilenkamp, F. (1994) Comparative mapping of recombinant proteins and glycoproteins by plasma desorption and matrix-assisted laser desorption/ionization mass spectrometry. *Analyt. Chem.* **66,** 2062–2070.
31. Papac, D., Wong, A., and Jones, A. (1996) Analysis of acidic oligosaccharides and glycopeptides by matrix-assisted laser desorption/ionization time-of-flight mass spectrometry. *Analyt. Chem.* **68,** 3215–3223.

32. Rush, R. S., Derby, P. L., Smith, D. M., Merry, C., Rogers, G., Rohde, M. F., and Katta, V. (1995) Microheterogeneity of erythropoietin carbohydrate structure. *Analyt. Chem.* **67,** 1442–1452.
33. Bloom, J. W., Madanat, M. S., and Ray, M. K. (1996) Cell-line and site specific comparative analysis of the *N*-linked oligosaccharides on human ICAM/-1DES454-532 by electrospray mass spectrometry. *Biochemistry* **35,** 1856–1864.
34. Schindler, P. A., Settineri, C. A., Collet, X., Fielding, C. J., and Burlingame, A. L. (1995) Site-specific detection and structural characterization of the glycosylation of human plasma proteins lecithin:cholesterol acyltransferase and apolipoprotein D using HPLC/electrospray mass spectrometry and sequential glycosidase digestion. *Protein Sci.* **4,** 791–801.
35. Medzihradszky, K. F., Maltby, D. A., Hall, S. C., Settineri, C. A., and Burlingame, A. L. (1994) Characterization of protein N-glycosylation by reverse-phase microbore liquid chromatography electrospray mass spectrometry, complimentary mobile phases and sequential exoglycosidase digestion. *J. Am. Soc. Mass Spectrom.* **5,** 350–358.
36. Ling, V., Guzzetta, A. W., Canova-Davis, E., Stults, J. T., Hancock, W. S., Covey, T. R., and Shushan, B. I. (1991) Characterization of the tryptic map of recombinant DNA derived tissue plasminogen activator by high performance liquid chromatography-electrospray ionization mass spectrometry. *Analyt. Chem.* **63,** 2909–2915.
37. Linsley, K. B., Chan, S. Y., Chan, S., Reinhold, B. B., Lisi, P. J., and Reinhold, V. N. (1994) Applications of electrospray mass-spectrometry to erythopoietin *N*-linked and *O*-linked glycans. *Analyt. Biochem.* **219,** 207–217.

127

Affinity Chromatography of Oligosaccharides and Glycopeptides with Immobilized Lectins

Kazuo Yamamoto, Tsutomu Tsuji, and Toshiaki Osawa

1. Introduction

Sugar moieties on the cell surface play one of the most important roles in cellular recognition. To elucidate the molecular mechanism of these cellular phenomena, assessment of the structure of sugar chains is indispensable. However, it is difficult to elucidate the structures of cell surface oligosaccharides owing to two technical problems. The first is the difficulty in fractionating various oligosaccharides heterogeneous in the number, type, and substitution patterns of outer sugar branches. The second problem is that very limited amounts of material can be available, which makes it difficult to perform detailed structural studies. Lectins are proteins with sugar binding activity. Each lectin binds specifically to a certain sugar sequence in oligosaccharides and glycopeptides. To overcome these problems, lectins can serve as very useful tools. Recently, many attempts have been made to fractionate oligosaccharides and glycopeptides on immobilized lectin columns. The use of a series of immobilized lectin columns, whose sugar binding specificities have been precisely elucidated, enables us to fractionate a very small amount of radioactive oligosaccharides or glycopeptides (~10 ng depending on the specific activity) into structurally distinct groups. In this chapter, we summarize the serial lectin–Sepharose affinity chromatographic technique for rapid, sensitive, and specific fractionation and analysis of asparagine-linked oligosaccharides of glycoproteins.

Structures of asparagine-linked oligosaccharides fall into three main categories termed high mannose type, complex type, and hybrid type *(1)*. They share the common core structure Manα1–3(Manα1–6)Manβ1–4GlcNAcβ1– 4GlcNAc-Asn but differ in their outer branches (**Fig. 1**). High mannose type oligosaccharides have two to six additional α-mannose residues to the core structure. A typical complex type oligosaccharide contains two to four outer branches with a sialyllactosamine sequence. Hybrid type structures have the features of both high mannose type and complex type oligosaccharides and most of them contain bisecting *N*-acetylglucosamine, which is linked β1–4 to the β-linked mannose residue of the core structure. Recently a novel type of carbohydrate chain, the so-called poly-*N*-acetyllactosamine type, has been described *(2–5)*. Its outer branches have a characteristic structure composed of N-acetyllactosamine repeating units. It may be classified as complex type; however, it

From: *The Protein Protocols Handbook, 2nd Edition*
Edited by: J. M. Walker © Humana Press Inc., Totowa, NJ

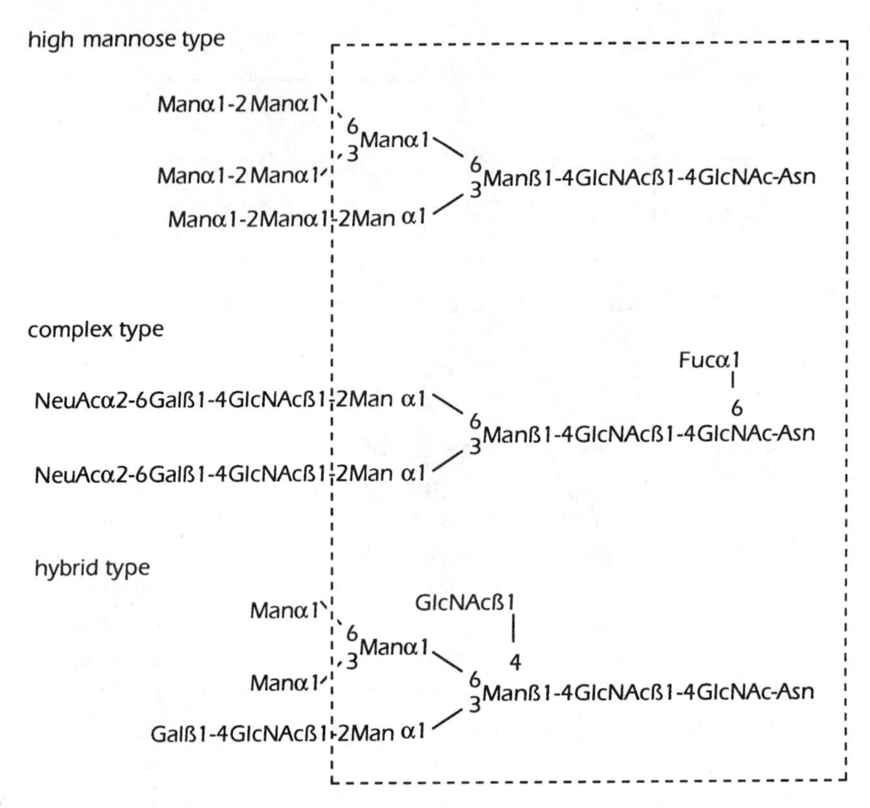

Fig. 1. Structures of major types of asparagine-linked oligosaccharides. The *boxed area* encloses the core structure common to all asparagine-linked structures.

is antigenically and functionally distinct from, standard complex type sugar chains *(4)*. Some of poly-*N*-acetyllactosamine type oligosaccharides have branched sequences containing Galβ1–4GlcNAcβ1–3(Galβ1–4GlcNAcβ1–6)Gal units *(2,3)*, which is the determinant of the I-antigen. Other novel complex type sugar chains having GalNAcβ1-4GlcNAc groups in their outer chain moieties have been found recently *(6,7)* and GalNAc residues are sometimes sulfated at the C4 position or sialylated at the C6 position.

Glycopeptides or oligosaccharides can be prepared from glycoproteins by enzymatic digestion or chemical methods as discussed in subsequent chapters in this book. The most widely used means for preparing glycopeptides is to completely digest material with pronase. Oligosaccharides can be prepared from glycoproteins or glycopeptides by treating samples with anhydrous hydrazine *(8)* or endoglycosidases. As the released oligosaccharides retain their reducing termini, they can be radiolabeled by reduction with NaB^3H_4 *(9)*. Primary amino groups of the peptide backbone of glycopeptides are labeled by acetylation with $[^3H]$- or $[^{14}C]$acetic anhydride *(10)*. Before employing columns of immobilized lectins for analyses, oligosaccharides or glycopeptides should be separated on a column of QAE- or DEAE-cellulose based on anionic charge derived from sialic acid, phosphate, or sulfate residues. Acidic oligosaccharides thus separated should be converted to neutral ones for simplifying the following separation. To simplify the presentation, the oligosaccharides discussed here do not contain sialic acid, phosphate, or sulfate residues, although these acidic residues, especially sialic acid, are found in many oligosaccharides. In most cases, the effect of these residues on the interaction of oligosaccharides with immobilized lectins is weak, but if documentation of this influence is available, it is mentioned in the appropriate sections. In this chapter we describe the general procedure of serial lectin affinity chromatography of glycopeptides and oligosaccharides using several well defined immobilized lectins.

2. Materials

1. Mono Q HR5/5, DEAE-Sephacel, Sephadex G-25 (Amersham Pharmacia Biotech, Uppsala, Sweden).
2. High-performance liquid chromatograph, two pumps, with detector capable of monitoring ultraviolet absorbance at 220 nm.
3. Neuraminidase: 1 U/mL of neuraminidase from *Streptococcus* sp. (Seikagaku Kogyo, Tokyo Japan) in 50 m*M* acetate buffer, pH 6.5.
4. Dowex 50W-X8 (50–100 mesh, H^+ form).
5. Bio-Gel P-4 minus 400 mesh (Bio-Rad, Richmond, CA).
6. HPLC mobile phase for Mono Q: A—2 m*M* Tris-HCl, pH 7.4; B—2 m*M* Tris-HCl, pH 7.4, 0.5 *M* NaCl.
7. HPLC mobile phase for Bio-Gel P-4: Distilled water.
8. HPLC standard for Bio-Gel P-4: Partial hydrolysate of chitin, which prepared according to Rupley *(11)*; 10 μg mixed with 50 μL of distilled water. Store frozen.
9. Concanavalin A, *Ricinus communis* lectin, wheat germ lectin, *Datura stramonium* lectin, *Maackia amurensis* leukoagglutinin, *Wistaria floribunda* lectin, *Allomyrina dichotoma* lectin, *Amaranthus caudatus* lectin (EY Laboratories, San Mateo, CA), *Phaseorus vulgaris* erythroagglutinin, *Phaseorus vulgaris* leukoagglutinin (Seikagaku Kogyo). Immobilized lectins were prepared at a concentration of 1–5 mg of lectin/mL of gel (*see* **Notes 1** and **2**) or obtained commercially (e.g., Amersham Pharmacia Biotech, EY Laboratories, Bio-Rad, Seikagaku Kogyo): *Galanthus nivalis* lectin, *Lens culinaris* lectin, *Pisum sativum* lectin, *Vicia fava* lectin, pokeweed mitogen, *Sambucus nigra* L lectin.
10. $[^3H]NaB^3H_4$: 3.7×10^9 Bq of $[^3H]NaBH_4$ (sp $1.9–5.6 \times 10^{11}$ Bq/mmol; NEN, Boston, MA) mixed with 2 mL of 10 m*M* NaOH; store at –80°C.
11. Tris-buffered saline (TBS): 10 m*M* Tris-HCl, pH 7.4, 0.15 *M* NaCl.
12. Lectin column buffer: 10 m*M* Tris-HCl, pH 7.4, 0.15 *M* NaCl, 1 m*M* $CaCl_2$, 1 m*M* $MnCl_2$ (*see* **Note 3**).
13. *N*-Acetylgalactosamine (Sigma, St. Louis, MO): 100 m*M* in lectin column buffer; store refrigerated.
14. Methyl-α-mannoside (Sigma): 100 m*M* in TBS; store refrigerated.
15. Methyl-α-glucoside (Sigma): 10 m*M* in TBS; store refrigerated.
16. Lactose (Sigma): 50 m*M* in TBS; store refrigerated.
17. *N*-Acetylglucosamine (Sigma): 200 m*M* in TBS; store refrigerated.

3. Methods

3.1. Separation of Acidic Sugar Chains on Mono Q HR5/5 or DEAE-Sephacel and Removal of Sialic Acids

3.1.1. Ion-Exchange Chromatography

1. Equilibrate the Mono Q HR5/5 or DEAE-Sephacel column with 2 m*M* Tris-HCl, pH 7.4, at a flow rate of 1 mL/min at room temperature.

2. Dissolve the oligosaccharides or the glycopeptides in 0.1 mL of 2 m*M* Tris-HCl, pH 7.4, and apply to the column.
3. Elute with 2 m*M* Tris-HCl, pH 7.4, for 10 min, then with a linear gradient (0–20 %) of 2 m*M* Tris-HCl, pH 7.4, 0.5 *M* NaCl for 60 min at a flow rate of 1 mL/min.
4. Neutral oligosaccharides are recovered in the pass-through fraction. Acidic monosialo-, disialo-, trisialo-, and tetrasialooligosaccharides are eluted out successively by the linear NaCl gradient.

3.1.2 Removal of Sialic Acid Residues

1. To 10–100 µg of oligosaccharides free of buffers or salts, add 100 µL of neuraminidase buffer and 100 µL of neuraminidase, and incubate at 37°C for 18 h.
2. Heat-inactivate the neuraminidase by immersion in a boiling water bath for 3 min.
3. Apply to the column of Dowex 50W-X8 (0.6 cm internal diameter × 2.5 cm), wash the column with 1 mL of distilled water, and concentrate the eluates under vacuum.

Alternatively, add 500 µL of 0.1 *M* HCl and heat at 80°C for 30 min, and dry up the sample using evaporator.

3.2. Separation of Poly-N-acetyllactosamine Type Sugar Chains from Other Types of Sugar Chains

Poly-*N*-acetyllactosamine type sugar chains vary as to the number of *N*-acetyllactosamine repeating units and the branching mode and the structural characterization of poly-*N*-acetyllactosamine type sugar chains has been quite difficult ***(12)***. This type of sugar chain has a higher molecular weight than other high mannose type, complex type or hybrid type chains. Poly-*N*-acetyllactosamine type sugar chains with a molecular mass > 4000 are excluded from the Bio-Gel P-4 column chromatography ***(2,13,14)*** and thus are easily separated from others.

1. Equilibrate two coupled columns (0.8 cm i.d. × 50 cm) of Bio-Gel P-4 in water at 55°C with a water jacket.
2. Elute the oligosaccharides at a flow rate of 0.3 mL/min and collect fractions of 0.5 mL. Monitor absorbance at 220 nm.
3. Collect poly-*N*-acetyllactosamine type oligosaccharides that are eluted at the void volume of the column. Other types of oligosaccharides included in the column are subjected to an additional separation (**Subheadings 3.3.–3.6.**) illustrated in **Fig. 2**. The specificity of the lectins used is summarized in **Fig. 3** and **Table 1**.

3.3. Separation of Complex Type Sugar Chains Containing GalNAcβ1–4GlcNAc Groups from Other Sugar Chains

Novel complex type oligosaccharides and glycopeptides with GalNAcβ1–4GlcNAcβ1–2 outer chains bind to *Wistaria floribunda* lectin (WFA)–Sepharose column ***(7)***.

1. Equilibrate the WFA–Sepharose column (0.6 cm i.d. ×5.0 cm) in lectin column buffer.
2. Dissolve oligosaccharides or glycopeptides in 0.5 mL of lectin column buffer, and apply to the column.
3. Elute (1.0-mL fractions) successively with three column volumes of lectin column buffer, and then with three column volumes of 100 m*M* *N*-acetylgalactosamine at flow rate 2.5 mL/h at room temperature.
4. Collect complex type sugar chains with GalNAcβ1–4GlcNAc outer chains, which are eluted after the addition of *N*-acetylgalactosamine.
5. Collect other type of sugar chains, which pass through the column.

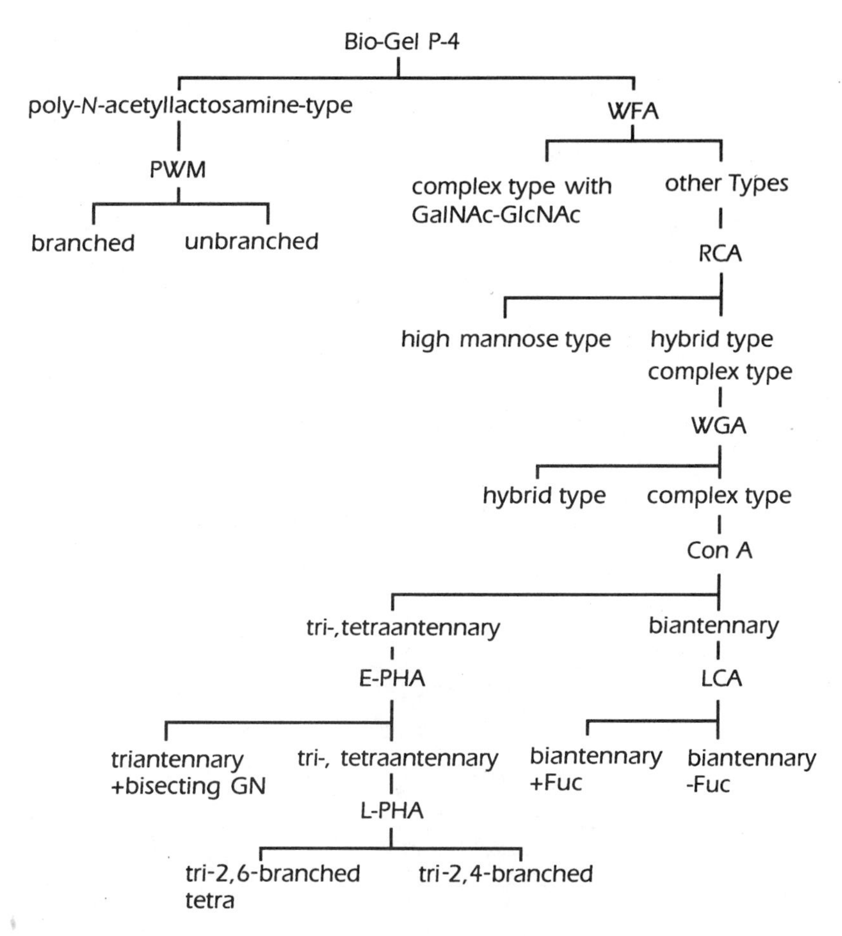

Fig. 2. Scheme of fractionation of asparagine-linked sugar chains by combining affinity chromatography on immobilized lectins.

3.4. Separation of High Mannose Type Sugar Chains from Complex Type and Hybrid Type Sugar Chains

3.4.1. Affinity Chromatography on Immobilized RCA

After the separation of high molecular weight poly-*N*-acetyllactosamine type oligosaccharides, a mixture of other three types of sugar chains can be separated on a column of *Ricinus communis* lectins (RCA) which recognizes the Galβ1–4GlcNAc sequence ***(15,16)***.

1. Equilibrate the RCA–Sepharose column (0.6 cm i.d. × 5.0 cm) in TBS.
2. Dissolve oligosaccharides or glycopeptides in 0.5 mL of TBS, and apply to the column.
3. Elute (1.0-mL fractions) successively with three column volumes of TBS, then with three column volumes of 50 m*M* lactose at a flow rate 2.5 mL/h at room temperature.
4. Bind both complex type and hybrid type sugar chains to the RCA–Sepharose column (*see* **Note 4**).
5. Collect high mannose type oligosaccharides, which pass through the column.
6. Purify the oligosaccharides or glycopeptides from salts and haptenic sugar by gel filtration on a Sephadex G-25 column (1.2 cm i.d. × 50 cm) equilibrated with distilled water.

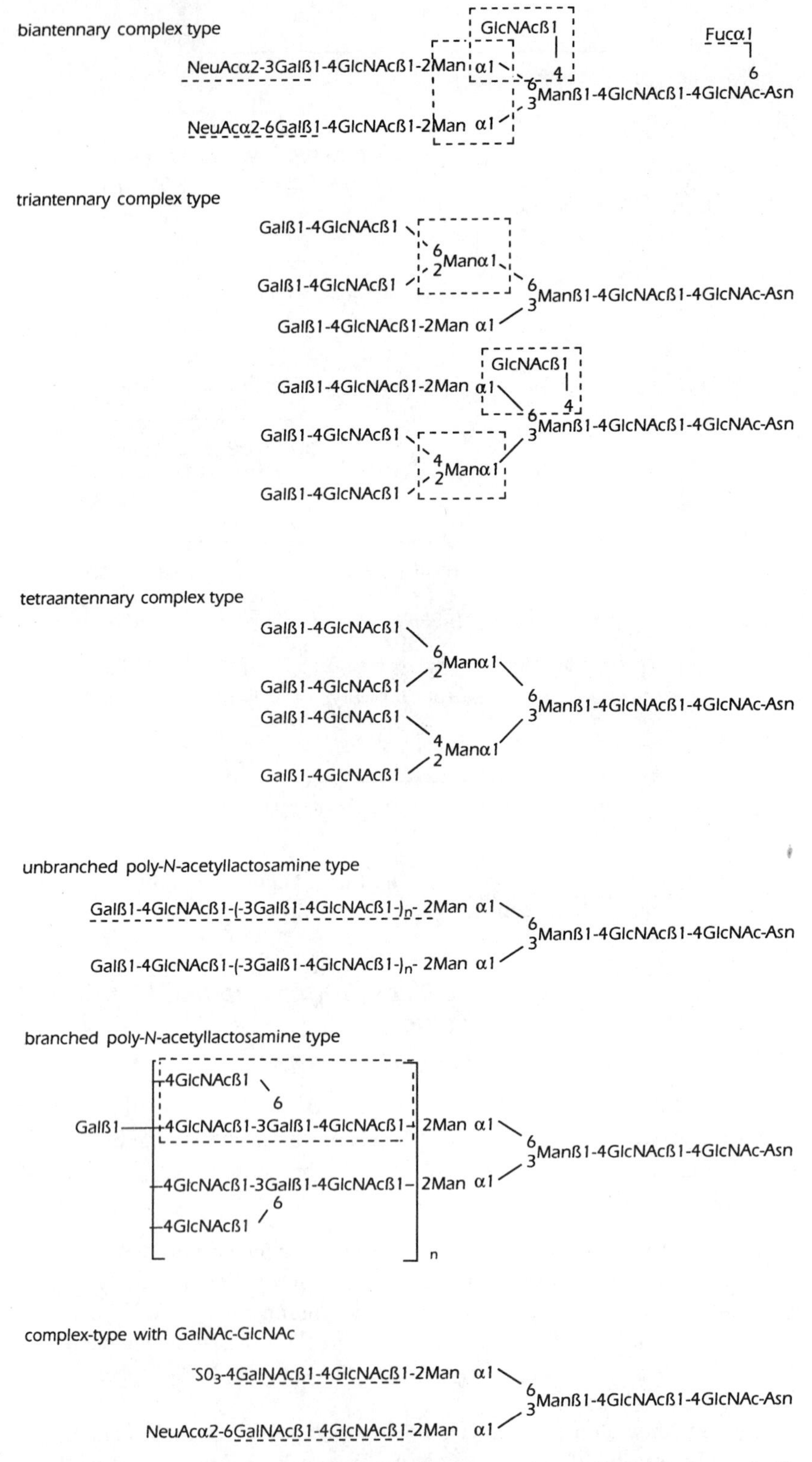

Fig. 3. Structures of several complex type oligosaccharides. The boxed area indicates the characteristic structures recognized by several immobilized lectins.

Table 1
Characteristic Structures Recognized by Several Immobilized Lectins

	RCA	GNA	WGA	ConA	LCA	PSA	VFA	E-PHA	L-PHA	DSA	WFA
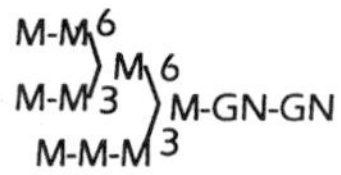 M-M 6 M 6 M-M 3 M-GN-GN M-M-M 3	-	R	-	+	-	-	-	-	-	-	-
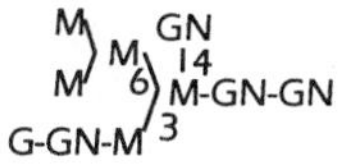 M GN M 14 M 6 M-GN-GN G-GN-M 3	+	-	R	+	-	-	-	-	-	-	-
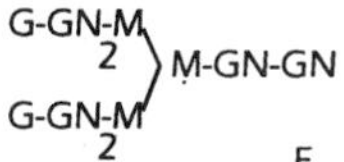 G-GN-M 2 M-GN-GN G-GN-M 2	+	-	-	+	-	-	-	-	-	-	-
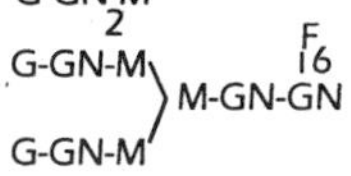 F G-GN-M 16 M-GN-GN G-GN-M	+	-	-	+	+	+	+	-	-	-	-
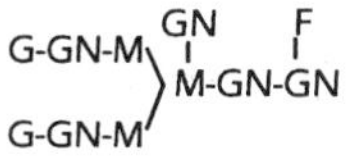 G-GN-M GN F M-GN-GN G-GN-M	+	-	-	+	-	-	-	R	-	-	-
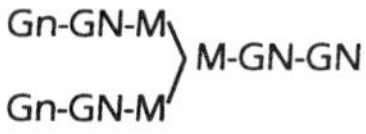 Gn-GN-M M-GN-GN Gn-GN-M	-	-	-	-	-	-	-	-	-	-	+
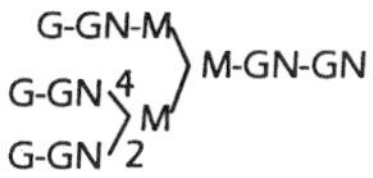 G-GN-M M-GN-GN G-GN 4 M G-GN 2	+	-	-	-	-	-	-	-	-	R	-
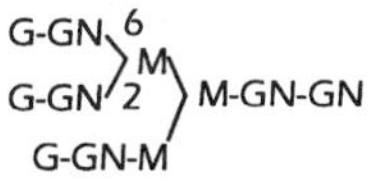 G-GN 6 M G-GN 2 M-GN-GN G-GN-M	+	-	-	-	-	-	-	-	R	+	-
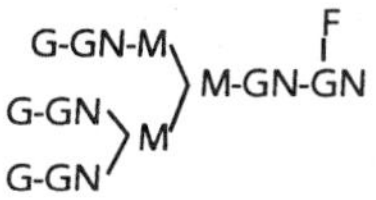 G-GN-M F M-GN-GN G-GN M G-GN	+	-	-	-	-	-	-	-	-	R	-
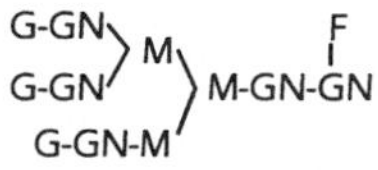 G-GN F M G-GN M-GN-GN G-GN-M	+	-	-	-	-	-	-	-	R	+	-
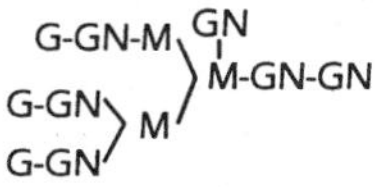 G-GN-M GN M-GN-GN G-GN M G-GN	+	-	-	-	-	-	-	R	-	R	-
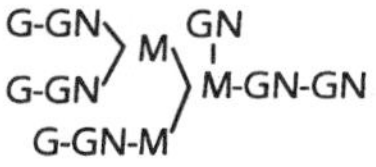 G-GN GN M G-GN M-GN-GN G-GN-M	+	-	-	-	-	-	-	-	R	N.D.	-
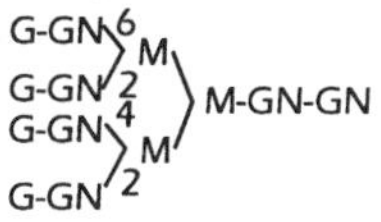 G-GN 6 M G-GN 2 G-GN 4 M-GN-GN M G-GN 2	+	-	-	-	-	-	-	-	R	+	-
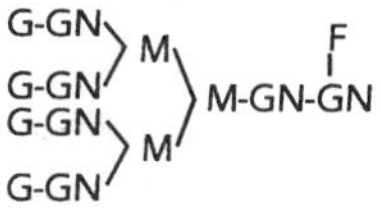 G-GN F M G-GN M-GN-GN G-GN M G-GN	+	-	-	-	-	-	-	-	R	+	-

+, bound; R, retarded; –, not bounds; N.D., not determined; G, galactose; M, mannose; F, fucose; GN, *N*-acetylglucosamine; Gn, *N*-acetylgalactosamine.

3.4.2. Affinity Chromatography on Immobilized Snowdrop Lectin

High mannose type glycopeptides, which carry Manα1–3Man units, are specifically retarded on the immobilized snowdrop *Galanthus nivalis* lectin (GNA) ***(17)***.

1. Equilibrate the GNA–Sepharose column (0.6 cm i.d. × 5.0 cm) in TBS.
2. Dissolve oligosaccharides or glycopeptides in 0.5 mL of TBS, and apply to the column.
3. Elute (0.5-mL fractions) successively with five column volumes of TBS, to collect sugar chains lacking Manα1–3Man units or hybrid type, which are not retarded.
4. Elute with three column volumes of 100 m*M* methyl-α-mannoside at a flow rate 2.5 mL/h at room temperature to obtain the specifically retarded high mannose type glycopeptides that carry Manα1–3Man units.

3.5. Separation of Hybrid Type Sugar Chains from Complex Type Sugar Chains

3.5.1. Affinity Chromatography on Immobilized WGA

Wheat germ agglutinin (WGA)–Sepharose has a high affinity for the hybrid type sugar chains. It is demonstrated that the sugar sequence GlcNAcβ1–4Manβ1–4GlcNAcβ1–4GlcNAc-Asn structure is essential for tight binding of glycopeptides to the WGA–Sepharose column ***(18)***.

1. Equilibrate the WGA–Sepharose column (0.6 cm i.d. × 5.0 cm) in TBS.
2. Dissolve glycopeptides in 0.5 mL of TBS and apply to the column.
3. Elute (0.5-mL fractions) successively with five column volumes of TBS.
4. Collect hybrid type glycopeptides with a bisecting *N*-acetylglucosamine residue, which are retarded on the WGA column.
5. Collect sugar chains having the typical complex type (and also high mannose type) sugar chains elute at the void volume of the column with TBS.

3.6. Separation of Complex Type Biantennary Sugar Chains

3.6.1. Affinity Chromatography on Immobilized Con A

Oligosaccharides and glycopeptides with tri- and tetraantennary complex type sugar chains pass through concanavalin A (Con A)–Sepharose whereas biantennary complex type, hybrid type and high mannose type sugar chains bind to the Con A and can be differentially eluted from the column ***(19,20)***.

1. Equilibrate the Con A–Sepharose column (0.6 cm i.d. × 5.0 cm) in lectin column buffer.
2. Pass the oligosaccharide mixture of the complex type chain from the WGA column through the Con A–Sepharose column.
3. Elute (1-mL fractions) successively with three column volumes of lectin column buffer.
4. Collect oligosaccharides with tri- and tetraantennary complex type sugar chains, which pass through a column. Complex type biantennary glycopeptides or oligosaccharides having bisecting GlcNAc also pass through the column.
5. Elute (1-mL fractions) successively with three column volumes of 10 m*M* methyl-α-glucoside and finally with three column volumes of 100 m*M* methyl-α-mannoside.
6. Collect complex type biantennary sugar chains, which are eluted after the addition of methyl-α-glucoside.
7. Collect high mannose type and hybrid type oligosaccharides or glycopeptides eluted after the addition of 100 m*M* methyl-α-mannoside.

3.6.2. Affinity Chromatography on Immobilized LCA, PSA, or VFA

The biantennary complex type sugar chains, bound to a Con A–Sepharose column and eluted with 10 m*M* methyl-α-glucoside, contains two types of oligosaccharides, which are separated on a column of lentil lectin (*Lens culinaris* ectin [LCA]) pea lectin (*Pisum sativum* lectin [PSA]) or fava lectin (*Vicia fava* ectin [VFA]) ***(21–23)***.

1. Equilibrate the LCA–, PSA–, or VFA–Sepharose column (0.6 cm i.d. × 5.0 cm) in lectin column buffer.
2. Pass the biantennary complex type sugar chains from the Con A column through a LCA–, PSA–, or VFA–Sepharose column.
3. Elute (1.0-mL fractions) successively with three column volumes of lectin column buffer, then with three column volumes of 100 m*M* methyl-α-mannoside at a flow rate 2.5 mL/h at room temperature.
4. Collect biantennary complex type sugar chains without fucose, which pass through the column.
5. Elute bound biantennary complex type sugar chains having a fucose residue attached to the innermost *N*-acetylglucosamine to the column.

3.6.3. Affinity Chromatography on Immobilized E-PHA

Complex type biantennary sugar chains having outer galactose residues and "bisecting" *N*-acetylglucosamine are retarded by *Phaseolus vulgaris* erythroagglutinin (E-PHA)–Sepharose ***(16,24)***.

1. Equilibrate the E-PHA–Sepharose column (0.6 cm i.d. × 5.0 cm) in lectin column buffer.
2. Apply the pass-through fraction from the Con A column on the E-PHA–Sepharose column.
3. Elute (0.5-mL fractions) successively with five column volumes of lectin column buffer at a flow rate 2.5 mL/h at room temperature.
4. Collect biantennary complex type sugar chains having the bisecting *N*-acetylglucosamine residue retarded on the E-PHA column (*see* **Note 6**). When the elution of the column is performed at 4°C, biantennary complex type oligosaccharide without bisecting *N*-acetylglucosamine is also retarded by the E-PHA–Sepharose column.

3.7. Separation of Complex Type Triantennary and Tetraantennary Sugar Chains

3.7.1. Affinity Chromatography on Immobilized E-PHA

E-PHA–Sepharose interacts with high affinity with triantennary (having 2,4-branched mannose) oligosaccharides or glycopeptides containing both an outer galactose residues and a bisecting *N*-acetylglucosamine residue ***(24)***.

1. Equilibrate the E-PHA–Sepharose column (0.6 cm i.d. × 5.0 cm) in lectin column buffer.
2. Apply the pass-through fraction from the Con A column on to the E-PHA–Sepharose column.
3. Elute (0.5-mL fractions) successively with five column volumes of lectin column buffer at a flow rate 2.5 mL/h at room temperature.
4. Collect retarded triantennary (having 2,4-branched mannose) oligosaccharides or glycopeptides containing both an outer galactose and a bisecting *N*-acetylglucosamine on the E-PHA column. Other tri- and tetraantennary oligosaccharides pass through the column (*see* **Note 7**).

3.7.2. Affinity Chromatography on Immobilized L-PHA

Phaseolus vulgaris leukoagglutinin (L-PHA), which is the isolectin of E-PHA, interacts with triantennary and tetraantennary complex type glycopeptides having an α-linked mannose residue substituted at positions C2 and C6 with Galβ1–4GlcNAc *(25)*.

1. Equilibrate the L-PHA–Sepharose column (0.6 cm i.d. × 5.0 cm) in lectin column buffer.
2. Apply the pass-through fraction from the Con A column on to the L-PHA–Sepharose column.
3. Elute (0.5-mL fractions) successively with five column volumes of lectin column buffer at a flow rate 2.5 mL/h at room temperature.
4. Collect retarded triantennary and tetraantennary complex type glycopeptides having both 2,6-branched α-mannose and outer galactose on the L-PHA column (*see* **Note 8**). Other tri- and tetraantennary oligosaccharides pass through the column.

3.7.3 Affinity Chromatography on Immobilized DSA

Datura stramonium lectin (DSA) shows high affinity with tri- and tetraantennary complex type oligosaccharides. Triantennary complex type oligosaccharides containing 2,4-substituted α-mannose are retarded by the DSA–Sepharose column. Triantennary and tetraantennary complex type oligosaccharides having an α-mannose residue substituted at the C2,6 position bind to the column and are eluted by the GlcNAc oligomer *(26,27)*.

1. Equilibrate the DSA–Sepharose column (0.6 cm i.d. × 5.0 cm) in TBS.
2. Apply the pass-through fraction from the Con A column on to the DSA–Sepharose column.
3. Elute (0.5-mL fractions) successively with three column volumes of TBS at a flow rate 2.5 mL/h at room temperature to obtain a retarded triantennary complex type sugar chain having 2,4-branched α-mannose on the DSA column.
4. Elute with three column volumes of 5 mg/mL *N*-acetylglucosamine oligomer at flow rate 2.5 mL/h at room temperature to obtain bound triantennary and tetraantennary complex type oligosaccharides having an α-mannose residue substituted at the C2,6 positions.

3.8. Separation of Poly-acetyllactosamine Type Sugar Chains

High molecular weight poly-*N*-acetyllactosamine type oligosaccharides are classified into two groups. One is branched poly-*N*-acetyllactoseminoglycan containing a Galβ1–4GlcNAcβ1–3(Galβ1–4GlcNAcβ1–6)Gal unit, and the other is a linear poly-N-acetyllactosamine structure that lacks galactose residues substituted at the C3,6 positions.

3.8.1. Affinity Chromatography on Immobilized PWM

Branched poly-*N*-acetyllactosamine type oligosaccharides can be separated by the use of a pokeweed mitogen (PWM)–Sepharose column *(28)*. Because the sugar sequence Galβ1–4GlcNAcβ1–6Gal firmly binds to the PWM–Sepharose column, the branched poly-*N*-acetyllactosamine chains can be retained by the column, while unbranched ones is recovered without any retardation *(29)* (*see* **Note 9**).

1. Equilibrate the PWM–Sepharose column (0.6 cm i.d. × 5.0 cm) in TBS.
2. Apply the poly-*N*-acetyllactosamine type sugar chains separated on Bio-Gel P-4 (*see* **Subheading 3.2.**) on to the PWM–Sepharose column.
3. Elute (1.0-mL fractions) successively with three column volumes of TBS, then with three column volumes of 0.1 *M* NaOH at a flow rate 2.5 mL/h at room temperature.

4. Collect unbranched poly-*N*-acetyllactosamine type sugar chains, which pass through the column.
5. Collect bound branched poly-*N*-acetyllactosamine type sugar chains.

3.8.2. Affinity Chromatography on Immobilized DSA

Immobilized DSA lectin interacts with high affinity with sugar chains having the linear, unbranched poly-*N*-acetyllactosamine sequence. For the binding to DSA-Sepharose, more than two intact *N*-acetyllactosamine repeating units may be essential ***(27)***.

1. Equilibrate the DSA–Sepharose column (0.6 cm i.d. × 5.0 cm) in TBS.
2. Apply the poly-*N*-acetyllactosamine type sugar chains separated on Bio-Gel P-4 (*see* **Subheading 3.2.**) on to the DSA–Sepharose column.
3. Elute (1.0-mL fractions) successively with three column volumes of TBS, then with three column volumes of 5 mg/mL of GlcNAc oligomer at a flow rate 2.5 mL/h at room temperature.
4. Collect branched poly-*N*-acetyllactosamine type sugar chains, which pass through the column, separated from unbranched poly-*N*-acetyllactosamine type sugar chains, that bind.

3.9. Separation of Sialylated Sugar Chains

The basic Galβ1–4GlcNAc sequence present in complex type sugar chains may contain sialic acids in α2,6 or α2,3 linkage to outer galactose residues.

3.9.1. Affinity Chromatography on Immobilized MAL

Maackia amurensis leukoagglutinin (MAL) ***(30,31)*** interacts with high affinity with complex type tri- and tetraantennary glycopeptides containing an outer sialic acid residue linked α2,3 to penultimate galactose. Glycopeptides containing sialic acid linked only α2,6 to galactose do not interacts detectably with the immobilized MAL (*see* **Note 10**).

1. Equilibrate the MAL–Sepharose column (0.6 cm i.d. ×5.0 cm) in lectin column buffer.
2. Apply the acidic oligosaccharides or glycopeptides separated on Mono Q HR5/5, or DEAE-Sephacel (*see* **Subheading 3.1.1., step 1**) on to the MAL–Sepharose column.
3. Elute (0.5-mL fractions) successively with five column volumes of lectin column buffer at a flow rate 2.5 mL/h at room temperature.
4. Collect glycopeptides or oligosaccharides containing α2,6-linked sialic acid(s), which pass through the column.
5. Collect retarded glycopeptides or oligosaccharides containing α2,3-linked sialic acid(s).

3.9.2. Affinity Chromatography on Immobilized Allo A

Allomyrina dichotoma lectin (allo A) ***(32,33)*** recognizes the other isomer of sialyllactosamine compared to MAL. Mono-, di-, and triantennary complex type oligosaccharides containing terminal sialic acid(s) in α2,6 linkage bind to allo A-Sepharose, while complex type sugar chains having isomeric α2,3-linked sialic acid(s) do not bind to the immobilized allo A.

1. Equilibrate the allo A-Sepharose column (0.6 cm i.d. × 5.0 cm) in TBS.
2. Apply the acidic oligosaccharides or glycopeptides separated on Mono Q HR5/5, or DEAE-Sephacel (*see* **Subheading 3.1.1., step 1**) on allo A-Sepharose column.
3. Elute (0.5-mL fractions) successively with three column volumes of TBS and then with three column volumes of 50 m*M* lactose at a flow rate 2.5 mL/h at room temperature.

4. Collect glycopeptides or oligosaccharides containing α2,3-linked sialic acid(s), which pass through the column.
5. Elute bound glycopeptides or oligosaccharides having α2,6-linked sialic acid(s) (*see* **Note 11**).

3.9.3. Affinity Chromatography on Immobilized SNA

Elderberry (*Sambucus nigra L.*) bark lectin (SNA) ***(34,35)*** shows the same sugar binding specificity as allo A. All types of oligosaccharides, which contain at least one NeuAcα2–6Gal unit in the molecule, bound firmly to the SNA–Sepharose.

1. Equilibrate the SNA–Sepharose column (0.6 cm i.d. × 5.0 cm) in TBS.
2. Apply the acidic oligosaccharides or glycopeptides separated on Mono Q HR5/5 or DEAE-Sephacel (*see* **Subheading 3.1.1., step 1**) on to the SNA–Sepharose column.
3. Elute (0.5-mL fractions) successively with three column volumes of TBS then with three column volumes of 50 m*M* lactose at a flow rate 2.5 mL/h at room temperature.
4. Collect glycopeptides or oligosaccharides containing α2,3-linked sialic acid(s), which pass through the column.
5. Elute bound glycopeptides or oligosaccharides having α2,6-linked sialic acid(s) in the 50 m*M* lactose eluant.

3.10. Summary

Various immobilized lectins can be used successfully for fractionation and for structural studies of asparagine-linked sugar chains of glycoproteins (*see* **Note 12**). This method needs <10 ng of a radiolabeled oligosaccharide or glycopeptide prepared from a glycoprotein by hydrazinolysis or by digestion with endo-β-*N*-acetylglucosaminidases or *N*-glycanases. The fractionation and the structural assessment through the use of immobilized lectins make the subsequent structural studies much easier.

4. Notes

1. During the coupling reactions, sugar binding sites of lectins must be protected by the addition of the specific haptenic sugars.
2. Immobilized lectin is stored at 4°C. In most cases, immobilized lectin is stable for several years.
3. Some lectins, especially legume lectins, need Ca and Mn ions for carbohydrate binding, so that the buffers used for the affinity chromatography on the lectin column must contain 1 m*M* $CaCl_2$ and $MnCl_2$.
4. Complex type or hybrid type oligosaccharides are retarded on a column of RCA–Sepharose rather than tightly bound when their sugar sequences are masked by sialic acids.
5. Intact *N*-acetylglucosamine and asparagine residues at the reducing end are required for tight binding of complex type oligosaccharides to an LCA–, PSA–, or VFA–Sepharose column.
6. High-affinity interaction with E-PHA–Sepharose is prevented if both outer galactose residues on a bisected sugar chain are substituted at position C6 by sialic acid.
7. Biantennary and triantennary complex type sugar chains having bisecting GlcNAc can be separated on a Bio-Gel P-4 column.
8. L-PHA–Sepharose does not retard the elution of sugar chains lacking outer galactose residues.
9. WGA can be used instead of PWM.
10. *Maackia amurensis* hemagglutinin (MAH), which is the isolectin of MAL, strongly binds to sialylated Ser/Thr-linked Galβ1–3GalNAc but not to sialylated Asn-linked sugar moieties ***(36)***.

11. Oligosaccharides without sialic acid(s) of mono-, di-, tri-, and tetraantennary complex type are retarded by the allo A lectin column.
12. More detailed reviews on the separation of oligosaccharides and glycopeptides by means of affinity chromatography on immobilized lectin columns have been published *(37–39)*.

References

1. Kornfeld, R. and Kornfeld, S. (1985) Assembly of asparagine-linked oligosaccharides. *Annu. Rev. Biochem.* **54,** 631–664.
2. Tsuji, T., Irimura, T., and Osawa, T. (1981) The carbohydrate moiety of Band 3 glycoprotein of human erythrocyte membrane. *J. Biol. Chem.* **256,** 10,497–10,502.
3. Fukuda, M., Dell, A., Oates, J. E., and Fukuda, M. N. (1984) Structure of branched lactosaminoglycan, the carbohydrate moiety of Band 3 isolated from adult human erythrocytes. *J. Biol. Chem.* **259,** 8260–8273.
4. Merkle, R. K. and Cummings, R. D. (1987) Relationship of the terminal sequences to the length of poly-N-acetyllactosamine chains in asparagine-linked oligosaccharides from the mouse lymphoma cell line BW5147. *J. Biol. Chem.* **282,** 8179–8189.
5. Fukuda, M. (1985) Cell surface glycoconjugates as onco-differentiation markers in hematopoietic cells. *Biochem. Biophys. Acta.* **780,** 119–150.
6. Green, E. D. and Baenziger, J. U. (1988) Asparagine-linked oligosaccharides on lutropin, follitropin, and thyrotropin: I. Structural elucidation of the sulfated and sialylated oligosaccharides on bovine, ovine, and human pituitary glycoprotein hormones. *J. Biol. Chem.* **263,** 25–35.
7. Nakata, N., Furukawa, K., Greenwalt, D. E., Sato, T., and Kobata, A. (1993) Structural study of the sugar chains of CD36 purified from bovine mammary epithelial cells: occurrence of novel hybrid type sugar chains containing the Neu5Acα2–6GalNAcβ1–4GlcNAc and the Manα1–2Manα1–3Manα1–6Man groups. *Biochemistry* **32,** 4369–4383.
8. Fukuda, M., Kondo, T., and Osawa, T. (1976) Studies on the hydrazinolysis of glycoproteins. Core structures of oligosaccharides obtained from porcine thyroglobulin and pineapple stem bromelain. *J. Biochem.* **80,** 1223–1232.
9. Takasaki, S. and Kobata, A. (1978) Microdetermination of sugar composition by radioisotope labeling. *Meth. Enzymol.* **50,** 50–54.
10. Tai, T., Yamashita, K., Ogata, M. A., Koide, N., Muramatsu, T., Iwashita, S., Inoue, Y., and Kobata, A. (1975) Structural studies of two ovalbumin glycopeptides in relation to the endo-β-*N*-acetylglucosaminidase specificity. *J. Biol Chem.* **250,** 8569–8575.
11. Rupley, J. A. (1964) The hydrolysis of chitin by concentrated hydrochloric acid, and the preparation of low-molecular-weight substrates for lysozyme. *Biochem. Biophys. Acta.* **83,** 245–255.
12. Krusius, T., Finne, J., and Rauvala, H. (1978) The poly(glycosyl) chains of glycoproteins. Characterization of a novel type of glycoprotein saccharides from human erythrocyte membrane. *Eur. J. Biochem.* **92,** 289–300.
13. Yamamoto, K., Tsuji, T., Tarutani, O., and Osawa, T. (1984) Structural changes of carbohydrate chains of human thyroglobulin accompanying malignant transformations of thyroid grands. *Eur. J. Biochem.* **143,** 133–144
14. Tsuji, T., Irimura, T., and Osawa, T. (1980) The carbohydrate moiety of Band-3 glycoprotein of human erythrocyte membranes. *Biochem. J.* **187,** 677–686.
15. Baenziger, J. U. and Fiete, D. (1979) Structural determinants of Ricinus communis agglutinin and toxin specificity for oligosaccharides. *J. Biol. Chem.* **254,** 9795–9799.

16. Irimura, T., Tsuji, T., Yamamoto, K., Tagami, S., and Osawa, T. (1981) Structure of a complex type sugar chain of human glycophorin A. *Biochemistry* **20,** 560–566.
17. Shibuya, N., Goldstein, I. J., Van Damme, E. J. M., and Peumans, W. J. (1988) Binding properties of a mannose-specific lectin from the snowdrop (*Galanthus nivalis*) bulb. *J. Biol. Chem.* **263,** 728–734.
18. Yamamoto, K., Tsuji, T., Matsumoto, I., and Osawa, T. (1981) Structural requirements for the binding of oligosaccharides and glycopeptides to immobilized wheat germ agglutinin *Biochemistry* **20,** 5894–5899.
19. Ogata, S., Muramatsu, T., and Kobata, A. (1975) Fractionation of glycopeptides by affinity column chromatography on concanavalin A–Sepharose. *J. Biochem.* **78,** 687–696.
20. Krusius, T., Finne, J., and Rauvala, H. (1976) The structural basis of the different affinities of two types of acidic N-glycosidic glycopeptides from concanavalin A–Sepharose. *FEBS Lett.* **71,** 117–120.
21. Kornfeld, K., Reitman, M. L., and Kornfeld, R. (1981) The carbohydrate-binding specificity of pea and lentil lectins. *J. Biol. Chem.* **256,** 6633–6640
22. Katagiri, Y., Yamamoto, K., Tsuji, T., and Osawa, T. (1984) Structural requirements for the binding of glycopeptides to immobilized vicia faba (fava) lectin. *Carbohydr. Res.* **129,** 257–265.
23. Yamamoto, K., Tsuji, T., and Osawa, T. (1982) Requirement of the core structure of a complex type glycopeptide for the binding to immobilized lentil- and pea-lectins. *Carbohydr. Res.* **110,** 283–289.
24. Yamashita, K., Hitoi, A., and Kobata, A. (1983) Structural determinants of Phaseolus vulgaris erythroagglutinating lectin for oligosaccharides. *J. Biol. Chem.* **258,** 14,753–14,755.
25. Cummings, R. D. and Kornfeld, S. (1982) Characterization of the structural determinants required for the high affinity interaction of asparagine-linked oligosaccharides with immobilized Phaseolus vulgaris leukoagglutinating and erythroagglutinating lectins. *J. Biol. Chem.* **257,** 11,230–11,234.
26. Cummings, R. D. and Kornfeld, S. (1984) The distribution of repeating [Galβ1,4GlcNAcβ1,3] sequences in asparagine-linked oligosaccharides of the mouse lymphoma cell line BW5147 and $PHA^R2.1$. *J. Biol. Chem.* **259,** 6253–6260.
27. Yamashita, K., Totani, K. T., Ohkura, Takasaki, S., Goldstein, I. J., and Kobata, A. (1987) Carbohydrate binding properties of complex type oligosaccharides on immobilized Datura stramonium lectin. *J. Biol. Chem.* **262,** 1602–1607.
28. Irimura, T. and Nicolson, G. L. (1983) Interaction of pokeweed mitogen with poly (N-acetyllactosamine) type carbohydrate chains. *Carbohydr. Res.* **120,** 187–195.
29. Kawashima, H., Sueyoshi, S., Li, H., Yamamoto, K., and Osawa, T. (1990) Carbohydrate binding specificities of several poly-N-acetyllactosamine-binding lectins. *Glycoconjugate J.* **7,** 323–334.
30. Wang, W.-C. and Cummings, R. D. (1988) The immobilized leukoagglutinin from the seeds of Maackia amurensis binds with high affinity to complex type Asn-linked oligosaccharides containing terminal sialic acid-linked a-2,3 to penultimate galactose residues. *J. Biol. Chem.* **263,** 4576–4585.
31. Kawaguchi, T., Matsumoto, I., and Osawa, T. (1974) Studies on hemagglutinins from Maackia amurensis seeds. *J. Biol. Chem.* **249,** 2786–2792.
32. Sueyoshi, S., Yamamoto, K., and Osawa, T. (1988) Carbohydrate binding specificity of a beetle (Allomyrina dichotoma) lectin. *J. Biochem.* **103,** 894–899.
33. Yamashita, K., Umetsu, K., Suzuki, T., Iwaki, Y., Endo, T., and Kobata, A. (1988) Carbohydrate binding specificity of immobilized Allomyrina dichotoma lectin II. *J. Biol. Chem.* **263,** 17,482–17,489.

34. Shibuya, N., Goldstein, I. J., Broekaert, W. F., Lubaki, M. N., Peeters, B., and Peumans. W. J. (1987) Fractionation of sialylated oligosaccharides, glycopeptides, and glycoproteins on immobilized elderberry (Sambucus nigra L.) bark lectin. *Arch. Biochem. Biophys.* **254,** 1–8.
35. Shibuya, N., Goldstein, I. J., Broekaert, W. F., Lubaki, M. N., Peeters, B., and Peumans, W. J. (1987) The elderberry (Sambucus nigra L.) bark lectin recognizes the Neu5Ac(α2–6)Gal/GalNAc sequence. *J. Biol. Chem.* **262,** 1596–1601.
36. Konami, Y., Yamamoto, K., Osawa, T., and Irimura, T. (1994) Strong affinity of Maackia amurensis hemagglutinin (MAH) for sialic acid-containing Ser/Thr-linked carbohydrate chains of N-terminal octapeptides from human glycophorin A. *FEBS Lett.* **342,** 334–338.
37. Osawa, T. and Tsuji, T. (1987) Fractionation and structural assessment of oligosaccharides and glycopeptides by use of immobilized lectins. *Ann. Rev. Biochem.* **56,** 21–42.
38. Osawa, T. (1989) Recent progress in the application of plant lectins to glycoprotein chemistry. *Pure & Appl. Chem.* **61,** 1283–1292.
39. Tsuji, T., Yamamoto, K., and Osawa, T. (1993) Affinity chromatography oligosaccharides and glycopeptides with immobilized lectins, in *Molecular Interaction in Bioseparations,* (Ngo, T.T., ed.), Plenum Press, New York, pp. 113–126.

Part VII

Antibody Techniques

128

Antibody Production

Robert Burns

1. Introduction

Antibodies are proteins that are produced by the immune systems of animals in response to foreign substances. The immune system has the ability to recognize material as nonself or foreign and mount a response to them. Substances that elicit this are known as immunogens or antigens, and one of the outcomes of the immune response is the production of antibodies that will recognize and bind to the eliciting substance.

The immune response results in the degradation of the antigen by cells called macrophages and its presentation in fragments to B lymphocytes that are found in the lymphatic tissues of the animal. On presentation of the antigen fragments the B lymphocytes mature and "learn" to produce antibody molecules that have a specific affinity to the antigen fragment that they have been presented with. This process gives rise to populations of B lymphocyte clones each of which produces antibody molecules to different locations known as epitopes on the target antigen *(1)*. As individual clones produce the antibody molecules and there are many of them, the resulting antibody mix in the blood is known as polyclonal and the fluid derived from the clotted blood is known as polyclonal antiserum.

Artificial exposure to antigens to produce antibodies in animals is known as immunization, and repeated exposure leads to a condition known as hyperimmunity, in which a significantly large proportion of the circulating antibodies in the animals blood will be directed toward the antigen of interest. Immunization gives rise to stable but quiescent populations of B lymphocytes known as memory cells that will respond to the presence of the antigen by dividing to increase their number and the level of circulating antibody in the blood.

The primary exposure to an antigen gives rise to a pentameric form of antibody known as immunoglobulin M (IgM). Subsequent exposure to the antigen causes a shift in the antibody type to the more stable immunoglobulin G (IgG). This shift in antibody type happens as affinity maturation of the B lymphocyte clones occurs.

Some antigens, particularly highly glycosylated proteins, do not elicit the production of memory B lymphocytes and also may not give rise to IgG antibodies even after repeated immunization. These are known as anamnestic antigens, and as no memory B lymphocytes are produced, the immune system always interprets exposure to them

From: *The Protein Protocols Handbook, 2nd Edition*
Edited by: J. M. Walker © Humana Press Inc., Totowa, NJ

as a primary one and mounts a primary IgM response. It is possible to produce IgG to anamnestic antigens by treating them to remove the carbohydrate moieties prior to immunization, but this may lead to antibodies that do not recognize the native protein structure.

Antibodies have two qualities known as avidity and affinity. Affinity is a measure of how well the binding site on the antibody molecule fits the epitope, and antibodies, which have high affinity also, have high specificity to the target epitope. Avidity is a measure of how strong the interaction between the epitope and the antibody molecule is.

For research purposes, antibodies to the antigen of interest should be expressed at fairly high levels in the sera of donor animals, should be of high affinity, and should predominantly be composed of IgG molecules, which are stable and easily purified. This can be achieved for most antigens providing they are not too highly glycosylated, have a molecular weight >5000, and are not toxic or induce immuneparesis in the donor animal.

Monoclonal antibodies are used very successfully in many areas of research and can either replace, complement, or be used in conjunction with immunoglobulins obtained from donor serum. Monoclonal antibodies are produced by hybridoma cell lines and can be grown in tissue culture in the laboratory. Hybridomas are recombinant cell lines produced from the fusion of B cell clones derived from the lymphatic tissue of donor animals and a myeloma cell line that imparts immortality to the cells *(2,3)*. As each hybridoma is descended from a single B cell clone the antibody expressed by it is of a single specificity and immunoglobulin type, and is thus termed a monoclonal antibody.

Each monoclonal antibody is monospecific and will recognize only one epitope on the antigen to which it has been raised. This may lead to practical problems if the epitope is not highly conserved on the native protein or where conformational changes may occur in the because of to shifts in pH or other environmental factors. Monoclonal antibodies are highly specific and will rarely if ever produce cross-reactions with other proteins. Polyclonal antibodies may cross-react with other closely related proteins where there are shared epitopes.

Polyclonal antisera contain a heterogeneic mixture of antibody molecules, many of which will recognize different epitopes on the protein, and their binding is much less likely to be affected by poorly conserved epitopes or changes in the protein shape.

Antisera are derived from individual animal bleeds and because of this are subject to batch variation. Individual animals can have very different immune responses to the same antigen, and individual bleeds from the same animal may vary in antibody content quite markedly. Monoclonal antibodies are produced from highly cloned cell lines that are stable and reliably produce a defined antibody product.

Polyclonal antibodies are generally less specific than monoclonals, which can be a disadvantage, as cross-reactivity may occur with nontarget proteins. Monoclonal antibodies can be too specific, as they will recognize only a single epitope that may vary on the protein of interest.

Both antibody types have their place in the research laboratory, and a careful evaluation of the required use should be undertaken before deciding which would be most applicable.

1.1. Donor Animals

Most polyclonal antibodies for research purposes are produced in domestic rabbits unless very large quantities are required, and then sheep, goats, donkeys, and even horses are used. Antibodies can also be produced in chickens; the antibodies are conveniently produced in the eggs. Rats and mice can also be used to produce antiserum but yield much smaller quantities of antibody owing to their relatively small size. Immunized mice can be used to produce 5–10 mL of polyclonal ascitic fluid by induction of ascites. Ascites is induced by the introduction of a nonsecretory myeloma cell line into the peritoneal cavity after priming with Pristane. Polyclonal antibodies are secreted into the peritoneum from the blood plasma of the animal and can achieve levels of 2–5 mg/mL. Ascitic fluid is aspirated from the peritoneum of the mouse when ascites has developed, indicated by bloating of the abdomen. The UK Home Office and other regulatory authorities now regard induction of ascites in mice as a moderate/severe procedure and this method is not normally used unless no alternative methods are available.

1.2. Adjuvants

Adjuvants are substances that increase the immune response to antigen by an animal. They may be simple chemicals such as alum, which adsorbs and aggregates proteins, increasing their effective molecular weight, or they may be specific immunestimulators such as derivatives of bacterial cell walls *(4)*. Saponins such as Quil-A derived from the tree *Quillaja saponaria* may also be used to increase the effective immunogenicity of the antigen. Ideally an adjuvant should not induce an antibody response to itself to ensure that antibodies generated are specific to the antigen of interest.

The most popular adjuvant, which has been used very successfully for many years, is Freund's adjuvant. It has two formulations, complete and incomplete, which are used for the primary and subsequent immunizations, respectively. Freund's incomplete is a mixture of 85% paraffin oil and 15% mannide monooleate. The complete formulation additionally contains 1 mg/mL of heat-killed *Mycobacterium tuberculosis*. In recent years the use of Freund's adjuvant has declined owing to animal welfare concerns and also the risk that it poses to workers, as it can cause localized soft tissue damage following accidental needlestick injuries. Injection preparations that contain Freund's adjuvant are also difficult to work with, as the resultant emulsion can be too thick to administer easily and it interacts with plastic syringes, preventing easy depression of the plunger.

Several adjuvant formulations are available that contain cell wall derivatives of bacteria without the intact organisms; they are much easier to administer and will achieve a similar immunostimulatory effect to Freund's without its attendant problems.

1.3. Legislation

Most countries have legislation governing the use of animals for all experimental work, and antibody production is no exception. Although the methods used are usually mild in terms of severity they must be undertaken with appropriate documentation and always by trained, authorized staff.

The legislation in terms of animal housing, immunization procedures, bleeding regimens, and choice of adjuvant vary widely according to local legislation, and it is imperative that advice be taken from the appropriate authorities prior to undertaking any antibody production work.

1.4. Antigens

The antigen chosen for an immunization program should be as close in structure and chemical identity to the target protein as possible. An exception to this is when synthetic peptides are produced to mimic parts of the native protein, an approach that is invaluable when the native antigen may be toxic to animals or nonimmunogenic.

The antigen should be soluble, stable at dilutions of approx 1 mg/mL, and capable of being administered in a liquid of close to physiological pH (6.5–7.5). The antigen should also be in as pure a form as practically possible to avoid the generation of antibodies to contaminating materials.

Many proteins are highly immunogenic in donor animals, particularly when the antigen is derived from a different species (*see* **Note 1**). Raising antibodies in the same species from which the antigen is derived from can be extremely difficult but can be overcome by conjugating the antigen to a carrier protein from another species prior to immunization. Carrier proteins such as hemoglobin, thyroglobulin, and keyole limpet hemocyanin are commonly used. Animals immunized with the conjugated form of the antigen will produce antibodies to both the protein of interest and the carrier protein. Apart from a lower specific antibody titer in the serum there should be no interference from the carrier protein antibodies.

1.5. Test Protocol

Ideally antiserum should be tested using the procedure in which it is to be used, as antibodies may perform well using one assay but not with another. This, however, is not always practical and so a number of tests can be carried out on antiserum to test its suitability for final use. The affinity of the antiserum to the antigen can be assayed by plate-trapped double-antibody sandwich enzyme-linked immunosorbent assay (DAS ELISA) *(5)*. The antigen is bound to a microtiter plate and then challenged with dilutions of the test antiserum. A secondary antispecies antibody enzyme conjugate is then added to the plate and will bind to any antibody molecules present. A chromogenic enzyme substrate is then added and the degree of color development indicates the quantity of antibody bound by the antigen.

Ouchterlony double-immunodiffusion can be used to observe the ability of the antiserum to produce immune complexes with the antigen in a semisolid matrix. This method can also be used to test cross-reactivity of the antiserum to other proteins closely related to the antigen. Radioimmunoassay and other related techniques can also be used to test the avidity of the antiserum to the antigen and will also give a measure of antibody titer.

In all the above tests preimmune antisera should be included to ensure that results obtained are a true reflection of antibodies produced by immunization and not due to nonspecific interactions.

2. Materials

1. At least two rabbits should be used for each polyclonal antibody production project, as they may have different immune responses to the antigen. They should be purchased from a reputable source and be parasite and disease free. New Zealand whites are often used but any of the domestic breeds will make acceptable donors (*see* **Note 2**).
2. The antigen should be in a buffer of pH 6.5–7.5 and be free of toxic additives (sodium azide is often added as a preservative). A concentration of 1 mg/mL is desirable but anything above 100 µg/mL is acceptable.
3. Complete and incomplete Freund's adjuvant or any of the propriety preparations containing purified bacterial cell wall components such as Titermax and RIBI.

3. Method

1. Prior to a course of immunizations a test bleed should be taken from the rabbits to provide a source of preimmune antiserum for each animal (*see* **step 5** below). It is usual to take only 2–3 mL for the test bleeds, which will yield 1–1.5 mL of serum. The blood should be allowed to clot at 4°C for 12 h and the serum gently aspirated from the tube.
2. The antigen should be mixed with the appropriate adjuvant according to the manufacturer's instructions to achieve a final volume of 0.5 mL/injection containing 50–500 µg of antigen (*see* **Note 3**). If Freund's adjuvant is to be used, the first injection only should contain the complete formulation and the incomplete one should be used for all subsequent immunizations.
3. The rabbit should be restrained and the antigen–adjuvant mixture injected into the thigh muscle. Alternate legs should be used for each injection (*see* **Note 4**).
4. The immunization should be repeated 14 d after the primary one and a test bleed taken 30 d after that.
5. If the antiserum shows that the desired immune status and antibody quality has been achieved then donor bleeds can be taken. The volume of blood collected and frequency of bleeding depends on animal welfare legislation and must be adhered to. Each bleed should be assayed individually, and once the antibody titer has started to fall a further immunization can be given, followed by donor bleeds, or the animal can be terminally anesthetized and exsanguinated by cardiac puncture or by severing the carotid artery.
6. Antiserum collected from rabbits can be stored for extended periods of time at 4°C but the addition of 0.02% sodium azide is recommended to prevent adventitious bacterial growth. Antiserum quite commonly has functional antibodies even after years of refrigerated storage, but storage at –20°C is recommended for long-term preservation.
7. Antiserum will often yield in excess of 5 mg/mL of antibody and can be purified to give the immunoglobulin fraction. This can be achieved with ammonium sulfate precipitation or by affinity chromatography using either the immobilized antigen or protein A. The antibody fraction can then be adjusted to 1 mg/mL, which is an ideal concentration both for its stability and for many practical applications. To purified antibody 0.02% sodium azide or some other preservative should be added to prevent the growth of adventitious organisms. Sodium azide can interfere with enzyme reactions and with photometric measurement, and this should be taken into account with regard to the final assay. Purified antibodies diluted to 1 mg/mL can be stored for long periods at 4°C with no loss of activity but for extended storage –20°C is recommended.

4. Notes

1. Some antigens will consistently fail to induce an antibody response in certain animals, and other species should be investigated as potential donors. In very rare occasions antigens

will not elicit an immune response in a range of species and the nature of the antigen will then have to be investigated with a view to modifying it to increase its immunogenicity.

2. Female rabbits are less aggressive, and although smaller and yielding smaller quantities of antiserum are preferable to male rabbits for antiserum production. Many biomedical facilities use communal floor pens for donor animals and female rabbits adapt better to this form of housing.
3. The use of excessive amounts of antigen in immunizations should be avoided, as this can lead to a poor immunological response. Swamping of the system can lead to selective deletion of the B cell clones of interest and a reduction in the specific antibody titer. High doses of antigen in the secondary and subsequent immunizations can cause anaphylactic shock and death of the donor animal.
4. It has been reported that increased stress levels in animals can depress the immune response, and appropriate measures should be taken to ensure that immunizations and bleeds are performed with the minimum of stress to the animals. General husbandry in terms of housing, noise levels, and other environmental factors should also be examined to ensure that animals for polyclonal production are maintained under suitable conditions.

References

1. Roitt, I., Brostoff, J., and Male, D. (1996) Immunology, C. V. Mosby, St. Louis, pp. 1.7–1.8.
2. Kennet, R. H., Denis, K. A., Tung, A. S., and Klinman, N. R. (1978) Hybrid plasmacytoma production: fusions with adult spleen cells, monoclonal spleen fragments, neonatal spleen cells and human spleen cells. *Curr. Top. Microbiol. Immunol.* **81,** *485.*
3. Kohler, G. and Milstein, C. (1975) Continuous cultures of fused cells secreting antibody of predefined specificity. *Nature* **256,** 495–497.
4. Harlow, E. and Lane, D. (1988) *Antibodies: A Laboratory Manual*, Cold Spring Harbor Laboratory Press, Cold Spring Harbor, NY, pp. 96–97.
5. Kemeny, D. M. and Chandler, S. (1988) *ELISA and Other Solid Phase Immunoassays*, John Wiley & Sons, pp. 1–29.

129

Production of Antibodies Using Proteins in Gel Bands

Sally Ann Amero, Tharappel C. James, and Sarah C. R. Elgin

1. Introduction

A number of methods for preparing proteins as antigens have been described *(1)*. These include solubilization of protein samples in buffered solutions (**ref.** *2* and *see* Chapter 128), solubilization of nitrocellulose filters to which proteins have been adsorbed (**ref.** *3* and *see* Chapter 130), and emulsification of protein bands in polyacrylamide gels for direct injections *(4–8)*. The latter technique can be used to immunize mice or rabbits for production of antisera or to immunize mice for production of monoclonal antibodies *(9–11)*. This approach is particularly advantageous when protein purification by other means is not practical, as in the case of proteins insoluble without detergent. A further advantage of this method is an enhancement of the immune response, since polyacrylamide helps to retain the antigen in the animal and so acts as an adjuvant *(7)*. The use of the protein directly in the gel band (without elution) is also helpful when only small amounts of protein are available. For instance, in this laboratory, we routinely immunize mice with 5–10 μg total protein using this method; we have not determined the lower limit of total protein that can be used to immunize rabbits. Since polyacrylamide is also highly immunogenic, however, it is necessary in some cases to affinity-purify the desired antibodies from the resulting antiserum or to produce hybridomas that can be screened selectively for the production of specific antibodies, to obtain the desired reagent.

2. Materials

1. Gel electrophoresis apparatus; acid-urea polyacrylamide gel or SDS-polyacrylamide gel.
2. Staining solution: 0.1% Coomassie brilliant blue-R (Sigma, St. Louis, MO, B-7920) in 50% (v/v) methanol/10% (v/v) acetic acid.
3. Destaining solution: 5%-(v/v) methanol/7% (v/v) acetic acid.
4. 2% (v/v) glutaraldehyde (Sigma G-6257).
5. Transilluminator.
6. Sharp razor blades.
7. Conical plastic centrifuge tubes and ethanol.
8. Lyophilizer and dry ice.
9. Plastic, disposable syringes (3- and 1-mL).
10. 18-gage needles.

From: *The Protein Protocols Handbook, 2nd Edition*
Edited by: J. M. Walker © Humana Press Inc., Totowa, NJ

11. Spatula and weighing paper.
12. Freund's complete and Freund's incomplete adjuvants (Gibco Laboratories, Grand Island, NY).
13. Phosphate-buffered saline solution (PBS): 50 m*M* sodium phosphate, pH 7.25/150 m*M* sodium chloride.
14. Microemulsifying needle, 18-g.
15. Female Balb-c mice, 7–8 wk old, or New Zealand white rabbits.

3. Method

1. Following electrophoresis (*see* **Note 1**), the gel is stained by gentle agitation in several volumes of staining solution for 30 min. The gel is partially destained by gentle agitation in several changes of destaining solution for 30–45 min. Proteins in the gel are then cross-linked by immersing the gel with gentle shaking in 2% glutaraldehyde for 45–60 min ***(12)***. This step minimizes loss of proteins during subsequent destaining steps and enhances the immunological response by polymerizing the proteins. The gel is then completely destained, usually overnight (*see* **Note 2**).
2. The gel is viewed on a transilluminator, and the bands of interest are cut out with a razor blade. The gel pieces are pushed to the bottom of a conical plastic centrifuge tube with a spatula and pulverized. The samples in the tubes are frozen in dry ice and lyophilized.
3. To prepare the dried polyacrylamide pieces for injection, a small portion of the dried material is lifted out of the tube with a spatula and placed on a small square of weighing paper. In dry climates it is useful to first wipe the outside of the tube with ethanol to reduce static electricity. The material is then gently tapped into the top of a 3-mL syringe to which is attached the microemulsifying needle (**Fig. 1A**). Keeping the syringe horizontal, 200 μL of PBS solution is carefully introduced to the barrel of the syringe, and the plunger is inserted. Next, 200 μL of Freund's adjuvant is drawn into a 1-mL tuberculin syringe and transferred into the needle end of a second 3-mL syringe (**Fig. 1B**). This syringe is then attached to the free end of the microemulsifying needle. The two plungers are pushed alternatively to mix the components of the two syringes (**Fig. 1C**). These will form an emulsion within 15 min; it is generally extremely difficult to mix the material any further.
4. This mixture is injected intraperitoneally or subcutaneously into a female Balb-c mouse, or subcutaneously into the back of the neck of a rabbit (*see* **refs. *13*** and ***14***). Since the emulsion is very viscous, it is best to use 18–g needles and to anesthesize the animals. For mice, subsequent injections are administered after 2 wk and after 3 more wk. If monoclonal antibodies are desired, the animals are sacrificed 3–4 d later, and the spleen cells are fused with myeloma cells (**ref. *13*** and *see* Chapter 159). The immunization schedule for rabbits calls for subsequent injections after 1 mo or when serum titers start to diminish. Antiserum is obtained from either tail bleeds or eye bleeds from the immunized mice, or from ear bleeds from immunized rabbits. The antibodies are assayed by any of the standard techniques (*see* Chapters 160 and 161).

4. Notes

1. We have produced antisera to protein bands in acetic acid–urea gels (*see* Chapter 16), Triton–acetic acid–urea gels ***(15,16)*** (*see* Chapter 17), or SDS–polyacrylamide gels (*see* Chapter 11). In our experience, antibodies produced to proteins in one denaturing gel system will crossreact to those same proteins fractionated in another denaturing gel system and will usually crossreact with the native protein. We have consistently obtained antibodies from animals immunized by these procedures.
2. It is extremely important that all glutaraldehyde be removed from the gel during the destaining washes, since any residual glutaraldehyde will be toxic to the animal. Residual glutaraldehyde can easily be detected by smell. It is equally important to remove all acetic

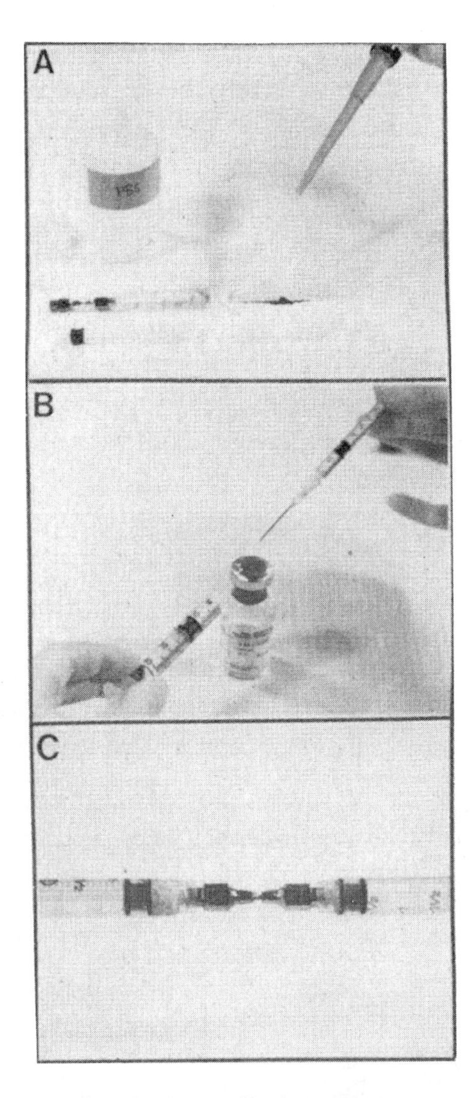

Fig. 1. Preparation of emulsion for immunizations. To prepare proteins in gel bands for injections, an emulsion of Freund's adjuvant and dried polyacrylamide pieces is prepared. **(A)** Dried polyacrylamide is resuspended in 200 μL of PBS solution in the barrel of a 3-mL syringe to which is attached a microemulsifying needle. **(B)** Freund's adjuvant is transferred into the barrel of a second 3-mL syringe. **(C)** An emulsion is formed by mixing the contents of the two syringes through the microemulsifying needle.

acid during lyophilization. Monoacrylamide is also toxic, whereas polyacrylamide is not. We do observe, however, that approx 50 mm^2 of polyacrylamide per injection is the maximum that a mouse can tolerate.

3. Freund's complete adjuvant is used for the initial immunization; Freund's incomplete adjuvant is used for all subsequent injections. The mycobacteria in complete adjuvant enhance the immune response by provoking a local inflammation. Additional doses of mycobacteria may be toxic.
4. High-titer antibodies have been produced from proteins in polyacrylamide gel by injecting the gel/protein mixture into the lumen of a perforated plastic golf ball implanted subcutane-

ously in rabbits *(17)*. This approach places less stress on the animal, as complete adjuvants need not be used, and bleeding is eliminated. The technique has also been used in rats.

References

1. Chase, M. W. (1967) Production of antiserum, in *Methods in Immunology and Immunochemistry,* vol. I (Williams, C. A. and Chase, M. W., eds.), Academic, New York, pp. 197–200.
2. Maurer, P. H. and Callahan, H. J. (1980) Proteins and polypeptides as antigens, in *Methods in Enzymology*, vol. 70 (Van Vunakis, H. and Langne, J. J., eds.), Academic, New York, pp. 49–70.
3. Knudson, K. A. (1985) Proteins transferred to nitrocellulose for use as immunogens. *Anal. Biochem.* **147,** 285–288.
4. Tjian, R., Stinchcomb, D., and Losick, R. (1974) Antibody directed against *Bacillus subtilis* σ factor purified by sodium dodecyl sulfate slab gel electrophoresis. *J. Biol. Chem.* **250,** 8824–8828.
5. Elgin, S. C. R., Silver, L. M., and Wu, C. E. C. (1977) The *in situ* distribution of drosophila non-histone chromosomal proteins, in *The Molecular Biology of the Mammalian Genetic Apparatus*, vol. 1 (Ts'o, P.O.P., ed.), North Holland, New York, pp. 127–141.
6. Silver, L. M. and Elgin, S. C. R (1978) Immunological analysis of protein distributions in *Drosophila* polytene chromosomes, in *The Cell Nucleus* (Busch, H., ed.), Academic, New York, pp. 215–262.
7. Bugler, B., Caizergucs-Ferrer, M., Bouche, G., Bourbon, H., and Alamric, F. (1982) Detection and localization of a class of proteins immunologically related to a 100-kDa nucleolar protein. *Eur. J. Biochem.* **128,** 475–480.
8. Wood, D. M. and Dunbar, B. S. (1981) Direct detection of two-cross-reactive antigens between porcine and rabbit zonae pellucidae by radioimmunoassay and immunoelectrophoresis. *J. Exp. Zool.* **217,** 423–433.
9. Howard, C. D., Abmayr, S. M., Shinefeld, L. A., Sato, V. L., and Elgin, S. C. I. (1981) Monoclonal antibodies against a specific nonhistone chromosomal protein of *Drosophila* associated with active genes. *J. Cell Biol.* **88,** 219–225.
10. Tracy, R. P., Katzmann, J. A., Kimlinger, T. K., Hurst, G. A., and Young, D. S. (1983) Development of monoclonal antibodies to proteins separated by two dimensional gel electrophoresis. *J. Immunol. Meth.* **65,** 97–107.
11. James, T. C. and Elgin, S. C. R. (1986) Identification of a nonhistone chromosomal protein associated with heterochromatin in *Drosophila melanogaster* and its gene. *Mol. Cell Biol.* **6,** 3862–3872.
12. Reichli, M. (1980) Use of glutaraldehyde as a coupling agent for proteins and peptides, in *Methods in Enzymology,* vol. 70 (Van Vunakis, H. and Langone, J. J., eds.), Academic, New York, pp. 159–165.
13. Campbell, A. M. (1984) *Monoclonal Antibody Technology.* Elsevier, New York.
14. Silver, L. M. and Elgin, S. C. R. (1977) Distribution patterns of three subfractions of *Drosophila* nonhistone chromosomal proteins: Possible correlations with gene activity. *Cell* **11,** 971–983.
15. Alfagame, C. R., Zweidler, A., Mahowald, A., and Cohen, L. H. (1974) Histones of *Drosophila* embryos. *J. Biol. Chem.* **249,** 3729–3736.
16. Cohen, L. H., Newrock, K. M., and Zweidler, A. (1975) Stage-specific switches in histone synthesis during embryogenesis of the sea urchin. *Science* **190,** 994–997.
17. Ried, J. L., Everad, J. D., Diani, J., Loescher, W. H., and Walker-Simmons, M. K. (1992) Production of polyclonal antibodies in rabbits is simplified using perforated plastic golf balls. *BioTechniques* **12,** 661–666.

130

Raising Highly Specific Polyclonal Antibodies Using Biocompatible Support-Bound Antigens

Monique Diano and André Le Bivic

1. Introduction

Highly specific antibodies directed against minor proteins, present in small amounts in biological fluids, or against insoluble cytoplasmic or membraneous proteins, are often difficult to obtain. The main reasons for this are the small amounts of protein available after the various classical purification processes and the low purity of the proteins.

In general, a crude or partially purified extract is electrophoresed on an SDS-polyacrylamide (SDS-PAGE) gel; then the protein band is lightly stained and cut out. In the simplest method, the acrylamide gel band is reduced to a pulp, mixed with Freund's adjuvant, and injected. Unfortunately, this technique is not always successful. Its failure can probably be attributed to factors such as the difficulty of disaggregating the acrylamide, the difficulty with which the protein diffuses from the gel, the presence of SDS in large quantities resulting in extensive tissue and cell damage, and finally, the toxicity of the acrylamide.

An alternative technique is to extract and concentrate the proteins from the gel by electroelution but this can lead to loss of material and low amounts of purified protein. Another technique is to transfer the separated protein from an SDS-PAGE gel to nitrocellulose. The protein-bearing nitrocellulose can be solubilized with dimethyl sulfoxide (DMSO), mixed with Freund's adjuvant, and injected into a rabbit. However, although rabbits readily tolerate DMSO, mice do not, thus making this method unsuitable for raising monoclonal antibodies.

To obtain highly specific antibodies the monoclonal approach has been considered as the best technique starting from a crude or partially purified immunogen. However, experiments have regularly demonstrated that the use of highly heterogeneous material for immunization never results in the isolation of clones producing antibodies directed against all the components of the mixture. Moreover, the restricted specificity of a monoclonal antibody that usually binds to a single epitope of the antigenic molecule is not always an advantage. For example, if the epitope is altered or modified (i.e., by fixative, Lowicryl embedding, or detergent), the binding of the monoclonal antibody might be compromised, or even abolished.

From: *The Protein Protocols Handbook, 2nd Edition*
Edited by: J. M. Walker © Humana Press Inc., Totowa, NJ

Because conventional polyclonal antisera are complex mixtures of a considerable number of clonal products, they are capable of binding to multiple antigenic determinants. Thus, the binding of polyclonal antisera is usually not altered by slight denaturation, structural changes, or microheterogeneity, making them suitable for a wide range of applications. However, to be effective, a polyclonal antiserum must be of the highest specificity and free of irrelevant antibodies directed against contaminating proteins, copurified with the protein of interest and/or the proteins of the bacterial cell wall present in the Freund's adjuvant. In some cases, the background generated by such irrelevant antibodies severely limits the use of polyclonal antibodies.

A simple technique for raising highly specific polyclonal antisera against minor or insoluble proteins would be of considerable value.

Here, we describe a method for producing polyclonal antibodies, which avoids both prolonged purification of antigenic proteins (with possible proteolytic degradation) and the addition of Freund's adjuvant and DMSO. Two-dimensional gel electrophoresis leads to the purification of the chosen protein in one single, short step. The resolution of this technique results in a very pure antigen, and consequently, in a very high specificity of the antibody obtained. It is a simple, rapid, and reproducible technique. Two-dimensional (2D) electrophoresis with ampholines for the isoelectric focusing (IEF) is still considered by most as time consuming and technically demanding. The introduction, however, of precast horizontal Ampholines or Immobilines IEF gels has greatly reduced this objection. Moreover, the development of 2D gel databases and the possibility to link it to DNA databases (Swiss-2D PAGE database) *(1)* further increases the value of using 2D electrophoresis to purify proteins of interest. Our technique was tested using two kinds of supports, nitrocellulose and polyvinylidene fluoride (PVDF), for the transfer of the protein. Both supports were tested for biocompatibility in rats and rabbits and were readily tolerated by the animals. Because PVDF is now used routinely for the analysis of proteins and peptides by mass spectrometry, it was of primary importance to test this as a prerequisite.

Even if 2D electrophoresis is a time-consuming technique it actually allows one to isolate several antigens in the same experiment, which is very important for applications such as maps of protein expression between wild-type and mutant cells or organisms or identification of proteins belonging to an isolated functional complex.

A polyclonal antibody, which by nature cannot be monospecific, can, if its titer is very high, behave like a monospecific antibody in comparison with the low titers of irrelevant antibodies in the same serum. Thus, this method is faster and performs better than other polyclonal antibody techniques while retaining all the advantages of polyclonal antibodies.

2. Materials

1. For 2D gels, materials are those described by O'Farrell *(2,3)* and Laemmli *(4)*. It should be noted that for IEF, acrylamide and *bis*-acrylamide must be of the highest level of purity, and urea must be ultrapure (enzyme grade) and stirred with Servolit MB-1 from Serva (0.5 g of Servolit for 30 g of urea dissolved in 50 mL of deionized water) as recommended by Görg *(5)*.
2. Ampholines with an appropriate pH range, for example, 5–8 or 3–9.

3. Transfer membranes: 0.45-μm BA 85 nitrocellulose membrane filters (from Schleicher and Schull GmBH, Kassel, Germany); 0.22-μm membranes can be used for low molecular weight antigenic proteins.
4. Transfer buffer: 20% methanol, 150 m*M* glycine, and 20 m*M* Tris, pH 8.3.
5. Phosphate-buffered saline (PBS), sterilized by passage through a 0.22-μm filter.
6. Ponceau red: 0.2% in 3% trichloroacetic acid.
7. Small scissors.
8. Sterile blood-collecting tubes, with 0.1 *M* sodium citrate, pH 6, at a final concentration of 3.2%.
9. Ultrasonication apparatus, with 100 W minimum output. We used a IOO-W ultrasonic disintegrator with a titanium exponential microprobe with a tip diameter of 3 mm (1/8 in.). The nominal frequency of the system is 20 kc/s, and the amplitude used is 12 p.

3. Methods

This is an immunization method in which nitrocellulose-bound protein purified by 2D electrophoresis is employed and in which *neither DMSO nor Freund's adjuvant* is used, in contrast to the method described by Knudsen *(6)*. It is equally applicable for soluble and membrane proteins.

3.1. Purification of Antigen

In brief, subcellular fractionation of the tissue is carried out to obtain the fraction containing the protein of interest. This is then subjected to separation in the first dimension by IEF with Ampholines according to O'Farrell's technique or by covalently bound Immobilines gradients (*see* Chapters 22 and 23). Immobilines allow a better resolution by generating narrow pH gradients (<1 pH unit), a larger loading capacity, and a greater tolerance to salt and buffer concentrations. At this point, it is important to obtain complete solubilization of the protein (*see* **Note 1** and Chapters 19 and 20).

Separation in the second dimension is achieved by using an SDS polyacrylamide gradient gel.

The proteins are then transferred from the gel to nitrocellulose (*7*, and *see* Chapters 39–42). It is important to work with gloves when handling nitrocellulose to avoid contamination with skin keratins.

3.2. Preparation of Antigen for Immunization

1. Immerse the nitrocellulose sheet in Ponceau red solution for 1–2 min, until deep staining is obtained, then destain the sheet slightly in running distilled water for easier detection of the spots. **Never let the nitrocellulose dry out.**
2. Carefully excise the spot corresponding to the antigenic protein. Excise inside the circumference of the spot to avoid contamination by contiguous proteins (*see* **Fig. 1**).
3. Immerse the nitrocellulose spot in PBS in an Eppendorf tube (1 mL size). The PBS bath should be repeated several times until the nitrocellulose is thoroughly destained. The last bath should have a volume of about 0.5 mL, adequate for the next step.
4. Cut the nitrocellulose into very small pieces with scissors. Then rinse the scissors into the tube with PBS to avoid any loss (*see* **Fig. 2**).
5. Macerate the nitrocellulose suspension by sonication. The volume of PBS must be proportional to the surface of nitrocellulose to be sonicated. For example, 70–80 μL of PBS is adequate for about 0.4 cm^2 of nitrocellulose (*see* **Notes 3** and **4**).

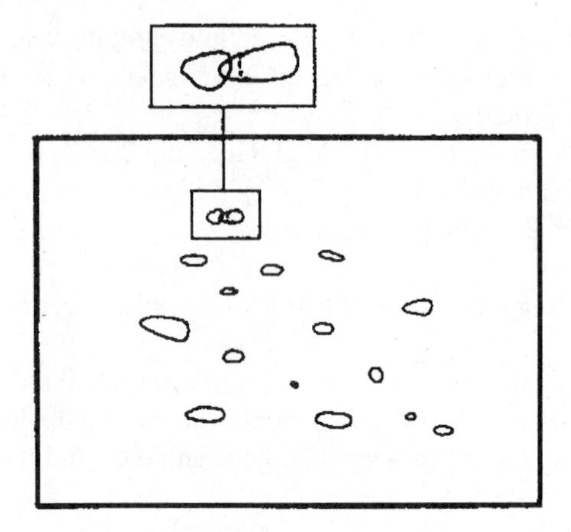

Fig. 1. Excision of the spot containing the antigen. Cut inside the circumference, for instance, along the dotted line for the right spot.

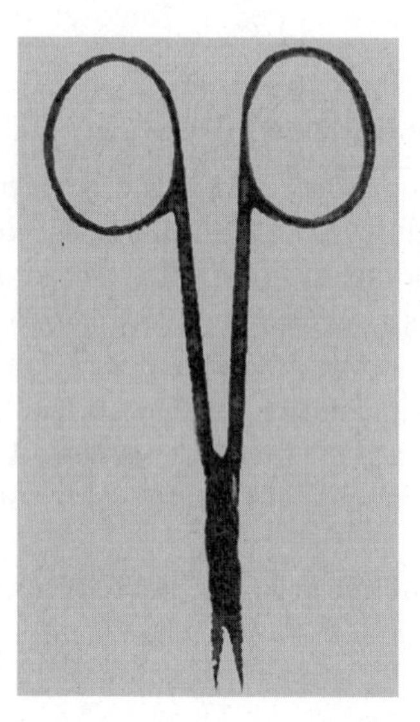

Fig. 2. Maceration of nitrocellulose.

6. After sonication, add about 1 mL of PBS to the nitrocellulose powder to dilute the mixture, and aliquot it in 500-, 350-, and 250-μL fractions and freeze these fractions at –80°C until use. Under these storage conditions, the aliquots may be used for immunization for up to 1 yr, or perhaps longer. Never store the nitrocellulose without buffer. Never use sodium azide because of its toxicity.

3.3. Immunization

1. Shave the backs of the rabbits or rats. Routinely inject two rabbits or two rats with the same antigen.
2. Thaw the 500-μL fraction for the first immunization and add 1.5–2 mL of PBS to reduce the concentration of nitrocellulose powder.
3. Inoculate the antigen, according to Vaitukaitis *(8)*, into 20 or more sites (Vaitukaitis injects at up to 40 sites). Inject subcutaneously, holding the skin between the thumb and

forefinger. Inject between 50 and 100 µL (a light swelling appears at the site of injection). As the needle is withdrawn, compress the skin gently. An 18-gauge hypodermic needle is routinely used, though a finer needle (e.g., 20- or 22-gauge) may also be used (*see* **Note 5**). Care should be taken over the last injection; generally, a little powder remains in the head of the needle. Thus, after the last injection, draw up 1 mL of PBS to rinse the needle, resuspend the remaining powder in the syringe, and position the syringe vertically to inject it.

4. At 3 or 4 wk after the first immunization, the first booster inoculation is given in the same way. The amount of protein injected is generally less, corresponding to two thirds of that of the first immunization.
5. At 10 d after the second immunization, bleed the rabbit (*see* **Note 6**). A few milliliters of blood is enough to check the immune response against a crude preparation of the injected antigenic protein. The antigen is revealed on a Western blot with the specific serum diluted at 1:500 and a horseradish peroxidase-conjugated or alkaline phosphatase-conjugated second antibody. We used 3,3'-diaminobenzidine tetrahydrochloride (DAB) for color development of peroxidase activity or the NBC-BCIP or the Western blue stabilizer (from Promega) for color development of phosphatase activity (*see* Chapter 48). If the protein is highly antigenic, the beginning of the immunological response is detectable.
6. At wk 2 after the second immunization, administer a third immunization in the same way as the first two, even if a positive response has been detected. If there was a positive response after the second immunization, one half of the amount of protein used for the original immunization is sufficient.
7. Bleed the rabbits 10 d after the third immunization and every week thereafter. At each bleeding, check the serum as after the second immunization, but the serum should be diluted at 1:4000 or 1:6000. Bleeding can be continued for as long as the antibody titer remains high (*see* **Note 7**). Another booster should be given when the antibody titer begins to decrease if it is necessary to obtain a very large volume of antiserum (*see* **Note 7**).
8. After bleeding, keep the blood at room temperature until it clots. Then collect the serum and centrifuge for 10 min at 3000*g* to eliminate microclots and lipids. Store aliquots at –22°C.

4. Notes

1. To ensure solubilization, the following techniques are useful:
   a. The concentration of urea in the mixture should be 9–9.5 *M*, that is, close to saturation.
   b. The protein mixture should be frozen and thawed at least 6×. Ampholines should be added only after the last thawing because freezing renders them inoperative.
   c. If the antigenic protein is very basic and outside the pH range of the Ampholines, it is always possible to carry out NEPHGE (nonequilibrium pH gradient electrophoresis) for the first dimension ***(9)***.
   d. A significant improvement using horizontal IEF is the possibility to load the sample at any place on the gel allowing to carry out in a single experiment both NEPHGE and IEF.
2. If the antigenic protein is present in small amounts in the homogenate, it is possible to save time by cutting out the part of the IEF gel where the protein is located and depositing several pieces of the first-dimension gel side by side on the second-dimension gel slab (*see* **Fig. 3**).
3. Careful attention should be paid to temperature during preparation of the antigen; always work between 2 and 4°C. Be particularly careful during sonication; wait 2–3 min between consecutive sonications. It helps to dip the eppendorf tube containing the nitrocellulose in liquid nitrogen until it is frozen and then to thaw it between each cycle of sonication.

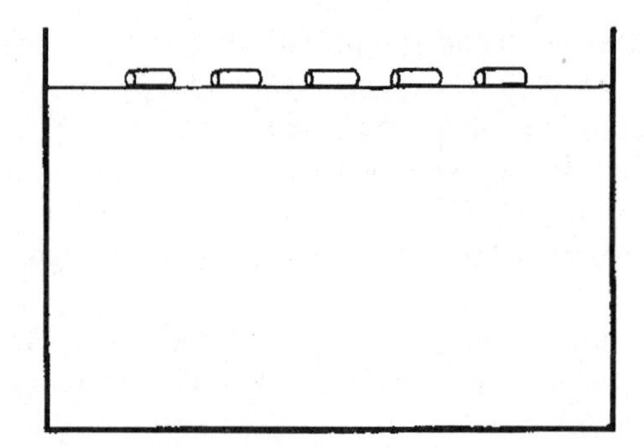

Fig. 3. Second dimension with several IEF gels. Several IEF gels are cut, 0.5 cm above and 0.5 cm below the isoelectric point of the protein of interest. They are placed side by side at the top of the second dimension slab gel. Thus, only one SDS gel is needed to collect several spots of interest.

4. This is a crucial point in the procedure. If too much PBS is added, the pieces of nitrocellulose will swirl around the probe and disintegration does not occur. In this case, the nitrocellulose pieces should be allowed to settle to the bottom of the tube before sonication and the excess buffer drawn off with a syringe or other suitable instrument (70–80 µL of PBS is sufficient for about 0.4 cm^2 of nitrocellulose). For these quantities, one or two 10-s cycles suffice to get powdered nitrocellulose. We mention the volume as a reference since the surface of nitrocellulose-bound antigen may vary. In every case the volume of PBS must be adjusted. We experienced that PVDF is more resistant to sonication than nitrocellulose and thus several cycles of freezing and thawing are necessary.
5. What is an appropriate amount of antigenic protein to inject? There is no absolute answer to this question. It depends both on the molecular weight of the protein and also on its antigenicity. It is well known that if the amount of antigen is very low (0.5–1 µg for the classic method with Freund's adjuvant), there is no antibody production; if the amount of antigen is very high (e.g., several hundred micrograms of highly antigenic protein), antibody production might also be blocked.

   It would appear that in our method, a lower amount of antigen is required for immunization; the nitrocellulose and PVDF act as if they progressively release the bound protein, and thus, the entire amount of protein is used progressively by the cellular machinery.

   Our experiments show that a range of 10–40 µg for the first immunization generally gives good results, although, in some cases, 5 µg of material is sufficient. The nitrocellulose powder has the additional effect of triggering an inflammatory process at the sites of injection, thus enhancing the immune response, as does Freund's adjuvant by means of the emulsion of the antigenic protein with the tubercular bacillus; macrophages abound in the inflamed areas.
6. It is perhaps worth noting that careful attention should be paid to the condition of the rabbit at time of bleeding. We bleed the rabbits at the lateral ear artery. When the rabbit is calm, 80–100 mL of blood may be taken. The best time for bleeding is early in the morning, and after the rabbit has drunk. Under these conditions, the serum is very clear. It is essential to operate in a quiet atmosphere. If the rabbit is nervous or under stress, the arteries are constricted so strongly that only a few drops of blood can be obtained. Note that to avoid clotting, the needle is impregnated with a sterile sodium citrate solution by drawing the solution into the syringe 3×. The atmosphere in the animal room should be also stressless for the rats to obtain enough blood.

7. When the effective concentration required corresponds to a dilution of 1 : 2000, the titer is decreasing. Serum has a high titer if one can employ a dilution >1 : 2000 and if there is a strong specific signal without any background.
8. We have also immunized mice with nitrocellulose-bound protein by intraperitoneal injection of the powder. This is convenient when time and material are very limited, since very little protein is needed to induce a response (3–5× less than for a rabbit) and since the time lag for the response is shorter (the second immunization was 2 wk after the first, and the third immunization, 10 d after the second). Mice have a high tolerance for the nitrocellulose powder. Unfortunately, the small amount of serum available is a limiting factor. This technique for immunizing mice can, of course, be used for the preparation of monoclonal antibodies. For people who need to raise antibodies against *Drosophila melanogaster* proteins, rats are better hosts than rabbits (mice too with the above restrictions). Nitrocellulose is injected on the top of the shoulder of the rats. The foot pad is also adequate as a site of priming injection for small quantities. Blood is collected from the tail.
9. Utilization of serum. The proper dilutions are determined. We routinely use 1 : 4000 for blots, 1 : 300–1 : 200 for immunofluorescence, and 1 : 50 for immunogold staining. Serum continues to recognize epitopes on tissue proteins after Lowicryl embedding. Labeling is highly specific, and gives a sharp image of *in situ* protein localization. *There is no need to purify the serum.* IgG purified from serum by whatever means usually gives poorer results than those obtained with diluted serum. Purification procedures often give rise to aggregates and denaturation, always increase the background, and result in loss of specific staining.
10. Bacterial antigenic protein. When antibodies are used in screening cDNA libraries in which the host is *Escherichia coli*, the antibodies produced against bacterial components of Freund's adjuvant may also recognize some *E. coli* components. An advantage of our technique is that it avoids the risk of producing antibodies against such extraneous components.
11. Is nitrocellulose antigenic? Some workers have been unable to achieve good results by immunization with nitrocellulose-bound protein. They reproducibly obtain antibodies directed against nitrocellulose. We found out that it is the result from injecting powdered nitrocellulose in Freund's adjuvant; using adjuvant actually increases the production of low-affinity IgM that binds nonspecifically to nitrocellulose. We have never observed this effect in our experiments when the technique described here was followed strictly. The same observations applied to PVDF membranes.
12. The purification step by 2D electrophoresis implies the use of denaturing conditions (SDS), and thus is not appropriate for obtaining antibodies directed against native structures. For that purpose, the protein should be transferred onto nitrocellulose after purification by classical nondenaturing methods and gel electrophoresis under nondenaturing conditions. However, it should be pointed out that, following the method of Dunn ***(10)***, it is possible partially to renature proteins with modifications of the composition of the transfer buffer.
13. Second dimension electrophoresis can be carried out with a first electrophoresis under native conditions, followed by a second electrophoresis under denaturing conditions, that is, with SDS. Because the resolution provided by a gradient is better, it should always be used in the 2D electrophoresis. Agarose may also be used as an electrophoresis support.
14. If only a limited amount of protein is available, and/or if the antigen is weakly immunogenic, another procedure may be used. The first immunization is given as a single injection, of up to 0.8 mL, into the popliteal lymphatic ganglion ***(10)***, using a 22-gauge needle, that is, the finest that can be used to inject nitrocellulose powder. In this case, the antigen is delivered immediately into the immune system. If necessary, both ganglions can receive an injection. The small size of the ganglions limits the injected volume. The even-

tual excess of antigen solution is then injected into the back of the rabbit, as described in Methods. The boosters are given in the classic manner, that is, in several subcutaneous injections into the rabbit's back. If the amount of protein available is even more limited, a guinea pig may be immunized (first immunization in the Iymphatic ganglion, and boosters, as usual).

15. The advantage of getting a high titer for the antibody of interest is that the amount of irrelevant antibodies is, by comparison, very low and, consequently, does not generate any background. Another advantage of using a crude serum with a high antibody titer is that this serum may be used without further purification to screen a cDNA expression library *(12)*.
16. The time required for transfer and the intensity used are dependent on the molecular weight and the nature of the protein to be transferred (hydrophilic or hydrophobic). During transfer, the electrophoresis tank may be cooled with tap water. We, however, prefer semidry transfer, which is time and buffer saving and cooling is not necessary.

References

1. Pennington, S. (1994) 2-D protein gel electrophoresis: an old method with future potential. *Trends Cell Biol.* **4,** 439–441.
2. O'Farrel, P. H. (1975) High resolution two-dimensional electrophoresis of proteins. *J. Biol. Chem.* **250,** 4007–4021.
3. O'Farrel, P. Z., Goodman, H. M., and O'Farrel, P. H. (1977) High resolution two-dimensional electrophoresis of basic as well as acidic proteins. *Cell* **12,** 1133–1142.
4. Laemmli, U. K. (1970) Cleavage of structural proteins during the assembly of the head of bacteriophage T4. *Nature (Lond.)* **227,** 680–685.
5. Görg, A. (1998) A laboratory manual. http://www.weihenstephan.de/blm/deg.
6. Knudsen, K. A. (1985) Proteins transferred to nitrocellulose for use as immunogens. *Analyt. Biochem.* **147,** 285–288.
7. Burnette, W. N. (1981) Electrophoretic transfer of proteins from sodium dodecylsulfate polyacrylamide gels to unmodified nitrocellulose and radiographic detection with antibody and radioiodinated protein A. *Analyt. Biochem.* **112,** 195–203.
8. Vaitukaitis, J., Robbins, J. B., Nieschlag, E., and Ross, G. T. (1971) A method for producing specific antisera with small doses of immunogen. *J. Clin. Endocrinol.* **33,** 98–988.
9. Dunn, M. J. and Patel, K. (1988) 2-D PAGE using flat bed IEF in the first dimension, in *Methods in Molecular Biology*, Vol. 3: *New Protein Techniques* (Walker, J. M., ed.), Humana Press, Totowa, NJ, pp. 217–232.
10. Dunn, S. D. (1986) Effects of the modification of transter buffer composition and the renaturation of proteins in gels on the recognition of proteins on Western blots by monoclonal antibodies. *Analyt. Biochem.* **157,** 144–153.
11. Sigel, M. B., Sinha, Y. N., and Vanderlaan, W. P. (1983) Production of antibodies by inoculation into lymph nodes. *Meth. Enzymol.* **93,** 3–12.
12. Preziosi, L., Michel, G. P. F., and Baratti, J. (1990) Characterisation of sucrose hydrolizing enzymes of *Zymomonas mobilis*. *Arch. Microbiol.* **153,** 181–186.

131

Production of Antisera Using Peptide Conjugates

Thomas E. Adrian

1. Introduction

Because an immunogen requires both an antigenic site and a T-cell receptor binding site, a minimum size is necessary *(1)*. Natural immunogens have a molecular weight >5000. Synthetic fragments of proteins may be able to bind to the surface of B cells, but do not stimulate an immune response. Such molecules are known as haptens. A hapten is an incomplete immunogen but can be made immunogenic by coupling to a suitable carrier molecule. A variety of different cross-linking agents are utilized for coupling of peptides to carrier proteins; examples of each type are covered in this chapter. In the case of larger fragments (approx 40 amino acids), it may be possible to stimulate an immune response by presenting the peptide together with a carrier such as polyvinylpyrrolidone without the need for conjugation. This method, which is also described in this chapter, has proven to be useful for the author for a number of different peptides. Unfortunately, however, the success of this method compared with the responses to conjugated peptides is largely a matter of trial and error, and peptide fragments of this size are expensive to produce.

2. Materials

2.1. Synthetic Peptides as Haptens

There is considerable advantage to be gained by using synthetic peptides as haptens. First, it is possible to raise region-specific antibodies directed perhaps to one end, or the active site of a protein. Second, it is possible to insert particular amino acids with specific side chains for coupling. For example, a peptide can be synthesized with an exã)a cysteine residue at one end of the molecule to enable coupling using sulfo-SMCC *(2)*. Alternatively, a tyrosine residue can be inserted that will enable specific coupling through the *bis*-diazotized benzidine reaction *(3)*. This latter approach is valuable because the same synthetic peptide can then serve as a radioligand for metabolic studies, for receptor binding, or for a radioimmunoassay, with the assurance that the antibodies raised will not be directed toward the tyrosine residue that will be iodinated *(3)*.

From: *The Protein Protocols Handbook, 2nd Edition*
Edited by: J. M. Walker © Humana Press Inc., Totowa, NJ

2.2. Protein Carriers

Factors governing the choice of the carrier include immunogenicity, solubility, and availability of functional groups for cross-linking. Although substances such as mucopolysaccharides, poly-L-lysine, and polyvinylpyrrolidone have been used as carriers, proteins are more widely used. Common protein carriers include serum albumin, ovalbumin, hemocyanin, and thyroglobulin. To find the very best immunogen it would be ideal to prepare conjugates with several different carriers with a range of hapten to carrier coupling ratios (*see* **Note 1**). The cost and time involved will usually make this impractical, however, and it is therefore necessary to carefully select the carrier most suitable for a particular antigen. In the classical hapten carrier system T lymphocytes recognize processed carrier determinates and cooperate with B cells which produce hapten-specific antibody response. Note that the amounts of carrier and peptide for conjugation are given in molar terms in the following conjugation protocols. This is necessary because of the wide variation in the molecular weights of potential carriers and neuropeptides to be coupled. The carrier protein represented by 100 nmol is approx 7 mg of bovine albumin, 4.5 mg of ovalbumin, 15 mg of γ-globulin, and 70 mg of thyroglobulin.

2.2.1. Bovine Serum Albumin

Because of its wide availability, high solubility, and relatively high number of coupling sites albumin is a popular choice as a carrier for weakly antigenic compounds. Albumin has a molecular weight of 67,000 and has 59 lysine residues providing primary amines useful for conjugation.

2.2.2. Ovalbumin

Ovalbumin (egg albumin) also has wide availability, as it is the primary protein constituent of egg white. This protein is smaller than serum albumin with a molecular weight of 45,000, but contains 20 lysine residues, 14 aspartic acid, and 33 glutamic acid residues for conjugation *(4)*. Ovalbumin exists as a single polypeptide chain with an isoelectric point of 4.6. Half of its 400 residues are hydrophobic. Caution should be exercised in handling of albumin, as it is denatured at temperatures above 56°C or even by vigorous shaking.

2.2.3. Hemocyanin

Keyhole limpit hemocyanin (KLH) is a copper-containing protein that belongs to a family of nonheme proteins found in arthropods and mollusca. KLH exists in five different aggregate states at neutral pH that will dissociate into subunits above pH 9.0 *(5)*. KLH is a valuable carrier protein because of its large molecular mass (approx 1×10^6 to 1×10^7) and numerous lysine groups for coupling. This property of dissociation at high pH can be utilized because it increases the availability of angiogenic sites, and this can produce improved antigenic responses *(5)*. The disadvantage of using KLH as a carrier protein is its poor water solubility. Although this makes the protein difficult to handle, it does not impair its immunogenicity.

2.2.4. Thyroglobulin

Thyroglobulin is another large molecular weight protein with a limited solubility. The advantage of thyroglobulin as a carrier comes from its large content of tyrosine residues which can be used for conjugation using the diazo reaction. The molecular weight of thyroglobulin is 670,000.

2.3. Coupling Agents

Chemical coupling agents or cross-linkers are used to conjugate small peptide haptens to large protein carriers. The most commonly used cross-linking agents have functional groups that couple to amino acid side chains of peptides (*see* **Table 1**). A host of different cross-linkers are commercially available, each with different characteristics regarding chemical groups that they are reactive toward, pH of coupling, solubility, and ability to be cleaved. The most comprehensive listing of coupling reagents is found in the Pierce catalog (Pierce, Rockford, IL; http://www.piercenet.com). Several things need to be taken into consideration when selecting a bifunctional coupling reagent. First is the selection of functional groups; this can be used to produce a specific type of conjugate. For example, if the only primary amine available is in the amino-(N)-terminal end of a peptide, then a coupling agent can be selected to specifically couple in this position, leaving the carboxyl (C)-terminal end of the peptide free and available as an antigenic site. Of course, a synthetic peptide can be engineered to contain an amino acid with a specific side chain available for coupling. This is usually placed at one end of the peptide, again making the other end available as an antigenic site. Second is the length of the cross bridge; the presence of a spacer arm may make the hapten more available and therefore produce a better immune response. Third is whether the cross-linking groups are the same (homobifunctional) or different (heterobifunctional). Once again this can alter the specificity of the coupling reaction. The last consideration is whether the coupling reaction is chemical or photochemical. For a good antigenic response it is necessary to maintain the native structure of the protein complex, and this can be achieved only using mild buffer conditions and near neutral pH. The reactive groups that can be targeted using cross-linkers include primary amines, sulfhydryls, carbonyl, and carboxylic acids (*see* **Note 2**). It is difficult to predict the proximity of protein–peptide interactions. The use of bifunctional reagents with spacer arms can prevent steric hindrance and make the hapten more available for producing a good immune response.

2.3.1. Carbodiimide

Carbodiimide condenses any free carboxyl group (nonamidated C-terminal aspartate or glutamate residue) or primary amino group (N-terminal or lysyl residue), to form a peptide bond (CO–NH). The most commonly used water-soluble carbodiimide is 1-ethyl-3-(3-dimethylaminopropyl)-carbodiimide hydrochloride (CDI or EDC). This coupling agent is very efficient and easy to use and usually couples at several alternative points on a peptide, giving rise to a variety of antigenic responses ***(6)***. Because of the unpredictable nature of the antibody responses, using this bifunctional agent the process has been termed "shotgun immunization." If it is necessary to raise antisera to a particular region of the peptide then this is not the method of choice. Furthermore, because the peptide bond linking the hapten to the carrier cannot rotate and holds the hapten physically close, steric hindrance is considerable.

Carbodiimide reacts with available carboxyl groups to form an active *O*-acylurea intermediate, which is unstable in aqueous solution, making it ineffective in two-step

Table 1
Bifunctional Cross-linking Agents Useful for Conjugation of Peptide Haptens to Protein Carriers for Immunization

Coupling method	Chemical name and formula	Reactive toward	Comments
CDC or EDC	1-Ethyl-3-[3-dimethylaminopropyl] -carbodiimide hydrochloride $CH_3{-}CH_2{-}N{=}C{=}N{-}(CH_2)_3{-}N(H^+Cl^-)(CH_3){-}CH_3$	Carboxyls or primary amines	Condenses any free carboxyl or primary amino groups (C- or N-terminus or side chain), rapid efficient, considerable steric hindrance, nonspecific coupling ("shotgun immunization").
Glutaraldehyde	glutaraldehyde $CHO{-}(CH_2)_3{-}CHO$	Primary amines	Forms bridge between primary amino groups (N-terminus or lysine residue, rapid, efficient, allows free rotation of hapten and thereby avoids steric hindrance.
Diazo	*BIS*-Diazotized benzidine $N{\equiv}N{-}C_6H_4{-}C_6H_4{-}N{\equiv}N$	Tyrosyl or histidyls	*BIS*-Diazonium salts bridge residues between hapten and carrier, very specific, overnight procedure, spacer arm holds hapten away from carrier, usually results in an excellent antigenic response.
Sulfo-SMCC	Sulfosuccinimidyl 4-[*N*-maleimidomethyl] -cyclohexane-1-carboxylate NaO_3S; N–O–C(=O)–; –CH_2–N	Primary amine and sulfhydryl	Contains a maleimide to react with a free sulfhydryl group and an NHS-ester group to react with a primary amino group, efficient and stable coupling, highly specific (e.g., coupling a synthetic peptide with a terminal cysteine residue.

conjugation procedures. This intermediate then reacts with the primary amine to form an amide derivative. Failure to immediately react with an amine results with hydrolysis of the intermediate. Furthermore, hydrolysis of CDI itself is a competing reaction during the coupling and is dependent on temperature 4-morphylinoethansulfonic acid (MES) can be used as an effective carbodiimide reaction buffer in place of water. Phosphate buffers reduce the efficiency of the CDI reaction, although this can be overcome by increasing the amount of CDI used to compensate for the reduction and efficiency. Loss of efficiency of the CDI reaction is even greater with Tris, glycine, and acetate buffers, and therefore use of these these should be avoided.

2.3.2. Glutaraldehyde

Glutaraldehyde links primary amino groups (either the N-terminal or lysyl residues) on both the peptide hapten and the carrier. This linkage allows free rotation of the hapten, which reduces possible steric hindrance which may otherwise block access to the immune system by the large carrier molecule.

2.3.3. Sulfo-succinimidyl 4-(N-maleimidomethyl) cyclohexane-1-carboxylate (sulfo-SMCC)

A peptide with a free sulfhydryl group, such as a synthetic peptide with a terminal cysteine residue, provides a highly specific conjugation site for reacting with sulfo-SMCC. This cross-linker contains a maleimide group that reacts with free sulfhydryl groups, along with an *N*-hydroxysuccinimidyl ester group that reacts with primary amines *(2)*. All peptide molecules coupled using this chemistry display the same basic antigenic conformation. They have a known and predictable orientation leaving the molecule free to interact with the immune system. This method can preserve the major epitopes on a peptide while enhancing the immune response. The water solubility of sulfo-SMCC, along with its enhanced maleimide stability, makes it a favorite for hapten carrier conjugation.

2.3.4. Bis-Diazotized Benzidine

Bis-diazonium salts bridge tyrosyl or hystidyl residues between the hapten and carrier. Overnight treatment of benzidine at 4°C with nitrous acid (hydrochloric acid and sodium nitrite) results in the two amino groups being diazotized. These two diazonium groups allow coupling at both ends of the molecule. Although limited in coupling points, diazotized benzidine provides a spacer arm holding the hapten away from the carrier and usually results in an excellent antigenic response.

2.4. Adjuvants

The adjuvant is important for inducing an inflammatory response. The author has had continuous success over two decades using Freund's adjuvant. Various synthetic adjuvants are available such as AdjuPrime (Pierce) and Ribi Adjuvant System (Ribi Immunochemicals) *(7)*. The author has had very limited success with the latter adjuvant, when used to immunize with several different small peptides. In contrast, responses with Freund's adjuvant run in parallel have always been good.

The conjugate is injected in the form of an emulsion made with Freund's adjuvant, which is a mixture of one part of detergent (Arlacel A, Sigma Chemicals) with four parts of *n*-hexadecane. This permits slow release of the coupled hapten into the circulation and may serve to protect labile antigens from degradation. Freund's adjuvant alone ("incomplete") causes an inflammatory response that stimulates antibody formation, and when made "complete" by addition of 1 mg/mL of heat-killed *Mycobacterium butyricum*, this response is further enhanced. It is convenient to purchase complete and incomplete Freund's adjuvant ready mixed (Sigma Chemicals or Calbiochem).

When preparing the emulsion, care should be taken to ensure that the oil remains in the continuous phase. Injection of aliquots of the aqueous conjugate solution into the oil via a fine bore needle, followed by repeated aspiration and ejection of the crude emulsion, will produce the required result. An alternative for making the emulsion is to use a homogenizer, although the generator should be retained specifically for this purpose to avoid subsequent peptide contamination. A simple test for the success of the preparation is to add a drop of the emulsion to the surface of water in a tube. The emulsion should stay in a single droplet without dispersing; confirming that it is immiscible and thus oil-phase continuous.

2.5. Choice of Animal for Immunization

Several factors need to be considered when choosing animal species for an immunization program, including cost, ease of handling and the volumes of antisera required (*see* **Note 3**). Small animals (such as rats and mice) have low blood volumes and present difficulties with bleeding. Large animals such as sheep or goats are expensive to house particularly over long periods. Rabbits or guinea pigs provide a near optimal solution, as they are relatively cheap to house and bleeding an ear vein or cardiac puncture in guinea pigs can provide between 10 and 30 mL of plasma from each bleed. For production of monoclonal antibodies immunization of mice is required.

3. Methods

3.1. Carbodiimide Procedure

1. Dissolve the peptide to be coupled (400 nmol) and protein carrier (100 nmol) in a small volume of water (<1 mL if possible) (*see* **Note 4**).
2. Add carbodiimide (200 mmol, 50 mg) to this solution.
3. Incubate the mixture overnight at 4°C (*see* **Notes 5** and **6**).

3.2. Glutaraldehyde Procedure

1. Dissolve the peptide (400 nmol) and carrier protein (100 nmol) in 1 mL of 0.1 *M* phosphate buffer, pH 7.5 (*see* **Note 4**).
2. Add glutaraldehyde (30 mmol, 1.5 mL of a 0.02 M solution) dropwise for 15 min.
3. Incubate the mixture overnight at 4°C (*see* **Notes 5** and **6**).

3.3. Sulfo-SMCC Procedure

1. Activate the carrier (100 nmol) by conjugating the active ester of sulfo-SMCC (2 μmol) via an amino group, in phosphate-buffered saline (PBS), pH 7.2 (with 5 m*M* EDTA) for 60 min at room temperature. This reaction results in the formation of an amide bond between the protein and the cross-linker with the release of sulfo-N-hydroxysuccinimidyl ester as a byproduct.
2. If desired the carrier protein can then isolated by gel filtration to remove excess reagent

using a gel such as Sephadex G25 (*see* **Note 5**). Desalt by eluting the column with PBS, collecting 0.5-mL fractions. Locate the protein peak using a protein assay (Bio-Rad micro method).

3. At this stage the purified carrier possesses modifications generated by the cross-linker, resulting in a number of maleimide groups projecting from its surface. The maleimide group of sulfo-SMCC is stable for several hours in solution at physiological pH. Therefore, even after activation and purification the greatest possible activity will still be left for conjugation with the peptide.
4. The maleimide group of sulfo-SMCC reacts at pH 7.0 with free sulfhydryls on the peptide to form a stable thioether bond. The peptide (100 nmol) with a free sulfhydryl group is incubated with the maleimide-activated carrier in 1 mL of PBS (with 5 m*M* EDTA) for 2 h (or overnight if more convenient) at 4°C (*see* **Note 6**).
5. Keep sulfo-SMCC away from moisture since it is subject to hydrolysis.

3.4. Diazo Procedure

1. Freshly prepare *bis*-diazotized benzidine on each occasion in the following manner: dissolve benzidine hydrochloride (80 mmol, 20.5 mg) in 10 mL of 0.18 *M* HCl and gently mix overnight with 1 mL of 0.16 *M* $NaNO_2$ (11 mg). This reaction should take place in an ice bath inside a cold room, and the temperature should never be allowed to rise above 4°C.
2. Dissolve the peptide (400 nmol) in 130 µL (1 µmol) of fresh *bis*-diazotized benzedine solution.
3. Add $NaHCO_3$ (40 µmol, 3.4 mg) followed immediately by the addition of the carrier predissolved in a minimal volume of aqueous solution (carrier protein 100 nmol) at 4°C.
4. Adjust the pH to 9.8 with NaOH using a microelectrode.
5. Incubate the mixture overnight at 4°C (*see* **Notes 5** and **6**).

3.5. Immunization Without Conjugation to Carrier Protein

1. This procedure, using polyvinylpyrrolidone as a noncovalent carrier, can be very valuable for peptide antigens with more than 40 amino acid residues. Good results have been obtained with 30–40 amino acid peptides, but it is unlikely to produce useful antisera with smaller antigens. As well as being quick and easy, it has the added advantage of retaining tertiary peptide structure.
2. Dissolve the peptide in aqueous solution containing a 100 *M* excess of polyvinylpyrrolidone (with respect to peptide concentration).
3. Emulsify the solution in Freund's adjuvant in the same manner as with peptide conjugates.

3.6. Making an Emulsion in Freund's Adjuvant

1. Dissolve conjugate in water (between 10 and 100 nmol of conjugated peptide/mL, 1 mL for each rabbit being immunized).
2. Make Freund's adjuvant by mixing one part of Arlacel A with four parts of *n*-hexadecane (allowing a little more than 1 mL per rabbit).
3. For primary injections only "complete" Freund's adjuvant is used. This is made by addition of 1 mg/mL of heat-killed mycobacteria. (Boosts are given in incomplete Freund's.)
4. When preparing the emulsion, care should be taken to ensure that the oil remains in the continuous phase. Injection of aliquots of the aqueous conjugate solution into the oil via a fine bore needle, followed by repeated aspiration and ejection of the crude emulsion, will produce the required result.
5. A simple test for the success of the preparation is to add a drop of the emulsion to the surface of water in a tube. The emulsion should stay in a single droplet without dispersing,

confirming that it is immiscible and thus oil-phase continuous.

3.7. Immunization Procedure

The emulsified conjugate can be administered in a variety of ways. For rabbits, the most frequently used are multiple (30–50) intradermal injections in the neck or back region, by four subcutaneous injections into each groin and axilla (0.5) *(8)*. The latter is the method we have successfully adopted for more than 20 yr. Injection into the foot-pads, which was at one time commonly employed, provides no advantage in terms of antibody response and should be avoided to prevent distress to the animals.

The procedure is as follows:

1. Bleed the animals and collect preimmune serum for later comparison with antisera produced by the immunization procedure.
2. Primary inoculations are given in complete Freund's adjuvant, 0.5 mL of emulsion into each groin and axilla.
3. Booster injections are given at 2–4-wk intervals, in the same manner but with incomplete Freund's adjuvant. The optimum is probably about 4 wk, but time constraints and cost may necessitate a shorter immunization schedule. With small synthetic haptens three to five, or even more boosts may be required to produce the desired high titer or high avidity antibody (*see* **Notes 7** and **8**).
4. After the first and subsequent boosts blood should be collected from an ear vein to test for the antibody titer and avidity.

3.8. Antibody Characterization for Radioimmunoassay

1. Serial dilutions of antisera are incubated with radiolabeled peptide under routine assay conditions to determine a working dilution.
2. The maximum displacement of radioactively labeled hormone from the antibody by the minimum amount of unlabeled peptide (the maximum displacement slope) is one of the main criteria for radioimmunoassay sensitivity.
3. Rapid screening for slope can be achieved by the addition of small amounts (usually about 10 fmol) of standard peptide to one set of a series of replicate antiserum dilutions set up to determine the antiserum titer. The amount of standard used should reflect the useful range (e.g., the concentration at which a hormone circulates).
4. Antibody heterogeneity may be due to use of nonhomogeneous antigens for immunization, polymerization or degradation of the hapten or carrier after coupling, or individual differences in the lymphocytic response to the antigen *(9)*.
5. Existence of heterogeneity can be revealed by Scatchard analysis. However, for high titer antisera there is frequently effectively only a single class of antibody that predominates in the reaction. Other populations of lower concentrations and avidities make insignificant contributions.
6. Specificity should be tested using related peptides.

4. Notes

1. The ratio of hapten reactive with protein is usually arranged to be in excess of 4:1 to achieve better antigenicity with respect to the hapten. Some authorities prefer ratios as high as 40:1 but in our experience this gives a lower affinity antibody response. This is presumably due to conjugation between hapten molecules rather between hapten and carrier. The hapten and protein carriers should be both present in high concentration to increase the efficiency of the cross-linking between the molecules.
2. Coupling of a hapten at a specific site gives more chance of governing which part of the

peptide becomes the antigenic determinant for the antibody, as the particular area of peptide where coupling occurs is likely to be hidden from immune surveyance.

3. Success in raising antisera is to some extent a hit and miss affair. Some workers have needed to immunize many animals to produce useful antisera, whereas three or four rabbits may produce the desired product in another immunization program. This depends, in part, on the antigenicity of the peptide and on the goals set for sensitivity and specificity. Some peptides are particularly susceptible to proteolysis (such as members of the vasoactive intestinal polypeptide family) or oxidation (such as cholecystokinin). In general, these less stable peptides make relatively poor antigens.
4. Usually the coupling agent is added after the hapten and carrier have been mixed together to minimize self-polymerization of either component.
5. Although not usually necessary, excess unreacted hapten and toxic byproducts may be removed by dialysis or gel permeation chromatography.
6. Quantification of the success of the coupling reaction is conveniently obtained by addition of a small amount of radioactively labeled hapten to the mixture prior to adding the coupling agent. Thus before and after the reaction small aliquots are removed and chromatographed. Small disposable columns contain Sephadex G-25 (Pharmacia) are ideal for this purpose. Proportion of radiation diluting in the high molecular weight position, together with the carrier, indicates the amount of coupling achieved.
7. If a particular animal has been boosted three or four times without producing detectable antibody, then the likelihood of it subsequently doing so is small and further effort is unprofitable.
8. On other occasions when animals do show a response but further boosting results in little improvement in avidity or titer, then variation in the coupling method for subsequent boosts can help. Changes including a different carrier protein or the cross-linking agent, or both, may result in production of a higher titer or more avid antisera.

References

1. Germain, R. (1986) The ins and outs of antigen processing and presentation. *Nature* **322,** 687–689.
2. Samoszuk, M. K., Petersen, A., Lo-Hsueh, M., and Rietveld, C. (1989) A peroxide-generating immunoconjugate directed to eosinophil peroxidase is cytotoxic to Hodgkin's disease cells *in vitro. Antibody Immunocon. Radiopharmaceut.* **2,** 37–46.
3. Adrian, T. E., Bacarese-Hamilton, A. J., and Bloom, S. R. (1985) Measurement of cholecystokinin octapeptide using a new specific radioimmunoassay. *Peptides* **6,** 11–16.
4. Harlow, E. and Lane, D. (1988) *Antibodies: A Laboratory Manual*, Cold Spring Harbor Laboratory Press, Cold Spring Harbor, NY, pp. 56–100.
5. Bartel, A. and Campbell, D. (1959) Some immunochemical differences between associated and dissociated hemocyanin. *Arch. Biochem. Biophys.* **82,** 2332–2336.
6. Bauminger, S. and Wilchek, D. (1980) The use of carbodiimides in the preparation of immunizing conjugates. *Methods Enzymol.* **70,** 151–159.
7. Chedid, L. and Lederer, E. (1978) Past, present and future of the synthetic immunoadjuvant MDP and its analogs. *Biochem. Pharmacol.* **27,** 2183–2186.
8. Vaitukaitis, J., Robbins, J. B., Nieschlag, E., and Ross, G. T. (1971) A method for producing specific antisera with small doses of immunogen. *J. Clin. Endocrinol. Metab.* **33,** 988–991.
9. Parker, C. W. (1976) *Radioimmunoassay of Biologically Active Compounds*, Prentice-Hall, Englewood Cliffs, NJ, pp. 36–67.

132

The Chloramine T Method for Radiolabeling Protein

Graham S. Bailey

Many different substances can be labeled by radioiodination. Such labeled molecules are of major importance in a variety of investigations, e.g., studies of intermediary metabolism, determinations of agonist and antagonist binding to receptors, quantitative measurements of physiologically active molecules in tissues and biological fluids, and so forth. In most of those studies, it is necessary to measure very low concentrations of the particular substance, and that in turn, implies that it is essential to produce a radioactively labeled tracer molecule of high specific radioactivity. Such tracers, particularly in the case of proteins, can often be conveniently produced by radioiodination.

Two γ-emitting radioisotopes of iodine are widely available, ^{125}I and ^{131}I. As γ-emitters, they can be counted directly in a well-type crystal scintillation counter (commonly referred to as a γ counter) without the need for sample preparation in direct contrast to β-emitting radionuclides, such as ^{3}H and ^{14}C. Furthermore, the count rate produced by 1 g atom of ^{125}I is approx 75 and 35,000 times greater than that produced by 1 g atom of ^{3}H and ^{14}C, respectively. In theory, the use of ^{131}I would result in a further sevenfold increase in specific radioactivity. However, the isotopic abundance of commercially available ^{131}I rarely exceeds 20% owing to contaminants of ^{127}I, and its half-life is only 8 d. In contrast, the isotopic abundance of ^{125}I on receipt in the laboratory is normally at least 90% and its half-life is 60 d. Also, the counting efficiency of a typical well-type crystal scintillation counter for ^{125}I is approximately twice that for ^{131}I. Thus, in most circumstances, ^{125}I is the radionuclide of choice for radioiodination.

Several different methods of radioiodination of proteins have been developed *(1,2)*. They differ, among other respects, in the nature of the oxidizing agent for converting $^{125}I^-$ into the reactive species $^{125}I_2$ or $^{125}I^+$. In the main, those reactive species substitute into tyrosine residues of the protein, but substitution into other residues, such as histidine, cysteine, and tryptophan, can occur in certain circumstances. It is important that the reaction conditions employed should lead on average to the incorporation of one radioactive iodine atom/molecule of protein. Greater incorporation can adversely affect the biological activity and antigenicity of the labeled protein.

The chloramine T method, developed by Hunter and Greenwood *(3)*, is probably the most widely used of all techniques of protein radioiodination, and is used extensively for the labeling of antibodies. It is a very simple method in which the radioactive iodide

From: *The Protein Protocols Handbook, 2nd Edition*
Edited by: J. M. Walker © Humana Press Inc., Totowa, NJ

is oxidized by chloramine T in aqueous solution. The oxidation is stopped after a brief period of time by addition of excess reductant. Unfortunately, some proteins are denatured under the relatively strong oxidizing conditions, so other methods of radioiodination that employ more gentle conditions have been devised, e.g., the lactoperoxidase method (*see* Chapter 133).

2. Materials

1. Na ^{125}I: 37 MBq (1 mCi) concentration 3.7 GBq/mL (100 mCi/mL).
2. Buffer A: 0.5 *M* sodium phosphate buffer, pH 7.4 (*see* **Note 1**).
3. Buffer B: 0.05 *M* sodium phosphate buffer, pH 7.4.
4. Buffer C: 0.01 *M* sodium phosphate buffer containing 1 *M* sodium chloride, 0.1% bovine serum albumin, and 1% potassium iodide, final pH 7.4.
5. Chloramine T solution: A 2 mg/mL solution in buffer B is made just prior to use (*see* **Note 2**).
6. Reductant: A 1 mg/mL solution of sodium metabisulfite in buffer C is made just prior to use.
7. Protein to be iodinated: A 0.5–2.5 mg/mL solution is made in buffer B.

3. Method

1. Into a small plastic test tube (1 × 5.5 cm) are added successively the protein to be iodinated (10 µL), radioactive iodide (5 µL), buffer A (50 µL), and chloramine T solution (25 µL) (*see* **Notes 3** and **4**).
2. After mixing by gentle shaking, the solution is allowed to stand for 30 s to allow radioiodination to take place (*see* **Note 5**).
3. Sodium metabisulfite solution (500 µL) is added to stop the radioiodination, and the resultant solution is mixed. It is then ready for purification as described in Chapter 136 and **Note 6**.

4. Notes

1. The pH optimum for the iodination of tyrosine residues of a protein by this method is about pH 7.4. Lower yields of iodinated protein are obtained at pH values below about 6.5 and above about 8.5. Indeed, above pH 8.5 the iodination of histidine residues appears to be favored.
2. If the protein is seriously damaged by the use of 50 µg of chloramine T, it may be worthwhile repeating the radioiodination using much less oxidant (10 µg or less). Obviously, the minimum amount of chloramine T that can be used will depend, among other factors, on the nature and amount of protein to be iodinated.
3. The total volume of the reaction should be as low as practically possible to achieve a rapid and efficient incorporation of the radioactive iodine into the protein.
4. It is normal to carry out the method at room temperature. However, if the protein is especially labile, it may be beneficial to run the procedure at a lower temperature and for a longer period of time.
5. Because of the small volumes of reactants that are employed, it is essential to ensure adequate mixing at the outset of the reaction. Inadequate mixing is one of the most common reasons for a poor yield of radioiodinated protein by this procedure.
6. It is possible to carry out this type of reaction using an insoluble derivative of the sodium salt of *N*-chloro-benzene sulfonamide as the oxidant. The insoluble oxidant is available commercially (Iodo-Beads, Pierce, Rockford, IL). It offers a number of advantages over the employment of soluble chloramine T. It produces a lower risk of oxidative damage to the protein, and the reaction is stopped simply by removing the beads from the reaction mixture, thus avoiding any damage caused by the reductant.

References

1. Bolton, A. E. (1985) *Radioiodination Techniques,* 2nd ed. Amersham International, Amersham, Bucks, UK.
2. Bailey, G. S. (1990) In vitro labeling of proteins, in *Radioisotopes in Biology* (Slater, R. J., ed.), IRL, Oxford, UK, pp. 191–205.
3. Hunter, W. M. and Greenwood, F. C. (1962) Preparation of iodine-131 labeled human growth hormone of high specific activity, *Nature* **194,** 495,496.

133

The Lactoperoxidase Method for Radiolabeling Protein

Graham S. Bailey

1. Introduction

This method, introduced by Marchalonis *(1)*, employs lactoperoxidase in the presence of a trace of hydrogen peroxide to oxidize the radioactive iodide $^{125}I^-$ to produce the reactive species $^{125}I_2$ or $^{125}I^+$. These reactive species substitute mainly into tyrosine residues of the protein, although substitution into other amino acid residues can occur under certain conditions. The oxidation can be stopped by simple dilution. Although the technique should result in less chance of denaturation of susceptible proteins than the chloramine T method, it is more technically demanding and is subject to a more marked variation in optimum reaction conditions.

2. Materials

1. $Na^{125}I$: 37 MBq (1 mCi) concentration 3.7 GBq/mL (100 mCi/mL).
2. Lactoperoxidase: available from various commercial sources. A stock solution of 10 mg/mL in 0.1 *M* sodium acetate buffer, pH 5.6, can be made and stored at –20°C in small aliquots. A working solution of 20 μg/mL is made by dilution in buffer just prior to use.
3. Buffer A: 0.1 *M* sodium acetate buffer, pH 5.6 (*see* **Note 1**).
4. Buffer B: 0.05 *M* sodium phosphate buffer containing 0.1% sodium azide, final pH 7.4.
5. Buffer C: 0.05 *M* sodium phosphate buffer containing 1 *M* sodium chloride 0.1% bovine serum albumin and 1% potassium iodide, final pH 7.4.
6. Hydrogen peroxide: A solution of 10 μg/mL is made by dilution just prior to use.
7. Protein to be iodinated: A 0.5–2.5 mg/mL solution is made in buffer A.

It is essential that none of the solutions except buffer B contain sodium azide as antibacterial agent, since it inhibits lactoperoxidase.

3. Method

1. Into a small plastic test tube (1 × 5.5 cm) are added, in turn the protein to be iodinated (5 μL), radioactive iodide (5 μL), lactoperoxidase solution (5 μL), and buffer A (45 μL).
2. The reaction is started by the addition of the hydrogen peroxide solution (10 μL) with mixing (*see* **Note 2**).
3. The reaction is stopped after 20 min (*see* **Note 3**) by the addition of buffer B (0.5 mL) with mixing.
4. After 5 min, buffer C (0.5 mL) is added with mixing. The solution is then ready for purification as described in Chapter 136 (*see* **Note 4**).

From: *The Protein Protocols Handbook, 2nd Edition*
Edited by: J. M. Walker © Humana Press Inc., Totowa, NJ

4. Notes

1. The exact nature of buffer A will depend on the properties of the protein to be radioiodinated. Proteins differ markedly in their pH optima for radioiodination by this method *(2)*. Obviously the pH to be used will also depend on the stability of the protein, and the optimum pH can be established by trial and error.
2. Other reaction conditions, such as amount of lactoperoxidase, amount and frequency of addition of hydrogen peroxide, and so forth, also markedly affect the yield and quality of the radioiodinated protein. Optimum conditions can be found by trial and error.
3. The longer the time of the incubation, the greater the risk of potential damage to the protein by the radioactive iodide. Thus, it is best to keep the time of exposure of the protein to the radioactive iodide as short as possible, but commensurate with a good yield of radioactive product.
4. Some of the lactoperoxidase itself may become radioiodinated, which may result in difficulties in purification if the enzyme is of a similar size to the protein being labeled. Thus, it is best to keep the ratio of the amount of protein being labeled to the amount of lactoperoxidase used as high as possible.

References

1. Marchalonis, J. J. (1969) An enzymic method for trace iodination of immunoglobulins and other proteins. *Biochem. J.* **113,** 299–305.
2. Morrison, M. and Bayse, G. S. (1970) Catalysis of iodination by lactoperoxidase. *Biochemistry* **9,** 2995–3000.

134

The Bolton and Hunter Method for Radiolabeling Protein

Graham S. Bailey

1. Introduction

This is an indirect method in which an acylating reagent (*N*-succinimidyl-3[4-hydroxyphenyl]propionate, the Bolton and Hunter reagent), commercially available in a radioiodinated form, is covalently coupled to the protein to be labeled *(1)*. The [^{125}I] Bolton and Hunter reagent reacts mostly with the side-chain amino groups of lysine residues to produce the labeled protein. It is the method of choice for radiolabeling proteins that lack tyrosine and histidine residues, or where reaction at those residues affects biological activity. It is particularly suitable for proteins that are sensitive to the oxidative procedures employed in other methods (*see* Chapters 132, 133, and 135).

2. Materials

1. [^{125}I] Bolton and Hunter reagent: 37 MBq (1 mCi), concentration 185 MBq/mL (5 mCi/mL) (*see* **Note 1**).
2. Buffer A: 0.1 *M* sodium borate buffer, pH 8.5.
3. Buffer B: 0.2 *M* glycine in 0.1 *M* sodium borate buffer, pH 8.5.
4. Buffer C: 0.05 *M* sodium phosphate buffer containing 0.25% gelatin.
5. Protein to be iodinated: A 0.5–2.5 mg/mL solution is made in buffer A.

It is essential that none of the solutions contain sodium azide or substances with free thiol or amino groups (apart from the protein to be labeled), since the Bolton and Hunter reagent will react with those compounds.

3. Method

1. The [^{125}I] Bolton and Hunter reagent (0.2 mL) is added to a small glass test tube (1 × 5.5 cm) and is evaporated to dryness under a stream of dry nitrogen.
2. All reactants are cooled in iced water (*see* **Note 2**).
3. The protein to be iodinated (10 µL) is added, and the tube is gently shaken for 15 min (*see* **Note 2**).
4. Buffer B (0.5 mL) is added (*see* **Note 3**). The solution is mixed and allowed to stand for 5 min.
5. Buffer C (0.5 mL) is added with mixing (*see* **Note 4**). The resultant solution is then ready for purification.

From: *The Protein Protocols Handbook, 2nd Edition*
Edited by: J. M. Walker © Humana Press Inc., Totowa, NJ

4. Notes

1. [^{125}I] Bolton and Hunter reagent is available from Amersham International (Little Chalfort, UK) and Dupont NEN (Stevenage, UK). The Amersham product is supplied in anhydrous benzene containing 0.2% dimethylformamide. Aliquots can be easily withdrawn from the vial. However, the Dupont NEN is supplied in dry benzene alone, and dry dimethylformamide (about 0.5% of the sample volume) must be added to the vial with gentle shaking to facilitate the removal of aliquots.
2. [^{125}I] Bolton and Hunter reagent is readily hydrolyzed in aqueous solution. Under the described conditions, its half-life is about 10 min.
3. Buffer B stops the reaction by providing an excess of amino groups (0.2 *M* glycine) for conjugation with the [^{125}I] Bolton and Hunter reagent. Thus, the carrier protein (0.25% gelatin) in buffer C does not become labeled.
4. This method of radioiodination has been used extensively, and various modifications of the described procedure, including time and temperature of the reaction, have been reported *(2)*. For example, it is possible first to acylate the protein with the Bolton and Hunter reagent, and then carry out radioactive labeling of the conjugate using the chloramine T method. However, in general, this procedure does not seem to offer any advantage over the method described here. Also, the time and temperature of the reaction can be altered to achieve optimal labeling.

References

1. Bolton, A. E. and Hunter, W. M. (1973) The labeling of protiens to high specific radioactivities by conjugation to a ^{125}I-containing acylating agent. *Biochem. J.* **133,** 529–538.
2. Langone, J. J. (1980) Radioiodination by use of the Bolton-Hunter and related reagents, in *Methods in Enzymology,* vol. 70 (Van Vunakis, H. and Langone, J. J., eds.), Academic, New York, pp. 221–247.

135

Preparation of ^{125}I-Labeled Peptides and Proteins with High Specific Activity Using IODO-GEN

J. Michael Conlon

1. Introduction

The reagent IODO-GEN (1,3,4, 6-tetrachloro-3α,6α-diphenylglycoluril) was first introduced by Fraker and Speck in 1978 *(1)* and rapidly found widespread use for the radioiodination of both peptides and proteins *(2)*. Although the usual application of the reagent in the laboratory is for introduction of an atom of ^{125}I, IODO-GEN has been used successfully for the preparation of proteins, particularly monoclonal antibodies, labeled with ^{131}I and ^{123}I for use in immunoscintigraphy and positron emission tomography *(3–5)*. This chapter addresses only the use of IODO-GEN to radiolabel pure peptides/proteins in aqueous solution, but iodination of proteins on the cell surface of a wide range of intact eukaryotic and prokaryotic cells *(6,7)* and on subcellular organelles *(8)* has been accomplished using the reagent. The advantages of IODO-GEN over alternative reagents previously used for radiolabeling such as chloramine T, lactoperoxidase/hydrogen peroxide, and ^{125}I-labeled Bolton–Hunter reagent (*N*-succinimidyl-3 [4-hydroxyphenyl] propionate) are that the reaction is rapid, technically simple to perform, and gives reproducibly high levels of incorporation of radioactivity with minimal oxidative damage to the protein.

The structure of IODO-GEN is shown in **Fig. 1**. The reagent is virtually insoluble in water and so a solution in dichloromethane or chloroform is used to prepare a thin film of material on the walls of the reaction vessel. Addition of an aqueous solution of Na^+ $^{125}I^-$ generates the reactive intermediate (I^+/I_3^+) that participates in electrophilic attack primarily at tyrosine but also at histidine residues in the peptide/protein. Reaction is terminated simply by transferring the contents of the reaction vessel to a clean tube, thereby obviating the need to add a reducing agent. In comparison to other radioiodination methods, particularly the use of chloramine T and lactoperoxidase, oxidative damage to sensitive residues in the peptide/protein (particularly methionine and tryptophan) is much less using IODO-GEN *(9)*. To minimize oxidative damage even further and to limit production of diiodotyrosyl derivatives, the strategy of trace labeling is recommended. This involves radioiodination of only approx 10% of the molecules and necessitates the separation, as completely as possible, of the radiolabeled

From: *The Protein Protocols Handbook, 2nd Edition*
Edited by: J. M. Walker © Humana Press Inc., Totowa, NJ

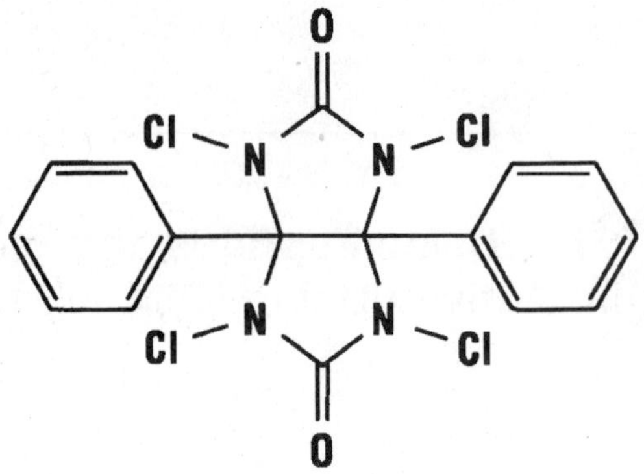

Fig. 1. The structure of IODO-GEN (1,3,4,6-tetrachloro-3α,6α-diphenylglycoluril).

peptide from the unreacted starting material in order to obtain a tracer of sufficiently high specific activity to be of use in radioimmunoassay, radioreceptor assay, or autoradiography. Reversed-phase high-performance liquid chromatography (RP-HPLC) combines rapidity and ease of operation with optimum separation of labeled and unlabeled peptide. The availability of wide-pore C_3 and C_4 columns generally permits good recoveries of labeled proteins with molecular mass (M_r) > 10,000. Techniques such as selective adsorption to diatomaceous materials (e.g., talc or microfine silica) and gel permeation chromatography give tracers of low specific activity, and ion-exchange chromatography is time consuming and results in a sample dilution that may be unacceptable.

2. Materials

2.1. Apparatus

No specialized equipment is required to carry out the iodination, but the reaction should be carried out in an efficient fume hood. Reaction takes place in a 1.5-mL natural-colored polypropylene microcentrifuge (Eppendorf tube) (*see* **Note 1**), immersed in an ice bath. A nitrogen or argon cylinder is required for removal of solvent. Liquids are dispensed with 10- and 100-μL pipets (e.g., Gilson Pipetman) that, because of inevitable contamination by radioactivity, should be dedicated to the reaction and stored in a designated area.

For HPLC, a system capable of generating reproducible linear two solvent gradients is required. Again, because of contamination by radioactivity, an injector, for example, Rheodyne Model 7125 with 1-mL sample loop; a 1-mL leak-free injection syringe, for example, Hamilton Gastight (cat. no. 1001); a 25 × 0.46 cm analytical reversed-phase column (*see* **Note 2** on column selection); and a fraction collector capable of collecting a minimum of 70 samples, for example, Frac 100 (Pharmacia, Uppsala, Sweden) should be dedicated to the radioiodination procedure. A UV detection system and chart recorder/integrator are not necessary unless it is important to demonstrate that the radiolabeled component has been completely separated from the starting material.

2.2. Chemicals

2.2.1. Iodination Reagents

1. IODO-GEN (Pierce, Rockford, IL): The reagent should be stored in a desiccator in the freezer and the bottle protected from light.
2. Dichloromethane (stabilized, HPLC grade): Redistillation is not required.
3. Carrier-free Na^+ $^{125}I^-$ in 0.1 *M* NaOH (3.7 GBq/mL; 100 mCi/mL, Amersham Pharmacia Biotech, Piscataway, NJ).
4. 0.2 *M* Disodium hydrogen phosphate–sodium dihydrogen phosphate buffer, pH 7.5 (*see* **Note 3**).

2.2.2. Chromatography

1. Solvent A: Add 1 mL of trifluoroacetic acid (Pierce HPLC/Spectro grade) to 1000 mL of water (*see* **Note 4**).
2. Solvent B: Mix 700 mL of acetonitrile (Fisher optima grade) with 300 mL of water and add 1 mL of trifluoroacetic acid. The solvents should be degassed, preferably by sparging with helium for 1 min, but passage through a filter is unnecessary.

3. Method

3. 1. Radioiodination

1. Dissolve 1.5 mg of IODO-GEN in 2 mL of dichloromethane.
2. Pipet 20 µL of the solution into a polypropylene tube and remove the solvent in a gentle stream of nitrogen or argon at room temperature. The aim is to produce a film of IODO-GEN on the wall of the tube. If the reagent has formed a visible clump, the tube should be discarded and a new tube prepared. According to the manufacturer's instructions (Pierce), the tubes can be stored in a vacuum desiccator for up to 2 mo, but it is recommended that a tube is prepared freshly for each reaction. The tube is set in an ice bath for 10 min prior to the iodination (*see* **Note 5**).
3. Dissolve 10 nmol of the peptide (*see* **Note 6**) in 0.2 *M* sodium phosphate buffer, pH 7.5 (100 µL) and pipet the solution into the chilled IODO-GEN-coated tube.
4. Add the Na^+ $^{125}I^-$ solution (5 µL; 0.5 mCi or 10 µL; 1 mCi depending on the quantity of tracer required).
5. Allow the reaction to proceed for between 1 and 20 min (*see* **Note 7**). The contents of the tube should be gently agitated by periodically tapping the side of the tube with a gloved finger.
6. Reaction is stopped by aspirating the contents of the tube into a solution of 0.1% (v/v) trifluoroacetic acid–water (500 µL) contained in a second polypropylene tube. The reaction vessel may be washed with a further 200 µL of trifluoroacetic acid–water and the washings combined with the reaction mixture.

3.2. Chromatography

1. Prior to carrying out the iodination, the column is equilibrated with 100% solvent A at a flow rate of 1.5 mL/min for at least 30 min. In the case of more hydrophobic peptides/proteins, the column is equilibrated with starting solvent containing up to 30% solvent B (21% acetonitrile).
2. The instrument is programmed to increase the proportion of solvent B from starting conditions to 70% (49 % acetonitrile) over 60 min using a linear gradient (*see* **Note 8**).

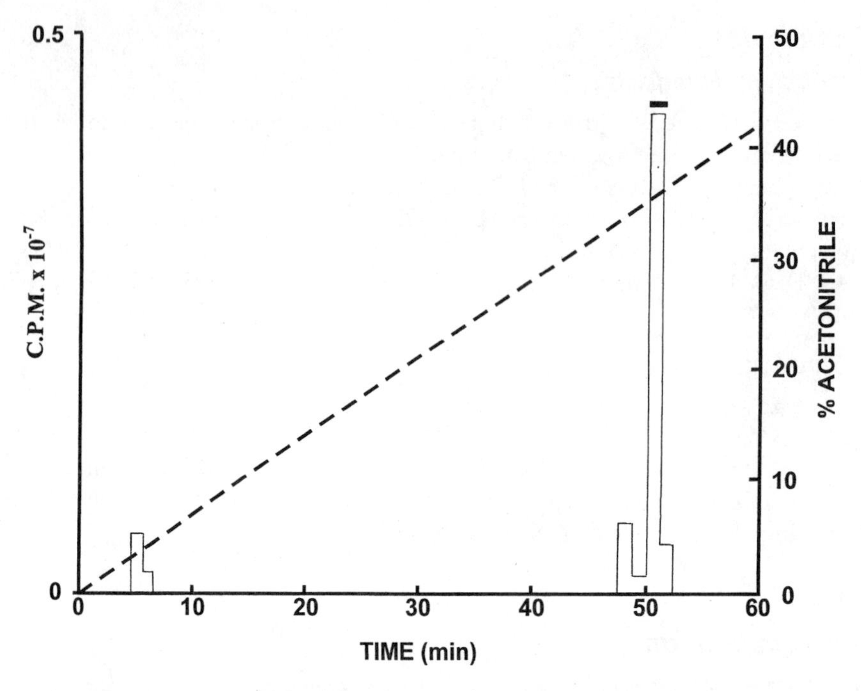

Fig. 2. RP-HPLC on a (0.46 × 25 cm) Vydac 218TP54 (C_{18}) column of the reaction mixture following incubation of 10 nmol of rat galanin with 0.5 mCi of Na ^{125}I in an IODOGEN-coated tube for 2 min at 0°C. Fractions (1 min) were collected and the fraction denoted by the *bar* contained tracer of high specific activity (74 TBq/mmol). The *dashed line* shows the concentration of acetonitrile in the eluting solvent.

3. The reaction mixture is injected onto the column (*see* **Note 9**). The fraction collector, programmed to collect 1-min fractions, is started and the linear elution gradient is begun. A total of 60 fractions are collected.
4. At the end of the chromatography, the column is irrigated with 100% solvent B for 30 min. The column may be stored in this solvent.
5. The radioactivity in aliquots (2 mL) of each fraction is counted in a gamma scintillation counter (*see* **Note 10** for optimum storage conditions of tracer).

The results of a typical radioiodination are illustrated in **Fig. 2**. The reaction mixture comprised 10 nmol of rat galanin and 0.5 mCi of Na^{+} $^{125}I^{-}$ and the reaction time was 2 min. Unreacted free iodide was eluted at the void volume of the column (fractions 5 and 6). The fraction denoted by the bar (tube 51) was of high specific activity (approx 74 TBq/mmol; 2000 Ci/mmol). The earlier eluting minor peak of radioactivity (tube 49) probably represented the diiodotyrosyl derivative. Before use in radioimmunoassay or radioreceptor studies, the quality of the tracer is assessed by incubating an aliquot (approx 20,000 cpm) with an excess of an antiserum raised against galanin (1:1000 dilution) in 0.1 *M* sodium phosphate buffer, pH 7.4 (final volume 300 µL) for 16 h at 4°C. Free and bound radioactivity are separated by addition of 100 µL of a 10 mg/mL solution of bovine γ-globulin (Sigma, St. Louis, MO) and 1 mL of 20% (w/v) solution of polyethylene glycol 6000 (Sigma, approx M_r 8000) in water followed by centrifugation (1600g for 30 min at 4°C). Under these conditions, >90% of the radioactivity is bound to antibody.

4. Notes

1. A systematic comparison of the properties of different IODO-GEN-coated surfaces has concluded that soda-lime glass produced the most rapid rate of oxidation of Na ^{125}I ***(10)***. However, irreversible binding of most peptides to the walls of polypropylene tubes is much less than to glass and so clear polypropylene Eppendorf tubes are routinely used in the author's laboratory. The use of borosilicate glass tubes ***(10)*** and polysytrene tubes is not recommended. Glass test tubes (12 mm × 75 mm) precoated with IODO-GEN (50 µg) are available from Pierce.
2. The choice of column is dictated by the nature of the radiolabeled peptide to be purified. For relatively small (M_r < 3000) peptides, good resolution and recoveries are generally obtained with (0.46 × 25 cm) narrow pore (80 Å), 5-µm particle size octadecylsilane (C_{18}) columns such as Supelcosil LC-18-DB (Supelco, Bellefonte, PA), Ultrasphere ODS (Beckman, Duarte, CA), or Spheri-5 RP-18 (Brownlee/Applied Biosystems, Foster City, CA). For purification of radiolabeled tracers of higher molecular mass (M_r > 3000), the use of columns containing wide-pore (300 Å) 5-µm particle size C_{18} packing materials is recommended. Suitable columns include Vydac 218TP54 (Separations Group, Hesperia, CA), Spherisorb wide-pore C_{18} (Phase Separations; U. K.), Waters Delta-Pak C_{18} (Millipore, Milford, MA), and Ultrapore C_{18} (Beckman). For purification of tracers of molecular mass >10,000, such as the pituitary glycoprotein hormones, sharper peaks and better recoveries of radioactivity may be obtained using wide-pore silica with propyl and butyl carbon loading. Suitable columns include Beckman Ultrapore C_3 and Vydac 214TP54 C_4 columns.
3. The reaction is not markedly pH sensitive and high incorporations of ^{125}I may be achieved in the pH range 6–9 ***(11)***.
4. Suitable water can be obtained using a Milli-Q purification system (Millipore) supplied with water that has been partially purified by single distillation or by treatment with a deionization resin.
5. A low reaction temperature is important in minimizing damage to the radiolabeled peptide/protein. For example, the use of IODO-GEN at room temperature to prepare ^{125}I-labeled human growth hormone resulted in the production of tracer containing a significant amount of the hormone in an aggregated form whereas only the radiolabeled monomer was produced when the reaction was carried out at 0°C ***(12)***.
6. As many peptides are relatively insoluble in buffers of neutral pH, it is recommended that the peptide first be dissolved in a minimum volume (approx 5 µL) of 0.1% (v/v) trifluoroacetic acid/water and the volume made up to 100 µL with 0.2 *M* sodium phosphate buffer, pH 7.5.
7. The optimum reaction time must be determined for each peptide and protein, but some general guidelines can be given. For small peptides (<15 amino acid residues) containing a tyrosine residue in a sterically unhindered region of the molecule, for example, the N- or C-terminal residue, the reaction proceeds rapidly and reaction times of between 0.5 and 2 min are generally sufficient. For larger peptides and proteins, it is often necessary to prolong the reaction time to between 10 and 15 min. Peptides that do not contain a tyrosine residue but possess a sterically unhindered histidine (e.g., neurokinin B ***[13]*** and secretin) may be iodinated using IODO-GEN, but longer reaction times (up to 20 min) may be required.

8. The relatively steep gradient (0 → 70% solvent B, equivalent to 0 → 49% acetonitrile) over 60 min is recommended as the initial elution conditions when preparing a radiolabel for the first time. Better separation of the tracer and the unlabeled peptide will be obtained using a shallower gradient. For example, relatively hydrophilic peptides such as $[Tyr^0]$bradykinin and $[Tyr^8]$substance P ***(14)*** may be purified using a gradient of 0 → 35% acetonitrile over 60 min, whereas hydrophobic peptides such as corticotropin-releasing hormone, urotensin-1, and neurokinin B may be purified using a gradient of 21 → 49% acetonitrile over 60 min. In a case where separation of radiolabeled peptide and starting material is incomplete, a label of higher specific activity may be obtained by prolonging the time of chromatography and collecting smaller fractions.
9. In some published protocols, the radiolabeled peptide or protein and unreacted $^{125}I^-$ are separated prior to RP-HPLC, for example, by adsorption on Sep-Pak C_{18} cartridges (Waters Associates, Milford, MA) or by gel-permeation chromatography on a Sephadex G-10 desalting column (Pharmacia). This procedure is not necessary and it is recommended that the reaction mixture is injected directly onto the HPLC column.
10. The stability of radiolabeled peptides and proteins varies dramatically, with useful lives ranging from a few days to more than 2 mo. Repeated freezing and thawing of the tracer is not recommended and so the HPLC fraction(s) containing the radiolabel should be aliquoted immediately, diluted with one volume ethanol or methanol, and stored at as low a temperature as possible (–70°C is preferred). The volume of the aliquot should be related to the size of a typical assay.
11. Although IODO-GEN is almost insoluble in water, its solubility in buffers containing detergent increases appreciably. Under these circumstances, oxidative damage to the peptide/protein may occur and the use of an alternative reagent IODO-BEADS (Pierce) should be considered. IODO-BEADS comprise the sodium salt of *N*-chloro-benzenesulfonamide immobilized on nonporous polystyrene beads ***(15)***. The reaction conditions and purification protocol using IODO-BEADS are the same as using IODO-GEN except that one or more of the beads are substituted for the film of IODO-GEN. High incorporations of radioactivity are observed even in the presence of detergents or chaotropic reagents, for example, urea.

References

1. Fraker, P. J. and Speck, J. C. (1978) Protein and cell membrane iodination with a sparingly soluble chloroamide 1,3,4,6-tetrachloro-3α,6α-diphenylglycoluril. *Biochem. Biophys. Res. Commun.* **80,** 849–857.
2. Salacinski, P. R. P., McLean, C., Sykes, J. E. C., Clement-Jones, V. V., and Lowry, P. J. (1981) Iodination of proteins, glycoproteins and peptides using a solid-phase oxidizing agent, 1,3,4,6-tetrachloro-3α,6α-diphenyl glycoluril (Iodogen). *Analyt. Biochem.* **117,** 136–146.
3. Fuchs, J., Gratz, K. F., Habild, G., Gielow, P., von-Schweinitz, D., and Baum, R. P. (1999) Immunoscintigraphy of xenotransplanted hepatoblastoma with iodine^{131}-labeled anti-alpha-fetoprotein monoclonal antibody. *J. Pediatr. Surg.* **34,** 1378–1384.
4. Richardson, A. P., Mountford, P. J., Baird, A. C., Heyderman, E., Richardson, T. C., and Coakley, A. J. (1986) An improved iodogen method of labeling antibodies with ^{123}I. *Nucl. Med. Commun.* **7,** 355–362.
5. Blauenstein, P., Locher, J. T., Seybold, K., Koprivova, H., Janoki, G. A., Macke, H. R., et al. (1995) Experience with the iodine-123 and technetium-99m labeled anti-granulocyte antibody Mab47: a comparison of labeling methods. *Eur. J. Nucl. Med.* **22,** 690–698.

6. Tuszynski, G. P., Knight, L. C., Kornecki, E., and Srivastava, S. (1983) Labeling of platelet surface proteins with ^{125}I by the iodogen method. *Analyt. Biochem.* **130,** 166–170.
7. Abath, F. G., Almeida, A. M., and Ferreira, L. C. (1992) Identification of surface-exposed *Yersinia pestis* proteins by radio-iodination and biotinylation. *J. Med. Microbiol.* **37,** 420–424.
8. Pandey, S. and Parnaik, V. K. (1989) Identification of specific polypeptides of the nuclear envelope by iodination of mouse liver nuclei. *Biochem J.* **261,** 733–738.
9. Thean, E. T. (1990) Comparison of specific radioactivities of human alpha-lactalbumin iodinated by three different methods. *Analyt. Biochem.* **188,** 330–334.
10. Boonkitticharoen, V. and Laohathai, K. (1990) Assessing performances of Iodogen-coated surfaces used for radioiodination of proteins. *Nucl. Med. Commun.* **11,** 295–304.
11. Saha, G. B., Whitten, J., and Go, R. T. (1989) Conditions of radioiodination with iodogen as oxidizing agent. *Int. J. Rad. Appl. Instrum. B* **16,** 431–435.
12. Mohammed-Ali, S. A., Salacinski, P. R., and Landon, J. (1981) Effect of temperature on the radioiodination of human growth hormone. *J. Immunoassay* **2,** 175–186.
13. Conlon, J. M. (1991) Measurement of neurokinin B by radioimmunoassay, in *Methods in Neurosciences*, Vol. 6 (Conn, P. M., ed.), Academic Press, San Diego, CA, pp. 221–231.
14. Conlon, J. M. (1991) Regionally-specific antisera to human β-preprotachykinin, in *Methods in Neurosciences*, Vol. 6 (Conn, P. M., ed.), Academic Press, San Diego, CA, pp. 221–231.
15. Markwell, M. A. K. (1982) A new solid-state reagent to iodinate proteins. *Analyt. Biochem.* **125,** 427–432.

136

Purification and Assessment of Quality of Radioiodinated Protein

Graham S. Bailey

1. Introduction

At the end of a radioiodination procedure, the reaction mixture will contain the labeled protein, unlabeled protein, radioiodide, mineral salts, enzyme (in the case of the lactoperoxidase method), and possibly some protein that has been damaged during the oxidation. For most uses of radioiodinated proteins, it is essential to have the labeled species as pure as possible. In theory, any of the many methods of purifying proteins can be employed *(1)*. However, the purification of the radioiodinated protein should be achieved as rapidly as possible. For that purpose, the most widely used of all separation techniques is gel filtration.

One of the most important parameters used to assess the quality of a purified labeled protein is its specific radioactivity, which is the amount of radioactivity incorporated/unit mass of protein. It can be calculated in terms of the total radioactivity employed, the amount of the iodination mixture transferred to the gel-filtration column, and the amount of radioactivity present in the labeled protein, in the damaged components, and in the residual radioiodine. However, in practice, the calculation does not usually take into account damaged and undamaged protein. The specific activity is thus calculated from the yield of the radioiodination procedure, the amount of radioiodide, and the amount of protein used, assuming that there are no significant losses of those two reactants. The yield of the reaction is simply the percentage incorporation of the radionuclide into the protein.

It is obviously important that the radioiodinated protein should as far as possible have the same properties as the unlabeled species. Thus, the behavior of both molecules can be checked on electrophoresis or ion-exchange chromatography. The ability of the two species to bind to specific antibodies can be assessed by radioimmunoassay.

This chapter will describe a protocol for the purification of a radiolabeled protein and an example of a calculation of specific radioactivity.

2. Materials

1. Sephadex G75 resin.
2. Buffer A: 0.05 *M* sodium phosphate containing 0.1% bovine serum albumin and 0.15 *M* sodium chloride, final pH 7.4.

From: *The Protein Protocols Handbook, 2nd Edition*
Edited by: J. M. Walker © Humana Press Inc., Totowa, NJ

3. Specific antiseum to the protein.
4. Buffer B: 0.1 *M* sodium phosphate buffer containing 0.15 *M* sodium chloride and 0.01% sodium azide, final pH 7.4.
5. γ-globulin solution: 1.4% bovine γ-globulins (Sigma [St. Louis, MO] G7516) in buffer B.
6. Polyethylene glycol/potassium iodide solution: 20% polyethylene glycol 6000 and 6.25% potassium iodide in buffer B.

3. Method

1. An aliquot (10 μL) of the mixture is retained for counting and the rest is applied to a column (1 × 20 cm) of Sephadex G75 resin (*see* **Notes 1** and **2**).
2. Elution is carried out at a flow rate of 20 mL/h, and fractions (0.6 mL) are automatically collected.
3. Aliquots (10 μL) are counted for radioactivity.
4. An elution profile of radioactivity against fraction number (for a typical profile, *see* **Fig. 1**) is plotted.
5. Immunoreactive protein is identified by reaction with specific antiserum to that protein in the following manner (**steps 6–13**) (*see* **Note 3**).
6. Aliquots (10 μL) of fractions making up the different peaks are diluted so that each gives 10,000 counts/min/100 μL.
7. Each diluted aliquot (100 μL) is incubated with the specific antiserum (100 μL) at 4°C for 4 h.
8. Protein bound to antibody and excess antibody are precipitated by the addition to each sample at 4°C of γ-globulin solution (200 μL) and polyethylene glycol/potassium iodide solution (1 mL) (*see* **Note 4**).
9. Each tube is vortexed and is allowed to stand at 4°C for 15 min.
10. Each tube is centrifuged at 4°C at 5000*g* for 30 min.
11. The supernatants are carefully removed by aspiration at a water pump, and the precipitates at the bottom of the tubes are counted for radioactivity.
12. Estimates of the yield of the radioiodination and specific radioactivity of the iodinated protein may then be made (*see* **Note 5**).
13. When the fractions containing radioiodinated protein have been identified, they are split into small aliquots that can be rapidly frozen or freeze-dried for storage (*see* **Note 6**).

4. Notes

1. Gel filtration by high-performance liquid chromatography (HPLC) provides a more rapid and efficient purification of iodinated protein than the conventional liquid chromatography described here, but it does entail the use of expensive columns and apparatus with the attendant danger of their contamination with radioactivity *(2)*.
2. A wide range of gel-filtration resins are available. In choosing a resin, the relative molecular masses (M_r) of the protein and other reactants and products must be borne in mind. Sephadex G-25 resin will separate the labeled protein from the low-mol-wt reagents, such as oxidants and reductants. However, if the labeled protein is contaminated with damaged protein (e.g., aggregated species), then a gel-filtration resin of higher porosity, such as Sephadex G-100, may produce a more efficient separation of the undamaged, mono-iodinated protein.
3. The occurrence of immunoreactive protein in more than one peak indicates the presence of polymeric or degradated forms of the iodinated protein. Ideally, iodinated protein should be present in a single, sharp, symmetrical peak. If this is not the case, then it is probably best to repeat the radioiodination under milder conditions or use a different method of iodination.

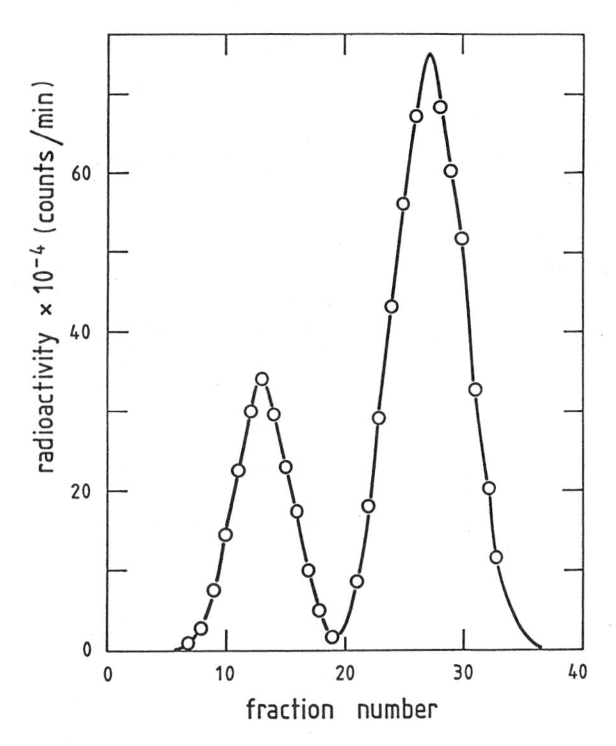

Fig. 1. Gel filtration of radioiodinated kallikrein of rat submandibular gland. The pure enzyme (10 µg) was iodinated with ^{125}I (18.5 MBq) by the chloramine T method. It was then purified on a column (1 × 20 cm) of Sephadex G-75 resin at a flow rate of 20 mL/h and collecting fractions of 0.6 mL. Aliquots (10 µL) of each fraction were measured for radioactivity. By radioimmunoassay, immunoreactive protein was found only in the first peak, and more than 90% of that radioactivity was bound by the antiserum to kallikrein from rat submandibular gland.

4. Polyethylene glycol produces precipitation of antibody and antibody-bound protein with little precipitation of unbound protein. Potassium iodide decreases the precipitation by polyethylene glycol of unbound protein *(3)*.
5. A typical example of the calculation of yield of iodination and specific radioactivity is as follows:

   Counts of 10 µL incubation mixture prior to gel filtration = 1,567,925 counts/10 s (1)

   Some of the radioiodinated protein is likely to bind to the reaction vessel and other surfaces, so it is best to calculate the radioactivity associated with the labeled protein in terms of the difference between total radioactivity applied to the column and the radioactivity associated with the unreacted iodide.

   Counts in 10-µL aliquots of iodide peak = 388,845 counts/10 s (2)

   Counts in 10-µL aliquots of protein peak
   = 1,567,925–388,845 counts/10 s = 1,179,080 counts/10 s (3)

   Yield of radioiodination reaction = % incorporation of ^{125}I into protein
   (1,179,080 × 100)/1,567,925 = 75.2% (4)

$$\text{Amount of radioactivity incorporated into protein} \\ = \% \text{ incorporation} \times \text{original radioactivity} \\ = 75.2\% \times 18.5 \text{ MBq} = 13.9 \text{ MBq} \quad (5)$$

$$\text{Specific radioactivity} = (\text{amount of radioactivity in protein/amount of protein used}) = \\ (13.9 \text{ MBq}/10 \text{ µg}) = (1.4 \text{ MBq/µg}) \quad (6)$$

6. Each aliquot should be thawed and used only once. Radioiodinated proteins differ markedly in their stability. Some can be stored for several wk (though it must be borne in mind that the half-life of ^{125}I is about 60 d), whereas others can only be kept for several days. If necessary, the labeled protein can be repurified by gel-filtration or ion-exchange chromatography prior to use.

References

1. Harris, E. L. V. and Angal, S. (1989) *Protein Purification Methods,* IRL, Oxford, UK.
2. Welling, G. W. and Welling-Webster, S. (1989) Size-exclusion HPLC of proteins, in *HPLC of Macromolecules* (Oliver, R. W. A., ed.), IRL, Oxford, UK, pp. 77–89.
3. Desbuquois, B. and Aurbach, G. D. (1971) Use of polyethyleneglycol to separate free and antibody-bound peptide hormones in radioimmunoassays. *J. Clin. Endocrin. Metab.* **33,** 732–738.

137

Purification of IgG by Precipitation with Sodium Sulfate or Ammonium Sulfate

Mark Page and Robin Thorpe

1. Introduction

Addition of appropriate amounts of salts, such as ammonium or sodium sulfate, causes precipitation of IgG *(1)* from all mammals, and can be used for serum, plasma, ascites fluid, and hybridoma culture supernatant. Although such IgG is usually contaminated with other proteins, the ease of these precipitation procedures coupled with the high yield of IgG has led to their wide use in producing enriched IgG preparations. They are suitable for many immunochemical procedures, e.g., production of immunoaffinity columns, and as a starting point for further purification. It is not suitable however for conjugating with radiolabels, enzymes, or biotin since the contaminating proteins will also be conjugated, thereby reducing the efficiency of the labeling and the quality of the reagent. The precipitated IgG is usually very stable, and such preparations are ideally suited for long-term storage or distribution and exchange between laboratories.

Ammonium sulfate precipitation is the most widely used and adaptable procedure, yielding a 40% pure preparation; sodium sulfate can give a purer preparation for some species, e.g., human and monkey.

2. Materials

2.1. Ammonium Sulfate Precipitation

1. Saturated ammonium sulfate solution: Add excess $(NH_4)_2SO_4$ to distilled water (about 950 g to 1 L), and stir overnight at room temperature. Chill at 4°C, and store at this temperature. This solution (in contact with solid salt) is stored at 4°C.
2. PBS: 0.14 *M* NaCl, 2.7 m*M* KCl, 1.5 m*M* KH_2PO_4, 8.1 m*M* Na_2HPO_4. Store at 4°C.

2.2. Sodium Sulfate Precipitation

This requires solid sodium sulfate.

3. Methods

3.1. Ammonium Sulfate Precipitation

1. Prepare saturated ammonium sulfate at least 24 h before the solution is required for fractionation. Store at 4°C.

From: *The Protein Protocols Handbook, 2nd Edition*
Edited by: J. M. Walker © Humana Press Inc., Totowa, NJ

2. Centrifuge serum or plasma for 20–30 min at 10,000g_{av} at 4°C. Discard the pellet (*see* **Note 1**).
3. Cool the serum or plasma to 4°C, and stir slowly. Add saturated ammonium sulfate solution dropwise to produce 35–45% final saturation (*see* **Note 2**). Alternatively, add solid ammonium sulfate to give the desired saturation (2.7 g of ammonium sulfate/10 mL of fluid = 45% saturation). Stir at 4°C for 1–4 h or overnight.
4. Centrifuge at 2000–4000g_{av} for 15–20 min at 4°C (alternatively for small volumes of 1–5 mL, microfuge for 1–2 min). Discard the supernatant, and drain the pellet (carefully invert the tube over a paper tissue).
5. Dissolve the precipitate in 10–20% of the original volume in PBS or other buffer by careful mixing with a spatula or drawing repeatedly into a wide-gage Pasteur pipet. When fully dispersed, add more buffer to give 25–50% of the original volume and dialyze against the required buffer (e.g., PBS) at 4°C overnight with two to three buffer changes. Alternatively, the precipitate can be stored at 4 or –20°C if not required immediately.

3.2. Sodium Sulfate Precipitation (see Note 3)

1. Centrifuge the serum or plasma at 10,000g_{av} for 20–30 min. Discard the pellet, warm the serum to 25°C, and stir.
2. Add solid Na_2SO_4 to produce an 18% w/v solution (i.e., add 1.8 g/10 mL), and stir at 25°C for 30 min to 1 h.
3. Centrifuge at 2000–4000g_{av} for 30 min at 25°C.
4. Discard the supernatant, and drain the pellet. Redissolve in the appropriate buffer as described for ammonium sulfate precipitation (**Subheading 3.1., step 5**).

4. Notes

1. If lipid contamination is excessive in ascites fluids, thereby compromising the salt precipitation, add silicone dioxide powder (15 mg/ mL) and centrifuge for 20 min at 2000g_{av} *(2)* before adding the ammonium or sodium salt.
2. The use of 35% ammonium sulfate will produce a pure IgG preparation, but will not precipitate all the IgG present in serum or plasma. Increasing saturation to 45% causes precipitation of nearly all IgG, but this will be contaminated with other proteins, including some albumin. Purification using $(NH_4)_2SO_4$ can be improved by repeating the precipitation, but this may cause some denaturation. Precipitation with 45% $(NH_4)_2SO_4$ is an ideal starting point for further purification steps, e.g., ion-exchange or affinity chromatography and FPLC purification (*see* Chapters 139 and 140).
3. Sodium sulfate may be used for precipitation of IgG instead of ammonium sulfate. The advantage of the sodium salt is that a purer preparation of IgG can be obtained, but this must be determined experimentally. The disadvantages are that yield may be reduced depending on the IgG characteristics of the starting material, IgG concentration, and composition. Fractionation must be carried out at a precise temperature (usually 25°C), since the solubility of Na_2SO_4 is very temperature dependent. Sodium sulfate is usually employed only for the purification of rabbit or human IgG.

References

1. Heide, K. and Schwick, H. G. (1978) Salt fractionation of immunoglobulins, in *Handbook of Experimental Immunology,* 3rd ed. (Weir, D. M., ed.), chap. 7. Blackwell Scientific, Oxford, UK.
2. Neoh, S. H., Gordon, C., Potter, A., and Zola, H. (1986) The purification of mouse MAb from ascitic fluid. *J. Immunol. Meth.* **91,** 231.

138

Purification of IgG Using Caprylic Acid

Mark Page and Robin Thorpe

1. Introduction

Caprylic (octanoic) acid can be used to purify mammalian IgG from serum, plasma, ascites fluid, and hybridoma culture supernatant by precipitation of non-IgG protein *(1)* (*see* **Note 1**). Other methods have been described where caprylic acid has been used to precipitate immunoglobulin depending on the concentration used. The concentration of caprylic acid required to purify IgG varies according to species (*see* **Subheading 3., step 2**). For MAb, it is usually necessary to determine experimentally the quantity required to produce the desired purity/yield. Generally, the product is of low to intermediate purity but this will depend on the starting material. Caprylic acid purified IgG preparations can be used for most immunochemical procedures, such as coating plates for antigen capture assays and preparation of immunoaffinity columns, but would not be suitable for conjugation with radioisotopes, enzymes, and biotin where contaminating proteins will reduce the specific activity.

2. Materials

1. 0.6 *M* sodium acetate buffer, pH 4.6. Adjust pH with 0.6 *M* acetic acid.
2. Caprylic acid (free acid).

3. Methods

1. Centrifuge the serum at 10,000g_{av} for 20–30 min. Discard the pellet and add twice the volume of 0.06 *M* sodium acetate buffer, pH 4.6.
2. Add caprylic acid dropwise while stirring at room temperature. For each 25 mL of serum, use the following amounts of caprylic acid: human and horse, 1.52 mL; goat, 2.0 mL; rabbit, 2.05 mL; cow, 1.7 mL. Stir for 30 min at room temperature.
3. Centrifuge at 4000g_{av} for 20–30 min. Retain the supernatant and discard the pellet. Dialyze against the required buffer (e.g., PBS) at 4°C overnight with two or three buffer changes.

4. Note

1. The method can be used before ammonium sulfate precipitation (*see* Chapter 137) to yield a product of higher purity.

Reference

1. Steinbuch, M. and Audran, R. (1969) The isolation of IgG from mammalian sera with the aid of caprylic acid. *Arch. Biochem. Biophys.* **134,** 279–284.

From: *The Protein Protocols Handbook, 2nd Edition*
Edited by: J. M. Walker © Humana Press Inc., Totowa, NJ

139

Purification of IgG Using DEAE-Sepharose Chromatography

Mark Page and Robin Thorpe

1. Introduction

IgG may be purified from serum by a simple one-step ion-exchange chromatography procedure. The method is widely used and works on the principle that IgG has a higher or more basic isoelectric point than most serum proteins. Therefore, if the pH is kept below the isoelectric point of most antibodies, the immunoglobulins do not bind to an anion exchanger and are separated from the majority of serum proteins bound to the column matrix. The high capacity of anion-exchange columns allows for large-scale purification of IgG from serum. The anion-exchange reactive group, diethylaminoethyl (DEAE) covalently linked to Sepharose (e.g., DEAE Sepharose CL-6B, Pharmacia, Uppsala, Sweden) is useful for this purpose. It is provided preswollen and ready for packing into a column, and is robust and has high binding capacity. Furthermore, it is relatively stable to changes in ionic strength and pH. Other matrices (e.g., DEAE cellulose) are provided as solids, and will therefore require preparation and equilibration *(1)*.

This procedure does not work well for murine IgG or preparations containing mouse or rat MAb, since these do not generally have the high pI values that IgGs of other species have. Other possible problems are that some immunoglobulins are unstable at low-ionic strength, e.g., mouse IgG_3, and precipitation may occur during the ion-exchange procedure. The product is of high purity (>90%) and can be used for most immunochemical procedures including conjugation with radioisotopes, enzymes, and so on, where pure IgG is required.

2. Materials

1. DEAE Sepharose CL-6B.
2. 0.07 *M* sodium phosphate buffer, pH 6.3.
3. 1 *M* NaCl.
4. Sodium azide.
5. Chromatography column (*see* **Note 1**).

3. Methods

1. Dialyze the serum (preferably ammonium sulfate fractionated; *see* Chapter 137) against 0.07 *M* sodium phosphate buffer, pH 6.3, exhaustively (at least two changes over a 24-h period) at a ratio of at least 1 vol of sample to 100 vol of buffer.

From: *The Protein Protocols Handbook, 2nd Edition*
Edited by: J. M. Walker © Humana Press Inc., Totowa, NJ

2. Apply the sample to the column, and wash the ion exchanger with 2 column volumes of sodium phosphate buffer. Collect the wash, which will contain IgG, and monitor the absorbance of the eluate at 280 nm (A_{280}). Stop collecting fractions when the A_{280} falls to baseline.
3. Regenerate the column by passing through 2–3 column volumes of phosphate buffer containing 1 *M* NaCl.
4. Wash thoroughly in phosphate buffer (2–3 column volumes), and store in buffer containing 0.1% NaN_3.
5. Pool the fractions from **step 2** and measure the A_{280} (*see* **Note 2**).

4. Notes

1. The column size will vary according to the user's requirements or the amount of antibody required. Matrix binding capacities are given by manufacturers and should be used as a guide.
2. The extinction coefficient ($E^{1\%}_{280}$) of human IgG is 13.6 (i.e., a 1 mg/mL solution has an A_{280} of 1.36). This can be used as an approximate value for IgGs from other sources.

Reference

1. Johnstone, A. and Thorpe, R. (1996) *Immunochemistry in Practice,* 3rd ed. Blackwell Scientific, Oxford, UK.

140

Purification of IgG Using Ion-Exchange HPLC

Carl Dolman, Mark Page, and Robin Thorpe

1. Introduction

Conventional ion-exchange chromatography separates molecules by adsorbing proteins onto the ion-exchange resins that are then selectively eluted by slowly increasing the ionic strength (this disrupts ionic interactions between the protein and column matrix competitively) or by altering the pH (the reactive groups on the proteins lose their charge). Anion-exchange groups (such as diethylaminoethyl; [DEAE]) covalently linked to a support matrix (such as Sepharose) can be used to purify IgG in which the pH of the mobile-phase buffer is raised above the p*I* or IgG, thus allowing most of the antibodies to bind to the DEAE matrix. Compare this method with that described in Chapter 139 in which the IgG passes through the column. The procedure can be carried out using a laboratory-prepared column that is washed and eluted under gravity *(1)*; however, high-performance liquid chromatography (HPLC) provides improved reproducibility (because the sophisticated pumps and accurate timers), speed (because of the small high-capacity columns), and increased resolution (because of the fine resins and control systems).

2. Materials

1. Anion-exchanger (e.g., Mono-Q HR 5/5 or HR 10/10, Pharmacia, Uppsala, Sweden.
2. Buffer A: 0.02 *M* triethanolamine, pH 7.7.
3. Buffer B: Buffer A containing 1 *M* NaCl.
4. 2 *M* NaOH.
5. Sodium azide.

3. Methods

1. Prepare serum by ammonium sulfate precipitation (45% saturation; *see* Chapter 137). Redissolve the precipitate in 0.02 *M* triethanolamine buffer, pH 7.7, and dialyze overnight against this buffer at 4°C. Filter the sample (*see* **Note 1**) before use (0.2 μm).
2. Assemble the HPLC system according to the manufacturer's instructions for use with the Mono-Q ion-exchange column.
3. Equilibrate the column with 0.2 *M* triethanolamine buffer, pH 7.7 (buffer A). Run a blank gradient from 0 to 100% buffer B (buffer A + 1 *M* NaCl). Use a flow rate of 4–6 mL/min for this and subsequent steps.

From: *The Protein Protocols Handbook, 2nd Edition*
Edited by: J. M. Walker © Humana Press Inc., Totowa, NJ

4. Load the sample depending on column size. Refer to the manufacturer's instruction for loading capacities of the columns.
5. Equilibrate the column with buffer A for at least 10 min.
6. Set the sensitivity in the UV monitor control unit, and zero the baseline.
7. Apply a salt gradient from 0 to 28% buffer B for about 30 min (*see* **Note 2**). Follow with 100% 1 *M* NaCl for 15 min to purge the column of remaining proteins.
8. Wash the Mono-Q ion-exchange with at least 3 column volumes each of 2 *M* NaOH followed by 2 *M* NaCl.
9. Store the Mono-Q ion-exchange column in distilled water containing 0.02% NaN_3.

4. Notes

1. It is essential that the sample and all buffers be filtered using 0.2-μm filters. All buffers must be degassed by vacuum pressure.
2. Using Mono-Q, IgG elutes between 10% and 25% buffer B, usually approx 15%. When IgG elutes at 25% (dependent on p*I*), then it will tend to coelute with albumin, which elutes at approx 27%. When this occurs, alternative purification methods should be employed.
3. Other anion-exchange media can be substituted for Mono-Q, for example, Anagel TSK DEAE (Anachem, Luton, UK).

Reference

1. Johnstone, A. and Thorpe, R. (1996) *Immunochemistry in Practice*, 3rd edition, Blackwell Scientific, Oxford, UK.

141

Purification of IgG by Precipitation with Polyethylene Glycol (PEG)

Mark Page and Robin Thorpe

1. Introduction

PEG precipitation works well for IgM, but is less efficient for IgG; salt precipitation methods are usually recommended for IgG. PEG precipitation may be preferred in multistep purifications that use ion-exchange columns, because the ionic strength of the Ig is not altered. Furthermore, it is a very mild procedure that usually results in little denaturation of antibody. This procedure is applicable to both polyclonal antisera and most MAb containing fluids.

2. Materials

1. PEG solution: 20% (w/v) PEG 6000 in PBS.
2. PBS: 0.14 *M* NaCl, 2.7 m*M* KCl, 1.5 m*M* KH_2PO_4, 8.1 m*M* Na_2HPO_4.

3. Methods

1. Cool a 20% w/v PEG 6000 solution to 4°C.
2. Prepare serum/ascitic fluid, and so forth, for fractionation by centrifugation at $10,000g_{av}$ for 20–30 min at 4°C. Discard the pellet. Cool to 4°C.
3. Slowly stir the antibody containing fluid, and add an equal volume of 20% PEG dropwise (*see* **Note 1**). Continue stirring for 20–30 min.
4. Centrifuge at $2000–4000g_{av}$ for 30 min at 4°C. Discard the supernatant (*see* **Note 2**), and drain the pellet. Resuspend in PBS or other buffer as described for ammonium sulfate precipitation (*see* Chapter 137).

4. Notes

1. Although the procedure works fairly well for most antibodies, it may produce a fairly heavy contamination with non-IgG proteins. If this is the case, reduce the concentration of PEG in 2% steps until the desired purification is achieved. Therefore, carry out a pilot-scale experiment before fractionating all of the sample.
2. PEG precipitation does not work for some antibodies. If the procedure is to be used for a valuable antibody for the first time, keep the supernatant in case precipitation has been inefficient.

From: *The Protein Protocols Handbook, 2nd Edition*
Edited by: J. M. Walker © Humana Press Inc., Totowa, NJ

142

Purification of IgG Using Protein A or Protein G

Mark Page and Robin Thorpe

1. Introduction

Some strains of *Staphylococcus aureus* synthesize protein A, a group-specific ligand that binds to the Fc region of IgG from many species *(1,2)*. Protein A does not bind all subclasses of IgG, e.g., human IgG_3, mouse IgG_3, sheep IgG_1, and some subclasses bind only weakly, e.g., mouse IgG_1. For some species, IgG does not bind to protein A at all, e.g., rat, chicken, goat, and some MAbs show abnormal affinity for the protein. These properties make the use of protein A for IgG purification limited in certain cases, although it can be used to an advantage in separating IgG subclasses from mouse serum *(3)*. Protein G (derived from groups C and G *Streptococci*) also binds to IgG Fc with some differences in species specificity from protein A. Protein G binds to IgG of most species, including rat and goat, and recognizes most subclasses (including human IgG_3 and mouse IgG_1), but has a lower binding capacity. Protein G also has a high affinity for albumin, although recombinant DNA forms now exist in which the albumin-binding site has been spliced out, and are therefore very useful for affinity chromatography. Other streptococcal immunoglobulin binding proteins are protein H (binds IgG Fc), protein B, which binds IgA and protein Arp, which binds IgG & IgA. These are not generally available for immunochemical use.

Another bacterial IgG-binding protein (protein L) has been identified *(4)*. Derived from *Peptostreptococcus magnus,* it binds to some κ (but not λ) chains. Furthermore, protein L binds to only some light-chain subtypes, although immunoglobulins from many species are recognized *(5,6)*.

Finally, hybrid molecules produced by recombinant DNA procedures, comprising the appropriate regions of IgG-binding proteins (e.g., protein L/G, protein L/A) also have considerable scope in immunochemical techniques. These proteins are therefore very useful in the purification of IgG by affinity chromatography. Columns are commercially available (MabTrap G II, Pharmacia, Uppsala, Sweden) or can be prepared in the laboratory. The product of this method is of high purity and is useful for most immunochemical procedures including affinity chromatography and conjugation with radioisotopes, enzymes, biotin, and so forth.

From: *The Protein Protocols Handbook, 2nd Edition*
Edited by: J. M. Walker © Humana Press Inc., Totowa, NJ

2. Materials

1. PBS: 0.14 *M* NaCl, 2.7 m*M* KCl, 1.5 m*M* KH_2PO_4, 8.1 m*M* Na_2HPO_4.
2. Sodium azide.
3. Dissociating buffer: 0.1 *M* glycine-HCl, pH 3.5. Adjust the pH with 2 *M* HCl.
4. 1 *M* Tris.
5. Binding buffer: Optimal binding performance occurs using a buffer system between pH 7.5 and 8.0. Suggested buffers include 0.1 *M* Tris-HCl, 0.15 *M* NaCl, pH 7.5; 0.05 *M* sodium borate, 0.15 *M* NaCl, pH 8.0; and 0.1 *M* sodium phosphate, 0.15 *M* NaCl, pH 7.5.
6. IgG preparation: serum, ascitic fluid, or hybridoma culture supernatant.
7. Protein A, G column.

3. Methods

Refer to Chapter 144 for CNBr activation of Sepharose and coupling of protein A, G.

1. Wash the column with an appropriate binding buffer.
2. Pre-elute the column with dissociating buffer, 0.1 *M* glycine-HCl, pH 3.5.
3. Equilibrate the column with binding buffer.
4. Prepare IgG sample: If the preparation is serum, plasma, or ascitic fluid, dilute it at least 1:1 in binding buffer and filter through 0.45-µm filter. Salt-fractionated preparations (*see* Chapter 137) do not require dilution, but the protein concentration should be adjusted to approx 1–5 mg/mL. Hybridoma culture supernatants do not require dilution.
5. Apply sample to column at no more than 10 mg IgG/2-mL column.
6. Wash the column with binding buffer until the absorbance at 280 nm is <0.02.
7. Dissociate the IgG-ligand interaction by eluting with dissociating buffer. Monitor the absorbance at 280 nm, and collect the protein peak. Neutralize immediately with alkali (e.g., 1 *M* Tris, unbuffered).
8. Wash the column with binding buffer until the pH returns to that of the binding buffer. Store the column in buffer containing at least 0.15 *M* NaCl and 0.1% sodium azide.
9. Dialyze the IgG preparation against a suitable buffer (e.g., PBS) to remove glycine/Tris.

References

1. Lindmark, R., Thorén-Tolling, K., and Sjöquist, J. (1983) Binding of immunoglobulins to protein A and immunoglobulin levels in mammalian sera. *J. Immunol. Meth.* **62,** 1–13.
2. Hermanson, G. T., Mallia, A. K., and Smith, P. K. (1992) *Immobilized Affinity Ligand Techniques.* Academic, San Diego, CA.
3. Ey, P. L., Prowse, S. J., and Jenkin, C. R. (1978) Isolation of pure IgG_1, IgG_{2a}, and IgG_{2b} immunoglobulins from mouse serum using protein A-sepharose. *Immunochemistry* **15,** 429–436.
4. Kerr, M. A., Loomes, L. M., and Thorpe, S. J. (1994) Purification and fragmentation of immunoglobulins, in *Immunochemistry Labfax* (Kerr, M. A. and Thorpe, R., eds.), Bios Scientific, Oxford, UK, pp. 83–114.
5. De Chateau, M., Nilson, B. H., Erntell, M., Myhre, E., Magnusson, C. G., Akerstrom, B., and Bjorck, L. (1993) On the interaction between protein L and immunoglobulins of various mammalian species. *Scand. J. Immunol.* **37,** 399–405.
6. Akerstrom, B., Nilson, B. H., Hoogenboom, H. R., and Bjorck, L. (1994) On the interaction between single chain Fv antibodies and bacterial immunoglobulin-binding proteins. *J. Immunol. Meth.* **177,** 151–163.

143

Analysis and Purification of IgG Using Size-Exclusion High Performance Liquid Chromatography (SE-HPLC)

Carl Dolman and Robin Thorpe

1. Introduction

The use of size-exclusion chromatography (SEC; also known as gel filtration chromatography) for purification of IgG has been described in protocol 131. SEC with dextran, agarose, and polyacrylamide soft gels was one of the first techniques developed for the purification of proteins. Soft gels, however, cannot tolerate high pressure and require long separation times. They also have limited stability to extremes of pH and to salt. These and other limitations of "conventional" size-exclusion chromatography are responsible for the problems with resolution often observed when analyzing/purifying IgG using such methods. However, many of these problems can be solved by using high performance liquid chromatography (HPLC) systems with appropriate size-exclusion columns.

Silica-based particles such as those used for many HPLC SEC columns are ideal for analysis and purification of IgG, as they are inert, have well-defined pore size, and because of their rigidity can be run at relatively high pressure, allowing reasonable flow rates and relatively quick separation times. Different pore sizes of the silica beads allow fractionation of different ranges of protein molecular weights. The column chosen should therefore have a pore size that allows the fractionation of proteins of molecular weights in the range 10,000–500,000, as this will separate aggregated, dimeric, monomeric, and fragmented immunoglobulins as discrete peaks (e.g., TSK G3000SW; *see* **Fig. 1**). Resolution of peaks is also dependent on column length, and in the past it has been necessary to join two columns to achieve optimal results. This of course doubles the time taken to run a sample. The alternative is to use a column with smaller bead dimensions, for example, the TSK G3000SW$_{XL}$, which has 5-μm beads compared to the 10-μm beads used for G3000SW columns.

As well as being much quicker than conventional size-exclusion chromatography, SE-HPLC is capable of much increased resolution, and the fully automated systems provide much greater reproducibility than "conventional" chromatography setups.

From: *The Protein Protocols Handbook, 2nd Edition*
Edited by: J. M. Walker © Humana Press Inc., Totowa, NJ

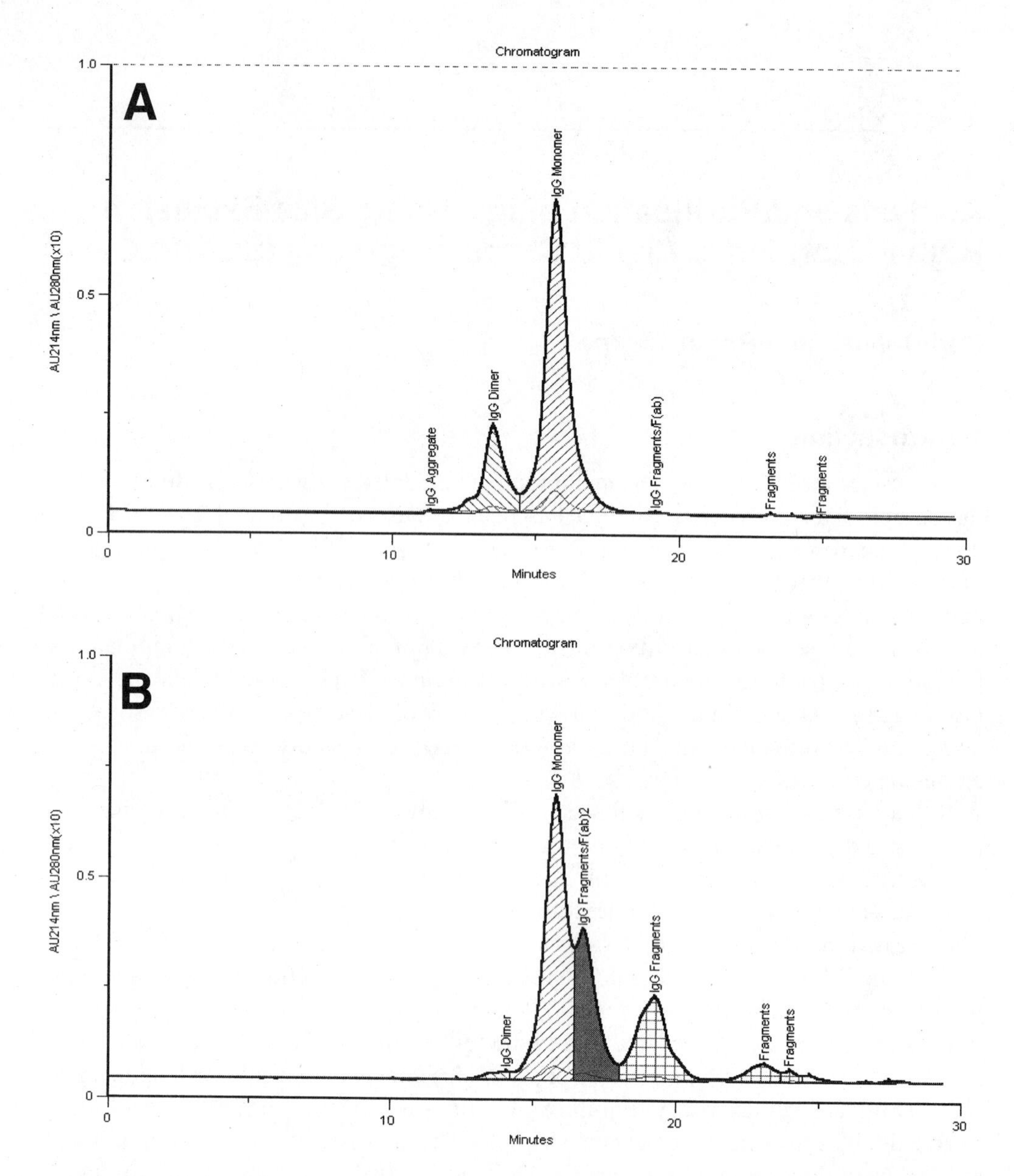

Fig. 1. Typical profiles for intact **(A)** and fragmented **(B)** human therapeutic immunoglobulin products.

2. Materials

1. The principal equipment used is a modular HPLC system (e.g., Gilson, Waters) comprising an autosampler, pump, UV-detector, and fraction collector.
2. The equipment is controlled using a Unipoint system controller or equivalent software.
3. TSK G3000SW$_{XL}$ 7.8 mm × 300 mm (or equivalent) column (supplied locally by Anachem, Luton, UK).
4. Mobile phase: 0.2 *M* Sodium phosphate, pH 6.0 + 0.1 *M* sodium sulfate, pH 6.0.

5. 18 Mohm deionized water.
6. 50% Methanol in deionized water.

3. Methods (*see* Note 1)

1. Assemble the HPLC components except the column according to the manufacturer's instructions. Equilibrate the system in the mobile phase (*see* **Note 2**).
2. Install the column (and guard column; *see* **Note 3**).
3. Equilibrate the column with mobile phase (*see* **Note 2**), making sure any bacteriostatic storage agent (*see* **Note 4**) is completely removed.
4. Monitor the baseline at 214 nm and/or 280 nm until stable (allow the UV-detector lamp up to 1 h to warm up).
5. Inject sample (*see* **Note 5**).
6. Monitor eluant at 214 nm and/or 280 nm and record data via computer software or integrator.
7. If purifying IgG, set the fraction collector to collect peaks using peak detection or collect 0.1–0.2 mL fractions.

4. Notes

1. All solutions should be 0.22-µm filtered and degassed under vacuum prior to use.
2. The flow rate should be 0.5 mL/min for 7.8 mm internal diameter (i.d.) columns (equal to a linear flow rate of 62.8 cm/h).
3. A guard column can be installed between the autosampler and the main SEC column to prolong the life of the main column, particularly when the sample purity is low.
4. Columns should be stored in the mobile phase with a bacteriostatic agent such as 0.02% azide. Ensure the column is saturated by passing several column volumes of this storage solution through it prior to removing the column from the system.
5. The sample volume is dependent on column size. Volumes above 100 µL will adversely affect sample resolution for 7.8 mm i.d. columns. For analytical purposes use 20 µL of a 1 mg/mL sample. For purification do not exceed a volume of 100 µL or a total load of 1 mg for 7.8 mm × 30 cm columns.

144

Purification of IgG Using Affinity Chromatography on Antigen-Ligand Columns

Mark Page and Robin Thorpe

1. Introduction

Affinity chromatography is a particularly powerful procedure, which can be used to purify IgG, subpopulations of IgG, or the antigen binding fraction of IgG present in serum/ascitic fluid/hybridoma culture supernatant. This technique requires the production of a solid matrix to which a ligand having either affinity for the relevant IgG or vice versa has been bound ***(1)***. Examples of ligands useful in this context are:

1. The antigen recognized by the IgG (for isolation of the antigen-specific fraction of the serum/ascitic fluid, and so forth).
2. IgG prepared from an anti-immunoglobulin serum, e.g., rabbit antihuman IgG serum or murine antihuman IgG MAb for the purification of human IgG (*see* **Note 1**).
3. IgG-binding proteins derived from bacteria, e.g., protein A (from *Staphylococcus aureus* Cowan 1 strain) or proteins G or C (from *Streptococcus* and *see* Chapter 142).

The methods for production of such immobilized ligands and for carrying out affinity-purification of IgG are essentially similar, regardless of which ligand is used. Sepharose 4B is probably the most widely used matrix for affinity chromatography, but other materials are available. Activation of Sepharose 4B is usually carried out by reaction with cyanogen bromide (CNBr); this can be carried out in the laboratory before coupling, or ready-activated lyophilized Sepharose can be purchased. The commercial product is obviously more convenient than "homemade" activated Sepharose, but it is more expensive and may be less active.

2. Materials

1. Sepharose 4B.
2. Sodium carbonate buffer: 0.5 *M* Na_2CO_3, pH 10.5. Adjust pH with 0.1 *M* NaOH.
3. Cyanogen bromide. **(Warning: CNBr is toxic and should be handled in a fume hood.)**
4. Sodium hydroxide: 1 *M;* 4 *M.*
5. Sodium citrate buffer: 0.1 *M* trisodium citrate, pH 6.5. Adjust pH with 0.1 *M* and citric acid.
6. Ligand solution: 2–10 mg/mL in 0.1 *M* sodium citrate buffer, pH 6.5.
7. Ethanolamine buffer: 2 *M* ethanolamine.
8. PBS: 0.14 *M* NaCl, 2.7 m*M* KCl, 1.5 m*M* KH_2PO_4, 8.1 m*M* Na_2HPO_4.

From: *The Protein Protocols Handbook, 2nd Edition*
Edited by: J. M. Walker © Humana Press Inc., Totowa, NJ

9. IgG preparation: serum, ascitic fluid, hybridoma culture supernatant.
10. PBS containing 0.1% sodium azide.
11. Disassociating buffer: 0.1 *M* glycine, pH 2.5. Adjust pH with 1 *M* HCl.
12. 1 *M* Tris-HCl, pH 8.8. Adjust pH with 1 *M* HCl.

3. Method

3.1. Activation of Sepharose with CNBr and Preparation of Immobilized Ligand

Activation of Sepharose with CNBr requires the availability of a fume hood and careful control of the pH of the reaction—failure to do this may lead to the production of dangerous quantities of HCN as well as compromising the quality of the activated Sepharose. CNBr is toxic and volatile. All equipment that has been in contact with CNBr and residual reagents should be soaked in 1 *M* NaOH overnight in a fume hood and washed before discarding/returning to the equipment pool. Manufacturers of ready activated Sepharose provide instructions for coupling (*see* **Note 2**).

1. Wash 10 mL (settled volume) of Sepharose 4B with 1 L of water by vacuum filtration. Resuspend in 18 mL of water (do not allow the Sepharose to dry out).
2. Add 2 mL of 0.5 *M* sodium carbonate buffer, pH 10.5, and stir slowly. Place in a fume hood and immerse the glass pH electrode in the solution.
3. **Carefully** weigh 1.5 g of CNBr into an air-tight container (Note: weigh in a fume hood; wear gloves)—remember to decontaminate equipment that has contacted CNBr in 1 *M* NaOH overnight.
4. Add the CNBr to the stirred Sepharose. Maintain the pH between 10.5 and 11.0 by dropwise addition of 4 *M* NaOH until the pH stabilizes and all the CNBr has dissolved. If the pH rises above 11.5, activation will be inefficient, and the Sepharose should be discarded.
5. Filter the slurry using a sintered glass or Buchner funnel, and wash the Sepharose with 2 L of cold 0.1 *M* sodium citrate buffer, pH 6.5—do not allow the Sepharose to dry out. Carefully discard the filtrate (use care: this contains CNBr).
5. Quickly add the filtered washed Sepharose to the ligand solution (2–10 mg/mL in 0.1 *M* sodium citrate, pH 6.5), and gently mix on a rotator ("windmill') at 4°C overnight (*see* **Note 3**).
7. Add 1 mL of 2 *M* ethanolamine solution, and mix at 4°C for a further 1 h—this blocks unreacted active groups.
8. Pack the Sepharose into a suitable chromatography column (e.g., a syringe barrel fitted with a sintered disk) and wash with 50 mL of PBS. Store at 4°C in PBS containing 0.1% sodium azide.

3.2. Sample Application and Elution

1. Wash the affinity column with PBS. "Pre-elute" with dissociating buffer, e.g., 0.1 *M* glycine-HCl, pH 2.5. Wash with PBS; check that the pH of the eluate is the same as the pH of the PBS (*see* **Note 4**).
2. Apply the sample (filtered through a 0.45-µm membrane) to the column. As a general rule, add an eqiuvalent amount (mole:mole) of IgG in the sample to that of the ligand coupled to the column. Close the column exit, and incubate at room temperature for 15–30 min (*see* **Note 5**).
3. Wash non-IgG material from the column with PBS; monitor the A_{280} as an indicator of protein content.

4. When the A_{280} reaches a low value (approx 0.02), disrupt the ligand–IgG interaction by eluting with dissociating buffer. Monitor the A_{280}, and collect the protein peak into tubes containing 1 *M* Tris-HCl, pH 8.8 (120 µL/1 mL fraction) to neutralize the acidic dissociating buffer.
5. Wash the column with PBS until the eluate is at pH 7.4. Store the column in PBS containing 0.1% azide. Dialyze the IgG preparation against a suitable buffer (e.g., PBS) to remove glycine/Tris.

4. Notes

1. The use of subclass-specific antibodies or MAb allows the immunoaffinity isolation of individual subclasses of IgG.
2. Coupling at pH 6.5 is less efficient than at higher pH, but is less likely to compromise the binding ability of immobilized ligands (especially antibodies).
3. Check the efficiency of coupling by measuring the A_{280} of the ligand before and after coupling. Usually at least 95% of the ligand is bound to the matrix.
4. Elution of bound substances is usually achieved by using a reagent that disrupts noncovalent bonds. These vary from "mild" procedures, such as the use of high salt or high or low pH, to more drastic agents, such as 8 *M* urea, 1% SDS or 5 *M* guanidine hydrochloride. Chaotropic agents, such as 3 *M* thiocyanate or pyrophosphate, may also be used. Usually an eluting agent is selected that is efficient, but does not appreciably denature the purified molecule; this is often a compromise between the two ideals. In view of this, highly avid polyclonal antisera obtained from hyperimmune animals are often not the best reagents for immunoaffinity purification, as it may be impossible to elute the IgG in a useful form. The 0.1 *M* glycine-HCl buffer, pH 2.5, will elute most IgG, but may denature some MAb. "Pre-elution" of the column with dissociating reagent just before affinity chromatography ensures that the isolated immunoglobulin is minimally contaminated with ligand.
5. The column will only bind to its capacity and therefore some IgG may not be bound; however, this can be saved and passed through the column again. The main problem is with back pressure or even blockage, but this can be reduced by diluting the sample by at least 50% with a suitable buffer (e.g., PBS). Incubation of the IgG containing sample with the ligand matrix is not always necessary, but this will allow maximal binding to occur. Alternatively, slowly recirculate the sample through the column, typically at <0.5 mL/min.

References

1. Hermanson, G. T., Mallia, A. K., and Smith, P. K. (1992) *Immobilized Affinity Ligand Techniques.* Academic, San Diego, CA.

145

Purification of IgG Using Thiophilic Chromatography

Mark Page and Robin Thorpe

1. Introduction

Immunoglobulins recognize sulphone groups in close proximity to a thioether group *(1)*, and, therefore, thiophilic adsorbents provide an additional chromatographic method for the purification of immunoglobulins that can be carried out under mild conditions preserving biological activity. A thiophilic gel is prepared by reducing divinylsulfone (coupled to Sepharose 4B) with β-mercaptoethanol. The product is of intermediate purity and would be useful for further processing, e.g., purification by ion-exchange, size exclusion, and/or affinity chromatography.

2. Materials

1. Sepharose 4B.
2. 0.5 *M* sodium carbonate.
3. Divinylsulfone.
4. Coupling buffer: 0.1 *M* sodium carbonate buffer, pH 9.0.
5. β-mercaptoethanol.
6. Binding buffer: 0.1 *M* Tris-HCl, pH 7.6, containing 0.5 *M* K_2SO_4.
7. IgG preparation: serum, ascitic fluid, or hybridoma culture supernatant.
8. 0.1 *M* ammonium bicarbonate.

3. Methods

Caution: Divinylsulfone is highly toxic and the column preparation procedures should be carried out in a well-ventilated fume cabinet.

1. Wash 100 mL of Sepharose 4B (settled volume) with 1 L of water by vacuum filtration.
2. Resuspend in 100 mL of 0.5 *M* sodium carbonate, and stir slowly.
3. Add 10 mL of divinylsulfone dropwise over a period of 15 min with constant stirring. After addition is complete, slowly stir the gel suspension for 1 h at room temperature.
4. Wash the activated gel thoroughly with water until the filtrate is no longer acidic (*see* **Note 1**).
5. Wash activated gel with 200 mL of coupling buffer using vacuum filtration and resuspend in 75 mL of coupling buffer.
6. In a well-ventilated fume cabinet, add 10 mL of β-mercaptoethanol to the gel suspension with constant stirring, and continue for 24 h at room temperature (*see* **Note 2**).

From: *The Protein Protocols Handbook, 2nd Edition*
Edited by: J. M. Walker © Humana Press Inc., Totowa, NJ

7. Filter and wash the gel thoroughly. The gel may be stored at 4°C in 0.02% sodium azide.
8. Pack 4 mL of the gel in a polypropylene column (10 × 1 cm) and equilibrate with 25 mL of binding buffer.
9. Perform chromatography at 4°C. Mix 1 mL of IgG containing sample with 2 mL of binding buffer, and load onto the column.
10. After the sample has entered the gel, wash non-IgG from the column with 20 mL of binding buffer; monitor the A_{280} as an indicator of protein content in the wash until the absorbance returns to background levels.
11. Elute the bound IgG with 0.1 *M* ammonium bicarbonate, and collect into 2-mL fractions. Monitor the protein content by absorbance at 280 nm, and pool the IgG containing fractions (i.e., those with protein absorbance peaks). Dialyze against an appropriate buffer (e.g., PBS) with several changes, and analyze by gel electrophoresis under reducing conditions (*see* Chapter 134).

4. Notes

1. The activated gel can be stored by washing thoroughly in acetone and kept as a suspension in acetone at 4°C.
2. Immobilized ligands prepared by the divinylsulfone method are unstable above pH 8.0.

References

1. Porath, J., Maisano, F., and Belew, M. (1985) Thiophilic adsorption—a new method for protein fractionation. *FEBS Lett.* **185,** 306–310.

146

Analysis of IgG Fractions by Electrophoresis

Mark Page and Robin Thorpe

1. Introduction (*see* Note 1)

After using a purification procedure it is necessary to obtain some index of purity obtained. One of the simplest methods for assessing purity of an IgG fraction is by sodium dodecyl sulfate-polyacrylamide gel electrophoresis (SDS-PAGE). Although "full-size" slab gels can be used with discontinuous buffer systems and stacking gels, the use of a "minigel" procedure, using a Tris-bicine buffer system *(1)*, rather than the classical Tris-glycine system, is quicker and easier, and gives improved resolution of immunoglobulin light-chains (these are usually smeared with the Tris-glycine system). Gel heights can be restricted to <10 cm and are perfectly adequate for assessing purity and monitoring column fractions.

2. Materials

1. Gel solution: 2.0 mL of 1 *M* Tris, 1 *M* bicine; 4.0 mL 50% w/v acrylamide containing 2.5% w/v *bis*-acrylamide; 0.4 mL 1.5% w/v ammonium persulfate; 0.2 mL 10% w/v SDS.
   Make up to 20 mL with distilled water.
2. Gel running buffer: 2.8 mL 1 *M* Tris, 1 *M* bicine; 1.4 mL 10% w/v SDS.
   Make up to 140 mL with distilled water.
3. Sample buffer: 1.0 g sucrose, 0.2 mL 1 *M* Tris, 1 *M* bicine, 1.0 mL 10% SDS, 0.25 mL 2-mercaptoethanol.
   Make up to 3 mL with distilled water, and add 0.001% w/v Bromophenol blue. Store at –20°C.
4. Coomassie blue R stain: Add coomassie brilliant blue R (0.025 g) to methanol (50 mL), and stir for 10 min. Add distilled water (45 mL) and glacial acetic acid (5 mL). Use within 1 mo.
5. Destain solution: glacial acetic acid (7.5 mL) and methanol (5 mL). Make up to 100 mL with distilled water.
5. *N,N,N',N'*-Tetramethylethylenediamine (TEMED).
7. Molecular-weight markers (mol-wt range 200,000–14,000).
8. Purified IgG or column fraction samples.

3. Methods

1. Prepare sample buffer.
2. Adjust antibody preparation to 1 mg/mL in 0.1 *M* Tris, 0.1 *M* bicine (*see* **Note 2**).

From: *The Protein Protocols Handbook, 2nd Edition*
Edited by: J. M. Walker © Humana Press Inc., Totowa, NJ

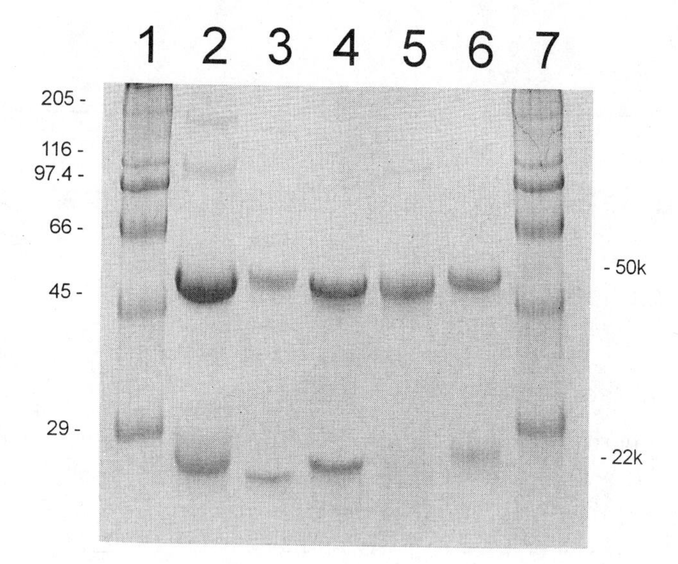

Fig. 1. SDS-PAGE minigel depicting purified IgG preparations derived from human serum (*lane 2*), mouse ascitic flude (*lanes 3* and *4*), rabbit serum (*lane 5*), and sheep serum (*lane 6*). Samples are elctrophoresed under reducing conditions and stained with Coomassie blue. All IgGs consist predominantly of two bands comprising heavy (50,000 M_r) and light (22,000 M_r) chains with no major contaminating proteins. Molecular-weight markers are shown in *lanes 1* and *7* and their molecular weights (in thousands) given on the left.

3. Mix the sample in the ratio 2:1 with sample buffer.
4. Heat at 100°C for 2–4 min.
5. Prepare gel solution and running buffer as described in **Subheading 2.**
6. Assemble gel mold according to manufacturer's instructions.
7. Add 30 µL TEMED to 10 mL of gel solution, pour this solution between the plates to fill the gap completely, and insert the comb in the top of the mold (there is no stacking gel with this system). Leave for 10 min for gel to polymerize.
8. Remove comb and clamp gel plates into the electrophoresis apparatus. Fill the anode and cathode reservoirs with running buffer.
9. Load the sample(s) (30–50 µL/track), and run an IgG reference standard and/or mol-wt markers in parallel.
10. Electrophorese at 150V for 1.5 h.
11. Remove gel from plates carefully, and stain with Coomassie blue R stain for 2 h (gently rocking) or overnight (stationary) (*see* **Note 3**).
12. Pour off the stain, and rinse briefly in tap water.
13. Add excess destain to the gel. A piece of sponge added during destaining absorbs excess stain. Leave until destaining is complete (usually overnight with gentle agitation).

4. Notes

1. The "minigel" is easily and quickly prepared consisting of a resolving (separating) gel only and takes approx 1.5 h to run once set up. The IgG sample is prepared for electrophoresis under reduced conditions, and is run in parallel with either a reference IgG preparation or with standard mol-wt markers. The heavy chains have a characteristic relative molecular weight of approx 50,000 and the light-chains a molecular weight of 22,000 (**Fig. 1**).

2. The extinction coefficient ($E^{1\%}_{280}$) of human IgG is 13.6 (i.e., a 1 mg/mL solution will have an A_{280} of 1.36).
3. Mark the gel uniquely prior to staining so that its orientation is known. By convention, a small triangle of the bottom left- or right-hand corner of the gel is sliced off.

References

1. Johnstone, A. and Thorpe, R. (1996) *Immunochemistry in Practice,* 3rd ed., Blackwell Scientific, Oxford, UK.

147

Purification of Immunoglobulin Y (IgY) from Chicken Eggs

Christopher R. Bird and Robin Thorpe

1. Introduction

Chickens produce an immunoglobulin G (IgG) homologue, sometimes referred to as IgY (to reflect the differences in the heavy chain domain compared with mammalian IgG), which can be conveniently isolated from the yolk of eggs *(1)*. The concentration of immunoglobulin in the egg yolk is roughly the same as in serum (10–15 mg/mL), and an average egg can yield approx 80–100 mg of immunoglobulin. Eggs can therefore provide an abundant source of polyclonal antibody that may be aquired non-invasively from eggs laid by immunized chickens *(2)*. The production of polyclonal antibodies in chickens provides other advantages over using conventional mammalian species. The mechanism of antibody production and organization of the avian immune system is quite different from that of mammals, and as chickens are phylogenetically distant from mammals because of their evolutionary divergence millions of years ago, it is possible to generate antibody responses to highly conserved proteins that do not easily elicit an immune response in mammalian species. Chicken immunoglobulins possess other characteristics that differ from mammalian antibodies that may offer advantages in certain immunological techniques. Most of the interactions via the Fc region in mammalian antibodies do not occur with chicken immunoglobulin. Chicken IgG does not activate mammalian complement systems, and does not react with mammalian rheumatoid factors, and neither protein A or protein G bind to chicken IgG. Although chicken antibodies can be substituted for mammalian antibodies in many techniques (and may even be advantageous) it may be necessary to optimize conditions in certain systems particularly with precipitation techniques where chicken IgG appears to be less efficient then conventional mammalian polyclonal antibodies.

To isolate chicken immunoglobulins from egg yolk the lipid first has to be removed. This can be achieved by dextran sulfate precipitation after which the chicken immunoglobulins can be purified by sodium sulfate precipitation.

2. Materials

1. Tris-buffered saline (TBS): 0.14 *M* NaCl in 10 m*M* Tris-HCl, pH 7.4.
2. 1 *M* Calcium chloride.
3. Dextran sulphate solution: 10% (w/v) in TBS.

From: *The Protein Protocols Handbook, 2nd Edition*
Edited by: J. M. Walker © Humana Press Inc., Totowa, NJ

4. Centrifuge.
5. Anhydrous sodium sulfate.
6. Saturated sodium sulfate solution: 36% (w/v) in water.

3. Methods

Break the eggshell and carefully separate the egg yolk from the white to minimize contamination of the yolk with egg white proteins. Place the yolk on a piece of filter paper in which a small hole has been cut. Position the yolk above the hole and pierce the yolk membrane with a needle and collect the yolk in a suitable container (e.g., 50-mL centrifuge tube). The yolk membrane and any remaining egg white will stick to the filter paper. Approximately 10 mL of yolk is obtained from an average sized egg.

1. Dilute the egg yolk with 4 vol of TBS.
2. Centrifuge at 2000–3000*g* for 20 min at room temperature with the brake off.
3. Remove the supernatant, and discard the membrane pellet and add 120 μL of dextran sulfate solution/mL of supernatant, mix well, and incubate at room temperature for 30 min.
4. Add 50 μL of 1 *M* $CaCl_2$/mL and mix well. Incubate for a further 30 min at room temperature.
5. Centrifuge at 2000–3000*g* for 30 min and collect the supernatant, which should be clear; if not, repeat **steps 3** and **4** using half the amounts of dextran sulfate and calcium chloride. The pellet can be washed once at this stage with 50 mL of TBS, and the supernatant combined with the first dextran sulfate supernatant obtained.
6. Adjust the volume of the pooled supernatants to 100 mL with TBS. Stir the supernatant and slowly add 20 g of sodium sulfate until completely dissolved and leave to stand for 30 min at room temperature.
7. Centrifuge for 20 min at 2000–3000*g* and discard the supernatant.
8. Redissolve the precipitate in 10 mL TBS and centrifuge for 20 min at 2000–3000***g***. Collect the supernatant and discard the pellet.
9. Stir the supernatant and slowly add 8 mL of 36% (w/v) sodium sulfate solution and leave to stand for 30 min at room temperature.
10. Centrifuge at 2000–3000*g* for 20 min, discard the supernatant, and redissolve the pellet in 5 mL of TBS. Dialyze and filter the purified immunoglobulin.

4. Notes

1. Eggs may be stored at 4°C until several have accumulated and then processed together. Alternatively separated egg yolk can be diluted with buffer containing preservative and stored at 4°C prior to purification.
2. 36% (w/v) Sodium sulfate is a supersaturated solution and requires heating to fully dissolve, after which the solution should be stored at 30–40°C before use. The entire procedure for IgG precipitation with sodium sulfate should be carried out at 20–25°C; otherwise the sodium sulfate may precipitate.
3. Purified chicken IgG can sometimes aggregate after freezing and thawing; therefore storing sterile solutions of purified chicken IgG at 4°C may be advisable; alternatively they can be lyophilized for longer term storage.
4. The chicken IgG heavy chain has a different mobility from mammalian IgG, which can be demonstrated by reducing sodium dodecyl sulfate-polyacrylamide gel electrophoresis (SDS-PAGE) (**Fig. 1**).

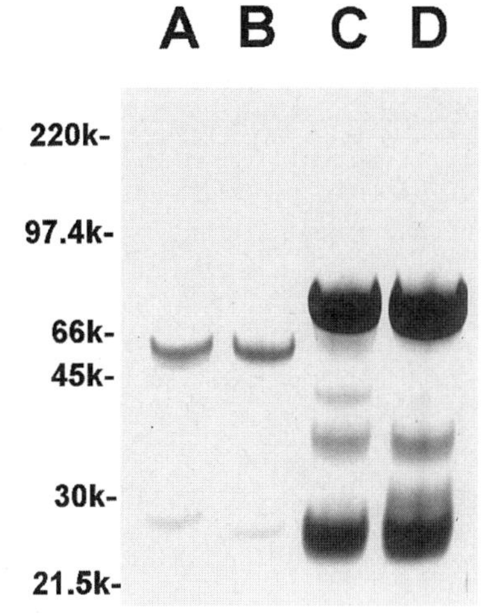

Fig. 1. Coomassie Blue stained SDS-polyacrylamide gel (10% total acrylamide) showing the different mobility of chicken IgG heavy chain compared to mammalian IgG. *Lanes: A*, protein G purified sheep IgG; *B*, therapeutic human IgG; *C* and *D*, two different preparations of purified chicken IgG. All samples were run reduced.

References

1. Jensenius, J. C., Andersen, I., Hau, J., Crone, M., and Koch, C. (1981) Eggs: conveniently packaged antibodies. Methods for purification of yolk IgG. *J. Immunol. Meth.* **46,** 63–68.
2. Jensenius, J. C. and Koch, C. (1997) Antibodies packaged in eggs, in *Immunochemistry: A Practical Approach*, Vol. 1 (Johnstone, A. P. and Turner, M. W., eds.), IRL Press, Oxford, UK, pp. 89–107.

148

Affinity Purification of Immunoglobulins Using Protein A Mimetic (PAM)

Giorgio Fassina, Giovanna Palombo, Antonio Verdoliva, and Menotti Ruvo

1. Introduction

While antibodies of the G class can be conveniently purified by affinity chromatography using immobilized protein A or G even at large scale, scaling up purification of IgM, IgA, and IgE and IgY still presents several problems, as specific and cost-effective ligands for these classes of immunoglobulins are not available. Protein A *(1)*, which is widely used for the affinity purification of antibodies from sera or cell culture supernatants, does not recognize immunoglobulins of the M, A, E, and Y classes well and is not used to capture and purify these immunoglobulins from crude sources. Moreover, these two proteins are obtained from microorganisms or genetically modified bacteria, which carries the risk of affecting the safety of the purified antibodies through the presence of contaminants such as viruses, pirogens, or DNA fragments. As a result, the availability of alternative ligands for the affinity purification of antibodies is highly important from an industrial aspect. After immobilization on solid supports, the mannan binding protein (MBP), an affinity ligand for IgM, provides affinity media useful for IgM isolation based on a temperature-dependent interaction of the ligand with the immunoglobulins *(2)*. The use of immobilized MBP for the purification of IgM is based on the adsorption in the presence of calcium at a temperature of 4°C, and the elution at room temperature of adsorbed immunoglobulins in the presence of ethylenediamineotetraacetic acid (EDTA). This ligand shows low binding affinity for IgG, but binds to bovine and human IgM with lower affinity than murine IgM. However, in addition to the complexity of MBP isolation, functional binding capacities of MBP columns are limited to 1 or 2 mg of IgM per milliliter of support. IgA, which is involved in the first specific defense against natural infection *(3)* and represents the second most abundant immunoglobulin in serum *(4)*, can be purified through the combination of different fractionation techniques such as ammonium sulfate precipitation, ion-exchange chromatography, and gel filtration *(5,6)*. All these procedures are time consuming, labor intensive, and are not compatible with industrial scaling up. Lectin jacalin, isolated from jackfruit seeds *(7)*, binds to IgA and can be conveniently used for the affinity purification of IgA from colostrum or serum *(8)*. However, several aspects

From: *The Protein Protocols Handbook, 2nd Edition*
Edited by: J. M. Walker © Humana Press Inc., Totowa, NJ

limit the use of this lectin for large-scale purification of monoclonal IgA from cell culture supernatants. First, jacalin is a biologically active lectin, being a potent T cell mitogen and a strong B cell polyclonal activator *(9)*, thus requiring a careful control for ligand leakage into the purified preparation. Second, jacalin binds to the carbohydrate moiety of IgA, and D-galactose is required to elute IgA from affinity columns, which is costly and impractical for large-scale operations.

Antibodies of the E class, which represent an extremely important class of immunoglobulins from a biological and clinical point of view, require complex and time-consuming isolation protocols that make their characterization very difficult. The main purification procedure is represented by immunoaffinity chromatography using anti-IgE antibodies immobilized on solid supports *(10,11)*. Even if selective enough for research application, scaling up immunoaffinity chromatography for preparative applications is very expensive and not easily accomplished. Other approaches for IgE purification include classical chromatographic protocols based on the combination of different sequential procedures such as salting out, affinity chromatography on lysine–Sepharose, ion-exchange, gel filtration, and immuno-affinity chromatography to remove interfering proteins *(12)*. Studies carried out with immobilized protein A show that this protein, known to recognize the immunoglubulin Fc region, does not bind to monoclonal IgE, but binds 12–14% of serum polyclonal IgE. Protein G binds to neither polyclonal nor monoclonal IgE *(13)*.

IgY is an important class of immunoglobulins obtained from chicken egg yolk that represents an economical source of polyclonal antibodies *(14)*. Despite the advantages in production of immunoglobulins from egg yolk in terms of efficiency *(15)*, immunogenicity against mammal proteins *(15)*, and applications in therapeutics and diagnostics *(16,17)*, only fewer than 2% of the total number of polyclonal antibodies produced worldwide and commercially available are raised in chickens. This low diffusion is related to difficulties in purifying IgY from egg yolk, particularly from the lipidic fraction, which represents the main contaminant. The complex isolation methods described in the literature *(18–20)* and the lack of ligands usable for the affinity purification of IgY make the isolation of this class of immunoglobulins a laborious process that cannot be scaled up easily for industrial applications.

A synthetic peptide ligand (PAM, Protein A Mimetic, TG19318) (*see* **Fig. 1**), derived from the synthesis and the screening of a combinatorial peptide library *(21)*, has been identified in our laboratory for its ability to recognize, as a protein A mimetic, the constant portion of immunoglobulins. Its applicability in affinity chromatography for the downstream processing of antibodies has been fully established in studies examining the specificity and selectivity for polyclonal and monoclonal antibodies derived from different sources. Ligand specificity is broader than for any existing ligand, as IgGs derived from human, cow, horse, pig, mouse, rat, rabbit, goat, and sheep sera *(21,22)*, as well as IgYs derived from egg yolk *(23)* have been efficiently purified on PAM affinity columns. The ligand proved useful not only for IgG purification, but also for IgM *(24)*, IgA *(25)*, and IgE *(26)* isolation from sera or cell culture supernatants (*see* **Table 1**). PAM ligand, a tetrameric tripeptide, can be produced at low cost by means of chemical solution phase or solid phase synthesis in large amounts, does not contain biological contaminants such as viruses, pyrogens, or DNA fragments,

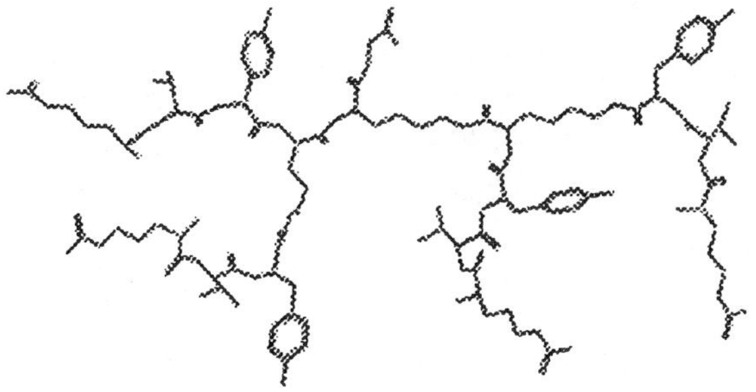

Fig. 1. Structure of PAM (TG19318) ($GKK_2[YTR]_4$).

as recombinant or extractive ligands such as protein A or G. In addition, the low toxicity of TG19318 and the low molecular weight of the resulting fragments considerably reduce the problems of contamination by leaked ligand, as is the case for protein A. Preliminary experiments suggest that the ligand is more stable to proteolytic digestion when coupled to solid supports, and the enzymatic activity normally found in crude feedstock derived from cell culture supernatants does not lead to noticeable loss of capacity. The ligand can be easily immobilized on preactivated solid supports, as the presence of four peptide chains departing from a central core, of which only a limited number are involved in the coupling with the solid phase, leaves the others fully available for the interaction. All the different supports tested so far maintain the ligand recognition properties for immunoglobulins, even if with different functional capacities. The affinity columns with immobilized PAM are not affected by the presence of denaturants, detergents, or other sanitation reagents commonly used for pirogen removal, and TG19318 columns can withstand a large array of harsh sanitizing agents with no capacity losses. Immunoglobulin adsorption on TG19318 affinity columns occurs with neutral buffers at low ionic strength and at room temperature. Elution of adsorbed immunoglobulins may be achieved under mild conditions not causing immunoglobulin denaturation, simply by changing the buffer pH to acid or alkaline conditions, with acetic acid (pH 3) or sodium bicarbonate (pH 9.0). Increasing the ionic strength of the dissociation buffer favors a more efficient elution of adsorbed antibodies.

Affinity interaction is strong enough to allow purification of antibodies directly from diluted supernatants in which the immunoglobulin concentration is very low, from 10 to 50 µg/mL. The main contaminant, albumin, is always efficiently removed in the purification step with any type of support tested for TG19318 immobilization. Column capacity depends on the type of support used for ligand immobilization, and may range from 10 to 25 mg of immunoglobulin/mL of support.

Purification of antibodies with clinical and therapeutic applications represents a key step in the validation process of these molecules, but the complex and the labor-intensive isolation procedures may alter the quality and the safety of the purified product. In this scenario, the availability of a synthetic ligand able to recognize antibodies independently from their class is of great importance from an industrial point of view, as it may simplify the isolation procedures, leading to lower production costs and reducing considerably the presence of biological contaminants in the purified antibody preparation.

Table 1
Immunoglobulins Specificity of PAM (TG 19318)

Ig class	Species	Source
IgG	Human	Serum
	Rabbit	Serum
	Cow	Milk
	Cow	Serum
	Sheep	Serum
	Rat	Serum
(IgG2b)	Rat	Cell supernatant
	Mouse	Serum
(IgG2a)	Mouse	Cell supernatant
(IgG1k)	Mouse	Cell supernatant
	Pig	Serum
	Horse	Serum
	Goat	Serum
IgM	Human	Serum
	Mouse	Serum
	Mouse	Ascites
	Mouse	Cell supernatant
IgA	Human	Serum
	Mouse	Serum
	Mouse	Cell supernatant
IgE	Rat	Ascites
IgY	Chicken	Egg yolk

2. Materials

2.1. Synthesis of PAM

1. Automated peptide synthesizer (Perkin-Elmer 431 A).
2. Resin Gly-hydroxymethylphenoxy (Gly-HMP).
3. 9-Fluorenyl-methoxycarbonyl-Lys(9-fluorenyl-methoxycarbonyl) (Fmoc-Lys[Fmoc]).
4. Fmoc-Arg(pentamethylchromane) (Fmoc-Arg[Pmc]).
5. Fmoc-Thr(*O-tert*-butyl) (Fmoc-Thr[OtBu]).
6. Fmoc-Tyr(OtBu).
7. *N*-Methyl-2-pyrrolidone.
8. Piperidine (20% in *N*-methyl-2-pyrrolidone).
9. 1 *M* Dicyclohexylcarbodiimide in *N*-methyl-2-pyrrolidone.
10. 1 *M* Hydroxybenzotriazole (HOBt) dissolved in *N*-methyl-2-pyrrolidone.
11. Methanol.
12. Dichloromethane.
13. Cleavage mixture: Trifluoroacetic acid (TFA)–phenol–water–ethanedithiol–thioanisol.
14. Ether.
15. High-performance liquid chromatography (HPLC) system.
16. Lichrospher RP-8 column (25 × 1 cm internal diameter [i.d.]).
17. Water–acetonitrile–TFA.

2.2. PAM Immobilization on Affinity Media

1. 0.5 *M* NaCl, 0.1 *M* Sodium bicarbonate, pH 8.5.
2. 0.1 *M* Tris-HCl, pH 8.5.
3. 0.1 *M* Acetic acid, 0.5 *M* NaCl, pH 4.0.
4. 0.1 *M* Tris-HCl, 0.5 *M* NaCl, pH 8.0.
5. NHS–Sepharose 4 Fast Flow (Pharmacia Biotech).
6. 3M Emphaze Biosupport Medium AB1 (Pierce).
7. HPLC system.
8. Lichrospher RP-8 column (25 × 1 cm i.d.).
9. Water–acetonitrile–TFA (95:5:0.1).

2.3. Affinity Purification on PAM Columns

1. HPLC/FPLC system.
2. 50–100 m*M* sodium phosphate, pH 7.0.
3. 25–50 m*M* bis-Tris buffer, pH 6.5.
4. 0.1 *M* Acetic acid.
5. 0.1 *M* Sodium bicarbonate, pH 8.5.
6. Protein A–Sepharose 4B (Pharmacia).

3. Methods

3.1. Synthesis of PAM

PAM can be produced in adequate amounts by solid-phase peptide synthesis on automatic peptide synthesizers, such as the Perkin-Elmer model 431A, software version 1.1, according to the synthesis procedure suggested by the manufacturer based on a consolidated methodology well known and widely reported in the literature.

1. Deprotect the Gly-HMP resin (0.1 mmol) by treatment with 3.0 mL of piperidine (20% in *N*-methyl-2-pyrrolidone) for 14 min, at room temperature with stirring.
2. Wash the resin 5× with 2.5 mL of *N*-methyl-2-pyrrolidone for 9 min under agitation at room temperature.
3. In the meantime, preactivate the amino acid residue in position 2 (Fmoc-Lys[Fmoc], 1 mmol) from the C-terminus (1 mmol) by incubation with 1 mL of 1 *M* HOBt dissolved in *N*-methyl-2-pyrrolidone and 1 mL of 1 *M* dicyclohexylcarbodiimide in *N*-methyl-2-pyrrolidone.
4. Incubate the activated amino acid with the resin for 51 min under constant agitation.
5. Wash the resin with *N*-methyl-2-pyrrolidone (four washes for 0.5 min with 2 mL).
6. Subject the resin to a further deprotection cycle with piperidine and a further coupling cycle with the next amino acid.
7. Repeat this sequential step procedure until all the amino acid residues are assembled. In detail, the following amino acid derivatives need to be used: Fmoc-Lys (Fmoc), Fmoc-Arg (Pmc), Fmoc-Thr (OtBu), and Fmoc-Tyr (OtBu).
8. Wash the resin with methanol, dichloromethane, and again with methanol and accurately dry the resin under vacuum for 12 h, after completion of synthesis cycles and removal of the N-terminal Fmoc group by piperidine treatment.
9. Detach protected peptide from resin by incubation of 100 mg of resin with 5 mL of a mixture of trifluoroacetic acid–phenol–water–ethandithiol–thioanisol 84:4:3:3:3 by vol for 2 h at room temperature under agitation.

10. Filter the resin using a sintered glass filter and reduce the filtrate in volume to few milliliters by vacuum evaporation and treat the residual liquid with 50 mL of cold ethyl ether.
11. Separate the precipitated peptidic material by centrifugation and resuspend the centrifuged material in 25 mL of water–acetonitrile–TFA 50:50:0.1, freeze, and lyophilize.
12. Purify the lyophilized material from contaminants by high-performance liquid chromatography (HPLC) using a Lichrospher RP-8 column (25 × 1 cm i.d.), equilibrated at a flow rate of 3 mL/min with water–acetonitrile–TFA 95/5/0.1, and eluting with a linear gradient of acetonitrile ranging from 5 to 80% in 55 min. Collect material corresponding to the main peak, freeze, and lyophilize.
13. Confirm chemical identity of PAM by determination of:
    a. Amino acid composition;
    b. N-terminal residue;
    c. Molecular weight by mass spectrometry.

3.2. PAM Immobilization on Affinity Media

3.2.1. PAM Immobilization on NHS-Sepharose 4 Fast Flow

1. Dissolve the peptide ligand in 0.1 *M* Na bicarbonate, 0.5 *M* NaCl, pH 8.5.
2. Collect the required amount of NHS-Sepharose 4 Fast Flow, remove the storage solution, and wash with 1 m*M* HCl.
3. Mix the PAM solution with gel suspension in a ratio 2:1 and leave the mixture to incubate for several hours at room temperature under gentle agitation, monitoring the extent of peptide incorporation by reverse phase (RP)-HPLC analysis of reaction mixture at different times (*see* **Note 1**).
4. Wash away excess ligand and incubate with 0.1 *M* Tris, pH 8.5, for 1 h at room temperature to deactivate residual active groups.
5. Wash peptide derivatized resin with at least three cycles of 0.1 *M* acetic acid, 0.5 *M* NaCl, pH 4.0, and 0.1 *M* Tris-HCl, 0.5 *M* NaCl, pH 8.0.
6. Store the product in 20% ethanol at 4°C.

3.2.2. PAM Immobilization on 3M Emphaze Biosupport Medium AB1

1. Dissolve the peptide ligand in 0.2 *M* Na bicarbonate, 0.6 *M* Na citrate, pH 8.0, at a concentration of 5–10 mg/mL.
2. Weigh out the required amount of 3M Emphaze and add the PAM solution directly to the dry beads.
3. Leave the mixture to incubate for several hours at room temperature under gentle agitation, monitoring the extent of peptide incorporation by RP-HPLC analysis of the reaction mixture at different times (*see* **Note 1**).
4. Wash away excess ligand with coupling buffer and incubate with 0.1 *M* Tris, pH 8.5, for 1 h at room temperature to deactivate residual active groups.
5. Wash peptide derivatized resin with at least three cycles of 0.1 *M* acetic acid, 0.5 *M* NaCl, pH 4.0, and 0.1 *M* Tris-HCl, 0.5 *M* NaCl, pH 8.
6. Store the product in 20% ethanol at 4°C.

3.3. Purification of Immunoglobulins on PAM Columns

3.3.1. Affinity Purification of IgG on the PAM-Emphaze Column

1. Dilute sera, ascitic fluids, or cell culture supernatants containing IgG 1:1 (v/v) with the column equilibration buffer, preferably 25–50 m*M* *bis*-Tris, pH 6.5, filtered through a 0.22-µm filter (Nalgene).

2. Load the sample on to the column equilibrated at a linear flow rate of 60 cm/h with 25–50 m*M* *bis*-Tris buffer, pH 6.5 (*see* **Notes 2** and **3**), monitoring the effluent by UV detection at 280 nm. Wash the column with the equilibration buffer until the UV absorbance returns to baseline.
3. Elute bound antibodies with 0.2 *M* acetic acid, pH 3.5 (*see* **Note 4**), wash the column with five volumes of 0.1 *M* acetic acid, and neutralize desorbed material immediately with 1.5 *M* Tris, pH 9.0.
4. Determine by sodium dodecyl sulfate-polyacrylamide gel electrophoresis (SDS-PAGE) and enzyme-linked immunosorbent assay (ELISA) purity and activity. The purity of adsorbed antibodies should be usually very high, ranging from 80% to 95%.
5. Store the column in 20% ethanol or 0.05% sodium azide (w/v) (*see* **Notes 5–7**).

3.3.2. Affinity Purification of IgM on the PAM-Sepharose 4 Fast Flow Column

3.3.2.1. Purification from Cell Culture Supernatants

Immobilized PAM is useful also for the capture of monoclonal IgM directly from crude cell supernatants (*see* **Note 8**) according to the following steps:

1. Load samples of crude cell culture supernatant obtained from stable hybridoma cell lines secreting murine IgM against specific antigens, even if containing a low concentration of IgM (10–100 mg/mL) on a PAM-affinity column equilibrated at a flow rate of 60 cm/h with 50–100 m*M* sodium phosphate, pH 7.0. Samples containing up to 5 mg of IgM may be loaded onto 1-mL bed volume columns. As before, it is recommended to dilute 1:1 (v/v) the samples with the elution buffer prior to application.
2. Wash the column with the equilibration buffer until complete removal of the unretained material is achieved, and then elute with 0.1 *M* acetic acid. Material desorbed by the acid treatment is collected and immediately neutralized.
3. Determine the protein content by the biuret method and IgM content by IgM specific ELISA assay (*see* **Note 9**).

3.3.2.2. Purification from Sera

IgM from sera can be purified by affinity chromatography on a PAM column after a preliminary IgG adsorption step on protein A–Sepharose 4B according to the protocol:

1. Load the serum sample on a protein A–Sepharose 4B affinity column equilibrated with 50 m*M* sodium phosphate, pH 7.0, at a flow rate of 60 cm/h.
2. Collect the column unretained material and dilute 1:1 (v/v) with 100 m*M* sodium phosphate, pH 7.0.
3. Load the protein A unretained fraction on the PAM column equilibrated at a flow rate of 60 cm/h with 50 m*M* sodium phosphate, pH 7.0. Wash the column and elute bound IgM as described previously in **Subheading 3.3.1.**
4. Collect fractions corresponding to the unbound and bound materials for SDS-PAGE analysis and ELISA determination of antibody recovery using an anti-IgM antibody conjugated to peroxidase for detection (*see* **Note 10**).

3.3.3. Affinity Purification of IgA on the PAM-Sepharose Fast Flow Column

3.3.3.1. Purification from Cell Culture Supernatants

Immunoglobulins of the A class secreted in cell culture supernatants derived from the cultivation of hybridoma can be conveniently purified on PAM columns equilibrated with 100 m*M* phosphate buffer, pH 7.0, at a flow rate of 60 cm/h.

1. Dilute sample containing up to 5 mg of IgA 1;1 (v/v) with 100 m*M* sodium phosphate, pH 7.0 and filter through a 0.22-mm filter and then load the sample onto the column.
2. Wash the column with loading buffer until the unbound material is completely removed.
3. Elute the adsorbed immunoglobulins with 0.1 *M* acetic acid and immediately neutralize with 1.5 *M* Tris, pH 9.0. Each fraction is checked for purity by SDS-PAGE and gel filtration analysis (*see* **Note 11**) and for IgA immunoreactivity using an ELISA assay (*see* **Note 12**).

3.3.3.2. Purification from Sera

Isolation of IgA from serum requires the prior removal of the IgG fraction. As in the case of IgM purification from sera, IgA-containing serum needs to be first fractionated on a protein A–Sepharose column, following conventional purification protocols.

1. Dilute 1:1 (v/v) the flow through material from protein A chromatography, lacking IgG and containing mainly IgA, IgM, and albumin, with 100 m*M* sodium phosphate, pH 7.0, and use directly for a subsequent fractionation on PAM columns.
2. Elute the bound fraction, after adsorption and column washing with 100 m*M* sodium phosphate, by a buffer change to 0.1 *M* acetic acid and immediately neutralize.

3.3.4. Affinity Purification of IgE on the PAM-Sepharose 4 Fast Flow Column

Monoclonal IgE obtained from the cultivation of stable hybridoma cell lines or contained in ascitic fluid can also be conveniently purified on PAM affinity columns.

1. Dilute samples containing up to 5 mg of IgE 1:1 with 100 m*M* sodium phosphate, pH 7.0, filter through a 0.22-mm filter, and then directly load onto a PAM column (1 mL bed volume) equilibrated at a flow rate of 1.0 mL/min with 100 m*M* sodium phosphate, pH 7.0, at room temperature.
2. Wash the column after sample loading with loading buffer to remove any unbound material.
3. Elute adsorbed immunoglobulins by a buffer change to 0.1 *M* acetic acid and immediately neutralize with 1.5 *M* Tris, pH 9.0. Each fraction should be checked for antibody reactivity by ELISA and for purity by SDS-PAGE electrophoresis. As in other cases, no traces of albumin are contaminating the purified IgE preparation. Immunoreactivity of IgE purified on PAM-columns can be determined by ELISA assay on polystyrene microtiter plates (*see* **Note 13**).

3.3.5. Affinity Purification of IgY on the PAM Emphaze Column

Y immunoglobulins from water-soluble yolk extract (*see* **Note 14**) can be isolated in a single-step protocol by affinity chromatography on a PAM column according to the following procedure:

1. Dialyze or dilute the sample 1:1 (v/v) with the starting buffer (25 m*M* *bis*-Tris, pH 6.5), filtered through a 0.45-mm filter (Nalgene).
2. Load the sample onto the column (1 mL bed volume) equilibrated at a flow rate of 1.0 mL/min with 25 m*M* *bis*-Tris, pH 6.5, monitoring the effluent by UV detection at 280 nm.
3. After elution of unbound material, change the eluent to 0.1 *M* acetic acid, pH 3.0, to elute bound material, and neutralize eluted IgY immediately with a few drops of 1 *M* Tris, pH 9.5.
4. Characterize collect fractions by SDS-PAGE, gel permeation, and radial immunodiffusion analysis to determine IgY recovery and purity, and by ELISA, to evaluate the immunoreactivity recovered after purification (*see* **Note 15**).

3.3.6. Sterilization of the PAM Matrix by Autoclaving

PAM resin can be sterilized by autoclaving according to the following procedure:

1. Centrifuge the resin to replace the storage buffer with 0.05 *M* Na phosphate, pH 7.0.
2. Autoclave the sample for up to 30 min at 120°C.
3. Repeat **step 1** to remove autoclaving buffer (*see* **Note 16**).
4. Store the matrix in 20% ethanol at 4°C.

4. Notes

1. PAM immobilization on preactivated solid supports occurs easily, with coupling yields generally between 80 and 95%. Recommended ligand density is between 6 and 10 mg/mL of support.
2. Optimal interaction of immunoglobulins to immobilized PAM occurs in the pH range 6.5–7.5. Compatible buffers are Tris, *bis*-Tris, and sodium phosphate. Phosphate-buffered saline (PBS) is not recommended because the high content of chloride ions interferes with binding. High salt concentrations reduce binding capacity.
3. The use of sodium phosphate as binding buffer, at concentrations from 100 to 200 m*M*, is suggested for samples containing high amounts of phospholipids.
4. Elution of adsorbed immunoglobulins can be performed using acetic acid or 0.1 *M* sodium bicarbonate, pH 8.5. Addition of sodium chloride to the elution buffer leads to recovery of antibodies in a more concentrated form.
5. PAM column sanitation is easily accomplished, as the ligand is stable to the vast majority of sanitizing agents and is not susceptible of denaturation.
6. Chemical stability of PAM is very high, and in the immobilized form is also sufficiently stable to enzymatic degradation. Columns can be reused for more than 40 purification cycles without appreciable loss of capacity.
7. Removal of adsorbed or precipitated proteins on the columns can be performed by repetitive washings with 0.1 *M* sodium hydroxide and 1 m*M* hydrochloridric acid. Check first supports compatibility with these eluents.
8. Binding affinity of PAM is higher for IgM than for IgG. Samples containing both immunoglobulins classes will be enriched in the IgM fractions.
9. Usually very high recovery (80%) is obtained. SDS-PAGE analysis of eluted fractions shows an excellent degree of purification, as no albumin traces are detected in the column-bound fraction, and all the material migrates at the expected molecular weight for IgM. Densitometric scanning of the purified fraction gel lane shows generally purity close to 95%. Column flowthrough material contains, on the other hand, the vast majority of albumin and the other contaminants. Extent of purification can be monitored also by gel filtration chromatography on calibrated columns. Gel filtration profiles of the affinity purified IgM validate SDS-PAGE data, indicating that a single affinity step on PAM columns is sufficient to remove albumin and capture and concentrate the IgM fraction. The effect of purification conditions on the maintenance of antibody antigen binding ability can be evaluated by ELISA assays on microtiter plates coated with the IgM corresponding antigen. For all cases tested, results indicate that the affinity fractionation step is mild and does not lead to loss of immunoreactivity, indicating that the purified antibody is fully active.
10. The vast majority of immunoreactivity (close to 80%) is generally found in the bound fraction, while only little activity is detected in the flow-through fraction. SDS-PAGE

analysis indicates that the column bound fraction contains mainly IgM (85% purity), with only trace amounts of IgG or other contaminating proteins. Only IgA are detected as minor contaminants, as this class of immunoglobulins, which is found in sera at very low concentrations, is also recognized by immobilized PAM. Immunoreactivity recovery of IgM from affinity purification can be checked using aliquots of crude material, unbound and bound fractions, directly coated on microtiter plates at the same concentration (10 μg/mL) in 0.1 *M* sodium carbonate buffer, pH 8.5, overnight at 4°C. After the plates are washed 5× with PBS, wells are then blocked with 100 μL of PBS containing 3% bovine serum albumin (BSA) for 2 h at room temperature, to prevent nonspecific adsorption of proteins. Plates are washed several times with PBS. IgM detection is performed by filling each well with 100 μL of an anti-IgM-peroxidase conjugate solution diluted 1:1000 with PBS containing 0.5% BSA, and incubating for 1 h at 37°C. Plates are then washed with PBS 5×, and developed with a chromogenic substrate solution consisting of 0.2 mg/mL ABTS in 0.1 *M* sodium citrate buffer, pH 5.0, containing 5 m*M* hydrogen peroxide. The absorbance at 405 nm of each sample is measured with a Model 2250 EIA Reader (Bio-Rad). If the antigen is available, recovery of immunoreactivity can be evaluated by immobilizing the antigen on microtiter plates, dissolved in 0.1 *M* sodium carbonate buffer, pH 8.5, overnight at 4°C. The plates are washed and saturated as described previously, and filled with crude, unbound and bound materials at the same concentration (10 μg/mL) diluted with PBS–0.5% BSA. The antibody detection and the development of the chromogenic reaction are then carried out as described in **Note 10**.

11. Determination by ELISA of IgA recovery indicates that the column retains 80% of the IgA immunoreactivity initially found in the sample. Gel electrophoretic analysis of the purified fraction indicates the absence of contaminating albumin; however, all the IgM originally present in the sample will be retained by the column. Detection of IgA immunoreactivity in the fractions derived from the affinity step can be accomplished by ELISA by immobilizing IgA containing samples on microtiter plates and detecting IgA with an anti-IgA antibody. SDS-PAGE analysis indicates that the majority of IgA in the sample is retained by the column, and only minute amounts of albumin are detected in the purified preparation. These results are confirmed by the gel filtration analysis, where the column bound fraction shows mainly the presence of IgA. ELISA determination of the IgA content of the column bound and unbound fractions after the purification step indicates that the majority (80–90%) of the initial immunoreactivity is retained by the column.
12. Aliquots of crude material, unbound and bound fractions (100 μL) are incubated on microtiter plates (Falcon 3912) in 0.1 *M* sodium carbonate buffer, pH 8.5, overnight at 4°C. After washing the plates 5× with PBS (50 m*M* phosphate, 150 m*M* sodium chloride), pH 7.5, plate wells are saturated with 100 μL of PBS containing 3% of BSA, for 2 h at room temperature, to prevent nonspecific protein adsorption. Plates are then washed with PBS several times. Detection of IgA antibody is performed by adding to each well 100 μL of an anti-IgA peroxidase conjugate solution (Sigma) diluted 1:1000 with PBS–0.5% BSA (PBS-B). The plates are incubated for 1 h at 37°C, washed with PBS-B containing 0.05% of Tween, then developed with a chromogenic substrate solution consisting of 0.2 mg/mL of ABTS in 0.1 *M* sodium citrate buffer, pH 5.0, containing 5 m*M* hydrogen peroxidase. The absorbance of each sample is measured with a Model 2250 EIA Reader (Bio-Rad).
13. Microtiter plates (Falcon 3912) are incubated with a 10 μg/mL solutions of crude sample, unbound and bound fractions (100 μL/well) in 0.1 *M* sodium carbonate buffer, pH 8.5, overnight at 4°C. After the plates are washed 5× with PBS (50 m*M* phosphate, 150 m*M* sodium chloride), pH 7.5, wells are saturated with 100 μL of a PBS solution containing 3% BSA, for 2 h at room temperature to block the uncoated plastic surface. The wells are

then washed with PBS and incubated with the biotinylated antigen (10 µg/mL) in PBS containing 0.5% BSA (PBS-B). After 1 h of incubation the plates are washed 5× with PBS containing 0.05% of Tween (PBS-T), then filled with 100 µL of a streptavidin peroxidase conjugate solution (Sigma) diluted 1:1000 with PBS–0.5% BSA. The plates are incubated for 1 h at 37°C, washed with PBS-T 5×, and then developed with a chromogenic substrate solution consisting of 0.2 mg/mL of ABTS in 0.1 *M* sodium citrate buffer, pH 5.0, containing 5 m*M* hydrogen peroxidase. The absorbance at 405 nm of each sample is measured with a Model 2250 EIA Reader (Bio-Rad). For antigen biotinylation, 2 mg of antigen, dissolved in 1 mL of 50 m*M* sodium phosphate buffer, pH 7.5, is treated with 200 µg of biotinamidocaproate *N*-hydroxysuccinimide ester dissolved in 20 µL of dimethyl sulfoxide (DMSO), under agitation at room temperature. After 2 h of incubation, 240 µL of a 1 *M* lysine solution is added to deactivate residual active groups, under stirring for 2 h. At the end the biotinylated antigen is extensively dialyzed against 50 m*M* sodium phosphate, pH 7.5, and used without any further treatment.

14. Water-soluble proteins are separated from the lipidic fraction of egg yolk by the water dilution method. Egg yolk is separated from the white, washed with distilled water to remove as much albumen as possible, diluted 1:9 with acidified distilled water, and incubated at least 6 h at 4°C. After incubation, sample is centrifuged at 10,000*g* for 20 min, the supernatant separated from pellet, filtered on 0.45-µm Nalgene filters, and loaded onto a PAM affinity column. Usually 10.4 mg of antibodies/mL of egg yolk are obtained with a purity close to 30%.
15. Radial immunodiffusion and SDS-PAGE analysis of eluted material show an excellent degree of purification in terms of recovery and purity, both very high and close to 98%. The gel filtration profile of purified IgY validates SDS-PAGE data, indicating that a single affinity step is sufficient to remove all contaminants and capture and concentrate the IgY fraction. The effect of purification conditions on the maintenance of antibody–antigen recognition can be evaluated by ELISA assays on microtiter plates coated with the IgY corresponding antigen. Results indicate that after the affinity fractionation step, antibodies were recovered fully active, with the majority of the immunoreactivity, about 99%, retrieved in the bound fraction.
16. Chemical and chromatographic stability of PAM resin after sterilization procedure can be tested by monitoring the release of the ligand from the resin by RP-HPLC analysis of the sterilizing buffer and measuring the Ig binding capacity of the matrix in column experiments. The investigations indicate that this treatment leads to significant loss of ligand (up to 5%), so that under process conditions the capacity of PAM columns to bind the immunoglobulins was quite reduced (up to 30%).

References

1. Fuglistaller, P. (1989) Comparison of immunoglobulin binding capacities and ligand leakage using eight different protein A affinity chromatography matrices. *J. Immunol. Meth.* **124,** 171–177.
2. Nevens, J. R., Mallia, A. K., Wendt, M. W., and Smith, P. K. (1992) Affinity chromatographic purification of immunoglobulin M antibodies utilizing immobilized mannan binding protein. *J. Chromatogr.* **597,** 247–256.
3. Tomasi, T. B. and Bienenstock, J. (1968) Secretory immunoglobulins. *Adv. Immunol.* **9,** 1.
4. Mestecky, J. R. and Kraus, F. W. (1971) Method of serum IgA isolation. *J. Immunol.* **107,** 605–607.
5. Waldam, R. H., Mach, J. P., Stella, M. M., and Rowe, D. S. (1970) Secretory IgA in human serum. *J. Immunol.* **105,** 43–47.

6. Khayam-Bashi, H., Blanken, R. M., and Schwartz, C. L. (1977) Chromatographic separation and purification of secretory IgA from human milk. *Prep. Biochem.* **7,** 225–241.
7. Roque-Barreira, M. R. and Campos-Nieto, A. (1985) Jacalin: an IgA-binding lectin. *J. Biol. Chem.* **134,** 1740–1743.
8. Kondoh, H., Kobayashi, K., and Hagiwara, K. (1987) A simple procedure for the isolation of human secretory IgA of IgA1 and IgA2 subclass by a jackfruit lectin, jacalin, affinity chromatography. *Mol. Immunol.* **24,** 1219–1222.
9. Bunn-Moreno, M. M. and Campos-Neto, A. (1981) Lectin(s) extracted from seeds of artocarpus integrifolia (jackfruit): potent and selective stimulator(s) of distinct human T and B cell functions. *J. Immunol.* **127,** 427–429.
10. Phillips, T. M., More, N. S., Queen, W. D., and Thompson, A. M. (1985) Isolation and quantification of serum IgE levels by high-performance immunoaffinity chromatography. *J. Chromatogr.* **327,** 205–211.
11. Lehrer, S. B. (1979) Isolation of IgE from normal mouse serum. *Immunology* **36,** 103–109.
12. Ikeyama, S., Nakagawa, S., Arakawa, M., Sugino, H., and Kakinuma, A. (1986) Purification and characterization of IgE produced by human myeloma cell line, U266. *Mol. Immunol.* **23,** 159–167.
13. Zola, H., Garland, L. G., Cox, H. C., and Adcock, J. J. (1978) Separation of IgE from IgG subclasses using staphylococcal protein A. *Int. Arch. Allergy Appl. Immunol.* **56,** 123–127.
14. Polson, A., von Wechmar, M. B., and van Regenmortel, M. H. V. (1980) Isolation of viral IgY antibodies fron yolks of immunized hens. *Immunol. Commun.* **9,** 475–493.
15. Gassmann, M., Thommes, P., Weiser, T., and Hubscher, U. (1990) Efficient production of chicken egg yolk antibodies against a conserved mammalian protein. *FASEB J.* **4,** 2528–2532.
16. Yang, J., Jin, Z., Yu, Q., Yang, T., Wang, H., and Liu, L. (1997) The selective recognition of antibody IgY for digestive cancers. *Chin. J. Biotechnol.* **13,** 85–90.
17. Terzolo, H. R., Sandoval, V. E., Caffer, M. I., Terragno, R., and Alcain, A. (1998) Agglutination of hen egg-yolk immunoglobulins (IgY) against *Salmonella enterica*, serovar enteritidis. *Rev. Argent. Microbiol.* **30,** 84–92.
18. Polson, A. (1990) Isolation of IgY from the yolks of eggs by a chloroform polyethylene glycol procedure. *Immunol. Invest.* **19,** 253–258.
19. Hansen, P., Scoble, J. A., Hanson, B., and Hoogenraad, N. J. (1998) Isolation and purification of immunoglobulins from chicken eggs using thiophilic interaction chromatography. *J. Immunol. Meth.* **215,** 1–7.
20. Hatta, H., Kim, M., and Yamamoto, T. (1990) A novel isolation method for hen yolk antibody, "IgY." *Agric. Biol. Chem.* **54,** 2531–2535.
21. Fassina, G., Verdoliva, A., Odierna, M. R., Ruvo, M., and Cassani, G. (1996) Protein A mimetic peptide ligand for affinity purification of antibodies. *J. Mol. Recogn.* **9,** 564.
22. Fassina, G., Verdoliva, A., Palombo, G., Ruvo, M., and Cassani, G. (1998) Immunoglobulin specificity of TG 19318: a novel synthetic ligand for antibody affinity purification. *J. Mol. Recogn.* **11,** 128–133.
23. Verdoliva, A., Basile, G., and Fassina, G. (2000) Affinity purification of immunoglobulins from chicken egg yolk using a new synthetic ligand. *J. Chromatogr. B.* **749,** 233–242.
24. Palombo, G., Verdoliva, A., and Fassina, G. (1998) Affinity purification of IgM using a novel synthetic ligand. *J. Chromatogr. B.* **715,** 137–145.
25. Palombo, G., De Falco, S., Tortora, M., Cassani, G., and Fassina, G. (1998) A synthetic ligand for IgA affinity purification. *J. Mol. Recogn.* **11,** 243–246.
26. Palombo, G., Rossi, M., Cassani, G., and Fassina, G. (1998) Affinity purification of mouse monoclonal IgE using a protein A mimetic ligand ***(TG 19318)*** immobilized on solid supports. *J. Mol. Recogn.* **11,** 247–249.

149

Detection of Serological Cross-Reactions by Western Cross-Blotting

Peter Hammerl, Arnulf Hartl, Johannes Freund, and Josef Thalhamer

1. Introduction

Antisera are frequently used tools for the characterization of proteins and peptides because they provide unique information about the structural features of antigens. Proteins that share structural similarities can be identified by serological cross-reactions. The Western blotting technique combines two steps that are characteristic for immunoaffinity chromatography *(1)*. One is the preparation of monospecific antibodies against a particular antigen and the second is the testing of their reactivity with the same or different antigens from any source of antigenic material.

As a rule, such experiments turn out to be elaborate and time-consuming procedures, consisting of many different steps; particularly because either a purified antigen or a monospecific antibody are basic requirements. In many cases, monospecific antibodies need to be purified from polyspecific antisera by immunoaffinity chromatography. However, only small amounts of monospecific antibody are required for analytical purposes. In such a case it may be sufficient to elute antibodies from selected bands off a Western blot. One protocol, for example, has been published by Beall and Mitchell *(2)*.

In contrast to experimental setups for the immunochemical analysis of one particular antigen, the method described in this chapter is especially designed for single-step analysis of cross-reactivities of multiple antigens, within the same or between different protein mixtures. The principle is to test antibodies that have bound to particular antigen bands of a Western blot against all antigen bands on a second blot. This is done by electrotransfer of antibodies from one Western blot to a second one, taking advantage of the dissociative effect of chaotropic ions on antigen–antibody complexes.

The strategy can be dissected into the following steps:

1. Two antigen mixtures are separated by sodium dodecyl sulfate-polyacrylamide gel electrophoresis (SDS-PAGE), each mixture on a separate gel. The samples are loaded onto the entire width of the gels. After electrophoresis, the proteins are blotted onto nitrocellulose (NC) paper.
2. One of the two blots (referred to as the "donor" blot in this text) is incubated with a polyspecific antiserum and then placed onto the second blot ("receptor" blot), upside

From: *The Protein Protocols Handbook, 2nd Edition*
Edited by: J. M. Walker © Humana Press Inc., Totowa, NJ

down, with the protein bands crossing the bands on the second blot. This assembly is wrapped in a dialysis membrane.
3. Antibodies are electrophoretically transferred from the donor blot to the receptor blot in the presence of NaSCN, which dissociates antigen–antibody complexes.
4. The donor blot is then discarded. The receptor blot, still wrapped in the dialysis membrane, is equilibrated with phosphate-buffered saline (PBS) to allow binding of the transferred antibodies to the protein bands.
5. Bound antibodies are detected by use of an enzyme-linked second antibody.

This method may be useful in a wide variety of serological studies. For example, it may help to identify structurally related molecules of different molecular weight within protein mixtures, for example, cellular extracts. It may provide information about the serological relationship of different antigen mixtures, and thereby help to investigate evolutionary distances. It may also find application in analyses of subunit composition of proteins and, in combination with peptide mapping, it can be a useful tool for epitope characterization.

2. Materials

2.1. SDS-PAGE

All solutions for SDS-PAGE should be prepared with chemicals of the highest purity available using double-distilled water.

1. Solution A (acrylamide): 29% (w/v) acrylamide, 1% (w/v) *bis*-acrylamide; store at 4°C, light sensitive, stable for approx 4 wk.
2. Solution B (separating gel buffer): 1.5 *M* Tris-HCl, pH 8.8; store at 4°C.
3. Solution C (stacking gel buffer): 0.5 *M* Tris-HCl, pH 6.8; store at 4°C.
4. SDS: 10% (w/v); store at room temperature.
5. Sample buffer: 62.5 m*M* Tris-HCl, pH 6.8, 20% (v/v) glycerol, 4% SDS, 1% (v/v) of a saturated aqueous solution of Bromophenol blue. Stored aliquots of 1 mL at –20°C may be thawed and frozen repeatedly.
6. Electrophoresis buffer: 25 m*M* Tris, 192 m*M* glycine, 0.1% (w/v) SDS. Do not adjust pH.
7. Dithiothreitol (DTT): 1 *M* Dithiotreitol. Stored aliquots of 1 mL at –20°C may be thawed and frozen repeatedly.
8. Iodoacetamide (IAA): 1 *M* IAA. Stored aliquots of 1 mL at –20°C may be thawed and frozen repeatedly.
9. *N,N,N',N'*-Tetramethylethylendiamine (TEMED): store at 4°C.
10. Ammonium persulfate (APS): 10% (w/v) APS. Store aliquots of 0.1 mL at –20°C.
11. *n*-Butanol (water saturated): Mix equal volumes of *n*-butanol and water and shake well. After separation of the two phases, the upper one is water-saturated butanol. Store at room temperature.

2.2. Electrotransfer

All buffers for this protocol contain methanol. Caution: methanol is toxic and volatile, and should be handled in a fume hood. Wear gloves and protective clothing and do not leave bottles open at the workbench.

1. Buffer A: 0.3 *M* Tris, 20% (v/v) methanol. Do not adjust pH. Store at 4°C.
2. Buffer B: 25 m*M* Tris, 20 % (v/v) methanol. Do not adjust pH. Store at 4°C.
3. Buffer C: 40 m*M* Aminocaproic acid, 25 m*M* Tris, 20% (v/v) methanol. Do not adjust pH. Store at 4°C.

2.3. Cross-Blot

1. Borate-Tween (BT): 50 m*M* $Na_2B_4O_7$, pH 9.3, 0.1 % (v/v) Tween 20. Store at 4°C.
2. SCN-borate-Tween (SBT): 1 *M* NaSCN, 50 m*M* $Na_2B_4O_7$, pH 9.3, 0.1 % (v/v) Tween 20. Store at 4°C.

2.4. Immunostaining

1. PBS: 8.1 m*M* Na_2HPO_4, 1.5 m*M* KH_2PO_4, 2.7 m*M* KCl, 140 m*M* NaCl. Store at 4°C.
2. PBS-Tween: 0.1% (v/v) Tween 20 in PBS. Store at 4°C.
3. PTS: 8.1 m*M* Na_2HPO_4, 1.5 m*M* KH_2PO_4, 2.7 m*M* KCl, 0.5 *M* NaCl, 0.1% (v/v) Tween 20. Store at 4°C.
4. 10 mg/mL Bovine serum albumin (BSA): Store aliquots of 10 mL at –20°C.
5. NC paper.
6. Horseradish peroxidase conjugated antibody with specificity for the immunoglobulin isotype of the test antiserum.
7. Chloronaphthol, solid. Store at –20°C. Caution: irritant to skin, eyes, and respiratory organs. Wear gloves, protective glasses, and clothing, especially when handling the solid substance (buffy crystals). May be purchased as tablets, which can be handled with lower risk.
8. 30% Hydrogen peroxide: Store at 4°C. Caution: strong oxidant. Corrosive. Avoid contact with eyes and skin.

2.5. Miscellaneous

1. Electrophoresis apparatus for SDS-PAGE: Mini-Protean (Bio-Rad, Richmond, CA) or equivalent.
2. Gradient mixer for 2 × 5 mL.
3. Semidry blotting apparatus.

3. Methods

Most conveniently, this protocol may be carried out according to the following time schedule:

Day 1: Casting two SDS gels.
Day 2: Casting the stacking gels.
Running SDS-PAGE.
Blotting of the gels onto NC.
Incubating one blot and of reference strips from both blots with antiserum.
Cross-blotting.
Reequilibration overnight.
Day 3: Incubation with second antibody.
Detection of spots by enzyme reaction.

This schedule proved convenient in our laboratory. Other schedules are possible, as time values for antibody incubations and washing steps given are minimum requirements from our experience and may be different with other materials.

3.1. SDS-PAGE

3.1.1. Casting Gradient Gels

Volumes are given for the Mini-Protean gel system (Bio-Rad) for gels of 83 × 55 × 1.5 mm.

1. Clean glass plates with detergent, rinse thoroughly with tap water followed by double-distilled (dd-) water, and let dry.
2. Assemble glass plates and prepare two gel solutions as follows:
   5% Acrylamide: 1.65 mL of solution A;
   2.50 mL of solution B;
   5.73 mL of dd-water;
   20% Acrylamide: 6.65 mL of solution A;
   2.50 mL of solution B;
   0.73 mL of dd-water.

   Degas under vacuum (e.g., in a sidearm flask) and add 100 µL of SDS.
3. Close the outlet valve and the valve connecting the two chambers of the gradient mixer. Pipet 3.6 mL of the 20% mix into the chamber of the gradient mixer that is connected to the outlet tubing, and then 3.6 mL of the 5% mix into the second chamber. To each chamber, add 1.8 µL of TEMED and 9 µL of APS and mix well.
4. Open the valves and cast the gel at a flow rate of not more than 2.5 mL/min. Control the flow rate either by hydrostatic pressure or by use of a peristaltic pump.
5. Immediately thereafter, rinse the gradient mixer with water and prepare for casting the second gel. Carefully overlay the casted gels with 1 mL of water-saturated butanol and polymerize for at least 3 h or overnight at room temperature. Do not move the gels while still fluid.

3.1.2. Sample Preparation

For gels 1.5 mm thick, approx 10–50 µg of protein in a volume of 20–50 µL/cm of gel width may be loaded. However, the optimal protein concentration may vary with the complexity of the sample.

1. Mix equal volumes of protein solution and sample buffer.
2. Add 1/20 volume of 1 *M* DTT and boil in a water bath for 5 min.
3. Alkylate free sulfhydryl residues on the proteins by adding 1/5 volume of 1 *M* IAA and incubating at 37°C for 30–60 min (*see* **Note 1**).

3.1.3. Casting Stacking Gels

1. Aspirate butanol from the polymerized gel and rinse the gel surface 3× with water. Let gels dry in an inverted position for 5–10 min (*see* **Notes 2** and **3**).
2. Meanwhile, prepare the stacking gel mix as follows:
   Stacking gel solution: 0.65 mL of solution A;
   1.25 mL of solution C;
   5.0 mL of dd-water.
3. Degas as described in **Subheading 3.1.** and add 50 µL of SDS, 50 µL of APS, and 5 µL of TEMED. Mix carefully and apply 1.3 mL onto each gel. Make sure there is 5 mm remaining from the stacking gel surface to the top of the glass plates for sample application. Overlay with water saturated butanol and polymerize for 10 min (*see* **Note 4**).

3.1.4. Running Electrophoresis

1. Assemble electrophoresis apparatus and fill with electrophoresis buffer.
2. Load protein samples onto gels and run at constant current, starting with 75 V for the Bio-Rad Mini-Protean system. Stop electrophoresis when bromophenol blue tracer dye has reached the bottom of the gel (*see* **Note 5**).

3.2. Electrotransfer onto NC

1. For each gel, cut one sheet of NC paper and 15 pieces of Whatman 3MM paper to fit the dimensions of the separating gel (*see* **Note 6**).
2. For each gel, prewet six sheets of the Whatman paper in buffer A, another six sheets in buffer C, and three sheets in buffer B. Prewet the single sheet of NC paper in buffer B.
3. Disassemble the electrophoresis unit, discard the stacking gel, and mount the blot in the following order, making sure that no air bubbles are trapped between individual stacks:
   a. Place the Whatman stack from buffer A onto the anode plate of the semidry blotting apparatus.
   b. Place the Whatman stack from buffer B on top and cover it with the NC sheet.
   c. Place the gel onto the NC paper, then cover it with the Whatman stack from buffer C.
4. Close the apparatus by mounting the cathode plate on top of the assembly and perform electrophoresis at a constant current of 1.4 mA/cm^2 for 2 h.

3.3. Blocking and Incubation of the Blots with Antiserum

All incubations are done on a laboratory shaker, either at room temperature for the time indicated or at 4°C overnight.

1. Disassemble blotting apparatus and incubate the blot, which is intended as the antibody source (i.e., the "donor" blot) in PBS-Tween at room temperature for 30 min. Incubate the second blot ("receptor" blot) in 1% BSA in PBS for 2 h.
2. Incubate the donor blot with antiserum in PBS-Tween with 1 mg/mL of BSA (*see* **Note 7**).
3. From the central area of both blots, cut square-shaped pieces fitted to the dimension of the separating gel. Make asymmetrical marks on edges of the square-shaped NC pieces to help remember the orientation of the antigen bands. Set aside the remaining pieces of the donor blot in PBS-Tween. Incubate the remaining pieces of the receptor blot with antiserum in PBS-Tween with 1 mg/mL of BSA. These margin pieces will serve as reference strips to identify spots on the cross blot.
4. Place donor blot onto a glass filter and wash with 20 mL of PTS in aliquots of approx 3 mL under vacuum using a sidearm flask. Incubate donor blot in PTS for 10 min on a laboratory shaker.
5. Repeat **step 4** twice (*see* **Note 8**).

3.4. Cross Blot

3.4.1. Electrotransfer of Antibodies

1. Prewet one stack of Whatman paper, 1 cm thick and sufficiently sized to cover the donor blot, in BT and a second, equally sized, stack in SBT.
2. Prewet a piece of dialysis membrane (with a pore size of approx 10 kDa) in BT, which is sufficiently sized to wrap the donor blot (i.e., at least twice the area of the donor blot).
3. Assemble the cross-blot as follows (**Fig. 1**):
   a. Place the Whatman stack soaked in BT onto the anode of the semidry blotting apparatus.
   b. Place the dialysis membrane on top, in such a way that one half covers the Whatman stack while the second half rests beside it on the anode plate.
   c. Place the receptor blot, with the protein bands upside, onto the part of the dialysis membrane, which covers the anodal Whatman stack.

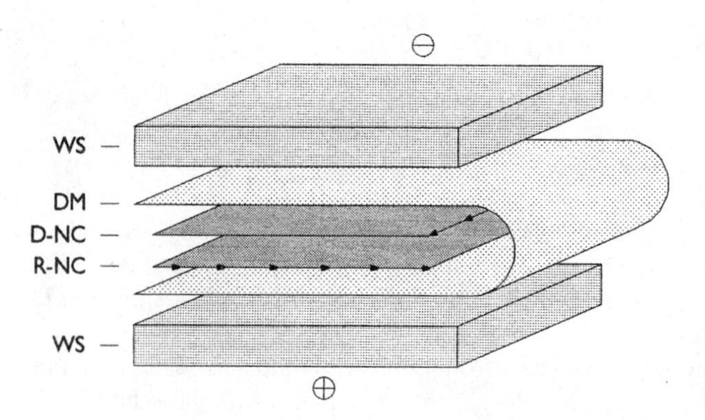

Fig. 1. Experimental setup for a cross-blot experiment: + and – indicate the position of the graphite electrodes of the semidry blotting apparatus. WS, stack of Whatman paper; DM, dialysis membrane; D-NC, donor blot NC sheet; R-NC, receptor blot NC sheet. The perpendicular orientation of the antigens bands on the blots to each other is indicated by *arrows*. The protein bands on the donor blot are faced downwards against the receptor blot, while the bands on the receptor blot are faced upwards against the donor blot.

   d. Place the donor blot, with the protein bands downward, onto the receptor blot in such a way that the bands on the donor blot are perpendicular to the bands on the receptor blot (*see* **Note 9**).
   e. Cover the donor blot with the dialysis membrane.
   f. Place the Whatman stack soaked with SBT onto the assembly and cover it with the cathodal plate of the semidry apparatus.
4. Perform electrophoresis at a constant current of 3 mA/cm^2 for 90 min (*see* **Note 10**).

3.4.2. Reequilibration of the Receptor Blot

1. Presoak two Whatman stacks of the same size as in **Subheading 3.4.1., step 1** in PTS.
2. **Important note:** During the following step, take extreme care not to slip the receptor blot against dialysis membrane! (*See* **Note 11**.)
   a. Disassemble the semidry apparatus.
   b. Remove the cathodal Whatman stack.
   c. Open the upper part of the dialysis membrane using a pair of forceps.
   d. Remove the donor blot and place into PBS-Tween.
   e. Close the dialysis membrane so that the receptor blot is completely wrapped in it.
   f. Place the receptor blot, wrapped in the dialysis membrane, between two Whatman stacks prewetted in PTS.
   g. Place this assembly into an appropriately sized tray and cover it with a glass plate and a weight of approx 50 g to stabilize it.
   h. Fill tray with additional PTS and incubate overnight without agitation.

3.4.3. Immunostaining

1. Remove the receptor blot from “reequilibration” assembly and wash all NC pieces (i.e., donor blot, receptor blot, and reference strips) in PBS-Tween 3 × 10 min on a laboratory shaker (*see* **Note 12**).
2. Incubate with enzyme-linked second antibody according to the manufacturer’s instructions (*see* data sheet) in PBS-Tween with 1 mg/mL of BSA.
3. Wash blots as described in **step 1**.

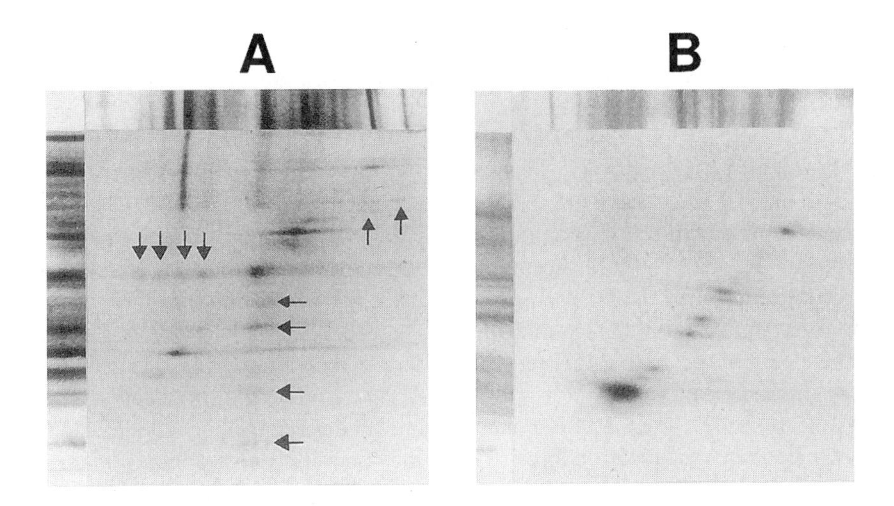

Fig. 2. Two examples of a cross-blot experiment carried out with cytoplasmic extracts of *E. coli* (taken from **ref. *1***). To the left of each cross-blot, a reference strip derived from the receptor blot and, on top of each cross-blot, a reference strip derived from the donor blot is included. **(A)** Homologous cross-blot of *E. coli* B wild-type strain antigens and a rabbit anti-*E. coli* antiserum showing numerous cross-reactions between proteins of different molecular weight (*arrows*). **(B)** Homologous cross-blot of *E. coli* CSH 57B antigens. The diagonal line of spots indicates a high specificity of the cross-reactions between proteins of the same molecular weight. No cross-reactions between proteins of different molecular weight could be detected in this experiment.

4. Wash twice for 5 min in PBS.
5. Meanwhile, prepare the staining solution as follows:
   a. Solubilize approx 10 mg of chloronaphthol in 1–3 mL of ethanol and mix with 50 mL of PBS under vigorous agitation.
   b. Incubate the solution for 5 min, filter through Whatman 3MM (or equivalent), and add 25 µL of 30% hydrogen peroxide.
6. Incubate NC papers in separate trays in staining solution until bands and spots become visible.
7. Stop the staining reaction by washing the blots in water and dry.
8. For documentation and interpretation of results, place the receptor blot and the reference strips together as shown in **Fig. 2** and photograph (*see* **Note 13**).

4. Notes

1. Samples for SDS-PAGE may be prepared in different ways. Depending on your requirements and intentions, the reduction of inter- and intramolecular disulfide bonds by DTT and alkylation of the resulting sulfhydryl groups may be omitted. However, be aware that remaining APS in the polymerized gel causes an oxidizing environment. This may result in protein oligomerization due to the formation of disulfide bonds between free SH- residues on individual molecules. In many cases, this may be the reason for irreproducible results in SDS-PAGE. For nonreduced samples, it may be beneficial to omit boiling of the sample. In this case, incubate at 37°C for 30 min before applying the sample onto gel.
2. Many types of electrophoresis equipment contain parts made of acrylic glass or equivalent. Avoid butanol or any other organic solvent coming into contact with such parts. This

can "corrode" the plastic surface. As a consequence, the transparency of these parts may be lost with time, which can cause difficulties in handling, (e.g., sample application).

3. While washing the gel surface, do not leave water on the gel for too long a time, as this would dilute the buffer and SDS concentration in the upper part of the gel.
4. The protocol for SDS-PAGE described here uses the discontinuous buffer system of Laemmli and colleagues *(3)*. For the band sharpening effect of this system a sharp increase in pH and buffer concentration between stacking and separating gel is essential. As soon as the stacking gel is casted, however, the two different buffer systems start to diffuse into each other, and consequently interfere with the beneficial effect of the system. Therefore, it is essential to minimize the time between casting the stacking gel and starting the electrophoresis. For this reason, we do not recommend casting the stacking gels before the samples are ready for application. The polymerization of the stacking gel is complete as soon as a second fluid phase is visible between the gel and the butanol phase.
5. For repeated runs, the electrophoresis buffer may be reused several times. After each run, mix the cathodal and the anodal buffers to restore the pH value. Reuse this buffer only as the anodal buffer in the following runs. For the cathode, use fresh buffer in every run. Reused buffer contains chloride ions from the separating gel of the previous run, which would eliminate the effect of the discontinuous buffer system.
6. Other blotting systems (e.g., in a tank blot module in 10 m*M* Na_2CO_3 or other buffer systems) are also suitable for this purpose. From our experience, the best results are obtained by semidry blotting, especially with respect to homogeneous transfer efficiency.
7. The working dilution of antiserum depends on parameters such as antibody titer and affinity, and therefore varies with the material used. For this reason, no suggestions on antiserum dilution and time of incubation are given in this protocol. It should be adjusted according to the experience of the investigator with the material at hand. However, conditions that ensure a maximum of specificity are a crucial requirement, especially for the cross-blotting method. Therefore, do not use too high a concentration of antiserum and add BSA at 1 mg/mL to minimize unspecific protein–protein interactions. If problems occur, increasing the ionic strength to 0.5 *M* NaCl may be helpful.
8. A rigorous washing procedure was included in the protocol to optimize the washing efficiency. Possibly, this is not necessary for all applications, depending on the materials used. Keep in mind that any antibody, which remains unspecifically attached to the donor blot, will increase background and unspecific signals on the receptor blot.
9. Make sure that the donor and receptor blot are cut of exactly equal size and placed onto each other accurately. This will facilitate the identification of individual spots in the final result. It may be helpful to mark the position of two NC sheets by penetrating them with a needle at three sites after they have been placed together.
10. The relatively long electrophoresis time used in the cross-blot step is intended to maximize the efficiency of antibody elution from the donor blot. However, some antibody species may not survive such a long exposure to 1 *M* NaSCN. This could be indicated by increased unspecific signals or low sensitivity in the final results. In such a case, try a shorter electrophoresis time or a lower concentration of NaSCN in the cathodal Whatman stack. Generally, it may be advantageous to minimize the electrophoresis time in this step so as not to exceed the buffer capacity in the Whatman stacks. This would cause a change in pH, which might affect the direction of antibody migration.
11. This is the most crucial step of the entire procedure. Right after the electrotransfer, the antibodies are not yet bound to the antigens on the receptor blot due to the presence of NaSCN. Instead, they are most likely trapped in the pores of the NC sheet and loosely attached on the surface of the dialysis membrane. Therefore, any movement of the recep-

tor sheet against the anodal surface of the dialysis membrane will result in an extreme decrease of the spot sharpness. The donor blot is removed in this step so that it does not compete with the antigens on the receptor blot during the subsequent reequilibration step. Given a sufficient sensitivity according to our experience with previous experiments, the removal of the donor blot may be omitted.

12. Alternate staining procedures are also possible, for example, second antibody conjugated to alkaline phosphatase and staining with nitroblue tetrazolium (NBT) and bromo-chloroindolyl phosphate (BCIP).
    As the staining intensity on the receptor blot is significantly lower than that on the reference strips, one may wish to employ a more sensitive detection system such as luminogenic enzyme substrates. However, some of these systems require the use of special membranes. As of now we have no experience to indicate that these membranes are also compatible with the cross blot procedure.
13. Interpretation of results: Only positive signals should be taken as a result. The failure of a band on the donor blot to react with a band on the receptor blot does not necessarily prove the absence of cross-reactivity. Such negative results could also be due to inefficient elution of an antibody from the donor blot or to denaturation of the eluted antibody. Likewise, the concentration of the antigen and/or antibody on the donor blot may be too low to give a signal on the receptor blot. As a rule, do not necessarily expect a signal on the cross-blot from donor blot bands which give only faint signals on the donor blot reference strips.
    Generally, it is sometimes difficult to judge the specificity of an antigen–antibody reaction. This is more likely for antibodies, which have been exposed to low pH values or chaotropic agents. With cross-blot results, be doubtful about donor blot bands, which give a signal with each band on the receptor blot. Also, receptor blot bands that react with each donor blot band should generally not be taken too seriously, unless special circumstances let you expect such a behavior.
    When working with different antigen mixtures on donor and receptor blots, respectively, include a homologous cross-blot as a control and reference. This is done by cross-blotting antibodies from a donor blot to a receptor blot containing the same antigens as the donor blot. Such homologous cross-blots have a "natural" internal reference because each antigen band on the donor blot is crossing "itself" on the receptor blot. In this way, the cross-blot yields a diagonal line of spots that are helpful in estimating both the sensitivity and the specificity you can expect with a particular band on the donor blot. Similarly, a homologous cross-blot may also be carried out with the receptor antigens. This will help to estimate the resistance of the receptor antigens against exposure to the cross-blot conditions.

References

1. Hammerl, P., Hartl, A., and Thalhamer, J. (1992) A method for the detection of serologically crossreacting antigens both within and between protein mixtures: the Western cross blot. *J. Immunol. Meth.* **151,** 299–306.
2. Beall, J. A. and Mitchell, J. F. (1986) Identification of a particular antigen from a parasite cDNA library using antibodies affinity purified from selected portions of Western blots. *J. Immunol. Meth.* **86,** 217–223.
3. Laemmli, E. K. (1970) Cleavage of structural proteins during the assembly of the head of bacteriophage T4. *Nature* **227,** 680–685.

150

Bacterial Expression, Purification, and Characterization of Single-Chain Antibodies

Sergey M. Kipriyanov

1. Introduction

In the past few years, some of the limitations of monoclonal antibodies as therapeutic and diagnostic agents have been addressed by genetic engineering. Such an approach is particularly suitable because of the domain structure of the antibody molecule, where functional domains carrying antigen-binding activities (Fabs or Fvs) or effector functions (Fc) can be exchanged between antibodies (**Fig. 1A**). Smaller antibody-derived molecules include enzymatically produced 50-kDa Fab fragments and engineered 25-kDa single-chain Fv (scFv) consisting of the heavy and light chain variable regions (V_H and V_L) connected by a flexible 14–24 amino acid long peptide linker *(1,2)* (**Fig. 1B**). Compared to IgG molecules, scFv exhibit significantly improved tumor specificity and intratumoral penetration *(3–5)*. However, the rapid blood clearance and monovalent nature of scFv fragments result in considerably lower quantitative tumor retention of these molecules *(3,6)*.

Recently, attention has focused on the generation of antibody molecules with molecular weights in the range of the renal threshold for the first-pass clearance. Construction of such molecules can be achieved by making single-chain constructs comprising four immunoglobulin variable (V_H and V_L) domains. Depending on the linker length and domain order the bivalent $(scFv)_2$ molecules *(7,8)* or single-chain diabody (scDb) *(9,10)* can be formed (**Fig. 1B**). These four-domain formats are also well suited for making bivalent bispecific molecules.

Unlike glycosylated whole antibodies, single-chain antibodies (SCAs) can be easily produced in bacterial cells as functional antigen binding molecules. There are two basic strategies to obtain recombinant antibody fragments from *Escherichia coli*. The first is to produce antibody proteins as cytoplasmic inclusion bodies followed by refolding in vitro. In this case the protein is expressed without a signal sequence under a strong promoter. The inclusion bodies contain the recombinant protein in a nonnative and nonactive conformation. To obtain functional antibody, the recombinant polypeptide chains have to be dissolved and folded into the right shape by using a laborious and time-consuming refolding procedure (for review *see [11]*). The second approach for

From: *The Protein Protocols Handbook, 2nd Edition*
Edited by: J. M. Walker © Humana Press Inc., Totowa, NJ

A Immunoglobulin G

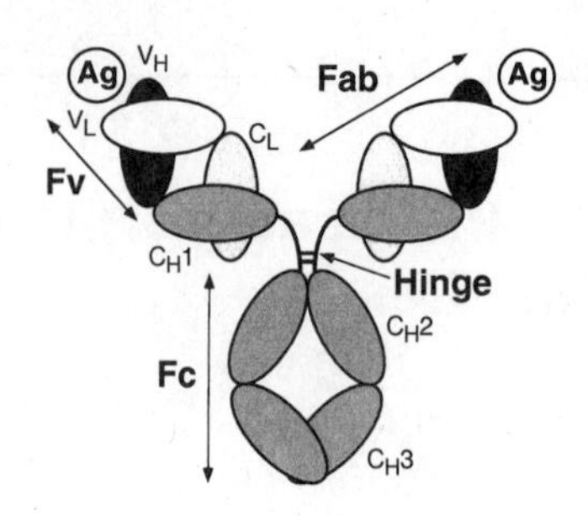

B Single-chain antibody fragments

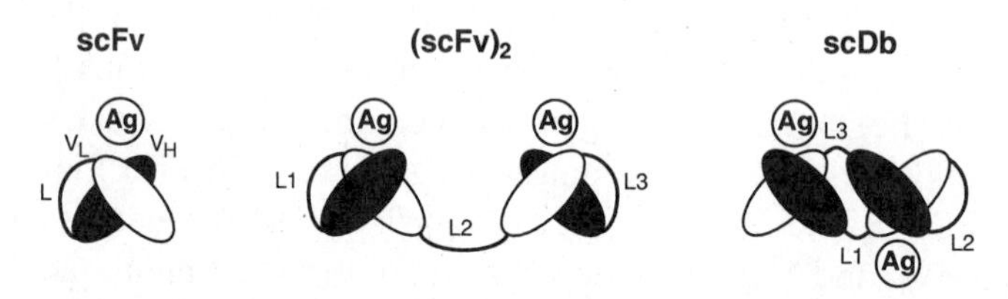

Fig. 1. Schematic representation of the domain structure of an immunoglobulin G (**A**) and recombinant single-chain antibody fragments (**B**). Antibody variable domains (V_H, V_L), peptide linkers (L), and antigen binding sites (Ag) of Fv modules are indicated.

obtaining functional antibody fragments is to imitate the situation in the eukaryotic cell for secreting a correctly folded antibody. In *E. coli*, the secretion machinery directs proteins carrying a specific signal sequence to the periplasm ***(12)***. The SCA fragments are usually correctly processed in the periplasm, contain intramolecular disulfide bonds, and are soluble. However, the high-level expression of a recombinant protein with a bacterial signal peptide in *E. coli* often results in the accumulation of insoluble antibody fragments after transport to the periplasm ***(13,14)***.

It is now recognized that aggregation in vivo is not a function of the solubility and stability of the native state of the protein, but of those of its folding intermediates in their particular environment ***(15,16)***. The degree of successful folding of antibody fragments in the bacterial periplasm appears to depend to a large extent on the primary sequence of the variable domains ***(17,18)***. The overexpression of some enzymes of the *E. coli* folding machinery such as cytoplasmic chaperonins GroES/L, periplasmic disulfide-isomerase DSbA, as well as periplasmic peptidylprolyl *cis,trans*-isomerases (PPIase) PpiA and SurA did not increase the yield of soluble antibody fragments ***(19–21)***. In contrast, the coexpression of either bacterial periplasmic protein Skp/OmpH or PPIase FkpA increased the functional yield of both phage-displayed and secreted scFv fragments ***(21,22)***. Modifications in bacterial growth and induction conditions can also increase the proportion of correctly folded soluble SCA. For example, lowering the bacterial growth temperature has been shown to decrease periplasmic aggregation and increase the yield of soluble antibody protein ***(15,23)***. In addition, the aggregation of recombinant antibody fragments in the *E. coli* periplasm can be reduced by growing the induced cells under osmotic stress in the presence of certain nonmetabolized additives such as sucrose ***(24,25)*** or sorbitol and glycine betaine ***(9)***. Moreover, inducing the synthesis of recombinant antibody fragments in bacteria under osmotic stress promotes the formation of domain-swapped SCA dimers ***(9)***.

SCA fragments produced in bacteria also provide new possibilities for protein purification by immunoaffinity chromatography. Their advantages include lower production costs, higher capacity for antigen on a weight basis, and better penetration in a small-pore separation matrix. Such recombinant immunosorbent proved to be useful for the one-step purification of a desired antigen from complex protein mixtures ***(26,27)***. Another interesting possible application is the purification or separation of toxic compounds, that cannot be used for immunization of animals, using antibodies selected from phage displayed antibody libraries.

In this chapter, generation of genetic constructs coding for SCA and the protocols for growing and inducing bacterial cells with or without nonmetabolized additives as well as for the purification of active SCA fragments from soluble periplasmic extracts are described. The purification scheme includes immobilized metal affinity chromatography (IMAC) as the main step for separating recombinant antibodies from bacterial proteins. In contrast to methods based on antigen affinity chromatography, it does not depend on the specificity of the particular SCA. The described procedure is useful for any SCA fragment that is secreted into the periplasm and has six histidine residues as a C-terminal tag. If the His-tagged protein is highly overexpressed in *E. coli*, a one-step IMAC purification can result in sufficiently pure material for most applications ***(28,29)***. However, if the protein of interest is present only as a small fraction, several contaminating bacterial proteins can bind to the IMAC column under the purification conditions and coelute (for a list of histidine-rich *E. coli* proteins *see* **ref. *30***). For further purification of antibody fragments from IMAC-eluted material, either antigen-affinity chromatography ***(14)***, thiophilic adsorption chromatography ***(31,32)***, or immunoaffinity purification using immobilized anti-His-tag monoclonal antibodies ***(30)*** has been used. Here you find a description of a relatively simple alternative procedure based on the separation of proteins by ion-exchange chromatography. This purification technique was tested for a number of SCA fragments and seems to be generally applicable.

2. Materials

2.1. Gene Assembly by PCR

1. Thermocycler PTC 150-16 (MJ Research, Watertown, MA).
2. *Vent* DNA polymerase (New England Biolabs, Beverly, MA).
3. 10× *Vent* buffer (New England Biolabs).
4. Sterile deionized water.
5. 100 m*M* dNTPs (New England Biolabs).
6. Bovine serum albumin (BSA), nonacetylated (10 mg/mL) (New England Biolabs).
7. Lambda *Bst*EII DNA molecular weight marker (New England Biolabs).
8. Agarose (FMC BioProducts, Rockland, ME).

9. 1× Tris-acetate electrophoresis buffer (1× TAE buffer): Prepare a stock solution of 50× TAE and dilute it 1:50 with water before use.
10. 50× TAE buffer: Dissolve 242 g of Trizma Base (Sigma-Aldrich Chemie GmbH, Steinheim, Germany) in distilled water. Add 57.1 mL of glacial acetic acid, 100 mL of 0.5 *M* EDTA, and water to a total volume of 1 L.
11. 0.5 *M* EDTA.
12. Ethidium bromide (10 mg/mL) (Merck, Darmstadt, Germany).

2.2. Cloning into the Expression Vector

1. Tabletop microcentrifuge.
2. QIAquick Gel Extraction Kit (Qiagen, Hilden, Germany).
3. QIAquick-spin PCR Purification Kit (Qiagen).
4. *Afl*II, *Bgl*II, *Eco*RI, *Eco*RV, *Hind*III, *Nco*I, *Nde*I, *Pvu*II, and *Xba*I restriction endonucleases (New England Biolabs).
5. 10× Restriction enzyme buffers (New England Biolabs).
6. Calf intestine alkaline phosphatase (CIP, New England Biolabs).
7. T4 DNA ligase (Stratagene).
8. 10× T4 DNA ligase buffer (Stratagene).
9. 3 *M* sodium acetate, pH 4.8.
10. Glycogen from mussels, molecular biology grade (20 µg/mL) (Boehringer GmbH).
11. Absolute ethanol.
12. 80% (v/v) ethanol.

2.3. Preparation of Bacterial Culture

1. 85-mm Petri dishes (Greiner, Frickenhausen, Germany).
2. 100-, 1000-, and 5000-mL Sterile glass Erlenmeyer flasks.
3. Thermostatic shaker (Infors GmbH, Einsbach, Germany).
4. Sorvall centrifuge with a set of fixed-angle rotors (Kendro Laboratory Products GmbH, Hanau, Germany).
5. Either *E. coli* K12 XL1-Blue (Stratagene, La Jolla, CA) or RV308 (Δ*lac*χ74*gal*ISII::OP308*strA*) *(33)* competent cells (*see* **Note 1**).
6. 2YT medium: 1 L contains 16 g of Bacto-Tryptone, 10 g of Bacto-Yeast extract, 5 g of NaCl, pH 7.5 (*see* **Note 2**).
7. $2YT_{GA}$: 2YT medium containing 0.1 g/L of ampicillin and 2% (w/v) glucose.
8. $2YT_{GA}$ agar plates. Media and agar plates are prepared according to standard protocols as described in **ref. *34***.
9. $2YT_{SA}$: 2YT medium containing 0.1 g/L of ampicillin and 0.4 *M* sucrose (*see* **Note 3**).
10. YTBS: 2YT medium containing 0.1 g/L of ampicillin, 1 *M* sorbitol, and 2.5 m*M* glycine betaine.
11. 100 m*M* solution of isopropyl-β-D-thiogalactopyranoside (IPTG). Store at –20°C.

2.4. Isolation of Recombinant Product from Soluble Periplasmic Fraction and Culture Medium

1. All-glass bacteria filter of pore size 10–16 µm (porosity 4) (Schott Glaswerke, Mainz, Germany).
2. Membrex TF filters of pore size 0.2 µm (MembraPure, Lörzweiler, Germany).
3. Amicon high-performance stirred ultrafiltration cell (Millipore GmbH, Eschborn, Germany).
4. Amicon YM10 membranes with a 10-kDa cutoff (Millipore).
5. Ammonium sulfate powder.

6. Magnetic stirrer.
7. Molecularporous dialysis tubes with a 12–14 kDa cutoff (Spectrum Laboratories Inc., Rancho Dominguez, CA).
8. 50 m*M* Tris-HCl, 20% sucrose, 1 m*M* EDTA, pH 8.0.
9. 50 m*M* Tris-HCl, 1 *M* NaCl, pH 7.0.
10. 50 m*M* Tris-HCl, 1 *M* NaCl, 50 m*M* imidazole, pH 7.0.
11. 50 m*M* Tris-HCl, 1 *M* NaCl, 250 m*M* imidazole, pH 7.0.
12. C16/20 column (Amersham Pharmacia Biotech, Freiburg, Germany).
13. Chelating Sepharose Fast Flow (Amersham Pharmacia Biotech).
14. 0.1 *M* $CuSO_4$.

2.5. Purification of SCA Fragments and Analysis of Molecular Forms

1. Mono S HR5/5 column (Amersham Pharmacia Biotech).
2. Mono Q HR5/5 column (Amersham Pharmacia Biotech).
3. Superdex 75 or Superdex 200 HR10/30 column (Amersham Pharmacia Biotech).
4. 50 m*M* Imidazole-HCl, pH 6.4. Filter (0.2 μm) and store at 4°C.
5. 50 m*M* Imidazole-HCl, 1 *M* NaCl, pH 6.4. Filter (0.2 μm) and store at 4°C.
6. 20 m*M* Tris-HCl, pH 8.0. Filter (0.2 μm) and store at 4°C.
7. 20 m*M* Tris-HCl, 1 *M* NaCl, pH 8.0. Filter (0.2 μm) and store at 4°C.
8. Phosphate-buffered saline (PBS): 15 m*M* sodium phosphate, 0.15 *M* NaCl, pH 7.4. Filter (0.2 μm) and store at 4°C.
9. PBSI: PBS containing 50 m*M* imidazole, pH 7.4. Filter (0.2 μm) and store at 4°C.
10. Biomax-10 Ultrafree-15 Centrifugal Filter Device (Millipore GmbH, Eschborn, Germany).
11. PD-10 prepacked disposable columns containing Sephadex G-25 (Amersham Pharmacia Biotech).
12. High and Low Molecular Weight Gel Filtration Calibration Kits (Amersham Pharmacia Biotech).
13. Bio-Rad Protein Assay Kit (Bio-Rad Laboratories GmbH, Munich, Germany).
14. 20% Human serum albumin (Immuno GmbH, Heidelberg, Germany).

2.6. Analysis of Antigen-Binding Activities by Enzyme-Linked Immunosorbent Assay (ELISA)

1. 96-Well Titertek polyvinylchloride ELISA microplates (Flow Laboratories).
2. 50 m*M* Sodium carbonate–bicarbonate buffer, pH 9.6. Store at 4°C.
3. PBS–Tween 20: 0.05% (v/v) Tween 20 in PBS. Store at room temperature.
4. 2% (w/v) Skimmed milk in PBS (milk–PBS). Store frozen at –20°C.
5. Mouse monoclonal antibody (mAb) 9E10 specific for the c-*myc* oncoprotein (Cambridge Research Biochemicals, Cambridge, UK).
6. Goat anti-mouse IgG conjugated to Horseradish peroxidase (HRP) (Jackson Immuno Research Laboratories, Inc., USA).
7. 3,3',5,5'-Tetramethylbenzidine (TMB) peroxidase substrate (Kirkegaard & Perry Laboratories Inc., USA).

3. Methods

3.1. Generation of Plasmids for Expression of SCAs

3.1.1. PCR Amplification

1. Perform the PCR amplification of DNA fragments in a total volume of 50 μL containing 50 ng of plasmid DNA, 25 pmol of each primer, 300 μ*M* dNTPs, 5 μL of 10× PCR buffer, 5 μg of BSA, and 1 U of *Vent* DNA polymerase.

2. Run 15–20 polymerase chain reaction (PCR) cycles on a thermocycler. The thermal cycle is 95°C for 1 min (denaturation), 57°C for 2 min (annealing), and 75°C for 2 min (extension). At the beginning of the first cycle incubate for 3 min at 95°C and at the end of the last cycle incubate for 5 min at 75°C.
3. Analyze the amplified DNA fragments by electrophoresis on a 1.5% agarose gel prestained with ethidium bromide.

3.1.2. Cloning into Expression Vector (see ***Note 4***)

1. Digest 10 μg of appropriate vector with suitable restriction endonucleases in the presence of alkaline phosphatase (CIP). Incubate at least 2 h at temperature recommended by the supplier.
2. Purify the PCR fragments and linearized vector by agarose gel electrophoresis followed by extraction using a QIAquick gel extraction kit.
3. Digest isolated PCR fragments with restriction endonucleases suitable for cloning into the vector of choice.
4. Remove stuffer fragments and purify the digested PCR products using the QIAquick-spin PCR purification kit.
5. Ligate the vector and insert using a molar ratio between 1:1 and 1:3. The reaction mixture consists of 50 ng of DNA, 1 U of T4 ligase, ligation buffer, and H_2O to a final volume of 10–20 μL. Incubate overnight at 16°C.
6. Precipitate the DNA by adding 1/10 volume of 3 *M* sodium acetate, 20 μg of glycogen, and 2.5× vol of absolute ethanol. Incubate for at least 3 h at –20°C. Sediment the precipitate by centrifugation for 15 min at 10,000*g* (minifuge). Wash the pellet 4× with 500 μL of 80% ethanol followed by centrifugation for 10 min at 10,000*g*. Allow the pellet to dry at room temperature. Dissolve the dry pellet in 5 μL of H_2O.
7. Use the products of one ligation reaction for the electroporation of 40 μL electrocompetent *E. coli* cells according to the supplier's protocol. Plate the bacteria on 2YT agar plates containing 0.1 g/L of ampicillin and 2% (w/v) glucose. Incubate overnight at 37°C.
8. Test individual colonies for the presence of the desired insert by plasmid minipreps (*see* **Note 5**).

3.2. Preparation of Bacterial Culture

1. Inoculate a few milliliters of $2YT_{GA}$ with an individual bacterial colony and let it grow overnight at 37°C when using *E. coli* XL1-Blue or at 26°C when using RV308.
2. Dilute an overnight bacterial culture 40× with fresh $2YT_{GA}$ and incubate at 37°C (XL1-Blue) or at 26°C (RV308) with vigorous shaking (180–220 rpm) until OD_{600} = 0.8–0.9.
3. Harvest bacteria by centrifugation at 1500*g* for 10 min and 20°C.
4. Resuspend the pelleted bacteria in the same volume of either fresh $2YT_{SA}$ or YTBS medium (*see* **Note 6**). Add IPTG to a final concentration of 0.2 m*M* (*see* **Note 7**) and incubate the bacterial culture for 14–16 h with shaking at room temperature (22–24°C).
5. Collect the cells by centrifugation at 6200*g* for 20 min and either discard the culture supernatant (RV308) or retain it and keep on ice (XL1-Blue) (*see* **Note 8**).

3.3. Isolation of Recombinant Product from Soluble Periplasmic Fraction and Culture Medium

1. Resuspend the pelleted bacteria in 5% of the initial volume of ice-cold 50 m*M* Tris-HCl, 20% sucrose, 1 m*M* EDTA, pH 8.0, and incubate on ice for 1 h with occasional stirring.
2. Centrifuge the cell suspension at 30,000*g* for 40 min at 4°C, and carefully collect the supernatant (soluble periplasmic extract). In case of using RV308, go to **step 5**. If using

XL1-Blue, combine the culture supernatant and the soluble periplasmic extract, clarify by additional centrifugation (30,000*g*, 4°C, 60 min), and pass through a glass filter of pore size 10–16 µm and then through a filter of pore size of 0.2 µm.

3. Either reduce the vol 10× by concentrating with Amicon ultrafiltration cell and YM 10 membrane (*see* **Note 9**) or concentrate the bispecific recombinant product by ammonium sulfate precipitation (*see* **Note 10**). In the latter case, place the beaker with the culture supernatant and the soluble periplasmic extract on a magnetic stirrer. Slowly add ammonium sulfate powder to a final concentration 70% of saturation (472 g/L of solution). Continue stirring for at least another 2 h at 4°C.
4. Collect the protein precipitate by centrifugation (30,000*g*, 4°C, 30 min) and dissolve it in 1/10 of the initial volume of 50 m*M* Tris-HCl, 1 *M* NaCl, pH 7.0.
5. Thoroughly dialyze the concentrated protein against 50 m*M* Tris-HCl, 1 *M* NaCl, pH 7.0 at 4°C. Clarify the dialyzed material by centrifugation (30,000*g*, 4°C, 60 min).
6. For IMAC, prepare a column of chelating Sepharose (1–2 mL of resin/L of flask culture), wash with 5 bed volumes of water. Charge the column with Cu^{2+} by loading 0.7 bed volume of 0.1 *M* $CuSO_4$ (*see* **Note 11**), wash the excess of ions with 10 bed volumes of water, and equilibrate with 3 volumes of 50 m*M* Tris-HCl, 1 *M* NaCl, pH 7.0 (*see* **Note 12**).
7. Pass the soluble periplasmic proteins over a chelating Sepharose column either by gravity flow or using a peristaltic pump. Wash the column with 10 bed volumes of start buffer (50 m*M* Tris-HCl, 1 *M* NaCl, pH 7.0) followed by start buffer containing 50 m*M* imidazole (*see* **Note 13**) until the absorbance (280 nm) of the effluent is minimal (20–30 column volumes). Perform all chromatography steps at 4°C.
8. Elute bound antibody fragments with a start buffer containing 250 m*M* imidazole (*see* **Note 14**).
9. Analyze the purity of eluted material by SDS-PAGE *(35)*.

3.4. Final Purification of SCA Fragments and Analysis of Molecular Forms

1. If the IMAC yields in homogeneous preparation of recombinant protein according to reducing SDS-PAGE, go to **step 6** (*see* **Note 15**). Otherwise, calculate the isoelectric point (p*I*) of your SCA on the basis of amino acid composition of antibody fragment (*see* **Note 16**).
2. Subject the protein material eluted from the IMAC column to buffer exchange either for 50 m*M* imidazole-HCl, pH 6.0–7.0, or 20 m*M* Tris-HCl, pH 8.0–8.5, using prepacked PD-10 columns (*see* **Note 17**). Remove the turbidity of protein solution by centrifugation (30,000*g*, 4°C, 30 min).
3. Load the protein solution either on a Mono Q or Monō S column equilibrated either with 20 m*M* Tris-HCl, pH 8.0–8.5, or 50 m*M* imidazole-HCl, pH 6.0–7.0, respectively. Wash the column with a least 10 volumes of the start buffer.
4. Elute the bound material using a linear 0–1 *M* NaCl gradient in the start buffer; collect 1-mL fractions.
5. Perform sodium dodecyl sulfate-polyacrylamide gel electrophoresis (SDS-PAGE) analysis of eluted fractions.
6. Pool the fractions containing pure recombinant antibodies. Determine the protein concentration (*see* **Note 18**).
7. Perform a buffer exchange for PBSI, pH 7.0–7.4, (*see* **Note 19**) and concentrate the purified antibody preparations up to 1.0–2.0 mg/mL using Ultrafree-15 centrifugal filter units.
8. Equilibrate a Superdex column with PBSI buffer; calibrate the column using high and low molecular weight gel filtration calibration kits (*see* **Note 20**).

9. For analytical size-exclusion chromatography, apply 50 µL of the concentrated preparation of SCA to a Superdex HR10/30 column. Perform gel filtration at 4°C, monitor the UV-absorption of effluent at 280 nm, and, if necessary, collect 0.5-mL fractions.
10. For long-term storage, stabilize purified antibody fragments by adding Human serum albumin (HSA) to a final concentration of 10 mg/mL. Store the sample at –80°C (*see* **Note 21**).

3.5. Analysis of Binding Properties of SCAs in ELISA

1. Coat 96-well ELISA microplates with 100 µL/well of a 10 µg/mL solution of antigen in 50 m*M* sodium carbonate–bicarbonate buffer, pH 9.6, overnight at 4°C.
2. Wash the wells 5× with PBS–Tween 20 and incubate with 200 µL of milk–PBS at room temperature for 2 h.
3. Load 100 µL of purified SCA at various dilutions in milk–PBS into the wells and incubate for another 2 h at room temperature.
4. Wash the wells 10× with PBS–Tween 20, add 100 µL of a 1:4000 dilution of mAb 9E10 in milk–PBS, and incubate the plates for 1 h at room temperature (*see* **Note 22**).
5. Wash the wells 10× with PBS–Tween 20, add 100 µL of 1:4000 dilution of goat anti-mouse IgG-peroxidase conjugate in milk–PBS, and incubate the plates for 1 h.
6. Wash the wells 10× with PBS–Tween 20, add 100 µL of a TMB peroxidase substrate, and incubate at room temperature for 20–40 min. Read the absorbance of the colored product at 655 nm on a Microplate Reader.

4. Notes

1. Both XL1-Blue and RV308 are suitable hosts for expression of antibody fragments in shake-flask bacterial cultures. XL1-Blue has the following advantages: electrocompetent bacterial cells are commercially available (Stratagene) and standard DNA isolation protocols yield in pure DNA preparations for restriction digests and sequencing. However, RV308 is more robust fast growing strain suitable for high-cell density fermentation ***(36)***. Moreover, unlike for XL1-Blue, no leakage of antibody fragments into the culture medium was observed for RV308.
2. LB (Luria Bertani) broth can also be used. However, we observed that the simple substitution of LB for somewhat richer 2YT medium gave an essential increase in the yield of soluble antibody molecules.
3. $2YT_{SA}$ medium is prepared directly before use by dissolving 137 g of sucrose powder in 1 L of sterile 2YT medium containing 0.1 g/L of ampicillin.
4. The protocols were established for vectors pHOG21 ***(25)*** and pSKK ***(9)*** designed for periplasmic expression of single recombinant product.
5. All DNA manipulations and transformation experiments are performed according to standard cloning protocols ***(34)***.
6. Direct induction without medium change is recommended for production of scFv and other SCA in predominantly monomeric form. However, the change of medium and induction of antibody synthesis in bacteria under osmotic stress significantly increases the yield of recombinant product, although these conditions promote the formation of domain-swapped dimers ***(9)***.
7. This concentration of IPTG was found to be optimal for vectors containing the *SCA* gene under the control of wt *lac* promoter/operator such as pHOG21 ***(25)*** or pSKK ***(9)***. Nevertheless, performing small-scale experiments to optimize the induction conditions is recommended for each vector.
8. For XL1-Blue either due to the leakiness of the outer membrane or due to the partial cell lysis, a significant fraction of antibody fragments is found in the culture medium. There-

fore, supernatant should also be used as a starting material for isolation of recombinant protein.

9. Other devices suitable for concentrating protein solutions with a cutoff of 10–12 kDa may also be used.
10. Ammonium sulfate precipitation is especially recommended for concentrating bispecific four-domain antibody fragments. This procedure was shown to be rather ineffective for precipitating monospecific scFv fragments ***(29)***.
11. IMAC can be also performed on Ni^{2+}-charged chelating Sepharose or Ni-NTA-Superflow resin (Qiagen, Hilden, Germany). However, the use of Cu^{2+} instead of Ni^{2+} is recommended for isolation of antibody fragments for clinical applications ***(28)***.
12. Tris buffer is usually not recommended for IMAC because of the presence of amines interacting with immobilized metal ions. However, we have found that such conditions do not influence the absorption of strong binders containing six histidines, while preventing nonspecific interactions of some *E. coli* proteins with the chelating Sepharose.
13. Unlike chelating Sepharose, the Ni-NTA columns should not be washed with buffers containing imidazole at concentrations higher than 20 m*M*.
14. To avoid the unnecessary dilution of eluted recombinant protein, collect 0.5–1.0-mL fractions and monitor the UV absorbance at 280.
15. The purity of the antibody fragments eluted from the IMAC column depends on the expression level of particular recombinant protein.
16. The isoelectric point of the protein can be calculated using a number of computer programs, for example, DNAid+1.8 Sequence Editor for Macintosh (F. Dardel and P. Bensoussan, Laboratoire de Biochimie, Ecole Polytechnique, Palaiseau, France). The calculated p*I* value gives you hint what ion exchange matrix and buffer system should be used.
17. For antibody molecules with p*I* values <7.0, we recommend using an anion exchanger such as a Mono Q with linear 0–1 *M* NaCl gradient in 20 m*M* Tris-HCl, pH 8.0–8.5. For proteins with p*I* values >7.0, the cation-exchange chromatography on a Mono S column with linear 0–1 *M* NaCl gradient in 50 m*M* imidazole-HCl buffer, pH 6.0–7.0 can be recommended. Moreover, we found that by exchanging the buffer after IMAC for 50 m*M* imidazole-HCl, pH 6.4–6.7, most of the contaminating bacterial proteins precipitate while the recombinant antibody fragments remain soluble ***(9)***.
18. For determination of protein concentrations, we recommend using a Bradford dye-binding assay because it is easy to use, sensitive and fast ***(37)***.
19. We recommend using PBSI buffer because, in our experience, PBS alone appeared to be unfavorable for the stability of some antibody fragments. The presence of imidazole stabilizes the scFv fragments. It was determined empirically that PBS with 50 m*M* imidazole, pH 7.0–7.5, is a suitable buffer for various antibody fragments kept at relatively high concentrations (2–3 mg/mL). Moreover, this buffer did not interfere with antigen binding and did not show any toxic effects after incubation with cultured cells or after injection into mice (intravenous injection of 200 µL) ***(9,38)***.
20. Size-exclusion chromatography on Superdex 75 allows separation of scFv monomers (M_r = 25–30 kDa) from dimers (diabody, M_r = 50–60 kDa) ***(14)***. Higher molecular forms will be eluted from this column in a void volume. In contrast, Superdex 200 enables analysis of larger SCA molecules ***(39)***. Accordingly, Superdex 200 should be calibrated with both the high and low molecular weight gel filtration calibration kits. In contrast, the low molecular weight gel filtration calibration kit is sufficient for calibrating the Superdex 75 column.
21. Alternatively SCA can be stabilized by adding BSA or fetal calf serum (FCS). HSA is recommended for antibody fragments developed for clinical applications. The recombi-

nant antibodies stabilized by albumin can be stored at –80°C for years without loss of activity. These preparations may be used for a number of biological assays such as ELISA, flow cytometry, and analyses of antitumor activity both in vitro and in vivo.

22. If different epitopes are used for immunodetection (non c-*myc*), the corresponding antibodies should be used.

References

1. Bird, R. E., Hardman, K. D., Jacobson, J. W., Johnson, S., Kaufman, B. M., Lee, S. M., et al. (1988) Single-chain antigen-binding proteins. *Science* **242,** 423–426.
2. Huston, J. S., Levinson, D., Mudgett Hunter, M., Tai, M. S., Novotny, J., Margolies, M. N., et al. (1988) Protein engineering of antibody binding sites: recovery of specific activity in an anti-digoxin single-chain Fv analogue produced in *Escherichia coli*. *Proc. Natl. Acad. Sci. USA* **85,** 5879–5883.
3. Milenic, D. E., Yokota, T., Filpula, D. R., Finkelman, M. A., Dodd, S. W., Wood, J. F., et al. (1991) Construction, binding properties, metabolism, and tumor targeting of a single-chain Fv derived from the pancarcinoma monoclonal antibody CC49. *Cancer Res.* **51,** 6363–6371.
4. Yokota, T., Milenic, D. E., Whitlow, M., and Schlom, J. (1992) Rapid tumor penetration of a single-chain Fv and comparison with other immunoglobulin forms. *Cancer Res.* **52,** 3402–3408.
5. Yokota, T., Milenic, D. E., Whitlow, M., Wood, J. F., Hubert, S. L., and Schlom, J. (1993) Microautoradiographic analysis of the normal organ distribution of radioiodinated single-chain Fv and other immunoglobulin forms. *Cancer Res.* **53,** 3776–3783.
6. Adams, G. P., McCartney, J. E., Tai, M. S., Oppermann, H., Huston, J. S., Stafford, W. F., et al. (1993) Highly specific *in vivo* tumor targeting by monovalent and divalent forms of 741F8 anti-c-*erb*B-2 single-chain Fv. *Cancer Res.* **53,** 4026–4034.
7. Gruber, M., Schodin, B. A., Wilson, E. R., and Kranz, D. M. (1994) Efficient tumor cell lysis mediated by a bispecific single-chain antibody expressed in *Escherichia coli*. *J. Immunol.* **152,** 5368–5374.
8. Kurucz, I., Titus, J. A., Jost, C. R., Jacobus, C. M., and Segal, D. M. (1995) Retargeting of CTL by an efficiently refolded bispecific single-chain Fv dimer produced in bacteria. *J. Immunol.* **154,** 4576–4582.
9. Kipriyanov, S. M., Moldenhauer, G., Schuhmacher, J., Cochlovius, B., Von der Lieth, C. W., Matys, E. R., and Little, M. (1999) Bispecific tandem diabody for tumor therapy with improved antigen binding and pharmacokinetics. *J. Mol. Biol.* **293,** 41–56.
10. Kontermann, R. E. and Müller, R. (1999) Intracellular and cell surface displayed single-chain diabodies. *J. Immunol. Meth.* **226,** 179–188.
11. Kipriyanov, S. M. and Little, M. (1999) Generation of recombinant antibodies. *Mol. Biotechnol.* **12,** 173–201.
12. Pugsley, A. P. (1993) The complete general secretory pathway in gram-negative bacteria. *Microbiol. Rev.* **57,** 50–108.
13. Whitlow, M. and Filpula, D. (1991) Single-chain Fv proteins and their fusion proteins. *Meth. Comp. Meth. Enzymol.* **2,** 97–105.
14. Kipriyanov, S. M., Dübel, S., Breitling, F., Kontermann, R. E., and Little, M. (1994) Recombinant single-chain Fv fragments carrying C-terminal cysteine residues: production of bivalent and biotinylated miniantibodies. *Mol. Immunol.* **31,** 1047–1058.
15. Plückthun, A. (1994) Antibodies from *Escherichia coli*, in *Handbook of Experimental Pharmacology*, Vol. 113: *The Pharmacology of Monoclonal Antibodies* (Rosenberg, M. and Moore, G. P., eds.), Springer-Verlag, Berlin, Heidelberg, pp. 269–315.

16. Hockney, R. C. (1994) Recent developments in heterologous protein production in *Escherichia coli. Trends Biotechnol.* **12,** 456–463.
17. Knappik, A. and Plückthun, A. (1995) Engineered turns of a recombinant antibody improve its in vivo folding. *Protein Eng.* **8,** 81–89.
18. Kipriyanov, S. M., Moldenhauer, G., Martin, A. C. R., Kupriyanova, O. A., and Little, M. (1997) Two amino acid mutations in an anti-human CD3 single-chain Fv antibody fragment that affect the yield on bacterial secretion but not the affinity. *Protein Eng.* **10,** 445–453.
19. Duenas, M., Vazquez, J., Ayala, M., Soderlind, E., Ohlin, M., Perez, L., et al. (1994) Intra- and extracellular expression of an scFv antibody fragment in *E. coli*: effect of bacterial strains and pathway engineering using GroES/L chaperonins. *BioTechniques* **16,** 476–477.
20. Knappik, A., Krebber, C., and Plückthun, A. (1993) The effect of folding catalysts on the in vivo folding process of different antibody fragments expressed in *Escherichia coli. Biotechnology* **11,** 77–83.
21. Bothmann, H. and Plückthun, A. (2000) The periplasmic *Escherichia coli* peptidylprolyl *cis,trans*-isomerase FkpA. I. Increased functional expression of antibody fragments with and without *cis*-prolines. *J. Biol. Chem.* **275,** 17,100–17,105.
22. Bothmann, H. and Plückthun, A. (1998) Selection for a periplasmic factor improving phage display and functional periplasmic expression. *Nat. Biotechnol.* **16,** 376–380.
23. Skerra, A. and Plückthun, A. (1991) Secretion and in vivo folding of the Fab fragment of the antibody McPC603 in *Escherichia coli*: influence of disulphides and *cis*-prolines. *Protein Eng.* **4,** 971–979.
24. Sawyer, J. R., Schlom, J., and Kashmiri, S. V. S. (1994) The effect of induction conditions on production of a soluble anti-tumor sFv in *Escherichia coli. Protein Eng.* **7,** 1401–1406.
25. Kipriyanov, S. M., Moldenhauer, G., and Little, M. (1997) High level production of soluble single-chain antibodies in small-scale *Escherichia coli* cultures. *J. Immunol. Meth.* **200,** 69–77.
26. Kipriyanov, S. M. and Little, M. (1997) Affinity purification of tagged recombinant proteins using immobilized single-chain Fv fragments. *Analyt. Biochem.* **244,** 189–191.
27. Arnold-Schild, D., Kleist, C., Welschof, M., Opelz, G., Rammensee, H. G., Schild, H., and Terness, P. (2000) One-step single-chain Fv recombinant antibody-based purification of gp96 for vaccine development. *Cancer Res.* **60,** 4175–4178.
28. Casey, J. L., Keep, P. A., Chester, K. A., Robson, L., Hawkins, R. E., and Begent, R. H. (1995) Purification of bacterially expressed single-chain Fv antibodies for clinical applications using metal chelate chromatography. *J. Immunol. Meth.* **179,** 105–116.
29. Kipriyanov, S. M., Moldenhauer, G., Strauss, G., and Little, M. (1998) Bispecific CD3 × CD19 diabody for T cell-mediated lysis of malignant human B cells. *Int. J. Cancer* **77,** 763–772.
30. Müller, K. M., Arndt, K. M., Bauer, K., and Plückthun, A. (1998) Tandem immobilized metal-ion affinity chromatography/immunoaffinity purification of His-tagged proteins—evaluation of two anti-His-tag monoclonal antibodies. *Analyt. Biochem.* **259,** 54–61.
31. Schulze, R. A., Kontermann, R. E., Queitsch, I., Dübel, S., and Bautz, E. K. (1994) Thiophilic adsorption chromatography of recombinant single-chain antibody fragments. *Analyt. Biochem.* **220,** 212–214.
32. Müller, K. M., Arndt, K. M., and Plückthun, A. (1998) A dimeric bispecific miniantibody combines two specificities with avidity. *FEBS Lett.* **432,** 45–49.
33. Maurer, R., Meyer, B., and Ptashne, M. (1980) Gene regulation at the right operator (O_R) bacteriophage λ. I. O_R3 and autogenous negative control by repressor. *J. Mol. Biol.* **139,** 147–161.
34. Sambrook, J., Fritsch, E. F., and Maniatis, T. (1989) *Molecular Cloning: A Laboratory Manual*, Cold Spring Harbor Laboratory, Cold Spring Harbor, NY.

35. Laemmli, U. K. (1970) Cleavage of structural proteins during the assembly of the head of bacteriophage T4. *Nature* **227,** 680–685.
36. Horn, U., Strittmatter, W., Krebber, A., Knupfer, U., Kujau, M., Wenderoth, R., et al. (1996) High volumetric yields of functional dimeric miniantibodies in *Escherichia coli,* using an optimized expression vector and high-cell-density fermentation under non-limited growth conditions. *Appl. Microbiol. Biotechnol.* **46,** 524–532.
37. Bradford, M. M. (1976) A rapid and sensitive method for the quantitation of microgram quantities of protein utilizing the principle of protein-dye binding. *Analyt. Biochem.* **72,** 248–254.
38. Cochlovius, B., Kipriyanov, S. M., Stassar, M. J., Schuhmacher, J., Benner, A., Moldenhauer, G., and Little, M. (2000) Cure of Burkitt's lymphoma in severe combined immunodeficiency mice by T cells, tetravalent CD3 × CD19 tandem diabody, and CD28 costimulation. *Cancer Res.* **60,** 4336–4341.
39. Le Gall, F., Kipriyanov, S. M., Moldenhauer, G., and Little, M. (1999) Di-, tri- and tetrameric single-chain Fv antibody fragments against human CD19: effect of valency on cell binding. *FEBS Lett.* **453,** 164–168.

151

Enzymatic Digestion of Monoclonal Antibodies

Sarah M. Andrew

1. Introduction

Originally, digestion of antibodies by proteolytic enzymes was used to study their structure. Many diverse structures can be obtained by fragmentation of the different classes of antibody with different enzymes, or by using the same enzyme and changing the conditions (**Fig. 1**). Not all the fragments obtained have significant binding activity; for example, in several studies by this author, Fv fragments obtained by digestion have been found to have lost their binding activity. Fragmentation of antibody is now usually carried out to introduce required properties (e.g., a decrease in molecular size), or to remove undesirable properties (e.g., nonspecific Fc receptor binding).

The most usual digestions carried out are:

1. Production of bivalent $F(ab')_2$ from mouse McAb IgG;
2. Production of univalent Fab from mouse McAb IgG;
3. Production of bivalent IgMs from mouse McAb IgM; and
4. Production of bivalent $F(ab')_{2\mu}$ from mouse McAb IgM.

$F(ab')_2$ and $F(ab')_{2\mu}$ are produced by digestion with pepsin and Fab is produced by digestion with papain. One useful fragmentation uses papain that has been preactivated with cysteine. This cleaves IgG_1 to produce $F(ab')_2$, and IgG_{2a} IgG_{2b} to produce Fab. It is a very stable fragmentation in which the times of incubation are not at all critical. The IgG-like subunit of IgM (IgMs) is the product of a mild reduction; this is most conveniently done using cysteine, which reduces the IgM and alkylates the subunit, thus preventing reassociation.

After digestion of the antibodies, it is necessary to purify the fragments for two reasons: to separate the fragment from any remaining intact antibody; and to separate the fragments of interest from other miscellaneous fragments produced by the process of digestion. Purification of IgG fragments by protein A affinity chromatography is possible if an intact Fc region remains after fragmentation, as it does in the case of undigested antibody and, sometimes, in digestions with pepsin. In general this method of purification is rough and ready and size exclusion chromatography is recommended as an additional step.

From: *The Protein Protocols Handbook, 2nd Edition*
Edited by: J. M. Walker © Humana Press Inc., Totowa, NJ

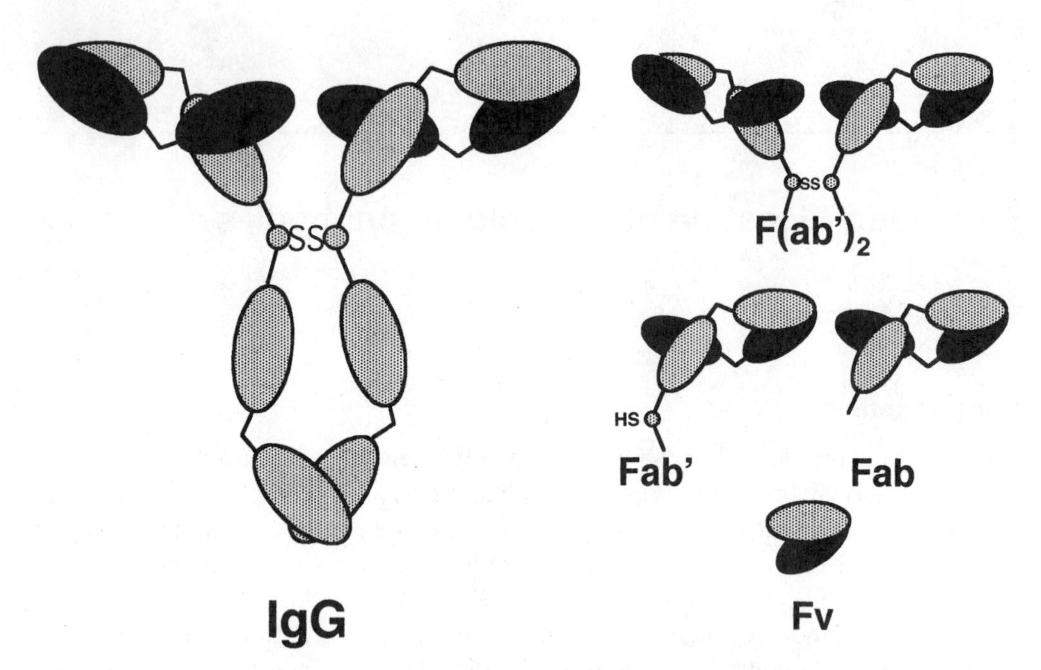

Fig. 1. Diagramatic representation of the major fragments of an IgG molecule that can be produced by enzymatic digestion, as described in the text.

2. Materials

2.1. Digestion with Pepsin

1. Purified solution of mouse McAb (IgG or IgM) at a concentration of ≥1 mg/mL.
2. 0.2 *M* Sodium acetate buffer brought to pH 4.0 with glacial acetic acid.
3. 0.2 *M* Sodium acetate buffer brought to pH 4.5 with glacial acetic acid.
4. 0.1 mg/mL pepsin (Sigma [St. Louis, MO] number P6887) in acetate buffer at appropriate pH (*see* **Subheadings 3.1.** and **3.2.**).
5. 2 *M* Tris base.
6. Phosphate-buffered saline (PBS) brought to pH 8.0 using NaOH.
7. Protein A-Sepharose CL-4B (Sigma number P 3391) swollen in PBS, pH 8.0, and packed into a 10 × 100 mm column, or a 1-mL HiTrap Protein A column (Sigma number 5-4838).
8. PBS.
9. A size exclusion column equivalent to 26 × 900 mm Sephacryl S-200 (Pharmacia, Uppsala, Sweden).
10. Centriprep 100 concentrator (Amicon, Beverly MA; Stonehouse, Glocestershire, UK).

2.2. Digestion with Papain

1. Purified solution of mouse IgG McAb at a concentration of ≥1 mg/mL.
2. PBS made to 0.02 *M* with respect to ethylenediaminetetra-acetic acid (EDTA) and 0.2 *M* with respect to cysteine.
3. Iodoacetamide crystals.
4. PBS brought to pH 8.0 using NaOH.
5. Protein A-Sepharose CL-4B (Sigma number P 3391) swollen in PBS, pH 8.0, and packed into 10 × 100 mm column, or a 1 mL HiTrap Protein A column (Sigma number 5-4838).
6. PBS.
7. A size exclusion column equivalent to 26 × 900 mm Sephacryl S-200 (Pharmacia) (optional).

2.3. Reduction and Alkylation with Cysteine

1. Purified solution of mouse IgM McAb at a concentration of ≥1 mg/mL in PBS.
2. 0.1 *M* cysteine stock solution in PBS (L-cysteine free base, Sigma number C7755).
3. Borate-buffered saline: 0.015 *M* sodium borate, 0.15 *M* sodium chloride, made to pH 8.5 with Sodium hydroxide.
4. A size exclusion column equivalent to 26 × 900 mm Sephacryl S-200 (Pharmacia).

3. Methods

3.1. Preparation of $F(ab')_2$ from IgG

The various subclasses of mouse IgG vary in their susceptibility to pepsin fragmentation. IgG_1 is quite resistant to digestion and and it is impossible to fragment IgG_{2b} to $F(ab')_2$ because it breaks down to Fab (*see* **Notes 1–3**). The method below is one that should work in most cases.

1. Dialyze the IgG against acetate buffer, pH 4.0, overnight at 4°C. Use any known amount of antibody between 1 and 20 mg.
2. Determine the concentration at A_{280}.
3. Add 0.1 mg/mL pepsin in acetate buffer, pH 4.0, to give an enzyme-to-antibody ratio of 1:20 (w:w).
4. Incubate for 6 h in a water bath at 37°C.
5. Stop the reaction by adding sufficient Tris base to bring the pH to roughly 8.0 (start by adding 50 µL Tris, mix, and test the pH with pH paper).
6. Dialyze the mixture against PBS, pH 8.0, overnight.
7. Equilibrate the protein A column with PBS, pH 8.0, and load the dialyzed mixture onto it 1 mL at a time. Collect the unbound fraction that contains the $F(ab')_2$ fragments.
8. If further purity is desired, the mixture should be concentrated to a volume of ≤3 mL and added to a precalibrated size exclusion column (26 × 900 mm Sephacryl S-200 or equivalent). At this stage regular PBS can be used to equilibrate the column and elute the fractions.
9. Collect 2.5-mL fractions over the molecular weight range of $F(ab')_2$ (110 kDa). The purity of the product can be assessed by sodium dodecyl sulfate-polyacrylamide gel electrophoresis (SDS-PAGE) in nonreducing conditions and in reducing conditions when a doublet at 25 kDa is seen.

3.2. Preparation of $F(ab')_{2\mu}$ from IgM (see Note 3)

1. Dialyze the IgM against acetate buffer, pH 4.5, overnight at 4°C.
2. Determine the concentration at A_{280} (*see* Chapter 1).
3. Add 0.1 mg/mL pepsin in acetate buffer, pH 4.5, to give an enzyme-to-antibody ratio of 1:20 (w:w).
4. Incubate for between 6 and 12 h in a water bath at 37°C.
5. Stop the reaction by adding sufficient Tris base to bring the pH to roughly 8.0 (start by adding 50 µL Tris, mix, and test the pH with pH paper).
6. Dialyze the mixture against PBS overnight, or concentrate and change the buffer using a Centriprep 100 concentrator.
7. Concentrate the mixture to a volume of ≤3 mL and add to a precalibrated size exclusion column (26 × 900 mm Sephacryl S-200 or equivalent). Elute fractions of between 1 and 2.5 mL. The molecular weight of $F(ab')_{2\mu}$ is 130 kDa. Assess purity on an 8% polyacrylamide gel under nonreducing conditions.

3.3. Preparation of Fab from IgG

Again, there is a variable susceptibility to the enzyme between the IgG subclasses. The method described is one that should give good results in most cases. If fragments are not obtained, variations in the concentration of enzyme or time of incubation can be tried. *See* **Notes 1**, **2**, and **4** for further information.

1. Use IgG in PBS at a concentration between 1 and 5 mg/mL.
2. Dissolve sufficient papain in an equal volume of digestion buffer to the volume of antibody solution to give a papain-to-antibody ratio of 1:20 (w:w).
3. Add the two equal solutions, one containing antibody and one containing papain together; mix thoroughly but gently.
4. Incubate for 4–6 h at 37°C.
5. Stop the reaction by adding crystalline iodoacetamide to make the mixture 0.03 *M* with respect to iodoacetamide, and dissolve by mixing gently.
6. Dialyze the mixture against PBS, pH 8.0, overnight at 4°C.
7. Equilibrate the protein A column in PBS, pH 8.0, then add the dialyzed mixture to it. Collect the unbound fraction. Wash the column with PBS, pH 8.0, to completely recover the Fab fragments.
8. Concentrate the Fab fragment mixture to a volume <5 mL.
9. Load the mixture containing the Fab fragments onto the size exclusion column equilibrated in PBS. Collect fractions corresponding to a molecular weight of 50 kDa.
10. Check the purity of the final product on a 10% SDS-polyacrylamide gel in nonreducing conditions.

3.4. Preparation of IgG Fragments Using Preactivated Papain

Bivalent $F(ab')_2$ fragments can be obtained from IgG_1 by this method. The protocol is also useful for producing monovalent Fab fragments from IgG_{2a} and IgG_{2b}. *See* **Note 4** for further information.

3.4.1. Preactivate the Papain

1. Make up a 2 mg/mL solution of papain in PBS with 0.05 *M* cysteine. Warm at 37°C for 30 min.
2. Equilibrate a PD10 column with acetate/EDTA buffer and apply the papain mixture.
3. Collect 10 × 1 mL fractions eluting with acetate/EDTA buffer. Assay the fractions at A_{280} and pool the two or three fractions containing protein.
4. Calculate the concentration of the preactivated papain using the following formula:

$$A_{280}/2.5 = \text{mg preactivated papain/mL} \quad (1)$$

3.4.2. Digest the IgG

1. Dialyse the IgG against acetate/EDTA buffer and determine the concentration after dialysis.
2. Mix the preactivated papain solution and the IgG solution in an enzyme-to-antibody ratio of 1:20.
3. Incubate for 6–18 h at 37°C.
4. Stop the reaction by adding crystalline iodoacetamide to a concentration of 0.03 *M*.
5. Dialyse the mixture against PBS, pH 8.0.
6. Purify the fragments as in **Subheading 3.3.**

3.5. Preparation of IgM Subunits from Pentameric IgM (see Note 5)

1. Add cysteine to IgM in PBS (up to 10 mg in 5 mL); make the solution 0.05 *M* with respect to cysteine.

2. Incubate the mixture for 2 h at 37°C.
3. Separate the fragments from intact antibody on a 26 × 900 mm size exclusion column or equivalent. The buffer in the column should be made 3 m*M* with respect to EDTA, if possible, because this prevents reassociation of the IgM fragments. Collect the fractions corresponding to a molecular weight of 180 kDa.

4. Notes

1. It is possible, but expensive, to buy kits for digestion of mouse antibody (Pierce, Warrington, UK). These kits work extremely well and can be a time-saving option. Each laboratory must consider whether the time saved in preparation is worth the extra expenditure on a kit. Fragmentation of antibodies from human and rabbit is described in **ref.** ***6***; generally the methods and enzymes used are similar.
2. The most common problems likely to occur when following these methods are that the antibody does not digest or that the molecule overdigests to produce small unrecognizable fragments. These can be overcome by varying the concentrations of enzyme, the times of digestion and, in the case of pepsin, the pH of the mixture ***(1)***. Generally, it is unwise to embark on fragmentation of IgG if the subclass of the antibody is not known. In digestion with both pepsin and papain the susceptibility to digestion varies with subclass. The order of susceptibility has been found to be $IgG_{2b} > IgG_3 > IgG_{2a} > IgG_1$ ***(2,3)***. Not all antibodies fall into this order (IgG_{2a} can be extremely sensitive to the action of papain in the presence of cysteine) and individual exceptions must be expected.
3. Digestion with pepsin has a great subclass variability. IgG_{2b} does not digest to $F(ab')_2$ fragments at all; the monovalent Fab/c (a single binding site and an intact Fc portion) is produced instead. This molecule has a very similar molecular weight to $F(ab')_2$; thus, is easy to imagine success with the fragmentation. The reason for the problem is thought to be an asymmetric glycosylation of the heavy chains in the molecule. All IgG subclasses can be further digested by pepsin to produce monovalent Fab fragments because there is a site of secondary cleavage on the NH_2-terminal side of the disulfide bonds. Further digestion with pepsin at a pH of 3.5 can produce Fv fragments after approx 3 h of incubation. There are reports of these having activity as antigen-binding fragments ***(4)***, but the personal experience of this author and colleagues suggests this is rare. It is also unfortunately true that it is difficult to produce active fragments from IgM. It has been suggested that IgM heavy chains can be truncated ***(5)***, but this has not been confirmed. The method given for IgM $F(ab')_{2\mu}$ will work, but one should not be too disappointed if the affinity is low.
4. On the whole, papain fragmentations work well and the timings of the incubations are not critical. Initially, care should be taken to mix the papain as it is in suspension; it will dissolve completely at the concentration given in the method. The methods using preactivated papain work extremely well and the incubation times are not at all critical.
5. The digestion to IgMs from IgM causes dissociation of the inter subunit disulfide bonds. It is possible to reduce the intrachain disulfide bonds on further reduction. This is why cysteine is the reducing agent chosen. Reduction by dithiothreitol or mercaptoethanol can be used, but more care is required with the incubations and a separate alkylation step is required.

References

1. Andrew, S. M. and Titus, J. A. (1997) Fragmentation of immunoglobulin G, in *Current Protocols in Immunology* (Coligan, J. E., Kruisbeek, A. M., Margulies, D. H., Shevach, E. M., and Strober, W., eds.), Wiley, New York, pp. 2.8.1–2.8.10.
2. Parham, P. (1986) Preparation and purification of active fragments from mouse monoclonal antibodies, in *Handbook of Experimental Immunology, Vol. 1: Immuno-chemistry* (Wier, D. M., ed.), Blackwell Scientific, London, UK, pp. 14.1–14.23.

3. Parham, P. (1983) On the fragmentation of monoclonal IgG_1, IgG_{2a} and IgG_{2b} from BALB/c mice. *J. Immunol.* **131,** 2895–2902.
4. Sharon, J. and Givol, D. (1976) Preparation of the Fv fragment from the mouse myeloma XPRC-25 immunoglobulin possessing anti-dinitrophenyl activity. *Biochemistry* **15,** 1591–1598.
5. Marks, R. and Bosma, M. J. (1985) Truncated μ (μ') chains in murine IgM: evidence that μ' chains lack variable regions. *J. Exp. Med.* **162,** 1862–1877.
6. Stanworth, D. R. and Turner, M. W. (1986) Immunochemical analysis of human and rabbit immunoglobulins and their subunits, in *Handbook of Experimental Immunology, Vol. 1: Immunochemistry* (Wier, D. M., ed.), Blackwell Scientific, London, UK, pp. 12.1–12.45.

152

How to Make Bispecific Antibodies

Ruth R. French

1. Introduction

This protocol describes the production of bispecific $F(ab')_2$ antibody derivatives (BsAbs) by the linking of two Fabэ fragments via their hinge region SH groups using the bifunctional crosslinker *o*-phenylenedimaleimide (*o*-PDM) as described by Glennie et al. *(1,2)*. The procedure is illustrated in **Fig. 1**. The first step is to obtain $F(ab')_2$ from the two parent IgG antibodies. Methods for digestion of IgG to $F(ab')_2$ are described in Chapter 151. Fab' fragments are then prepared from the two $F(ab')_2$ species by reduction with thiol, thus exposing free SH groups at the hinge region (three SH-groups for mouse IgG1 and IgG2a antibodies) (*see* **Note 1**). One of the Fabэ species (Fab'-A) is selected for alkylation with *o*-PDM. Because *o*-PDM has a strong tendency to crosslink adjacent intramolecular SH-groups, two of the three hinge SH-groups will probably be linked together, leaving a single reactive maleimide group available for conjugation (**Fig. 1**; *see* **Note 2**). Excess *o*-PDM is then removed by column chromatography, and the Fab'-A(mal) is mixed with the second reduced Fab' (Fab'-B) under conditions favoring the crosslinking of the maleimide and SH groups. When equal amounts of the two parent Fab' species are used, the major product is bispecific $F(ab')_2$, resulting from the reaction of one Fab'-A(mal) with one of the SH groups at the hinge of Fab'-B. Increasing the proportion of Fab'-A(mal) in the reaction mixture results in a significant amount of $F(ab')_3$ product by the reaction of two molecules of Fab'-A(mal) with two free SH-groups at the hinge of a single Fab'-B molecule (*see* **Note 3**). The remaining free SH groups on Fab'-B are alkylated, and the $F(ab')_2$ bispecific antibody product (Fab'-A × Fab'-B) is separated by gel filtration chromatography. Each stage of the procedure is checked by HPLC.

Using this method, well-defined derivatives are produced with good yield, and the products are easily isolated; starting with 10 mg each of two parent $F(ab')_2$ species, expect to obtain 5–10 mg BsAb. The derivatives can be produced at relatively low cost, and quickly. It is possible to obtain the BsAb product from the parent IgG in five working days. The protocols can be scaled up to produce larger amounts (100–150 mg) of derivative for therapeutic applications if required. However, bear in mind that the derivatives produced by this procedure are almost always contaminated with trace amounts of parent IgG antibody or Fc fragments, which are coharvested with the parent

From: *The Protein Protocols Handbook, 2nd Edition*
Edited by: J. M. Walker © Humana Press Inc., Totowa, NJ

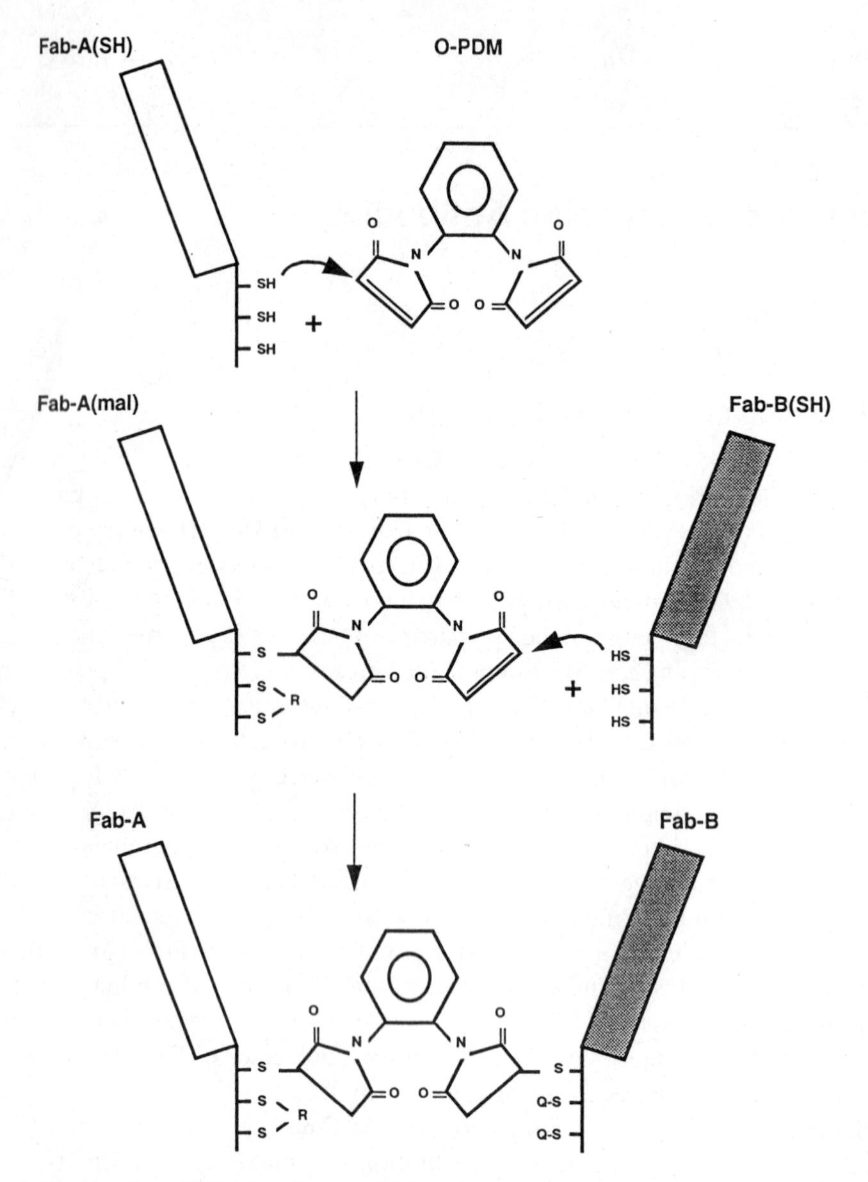

Fig. 1. Preparation of BsAb using *o*-PDM as crosslinker. The F(ab')$_2$ BsAb illustrated is produced from Fab' fragments derived from mouse IgG1 or IgG2a antibody. Two adjacent hinge SH-groups of Fab'-A are cross-linked by *o*-PDM (R, *o*-phenylenedisuccinimidyl linkage), leaving one with a free maleimide group for cross-linking with an SH-group at the hinge of Fab'-B. Unconjugated SH-groups at the Fab'-B hinge are blocked by alkylation (Q, carboxyamidomethyl). Increasing the ratio of Fab'-A(mal) to Fab'-B(SH) will favor the production of bispecific F(ab')$_3$, in which two molecules of Fab'-A(mal) are linked to one molecule of Fab'-B.

F(ab')$_2$ and the final product. If the presence of Fc is likely to be a problem, preparations can be tested for Fc by enzyme-linked immunosorbent assay (ELISA), and, if necessary, Fc removed by immunoaffinity chromatography ***(2)***.

2. Materials

2.1. Reagents

1. 2 *M* TE8: 2 *M* Tris-HCl, pH 8.0, 100 m*M* ethylenediaminetetra-acetic acid (EDTA). Prepare 0.2 *M* TE8 from 2 *M* stock.
2. $F(ab')_2$ reducing solution: 220 m*M* 2-ME, 1 m*M* EDTA. Make up 10 mL. Use a fume cupboard.
3. Sephadex G25 (Pharmacia, Uppsala, Sweden) and Ultragel AcA44 (Biosepra S. A., Villeneuve la Garenne, France) gel filtration media.
4. G25 column buffer (50 m*M* AE); 3.35 g sodium acetate, 526 µL glacial acetic acid, 0.186 g EDTA, made up to 1 L. Degas before use under vacuum or using nitrogen.
5. High performance liquid chromatography (HPLC) buffer (0.2 *M* phosphate, pH 7.0): Add 0.2 *M* Na_2HPO_4 to 0.2 *M* NaH_2PO_4 to obtain the required pH.
6. *o*-PDM/DMF for Fab'(SH) alkylation: 12 m*M* *o*-PDM in dimethylformamide. Make up just prior to use. Chill in a methylated spirit/ice bath. Caution: *o*-PDM is toxic and should be handled with care.
7. NTE8, 1 *M*: 1 *M* NaCl, 0.2 *M* Tris-HCl, pH 8.0, 10 m*M* EDTA.
8. Iodoacetamide: 250 m*M* in 0.2 *M* TE8 and 50 m*M* in 1 *M* NTE8.

2.2. Chromatography Equipment

1. Two chromatography columns packed with Sephadex G25 (*see* **Subheading 2.1.**) and equilibrated and run in 50 m*M* AE are required. The first (column 1) should be 1.6 cm in diameter, packed to a height of 25 cm with gel, and pumped at approx 60 mL/h. The second (column 2) should be 2.6 cm in diameter, packed to a height of 20 cm with gel, and pumped at approx 200 mL/h. The columns must be fitted with two end-flow adaptors and water jackets to allow chilling throughout the procedure. Pharmacia K Series columns are ideal.
2. Two larger columns packed with polyacrylamide agarose gel (Ultragel AcA44; *see* **Subheading 2.1.**) and run in 0.2 *M* TE8 are used for the size exclusion chromatography of the BsAb products. These should be 2.6 cm in diameter, and packed to a height of 80 cm with gel. The two columns should be joined in series using Teflon capillary tubing and pumped at approx 30 mL/h. Chilling is not required at this stage of the preparation, and the columns can be run at room temperature.
3. Two peristaltic pumps capable of rates between 15 and 200 mL/h for column chromatography.
4. Chiller/circulator to cool columns. A polystyrene box containing water and crushed ice and a submersible garden pond pump (rate approx 10 L/min) can be used as an alternative to a commercial chiller.
5. UV monitor, chart recorder, and fraction collector.
6. Amicon stirred concentration cell (Series 8000, 50 or 200 mL) with a 10,000 M_r cutoff filter for concentration of products.
7. HPLC system fitted with Zorbex Bio series GF250 column (Du Pont Company, Wilmington, DE) or equivalent gel-permeation column capable of fractionation up to approx 250,000 M_r.

3. Methods

3.1. Preparation of Bispecific $F(ab')_2$ Derivatives

The method described here is for the preparation of $F(ab')_2$ BsAb starting with 5–20 mg of each parent $F(ab')_2$ to obtain 1–8 mg of BsAb product.

1. Use equal amounts of $F(ab')_2$ from the two parent antibodies. The $F(ab')_2$ should be in 0.2 *M* TE8 at 5–12 mg/mL in a final volume of 1–3 mL. Keep a 50 µL sample of both $F(ab')_2$ preparations for HPLC analysis (*see* **Subheading 3.3.**).
2. Reduce both parent $F(ab')_2$ preparations to Fab'(SH) using 1/10 vol $F(ab')_2$ reducing solution (final concentration 20 m*M* 2-ME). Incubate at 30°C for 30 min and then keep on ice. Maintain the tempterature at 0–5°C for the rest of the procedure unless stated otherwise.
3. Select the species to be maleimidated (Fab'-A[SH]) (*see* **Note 3**). Remove 2-ME by passing through the smaller Sephadex G25 column (column 1). Collect the protein peak, which elutes after approx 8–10 min, in a graduated glass tube in an ice bath (*see* **Note 4**). Take a 45 µL sample from the top of the peak for HPLC analysis (*see* **Subheading 3.3.**). Keep the column running to completely elute 2-ME, which runs as a small secondary peak.
4. When the chart recorder has returned to baseline, load the second Fab'(SH) species [Fab'-B(SH)] onto the column, and separate as for Fab'-A(SH), again taking a sample for HPLC analysis (*see* **Subheading 3.3.**).
5. After the Fab'-B(SH) has been loaded onto the G25 column, the Fab'-A(SH) partner can be maleimidated. Rapidly add a 1/2 vol (normally 4–5 mL) of cold *o*-PDM/DMF to the Fab'-A(SH), seal the tube with Parafilm or similar, and mix by inverting two to three times (*see* **Note 5**). Stand in an ice bath for 30 min.
6. When the Fab'-B(SH) has been collected, connect the larger Sephadex G25 column (column 2) to the chart recorder. After the 30 min incubation, load the Fab'-A(SH)/*o*-PDM/DMF mixture onto this column. Collect the Fab'-A(mal) protein peak (elutes after 8–10 min) (*see* **Note 6**).
7. Pool the Fab'-A(mal) and the Fab'-B(SH). Immediately concentrate in a stirred Amicon concentration cell to around 5 mL, and then transfer to a tube for overnight incubation at 4°C (*see* **Note 7**).
8. During conjugation, in addition to the required BsAb, disulfide bonded homodimers may also form. To eliminate these, after overnight incubation add 1/10 volume 1 *M* NTE8 to the mixture to increase the pH, and then 1/10 vol $F(ab')_2$ reducing solution to reduce the homodimer disulfide bonds. Incubate at 30°C for 30 min.
9. Alkylate to block sulphydryl groups by the addition of 1/10 vol 250 m*M* iodoacetamide in 0.2 *M* TE8 (*see* **Note 8**). Check the composition of the mixture by HPLC (*see* **Subheading 3.3.**).
10. Separate the products on two AcA44 columns run in series. Collect 10–15 min fractions. A typical elution profile is shown in **Fig. 2**.
11. Pool the fractions containing the BsAb product. To minimize contamination, only take the middle two-thirds of the peak. Concentrate and dialyze into appropriate buffer.
12. If required, check the final product by HPLC (*see* **Subheading 3.3.**).

3.2. Preparation of Bispecific $F(ab')_3$ Derivatives

This is as for the preparation of bispecific $F(ab')_2$ except that the ratio of Fab'(mal) to Fab'(SH) is increased from 1:1 to 2:1 or greater. Therefore, start with at least twice as much of the $F(ab')_2$ which is to provide two arms of the $F(ab')_3$ product.

3.3. HPLC Monitoring

For rapid analysis of products during the preparation, an HPLC system is used as described in **Subheading 2.2.** This will resolve IgG, $F(ab')_2$, and Fab' sized molecules in approx 20 min, and can be performed while the preparation is in progress. The parent $F(ab')_2$ and the alkylated reaction mixture can be loaded directly onto the column

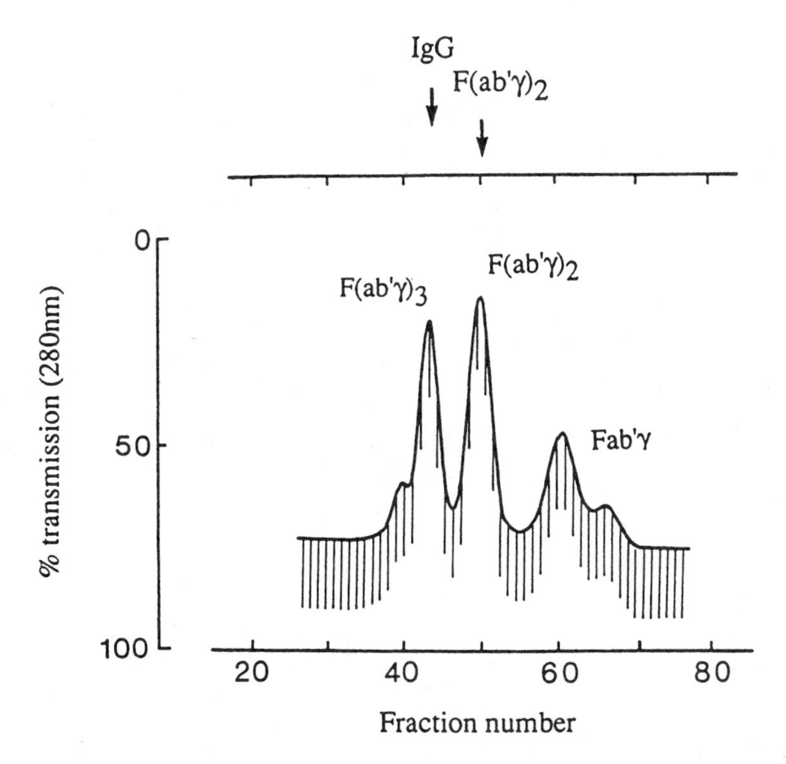

Fig. 2. Chromatography profile showing the separation of parent Fab' and bispecific F(ab')$_2$ and F(ab')$_3$ products on AcA44 columns. In this case, Fab'-A(SH) and Fab'-B(mal) were mixed at a ratio of 2:1 to increase the formation of bispecific F(ab')$_3$. The unreacted Fab' fragments and the F(ab')$_2$ and F(ab')$_3$ products are indicated. The *arrows* show the points at which protein standards eluted from the same columns.

and the eluted product monitored at 280 nm. However, we have found that F(ab')SH rapidly reoxidizes back to F(ab')$_2$ while on the column. This can be overcome by alkylating the free SH-groups by the addition of 5 µL 50 m*M* iodoacetamine in 1 *M* NTE8 to the 45-µL sample from the G25 column.

Fab' will elute from the column later than F(ab')$_2$ resulting in a shift in the position of the peak on reduction. In most cases >95% of the F(ab')$_2$ is reduced. Following alkylation and overnight incubation, the reaction mixture typically elutes from HPLC as a triplet, containing a mixture of alkylated Fabэ and Fab'(mal), which elute in a similar postion to Fab'(SH), bispecific F(ab')$_2$ product, which elutes similarly to the parent F(ab')$_2$, and a smaller amount of bispecific F(ab')$_3$, which elutes similarly to IgG.

4. Notes

1. Two SH groups may also be produced by the reduction of the heavy/light chain disulfide bond. However, under the conditions used, this bond is not fully reduced, and any SH groups that are produced are less likely to be available for conjugation ***(1,2)***. This procedure relies on one maleimidated hinge SH-group remaining free for conjugation after the intramolecular cross-linking of adjacent SH-groups with *o*-PDM. It follows that the Fab' species chosen to be maleimidated must be derived from IgG with an odd number of hinge

region disulfide bonds. Of the mouse IgG subclasses, IgG1 and IgG2a (three bonds), and IgG3 (one bond) qualify, whereas IgG2b (four bonds) does not. $F(ab')_2$ derived from rabbit Ig (one bond) and rat IgG1 (three bonds)can also be employed. However, rat IgG2a and IgG2c both have two, and rat IgG2b has four and so cannot be used as the maleimidated partner.

2. If $F(ab')_3$ derivatives are required, the number of SH-groups at the hinge of the unmaleimidated partner should be at least two and preferably three, because this determines the number of Fab'(mal) arms that can be conjugated.
3. We have found that a few antibodies give consistently low yields of BsAb when used as the maleimidated partner. If large quantities of a derivative are required, it is worthwhile performing small scale pilot preparations to determine which maleimidated partner gives the optimal yield.
4. It is very important to avoid contamination of the Fab'-A(SH) with 2-ME. In order to minimize the risk, stop collecting when the recorder has returned two-thirds of the way back to the baseline. In order to ensure that Fab'-B(SH) is not contaminated with 2-ME left over from the first run, make sure that it is not loaded until the chart recorder has returned to the baseline.
5. Sometimes the mixture becomes slightly cloudy.
6. To avoid contamination with *o*-PDM/DMF, which elutes as a second large peak, stop collecting when the chart recorder has returned halfway to baseline.
7. To avoid loss of product, slightly over concentrate, then wash the cell with a small volume of chilled buffer.
8. It is very important to add excess iodoacetamide at this stage; otherwise the BsAb derivative can precipitate.

References

1. Glennie, M. J., McBride, H. M., Worth, A. T., and Stevenson, G. T. (1987) Preparation and performance of bispecific $F(ab'\gamma)_2$ antibody containing thioether-linked Fab'γ fragments. *J. Immunol.* **139,** 2367–2375.
2. Glennie, M. J., Tutt, A. L., and Greenman, J. (1993) Preparation of multispecific $F(ab')_2$ and $F(ab')_3$ antibody derivatives, in *Tumour Immunobiology, A Practical Approach* (Gallagher, G., Rees, R. C., and Reynolds, C. W., eds.), Oxford University Press, Oxford, UK, pp. 225–244.

153

Phage Display

Biopanning on Purified Proteins and Proteins Expressed in Whole Cell Membranes

George K. Ehrlich, Wolfgang Berthold, and Pascal Bailon

1. Introduction

Phage display technology ***(1)*** is rapidly evolving as a biomolecular tool with applications in the discovery of ligands for affinity chromatography ***(2–10)*** and drugs ***(11–17)***, in the study of protein–protein interactions ***(18–21)***, and in epitope mapping ***(22–26)***, among others. This technology relies on the utilization of phage display libraries (*see* **Notes 1** and **2**) in a screening process known as biopanning ***(27)***.

In biopanning, the phage display library is incubated with a target molecule. The library can be incubated directly with an immobilized target (as in a purified protein attached to a solid support or a protein expressed in a whole cell surface membrane) or preincubated with a target, prior to capture on a solid support (as in Streptavidin capture). As in affinity chromatography, noninteracting peptides/proteins are washed away and then interacting peptides/proteins are eluted specifically or nonspecifically. These interacting phage display peptides/proteins can be amplified by bacterial infection to increase their copy number. This screening/amplification process can be repeated as necessary to obtain higher affinity phage display peptides/proteins. The desired sequences are obtained by DNA sequencing of isolated phage DNA (**Fig. 1**).

In this chapter, methodologies to generate targets for phage display and to biopan against generated targets with phage display peptide libraries are described (*see* **Note 3**).

2. Materials

2.1. Stock Solutions

1. Phosphate-buffered saline (PBS): 137 m*M* NaCl, 3 m*M* KCl, 8 m*M* Na_2HPO_4, 1.5 m*M* KH_2PO_4, pH 7.5.
2. DMEM: Dulbecco's modified Eagle's medium with 4 m*M* L-glutamine and 4.5 g/L of glucose (Cellgro™, Mediatech, Herndon, VA).
3. Immobilization buffer: 0.1 *M* $NaHCO_3$, pH 8.6.
4. Blocking buffer: Immobilization buffer containing 5 mg/mL of Bovine serum albumin (BSA) (or target) and 0.02% sodium azide (NaN_3).

From: *The Protein Protocols Handbook, 2nd Edition*
Edited by: J. M. Walker © Humana Press Inc., Totowa, NJ

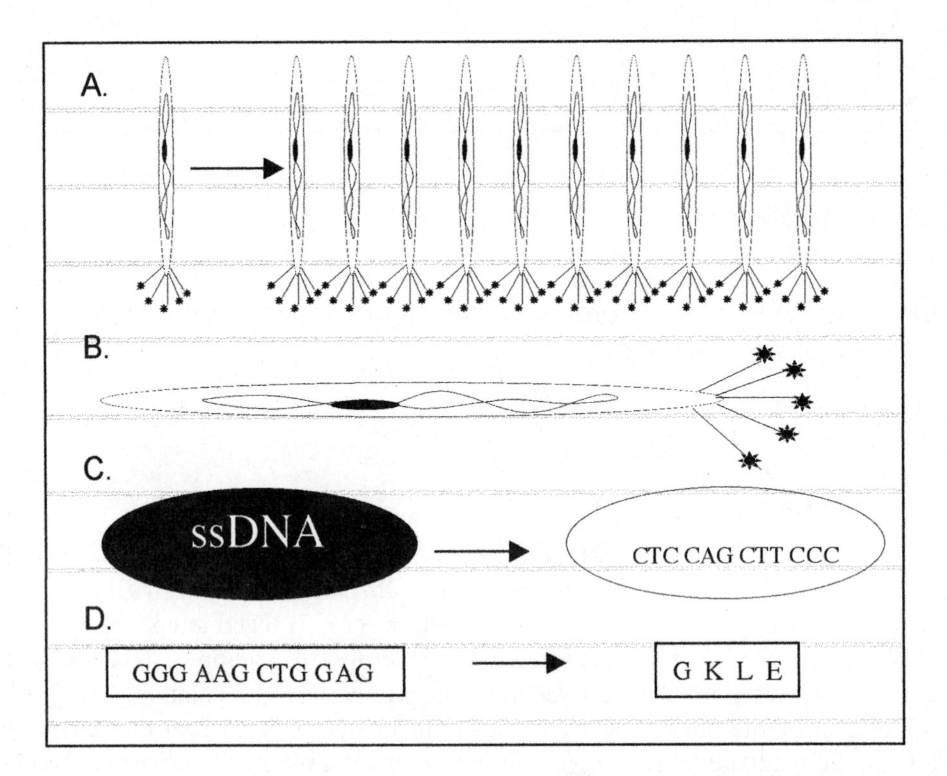

Fig. 1. Following elution of interacting peptides peptides/proteins in phage display, their copy number can be increased by bacterial infection, as these peptides/proteins are displayed on bacteriophage **(A)**. Interacting phage displayed peptides/proteins can then be isolated **(B)** in order to purify and sequence their coding ssDNA **(C)**. In turn, these codons can be translated to determine the amino acid sequences of the interacting phage-derived peptides/proteins **(D)**.

5. Tris buffer: 50 m*M* Tris-HCl, 150 m*M* NaCl, pH 7.5.
6. Biopan buffer: Tris buffer with 0.1% (v/v) Tween 20.
7. Elution buffer: 0.2 *M* glycine-HCl, pH 2.2, with 1 mg/mL of BSA.
8. PEG: 20% (w/v) Polyethylene glycol-8000 (Sigma), 2.5 *M* NaCl. Autoclaved and stored at room temperature.

2.2. Phage Display on Purified Proteins

2.2.1. Target Generation

2.2.1.1. Expression and Purification of Protein Target

Protein targets for phage display are typically produced using recombinant protein technology *(28)*. All starting materials used here were prepared at Hoffmann-LaRoche and supplied as either *E. coli* cell paste or cell culture medium by the fermentation group and purified in our laboratories (Biopharmaceutical R&D, Hoffmann-LaRoche, Nutley, NJ).

2.2.1.2. Biotinylation for Streptavidin Capture

1. EZ-Link™ Sulfo-NHS-Biotinylation Kit (Pierce, Rockford, IL) containing 25 mg of sulfo-NHS-biotin, one BupH™ pack, 10 mL D-Salt™ dextran desalting column MWCO

5000, 1 mL of 2-hydroxyazo-benzene-4′-carboxylic acid (HABA, 10 m*M* in approx 0.01 *N* NaOH), and 10 mg of Avidin.
2. Protein for biotinylation.
3. PBS.
4. 1.5-mL Microcentrifuge tubes (10 × 38 mm).
5. Round-bottom polypropylene test tubes, 5 or 13 mL (75 × 12 mm or 95 × 16.8 mm).
6. UV-visible spectrophotometer.

2.2.2. Biopanning Against Purified Immobilized Target

2.2.2.1. Target Immobilization

1. 60 × 15 mm Petri dishes or 96-well microtiter plates.
2. Target protein: 1.5 mL (100 µg/mL of immobilization buffer) for a Petri dish (one round of biopanning) or 0.15 mL/well (96-well microtiter plate).
3. Sterile needle (18-gauge).
4. Immobilization buffer.
5. Blocking buffer.
6. Biopan buffer.

2.2.2.2. Affinity Selection

1. Phage display library: 1–2 × 10^{11} phage/round.
2. Plates containing immobilized target.
3. Target ligand: 0.1–1 m*M* ligand for each elution.
4. Target protein: 100 µg for each elution.
5. Biopan buffer.
6. Elution buffer.
7. 1 *M* Tris-HCl, pH 9.1.

2.2.3. Biopanning Against Biotinylated Target

2.2.3.1. Immobilization of Streptavidin

1. 60 × 15 mm Petri dishes or 96-well microtiter plates.
2. Sterile needle (18-gauge).
3. Immobilization buffer: 0.1 *M* $NaHCO_3$, pH 8.6.
4. Blocking buffer: containing 0.1 µg of Streptavidin/mL (to complex any residual biotin in BSA).
5. Biopan buffer.
6. Streptavidin: 100 µg/mL of immobilization buffer (1.5 mL/dish [one round of biopanning] or 0.15 mL/well [96-well microtiter plate]).

2.2.3.2. Affinity Selection

1. Phage display library: 1–2 × 10^{11} phage/round.
2. Plates containing immobilized Streptavidin.
3. Target ligand: 0.1–1 m*M* ligand/elution.
4. Target protein: 100 µg/elution.
5. Biopan buffer.
6. Elution buffer.
7. 1 *M* Tris-HCl, pH 9.1.

2.3. Phage Display on Proteins Expressed in Whole Cells

2.3.1. Target Generation

2.3.1.1. Infection for Biopanning on Insect Cells

1. Cells for infections (*see* **Note 4**).
2. Cell incubator.

3. Laminar flow hood.
4. Baculovirus containing the gene for the desired receptor.
5. EX-CELL™ 401 medium (JRH Biosciences, Lenexa, KS).
6. Fetal bovine serum.
7. 35 × 10 mm Six-well tissue culture plates.
8. 15-mL Conical-bottom polypropylene test tubes with cap (17 × 100 mm).

2.3.1.2. Transfection for Biopanning on Mammalian Cells

1. Cells for transfections (*see* **Note 5**).
2. Cell incubator.
3. Laminar flow hood.
4. Desired cDNA in mammalian cell expression vector.
5. Lipofectamine™ (Gibco-BRL, Life Technologies, Gaithersburg, MD).
6. DMEM.
7. Gibco-BRL Penicillin–Streptomycin, liquid (10,000 IU of Penicillin/10,000 µg/mL of Streptomycin, Life Technologies).
8. Complete DMEM: DMEM supplemented with 10% fetal bovine serum, 100 U/mL of Penicillin, 100 µg/mL of Streptomycin.
9. 35 × 10 mm Six-well tissue culture plates.
10. 15-mL Conical-bottom polypropylene test tubes with cap (17 × 100 mm).

2.3.2. Affinity Selection by Biopanning Against Control (Uninfected) and Infected/Transfected Whole Cells

1. Infected/transfected and uninfected (control) cells.
2. Grace's insect cell medium.
3. Nonfat dry milk.
4. Phage display library: 1–2 × 10^{11} phage/round.
5. 15-mL Conical-bottom polypropylene test tubes with cap (17 × 100 mm).
6. Urea elution buffer: 6 *M* urea, pH 3.0.
7. 2 *M* Tris base to neutralize pH.
8. HEPES buffer: DMEM, 2% nonfat dry milk, 20 m*M* HEPES (Sigma), pH 7.2.

2.4. General Molecular Techniques

2.4.1. Phage Titering

1. Phage (input or output) from library.
2. Bacterial strain: XL1-Blue (Stratagene, La Jolla, CA).
3. Tetracycline (Sigma, St. Louis, MO), 12.5 mg/mL stock solution in 50% ethanol. Add to a final concentration of 12.5 µg/mL.
4. LB medium: 10 g of Bacto-Tryptone, 5 g of yeast extract, 5 g of yeast, 5 g of NaCl, in 1 L, pH 7.0. Sterilize by autoclaving.
5. 95 × 15-mm Petri dishes.
6. LB plates: LB medium, 15 g/L of agar. Autoclave medium. Allow to cool (<70°C). Add Tetracycline (12.5 µg/mL of medium). Pour medium into plates. Allow coated plates, partially covered, to cool to room temperature. Cover, invert, and store coated plates at 4°C. Warm plates slowly to 37°C before titering.
7. 250-mL Triple baffle shake flasks (Bellco Glass, Inc., Vineland, NJ).
8. Shaking water bath for 37°C and 45°C incubations.
9. UV-visible spectrophotometer.
10. Microwave oven.

11. Agarose top: 10 g of Bacto-Tryptone, 5 g of yeast extract, 5 g of NaCl, 1 g of $MgCl_2 \cdot 6H_2O$, 7 g of agarose/L. Autoclave. Save in 200-mL aliquots and store at room temperature. Melt Agarose Top in a microwave oven prior to use.
12. 17 × 100 mm Sterile polypropylene culture tubes.

2.4.2. Phage Amplification

1. Bacterial strain: XL1-Blue.
2. LB broth.
3. Tetracycline.
4. 250-mL Triple baffle shake flasks.
5. Eluate.
6. 30-mL Polypropylene centrifuge tubes (Rochester, NY).
7. 1.5-mL Microcentrifuge tubes.
8. PEG.
9. Tris buffer (*see* **Subheading 2.1.**).
10. NaN_3.

2.4.3. Plaque Amplification and Purification of Single-Stranded Bacteriophage M13 DNA

1. QIAprep Spin M13 Kit (QIAGEN, Chatsworth, CA) containing QIAprep spin columns, M13 precipitation buffer (MP buffer), M13 lysis and binding buffer (MLB buffer, concentrated PE buffer, and 2-mL collection tubes.
2. Bacterial strain: XL1-Blue.
3. LB broth.
4. Tetracycline.
5. 250-mL Triple baffle shake flasks.
6. Plates containing selected phage.
7. Pasteur pipets (Fisher Scientific).
8. 12 × 75 mm Sterile polypropylene culture tubes.
9. 1.5-mL Microcentrifuge tubes.
10. 1 m*M* Tris-HCl, pH 8.5.

3. Methods

3.1. Phage Display on Purified Proteins

3.1.1. Target Generation

3.1.1.1. Expression and Purification of Protein Target

See **Subheading 2.2.2.1.**

3.1.1.2. Biotinylation for Streptavidin Capture

An efficient coupling chemistry for target biotinylation is via reaction of primary amine on target protein with the activated *N*-hydroxysulfosuccinimide ester of biotin (EZ-Link Sulfo-NHS-Biotinylation Kit Instructions, Pierce, USA).

1. Add 2 mg of target protein in 1 mL of PBS to a 20-fold molar excess of Sulfo-NHS-Biotin.
2. Incubate on ice for 2 h.
3. Meanwhile, equilibrate dextran desalting column (mol wt cutoff 5000) with 30 mL of PBS.
4. Load sample onto gel.

5. Separate biotin from biotinylated protein with PBS.
6. Collect 1-mL fractions.
7. Measure protein absorbance at 280 nm or by protein assay.
8. Pool fractions containing protein.
9. Store biotinylated protein at 4°C until ready for use.
10. Use the HABA method to determine amount of biotinylation (see kit instructions).

3.1.2. Biopanning Against Purified Immobilized Target (see **Note 7**)

3.1.2.1. Target Immobilization

The following method can be used to immobilize target proteins by hydrophobic adsorption on polystyrene plates whereby the target is coated onto the plate and then excess target is removed.

1. Add 1.5 mL (150 μL of 100 μg target/mL) of immobilization buffer to each 65 × 15-mm Petri dish (96-well microtiter plate).
2. Swirl plate repeatedly to ensure that plate is completely coated with target solution.
3. Incubate plate(s) overnight with gentle shaking in a container containing wet paper towels at 4°C.
4. Carefully remove target solution by aspiration through a sterile needle.
5. Fill each plate (well) completely with a sterile solution of blocking buffer containing 5 mg/mL of BSA or target (if available) and 0.02% NaN_3.
6. Incubate for 1 h at 4°C.
7. Remove solution by aspiration as described in **step 4**.
8. Wash plates (wells) by rinsing with biopan buffer.
9. Repeat wash step 5×.

3.1.2.2. Affinity Selection

1. Add 1–2 × 10^{11} phage to 1 mL (100 μL) of biopan buffer/plate (well).
2. Pipet phage solution onto plates (wells).
3. Incubate for 60 min at room temperature with gentle rocking.
4. Carefully remove phage solution from plates (wells).
5. Wash plates (wells) 10× with biopan buffer.
6. Elute bound phage specifically (or nonspecifically) by adding 1 mL (100 μL) of target ligand (0.1–1 m*M*) or 100 μg of free target in biopan buffer (elution buffer) to plates (wells).
7. Incubate with gentle rocking for 60 min (10 min) at room temperature.
8. Collect eluates in microcentrifuge tubes.
9. Neutralize nonspecific eluate with 150 μL (15 μL) of 1 *M* Tris-HCl, pH 9.1.
10. Titer 1 μL of eluate (*see* **Subheading 3.3.1.**).
11. Amplify rest of eluate or store at 4°C (*see* **Subheading 3.3.2.**).
12. Repeat additional rounds of biopanning using the same pfu (1–2 × 10^{11}) as described in **step 1**.
13. Pick plaques from the last round (unamplified) phage titer for amplification and isolation of single-stranded bacteriophage M13 DNA (*see* **Subheading 3.3.3.**).

3.1.3. Biopanning Against Biotinylated Target

The described protocol is a modification of that described by Smith and Scott *(27)*.

3.1.3.1. Immobilization of Streptavidin

Immobilize Streptavidin as described in **Subheading 3.1.2.1.** Add 0.1 μg/mL of Streptavidin to the blocking buffer to complex any biotin that may be contained in the BSA.

3.1.3.2. Affinity Selection

1. Incubate phage display library (1–2 × 10^{11} pfu) with 0.1 μg of biotinylated target (*see* **Subheading 3.1.1.2.**) in 500 mL of biopan buffer overnight at 4°C.
2. Add phage display library–target mixture to Streptavidin coated plates.
3. Incubate at room temperature for 10 min.
4. Add 0.2 m*M* biotin in biopan buffer (500 μL) to displace any Streptavidin-binding phage.
5. Incubate an additional 5 min.
6. Proceed from **step 4** in **Subheading 3.1.2.2.** (*see* **Note 8**).

3.2. Phage Display on Proteins Expressed in Whole Cells

3.2.1. Target Generation

3.2.1.1. Infection for Biopanning on Insect Cells

1. Seed 1–2 × 10^6 cells in 1 mL of EX-CELL™ 401 medium containing 1% Fetal bovine serum into 35-mm plates (one well/infection).
2. Let cells adhere for 1 h at 27°C.
3. Add 0.5–5 MOI Baculovirus (for target and control) to medium.
4. Grow for 48–72 h at 27°C.
5. Harvest Baculovirus-infected cells into 15-mL polypropylene tube(s).

3.2.1.2. Transfection for Biopanning on Mammalian Cells

Transient transfections can be performed using Lipofectamine™ to make a complex with the target cDNA and incubating the complex with the desired cell line. Satisfactory results can be obtained using target cDNA in pCDNA3 (Invitrogen) ***(29)***.

1. Seed 1–3 × 10^5 cells/well in 2 mL of complete DMEM, in a six-well tissue culture plate.
2. Grow cells at 37°C in an atmosphere of 5% CO_2–95% air to 80% confluence.
3. Use 3 μg of DNA and 8 mL of Lipofectamine™ for each transfection.
4. Dilute DNA and Lipofectamine™ separately in serum- and antibiotic-free DMEM (100 μL/transfection).
5. Combine solutions with gentle mixing.
6. Incubate at room temperature for 30 min.
7. Meanwhile, wash cells with 2 mL of serum- and antibiotic-free DMEM.
8. Add 0.8 mL of serum- and antibiotic-free DMEM, for each transfection, to the DNA–lipid mixture.
9. Mix gently.
10. Carefully place DNA–lipid mixture over rinsed cells (1 mL total volume).
11. Incubate complex mixture with cells at 37°C in an atmosphere of 5% CO_2–95% air for 5 h.
12. Add 1 mL of DMEM, 20% fetal bovine serum to each transfection mixture after 5 h incubation.
13. Incubate complex mixture at 37°C in an atmosphere of 5% CO_2–95% air for 18 h.
14. Replace medium with 2 mL of complete DMEM.
15. Collect cells after an additional 24-h incubation at 37°C in an atmosphere of 5% CO_2–95% air.

3.2.2. Affinity Selection by Biopanning Against Control (Uninfected) and Infected/Transfected Whole Cells

There are at least two published protocols that describe the use of phage display targeting proteins expressed as cell surface receptors in whole cells ***(30,31)***. The following method is an adaptation of that described by Goodson et al. ***(30)***, who discovered novel urokinase receptor antagonists with phage display (*see* **Note 9**).

1. Resuspend control insect cells (10^6) in 0.5 mL of Grace's insect cell medium–2% nonfat dry milk containing $1–2 \times 10^{11}$ bacteriophage library.
2. Incubate cells with bacteriophage library for 30 min at room temperature.
3. Sediment cells in a centrifuge at 1000*g* for 10 min.
4. Carefully remove supernatant containing bacteriophage.
5. Incubate supernatant with receptor bearing insect cells (10^6) for 30 min with gentle agitation.
6. Sediment cells in a centrifuge at 1000*g* for 5 min.
7. Discard supernatant.
8. Wash cells with 10 mL of Grace's insect cell medium–2% nonfat dry milk.
9. Sediment cells in a centrifuge at 1000*g* for 5 min.
10. Discard supernatant.
11. Repeat **steps 8–10** four more times.
12. Elute interacting phage with 0.5 mL of 6 *M* urea, pH 3, for 15 min at room temperature.
13. Neutralize eluate by adding 10 µL of 2 *M* Tris base.
14. Titer 1 µL (*see* **Subheading 3.3.1.**).
15. Amplify rest of eluate or store at 4°C (*see* **Subheading 3.3.2.**).
16. Resuspend receptor transfected mammalian cells (10^6) in 0.5 mL of HEPES buffer.
17. Add amplified eluate ($1–2 \times 10^{11}$ pfu).
18. Incubate at room temperature for 30 min.
19. Sediment cells in a centrifuge at 1000*g* for 5 min.
20. Discard supernatant.
21. Wash cells with 10 mL of HEPES buffer.
22. Sediment cells in a centrifuge at 1000*g* for 5 min.
23. Discard supernatant.
24. Repeat **steps 21–23** four more times.
25. Elute interacting phage with 0.5 mL of 6 *M* urea, pH 3.0, for 15 min at room temperature.
26. Neutralize eluate by adding 10 µL of 2 *M* Tris base.
27. Titer 1 µL (*see* **Subheading 3.3.1.**).
28. Amplify the rest of the eluate or store at 4°C (*see* **Subheading 3.3.2.**).
29. Repeat **steps 5–14** using an input of 1×10^{11} pfu.
30. Pick plaques for amplification and isolation of single stranded bacteriophage M13 DNA (*see* **Subheading 3.3.3.**).

3.3. General Molecular Techniques

3.3.1. Phage Titering

Phage titering is a way of calculating the input and output of phage particles in biopanning. This measurement is determined experimentally by infecting *E. coli* in a petri dish and then evaluated by quantifying the number of plaque-forming units (pfu) left overnight on the bacterial lawn.

1. Streak out XL1-Blue cells onto LB plates containing Tetracycline.
2. Invert plates.
3. Incubate at 37°C for 24 h.
4. Store (approx 1 mo) wrapped in parafilm at 4°C until needed.
5. Set up overnight culture by inoculating a single colony of XL1-Blue in LB broth (20 mL) containing Tetracycline in a triple baffle shake flask (*see* **Note 6**).
6. Incubate at 37°C with vigorous shaking.
7. Dilute overnight culture 1:100 in 20 mL of LB broth containing Tetracycline in another shake flask.

8. Shake vigorously at 37°C until mid-log phase (OD_{600} approx 0.5).
9. Set up for titering while cells are growing.
10. Melt Agarose Top in microwave oven (3 min on high setting).
11. Add 3 mL of Agarose Top into sterile culture tubes (one tube/dilution) in a 45°C water bath.
12. Prewarm LB plates (one plate/dilution) at 37°C until needed.
13. Prepare 10-fold serial dilutions (10^8–10^{11} for amplified eluates and phage libraries and 10^1–10^4 for unamplified eluates).
14. Dispense 200 µL of culture at mid-log phase into microcentrifuge tubes (one tube/dilution).
15. Add 10 µL of each dilution to microcentrifuge tubes containing culture.
16. Mix on a vortex mixer.
17. Incubate at room temperature for 5 min.
18. Add infected cells to preequilibrated culture tubes containing Agarose Top.
19. Vortex-mix rapidly.
20. Pour onto prewarmed LB plate.
21. Spread Agarose Top evenly to cover surface of plate.
22. Cool plates at room temperature for 5 min with cover slightly off.
23. Cover plates completely.
24. Incubate inverted plates overnight at 37°C.
25. Count plates having approx 10^2 plaques. Multiply number of plaques with dilution factor to determine the pfu/10 µL.

3.3.2. Phage Amplification

1. Prepare an overnight culture by inoculating a single colony of XL1-Blue in LB broth containing 12.5 µg of Tetracycline (20 mL) in a 250-mL shake flask.
2. Incubate at 37°C with vigorous shaking.
3. Add eluate to a 1:100 dilution of the overnight culture of XL1-Blue in LB broth (20 mL) containing Tetracycline in another shake flask.
4. Incubate culture with vigorous shaking for 4.5 h at 37°C.
5. Spin culture in a centrifuge tube for 10 min at 12,000*g* at 4°C.
6. Recentrifuge supernatant in another centrifuge tube.
7. Pipet supernatant into another centrifuge tube.
8. Add 1/6 vol of PEG to precipitate phage at 4°C overnight.
9. Sediment precipitated phage by centrifugation at 12,000*g* at 4°C for 15 min.
10. Discard supernatant.
11. Recentrifuge for 1 min.
12. Remove residual supernatant.
13. Suspend pellet in 1 mL of Tris.
14. Place suspension in a microcentrifuge tube.
15. Centrifuge at high speed in a microcentrifuge for 5 min at 4°C.
16. Reprecipitate phage with 1/6 vol of PEG in a new tube.
17. Incubate on ice for 60 min.
18. Sediment phage in microcentrifuge at high speed for 10 min at 4°C.
19. Remove supernatant.
20. Recentrifuge.
21. Discard any residual supernatant.
22. Suspend pellet in 200 µL of Tris buffer containing 0.02% NaN_3.
23. Sediment insoluble matter by centrifugation for 1 min at high speed in a microcentrifuge.
24. Transfer amplified eluate in supernatant to new tube.
25. Titer amplified eluate (*see* **Subheading 3.3.1.**).

3.3.3. Plaque Amplification and Purification of Single-Stranded Bacteriophage M13 DNA

Single-stranded bacteriophage M13 DNA from isolated plaques can be purified using the QIAprep Spin M13 Kit (QIAGEN, Chatsworth, CA). Satisfactory results were also obtained by using phenol–chloroform extraction *(32)*.

1. Set up an overnight culture by inoculating a single colony of XL1-Blue in LB broth containing Tetracycline (20 mL) in 250-mL shake flasks.
2. Incubate at 37°C with vigorous shaking.
3. Select 10 colonies from the desired plates using a pasteur pipet.
4. Place each colony in a separate sterile culture tube containing 1 mL of a 1:100 diluted culture of XL1-Blue containing Tetracycline.
5. Incubate tubes with vigorous shaking for approx 4.5 h.
6. Transfer cultures to a 1.5-mL microcentrifuge tube.
7. Centrifuge in a microcentrifuge at high speed for 10 min.
8. Transfer the supernatant to a fresh 1.5-mL microcentrifuge tube.
9. Centrifuge in a microcentrifuge at high speed for 8 min.
10. Transfer the supernatant to another 1.5-mL microcentrifuge tube.
11. Use 750 µL of supernatant for obtaining single-stranded DNA.
12. Save the rest for phage stock at 4°C.
13. Add 7.5 µL of MP buffer (1:100 volume) to the 750 µL of supernatant to initiate phage precipitation.
14. Vortex-mix.
15. Incubate at room temperature for 2 min.
16. Add the precipitated phage sample to a QIAprep spin column in a 2-mL collection tube.
17. Centrifuge the sample in a microcentrifuge at 8000 rpm for 15 s to recover precipitated phage on a silica membrane.
18. Discard the flow-through from the collection tube.
19. Add the rest of the sample (57.5 µL) to the spin column.
20. Repeat **steps 17** and **18**.
21. Add 0.7 mL of MLB buffer to the spin column for phage lysis and subsequent binding of M13 DNA to the column.
22. Centrifuge the sample in a microcentrifuge at 8000 rpm for 15 s.
23. Add another 0.7 mL of MLB buffer to the spin column.
24. Incubate for 1 min at room temperature for complete phage lysis.
25. Centrifuge the column at 8000 rpm for 15 s.
26. Add 0.7 mL of diluted PE buffer (5 parts EtOH/part PE buffer concentrate).
27. Centrifuge the column at 8000 rpm for 15 s to remove excess salt.
28. Remove PE buffer from the collection tube.
29. Recentrifuge the column at 8000 rpm for 15 s to avoid carryover of residual EtOH.
30. Place the spin column into a 1.5-mL microcentrifuge tube.
31. Add 100 µL of 1 m*M* Tris-HCl, pH 8.5, to the column membrane.
32. Incubate for 10 min.
33. Centrifuge the column at 8000 rpm for 30 s to elute single-stranded M13 DNA.
34. Sequence DNA (*see* **Note 10**).
35. Translate the DNA sequence from the reduced genetic code (32 codons) of the libraries randomized region to obtain the amino acid sequence of the affinity selected phage display peptide (*see* **Note 11**).
36. Determine substrate specificity for binding clones (*see* **Note 12**).

4. Notes

1. Filamentous phage, such as f1, fd, and M13, are a group of highly homologous bacteriophage that are capable of infecting a number of Gram-negative bacteria. The infectious nature of these phage are determined by their ability to recognize the F pilus of their host bacteria. This particular biorecognition is effected by one (pIII) of five coat proteins at its amino-terminus. The minor coat protein, pIII, consists of 406 amino acids and exists as five copies oriented on one tip of the cylindrical phage particle. Interestingly, only one copy of pIII is necessary for phage infectivity.
2. Phage display libraries consist of peptides/proteins individually expressed on the surface of filamentous M13 bacteriophage. The expressed peptides/proteins are typically engineered by inserting foreign DNA fragments encoding for the desired peptide/protein into the 5' region of gene III. The insertion allows the peptide/protein product to be fused to the amino-terminus of the minor coat protein pIII. The peptides/proteins can be variable in length (5–1000 amino acids) and number (tens to billions of copies). Moreover, the peptides/proteins can be identified because the DNA encoding them resides inside the phage virion.
3. Current practical limitations on the size of phage display libraries (approx 10^{10} permutations) have placed constraints on the exploration of total molecular diversity ($20^{\text{length of peptide/protein}}$ = no. of possible permutations). For this reason, some phage display libraries have been engineered to minimize the need for ultimate molecular diversity by being strategic in design with respect to structure (as in microprotein scaffolds in constrained peptide libraries) and sequence (as in conserved amino acids in variable domains of antibodies in F_v, single chain F_v, and FAb peptide libraries). Although any of the described phage display libraries can be used for biopanning, the three phage display libraries (7-mer, 9-mer [33], and 12-mer) used in our studies were linear and numbered about 10^9 in molecular diversity. The commercial use of phage display libraries may require a licensing agreement from its proprietor ***(34)***.
4. The insect cells used in our studies were Hi Five™ Cells (BTI-TN-5B1-4, Invitrogen, Carlsbad, CA). However, successful results were reported using Sf9 Cells ***(30,31)***.
5. The mammalian cells used in our studies were human embryonic kidney 293 (HEK293, cat. no. 45504, ATCC, Manassas, VA). However, successful results were reported using Chinese Hamster Ovary cells ***(35)***, COS-1 ***(31)***, and COS-7 monkey kidney cells ***(30)***.
6. Be sure to set up an overnight culture the day before phage titering is to be performed. If not, inoculate 5–10 mL of LB broth containing 12.5 µg of Tetracycline/mL of broth with a single colony of XL1-Blue the morning titering is to be performed. Let it grow until mid-log phase $OD_{600\sim}$ approx 0.5). Proceed as described from **step 9, Subheading 3.3.1.**
7. The methods described here were performed essentially following the instructions provided for the Ph.D-7 and Ph.D.-12 Phage Display Library Kit (New England Biolabs, USA). However, conditions (buffers, input phage, incubation/elution time and temperature, number of washes, number rounds of biopanning, etc.) may need to be optimized for ideal results.
8. The key to successful biopanning using Streptavidin capture is to avoid the binding of phage to Streptavidin by competition with biotin. However, phage-derived peptide sequences containing the HPQ amino acid motif can be presumed to be Streptavidin binders ***(36)***.
9. Expression of cell surface receptors can be examined by immunocytochemistry on whole cells and/or immunoblots on membrane homogenates ***(29)***.
10. All sequencing was performed on an Applied Biosystems 373A or 377 automated DNA sequencer using the Sanger method with fluorescently labeled dideoxy terminator chemistry and ThermoSequenase enzyme (Amersham Pharmacia Biotech).

11. The 32 codons encoding the possible 20 amino acids are (1) TTT = F (phenylalanine), (2) TTG = L (leucine), (3) TCT = S (serine), (4) TCG = S, (5) TAT = Y (tyrosine), (6) TAG = Q (glutamine), (7) TGT = C (cysteine), (8) TGG = W (tryptophan), (9) CTT = L (leucine), (10) CTG = L, (11) CCT = P (proline), (12) CCG = P, (13) CAT = H (histidine), (14) CAG = Q (glutamine), (15) CGT = R (arginine), (16) CGG = R, (17) ATT = I (isoleucine), (18) ATG = M (methionine), (19) ACT = T (threonine), (20) ACG = T, (21) AAT = N (asparagine), (22) AAG = K (lysine), (23) AGT = S, (24) AGG = R (arginine), (25) GTT = V (valine), (26) GTG = V, (27) GCT = A (alanine), (28) GCG = A, (29) GAT = D (aspartic acid), (30) GAG = E (glutamic acid), (31) GGT = G (glycine), and (32) GGG = G.
12. Biopanning is an ideal way of selecting for phage displayed binding peptides. However, it is important that substrate specificity for selected phage is determined, as not all phage are always binding phage. Some of these nonspecific binding phage might be plastic binders *(37)* or they might be favored by natural selection (*see* **Note 13**) during phage amplication. Although phage selectivity can be built directly into the target/phage interaction *(38)*, the most common and fastest way of determining substrate specificity is by enzyme-linked immunosorbent assay (ELISA) using an antiphage antibody *(27,39)*.
13. The "quickscreen" has been employed to avoid this problem, as no phage amplification occurs between rounds of biopanning, that is, recovered phage from the first round is used for the second round upon elution, and so forth *(5)*. In this method, stepwise pH elution can be used to select for high-affinity binding phage *(5)*.

Acknowledgment

We appreciate the assistance of G. Katarina Luhr (Karolinska Institute, Stockholm, Sweden) in rendering **Fig. 1**.

References

1. Smith, G. P. (1985) Filamentous fusion phage: novel expression vectors that display cloned antigens on the virion surface. *Science* **228,** 1315–1317.
2. Braisted, A. C. and Wells, J. A. (1996) Minimizing a binding domain from protein A. *Proc. Natl. Acad. Sci. USA* **93,** 5688–5692.
3. Murray, A., Sekowski, M., Spencer, D. I., Denton, G., and Price, M. R. (1997) Purification of monoclonal antibodies by epitope and mimotope affinity chromatography. *J. Chromatogr. A* **782,** 49–54.
4. Nord, K., Gunneriusson, E., Ringdahl, J., Stahl, S., Uhlen, M., and Nygren, P. A. (1997) Binding proteins selected from combinatorial libraries of an alpha-helical bacterial receptor domain. *Nat. Biotechnol.* **15,** 772–777.
5. Ley, C. A. (1997) Custom affinity ligands from phage display for large-scale affinity purification, in *IBC International Conference on Display Technologies*, Lake Tahoe, Nevada.
6. Maclennan, J. (1997) The generation of process suitable, rugged, targeted affinity ligands using phage display technology, in *Twelfth Symposium on Affinity Interactions: Fundamentals and Applications of Biomolecular Recognition.* Abstract L30, Kalmar, Sweden.
7. Ehrlich, G. K. and Bailon, P. (1998) Identification of peptides that bind to the constant region of a humanized IgG1 monoclonal antibody using phage display. *J. Mol. Recogn.* **11,** 121–125.
8. Ringdahl, J., Gunneriusson, E., Gronlund, H., Uhlen, M., and Nygren, P.-A. (1999) Selection of a robust affinity ligand for IgA from a combinatorial protein library displayed on phage, in *Thirteenth International Symposium on Affinity Technology and BioRecognition*, Compiegne, France, P60.

9. Sengupta J., Sinha P., Mukhopadhyay, C., and Ray, P. K. (1999) Molecular modeling and experimental approaches toward designing a minimalist protein having Fc-binding activity of Staphylococcal protein A. *Biochem. Biophys. Res. Commun.* **256,** 6–12.
10. Ehrlich, G. K., Bailon, P., and Berthold, W. (2000) Phage display technology. Identification of peptides as model ligands for affinity chromatography. *Meth. Mol. Biol.* **147,** 209–220.
11. Doorbar, J. and Winter, G. (1994) Isolation of a peptide antagonist to the thrombin receptor using phage display. *J. Mol. Biol.* **244,** 361–369.
12. Nagai, Y., Tucker, T., Ren, H., Kenan, D. J., Henderson, B. S., Keene, J. D., et al. (2000) Inhibition of polyglutamine protein aggregation and cell death by novel peptides identified by phage display screening. *J. Biol. Chem.* **275,** 10,437–10,442.
13. Binetruy-Tournaire, R., Demangel, C., Malavaud, B., Vassy, R., Rouyre, S., Kraemer, M., et al. (2000) Identification of a peptide blocking vascular endothelial growth factor (VEGF)-mediated angiogenesis. *EMBO J.* **19,** 1525–1533.
14. Szardenings, M., Muceniece, R., Mutule, I., Mutulis, F., and Wikberg, J. E. (2000) New highly specific agonistic peptides for human melanocortin MC(1) receptor. *Peptides* **21,** 239–243.
15. Beck, Z. Q., Hervio, L., Dawson, P. E., Elder, J. H., and Madison, E. L. (2000) Identification of efficiently cleaved substrates for HIV-1 protease using a phage display library and use in inhibitor development. *Virology* **274,** 391–401.
16. Hammond, F., Cavanagh, A., Morton, H., Hillyard, N., Papaioannou, A., Clark, M., et al. (2000) Isolation of antibodies which neutralize the activity of early pregnancy factor. *J. Immunol. Meth.* **244,** 175–184.
17. Mummert, M. E., Mohamadzadeh, M., Mummert, D. I., Mizumoto, N., and Takashima, A. (2000) Development of a peptide inhibitor of hyaluronan-mediated leukocyte trafficking. *J. Exp. Med.* **192,** 769–780.
18. Hong, S. S. and Boulanger, P. (1995) Protein ligands of the human adenovirus type 2 outer capsid identified by biopanning of a phage-displayed peptide library on separate domains of wild-type and mutant penton capsomers. *EMBO J.* **14,** 4714–4727.
19. Stoop, A. A., Jespers, L., Lasters, I., Eldering, E., and Pannekoek, H. (2000) High-density mutagenesis by combined DNA shuffling and phage display to assign essential amino acid residues in protein–protein interactions: application to study structure-function of plasminogen activation inhibitor 1 (PAI-I). *J. Mol. Biol.* **301,** 1135–1147.
20. Cochran, A. G. (2000) Antagonists of protein–protein interactions. *Chem. Biol.* **7,** R85–R94.
21. Huang, W., Zhang, Z., and Palzkill, T. (2000) Design of potent beta-lactamase inhibitors by phage display of beta-lactamase inhibitory protein. *J. Biol. Chem.* **275,** 14,964–14,968.
22. Cwirla, S. E., Peters, E. A., Barrett, R. W., and Dower, W. J. (1990) Peptides on phage: a vast library of peptides for identifying ligands. *Proc. Natl. Acad. Sci. USA* **87,** 6378–6382.
23. Scott, J. K. and Smith, G. P. (1990) Searching for peptide ligands with an epitope library. *Science* **249,** 386–390.
24. Grabowska, A. M., Jennings, R., Laing, P., Darsley, M., Jameson, C. L., Swift, L., and Irving, W. L. (2000) Immunization with phage displaying peptides representing single epitopes of the glycoprotein G can give rise to partial protective immunity to HSV-2. *Virology* **269,** 47–53.
25. Bentley, L., Fehrsen, J., Jordaan, F., Huismans, H., and du Plessis, D. H. (2000) Identification of antigenic regions on VP2 of African horsesickness virus serotype 3 by using phage-displayed epitope libraries. *J. Gen. Virol.* **81,** 993–1000.
26. Gazarian, K. G., Gazarian, T. G., Solis, C. F., Hernandez, R., Shoemaker, C. B., and Laclette, J. P. (2000) Epitope mapping on N-terminal region of taenia solium paramyosin. *Immunol. Lett.* **72,** 191–195.

27. Smith, G. P. and Scott, J. K. (1993) Libraries of peptides and proteins displayed on filamentous phage. *Meth. Enzymol.* **217,** 228–257.
28. Sambrook, J., Fritsch, E., and Maniatis, T. (1989) *Molecular Cloning: A Laboratory Manual.* Cold Spring Harbor Laboratory Press, Cold Spring Harbor, NY, PC24.
29. Ehrlich, G. K., Andria, M. L., Zheng, X., Kieffer, B., Gioannini, T. L., Hiller, J. M., et al. (1998) Functional significance of cysteine residues in the delta opioid receptor studied by site-directed mutagenesis. *Can. J. Physiol. Pharmocol.* **76,** 269–277.
30. Goodson, R. J., Doyle, M. V., Kaufman, S. E., and Rosenberg, S. (1994) High-affinity urokinase receptor antagonists identified with bacteriophage peptide display. *Proc. Natl. Acad. Sci. USA* **91,** 7129–7133.
31. Szardenings, M., Tornroth, S., Mutulis, F., Muceniece, R., Keinanen, K., Kuusinen, A., and Wikberg, J. E. S. (1997) Phage display selection on whole cells yields a peptide specific for melanocortin receptor 1. *J. Biol. Chem.* **272,** 27,943–27,948.
32. Ehrlich, G. K., Berthold, W., and Bailon, P. (2000) Phage display technology. Affinity selection by biopanning. *Meth. Mol. Biol.* **147,** 195–208.
33. Hammer, J., Takacs, B., and Sinigaglia, F. (1992) Identification of a motif for HLA-DR1 binding peptides using M13 display libraries. *J. Exp. Med.* **176,** 1007–1013.
34. Glaser, V. (1997) Conflicts brewing as phage display gets complex. *Nat. Biotechnol.* **15,** 506.
35. Cabilly, S., Heldman, J., Heldman, E., and Katchalski-Katzir, E. (1998) The use of combinatorial libraries to identify ligands that interact with surface receptors in living cells, in *Methods in Molecular Biology*, Vol. 87: *Combinatorial Peptide Library Protocols* (Cabilly, S., ed.), Humana Press, Totowa, NJ, pp. 175–183.
36. Giebel, L. B., Cass, R. T., Milligan, D. L., Young, D. C., Arze, R., and Johnson, C. R. (1995) Screening of cyclic peptide phage libraries identifies ligands that bind Streptavidin with high affinity. *Biochemistry* **34,** 15,430–15,435.
37. Adey, N. B., Mataragnon, A. H., Rider, J. E., Carter, J. M., and Kay, B. K. (1995) Characterization of phage that bind plastic from phage-displayed random peptide libraries. *Gene* **156,** 27–31.
38. Spada S., Krebber C., and Pluckthun A. (1997) Selectively infective phages (SIP). *Biol. Chem.* **378,** 445–456.
39. Sparks, A. B., Adey, N. B., Cwirla, S., and Kay, B. K. (1996) Screening phage-displayed random peptide libraries, in *Phage Display of Peptides and Proteins* (Kay, B. K., Winter, J., and McCafferty, J., eds.), Academic Press, San Diego, CA, pp. 227–253.

154

Screening of Phage Displayed Antibody Libraries

Heinz Dörsam, Michael Braunagel, Christian Kleist, Daniel Moynet, and Martin Welschof

1. Introduction

Human monoclonal antibodies are more suitable than monoclonal antibodies of animal origin for clinical applications because of lower hypersensitivity reactions, less formation of circulating immune complexes, and lower anti-immunoglobulin responses. The classical production of human monoclonal antibodies via the hybridoma technique or Epstein–Barr virus (EBV) transformation is limited by the instability of cell lines, low antibody production, and the problems of immunizing humans with certain antigens *(1,2)*. A promising alternative is the production of human recombinant antibodies *(3)*. Recombinant DNA technology has made it possible to clone human antibody genes in vectors and to generate antibody expression libraries *(4–7)*. One approach has been to amplify and recombine the IgG repertoire of an "immunized" donor. This has been used to isolate several antibodies that were related to diseases *(8–10)*. To obtain more universal antibody libraries the naive IgM repertoire of several "unimmunized" donors were pooled *(11–13)*. The complexity of the combinatorial libraries has been further increased by creating the so-called "semisynthetic" antibody libraries *(14–16)*.

To prepare antibody DNA from peripheral lymphocytes, spleen lymphocytes, or B-cell lines, mRNA is first isolated by standard methods. After preparation of the first strand of cDNA the Fv- or Fab-encoding regions are amplified using the polymerase chain reaction (PCR) and a set of primers homologous to the variable region of the heavy (μ,γ) and light chains (κ,λ) *(17–21)*. The PCR products are randomly combined in an appropriate expression vector. For extra stability, the V_L and V_H domains of Fv fragments are often joined with a peptide linker *(22,23)*. The larger Fab fragment contains the V_L–C_L and V_H–C_{H1} segments linked by disulfide bonds. To facilitate the screening of these scFv- or Fab-antibody libraries, phagemid pIII display vectors are commonly used *(24–26)*. These vectors contain a phage intergenic region to provide a packaging signal. The expression of the pIII-antibody fusion protein is regulated by a bacterial promoter under the control of a *lac* operator.

To display the antibody fragments on the phage surface, the phagemid must be packaged with proteins supplied by helper phages. First, *Escherichia coli* is transformed

From: *The Protein Protocols Handbook, 2nd Edition*
Edited by: J. M. Walker © Humana Press Inc., Totowa, NJ

with the phagemid vector to which the PCR products have been ligated. The phagemid-containing bacteria are infected with a helper phage such as M13KO7 to yield recombinant phage particles that display scFv or Fab fragments fused to the pIII protein. Phage displaying antibody fragments that bind to a specific antigen can be enriched from a large library of phages by panning over the antigen. Nonspecific phages are removed during the washing procedure, following in which the remaining antigen-specific phages are eluted and used to reinfect exponentially growing *E. coli*. The specificity of the enriched phage particles can be confirmed in an enzyme-linked immunosorbent assay (ELISA). The entire procedure, which is described in this chapter, is outlined on a flow chart (**Fig. 1**).

2. Materials

2.1. Phage Rescue, Panning, Reinfection, Small-Scale Phage Rescue

1. Sterile glass 100-mL and 1000-mL Erlenmeyer flasks.
2. Petri dishes, 85-mm and 145-mm diameter (Greiner, Frickenhausen, Germany).
3. Maxisorb® Immunotubes and caps (Nunc, Roskilde, Denmark).
4. Sterile polypropylene 96-deep-well microtiter plate (Beckmann, München, Germany) and sealing device (Beckman, Biomek™, cat. no. 538619).
5. End over end shaker.
6. Tabletop microcentrifuge.
7. Clinical centrifuge with swinging-bucket rotor.
8. IEC refrigerated centrifuge (or equivalent) with adaptors for microtiter plates.
9. ELISA plate shaker (e.g., IKA MTS-2).
10. Helper phage M13KO7-derived VCSM13 (Stratagene cloning systems, La Jolla, CA).
11. XL-1 Blue bacteria (Stratagene cloning systems, La Jolla, CA).
12. 2 *M* Glucose, sterile filtered.
13. (5 mg/mL) Ampicillin, sterile filtered.
14. (10 mg/mL) Kanamycin, sterile filtered.
15. M9-minimal medium agar plates: To 15 g of Bacto-agar add water to 750 mL and autoclave. After the solution has cooled to 55°C, add 200 mL of sterile 5× M9 salts, 50 mL of sterile 2 *M* glucose, and 10 mL of sterile-filtered 100× supplement. Pour plates quickly.
16. 5× M9 salts: To 37.4 g of $Na_2HPO_4 \cdot 2H_2O$, 11.75 g of $NaH_2PO_4 \cdot H_2O$, 2.5 g of NaCl, and 5 g of NH_4Cl, add distilled water to 1 L and autoclave.
17. 100× Supplement: To 1 mL of 1 *M* $MgSO_4 \cdot 7H_2O$, 1 mL of 100 m*M* $CaCl_2 \cdot 2\ H_2O$, 30 µL of 100 m*M* $Fe(III)Cl_3 \cdot 6H_2O$ and 60 µL of 500 m*M* thiamin, add distilled water to 10 mL and sterile filter.
18. SOB-GA medium agar plates: To 15 g of Bacto-agar, 20 g of Bacto-tryptone, 5 g of Bacto-yeast extract, and 0.5 g of NaCl, add distilled water to 920 mL and autoclave. After the medium has cooled to 55°C, add 10 mL of sterile $MgCl_2$, 50 mL of sterile 2 *M* glucose, and 20 mL of sterile filtered ampicillin (5 mg/mL). Pour plates quickly.
19. 2× YT Medium: Dissolve 15 g of Bacto-agar, 17 g of Bacto-tryptone, 10 g of Bacto-yeast extract and 5 g of NaCl in 800 mL of distilled water, adjust the pH of the medium to 7.5 with 1 *M* NaOH, add distilled water to 1 L total volume and autoclave.
20. 2× YT-GA Medium: 2× YT medium containing 100 µg/mL of Ampicillin and 100 m*M* glucose.
21. 2× YT-AK Medium: 2× YT medium containing 100 µg/mL of Ampicillin and 50 µg/mL of Kanamycin.

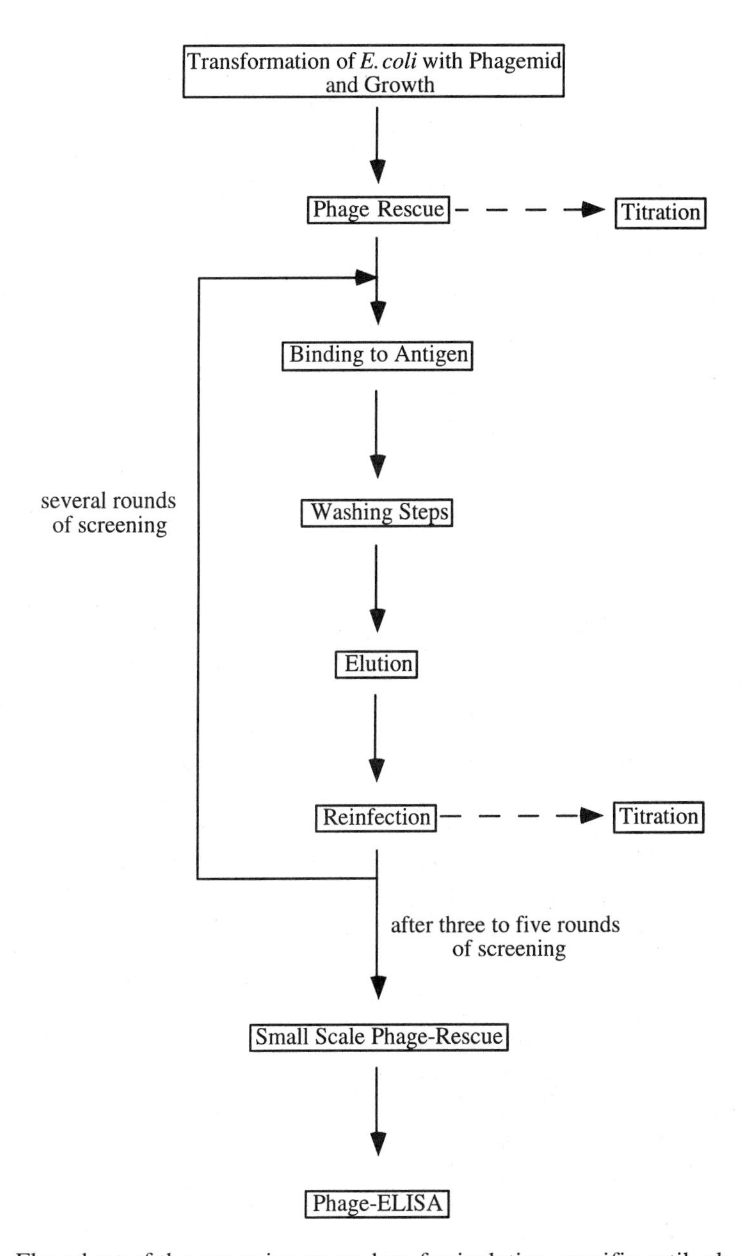

Fig. 1. Flowchart of the screening procedure for isolating specific antibody clones.

22. Sterile glycerol.
23. PEG–NaCl: To 200 g of polyethylene glycol 6000 and 146.1 g of NaCl add distilled water to 1 L, heat to dissolve, and autoclave.
24. Phage dilution buffer: 10 m*M* Tris-HCl, pH 7.5, 20 m*M* NaCl, 2 m*M* EDTA.
25. Coating buffer: 50 m*M* Na_2CO_3, pH 9.6.
26. PBS: 50 m*M* phosphate buffer, pH 7.2, 140 m*M* NaCl.
27. PBST: PBS, 0.1% (v/v) Tween 20.
28. Blocking buffer: 2% (w/v) Skimmed milk powder in PBS; use immediately.

29. 100 m*M* Triethylamine.
30. 1 *M* Tris-HCl, pH 7.4.

2.2. Phage ELISA

1. Anti-M13-antibody: Rabbit-anti-M13-sera, diluted 1 : 1000 in blocking buffer (Stratagene, cat. no. 240210).
2. Second antibody: Goat-anti-rabbit IgG, horseradish peroxidase (HRP)-conjugated (Jackson Immuno Research Laboratories, Inc., West Grove PA, USA, cat. no. 111-095-144).
3. Enzyme substrate: TMBKit from Kierkegaard-Perry Labs, Gaithersburg, Maryland, USA (cat. no. 507600).
4. ELISA plates: Bibby-Dunn (Asbach, Germany), polyvinyl chloride (PVC) activated, flat bottom (cat. no. 77-173-05).
5. PBS: 50 m*M* phosphate buffer, pH 7.2, 140 m*M* NaCl.
6. PBST: PBS, 0.1% (v/v) Tween 20.
7. Blocking buffer: 2% (w/v) Skimmed milk powder in PBS; store at –20°C.
8. Coating buffer: 50 m*M* Na_2CO_3, pH 9.6.
9. *E. coli*-lysate-coated nitocellulose:

   Pellet a 50-mL overnight culture of *E. coli* at 6000*g* for 10 min. Discard the supernatant.
   Resuspend the pellet in 10 mL of PBS and lyse by sonification.
   Incubate a sheet of nitrocellulose (approx 50 cm^2) with the cell lysate for 1 h at room temperature (RT) on a shaker.
   Wash the nitrocellulose 3 × 15 min in PBST.
   Cut the nitrocellulose sheet into 1-cm^2 pieces. Store them at –20°C.
   One piece of the *E. coli*-lysate-coated nitrocellulose is enough to preabsorb up to 20 mL of diluted antibody.

3. Methods

3.1. Phage Rescue

1. Plate each transformation of the ligated DNA library onto three 145-mm diameter SOB-GA agar plates and incubate overnight at 30°C (*see* **Notes 1** and **2**).
2. Add 10 mL if 2× YT-GA medium to each plate and scrape bacteria into the medium with a sterile glass spreader. Pool cells from all three plates and measure absorbance at 600 nm (A_{600}) in an appropriate dilution.
3. Inoculate 400 mL of 2× YT-GA medium with bacteria from the plate to A_{600} of 0.025 (*see* **Note 3**).
4. To the remaining bacteria harvested from the plates add glycerol to a final concentration of 20% (v/v) and freeze in appropriate aliquots at –70°C. These are the library stock cultures and should be handled with care (*see* **Note 4**).
5. Grow the inoculated culture at 37°C with shaking at 280 rpm until an A_{600} of 0.1 is reached.
6. Add M13KO7 helper phage at multiplicity of infection of 20 and mix gently (*see* **Notes 5** and **6**).
7. Incubate for 15 min at 37°C without shaking, then for 45 min at 37°C with shaking at 260 rpm.
8. Pellet infected bacteria for 10 min at 1500*g*.
9. Gently resuspend the bacteria in 400 mL of 2× YT-AK medium **(no glucose)** (*see* **Note 7**).
10. Grow at 37°C with shaking at 280 rpm for 5–8 h (*see* **Note 8**).
11. Take 1 mL of the culture and centrifuge in a microcentrifuge ≥ 13000*g* for 2 min; use supernatant for determination of number of infectious particles (titration) (*see* **Notes 9** and **10**).

12. Centrifuge the rest of the culture at ≥ 6000*g* for 15 min at 4°C to pellet the cells. From now on keep everything on ice!
13. Add 1/5 volume of ice-cold PEG solution to the supernatant and place on ice for at least 30 min.
14. Centrifuge at 10,000*g* for 20 min at 4°C (*see* **Note 11**).
15. Discard the supernatant and invert the tube over a clean paper towel to remove remaining buffer.
16. Resuspend the pellet in 4 mL of ice-cold phage dilution buffer (*see* **Note 12**).
17. Transfer the resuspended phages to 1.5-mL tubes and centrifuge in a microcentrifuge for 5 min at ≥ 13000*g* at 4°C to pellet cellular debris.
18. Transfer supernatant to fresh tubes and place aliquots on ice. Take a 10-µL aliquot and determine the number of infectious particles (titration of phages) (*see* **Note 9**).
20. Proceed with panning to select for antigen-positive recombinant phage antibodies (*see* **Note 13**).

3.2. Panning to Select for Antigen Binding Clones

1. Coat one Maxi Sorb® Immunotube with appropriate amounts of your protein diluted in 50 m*M* Na_2CO_3, pH 9.6 buffer or PBS. Coating can be performed for 2 h at RT or overnight at 4°C by rotating end over end (*see* **Note 14**). Always incubate one tube only with coating buffer but without antigen. This will be the negative control (*see* **Note 15**).
2. For each Immunotube inoculate 20 mL of 2× YT-G medium with a colony from a minimal medium plate and incubate at 37°C with shaking at 280 rpm until the culture reaches an A_{600} of 0.4. These bacteria are needed for reinfection with the enriched phage clones (*see* **Subheading 3.3.**).
3. Wash Immunotubes 3× with 2 mL of PBS. For washing, add 2 mL of PBS to the tube, close it, and vortex-mix for 2 s. Empty the tube completely after each wash (*see* **Note 16**).
4. Block with blocking buffer at RT by rotating end over end for 3 h (*see* **Note 30**).
5. Wash Immunotubes 3× with PBS as described in **step 3**.
6. Dilute 1×10^{11}–1×10^{12} of the recombinant phages in 5 mL of blocking buffer and incubate for 15 min at RT (*see* **Notes 17** and **18**).
7. Add the preincubated phage dilution to the antigen coated tube and incubate at RT by rotating end over end for 30 min, then leave it undisturbed for another 1.5 h.
8. Wash the tube 20× with PBS containing 0.1% Tween 20 and 20× with PBS as described in **step 3**.
9. To elute the bound phage antibodies add 1 mL of 100 m*M* triethylamine to the tube and agitate gently for 5 min. Make sure that all parts of the tube are rinsed. Neutralize phage solution with 1 mL of Tris-HCl, pH 7.5, and place it immediately on ice (*see* **Note 19**).

3.3. Reinfection of E. coli with Enriched Phage Clones

1. Each sample of phages eluted from an individual Immunotube is added to 20 mL of exponentially growing XL-1 Blue or TG1 cells. These bacteria are prepared by transferring a colony from a minimal medium plate to 20 mL of 2× YT medium and incubating at 37°C with shaking at 280 rpm until the culture reaches an A_{600} of 0.4.
2. Mix phages and bacteria by gently agitating the culture (*see* **Note 20**).
3. Leave the culture undisturbed for 15 min at 37°C, then shake at 260 rpm for 45 min at 37°C.
4. Remove 100 µL of the 20-mL cell suspension. Prepare 10-fold dilutions of the cell suspension in 2× YT-GA medium and plate onto separate SOB-GA agar plates (titer plates) (*see* **Note 21**). Incubate titer plates overnight at 30°C.

5. Centrifuge the reinfected cell suspension for 5 min at 3400*g*.
6. Discard supernatant, resuspend the bacteria immediately in 1 mL of SOB-GA medium, and plate them onto three SOB-GA medium 145-mm diameter agar plates. When dry, invert the plates and incubate overnight at 30°C (*see* **Note 22**).
7. The next day resuspend the bacteria in 2× YT-GA medium as described in **Subheading 3.1., step 2**, and proceed with phage rescue. Plates can also be sealed and stored for up to 2 wk at 4°C before rescue (*see* **Note 23**).
8. For analyzing individual clones, take isolated colonies from your titer plates and proceed as described in **Subheading 3.4.** (*see* **Note 24**).

3.4. Small-Scale Phage Rescue

1. Add 500 µL of 2× YT-GA medium to each well of a sterile polypropylene 96-deep-well microtiter plate (*see* **Note 25**).
2. Transfer individual isolated colonies (e.g., from your titer plates) to separate wells using sterile toothpicks or pipet tips.
3. Incubate overnight (18 h) at 37°C at 1100 rpm on an IKA MTS-2 ELISA shaker. This is the master plate (*see* **Notes 26** and **27**).
4. Transfer 5 µL of saturated culture from each well of the master plate to the corresponding well of a new plate containing 500 µL of 2× YT-GA medium.
5. Incubate at 1100 rpm on an IKA ELISA shaker for 3 h at 37°C if using XL-1 Blue or 1.5 h if using TG1 cells.
6. Add 1×10^{10} M13KO7 helper phages to each well.
7. Place the microtiter plates onto the ELISA shaker, mix at 1100 rpm for 5 s, and leave them undisturbed for 30 min at 37°C.
8. Incubate with shaking (1100 rpm) at 37°C for 1 h.
9. Centrifuge plates at 350*g* for 10 min at RT.
10. Carefully remove the supernatant and discard in an appropriate waste container.
11. Add 500 µL of prewarmed (RT or 37°C) 2× YT-AK medium **(no glucose)** to each well.
12. Incubate at 37°C with shaking at 1100 rpm for 5–8 h (*see* **Note 8**).
13. Centrifuge 20 min at 1000*g* at 4°C to pellet the cells.
14. Transfer 400 µL of the supernatant to a polypropylene 96-deep-well microtiter plate precooled to 4°C.
15. Seal the plate and store on ice or 4°C until required for the phage ELISA (*see* **Note 13**).

3.5. Phage ELISA

1. Add 1 µg of antigen to 100 µL of coating buffer per well. Incubate overnight at 4°C or 4 h at RT. Discard the coating solution and wash 1× with 200 µL of PBS (*see* **Notes 28** and **29**).
2. Each well of the microwell titer plate is blocked with 200 µL of blocking buffer. Incubate 2 h at RT. Discard the blocking buffer and wash 3× with PBS (*see* **Note 30**).
3. Add 100 µL of rescued phagemids (containing 10^8–10^9 particles) to each well. Incubate for 2 h at RT. Discard the phage solution and wash 5× with PBST (*see* **Note 31**).
4. Add an appropriate amount of anti-M13 antibody diluted in 100 µL of blocking buffer. Incubate for 2 h at RT. Discard the phage solution and wash 6× with PBST (*see* **Note 32**).
5. Add an appropriate amount of "second antibody" diluted in 100 µL of blocking buffer. Incubate for 2 h at RT. Discard the phage solution and wash 5× with PBST (*see* **Note 33**).
6. Add an appropriate amount of "enzyme substrate" in a 100-µL solution and quantify the reaction with an ELISA reader (*see* **Notes 34** and **35**).

4. Notes

1. The protocols described were established using the phagemid vector pSEX81. A detailed protocol regarding the cloning of an antibody repertoire into pSEX 81 with the title *Production of a Human Antibody Library in the Phage-Display Vector pSEX81* is provided in *Methods in Molecular Medicine*, Vol. 13: *Molecular Diagnosis of Infectious Diseases*, edited by U. Reischl, Humana Press, Totowa, NJ.
2. Both XL-1 Blue and TG1 bacteria are suitable hosts for phagemid libraries, whereby TG1 grows faster than XL-1 Blue. Plating transformed bacteria onto SOB-GA plates reduces the risk of losing slower growing clones.
3. For performing phage rescue on a smaller or larger scale adjust the volume of the media and the number of M13KO7 helper phages. It is important to use optimal growth conditions for the bacteria at all times. It is therefore important to use large flasks for better aeration.
4. Phage rescue can also be started from an overnight culture or with bacteria from a glycerol stock. In the latter case, the concentration of glycerol in the culture should not exceed 2% to avoid inhibition of bacterial growth.
5. Do not handle the cells too vigorously before and during superinfection with helper phage, as this may decrease their capacity for phage infection.
6. One milliliter of XL-1 Blue at A_{600} of 0.1 corresponds to 5×10^8 bacteria. To obtain a multiplicity of infection of 20, add 1×10^{10} phages to each milliliter of the culture. One milliliter of TG1 at A_{600} of 0.1 corresponds to 1×10^8 bacteria, so 2×10^9 phages/mL are required.
7. It is very important to use glucose-free medium at this step, as the presence of glucose will inhibit expression of the antibody genes.
8. Extended incubation increases the number of rescued phages (infectious particles), but decreases the ratio of functional recombinant antibody to phage particles. The binding capacity of the rescued phage antibodies is lost mainly owing to protease activity in the culture.
9. Determination of the number of infectious particles defined as colony-forming units (cfu) can be performed according to standard methods (*see* **ref. *25***).
10. The number of infectious particles in the supernatant is usually about 1×10^{10} to 1×10^{11} cfu/mL. If it is less than 1×10^9 cfu/mL, the rescue should be repeated.
11. Take care, as the pellet may be hardly visible.
12. Rinse that side of the tube where the pellet is to be expected with 1 mL of buffer several times, pool in a polypropylene tube, and repeat this procedure 4×.
13. Panning should be performed as soon as possible following rescue, as some phage displayed recombinant antibody preparations may be unstable. It is possible to freeze rescued phage antibodies at –20°C, although it is presently unknown wether some antibodies might possibly be denatured during freezing. In any case, repeated freezing and thawing should be avoided.
14. The optimal conditions for coating may differ for each antigen. Small-scale trials should therefore first be carried out. The present protocol is suitable for protein antigens. We use 200 µg of protein per tube if sufficient amounts are available, but the optimal amounts may vary. Our suggestion is to make an empiricall estimation of the amount of antigen necessary to coat a well of an ELISA plate. This amount can then be scaled up to the volume of the Immunotubes (5 mL). The suitability of the plastic support for immobilization of the particular antigen should also be determined prior to panning. If possible, an immunoassay should be established prior to panning to ensure that under the conditions used for coating the antigen binds to the tube wall.

15. A negative control is always necessary to verify that an enrichment is due to binding of the recombinant phage antibodies to the antigen rather than to components of the blocking buffer. The tube used as negative control should be treated exactly like the antigen coated tubes.
16. The tube can be closed with Parafilm® or with the caps provided by the supplier of the tubes.
17. The amount of rescued phagemid antibodies used for panning is dependent on the complexity of the library. Make sure that the number of phages is at least 10^3–1×10^4 times more than the complexity of the library.
18. To reduce nonspecific, hydrophobic protein–protein interactions between native M13 phage proteins and some antigens, Triton X-100 may be added to a final concentration of 0.1%.
19. Incubation with triethylamine reduces the infectivity of the recombinant phage antibodies. Do not exceed an incubation time of more than 5–10 min.
20. Handle the cells gently before and during infection with eluted phages.
21. This titration is to determine the number of infectious phagemid particles eluted from the tube after panning. The number should be low after first round of panning and then increase after subsequent rounds as specific clones are enriched.
22. It is very important that the agar plates are well dried before plating the bacteria.
23. If there are only a few colonies on the plates, extend the incubation at 30°C until the colonies are fairly large. Resuspend the cells in a smaller volume.
24. After the third round of panning and reinfection, its possible to start analyzing single clones.
25. For the small-scale rescue of phagemids, we recommend the use of special polypropylene 96-deep-well microtiter plates (volume per well = 1 mL). However, these plates may require a special adaptor for centrifugation.
26. If another shaker is used, speed may have to be decreased to prevent spillage of the medium into adjacent wells.
27. The remaining master plate may be stored at 4°C for up to 2 wk. To store it longer, add sterile glycerol to the cultures to a final concentration of 20%, seal the plate and mix by inverting several times. Store at –70°C.
28. All incubation steps should be carried out in a wet chamber to prevent evaporation or contamination. For example, the microwell plate can be placed in a closed box containing moistened tissue paper.
29. These conditions work for most protein antigens or haptens bound to a protein carrier. The optimal amount of antigen may vary. Nonprotein antigens (e.g., lipopolysaccharides) may require a different buffer. In this case, refer to the buffers used in a conventional ELISA.
30. 2% Skimmed milk in PBS works very well as a blocking buffer. Some commercially available blocking buffers (e.g., Pierce Superblock™) give comparable results and should be used according to the manufacturers instructions. If the solid support is coated with a nonprotein antigen, other blocking buffers may be required. In this case, refer to the buffers used in a conventional ELISA. Take note that collagen, FCS or gelatin do not work well as blocking agents owing to their interaction with phagemid particles.
31. Both supernatant or PEG-precipitated phagemids may be used in phage ELISA. PEG precipitation lowers the background and removes contaminants interferring with the binding.
32. The anti-M13 antibody is a polyclonal serum raised against M13 or fd phages. A commercial source is listed in **Subheading 2.** The serum should be diluted according to the manufacturer's instructions in the same solution used for blocking as described in **step 2**. The serum used here contains a high titer of antibodies against *E. coli* proteins and therefore cross-reacts with proteins in the phage supernatant. To avoid this, the serum must be preabsorbed. Incubate the diluted serum overnight with a piece of *E. coli*-lysate-coated nitocellulose at 4°C. This can be done in parallel with the coating (**Subheading 3.5., step 1**).

33. The "second antibody" is a serum raised against the anti-M13 antibody linked to HRP. A commercial source is listed in **Subheading 2.** The serum should be diluted according to the supplier.
34. "Enzyme substrate" is a substrate for HRP. A commercial source is listed in **Subheading 2.**
35. The detection limit of the phage ELISA is about 10^5 bound phages/well.

References

1. Carson, D. A. and Freimark, B. D. (1986) Human lymphocyte hybridomas and monoclonal antibodies. *Adv. Immunol.* **38,** 275.
2. Glassy, M. C. and Dillman, R. O. (1988) Molecular biotherapy with human monoclonal antibodies. *Mol. Biother.* **1,** 7.
3. Winter, G. and Milstein, C. (1991) Man made antibodies. *Nature* **349,** 293.
4. McCafferty, J., Griffiths, A. D., Winter, G., and Chiswell, D. J. (1990) Phage antibodies: filamentous phage displaying antibody variable domains. *Nature* **348,** 552.
5. Mullinax, R. L., Gross, E. A., Amberg, J. R., Hay, B. N., Hogrefe, H. H., Kubitz, M. M., et al. (1990) Identification of human antibody fragment clones specific for tetanus toxoid in bacteriophage lambda immunoexpression library. *Proc. Natl. Acad. Sci. USA* **87,** 8095.
6. Marks, J. D., Hoogenboom, H. R., Bonnert, T. P., McGafferty, J., Griffiths, A. D., and Winter, G. (1991) By-passing immunization. Human antibodies from V-gene libraries displayed on phage. *J. Mol. Biol.* **222,** 581.
7. Barbas, C. F., Rosenblum, J. S., and Lerner, R. A. (1993) Direct selection of antibodies that coordinate metals from semisynthetic combinatorial libraries. *Proc. Natl. Acad. Sci. USA* **90,** 6385.
8. Barbas, C. F., Collet, T. A., Amberg, W., Roben, P., Binley, J. M., Hoekstra, D., et al. (1993) Molecular profile of an antibody response to HIV-1 as probed by recombinatorial libraries. *J. Mol. Biol.* **230,** 812.
9. Rapoport, B., Portolano, S., and McLachlan, S. M. (1995) Combinatorial libraries: new insights into human organ-specific autoantibodies. *Immunol. Today* **16,** 43.
10. Welschof, M., Terness, P., Kipriyanov, S., Stanescu, D., Breitling, F., Dörsam, H., et al. (1997) The antigen-binding domain of a human IgG-anti-F(ab´)2 autoantibody. *Proc. Natl. Acad. Sci. USA* **94,** 1902.
11. Marks, J. D., Hoogenboom, H. R., Bonnert, T. P., McGafferty, J., Griffiths, A. D., and Winter, G. (1991) By-passing immunization. Human antibodies from V-gene libraries displayed on phage. *J. Mol. Biol.* **222,** 581.
12. Griffiths, A. D., Malmquist, M., Marks, J. D., Bye, J. M., Embleton, M. J., McCafferty, J., et al. (1993) Human anti-self antibodies with high specificity from phage display libraries. *EMBO J.* **12,** 725.
13. Little, M., Welschof, M., Braunagel, M., Hermes, I., Christ, C., Keller, A., et al. (1999) Generation of a large complex antibody library from multiple donors. *J. Immunol. Meth.* **231,** 3.
14. Hoogenboom, H. R. and Winter, G. (1992) By-passing immunization. Human antibodies from synthetic repertoires of germline V_H gene segments rearranged in vitro. *J. Mol. Biol.* **227,** 381.
15. Barbas, C. F., Bain, J. D., Hoekstra, D. M., and Lerner, R. A. (1992) Semisynthetic combinatorial antibody libraries: a chemical solution to the diversity problem. *Proc. Natl. Acad. Sci. USA* **89,** 4457.
16. Hayashi, N., Welschof, M., Zewe, M., Braunagel, M., Dübel, S., Breitling, F., and Little, M. (1994) Simultaneous mutagenesis of antibody CDR regions by overlap extension and PCR. *BioTechniques* **17,** 310.

17. Larrick, J. W., Danielsson, L., Brenner, C. A., Wallace, E. F., Abrahamson, M., Fry, K. E., and Borrebaeck, C. A. K. (1989) Polymerase chain reaction using mixed primers: cloning of human monoclonal antibody variable region genes from single hybridoma cells. *BioTechnology* **7,** 934.
18. Marks, J. D., Tristem, M., Karpas, A., and Winter, G. (1991) Oligonucleotide primers for polymerase chain reaction amplification of human immunoglobulin variable genes and design of family-specific oligonucleotide probes. *Eur. J. Immunol.* **21,** 985.
19. Campbell, M. J., Zelenetz, A. D., Levy, S., and Levy, R. (1992) Use of family specific leader region primers for PCR amplification of the human heavy chain variable region gene repertoire. *Mol. Immunol.* **29,** 193.
20. Barbas, C. F., Björling, E., Chiodi, F., Dunlop, N., Cababa, D., Jones, T. M., et al. (1992) Recombinant human Fab fragments neutralize human type I immunodeficiency virus in vitro. *Proc. Natl. Acad. Sci. USA* **89,** 9339.
21. Welschof, M., Terness, P., Kolbinger, F., Zewe, M., Dübel, S., Dörsam, H., et al. (1995) Amino acid sequence based PCR primers for the amplification of human heavy and light chain immunoglobulin variable region genes. *J. Immunol. Meth.* **179,** 203.
22. Huston, J. S., Levinson, D., Mudgett-Hunter, M., Tai, M.-S., Novotny, J., Margolies, M. N., et al. (1988) Protein engineering of antibody binding sites: recovery of specific activity in an anti-digoxin single-chain Fv analouge produced in *Escherichia coli. Proc. Natl. Acad. Sci. USA* **85,** 5879.
23. Bird, R. E., Hardman, K. D., Jacobson, J. W., Johnson, S., Kaufman, B. M., Lee, S.-M., et al. (1988) Single-chain antigen-binding proteins. *Science* **242,** 423.
24. Breitling, F., Dübel, S., Seehaus, T., Klewinghaus, I., and Little, M. (1991) A surface expression vector for antibody screening. *Gene* **104,** 147.
25. Hoogenboom, H. R., Griffiths, A. D., Johnson, K. S., Chiswell, D. J., Hudson, P., and Winter, G. (1991) Multi-subunit proteins on the surface of filamentous phage: methodologies for displaying antibody (Fab) heavy and light chains. *Nucl. Acids Res.* **19,** 4133.
26. Barbas, C. F., Kang, A. K., Lerner, R. A., and Benkovic, S. J. (1991) Assembly of combinatorial antibody libraries on phage surfaces: the gene III site. *Proc. Natl. Acad. Sci. USA* **88,** 7978.
27. Sambrook, J., Fritsch, E. F., and Maniatis, T. (1989) *Molecular Cloning: A Laboratory Manual* (Nolan, C., ed.), Cold Spring Harbor Laboratory Press, Cold Spring Harbor, NY.

155

Antigen Measurement Using ELISA

William Jordan

1. Introduction

The initial description of the enzyme-linked immunosorbent assay (ELISA) almost 30 yr ago *(1)* marked a technological advance that has had an immense impact in both clinical diagnostic and basic scientific applications. This assay represents a simple and sensitive technique for specific, quantitative detection of molecules to which an antibody is available. Although there are a huge number of variations based on the original ELISA principle, this chapter focuses on perhaps the two most useful and routinely performed: (1) the indirect sandwich ELISA—providing high sensitivity and specificity and (2) the basic direct ELISA—useful when only one antibody to the sample antigen is available.

1.1. The Direct ELISA

During the indirect sandwich ELISA (**Fig. 1A**), an antibody specific for the substance to be measured is first coated onto a high-capacity protein binding microtiter plate. Any vacant binding sites on the plate are then blocked with the use of an irrelevant protein such as fetal calf serum (FCS) or Bovine serum albumin (BSA). The samples, standards, and controls are then incubated on the plate, binding to the capture antibody. The bound sample can be detected using a secondary antibody recognizing a different epitope on the sample molecule, thus creating the "sandwich." The detection antibody is commonly directly conjugated to biotin, allowing an amplification procedure to be carried out with the use of streptavidin bound to the enzyme horseradish peroxidase (HRP). As streptavidin is a tetrameric protein, binding four biotin molecules, the threshold of detection is greatly enhanced. The addition of a suitable substrate such as 3,3',5,5'-tetramethylbenzidine (TMB) allows a colormetric reaction to occur in the presence of the HRP that can be read on a specrophotometer with the resulting optical density (OD) relating directly to the amount of antigen present within the sample.

1.2. The Indirect ELISA

In some cases, however, only one specific antibody may be available, and in such a small quantity that directly conjugating it to biotin would be impractical. In this situation a direct ELISA should be used. During the direct ELISA the sample itself is coated

From: *The Protein Protocols Handbook, 2nd Edition*
Edited by: J. M. Walker © Humana Press Inc., Totowa, NJ

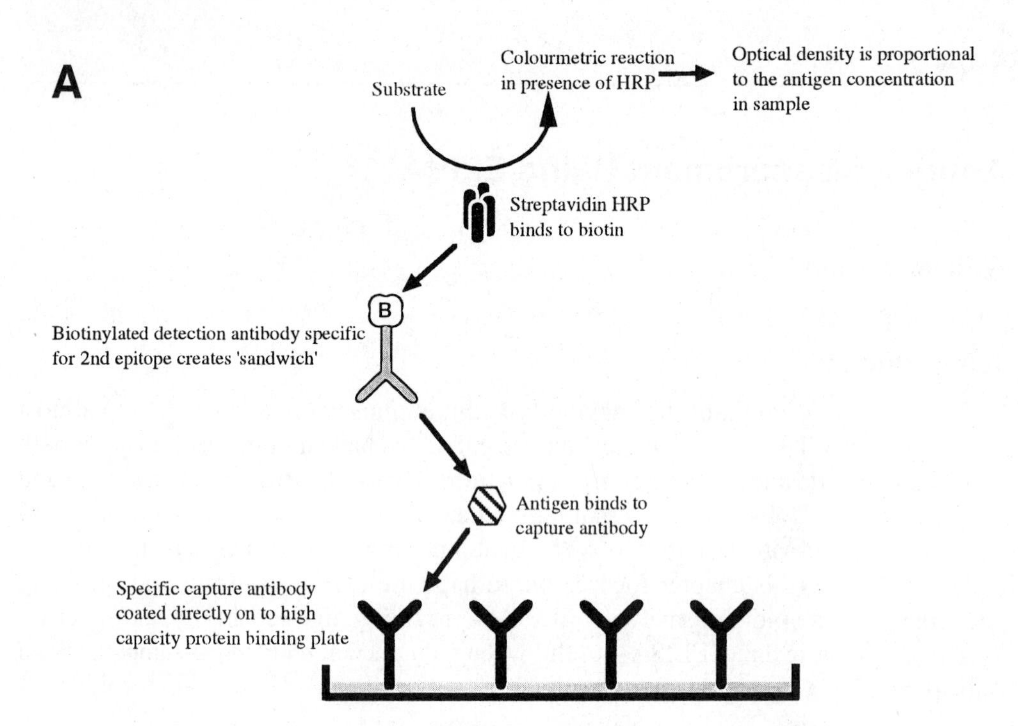

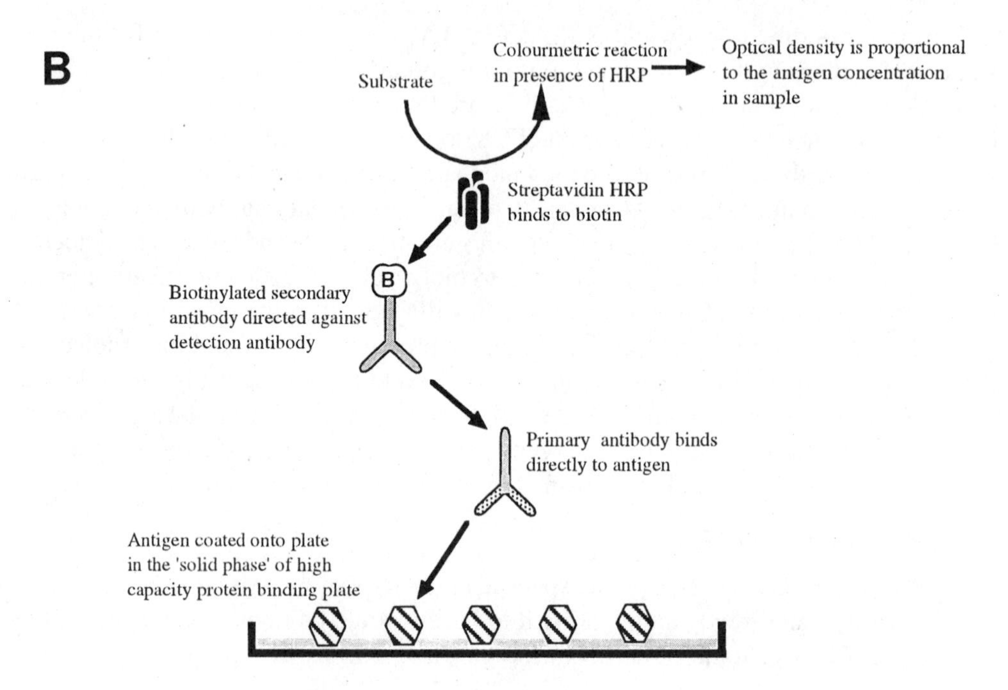

Fig. 1. The indirect sandwich ELISA.

directly onto the microtiter plate and is then detected using the specific antibody. If this antibody is biotinylated, then the procedure can then proceed as in the indirect ELISA; if not, then a secondary biotinylated antibody directed against the species of the detecting antibody itself can be used. **Figure 1B** demonstrates this technique.

To set up a reliable and durable ELISA it is essential to first optimize a number of the parameters mentioned previously. The level of optimization will, of course, depend on exactly what is required from the assay. In some cases a simple "yes-or-no" answer is desired and a simple standard procedure may be sufficient. If, however, high sensitivity is the aim with accurate quantatation of the molecule in question, then carefully setting up the optimal conditions in advance will pay dividends and save a great deal of time in the long term.

2. Materials

1. Antibodies (*see* **Note 1**): For the indirect ELISA, antibody pairs can often be bought commercially and consist of a capture antibody and a biotinylated detection antibody. Both antibodies are specific for the molecule in question, with the detection antibody being directly biotinylated, and able to recognize the sample molecule when it is bound to the capture antibody on the plate. For the direct ELISA, one specific detection antibody is required, preferably biotinylated, but if not a secondary biotinylated antibody specific for the detection antibody is also needed. Aliquot and freeze at –20°C or lower in small, usable quantities with a carrier protein at such as 10% FCS or 1% BSA.
2. Blocking buffer: Phosphate-buffered saline (PBS), pH 7.4, supplemented with either 1% fatty acid free BSA or filtered 10% FCS. If measuring in tissue culture samples, substitute the FCS with protein representing that within the culture conditions, that is, 10% FCS; in effect the sample will have been preabsorbed with this during culture and thus any nonspecific binding to the blocking buffer is avoided.
3. Carbonate coating buffer: 8.41 g Na_2HCO_3 in 1 L freshly made PBS. Dissolve and adjust to desired pH with HCl or NaOH. Store for no more than 1 mo at 4°C (*see* **Note 2**).
4. High-capacity protein binding 96-well microtiter plates: There are a large number of suitable makes including Maxisorp (Nunc), Immunoware (Pierce), Immunlon II (Dynatech), and Costar (*see* **Note 3**).
5. PBS–Tween–10% FCS: Add 0.05 mL of Tween 20 to 90 mL of PBS, pH 7.4, and 10 mL of filtered FCS. Make up fresh as required.
6. Plate sealers or cling film wrap.
7. Plate washing apparatus: Adequate washing is a vital element of achieving a successful ELISA. Although a number of automatic plate washers are available, they are expensive and the use of a wash bottle with good pressure is perfectly suitable, although a little more time consuming.
8. Samples/standards: Standards of known amounts are required for positive controls, and for estimation of levels within samples (via comparison to a titration of known amounts). All standards should be diluted in medium as near as, or identical to that of the sample solution. Standards are best obtained from a reliable commercial source having been mass calibrated and should be frozen in small, concentrated aliquots at –20°C or lower. Repeated freeze–thaw cycles must be avoided.
9. Spectrophotometer: Any suitable microplate reader able to measure absorbance at the appropriate wavelength.
10. Stop solution: 0.5 *M* H_2SO_4.

11. Streptavidin HRP (*see* **Note 4**): Use in accordance to manufacturer's instructions. Sources of streptavidin HRP include Sigma, Biosource, Pierce, and Zymed.
12. Substrate: One-step TMB (Zymed). Although many other substrates are available for HRP, TMB has high sensitivity with a quick development time. OD can be monitored at 650 nm while the color develops, then at 450 nm when the reaction is stopped with H_2SO_4. TMB may also be obtained in a lyophilized state and made up fresh with hydrogen peroxidase.
13. Washing buffer: Add 0.5 mL of Tween 20 to 1 L of PBS, pH 7.4. Make up fresh as required.

3. Methods

3.1. Basic Sandwich ELISA Protocol

1. Dilute the capture antibody to 1 μg/mL in coating buffer, pH 9.5. Add 100 μL to each well of a high-capacity protein binding 96-well microtiter plate.
2. Seal the plate to avoid evaporation and incubate overnight (12–18 h) at 2–8°C.
3. Wash plate: Discard unbound antibody by inverting and flicking the plate over a sink. Fill each well with washing buffer and leave for at least 10 s before discarding once more. Ensure all liquid has been removed between each wash by repeatedly tapping the plate onto clean paper towels. Repeat 3×.
4. Add 200 μL of blocking buffer to each well. Seal plate and incubate for at least 2 h at room temperature.
5. Discard blocking buffer. Wash plate 3× as described in **step 3**.
6. Prepare a titration series of known standards (e.g., in 1.5-mL Eppendorf tubes) diluted in a matrix representing that of the samples (e.g., culture medium or human serum). Include a negative control such as culture medium only. Transfer samples and antigen standards to the ELISA plate in duplicate at 100 μL/well. Seal plate and incubate at room temperature for at least 2 h, or overnight at 4°C for increased sensitivity.
7. Wash plate 3× as described in **step 3**.
8. Dilute biotinylated detection antibody to 1 μg/mL in PBS, pH 7.4. Add 50 μL/well. Incubate at room temperature for 2 h. If problems with nonspecific binding of the biotinylated antibody to the plate occur, dilute the antibody in PBS–Tween–10% FCS rather than just PBS.
9. Wash plate 3× as described in **step 3**.
10. Dilute streptavidin HRP according to manufacturer's instructions. Add 100 μL/well. Incubate at room temperature for 30 min. Again, if problems with nonspecific binding of the biotinylated antibody to the plate occur, dilute the antibody in PBS-Tween-10% FCS rather than just PBS.
11. Wash plate 4× as described in **step 3**.
12. Add 100 μL/well of "one step" TMB. Allow color to develop for 10–60 min (20 min is usually sufficient). OD may be monitored at this stage at 650 nm as the color develops.
13. Add 100 μL of 0.5 *M* H_2SO_4 to each well to stop the reaction. Read OD at 450 nm.
14. Estimate amount of antigen within samples by comparing ODs to those of known standards.

3.2. Direct ELISA Protocol

1. Make serial dilutions of samples from neat to 1:16 in coating buffer, pH 9.5. Add 100 μL to each well of a high-capacity protein binding microtiter plate. Also set up wells with antigen standards and a negative control diluted in the same coating buffer.
2. Seal the plate with an acetate plate sealer or cling film wrap to avoid evaporation and incubate overnight (12–18 h) at 2–8°C.
3. Wash plate: Discard unbound antibody by inverting and flicking the plate over a sink. Fill each well with washing buffer leave for at least 10 s and discard once more. Ensure all

liquid has been removed between each wash by repeatedly tapping the plate onto clean paper towels. Repeat 3×.
4. Add 200 µL of blocking buffer to each well. Seal plate and incubate for at least 2 h at room temperature.
5. Discard blocking buffer. Wash plate 3× as described in **step 3**.
6. Dilute detection antibody to 2 µg/mL in PBS. Add 50 µL/well. Incubate at room temperature for 2 h. If the antibody is biotinylated proceed to **step 9**.
7. Wash plate 3× as described in **step 3**.
8. Add secondary antibody (100 µL, 2 µg/mL) specific for the detection antibody (i.e., if the detection antibody is mouse, use biotinylated rabbit anti-mouse Ig).
9. Wash plate 3× as described in **step 3**.
10. Dilute streptavidin HRP according to the manufacturer's instructions. Add 100 µL/well. Incubate at room temperature for 30 min.
11. Wash plate 4× as described in **step 3**.
12. Add 100 µL/well of "one-step" TMB. Allow color to develop for 10–60 min (20 min is usually sufficient). OD may be monitored at this stage at 650 nm, as the color develops.
13. Add 100 µL of 0.5 *M* H_2SO_4 to each well to stop the reaction. Read OD at 450 nm.
14. Estimate amount of antigen within samples by comparing ODs to those of known standards of sample.

3.3. Optimization of ELISA

At this stage the ELISA may be more than adequate, although optimization of the following parameters is usually required:

1. pH of carbonate coating buffer (pH 7.0–pH 10.0). Also try PBS at pH 7.4.
2. Concentration of capture antibody (0.5–10.0 µg/mL).
3. Concentration of secondary antibody (indirect ELISA, 0.5–4.0 µg/mL).
4. Concentration of biotinylated detection antibody (0.1–2.0 µg/mL).
5. Concentration of streptavidin–HRP conjugate (usually between 1:1000 and 1:10,000).

4. Notes

1. The quality of the antibodies used is perhaps the most important aspect in setting up a good ELISA. Antibodies need to have a high affinity for the sample to be measured, reducing the chance of being "washed off" during the assay. The indirect ELISA relies on two specific antibodies and thus the increased specificity increases sensitivity of the assay by reducing background.
2. The pH of the coating buffer can have a great effect on the amount of antibody that will bind to the plate and thus to the ELISA as a whole. Basically, a higher pH will result in more antibody binding but may have a detrimental effect on its immunoreactivity. Thus a pH must be found that is suitable for the antibody in question, and this can vary dramatically. When beginning optimization of the assay, test a range of carbonate buffers from pH 7.0 to 10.0 as well as PBS, pH 7.4. We usually find a carbonate buffer pH of 9.5 gives good results. In some cases we have found commercially available coating antibodies recommended by the manufacturer to be adsorbed onto the plate at pH 7.4 to be far more effective at higher pH values, improving the sensitivity by up to 10-fold. This can allow the coating concentration to be vastly reduced, creating an extremely cost-effective assay.
3. There can be a significant difference between the protein binding capability of different makes, and even batches of microtiter plates. Some appear to be extremely good for binding antibodies, whereas others more useful for other proteins. The only real way to choose

a suitable plate is by trial and error, or using a recommended make known to bind the protein you intend to coat.

4. Although the HRP–TMB system represents a good, reliable, and sensitive combination, HRP has a number of alternative substrates that can be used *(2)*. These include o-phenylene diamine (OPD) or 2,2′-azino-di(3-ethylbenzthiazoline-6-sulfonic acid (ABTS). There are also a number of options for the enzyme used other than HRP, such as alkaline phosphatase (AP) which can be used in combination with the substrate *p*-nitrophenyl phosphate (p-NPP) or phenolphthalein monophosphate (PMP). The choice of enzyme–substrate system depends on a number of factors including price, sensitivity, and whether a filter is available for the substrate specific wavelength to be measured.
5. The use of biotinylated secondary antibodies (or detection antibody in the indirect sandwich ELISA) in conjunction with enzyme-conjugated streptavidin (or avidin, extravidin, etc.) both increase sensitivity and save time in that a further step is eliminated from the assay and therefore another step of optimization is eliminated.
6. If the major aim of the ELISA is to obtain quantitation of substances present in extremely low concentrations a number of adaptations to the technique can be used. Such techniques often use AP enzyme systems which can be utilized, for example, to lock into a circular redox cycle producing an end product such as red formazan which is greatly amplified in comparison to standard amplification methods *(3)*. Chemiluminescent amplified ELISA principles have also been shown to give very high sensitivity *(4)* and can be optimized to measure as little as 1 zeptomol (about 350 molecules!) of AP *(5)*. Although extremely sensitive, such techniques are time consuming to set up and optimize, and are far more expensive than the simple colormetric ELISAs described in this chapter.
7. In some cases, molecules present in a sample are masked by the solution that they are in. This problem can sometimes be solved by diluting the samples in PBS–Tween–10% FCS. If this is performed, remember to make similar adjustments to the solution used for the standards. Possible interference molecules within samples such as soluble receptors for the antigen can also cause a problem. Commercially available matched antibody pairs for molecules should have been pretested and guaranteed against being affected by such problems.

References

1. Engvall, J. R. and Perlamann, P. (1971) Enzyme-linked immunosorbent assay (ELISA). Quantitative assay of immunoglobulin G. *Immunochemistry* **8,** 871–874.
2. Roberts, I. M., Solomon, S. E., Brusco, O. A., Goldberg, W., and Bernstein, J. J. (1991) A comparison of the sensitivity and specificity of enzyme immunoassays and time-resolved fluoromimmunoassay. *J. Immunol. Meth.* **143,** 49–56.
3. Self, C. H. (1985) Enzyme amplification—a general method applied to provide an immunoassisted assay for placental alkaline phosphatase. *J. Immunol. Meth.* **76,** 389–393.
4. Bronstein, I., Voyta, J. C., Thorpe, G. H., Kricka, L. J., and Armstrong, G. (1989) Chemiluminescent assay of alkaline phosphatase applied in an ultrasensative enzyme immunoassay of thyrotropin. *Clin. Chem.* **35,** 1441–1446.
5. Cook, D. B. and Self, C. H. (1993) Determination of one-thousandth of an attomole (1 zeptomole) of alkaline phosphatase: application in an immunoassay of proinsulin. *Clin. Chem.* **39,** 965–971.

156

Enhanced Chemiluminescence Immunoassay

Richard A. W. Stott

1. Introduction

Chemiluminescence results from reactions with a very high energy yield, which produce a potentially fluorescent product molecule; reaction energy passed to the product may result in an excited state and subsequent production of a single photon of light. The light yield is usually low, but can approach one photon per molecule in bioluminescent reactions catalyzed by dedicated enzymes.

A wide variety of immunoassay systems that use chemiluminescence or bioluminescence have been developed with the aim of detecting low concentrations of biologically active molecules. Even when emission efficiencies are <1%, chemiluminescence is a sensitive label detection method compared with isotopic methods in which very large numbers of molecules must be present for each detected disintegration (e.g., about 1×10^7 atoms of ^{125}I give 1 count/s). The production of light against a low background permits detection of small numbers of reacting molecules by measuring total light output. Luminescent emissions can be measured over a range of at least six orders of magnitude by all but the simplest luminometers. This is in marked contrast to fluorescent or spectrophotometric detection of reaction products, where sensitivity and instrument linear range are limited by the stability of light sources and wavelength selection. For example, a good spectrophotometer may achieve a linear range slightly greater than three orders of magnitude.

Directly labeled chemiluminescent systems produce <1 photon/label and require complex chemical synthesis to produce each new labeled molecule *(1,2)*. In contrast, chemiluminescent detection of enzyme labels combines the advantages of a high specific activity label with the convenience of relatively simple coupling chemistries, which use commercial reagents. A number of enzyme labels can be detected via chemiluminescent or bioluminescent reactions, including β-galactosidase *(3)*, alkaline phosphatase *(4)*, peroxidase, luciferin and a variety of enzymes indirectly linked via production of ATP or NADH *(5)*. Systems that use alkaline phosphatase and peroxidase have the additional advantage of commercial availability of a wide range of labeled molecules and complete assay kits suitable for adaptation to luminescent detection.

Enhanced chemiluminescence is based on the reaction of luminol (3-aminophthalhydrazide) with an oxidizing agent, such as hydrogen peroxide or sodium perbo-

From: *The Protein Protocols Handbook, 2nd Edition*
Edited by: J. M. Walker © Humana Press Inc., Totowa, NJ

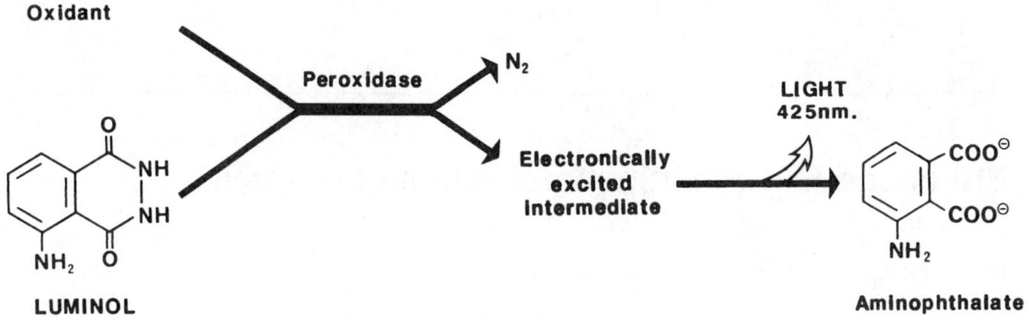

Oxidant

Peroxidase N_2

LIGHT 425nm.

Electronically excited intermediate

LUMINOL

Aminophthalate

Fig 1. Chemiluminescent oxidation of luminol by peroxidase.

rate. This reaction is catalyzed by metal ions at high pH, resulting in emission of blue light (emission peak about 425 nm). At lower pH, the reaction is catalyzed by heme-containing enzymes, such as horseradish peroxidase, catalase, cytochrome C, and hemoglobin (**Fig. 1**). However, the light output is low with a half-life of a few seconds. The presence of any one of a series of "enhancer molecules" increases the light emission from horseradish peroxidase by 1000-fold or more, and alters the kinetics so that a steady glow is produced lasting several hours *(6,7)*. Microparticles, plastic beads, plastic tubes, microtiter plates, membranes, and plastic pins have all been used successfully as solid supports in a wide range of enhanced chemiluminescence assays, including competitive immunoassays, immunometric assays, and RNA and DNA binding assays *(6,8,9)*.

In common with all other sensitive detection systems, maintenance of the label enzyme in its active state is important. The precautions detailed in **Notes 1–3** should be observed to maximize the sensitivity achieved. Reagents for enhanced chemiluminescence can be prepared in the laboratory or are available commercially (*see* **Note 4**). The purity of the substrate solution is important in achieving maximum sensitivity. Therefore, the precautions detailed in **Notes 5–7** should be followed if preparing substrate solutions. The free base form of luminol undergoes rearrangement to a mixture of luminol and a series of contaminants. Therefore, luminol should be purified by recrystallization as the sodium salt before use (*see* **Note 8**).

1.1. Light Measurement Instruments (Luminometers)

Commercial luminometers range from low-cost manual single tube instruments to fully automated high-capacity machines and have been reviewed previously *(10)*. However, application-specific requirements are rarely discussed, and the first-time user will require some guidance in matching an instrument to the chemistry or chemistries to be used.

There is normally no requirement for wavelength selection, because very few reactions produce significant light output. Therefore, a simple luminometer can consist of a detector and some means of presenting a sample or samples in a light-tight compartment. There may also be a system for adding reagents to the sample while in the chamber. The detector is usually a photomultiplier tube for sensitive instruments, but can be a photodiode in a portable instrument *(7,11)* or photographic film, if a semiquantitative result is sufficient *(2,12)*.

Table 1
Carryover from a Single Glowing Well in a Rigid[a] Black Microtiter Plate Measured Using a Prototype Luminometer

0.02%	0.01%	0.007%	ND	ND
0.04%	0.035%	0.018%	0.005%	ND
0.71%	0.075%	0.034%	0.007%	ND
Source	0.1%	0.03%	0.007%	ND
0.045%	0.034%	0.025%	0.006%	ND
0.03%	0.025%	0.013%	0.005%	ND
0.009%	0.009%	0.006%	ND	ND

[a]Data represent the mean of several readings obtained for empty wells expressed as a percentage of the mean light output of the source well. Positions marked ND gave readings that were not significantly different from the photomultiplier background.

Luminometers designed for use with short-lived reactions have complex high-precision reagent injection systems. There may also be a short measurement prior to reagent injection to correct for background owing to light leaks, phosphorescence, and scintillation from sample tubes. Photon-counting and cooled detectors may also be used to achieve maximal light-detection sensitivity. None of these features is essential for use with enhanced chemiluminescence.

Relatively high light intensities and prolonged emission are produced by enhanced chemiluminescence detection of peroxidase and the substituted dioxetane-based detection reactions for alkaline phosphatase or β-galactosidase *(3,4)*. These reactions require a short stabilizing time before reading and are conveniently performed by adding the reagents before the sample reaches the measuring position. This can only be done if any preinjection blank measurement can be disabled. It is also practical to handle large numbers of samples using a timed reagent addition outside the instrument, completely eliminating the need for automatic reagent handling. High light output can lead to "pulse pileup" in photon-counting electronics, resulting in nonlinearity and eventually zero apparent signal *(13)*. Linearity can be improved via mathematical correction for the dead time or insertion of a neutral density filter.

Although light output from individual transparent microtiter wells can be measured in a luminometer designed for tubes, it is more convenient to use opaque microtiter plate wells and one of the purpose-designed readers. Prolonged light output makes it possible to start the reaction outside the instrument in the same pattern as the light emission is read. However, the presence of other glowing wells introduces the possibility of light carryover into the well being read. Carryover is important if the dynamic range of light emissions from a plate is expected to be higher than three orders of magnitude; in this case, there is a risk of false-positive results owing to a fraction of a percent of the emission from a very high sample being transmitted to a low one as much as three positions away (**Table 1**).

Both white and black microtiter plates are available in single-well, strip, and plate formats. The plastic is made opaque by incorporating colored particles into transparent plastic. Therefore, some light can pass through the plastic. The light transmission differs considerably between formats and individual manufacturers. However the greatest

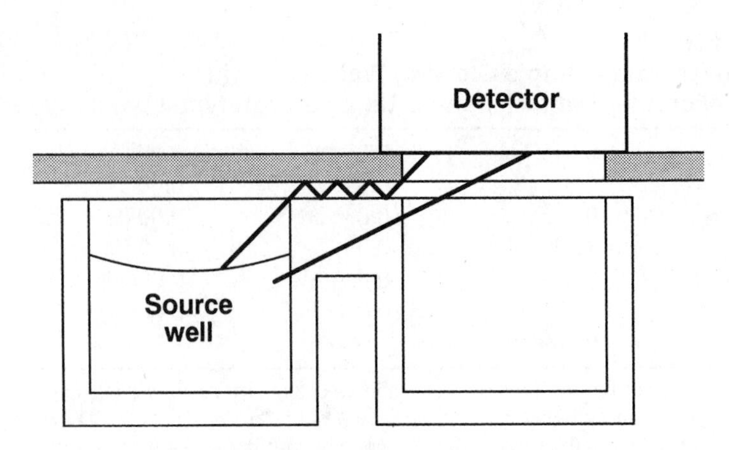

Fig. 2. Origin of light carryover between wells of an "opaque" microtiter plate.

carryover occurs via external reflection from the shiny top surface (**Fig. 2**). Carryover is least for black plastic, although the loss of reflection within the well also reduces the signal available to the detector. Multiwell strips have higher carryover along the strip than between adjacent strips owing to the plastic web that links the wells. Similarly, individual wells have lower carryover than joined ones.

Carryover varies considerably between instruments, but it is particularly low if the instrument has an antireflection mask between the plate and detector. Individual instruments should be assessed for carryover using the type of plates that will be used with it. The pattern of carryover should be determined by reading light output from all wells of a plate containing a single glowing well with a light output, which represents the highest expected from the assay. The location of this well may affect the results, and carryover should be assessed using each corner well and one close to the center of the plate.

2. Materials

1. High-quality deionized water (*see* **Note 3**).
2. Luminol stock solution: 1.25 m*M* luminol in 0.1 *M* Tris-HCl, pH 8.6. Store at 4°C in the dark. Make up fresh each week. Luminol should be recrystallized as the sodium salt before use (*see* **Note 8**).
3. Hydrogen peroxide: 30% (w/v). Store at 4°C.
4. p-Iodophenol stock solution: p-iodophenol, 1 mg/mL in dimethyl sulfoxide (DMSO). Make up fresh each day.
5. Microfluor "B" microtiter plates (Dynatech Laboratories, Chantilly, VA).
6. Coating buffer: 0.1 *M* glycine, pH 8.8. Adjust pH using NaOH. Store at 4°C, and make up fresh each week.
7. Rabbit anti-α fetoprotein (AFP) (Dako [Glostrup, Denmark] cat. no. A0008).
8. Phosphate-buffered saline (PBS), pH 7.2: 0.14 *M* NaCl, 2.7 m*M* KCl, 1.5 m*M* KH_2PO_4, and 8.1 m*M* Na_2HPO_4. Store at 4°C, and make fresh each week.
9. Blocking solution: PBS containing Bovine serum albumin (BSA) 0.1% (w/v). Make up fresh each day.
10. PBS-Tween: PBS containing Tween-20, 0.05% (v/v). Make an additional batch using high-quality deionized water (*see* **Note 3**). Store at 4°C, and make fresh each week.
11. Assay diluent: PBS-Tween containing BSA 0.5% (w/v). Make up fresh each day.
12. Working standards concentration range 0–800 ng/mL made fresh for each assay batch by serial dilution of stock standard using normal human serum. Stock standard (1600 ng/mL)

is made by diluting AFP standard serum (Dako cat. no. X900 or Boehring Diagnostics [Westwood, MA] cat. no. OTOD 02/03) in human serum containing 0.05% (w/v) sodium azide. Store frozen as 1-mL aliquots.

13. Peroxidase-conjugated anti-AFP (Dako cat. no. P128) diluted 1/1000 in assay diluent. Make up fresh for each assay batch.
14. Working enhanced chemiluminescence substrate solution (*see* **Note 9**)—either:
    a. p-iodophenol-enhanced substrate: 1.25 m*M* luminol, 4 μ*M* p-iodophenol, 2.7 m*M* H_2O_2). Add 15 μL of stock hydrogen peroxide and 40 mL of p-iodophenol stock solution to 50 mL of stock substrate: solution,, and mix well. Make up daily using high-quality deionized water (*see* **Note 3**), and store in the dark when not in use, or
    b. p-Hydroxycinnamic acid-enhanced substrate (1.25 m*M* luminol, 30 μ*M* p-hydroxycinnamic acid, 2.7 m*M* H_2O_2). Add 15 mL of stock hydrogen peroxide and 1 mg of p-hydroxycinnamic acid (alternative names: p-coumaric acid or 4-hydroxycinnamic acid, e.g., Aldrich [Gillingham, UK] cat. no. 2,320-7) to 50 mL of stock substrate solution. Mix for 30 min before using. Make up daily using high-quality deionized water (*see* **Note 3**), and store in the dark when not in use.

3. Method

1. Coat the wells of a microtiter plate with 100 mL of anti-AFP (1/1000 in coating buffer). Allow the protein to bind for either 2 h at room temperature or overnight at 4°C (*see* **Note 10**).
2. Empty the wells, and wash off any unbound antibody by filling each well with PBS and shaking the plate to re-empty. Repeat the wash and block unbound sites by incubating with 200 mL/well of blocking solution for 30 min at room temperature.
3. Empty the wells, and wash the plate twice (as described in **step 2**) with PBS-Tween to remove any unbound albumin.
4. Prepare 1/20 dilutions of samples, standards, and quality control specimens using assay diluent, and add 150 μL to each of the microtiter plate wells. Cover the plate with plastic film, and incubate at 37°C for 1 h.
5. Empty the wells, and wash the plate three times with PBS-Tween. Shake the plate over a sink to ensure complete removal of wash solution. Add 150 μL of working conjugate to each well, cover, and incubate at 37°C for 1 h.
6. Empty the wells, and wash the plate three times with PBS-Tween made using high-quality deionized water (*see* **Note 3**).
7. Add 150 μL of enhanced chemiluminescent substrate to each well in the same order and preferably with the same timing as used by the plate reader. Allow at least 2 min for the light output to stabilize before reading the plate.
8. Obtain unknown specimen results either by reading off a plotted calibration curve or use a computer program to calculate from a fitted curve.

4. Notes

1. Peroxidase is inactivated by anions and certain antimicrobial agents, including azide, cyanide, and thiomersal. Antimicrobial agents may be present in concentrated enzyme label solutions and assay buffers at typically active concentrations, but must not be present in wash solutions or substrate. The latter reagents must be freshly made each day from concentrated stocks.
2. Powerful oxidizing or reducing agents may interfere with any peroxidase detection reaction by inactivating peroxidase, oxidizing the substrates or reducing the oxidants in the reagents. There are potentially many of these in the laboratory environment, including chlorine in water, disinfectants, paper dust, laboratory coats, skin, and so forth. Care should be taken to avoid contamination of individual assay wells or equipment.

3. For best possible sensitivity, the final assay wash (**step 6**) and all substrate reagents should be made up in water of the highest possible purity. Trace contamination with bacteria, algae, organic compounds, and chlorine is a particular problem. Laboratory-grade distilled or reverse osmosis water should be further treated using a deionization cartridge. The water plant must be well maintained to avoid bacterial growth in deionization columns, plumbing, and storage tanks. Alternatively, commercial HPLC-grade water has been found to be satisfactory.
4. Amerlite signal reagent (Johnson and Johnson Clinical Diagnostics) is supplied as separate bottles of buffer and substrate tablets. One "A" and one "B" substrate tablet (cat. no. LAN.4401) are dissolved in each bottle of substrate buffer (cat. no. LAN. 4402) prior to use. This reagent is stable for a day at room temperature providing it is kept in the dark glass bottle.
5. In order to ensure stable concentrations, the anhydrous form of sodium luminol is preferred molecular weight 199.1. This should be stored over silica gel in the dark. Luminol solutions should be stored in the dark at 4°C. Stock solutions must be made up at least weekly and working substrate daily.
6. DMSO is a colorless, odorless compound. However, it is hygroscopic and acquires an onion-like smell. DMSO in this state has been found to be inhibitory in the enhanced chemiluminescent reaction. Therefore, the highest available grade should be purchased in small amounts and carefully stored to minimize water uptake.
7. Hydrogen peroxide is a powerful oxidant, but gradually loses activity. The highest available grade should be purchased in small amounts, and stored at 4°C.
8. Luminol is available as the free base under alternative chemical names from several chemical suppliers, including:
   a. 5-Amino-2,3-dihydro-1,4-phthalazinedione (Sigma [St. Louis, MO], cat. no. A 8511P).
   b. 3-Aminophthalhydrazide (Aldrich cat. no. 12,307-2).

   There is considerable batch to batch variability in commercial luminol. The purity of the original material is only important in determining how many recrystallization steps are required. Recrystallization gives a consistently high activity product and may be performed as follows (for further details, *see* **ref. 7**):
   a. Dissolve luminol in 5% (w/v) sodium hydroxide at room temperature until close to saturation (about 200 g/L). The color of this solution will depend on the original luminol. Cool the solution in an ice bath and allow to crystallize for 4 h. Recover the sodium luminol crystals using suction filtration on a glass fiber filter disk (Whatman GFA or similar), and wash using a small volume of ice-cold 5% sodium hydroxide. The crystals should be white or only slightly discolored. Sodium luminol should be recrystallized at least twice after a white product is obtained.
   b. Dissolve the sodium luminol in a minimum volume of 5% sodium hydroxide at room temperature. Allow to crystallize on an ice bath for 18 h. A refrigerator can be used as an alternative, although crystallization may be slower and yield may be low. Recover the crystals by filtration as above.

      The initial crystalline form is the hexa-hydrate, which converts to an anhydrous powder on drying over silica gel. Sodium luminol is stable to heat (melting point >400°C), but Sodium luminol has recently become available from Sigma (cat. no. A 4685). The author has no experience with this product.
9. The substrate solutions detailed here (**Subheading 2., item 14**) are essentially interchangeable with no sensitivity advantage for either. Both systems are optimized for a reasonably steady light output at a peroxidase concentrations typically encountered in immunoassays. Use of final peroxidase activities that are markedly higher will result in declining

light output owing to substrate exhaustion. This cannot be avoided by alteration of the reaction conditions. Slight reduction in enhancer concentration may give more stable light output for assays with atypically low peroxidase activity.

10. For best results, plates should be coated with the IgG fraction of an antiserum, this can be conveniently prepared using caprylic acid precipitation (*see* Chapter 138 and **ref. *15***). Where this is not possible, indirect capture may be used, such as antispecies antibody on the plate, or the streptavidin–biotin system. Any indirect capture system must be compatible with the final label, e.g., labeled antigen or different species antisera with no cross-reaction with the indirect coating antibody.
11. Safety data (from **ref. *16***):
    a. Luminol (commercial-grade): Irritating to eyes, respiratory system, and skin. No specific information is available for pure luminol or for the sodium salt.
    b. Hydrogen peroxide: Contact with combustible materials may cause fire. Causes burns. Keep in a cool place. After contact with skin, wash immediately with plenty of water.
    c. DMSO: Irritant to eyes, skin, and respiratory system. Harmful by inhalation, skin contact, and if swallowed. May cause sensitization by inhalation or skin contact.
    d. *p*-Iodophenol: Irritant to eyes, skin, and respiratory system.
    e. *p*-Hydroxycinnamic acid: Irritant to eyes, skin, and respiratory system.

References

1. Zomer, G., Stavenuiter, J. F. C., Van Den Berg, R. H., and Jansen, E. H. J. M. (1991) Synthesis, chemiluminescence and stability of acridinium ester labeled compounds, in *Luminescence Techniques in Chemical and Biochemical Analysis* (Bayens, W. R. G., De Kekeleire, D., and Korkidis, K., eds.), Dekker, New York, pp. 505–521.
2. Jansen, E. H. J. M., Zomer, G., and Van Peteghem, C. H. (1991) Chemiluminescence immunoassays in vetinary and food analysis, in *Luminescence Techniques in Chemical and Biochemical Analysis* (Bayens, W. R. G., De Kekeleire, D., and Korkidis, K., eds.), Dekker, New York, pp. 477–504.
3. Bronstein, I. and McGrath, P. (1989) Chemiluminescence lights up. *Nature* **333,** 599,600.
4. Bronstein, I., Voyta, J. C., Thorpe, G. H. G., Kricka, L. J., and Armstrong, G. (1989) Chemiluminescent assay of alkaline phosphatase applied in an ultrasensitive enzyme immunoassay of thyrotropin. *Clin. Chem.* **35,** 1441–1446.
5. Bronstein, I. and Kricka, L. J. (1989) Clinical applications of luminescent assays for enzymes and enzyme labels. *J. Clin. Lab. Analyt.* **3,** 316–322.
6. Thorpe, G. H. G. and Kricka, L. J. (1986) Enhanced chemiluminescent reactions catalyzed by horseradish peroxidase. *Meth. Enzymol.* **133,** 331–354.
7. Kricka, L. J., Stott, R. A. W., and Thorpe, G. H. G. (1991) Enhanced chemiluminescent detection of Horseradish peroxidase labels in ligand binder assays, in *Luminescence Techniques in Chemical and Biochemical Analysis* (Bayens, W. R. G., De Kekeleire, D., and Korkidis, K., eds.), Dekker, New York, pp. 599–635.
8. Thorpe, G. H. G., Stott, R. A. W., Sankolli, G. M., Catty, D., Raykundalia, C., Roda, A. and Kricka, L. J. (1987) Solid supports and photodetectors in enhanced chemiluminescent immunoassays, in *Bioluminescence and Chemiluminescence. New Perspectives* (Scholmerich, J., Andreesen, R., Kapp, M., and Woods, W. G., eds.), Wiley, Chichester, UK, pp. 209–213.
9. Matthews, J. A., Batki, A., Hynds, C., and Kricka, L. J. (1985) Enhanced chemiluminescent method for the detection of DNA dot-hybridization assays. *Analyt. Biochem.* **151,** 205–209.

10. Stanley, P. E. (1992) A survey of more than 90 commercially available luminometers and imaging devices for low light measurements of chemiluminescence and bioluminescence, including instruments for manual, automatic and specialized operation, for HPLC, LC, GLC, and microtiter plates. Part 1 Descriptions. *J. Biolum. Chemilum.* **7,** 77–108.
11. Marks, K., Killeen, P. R., Goundry, J., Gibbons, J. E. C., and Bunce, R. A. (1987) A portable silicon photodiode luminometer. *J. Biolum. Chemilum.* **1,** 173–179.
12. Kricka, L. J. and Thorpe, G. H. G. (1986) Photographic detection of chemiluminescent and bioluminescent reactions. *Meth. Enzymol.* **133,** 404–420.
13. Stott, R. A. W., Moseley, S. B., Williams, L. A., Thorpe, G. H. G., and Kricka, L. J. (1987) Enhanced chemiluminescent quantitation of horseradish peroxidase labels in commercial EIA kits using a modified Berthold LB950T, in *Bioluminescence and Chemiluminescence. New Perspectives* (Scholmerich, J., Andreesen, R., Kapp, M., and Woods, W. G., eds.), Wiley, Chichester, UK, pp. 249–252.
14. Stott, R. A. W. and Kricka, L. J. (1987) Purification of luminol for use in enhanced chemiluminescence immunoassay, in *Bioluminescence and Chemiluminescence. New Perspectives* (Scholmerich, J., Andreesen, R., Kapp, M., and Woods, W. G., eds.), Wiley, Chichester, UK, pp. 237–240.
15. Steinbuch, M. and Audran, R. (1969) The isolation of IgG from mammalian sera with the aid of caprylic acid. *Arch. Biochem. Biophys.* **134,** 279–284.
16. Lenga, R. E. and Votoupal, K. L. (eds.) (1993) The Sigma-Aldrich Library of Regulatory and Safety Data. Sigma-Aldrich Corporation.

157

Immunoprecipitation

Kari Johansen and Lennart Svensson

1. Introduction

Immunoprecipitation allows the investigator to detect and quantitate antigens in a mixture of proteins or characterize a specific antibody response to already well-characterized proteins. Addition of antibodies to proteins, usually radiolabeled, allows formation of antigen–antibody complexes. After separation from contaminating proteins, the complexes are disassociated and the proteins of interest are separated by sodium dodecyl sulphate-polyacrylamide gel electrophoresis (SDS-PAGE). Size and quantity of proteins may then be analyzed either by autoradiography or a gel scanning procedure. Immunoprecipitation is extremely sensitive and may detect very small amounts of radiolabeled protein (detection level ~100 pg protein or 100 cpm/protein). Unlabeled proteins may be used if other sensitive detection methods are utilized, e.g., enzymatic activity assays or Western blotting. The advantage of the immunoprecipitation technique vs immunoblotting is the possibility to analyze the immune response to proteins expressed in their native conformation. Radioimmunoprecipitation assay (RIPA) is used routinely for the detection of viral proteins, characterization of monoclonal and polyclonal antibody preparations, and determination of the specificity of the immune response to various pathogens ***(1–3)***.

The major steps in immunoprecipitation are:

1. Labeling of proteins expressed by mammalian cells;
2. Lysis of the cells;
3. Addition of antibodies to lysed cells and formation of antigen–antibody complexes;
4. Purification of the specific immune complexes; and
5. Analysis of the immunoprecipitated proteins.

1.1. Metabolic Labeling of Proteins

Several techniques are available for labeling proteins. Usually, radiolabeled essential amino acids, such as ^{35}S-labeled methionine and ^{35}S-labeled cysteine, are used to label newly synthesized proteins expressed in mammalian cells, either naturally or after transfection or infection. ^{35}S emits a weak β-radiation easily detected either by autoradiography or gel scanning and has a half-life of 87.1 d. Although most strains of yeasts and prokaryotes synthesize methionine and cysteine ***(1)***, ^{35}S-labeled sulphate is used as the primary metabolic precursor to label proteins in these organisms.

From: *The Protein Protocols Handbook, 2nd Edition*
Edited by: J. M. Walker © Humana Press Inc., Totowa, NJ

Table 1
Detergents Often Used in Lysis Buffer

Detergent	Ability to disperse protein aggregates	Denatures proteins	Working concentrations
Ionic detergents			
Deoxycholic acid, sodium salt	High	No	0.1–10 mg/mg membrane lipid
SDS	High	Yes	>10 mg/mg protein
Zwitterionic detergents			
CHAPS	High	No	6.5–13 m*M*
Nonionic detergents			
Nonidet P40 (NP-40)	Low	No	1–10 m*M*/L
Triton® X-100	Low	No	1–5 m*M*
n-Octylglucoside	Low	No	46 m*M*

Other possibilities for labeling are biotinylation or iodination of, e.g., surface proteins or secreted proteins. The biotinylated proteins may be immunoprecipitated by avidin bound to Agarose beads (Vector, Burlingame, CA), which presents as an interesting alternative to the use of radioactive isotopes.

1.2. Lysis of Cells

Lysis of the cells is the most crucial part of immunoprecipitation and several techniques may be utilized. The aim is usually to solubilize the target antigen in an immunoreactive, undegraded, and biologically active form. When choosing a lysis method, a good strategy is to start with a crude lysis buffer and if the antigen is released, then step back, remove and alter the buffer composition until conditions have been optimized. Variables that have been found to influence the efficiency of solubilization are the ionic strength; salt concentration and pH of the lysis buffer; the presence of divalent cations, cofactors, and stabilizing ligands, and the concentration and type of detergent used (ionic, zwitterionic, or nonionic). Among the common detergents, SDS denatures proteins, whereas most other detergents do not. Antigen release may be tested by immunoblotting of the lysate and the remaining cell debris. In general, detergent lysis with a single or triple detergent lysis buffer is usually advised as a start. Repeated freezing and thawing of lysates may lead to excessive proteolytic degradation.

Most soluble nuclear and cytoplasmic proteins may be solubilized by lysis buffers containing the nonionic detergents Nonidet P-40 (NP-40), Triton X-100 or the zwitterionic detergent 3-([3-cholamidopropyl]dimethylammonia)-1-propane-sulfonate (CHAPS) (Boehringer-Mannheim, Mannheim, Germany) containing a relatively high salt concentration (0.5 *M* NaCl or 0.6 *M* KCl), low salt concentration, or no salt (**Table 1**). Possible ranges for the lysis buffers include nonionic detergent concentrations between 0.1 and 2%, ionic detergent concentrations between 0.01 and 0.5%, salt concentrations between 0 and 1 *M*, divalent cation concentrations between 0 and 10 m*M*, EDTA concentrations between 0 and 5 m*M*, and pH values between 6.0 and 9.0. The addition of RNases and DNases as well as protease inhibitors may protect the target antigen (**Table 2**). The susceptibility to proteases varies greatly with cell-surface

Table 2
Protease Inhibitors Often Used in Lysis Buffers

Inhibitor	Specificity	Solubility/stability	Starting concentrations
Antipain-dihydrochloride	Trypsin, papain, plasmin	Soluble in H_2O, methanol DMSO. Dilute solutions stable 1 mo at –20°C	50 µg/mL
Aprotinin	Plasmin, trypsin, chymotrypsin	Soluble in H_2O, aqueous buffers. Dilute solutions stable 6 mo at –20°C	0.06–2 µg/mL
Chymostatin	α, β, γ, δ-Chymotrypsin	Soluble in glacial acetic acid, DMSO. Dilute solutions stable 1 mo at –20°C	0.1 ng/mL
E-64	Papain, cathepsin B, L	Soluble in a 1:1 mixture (v/v) of ethanol and water	0.5-10 µg/mL
Leupeptin	Trypsin, papain, plasmin, cathepsin B	Soluble in H_2O. Dilute solutions stable 6 mo at –20°C	0.5 µg/mL
PMSF	Chymotrypsin, trypsin, thrombin, papain	Soluble in 2-propanol Stock solutions >100 m*M* PMSF in 100% 2-propanol stable at room temperature >1 yr	17–170 µg/mL (0.1–1 m*M*)

proteins and secreted proteins generally being more resistant than cytoplasmic proteins. Therefore, it is advised to work on ice and include a protease inhibitor in the lysis buffer. Proteases are divided into five classes according to their mechanism of catalysis: serine proteases, cysteine proteases, aspartic proteases, metalloproteases, as well as enzymes with unknown reaction sites *(4)*. Among the two most commonly used protease inhibitors, aprotinin is a serine protease inhibitor and phenylmethylsulfonyl fluoride (PMSF) is a serine and cysteine protease inhibitor *(5)*.

Unlike other compartments in the cell where protein folding occurs, the endoplasmic reticulum is oxidizing and therefore promotes the formation of disulfide bonds. Although the cytosol of intact cells is reduced, and therefore prevents formation of disulfide bonds, disruption of cells without previous treatment with alkylating agents such as *N*-ethylmaleimide (NEM), will incorrectly introduce disulfide bonds in cytosolic proteins *(6)*. To avoid artificial disulfide bond formation during cell lysis, cells should be washed with ice-cold phosphate-buffered saline (PBS) containing NEM and lysed in lysis buffer containing NEM.

Table 3
Specificity of Protein A, Protein G, and Jacalin for Immunoglobulins from Various Species[a,b]

Antibodies	Protein A	Protein G type 2	Jacalin
Human IgG 1	++	++	–
Human IgG 2	++	++	–
Human IgG 3	–	++	–
Human IgG 4	++	++	–
Human IgA 1	+	–	++
Human IgA 2	+	–	?
Rabbit	++	++	
Cow	+	++	
Horse	–	++	
Goat	+	++	
Guinea pig	++	++	
Sheep	–	++	
Dog	++	++	
Pig	++	++	
Mouse IgG	+	++	
Mouse IgA	?	?	–
Rat	–	+	

[a]Reproduced with the permission of Pharmacia Biotech from the data file for GammaBind™ G, type 2.

[b]Binding capacity: ++, strong; +, intermediate; –, weak.

1.3. Formation of Antigen–Antibody Complexes

1.3.1. Direct Immunoprecipitation

The antibody or antigen preparation to be tested by immunoprecipitation is now allowed to react with aliquots of the cell lysate or a specific antibody. The antibody–antigen complex is then precipitated by one of several methods. Depending on the antibody to be precipitated, staphylococcal protein A.

Protein A consists of a single polypeptide chain with two accessible high-affinity binding sites for the Fc region of IgG and secondary binding sites for the Fab region. The protein A antibody complexes disassociate at low pH, boiling, and/or the addition of reducing reagents. Protein A binds a relatively broad spectrum of antibodies and antibody subclasses from different species (**Table 3**). One milliliter of protein A-Sepharose beads can adsorb 10–20 mg of IgG. The specificity of antibody capture applications can be manipulated by adjusting the pH and ionic strength of the binding and elution buffers ***(7–9)***.

Native and recombinantly engineered protein G bind specifically to more species of immunoglobulins than does protein A (**Table 3**). The recombinant form does not crossreact with IgM, IgE, IgA, IgD, or serum albumin as protein A may do ***(10–12)***.

Jacalin, purified from the jack-fruit *Artrocarpus integrifolia*, bound to Agarose beads, has recently been introduced for the purification of human monomeric immunoglobulin A, subclass 1 (**Table 3**) ***(13,14)***. Jacalin is a lectin composed of four identical subunits of approx 10,000 Da each. This glycoprotein appears to bind only *O*-glycosidically linked oligosaccharides, preferring the structure galactosyl (β-1,3) *N*-acetylgalactoseamine. The binding capacity of the jacalin-agarose beads is 3–4 mg monomeric IgA/mL of gel (Vector). Immunoglobulin A may be eluted in a biologically active form by 0.8 *M* D-galactose in 175 m*M* Tris-HCl (pH 7.5).

1.3.2. Indirect Immunoprecipitation

If your antibody does not sufficiently bind in any of the aforementioned systems, another possibility is to utilize an anti-immunoglobulin antibody already bound to, e.g., protein A or G.

1.4. SDS-PAGE

The immunoprecipitated proteins may be analyzed by electrophoresis in polyacrylamide gels either under reducing or nonreducing conditions ***(15,16)***. SDS, a strong anionic detergent, is often used in combination with a reducing agent and heat to disassociate the proteins before loading on to the gel. SDS bound to denatured proteins will give the protein a negative charge. The amount of SDS bound to proteins is usually proportional to the size of the polypeptide so that the formed complexes of SDS-polypeptide will migrate through the gel according to size. Mol-wt markers included in each run will help to calculate the estimated molecular weights of unknown polypeptides. *N*- and *O*-linked glycosylation will decrease the electrophoretic mobility through the gel. If proteins are separated under nonreducing conditions the electrophoretic mobility will be changed as compared to separation under reducing conditions.

Polyacrylamide gels are chains of polymerized acrylamide that are crosslinked by *N,N'*-methylenebisacrylamide to form pores through which the polypeptides must pass. By varying the concentration of polyacrylamide the effective range of separation of polypeptides may be changed (**Table 4**).

Initially introduced by Ornstein and Davis ***(17,18)***, discontiunous buffer systems are commonly used. The buffer in the reservoirs differs in pH and ionic strength from the buffer used to cast the gel. SDS added in all buffers and the gel, introduced initially by Leammli ***(19)***, will bind to all denatured proteins and give them a negative charge. Full details of SDS-PAGE are given in Chapter 11.

1.5. Detection and Analysis of Immunoprecipitated Proteins

Although gel scanners are slowly moving into many research laboratories, most researchers are still using autoradiography because of the high costs of gel scanners. The great advantage of gel scanners is the shortening of the time before results are obtained. With the gel scanner, results are available immediately while autoradiography results are available within days or weeks. In autoradiography, permanent images are produced on photographic film applied to a dried gel. The β-particles emitted from the labeled proteins in the gel will interact with the silver halide crystals in the emulsion of the film. The film exposure should take place in a light-proof cassette at –70°C in order to stabilize the silver atoms and ions that form the image of the radioactive source.

The autoradiographic images may be amplified by fluorescent chemicals that emit photons when they encounter a single quantum of radiation and this may increase the

Table 4
Polyacrylamide Gel Mixtures

Separation gel[a]	6%	7.5%	10%	11%	15%
40% Acrylamide	6 mL	7.42 mL	10 mL	11 mL	15 mL
2% *bis*	3.24 mL	4 mL	5.3 mL	5.88 mL	10 mL
8X 1 *M* Tris (pH 8.8)	4.9 mL	4.9 mL	4.9 mL	4.9 mL	4.9 mL
dH_2O	25.22 mL	22.77 mL	19 mL	17.5 mL	9.48 mL
10% Amps	230 µL	230 µL	138 µL	138 µL	138 µL
10% SDS	0.4 mL	0.4 mL	0.4 mL	0.4 mL	0.4 mL
TEMED	33 µL	33 µL	14 µL	14 µL	14 µL
Stacking gel[b]	**3.5%**	**4.5%**			
40% Acrylamide	1.3 mL	1.68 mL			
2% *bis*	1 mL	0.9 mL			
8X 1 *M* Tris (pH 6.8)	1.9 mL	1.9 mL			
dH20	11 mL	10.5 mL			
10% Amps	110 µL	100 µL			
10% SDS	150 µL	150 µL			
TEMED	20 µL	20 µL			

[a]Total volume 40 mL.
[b]Total volume 15 mL.

detection level 10-fold. Several commercial preparations are available but in most cases 1 *M* sodium salicylate (pH 6.0) treatment for 30 min will serve the purpose.

2. Materials

2.1. Metabolic Labeling of Proteins

1. Monolayer cultures approximately subconfluent to confluent or suspension cultures.
2. Methionine-free and/or cysteine-free medium.
3. ^{35}S methionine and/or ^{35}S cysteine. The rate of synthesis and the half-life of the protein of interest as well as the number of cells to be labeled affect the time required as well as the intensity of labeling. Labeling for 2–4 h with 100–400 µCi of ^{35}S-labeled amino acids is commonly used, preferably at the time of maximum protein synthesis in the cells (*see* **Note 1**).

2.2. Lysis of Cells

Choose one of the following lysis buffers ***(1–6)*** (*see* **Note 3**):

1. Triple-detergent lysis buffer: 50 m*M* Tris-HCl (pH 8.0), 150 m*M* NaCl, 0.02% sodium azide, 0.1% SDS, 1% NP-40, 0.5% sodium deoxycholate.
2. Single-detergent lysis buffer: 50 m*M* Tris-HCl (pH 8.0), 150 m*M* NaCl, 0.02% sodium azide, 1% Triton X-100 or NP-40.
3. High salt lysis buffer I: 50 m*M* HEPES (pH 7.0), 500 m*M* NaCl, 1% NP-40.
4. High salt lysis buffer II: 10 m*M* Tris-HCl (pH 7.8), 150 m*M* NaCl, 600 m*M* KCl, 5 m*M* EDTA, 2% Triton X-100.
5. No salt lysis buffer: 50 m*M* HEPES (pH 7.0), 1% NP-40.

6. Very gentle single detergent buffer: 50 m*M* HEPES (pH 7.5), 200 m*M* NaCl, 2% CHAPS.
7. Protease inhibitors are often included in lysis buffers. Aprotinin at a concentration of 1 μg/mL and PMSF at a concentration of 100 μg/mL are regularly used (*see* **Note 5**). Many other protease inhibitors are available and the need for additional inhibitors must be established in each new system. For further information, *see* **Table 2**.
8. To avoid artificial disulfide bond formation during cell lysis cells should be washed with ice-cold PBS containing 20 m*M* NEM and lysed in lysis buffer containing 20 m*M* NEM.

2.3. Formation of Antigen–Antibody Complexes

1. Radiolabeled cell lysate (*see* **Note 6**).
2. Monoclonal or polyclonal antibody for immunoprecipitation.
3. Protein A or G bound to Sepharose beads (Pharmacia Biotech, Piscataway, NJ) or Jacalin-Agarose (Vector).
4. RIPA-buffer: 2% (v/v) Triton X-100, 150 m*M* NaCl, 600 m*M* KCl, 5 m*M* disodium EDTA, 3 m*M* PMSF, 1 μg/mL aprotinin, and 20 m*M* NEM in 10 m*M* Tris-HCl, pH 7.8.
5. Low-salt washing buffer: 10 m*M* Tris-HCl (pH 8.0), 150 m*M* NaCl.
6. Reducing sample buffer: 3% SDS, 3% 2-mercaptoethanol, 0.1% EDTA, 10% (v/v) glycerol in 62 m*M* Tris-phosphate, 0.02% bromphenol blue, pH 6.8. A 2X stock solution can be stored at room temperature.
7. Nonreducing sample buffer: 3% SDS, 0.1% EDTA, 10% (v/v) glycerol in 62 m*M* Tris-phosphate, 0.02% bromphenol blue, pH 6.8.

2.4. SDS-PAGE

See Chapter 11.

2.5. Detection and Analysis of Immunoprecipitated Proteins

1. Fix 10% (v/v) glacial acetic acid and 35% (v/v) methanol in deionized water.
2. 1 *M* Sodium salicylate (pH 6.0) in deionized water (Sigma, St. Louis, MO) (*see* **Note 10**).
3. Whatman 3-mm paper.
4. Gel dryer (*see* **Notes 11** and **12**).
5. Light proof cassette (*see* **Note 13**).
6. X-ray film (Kodak Xomat, Eastman Kodak, Rochester, NY).
7. Developer (X-ray developer LX 24, Eastman Kodak).
8. Fixative for X-ray film (X-ray, Eastman Kodak).
9. Dark room.

3. Methods

3.1. Metabolic Labeling of Proteins

1. Wash monolayers twice with PBS or wash and centrifuge cells in suspension twice, and add methionine and/or cysteine deficient medium (without fetal calf serum) prewarmed to the appropriate temperature.
2. Incubate the cells for 20–60 min to deplete the intracellular pools of methionine and/or cysteine.
3. Replace the amino acid deficient medium with methionine and/or cysteine free medium including ^{35}S-labeled amino acids. Incubate for the desired period of time (*see* **Note 1**). Keep the volume of medium down to increase the concentration of radiolabeled amino acids. Suggested volumes for adherent cells are: 1–2 mL for a 25-cm^2 flask, 1–2 mL for a 90-mm Petri dish, 250–500 μL for a 60-mm Petri dish, 100–200 μL for a 30-mm Petri dish. Cells growing in suspension should be resuspended at a concentration of 10^7/mL (*see* **Note 2**).

4. If the antigen of interest is a secreted antigen, the radioactive supernatant is saved for immunoprecipitation. If the antigen of interest accumulates intracellularly, the radioactive supernatant is discarded in the radioactive waste.

3.2. Lysis of Cells

1. Wash the cells twice with ice cold PBS containing 20 m*M* NEM, drain the last PBS with a Pasteur pipet and add the lysis buffer of choice (2–3 mL to a 25-cm^2 flask) (*see* **Note 3**). Let monolayers solubilize for 30 min on ice.
2. To clear the lysate from cell debris, centrifuge for 10 min at 12,000*g* in a microfuge. Before storage of the labeled antigen at –70°C, aliquot the antigen to avoid repeated freezing and thawing (*see* **Note 4**).
3. Check efficiency of metabolic labeling by running an SDS-PAGE and autoradiography.

3.3. Formation of Antigen–Antibody Complexes

1. Mix appropriate amounts of the cell lysate (usually 5–100 μL) with the monoclonal or polyclonal antibody and dilute to 500 μL in RIPA buffer and incubate at 4°C overnight. Aim at a sufficient amount of antibody to precipitate all of the target antigen. Several factors, such as concentration of antigen and titer and avidity of the antibody, will affect the amount of antibody to be used. Start the immunoprecipitation by titrating the antibody against a fixed amount of target antigen.

   Complete immunoprecipitation is usually obtained by 0.5 μL–5 μL of polyclonal antiserum, 5–100 μL of hybridoma tissue culture medium, or 0.1–1.0 μL of ascitic fluid (*see* **Note 6**).
2. If your antibody does not bind efficiently to protein A, G, or jacalin, add a second antibody that is directed against your primary antibody and binds strongly to one of these proteins, and incubate at 4°C for 1 h. The amount of anti-immunoglobulin antibody must be titrated against a fixed amount of antigen–primary antibody complex to exclude reactivity between secondary antibody and target antigen.
3. Add 25–100 μL of protein A-/protein G-Sepharose beads or jacalin-agarose beads to the antigen–antibody mixture and incubate on a rocker for 1 h at 4°C or at room temperature.
4. Centrifuge the newly formed immune complexes at 12,000*g* for 30 s and remove the supernatant. Wash the complexes to remove nonspecifically adsorbed proteins at least four times with 1 mL of RIPA buffer and resuspend the beads with careful vortexing between washes. The last wash should always be performed with the low-salt washing buffer. Take care to remove the last traces of the final wash.
5. To disassociate the immune complexes, add 40 μL of sample buffer and incubate at 100°C for 2–3 min (*see* **Note 7**). Centrifuge the samples for 20 s at 12,000*g* in a microfuge and save the supernatants for SDS-PAGE. Samples can be frozen at –20°C before analysis by SDS-PAGE.

3.4. SDS-PAGE

Prepare an SDS-PAGE gel as described in Chapter 11, **Subheading 3.** Following the run, fix the gel.

3.5. Detection and Analysis of Immunoprecipitated Proteins

Autoradiography:

1. After washing the fixed gel twice in deionized water soak the gel in 50–100 mL of 1 *M* sodium salicylate in deionized water for 30 min on a rocker (*see* **Note 10**).

2. Dry the gel onto a 3-mm Whatman paper, presoaked in water, in a gel dryer at 60°C for 2–6 h (*see* **Note 11**).
3. Place the gel together with an X-ray film in a light-excluding X-ray film casette at –70°C and expose the film for an appropriate time (days–weeks) (*see* **Note 12**).
4. Develop the film in an automatic X-ray film processor or manually for 5 min each as follows (*see* **Note 13**): developer, water bath, fixative, and running water.

4. Notes

4.1. Metabolic Labeling of Proteins

1. When radiolabeling cells for longer than 6 h, all the radiolabeled amino acids may be consumed. It is therefore sometimes necessary to add unlabeled amino acids, i.e., methionine and/or cysteine as shortage of amino acids may cause an interruption in protein synthesis.
2. If small volumes are used during labeling, keep the flask on a slow rocker or shake the dishes every 15 min to ensure that the cells do not dry.

4.2. Lysis of Cells

3. To optimize the extracting conditions, try a stronger lysis buffer on the centrifuged cell debris than the buffer you initially used for lysis of the cells. Then test by immunoblotting whether or not most of the labeled proteins were extracted with the lysis buffer initially used.
4. To remove aggregates of cytoskeleton elements after thawing of lysate, centrifuge samples at 12.000*g* for 5 min.
5. **Caution:** PMSF is extremely destructive to mucous membranes of the respiratory tract, the eyes and skin. Also, the other protease inhibitors are toxic.

4.3. Formation of Immune Complexes

6. To avoid nonspecific binding of complexes to the tube wall, use good quality tubes (e.g., Eppendorf, Hamburg, Germany) or pre-coat the tubes with 0.5% bovine serum albumin for 15 min.
7. Before heating samples to 100°C, use a needle to make a small hole in the tube cap. This will prevent the building of excess pressure in the tube and the tube cap will remain closed during the heating step.

4.4. SDS-PAGE

8. **Caution:** Acrylamide and *bis* acrylamide are neurotoxic and may be absorbed through the skin. Polyacrylamide is considered nontoxic but may contain unpolymerized material.
9. When preparing and handling gels, use gloves to avoid exposure to unpolymerized polyacrylamide and radioisotopes.

4.5. Detection and Analysis of Immunoprecipitated Proteins

10. **Caution:** Salicylate may elicit allergic reactions and is readily absorbed through the skin. As an alternative use commercial fluorescents available from several companies.
11. Shrinkage, distortion, and cracking of the gel are common problems encountered when trying to dry gels. To avoid shrinkage and distorsion, dry the gel onto a 3-mm Whatman paper (presoaked in water). Make sure there are no air bubbles between the gel and the paper before starting the gel dryer. To avoid cracking do not turn off the gel dryer or break the vacuum before the gel is completely dry. If possible, use thin gels (0.75 mm), since cracking is more common with thicker gels containing larger amounts of polyacrylamide.

12. Take care to prewarm the cassette to room temperature for 15 min before developing the film, since moisture inside the cassette will destroy the emulsion.
13. The new time-saving gel scanners provide an alternative to autoradiography.

References

1. Sambrook, J., Fritsch E. F., and Maniatis, T. (eds.) (1989) *Molecular Cloning: A Laboratory Manual* (2nd ed.), Cold Spring Harbor Laboratory Press, Cold Spring Harbor, NY.
2. Harlow, E. and Lane, D. (1988) *Antibodies: A Laboratory Manual.* Cold Spring Harbor Laboratory Press, Cold Spring Harbor, NY.
3. Burleson, F. G., Chambers, T. M., and Wiedbrauk, D. L. (eds.) (1992) *Virology: A Laboratory Manual.* Academic, San Diego, CA.
4. Barrett, A. J. and Salvesen, G. (eds.) (1986) *Proteinase Inhibitors. Research Monographs in Cell and Tissue Physiology*, vol. 12. Elsevier, Amsterdam, The Netherlands.
5. James, G. T. (1978) Inactivation of the protease inhibitor phenylmethylsulfonyl fluoride in buffers. *Analyt. Biochem.* **86,** 574–579.
6. Hammond, C. and Helenius, A. (1994) Quality control in the secretory pathway of a misfolded viral membrane glycoprotein involves cycling between the ER, intermediate compartment, and Golgi apparatus. *J. Cell Biol.* **126,** 41–52.
7. Hjelm, H., Hjelm, K., and Sjöquist, J. (1972) Protein A from *Staphylococcus aureus.* Its isolation by affinity chromatography and its use as an immunosorbent for isolation of immunoglobulins. *FEBS Lett.* **28,** 73–76.
8. Sjödahl, J. (1977) Structural studies on the four repetitive Fc-binding regions in protein A from *Staphylococcus aureus. Eur. J. Biochem.* **78,** 471–490.
9. Goudswaard, J., van der Donk, J. A., Noordzij, A., van Dam, R. H., and Vaerman, J.-P. (1978) Protein A reactivity of various mammalian immunoglobulins. *Scand. J. Immunol.* **8,** 21–28.
10. Björck, L. and Kronvall, G. (1984) Purification and some properties of streptococcal protein G: a novel IgG-binding reagent. *J. Immunol.* **133,** 969–974.
11. Åkerström, B., Brodin, T., Reis, K., and Björck, L. (1985) Protein G: a powerful tool for binding and detection of monoclonal and polyclonal antibodies. *J. Immunol.* **135,** 2589–2592.
12. Fahnestock, S. R., Alexander, P., Nagle, J., and Filpula, D. (1986) Gene for an immunoglobulin-binding protein from a group G streptococcus. *J. Bacteriol.* **167(3),** 870–880.
13. Roque-Barreira, M. C. and Campos-Neto, A. (1985) Jacalin: an IgA-binding lectin. *J. Immunol.* **134,** 1740–1743.
14. Johansen, K., Granqvist, L., Karlén, K., Stintzing, G., Uhnoo, I., and Svensson, L. (1994) Serum IgA immune response to individual rotavirus polypeptides in young children with rotavirus infection. *Arch. Virol.* **138,** 247–259.
15. Studier, F. W. (1973) Analysis of bacteriophage T7 early RNAs and proteins on slab gels. *J. Mol. Biol.* **79,** 237–248.
16. Hames, B. D. and Rickwood, D. (eds.) (1981) *Gel Electrophoresis of Proteins: A Practical Approach.* IRL, Oxford, UK.
17. Ornstein, L. (1964) Disc electrophoresis-I. Background and theory. *Ann. NY Acad. Sci.* **121,** 321–349.
18. Davis, B. J. (1964) Disc-electrophoresis II. Method and application to human serum proteins. *Ann. NY Acad. Sci.* **121,** 404–427.
19. Laemmli, U. K. (1970) Cleavage of structural proteins during the assembly of the head of bacteriophage T4. *Nature* **227,** 680–685.

Part VIII

Monoclonal Antibodies

158

Immunogen Preparation and Immunization Procedures for Rats and Mice

Mark Page and Robin Thorpe

1. Introduction

A high-titer antibody response usually requires use of an adjuvant for the first (priming) immunization. For most purposes, the immunogen is prepared by emulsification in a mineral oil containing heat-killed mycobacterium (Freund's complete adjuvant—FCA). The emulsion ensures that the antigen is released slowly into the animal's circulation, and the bacteria stimulate the animal's T-helper cell arm of the immune system. Further booster (secondary) immunizations are almost always necessary for production of high antibody levels, and these are given either in phosphate-buffered saline (PBS) or as an oil emulsion (bacteria are not normally included in the boosting injections; a suitable oil adjuvant is Freund's incomplete adjuvant—FIA). A large number of alternative adjuvants are available, but FCA/FIA (for priming and boosting respectively) usually produces maximal immune responses. However, FCA in particular can produce adverse effects in some cases and is not normally recommended for use in humans or primates Alum adjuvants are often chosen as an alternative and can be used in humans. Immunization with substances with molecular weights <3000 (such as peptides) are not normally immunogenic and will require conjugation to a carrier protein (*see* **ref. *1*** and Chapters 117–119), such as purified protein derivative (PPD) or keyhole limpet hemocyanin (KLH).

2. Materials

1. Freund's complete adjuvant (FCA).
2. Freund's incomplete adjuvant (FIA).
3. Phosphate-buffered saline (PBS).
4. Immunogen preparation.
5. 2-mL glass Luer lock syringes (two).
6. Syringe coupler, Luer lock with female inlet and outlet ports (Sigma, St. Louis, MO).
7. Rat/mouse.

3. Methods

3.1. Immunogen Preparation (see Notes 1–5)

1. Dilute immunogen in physiological buffer (e.g., PBS without sodium azide) so that the preparation will contain 10–100 μg of protein in approx 300–600 μL/animal (Note: the

From: *The Protein Protocols Handbook, 2nd Edition*
Edited by: J. M. Walker © Humana Press Inc., Totowa, NJ

final quantity will be 5–50 μg, since the preparation will be diluted with an equal volume of adjuvant). Mix the immunogen solution with an equal volume of FCA and draw up into a glass syringe or prepare directly in the syringe barrel.
2. Remove excess air from the syringe barrel, and connect to a second glass syringe with its plunger fully depressed via a double-hub connector.
3. Ensure the connections are tight and not leaking, and transfer the oil and immunogen solutions from one syringe to the other. Continue this action until the mixture is fully emulsified when it should appear as a creamy, white thick liquid.
4. Transfer emulsion into one of the glass syringes, remove from double-hub connector and empty syringe, and fit a small-diameter needle (the size of which will depend on the animal to be immunized and route of immunization).

3.2. Immunization Procedure

1. Prime mice or rats by immunizing with immunogen subcutaneously on the flanks and neck (0.1 mL/site, 3–5 sites). Do his by raising the skin between thumb and forefinger, and inserting the needle into the raised area at a shallow angle. A short narrow-diameter needle is preferred (0.4 × 27 mm) to avoid injection into the deeper body layers/cavities. The result should be a discrete lump under the skin.
2. Boost intraperitoneally after 14–28 d using the immunogen prepared in PBS via a short narrow-diameter needle. Administer the immunogen at one site in no more than 0.5 mL using the same total dose as that used for priming (usually 5–50 μg).
3. Three days after the booster immunization, withdraw blood from the tail vein with a needle (0.4 mm) and syringe, and use this as a positive control in screening assays during hybridoma production and as a check on the success of the immunization. Sacrifice mouse, and remove spleen aseptically.

4. Notes

1. Shake the complete adjuvant before use to disperse the Mycobacterium particles fully.
2. The emulsion is very difficult to recover completely from the walls of vessels, and so forth; therefore, it is inevitable that some will be lost during preparation. To minimize this, prepare the emulsion in glass syringes, one of which can be used for the immunization. If possible, prepare slightly more than required to compensate for losses, which normally amount to around 10%.
3. If Luer connectors are not available, the emulsion can be prepared by vigorous shaking or mixing using a whirlimixer. This is less efficient at producing an emulsion and requires larger volumes of immunogen.
4. When using connected syringes, keep the hub connector as short as possible to avoid emulsion loss.
5. Use glass syringes to prepare the emulsion since the rubber seals of plastic syringes are not compatible with the mineral oil of the adjuvant.

Reference

1. Thorpe, R. (1994) Producing antibodies, in *Immunochemistry Labfax* (Kerr, M. A. and Thorpe, R., eds.), Bios Scientific, Oxford, UK, pp. 63–81.

159

Hybridoma Production

Mark Page and Robin Thorpe

1. Introduction

Köhler and Milstein *(1)* introduced technology for the production of MAb in vitro by the construction of hybridomas. These hybridomas are formed by the fusion of neoplastic B-cells (normally a B-cell line derived from a tumor) with spleen cells from an immune animal. Cells can be induced to fuse by mixing them together at high density in the presence of polyethylene glycol (some viruses also induce cell fusion). The efficiency of fusion is usually fairly low, but hybridomas can be selected for if the parent neoplastic cell line is conditioned to die in selective medium. The tumor cells are killed by the selective medium, normal nonfused spleen cells die after a period in culture, but hybridomas inherit the ability to survive in the selective medium from the normal parent cell. Usually, medium containing hypoxanthine, aminopterin, and thymidine (HAT) is used. There are two pathways available to the cell for synthesis of nucleic acid: (1) *de novo* synthesis and (2) synthesis by salvaging nucleotides produced by breakdown of nucleic acid. Aminopterin (and thus HAT medium) inhibits *de novo* synthesis of nucleic acid, but this is not lethal for normal cells, since the salvage pathway can still function (the hypoxanthine and thymidine present in HAT medium ensures that there is no deficiency of nucleotides). However, the enzyme hypoxanthine-guanine phosphoribosyl transferase (HGPRTase) is essential for the operation of the salvage pathway, so if the tumor cell line is deficient in this enzyme, it will be unable to synthesize nucleic acid and die. HGPRTase-deficient cell lines are produced by selection in medium containing 8-azaguanine. Cell possessing HGPRTase incorporate the 8-azaguanine into their DNA and die, whereas HGPRTase-deficient cells survive.

If individual hybridomas are isolated by cloning, it is possible to produce large quantities of the secreted MAb (all the immunoglobulin secreted by a clone has identical antigen-binding specificities, allotype, heavy chain subclass, and so forth).

Hybridoma technology requires facilities for, and knowledge of, tissue culture, freezing and storing viable cells, and methods of screening for antibody secretion. It is advisable to learn the cell culture, fusion, and cloning techniques from an experienced worker rather than trying to set them up in isolation. The exact methodology for hybridoma production varies considerably between laboratories; one such protocol is given below.

From: *The Protein Protocols Handbook, 2nd Edition*
Edited by: J. M. Walker © Humana Press Inc., Totowa, NJ

2. Materials

1. Media: RPMI-1640 or DMEM. Add penicillin (60 mg/L) and streptomycin (50 mg/L) to medium.
2. Selective media supplements: These reagents contain hypoxanthine (H), aminopterin (A), and thymidine (T), and are utilized for the selective growth of hybridomas. HAT supplement and HT supplement are both supplied in 50X concentrated stock solutions.
3. Polyethylene glycol 1500 (PEG 1500) in 7.5 m*M* HEPES (PEG 50% w/v).
4. Fetal bovine serum (FBS) (*see* **Note 1**).
5. NSO myeloma cells: The NSO cell line is a mouse myeloma line that does not synthesize or secrete immunoglobulin. Grow in 3–5% FBS-RPMI at a density of $2–5 \times 10^5$ cell/mL. Approximately $3–5 \times 10^7$ cells are required for a mouse spleen fusion (*see* **Note 2**).
6. Dimethylsulfoxide (DMSO).

3. Method

Cells must be grown and processed using aseptic conditions.

3.1. Fusion Protocol

1. Warm stock solutions of medium, FBS, PEG, and 50X HAT by placing stock bottles in a water bath set at 37°C.
2. Estimate cell numbers of myeloma cells (NSO) using a hemocytometer. The cells should be in log growth at $2–4 \times 10^5$/mL.
3. Remove the spleen from an immunized mouse (*see* Chapter 158) under sterile conditions. If rats are used, then the Y3 myeloma cell line should be used as fusion partner.
4. Transfer the spleen to a glass homogenizer, and add 15 mL of warmed RPMI. Push the pestle down the homogenizer, twist four times, and then pull the pestle up. Repeat three times. Alternatively, transfer the spleen to a Petri dish containing 15 mL RPMI. Hold the spleen at one end with forceps, and pierce the other end of the spleen with a bent needle attached to a 5-mL syringe. Tease out the cells by stroking the spleen with the bent needle, and then pass the cell suspension three to four times through the needle and syringe.
5. Transfer the cells to a 50-mL tube, add warm RPMI, and centrifuge for 5 min at 300*g*. At the same time, pellet the NSO cells by centrifugation.
6. Decant the supernatant. Assume the mouse spleen contains about 10^8 cells (a rat spleen contains 2×10^8 cells).
7. Mix the NSO and spleen cells together at a ratio of 1 NSO cell:2 spleen cells in 50 mL RPMI.
8. Centrifuge and decant the supernatant. Remove the last drop of supernatant with a pipet.
9. Add 1 mL 50% of PEG over 1 min (30 s for rat). Stir gently with the pipet while adding the PEG.
10. Leave for 1 min.
11. a. Add 1 mL RPMI over 1 min while stirring gently.
    b. Add 2 mL RPMI over 1 min while stirring gently.
    c. Add 5 mL RPMI over 1 min while stirring gently.
    d. Add 10 mL RPMI over 1 min while stirring gently.
    Top up the tube with RPMI and spin at 300*g* for 5 min.
12. Resuspend in 200 mL HAT medium (4 mL 50X HAT + 196 mL 15% FBS/RPMI).
13. Distribute into 8×24-well culture plates, 1 mL/well.
14. Feed with HAT medium every 6–8 d—remove half of the medium from each well, and replace with an equal amount of fresh HAT medium. Check for hybridoma growth and tumor cell death using an inverted microscope.

15. After 10–14 d, feed the wells containing hybridomas with warm HT medium (2 mL 50X HT + 98 mL 15% FBS/RPMI).
16. Screen the wells containing hybridomas for antibody secretion (*see* Chapters 160 and 161) when the medium becomes yellow-orange (acid) and hybridomas are nearly confluent.
17. Expand positive wells into further plates or bottles (after a further 5–8 d, the cells will grow in medium containing 10% v/v FBS). Clone or cryopreserve in liquid nitrogen.
18. To cryopreserve, resuspend about 10^7 hybridomas in 0.5–1.0 mL of medium containing 20% v/v FBS and 10% v/v DMSO in cryotubes. The freezing mixture should be cooled to approx 4°C during use by storing the solution in a refrigerator and placing the cryotubes in ice water. Slowly freeze the cells in nitrogen vapor for at least 2 h, followed by storage in liquid nitrogen (*see* **Note 3**). Hybridomas are thawed rapidly in a 37°C water bath, washed in ~50 mL medium to remove DMSO, spun down, and resuspended in fresh medium/FBS.

3.2. Cloning (see Note 4)

1. Weigh out 0.5 g agar into a 100-mL autoclave bottle.
2. Add 10 mL distilled water, and autoclave for 15 min.
3. Transfer to a water bath set at 50–55°C, and mix with 90 mL warmed medium (15% FBS/RPMI at 50°C).
4. Add 12–14 mL of the agar solution to each Petri dish, and allow to set for 20 min at room temperature (make up three dishes for each line to be cloned).
5. Suspend hybridomas in a well or flask, and transfer 2–4 × 10^3 cells in 0.2 mL of medium to one well of a fresh 24-well plate. Add 0.8 mL medium to this and make two more serial 1:5 dilutions (i.e., 0.2 + 0.8 mL medium). Discard 0.2 mL from the final well.
6. Allow the agar solution to cool to about 45–47°C, and add 1 mL of the agar to each well. Mix and transfer the contents of each well into separate agar containing Petri dishes (avoiding introducing bubbles). Incubate at 37°C in a CO_2 incubator.
7. Check daily for cell growth with an inverted microscope, and discard plates containing overgrown or no cells. When clones grow into colonies of 20–100 cells, pick them individually from the plate using a Pasteur pipet, and transfer into 1 mL of medium in separate wells in a 24-well plate.
8. Feed every 2–4 d, and when cells are 75–100% confluent, test for antibody secretion (*see* Chapters 160 and 161) and expand positive wells. Hybridomas may be weaned onto 10% FBS/RPMI at this stage.
9. Repeat the cloning procedure.

4. Notes

1. Test several batches of FBS before ordering a large quantity. Check for toxicity and sterility, and if possible, compare with a previous batch of FBS that is known to support hybridoma growth. Alternatives to FBS are available, such as CLEX, Dextran Products, Canada.
2. To reduce the possibility of contamination of the NSO line with mycoplasma, bacteria, fungi, yeast, or virus, always:
   a. Use a separate bottle of medium for NSO.
   b. Do not manipulate other cell lines in the hood at the same time as NSO.
3. Do not leave cells in DMSO at room temperature since they will die.
4. If hybridomas are in HAT medium, carry out cloning in HT medium.

Reference

1. Köhler, G. and Milstein, C. (1975) Continuous cultures of fused cells secreting antibody of predefined specificity. *Nature* **256,** 495.

160

Screening Hybridoma Culture Supernatants Using Solid-Phase Radiobinding Assay

Mark Page and Robin Thorpe

1. Introduction

A large number of hybridoma culture supernatants (up to 200) need to be screened for antibodies at one time. The assay must be reliable so that it can accurately identify positive lines, and it must be relatively quick so that the positive lines, which are 75–100% confluent, can be fed and expanded as soon as possible after the assay results are known. Solid-phase binding assays are appropriate for this purpose and are commonly used for detection of antibodies directed against soluble antigens *(1)*. The method involves immobilizing the antigen of choice onto a solid phase by electrostatic interaction between the protein and plastic support. Hybridoma supernatants are added to the solid phase (usually a 96-well format) in which positive antibodies bind to the antigen. Detection of the bound antibodies is then achieved by addition of an antimouse immunoglobulin labeled with radioactivity (usually ^{125}I) and the radioactivity counted in a γ counter.

2. Materials

1. Antigen: 0.5–5 µg/mL in phosphate-buffered saline (PBS) with 0.02% sodium azide.
2. PBS: 0.14 *M* NaCl, 2.7 m*M* KCl, 1.50 m*M* KH_2PO_4, 8.1 m*M* Na_2HPO_4.
3. Blocking buffer: PBS containing either 5% v/v pig serum, 5% w/v dried milk powder, or 3% w/v hemoglobin (*see* **Note 1**).
4. Hybridoma culture supernatants.
5. Negative control: irrelevant supernatant or culture medium.
6. Positive control: serum from immunized mouse from which spleen for hybridoma production was derived.
7. Antimouse/rat immunoglobulin, ^{125}I-labeled (100 µCi/5 µg protein).

3. Method

1. Pipet 50 µL of antigen solution into each well of a 96-well microtiter plate and incubate at 4°C overnight. Such plates can be stored at 4°C for several weeks. Seal plates with cling film to prevent the plates from drying out (*see* **Note 2**).
2. Remove antigen solution from wells by either a Pasteur pipet (cover the tip with a small piece of plastic tubing to prevent scratching the antigen-coated wells) or by shaking the contents from the plate in one quick movement (*see* **Note 3**).

From: *The Protein Protocols Handbook, 2nd Edition*
Edited by: J. M. Walker © Humana Press Inc., Totowa, NJ

3. Wash the plate by filling the wells with PBS and rapidly discarding the contents. Repeat this twice more, tap the plate dry, then fill the wells with blocking buffer, and incubate at room temperature for 30 min to 1 h. Wash three more times with PBS.
4. Pipet 100 µL of neat hybridoma supernatant into the wells, and incubate for approx 3 h at room temperature. Include the negative and positive controls.
5. Wash with blocking buffer three times.
6. Dilute the ^{125}I-labeled antimouse immunoglobulin in blocking buffer to give 1×10^6 cpm/mL. Add 100% µL to each well, and incubate for 1 h at room temperature.
7. Wash with blocking buffer.
8. Cut out individual wells using scissors or a hot nichrome wire plate cutter, and determine the radioactivity bound using a γ counter. A positive result would have counts that are four to five times greater than the background. Normally, the background counts would be around 100–200 cpm with a positive value of at least 1000 cpm (*see* **Note 4**).

4. Notes

1. Filter the milk and hemoglobin solutions coarsely through an absorbent cloth (e.g., Kimnet, Kimberley Clark) to remove lumps of undissolved powder.
2. Antigens (diluted in distilled water) can be dried onto the plates by incubation in a warm room (37°C) without sealing. This method is usually preferred if the antigen is a peptide improving binding to the plastic; however, the background signal may be increased.
3. The antigen solution may be reused several times to coat additional plates, since only a small proportion of the protein adheres to the plastic.
4. Counting radioactivity in a 96-well format can also be performed by using plates designed for use in scintillation counters. The assay is performed as described, except that in the final step, a scintillation fluid (designed to scintillate with ^{125}I, such as Microscint 20, Packard Instrument Co.) is added to each well, and the scintillation counted in a purpose-built machine (e.g., Topcount, Packard Instrument Co.).

Reference

1. Johnstone, A. and Thorpe, R. (1996) *Immunochemistry in Practice,* 3rd ed. Blackwell Scientific, Oxford, UK.

161

Screening Hybridoma Culture Supernatants Using ELISA

Mark Page and Robin Thorpe

1. Introduction

Enzyme-linked immunosorbent assay (ELISA) is a widely used method for the detection of antibody and is appropriate for use for screening hybridoma supernatants *(1)*. As with radiobinding assays, it is a solid-phase binding assay that is quick, reliable, and accurate. The method is often preferred to radioactive assays, since the handling and disposal of radioisotopes is avoided and the enzyme conjugates are more stable than radioiodinated proteins. However, most of the substrates for the enzyme reactions are carcinogenic or toxic and, hence, require handling with care.

Enzymes are selected that show simple kinetics, and can be assayed by a simple procedure (normally spectrophotometric). Cheapness, availability, and stability of substrate are also important considerations. For these reasons, the most commonly used enzymes are alkaline phosphatase, β-D-galactosidase, and horseradish peroxidase.

2. Materials

1. PBS-Tween: Phosphate-buffered saline (PBS) (0.14 *M* NaCl, 2.7 m*M* KCl, 1.5 m*M* KH_2PO_4, 8.1 m*M* Na_2HPO_4) containing 0.05% Tween-20.
2. Antigen: 1–5 μg/mL in PBS.
3. Blocking buffers: PBS-Tween containing either 5% v/v pig serum or 5% w/v dried milk powder (*see* **Note 1**).
4. Hybridoma culture supernatants.
5. Alkaline phosphatase conjugated antimouse immunoglobulin.
6. Carbonate buffer: 0.05 *M* sodium carbonate, pH 9.6.
7. *p*-Nitrophenyl phosphate, disodium hexahydrate: 1 mg/mL in carbonate buffer containing 0.5 m*M* magnesium chloride.
8. 1 *M* sodium hydroxide.

3. Method

1. Pipet 50 μL of antigen solution into each well of a 96-well plate, and incubate overnight at 4°C. Such plates can be stored for 4°C for several weeks. Seal plates with cling film to prevent the plates from drying out (*see* **Note 2**).
2. Remove antigen solution from wells by either a Pasteur pipet (cover the tip with a small piece of plastic tubing to prevent scratching the antigen-coated wells) or by shaking the contents from the plate in one quick movement (*see* **Note 3**).

From: *The Protein Protocols Handbook, 2nd Edition*
Edited by: J. M. Walker © Humana Press Inc., Totowa, NJ

3. Wash the plate by filling the wells with PBS and rapidly discarding the contents. Repeat this twice more, then fill the wells with blocking buffer, and incubate at room temperature for 30 min to 1 h. Wash three more times with PBS.
4. Pipet 100 μL of neat hybridoma supernatant into the wells, and incubate for approx 3 h at room temperature. Include the negative and positive controls (*see* **Note 4**).
5. Wash with blocking buffer three times.
6. Prepare 1:1000 dilution of alkaline phosphatase-conjugated antimouse immunoglobulin in PBS-Tween, and add 200 μL of this to each well. Cover, and incubate at room temperature for 2 h.
7. Shake-off conjugate into sink and wash three times with PBS.
8. Add 200 μL *p*-nitrophenyl phosphate solution to each well, and incubate at room temperature for 20–30 min.
9. Add 50 μL sodium hydroxide to each well, mix, and read the absorbance of each well at 405 nm (*see* **Note 5**). Typical background readings for absorbance should be less than 0.2 with positive readings usually three or more standard deviations above the average background value.

4. Notes

1. Filter the milk solution coarsely through an absorbent cloth (e.g., Kimnet, Kimberley Clark) to remove lumps of undissolved powder.
2. Antigens (diluted in distilled water) can be dried onto the plates by incubation in a warm room (37°C) without sealing. This method is usually preferred if the antigen is a peptide improving binding to the plastic; however, the background signal may be increased.
3. The antigen solution may be reused several times to coat additional plates, since only a small proportion of the protein adheres to the plastic.
4. All assays should be carried out in duplicate or triplicate.
5. Purpose-built plate readers provide a very rapid and convenient means of determining the absorbance.

Reference

1. Johnstone, A. and Thorpe, R. (1996) *Immunochemistry in Practice,* 3rd ed. Blackwell Science, Oxford, UK.

162

Growth and Purification of Murine Monoclonal Antibodies

Mark Page and Robin Thorpe

1. Introduction

Once a hybridoma line has been selected and cloned, it can be expanded and seed stocks cryopreserved for future use. Relatively large amounts of purified MAb may also be required. There are a variety of procedures for this that ensure the establishment of a stable cell line secreting high levels of specific immunoglobulin. High concentrations of antibody can be generated by growing the line in the peritoneal cavity of mice/rats of the same strain as the tumor cell line donor and spleen cell donor. Antibody is secreted into the ascitic fluid formed within the cavity at a concentration up to 10 mg/mL. However, the ascites will contain immunoglobulins derived from the recipient animal that can be removed by affinity chromatography if desired. Several in vitro culture methods using hollow fibres or dialysis tubing *(1)* have been developed and are commercially available; this avoids the use of recipient mice/rats and contamination by host immunoglobulins, although contamination with culture medium-derived proteins may be a problem.

2. Materials

1. RPMI-1640/DMEM medium.
2. Fetal bovine serum (FBS).
3. Phosphate-buffered saline (PBS): 0.14 *M* NaCl, 2.7 m*M* KCl, 1.50 m*M* KH_2PO_4, 8.1 m*M* Na_2HPO_4.
4. HT medium: 1 mL 50X HT supplement in 50 mL 10% FBS/RPMI.
5. HT supplement (Gibco, Paisley, UK).

3. Method

3.1. Growth of Hybridomas

1. After cloning, grow hybridomas in 1 mL HT medium in separate wells of a 24-well plate. Check growth every day and feed with 0.5–1 mL HT medium if supernatant turns yellow (*see* **Note 1**).
2. When cells are 75–100% confluent (*see* **Note 2**), expand into two further wells with fresh HT medium. When these are confluent, transfer into a 25-cm^3 flask and feed with approx 10 mL medium (*see* **Note 3**). Subsequently, cell lines should be weaned off HT

From: *The Protein Protocols Handbook, 2nd Edition*
Edited by: J. M. Walker © Humana Press Inc., Totowa, NJ

medium by increasing the dilution of HT medium stepwise with normal medium lacking the HT supplement, eventually culturing the cells in normal medium. Do this by reducing the amount of HT medium by 25–50% at each feed for expansion of the hybridomas. For example, reduce the amount of HT from 100 to 75 to 50 to 25 to 0% replacing with normal medium.

3. The FBS may be reduced to 5% or lower if hybridoma growth is strong.

3.2. Purification of MAb

MAb purification can be carried out by a number of methods as described in Chapters 137–148 and 163.

4. Notes

1. Phenol red is included in the medium as a visual indicator of pH. A yellow supernatant indicates acid conditions as a result of dissolved CO_2 (carbonic acid) derived from active cell growth (respiration). Normally, a yellow supernatant will indicate that the cells should be split (passaged) because the active cell growth will have exhausted the medium of nutrients and hence will not sustain cell viability.
2. Cells are confluent when the bottom of the flask/well is completely covered. Usually, the cells will require feeding or expanding at this stage (*see* **Note 1**).
3. To prevent inadvertant loss of precious cell lines, it may be prudent to freeze cloned lines before expansion into flasks.

Reference

1. Pannell, R. and Milstein, C. (1992) An oscillating bubble chamber for laboratory scale production of MAb as an alternative to ascitic tumors. *J. Immunol. Meth.* **146,** 43–48.

163

Affinity Purification Techniques for Monoclonal Antibodies

Alexander Schwarz

1. Introduction

Monoclonal antibodies have many applications in biotechnology such as immunoaffinity chromatography, immunodiagnostics, immunotherapy, drug targeting, biosensors, and many others. For all these purposes homogeneous antibody preparation are needed. Affinity chromatography, which relies on the specific interaction between an immobilized ligand and a particular molecule sought to be purified, is a well-known technique to purify proteins from solution. Three different affinity techniques for the one-step purification of antibodies—protein A, thiophilic adsorption, and immobilized metal affinity chromatography—are described in this chapter.

1.1. Protein A Chromatography

Protein A is a cell wall component of *Staphylococcus aureas*. Protein A consists of a single polypeptide chain in the form of a cylinder, which contains five highly homologous antibody binding domains. The binding site for protein A is located on the Fc portion of antibodies of the immunoglobulin G (IgG) class *(1)*. Binding occurs through an induced hydrophobic fit and is promoted by addition of salts such as sodium citrate or sodium sulfate. At the center of the Fc binding site as well as on protein A reside histidine residues. At alkaline pH, these residues are uncharged and hydrophobic, strengthening the interaction between protein A and the antibody. As the pH is shifted to acidic values, these residues become charged and repel each other. Differences in the pH-dependent elution properties (**Table 1**) are seen between antibodies from different classes as well as different species due to minor differences in the binding sites. These differences can be successfully exploited in the separation of contaminating bovine IgG from mouse IgG, as shown in **Subheading 1.3.**

The major attraction in using protein A is its simplicity. The supernatant is adsorbed onto a protein A gel, the gel is washed, and the antibody is eluted at an acidic pH. The antibody recovered has a purity of >90%, often with full recovery of biological activity. The method described in **Subheading 3.1.** provides a more detailed description of the purification for high-affinity antibodies.

From: *The Protein Protocols Handbook, 2nd Edition*
Edited by: J. M. Walker © Humana Press Inc., Totowa, NJ

Table 1
Affinity of Protein A for IgG for Different Species and Subclasses

IgG species/subclass	Affinity	Binding pH	Elution pH
Human IgG1	High	7.5	3
Human IgG2	High	7.5	3
Human IgG3	Moderate	8	4–5
Human IgG4	High	7.5	3
Mouse IgG1	Low	8.5	5–6
Mouse IgG2a	Moderate	8	4–5
Mouse IgG2b	High	7.5	3
Mouse IgG3	Moderate	8	4–5
Rat IgG1	Low	8.5	5–6
Rat IgG2a	None–low		
Rat IgG2b	Low	8.5	5–6

The affinity between protein A and the antibody is due mainly to hydrophobic interactions at the binding sites *(1)*. The interaction can be strengthened by the inclusion of higher concentrations of chaotropic salts such as sodium citrate or sodium sulfate. Although the addition of chaotropic salts is unnecessary in the case of high-affinity antibodies, it allows weakly binding antibodies such as mouse IgG1 to strongly interact with protein A. This purification scheme is outlined in **Subheading 3.2.** The different affinities resulting from minor differences in the Fc region can also be utilized for the separation of subclasses or the separation of different species in hybridoma supernatant using a pH gradient. However, this is possible only if the difference in affinity is large enough like in the case of the separation of mouse IgG2b from bovine IgG. If fetal calf serum supplemented growth media is used, bovine IgG will contaminate the monoclonal antibody if the method described in **Subheading 3.1.** or **3.2.** is used. An improved protocol utilizing the different affinities is described in **Subheading 3.3.**

1.2. Thiophilic Adsorption Chromatography

The term thiophilic adsorption chromatography was coined for gels, which contain low molecular weight sulfur-containing ligands such as divinyl sulfone structures or mercapto-heterocyclic structures *(2,3)*. The precise binding mechanism of proteins to these gels is not well understood. However, thiophilic chromatography can be regarded as a variation of hydrophobic interaction chromatography inasmuch as chaotropic salts need to be added to facilitate binding of antibodies to the gel. Therefore, the interaction between the ligands and the antibody is likely to be mediated through accessible aromatic groups on the surface of the antibody.

The major advantage of these thiophilic gels over protein A gels is that they bind **all** antibodies with sufficiently high capacity and very little discrimination between subclasses or species. Furthermore, elution conditions are much milder, which might be beneficial if the harsh elution conditions required for strongly binding antibodies to protein A lead to their denaturation. Thiophilic gels also purify chicken IgY *(4)*. A general method is provided in **Subheading 3.4.**

The major disadvantage of thiophilic gels is the requirement to add high concentrations of chaotropic salts to the adsorption buffer to facilitate binding. This requirement is not a problem for small-scale purifications at the bench, but it is unattractive at larger scale.

Recently, a major improvement of thiophilic chromatography came about with the utilization of very high ligand concentrations *(9)*. By immobilizing mercapto-heterocyclics such as 4-mercaptoethyl pyridine at very high concentrations, the need to add chaotropic salts such as sodium sulfate was abolished. All antibodies investigated are adsorbed at neutral pH, independent of the salt concentration. At acidic pH values, the ligand becomes charged and through charge–charge repulsion, the antibody is eluted. This variation of thiophilic adsorption chromatography is highly reminiscent of protein A chromatography in its simplicity, but with the added advantage that it is nonspecific for antibodies. Purities and recoveries obtained range from 90% for antibodies from protein-free cell culture supernatant to approx 75% from ascitic fluid. The method is outlined in **Subheading 3.5.**

1.3. Immobilized Metal Affinity Chromatography

Immobilized metal affinity chromatography (IMAC) is a general term for a variety of different immobilization chemistries and metals utilized *(10)*. The most commonly used gel for IMAC is nickel-loaded iminodiacetic acid (Ni-IDA) gel, used, for example, in kits for the separation of His-tail modified proteins. Under slightly alkaline conditions, the interaction between the immobilized nickel and proteins is strongest with accessible histidine residues. Ni-IDA binds to the Fc portion of the antibody *(11)*, and similar to thiophilic adsorption chromatography, it binds all antibodies without discrimination between subclasses or species *(11,12)*. Depending on the growth medium used, some contamination is apparent, namely transferrin and traces of albumin. The purity of monoclonal antibodies purified out of ascitic fluid is generally lower than out of hybridoma cell culture supernatant. As described for protein A chromatography, a shift in pH to acidic values leads to the generation of charges on the histidine residues and consequent elution of the protein bound. Alternatively, competitive elution with either imidazole or EDTA yields antibody in good purity. However, the elution with EDTA delivers the antibody with the metal still bound to the protein, while imidazole adsorbs at 280 nm, interfering with UV detection. A general protocol is outlined.

2. Materials

2.1. General Purification of Human IgG, Humanized IgG, and Mouse IgG2a and Mouse IgG2b Using Protein A

1. Buffer A: 50 m*M* Tris-HCl, pH 7.5.
2. Buffer B: 50 m*M* Tris-HCl, 500 m*M* NaCl, pH 8.0.
3. Buffer C: 100 m*M* acetate buffer, 50 m*M* NaCl, pH 3.0.
4. Buffer D: 10 m*M* NaOH.

2.2. General Purification of Mouse IgG1, Rat IgG1, and Rat IgG2b Using Protein A

1. Buffer A: 500 m*M* sodium citrate, ph 8–8.4.
2. Buffer B: 100 m*M* sodium acetate, 50 m*M* NaCl, pH 4.0.
3. Buffer C: 10 m*M* NaOH.

2.3. General Purification of IgG with Removal of Bovine IgG Using Protein A

1. Buffer A: 50 m*M* Tris-HCl, pH 7.5.
2. Buffer B: 50 m*M* citrate, pH 5.0.
3. Buffer C: 50 m*M* citrate, 50 m*M* NaCl, pH 3.0.
4. Buffer D: 10 m*M* NaOH.

2.4. Thiophilic Purification of IgG

1. Buffer A: 50 m*M* Tris-HCl, 500 m*M* sodium sulfate, pH 7.5.
2. Buffer B: 50 m*M* sodium acetate, pH 5.0.
3. Buffer C: 10 m*M* NaOH.

2.5. General Purification of IgG Using MEP-HyperCell

1. Buffer A: 50 m*M* Tris-HCl, pH 7.5.
2. Buffer B: 100 m*M* acetate buffer, 50 m*M* NaCl, pH 3.5.
3. Buffer D: 10 m*M* NaOH.

2.6. Immobilized Metal Affinity Purification of IgG

1. Buffer A: 50 m*M* sodium phosphate, 500 m*M* sodium chloride, pH 8.0.
2. Buffer B: 50 m*M* sodium acetate, 50 m*M* sodium chloride, pH 4.5.
3. Buffer C: 200 m*M* sodium phosphate, 200 m*M* sodium chloride, pH 8.0.
4. Buffer D: 50 m*M* Tris-HCl, 500 m*M* sodium chloride, 50 m*M* EDTA, pH 8.0.
5. Buffer E: Buffer B, 50 m*M* nickel chloride.

3. Method

3.1. General Purification of Human IgG, Humanized IgG, and Mouse IgG2a and Mouse IgG2b Using Protein A (see Note 1)

This protocol is designed for the purification of high-affinity IgG monoclonal antibodies from hybridoma cell culture supernatant and ascitic fluid. For lower affinity monoclonals, refer to **Subheading 3.2.** or the other purification methods described below.

For cell culture supernatants that contain fetal calf serum and the separation of contaminating bovine IgG from the monoclonal antibody, please see **Subheading 3.3.**

1. Bring all materials to room temperature.
2. Equilibrate a 10-mL protein A gel with 5 column volumes of buffer A.
3. Load supernatant onto protein A–gel column at 5 mL/min.
4. Load up to 20 mg of antibody/mL of gel.
5. Wash with 10 column volumes of buffer B.
6. Elute antibody with 5 column volumes of buffer C.
7. Reequilibrate column with 2 column volumes of buffer A.
8. Wash with 5 column volumes of buffer E followed by 10 column volumes of buffer A.

The column is ready for the next chromatography run.

3.2. General Purification of Mouse IgG1, Rat IgG1, and Rat IgG2b Using Protein A (see *Note 2*)

For cell culture supernatants that contain fetal calf serum and the separation of contaminating bovine IgG from the monoclonal antibody, please *see* **Subheading 3.3.**

1. Bring all materials to room temperature.
2. Dilute supernatant 1:1 with 1000 m*M* sodium citrate solution. Filter the resulting solution through a 0.2-μm filter (**important**).
3. Equilibrate a 10-mL protein A gel with 5 column volumes of buffer A.
4. Load supernatant onto the protein A–gel column at 5 mL/min.
5. Load up to 12 mg of antibody/mL of gel.
6. Wash with buffer A until the UV baseline is reached.
7. Elute antibody with 10 column volumes of buffer B.
8. Reequilibrate column with 2 column volumes of buffer A.
9. Wash with 5 column volumes of buffer C followed by 10 column volumes of buffer A.

The column is ready for the next chromatography run.

3.3. General Purification of IgG with Removal of Bovine IgG Using Protein A (see *Note 3*)

This method is designed for the purification of human, humanized, and mouse IgG2a and IgG2b monoclonal antibodies from hybridoma cell culture supernatant containing fetal calf serum. It does not work well with weakly binding antibodies such as mouse IgG1.

1. Bring all materials to room temperature.
2. Equilibrate a 10-mL protein A gel with 5 column volumes of buffer A.
3. Load supernatant onto the protein A–gel column at 5 mL/min.
4. Load up to 20 mg of antibody/mL of gel.
5. Wash with 5 column volumes of buffer A.
6. Wash with 15 column volumes of buffer B.
7. Elute antibody with 10 column volumes of buffer C.
8. Wash with 5 column volumes of buffer E followed by 10 column volumes of buffer A.

The column is ready for the next chromatography run.

3.4. Thiophilic Purification of IgG (see *Note 4*)

This protocol is designed for the purification of all IgG from hybridoma cell culture supernatant and ascitic fluids. For cell culture supernatants that contain more than 5% fetal calf serum, the monoclonal antibody will be contaminated with bovine IgG.

1. Bring all materials to room temperature.
2. Dilute supernatant 1:1 with 1000 m*M* sodium sulfate solution.
3. Filter the resulting solution through a 0.2-μm filter (**important**).
4. Equilibrate thiophilic gel with 5 column volumes of buffer A.
5. Load supernatant onto the thiophilic gel column at 5 mL/min.
6. Load up to 10 mg of antibody/mL of gel.
7. Wash with buffer A until the UV baseline is reached.
8. Elute antibody with 10 column volumes of buffer B.
9. Wash with 5 column volumes of buffer C followed by 10 column volumes of buffer A.

The column is ready for the next chromatography run.

3.5. General Purification of IgG Using MEP-HyperCell (see Note 5)

This protocol is designed for the purification of all IgG from hybridoma cell culture supernatant and ascitic fluids. For cell culture supernatants that contain more than 5% fetal calf serum, the purity is relatively low at approx 50–65%.

1. Bring all materials to room temperature.
2. Equilibrate a 10-mL MEP-HyperCell gel with 5 column volumes of buffer A.
3. Load supernatant onto the column at 5 mL/min.
4. Load up to 20 mg of antibody/mL of gel.
5. Wash with 10 column volumes of buffer A.
6. Elute antibody with 8 column volumes of buffer B.
7. Reequilibrate the column with 2 column volumes of buffer A.
8. Wash with 8 column volumes of buffer E followed by 10 column volumes of buffer A.

The column is ready for the next chromatography run.

3.6. Immobilized Metal Affinity Purification of IgG (see Note 6)

This protocol is designed for the purification of all IgG from hybridoma cell culture supernatant and ascitic fluids. For cell culture supernatants that contain more than 5% fetal calf serum, the monoclonal antibody will be contaminated with bovine IgG.

1. Bring all materials to room temperature.
2. Dilute three parts of supernatant with one part of buffer C.
3. Equilibrate a 10-mL IMAC gel with 5 column volumes of buffer B.
4. Load the column with buffer E until the column is completely colored.
5. Wash with 5 column volumes of buffer B.
6. Equilibrate with 5 column volumes of buffer A.
7. Load supernatant onto the IMAC gel column at 5 mL/min.
8. Load up to 10 mg of antibody/mL of gel.
9. Wash with buffer A for 15 column volumes until the UV baseline is reached.
10. Elute antibody with 10 column volumes of buffer B.
11. Strip column with 5 column volumes of buffer D, followed by 5 column volumes of buffer B.

The column is ready for the next chromatography run.

3.7. Concluding Remarks

All the protocols provided in the yield antibodies in good yield and good purity. If not further specified in the accompanying notes, the protocols are usable with virtually all commercially available gels. Difficulties in using a protein A gel can be overcome by using a different affinity column such as a thiophilic gel or an IMAC gel. In most protocols provided, elution is accomplished at acidic pH values. Therefore, further purification can be accomplished by adsorbing the eluate onto a cation-exchange gel. Most of the impurities will flow through and, when using gradient elution, the antibody is generally the first protein to elute from the column.

4. Notes

1. This general protocol works well with high-affinity antibodies and is independent of the particular protein A gel used. The purity of the antibody recovered can be increased by first dialyzing the supernatant or the ascitic fluid against buffer A. Sometimes, phenol red

is difficult to remove and the washing step has to be increased accordingly until no red color can be detected on the gel.

2. No pH adjustments have to be made for the sodium citrate solution as the pH will be between 8.0 and 8.5. It is very important to filter the resulting solution after mixing of the supernatant and the sodium citrate solution. In the case of ascitic fluid, a precipitate is clearly visible, while in the case of hybridoma cell culture supernatant precipitating material might not be visible. The purity is somewhat lower than in the case of high-affinity antibodies, but it is very acceptable in the 85–90 % range.
3. The molarity of the citrate buffer used in the intermediate wash is slightly dependent on the protein A gel used. It is possible to wash the protein A gel from Pharmacia with 100 m*M* citrate buffer, pH 5.0, while the higher molarity would start to elute antibody from a protein A gel from BioSepra. A good starting point is the 50 m*M* citrate buffer outlined above, and if no loss of the desired monoclonal antibody is detected in the wash fraction, higher buffer concentrations can be employed.
4. There are only a few thiophilic gels commercially available. Of these, the thiophilic gel from E. Merck is by far the best. The most important precaution in using thiophilic gels is the need to filter the supernatant after adjustment to 500 m*M* sodium sulfate. Other antibody molecules like IgA ***(5)***, bispecific antibodies ***(6)***, and single-chain antibody fragments have been purified using thiophilic gels ***(7,8)***.
5. All antibodies, independent of species and subclass, can be purified with this gel. No separation of subclasses can be achieved, however. It is highly likely that similarly to the thiophilic gels described above, other antibody classes such as IgA, IgE, IgM, and chicken IgY can be purified using MEP-HyperCell. The MEP-HyperCell resin is available from BioSepra Inc., a unit of Ciphergen, Inc.
6. Generally, all chelating and amine-containing reagents such as citrate and Tris buffers should be avoided in the feedstocks and buffers used, as these buffers will strip the column of the immobilized metal. If possible, using a gradient to elute the proteins results in better purity of the antibody sought. Three different metals can be used to purify antibody, namely copper, nickel, and cobalt. A good starting metal of any purification using IMAC is nickel. Other metals that work in the purification of antibodies are cobalt and copper. The use of zinc does not yield any antibody. The addition of a His-tag allows the specific adsorption and elution of single-chain antibodies and antibody fragments ***(13,14)***.

References

1. Diesenhofer, J. (1981) Crystallographic refinement and atomic models of a human Fc fragment and its complex with fragment B of protein A from *Staphylococcus aureus* at 2.9 and 2.8 Å resolution. *Biochemistry* **20,** 2361–2370.
2. Porath, J., Maisano, F., and Belew, M. (1985) Thiophilic adsorption—a new method for protein fractionation. *FEBS Lett.* **185,** 306–310.
3. Oscarsson, S. and Porath, J. (1990) Protein chromatography with pyridine- and alkylthioester based Agarose adsorbents. *J. Chromatogr.* **499,** 235–247.
4. Hansen, P., Scoble, J. A., Hanson, B., and Hoogenraad, N. J. (1998) Isolation and purification of immunoglobulins from chicken eggs using thiophilic interaction chromatography. *J. Immunol. Meth.* **215,** 1–7.
5. Leibl, H., Tomasits, R., and Mannhalter, J. W. (1995) Isolation of human serum IgA using thiophilic adsorption chromatography. *Protein Exp. Purif.* **6,** 408–410.
6. Kreutz, F. T., Wishart, D. S., and Suresh, M. R. (1998) Efficient bispecific monoclonal antibody purification using gradient thiophilic affinity chromatography. *J. Chromatogr. B Biomed. Sci. Appl.* **714,** 161–170.

7. Schulze, R. A., Kontermann, R. E., Queitsch, I., Dubel, S., and Bautz, E. K. (1994) Thiophilic adsorption chromatography of recombinant single-chain antibody fragments. *Analyt. Biochem.* **220,** 212–214.
8. Fiedler, M. and Skerra, A. (1999) Use of thiophilic adsorption chromatography for the one-step purification of a bacterially produced antibody F(ab) fragment without the need for an affinity tag. *Protein Exp. Purif.* **17,** 421–427.
9. Burton, S. C. and Harding, D. R. (1998) Hydrophobic charge induction chromatography: salt independent protein adsorption and facile elution with aqueous buffers. *J. Chromatogr. A* **814,** 71–81.
10. Porath, J. and Olin, B. (1983) Immobilized metal ion affinity adsorption and immobilized metal ion affinity chromatography of biomaterials. *Biochemistry* **22,** 1621–1630.
11. Hale, J. E. and Beidler, D. E. (1994) Purification of humanized murine and murine monoclonal antibodies using immobilized metal-affinity chromatography. *Analyt. Biochem.* **222,** 29–33.
12. Al-Mashikhi, S. A., Li-Chan, E., and Nakai, S. (1998) Separation of immunoglobulins and lactoferrin from cheese whey by chelating chromatography. *J. Dairy Sci.* **71,** 1747–1755.
13. Laroche-Traineau, J., Clofent-Sanchez, G., and Santarelli, X. (2000) Three-step purification of bacterially expressed human single-chain Fv antibodies for clinical applications. *J. Chromatogr. B Biomed. Sci. Appl.* **737,** 107–117.
14. Kipriyanov, S. M., Moldenhauer, G., and Little, M. (1997) High level production of soluble single chain antibodies in small-scale *Escherichia coli* cultures. *J. Immunol. Meth.* **200,** 69–77.

164

A Rapid Method for Generating Large Numbers of High-Affinity Monoclonal Antibodies from a Single Mouse

Nguyen Thi Man and Glenn E. Morris

1. Introduction

Since the first description by Kohler and Milstein *(1)*, many variations on this method for the production of monoclonal antibodies (MAb) have appeared (e.g., *see* **refs. *2–4*** and Chapter 162), and it may seem superfluous to add another. The variation we describe here, however, includes a number of refinements that enable rapid (6–10 wk) production from a single spleen of large numbers ***(20–30)*** of cloned, established hybridoma lines producing antibodies of high-affinity. We have applied this method to recombinant fusion proteins containing fragments of the muscular dystrophy protein dystrophin *(5,6)*, and dystrophin-related proteins *(7)*, to hepatitis B surface antigen *(8)* and to the enzyme creatine kinase (CK) *(9)*, and have used the MAb thus produced for immunodiagnosis, epitope mapping, and studies of protein structure and function *(5–12)*. Epitopes shared with other proteins are common, so availability of several MAb against different epitopes on a protein can be important in ensuring the desired specificity in immunolocalization and Western blotting studies *(7)*.

In the standard Kohler and Milstein method, Balb/c mice are immunized with the antigen over a period of 2–3 mo, and spleen cells are then fused with mouse myelomas using polyethylene glycol (PEG) to immortalize the B-lymphocytes secreting specific antibodies. The hybrid cells, or hybridomas, are then selected using medium containing hypoxanthine, aminopterin, and thymidine (HAT medium) *(13)*. All unfused myelomas are killed by HAT medium, and unfused spleen cells gradually disappear in culture; only hybridomas survive.

Special features of the detailed method to be described here include:

1. A culture medium for rapid hybridoma growth without feeder layers;
2. Screening early to select high-affinity antibodies;
3. Cloning without delay to encourage hybridoma growth and survival; and
4. The use of round-bottomed microwell plates to enable rapid monitoring of colony growth with the naked eye.

In typical fusion experiments, cells are distributed over eight microwell plates (768 wells) and 200–700 wells show colony growth. Of these hybridomas, up to 50%

From: *The Protein Protocols Handbook, 2nd Edition*
Edited by: J. M. Walker © Humana Press Inc., Totowa, NJ

Table 1
Examples of Colony and MAb Yields Using This Method

Antigen	Wells with growth	Wells binding to antigen strongly	Wells with desired affinity/specificity	Final no. of MAb
108-kDa rod fragment of dystrophin in fusion protein *(5)*	587	136	27	16
59-kDa rod fragment of dystrophin in fusion protein *(6)*	669	147	30	25
55-kDa C-terminal fragment of dystrophin in fusion protein *(6)*	700+	42	20	17
37-kDa C-terminal fragment of a dystrophin-related protein in fusion protein *(7)*	186	54	24	19

(100–300) produce antibodies that show some antigen binding, though many may be too weak to be useful. After further screening for desired affinity and specificities, we normally select 25–30 as a convenient number to clone. Only occasionally, in our experience, are two or more MAb produced that are indistinguishable from each other when their epitope specificities are later examined in detail. **Table 1** illustrates the results obtained from four recent fusions by this method.

In the early stages after fusion, a particularly good cell-culture system is needed to promote rapid hybridoma growth in HAT medium. Horse serum, selected for high cloning efficiency, is a much better growth promoter than most batches of fetal calf serum, and HAT medium is prepared using myeloma-conditioned medium and human endothelial culture supernatant, a hybridoma growth promoter. Under these conditions, colonies are visible with the naked eye 7–10 d after fusion, and screening should start immediately. Even unimmunized mouse spleens will produce large numbers of colonies, so the use of hyperimmunized mice, with high serum titers in the screening assay, is essential for generating a high proportion of positive colonies. The screening assays are also of critical importance, because they determine the characteristics of the MAb produced. Both these points are illustrated in **Fig. 1**. CK, a fairly typical "globular" enzyme, is denatured and inactivated when attached directly to ELISA plastic, but it can be captured onto ELISA plates in its native form by using Ig from a polyclonal antiserum *(9)*. When mice were immunized with untreated CK, antibody titers against denatured CK were high, whereas titers of antibody against native CK were insufficient to produce a plateau in the capture ELISA, even at high serum concentrations (**Fig. 1A**). Seven fusions from such mice (over 1000 colonies screened) produced only two, low-affinity MAb specific for native CK. In contrast, when we immunized with

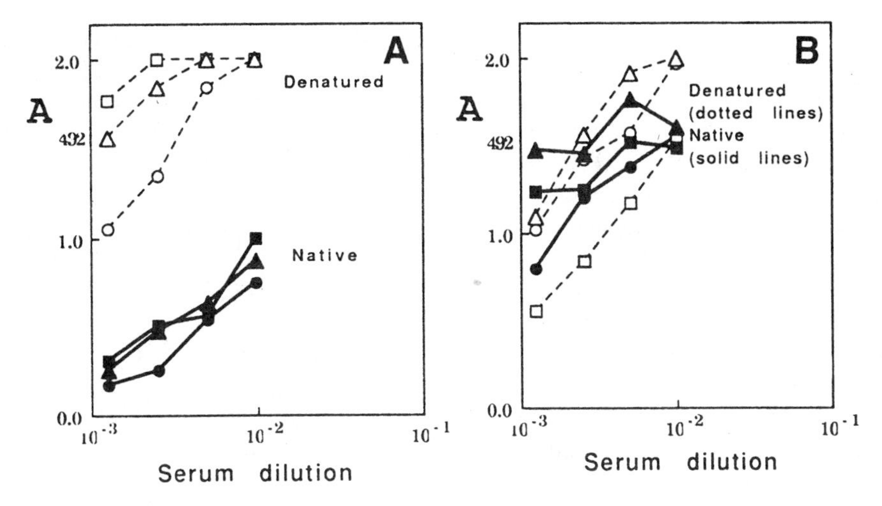

Fig. 1. Effect of aggregation of CK on the immune response in Balb/c mice. Test sera were taken from three mice immunized with untreated CK **(A)**: circles, squares, and triangles represent individual mice) or with glutaraldehyde-aggregated CK **(B)**, and serial dilutions were used as primary antibody in either direct ELISA for denatured CK (dotted lines) or sandwich ELISA for native CK (solid lines). (Reproduced with permission from **ref.** ***9***.)

CK aggregated by glutaraldehyde treatment (**Fig. 1B**), titers were higher for native CK antibodies than for those against denatured CK. In a fusion performed with one of these mice, only 2% of the fusion wells were positive in the direct ELISA (denatured CK), but 13% recognized only native CK in the capture ELISA *(9)*. This illustrates the importance of having both high-titer mice and a suitable screen.

Finally, although the method is rapid and efficient, it is also labor-intensive, especially during screening and cloning. When embarking on the method, one should not be daunted by the prospect of having over 70 culture plates in the incubator simultaneously by the time that the later cloning stages of a single experiment are reached.

2. Materials

1. One or two vertical laminar flow sterile hoods (e.g., Gelaire BSB4).
2. 37°C CO_2 incubator (e.g., LEEC, Nottingham, UK; *see* **Note 1**).
3. Transtar 96 (Costar, Cambridge, MA) for plating fusions and removing hybridoma supernatants.
4. Round-bottomed 96-well tissue-culture microwell plates (NUNC/Gibco, Life Technologies, Gaithersburg, MD, cat. no. 1–63320) (*see* **Note 2**).
5. Sterile tissue-culture flasks 50 mL (NUNC/Gibco cat. no. 1–63371), 260 mL (NUNC/Gibco cat. no. 1-53732), and 24-well plates (NUNC/Gibco cat. no. 1-43982), 1.8-mL cryotubes (NUNC/Gibco cat. no. 368632), and sterile 30-mL universal bottles.
6. Freund's adjuvant, complete (Sigma [St. Louis, MO] no. F-4258) stored at 4°C.
7. Freund's adjuvant, incomplete (Sigma no. F-5506) stored at 4°C.
8. Pristane (2,6,10,14 tetramethyl pentadecane; Aldrich Chemical Co. [Milwaukee, WI], no. T2280–2) stored at room temperature.
9. 500X HT: 6.8 mg/mL hypoxanthine (Sigma no. H-9377) and 1.95 mg/mL thymidine (Sigma no. T-9250) (*see* **Note 3**). Filter to sterilize. Store in 5-mL aliquots in sterile plastic universals at –20°C (stable for a few years).

10. 1000X Aminopterin (Sigma no. A-2255): 4.4 mg are dissolved in 25 mL of water (*see* **Note 3**). Since the chemical is toxic and teratogenic, it must be weighed with appropriate containment and operator protection. Filter to sterilize. Store in 2-mL aliquots at –20°C, wrapped in aluminum foil to protect from light (stable for years).
11. 50% polyethylene glycol (PEG), mol wt 1500 (BDH-Marck, Poole, Dorset, UK, Prod. 29575), is made up by weighing 10 g of PEG in a glass universal and autoclaving. Although the solution is still hot (about 60°C), add 10 mL of DMEM/25 m*M* HEPES prewarmed in a 37°C water bath. Mix well, and store in 5-mL aliquots at room temperature in the dark; stable at least for 6 mo. The pH should be slightly alkaline (as judged by phenol red) when used for fusion.
12. DMEM/25 m*M* HEPES. (Gibco cat. no. 041–2320H). This is used as a "physiological saline," not for growing cells.
13. DMEM/20% HS is prepared by adding to each 100 mL of 1X DMEM (Gibco cat. no. 041-1885): 1.3 mL L-glutamine (200 m*M*) (Gibco cat. no. 043-5030H), 1.3 mL sodium pyruvate (100 m*M*) (Gibco cat. no. 066–01840E), 1.3 mL penicillin-streptomycin (Gibco cat. no. 043-5140H), 0.6 mL nystatin (Gibco cat. no. 043-5340H), 1.3 mL nonessential amino acids (MEM) (Gibco cat. no. 043-1140H), and 25 mL selected horse serum (HS). For routine growth and maintenance of both myelomas and established hybridomas, serum is reduced to 5%.
14. Cloning medium: DMEM/20% HS supplemented with 1X HT and 5% human endothelial culture supernatant (HECS) (Costar no. M712).
15. HAT medium: Myeloma-conditioned DMEM/20% HS with 1X HAT and 5% HECS.
16. Medium for feeding the fusion: DMEM/20% HS with 5% HECS and 5X HT.
17. Medium for freezing down cells: 92.5% HS/7.5% dimethyl sulfoxide (DMSO) (Sigma no. D-5879).
18. Trypan blue (0.1%) (Sigma no. T6146) in PBS.

3. Methods

3.1. Immunization of Balb/c Mice

1. Purify the protein antigen. With purified antigen, screening is faster and easier, and the yield of MAb is higher, but proteins of 50% purity or less on SDS-PAGE may still give good results.
2. Each of three Balb/c mice (6–8 wk old) is given an ip injection of 50–100 µg of antigen in 0.1 mL PBS emulsified with an equal volume of Freund's complete adjuvant by sucking up and down many times into a disposable sterile plastic 2-mL syringe with a sterile 21-gage 1–1/2 in. no. 2 needle (*see* **Note 4**).
3. Four weeks later, an ip boost of 50–100 µg antigen in Freund's incomplete adjuvant is given.
4. Ten days later, blood (ca. 0.1 mL) is taken from the tail vein, allowed to clot, and centrifuged at 10,000 rpm for 5 min. The serum is tested by the method to be used for screening hybridomas. If the titers are high in this assay (*see* **Fig. 1**), the mice should be left to rest for an additional month before fusion.
5. The best mouse is boosted with 50–100 µg of antigen in PBS, ip, and/or iv in the tail vein, 4, 3, and 2 d before fusion.

3.2. Selection of Horse Serum

1. Most suppliers will provide test samples of different serum batches and keep larger amounts on reserve. All sera are heat-inactivated at 56°C for 30 min before use, though this is not known to be essential. Thaw serum in 37°C water bath and swirl to mix before incubating in a 56°C water bath for 30 min with occasional swirling.

2. Prepare cells of any hybridoma or myeloma line at 1000 cells/mL in DMEM and dilute to 40 cells/mL, 20 cells/mL, 10 cells/mL in DMEM/20% HS of different batches (cloning of established cell lines does not require feeder layers or HT and HECS). For each test serum, plate one 96-well round-bottomed microplate with four rows at four cells in 0.1 mL/well, four rows at two cells in 0.1 mL/well, and four rows at one cell in 0.1 mL/well.
3. From the fifth day onward, note the number of colonies at each cell density for each test serum, as well as the size of the clones. After 2 wk, choose the batch of serum with the highest number and the largest clones. Serum may be stored at –70°C for at least 1–2 yr without loss of activity.

3.3. Growth and Maintenance of Myelomas

NSO/1 *(2)* and SP2/O ***(14)*** myeloma lines grow very fast and do not synthesize immunoglobulins They should be kept in logarithmic growth in DMEM/5% HS for at least a wk before fusion by diluting them to 5×10^4/mL when the cell density reaches 4 or 5×10^5/mL. If thawing myelomas from liquid nitrogen, it is advisable to begin at least 2–3 wk before fusion (*see* **Note 5**). We have never found it necessary to treat lines with azaguanine to maintain their aminopterin sensitivity, which can be checked by plating in HAT medium at 100 cells/well.

3.4. Fusion with Spleen Cells

1. All dissection instruments are sterilized by dry heat (160°C for 5 h), autoclaving (120°C for 30 min) or dipping in 70% ethanol, and flaming in a Bunsen burner. Two separate fusions are done with each spleen as a precaution against accidents.
2. Two days before fusion, set up 2 × 260-mL flasks with 40 mL each DMEM 20% HS and 1×10^5 myeloma cells/mL (NSO/1 or Sp2/O).
3. On the day of fusion, place in a 37°C water bath 100 mL of DMEM/25 m*M* HEPES, 10 mL of DMEM/20% HS, and 1 bottle of PEG1500.
4. Count the myeloma cells in a hemocytometer after mixing an aliquot with an equal volume of 0.1% Trypan blue in PBS; cell density should be 4–6 × 10^5 with viability 100% and no evidence of contamination.
5. Transfer the myeloma cells to four universals and centrifuge for 7 min at 300*g* (MSE 4 L). Remove the supernatants with a 10-mL pipet and retain them for plating the fusion later (myeloma-conditioned medium). Resuspend each pellet in 5 mL of DMEM/HEPES, and combine in two universals (2 × 10 mL).
6. The immunized mouse is killed by cervical dislocation outside the culture room and completely immersed in 70% ethanol (15–30 s). Subsequent procedures are performed in sterile hoods. The mouse is pinned onto a "sterile" surface (polystyrene covered with aluminum foil and swabbed with 70% ethanol), and a small incision in the abdominal skin is made with scissors. The peritoneum is then exposed by tearing and washed with 70% ethanol before opening it and pinning it aside. The spleen is lifted out, removing the attached pancreas with scissors, and placed in a sterile Petri dish (*see* **Note 6**).
7. After removing as much surrounding tissue as possible, the spleen is placed in 10 mL of DMEM/25 m*M* HEPES and held at one end with forceps while making deep longitudinal cuts with a sterile, curved scalpel to release the lymphocytes and red blood cells. This process is completed by scraping with a Pasteur pipet sealed at the end until only connective tissue is left (2–3 min).
8. The spleen cell suspension is transferred to a 25-mL universal, with pipeting up and down five to six times to complete the dispersal. Large lumps are allowed to settle for 1 min before transferring the supernatant to a second universal and centrifuging for 7 min at

300*g*. Resuspend the pellet in 5 mL of DMEM/25 m*M* HEPES, and keep at room temperature. Usually about 1×10^8 cells/spleen are obtained (*see* **Note 7**).

9. Add 2.5 mL spleen cell suspension to each 10 mL of myeloma cell suspension, and mix gently. Centrifuge at 300*g* for 7 min at 20°C. Remove supernatants with a pipet, and resuspend pellets in 10 mL DMEM/25 m*M* HEPES using a pipet. Centrifuge at 300*g* for 7 min. Remove the supernatant completely from one pellet (use a Pasteur pipet for the last traces), and then loosen the pellet by tapping the universal gently (*see* **Note 8**).
10. Remove the DMEM/25 m*M* HEPES and the 50% PEG from the water bath just before use. Take 1 mL of PEG in a pipet, and add dropwise to the cell pellet over a period of 1 min, mixing between each drop by shaking gently in the hand. Continue to shake gently for another minute. Add 10 mL DMEM/25 m*M* HEPES dropwise with gentle mixing, 1 mL during the first minute, 2 mL during the second, and 3.5 mL during the third and fourth minutes. Centrifuge at 300*g* for 7 min. (This procedure can be repeated with the second cell pellet, while the first is spinning.)
11. Remove supernatant, and resuspend pellet gently in 5 mL of DMEM/20% HS. Place both 5-mL cell suspensions in their universals in the CO_2 incubator for 1–3 h.
12. During this time, take the ca. 80 mL of myeloma-conditioned medium and add 4.5 mL of HECS, 90 µL of aminopterin, and 180 µL of hypoxanthine/thymidine solutions. Filter 2×40 mL through 22-µm filters (47 mm; Millipore [Bedford, MA]) to resterilize, and remove any remaining myeloma cells.
13. To each 40 mL, add the 5-mL fusion mixture from the CO_2 incubator and distribute in 96-well microtiter plates (4 plates/fusion; 100 µL/well) using a Transtar 96 or a plugged Pasteur pipet (3 drops/well) (*see* **Note 9**). Put the plates from each fusion in a separate lunch box in the CO_2 incubator and leave for 3 d.
14. On d 4, add to each well 80 µL of DMEM/20% HS supplemented with HECS and 5X HT. Replace in the CO_2 incubator as quickly as possible.
15. By d 10–14, a high proportion of the wells should have a clear, white, central colony of cells, easily visible with the naked eye. If you want to select for high-affinity MAb (and you would be well advised to do so, unless you have some very special objectives), do not delay screening. If you are using a sensitive screening method, such as ELISA, immunofluorescence, or Western blotting, you will detect high-affinity antibodies from even the very small colonies. Do not wait for the medium to turn yellow, or your cells may start to die (if you are using a less sensitive screen, such as an inhibition assay, you *may* need to wait longer). A high proportion of the 768 wells (50–100%) should have colony growth by 14 d. If it is <10%, try reducing the PEG concentration. If you regularly obtain over 70%, use 6–8 plates/fusion instead of 4. There should be no microbial contamination (*see* **Note 10**).

3.5. Screening

It would be quite wrong to describe one screening method in detail, since this must be chosen to suit the purpose for which MAb are required. We recommend ELISA as the simplest technique for screening 8×96 wells in 1 d, but for coating ELISA plates directly, 10–50 µg of reasonably pure antigen are required. Alternatively, plates can be coated with a capture antibody; we have used this approach to capture CK in its native conformation *(9)* and to capture hepatitis B surface antigen from human plasma *(8)*. The availability of multichannel miniblotters makes direct screening by Western blotting feasible, and with two miniblotters, 224 microwells can be screened in one long day. Immunohistochemistry on cultured cells grown in microwells is also feasible, though we have not used it ourselves. A Transtar 96 apparatus is a rapid and sterile way to remove 10–50 µL of culture supernatant from all 96 wells of a microwell plate. The culture plates have to be open inside the hood for only a few seconds. They should always be wiped with paper tissue soaked in 70% ethanol before opening. Screening the whole plate, including wells without cell growth, by ELISA avoids possible errors when trying to keep track of individual wells. If ELISA-positive wells are to be screened further (e.g., by Western blotting using a multichannel miniblotter), this first 50 µL of supernatant may be diluted into 100 µL or more of PBS; multiple sampling of fusion wells should be minimized to avoid contamination. We usually screen by ELISA on d 10, carry out further screening of ELISA-positive wells on d 11 and 12, and clone about 30 of the best wells on d 12, 13, and 14.

3.6. Cloning by Limiting Dilution

1. Using a plugged Pasteur pipet, transfer the positive clone to a 25-well plate containing 0.5 mL of cloning medium (*see* **Note 11**).
2. Take an aliquot of these cells, dilute with an equal volume of 0.1% trypan blue in PBS, and count cells on at least two chambers of a hemocytometer.
3. Prepare 6 mL of 160 cells/mL and perform serial dilutions to 40 cells/mL and 10 cells/mL. Plate four rows at 16 cells in 0.1 mL/well, four rows at four cells in 0.1 mL/well, and four rows at one cell in 0.1 mL/well. Three drops from a plugged Pasteur pipet are about 0.1 mL.
4. Eight to ten days after plating, clones at 1 cell/well are visible, and screening can start immediately. It is best to screen at least 16–24 wells from each cloning plate. As soon as positive clones have been identified, they should be cloned a second time, but at 4, 1, and 0.5 cells/well instead of 16, 4, and 1.
5. When these plates are screened again after another 8–10 d, at least one dilution should have <50% of the wells with growth, and *all wells with growth* should be positive in the screening. If any colonies are negative, cloning should be repeated until all wells are positive (*see* **Note 12**).

3.7. Preservation of Hybridoma Lines

1. Hybridoma clones should be expanded slowly from microwells, first into 0.5 mL in 25-well plates, feeding to 1 or 2 mL before transferring to a 50-mL flask in 5 mL of medium.
2. To preserve cell lines indefinitely, centrifuge 1–3 × 10^6 growing cells in a universal, remove all supernatant, suspend cell pellet in 0.3–0.5 mL of ice-cold 7.5% DMSO/HS, and transfer the suspension into a 1.8-mL cryotube.
3. Surround the cryotube with 1 in of polystyrene, and place in a –70°C freezer. Transfer it the next day to a liquid nitrogen container.
4. It is advisable to freeze several vials of the same line and thaw one after a week to ensure viability, while maintaining the cells in culture.
5. To recover cells from liquid nitrogen, thaw the cryovial quickly in a 37°C water bath until only a tiny piece of ice is visible. Transfer the cell suspension immediately to 10 mL of ice-cold DMEM/20% HS. Centrifuge at 300*g* for 7 min, remove supernatant completely, and resuspend the cell pellet in 3–5 mL of cloning medium. Transfer the cell suspension into a 50-mL flask, reserving about 0.1 mL to count viable cells, and place flask in the CO_2 incubator. If the viable cell count is <1 × 10^4/mL, cloning by limiting dilution should be carried out immediately (*see* **Note 13**).

3.8. Antibody Production

1. Antibody can be generated as culture supernatant by starting a flask culture at 1×10^5/mL in DMEM/20% HS and, when the cell density reaches $4–6 \times 10^5$/mL, diluting to 1×10^5/mL. The concentration of horse serum is gradually reduced to 10% and then 5% in the process of reaching the volume required (*see* **Note 14**). The cells are then left to grow undisturbed until they are all dead (*see* **Note 15**). Culture supernatant can then be harvested by centrifugation, and stored in aliquots at –20°C until required. For routine use, keep an aliquot of 1–2 mL with 0.1% sodium azide at 4°C
2. For ascites fluid, an adult Balb/c mouse is primed with 0.2 mL of pristane ip.
3. After 7–10 d, $1–3 \times 10^6$ hybridoma cells in logarithmic growth (100% viability) are centrifuged at 300*g* for 7 min at 20°C and washed once with DMEM/25 m*M* HEPES to remove all serum. The cell pellet is resuspended in 0.2 mL of DMEM/HEPES, and injected ip into a primed mouse using a 21-gage no. 2 needle (green).
4. Ascites development becomes evident by external examination within 7–14 d and mice must be examined twice a day over this period. Mice are killed by cervical dislocation as soon as they show signs of discomfort (advice should be sought and followed). The peritoneum is exposed as in **Subheading 3.4., step 6**, and a 21-gage needle on a 5-mL syringe is inserted about 1 cm so that the tip remains visible through the peritoneum and fluid can be withdrawn without blockage. Last traces of fluid are removed after opening the cavity with scissors. The final volume obtained is usually 2–3 mL, though blood from the heart and thoracic cavity can also be collected (about 0.5 mL) and processed separately as a source of antibody.
5. Fluids are allowed to clot and then centrifuged at 3000*g* for 20 min. Store in aliquots of 200 µL at –20°C or preferably –70°C.

4. Notes

1. Fusions from a single mouse spleen aimed at producing 20–30 cloned hybridoma lines by the method described here can eventually fill a standard size bench-top incubator completely. The floor of the incubator is filled with autoclaved, double-distilled water. Do not use the heater on the inner glass door; a soaking wet door is a good check for 100% humidity. The interior of the incubator must be kept as free of contamination as possible; this may necessitate removing the contents, changing the water, and wiping the interior with 70% ethanol on several occasions during hybridoma growth and cloning. Even so, we use large plastic lunch boxes to insulate unsealed culture vessels (e.g., microwell plates) from the turbulent incubator environment (a fan is desirable for uniform temperature and humidity). Holes plugged with cotton wool at each end of the lunch box to admit CO_2 are optional; an alternative is to leave the lids ajar for 1–2 min to equilibrate before closing them. They can be wiped regularly, inside and out with 70% ethanol. Never use toxic chemicals (e.g., bleach) to decontaminate an incubator. Never handle culture plates without gloves.
2. Cells roll to the bottom giving a white colony in the exact center of each well, visible even at early stages; flat-bottomed plates have initially transparent colonies, often at the side of the well and easily overlooked. Use of round-bottomed plates makes it very easy to monitor colony growth after fusion without using a microscope.
3. The mixture does not dissolve by itself, so add 1 *M* NaOH dropwise while stirring until the solution goes clear.
4. Avoid syringes with rubber plunger tips; when the emulsion thickens, they come off the plunger.

5. The myeloma cells should look uniform and healthy by this time with no sign of cell debris.
6. We normally perform all steps up to this point in a separate sterile hood. The mouse is nonsterile externally, and particular care is taken to avoid hairs; the outer skin is torn, rather than cut, for this reason. Mice should not be introduced into the hood used for cell culture.
7. Some protocols remove red blood cells by differential lysis at this point, but we have not found it necessary.
8. The final concentration of PEG during fusion is thought to be critical for the yield of colonies. If initial yields are low, try adding increasing (but small) amounts of DMEM/25 m*M* HEPES to the pellet before adding PEG.
9. Never use unplugged micropipet tips for adding to culture plates; sterile tips can be used for *removing* culture supernatant only.
10. Bacterial and yeast contaminations rarely occur and are usually the result of a major failure in sterile technique (e.g., inadequate sterilization of culture medium or glassware). Sources of any sporadic fungal contamination should be tracked down and eliminated.
11. Some protocols recommend freezing down uncloned or partly cloned cells as a precautionary measure. Following our rapid cloning protocol, however, the first round of cloning and screening is often complete before the original colony is sufficiently expanded for freezing. It is certainly advisable to keep the original culture alive, however, by adding 0.1 mL of cloning medium back to the fusion well and by feeding the 24-well culture, if necessary. Cloning should not be regarded as an ordeal suitable only for healthy cells, but rather as a means of invigorating a failing culture. We always clone twice for this reason, even if the line is evidently clonal after the first round. *As a general maxim*: if in doubt or trouble, clone immediately.
12. There are rare exceptions to the general principle that cloning should be continued until all colonies are positive in the screens. We were once performing an initial screen by ELISA and then testing ELISA-positive wells by immunofluorescence microscopy (IMF) on muscle sections. After cloning one well that was positive in both assays, we found that only half the clones were ELISA-positive and very few of these were also IMF-positive. We recloned wells that were positive in both assays again with the same result. We thought we had come across our first "unstable" hybridoma, but by chance we tested ELISA-negative wells in IMF and found they were all IMF-positive. Only then did we realize that the original fusion well had contained two different hybridomas, one ELISA-positive and one IMF-positive. The purpose of cloning is to separate such lines, but by selecting wells that were positive in both assays (and hence still had two clones), we had been systematically defeating this objective.
13. When very few cells survive after being kept in liquid nitrogen, "cloning" by limiting dilution at about 100 total cells/well may be the only way to recover the cell line.
14. Cells may also be collected by centrifugation and resuspended in 5% fetal calf serum at this stage, if a culture supernatant with low levels of nonmouse Ig is required.
15. Most MAb are stable for long periods in sterile culture medium, but there are undoubtedly some that lose activity rapidly even at 4°C. These are perhaps best avoided, but, if required, bulk culture would have to be monitored regularly and supernatants harvested when their antibody activity is still high.

Acknowledgment

We thank C. J. Chesterton (King's College, London) for sharing with us his enthusiasm for, and experience of, hybridoma technology in 1981.

Further Reading

1. Langone, J. J. and Van Vunakis, H. (eds.) (1986) Methods in enzymology, in *Immunochemical Techniques,* vol. 121, part I. Academic, New York.
2. Goding, J. W. (1986) *Monoclonal Antibodies: Principles and Practice.* Academic, New York.

References

1. Kohler, G. and Milstein, C. (1975) Continuous cultures of fused cells secreting antibody of predefined specificity. *Nature* **256,** 495–497.
2. Galfre, G. and Milstein, C. (1981) Preparation of Monoclonal antibodies: strategies and procedures. *Meth. Enzymol.* **73,** 3–46.
3. Fazekas de St. Groth, S. and Scheidegger, D. (1980) Production of monoclonal antibodies: strategy and tactics. *J. Immunol. Meth.* **35,** 1–21.
4. Zola, H. and Brooks, D. (1982) Techniques for the production and characterization of monoclonal hybridoma antibodies, in *Monoclonal Antibodies: Techniques and Applications* (Hurrell, J. G., ed.), CRC, FL, pp. 1–57.
5. Nguyen thi Man, Cartwright, A. J., Morris, G. E., Love, D. R., Bloomfield, J. F., and Davies, K. E. (1990) Monoclonal antibodies against defined regions of the muscular dystrophy protein, dystrophin. *FEBS Lett.* **262,** 237–240.
6. Ellis, J. M., Nguyen thi Man, Morris, G. E., Ginjaar, I. B., Moorman, A. F. M., and van Ommen, G.-J. B. (1990) Specifity of dystrophin analysis improved with monoclonal antibodies. *Lancet* **336,** 881,882
7. Nguyen thi Man, Ellis, J. M., Love, D. R., Davies, K. E., Gatter, K. C., Dickson, G., and Morris, G. E. (1991) Localization of the DMDL-gene-encoded dystrophin-related protein using a panel of 19 monoclonal antibodies. Presence at neuromuscular junctions, in the sarcolemma of dystrophic skeletal muscle, in vascular and other smooth muscles and in proliferating brain cell lines. *J. Cell Biol.* **115,** 1695–1700.
8. Le Thiet Thanh, Nguyen thi Man, Buu Mat Phan, Ngoc Tran Nguyen, thi Vinh Ha, and Morris, G. E. (1991) Structural relationships between hepatitis B surface antigen in human plasma and dimers of recombinant vaccine: a monoclonal antibody study. *Virus Res.* **21,** 141–154.
9. Nguyen thi Man, Cartwright, A. J., Andrews, K. M., and Morris, G. E. (1989) Treatment of human muscle creatine kinase with glutaraldehyde preferentially increases the immunogenicity of the native conformation and permits production of high-affinity monoclonal antibodies which recognize two distinct surface epitopes. *J. Immunol. Meth.* **125,** 251–259.
10. Nguyen thi Man, Cartwright, A. J., Osborne, M., and Morris, G. E. (1991) Structural changes in the C-terminal region of human brain creatine kinase studied with monoclonal antibodies. *Biochim. Biophys. Acta* **1076,** 245–251.
11. Morris, G. E. and Cartwright, A. J. (1990) Monoclonal antibody studies suggest a catalytic site at the interface between domains in creatine kinase. *Biochim. Biophys. Acta* **1039,** 318–322.
12. Sedgwick, S. G., Nguyen thi Man, Ellis, J. M., Crowne, H., and Morris, G. E., (1991) Rapid mapping by transposon mutagenesis of epitopes on the muscular dystrophy protein, dystrophin. *Nucleic Acids Research* **19,** 5889–5894.
13. Littlefield, J. W. (1964) Selection of hybrids from matings of fibroblasts in vitro and their presumed recombinants. *Science* **145,** 709,710.
14. Shulman, M., Wilde, C. D., and Kohler, G. (1978) A better cell line for making hybridomas secreting specific antibodies. *Nature* **276,** 269,270.

Index

W

X

Z